D0906616

GEODESY

GEODESY

BY

G. BOMFORD

FOURTH EDITION

CLARENDON PRESS · OXFORD
1980

Oxford University Press, Walton Street, Oxford OX2 6DP

GLASGOW NEW YORK TORONTO MELBOURNE WELLINGTON
CAPE TOWN IBADAN NAIROBI DAR ES SALAAM LUSAKA ADDIS ABABA
DELHI BOMBAY CALCUTTA MADRAS KARACHI LAHORE DACCA
KUALA LUMPUR SINGAPORE HONG KONG TOKYO

First edition 1952
Second edition 1962
Third edition 1971
Fourth edition 1980

Published in the United States of America by Oxford University Press Inc., New York.

British Library Cataloguing in Publication Data

Bomford, Guy
Geodesy. – 4th ed.
1. Geodesy
I. Title
526 QB281 80-40089

ISBN 0-19-851946-X

Filmset by Universities Press, Belfast
Printed in Great Britain by
Lowe & Brydone Printers Ltd, Thetford, Norfolk

PREFACE TO THE FOURTH EDITION

THE definition of geodesy. The literal meaning of 'geodesy' is 'dividing the earth', and its first object is to provide an accurate geometrical framework for the control of topographical and other surveys. From this point of view geodesy has been taken to include:

(*a*) Primary or zero-order triangulation, trilateration, and traverse. Chapters 1 and 2.

(*b*) The measurement of height above sea-level by triangulation or spirit levelling. Chapter 3.

(*c*) Astronomical observations of latitude, longitude, and azimuth to locate the origins of surveys, and to control their direction. Chapter 4.

(*d*) Crustal movements. To detect changes in the relative positions of points on the ground, and in their heights above sea-level. Chapter 6, Section 12.

This, however, has not been the end of the subject. Circumstances have caused geodesy to overlap to some extent with what might reasonably be described as geophysics. Triangulation cannot be computed without a knowledge of the figure of the earth, and from very early days geodesy has included astronomical observations of latitude and longitude, not only to locate detached survey systems, but to enable triangulation to give the length of the degree of latitude or longitude in different parts of the earth, and so to determine the earth's figure. An alternative approach to the same subject has been via the variation of gravity between equator and pole, as measured by timing the swing of a pendulum. But these two operations, the measurement of the direction and intensity of gravity, have led to more than the determination of the axes of a spheroidal earth. They have revealed the presence of irregularities in the earth's figure and gravitation which have constituted an important guide to its internal composition. It is difficult to be precise about the dividing line between geodesy and geophysics, but for the present purpose geodesy is held to include:

(*e*) Observation of the direction of gravity by astronomical observations for latitude and longitude.

(*f*) Observation of the intensity of gravity by the pendulum or other apparatus.

(*g*) The use of (*e*) and (*f*) to deduce the exact form of the earth's sea-level equipotential surface, the geoid, and of the external equipotential surfaces at all heights, Chapter 4, Section 6 and Chapter 6, Sections 6 and 10. Deductions about the density, temperature, and constitution of the earth's interior, which may be based on the irregularities in the geopotential, are matters for geologists and geophysicists, although it may be proper for geodesists to take some interest in their findings.

Other subjects which are to some extent shared with other disciplines are:

(*h*) Polar motion, § 4.09. Shared with geophysics and astronomy.
(*i*) Earth tides, Chapter 6, Section 13. Shared with geophysics.
(*j*) The separation between the geoid and mean sea-level. Chapter 3, Section 4. Shared with oceanography.
(*k*) Engineering surveys, miniature geodesy, where geodetic accuracy is required over a small area for the laying out of special engineering works. This is mentioned in § 1.13, but otherwise it is not specifically dealt with here. Shared with engineering.
(*l*) Finally there is satellite geodesy, which is now spreading its beneficial influence over all other branches of the subject.

Satellite geodesy, Chapter 7, has made great advances during the last 10 years. With the completion of the BC-4 Pageos programme the method of astro-triangulation, Chapter 7, Section 4, may have become obsolete. At the moment its place has been taken by the doppler Transit system, Chapter 7, Section 7. The great advantage of the doppler method is that the satellite ephemerides are prepared by US agencies which control a network of tracking stations, and then any other organization can obtain geocentric fixes in any weather, anywhere in the world, at any time, by sending a one- or two-man team with an easily portable instrument to spend a few days at the required site. Some patience may be required before the computations are completed, but no special field organization is required. The fix may (hopefully) be good to ±5 metres relative to the geocentre, while the relative positions of points 100 or 200 km apart, if observed simultaneously by a pair of instruments, are probably good to ±one metre. If these accuracies can in fact be confidently achieved, the impact on conventional geodesy will be very great, as follows.

(*a*) All detached survey datums, such as the European datum, North American datum, etc., can be converted to a single geocentric system, see § 7.77. This has already given better results than could be attained by any practical conventional means.
(*b*) In extensive existing systems such as the Europe–India–Singapore or the Europe–Egypt–Cape systems, over distances of a few thousand km or more, intermediate satellite fixes are likely to be more accurate *inter se* than the best conventional methods. See § 7.78.
(*c*) Satellites give fixes in all three dimensions, so long lines of geoid section (Chapter 4, Section 6 and Chapter 6, Section 10) can be controlled by satellite fixes, and ocean gaps can be crossed.

(*d*) If satellite fixes 200 km apart are good to ±1 m, their mutual azimuth is good to ±1″ (second of arc), and the necessity for Laplace control observations, § 1.03 and § 2.05, becomes questionable.

(*e*) With the separation of the geoid and the spheroid determined by the combination of satellite fixes and spirit levelling, the value of using Stokes's integral, § 6.30, for this purpose will be reduced.

(*f*) Satellites have already given a good generalized analysis of the earth's external gravity field.

At the moment (1978) this list perhaps contains an element of optimism. The accuracies quoted are expected but not quite firmly established. It is certainly much too early to dismiss all non-satellite work as obsolete. One reason for this is that at present it is only over long distances that satellite methods are the more accurate. All the finer details of topographical control and of gravity and geoid surveys must still be provided in the traditional way. Further, although satellite methods might now replace conventional methods in an extensive new survey, such as a framework for all Africa, existing geodetic frameworks are all of the pre-satellite era. Their methods, and their strengths and weaknesses, need to be known and studied until these frameworks have been replaced by something better.

Looking a little further into the future, lunar lasers, § 7.45, and Very Long Baseline Interferometry (VLBI), Chapter 7, Section 9, give promise of much higher accuracy, but do not appear likely to offer the convenience of doppler. The Global Positioning System (GPS), § 7.44, which it is hoped to operate from 1984, is expected to be an improvement on the present doppler system both in convenience and in accuracy.

Changes in the present edition. The treatment of satellites, Chapter 7, is naturally much expanded, but it is impossible to deal with the subject fully in 100 pages: ten times as much would be insufficient. This chapter is not addressed to those who have already specialized in the subject, except to underline its uses for geodesy.

Of the other chapters, apart from thorough revision and amendment in matters of detail, Chapters 1 and 2 contain relatively little that is entirely new. Care has been taken to eliminate anything which has become obsolete, for which reference can be made to the previous edition.

Chapter 3, Section 4 has been enlarged in its treatment of the varying separation between mean sea-level and an equipotential surface.

In Chapter 4, Sections 1, 3, and 4, which deal with time and longitude, have mostly been rewritten. The most promising modern instruments for this purpose appear to be small portable zenith cameras. The section on gyro-theodolites has been moved to this chapter.

In Chapter 5, Gravity, instruments for measuring the absolute value of g have become more accurate and more portable, to the extent that the pendulum may be becoming obsolete. The completion of the International Gravity Standardization Net (1971) has been a great step forward in the unification of gravity surveys made by different agencies.

In Chapter 6, Physical Geodesy, the arrangement differs little from that of the previous Chapter 7, but its revision has involved a higher proportion of rewriting than has been required in the earlier chapters.

Of the Appendices, Appendix D (Theory of errors) has been substantially enlarged, but changes in the others are relatively small.

In the Bibliography about 250 new items have been included, and 160 older items removed.

SI units have been adopted in principle, but there are three points of difficulty as below.

(*a*) The gal ($=1\ \text{cm s}^{-2}$ or $0.01\ \text{m s}^{-2}$) and mGal are universally used in geodesy, but are not SI units. The SI system admits their retention for a limited period, but they seem unlikely to survive for another ten years. At the expense of what is probably irritating repetition, figures and formulae have been given in both units. See § 5.01.

(*b*) The bar ($=10^5$ Pa) is in a similar position, but is perhaps more firmly entrenched than the gal. Figures and formulae have generally been given in both units, but where this becomes unduly monotonous, preference has been given to the bar and mbar.

(*c*) Potential. The International Association of Geodesy introduced the Geopotential unit (GPU) some years ago as a unit of 'dynamic height', § 3.00. $1\ \text{GPU} = 1\ \text{kGal metre} = 10\ \text{m}^2\,\text{s}^{-2}$. It has the convenience that 'heights' expressed in GPU differ numerically by about 2 per cent from their values expressed in metres. The SI system does not permit its use, but in the limited context of Chapter 3, it has been retained. It is to be hoped that the IAG will presently make some recommendation in the matter. In § 3.28 reference is made to the rather similar dynamic centimetre, as used by oceanographers ($1\ \text{dyn cm} = 10^{-2}\ \text{GPU} = 10^{-1}\ \text{m}^2\,\text{s}^{-2}$). The position is explained where it occurs.

Acknowledgements. I have to thank many who have helped me with information and advice, especially Dr A. R. Robbins (geodetic astronomy and other matters), Prof. V. Ashkenazi (theory of errors and satellite doppler), and also Mr G. Gebel, (satellite doppler and connected matters), Mr T. Vincenty and Mr B. R. Bowring (triangulation computations). I also have to thank many organizations which have sent me long series of their publications, notably the Smithsonian Astrophysical Observatory, the Goddard Space Flight Centre, the Ohio State University, the Royal Aircraft Establishment (Farnborough), the International Association of Geodesy, the Deutsche Geodätische Kommission, and the Geodetic

Institutes of many countries, as well as numerous individuals who have sent me copies of their personal publications.

General references. The following cover considerable parts of the subject: [76], [155], [262], [270], [273], [315], [316], [386], and [620].

Sutton Courtenay G.B.
December 1978

CONTENTS

CHAPTER 1. TRIANGULATION, TRAVERSE, AND TRILATERATION (FIELD WORK)

Section 1. Methods and layout	1
2. Angle measurement	20
3. Measurement of distances	36

CHAPTER 2. COMPUTATION OF TRIANGULATION, TRAVERSE, AND TRILATERATION

Section 1. The geodetic reference spheroid	92
2. Computation of a single triangulation chain or traverse	107
3. Computation by variation of coordinates	126
4. Adjustment by conditions	144
5. The adjustment of a large geodetic framework	153
6. Estimates of accuracy	163
7. Change of spheroid	177
8. Computation in plane rectangular coordinates	182

CHAPTER 3. HEIGHTS ABOVE SEA-LEVEL

Section 1. Introductory	197
2. Spirit levelling	203
3. Heights by vertical angles, and refraction	228
4. Mean sea-level and the tides	243

CHAPTER 4. GEODETIC ASTRONOMY

Section 1. Introductory	254
2. Latitude	286
3. Longitude. Meridian transits	299
4. Time and latitude. Simultaneous observations	316
5. Azimuth	330
6. Geoidal sections	342
7. Gyro-theodolites	347

CHAPTER 5. GRAVITY OBSERVATIONS

Section 1. Introductory and absolute values of gravity	356
2. Pendulum observations	364
3. Gravimeters	373

CHAPTER 6. PHYSICAL GEODESY

Introduction	386
Section 1. The earth's internal structure	389
2. The attraction of standard bodies	396
3. Green's and Clairaut's theorems	402

4. The variation of attraction and gravity with height 420
5. Reference spheroids and standard gravity formulae 425
6. Stokes's integral 430
7. Masses outside the geoid. Gravity anomalies 442
8. Stokes's integral applied to the ground surface 456
9. The deviation of the vertical 464
10. Geoid surveys 470
11. Gravity as a guide to internal densities 478
12. Crustal movements 491
13. Earth tides 497

CHAPTER 7. ARTIFICIAL SATELLITES
Section 1. Introductory 504
2. Atmospheric refraction 516
3. Zero-order and resonant harmonics 528
4. Satellite photography. Astro-triangulation 537
5. Satellite photography. Dynamic use 562
6. Measurement of satellite distances 575
7. Doppler 584
8. Satellite altimetry 615
9. Minitrack and very long baseline interferometry 620
10. Synthesis. Geodetic results 627

APPENDIX A. The geometry of the spheroid 643
B. Matrix algebra 650
C. Cartesian coordinates in three dimensions 678
D. Theory of errors 691
E. Vector algebra 740
F. Complex numbers and conformal mapping 757
G. Modulated waves and tellurometer ground swing 764
H. Spherical harmonics 775
I. Rotating axes. Coriolis force 790
J. Gravity reduction tables 794

BIBLIOGRAPHY 800

INDEX 829

1

TRIANGULATION, TRAVERSE, AND TRILATERATION (FIELD WORK)

SECTION 1. METHODS AND LAYOUT

1.00. Objects and definitions

THE objects of a geodetic framework are as follows.

(*a*) To constitute the main framework on which less precise observations may be based, which in turn may form a basis for topographical and cadastral maps.

(*b*) In combination with observations of latitude, longitude, and gravity to assist in determining the size and shape of the earth, and the form of its external equipotential surfaces.

(*c*) To detect and record movements of the earth's crust.

Of these objects the first is a practical necessity. When it is at all possible the surveyor likes to work from the whole to the part. He does not like to start making local surveys of disconnected areas, hoping that it may later be possible to join them together satisfactorily. A small island of a few hundred square kilometres can of course be accurately surveyed without a primary framework, but any continuous mapping of a large area without one will result in confusion. It is often difficult to justify expenditure on the purely scientific aspects of geodesy, and it is fortunate that the most expensive geodetic process has been justifiable as necessary to the surveyor.

The main framework of a country, if it conforms to the usual standards, has been described as *primary* or *first order*, while triangles or traverses fixing the numerous points required for detailed survey constitute the *tertiary, third order*, or *topographical* framework. *Secondary* may intervene. Some modern methods such as special traverses, § 1.06 (*c*), or satellite surveys, § 1.09, more accurate than classical triangulation, may in future be described as of *zero* order, and will control the first-order framework.

The zero-order and primary framework, together with such secondary as may be suitable for extending the main framework or contributing to its strength, is known as *geodetic*. About half the world's habitable land areas are now covered by a good geodetic framework.

1.01. Accuracy required

The geodetic framework is generally made as accurate as is reasonably possible. For making topographical maps and property surveys an error of scale of as much as one part in 5000 may be admissible, but a much higher over-all accuracy must be achieved in the framework if this error is never to be locally exceeded, especially if the work is to be used for mapping purposes before it has been adjusted as a whole. With this in mind it is not unreasonable to aim at determining the sides of all main triangles and traverse legs with an error which will seldom exceed one part in 50 000 or 20 parts per million (20 ppm).

The principal scientific objective is to measure the departures of the earth's sea-level figure from the exact spheroidal form, such departures seldom amounting to more than ±60 metres, or 10 ppm of the earth's radius. To delineate them approximately demands an over-all accuracy of 2 ppm in surveys of continental extent, and this in turn demands a random accuracy of (say) 20 ppm in each triangle or traverse leg. At present (1978) the question now is to what extent satellite fixes can assist classical methods, or indeed take over the task from them.

In primary triangulation the average triangular error has generally been between 0″.3 and 1″.0. A widely accepted formula for the desirable standard (root-mean-square) error in the relative position of two geodetic points S km apart, after adjustment, is one part in $100\,000\sqrt{(S/30)}$, or $55S^{-\frac{1}{2}}$ parts per million (ppm), [74].

One way of deciding on what accuracy is necessary in the geodetic framework is to hope that no ordinary user will be able to detect, or suffer from, its errors. It is relevant to remark that many users of geodetic frameworks (topographical, cadastral, and engineering surveyors) have EDM instruments at their disposal whose accuracy is better than one part in 50 000, except in very short lines. It is a matter of some inconvenience if their surveys are locally more accurate than the national framework to which their work is supposed to be adjusted. The decision lies with the government, or other authority, that pays for the work. The task of the geodesist is to do the best he can with the funds provided, and to advise what gain or loss may arise from an increase or decrease in their amount.

The adjustment of the geodetic framework, and later that of all subsidiary control, is most important. Until the work has been adjusted the prescribed standard errors are locally liable to be grossly exceeded. During the last 20 years, adjustment difficulties have largely been overcome, but adjustment necessarily has to be postponed until at least the greater part of the observations concerned have been completed.

1.02. The layout of a geodetic framework

This may take several forms, as below.

(*a*) Classical triangulation. §§ 1.03–1.05.
(*b*) A system of traverses, § 1.06.
(*c*) Trilateration, § 1.07.
(*d*) Satellite fixes, §§ 1.08 and 1.09.

These different types of framework can of course be used in combination.

1.03. Classical triangulation

(*a*) The layout of a triangulation system may either be a continuous net as in Fig. 1.3, or a system of chains as in Fig. 1.1. The advantages of a net are that it makes less demand on the topographical triangulators, and that provided the adjustment can be satisfactorily carried out (Chapter 2, Sections 3, 4, and 5) it gives the more accurate result.

The advantages of a system of chains are that except in a very small area it costs less, and that it is easier to adjust. It is also quicker to observe and compute, so that adjusted results may be obtainable before much topographical work has been carried out on a temporary basis.

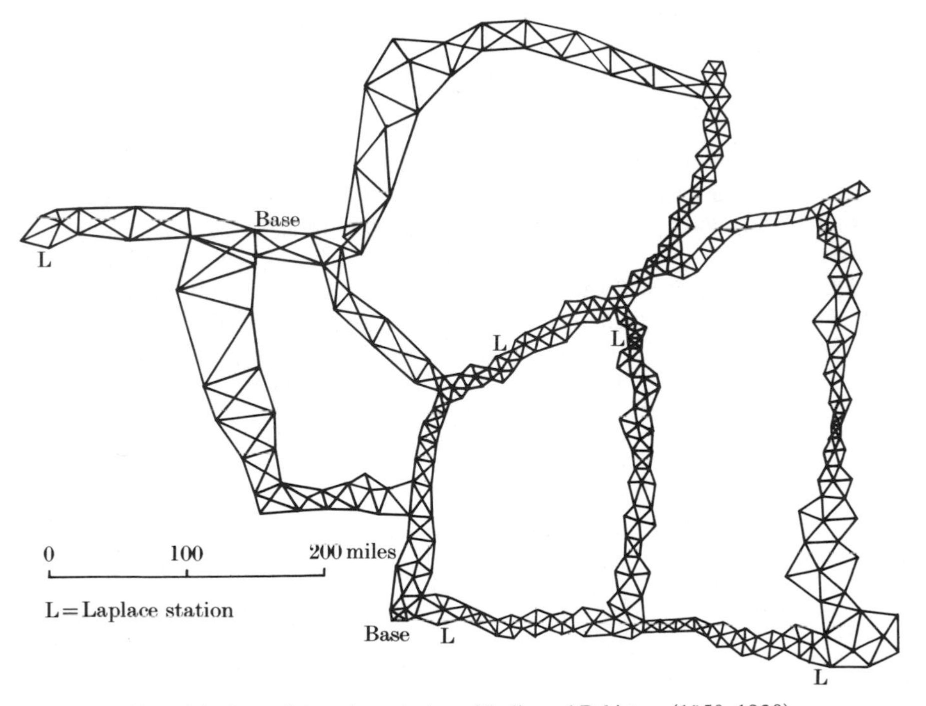

FIG. 1.1. Part of the triangulation of India and Pakistan (1850–1930).

These advantages apply with added force to a system of EDM traverses.

The weakness of triangulation is that it tends to accumulate errors of scale and azimuth, since each side derives its scale and azimuth from the preceding side. Consequently, although a single base might in principle suffice for computation, the accumulation of error must be controlled by other bases, and the error in azimuth must be controlled by astronomical observations for azimuth (and longitude).† Stations at which azimuth is so controlled are known as *Laplace stations*, § 2.05.

(*b*) *The system of chains* was originally adopted by most large survey organizations, and still forms the basis of most of the world's mapping. Apart from all other considerations, a large continuous net was impossible to adjust until after 1950. In Europe, for instance, the 1950 adjustment was mostly based on chains, even where nets existed. The readjustment, for which preparations began in 1962, is still not quite complete in 1978. The system of chains is in principle obsolete, but some description of it is still necessary.

(*c*) *Spacing of chains, bases, and Laplace stations. Historical.* The adopted intervals between chains have depended on the following.

(i) The possibility of simultaneous adjustment. The accuracy of primary observation is lost if it cannot be incorporated in the adjustment, either because it cannot be finished in time or because its inclusion would make the adjustment too complicated.

(ii) The economics of secondary or tertiary breakdown.

(iii) The geography of the region.

(iv) The location of suitable sites (in the past) for invar bases.

In India, where the aim was to produce a 1:63360 map, the geodetic chains form rectangles of about 150–200 km by 250–500 km, and the tertiary triangulation has generally filled the enclosed areas with errors of less than 15 metres without any elaborate system of secondary chains. In the United States, primary chains were originally rather less widely spaced in the east, and more widely in the west, but the intervals have since been reduced to little more than 100 by 100 km.

In the nineteenth century, bases and Laplace azimuths were difficult to observe. The intervals between them were therefore large, and great care was taken over the intervening triangulation. In India bases were 700–1200 km apart, while no Laplace stations were used in the adjustment of

† A simple astronomical azimuth is spoiled by the fact that the local vertical, on which the levelling of the instrument depends, does not coincide with the normal to the spheroid of reference on which the computations are based. Astronomical longitude observations determine the angle between the local vertical and the spheroidal normal, and so enable the azimuth observation to be corrected. See §§ 2.05 and 4.42. An azimuth which has been so corrected is known as a *Laplace azimuth*.

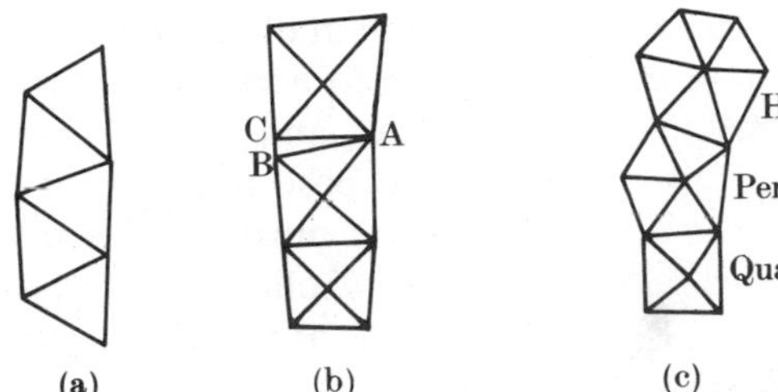

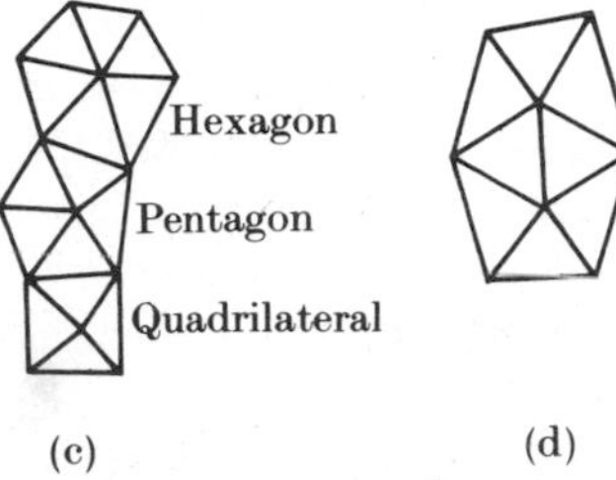

FIG. 1.2. (a) Simple triangles. (b) Braced quadrilaterals. (c) Centred figures. (d) Double-centred figures.

1880. In the United States, where adjustment was postponed until 1925–30, the intervals between bases averaged 400 km, and between Laplace stations 250 km. At that time these were reasonable figures, but by modern standards they were much too sparse.

A chain of triangles may consist of *figures*† of many different types, such as simple triangles, Fig. 1.2(a), braced quadrilaterals, Fig. 1.2(b), centred quadrilaterals, pentagons, etc., Fig. 1.2(c), or more complex figures, Fig. 1.2(d). Even more complex figures may occur at the junction of two chains or for the extension of a base, as in Figs. 1.1 and 2.17.

In hilly country braced quadrilaterals are generally the best, although centred figures are qute satisfactory. In flat country the long diagonals of quadrilaterals are difficult to observe, and centred figures have been preferred. Occasional simple triangles are admissible where difficult to avoid.

Ideally all figures should be regular, or perhaps a little elongated along the chain. Irregularity causes more or less rapid increase in scale error, which will demand closer base control. The point is that no angle should be small which will fall opposite the known side in the course of computation, by the strongest route, from either end of the chain. For if it does an unstable cosecant becomes involved. It is worth noting, however, that an acute angle such as CAB in Fig. 1.2(b) does no great harm, since in the computations it is never opposite the known side.

In most countries the essential angles of simple triangles and of centred figures have been kept above 40°, and those of quadrilaterals above 35°.

Unvisited pivot stations, eccentric stations, and resections of all kinds have generally been avoided in geodetic triangulation, although eccentric stations may be unavoidable when high towers or beacons are used. The three angles of all triangles have ordinarily been observed.

(*d*) *Continuous nets.* A triangulation net must at least comprise a continuous net of simple triangles, like that in Fig. 1.3, and the three

† A *figure* is a group of triangles such that any figure has one side, and only one, common to each of the preceding and following figures.

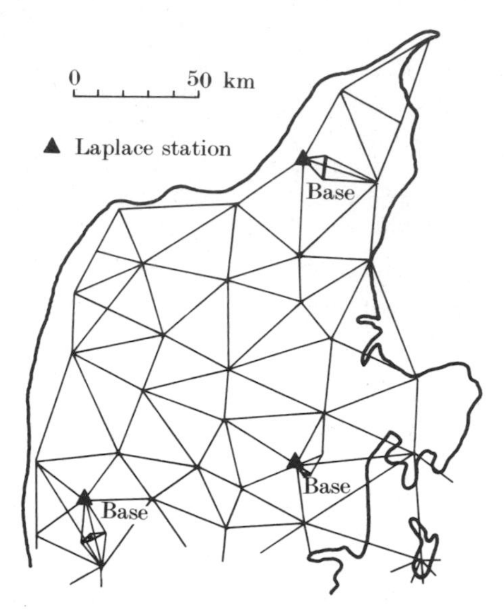

FIG. 1.3. Part of the triangulation of Denmark, as in 1930. The length of each line has since been measured by EDM.

angles of all the triangles should in general be observed. In addition, extra lines may be observed, either a few as in Fig. 1.3 or many. At every station, observations may well be made to all others visible from it. Extra lines give extra strength provided that they are not unduly long.

It is, of course, desirable for triangles to be equilateral. Variations of side length will be unavoidable as between flat and hilly country, but abrupt changes should be avoided as far as possible.

1.04. Length of triangulation lines

In hills a good length is about 50 km, although lines as long as 150 km have been observed. In plains they may have to average as little as 15 km if high towers are difficult.

Long lines make for rapid progress, but poor weather may cause prolonged delays, and unless helicopters or radio-telephones are available, misunderstandings between observers and lamp men may be difficult to remedy. Long lines make for maximum accuracy if light can be trusted to curve only in a vertical plane, but its failure to do so is one of the most serious sources of error, and triangles with unusually long sides are not likely to be the most accurate. See [436], p. 36. Long sides also increase the inaccuracy of base extensions. Shorter sides make things easier for the tertiary triangulators, but this point should not be given undue weight, since they can always connect to a long primary side by running a chain of small triangles along it.

Grazing lines. The decrease of air-density with height causes light to be curved in a vertical plane. Lateral refraction, or curvature in a horizontal plane, will similarly occur when conditions differ on the two sides of a line, and a line grazing close to the ground, particularly to ground sloping across the line, will be liable to such disturbance. No permissible tolerance can be quoted. It can only be said that grazes should be avoided as far as possible. To observe lines which are only clear at the hours of high refraction is to risk trouble, although such lines have been satisfactorily observed. See also §§ 1.19 and 3.22.

In hills serious grazes can generally be avoided, but in flat country they cannot, and their effects can only be minimized by short lines and high towers. In any case accuracy is likely to fall off, and closer base and Laplace control may be necessary. In modern work EDM traverse is of course the obvious substitute for triangulation in flat ground. Lines with sea on one side and land on the other for long distances, or between desert and vegetation, are similarly susceptible to lateral refraction, even though well clear of the ground.

If possible, triangulation stations should be on the highest point of a hill, and preferably on the highest hill of a group. Where features are very large, the latter may be impracticable, but as a general rule a primary station should be on the highest point within 5 or 10 km. This will minimize the risk of grazing lines.

1.05. The strengthening of old triangulation

The spacing of invar baselines in classical frameworks was far too great. Suitable sites were difficult to find, or not existent, and the extension of a short baseline to a triangulation side of normal length was a bad source of error.

EDM instruments make it easy to measure an unlimited number of long triangulation sides, and any new triangulation that may be observed will, of course, take advantage of that. In principle, existing triangulation can be strengthened by the introduction of numerous EDM lines, but unless they are extremely numerous it is essential to prove that the old station marks have not moved by as much as a few centimetres, since the measurement of an isolated line whose ends have moved will impose a common error of scale on a large area of the old work. Reasonable proof, or disproof, of stability may be got by measuring three or more lines radiating from a central station, or by measuring all the lines of an old figure. But if this is to be done at all frequently it may be better, and little more expensive, to run an EDM traverse along one or two selected lines passing right through the old triangulation. In some countries every old triangulation line has been remeasured by EDM.

Improvement of scale in this way is to some extent wasted effort unless new Laplace stations are also observed, with the same proof of stability, at suitable intervals.

The question of how many EDM lines and Laplace azimuths can usefully be added to existing triangulation is one that can best be answered by strength analysis. See further in § 2.48.

1.06. Traverses, using electromagnetic distance measurement (EDM)†

(*a*) A system of EDM traverses has superseded triangulation as the recommended form of primary framework in large countries where none yet exists. Compared with triangulation, its advantages are as follows.

(i) The ease with which a single line of easily accessible stations can be selected, compared with the difficulty of laying out a chain of well-conditioned triangulation figures.

(ii) Similarly, the ease of organizing observations to only two stations at a time, all on the same line of communication, instead of to six or more stations scattered over an area.

(iii) A radio-telephone is part of the equipment.

(iv) In a traverse there is no accumulation of scale error.

(v) A system of traverses is easier to adjust than a net or system of triangulation chains.

(vi) The speed with which a traverse framework can be completed decreases the amount of topographical work which has to be provisionally published before the primary framework has been adjusted.

The *spacing* between primary traverse lines will be governed by the considerations listed in § 1.03 (*c*) (i) to (iii). If similar traverses are used for the secondary breakdown, as they may well be, the spacing of those described as primary will depend mostly on how long the adjustment can be postponed, and on how complicated a system can be adjusted simultaneously. Fig. 1.4 shows the primary traverse framework of Australia as adjusted in 1966.

The *lengths of traverse legs* will, of course, depend on the ground. In flat country they may average 10–20 km, using 8- to 15-metre portable observing towers. In hills it may be convenient to average not more than 40 km.

† EDM is electromagnetic distance measurement, in general. MDM is microwave distance measurement, as has been associated with the name Tellurometer, with wavelengths of between, say 10 cm and 8 mm. EODM is electro-optical distance measurement, using light (including infra-red). Associated with the name Geodimeter, but not confined to it. In general it is more accurate than MDM, see §§ 1.47 and 2.44.

FIG. 1.4. The MDM tellurometer traverses of Australia, 1966. Triangulation chains are not shown.

Laplace stations. It is rational to specify that the freedom from accumulation of scale error must be matched by similar freedom from azimuth error, as can be ensured by the provision of frequent Laplace stations. A Laplace azimuth at every alternate traverse station will give a direct determination of the azimuth of every traverse leg, and this is a reasonable aim, although it is no disaster if bad weather causes some azimuths to be lost.

In countries where the weather makes regular astronomical observations difficult, a Laplace azimuth at every fourth or even sixth station may have to suffice, provided the traverse angles are observed with extra accuracy. But where the sky is generally clear, nothing prevents the inclusion of Polaris or some other suitable star in every round of traverse angles, and the longitude observations are the only additional cost. The more numerous the Laplace azimuths are, the less accurate does each have to be, provided that systematic error is avoided. The necessary accuracy of the traverse angles is also reduced. See [101]. If latitude is observed as well as longitude, Laplace observations at every station will provide the geoid profile, § 4.50, which (at least along selected lines) is necessary in order that all measured distances may be reduced to spheroid level, § 1.30 (*f*).

For recommended EDM programmes see §§ 1.39 and 1.43; for horizontal angles see § 1.18; and for the Laplace azimuths see Chapter 4, Section 5.

Current practice is that traverse lines should roughly form squares or rectangles, as has been usual for triangulation chains, but there is no obvious reason why they should not be laid out in the form of a net of roughly equilateral triangles. If traverses were approximately straight, this would give the net some stiffness quite apart from Laplace azimuths and accurate traverse angles. The ratio of distance traversed to area covered, coupled with a specified maximum diatance between any internal point and the nearest traverse, is the same for the two layouts.

(*b*) *MDM traverses.* In Australia the primary framework (of 1966) consists almost entirely of a net of MDM (tellurometer) traverses, as shown in Fig. 1.4. For the routine of observation, see § 1.46, and for accuracy achieved see §§ 2.44 and 2.47. For the relative advantages of EODM and MDM, see § 1.47. The laser geodimeter was not available when this survey was observed. The geodetic framework of Saudi Arabia is similar, [47].

Laplace azimuths were observed on average at every fourth station.

(*c*) *Zero-order EODM* (*Geodimeter*) *traverse.* Figs. 1.5 and 1.6 illustrate a system of traverse now being adopted by the USC & GS as zero-order control, [405] and [406].

Traverse legs 8–15 km long are measured by geodimeter. As a check, each leg is duplicated, alternate stations A, D, G,... being common to both measures, while the intervening stations BC, EF,... are in pairs separated by about 25 metres. Traverse angles are measured along both sets of lines at all stations, including both B and C, etc.

The short distances BC, EF,... are taped, and the angles ABC, DEF,... are also measured, perhaps only to the nearest 10″, because of the short lengths of BC, EF,... . The values of BC and the angle ABC, taken with any approximate values of AB and BD, then give accurate values for the angles BAC and BDC, and also of the differences AB minus AC and BD minus CD. The traverse angles at A, D, G,... and all the geodimeter distances are thus duplicated and if the comparison is bad the observations can be repeated, or the stations can be re-sited for better meteorological conditions.

Azimuths, latitudes, and longitudes are observed at all the single stations A, D, G,... and provide Laplace azimuths, and deviations of the vertical every 16–30 km for the geoid profile.

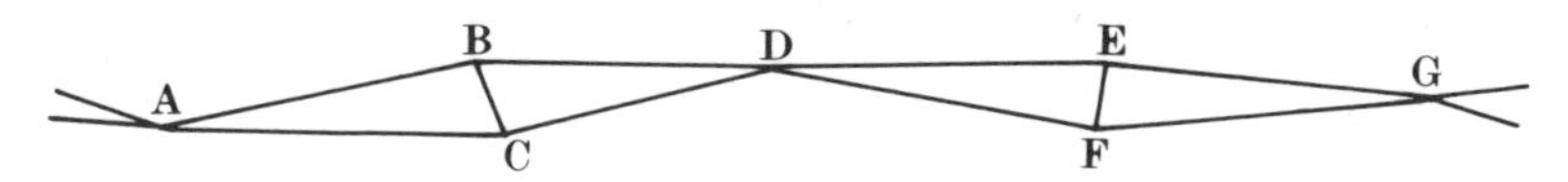

FIG. 1.5. Zero-order geodimeter traverse. The lengths of BC and EF are exaggerated.

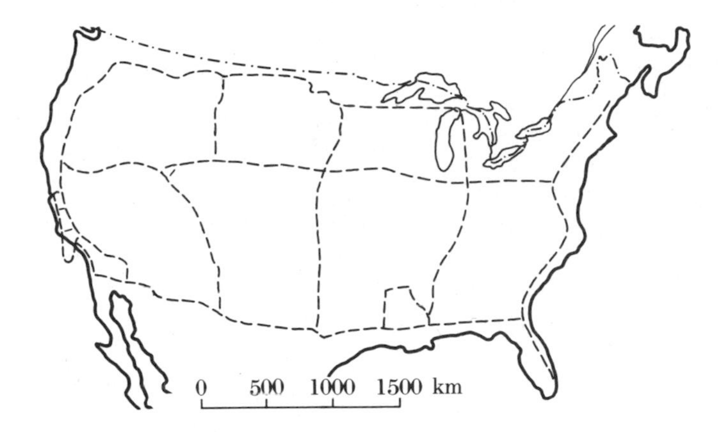

FIG. 1.6. United States. Zero-order traverses, (1974).

[405] gives detailed specifications and some results. Accuracy over such distances as 1000 km is believed to be ±1 ppm (s.e.). See § 2.44(*b*) and (*c*)(iv).

[455] describes a 890 km traverse in Finland, observed with a model 8-laser geodimeter for scaling the stellar triangulation, § 7.29. It was nearly straight with 25 legs. The over-all error of length is estimated as 0.135 ppm.

If the accuracy of the geodimeter should presently reach as small a figure as ±0.3 ppm, far exceeding the accuracy of angular measurement, a very strong framework will be obtained by laying out nearly straight traverses to form a traverse trilateration, as suggested in § 1.06 (*a*), whose form will be insensitive to any angular errors, while its orientation will be given by the mean of all the Laplace azimuths included in the system.

1.07. Trilateration

(*a*) *EDM trilateration.* The ease and accuracy of EDM, especially of the laser geodimeter, invites the substitution of trilateration for triangulation. A continuous net of EODM lines would indeed be very strong by all modern standards. So far as its internal control is concerned, it would only derive appreciable extra strength from Laplace stations if it extended over very large areas, although in the absence of satellite control it would need Laplace azimuths, possibly a large number of them, to give a correct orientation to the net as a whole.

If it is practicable to adjust it as a whole, an EODM trilateration is probably the most accurate form of framework.

Trilateration is not recommended as a substitute for triangulation if a system of chains is required. A narrow trilateration chain rapidly looses direction and requires close azimuth control.

(*b*) *Shoran and Hiran trilateration.* Between 1945 and 1960 Shoran and Hiran were systems of radar trilateration between ground stations up to 500 or even 800 km apart, observations being made from the ground to an aircraft flying across the middle point of the line. Accuracy was of the order 1 in 10^5. This system enables latitude and longitude to be carried across seas of up to 800 km in width as in Fig. 1.7, but the third coordinate (height above the spheroid) has to be determined by other means. It has been widely used over undeveloped country in northern Canada, and as a connection between Europe and North America via Greenland, and around the Caribbean Sea and adjacent parts of South America. It has now been superseded by satellite observations whose accuracy and convenience are greater, and which also provide the third coordinate.

For a general description of Shoran and Hiran see [113], 3 Edn, pp. 94–102. For more details see [499]. For reports on the Norway–Greenland tie see [28].

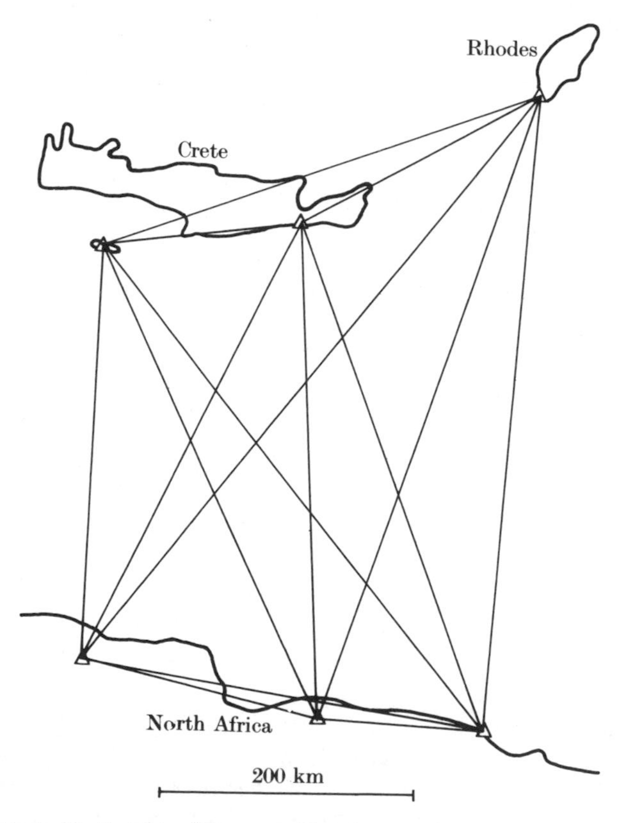

Fig. 1.7. Crete to North Africa Hiran net. The three stations on each side of the sea gap are also mutually joined by conventional triangulation.

1.08. International and intercontinental connections

Every effort must be made to connect with all adjacent primary frameworks. Conventional connections across land can hardly ever be impossible, except for political reasons. It may be assumed that every separate survey contains its own baselines and Laplace stations, so a connection between two surveys only demands (as a minimum) the existence of one common station whose three coordinates (x, y, z, or latitude, longitude, and height above the spheroid) have been determined by both parties. See §§ 2.50–2.53.

Connections across the sea or other obstacle may be carried out by any of the following methods.

(*a*) A sea crossing between points which are high enough to be intervisible may be made by one or more lines whose azimuths and lengths are directly measured, provided the distance does not exceed the extreme range of the EDM employed. With the tellurometer this is about 120 km, but the distance can be increased by carrying two tellurometers in a ship which passes back and forth across the middle of the line, making frequent measures of distance to the points on either side. A minimum sum can then be used for the derivation of the total distance.

(*b*) By Shoran nets, § 1.07 (*b*).

(*c*) By theodolite observations to flares dropped from an aircraft [525], or by astro-triangulation using balloons, § 7.29, [321], and [51].

(*d*) By satellite observations.

So far as intercontinental connections are concerned, satellite observations have already practically solved the problem. They have the great advantage that some satellite methods, or combinations of methods, give the positions of survey stations relative to the earth's centre of mass, the geocentre. The interconnection of national frameworks by classical methods is still of course a necessary task, but as satellite methods become more accurate, they will be used for the connection of progressively smaller areas of survey, and even for the internal control of individual national frameworks, as in § 1.09.

See §§ 7.77 and 7.78, and Tables 7.5 and 7.6, and the references there quoted for the unification of national and continental surveys, as it stands in 1977. See §§ 2.50–2.54 for formulae for converting surveys from one national datum to another.

1.09. Satellite fixes as a geodetic framework

It is now beginning to be possible to use satellite fixes as the main internal primary framework of national surveys of quite moderate (not too small) size. The system which (in 1978) appears to be the most suitable for this task is the doppler system, described in §§ 7.46–7.63.

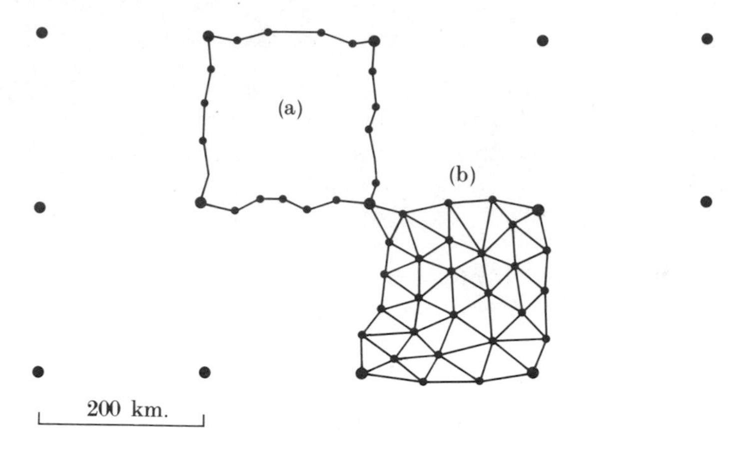

FIG. 1.8. Doppler control points. Connected by EDM traverse in (a). Infilled by EDM trilateration in (b).

This system gives the geocentric coordinates of any ground point by a few days observations with a relatively inexpensive portable instrument, operated by one or two men, quite independently of any other organization except for the basic data (satellite ephemeris) supplied by the satellite tracking organization based on the United States. From 1984 the doppler system is likely to be replaced by the *global positioning system* (Navstar), § 7.44.

It appears probable that doppler fixes made at the corners of (say) a 200×200 km square, as in Fig. 1.8, will give positions in a geocentric framework, which will be relatively correct to within a standard deviation (s.d.) of ± 1 metre. This is equivalent to an accuracy of 1 part in 200 000 in their differences of latitude or longitude, or of 1″ in the mutual azimuth. This may be said to match current geodetic standards over the same distances. It is further to be hoped that the errors in adjacent 200×200 km squares will be independent, so that higher proportional accuracy may be attainable over long distances.†

If it is presently really established that Doppler or other satellite fixes 200 km apart are relatively correct to ± 1 metre, they might be accepted as correct for most practical purposes without further adjustment. Occasional less accurate fixes would be detected by the secondary breakdown within each 200-km square. Alternatively, in a country which was not too large, the secondary breakdown and the doppler fixes could be contained in a single adjustment, in which the doppler fixes would be given high, but

† In 1978 there is still some doubt, § 7.63 (*d*), about the scale and orientation of the geocentric framework in which the doppler satellites' ephemeris is given. Doubts on this account may amount to 1 ppm or 0″.2 over long distances.

not infinite, weight. In either case there would seldom be need for Laplace azimuths, nor generally for geoid sections (§ 4.50) between the doppler fixes.

Fig 1.8 suggests two methods for basing a primary framework on doppler fixes. In Fig. 1.8(a) the doppler fixes are connected by primary traverses which are adjusted on to or with the doppler. The infilling of each 200-km square is then isolated from that of adjacent squares and can be adjusted separately by itself. In Fig. 1.8(b) each 200-km square is filled in by EDM trilateration. This may be very strong, but near points in adjacent squares may not be adequately compatible with each other, unless it has been possible to adjust all the squares together simultaneously

1.10. Secondary breakdown

§ 1.00 defines the secondary framework as being of geodetic significance if it contributes to the strength of the primary framework. For this the following conditions are necessary.

(*a*) An adequate standard of accuracy. Most EDM lines would pass this test.

(*b*) It must be completed in time to be included in the adjustment of the primary framework.

(*c*) It must be possible to include it without the adjustment becoming unduly complicated. Modern computers have a great capacity, but there is a limit. If it were possible to include all the so-called secondary work, there would then be the question whether tertiary EDM lines should be included. The process has to stop somewhere.

A possible, but rather strict, rule for the required accuracy of secondary breakdown, is that the standard error of any secondary station relative to the nearest primary should be the same in linear units (not ppm) as the standard error of one primary station relative to adjacent primary stations. This results in the latitudes and longitudes of secondary stations being about as well determined as those of the adjacent primaries. On the shorter distances involved a reduced accuracy in ppm or seconds of arc would of course be allowable.

The methods employed for secondary breakdown may be (*a*) triangulation, or (*b*) EDM trilateration, or (*c*) EDM traverses, or (*d*) aerodist, see § 1.48. When it is impossible or inconvenient to adjust the secondary concurrently with the primary, it has to be adjusted later with all the connected primary stations held fixed. There is then some risk of unwelcome relative error between near stations in separately adjusted secondary blocks. This risk constitutes some justification of the high standards of accuracy usually specified.

Unvisited points. Geodetic triangulation or traverse parties cannot usefully spend their time fixing unvisited points as an aid to mapping, but it may be worthwhile for them to fix some of the following. (*a*) Existing topographical stations or points. (*b*) A number of sharply defined points, if any such are likely to be more permanent than the best stations. (*c*) Distant peaks of special geographical interest, if they are too distant to be fixed by topographical methods. (*d*) Easily accessible sites at which astronomical observations for the deviation of the vertical may be made.

1.11. Reconnaissance

Careful reconnaissance is essential before observations begin, although it is much easier for a traverse than for triangulation or trilateration. It may take three forms.

(*a*) *Examination of maps*, if they exist. In easy country it may be possible to decide on the layout with 95 per cent certainty without visiting the ground. The following rule is a useful test for intervisibility. If A and B are two stations of heights h_1 and h_2 metres, the height above sea-level, h, of a point on the ray of light between them at a distance of d_1 km from A and d_2 from B, allowing for earth's curvature and normal refraction ($k = 0.075$), is given by

$$h = h_1 + (h_2 - h_1)\frac{d_1}{d_1 + d_2} - 0.067 d_1 d_2 \text{ metres.} \tag{1.1}$$

In foot-mile units the factor 0.067 becomes 0.57. For microwaves the term $0.067 d_1 d_2$ (metres) becomes $0.059 d_1 d_2$, corresponding to $k = 0.125$, § 1.35 (g).

(*b*) *Aerial reconnaissance.* This is the natural method in difficult country. A first reconnaissance can be made by light aeroplane, but if it is necessary to land on or near proposed stations to test intervisibility or to ascertain access routes a helicopter will be required.

(*c*) *Ground reconnaissance.* In unsurveyed and inaccessible country, without aircraft, the ground reconnaissance of a triangulation chain has sometimes been a difficult matter. A traverse line is much easier provided it follows a passable route reasonably closely. See §§ 1.37 and 1.44 for the characteristics of a good tellurometer line, and § 1.37 for the geodimeter. The demands of the EDM and of the traverse angles may conflict.

(*d*) *Advanced party.* Especially when the reconnaissance has been by map or air, triangulators have to be preceded by an advanced party to clear trees, build stations, and post lamp or helio squads. This may not be necessary for a traverse; see § 1.46. Stations must be constructed to last for 100 years.

(*e*) *Connection with old work.* If triangulation is to derive scale or azimuth from an existing pair of stations, they must not only be accurately identified, but there must be certainty that they have not moved. A shift of 0.3 m in a 30-km side would be regrettable, and a few centimetres is the standard to be aimed at. Hidden witness marks provide some check, but not necessarily against earthquakes, the bodily sliding of hill tops, or against creep in alluvial areas. The usual check is to connect with three old stations, instead of the minimum two, and to re-observe two of the angles formed by them. If these agree with their old values within a second or less, serious error is unlikely. But when several decades have elapsed since the old stations were fixed, it is essential to include measured lines and Laplace stations in the new work, and to compute in terms of them. Since the introduction of EDM, measured lines are sure to be included in the new work in any case.

1.12 Station building

Essentials of good construction are as follows.

(*a*) A distinctive, and so far as possible indestructible, mark at or below ground level. This may be an inscribed brass plug, but where brass is liable to be stolen, a mark cut on rock or a large stone is better. Two marks should be provided, one visible on the surface and one buried vertically below.

(*b*) Two or three hidden witness marks, whose distances and bearings from the centre and from each other are recorded. They may be of brass or stone, differently inscribed, or they may be glass bottles.

(*c*) Provision for plumbing over the ground-level mark. Tilting of a tower or pillar may shift an upper mark laterally.

(*d*) Except when the instrument tripod stands directly on solid rock, the structure supporting it must make no direct contact with that on which the observer stands. See Fig. 1.9.

(*e*) An opaque beacon, if used, must be such that no conditions of light or shade will result in its being intersected off centre. Fig. 1.10 shows a good design. The two vanes are at right angles in plan, and each is either all shaded or all in sunlight. The lower vane must be well separated from the upper, so that shade cannot fall on it.

All survey organizations have their own designs for survey pillars and marks, based on local requirements and materials.

Towers. Theodolites, tellurometers, and signals can if necessary be raised to a height by any of the following means.

(*a*) On existing structures such as church towers. Convenient but generally bad, as it is difficult to plumb down to ground level.

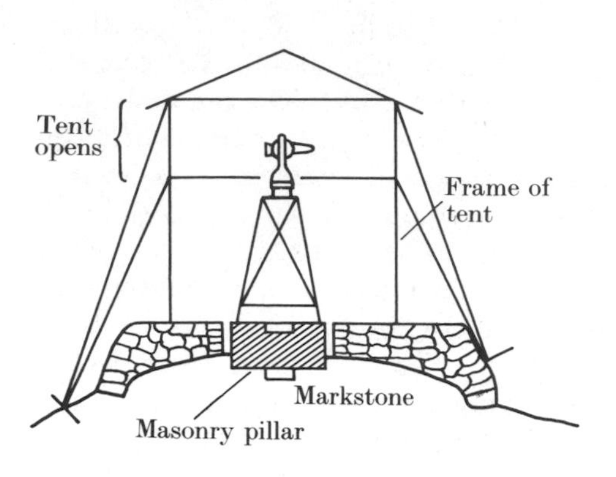

FIG. 1.9. Hill station and tent. The braced tripod may be replaced by a concrete pillar, with holes to permit plumbing over the upper mark.

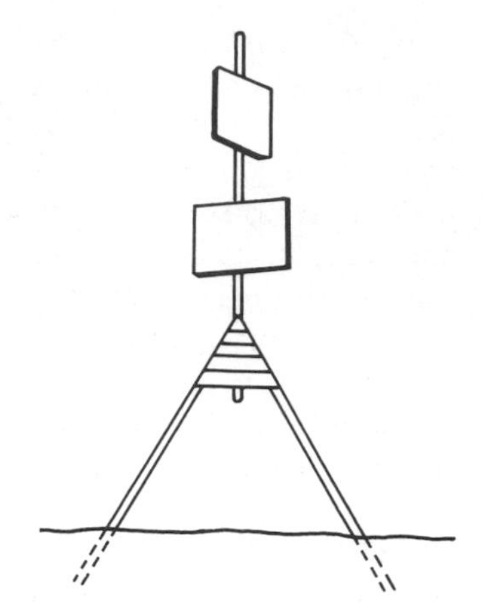

FIG. 1.10. Opaque beacon.

(*b*) Brick or concrete towers, as used in India in the nineteenth century, but now expensive. An inner tower of about 1-metre diameter at the top, with a central 15-cm hole for plumbing, supports the theodolite, while an independent outer tower supports the observer and tent.

(*c*) Improvised wooden structures. [394] describes a tower on which the instrument can be raised 7 metres on a 6-cm pipe with steadying guys, the observer's stand being improvised from local timber.

(*d*) Portable steel towers, which can be moved and erected many times. The *Bilby tower*, [90], can raise the observer and lamp to a height of 30 or even 40 metres, with a beacon 3 metres higher. Weight of a 30-metre tower is 3000 kg, maximum weight of any piece is 20 kg, and the longest piece is 7 metres. Five men can erect in five hours.

The *Lambert tower* [538] consists of two concentric tubes, in 4.5-m (15 foot) lengths, which can be built up to a height of 18 m. The outer tube is about 0.3 m in diameter and is supported by guys. It carries a platform on which the observer stands. The inner tube of diameter 0.1 m ($3\frac{1}{2}$ inches) supports the theodolite or EMD instrument. It is laterally supported by the outer tube by constraints which do not transmit torque. The assembled 18-m tower weighs 550 kg (1200 pounds), transportable as five loads each of 120 kg.

For astronomical azimuth observations, or if ground stations are much elevated above or below the horizon, extra stability is necessary, since varying dislevelment will cause errors as in § 1.15 (vii). It is doubtful whether work of this kind can be done with primary accuracy on high towers.

1.13. Engineering geodesy

Some special engineering surveys such as for the layout of synchrotrons and radio telescopes, or for monitoring the movements of dams or of the strains in the keels and decks of large ships while under construction, demand geodetic accuracy over relative short distances. *Miniature geodesy* may be defined to comprise measurements where an accuracy better than perhaps 10 ppm, or 2 seconds of arc, is required over distances of less than 1 km. These limits are naturally ill-defined.

The most striking differences between miniature geodesy and normal geodesy are as in (*a*) to (*f*) below.

(*a*) Miniature geodesy requires exceptionally precise centring, perhaps to an accuracy of 0.1 mm. Instruments and targets need to be accurately interchangeable.

(*b*) The fine readings required for distance measurements, and the requirement of accurate calibration or the elimination of zero errors.

(*c*) The special role of alignments in both the horizontal and vertical planes.

(*d*) The problem of transferring azimuth down vertical shafts and steep slopes.

(*e*) The use of hydraulic systems as a basis for heights.

(*f*) The use of rectangular coordinates for computations. This is clearly convenient for the recording of horizontal x and y, but the level surface to which spirit levelling and hydraulic heights refer is curved, and necessarily departs† from the x, y-plane.

Miniature geodesy employs many of the instruments and practices of normal geodesy, but its special techniques are not further described here.

References for engineering geodesy. [181] and [484]. Many papers are included in the proceedings of the International Federation of Surveyors, published by the Bureau of the FIG every four years (1973, 1977 etc.)

General references for Section 1

Examples of national primary frameworks are to be seen in the National Reports presented every four years to the International Association of Geodesy. Narrative descriptions of reconnaissance, station building, and the routine of observation are to be found in the annual or other reports of many survey departments, and also in [41], [47], [97], [292], [293], [295], [436], and [589].

† The separation is conventionally $l^2/2R$, where l is distance from the origin (where the plane is tangent to the sphere) and R is the earth's radius. It is 8 cm if $l = 1$ km. The actual datum surface is the geoid whose curvature is variable, but the spherical approximation will generally suffice.

SECTION 2. ANGLE MEASUREMENT

1.14. Theodolites

The design of geodetic theodolites has not changed notably in the last 10 years. The most suitable instruments are still those, such as the Wild T3 or Kern DKM 3, which have a horizontal circle of about 140-mm diameter, and whose circle readings can be recorded to 1″ with estimation to 0.1″. The large Wild T4, with a 250-mm circle is principally used for astronomical observations, § 4.20.

The distinctive features of these instruments are their small size and the optical system by which opposite sides of the circle are brought into coincidence† by a micrometer which can be read as the observer stands in position to look into the telescope.

Details of construction, and of the methods of reading the circles, can be seen in the makers' instruction books. They are not illustrated here.

Horizontal angles measured by the old 36-, 24-, or 12-inch micrometer theodolites between 1800 and 1950 are not necessarily inferior to those measured by modern instruments. In fact, the long time required to complete observations with them probably resulted in better elimination of refraction errors than modern observers have time for.

Stands. If the legs of the tripod ordinarily provided are properly spread, they cannot stand on an isolated pillar, § 1.12, of diameter small enough for the observer to be able to avoid standing on it too. The remedy is either to use a braced wooden stand such as is shown in outline in Fig. 1.9, or to build concrete pillars on which the theodolite can stand directly. [589], pp. 13–17 describes an improvised design. On suitable hard ground work has been done with tripods supported on wooden pegs or metal pipes, driven into the ground [100].

For the use of theodolites for astronomical observations see § 4.20. In this section §§ 1.17–1.25 refer to horizontal angles only, and § 1.26 to vertical angles.

1.15. Theodolite adjustments

Theodolites require adjustments as below.

(*a*) Collimation in azimuth. The line of sight must be approximately perpendicular to the horizontal axis.

(*b*) Adjustment of the upper bubble.

(*c*) The horizontal axis must be approximately perpendicular to the vertical axis.

(*d*) Adjustment of the lower bubble.

† Thereby eliminating error due to eccentricity of the divided circle.

(*e*) Determination of the value of one division of each bubble, if the instrument is to be used for astronomical work.

(*f*) Adjustment of the optical plummet.

(*g*) Verticality of the cross-wires.

(*h*) Micrometer run.

(*i*) Adjustment of the striding level, if required for astronomical work. See §§ 4.12 and 4.26.

These are more or less permanent adjustments. The first three only need attention if abnormal differences occur between face right (FR) and face left (FL). Items (*d*), (*f*), (*g*), and (*h*) are easily tested at each station, but should seldom need change.

The following are necessary at every station or more frequently.

(*j*) Examination for stiff bearings and loose joints.

(*k*) Centring over the station mark.

(*l*) Levelling, i.e. getting the vertical axis vertical.

(*m*) Clear vision of the cross-wires, by focusing the eye-piece.

(*n*) Clear vision of the object and absence of parallax, by focusing the object glass.

(*o*) Clear vision in the circle-reading microscope.

All these adjustments are made in a similar way to those of smaller theodolites, but the following points may be noted.

(i) *Collimation in azimuth.* If the line joining the optical centre of the object glass to the intersection of the cross-wires is off perpendicular to the transit axis by θ'', the error in the direction of an object at elevation α is $\theta'' \sec \alpha$, and in an angle it is $\theta''(\sec \alpha_1 - \sec \alpha_2)$. It is cancelled by change of face, and $\theta = 10''$ is quite harmless.† Rigidity of the various parts of the instrument is more important than perfect adjustment of collimation. This applies also to adjustment of the upper bubble and to dislevelment of the transit axis.

(ii) *Adjustment of the upper bubble,* sometimes known as *collimation in altitude.* The constant error of an observed elevation on one face is the so-called collimation error in altitude. It is cancelled by change of face, and adjustment is only required if differences between FR and FL are annoying, or if the line of sight is seriously off the mechanical axis of the telescope.

† The *line of collimation* is the perpendicular from the optical centre of the object glass to the horizontal axis, the optical centre being the point through which light passes undeviated. The optical axis of the lens must, of course, approximately coincide with the line of collimation, but this is a makers' adjustment. The *collimation error* is the angle θ. The line joining the cross-wires to the optical centre of the object glass is the *line of sight*.

(iii) *Horizontal (or transit) axis.* Some modern theodolites have no provision for this adjustment, and it cannot go wrong unless serious damage is done elsewhere.

(iv) *Bubble scale* values, if required, may be got by calibration against the theodolite's own circle, or from a *bubble tester* at HQ. See § 4.12. The lower bubble should be adjusted so that it lies within one or two divisions of the centre of its run when the axis is vertical.

(v) *Micrometer run.* Optical micrometers are less liable to go out of adjustment than the micrometers of old-fashioned theodolites, but the run from 0″ to 120″, or whatever it may be, must be tested. If it is wrong, its adjustment in the field may not be easy. The effect of errors of run is much reduced by the programme of observation, § 1.18. If the error is more than (say) 2″ an increased number of zero settings may be necessary.

(vi) *Stiff bearings and loose joints.* An important test. Clamp the horizontal circle, intersect some sharp object with the vertical wire, and while looking through the telescope take hold of the tribrach with both hands and try to twist it. The wire will move off the object, but should return most of the way when the twist is relaxed. Only experience can say what pressure is reasonable. Loose foot-screws, or the joints between wood and metal in the stand, are the usual sources of trouble. Foot-screws should be tight in their bearings, and not unscrewed too far. Then, more gently, repeat with the twist applied to the upper part of the theodolite, and finally by applying gentle side pressure to the eye-end of the telescope.

Any unusual stiffness of either the horizontal or vertical axis is of course serious.

(vii) *Levelling.* A levelling error of θ'' will introduce an error of between + and $-\theta'' \tan\beta$ in the horizontal direction of an object of elevation β.† A few seconds off level thus seldom matter except in astronomical work, but more should not be allowed. Change of face does not cancel the error. See also § 4.42 (*c*).

(viii) *Focus.* Eye-piece focus, which is adjusted first, depends on the observer's eye or spectacles. Adjustment of the object glass for clear vision of the object should then eliminate parallax. If the eye persists in preferring a focus which leaves parallax present, test the eye-piece focus again. If the two cannot be reconciled, absence of parallax is generally more important than the clearest vision.

† Let ψ be the inclination of the horizontal axis, positive when its left-hand end is too high. Then $\theta \tan\beta$ (with the correct sign) is to be added to the (clockwise) circle reading of the object. Depressions are negative β.

1.16. Lamps and heliotropes

Signals in current use are as follows.

(*a*) *Heliotropes.* On a bright and clear day a 15-cm helio can be seen at 50 km, but 20 or even 30 cm are needed for very long lines. The sun's diameter being only 30 minutes, a 15-minute error of alignment causes loss of brightness, although the sky near the sun is often bright enough to show when the error is much larger. The helio must thus be carefully aligned in the first instance, and periodically checked, while the mirror needs adjustment every half-minute or so to allow for the sun's movement. A second helio is needed to reflect the sun into the first when its direction is unfavourable. Helios are good objects when the sun shines regularly, and provided they are conscientiously attended to. Unless all lines are very long, some provision for reducing the light will be required, either by stops on the helio, or as in § 1.27 (*f*).

(*b*) *Electric lamps.* A typical make has a 6-watt, 6-volt bulb (1 amp) with a small filament placed in front of a 15 cm reflector to give a concentrated beam (3° divergence). It is lit by an 85 amp-hour accumulator. Lamps must be accurately pointed in the right direction or the light may be intersected off centre.

(*c*) *Opaque beacons.* These require no attention, and may be visible when helios are not, but they are apt to be invisible on long lines, or with a background other than the sky. 'Phase error' is also a danger; see § 1.12 (*e*).

Electric lamps are recommended by night, or in poor daylight on lines that are not too long; 20-cm helios or larger on long lines and 15-cm on short, in bright sunlight. A standing opaque beacon may be useful in addition.

In triangulation a detailed programme is an essential preliminary. The layout having been decided on as in § 1.11, the observer must make out a programme showing day by day the location of the advanced party, of each helio of lamp squad, and of the observing party. Attention must, of course, be given to where each finds its supplies and transport, and how orders are to be communicated. Dates given should be those to be hoped for if all goes well. Bad weather may cause delay, but that will not upset the serial order in which all moves will take place, and the actual dates of moves are regulated by signals at the completion of the observations at each station. Where supplies and transport are easy and with lamp men who can find their own way about, the programme presents no special problems, but in uninhabited country, or with illiterate lamp men, it calls for very careful thought, and the whole success of the work depends on the care with which it has been made.

If lamp men can read Morse, orders can be sent when plans have to be changed, but otherwise only a few distinctive signals can be sent, such as to call for the light, to report completion of observations at a station, and to instruct the lamp man to move to his next pre-arranged point. Lamp men must be able to signal "Seen and understood".

If lines are long and visibility poor, the advance party may have to lay out aiming marks in the correct directions, so that helios and lamps can be aligned correctly when the stations cannot be seen.

1.17. Methods of observation

In *triangulation* horizontal angles can be observed on several systems.

(*a*) *The method of rounds.* Starting with the intersection of one selected station, a flank station if there is one, the theodolite is swung in turn on to each of the others, right round the horizon until the first has been intersected again. This constitutes one round.

(*b*) *The method of directions.* One station is selected as a reference mark (RM), preferably that most likely to be continuouly visible. Angles are then independently measured between this RM and every other station. For a given degree of precision this involves nearly twice as much work as the method of rounds, but if stations are only intermittently visible it may be the only practicable system. If no station is fairly continuously visible, two stations may be selected as alternative RMs. The angle between them must, of course, be strongly measured.

(*c*) *The method of angles.* The angles between adjacent stations in a round are independently measured. This also involves nearly twice as much work as method (*a*), in spite of which large angles (the sum of two or three smaller) may be less precisely measured, since they have to be deduced by summing the independently measured smaller ones. But see § 2.43 (*e*).

(*d*) *Schreiber's method.*† The angle between each station and every other is independently measured. This again involves double work, but there is some choice in the order in which observations are made, and work can proceed in worse weather than is required for rounds. It has been much used in Europe and Africa. [123] gives details and suggests a modification.

If conditions admit, the method of rounds is advised. Directions or Schreiber's method may be used when reasonably uninterrupted rounds are impossible. Independent angles are only advised for additional measures of any angle for which extra accuracy may be required. Also see end of § 1.19. The system of measuring angles by *repetition*, swinging the

† This is not to be confused with Schreiber's method of forming normal equations in § 2.25.

lower and upper plates backwards and forwards, is not generally recommended.

The closing of rounds. When there are more than two or three directions in a round of angles, it is advisable to close the round on the starting station, although some authorities only specify closure at a central (as opposed to a flank) station. If a misclosure is abnormally large, the round should be rejected and repeated.

Broken rounds. A helio or lamp is often not visible when required, and it may have to be missed out. Then, when it shows again, the missing direction can be made good by observing the angle between it and *one* other station, preferably adjacent to it. Combining this measure with the circle reading of the other station in the broken round then enables an entry to be made (in brackets or otherwise distinguished) in the abstract form, to replace the missing direction. This practice ensures that there will be no trouble over station adjustment, § 1.23.

Traverses. In a traverse there will generally be only two directions to be observed. Closing the round will add 50 per cent to the number of observations, and there is little advantage in measuring one direction twice as often as the other. But when observing (say) at B to A and C, the theodolite should swing clockwise from A to C as often as from C to A, and the same anti-clockwise. The angles on both sides of the traverse will thus be equally strongly measured. If they do not total 360° the error will be visible, and can be adjusted.

1.18. Number of measures of each angle

A *pointing* comprises a single intersection of an object and the reading of the circle. A *measure* of an angle or a direction is obtained by subtracting the circle reading of one object from that of another. Such readings may involve the means of several pointings. A *set* is a pair of measures, FL and FR. A *zero* comprises all the measures taken on one position of the graduated circle.

(*a*) *Pointings.* With some theodolites pointings are intended to be made in pairs. One pair may be enough, but in bad conditions morc may bc needed. A good rule is that if the pair differ by more than 2″ (1″ as read in a Wild), another pair is needed. Then if the over-all range of the four exceeds 4″ (2″ as read), a third pair should be made. And so on, except that it would probably then be better to wait for better conditions. With other instruments two or three pointings may be made as a normal minimum, with a similar rule for repetitions.

(*b*) *Measures and sets.* An equal number of measures must, of course, be made on each face. If a theodolite is poorly divided the programme should be one set on each of many zeros, but with good modern theodolites two or three sets per zero may be preferred, unless the micrometer run is wrong. An alternative is to take 2 or 3 FL on one zero followed by 2 or 3 FR on the next.

(*c*) *Zeros.* In primary triangulation 10 to 16 zeros are generally observed, with 1, 2, or 3 sets on each, or twice as many zeros with FL *or*† FR on each. In India the rule has been three sets on each of 10 zeros: in the United States [229] one set on 16: in the Sudan one set on 16, sometimes duplicated. However accurate the instrument may be, a protracted programme is necessary to reduce lateral refraction error.

The referring mark, or one of the flank stations, is selected as the point on which the zero is to be set. The degrees of FL setting on this point should then be regularly spaced through 180°, and the minutes and seconds should be spaced through the range of the micrometer to minimize the effects of error of run.‡ It suffices to set zero within 15″ of the prescribed figure unless the run error is large, in which case better elimination will be got by setting within 3″ or 4″.

While the above represents the usual custom, it can be argued that 11 or 13 zeros are better. The instrument maker may have prepared his master circle by striking arcs of 45°, which have been mechanically subdivided. If so, errors of division might be periodic with 45° as a period, and 12 or 16 zeros at intervals of one-third or one-quarter of 45° will be ineffective. Similarly with the minutes and seconds. [589], p. 37, mentions a theodolite whose alternate 10′ intervals were large and small, so that error would only have been eliminated if micrometer settings had been spaced over the double interval. The point may seldom be of much consequence, but unless the exact method used for dividing the circle is known, a safe rule seems to be to space zeros at 180° ÷ 11, 13, or 17.

Traverses. In a traverse, where Laplace stations may be very frequent, the necessary accuracy of the traverse angles may be less than has been usual for triangulation. The time taken to complete one zero is also much reduced, so in practice the number of zeros may be about the same. See § 1.19. On the other hand, if Laplace stations were as far apart as has been usual in triangulation, such as 300 km, the azimuth accuracy of a primary triangulation chain would only be equalled if several days were spent on each traverse angle, since from this point of view a triangulation chain is the equivalent of three or four traverses. At the other extreme, if

† In some theodolites, such as the Kern DKMs, there are two concentric divided circles, whose images are made to coincide by the micrometer. Then FL and FR readings do not have the same graduation errors, and there is no advantage in changing zero between them.

‡ [589], pp. 34–6 advises that the minutes and seconds of the zero settings should be $D/2n$, $3D/2n, \ldots$, i.e. equally spaced through the interval D between circle divisions, but starting with $D/2n$ instead of 00′ 00″, if the number of zeros n is even. If n is odd, the first zero should be 00′ 00″.

Laplace azimuths are observed at every traverse station, the traverse angles (as such) do not have to be measured at all.

In the United States, the angles of zero-order traverses, § 1.06 (*c*), are measured on 16 zeros on each of two nights, with a third night if the first two differ by 1″.

1.19. Time of day

With good instruments, properly used, the worst source of error is believed to be the refraction of light out of the vertical plane. The ordinary 'jumping' which occurs on hot days is met by increasing the number of pointings, but refraction may incline more to one side than the other, and repetition then does little good. This is especially possible if the line grazes along a hillside or over flattish but sloping ground. The lateral refraction is then likely to be maximum during the heat of the day, when the air near the ground will be overheated. It should be small an hour or two after sunrise and again in the evening. And at night when the ground is cooler than the air, its sign may be reversed. The ideal is therefore to divide about equally between these four periods, omitting the afternoon if the jumping is excessive, § 1.18 (*a*). With short lines well clear of the ground, or if the diurnal temperature range is small, less care is necessary.† See § 3.22.

At night, lights may appear good and steady, but they may slowly drift and give worse results than apparently bad lights at midday.

In the United States most work is now done at night, with satisfactory results, although what is advised above is thought better in dry tropical conditions, and if the programme is a long one, day work saves inconvenience and time. In Australia traverse angles are observed in the evening twilight.

If the methods of directions or angles are used, the pointings to each station must be distributed through the whole period of work, and not concentrated at one or a few periods. The method of rounds automatically secures this. Subject to exceptions, it may be said that the accuracy with which an angle is measured depends on the length of time over which its observation is evenly spread.

1.20. Abstract of observations

In the angle book, pointings are meaned and subtractions are made to give measures of directions or angles, a *direction* being the angle made with the station selected as RM for the zero settings. It does not much matter whether angles or directions are abstracted, but directions are

† See [105] for an analysis of angles read at different hours. The 1934 report includes 6 angles in which the means of 20 afternoon measures were wrong by 3″ to 6″. In five the night errors were of opposite sign. Also see [229], pp. 135–6.

advised as being simpler for subsequent small corrections, §§ 1.22 and 2.09.

As observations are made, angles or directions are copied on to an *abstract form*, which separately records each measure, those on each zero being grouped together under sub-heads FL and FR. This abstract ensures the observation of the full programme of measures and zeros, it shows up unfilled gaps in broken rounds, and reveals blunders and inaccuracies. Zero means are taken out, and thence the final observed value, subject only to correction for eccentric observations or station adjustment, § 1.23.

When angles have been observed in closed rounds, § 1.17 (*a*), two courses are possible.

(*a*) The accepted reading of the RM, to be used for getting directions, may be the mean of its opening and closing values.

(*b*) Directions may be got from the opening value only, and the closing readings may be used to give a closing direction for the RM, nearly but not quite equal to zero.

The former course saves some trouble, but the latter will reveal any systematic drag of the lower plate, such as may tend to make all angles too small, and may perhaps be preferred. See further in § 1.23.

For rejections and repetitions see § 1.25.

The abstract should be prepared in duplicate, and one copy should be sent to a safe place as soon as possible.

In *traverses* the abstract is much more simple than in triangulation. So much so that it is probably only necessary to prepare a single copy for insurance against loss of the angle books.

1.21. Witness marks and descriptions of stations

Azimuths of witness marks and horizontal distances to the nearest centimetre are required. The best procedure is as follows. Observe one round FR and one FL, horizontal and vertical, including one station and all the witness marks. Then turning on to each mark again, measure the sloping distance from the transit axis to each mark with a steel tape. The vertical angle enables this to be reduced to horizontal. The marks should not be so close as to make fairly sharp focus impossible.

The angle book should contain descriptions with special reference to the following.

(*a*) The general whereabouts, so that the site can be found again.

(*b*) Details of construction, especially of the station marks and the vertical distances between them.

(*c*) The type of signal used, helio, lamp, or beacon, with a diagram of

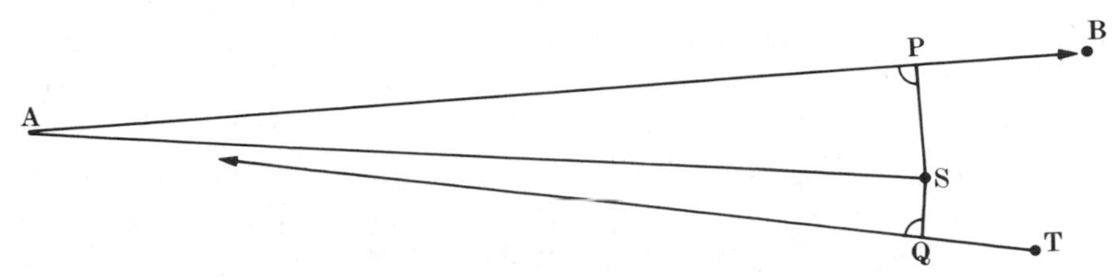

FIG. 1.11. Eccentric observations. AB = many kilometres. SP = some metres.

the latter if not of standard type. Their heights, and that of the transit axis, to the nearest centimetre above one of the station marks.

(*d*) Descriptions of the witness marks, with their distances and azimuths to the nearest centimetre and minute.

1.22. Eccentric observations†

Some or all of the observations from A to a station S may have been made to a signal B which is not vertically over the point at which observations have been made. Or, observations at a station may have been made at two slightly separated points S and T. The procedure is then as follows.

(*a*) Decide and record which point is to be the accepted position of the station: S in Fig. 1.11, but the points marked B or T could have been accepted.

(*b*) Measure the horizontal distance from S to any beacon B or other observation point T. Perhaps as for witness marks in § 1.21.

(*c*) If S is an observation point, include B and T in a round of angles with any distant station such as A.

(*d*) If S is not an observation point, include it and B in a round of angles at T.

(*e*) Then the correction to directions observed to B from A is SP/AS, and the correction to a direction observed to A at T is SQ/AS. The sign of the correction is best determined from a diagram.

(*f*) If the correction is to be correct to 0.1″, which may not be necessary if AS is short, the distances SB and ST must be correct to within 1 cm per 20 km of AS, which may be difficult unless they are short and nearly horizontal. Similar accuracy is required when pointing on B or T for the angles ASB, etc. If the corrections to the directions AB or TA amount to n'', AS must be known to one part in $10n$, either from the map or from a preliminary solution of triangles. Corrections of more than (say) 60″ are likely to introduce serious error. It is always best to have no eccentric stations.

(*g*) A diagram should be recorded in the angle book. Corrections

† Sometimes described as satellite observations, but this now invites confusion with earth satellites.

should be computed, at least with preliminary values of AS, etc., before leaving the station while signs are still visible to the eye, and the corrections should be applied to the general means in the abstract form.

In *EDM lines*, especially when using microwaves, the distance may have been measured from eccentric stations. The routine reduction to spheroid given in § 1.45 will then give the distance between spheroidal points corresponding to the eccentric stations, provided their heights have been used in the formulae of § 1.45. The spheroidal separation of the principal marks can then be deduced from that of the eccentric marks as in (*a*) to (*g*) above.†

1.23. Station adjustment

(*a*) If angles have been independently‡ measured all round the horizon at any station, they will not exactly total 360°, and the misclosure needs to be distributed between them, either equally or in inverse proportion to their weights. Weights may often be regarded as equal, but there may be other considerations, such as differing numbers of measures. Angles are seldom measured independently, and this situation will seldom arise.

(*b*) If observations are made in rounds, and if broken rounds are dealt with as in § 1.17, and if the accepted reading of the RM is the mean of first and last, there will be no misclosures. But if the closing reading of the RM has been separately abstracted, the misclosure will need to be distributed among the angles of the round.

(*c*) If the method of directions has been used with a single RM, there will be no misclosure to adjust. If two RMs have had to be used, adjustment can be done by common sense, or perhaps rather better by least squares as below.

In Fig. 1.12 suppose that stations 2, 3, 4, and 5 have been observed with station 1 as RM on some zeros, and that 2, 4, and 5 have also been observed with 3 as RM. Then as in § D.14 observation equations are

$$\begin{aligned}
D_2 \qquad\qquad\qquad\qquad &= k_1 \\
D_3 \qquad\qquad\qquad &= k_2 \\
D_4 \qquad\qquad &= k_3 \\
D_5 &= k_4 \\
-D_3+D_4 \qquad\qquad &= k_5 \\
-D_3 \qquad\qquad +D_5 &= k_6 \\
D_2-D_3 \qquad\qquad\qquad &= k_7,
\end{aligned} \tag{1.2}$$

† If the stations are at great altitudes, and if the eccentric stations are rather distant from the principal stations, a reduction of 1 ppm per 6.4 m of altitude will be required in distances such as ST, but this should very seldom be of consequence.

‡ i.e. as in § 1.17 (*c*), not in rounds.

where the D's are the required adjusted directions from 1 as RM, k_1 to k_4 are recorded mean angles from 1, and k_5 to k_7 are recorded means from 3. The weights may be the number of zeros on which each k has been observed, or anything else more appropriate. We have seven equations for four unknowns, and the most probable D's are computed as in § D.14.

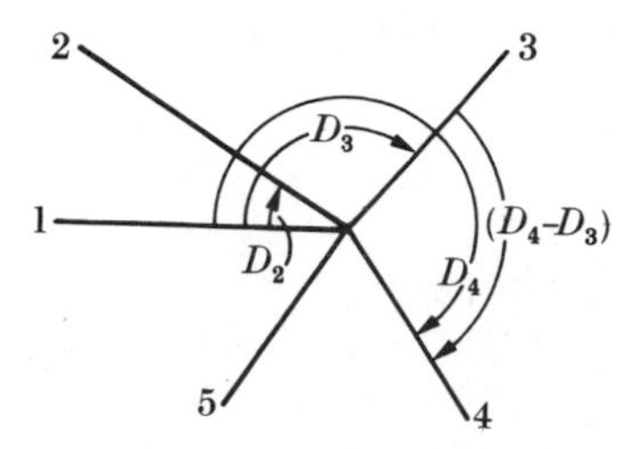

FIG. 1.12. Directions observed from two RM's, 1 and 3.

(d) Another situation is that the angles at a station may have been measured (in rounds or otherwise) in different combinations in different years or by different instruments. In Fig. 1.13 suppose that the angles marked 1 to 6 have been measured independently, and let $x_1, \ldots, x_6$ be the errors of these measures. Then the x's must be related by the conditions

$$\begin{aligned} x_1+x_2+x_3+x_4 &= e_1, \\ x_5-(x_1+x_2+x_3) &= e_2, \\ x_6-(x_2+x_3) \quad &= e_3, \end{aligned} \tag{1.3}$$

where e_1, e_2, and e_3 are the known small misclosures. The weights of the observations must be assessed, or regarded as equal, and the most probable values of $x_1, \ldots, x_6$ are then given by the routine of § D.16.

If the triangulation is being computed by the method of variation of coordinates, it is possible, although generally very inconvenient, to avoid station adjustment by forming separate observation equations for each separately observed angle or direction.

(e) Examples of station adjustment are given in [466], pp. 79–121. There is much to be said for avoiding anything more complicated than a simple distribution of closing error. See [123] for an example of station adjustment after Schreiber's method of observation.

(f) In *traversing*, adjustment as in (c) or (d) is unlikely to be necessary except possibly at the junctions of several lines.

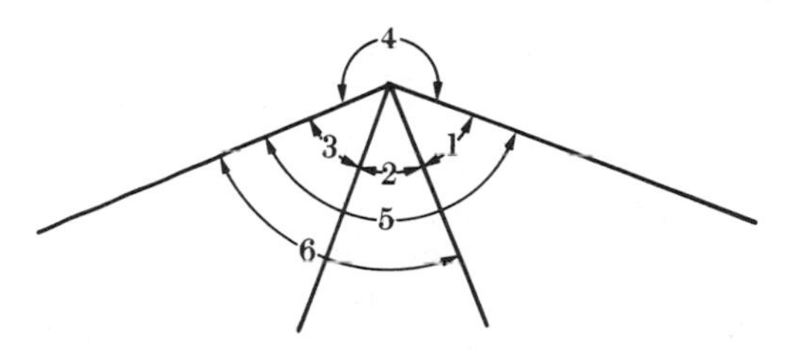

FIG. 1.13. Angles measured in different combinations.

1.24. Sources of errors in angles

The principal sources of error in a horizontal angle are as follows.

(*a*) Lateral refraction.

(*b*) The centring of the theodolite, helio, or lamp with short lines or high towers.

(*c*) Stiff axes or loose joints.

(*d*) Dislevelment (of the cross level), or the omission of formula (2.5), in lines which are steeply elevated or depressed below the horizon.

(*e*) Unsteadiness on a tower.

(*f*) Occasionally, unsuspected obscure errors like those of § 1.27 (*h*) or (*k*).

(g) Arithmetical mistakes in abstracting or taking means.

Lateral refraction is believed to be the most serious source of error in current practice. See §§ 1.04, 1.19, and 3.22. Neither graduation error of the circle, nor random errors in pointing or in circle readings are likely to do significant harm in a 10-zero programme observed as in § 1.18.

Errors of adjacent small angles are likely to be somewhat positively correlated in so far as they may be due to causes (*a*,) (*b*), (*c*), or (*d*) listed above.

1.25. Rejections and repetitions

An obvious blunder, such as a mistake in the degrees or minutes, can be corrected in the record. Bad misclosures of rounds, or recorded accidents, are, of course, dealt with by rejection and repetition.

Unusually discrepant pointings should be dealt with as in § 1.18 (*a*).

If a zero mean, or a group of several means, is widely discrepant, lateral refraction is the most probable cause. The best course is then to repeat that series of zeros at the same time next day. If normal values result, they are meaned with the first values, and the anomaly will be halved. If the wide values recur, rejection is not necessarily indicated, since the apparently bad observations may be useful items in a periodic variation. See § 1.19. If there is doubt whether to reject a whole time of day, such as the afternoon, consideration may as a last resort be given to triangular errors or misclosures on Laplace stations. This may help to give the most probable values of the angles, but it will falsify any estimate of accuracy based on the resulting misclosures. The rejection and the reasons for it should be prominently recorded.

Triangular error. The sum of the angles of a triangle should be 180° plus the spherical excess, § 2.12. This should be approximately computed, and as soon as three angles of any triangle have been measured, their sum should be compared with the correct value. In primary triangulation the error should average less than 1 second, and an error of more than 2 seconds is generally looked on as rather a serious weakness. If the error in any triangle is abnormally large there is no easy remedy, except possibly checking the arithmetic, since to revisit stations will completely upset the programme. Whether to reobserve or not is generally best decided after the cause has been traced and put right.

In general it is seldom good practice for a superior authority to order that observations of angles (or of anything else) should be rejected if their discrepancies from a general mean should exceed a certain limit. The proper criterion for the rejection of observations is the range of variation actually being found in the current working conditions. An observation which is otherwise apparently good should only be rejected if it is badly outside that range. If discrepancies are unusually large the programme may be increased, or work may be suspended, but it is of course better to locate and remedy the source of the trouble. Also see § D.12.

1.26. Vertical angles

(*a*) *Accuracy.* Less accurate observations are required than for horizontal angles, since atmospheric refraction makes high accuracy impossible in any case, and since the final height control will be by spirit levelling. Nevertheless it may be many years before spirit levelling is available, and it will then be very inconvenient if the contouring of maps based on the triangulation is upset. The aim should therefore be to measure vertical angles as accurately as refraction permits, and to hope that height errors will nowhere accumulate to more than 2 or 3 metres.

Height can be carried through a triangulation chain by a single line of good reciprocal vertical angles. The omission of some vertical angles, especially over long diagonals, is therefore less serious than a loss of horizontal angles, and when they are hard to get there is no need to delay indefinitely for them all. If the full programme is impossible, the observer should try to get at least a series of simple triangles, all of whose sides have been reciprocally observed at the right time of day, as they provide a set of triangular height closures from which accuracy can be assessed. See § 3.15.

In a traverse, reciprocal vertical angles are necessary between all stations, unless spirit levelling is soon likely to be available. Slopes are in any case required for the reduction of measured distances to horizontal, although in flat lines it is possible to do that with barometric heights; see § 1.46 (*b*).

(*b*) *Method of observation.* Vertical angles must be measured in pairs, FL and FR, to cancel bubble adjustment error. Three sets should suffice, but if the station is occupied for more than one day, a couple of sets should be observed each day up to a maximum of two or three. The upper bubble should be centred at each pointing. It is not ordinarily necessary to read the ends of the bubble and to apply a bubble correction.

A strange instrument must be examined to discover the rule for converting circle readings into elevations or depressions. Recorded angles should be converted and entered as E or D before leaving the station.

(*c*) *Time of day*. Vertical angles should be observed at between 12.00 and 15.00 hours local mean time, when refraction is not only least, but (more important) most uniform from place to place and from day to day. Reciprocal angles simultaneously observed are better than observations made on different days, but in triangulation they may be difficult to arrange. If afternoon angles are impossible, simultaneous reciprocal observations are the best substitute, and these may be better at night than near sunrise or sunset. How satisfactory they may be will depend on how uniform conditions are along the line.

Rules may more readily be relaxed in short lines and in cool cloudy weather.

It has sometimes been suggested that vertical angles are best observed soon after dawn, when refraction is unlikely to be abnormally high or low. The answer to this is that dawn is the time of greatest change, when the temperature gradient and the coefficient of refraction are at their least predictable. Equality of refraction at the two ends of a line is best secured at the relatively stable time of minimum refraction.

If vertical angles are required to distant peaks which may be cloud-covered after the early morning, a series of observations may be made from an hour or two after sunrise until as late as possible, temperatures being recorded and continued until 15.00 hours. A value for the minimum vertical angle can then be obtained on the assumption that temperature and refraction angle vary linearly in such a line; see § 3.20 (*a*).

In a traverse, afternoon reciprocal angles should be easy to arrange; see § 1.46.

The shade temperature should be recorded to the nearest degree when vertical angles are observed, and the pressure to the nearest 5 or 10 millibars. Humidity is not required.

(*d*) At each pointing, the telescope bubble must either be centred, or it should be read with a view to a correction being applied to the circle reading. If the bubble divisions are numbered from the centre outwards (as is usual), the correction to elevations (with the correct sign) is

$$(d/n)\left\{\sum(\text{Object end readings})-\sum(\text{Eye end readings})\right\}. \qquad (1.4)$$

where n is the total number of readings (an O and an E each counting as one), d is the change (in seconds) of inclination when one end of the bubble moves through one division of the scale, and elevations are positive. For the determination of d see § 4.12.

1.27. Miscellaneous advice to observers

(*a*) During any round the theodolite must move continuously in one direction. If a station is accidentally overshot by more than the slow-motion can correct, the telescope should be swung on right round the circle until it comes up to the station again. A good rule is 'face left, swing left' and vice versa. After any change of face or zero, or other reversal of swing, the telescope should be turned at least 360° in the new direction before a mark is intersected.

(*b*) Final movements of a slow-motion screw or micrometer are probably best made against the spring, although the point is generally unimportant, except from the point of view of (*h*) below.

(*c*) Intersections should always be made with the same point of the vertical wire, close to but just off the horizontal wire, e.g. just below on FL and above on FR. Similarly intersections for vertical angles should be made at one point on the horizontal wire.

(*d*) Vertical wires are sometimes made half single and half double. Except for opaque marks the latter is usually better, but the same half must be used for all the objects in one round.

(*e*) Electric illumination of the circles is better than natural, even by day.

(*f*) It is desirable that all helios and lights in a round should be of approximately equal brilliance and apparent size, as some observers may have a personal bisection error depending on brilliance; see [472] and [485], where a reversible eye-piece prism is suggested as a remedy. If helios or lamps are too bright they can be dimmed by suitable stops, or by a small muslin fan held in front of the object glass, or sometimes by increasing the telescope illumination. A red or yellow filter on the eye-piece has been advised as an aid to clear vision. Opinions differ, but a filter is only likely to help with opaque objects or when reduction of brilliance is all that is needed.

(*g*) The eye must always look straight into an eye-piece or micrometer, to avoid parallax.

(*h*) Slow-motion clamping should be light. The vertical circle should be clamped while horizontal angles are being measured, but sub-paragraph (*b*) should not be overlooked, as theodolites have existed in which reversal of the vertical slow-motion causes a sharp jump in the horizontal pointing. Care should be taken to turn all screws with a pure rotary motion about a horizontal axis, without torque about the vertical axis.

(*i*) The quicker a round is finished, the less the risk of the stand twisting. Readings should be made briskly without undue pause for thought or verification. An occasional blunder, although unnecessary, is harmless, § 1.25. It is only on the intersection of the mark in bad conditions that one may have to dwell for some time to get a good bisection of the range through which it is jumping.

(*j*) Avoid frequent change of collimation or focus. Aim at getting the theodolite into good adjustment, and then leaving it alone. Footscrew levelling must, of course, be attended to as soon as it goes wrong, but not in the middle of a round.

(*k*) When one-third of the zeros at any station have been done, the whole instrument, foot-screws and all, but not the tripod or stand, should be turned through 120°, and again after another third. This is to cancel the effect of possible strain in the axis bearing [473]. [590], pp. xiiD–xixD, gets near to making a similar recommendation.

(*l*) Keep the sun off the instrument stand. An umbrella may suffice, but in some countries a tent is advised. Fig. 1.9 shows a suitable design, 2×2 m. The roof takes off for star work, and the upper parts of the sides can be opened as required.

(*m*) See that heat rising from camp-fires, even though extinguished, does not cause lateral refraction.

(*n*) The employment of a recorder is advised. He saves the observer's eye from constant change of focus, keeps up the abstract, and is useful in many ways.

(*o*) The time should be recorded every hour or so, and also the change over from helios to lamps.

General references for theodolite observations. [41], [229], [293], [436], and [589].

SECTION 3. MEASUREMENT OF DISTANCES

1.28. Introductory and notation

Between 1900 and 1950 geodetic base lines were measured by invar wires or tapes, typically 24 m long, hung under tension in catenary, and stepped along over a distance of 7 to 15 km of necessarily flat ground. The length of the wires was determined by comparison with a standard metre bar, which via a few intermediary comparisons had been standardized in terms of the international metre, which was kept at Sèvres, France. Invar baselines are described in § 1.30.

Since about 1950 measurement by invar wires has been superseded by the much more convenient system of measurement by EDM. In principle the method is that a modulated beam of light or microwaves of accurately known modulation frequency, is emitted from one end of the line which is to be measured. It is reflected or transponded back by an instrument at the other end, and the phase difference of the returning modulation is recorded. The double distance is then a whole number (M) of modulation wavelengths plus the phase difference θ. M is given by measures of the phase difference on other frequencies, and the distance is

$$\begin{aligned} D &= \tfrac{1}{2}(M+\theta/2\pi)(\text{modulation wavelength}) \\ &= \tfrac{1}{2}v(M+\theta/2\pi) \div (\text{modulation frequency}), \end{aligned} \tag{1.5}$$

where v is the velocity with which the modulation is transmitted, and θ is measured in radians.†

An EDM measurement is thus dependent on a knowledge of the velocity of light through the atmosphere, together with a frequency meter to record the modulation frequency in Hz.‡

The electronic circuits inside the instruments are not here regarded as part of geodesy, and are not described. Geodesy is regarded as beginning when the waves leave the instrument and ceasing when they return. The observation programme with different types of instrument is only described so far as is necessary to illustrate general principles. The detailed instructions for switching and operating the instruments will be found in the makers' handbooks for different models.

Accuracy is much dependent on the refractive index of the wave in the atmosphere through which it passes. The normal state of the atmosphere, from this point of view, is described in § 1.35, and the possible irregularities in § 1.36. § 1.37 describes the extent to which meteorological observations at the two ends of a line may give the average value of the refractive index along it, and the average curvature of its path. § 1.38 deals with the reflection of microwaves, and its application to tellurometer ground swing is in § 1.44 and Appendix G. The routine reduction of measured distances is in § 1.45.

The accuracy aimed at is 1 ppm, and sources of error a few times smaller are regarded as negligible. Random errors of ±5 ppm may often be harmless, since in a traverse of 25 legs they will have no more effect than a systematic error of 1 ppm.

The wavelength of light Λ is measured in micrometres ($1\ \mu\text{m} = 10^{-6}$ m), formerly known as microns. It used also to be measured in ångströms ($1\ \text{Å} = 10^{-10}$ m). It varies between about 0.40 μm (violet) and 0.72 μm (red). Electromagnetic waves of lengths between about 1 m and 1 mm are known as microwaves.

The following notation is used in §§ 1.28–1.49. SI units are used, except that pressures are primarily given in bars, but are generally duplicated by figures in Pa. Temperatures are in °C or K as stated.

† The angular velocity of wave motion measured in radians equals its frequency in Hz multiplied by 2π. In these circumstances 1 Hz is equivalent to 2π rad s^{-1}.

‡ 1 Hz = 1 hertz = 1 cycle per second. 1 kilohertz (kHz), 1 megahertz (MHz), and 1 gigahertz (GHz) signify 10^3, 10^6, and 10^9 cycles per second respectively, [53].

General and geometrical

c = velocity of light in vacuum
v = phase velocity of light or microwaves in air.
v_G = group velocity.
Λ = wavelength of light.
n_0 = refractive index of light or microwaves at 0 °C and 1013.25 mbar.
n = refractive index in other conditions.
n_G = group refractive index, (1.13)–(1.15).
$N = (n-1)10^6$.
$M = N + 0.16$ per metre of height, (1.47). Also number of wavelengths in (1.5).
β = vertical angle. Also angle of reflection in Fig. 1.20.
D = distance along curved path. Not in (1.50).
C = length of path chord. Not in (1.17).
L = spheroidal distance.
C_s = length of spheroidal chord.
s = variable distance along path or chord.
θ = Any angle.
σ = radius of curvature of path.
R = earth's radius (6370 km), when this approximation suffices.
$K = R/\sigma$.
R_α = radius of curvature of spheroid in prescribed latitude and azimuth, (1.6).
$1/R' = 1/R - 1/\sigma$. $R' \approx 7400$ km for light and 8500 km for microwaves.
$k = R/2\sigma = \frac{1}{2}K$.
a_P = reflection factor from a plane surface.
$a = a_P \times D$ = reflection factor from a sphere, (1.50).
h = height above sea-level or ground, as stated.
$h' = h - L^2/8R'$ in § 1.38.
p, q = elements of standard error, (1.55) and (1.57).
s.e. = standard error. ±indicates s.e. unless otherwise stated.

Meteorological

P = total pressure.
T = temperature in °C or K as stated.
e = partial pressure of water vapour.
e' = saturation pressure of water vapour.
$\bar{P}, \bar{T}, \bar{e}, \bar{n}$ = average values along a measured line.
P_m, T_m, e_m, n_m = mean of values at the two ends of a line. Except T_m in (1.27).
ρ = density.
χ = concentration of water vapour. Mass/volume.
c_d = gas constant for dry air, (1.21).
c_w = gas constant for water vapour, (1.22).
α = coefficient of expansion of dry air = 1/273.16.
Γ = adiabatic temperature gradient of dry air = −0.00986 °C/m.
λ = normal average temperature gradient of air = −0.0055 °C/m.
κ = eddy conductivity. Generally in metre² per hour in § 1.36.
t = time. In hours from midnight in § 1.36.
$h' = (24\kappa/\pi)^{\frac{1}{2}}$. See § 1.36 (*a*). Different in § 1.38.

Electromagnetic

l_C = wavelength of microwave carrier.
l_M = wavelength of microwave modulation
f_C = Frequency of carrier in Hz $= v/l_C$.

f_M = frequency of modulation = v/l_M.
w = angular velocity of carrier = $2\pi f_C$ rad s^{-1}.
W = angular velocity of modulation = $2\pi f_M$ rad s^{-1}.
T = travel time over the direct route (single distance) = D/v.
Q = excess travel time of a reflected wave (single distance).

Meteorological units

1 millibar (mbar) is defined to be 1000 dyn/cm^2 = 10^2 Pa.†
Standard sea-level pressure = 1013.25 mbar = 101.325 kPa.
Absolute zero temperature = 0 K = −273.16 °C.

Abbreviations

SI = International system (of units).
EDM = Electromagnetic distance measurement.
EODM = Electro-optical distance measurement. Including infra-red.
MDM = Microwave distance measurement.
ppm = parts per million.

1.29. Standards of length

Between 1889 and 1960 the international standard of length was a metre bar kept at Sèvres. Similar bars, which have been accurately compared with it, exist in other countries. Since 1960 the metre has been defined, conformably with the length of the bar at that date, in terms of the wavelength of a spectral line of the krypton-86 atom, so that the metre is now independent of any possible change in the length of the bar.

The SI definition of the metre is 1 m = 1 650 763.73 wavelengths, in vacuum, of the radiation corresponding to the transition between the levels $2p_{10}$ and $5d_5$ of the krypton-86 atom.

Invar wires used for base measurement since 1900 will have been indirectly compared with the standard metre, but the currently accepted triangulation framework of some countries may still depend on earlier baselines whose unit of length is rather remotely connected with the metre. In some countries there is the further complication that metre units have been converted to feet by different foot–metre ratios in different countries. These problems will disappear with time as EDM lines take over the work of invar bases, and become incorporated in new adjustments. For a summary of the positions in Great Britain, the United States, and India see [113], 3 Edn, pp. 35–37.

For EDM the definition of the metre enters (1.7) through the velocity v whose accepted value will be recorded in SI metres per SI second. So if the frequency is recorded in SI seconds, the distance will be in SI metres. We are not here concerned with the methods‡ of determining the velocity of light, but the basic velocity in vacuum is currently given by c = 299 792 458 ± 1.2 ms^{-1}, [53] and [346], a figure which is accurate enough

† 1 Pa = 1 pascal = 1 m^{-1} kg s^{-2}. 1 bar = 10^5 Pa. [53].

‡ An obvious method is to measure a well calibrated invar baseline with a geodimeter, but more refined laboratory methods are available.

for present geodetic purposes.† See §§ 1.32–1.36 for the velocity of light and microwaves in the atmosphere.

1.30. Invar baselines

The routine of measuring invar baselines is described in [113], 3 Edn, where numerous references are given, and the 2 Edn also describes the method of standardizing the wires. In the present paragraph their description is limited to matters affecting the utility of existing baselines.

(*a*) *Properties of invar.* Invar is an iron alloy containing 36 per cent nickel. Its peculiarity is its low temperature coefficient, which may vary between +0.5 and −0.2 ppm per °C for invar wires, or sometimes as high as +2 ppm for bars. Other peculiarities result as below.

(i) The temperature coefficient varies with the proportions of the alloy, and also according to the thermal and mechanical history of any particular wire. It must be separately determined for each wire, and may change.

(ii) Length slowly increases with time, possibly by several parts per million per year when a wire is young, but less later. Wires should not be used until found to be reasonably stable. They must always be compared with a stable standard not more than a year before and after use.

(iii) Winding a 1.65-mm diameter wire off its 50-cm drum and returning it has been found to lengthen it by an amount given as 0.17 ppm, while the measurement of 100 bays of base has tended to shorten it by a similar amount, [115] and [116], but such figures are not necessarily applicable to all wires.

(iv) Length does not depend uniquely on temperature. If temperature is reduced, the normal contraction (if the coefficient is positive) takes place at once, but it will be followed by a slow elongation amounting to perhaps 1 ppm: and after a rise vice versa. This probably makes 1 ppm the limit of certain precision, [242].

(v) Damage not only changes the length of a wire, but may induce instability persisting for a year or more. A damaged wire cannot be trusted until repeated tests have proved it stable, [106], pp. 36–8. Invar wires must be carefully handled if they are to hold their lengths during a base measurement to within 1 or 2 ppm.

(*b*) *Baselines* 1900–55. The routine has been to measure baselines by suspending 24-m invar tapes or wires under tension in catenary between

† The metre and the second being defined, the velocity of light is determined by experiment. It appears likely that during the next decade some suitable figure will be accepted for the velocity of light in vacuum as a base unit, and that experiment will determine the length of the metre, which will then cease to be a base unit.

suitable tripods, to cover a total length of between (say) 7 and 15 km. The line must be fairly flat and cleared of all obstacles. It should be straight, although one or two changes of direction of a few degrees are permissible provided that the ends of each section are intervisible, and also the two ends of the whole base.

The measurement has generally been carried out once in each direction, preferably by two wires, used simultaneously, in each direction; four wires in all. The wires should have been standardized in a comparator both before and after field work, and the changes should be on record. The interval between these two comparisons should not have been more than a few months, with not more than three or four baselines measured during the interval.

The avoidance of wind has been important. Exposure of a wire to a cross-wind of 3.5 m s^{-1} may introduce an error of 0.5 ppm, [552]. For the use of screens see [435]. Tapes are more sensitive to wind than wires are.

(*c*) *Extension.* The longest baseline is much shorter than the average side length of most geodetic triangulation, to which it has had to be connected by a geometrically complicated extension figure, such as that shown in Fig. 2.17. This extension figure has generally been liable to introduce more error than the measurement of the baseline itself, and the necessity for a reasonably strong extension figure has added much to the difficulty of finding good baseline sites.

(*d*) *Accuracy.* The accord of different wire values of each kilometre section will usually have suggested a standard error of ±0.3 or 0.5 ppm, but a true value is only obtainable by considering the errors possibly arising from standardization and changes in wire lengths, the effect of wind and pulley friction, ignorance of the separation between geoid and spheroid, see (*f*) below, and from the corrections for slope and temperature if they should be extreme. It will be difficult to say of any invar base that it cannot be wrong by 1 or 2 ppm.

The extension. The accuracy of a well-measured base is likely to be largely lost in the extension, although care taken to avoid systematic errors in the base measurement, which may be shared by other bases, is not at all wasted. For estimating the accuracy of base extensions see § 2.44 (*e*).

(*e*) *Standardization of invar wires by the Vaisäla Comparator.* With this apparatus, distance is measured by the interference of light. Accuracy is high, such as 0.1 ppm, but atmospheric conditions must be good, and the instrument mountings must be stable. It cannot be used for extensive field work, but it provides a base of up to 864 metres in length on which invar wires and their straining gear can be calibrated in field conditions with more realistic accuracy than may be obtainable in the laboratory. It is of course essential to prove the stability of the terminals of a Vaisäla base by occasional remeasurement. The Nummela base in Finland changed by 0.5 mm (0.6 ppm) between 1947 and 1958.

Vaisäla baselines have been measured at Nummela, Finland (864 m); at Buenos Aires (480 m); At Loenermark, Netherlands (576 m); near Munich (864 m); at Ohio State University (500 m); and near Lisbon (480 m). The method of observation is described in [278]. Shorter reports are in [260], [279], and [353]. [354] summarizes the lengths of these lines, and their variations when repeat observations have been made.

(*f*) *Reduction to spheroid level.* An elevated base must clearly be reduced by (length)$\times h/(R+h)$, to reduce it to 'sea-level', where h is the height and R is the earth's radius.

(i) An error of 1 ppm results from an error of 6.38 m in h.

(ii) Computations all refer to a reference spheroid, §§ 2.02 and 2.08, and h must be measured above it, not above the mean sea-level or geoid. In a small country the difference may be small, but in a continental area it may be large, especially if an old and ill-fitting spheroid is in use.† Unfortunately spirit levelling and triangulation give heights above MSL, and the separation between MSL and the spheroid can only be obtained from extensive special surveys, as in §§ 4.50–4.52. An error of 6 m in the separation results in an error of 1 ppm in the base.

(iii) Strictly, the radius R should be the radius of curvature of the spheroid in the latitude and azimuth concerned, as given by

$$R_\alpha = \rho\nu/(\rho \sin^2\alpha + \nu \cos^2\alpha). \quad \text{See (A.67)} \qquad (1.6)$$

This can hardly differ from 6.38×10^6 m by $\frac{1}{2}$ per cent, and this figure suffices for any base less than 500 m above sea-level.

(iv) If the height of the base is not constant, the accepted h should be its average height above the spheroid along the line of the base. But see § 1.45 for the reduction of EDM lines.

General references for invar base measurement. [113], 3 Edn, [229], [294], [435], [436], and [589].

1.31. Atmospheric refraction

The effect of the atmosphere on the curvature of the path of light is well known. Below the path the air is more dense, and its refractive index is greater, than it is above the path. The gradient of the refractive index causes the curvature. Similarly, the velocity of light, or of microwaves, in

† In India and Burma the Mergui base, 3 m above sea-level, stands 100 m above Everest's spheroid, and acceptance of the MSL height would put it wrong by 1 in 60 000. See [110], Appendix 1. This may be an exceptional case, but [96], p. 183, quotes 300 m as the separation between the geoid in Far Eastern Asia and Bessel's spheroid as oriented at Pulkova, in European Russia.

the atmosphere depends on the refractive index of the air along the path. One or other, or both, of these two effects introduces trouble into almost every observation which the surveyor makes. It might perhaps be reasonable to devote a chapter exclusively to refraction, bringing all its manifestations together in one place, but actually the formulae which give the practical corrections to different observations vary so much between one situation and another that it is more convenient to give them separately in the different chapters in which the observations are described.

The present chapter gives the general theory in §§ 1.32 and 1.33, with its application to EDM between ground points in §§ 1.35–1.37. Other aspects of refraction are dealt with in the following paragraphs.

(*a*) Levelling, § 3.06 (*b*) (iv).

(*b*) Vertical angles between ground points, §§ 3.17–3.21.

(*c*) Horizontal angles, § 3.22.

(*d*) Altitudes of stars, celestial refraction, § 4.11.

(*e*) Parallactic refraction, as between a satellite and a star behind it, § 7.07.

(*f*) Distances between ground points and satellites, including the ionospheric refraction of radar and microwaves, §§ 7.08–7.10.

(*g*) Effect of the curvature of the path on the distance from the ground to a satellite, § 7.08 (*d*) and § 7.10 (*f*).

1.32. The velocity of light

The length of a wave is given by

$$\text{wavelength} = \text{velocity} \div \text{frequency, or } l = v/f. \tag{1.7}$$

For light, v is approximately 3×10^8 metres per second. So if $l = 10$ cm, $f \approx 3$ GHz.

The wave or *phase velocity* of light or other electromagnetic waves through any medium such as air, is given by

$$v = c/n, \tag{1.8}$$

where c is the velocity in vacuum, § 1.29, the same for light and all other such waves, and n is the refractive index appropriate to the wavelength and medium concerned.

The refractive index depends on wavelength in a rather irregular way. In the neighbourhood of strong absorption lines the variation (or *dispersion*) is considerable, but elsewhere it can be neglected. For light waves the dispersion must be allowed for as in (1.9) and (1.15). For microwave instruments the wavelengths can be, and generally are, so chosen that the dispersion is negligible. But see 8-mm microwaves in § 1.33.

The remainder of this paragraph refers to light. For microwaves see § 1.33.

The refractive index of light, but not of microwaves, in air is given [83] by

$$(n_0-1)10^6 = N_0 = 287.604 + 1.6288/\Lambda^2 + 0.0136/\Lambda^4 = A + B/\Lambda^2 + C/\Lambda^4 \tag{1.9}$$

defining A, B, and C, where n_0 is the index in dry air 0 °C and 1013.25 mbar, with 0.03 per cent of CO_2, and Λ is the wavelength in micrometres. So n_0 is 1.000 2908 when $\Lambda = 0.72$ (red), 1.000 2924 when $\Lambda = 0.589$ (sodium), and 1.000 2982 when $\Lambda = 0.40$ (violet), and the velocity of red light is greater than that of violet.

The index in other atmospheric conditions is given [83] by

$$\begin{aligned}(n-1) &= \frac{(n_0-1)}{\alpha T}\cdot\frac{P}{1013.25} - \frac{0.000\,000\,041e}{\alpha T}\\ &= 0.2696(n_0-1)(P/T) - 11.2(e/T)\,10^{-6}\\ &\approx 0.2696(n_0-1)(1/T)(P-0.14e),\end{aligned} \tag{1.10}$$

where T = temperature in K, P = total pressure in mbar, or in units of 100 Pa, e = partial pressure of water vapour in the same units as P, α = coefficient of expansion of air, 0.003661 or 1/273.16.

If $\Lambda = 0.72\ \mu$m (red), (1.10) becomes

$$(n-1) \approx 78.6(P/T)10^{-6} \text{ approximately,} \tag{1.11}$$

and for $\Lambda = 0.59$ or 0.40 the factor is 79.1 and 80.6 respectively.

In numerical work it is convenient to write $N = (n-1)10^6$. The magnitude of N is then of the order 300, and one unit of N corresponds to 0.000 001 in n, or to 1 ppm in n, or to 1 ppm in v with opposite sign.

Group velocity. The formula $v = c/n$ gives the velocity of a single pure wave such as can be represented by a sine curve. If two or more such waves, of slightly different wavelengths and consequently different velocities, are superposed as in Fig. G.1, the resulting modulated waveform will travel with a different velocity known as the group velocity v_G, see § G.11, where

$$\begin{aligned} v_G &= v - \Lambda(\mathrm{d}v/\mathrm{d}\Lambda) \\ &= v - \Lambda(\mathrm{d}v/\mathrm{d}n)(\mathrm{d}n/\mathrm{d}\Lambda)\\ &= v - \Lambda c(2B/\Lambda^3 + 4C/\Lambda^5)(P/1013.2\alpha T), \text{ using (1.8), (1.9), (1.10)}\\ &= v - c(3.26/\Lambda^2 + 0.054/\Lambda^4)(P/1013.2\alpha T)10^{-6}.\end{aligned} \tag{1.12}$$

Whence at 0 °C and 1013.25 mbar

$$(v_G - v)/c = -22.5 \text{ ppm if } \Lambda = 0.40\ \mu\text{m (violet)}$$

or -11.3 ppm if $\Lambda = 0.55$

or -6.8 ppm if $\Lambda = 0.70$ (red).

In comparison with (1.8) we may write

$$v_G = c/n_G, \tag{1.13}$$

defining n_G, the *group refractive index*. Note that n_G has no particular connection with the refraction of light. Then substituting (1.8) and (1.13) in (1.12) gives

$$n_G = n - \Lambda(dn/d\Lambda) \tag{1.14}$$

and at 0 °C and 1013.25 mbar,

$$N_{G0} = A + 3B/\Lambda^2 + 5C/\Lambda^4, \tag{1.15}$$

where A, B, and C are as in (1.9). For $\Lambda = 0.72$, 0.694 (ruby laser), 0.59, and 0.40 μm, $N_{G0} = 297.2$, 298.0, 302.0, and 319.2 respectively.

For other conditions (1.10) applies with n_G and n_{G0} instead of n and n_0.

The following summarizes the situations in which the group velocity or the phase velocity respectively should be used.

(*a*) There is no distinction between v_G and v when there is no dispersion, i.e.
- (i) for all waves in vacuum,
- and (ii) in general for microwaves in air, but see (*c*)(iii) below.

(*b*) There is no question of using v_G or n_G
- (i) when computing astronomical aberration, § 4.04 (*d*), because of (*a*) (i) above;
- (ii) when computing the refraction or curvature of light or microwaves;
- (iii) when using (1.7) to relate the wavelength and the frequency of light used to define the standard of length.

(*c*) The group velocity, and n_G, must be used in the following circumstances.
- (i) For the velocity of any kind of modulation or combination of light waves, including lasers, but generally not microwaves. But if the range of wavelengths involved is not small enough for $\Lambda(dv/d\Lambda)$ in (1.12) to have a substantially unique value, the modulation wave form will be varying, and its velocity will not be precisely definable.
- (ii) For the velocity of the beginning and end of a pulse or trail of light, including lasers.
- (iii) For microwaves in the ionosphere, § 7.10.

Differentiating (1.10) and (1.14) shows that in average sea-level conditions an increase of one N-unit, and a decrease of 1 ppm in the group velocity of light results from changes of

(i) $-0.02\ \mu$m in Λ, if $\Lambda = 0.6\ \mu$m, or
(ii) +3.4 mbar (340 Pa) in the total pressure, or
(iii) −26 mbar (2600 Pa) in the water vapour pressure, or
(iv) −1 °C in the temperature.

At greater altitudes with lower pressures the tolerances given in (i) and (iv) will be increased in proportion to P, and at lower temperatures (ii) will be decreased in proportion to T. The effect of humidity on the velocity of light (not of microwaves) is practically negligible in nearly all circumstances.

1.33. The velocity of microwaves

The velocity of microwaves differs from that of light as follows.

(i) It is almost independent of wavelength, and consequently there is generally no question of a group velocity differing from the phase velocity. But see 8-mm waves, below, and § 7.10(a) for the ionosphere.

(ii) It is much more affected by humidity.

For 12.5-mm microwaves the refractive index is given [197] by†

$$(n-1)10^6 = N = 77.6\frac{P}{T} - 13\frac{e}{T} + \frac{3.7e\times 10^5}{T^2}, \tag{1.16}$$

where P and e are measured in mbar, and T in K. In dry air ($e = 0$), at 1013.25 mbar and 273 K, $n_0 = 1.0002877$, and is substantially the same as the value given by (1.9) with Λ large. This formula is also applicable to 10-cm and 3-cm microwaves, and it is almost correct for 8 mm, but see below.

If P and e are measured in Pa the factors 77.6, 13, and 3.7×10^5 are changed to 0.776, 0.13, and 3.7×10^3.

The velocity is given by $v = c/n$ as in (1.8).

In average sea-level conditions with $P = 1000$ mbar (10^5 Pa), $e =$ 10 mbar (10^3 Pa), and $T = 290$ K, a decrease of 1 ppm in the velocity of microwaves will result from changes of

(i) +3.6 mbar (360 Pa) in P.
(ii) +0.2 mbar (20 Pa) in e. Measuring e is a serious difficulty.
(iii) −0.8 °C in T.

† The presence of CO_2 increases n by $(48/T)10^{-6}$ per millibar of partial pressure of CO_2. This is about 0.05×10^{-6} for the usual 0.03 per cent of CO_2, and it may be ignored. [34] puts (1.16) into the form $(n-1) = A(P-e) + Be$, where A and B depend only on T and are tabulated.

At a height of 5000 m with $P = 580$ mbar, $e = 2$ mbar, and $T = 275$ K, these figures are +3.5 mbar in P, +0.2 mbar in e, and −1.5 °C in T.

8-mm microwaves. The velocity of 8-mm microwaves in dry air is slightly affected by an absorption line of oxygen at 5 mm, and the effect of water vapour is also slightly abnormal. The refractive index of dry air and of water vapour for 3-cm, 12.5-mm, and 4-mm waves is given in [212], p. 834. From those figures [213] estimates that on account of the oxygen line intervening between 12.5 and 4 mm, n of dry air at 8 mm will be greater than that given by (1.16) by rather less than 0.1 ppm, and that the abnormal affect of water vapour (13 mbar) will be similar. The total effect on n_G is then estimated as +0.5 ppm, so the group velocity of 8-mm microwaves is estimated to be 0.5 ppm less than the velocity derived from (1.16) which is used for 10- and 3-cm waves. The difference between n_G and n arises mostly from the oxygen line and is not much dependent on the actual water-vapour pressure.

The velocity of 8-mm waves, unlike that of 10- and 3-cm waves, is affected by falling rain. [141], p. 193 states that the n of 8-mm waves is increased (and v decreased) by $(\frac{1}{2}\pi)(0.085)R^{0.84}$ ppm, where R is the rate of rainfall in millmetres per hour. If $R = 16$ mm, a heavy rate, this correction is −1.3 ppm.

1.34. Meteorological observations

§§ 1.32 and 1.33 show that to determine the velocity of light or microwaves to 1 ppm, the average pressure along the path must be determined to 3.5 mbar, the temperature to 1 °C, and (for microwaves only) the water vapour pressure to 0.2 mbar. Difficulties are

(i) instrumental, see below, and
(ii) obtaining an average value from observations made at a few particular places. See § 1.37.

(*a*) *Pressure.* With a good aneroid there is no difficulty in measuring pressure correct to 3 mbar (300 Pa), provided the calibration is occasionally checked at the pressures expected in the field as well as for sea-level.

(*b*) *Temperature* must be measured by an aspiration thermometer screened from radiation from the sun, sky, and ground, by night as well as by day. Errors of measurement need not amount to 1 °C. [483] describes dry and wet bulb thermometers for use on masts. In hot sunlight the temperature at (say) 1.5 metres above the ground may fluctuate, falling some degrees with a gust of convectional wind every few minutes. In such circumstances the lowest values are likely to be the most typical of air temperatures at (say) 30 metres above the ground. See [97], p. 319.

(c) *Water vapour pressure*, e, is measured by the difference between wet and dry bulb thermometers. [34] gives

$$e = e' - C(t_d - t_w)P \qquad (1.17)\dagger$$

where e' is the saturated pressure at the wet bulb temperature. [25] and [34] give tables of e'. It is 6, 12, 23, and 41 mbar at 0, 10, 20, and 30 °C respectively. For electronic computing it may be more convenient to programme the Goff-Gratch formula on which the tables are based, [29], p. 350. e, e', and P must be in the same units. $C = 0.66 \times 10^{-3}$ if the wet bulb is water covered, and 0.57×10^{-3} if it is iced. t_d and t_w are the dry and wet bulb temperatures in °C, as given by a forced draft (Assman) hygrometer giving an air velocity of 3 m/s over the bulbs. The bulbs must be screened by radiation shields.

[34] gives convenient tables for computing e.

At $t_w = 0$, 10, 20, and 30° respectively, errors of 0.20, 0.14, 0.10, and 0.07 °C in $(t_d - t_w)$ may result in errors of 0.2 mbar (20 Pa) in e, corresponding to 1 ppm in the velocity of microwaves.

[34] advises the following precautions when reading the hygrometer, and considers that errors of more than 0.2 mbar (20 Pa) should then be unusual.

(i) Avoid breathing on the air intake, especially below 0 °C. A transparent screen may help.

(ii) Repeated readings must confirm that the wet bulb has reached its equilibrium position.

(iii) Stand clear of the thermometers between readings.

(iv) The wet bulb sleeve must be clean and in good thermal contact with the thermometer, tied with cotton above the bulb and 6 mm below it, so as not to constrain it.

(v) After reading, examine the wet bulb from below, with the radiation shield in place, to check that no bead of water bridges the gap between bulb and shield.

(vi) Verify that there is no moisture or ice on the inside of the shield or on the dry bulb.

(vii) The sleeve must be wetted with the shield removed.

(viii) Use distilled water or rain or melted ice.

(ix) Below 0 °C see whether the wet bulb is covered with ice or with super-cooled water, for C in (1.17). If in doubt, get it to freeze by touching it with a piece of ice or cold metal.

(x) Below −24 °C [34] advises that e is best got by assuming that $e = 0.6e'$.

† In [113], 3 Edn, p. 54 the figure 1006 in (1.37) is wrong. It should have been 755. The figures here given for C in (1.17) include the factor 1/755.

1.35. The atmosphere. Normal conditions

(*a*) *Objects.* Knowledge of the variations of P, T, and e, and thence of n_G is required in connection with EDM for two purposes.

(i) When P, T, and e vary along the path, the average velocity will be given by $v = c/\bar{n}_G$, where $\bar{n}_G$ is the average value of n_G. In practice, observations can generally only be made at the two ends of the line, and n_G is taken to be the mean of its end values. This assumption may not be a good one, and although it is seldom possible to do better, it is necessary to consider the size of the possible error, and what circumstances are least favourable.

Except in a horizontal line $\bar{n}_G$ is not quite the same as n_G computed with $\bar{P}$, $\bar{T}$, and $\bar{e}$, the average values of P, T, and e, but when considering the effect of meteorological irregularities, it is convenient to consider the possible differences between $\bar{P}$, $\bar{T}$, $\bar{e}$, and P_m, T_m, e_m, the means of the two end values.

(ii) The radius of curvature of the path, σ, and thence the course of the path, is also required (*a*) because the values of P, T, e, and n at any point depend on its height above ground or sea-level, and (*b*) because the measured length D is longer than the chord C, from which the corresponding spheroidal length can be computed, § 1.45.

It is shown in § 3.17 that

$$\frac{1}{\sigma} = -\frac{1}{n}\frac{dn}{dh}\cos\beta, \tag{1.18}$$

where β is the slope of the line, n is the phase index, h is height (positive upwards), and the negative sign makes σ positive when dn/dh is negative and the path is above the chord, as is most usual. In the present context $1/n$ and $\cos\beta$ can be taken as unity, and

$$\frac{1}{\sigma} = -\frac{dn}{dh}. \tag{1.19}$$

It is convenient to write

$$K = R/\sigma, \quad \text{where } R \text{ is the earth's radius.} \tag{1.20}$$

If σ is constant, $\tfrac{1}{2}K$ then equals k, the surveyor's *coefficient of refraction* as in § 3.18. Note that microwaves do not have the same K or k as light.

(*b*) *Water in the atmosphere.* The lower atmosphere consists of a mixture of dry air of very uniform composition and of average molecular weight 28.9, with a variable amount of water vapour of molecular weight 18.0. In any particular volume of air the pressure, temperature, and density ρ of these two constituents are independently related by the perfect gas law

$$(P-e)=c_d\rho_d T \tag{1.21}$$

and

$$e=c_w\rho_w T, \tag{1.22}$$

where P is the total pressure, e is the partial pressure of water vapour, T is in K necessarily the same for both, and suffixes d and w refer to dry air and water vapour respectively. $c_d=2.87\times10^2$, and $c_w=4.61\times10^2\ \mathrm{m^2\,s^{-2}\,K^{-1}}$, [547], p. 2.

It follows that

$$\frac{\rho_w}{\rho_d}=\frac{2.87}{4.61}\frac{e}{P-e}=0.622e/(P-e). \tag{1.23}$$

So in any particular volume the ratio of the two pressures is not directly affected by changes of temperature, and can only significantly change if water vapour is added or subtracted by evaporation, condensation, or mixing, [547], p. 5.

The density ρ of damp air is given by

$$P=c_d\rho T(1+0.378e/P),$$

whence

$$\rho=\frac{0.3483P}{T(1+0.378e/P)}\ \mathrm{kg\,m^{-3}}\quad \text{if } P \text{ and } e \text{ are in mbar}$$

$$=\frac{0.3483\times10^{-2}P}{T(1+0.378e/P)}\ \mathrm{kg\,m^{-3}}\quad \text{if } P \text{ and } e \text{ are in Pa.} \tag{1.24}$$

T is in K.

If $P=1000$ mbar (10^5 Pa), and $T=273.16$ K, and $e=0$, $\rho=1.276\ \mathrm{kg\,m^{-3}}=1.276\ 10^{-3}\ \mathrm{g\,cm^{-3}}$.

(*c*) Total pressure decreases with height. Considering the equilibrium of a metre cube of air

$$\frac{dP}{dh}=-\rho g=-\frac{0.3483\times10^{-2}P}{T(1+0.378e/P)}\times\frac{g}{g_{45}}\times g_{45}\ \mathrm{Pa\,m^{-1}}$$

introducing g_{45} ($9.806\ \mathrm{m\,s^{-2}}$) for convenience.

Whence

$$\frac{dP}{dh}=-\frac{0.0342P}{T(1+0.378e/P)}\times\frac{g}{g_{45}} \tag{1.25}$$

$$=-0.0342(P/T)\ \text{within 1 per cent.} \tag{1.26}$$

In (1.25) and (1.26) the gradient is in Pa m^{-1} if P and e are in Pa, or in mbar m^{-1} if they are in mbar.

Integrating (1.25)

$$P = P_0 \exp\left\{\frac{-0.0342h}{(g_{45}/g_m)T_m(1+0.378e/P)_m}\right\}, \tag{1.27}$$

where P_0 and e are sea-level pressures in either Pa or mbar, h is in metres, T is in K, and suffix m indicates mean values between sea-level and height h.

(*d*) *Temperature* ordinarily decreases with height, but inversions occur and the temperature gradient is far less predictable than the pressure gradient. The adiabatic gradient for dry air is −0.00986 °C/metre, or close to −1 °C/100 metre, but at a height of more than a few hundred metres above the ground the more usual average value is

$$(\mathrm{d}T/\mathrm{d}h) = -0.0055\ ^\circ\text{C/metre}. \tag{1.28}$$

See § 1.36 for departures from this normal value.

(*e*) *Water vapour pressure* also tends to decrease with height, but even less regularly than T. [60], p. 63 gives as a standard

$$e = e_0 \exp(-h/2700\ \text{metres}), \tag{1.29}$$

where e_0 is an average sea-level value at the time and locality. As an alternative, [139], p. 328 for temperate latitudes gives

$$e = 10\ \text{mbar} - 1\ \text{mbar per 300 metres of } h. \tag{1.30}$$

Neither formula pretends to be more than an approximation to average conditions, and no useful formula can be given for $\mathrm{d}e/\mathrm{d}h$, although if a standard is required (1.29) gives

$$\frac{\mathrm{d}e}{\mathrm{d}h} = -\frac{e_0}{2700}\exp(-h/2700)\ \text{mbar/metre}. \tag{1.31}$$

(*f*) *Standard curvature. Light.* Differentiating (1.10), and ignoring e gives

$$\frac{\mathrm{d}n}{\mathrm{d}h} = (n_0-1)(0.269)\left(\frac{1}{T}\frac{\mathrm{d}P}{\mathrm{d}h} - \frac{P}{T^2}\frac{\mathrm{d}T}{\mathrm{d}h}\right)\text{metres}^{-1}, \tag{1.32}$$

P being in mbar and T in K. Using (1.26) and (1.28) for $\mathrm{d}P/\mathrm{d}h$ and $\mathrm{d}T/\mathrm{d}h$, and putting $(n_0-1) = 0.000293$ gives (for metres, millibars, and K)

$$\frac{\mathrm{d}n}{\mathrm{d}h} \times 10^6 = \frac{\mathrm{d}N}{\mathrm{d}h} = -2.26\frac{P}{T^2}. \tag{1.33}$$

The first three lines of Table 1.1 give the values of dN/dh and σ/R, from (1.19), corresponding to typical values of P and T at different heights. An average value is

$$\frac{dN}{dh} = -0.023\ \text{m}^{-1}; \quad \text{and} \quad \text{from (1.19) } \sigma \text{ (for light)} = 7R, \tag{1.34}$$

corresponding to $k = 0.072$. See § 3.17 for more details.

(g) *Standard curvature. Microwaves.* Differentiating (1.16), and neglecting two small terms, gives

$$\frac{dn}{dh} \times 10^6 = \frac{dN}{dh} = \frac{77.6}{T}\frac{dP}{dh} - \frac{1}{T^2}(77.6P + 740\,000e/T)\frac{dT}{dh} + \frac{370\,000}{T^2}\frac{de}{dh}. \tag{1.35}$$

Using (1.25) and (1.28),

$$\frac{dN}{dh} = -2.22\frac{P}{T^2} + 4100\frac{e}{T^3} + \frac{370\,000}{T^2}\frac{de}{dh}, \tag{1.36}$$

in metres, millibars, and K units. Multiply the right-hand side by 10^{-2} if P and e are measured in Pa. In this formula the term in de/dh is important and may be dominant.

The first three lines of Table 1.2 give the values of dN/dh and of σ/R at different heights, firstly with $e = \frac{1}{2}e'$ and secondly with $e = e'$ (saturated). The value of de/dh is taken from (1.31). The values of σ/R are very variable, but the result

$$\frac{dN}{dh} = -0.040\ \text{m}^{-1}, \quad \text{and} \quad \sigma \text{ (for microwaves)} = 4R, \tag{1.37}$$

is the value generally accepted as standard, corresponding to $K = 0.25$ and $k = 0.125$. Important abnormalities frequently occur; see §§ 1.36 and 1.37. At great heights σ will be greater, less doubtful, and more equal to that of light, such as equal to $5R$, $7R$, and $8R$ at 3, 4, and 6 km respectively.

[510] gives tables for computing σ for both light and microwaves, given the meteorological data.

(*h*) *Curvature correction.* Referring to § 1.35 (*a*) (ii), the curvature of the path ACB in Fig. 1.14 raises the mid-point C above the chord by a distance $\text{CD} = D^2/8\sigma$, while the earth's curvature raises the surfaces of (presumed) equal n by a distance $\text{C'D'} = D^2/8R$. At C the refractive index n will therefore be greater than the mean of its values at A and B by

$$-\frac{D^2}{8}\left(\frac{1}{R} - \frac{1}{\sigma}\right)\frac{dn}{dh} = \frac{D^2}{8R^2}(K - K^2), \tag{1.38†}$$

† In (1.38), if light is being considered dn/dh must be the gradient of the group index n_G, while $1/\sigma$ equals the gradient of the phase index, but in this context the distinction is immaterial.

TABLE 1.1†
Curvature of light

h/m	P/mbar	T/K	Time	$\mathrm{d}T/\mathrm{d}h$ ÷(°C/m)	$\mathrm{d}N/\mathrm{d}h$ ÷m	$k=R/2\sigma$ § 3.18	σ/R	Note
0	1000	300	Afternoon	−0.0055	−0.025	+0.080	+6.2	(i)
1500	840	285	Any time	−0.0055	−0.023	+0.072	+7.0	(ii)
3000	700	275	Any time	−0.0055	−0.021	+0.066	+7.6	(iii)
0	1000	300	Midday	−0.033	0	0.0	∞	(iv)
0	1000	300	Midday	−0.07	+0.032	−0.10	−5.0	(v)
0	1000	270	Night	+0.11	−0.157	+0.50	+1.0	(vi)
0	1000	270	Night	+0.20	−0.254	+0.81	+0.62	(vii)

Notes

(i) Between hills near sea-level. In the afternoon, or all day if cool and cloudy with moderate wind.
(ii) Between hills well clear of ground. All day.
(iii) Between mountains. All day.
(iv) About 10 metres above bare plains. Midday, still and clear.
(v) As (iv) but more extreme. The path is convex towards the ground. Probably some heat mirage.
(vi) About 10 metres above bare plains. Night. Still and clear. $\sigma = R$.
(vii) As (vi) but more extreme. k's of 1.2 have been recorded. [591] Appendix III.

Cold air above warm water will give a curvature of the same kind as (iv), and hot air above cold water will have an effect similar to (vi).

TABLE 1.2
Curvature of microwaves
$\mathrm{d}T/\mathrm{d}h = -0.0055$ °C/m. $\mathrm{d}e/\mathrm{d}h$ *in first three lines is from* (1.29)

			$e=\frac{1}{2}e'$				$e=e'$			
h/m	P/mbar	T/K	e/mbar	$\mathrm{d}e/\mathrm{d}h$ ÷(mbar/m)	$\mathrm{d}N/\mathrm{d}h$ ÷m	σ/R	e/mbar	$\mathrm{d}e/\mathrm{d}h$ ÷(mbar/m)	$\mathrm{d}N/\mathrm{d}h$ ÷m	σ/R
0	1000	300	17.9	−0.0066	−0.050	+3.1	35.8	−0.0132	−0.075	+2.1
1500	840	285	7.0	−0.0038	−0.039	+4.0	14.0	−0.0076	−0.056	+2.8
3000	700	275	3.6	−0.0022	−0.031	+5.1	7.1	−0.0044	−0.041	+3.8
2	1000	295					26.0	−0.066	−0.31	+0.5
2	1000	295					26.0	−0.031	−0.155	+1.0

Notes. The first three items are for normal P and T at different heights (a) with e rather dry ($=\frac{1}{2}e'$), and (b) very damp ($e=e'$).
The last two items are for low lines at 2 metres above damp grass in daylight in summer in Europe; see § 1.37 (e).

† In this and other tables, a column heading such as h/m or $h \div$m, etc. implies that the figures in that column are pure numbers, namely the ratio of the physical quantity h to the unit of length m. See [53] pp. 6–7.

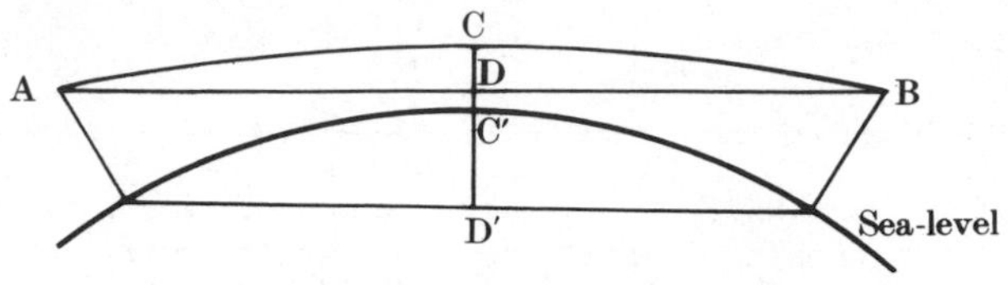

FIG. 1.14. AB = D km. Radius of path ACB = σ.

using (1.19) and (1.20). The average difference along AB will be two-thirds as much, so on this account

$$\begin{aligned}\bar{n}-n_m &= (D^2/12R^2)(K-K^2)10^6 \text{ ppm}\\ &= (\mathrm{D/R})^2 10^4 \text{ ppm if } \sigma = 7R \text{ for light}\\ &= 1.6(D/R)^2 10^4 \text{ ppm if } \sigma = 4R \text{ for microwaves.}\end{aligned} \tag{1.39}$$

If this is ignored when computing D, the preliminary value of D, which varies as $1/n$, will require a correction of $-(D^3/12R^2)(K-K^2)$. See further in § 1.45.

(*i*) *Exponential pressure correction.* If the two ends of a line are at very unequal heights h_1 and h_2 metres, the roughly exponential variation of P with h will cause $\bar{P}$ and P_m to be unequal. If P_1 is the pressure at h_1

$$\bar{P}-P_m \text{ (on this account)} = -\frac{P_1}{12}(h_2-h_1)^2/(8000)^2 \text{ mbar.} \tag{1.40}$$

If $h_2-h_1 = 1000$ m, this is only 1 mbar, corresponding to 0.3 ppm.

This is of no consequence. In a steep line $\bar{n}$ would in any case necessarily be computed from independently estimated values of P, T, and e at different heights rather than from $\bar{P}$, $\bar{T}$, and $\bar{e}$.

1.36. The atmosphere. Irregular conditions

(*a*) *Temperature.* The temperature of the air within 500 metres of the ground is much influenced by the temperature of the ground over which it has been passing. Direct heating of the air by the sun, and its cooling by radiation, are less important.

Over land. Ground is heated by sunlight and chilled by exposure to a clear night sky. Its changes of temperature are communicated to the adjacent air, and thence diffused upwards. The equation of heat transfer by turbulent air is

$$\frac{\partial T}{\partial t} = \frac{\partial}{\partial h}\left[\kappa\left(\frac{\partial T}{\partial h}+\Gamma\right)\right], \tag{1.41}$$

where T is temperature, t is time, h is height, Γ is the adiabatic gradient, and κ is the conductivity, of dimensions l^2t^{-1}. If κ is constant

$$\frac{\partial T}{\partial t} = \kappa \frac{\partial^2 T}{\partial h^2}. \tag{1.42}$$

The conductivity of still air is extremely small, but when heat is carried upwards by eddies the *eddy conductivity* is very much greater. Given T at ground level as a function of t, and given κ, T at any height and time is known. An approximation for T at ground level is

$$T_{(0,t)} = T_{(0,m)} + A\cos(\tfrac{1}{12}\pi)(t - t_0), \tag{1.43}$$

where $T_{(0,m)}$ is the mean of the day, $2A$ is the diurnal temperature range, t is time in hours from midnight, and t_0 hours is the time of maximum T_0, suffix 0 indicating ground level. Then if κ is constant, $T_{(h,t)}$ at height h and time t is given by

$$T_{(h,t)} = T_{(0,m)} + \lambda h + Ae^{-h/h'}\cos[(\tfrac{1}{12}\pi)(t - t_0) - h/h'], \tag{1.44}$$

where λ is the gradient of the diurnal mean T, such as -0.0055 °C/m, and $h' = (24\kappa/\pi)^{\frac{1}{2}}$ if κ is in metre^2 per hour units. In plain language, (1.44) says that the daily temperature range is reduced to 1/2.7 of its surface value at height h' above the ground, and that the daily maximum temperature at this height occurs $12/\pi$ hours after the ground-level maximum, with lag at other heights in proportion to h.

This simple theory is complicated by the fact that κ and h' depend on the vigour of the air's turbulence, and are not constant† with either height or time. They will be small close to the ground, where eddies are necessarily small, increasing at greater heights. [546] suggests that up to 100 metres above the ground κ is proportional to $h^{1.6}$ or $h^{1.8}$. It may reach a maximum at about 250 m, [139], p. 197. It is also clear that κ will be smaller in still air than in a strong wind, and when dT/dh is positive (temperature inversion) than when it is negative.‡ These last two considerations may cause κ to be smaller at night than in the afternoon.

Fig 1.15 shows T at different times of day at heights of between 1.2 and 87 m in England in spring. Between 10.00 and 16.00 hours the average gradient above 10 m is very closely adiabatic. Fig. 1.16 shows a typical variation of diurnal maximum and minimum temperatures at heights of up to 1000 m.

[546], p. 40 gives an approximate solution of (1.41) when κ is proportional to h^c, which suggests that the time-lag of maximum temperature

† If κ is not constant, h' ceases to be the height at which the temperature range is decreased by 1/2.7, but it is still convenient to think in terms of a variable height h' which is defined as equal to $(24\kappa/\pi)^{\frac{1}{2}}$, where κ has its variable value.

‡ When the gradient much exceeds the adiabatic gradient of 0.01 °C/m there is liable to be direct upward convection, as may often be seen in dusty countries.

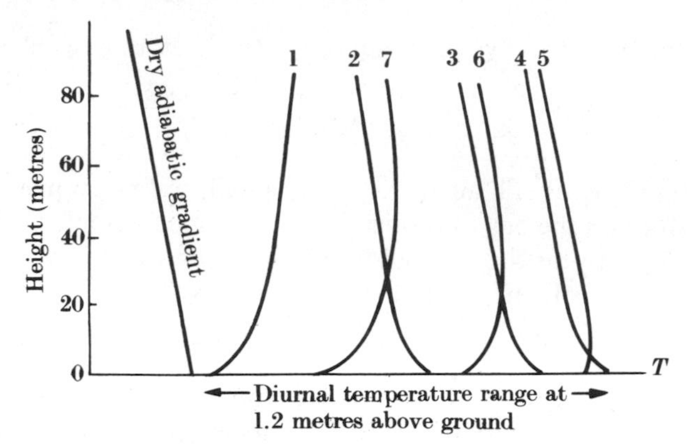

FIG. 1.15. Variation of temperature with height in the first 80 metres above the ground. Curve (1) = sunrise, (2) = 10.00 hours, (3) = 3 hours before maximum temperature, (4) = 1 hour before maximum, (5) = 1 hour after maximum, (6) = sunset, (7) = 22.00 hours. England, spring. Based on [547], Fig. 21.

may vary as $h^{(1-\frac{1}{2}c)}$ instead of as h. If $c = 1.6$ as above, the lag may then vary as $h^{0.2}$, so after rapidly reaching a value of (say) 1 hour at h = (say) 10 m, [547], p. 206, it may not increase to 2 hours until a height of 320 m. This is in better accord with observation than the solution (1.44). Above 1000 m the diurnal change of T due to radiation exceeds that brought up from the ground, and the time of maximum will tend towards midday.

Making empirical allowance for the effect of radiation, and for a slow rate of increase of time lag with height, [387] gives

$$T_{(h,t)} = T_{(0,m)} + \lambda h + e^{-h/h'} f(t - h/V) + \phi(t), \tag{1.45}$$

where $T_{(0,m)}$, λ, h, and h' are as in (1.43) and (1.44). $\phi(t)$ is the estimated

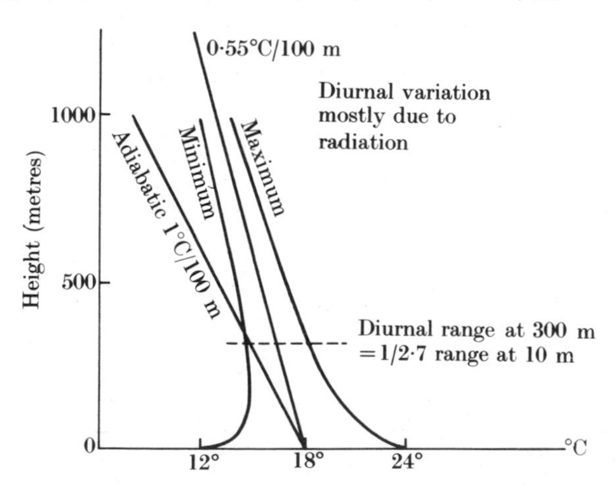

FIG. 1.16. Typical variation of diurnal maximum and minimum temperatures with height.

effect of radiation, namely $\frac{1}{8}(T_{(0,t)}-T_{(0,m)})$, constant at all heights, suffix 0 referring to ground level. $f(t)=\frac{7}{8}(T_{(0,t)}-T_{(0,m)})$. For $f(t-h/V)$ write $(t-h/V)$ for t. V is the velocity with which the diurnal maximum temperature is transmitted upwards, estimated to be 500 metres/hour. So the lag is h/V hours.

This formula gave good agreement with the observed diurnal variation of vertical angles between triangulation stations in France.

Values of κ and h' may be obtained in several ways, as below.

(i) By measuring the temperature changes at two heights.

(ii) By the differences in wind direction and velocity at different heights.

(iii) By the diurnal variation of atmospheric refraction in vertical angles.

Table 1.3 gives some recorded values of κ and h'. Their increase with height is apparent.

TABLE 1.3

Values of eddy conductivity κ in metre² per hour, and of $h'=\sqrt{(24\kappa/\pi)}$ metres

Place	Season	(Height range)/m	κ†÷ (m^2/hour)	h'/m‡	Method	Reference
Porton, England	Mar., June and Dec.	0.025–0.30	0.8	2.4	(i)	[547] quoting [88]
		0.3–1.2	16	11		
		1.2–7.1	320	49		
		7–17	1800	117		
Punjab, India	Cold	7–12	400	55	(iii)	[591], App. III
Leafield, England	Spring	1.2–87	3800	170	(i)	Fig. 1.15
Greenland, Ice	June	1–2.6	550	65	(iii)	[456]
Boston, USA Sea		0–6	52	20	(i)	[139], p. 197
		0–15	250	43		
		0–30	650	71		
		0–60	1360	103		
		0–90	2400	136		
Newfoundland. Sea	May–Aug.	0–about 500	1050	90	(i)	[554]
Salisbury, England		0–650	18 000	370	(ii)	[554]
Eiffel Tower, Paris	Jan.	18–302	15 000	340	(i)	[554]
	Aug.	18–302	64 000	700	(i)	[554]
Trappes, France	Hot years	167–1000	33 000	500	(i)	[387], p. 42
	Wet years	167–1000	14 300	330	(i)	[387], p. 42
South Africa	All year	0–4500	13 400	320	(i)	[20]
Various. Lat. 26–37°	Summer	0–4500	19 000	380	(i)	[20]

† κ is generally quoted in centimetres² per second. Multiply by 0.36 to give metres² per hour.

‡ If the range of h in column 3 is not grossly dissimilar to h', the latter correctly gives the height at which the diurnal temperature range at the lower level is multiplied by 1/2.7. The figure 1.2 m should probably be substituted for 0, where the latter is entered in column 3, after the first two lines.

Temperature over water. The temperature of air in contact with the sea will be the same as that of the surface water, with little or no diurnal variation. If the air has travelled far over water of uniform temperature, its vertical temperature gradient is likely to be about −0.0055 °C/m. On the other hand, if it has recently come from over land, so that its temperature is likely to differ from that of the sea, the gradient will be abnormal. A change of temperature will be transmitted upwards by eddy conductivity in accord with (1.41), but in this situation the solution (1.44) is not appropriate, because the gain or loss of heat is not diurnally periodic. See under *Water vapour*, and Fig. 1.17 below.

(*b*) *Water-vapour pressure.* The water content of a particular volume of air can be changed by any of the following.

(i) Evaporation from a water or ice surface, or from damp land, and consequent eddy diffusion.
(ii) Evaporation from rain.
(iii) Mixing with other air.
(iv) Condensation.

Air in contact with a water surface has the same temperature as the water, and is saturated at that temperature. If the air above it contains less water vapour, and is unsaturated, the vapour will be distributed upwards in accordance with

$$\frac{\partial \chi}{\partial t} = \frac{\partial}{\partial h}\left(\kappa \frac{\partial \chi}{\partial h}\right), \tag{1.46}$$

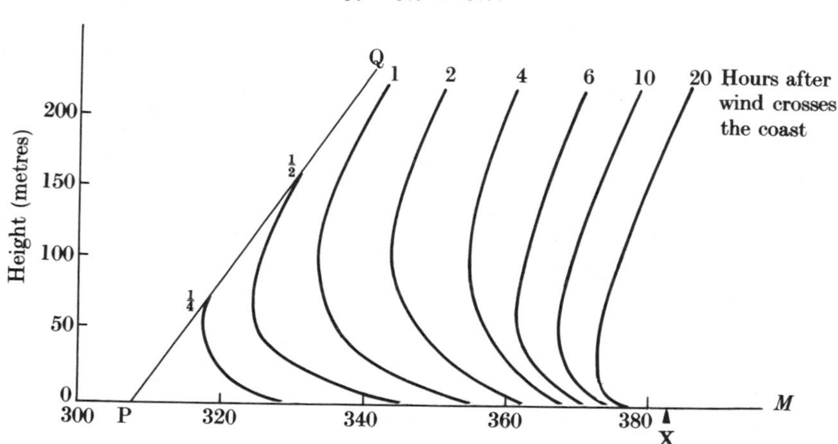

FIG. 1.17. Values of M for microwaves over the sea. From[139].

$M = N + (0.16$ per metre of $h)$.

All the curves pass through the point X. $PX = 74$ N-units.

At any height the horizontal distance between any curve and the line PQ equals the number of units by which N at that height has changed after the time interval concerned. See §1.36 (*b*).

where χ is the concentration (mass/volume) of water vapour, and κ is substantially the same† as the κ of (1.41). A dry land wind blowing over the sea thus becomes saturated at the surface immediately after passing the coast, and the increase of water content extends upwards to greater heights as the air passes further over the sea.

Fig. 1.17, from [139], p. 198, shows the changes in the index of refraction (for microwaves) in an initially warm (32 °C) and dry ($e =$ 12.3 mbar) land wind passing over sea at $T = 22$ °C to which corresponds $e' = 26.5$ mbar. The diagram covers the first 20 hours after passing the coast. The index actually plotted is

$$M = N + (h \times 10^6)/(\text{earth's radius}) \\ = N + 0.16 \text{ per metre of height.‡} \tag{1.47}$$

If δN is the initial difference of N between the dry wind and the air in contact with the sea, the diagram shows the proportion of δN reaching different heights after different intervals of time. Thus after a quarter of an hour 6 units out of 74 have reached 30 metres, but none have reached 60 metres.

Temperature differences between the water and the air from the land will be transmitted upwards in the same way, in the sense that the proportions of initial δT, δe, and δN reaching a certain height after a certain interval of time are likely to be the same. In fact, [139], p. 198,

$$\frac{T_L - T}{T_{0L} - T_w} = \frac{e_L - e}{e_{0L} - e_w} = 1 - E\left(\frac{h}{\sqrt{(4\kappa t)}}\right), \tag{1.48}$$

where $E(\xi) = \dfrac{2}{\sqrt{\pi}} \displaystyle\int_0^\xi e^{-x^2}\,dx$, which is tabulated in most compilations,

T and e refer to height h at time t after crossing the coast,
T_w and e_w refer to air at sea-level on the sea,
T_L and e_L refer to height h over the coastline,
T_{0L} and e_{0L} refer to sea-level on the coastline.

Note that κ must be in the same units as h and t.

(*c*) *Irregularities at greater heights.* Abrupt changes of T and e often occur when a body of hot dry air overlies colder and more humid air. There may be a sudden upward increase of temperature of a few °C. [139], pp. 20–4 and [371] illustrate cases where there are abrupt decreases of 20 or more units in N (for microwaves) at various heights of

† [547], p. 319. The two κ's become unequal when the gradient is a large negative.

‡ The term 0.16 m^{-1} has been arbitrarily included in order to give the baseline PQ a slope which keeps the curves well separated.

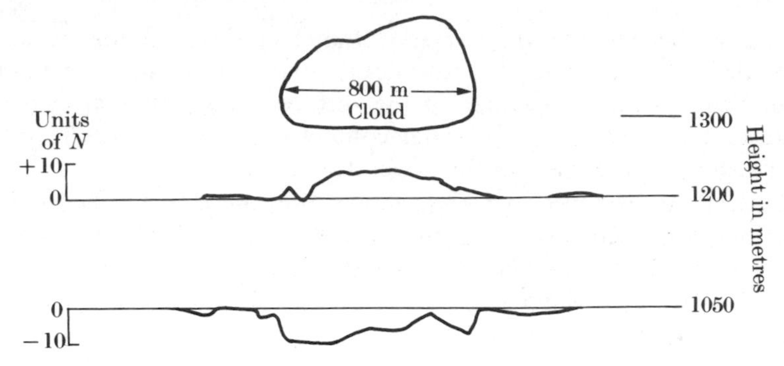

FIG. 1.18. Variations of N under cumulus cloud. From [497].

between 100 and 2000 metres, and [139], p. 203 shows a temperature inversion of 13 °C at 1000 m.

Fig. 1.18, from [497], shows an increase of 8 units of N for microwaves immediately under the base of cumulus cloud at 1200 metres compared with the same height not underlying cloud, while at the same time there was a relative decrease of 10 units under the cloud 150 metres lower down. The associated temperature variations were only about 1 °C.

1.37. Average refractive index and curvature along a line

Continuing from § 1.35 (*a*) (i), this paragraph summarizes the possible errors arising from the assumption that $\bar{P}$, $\bar{T}$, $\bar{e}$, and $\bar{n}$ are equal to P_m, T_m, e_m, and n_m, in the variable conditions described in § 1.36.

(*a*) *Total pressure.* The assumption that $\bar{P} = P_m$ in a line of under 100 km, after correction for curvature, is very unlikely to be wrong by 3 mbar (1 ppm), as may be seen from typical daily weather charts.

(*b*) *Temperature.* In lines between high hills errors of several degrees may arise if the line passes across a front between bodies of air of different origins, but any such errors made on one day are unlikely to be repeated on the next. If the hills are sharp, and if most of the line is 500 m or more above the ground, the diurnal variation of temperature at the end stations should not much differ from the mean along the line. The diurnal range at the stations can be measured, and if it is much greater than Fig. 1.16 suggests, anxiety may be felt. Error may be lessened by observing near the time when the temperature equals the mean of the day.

In flat ground, lines between towers will be short, and temperature errors should be small, provided the character of the ground is uniform. But if stations are on arid desert, while most of the line is over forest, cultivation or water, errors of as much as 5 °C may occur at night or in the afternoon. As before, the time when T equals the daily mean will be the best.

If a line passes at an altitude of 100 or 200 m above ground or water which is so variable as to require special treatment, ground-level T can be observed throughout the day at selected places, and Figs. 1.15 and 1.16 can be used to estimate the temperature on the path. Such estimates are possibly best made between 10.00 and 14.00 hours when dT/dh may be a uniform -0.01 °C/m. At night they will be less satisfactory.

In the US zero-order geodimeter traverses, § 1.06 (*c*), in non-mountainous country the temperature at the mid-point of the path is measured by balloon, except in windy weather when that is diffcult and also less necessary. This measure is given double weight in combination with the two end measures.

Large errors due to wrongly estimated $\bar{T}$ may be detected and their effects reduced, if lines are measured once by day, when the ground is hotter than the air, and once by night, but this is not possible with the (non-laser) geodimeter.

Errors may be very large within two metres of the ground, and no considerable fraction of the path should be so low. If the path and the thermometers can be kept above 10 or 15 metres, so much the better. See also § 1.34 (*b*).

[509] suggests that in a line with a large, known, height difference, $\bar{T}$ might be estimated from measures of P at the two ends using the barometric formula (1.27) with P_1 and P_2 in place of P_0 and P, and $(h_2 - h_1)$ for h. Errors of 1 °C in $\bar{T}$ would be caused by errors of 0.1 mbar in either pressure, or of 0.7 metres in $(h_2 - h_1)$, if $(h_2 - h_1) = 200$ m. Very good barometers would be necessary, and isobaric charts for any but short lines.

Simultaneous use of two colours. For the use of 2-colour EODM to eliminate temperature errors, see § 1.40.

Favourable conditions for the estimation of $\bar{T}$ are as follows.

(i) Short lines

(ii) Cloudy skies and moderate wind with small variations of T.

(iii) High lines between steep hills, provided there are no large temperature inversions.

(iv) Low lines above very flat ground of uniform character, preferably forested, avoiding heights of less than 2 or better 10 metres.

(v) Observations made near the time when T equals the mean of the day.

Temperature is the critical factor in estimating the velocity of light.

Apart from the possibility of using two colours, there is clearly liable to be an error of 1 °C in almost any line, but errors of 3 °C should be avoidable except in recognizably bad circumstances. It may be hoped that the systematic average error in a number of lines will not exceed 1 °C.

(*c*) *Water-vapour pressure and n for microwaves.* The estimation of $\bar{e}$ is the critical factor in determining the velocity of microwaves, since an error of 1 ppm arises from one of only 0.2 mbar in $\bar{e}$.

Examples quoted in § 1.36 (*c*) show that at quite high altitudes, such as 1000 m, the refractive index may differ by 20 *N*-units (ppm) between two adjacent bodies of air, and Fig. 1.17 shows that at low levels over water it may vary by 50 *N*-units in the lowest 10 metres, and by 10 units in the next 30 metres. These variations are due to *e* rather than to T. Ordinarily, of course, the average will not differ from the mean of the ends by these amounts, but it is not impossible for at least one end to be quite untypical of most of the line.

In spite of these doubts, microwave instruments undoubtedly do generally produce good results. Assuming that *e* is the main source of the errors there quoted, it seems that $\bar{n}$ is commonly estimated with an s.e. of $\pm 5 \times 10^{-6}$ in a single measure, which is equivalent to an s.e. of ± 1 mbar in $\bar{e}$. But it must be realized that in circumstances which are both unfavourable and unfortunate, errors five times as great are not impossible.

In Australia, which is generally a favourable country for microwaves, it is specified that the values of *e* at the two ends of a horizontal line should agree within 1.6 mbar (8 *N*-units), and meeting this specification seldom causes difficulty.

In general [312] it appears that EODM distances (made at night with a non-laser) tend to average about 2 ppm greater than MDM (made by day). Errors in the estimated temperatures by day and by night may contribute, but the most probable source of error is the estimated values of *e* for the MDM. The EODM values are certainly to be preferred. See § 2.44.

Unfavourable meteorological circumstances for the determination of the velocity of microwaves are as follows.

(i) A mixture of land and sea, or of desert and irrigation.
(ii) A low line over sea, when the wind is blowing off the land.
(iii) Lines grazing within 2 metres of water or damp land are liable to extraordinary error. A clearance of at least 10 metres is most desirable.
(iv) Low lines, at or near the time when dew is falling or evaporating.
(v) High temperatures.
(vi) Long lines.
(vii) A climate characterized by alternations of wet and dry air masses is bad for high-altitude lines.
(viii) Lines passing close below cumulus cloud base, or humidity measures made in the same situation.

The following are relatively favourable circumstances.

(ix) A line high above land, distant from any sea up-wind. But high lines may be bad for ground-swing, § 1.40.

(x) A low line (but above 10 metres) over uniform, flat, arid land far from the sea or other sources of moisture.

(*d*) *Curvature of light*, dn/dh. The curvature of light affects the vertical angles of triangulation, and is more fully dealt with in §§ 3.17–3.20. Very abnormal curvatures may occur near the ground, as may be deduced from the variations of T quoted in § 1.36 (*a*), or from the observation of reciprocal vertical angles.† Table 1.1, items (iv) to (vii), show how the radius of curvature σ, which is normally $7R$, may vary between $\frac{1}{2}R$ over land at night, through ∞, to $-5R$ (convex to the earth) over land in the afternoon, When $\sigma = R$, as may occur over flat ground at night or over cold water by day, a horizontal light ray will follow the curve of the earth so long as the conditions persist.

Fortunately, for EDM the corrections depending on σ are small, especially for short lines such as may lie wholly within the most disturbed levels of the air. Thus if $D = 20$ km, and $\sigma = \frac{1}{2}R$ with $K = 2$, (1.39) gives a correction of 1.7 ppm, while the conventional $\sigma = 7R$ gives sensibly zero.

The most unfavourable line for the geodimeter, from this point of view, would be a very low line over a cold sea, possibly giving a greater range than would have been expected. In such a line the observation of vertical angles will determine σ and avoid risk of error.

(*e*) *Curvature of microwaves*. dn/dh. [282] summarizes the results of many observations of microwave path curvature over flat, damp fields in England and Germany. At heights of 40 m above the ground the figure $\sigma = 4R$ is well confirmed, but at 2 metres σ may equal R or even $\frac{1}{2}R$ as a monthly mean during daylight in summer. Table 1.2, last two items, shows values of de/dh which can produce these curvatures. Such low lines are necessarily short, and serious errors of distance due to curvature are unlikely. With microwaves, uncertainty about e itself is likely to be much more serious than doubt about de/dh and σ. It is fortunate that de/dh and dT/dh do not both tend to give small values of σ at the same time.

If a microwave passes horizontally into a temperature inversion or into a level at which de/dh is strongly negative, its curvature may become equal to that of the earth. For some distance it will be trapped at that level, and its path will depart seriously from the normal curve.

The most unfavourable situation for the curvature of microwaves will be as in Fig. 1.19, where a very long line between hills A and D is only made possible by abnormal curvature in the centre section BC, where the

† Formula (3.53), in which ψ_P and ψ_Q may be neglected for the present purpose. Note that the vertical angles of light do not give the curvature of microwaves.

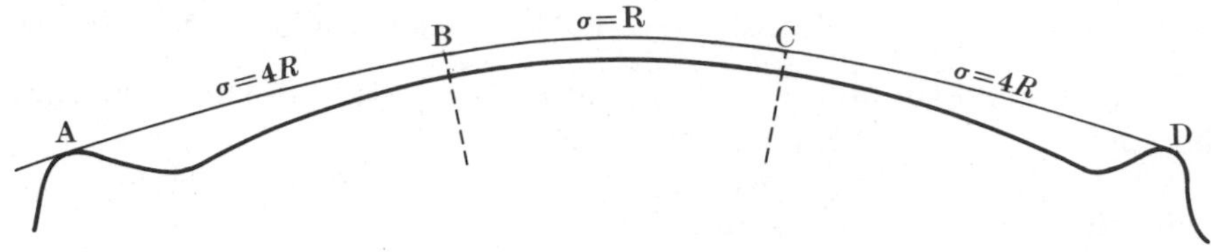

FIG. 1.19. Between B and C the path is a few metres above water.

path grazes low over damp ground or water. In most of the sections AB and CD σ will be the usual $4R$, while in BC it will be approximately R.† The curvature correction of a line which is observable at an unexpectedly great distance will best be computed on the assumption that it is curved as in Fig. 1.19.

At high altitudes, such as 1000 or 2000 m, abrupt changes in e may occur in a thin horizontal layer between two different air masses. The resulting $\mathrm{d}e/\mathrm{d}h$ may be more negative than -0.031 mbar/m, which as in Table 1.2 last item gives $\sigma = R$. Such a situation, like that at sea-level referred to above, constitutes a *duct*, in which microwaves may remain trapped following the curve of the earth. A duct will also result, with a normal e-gradient, in a temperature inversion in which $\mathrm{d}T/\mathrm{d}h$ exceeds $+0.1$ °C/metre.

General references for meteorology. [34], [139], [546], and [547].

1.38. The reflection of microwaves

Unwanted reflections of light cause no inaccuracy with EODM. The light beam is narrow, and most natural surfaces like trees and grass are bad reflectors. Trouble may occur in a EODM line grazing low over water or snow, but it will result in inability to make a satisfactory reading, rather in a reading which is wrong but apparently good.

With microwaves the position is different, and unwanted reflections of considerable strength may reach the remote instrument and there produce readings which are wrong but of apparently perfect quality, § 1.40.

A reflected microwave such as ACB in Fig. 1.20 differs from the direct wave AB as follows.

(i) Its amplitude, as received at B, is reduced by a reflection factor a, between 0 and 1. [462] gives the following formula for a for the reflection of microwaves from smooth water.

$a = a_P D$, where a_P is for reflection in a plane surface, and D allows for divergence on reflection from a curved surface. This D has no connection

† At some small height above the water surface there will be a level at which σ is a little greater than R, with smaller σ below, and greater above. This is the path which will be followed by microwaves going from B to C.

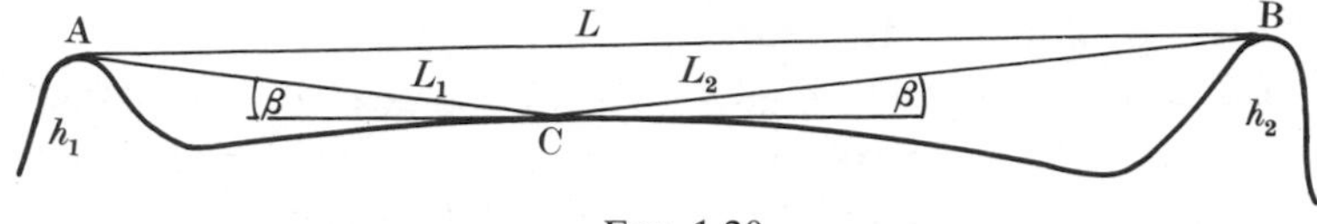

FIG. 1.20.

with the path length D.

$$a_P = (1+17.2 \sin \beta + 79 \sin^2\beta)^{-\frac{1}{2}}, \tag{1.49}$$

where β as in Fig. 1.20 is the actual reflection angle, allowing for the curvature of the path and of the earth.

$$D = [1+2L_1L_2/R'\{h_1-(L_1^2/2R')+h_2-(L_2^2/2R')\}]^{-\frac{1}{2}}, \tag{1.50}$$

where h_1 and h_2 are the heights of the stations, and L_1 and L_2 are the two sections of the total distance L.

$1/R' = (1/R)+(dn/dh) =$ conventionally $(1/R)-(1/4R)$, whence

$$R' = 8500 \text{ km}. \tag{1.51}$$

From this, a_P will exceed 0.9 if $\sin \beta$ is less than 0.01, as will often be the case. In (1.50) if $L_1 = L_2 = \frac{1}{2}L$, and if $h_1 = h_2 = h$,

$$D = (1+L^2/4R'h')^{-\frac{1}{2}}, \tag{1.52}$$

where $h' = h - L^2/8R'$. If $L = 50$ km and h' is as little as 10 m, $D = 0.3$, while if $L = 50$ km and $h' = 100$ m, $D = 0.75$. [462] gives a routine for computing β and also the excess path ACB $-$ AB.

Reflections from a rough surface will be weaker, a surface being 'rough' if, [462], the height of its irregularities exceeds (1/8)(their wavelength)$\times$ cosec β.

As an approximation [617] gives $a = 0.3$, 0.6, and 0.9 for small reflection angles over dry land, swamp, and water respectively.

(ii) On reflection the phase is changed, [462], by

$$\pi - 2.5 \sin \beta. \tag{1.53}$$

When β is small, this is nearly equal to π. If a is nearly 1, and if AB and ACB are of sufficiently nearly equal length, the combined result may be that the reflection extinguishes the direct ray.

(iii) The reflection will be more or less polarized.

(iv) The longer path of the reflected ray will result in its arriving at B out of phase with the direct ray by something additional to (1.53).

The change of phase (iv) is the source of trouble. The result of combining the modulated reflected ray with the modulated direct ray is not a double-peaked modulation but an apparently normal modulation with a change in its phase, and consequent error in the deduced distance. See § 1.43(b).

1.39. EODM. The geodimeter

This section deals with the long-range optical instruments typified by (but not confined to) the series of Aga geodimeters.† Experiments with two-colour lasers are referred to in § 1.40, short-range instruments in § 1.41, and the mekometer in § 1.42.

(*a*) *General description.* The geodimeter (or other similar instrument) is set up at one end of the line to be measured, and a reflector consisting of a number of corner cubes is set up at the other. Within the geodimeter, monochromatic light from an intense source is amplitude-modulated by a

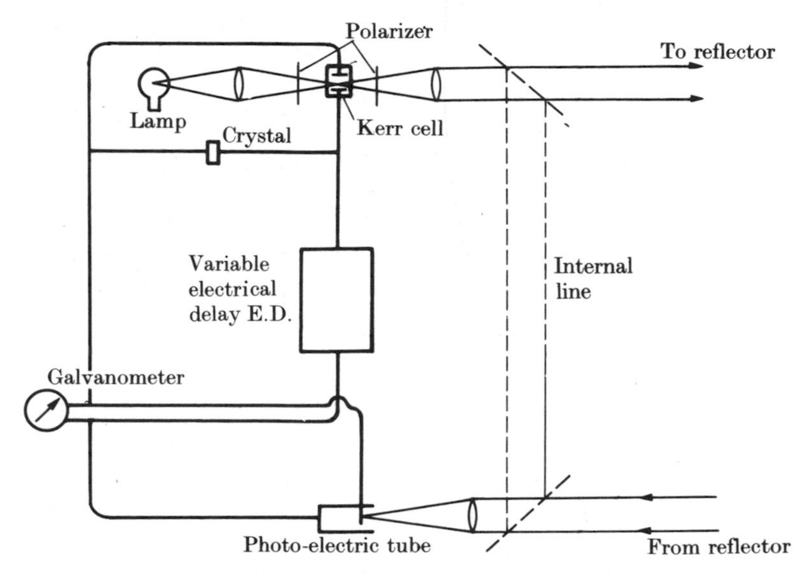

FIG. 1.21. Geodimeter. The lines to and from the reflector are arranged so as to be coaxial. The internal direct line is switched on for calibration after every measure of the reflected line.

Kerr cell controlled by a crystal oscillator, and is directed on to the distant reflector. On its return, the light is focused on to a photo cell whose output will in general be out of phase with the modulation leaving the Kerr cell, by an amount depending on the length of the line being measured. The output of the photo cell, suitably amplified, and the input voltage of the Kerr cell are applied to a galvanometer so that they will produce a null reading if they are in phase. A variable electrical time dalay (ED) is introduced into the voltage from the Kerr cell, so that if the two are not in phase, a null reading can be got by varying this delay.

† The name geodimeter for EODM as opposed to tellurometer for MDM is no longer apt, as the manufacturers of tellurometers now also produce EODM instruments, and name them as tellurometers, e.g. their MA 100.

If l_M is the modulation wavelength, null readings will be obtained when the distance D is given by

$$D = (ED \times v) + N(l_M/4) + Z, \tag{1.54}$$

where ED is the variable delay which gives a null reading, Z is a calibration correction, v is the velocity of the light, as given by (1.8), (1.10), and (1.15). N is any whole number, and $l_M = v \div$ modulation frequency. In other words, the distance D is a whole number of quarter modulation wavelengths, plus the distance corresponding to ED plus a calibration correction.

With a crystal frequency of 30 MHz, l_M is about 10 metres, and the quarter wavelength is 2.5 m. So the variable delay covers this range and measures the fraction of 2.5 m that has to be added to the whole number to give the correct distance. A second frequency, differing from the first by 2.5 per cent, enables the whole number to be determined if the distance is known to the nearest 100 m, and a third frequency resolves doubts of up to 2000 m. A fourth frequency may be added.

Any internal time lags or zero error in the electrical delay are determined after each measure of ED, by switching light from the Kerr cell directly across to the photo cell through a short internal route of known length.† See (g) below.

(*b*) *Modulation by Kerr cell.* Light from the lamp is passed through two crossed polarizers, a combination which by itself allows no light to pass. But between them is placed a Kerr cell, consisting of two metal plates separated by a layer of nitro-benzene through which the light passes. When a voltage is applied to the plates of the cell, the nitro-benzene modifies the plane polarization induced by the first polarizer, so that some light is now able to pass through the second. The intensity of the light so passing depends on the voltage applied to the plates, and amplitude modulation results.

(*c*) *The crystals.* Frequency varies with temperature but the temperature of the cyrstals is very adequately thermostatically controlled, and this is not a probable source of significant error. The frequency should be checked periodically, especially when the crystal is new. Changes such as 1 ppm per year may be recorded. See (*h*) (ii) below.

(*d*) *The light source.* Some early geodimeter models used a tungsten filament lamp, but the current model 6 gets a longer range with a mercury vapour lamp. The latter requires a petrol generator. It is not necessary for the out-going light to be monochromatic, since a colour filter is placed in front of the receiving element to give the same effect. An error of about

†This supersedes the small computation associated with every measure in the older instruments described in [113], 3 Edn, p. 77.

150 Å (0.015 μm) in the reputed wavelength of the light results in an error of 1 ppm in the group refractive index, and hence in the velocity, as given by (1.15).

In geodimeter models 8, 6BL, and 600 a laser source is used, see (*j*) below.

(*e*) *Reflectors.* A corner cube consists of three mutually perpendicular reflecting planes, and ordinarily takes the form of a glass cube cut diagonally. Light shining into the cube is reflected back to its source. The cube need not be accurately aligned, and an error of even 10 degrees is harmless.

The velocity of light through glass is 1/(1.57) times its velocity through air. It follows that the point on a corner cube assembly which should be plumbed over the station mark is behind the apex of the cube by a distance of $0.57d$ where d is the thickness of the cube from its front to its apex.

The number and size of the corner cubes required depends on the distance, the weather, and time of day, and on the light source. As many as 54 have been used.

(*f*) *Instrument statistics.* Table 1.4 gives some statistics for the Aga models 2, 6, 8, and 600, of which the first has long been obsolete. See (*j*)

TABLE 1.4
Geodimeter statistics

Model	2	6†	8 Laser	600 Laser
Maximum range. Clear, dark night/km	35–50	25	55	40
Maximum range. Clear day/km	6	5–10	55	40
Approximate crystal frequency/MHz	10 10.05 31.468	29.970 30.045 31.468	29.970 30.045 31.468	30 0.750
Instrument weight/kg	100	} 26	20	15
Transport case weight/kg	46			8.5
Tripod weight/kg	—	7	} 30	
Power unit weight/kg	24	20		3–10
Power source‡	P	P	B	B
Accuracy. p/mm	22	10	6	1–10§
For q/ppm see (*i*) below.				

† With mercury vapour lamp.
‡ B = battery. P = petrol generator.
§ 10 mm for 1-minute programme. 1 mm for 20–30 minutes.

for some description of the laser models. Such statistics as these are liable to get out of date very rapidly.

(g) *Routine of operation.* Details vary with different instruments, and the makers' instructions will always be clear. In general, readings are made in pairs of R and C, where R is the null reading for light returning from the far end of the line, and C is the reading when the light passes straight through the instrument, as in (a) above. These pairs of readings are made in sets of four corresponding to combinations of phase settings within the instrument. The mean of the four R's minus the mean of the C's then gives one value of the delay ED for (1.54).

To obtain the correct number of full quarter-wavelengths in (1.54), three or four different modulation frequencies are provided, as given in Table 1.4. The routine outlined above is repeated on the other two, or three, frequencies. The instrument automatically records the number of quarter-wavelengths, and the mean of all three frequencies is accepted for the value of ED.

The number of repetitions must depend on the circumstances and the accuracy required. Two sets of the four phase readings on all the three or four frequencies may be considered as sufficient for a single day (or night). To reduce atmospheric uncertainties a second day or night may be specified, with a repetition if the two differ unduly.

Temperatures and pressures are recorded at the start and finish of the measure on each frequency, at both ends of the line, while humidity need only be recorded at the start and finish of the day (or night). Correction for the velocity of light, in the actual atmospheric conditions, is of course made later.

Note that, in contrast to the tellurometer, each of the three frequencies is used to given an accurate measure of the distance, and the results of all three are meaned with equal weight.

(h) *Calibration* is required to eliminate three sources of error, as below.

(i) *An additive constant*, or zero error, Z in (1.54). Plumbing errors at either end of the line, and allowance for the velocity of light through glass, (e) above, could come under this heading if they were truly constant, and not due to random carelessness. The value of Z can be determined by using the instrument to measure a short known distance, but see cyclic error in (iii) below.

(ii) *Crystal frequency*, see § 1.39 (c). An error in the supposed frequency of the crystal constitutes a simple error of scale in all lines measured with the error unchanged. In principle it may be determined by the measurement of a known long distance, provided other errors have been eliminated: zero error, cyclic error, and atmospheric errors. But for calibration correct to 1 ppm the base must be at least 1.5×10^6 times the standard error of reading at each end, i.e. at least 10 km.

Alternatively, and better, the frequency can be calibrated by a frequency meter, provided the frequency of the meter is known to (say) 1 ppm, and is itself periodically tested. Frequency calibration while at work in the field may be thought to be desirable at (say) monthly intervals (for 1 ppm accuracy), unless experience with any particular crystal (or type of crystal) shows it to be unnecessary, or more frequently necessary. The seconds of radar time signals constitute a reliable measure of SI seconds, and a local crystal can in principle be calibrated by two comparisons with a time signal (correct to 0.001 s) at an interval of several hours. The necessary apparatus is included in the Chronocord, § 4.31, provided a radio receiver is available.

For long-range geodetic work it is important that the design of the instrument should incorporate a socket to which a tester can be connected without opening up the instrument.

(iii) *Cyclic error.* In some instruments there occurs a cyclic error depending on the fraction of the modulation wavelength involved in the distance measured. This is similar to the trouble in the tellurometer, described in § 1.43 (*c*) (ii). It can be corrected by multiple zero calibration measurements to give the variation of the readings ED when a target at some quite small distance is moved in short steps along a *graduated bar* about half a modulation wavelength long.

The *Nottingham baseline* [71]. This is designed for the calibration of instruments with ranges of not more than 2 km. It is a base 820 m long, which contains five intermediate pillars. It has been measured by mekometer (which has no cyclic error), § 1.42, and also by a 70-m Hewlett-Packhard laser interferometer. Periodic remeasurement is of course necessary. The intermediate pillars are spaced unequally so as to produce 21 selected distances, such that several values of $\mathrm{ED} \times v$ in (1.54) are spaced over each of the half-cycles of 10 m, 2 m, and 30 cm, which are most usually found.

Calibration over one short distance, and also over the whole 820 m, would give the zero error and also the crystal frequency (to a few ppm), in the absence of cyclic error. The latter can be partly eliminated by including a selection of other distances giving $\mathrm{ED} \times v$ spaced over the period of the instrument concerned, but direct determination by a graduated bar, as in (iii) above, is also advised.

(*i*) *Accuracy.* The two principal sources of error are as follows.

(i) Calibration error, p mm, which is independent of the distance measured and

(ii) Refractive index error, q ppm of the distance. This mostly arises from doubt in the average temperature $\bar{T}$.

The combined error from all causes can be expressed as $\pm p$ mm $\pm q$ ppm, where p and q are standard errors (68 per cent probability). Then the estimated s.e. in a distance of D km will be

$$(p^2+q^2D^2)^{\frac{1}{2}} \text{ mm.} \tag{1.55}$$

For p Table 1.4 gives the figures quoted by the makers. These are generally found to be correct. They depend on the design of the instrument. They are naturally small for instruments intended for use over short distances, such as those described in § 1.41. For q the figure mostly depends on the accuracy with which the average temperature along the line can be estimated, although frequency calibration errors, if neglected, can of course be serious. A commonly accepted figure is 1 or 2 ppm for long-range EODM instruments.

It has been suggested, but not proved, that the customary measurement of temperature close to the ground may result in the average of any large sample of geodimeter measures being too small by something of the order of 1 ppm. See § 2.44 (*c*)(iv).

(*j*) *Laser geodimeters.* In the models 8, 6BL, and 600 the light source is a helium–neon laser with a wavelength of 0.6328 μm. In models 6BL and 600 the modulation is by a Kerr cell as in Model 6 and other geodimeters, but in Model 8 it is by means of a crystal (KDP) Pockels cell. Model 8 has been much used, but 600 is a later model. The rather smaller range of 600, 40 km instead of 55, is of little disadvantage since in many circumstances atmospheric considerations make longer ranges undesirable.

The great advantages of laser light over ordinary light are that in daylight the range is the same as at night, and that the reduced power consumption can be met by a battery instead of a petrol generator.

1.40. EODM. Two-colour laser geodimeters

(*a*) In principle, $\bar{N}$ (for light) can be obtained without any temperature measurements, if the path length is simultaneously measured with light of two different colours. Formula (1.15) shows a variation of about 18×10^{-6} in the group index n_G between wavelengths of 0.4 and 0.6 μm, which is about 6 per cent of the normal value of (n_0-1), namely of 300×10^{-6}. An error in either of the two measurements will then introduce an error of 1/0.06 (=17) times as great in the deduced distance. So if the distance is to be correct to 1 ppm, the difference between the two colours must be measured correctly to 0.06 ppm.

It may be noticed that in the present context (of distance measurement) the 6 per cent variation between the two colours is between their group velocities, (1.15), in contrast to § 3.21, where the separation between the two vertical angles depends on the phase index, (1.9), and the situation there is consequently three times less favourable for the same pair of wavelengths, with an adverse factor of about 50 instead of 17.

For an accuracy of 1 ppm, using light, e in (1.10) is not quite negligible. For higher accuracies, if otherwise obtainable, $\bar{e}$ can in principle be measured by the inclusion of a simultaneous 3-cm microwave with the two light wavelengths. It may be necessary to remember that different wavelengths have different paths with slightly different curvature corrections and atmospheric conditions.

(*b*) No multicoloured instruments are yet (1977) commercially available, but two experimental models are reported.

(i) *Georan*, [524]. This is a light-weight two-colour instrument, using the 0.458 (blue) and 0.514-m (red) lines of a pulsed argon ion laser. The polarization of each pulse is modulated at 500 MHz by a KDP crystal. The difference between these two wavelengths is rather small, with only 5.2×10^{-6} between their group indices, and the adverse factor of 17 quoted above is here about 55. This was accepted in the design, because the argon ion laser is currently the only light-weight two-colour laser available.

The instrument weighs 25 kg and can be mounted on a tripod. The power supply (12 V, 45 W) weighs 10 kg.

The anticipated accuracy is 0.7 ppm at 10 km, and 0.2 ppm at 30 km, the maximum range. Careful measurements of the water vapour pressure are necessary for this accuracy.

(ii) A two-colour laser, combined with a 3-cm microwave to correct for the water vapour, is described in [298]. The wavelengths are 0.6329 μm (He–Ne gas laser) and 0.4416 μm (He–Cd metal vapour laser).

The instrument is described as portable, and its accuracy has been estimated as about 0.1 ppm. It has been used to measure the strains in lines 3 to 9 km long in a seismic area.

1.41. EODM. Short-range instruments

There is a demand for EDM instruments to measure ranges of up to one or two km for engineering, cadastral, and topographical purposes, and a great variety of such instruments now exists. Strictly, most of the work they are used for is no part of geodesy, but they do achieve geodetic accuracy.

These instruments are small. Many will either fit on top of a small theodolite, or at least will use the same tribach. The emphasis is on light weight and speed of use, with fine resolution and centring.

The light source of nearly all short-range EDM instruments is a GaAs semi-conductor diode, which emits infra-red light of wavelength within 0.05 of 0.90 μm. As with other light, the refractive index for this wavelength is little affected by water vapour. The group refractive index in ordinary conditions will be about 1.000280 and is certified for each instrument. It varies with temperature and pressure as in (1.10), 1 ppm per 1 °C or 3.5 mbar.

The great utility of the GaAs diode is that its output of light is closely proportional to the input current, with a very small time lag. The beam can thus be amplitude modulated by simply applying an alternating current of the required frequency.

Most instruments give an automatic read-out, with an expected accuracy of about ±5 mm or less plus 10 ppm of the distance. This rather high figure of 10 ppm is due to the automatic read-out, which is designed to work with very little information about atmospheric conditions. The operator is given a choice of 5 or 10 values for the refractive index.

Instruments of European make which are typical of those described here, with ranges up to about 2 km, are currently the Aga 12, the Tellurometer MA 100, the Wild Dl 3S, and Kern DM 2000.

1.42. EODM. The mekometer

The mekometer, [214] and [124], using 0.48 μm light, is designed to give high accuracy over distances of up to 3 km, by day or by night. The current model III is known as the Kern ME 3000.

The light waves are modulated to 500 MHz by means of a quarter-wave coaxial cavity resonator, the frequency of which is related to a standard cavity made of fused quartz and is compensated for temperature change. They are reflected from the other end of the line, and as in other EDM instruments the phases of the outgoing and returning waves are compared. The use of the cavity resonator results in the modulation wavelength being determined by the physical dimensions of the resonator, instead of the usual dependence on a frequency standard. To this extent the measures are independent of the velocity of light, but they are, of course, dependent on the temperature and pressure of the air inside the cavity being the same as, or knowably different from, the average along the path.

The resolution is very fine, to 1 mm in Mark II and to 0.1 mm in Mark III at distances of up to 1 km. There is reported to be no cyclic zero error of as much as 0.1 mm. In short lines the temperature error may be expected to be less than in typical geodimeter lines of 20 or 30 km, and in the laboratory conditions in which some engineering apparatus is now constructed the temperature errors may be relied on to be less than 1 ppm.

The weight of the Mark III is 5.5 kg for the instrument excluding the stand, plus 5.5 kg for the power unit.

The accuracy is reported [215], as ±0.2 mm ±3 ppm of the distance, the latter figure being appropriate if no external temperature measures are made and if the temperature of the cavity is assumed to be the same as the average along the line. In stable atmospheric conditions a lower figure may well be expected.

Also see [407] which gives practical advice about precautions required for the best results.

1.43. MDM. The tellurometer

(a) *General description.* Microwave tellurometers† and instruments like them are used in pairs of two identical instruments, one at each end of the line to be measured. One, called the Master, transmits modulated microwaves, which are returned by the other, called the Remote, and the phase difference of the returning modulation gives the transmission time over the double distance. The following details refer to the 3-cm model. Other models operate on 10-cm or 8-mm microwaves, with corresponding differences in other figures given below.

The master emits approximately 3-cm microwaves, corresponding to a frequency of between 10.025 and 10.45 GHz, which are frequency modulated at 7.5 MHz (40-m wavelength); this is referred to as the A-pattern modulation. The remote instrument, which is sensitive only to frequency modulation,‡ receives the incoming signal and as a matter of electronic convenience retransmits it on a carrier wavelength with a frequency 33 MHz greater. The master then records the phase difference, normally as so many metres and decimals of a metre corresponding to some standard value of the velocity.

The remaining figures in the total distance remain to be determined, by the provision of other modulations which can be switched on when required. Thus the B-pattern may give the distance in tens and units of km, the C-pattern in km and tenths, D in hundreds and tens of metres, and E to tens and single metres. In each case the second figure may be in error by 2 or 3 units, but the next pattern gives that figure correctly.

The pattern frequencies are formed by crystals which are thermostatically controlled. The A-pattern crystal must be correct to 1 ppm or better, and must be periodically calibrated, but the exact accuracy of the others is not important. Ambiguity in the coarse readings will only occur if they are wrong by about 5 or 10 ppm.

To eliminate possible internal maladjustments, A pattern readings are made in sets of four, referred to as +A and −A, either forward or reverse. The mean of each set of four is recorded.

Operating with microwaves, tellurometers require a line which is clear

† The generic name tellurometer has been used to include all long-range instruments using microwaves, which are similar to the tellurometer originally developed by T. L. Wadley in 1956 [585].

‡ So it ignores any amplitude modulation which may result from ground reflections, see Appendix G.

of the ground and dense vegetation, but are not troubled by sunlight. The 10-cm and 3-cm models can operate in haze, fog, dust, or cloud, but rain or falling snow are not recommended. Communication between Master and Remote is provided by telephones built into the instruments, using the same microwaves as are used for measurement. In ordinary circumstances the 10- and 3-cm instruments can be relied on to work at a range of 50 km, and ranges of 150 km have been achieved. The 8-mm model has a rather shorter range. [248] gives figures for its range in different densities of rain and cloud.

Tellurometers can be transported by back pack, and can be operated from portable towers. Power can be supplied by batteries.

(*b*) *Ground swing.* Reflections from the ground or from water or other objects cause an error which may be expected to vary cyclicly if the carrier frequency is changed. Provision is made for the frequency to be varied through a range of 10 per cent (in the 10-cm models, less in the 3-cm and 8-mm models), leaving the modulation frequency unchanged. Measures are generally made on 20 different frequencies (*cavity tunings*) equally spaced through this range. The apparent variation in the measured distance is known as *ground swing.* See further in (*f*), below, and § 1.44.

(*c*) *Plumbing and zero correction.* The point in each instrument to which the measured distance refers is a dipole placed at the focus of the parabolic reflector, the measured distance being via the two reflectors. The point which should, in principle, be plumbed over the station mark is therefore the directrix plane of the reflector. Actually, as the result of calibration over short known distances, a calibration constant is determined which must be added to distances measured with any particular pair of instruments† when normally plumbed, each member of the pair acting as master for half the measures.

This calibration 'constant' has been a source of difficulty, since cyclic changes are apt to occur in it from two distinct causes, as in (i) and (ii) below.

(i) In some 10- and 3-cm instruments the instrumental error varies with the carrier frequency through a range of a few centimetres, on account of internal reflections producing effects similar to ground swing. The usual changes of cavity tuning as in (*b*) above eliminate the error, so the usual programme of cavity tunings should be observed, both in the field and when calibrating, whether ground swing is to be feared or not. Instruments designed since 1965 may be free from this trouble, and in 8-mm instruments the effect is in any case much reduced.

† Each pair of master and remote is to this extent inseparable. If they work as a pair, they must be calibrated as a pair. Of course, three or more instruments can be used in all possible pairings, provided the appropriate calibrations have been carried out.

(ii) In some MRA 3 (3-cm) tellurometers, the zero error has also been found to depend cyclicly on the A-pattern reading, namely the fraction of the modulation wavelength comprised in the measured distance. It may take the form

$$\text{error} = b_0 + b_2 \sin(4\pi A/100), \tag{1.56}$$

where A is the reading on a scale of 100 divisions covering the modulation wavelength. See [193] with many references giving $b_2 = 2$ to 8 cm. A remedy is to calibrate over a number of distances equally spaced through the A-pattern modulation wavelength, as described for the geodimeter in § 1.39 (h). It has been suggested that the trouble may lie in the phase resolvers of some instruments, and not in others, but the cause is doubtful.

If calibration is done over a flat surface, ground swing may by mischance assume the abnormal form of Fig. 1.22(e). [355], p. 28 remarks that this happens if the excess path is such that $\cos wQ = 1$. See § G.10. If the calibration distance is L and the heights of the two instruments are h, this will occur when $h^2 = (\frac{1}{2}Ll_c n)$, where n is 0, 1, 2 etc. For example, if L is 100 metres and l_c is 3 cm, these critical values of h will be 0, 1.2 m, 1.7 m, etc., and in these circumstances the instrument height $h = 0.5$ m is advised.

(d) *Frequency calibration* of the crystals controlling the modulation frequency must be carried out periodically, as for those of the geodimeter, § 1.39 (h).

(e) *The routine of operation.* The routine involves a complicated sequence of testing and switching, which varies with different makes and models. It is necessarily given in full in the makers' instruction manuals. The following is a summary of what has to be done.

(i) Set up the instrument, plumbed over the mark and approximately directed towards the other end of the line. The beam is wide, and a rough bearing will suffice for getting signals. It will previously have been agreed which instruments is to act first as Master.

(ii) Switch on the crystal thermostat.

(iii) As soon as the crystal is warm carry out routine checks and establish telephonic communication.

(iv) Perfect the pointing of the two instruments to give maximum signal strength.

(v) Record coarse pattern readings, B, C, D, and E.

(vi) Record barometer and wet and dry bulb thermometers, as in § 1.34.

(vii) Record about twenty sets of four A-pattern readings, equally spaced through the range of carrier frequencies provided.

(viii) Repeat meteorological readings.
(ix) Repeat coarse pattern B, C, D, and E.

All this may take one hour, and constitutes one *measure* of the line. It is customary to make four such measures, half with one instrument acting as Master, and half with the other. As a precaution against abnormal meteorological conditions, half the measures should be made on one day and half on another, preferably at different times of day. See also § 1.46.

(*f*) *Computation of path distance.* The twenty sets comprising each measure can be plotted graphically† against cavity tuning to exhibit the *ground swing curve*, and if the curve is satisfactory the mean reading should be estimated as in § 1.44. This gives the accepted A-pattern reading, which in combination with the coarse patterns gives the travel time, or a preliminary figure for the distance.

Given the mean meteorological readings,‡ (1.16) then gives the true path distance. The calibration constant must of course be added, and also any recorded plumbing off-sets. For the final reduction to spheroidal distance see § 1.45.

Tables, [34], facilitate the computation of velocity from the meteorological data.

(*g*) *Instrument statistics.* Currently (1978) there have been three main types of tellurometer and similar MDM instruments, working on 10-cm, 3-cm, and 8-mm wavelengths. The 3-cm is the most usual. Some details are given in Table 1.5.

TABLE 1.5
Tellurometer statistics

Model	MRA 2	MRA 101	MRA 4	MRA 5
Carrier wavelength/cm	10	3	0.8	3
Carrier frequencies/MHz	2800 to 3200	10 025 to 10 450	34 500 to 35 100	10 000 to 10 500
Modulation frequency/MHz	10	7.5	75	7.5
Modulation wavelength/m	30	40	4	40
Weight for back pack/kg	20	12	24	18
plus tripod/kg	7	7	7	7
plus battery/kg	20	10	10	10
Power required/W	75	40	40	40
Accuracy. $\pm p$/mm	30	20	3	10
For q see (*h*) below.				

Frequencies and wavelengths given are approximate.

Range of all models in good conditions exceeds 50 km, probably 80. 150 km has been observed.

† This graphical plotting will give the best result in bad circumstances, but if conditions are known to be favourable, and if the various measures are seen to be normally consistent, a simple mean may be expected to suffice.

‡ At this stage curvature corrections, § 1.35 (*a*) (ii) (*a*) are ignored. They are included later, see § 1.45.

(*h*) *Accuracy.* The principal sources of error are

(i) Cyclic zero error, if present. See (*c*) above.
(ii) Ground swing, except with the 8-mm wavelength.
(iii) Vapour pressure.
(iv) Temperature. Generally less serious.

In short lines zero error may be the most serious, and in long lines the vapour pressure, especially as liability to large ground swing error will be apparent from the ground swing curves, and may be avoided by immediate rejection.

As for the geodimeter the total expected error in a line of D km may be expressed in the form

$$\text{s.e.} = \pm p \pm qD \text{ mm} = \pm (p^2 + q^2 D^2)^{\frac{1}{2}} \text{ mm}. \tag{1.57}$$

Values of p for instruments using different wavelengths are given in Table 1.5. It is difficult to give reasonable values for q. With good calibration (especially frequency calibration), and disregarding lines with evidently bad ground swing, a figure such as ±3 ppm might be quoted if it were not for the water vapour problem. But it must be recognized that irregularities in water vapour pressure may produce errors much in excess of the two or three times the s.e. which are ordinarily taken as the maximum to be feared. § 1.37 (*c*) suggests an s.e. of ±5 ppm, with occasional errors of up to five times as much. Smaller figures may reasonably be quoted for dry countries, or for lines of less than (say) 10 or 20 km.

If calibration is carried out infrequently, or in non-typical conditions, calibration errors may be systematic over a number of consecutive traverse lines, making them more serious than they are in a single line. Meteorological errors are clearly unlikely to be systematic to any great extent, but it is not possible to be certain that observations of T and e made at the ends of typical lines, near ground level, may not systematically differ from the path mean by the equivalent of 2 or 3 ppm. See § 2.44 (*c*) (iii).

(*i*) *Favourable conditions* for T are listed in § 1.37 (*b*), for e in § 1.37 (*c*), and for ground swing in § 1.44. The last two are sometimes in conflict. Keeping both considerations in mind, it may be said that the following are generally good lines.

(i) Lines pasing 10 or 15 metres above uniform, flat, dry ground, preferably tree-covered or in cloudy weather with some wind.

(ii) Lines between hills, provided that either the foreground or intervening crests screen reflections with an excess path (§ 1.44) of more than 2 or 3 metres.

From both points of view it is undesirable that the line should pass close above the sea or other extensive water. See also § 1.37 (*e*) for curvature.

(*i*) *Characteristics of different wavelengths.* The effect of shorter wavelength is as follows.

(i) Less power to penetrate haze and cloud, etc.

(ii) A narrower beam. This may be inconvenient at long ranges, but it helps to reduce reflections.

(iii) Smaller ground swing. See § 1.44. With 8 mm, ground swing is almost entirely eliminated. But meteorological troubles, which on long lines are the serious thing, remain unchanged.

(iv) Better resolution and smaller zero errors.

(v) With 8 mm, velocity is slightly affected by absorption lines. See § 1.33.

To summarize, 8 mm is substantially more accurate than 10 cm or 3 cm for lines of up to (say) 10 km. For ordinary geodetic lines of over 10 km, 3 cm seems to be the best microwave.

For comparison with the geodimeter, see § 1.47.

1.44. Tellurometer ground swing (continuing from §§ 1.38 and 1.43 (*b*))

The theory of the combination of the direct wave with a single strong reflected wave is given in Appendix G. If the reflected wave is not too strong, with a reflection factor of less than (say) 0·3, the error in the deduced distance is approximately given by (G.26) and (G.27) as

$$\Delta = -(al_M/2\pi)\sin(WQ)\cos(wQ) \tag{1.58}$$

$$= -a\left(\frac{l_M}{2\pi}\right)\sin\left(\frac{2\pi d}{l_M}\right)\cos\left(\frac{2\pi d}{l_c}\right), \tag{1.59}$$

where l_M and l_c are the modulation and carrier wavelengths,

d is the excess path of the reflected waves, ACB—AB in Fig. 1.20,

$w/2\pi$ and $W/2\pi$ are the carrier and modulation frequencies,

a is the reflection factor.

Formula (1.59) shows that for small values of a

(i) The error of distance cannot exceed $al_M/2\pi$. (1.60)

(ii) The error is proportional to $\cos(2\pi d/l_c)$, so if a series of measures is made with w varying through such a range as will make d/l_c vary through one or more exact multiples of unity, the mean error will be zero.

In a 10-cm instrument, if the 0.1-metre carrier wavelength is varied through 10 per cent, the term $2\pi d/(0.1)$ will be varied through $2\pi d$, so the ground swing curve will contain as many full cycles as there are metres in the excess path d, and there will be at least one full cycle if $d>1$ metre. When $d=\frac{1}{2}$, $1\frac{1}{2}$, or 5 metres the theoretical curve takes the forms of Figs. 1.22(a), (b), and (c).

In the same circumstances (10-cm carrier and a small a) the semi-range of the theoretical swing will be

$$\frac{30a}{2\pi}\sin\frac{2\pi d}{30} \tag{1.61}$$

$$\approx 5a\sin(d/5) \text{ metres}$$

$$\approx ad \quad \text{if } d<3 \text{ or } 4 \text{ m}. \tag{1.62}$$

From (1.61), the semi-range will be zero if $d=0, 15, 30$, etc., metres, and it will be a maximum of $\pm 5a$ metres if $d=7\frac{1}{2}$, $22\frac{1}{2}$, etc., metres.

With a 3-cm carrier and 40-m modulation, (1.61) and (1.62) are little changed, $6\frac{1}{2}$ being substituted for 5, and with the 4 per cent variation a

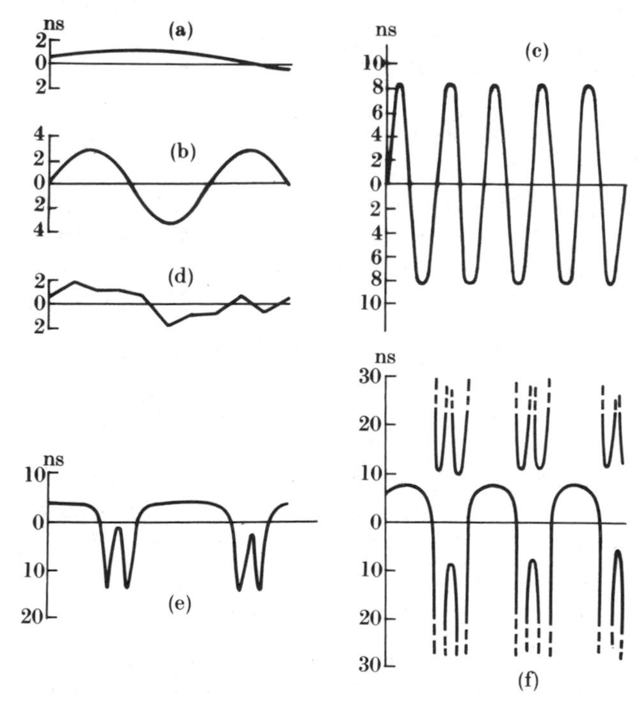

FIG. 1.22. Tellurometer ground swing, 10-cm microwaves. In units of nano-seconds for the double path, or of 0.15 metres in the single distance. (a) $a=0.3$, $d=\frac{1}{2}$ m. (b) $a=0.3$, $d=1\frac{1}{2}$ m. (c) $a=0.3$, $d=5$ m. (a), (b), and (c) are theoretical errors computed by (1.61). (d) Typical, satisfactory swing curve over land. (e) Theoretical curve for $a=0.9$, $d=1.4$ m, for reflection from water, formula (G.25). (f) Theoretical curve for strong water reflection ($d=2.1$ m) combined with another reflection, from [617].

full cycle will be got if $d > 0.75$ m instead of 1.0 m. The factor a may be expected to be rather smaller for ground (but not smooth water) reflections, and 3-cm ground swings are generally expected to be less than those of 10-cm waves.

With an 8-mm carrier and 4-metre modulation, the figures 5 in (1.62) become 0.65, and the approximation that the semi-range $= ad$ is only valid if $d < 0.4$ m. The maximum error is $0.65a$ m, and with the 2.0 per cent frequency variation a full cycle is got if $d > 0.4$ m. For reflections from the ground, a may also be expected to be further reduced. Ground swing is thus nearly eliminated, but the usual 20 sets of A-pattern readings should nevertheless be made.

If a is not small, the curve resulting from a single strong reflection is given by (G.25) in Appendix G, as shown in Fig. 1.22(e) for $a = 0.9$ and $d = 1.4$ metres.

In practice it is not to be expected that the ground swing curve will take the ideal form, since it will be affected by varying errors from meteorological or other sources, and the reflections may be multiple and of varying strength. The effect of many weak reflections is likely to produce many small ups and downs in the curve, with a mean error of near zero, [617], as shown in Fig. 1.22(d), but a combination of one strong reflection with another of only moderate strength may produce a very bad situation as in Fig. 1.22(f), from [617].

The ground swing curve for any measure having been plotted, an estimate of the best value must be made. Ordinarily the simple mean of all the 20 readings cannot be improved on, but exceptions may be as follows.

(i) If a very regular curve visibly comprises an unequal number of maxima and minima, as in Fig. 1.22(b), this may be allowed for.

(ii) The single strong reflection of Fig. 1.22(e) may be treated as in § G.10.

If $d < 1$ metre, so that the curve comprises less than a full cycle, it will be difficult to estimate the mean, especially as the curve will never be as smooth as Fig. 1.22(a). But if $d < 0.5$ metre and $a < 0.4$, the error is limited to 0.2 metres by (1.62), with the probability of its being less. Values of d between 0.5 and 1 metre are thus best avoided with 10-cm instruments, and of between 0.3 and 0.7 m with the 3-cm.

An important use of the ground swing curve is to show whether the measure is satisfactory or not. With 10- or 3-cm it is usual to accept a curve whose total range, maximum to minimum, is less than 6 ns in the double distance (0.9 metres in the single distance), provided it shows some ups and downs, rather than a continuous slope from one end to the other. [617] reports that only 8 per cent of 496 lines in Canada, of average length 26 km, exceeded this limit as the mean range of four measures, so it is not inconveniently strict. A criterion of 4 ns has been suggested, but its acceptance is not always possible. A curve whose form corresponds to a known d and an appropriate a, even though the swing is large, is of course more readily acceptable than one with large variations of inexplicable origin. Another criterion is that a measure may be accepted if its curve suggests that the mean can be correctly drawn to within 6 ppm of the length of the line, with a 50 per cent probability.

If a swing curve is unsatisfactory, simple repetition is unlikely to produce anything much better. The best course of action is then to locate the source of the reflection, and to screen it from one or other instrument, by one of the following methods.

(i) Tilting the beam upwards and working on its bottom fringe. This may result in weak signals.

(ii) Screening the reflection by having the beam graze over a wall or screen in front of one instrument. This may result in reflections from the wall, and is not advised.

(iii) Lowering the instrument, or setting it back behind the crest of the hill, so that the beam grazes along the foreground. This will often be effective, provided the ground is free from rocks and such-like excresences which may send reflections back into the instrument.

Reflections can also be caused by near objects, such as steel or wooden members of the supporting tower [617], vehicles, walls, or even trees.

To summarize, unfavourable conditions for tellurometer ground swing are as follows.

(i) Reflections from water, except with a very small d. This situation may be even worse if there is another reflection as well.

(ii) Reflections from ground with a d of more than 2 or 3 metres, or with such a d as will give between half a cycle and a full cycle.

1.45. Computation of spheroidal distance

The arc AB, Fig. 1.23, having been measured, the following corrections must be applied to give the spheriodal arc HK.

(a) When computing the velocity of transmission, the effect of the curvature of the path and of the earth on the refractive index will probably have been ignored, as is recommended in § 1.43 (f). If so, the correction required (1.39) is

$$-(D^3/12R^2)(K-K^2), \tag{1.63}$$

where $K = R/\sigma$. See further in (f) below.

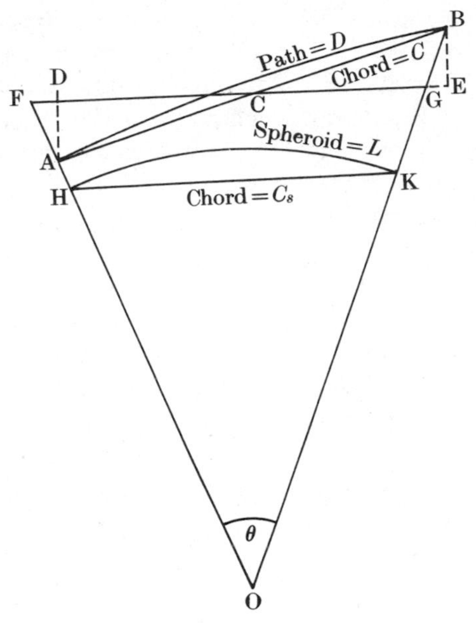

FIG. 1.23. $AH = h_A$. $BK = h_B$. $HF = KG = \frac{1}{2}(h_A + h_B)$. AC = CB. $HO = R_\alpha$,

(*b*) Chord AB − arcAB $= C - D = -D^3/24\sigma^2 = -(D^3/24R^2)(K^2)$. See (*f*). (1.64)

(*c*) *Slope correction.* Given $(h_B - h_A) = \delta h$, the chord distance FG at height $\frac{1}{2}(h_B + h_A)$ is given by

$$FG = DE = (C^2 - \delta h^2)^{\frac{1}{2}}.$$

So
$$FG - \text{chord } AB = -(\delta h^2/2D) - (\delta h^4/8D^3), \tag{1.65}$$

the distinction between C and D here being immaterial. The term δh^4 is seldom necessary. The accuracy required for δh is given by the rule that error in L = (error in δh)($\delta h/L$). In principle δh is the difference of spheroidal heights, not geoidal as given by spirit levelling. The difference between these two values of δh is very unlikely to be as much as 5 m in 50 km, and will seldom be of consequence. But, see also (*d*) below, if an accuracy of 1 ppm in spheroidal distances is to be attained in areas where the geoid is irregular and values of δh are large, the form of the geoid will have to be determined.

If the slope of AB is more than about 5°, it is better to compute the chord at the height of the *lower* station, A, by

$$\text{chord at } h_A = C \cos \beta \sec \tfrac{1}{2}\theta, \tag{1.66}$$

where $\sin\frac{1}{2}\theta = \frac{1}{2}$(chord at h_A)/$(R_\alpha + h_A)$†, and β is the spheroidal angle of depression at the higher station, (B), freed from refraction and from deviation of the vertical at B.‡ An error of $\delta\beta$ radians in β causes an error of $C \times \delta\beta \sin\beta \sec\frac{1}{2}\theta$ in the chord, so if $\beta = 5°$ the refraction and deviation of the vertical must be estimated to within 2″ for 1 ppm accuracy.

(*d*) *Reduction to spheroid level*

Spheroid chord HK$(=C_s)$ − mean-level chord FG $= -(\text{FG})h_m/(R_\alpha + h_m)$, (1.67)

where $h_m = \frac{1}{2}(h_A + h_B)$, and R_α is the spheroidal radius of curvature in the latitude and azimuth concerned, as given by (1.6). R_α can be taken as $6{\cdot}38 \times 10^6$ metres if $h_m < 500$ m.

If (1.66) has been used, substitute h_A (the lower) for h_m in (1.67).

As in § 1.30 (*f*), heights used in (1.67) must be heights above the spheroid, namely ordinary heights above the geoid plus heights of the geoid above the spheroid. An error of 1 ppm in L results from an error of 6 m in h_m.

(*e*) Spheroid arc HK $(=L)$ − chord HK $= D^3/24R^2$. (1.68)

See (*f*) below.

(*f*) It is convenient to combine the three curvature corrections (1.63), (1.64), and (1.68) to give the single correction

$$+(D^3/24R^2)(1-K)^2 \tag{1.69}$$

= (conventionally) $D^3/33R^2$ for light with $K = 1/7$

or $D^3/43R^2$ for microwaves with $K = 1/4$.

(*g*) *Summary*

$$L - D = -\left(\frac{\delta h^2}{2D} + \frac{\delta h^4}{8D^3}\right) - \frac{h_m\{D - (\delta h^2/2D) - (\delta h^4/8D^3)\}}{R_\alpha + h_m} + \frac{D^3}{(33 \text{ or } 43)R^2}. \tag{1.70}$$

(*h*) As an alternative to (*c*) and (*d*) above, [251], gives

$$C - C_s = \frac{(h_A - h_B)^2}{C + C_s} + \frac{(h_A + h_B)}{R_\alpha(C + C_s)} C_s^2 + \frac{h_A h_B}{R_\alpha^2(C + C_s)} C_s^2, \tag{1.71}$$

in which the last term is likely to be negligible. In the denominators C and C_s can generally be taken as equal to D, but a second approximation

† This can be computed using a preliminary value of the chord at h_A.

‡ The observed vertical angle at B is with reference to the geoidal vertical. Here we require it with reference to the spheroidal normal. If the difference is material it can only be obtained by astronomical observations.

for C_s may be necessary. This formula is convenient for electronic computing. The various curvature corrections (a), (b), and (e) are finally included by adding $C_s^3/(33 \text{ or } 43R^2)$ to C_s.

(i) *Record of distances between station marks.* It may be convenient to keep a permanent record of distances between station marks on the ground, as well as of the spheroidal distances. The latter depend on the accepted values of spheroidal height, which may change with the years, and old values may not be easy to trace. For this purpose, [581 B],

$$D_m = \{D^2-(h_2+\delta h_2-h_1-\delta h_1)^2+(h_2-h_1)^2\}^{\frac{1}{2}}-\tfrac{1}{2}(\delta h_1+\delta h_2)(D/R),$$
$$\approx \{D^2-2(h_2-h_1)(\delta h_2-\delta h_1)\}^{\frac{1}{2}}-\tfrac{1}{2}(\delta h_1+\delta h_2)(D/R), \text{ neglecting } \delta h_1^2 \text{ etc,} \tag{1.72}$$

where D_m is the distance between marks, D is the distance between instruments, h_1 and h_2 are the heights of the marks above the spheroid, δh_1 and δh_2 are the heights of the instruments above the marks, and R is any value for the radius of the earth. Small changes in the values of h_1 and h_2 have little effect on the result.

If eccentric stations have been used, see § 1.22.

1.46. Routine for tellurometer or laser geodimeter traverse

See § 1.06 for the general specification, § 1.11 for the reconnaissance, § 1.39 or § 1.43 for distance measurement, §§ 1.17 and 1.18 for traverse angles, § 1.34 for meteorological observations, and Chapter 4, section 5 for the azimuths. Descriptions of tellurometer traversing in Australia, as in [97], are outlined below.

(a) *Organization.* It is assumed that the reconnaissance has been completed, so that the selected stations provide acceptable clear lines. The observing party is divided into three detachments of at least two men (observer and recorder) and a tellurometer. In the most usual system of work these three detachments occupy three consecutive traverse stations, and move forward one station more or less simultaneously, keeping in the same order. In ideal circumstances with motor or helicopter transport between station sites, all three detachments can move forward once a day, and can produce an out-turn of (say) five lines a week, although that can hardly be kept up indefinitely. In less ideal circumstances, alternate days may be occupied by moving camp, and the number in each detachment may be increased by transport or tower-erecting personnel.†

The duties of the three detachments may be as follows. *Fore detachment.* Proving and clearing the fore station on arrival; emplacing the

† Air transport demands considerable base organization to deliver petrol and stores to convenient places.

station mark in concrete, and the witness marks; tellurometer measurements to the centre station; showing a helio or lamp to the centre. *Centre detachment.* Tellurometer measures to fore and back; horizontal angles; vertical angles to rear; showing a helio or light to rear for vertical angles; azimuth and longitude observations. *Rear detachment.* Tellurometer measures to centre; vertical angles to centre (simultaneous); showing helio or lamp to centre; latitude observations; completion of station monument.

If transport is by light helicopter, the time-table of the centre detachment each day may be

08.00 Rear tellurometer.
09.00 One member moves forward by air with half the equipment.
11.00 Second member moves with remaining equipment.
12.00 Vertical angles (rearwards) and fix witness marks.
13.00 Forward tellurometer.
17.00 Horizontal angles. Sixteen zeros.
18.30 Azimuth, Eight zeros on Polaris or σ Oct, and both fore and back stations.
19.30 Longitude until 20.30. Six pairs of timed altitudes.

The other detachments conform. In this programme the two tellurometer measures of any line are separated by 40 hours, giving maximum opportunity for meteorological conditions to change.

Alternatively, if each detachment can include a trained surveyor and the necessary instruments, everything is made easier by the rear party moving up to the front every day.

(*b*) *Heights.* Barometric height differences may suffice for the slope corrections if the lines are flat, such as $\frac{1}{2}°$ or less, but vertical angles provide better heights for topographical control, and also (together with astronomical stations for geoid section) for the reduction of the traverse legs to spheroid level.

(*c*) *Short lines.* There is no great objection to an occasional line being as short as even 1 km, if it gets over some difficulty. Such a line introduces little extra error into the distances, and the azimuth weakness normally associated with short traverse lines can be eliminated by close Laplace control.

(*d*) *Field computations.* Preliminary computations should be made in the field as a check on blunders or other unusual sources of error. The meteorological observations made at the two ends of each line should be examined for unusually large differences. Tellurometer distances should be computed as in § 1.43 (*f*), including the meteorological corrections, so that the two measures can be compared. On average they should differ by not more than 6 ppm. A difference of 10 ppm in a line of over 10 km invites repetition. Care must be taken to avoid a repetition of error in interpreting the coarse readings. It is important that there should be no blunders in recording the positions of any eccentric stations, such as may be necessary for reducing ground swing, and these should be separately measured by both the fore and centre detachments.

Horizontal angles are ultimately checked by closure on Laplace stations. In the field it is only possible to see that the several zeros agree normally well, making especially sure of the degrees and minutes.

1.47. Light versus microwaves

The relative advantages of using light as in the geodimeter, and microwaves as in the tellurometer, are as below.

In favour of light

(*a*) Greater accuracy. Ground swing does not occur, while zero errors, coarse resolution, and error due to water vapour pressure are much reduced. The first three can be eliminated in the 8-mm tellurometer, but not the last, which is perhaps the most serious.

(*b*) Only a single instrument and operator are required. But the reflectors do of course have to be emplaced and removed.

(*c*) The possibility of using two colours as in § 1.40 to eliminate temperature error.

(*d*) No interference with telecommunications.

In favour of microwaves

(*e*) Longer range, but the range of laser geodimeters is ordinarily sufficient, and ultra-long ranges aggravate meteorological errors.

(*f*) MDM operates through cloud and haze (except 8 mm), but in a traverse clear air is in any case required for the angles.

(*g*) The wider beam helps initial alignment.

Summary. EODM must be used if the highest accuracy is required, or if very short ranges are involved. MDM must be used if delay from bad visibility is to be expected, and cannot be tolerated.

1.48. Aerodist

This is an airborne equipment very similar to the tellurometer, designed to give the distance of an aircraft from two or three ground stations. The carrier wave is of between 1200 and 1470 MHz (about 25 cm), modulated to 1.5 MHz. It is transmitted from the aircraft and returned from ground transponders which are very similar to tellurometers. The ranges of the ground stations are simultaneously recorded by a pen recorder in the aircraft, the coarse ranges being determined by automatic switching. One operator is required in the aircraft, and one at each ground station. The master weighs 15 kg and requires 10 amperes at 28 volts. [193], pp. 83–4.

To measure the distance between two ground stations, up to 400 km apart, the aircraft flies across the line at approximately its middle point, and the minimum sum of the two distances is recorded.

Reading sensitivity is claimed to be about ±1 metre (probable), the critical item being the reading of the graphical record. Errors in meteorological corrections will be similar to those of the tellurometer, aggravated by the great length of the line, but insensitive to abnormalities close to the ground. The effect of an error δh in the height of the aircraft is to increase the apparent horizontal distance to a ground point by approximately $\delta h \sin \beta$, where β is the angle of depression at the aircraft. No provision is made for reducing the effect of reflections (ground swing) by varying the carrier wavelength, but the movement of the aircraft produces changes in the excess path, which have the same effect, as is apparent from (1.59) in which l_c is 20 or 25 cm, and a full ground swing cycle will be produced if the excess path varies through the same amount.

[364] reports an s.e. of ±3 or 4 metres in a single line-crossing measure of a distance of between 80 and 150 km, and expects an s.e. of ±1 metre (or about 10 ppm) in the mean of 9 measures.

For secondary breakdown, the line-crossing method need not be used. The aircraft can be fixed by distances to two fixed points, while the distance to a third is simultaneously recorded. The third point is then fixed by one or two more such distances from different directions.

Aerodist has been a useful equipment for secondary breakdown.

1.49. Lasers

A laser is a device for generating a beam of light in a particular way, which gives it the following characteristics.

(*a*) It is extremely monochromatic. Its whole energy thus has a nearly identical group velocity (1.12), so that a pulse or other wave pattern can proceed to a great distance without losing its sharpness or changing its form. Interferometry is possible at relatively great distances (some hundreds of metres), and frequency modulation is possible. In contrast, if ordinary light could be made equally monochromatic by colour filters, the filters would have to absorb nearly all the energy.

(*b*) It is emitted from a source of finite size as a nearly parallel beam, so that the whole energy from a finite source can be concentrated and transmitted as nearly parallel light, instead of only the energy from a point source as with ordinary light.

(*c*) As the result of (*a*) and (*b*) the narrow beam of a laser is very intense. Its light is described as *coherent*.

Laser beams are produced either as short pulses, each lasting for perhaps 10^{-7} seconds, of great power during that short period, or as continuous waves of relatively slight power. There are three main methods for their production.

(*a*) The 'solid state' method, generally using either a ruby crystal, or neodymium in glass, or calcium tungstate.
(*b*) The 'semiconductor' method, using gallium arsenide.
(*c*) The 'gas laser' method, using a mixture of helium and neon, or some other gas.

In each case the active medium is placed inside a Fabry–Perot resonator, which permits the build-up of light at selected wavelengths only. To sharpen a pulse it can be triggered off at maximum energy by a process known as 'Q-switching'. See [180] for an intelligible explanation of current methods of producing laser beams.

Table 1.6, from [180] amplified, gives the characteristics of four lasers which have been much used.

The phase and group refractive indexes and velocities of laser beams are the same as those of ordinary light, (1.9)–(1.15), with the same error of 1 ppm corresponding to an error of 1 °C, and with the influence of water vapour very small. Pulses and modulation patterns travel with the group velocity. As with ordinary light, laser range measurements to satellites are not troubled by ionospheric refraction.

Lasers can be dangerous to the human eye. See [185], pp. 74–5 for permissible safety limits. These limits may vary in different countries, and may be changed from year to year.

TABLE 1.6
Laser characteristics

Method	Solid-state		Semi-conducting	Gas
Substance	Ruby	Neodymium–glass	Ga–As	He–Ne
Wavelength/μm	0.6943†	1.0600‡	0.8400 approx.	0.6328 or 1.1530
Pulse or continuous	Pulse	Either	Either	Continuous
Maximum pulses per minute	A few	Over 1000	Over 1000	—
Peak power, pulsed	Many megawatts§		100 watts	—
Continuous power	Insignificant	Insignificant	A few watts	Tens of milliwatts

† At room temperature. 0.6934 μm at low temperatures.
‡ Outside the visible spectrum. May vary through 20 Å.
§ For a few pulses per minute. Reduced in proportion for more.

1.50. Length ratios

Formulae (1.55) and (1.57) give the expected error of EDM lines in the form $(p^2+q^2D^2)^{\frac{1}{2}}$. Except on short lines the second term is preponderant. Its most natural cause is error in estimating the mean meterological correction along the line. The error will result from ignorance of humidity in MDM, and of temperature in EODM. In two-colour EODM the q term should be unimportant.

In most EDM lines there is a short length near the ground at each end, where temperature conditions will have been recorded, and there will be a much larger distance in the middle where there are no direct measurements. In many lines this section will be well above the ground. Then at any particular time its temperature at a given height may be fairly constant over a wide area, but the available ground measurements will not typify it. The error in the computed distance will vary between day and night, and from one day to another, but there is a reasonable hope that the error (expressed in ppm) will be much the same for several of the lines made at nearly the same time (within half an hour) from any one (suitable) station. The ratio of the lengths of two such lines may be expected to be substantially more accurate than the computed length of each.

There is a certain amount of evidence to show that in some circumstances this is actually the case. Thus, [146] quotes lines in which the recorded EODM (single-colour) distances changed by up to 2 ppm between 18.00 and 22.00 hours on a certain day, while the length ratios of three lines changed by only 0.2 ppm. The three lines were 32, 39, and 74 km in length, in directions 0°, 45°, and 180°.

If the side ratios of a triangle ABC are measured at A, B, and C, on different days, the product (AB/AC) (CA/CB) (BC/BA) should of course equal unity, and its failure is a measure of the errors of the ratios. [495], Table 6 gives four sets of such products determined by EODM on different occasions for the sides of four triangles. The average of the 16 misclosures from unity is 1.6 ppm, which suggests an s.e. of 1.2 ppm for each single side ratio. Further, the angles of each triangle were determined from the side ratios with an average s.e. of 0″.27, while the values determined from the directly measured distances averaged 0″.80, [495], Table 7.

While this system of measurement appears to give the shapes of triangles or nets with unusual accuracy in favourable conditions, there is the difficulty that the scale remains undetermined. It can of course be deduced by accepting a mean from the original EDM observations, leaving the over-all scale the same as it would have been if ratios had not been used. Alternatively, the scale may be got from what may appear to have been the most satisfactory lines, or in a large net perhaps from distant satellite fixes.

The relative weighting of the three ratios between the sides of a triangle is a matter of some consequence. See [169] and [75].

The quasi-simultaneous observation of every side ratio in a net, may add substantially to the amount of travelling and reoccupation of stations and beacons.

General references for Chapter 1, Section 3.
For meteorological conditions. [60], [139], [546], [547], [554], and [555].
For EDM measurements. [138], [157], [355], and [373].

2,

COMPUTATION OF TRIANGULATION, TRAVERSE, AND TRILATERATION

SECTION 1. THE GEODETIC REFERENCE SPHEROID

2.00. Notation

THE following symbols are used in Chapter 2, Sections 1 and 2, and Appendix A. So far as possible the more common such as a, b, e, f, ϕ, λ, ρ, ν, ξ, and η are used in the same sense elsewhere. See §§ 2.19, 2.42, 2.50, and 2.55 for the notation used in Chapter 2, Sections 3, 6, 7, and 8.

a, b, e, f = Spheroid's semi-axes, eccentricity $(1/a)\surd(a^2-b^2)$ and flattening $(a-b)/a$.
$\epsilon = e^2/(1-e^2) \approx 1/150$. Often written $(e')^2$. Also ϵ = triangular error.
ϕ = Latitude. North positive.
λ = Longitude, east of Greenwich.
u = Reduced latitude. Also $= 1/w$.
ω = Longitude measured east from origin.
ρ = Radius of curvature in meridian.
ν = Radius of curvature in prime vertical.
R_α = Radius of curvature in azimuth α.
$K = 1/\rho\nu = 1/r^2$.
$K_m = \frac{1}{3}(K_1+K_2+K_3)$
$r = \surd(\rho v)$.
$1/r_m = (1/\surd 3)\surd(1/r_1^2+1/r_2^2+1/r_3^2) = \surd K_m$,

(For ϕ to $1/r_m$:) Suffix 1 refers to point P_1, etc. Suffix m refers to a mid-point or mean value.

R = Earth's radius in small terms where exact definition is immaterial.
$\Delta\phi, \Delta\lambda = (\phi_2-\phi_1), (\lambda_2-\lambda_1)$.
$\phi_m = \frac{1}{2}(\phi_2+\phi_1)$
A_{12} = Azimuth at P_1 of normal section containing P_2.
α_{12} = Azimuth at P_1 of geodesic P_1P_2.
A_{21} = Azimuth at P_2 of normal section containing P_1.

(For A_{12}, α_{12}, A_{21}:) Measured clockwise from north. Suffix 12 often omitted when no confusion is possible.

$\Delta A = A_{21} - 180° - A_{12}$.
L_G = Distance P_1P_2 in linear units, along geodesic.
L = Distance P_1P_2 in linear units, along normal section.
$\sigma = L/R$, in small terms.
h = Height above spheroid.† Suffix 1 refers to p_1, etc.
β = Observed angle of elevation above horizontal. Positive elevation, negative depression.
E = Spherical excess of spheroidal or spherical triangle.
w, u = Weight of an observation, and its reciprocal. u is also reduced latitude.
ϵ = Triangular error, or error in other condition equation. Also $\epsilon = e^2/(1-e^2)$.

† In other chapters, and in Section 7 of this chapter, h is height above the geoid, and the (less common) height above the spheroid is h_s or H.

ξ = Deviation of vertical in meridian.
η = Deviation in prime vertical. } For signs see § 2.05.
ζ = Component of deviation in azimuth $A+90°$. See §2.09 (*b*).
N_0 = height of geoid above spheroid at the origin. Elsewhere without suffix.
p_1, p_2 = Points on the surface of the ground.
P_1, P_2 = Points on the surface of the spheroid.

2.01. Reference systems

To compute and record the positions of points on or above the earth's surface, some coordinate system is necessary. Such a reference system may take many forms as below, of which sometimes one and sometimes another may be the most convenient.†

(*a*) Cartesian coordinates x, y, z. The z-axis will naturally be parallel to the earth's axis of rotation,‡ and the x- and y-axes will rotate with the earth, the x-axis being parallel to the Greenwich meridian. Ideally the origin should be at the earth's centre of gravity, but that presents the same difficulty as arises with the spheroidal reference system, § 2.02 (*c*) (iii). Computations in cartesian coordinates are described in Appendix C.

(*b*) Geocentric polar coordinates r, θ, λ, equivalent to the above. Not often used.

(*c*) Spheroidal coordinates. Latitude and longitude on an arbitrary reference spheroid as defined in §§ 2.02, 2.03, and 2.06, there being an arbitrary one-to-one correspondence between points on the ground and points on the reference spheroid. The third coordinate (height) may be measured above the spheroid, as is natural, but for cartographical purposes heights are necessarily recorded above the geoid (sea-level). See §§ 3.00 and 1.30(*f*), footnote.

(*d*) Plane coordinates. It is sometimes convenient to record the position of points by rectangular coordinates x and y on a plane 'projection' of the spheroid. See §§ 2.55–2.64.

(*e*) Instead of recording the height coordinate in metres above the geoid, it is possible to specify the position of a point by latitude, longitude, and the gravitational potential. See § 3.00.

(*f*) Astronomical observations for latitude and longitude are measured

† All these systems are fixed in relation to the solid earth, so that the coordinates of points do not in general change with time. Local earth movements do, of course, occur, such as may be due to earthquakes or landslips, and these properly cause coordinates to change. Widespread tectonic movements also occur, §§ 6.71 to 6.75, but so slowly that they are very seldom measurable. Earth tides, §§ 6.76 to 6.84, cause a very small periodic deformation which is not detectable by ordinary survey methods, and which causes no appreciable change in the relative positions of near points.

‡ The mean polar axis rather than the instantaneous axis, so as to avoid periodic changes in the coordinates. See §§ 2.04 and 4.09. The word 'mean' has reference to polar motion not to nutation.

with reference to the direction of gravity at the point of observation, which in general is not parallel to the spheroidal normal at the corresponding point on the spheroid. Such observations are of great interest and value, but lines of equal astronomical latitude and equal longitude do not constitute a useful coordinate system.

The usual reference system for geodetic computations is that of (*c*), in which the position of a point on the ground is described by the latitude and longitude of the corresponding point on the spheroid, and its height by its height above the geoid. Computations on this system are described in detail in the remainder of this chapter, except Section 8. Heights are referred to in Chapter 3. To complete the record it is, of course, necessary to determine the separation of geoid and spheroid as described in §§ 4.50–4.52, or otherwise.

Cartesian coordinates are more convenient for computing observations to artificial satellites, see Chapter 7, and may also be used for other purposes as, for example, in § 2.51.

Plane coordinates are generally only used for cadastral, military, engineering, or sometimes topographical surveys, but their derivation is commonly regarded as a geodetic problem, §§ 2.55–2.64.

2.02. Geoid and spheroid

We have to consider the three surfaces.

(*a*) The surface of the solid earth. This is roughly an oblate spheroid with semi-axes of 6378 and 6357 km, but it may locally depart from the spheroidal shape by some kilometres.

(*b*) The mean sea-level surface or *geoid*, which is much more nearly spheroidal. The geoid may be described as a surface coinciding with mean sea-level in the oceans, and lying under the land at the level to which the sea would reach if admitted by small frictionless channels. More precisely it is that equipotential surface of the earth's attraction and rotation which, on average, coincides with mean sea-level in the open ocean.† If the solid earth was itself a perfect spheroid without internal anomalites of density, the geoid would be very nearly exactly spheroidal, but irregularities in shape and density cause the geoid to depart from the spheroidal form by amounts of possibly 100 metres, at inclinations of generally a few seconds but possibly amounting to as much as 1 minute in mountainous country.

The geoid is a physical reality. At sea-level the direction of gravity and

† Ambiguity due to mean sea-level not being exactly an equipotential surface or to periodic changes in the form of the geoid due to earth tides, will not amount to more than 1 or 2 metres. Except when studied for their own interest, these phenomena are generally ignored. See §§ 3.08 and 3.28.

the axis of a level theodolite are perpendicular to it, and the process of spirit levelling measures heights above it. Astronomical observations made with instruments whose axes are set perpendicular to the geoid could be used to define 'meridians' and 'parallels' upon it, but these would not be a suitable basis for computing triangulation, because the irregular form of the geoid would cause them to be irregularly spaced. Two such 'astronomical parallels' 10 000 metres apart in one longitude might be only 9900 metres apart at a distance of a few kilometres.

(*c*) The *reference spheroid*. The geoid being unsuitable, the position of points on the earth's surface must be expressed by coordinates on an arbitrarily defined geometrical figure. The adopted figure is a spheroid (of revolution) with axes approximating to those of the geoid as closely as possible. Its meridians and parallels constitute the reference system.

The definition of a spheroidal reference system involves the choice of eight independent constants, as follows.

(i) The minor axis is defined to be parallel to the earth's mean polar axis, the CIO, a direction fixed in relation to the solid earth, § 4.09. This is the equivalent of two constants.†

(ii) Lengths are assigned to the major and minor axes, or to the major axis and the flattening. Two more constants. Many different values have been used, some of which are given in Chapter 6, Table 6.5. International values have been agreed on, but in most countries past history continues to dictate the adoption of others. The International Spheroid of 1924 (6 378 388 and 1/297) no longer has special significance. Recommended standard values in 1967 were 6 378 160 and 1/298.247.

(iii) The definition of the centre of the spheroid involves another three constants. Ideally it should be at the earth's centre of gravity, but this has only recently become possible. In the past, the surveyor standing on his triangulation station has had no precise knowledge of the direction and distance of the centre of gravity, and he has not been able to relate his measures, or the definition of his spheroid, to it. The system adopted in the past, and still generally in common use, is described in § 2.06 (a).

(iv) The zero of longitude is taken to be the 'Greenwich meridian', or the Conventional Zero Meridian (CZM), as defined by the Bureau International de l'Heure (BIH), see § 4.10 (*d*). This is equivalent to the eighth constant.

† The use of (2.1), (2.2), and (2.3) to give the deviation of the vertical implies that the minor axis of the spheroid is parallel to the axis of rotation. If astronomical observations have been reduced to the CIO by (4.25)–(4.28) the axis of the spheroid will have been placed parallel to the mean axis defined by the CIO. This has not generally been done, and small errors result, but unless all observations have consistently been reduced to some other pole, it is best to regard the axis as parallel to the CIO, and to admit the existence of errors. It is believed that no substantial survey, except possibly the Australian 1966 adjustment, has ever been consistently related to any other mean axis.

(*d*) *One-to-one correspondence between ground points and points on the spheroid.* One other matter requires arbitrary definition. The meridians and parallels of the spheroid describe the position of a point on the spheroid, but it remains to define what point on the spheroid is to correspond to a point on the earth's surface some distance above it. The definition adopted, Helmert's [270], is that if p is a point on the ground and P is the corresponding point on the spheroid, p lies on the normal to the spheroid at P. This definition provides the simplest formulae for computation.†

The word *normal* is used for the line perpendicular to the spheroid at any point, while the *vertical* is the direction of gravity, a curve perpendicular to the geoid and other equipotential surfaces at the points where it cuts them. The word *horizontal* is applied only to the plane which is tangent to the ground-level equipotential surface. The vertical at the point is, of course, perpendicular to it.

2.03. Spheroidal latitude, longitude, and azimuth

The spheroidal or *geodetic latitude* of a point P on a spheroid is defined to be the angle between the normal at P and the plane of the spheroidal equator. In Fig. 2.1 it is 90° − YQP.‡ The *geodetic longitude* of P is the angle AOB between the meridian plane through P and an arbitrarily defined zero meridian plane, such as YAY′ in Fig. 2.1. This zero meridian plane might be defined by a point on the spheroid, but it is actually defined as in § 4.10 (*d*).

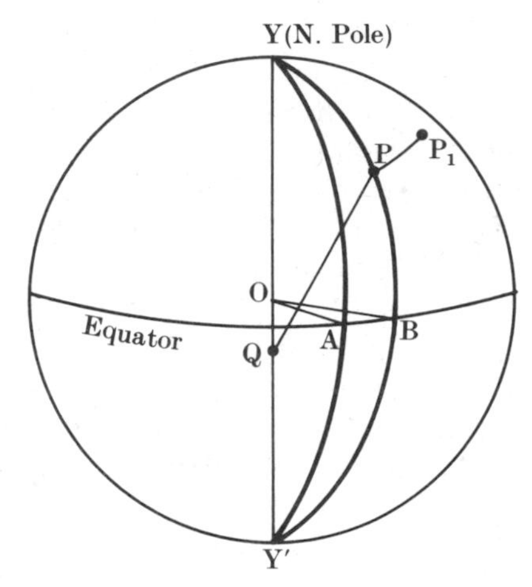

FIG. 2.1. The spheroid.

At P the geodetic azimuth YPP_1 of (the normal section containing)‖ another point P_1 is the angle between two planes, both containing the spheroidal normal at P, one of which contains the north pole of the spheroid, and the other the point P_1.

† It has been proposed that p should not lie on the straight normal at P, but on a curved 'normal'. See [113], 2 Edn, p. 86 first footnote. The difference is trivial, and there is no apparent merit in this alternative. It is unlikely to have ever been used. There is also Pizzetti's system, briefly described in [262], p. 180, but it is more complicated than Helmert's and has no advantage as a reference system.

‡ The use of YY′ for the polar axis does not imply connection with the *Y* in cartesian coordinates *X*, *Y*, *Z*. Z is preoccupied for *zenith* in Fig. 2.3 and P for points on the spheroid in general.

‖ See § 2.08.

2.04. Astronomical latitude, longitude, and azimuth

Astronomical observations are made with reference to the vertical at the point of observation. The *astronomical latitude* is then the inclination of this vertical to the plane of the equator, and the *astronomical longitude* is the angle between two planes, both containing (or parallel† to) the axis of rotation, one being also parallel to the vertical at the point of observation, and the other being the arbitrarily defined 'Greenwich' zero meridian plane, the same plane as that used for geodetic longitudes (§ 2.03). See § 4.10(*d*) for its precise definition.

Similarly, the *astronomical azimuth* from p to a point p_1 is the angle between two planes, both containing the vertical at p, one of which (the meridian) contains the celestial north pole (i.e. it is parallel to the axis of rotation), and the other contains p_1. Note that p_1 is a point on the ground. The azimuth of P_1, the corresponding point on the spheroid, will differ by the small correction described in § 2.09 (*c*).

The short definitions given above are ambiguous, and must be amplified in two respects. Firstly, actual observations refer to the instantaneous pole, defined by the instantaneous axis of rotation, but it is convenient to reduce them to the mean pole‡ defined by the same mean axis of rotation as that to which the minor axis of the spheroid is defined to be parallel, § 2.02 (*c*) (i). The angle between the instantaneous and mean poles is small, less than 0.″6, and has often been neglected, but we here define astronomical latitude, longitude, and azimuth as the values obtained *after reduction to mean pole.* See § 4.09 for the definition of the mean pole, and for the corrections required by different types of astronomical observation.

The second ambiguity comes from the fact that the ground-level equipotential surfaces (to which verticals are perpendicular) are not in general parallel to the geoid below them. Simple theory, assuming a homogeneous earth, indicates an inclination, in meridian only, of 0″17 sin 2ϕ at a height of 1000 metres (§ 6.24), but on a steep mountain side (not a typical astronomical site) it may be 5″ or more. Three possible definitions arc therefore

(*a*) The ground-level value as observed.

(*b*) The geoid-level value. This can be calculated, laboriously, if the form of the ground and the density of the rock are adequately known, as in general they are, since 10 per cent is ample accuracy.

(*c*) The ground-level value of latitude corrected to geoid level by the

† These planes need not contain the axis of rotation, which is not easy to define and is impossible to locate, § 4.09. It suffices that they should be parallel to it. Its direction can be defined without ambiguity.

‡ Meaned with respect to polar motion, not to nutation.

conventional $-0''.17 \sin 2\phi$ per 1000 m. In this case longitude and azimuth are not affected.

Of these three the first is what is actually observed, and is what is required for most purposes, such as the correction of horizontal angles in § 2.09 (*b*), or for vertical angles carrying forward spheroidal heights as in § 6.59. Unless otherwise stated, astronomical latitudes, longitudes, and azimuths are defined to be the *ground-level values reduced only to mean pole.* See § 4.51 (*e*) for a situation in which geoid-level values are, in principle, to be preferred.

2.05. Deviation of the vertical and Laplace's equation

The deviation of the vertical at any ground point p is the angle between the vertical and the spheroidal normal. It is the angle between the tangent plane of the spheroid and that of the equipotential surface at the corresponding ground point. Except when otherwise specified, ground-level, not geoid-level, values are implied. It is recorded in terms of its two components, ξ in the meridian and η† at right angles to it, with signs as below. The deviation depends on both the arbitrary definition of the spheroid and on the actual form of the geoid as brought about by the earth's irregularities of form and density. Thus a deviation of the downward vertical to the south-west, for example, may be due to some excess of mass in that direction, but it may also arise from the arbitrary definition of the deviations at the origin or from a general misfit between geoid and spheroid. Fig. 2.2 illustrates deviation in meridian.

Fig. 2.3 shows the celestial sphere, points on which represent the directions of straight lines in space. YY′ is the mean axis of rotation, YA is the zero meridian for both astronomical and geodetic longitudes, Z_A represents the direction of the vertical at a ground point p, and Z_G the normal at the corresponding spheroid point P, on which p lies. $Z_A X$ and XZ_G equal ξ and η, the components of the deviation. These are conventionally reckoned positive when the downward vertical is deviated to the south or west, respectively, of the inward normal, or when the geoid is rising to the south or west relative to the spheroid.

Deviation in meridian. In Fig. 2.3 the astronomical latitude of p is $90°-YZ_A$, and the geodetic latitude is $90°-YZ_G$. $Z_G X$ is perpendicular to YZ_A, and since ξ and η are small quantities $B'Z_G = BZ_A - \xi$. Whence

$$\xi = \text{Astro latitude} - \text{Geod latitude}. \tag{2.1}$$

This is immediately apparent in Fig. 2.2 also. Note that the identity (2.1),

† In some countries η has been used for the deviation in meridian and ξ for the other, as in the 1952 edition of this book. The present convention has been agreed to internationally, and it is to be hoped that the other will cease to be used.

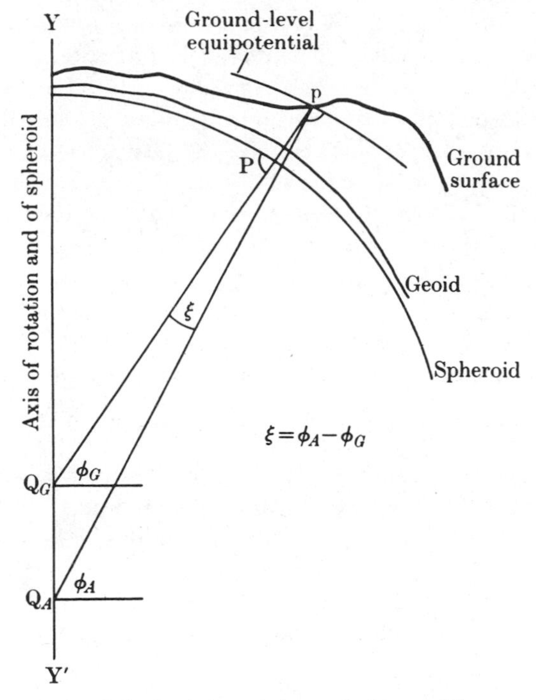

FIG. 2.2. Deviation of the vertical in meridian.

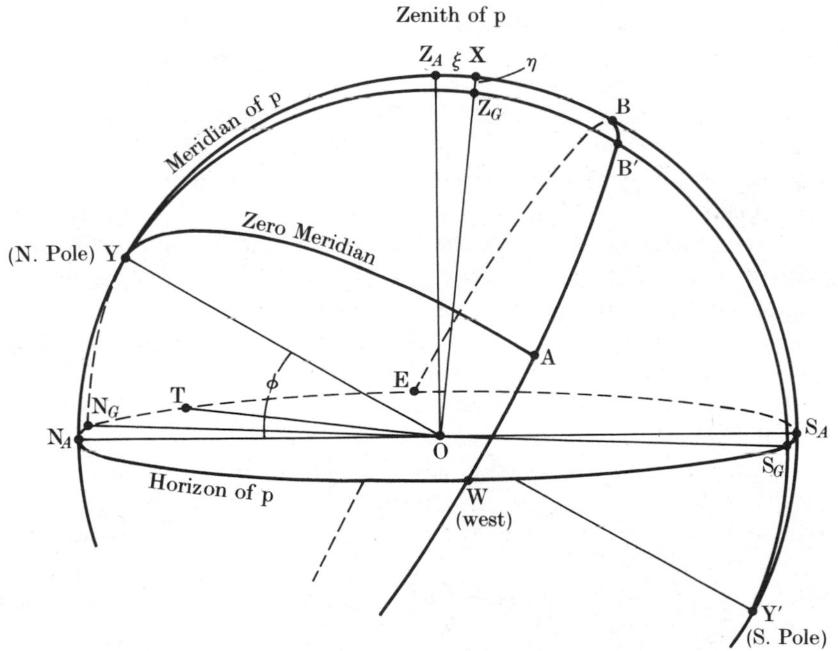

FIG. 2.3. The celestial sphere.

and also (2.2) and (2.3) below, depends on the minor axis of the spheroid being defined to be parallel to the (mean) axis of rotation.

Deviation in prime vertical. η is similarly related to the difference between the geodetic and astronomical longitudes, for at p the astronomical longitude minus the geodetic is AB − AB′, which equals $XZ_G \sec \phi$, whence

$$\eta = (\text{Astro long} - \text{Geod long})\cos \phi. \tag{2.2}†$$

Azimuth. Let the straight line joining p to some point on its horizon, whose azimuth is astronomically observed, cut the celestial sphere at T. Then the astronomical azimuth of T is $N_A T$, and the geodetic azimuth is $N_G T$. The difference $N_A N_G = N_A Y N_G \sin \phi = BB' \sin \phi = \eta \tan \phi$, or

$$\eta = (\text{Astro azimuth} - \text{Geod azimuth})\cot \phi. \tag{2.3}$$

Laplace's equation. Equations (2.2) and (2.3) show that η can be deduced from the difference of either the astro and geodetic longitudes or of astro and geodetic azimuths. the two values must be the same. Therefore

$$(\text{Astro} - \text{Geod longitude}) \cos \phi = (\text{Astro} - \text{Geod azimuth})\cot \phi.$$

Whence

$$\begin{aligned} \text{Geod azimuth} &= \text{Astro azimuth} - \eta \tan \phi \\ &= \text{Astro azimuth} - (\text{Astro along} - \text{Geod long})\sin \phi. \end{aligned} \tag{2.4}$$

This is a most important result. It enables the geodetic azimuth at any geodetically fixed station to be determined from a combination of astronomical azimuth and longitude observations. Stations at which this equation can be formed are known as Laplace stations. They control the geodetic azimuths of a triangulation system in the same way as bases control its scale. They are, of course, of equal or greater importance in geodetic transverse and trilateration.

When ϕ is small, $\eta \tan \phi$ in (2.4) is necessarily small too. In the neighbourhood of the equator it becomes possible to write geodetic azimuth = astro azimuth with relatively little error.

A point which requires consideration is that the geodetic longitude must be adequately known, combined with the fact that at great distances from the origin azimuth error may introduce significant longitude error. In computations carried out by the method of variation of coordinates this

† We note that (2.2) is only correct if the zero meridian is the same for both astronomical and geodetic longitudes, see § 2.07.

matter is very easily allowed for, see § 2.23(*f*) and (*g*), but in the older methods of computation some sort of successive approximation may be necessary in exceptional cases. See § 2.40 (*a*).

2.06. Definition of the centre of the spheroid

(*a*) See § 2.02 (*c*) (iii). The spheroid has been located in relation to the ground survey stations as follows.

At one station, known as the *origin*, the surveyor arbitrarily defines the height of the ground above the spheroid,† the geodetic latitude, and the geodetic azimuth of another survey station. For the height he may adopt that given by spirit levelling, in which case the spheroid and geoid will coincide beneath the station. For the latitude and azimuth he may adopt observed astronomical values, and the spheroid beneath the station will then be parallel to the horizontal plane at the station. Or, with the intention of securing general agreement between astronomical and geodetic coordinates over an area, the defined geodetic values may differ from the astronomical. Then the deviations of the vertical at the origin, ξ_0 and η_0 are given by (2.1) and (2.3).

The surveyor will also need to define his geodetic longitude, and in combination with the observed astronomical longitude this will give η_0 by (2.2). But Laplace's equation (2.4) must be satisfied, and the two values of η_0 must be the same, so there is an arbitrary choice of either geodetic azimuth or of geodetic longitude, but not of both.‡

In a new independent modern survey the accepted latitude, longitude and height of the origin will be as given by satellite observations.

The defined geodetic latitude, longitude, and azimuth enable computations of coordinates to be carried out, and the defined height above the spheroid enables the opening base to be reduced to spheroid level. These definitions also relate the centre of the spheroid to the ground point and the vertical at the origin. For the direction of the spheroidal normal (PQ in Fig. 2.1) relative to the vertical is defined by ξ_0 and η_0, and its length can be calculated from the axes of the spheroid, formula (A.63), to which must be added the height of the ground above the spheroid. This locates Q. The distance OQ is similarly given by (A.60) and its direction is toward the celestial pole, which the surveyor can see. If pressed to do so, the surveyor can point a telescope at the centre of his spheroid, and can state its distance.

† In the past the necessity for defining the spheroidal height at the origin has often been overlooked, but there has been a base nearby, which has been reduced to sea-level through its spirit-levelled height, and that length has been used for computations on the spheroid. The height of the ground above the spheroid at that point has thus been implicitly defined to be the same as its height above the geoid.

‡ This point has sometimes been overlooked. See end of § 2.06 (*b*) (iii).

The origin of the coordinates derived from many systems of satellite observation necessarily coincide with the earth's centre of mass, the geocentre. The modern aim is to convert all non-geocentric coordinate systems to the geocentre, and so to combine all local surveys to one single system. See §7.77.

(*b*) *Error in* ϕ, λ *and A at the origin.* The observed values of the astronomical latitude, longitude, and azimuth at the origin will inevitably contain small errors, and the adopted arbitrary values of geodetic longitude and azimuth, taken with the correct astronomical values, may fail to satisfy Laplace's equation by small or (by oversight) large amounts. If such errors are detected, the following courses are possible.

(i) An error in the accepted astronomical latitude at the origin affects ξ_0 and thence the position of the defined spheroid centre in relation to the vertical. No other change or inconsistency arises.

(ii) An error in the astronomical longitude similarly affects η_0 and the position of the spheroid centre, and in addition results in the non-satisfaction of Laplace's equation unless either the geodetic longitude or azimuth is changed also. To change the geodetic azimuth involves complete recomputation, but there is no great harm in having to recognize that the value which has been used for computation is *in error* by any small amount up to (say) 1″. At every other Laplace station the accepted geodetic azimuth is liable to be wrong by as much.

(iii) If an error in the astronomical longitude, or the inadvertent non-satisfaction of Laplace's equation at the origin, demands a change of more than about 1″ in the geodetic azimuth there, it will be more convenient to leave the azimuth alone and to admit error in the accepted geodetic longitude. This will equally affect all geodetic longitudes throughout the area of survey. It is a nuisance, but in practice the error of longitude can be ignored except when forming Laplace's equation or deducing η_0 from (2.2). See [231], p. 162 and [110], p. 105 for an error of +3″.16 in all Indian longitudes, due to oversight in defining geodetic longitude and azimuth at the origin.

'*Spheroid*' *and* '*Datum*'. The word 'spheroid' can be used either with reference to the axes and flattening of the figure or to the three arbitrary constants at the origin as well. The word 'datum'† can also be used to refer only to the latter, or to the dimensions of the spheroid as well. Confusion could perhaps be avoided if the word 'origin' was used in connection with the three elements by themselves. For example, the International spheroid (a and f), located by the Potsdam 1950 origin (ξ_0, η_0, N_0), constitutes the European 1950‡ Datum.

† Datums in the plural.

‡ When triangulation or other frameworks have been adjusted at various dates, the date of any such adjustment may be included in the words by which the reference system is described.

See [49], Vol i, pp. 29–44 for details of many current local and continental datums.

2.07. Geodetic longitude

As stated in § 2.04, the zero of both astronomical and geodetic longitudes ordinarily is, and should be, the same 'Greenwich' zero meridian plane. The point needs some amplification.

The 'Greenwich' meridian was originally a plane containing, or parallel to, the (mean) axis of rotation and also parallel to the vertical at the site of the Greenwich meridian transit telescope. As now defined it has much less connection with Greenwich, but so far as possible the new definition has been made to give the same plane as the old one.

Fig. 2.4 shows the spheroid viewed from a point on the prolongation of its minor axis above the north pole. The point p is any geodetic survey station, K is the origin of the system of which p is a member, and 'Greenwich' is the site of the Greenwich transit. The figure shows a section of the ground level equipotential surface, and the spheroidal parallel, at each place.

At K AS is the vertical, and $A'KA = \eta_0 \sec \phi_0$, where η_0 is the arbitrarily accepted deviation at K, defining the direction of A′O the spheroidal normal, which in turn locates O on the minor axis since in the diagram $KO = (\nu_0 + \text{spheroidal height}) \cos \phi_0$.

At p BS is the vertical, in general not passing through O, and B′O is the spheroidal normal, the angle A′OB′ (=geodetic $\lambda_p - \lambda_K$) having been determined by triangulation computations.

The zero longitude plane is OC″ defined as in § 4.10 (*d*). So far as has been possible this has been defined to be parallel to the old Greenwich meridian plane CQ which was parallel to the vertical at Greenwich. The spheroidal normal at Greenwich is C′O, not in general coinciding with CQ, and the geodetic longitude of the Greenwich transit site is not in general zero.‡

The angle ATC″ is the astronomical longitude of the origin K.

Then the geodetic longitude of p, as given by computation is

$$\begin{aligned} &\lambda_{GK} + (\lambda_{Gp} - \lambda_{GK}) \\ &= (\lambda_{AK} - \eta_0 \sec\phi) + (\lambda_{Gp} - \lambda_{GK}) \\ &= A'OC'' + B'OA' \\ &= B'OC'', \end{aligned}$$

‡ In the geodetic survey of Great Britain the 1968 geodetic longitude of the Greenwich transit site, or the 'meridian mark', is 0″.418 E. It is the vertical there that is (or used to be) parallel to the meridian plane.

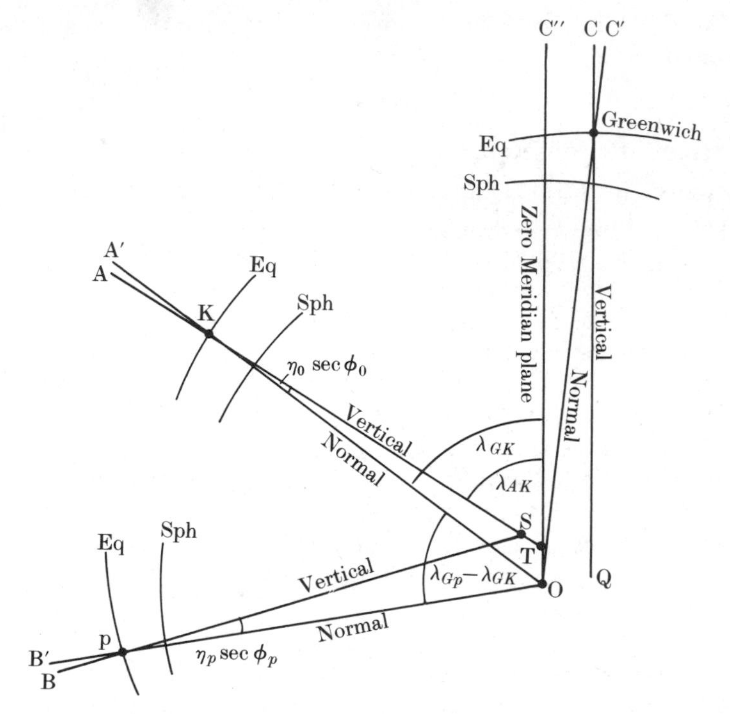

FIG. 2.4. The spheroid seen from a point on its minor axis above the north pole. O = spheroid axis. Eq = ground level equipotential surfaces. Sph = spheroidal parallels. Suffix G = geodetic. Suffix A = astronomical. Suffix 0 refers to origin K.

showing that the 'Greenwich' zero meridian plane is also the origin of geodetic longitudes, if (2.2) is to be true.

2.08. The projection method of computation

The basis of computation on the defined spheroid is as follows.

(i) Angles observed between ground points, with the theodolite levelled so that the axis lies in the vertical, must receive two small corrections to reduce them to the angles between planes containing the spheroidal normal at the station and the spheroid points corresponding to the points observed. In principle this requires a knowledge of the deviation of the vertical at every station, but actually the effect of the deviation is generally negligible. See § 2.09 (b) and (c). The line in which a plane containing the normal intersects the surface of the spheroid is known as a *normal section.*

(ii) Distances measured between ground points must be corrected to give distances between corresponding spheroid points. See § 1.30(f) for the reduction of invar bases, and § 1.45 for lines measured by EDM. It is important to note that the reduction must be through heights above the spheroid, not through heights above the geoid, as given by spirit levelling. This involves an adequate network of geoid profiles over the area of survey. See § 4.50.

Given angles between spheroid points, the distance and azimuth of the normal section between at least one pair of such points, and the spheroidal coordinates of one point, the computation of the latitudes and longitudes of the other points is in principle straightforward. See §§ 2.11–2.14.

Computations carried out in this way are described as being done by *Helmert's projection method*, [270]. Given errorless observations, and the necessary deviations of the vertical and geoid profiles, the computation process is free from any sort of approximation.

Computations which ignore significant deviations of the vertical at theodolite stations, and the separation of the geoid and the spheroid in the reduction of measured distances, have sometimes been described as having been carried out by the *development method*, but this is a misnomer. It is not a method at all, it is simply the neglecting of certain corrections required by the projection method. It results in more or less serious errors according to circumstances.

Accuracy of computations. In the field, angles can be measured with errors which may average as little as some tenths of a second, and distances to a few parts per million or less. It is a pity to spoil good field work by unnecessary rounding errors in computations, and in the best geodetic computations it is proper to record and compute distances to 0.1 ppm, while angles, directions, and azimuths may be recorded to 0".1, or perhaps to 0".01 with a good deal of tolerance about the accuracy of the last figure. Latitudes and longitudes may be computed to 0".001, equivalent to 3 cm on the ground, or to 0".0001 where geodetic points are close together. Engineering works, where 0.1 mm may be significant, will generally be computed in plane coordinates.

2.09. Reduction of observed directions

(*a*) If at a station p_1, h_1 metres above the corresponding spheroidal point P_1, observations are made to another station p_2 at height h_2, the direction actually observed is that of the plane containing the vertical at p_1 and the point p_2. This direction must then be corrected to that of the plane containing the normal p_1P_1 and the spheroidal point P_2. It is to be noted that the normals at P_1 and P_2 will not generally be coplanar. Corrections arise from two causes, namely this skewness of the normals, and the fact that the vertical at p_1 is inclined to the normal p_1P_1 by the deviation of the vertical.

(*b*) *Correction for deviation.* In Fig. 2.5 p_1P_1 is the normal, and p_1P_1' the vertical. p_1P_1'' resolves the deviation into two components, ζ $(=\xi \sin A - \eta \cos A)$ being that at right angles to p_1p_2. Let $p_2t''P_2''$ and $p_2t'P_2'$ be drawn parallel to p_1P_1'' and p_1P_1' respectively. t′ and t″ are on the horizon plane of p_1, p_2p_1t' $(=\beta)$ being the vertical angle from p_1 to p_2. Then $t't'' = \zeta \times p_1t' \tan \beta$, and the required correction

$$t'p_1t'' \text{ is } -\zeta \tan \beta. \tag{2.5}$$

This correction can be important. In an extreme case ζ can be 60″ and β 5°. The correction would then be $5\frac{1}{2}''$. More often ζ may be 20″ and β 1°, giving a correction of 0″.3. In the primary triangulation of India an angle does actually exist whose correction is $4\frac{1}{2}''$. The correction cannot be applied unless the deviation has been observed, and it follows that it may be necessary to observe astronomical latitude and longitude (or azimuth) at all primary stations in mountainous country. See § 6.59 for an alternative method of finding the deviation in mountains.

(*c*) *Correction for skew normals.* In Fig. 2.6 p_2, P_1, and P_2 are as above and the normals cut the polar axis at Q and R. QP_2 is joined and produced to p_2' so that $P_2p_2' = h_2$. P_1C is a parallel of latitude, so that $CP_2 = \Delta\phi$. Then the required correction to the observed direction p_1p_2 is $p_2p_2'(\sin A_{21})/L$, the $\sin A_{21}$ being due to the fact that p_2' lies in the meridian of p_2. Let $QP_2R = \psi$. Join QC and let it meet P_2R at D. Then $DP_2 \approx \rho_m$, since D is the centre of curvature of CP_2, being the intersection of the normals at C and P_2. Whence $DQ = \nu_1 - \rho_m$.

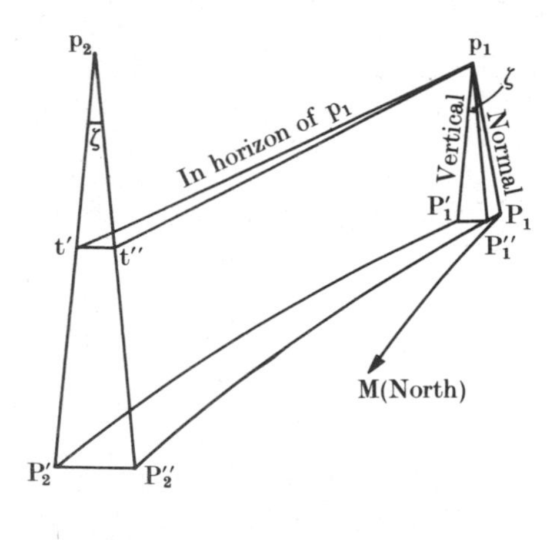

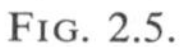

FIG. 2.5.

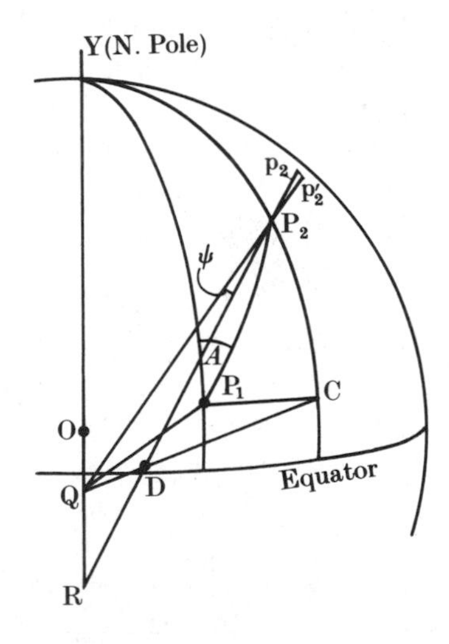

FIG. 2.6.

Then from the triangle DQP_2,

$$\psi = \Delta\phi(\nu_1 - \rho_m)/R$$
$$= \Delta\phi \cdot \epsilon \cos^2\phi, \text{ from (A.62) and (A.63),}$$
$$= \epsilon(L/R)\cos A \cos^2\phi, \qquad (2.6)$$

and ignoring the difference between A and A_{21}, the required correction is

$$\frac{h_2\epsilon}{2R}\sin 2A \cos^2\phi, \qquad (2.7)$$

or $0''.11$ $(h_2/1000)\sin 2A \cos^2\phi$, if h_2 is in metres. Positive if p_2 is south-west or north-east of p_1.

This is a small correction, not more than $0''.3$ if $h_2 = 3000$ metres. Except in mountainous country it can reasonably be ignored, and smaller terms are always negligible.

(*d*) *Corrections to an angle.* If angles, rather than directions, have been abstracted (§ 1.20) the correction to an angle is that of the direction with greater azimuth minus that of the direction with smaller azimuth. (Except when azimuth 0° intervenes between the two arms of the angle.) If directions have been abstracted, they can be corrected as above, and corrected directions can then be subtracted from each other to give corrected values of the angles.

2.10. Geodetic tables

Computations such as those of §§ 2.14 and 2.15 are usually expressed in terms of ρ, the radius of curvature of the spheroidal meridian, and of ν, the *normal terminated by the minor axis*, PQ in Fig. 2.1. These vary with the latitude, and also depend on the axis and flattening of the adopted spheroid. See § A.03. Tables for the correct spheroid must of course be used.

When formulae are being computed electronically, no tables of ρ and v are required, since it is easier to compute them when necessary, using (A.61)–(A.63), than to refer to a table placed in the store.

Section 2. Computation of a Single Triangulation Chain or Traverse

2.11. Outline

This section describes the computation of a single triangulation chain or very simple network by the methods used with logarithms or desk computers before the introduction of electronic computers.† Using the latter, a much more suitable method is that of Variation of Coordinates, which is described in Section 3, but the old methods cannot be ignored, and their description provides an opportunity for recording some basic formulae in §§ 2.14 and 2.15, and for giving details of a simple example of adjustment by least squares. The student is advised to study this section, even though he may never have to compute by the methods described.

For the computation of a single traverse see § 2.16.

It is required to compute a chain of triangulation as described in § 1.03 and shown in Figs. 1.1 and 2.8, consisting of a series of simple independent figures as defined in § 1.03. It is assumed to emanate from a side of known length and azimuth, with known latitude and longitude at one end. All, or at least sufficient, angles have been observed. The angles and the known base have been reduced to the spheroid as described in §§ 2.08(ii) and 2.09.

2.12. Solution of triangles. Geodesics

Given one side and the three angles of a spherical (not spheroidal) triangle $P_1P_2P_3$, whose angles are I, II, and III, and whose sides are L_1, L_2, and L_3, L_1 being opposite P_1, it is well known that

$$L_1/\sin \mathrm{I}_P = L_2/\sin \mathrm{II}_P = L_3/\sin \mathrm{III}_P, \tag{2.8}$$

where $\mathrm{I}-\mathrm{I}_P = \mathrm{II}-\mathrm{II}_P = \mathrm{III}-\mathrm{III}_P = E/3$, and where E is the spherical excess, namely area/(radius)^2 radians. This is Legendre's theorem, which is usually quoted as stating that the angles of a plane triangle, whose sides are equal to those of a certain spherical triangle, are equal to the angles of the spherical triangle each diminished by one-third of the spherical excess.

On the spheroid this is also very nearly true if the sides are less than 150 km. For the spherical excess the earth's radius should be taken as $\surd(\rho\nu)$ at the centre of the triangle. So

$$E = (\text{area} \times \text{cosec } 1'')/(\rho\nu)_m \text{ seconds.} \tag{2.9}$$

But the point requires investigation as below.

In Fig. 2.6 the plane containing P_2 and the normal at P_1 cuts the spheroid in the normal section P_1P_2 shown in Fig. 2.7 as $P_1\alpha P_2$. For reciprocal observations at P_2 the plane containing P_1 and the normal at P_2 cuts the spheroid in a normal section $P_2\gamma P_1$ which does not coincide with $P_1\alpha P_2$. The separation between them is very slight, but the six curves marked with arrows in Fig. 2.7 cannot be described as a triangle. One

† These methods may also be suitable for simple frameworks using the hand or small desk electronic computers, which are now (1978) in common use.

unique line between each pair of stations must therefore be defined, which shall be the side of the spheroidal triangle. The natural line to choose is the *geodesic* or shortest line on the spheroid joining the two points. It is shown as $P_1\beta P_2$, and it generally lies between the two normal sections as shown.

The correction to the normal section P_1P_2 is $\beta P_1\alpha$, and is

$$-(\epsilon/12)(L/R)^2 \sin 2A_{12} \cos^2 \phi_1 + (\epsilon/48)(L/R)^3 \sin A_{12} \sin 2\phi_1, \quad (2.10)$$

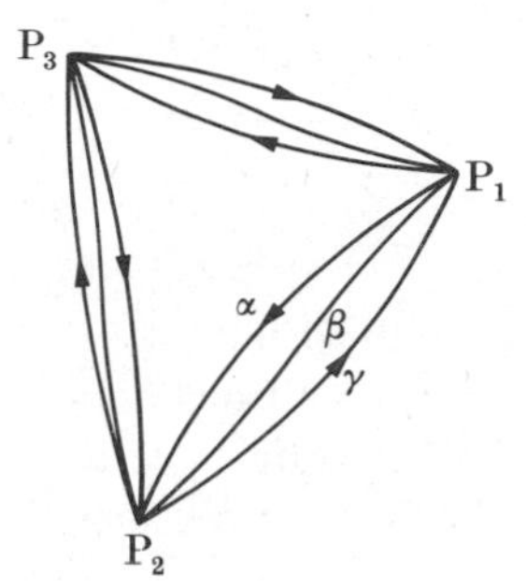

FIG. 2.7. The geodesic triangle. Arrows mark the normal sections. Intermediate lines with reversed curvature are geodesics.

or $-0''.028\ (L/100)^2 \sin 2A \cos^2\phi$, where L is in kilometres. Positive if P_2 is north-west or south-east of P_1. See [113], 1 and 2 Edns, §§ 8.06–8.10, [155], pp. 124–30, [270], and [563]. This correction is minute, less than $0''.03$ if $L = 100$ km, and less than $0''.07$ if $L = 150$ km. In any terrestrial triangle whose angles can be observed it can be neglected, and the geodesic and the two normal sections can be regarded as identical. Neglected terms in $(L/R)^4$ would be serious if $L > 2000$ or 3000 km.

The angle $\alpha P_1 \gamma$ approximately equals three times $\alpha P_1 \beta$. The geodesic divides the angle between the two sections in the ratio 2 to 1, unless the azimuth A is such that the second term of (2.10) is an appreciable fraction of the whole, in which case the geodesic and both sections will nearly coincide.

Continuing, on the supposition that these minutiae are not negligible, the formula for the spherical (or spheroidal) excess of a large triangle is

$$E = (\Delta_p/r_m^2)(1 + m^2/8R^2), \quad (2.11)$$

where Δ_p is the area of a plane traingle with sides L_1, L_2, L_3 the same as those of the spheroidal triangle, $1/r_m^2 = \frac{1}{3}(1/\rho_1\nu_1 + 1/\rho_2\nu_2 + 1/\rho_3\nu_3)$, $3m^2 = L_1^2 + L_2^2 + L_3^2$, and R is any reasonable value of the earth's radius.

For a triangle with 150-km sides E is $50''$ and the term in $m^2/8R^2$ is only $0''.005$. The formula $E = \Delta/\rho_m\nu_m$ is then adequate, and the area can often be taken from a map. On the other hand in a triangle with 800-km sides E is about $1500''$, and the small term is $3''$. In such a triangle (if one was ever observed) an approximate solution would give m, and (2.11) could be used in full.

And for the solution of triangles as in (2.8), instead of $\mathrm{I} - \mathrm{I}_p$, etc. $= E/3$, we have

$$\mathrm{I} - \mathrm{I}_p = E/3 + (E/60\,R^2)(m^2 - L_1^2) + (ER^2/12)(K_1 - K_m), \text{ etc.} \quad (2.12)$$

where E, R, and m are as in (2.11), $K_1 = 1/\rho_1\nu_1$, etc., and

$$K_m = 1/r_m^2 = \tfrac{1}{3}(K_1 + K_2 + K_3).$$

See [113], 2 Edn. §§ 8.13 and 8.14, [270], § 8.4, and [563]. The second and third terms of (2.12) are less than 0″.05 if sides are less than 500 km, and for all ordinary purposes $\mathrm{I} - \mathrm{I}_p = E/3$ as in (2.08).

The difference between the length of the geodesic, L_G, and either of the normal sections, L, is even more negligible.

$$L = L_G\{1 + (\epsilon^2 L^4/360\ R^4)\sin^2 2A\ \cos^4\phi\}, \qquad (2.13)†$$

where $\epsilon^2 = e^4/(1-e^2)^2 \approx 1/22\,000$. See [113], 2 Edn, § 8.11, and [563]. The difference is less than (1/150) ppm when $L = 3000$ km.

The geodesic departs laterally from the normal section by rather less insignificant amounts. If a long line between two fixed points is used to define a boundary it would naturally be the geodesic rather than one of the two normal sections. In the centre of such a line the geodesic is midway between the normal sections, at a distance from each of $(L/2)(\alpha \mathrm{P}_1 \beta)$ sin 1″, or less than $2\frac{1}{2}$ cm in a line of 150 km.‡

In 1950, when Shoran trilateration with sides of 800 km was beginning to be observed, a full investigation of the solution of triangles with sides of 1000 km or more seemed necessary. Now (1978) it is evident that all such computations will be done by variation of coordinates, in which this complication does not arise. It appears that the geodesic is generally only likely to be required for the solution of triangles whose sides or angles can be directly measured, but the sides of such triangles seldom exceed 150 km, at which distances the distinction between geodesic and normal section is immaterial. It is of course proper to record formulae for greater distances in case unusual problems arise.

When computation is to be by the method of variation of coordinates, §§ 2.18–2.24, there is no need to consider the geodesic at all. Normal section angles are observed, and normal section formulae can be used for the computations.

† Further terms are, [121], 1974.

$$-(L^6/576\ R^5)\epsilon^2 \cos^2\phi \sin 2\phi \sin A \sin 2A + (L^7/3584R^6)\epsilon^2 \sin^2 2\phi \sin^2 A - -(L^7/7560R^6)\epsilon^2 \cos^4\phi \sin^2 2A. \qquad (2.13\text{A})$$

There is a difference between the length of the section containing the normal at P_1 and the other containing the normal at P_2. These smaller terms are correct if A and ϕ at P_1 are used in (2.13) for the section containing the normal at P_1, and the A and ϕ of P_2 for the reverse section.

‡ Regarding the separation as parabolic, the separation of the two normal sections at a distance of $L/2$ is $(2/3)(\alpha \mathrm{P}_1 \gamma)(L/2) = (L)(\alpha \mathrm{P}_1 \beta)$.

2.13. Figural adjustments

(*a*) *Conditions.* Every figure, § 1.03(*c*), except a single triangle with only two angles observed, is likely to contain some redundant observations, and before the computations can be carried further, values of every observed angle must be obtained such as will produce zero triangular errors and satisfy other such conditions. The object is to secure self-consistency, and to obtain improved values of the angles. It is assumed that station adjustment, if required, and the reduction of angles and measured distances to the spheroid, have been carried out as in §§ 2.08(ii) and 2.09. There are four kinds of conditions which the observations ought to satisfy, and the aim is to find minimum corrections which will cause them to be satisfied.

(i) *Triangular conditions.* The three angles of every triangle must add up to $180° + E$.

(ii) *Side conditions.* In figures more complex than simple triangles, there are relations between the side lengths. Thus in ABCDE of Fig. 2.8, starting with OA one can compute round the polygon, and deduce OA again. Whence $\sin 1 \cdot \sin 3 \cdot \sin 5 \cdot \sin 7 \cdot \sin 9$ should equal $\sin 2 \cdot \sin 4 \cdot \sin 6 \cdot \sin 8 \cdot \sin 10$. Similarly in EFGD $\sin(2+3) \cdot \sin 5 \cdot \sin 7$ should equal $\sin 4 \cdot \sin(6+7) \cdot \sin 2$, or computing round the intersection of the diagonals, as is proper even though no station exists there†, $\sin 1 \cdot \sin 3 \cdot \sin 5 \cdot \sin 7$ should equal $\sin 2 \cdot \sin 4 \cdot \sin 6 \cdot \sin 8$. But if one of these conditions is satisfied so will be the other, and both must not be written down.

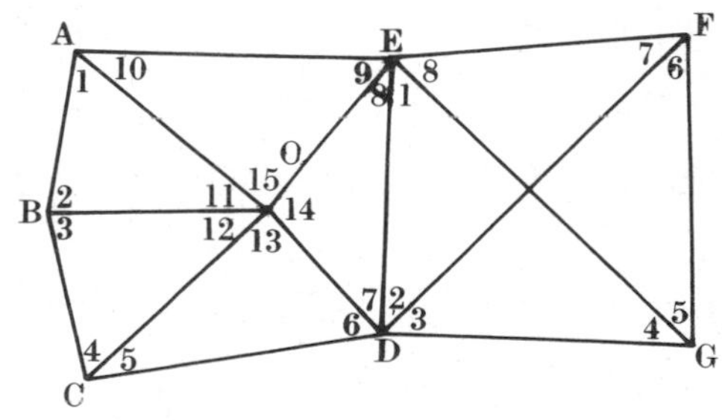

FIG. 2.8.

(iii) *Central conditions.* Where angles comprising the whole horizon enter the computations, their sum must be held to 360°. Thus in ABCDE $11+12+13+14+15$ should be 360°, and the fact that station adjustment may have satisfied this condition does not make a central equation unnecessary, since the condition must be maintained, in spite of changes introduced by other conditions. This is not necessary at stations such as

† The figure is spheroidal, not plane, and the diagonals may not exactly intersect. The resulting error is immaterial in figures small enough for their stations to be intervisible.

A, however, in the form that

$$1+10+(\text{external BAE}) = 360°,$$

because corrections to 1 and 10 will be got from the adjustment, and BAE, which does not enter into any other equation, can only be deduced from them.†

(iv) *Measured side ratio.* If AB and BC are measured lengths, presumed errorless, their ratio must be preserved by an equation such as that AB · sin 1 · sin 12 should equal BC · sin 4 · sin 11. For the inclusion of fallible measured distances see § 2.31 (*a*).

(*b*) *The form and number of condition equations.* Every condition equation is a relation between the unknown errors of the observed angles.‡ If x_n is the error of the angle n, the triangle equation of AOB gives

$$x_1+x_2+x_{11} = 1+2+11-(180°+E) = \epsilon_{\text{AOB}}, \tag{2.14}$$

ϵ_{AOB} being the triangular error.

Similarly the central equation gives

$$x_{11}+x_{12}+x_{13}+x_{14}+x_{15} = (11+12+13+14+15)-360°, \tag{2.15}$$

a known quantity.

For a side equation we have

$$\ln\sin(1-x_1)+\ln\sin(3-x_3)+\ldots = \ln\sin(2-x_2)+\ln\sin(4-x_4)+\ldots,\|$$

where 1, 2, 3, etc., are the observed angles,

or $$(\ln\sin 1 - x_1\cot 1)+\ldots = (\ln\sin 2 - x_2\cot 2)+\ldots.$$

We have

$$\ln\sin 1+\ln\sin 3+\ldots = \ln\sin 2+\ln\sin 4+\ldots+\epsilon_s, \tag{2.16}$$

where ϵ_s is the error of the logarithmic side closure. Whence, subtracting,

$$x_1\cot 1+x_3\cot 3+\ldots = x_2\cot 2+x_4\cot 4+\ldots+\epsilon_s, \tag{2.17}$$

All the condition equations thus give rise to equations of the form

† If BAE has been separately measured, not as part of a round, but independently, its measured value will have influenced the values allotted to 1 and 10 (and their weights) in the result of the station adjustment.

‡ i.e. after station adjustment and reduction to spheroid. In (2.16) one-third of the appropriate spherical excess should be deducted from each angle, although the same value of ϵ_B is likely to be got, whether this is done or not.

‖ The side equation is not expressed logarithmically in the expectation that computations will be done by logarithms, but because it is the easiest way of deriving (2.17).

$$\left.\begin{aligned} a_1x_1+a_2x_2+\ldots a_tx_t &= \epsilon_a \\ b_1x_1+b_2x_2+\ldots b_tx_t &= \epsilon_b \\ \cdot\quad\cdot\quad\cdot\quad\cdot\quad\cdot\quad &\cdot\quad\cdot \\ n_1x_1+n_2x_2+\ldots n_tx_t &= \epsilon_n \end{aligned}\right\}. \tag{2.18}$$

The coefficients $a_1, a_2, \ldots, b_1, b_2, \ldots$ are mostly zero; others are unity, if weights are equal, see below; and in side equations they are cotangents.

The errors x and the closing errors ϵ will be in seconds, except that in the side equations as they stand the x's are in radians and the ϵ's are fractions such as (discrepancy in OA÷OA). If the ϵ's are recorded in ppm and the x's in seconds of arc, the cotangent terms in (2.17) must be multiplied by $10^6 \times \sin 1''$ or 4.8481. Alternatively, in old computations, if ϵ is in units of the sixth decimal of common logs (one or two decimals of the unit being retained), the cotangents must be replaced by the change per 1″ in the sixth decimal of the log sine.†

Sufficient relevant equations must be formed, but no redundant ones. The conditions for common figures are listed below, but for more complex figures see § 2.32.

Simple triangle:	One triangular.
Braced quadrilateral:	Three triangular and one side.‡
Centred n-sided polygon:	n triangular, one side, and one central.

The above is if all angles have been observed. For any one missing reduce the number of triangular conditions by one. If a redundant equation is included, or if one equation is identically obtainable from one or more of the others, the fault will become apparent in the solution of the normal equations, where sooner or later two of the equations will become identical, and one of the λ's in (2.19) will be indeterminate.

(*c*) *Weights.* Before going further, weights may be assigned to the observed angles. The weight can be defined as equal to, or proportional to, the inverse square of the standard error, and the mean of two independent and equally reliable measures has twice the weight of a single one. Let w_α be the weight of an angle α, and let $u_\alpha = 1/w_\alpha =$ the square of the s.e.‖

There is much to be said for giving equal weights to all the angles of

† The 6th decimal is preferable to the 7th or 8th as the ϵ's in the side equations will then be of the same order of magnitude as those in the angle equations.

‡ There are four triangles, but satisfying three automatically satisfies the fourth.

‖ At this stage the s.e.'s of angles may be unknown, but it will suffice if the allotted weights are made proportional to the inverse squares of the s.e.'s. The relative reliability of angles can be assessed comparatively easily.

any one figure if they have been observed with the same programme in similar circumstances, but extra weight can be given to any which have been more fully observed for reasons other than exceptional internal discrepancy, and reduced weight to any which seem exceptionally doubtful. Very small angles may perhaps be given extra weight, § 2.43 (*e*). The allotting of weights based on the scatter of the ordinary programme of 10 or 16 zeros is not recommended as a regular routine.

(*d*) *Solving condition equations.* In a simple triangle the error is distributed between the angles is proportion to their u's, the correction to an angle α being $u_\alpha \epsilon/(u_\alpha + u_\beta + u_\gamma)$.

In other figures (2.18) consists of n equations for t unknowns, where $t > n$, and their solution has to make $x_1^2/u_1 + x_2^2/u_2 + \ldots$ a minimum. The routine procedure is to form n *normal equations* for n *correlates* $\lambda_1, \ldots, \lambda_n$ as below. See § D.16, and [466], of which the last gives simple examples.

$$\left.\begin{array}{l} [aau]\lambda_a + [abu]\lambda_b + \ldots [anu]\lambda_n = \epsilon_a \\ [bau]\lambda_a + [bbu]\lambda_b + \ldots [bnu]\lambda_n = \epsilon_b \\ \quad\cdot \quad\quad \cdot \quad\quad \cdot \quad\quad \cdot \\ \quad\cdot \quad\quad \cdot \quad\quad \cdot \quad\quad \cdot \\ \quad\cdot \quad\quad \cdot \quad\quad \cdot \quad\quad \cdot \\ [nau]\lambda_a + [nbu]\lambda_b + \ldots [nnu]\lambda_n = \epsilon_n \end{array}\right\}, \tag{2.19}$$

where $[aau]$ means $a_1 a_1 u_1 + a_2 a_2 u_2 + \ldots a_t a_t u_t$,

and $[bau] = [abu]$ means $a_1 b_1 u_1 + a_2 b_2 u_2 + \ldots a_t b_t u_t$.

The a's, b's, etc., will be 0 or 1, or in side equations numbers such as (e.g.) 3.21 or 0.56. Two significant figures suffice for the u's, and it is convenient to have them vary between about 0.10 and 5.0, as can generally be got by multiplying all by some power of 10 if necessary. The totals $[aau]$, etc., should then be formed exactly. Note that those below a diagonal drawn from top left to bottom right are identical with those above, and so need not be recorded. Check arithmetic as follows. In (2.18) let $a_1 + b_1 + \ldots n_1 = \sigma_1$, etc., and compute $[a\sigma u]$, $[b\sigma u]$, etc., where $[a\sigma u] = a_1\sigma_1 u_1 + \ldots a_t\sigma_t u_t$. Then in (2.19) $[aau] + \ldots [anu]$, the sum of the coefficients in the first equation, should equal $[a\sigma u]$, and so on. Also $[\sigma\sigma u]$ should equal $[a\sigma u] + [b\sigma u] + \ldots [n\sigma u]$.

To solve the normal equations (2.19), Gauss's classic routine is as follows. Multiply the first by $[abu]/[aau]$, and subtract the result from the second. Then multiply the first by $[acu]/[aau]$ and subtract from the third, and so on. This gives a set of $(n-1)$ equations from which λ_a has been eliminated. The symmetry about the diagonal is maintained. Proceeding as above with this set then eliminates λ_b, and so on until there remains a single one-term equation for λ_n. Substituting it in the equation for λ_m and λ_n then gives λ_m, and so on back to λ_a.†

The normal equations having been solved, the x's are given by

$$\left.\begin{aligned} x_1 &= u_1(\lambda_a a_1 + \lambda_b b_1 + \ldots \lambda_n n_1) \\ x_2 &= u_2(\lambda_a a_2 + \lambda_b b_2 + \ldots \lambda_n n_2) \\ &\quad \text{etc.} \end{aligned}\right\} \tag{2.20}$$

Check by the rule that $x_1^2/u_1 + x_2^2/u_2 + \ldots x_t^2/u_t$ should equal

$$\epsilon_a\lambda_a + \epsilon_b\lambda_b + \ldots \epsilon_n\lambda_n.$$

The final values of the angles are the observed angles corrected as in § 2.09 minus the appropriate x's. For final check these should be seen to satisfy the original conditions.

The work in solving the normals varies as the cube of their number. Detailed advice and examples are given in [466].

In any but the simplest cases the work of forming and solving the normal equations, and of obtaining and checking the unknowns, will be done by a standard programme on an electronic computer, using methods which differ in detail from the above, see §§ B.14 and B.15. The method of variation of coordinates, § 2.18 is likely to be preferred.

(*e*) *Directions or angles.* In the method outlined above, the unknown errors are those of the observed angles. Alternatively it is possible to solve for the errors of the observed directions. The procedure is then similar, every unknown angular error x in (2.18) being replaced by the difference between the errors of the two directions concerned. The method of angles is the simpler, and is thought to give slightly the more probable results. See § 2.22.

2.14. Computation of coordinates

(*a*) *Distances and azimuths.* The lengths of all sides of triangles are given by successive use of (2.8), the 'plane' angles I_p, etc., now being replaced by $I_p - x$, where x is the error given by (2.20). Where there are two routes through any figure, the solution of triangles should be carried through both, and the resulting values of the terminal side should agree.

† No firm rule can be given for the number of figures to keep when solving (2.19). If a's, b's, etc., and u's are as advised above, 5 decimals ought to be ample, but more may be needed if the equations are very numerous or unstable. [466], pp. 57–8, gives some advice.

When the λ's have been found, substitute in (2.19), and if each equation is not satisfied within about 0".03 a second approximation may be made. For let the true λ_a be $\lambda'_a + \Delta\lambda_a$, etc., where λ'_a is the value got from the first solution, and let the substitution of λ'_a, etc., in (2.19) give $\epsilon_a - \Delta\epsilon_a$, etc. Then (2.19) rewritten with $\Delta\lambda$'s for λ's and $\Delta\epsilon$'s for ϵ's constitutes a set of equations for the $\Delta\lambda$'s, which can be solved with little new arithmetic. A further approximation can be made if necessary.

For azimuths, if A_{10} is the known (normal section) azimuth from P_1 to P_0, the forward azimuth A_{12} from P_1 to P_2 is given by adding or subtracting the observed angle $P_0P_1P_2$ (corrected as in § 2.09) less the appropriate error x. Note that $E/3$ must not be subtracted. The spherical excess only enters into the computation of the sides.

The reverse azimuth from P_2 to P_1 is given by the formulae which follow.

(*b*) *Latitudes and longitudes.* Given ϕ_1, λ_1, A_{12} or α_{12}, and L_{12}, many formulae exist for finding ϕ_2, λ_2 and the reverse azimuth A_{21} or α_{21}, and since visibility ordinarily limits L_{12} to 150 km or so, most formulae in common use are amply accurate. In the past, much attention has been given to deriving formulae which will be accurate over great distances, but the importance of the problem has diminished, for the reasons given at the end of § 2.12. The following sub-paragraphs give a selection.

Formulae for the reverse problem § 2.15, to determine the distance and mutual azimuths between points of known ϕ and λ, are now more important than the direct formulae. They are currently used, for lines of up to about 500 km.† There is no apparent use for the surface, spheroidal, length of lines longer than this.

Most of the formulae quoted or referred to here and in § 2.15 use the normal section azimuth A. If a formula using the geodesic azimuth α is employed, observed angles and azimuths must be corrected by (2.10), if the correction is appreciable.

These formulae are of two kinds. (i) Those such as (*f*) and (*g*) which give the difference $\phi_2 - \phi_1$, and whose results are consequently correct to 1 ppm of L or better (within the distance to which the formula will give it) if 7-figure trig functions are used. An 8th figure may possibly be advised for safety, but they are referred to below as '7-figure'. And (ii) those which give sin or tan ϕ_2, so that an error of 0".001, which is about 3 cm on the ground, may arise from an error of 2 in the 9th decimal of the sine (at 60°), or of 4 in the tangent (of small angles). So 8-figure tables are not good enough in short lines or when the answer comes in a part of the table where differences are small, and they are referred to below as '9-figure'. For non-electronic computing there has been obvious advantage in using the former type. Fewer figures may, of course, always be used in the smaller terms.

The selection of formulae given below, and in § 2.15 necessarily excludes many good and widely used formulae. The choice between different formulae depends on the lengths of the lines for which they are intended, on the accuracy required, and on the computer for which they are to be programmed. Any organization which has got its own formula

† See § 2.39 (*a*), where traverse lines of up to 500 km in length have been treated as single units for adjustment by variation of coordinates.

programmed, and is satisfied with its accuracy, will probably do well to avoid change. [33] lists the errors in 32 reverse formulae at distances of 50 and 500 miles. [32] similarly lists the errors of distance in 22 formulae, and of azimuth in 10, over distances of 500, 1000, 3000, and 6000 miles.†

(*c*) *Rudoe's formula.* '9-figure'. [503].‡ For use as a standard of comparison. If sufficient figures are used in the computation, this formula is correct to fractions of a millimetre at all distances. All the formulae are closed except the expansion for $u_2'-u_1'$ in (v), which is carried to terms in ϵ^3, of magnitude 0".0005. The first omitted term will be more than 100 times smaller. The principle is to compute the axes and eccentricity of the ellipse in which the normal section cuts the spheroid: to get the 'reduced latitude' (A.52) of P_2 in this ellipse: and thence to return to the spheroid. It is not intended for general use with short lines.

(i) $C^2 = \cos^2\phi_1 \cos^2 A + \sin^2\phi_1$. Here $A = A_{12}$.

(ii) $\epsilon_0 = C^2\epsilon$. ϵ_0 is ϵ of ellipse of normal section.

(iii) $b_0 = \{\nu_1\sqrt{(1+\epsilon\cos^2\phi_1\cos^2 A)}\}\div(1+\epsilon_0)$. b_0 is minor axis of above.

(iv) $\tan u_1' = \dfrac{\tan\phi_1}{\cos A\sqrt{(1+\epsilon_0)}}$. u_1' is reduced lat of P_1 in above.

(v)
$$\begin{aligned} u_2'-u_1' = (\sigma_2-\sigma_1)+2\gamma_2 \sin(\sigma_2-\sigma_1)\cos(\sigma_2+\sigma_1)+ \\ +2\gamma_4 \sin 2(\sigma_2-\sigma_1)\cos 2(\sigma_2+\sigma_1)+ \\ +2\gamma_6 \sin 3(\sigma_2-\sigma_1)\cos 3(\sigma_2+\sigma_1), \end{aligned}$$

where $(\sigma_2-\sigma_1) = L\gamma_0/b_0$,

$$\gamma_0 = 1-\tfrac{1}{4}\epsilon_0+\tfrac{7}{64}\epsilon_0^2-\tfrac{15}{258}\epsilon_0^3,$$

$$\gamma_2 = \tfrac{1}{8}\epsilon_0-\tfrac{1}{16}\epsilon_0^2+\tfrac{71}{2048}\epsilon_0^3,$$

$$\gamma_4 = \tfrac{5}{256}\epsilon_0^2-\tfrac{5}{256}\epsilon_0^3,$$

$$\gamma_6 = \tfrac{29}{6144}\epsilon_0^3,$$

$$2\sigma_1 = 2u_1'-(\tfrac{1}{4}\epsilon_0-\tfrac{1}{8}\epsilon_0^2)\sin 2u_1'-\tfrac{1}{128}\epsilon_0^2\sin 4u_1',$$

$$\sigma_1+\sigma_2 = 2\sigma_1+(\sigma_2-\sigma_1)$$

(vi) $\sin u_2 = \dfrac{b_0 C}{b}\sin u_2' - \left(\dfrac{\epsilon-\epsilon_0}{1+\epsilon_0}\right)\sin u_1$. (Back to the spheroid),

where $\sin u_1 = \tan\phi_1 \div \sqrt{(1+\epsilon+\tan^2\phi_1)}$.

(vii) Latitude.

$$\sin\phi_2 = \sin u_2 \div \sqrt{(1-e^2\cos^2 u_2)}.$$

† Large errors are incorrectly listed for Rudoe's formula at 500 miles in [32], and in one line in [33] at 6000 miles.

‡ Not published. The order of items (i) to (vii) gives all the geodesy necessary for a proof: all else is straightforward mathematics.

(viii) Longitude. $\cos(\mu+\Delta\lambda)\dagger=\dfrac{a_0 \cos u_2'}{a \cos u_2}$,

where $\tan \mu=\sin \phi_1 \tan A.$

and $a_0=b_0\sqrt{(1+\epsilon_0)}.$

For the reverse azimuth use (2.23), or alternatively use Dalby's theorem [153], pp. 235–6, which states

$$A_{12}-A_{21}=B_{12}-B_{21}+\tfrac{1}{4}e^4(\lambda_2-\lambda_1)(\phi_2-\phi_1)^2 \cos^4\phi_1 \sin \phi_1, \quad (2.21)$$

where B_{12} and B_{21} are the reciprocal azimuths of points $\phi_1\lambda_1$ and $\phi_2\lambda_2$ on a sphere. With $L<800$ km the small term $<0''.0005$ and is negligible. For $(B_{12}-B_{21})$ the formula is:

$$\cot \tfrac{1}{2}(B_{12}-B_{21})=\tan \tfrac{1}{2}(\lambda_2-\lambda_1)\sin \tfrac{1}{2}(\phi_2+\phi_1)\sec \tfrac{1}{2}(\phi_2-\phi_1). \quad (2.22)$$

As an alternative to (vi), (vii), and (viii) of Rudoe's formula, [121], 1978, gives the following.

(vi)‡ $x_2=a_0 \cos u_2' \cos \mu+b_0 \sin u_2' \sin \mu \cos \phi_1 \sin A$
$\quad +\delta \sin \phi_1 \sin A$

$y_2=-a_0 \cos u_2' \sin \mu+b_0 \sin u_2' \cos \mu \cos \phi_1 \sin A+\delta \cos A$

$z_2=b_0 C \sin u_2'-(\delta \cos \phi_1 \sin A)/(1+\epsilon)$

where $\tan \mu=\tan A \sin \phi_1,$

$$a_0=b_0(1+\epsilon_0)^{\frac{1}{2}},$$

$$\delta=(e^2 \nu_1 \sin \phi_1 \cos \phi_1 \sin A)/(1-e^2 \cos^2\phi_1 \sin^2 A).$$

(vii) Latitude. $\tan \phi_2=(1+\epsilon)z_2/(x_2^2+y_2^2)^{\frac{1}{2}}.$

(viii) Longitude. $\tan \Delta\lambda=y_2/x_2.$

(*d*) *Clarke's best formula.* '9-figure'. [155], pp. 268–270 and 275. Correct to (1/25) ppm at 800 km. For lines between 100 and 1200 km. [486] gives a few more terms by which accuracy is increased to (1/180) ppm at 1600 km.

(i) $r_2'=-\epsilon \cos^2\phi_1 \cos^2 A.$

(ii) $r_3'=3\epsilon(1-r_2')\cos \phi_1 \sin \phi_1 \cos A.$

(iii) $\theta=\dfrac{L}{\nu_1}-\dfrac{r_2'(1+r_2')}{6}\left(\dfrac{L}{\nu_1}\right)^3-\dfrac{r_3'(1+3r_2')}{24}\left(\dfrac{L}{\nu_1}\right)^4.$

† If $(\mu+\Delta\lambda)$ is small, use $\sin(\mu+\Delta\lambda)=\Lambda_{12} \sin \mu$,

where $\Lambda_{12}=\dfrac{\tan u_2}{(1+\epsilon)\tan u_1}+e^2\dfrac{\cos u_1}{\cos u_2}.$

‡ Here the axis of x is taken to be in the meridian plane of point P_1.

(iv) $\dfrac{\nu_1}{r} = 1 - \dfrac{r_2'}{2}\theta^2 - \dfrac{r_3'}{6}\theta^3.$

(v)† $\sin\psi = \sin\phi_1 \cos\theta + \cos\phi_1 \cos A \sin\theta.$

Longitude (vi) $\sin\Delta\lambda = \sin A \sin\theta \sec\psi.$

Latitude (vii) $\tan\phi_2 = (1+\epsilon)\left\{1 - e^2\left(\dfrac{\nu_1}{r}\right)\dfrac{\sin\phi_1}{\sin\psi}\right\}\tan\psi.$

Azimuth (viii) Use (2.23) or (2.21).

(e) *Clarke's approximate formula.* '7-figure'. [155], p. 273. Given by A. R. Clarke for $L < 150$ km and correct to 1 ppm. [465] lists innumerable smaller terms by which accuracy can be increased to (1/25) ppm at 800 km.

(i) $p = \dfrac{L^2 \sin A \cos A}{2\rho_X \nu_X \sin 1''}, \quad q = p \tan A \tan\phi_X.$

Latitude (ii) $\phi_2 - \phi_1 = \dfrac{L\cos(A - \frac{2}{3}p)}{\rho_Y \sin 1''} - q.$

Longitude (iii) $\lambda_2 - \lambda_1 = \dfrac{L\sin(A - \frac{1}{3}p)}{\nu_X \sin 1''}\sec(\phi_2 + \tfrac{1}{3}q).$

Azimuth (iv) $A_{21} = 180 + A_{12} + (\lambda_2 - \lambda_1)\sin(\phi_2 + \tfrac{2}{3}q) - p.$

Suffix X refers to $\phi_X = \phi_1 + \dfrac{L\cos A}{\rho_1 \sin 1''}.$

Suffix Y refers to $\phi_Y = \frac{1}{2}(\phi_1 + \phi_X).$

(f) *Topographical formula.* '5-figure'. This well-known formula gives approximate positions over lines of 50–60 km for variation of coordinates. It is correct to about 0.3 m at 40 km.

$$\Delta A'' = \Delta\lambda'' \sin\phi_m,$$

$$\Delta\phi'' = (L/\rho_m)\cos(A + \tfrac{1}{2}\Delta A)\operatorname{cosec} 1'',$$

$$\Delta\lambda'' = (L/\nu_m)\sin(A + \tfrac{1}{2}\Delta A)\sec(\phi + \tfrac{1}{2}\Delta\phi)\operatorname{cosec} 1''.$$

Suffix m is for the mean latitude, estimated to $1'$ by any means. In the first line estimate $\Delta\lambda$ by any means, and repeat if the last line shows that the estimate is not good enough.

(g) *Puissant's (USC & GS) formula.* '7-figure'. [21] and [370]. Correct

† In high latitudes, instead of (v) and (vi) in Clarke's best formula, use

$$\cot\Delta\lambda = (\cos\phi_1 \cot\theta - \sin\phi_1 \cos A)/\sin A,$$

$$\cos\psi = \sin\theta \sin A \operatorname{cosec}\Delta\lambda.$$

to 1 ppm at 80 or 100 km, beyond which it rapidly goes wrong. (40 ppm at 250 km in latitude 60°). Within its limits it has been very convenient for desk (not electronic) computation, provided tables for five functions B to F are available. It is given in full in [113], 3 Edn, p. 134.

(*h*) *Levallois and Dupuy*. '9-figure'. [388]. Expansions are in powers of ϵ, and accuracy is substantially perfect. Geodesic azimuths are used.

[275], pp. 381–3 gives a clear summary, in a form suitable for desk machine computation.

(*i*) *de Graaff-Hunter's formula*. '9-figure'. [113], 2 Edn, pp. 106 and 505–6. Correct to 1 ppm at 500 km. The method is of interest. $\Delta\phi$, $\Delta\lambda$, and $\Delta\alpha$ are computed for the spheroid in the form of series in which some terms contain ϵ and some do not. The terms not involving ϵ give the result for a sphere of radius ν_1, and can be replaced by the ordinary closed formulae of spherical trigonometry, to which must be added the small terms in ϵ.

2.15. Reverse formulae

The mutual distance and azimuths of two points of given ϕ and λ are required (accurately) for computation by variation of coordinates.

(*a*) *Cunningham's azimuth formula* [172]. This is a closed formula, which is exact at all distances if computed with sufficient figures, but the mutual azimuth of near points is of course unstable in the sense that the usual amount of inaccuracy in the data will produce an abnormal error in the computed azimuth. Cunningham's formula is

$$\cot A_{12} = (\Lambda_{12} - \cos \Delta\lambda)\sin \phi_1 \operatorname{cosec} \Delta\lambda,$$

where

$$\Lambda_{12} = \frac{\tan \phi_2}{(1+\epsilon)\tan \phi_1} + e^2\sqrt{\frac{(1+\epsilon)+\tan^2\phi_2}{(1+\epsilon)+\tan^2\phi_1}}$$

$$= \frac{\tan \phi_2}{(1+\epsilon)\tan \phi_1} + e^2\frac{\nu_1 \cos \phi_1}{\nu_2 \cos \phi_2}. \tag{2.23}$$

A_{21} may be got from the same by transposing suffixes 1 and 2, or from (2.21), or from $\nu \cos \phi \sin \alpha = \text{constant}$ together with (2.10) if the distance is not too great for the accuracy of (2.10).

As given above, Λ_{12} is indeterminate if ϕ_2 and $\phi_1 = 0$, but there is no instability in $\Lambda_{12} \sin \phi_1$, which is what is required for cot A_{12}, and which is firmly near zero in these circumstances. (i) and (iii) of (*c*) below give an alternative arrangement.

(*b*) *Rudoe's reverse formula*. [503]. This is the reverse of § 2.14 (*c*), and gives normal section distances correct to a fraction of a millimetre if sufficient figures are used. For use as a standard of comparison.

A_{12} must first be got from (2.23). Then compute b_0 and ϵ_0, the minor axis and ϵ of the ellipse in which the normal section at P_1 cuts the spheroid, from

(i) $b_0 = \{\nu_1/(1+\epsilon_0)\}\sqrt{(1+\epsilon\cos^2\phi_1\cos^2 A_{12})}$,

(ii) $\epsilon_0 = \epsilon(\cos^2\phi_1\cos^2 A_{12} + \sin^2\phi_1)$.

If u'_1 and u'_2 are reduced latitudes in this ellipse,

(iii)† $\tan u'_1 = \dfrac{\tan\phi_1}{\cos A_{12}\sqrt{(1+\epsilon_0)}}$,

(iv) $\tan u'_2 = \dfrac{\nu_1\sin\phi_1 + (1+\epsilon_0)(z_2 - z_1)}{(x_2\cos A_{12} - y_2\sin\phi_1\sin A_{12})\sqrt{(1+\epsilon_0)}}$,

where
$$x_2 = \nu_2\cos\phi_2\cos\Delta\lambda,$$
$$y_2 = \nu_2\cos\phi_2\sin\Delta\lambda,$$
$$z_2 = \nu_2(1-e^2)\sin\phi_2,$$
$$z_1 = \nu_1(1-e^2)\sin\phi_1.$$

The axis of x is here taken to be in the meridian plane of P_1.

Then the length of the normal section (from P_1) is given by

(v) $L = b_0\{c_0(u'_2 - u'_1) + c_2(\sin 2u'_2 - \sin 2u'_1) + c_4(\sin 4u'_2 - \sin 4u'_1) + \ldots\}$,

where
$$c_0 = 1 + \tfrac{1}{4}\epsilon_0 - \tfrac{3}{64}\epsilon_0^2 + \tfrac{5}{256}\epsilon_0^3,$$
$$c_2 = -\tfrac{1}{8}\epsilon_0 + \tfrac{1}{32}\epsilon_0^2 - \tfrac{15}{1024}\epsilon_0^3,$$
$$c_4 = -\tfrac{1}{256}\epsilon_0^2 + \tfrac{3}{1024}\epsilon_0^3.$$
$$c_6 = -\tfrac{1}{3072}\epsilon_0^3$$

(vi) For A_{21} use (2.23).

In this formula there is an apparent instability in (iii) if $\tan\phi_1$ and $\cos A_{12}$ are both zero, as occurs when both points are very close to the equator. The trouble occurs because in this situation the normal section is nearly circular, and the major axis from which u's are measured is indeterminate. This does not matter provided some single point (reasonably accurately located) is used for both u'_1 and u'_2. This is secured by computing with sufficient significant digits, using floating point if necessary.‡ [581B]. In these circumstances $\cos A_{12}$ must be obtained from $\cos A_{12} = (1+\tan^2 A_{12})^{-\frac{1}{2}}$, and must not be taken from tables of $\cos A_{12}$.

† If A_{12} is near 90° or 270°, $\cos A_{12}$ in (iii) and (iv) must be determined from the value of $\cot A_{12}$ given by (2.23), not from the value of A_{12} itself which may have been got from $\cot A_{12}$.

‡ In the extreme case where both points lie exactly on the equator, L is equal to $\alpha\Delta\lambda\sin 1''$.

(*c*) *Robbins's formula.* [486]. Based on Clarke's best formula, reversed. Correct to (1/100) ppm at 1600 km. Errors of 20 m at 5000 km. Some small terms in (vii) may be omitted in short lines.

(i)† $\tan \psi_2 = (1-e^2)\tan \phi_2 + e^2 \dfrac{\nu_1 \sin \phi_1}{\nu_2 \cos \phi_2}$,

(ii) $\delta\phi_2 = \phi_2 - \psi_2$,

(iii) $\cot A_{12} = (\cos \phi_1 \tan \psi_2 - \sin \phi_1 \cos \Delta\lambda)\operatorname{cosec} \Delta\lambda$,

(iv) $\cot A'_{21} = (\sin \psi_2 \cos \Delta\lambda - \cos \psi_2 \tan \phi_1)\operatorname{cosec} \Delta\lambda$,

(v) $\sin \sigma = \sin \Delta\lambda \cos \psi_2 \operatorname{cosec} A_{12}$
$= -\sin \Delta\lambda \cos \phi_1 \operatorname{cosec} A'_{21}$ for check,

(vi) $A_{21} = A'_{21} - \delta\phi_2 \sin A'_{21} \tan \tfrac{1}{2}\sigma$,

(vii)
$$L = \nu_1\sigma\left[1 - \frac{\sigma^2}{6} h^2(1-h^2) + \frac{\sigma^3}{8} gh(1-2h^2) + \right.$$
$$\left. + \frac{\sigma^4}{120}\{h^2(4-7h^2) - 3g^2(1-7h^2)\} - \frac{\sigma^5}{48} gh\right],$$

where $g^2 = \epsilon \sin^2\phi_1$ and $h^2 = \epsilon \cos^2\phi_1 \cos^2 A_{12}$.

When A_{12} is near 0° or 180°, get ($\sin \Delta\lambda \operatorname{cosec} A_{12}$) for (v) from (iii) written as $\sin \Delta\lambda \operatorname{cosec} A_{12} = (\cos \phi_1 \tan \psi_2 - \sin \phi_1 \cos \Delta\lambda)\sec A_{12}$ and similarly from (iv) when A'_{21} is near 0° or 180°.

A_{21} may be got from (i) and (iii) with suffixes 1 and 2 interchanged, instead of from (iv) and (vi). The second expression for $\sin \sigma$ in (v) cannot then be used. (i) and (iii) are essentially the same as (2.23).

(*d*) *Chord-to-arc method*

(i) Let ϕ_1 and λ_1, ϕ_2 and λ_2 be converted to (x_1, y_1, z_1) and (x_2, y_2, z_2) by the formulae

$$x = \nu \cos \phi \cos \lambda$$
$$y = \nu \cos \phi \sin \lambda$$
$$z = (1-e^2)\nu \sin \phi$$

(ii) The chord $C = \{(x_2-x_1)^2 + (y_2-y_1)^2 + (z_2-z_1)^2\}^{\frac{1}{2}}$

(iii) For the azimuth use Cunningham's formula (2.23).

(iv) The difference between the geodesic arc and the chord, $L_G - C$, which is given in (1.68) as $D^3/24\,R^2$, can be given more precisely by variety of formulae, such as (A) and (B) below.

(A) Let $R_1 = \nu_1(1+\epsilon \cos^2 A_{12} \cos^2\phi_1)^{-1}$

and let $R_2 = \nu_2(1+\epsilon \cos^2 A_{21} \cos^2\phi_2)^{-1}$.

Then $L_G = 2\,R \sin^{-1}$ (C/2R), where $R = \tfrac{1}{2}(R_1+R_2)$.

From [581B], which gives a maximum error of 1×10^{-5} m in a number of 160 km test lines. See also [155], p. 109.

† In high latitudes write $\tan \psi_2 = (1-e^2+e^2\,(\nu_1 \sin \phi_1/\nu_2 \sin \phi_2)) \tan \phi_2$.

(B) [563], pp. 19–20, gives

$$L_G - C = C\left\{\frac{C^2}{24\nu_1^2} + \frac{C^2\epsilon\cos^2\phi_1\cos^2 A_{12}}{12\,\nu_1^2} - \frac{C^3\epsilon\sin 2\phi_1\cos A_{12}}{16\nu_1^3} + \text{terms in } \epsilon^2(C/\nu_1)^2 \text{ and } \epsilon(C/\nu_1)^4\right\}.$$

Writing $C/\nu_1 = 1/40$ for lines of 160 km, and $\epsilon \approx 1/150$, the maximum magnitudes of the two terms following $C^2/24\nu_1^2$ are 0.3 and 0.007 ppm respectively, so the formula

$$L - C = L_G - C = C\left\{\frac{C^2}{24\nu_1^2} + \frac{C^2\epsilon\cos^2\phi_1\cos^2 A_{12}}{12\nu_1^2}\right\}$$

can be used for any intervisible line between ground stations with an error of less than 0.1 ppm.

(*e*) *Topographical formula.* Correct to about 0.3 metres in L and to 1″ in A if $L < 40$ km.

$$\Delta A'' = \Delta\lambda'' \sin\phi_m,$$

$$\tan(A + \tfrac{1}{2}\Delta A) = \frac{\nu_m}{\rho_m}\cdot\frac{\Delta\lambda''\cos\phi_m,}{\Delta\phi''},$$

or

$$\cot(A + \tfrac{1}{2}\Delta A) = \frac{\rho_m}{\nu_m}\cdot\frac{\Delta\phi''}{\Delta\lambda''\cos\phi_m},$$

$$L = \rho_m\Delta\phi''\sec(A + \tfrac{1}{2}\Delta A)\sin 1'',$$

or

$$L = \nu_m\Delta\lambda''\cos\phi_m\,\mathrm{cosec}(A + \tfrac{1}{2}\Delta A)\sin 1''.$$

(*f*) *Vincenty's formula.* [581]. This is a pair of formulae for direct and reverse computations, devised to minimize the amount of core which the programs will occupy in the computer. The equations are in nested form. The program is intended to give sub-millimetre accuracy over distances of nearly 20 000 km for the geodesic.

(g) *USC&GS formula. Gauss's mid-latitude* [370]. Correct to 1 ppm at 100 km. Convenient for desk computation provided tables for functions A', B, and F are available. Not suitable for electronic computing. It is given in full in [113], 3 Edn, p. 137.

(*h*) *Helmert's formula.* [270] i. 5 § 13, and [32],† p. 34, give a formula for the length of the geodesic, which by reiteration can give any required accuracy over any distance.

† In [32] the reduced latitude u, given by $\tan u = (b/a)\tan\phi$, is referred to as the parametric latitude.

(*i*) *Hyperbolic formula.* [125] and [113], 2 Edn., § 3.13 (*d*), give a very simple formula for the geodesic distance and azimuth using hyperbolic functions, which an electronic computer computes as easily as trigonometrical functions. It is correct to 1 ppm at 1000 km, using 8-figure tables.

(*j*) *Sharma's formula.* [523] and [113], 3 Edn., p. 138. Using the property that along a geodesic $(\nu/a)\cos\phi\sin\alpha$ equals a constant k, a formula is given for the length of a geodesic in terms of k and the latitudes of its two ends. The length of any geodesic can then be computed by dividing it into sections with equal latitude differences.

2.16. Computation of a single traverse

(*a*) It is required to compute a traverse, as in Fig. 2.9, given ϕ and λ at P_0, and with Laplace azimuths at P_0 and other stations at varying intervals. Traverse angles are, of course, corrected by (2.5) and (2.7), if these corrections are appreciable, and measured distances are reduced to the spheroid. There is no question of any deduction for spherical excess. Computation by variation of coordinates, which is described in Section 3, is not the natural procedure, but if an available computer has been programmed for computing triangulation by that method, it will probably be convenient to use it for traverses also.

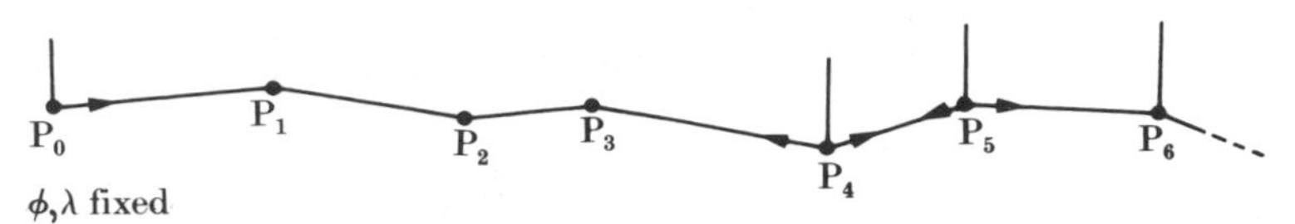

FIG. 2.9. Traverse with Laplace aximuths at P_0, P_4, and P_5.

Laplace azimuths cannot be computed unless their geodetic longitudes are known. This involves some kind of successive approximation as described in § 2.40 (*a*), and in (*b*) below. The method of variation of coordinates does this very conveniently, § 2.23 (*f*) and (*g*).

(*b*) *Errorless Laplace azimuths.* If, as is not ordinarily advised, the Laplace azimuths are to be treated as infallible, it is only necessary to adjust the horizontal angles between consecutive Laplace stations. Starting from P_0, preliminary computations of longitude and reverse azimuths will give a value for the azimuth of P_4P_3 in Fig. 2.9, and if this does not agree with the Laplace value the discrepancy is distributed through the angles at P_1P_2, and P_3, with such relative weighting as may be proper. The traverse can then be rigorously computed from P_0 to P_4, but if the resulting longitude of P_4 differs significantly from that given by the preliminary computations, the process may have to be repeated.†

† Unless Laplace stations are very far apart, this should not cause trouble. An average error of 3″ of azimuth over a distance of 150 km between Laplace stations (both large figures) will cause an error of less than 2.2 metres at the far end, which changes the Laplace azimuth by 0″07 tan ϕ, which is trivial except in high latitudes.

A pair of reciprocal Laplace azimuths, such as between P_4 and P_5, cannot both be treated as infallible. Accepting P_4P_5, the reverse azimuth P_5P_4 must be computed and compared with its Laplace value. Any discrepancy must then be divided between the Laplace values of P_4P_5 and P_5P_4, before they can be treated as infallible.

(*c*) *Fallible azimuths.* Let x_{01}, x_{43}, and x_{45}, etc. be the required errors of the Laplace azimuths P_0P_1, P_4P_3, and P_4P_5, etc.: and let x_{02}, x_{13}, etc., be the errors of the traverse angles† $P_0P_1P_2$, $P_1P_2P_3$, etc. Then

$$\left.\begin{aligned} x_{01}+x_{02}+x_{13}+x_{24}-x_{43} &= \epsilon_{43} \\ x_{45}-x_{54} &= \epsilon_{54} \\ x_{56}+x_{57}+x_{68}+\ldots &= \text{etc.} \end{aligned}\right\}, \tag{2.24}$$

where ϵ_{43} is (Laplace azimuth P_4P_3) minus (azimuth P_4P_3 as brought forward from P_0).

In (2.24) no x has been allotted to the traverse angles, such as $P_3P_4P_5$, at the Laplace stations, and the equations (2.24) are consequently independent. The x's are then immediately got by dividing the ϵ's of each equation among its x's according to the latter's estimated weights.‡

It may, however, happen that the traverse angles at Laplace stations have been independently measured, as they would have been if the observation programme at (say) P_4 was in closed rounds $P_3-\text{star}-P_5-P_3$. In this case there are additional equations

$$\left.\begin{aligned} -x_{43}+x_{45}-x_{35} &= 0,\| \\ -x_{54}+x_{56}-x_{46} &= 0, \text{ etc.} \end{aligned}\right\} \tag{2.25}$$

The combined equations (2.24) and (2.25) now have to be solved by least squares, using the routine described in § 2.13 (*d*).

With fallible Laplace azimuths, reciprocal Laplace stations require no special treatment.

[99] describes a simple routine for computing the azimuths of a traverse, which is to be held to predetermined azimuths at its two ends, but which contains other (fallible) Laplace azimuths. All Laplace azimuths and traverse angles are assumed to have equal weight.

See § 2.32 (*k*) for the adjustment of a traverse on to fixed ϕ and λ at its two ends.

2.17. Miscellaneous computation notes

(*a*) In § 2.14 and 2.15 formulae are given in basic form, and devices for simplifying computation may sometimes be found in the sources quoted.

† The angles on the left-hand side of the traverse, starting from P_6.

‡ The weight of a Laplace azimuth is not necessarily equal to that of a traverse angle.

‖ Zero if the misclosure of rounds has been distributed.

Watch must always be kept for instability, such as occurs when a small angle is deduced from its cosine, and also for the necessity for using second differences, or some alternative, such as used to be necessary when taking the log sine of a small angle.

(*b*) In desk computations coordinates should always be computed from two known points. Agreement then checks all work done under §§ 2.12 and 2.14, but not the field abstract, nor § 2.09, nor the final compilation of results, where mistakes can only be avoided by computing in duplicate, and then only if computers are experienced and agree without frequent comparison. Where work is self-checking, duplicate computations are not essential, although they may be economical, as the cause of failure to check may take long to trace. Work should be done on standard printed forms if possible.

Surveyors should make it a rule that they will not give field data to a computer until they have been examined and checked. It is proper than observers should themselves abstract and examine means, an action which draws their attention to abnormal scatter or other irregularity. It is essential that triangular errors and the like should be examined before computations go further.

General references for Sections 1 and 2. [466], [484], and [563].

SECTION 3. COMPUTATION BY VARIATION OF COORDINATES

2.18. General description

Triangulation, trilateration, and traverse, or any combination of them, can be computed and simultaneously adjusted by the method of *Variation of Coordinates.* This method is very suitable for use with an electronic computer, and in any large organization it is likely to supersede the method described in Section 2. The computer may be programmed for the largest system that it can conveniently take, and can then be used for all such computations great or small. Alternatively, smaller computations can be handled on small computers, and the use of really large ones can be reserved for major framework adjustments.

The method of variation of coordinates is as follows. Approximate provisional coordinates ϕ, λ are first computed for all stations by any means. Approximate formulae such as § 2.14 (*f*) with 5-figure tables, will amply suffice, and all redundant observations may be neglected. Adequate coordinates may already have been computed for other purposes. For the desirable accuracy see § 2.21. Accepting these coordinates, the mutual azimuths of all lines whose directions have been observed, and the lengths of all lines whose lengths have been measured, are then accurately computed to perhaps 0.1 ppm accuracy, using a formula such as one of those listed in § 2.15; but not § 2.15 (*e*) or (g).

If the provisional coordinates and all the observations were correct, measured† angles (O) would agree with differences (C) of computed distances. Actually they will not do so, and the problem is to find corrections $\delta\phi$, $\delta\lambda$ to the provisional coordinates of each station, such as will minimize the sum of the squares of the weighted differences (the v's of § 2.23) between the observed angles and distances and the values computable from the corrected coordinates.

This is a case of adjusting by the parametric method without explicitly stated conditions, as described in § D.14. In the notation of (D.42) the x's are the unknown $\delta\phi$, $\delta\lambda$ at each station, the k's are the differences (observed minus provisionally computed), and the coefficients a, b, etc., are the effect on the computed mutual azimuth or distance, δA or δL, of unit values of $\delta\phi$ or $\delta\lambda$ at each end of the line, namely $\mathrm{d}A_{12}/\mathrm{d}\phi_1$, $\mathrm{d}A_{12}/\mathrm{d}\phi_2 \ldots \mathrm{d}L/\mathrm{d}\phi_1 \ldots\ldots$, etc.

The *observation equations*, in which the observations are compared with the provisionally computed values, take many different forms, for angles, directions, or distances, with modifications if (for instance) a particular latitude, longitude, azimuth, or distance is to be held fixed. These are given in § 2.23.

This method of computation produces fully adjusted results. The reconciliation of triangular errors, and the adjustment of triangulation figures as in § 2.13, calls for no special treatment.‡ Within the limits of any piece of work which is dealt with simultaneously, adjustment on to bases and Laplace stations, and the adjustment of circuit closures, is also disposed of. In this respect, the method is to be contrasted with that of Section 2, which calls for subsequent adjustment as in Section 4.

The advantages of the method of variation of coordinates are very great, as follows.

(*a*) As above, no subsequent adjustment is necessary. This avoids

(i) the difficulty of correctly forming the conditions in large systems, § 2.32;

(ii) special trouble in forming the condition equations for circuit closures in latitude and longitude, § 2.31 (*c*) (iii).

† Observed directions and angles should first be corrected as in § 2.09, but normal section azimuths should not be corrected to geodesic, provided the computed distances and azimuths are for normal section. Theodolite rounds must be reconciled by station adjustment or otherwise, § 1.23. Measured distances must be reduced to spheroid level, §§ 1.30 and 1.45. If rectangular coordinates are used, observed directions, angles, and distances must receive the arc-to-chord and scale corrections of § 2.59.

‡ But see § 2.29. Closing errors should be recorded and examined as checks on gross errors.

(*b*) One computer program deals with all combinations of observed angles, azimuths, and distances, whether regarded as fallible or errorless.

(c) It avoids the difficulty about Laplace azimuths changing with revised values of longitude. See § 2.23 (*f*) and (*g*).

(*d*) The normal equations, although possibly more numerous than if the method of conditions is used, generally cover a narrower bandwidth, and are easier to solve. See § B.14 (*f*) and (g).

(*e*) Additional observations can afterwards be incorporated, to produce a revised adjustment, with the minimum of trouble.

Simple worked examples are given in [466].

See § C.02 for the method of variation of coordinates in cartesian coordinates *x*, *y*, *z* in three dimensions.

2.19. Notation

$\delta\phi$, $\delta\lambda$, δA, δL = changes in ϕ, λ, and A in seconds of arc, and in L in metres.
$a, b, c, d = \partial A_{12}/\partial\phi_1, \partial A_{12}/\partial\lambda_1, \partial A_{12}/\partial\phi_2, \partial A_{12}/\partial\lambda_2 = (\rho_1/L)\sin A_{12}$, etc., as in (2.27).
$p, q, r, s = \partial L/\partial\phi_1$, etc. $= -\rho_1 \cos A_{12} \sin 1''$, etc., as in (2.27),
E, N = rectangular coordinates, easting and northing.
O = an observed value of azimuth, length, or angle.
C = a value computed from provisional ϕ, λ.
σ = the estimated standard error of an observed quantity.
w = the weight of an observation = $1/(\text{its } \sigma^2)$.
n = the serial number of a station.
x = an unknown optimum correction $\delta\phi$ or $\delta\lambda$.
Z = the unknown azimuth of the zero of observed directions.
N^{-1} = the inverse of the matrix of the normal equations.
α_{nm} = the element in row n and column m of N^{-1}.
v = the residual, (change in C resulting from $\delta\phi$, $\delta\lambda$) $-$ (O $-$ C).
RM = referring mark.

2.20. Differential formulae

In rectangular coordinates the formulae for δA and δL are very simple.

$$\delta A_{12} = (1/L \sin 1'')(-\cos A_{12}\delta E_1 + \sin A_{12}\delta N_1 + \cos A_{12}\delta E_2 - \sin A_{12}\delta N_2),$$

$$\delta L = -\sin A_{12}\delta E_1 - \cos A_{12}\delta N_1 + \sin A_{12}\delta E_2 + \cos A_{12}\delta N_2. \quad (2.26)$$

The correction to an angle 213 is $(\delta A_{13} - \delta A_{12})$, rotation from 2 to 3 through the angle being clockwise.

For computation in ϕ, λ [449] gives

$$\delta A_{12} = (\rho_1/L)\sin A_{12}\delta\phi_1 + (\rho_2/L)\sin A_{21}\delta\phi_2 + (\nu_2/L)\cos\phi_2 \cos A_{21}(\delta\lambda_1 - \delta\lambda_2),$$

$$\delta L = \{-\rho_1 \cos A_{12}\delta\phi_1 - \rho_2 \cos A_{21}\delta\phi_2 + \nu_2 \cos\phi_2 \sin A_{21}(\delta\lambda_1 - \delta\lambda_2)\}\sin 1'', \quad (2.27)$$

where δA, $\delta\phi$, and $\delta\lambda$ are in seconds of arc, and δL is linear units. The values used for sin A_{12}, etc., ρ, and ν come from the provisional computations, which should be sufficiently accurate, see § 2.21.

2.21. Accuracy required in provisional coordinates

In (2.27) ρ and ν are insensitive to error in ϕ and λ, and the chief source of error, especially in short lines, lies in the errors of sin A, cos A, and L caused by large relative errors in the provisional coordinates at the ends of a line.

For simplicity considering rectangular coordinates (2.26), let the relative error in the provisional coordinates at two adjacent stations be d (metres) in azimuth β. Formula (2.26) assumes that d is small. If it is large, a correct formula is

$$\tan \delta A = -\{d \sin (A-\beta)\} \div \{L + d \cos(A-\beta)\}, \tag{2.28}$$

with a similar formula for δL. Then the errors of (2.26) are

$$\text{in } \delta A,\ \epsilon_A = -(d^2/L^2)\sin(A-\beta)\cos(A-\beta)\operatorname{cosec} 1'' \tag{2.29}$$

$$\text{maximum} = (d^2/2L^2)\operatorname{cosec} 1,''$$

$$\text{and in } \delta L,\ \epsilon_L = (d^2/2L)\sin^2(A-\beta) \tag{2.30}$$

$$\text{maximum} = (d^2/2L).$$

For $\epsilon_A < 0''.01$, permissible values of the relative error d are 3, 15, and 62 m when $L = 10$, 50, and 200 km respectively. And for $\epsilon_L < 0.01$ metres, the corresponding permissible values of d are 14, 32, and 64 m, and 100 m when $L = 500$ km. An error of 1 part in 4000 in the provisional length of a line, or of 1 minute in its azimuth, will always be harmless.

If there is any risk that the first provisional coordinates are not sufficiently accurate, a second approximation must be made. The values of $\delta\phi$ and $\delta\lambda$ given by the first solution are added to the first provisional coordinates to give second provisional coordinates. New accurate mutual azimuths and distances must then be computed, and observation equations formed with new right-hand sides, and solved to give additional $\delta\phi$ and $\delta\lambda$'s.

When a computer is used, a second approximation, and more if indicated, should be made as a routine check, see § 2.29. The accuracy of the provisional coordinates, and of the coefficients $(\rho/L)\sin A_{12}$, etc. in (2.27) is then almost immaterial. The provisional coordinates can be taken from a map if necessary, with errors of even as much as $L/10$. In such a case several iterations will of course be necessary.

2.22. Angles or directions

Rounds of angles can give rise to observation equations of two different kinds.

(*a*) *Angles.* The observed value of an angle 213 (O) may be compared with the value (C) given by the computed normal azimuths 12 and 13. The resulting observation equation is then in the form of (2.34).

(*b*) *Directions.* Theodolite observations recorded as in § 1.20 give directions, namely the angle between each station in the round and an arbitrary RM, such as a flank station. Then for every line

$$\text{True azimuth} = (\text{Recorded direction}) + Z - \text{Error}, \qquad (2.31)$$

where Z is an unknown which is constant for all the directions concurrently observed at the same station. As a matter of convenience Z may be made small,† but like the errors of the provisional latitudes and longitudes it does not have to be minimized. Observation equations take the form of (2.35).

Angle equations involve two unknowns $\delta\phi$ and $\delta\lambda$ at each station, while directions involve three, $\delta\phi$, $\delta\lambda$, and Z. Which system is the better depends on which gives the more probable results, and also on which is the more convenient.

The essential difference between the two systems is that for observations of apparently uniform quality the method of angles will give equal weights to every undivided angle at a station, such as α, β, and γ in Fig. 2.10, which implies reduced weight for combinations such as $\alpha+\beta$ or $\alpha+\beta+\gamma$. The method of directions, on the other hand, will give the same weight to single angles and to all combinations of them. In other words, the method of angles treats adjacent angles as independent, while the other treats directions as independent. At first sight all pointings and directions may be thought to be independent and equally reliable, and since every angle great or small is the difference of two directions, equal weight may seem proper for them. On the other hand, for the reasons given in § 2.43 (*e*), large angles such as $(\alpha+\beta+\gamma)$ or δ in Fig. 2.10 are in fact believed to be less accurate than smaller angles such as α, β, or γ to the extent that the s.e. of a 180° angle tends to be about $\sqrt{3}$ times that of a 60°‡ angle. This, perhaps fortuitously, is what results from treating α, β, and γ as independent.‖ Consequently, although there is no general agreement on the point, the method of independent angles is here

† By recording directions from approximate north instead of from an arbitrary RM. If $Z<1'$, directions can be used for A in $(\rho/L)\sin A$, etc.

‡ An angle $180°+\theta$ has the same standard error as $180°-\theta$. For this purpose, no angle is greater than 180°.

‖ [8] shows that the use of either angles or directions gives identically the same results if adjacent angles, instead of being treated as independent, are given the correlations which they would have if directions were independent. As should be obvious. The differing results of a small computation by the usual two methods, independent angles and independent directions, are also given.

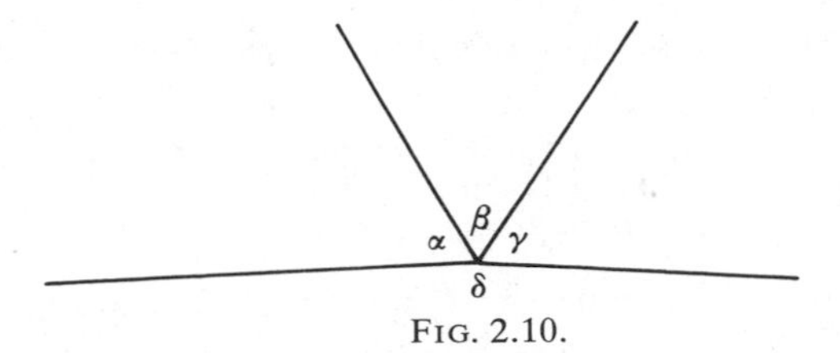

FIG. 2.10.

regarded as slightly the better on grounds of probability. It is fortunate that § 2.25, Table 2.1, also finds it the more convenient.

(*c*) *Exterior angles.* A related problem is whether equations should be formed with the large (typically 180°) angles on the exterior of a system, like 234 in Fig. 2.11. Let the average correction to all the (60°) angles of the system be $\bar{x}$. The magnitude of the expected correction to the sum of the three angles 238+839+934 will then be $\bar{x}\sqrt{3}$, and if the exterior angle 234 is excluded, its correction will automatically be the same with opposite sign.† On the other hand, if the exterior 234 is given an observation equation with the same weight as the other angles, it may expect a correction no greater than each of theirs. Here also the argument of § 2.43 (*e*) suggests that the larger correction is likely to be more appropriate, and it is therefore advised that observation equations should not usually be formed for exterior angles.‡ § 2.25 suggests the same conclusion on grounds of convenience.

One who believes that the s.e. of a 180° angle tends to be the same as that of a 60° angle will prefer to adjust by directions, and will do so unless it involves him in excessive extra cost. On the other hand, one who believes that the s.e. of a 180° angle is substantially the greater will prefer to adjust by angles (without exteriors), and will find it economical to do so.

In a traverse only one of the two angles (totalling 360°) at each traverse station should be used. If both are used, significant mis-weighting, as compared with distances, may result. See § 2.43 (g).

2.23. Observation equations

(2.27) may be written

$$\begin{aligned} \delta A_{12} &= a\delta\phi_1 + b\delta\lambda_1 + c\delta\phi_2 + d\delta\lambda_2, \\ \delta L_{12} &= p\delta\phi_1 + q\delta\lambda_1 + r\delta\phi_2 + s\delta\lambda_2, \end{aligned} \tag{2.32}$$

where $a = (\rho_1/L)\sin A_{12}$, $p = -\rho_1 \cos A_{12} \sin 1''$, etc., $d \approx -b$, and $s \approx -q$.

† The total of the round will have been previously adjusted to 360°, § 1.23.

‡ The same result can be got, with unnecessary extra trouble, by including an equation for the exterior angle and giving it reduced weight. If exterior angles are included with the same weights as are given to 60° angles, the results are likely to be very similar to those given by the method of directions.

Then the observation equations may take different forms as in (*a*)–(h) below, the more common forms being (*a*), (*b*), (*d*), and (g).

(*a*) If the *distance* L_{12} has been fallibly measured, the equation is

$$p\delta\phi_1 + q\delta\lambda_1 + r\delta\phi_2 + s\delta\lambda_2 = (\text{O}-\text{C}) + v, \tag{2.33}$$

where (O—C) is the observed† value of L minus that computed from the provisional coordinates, and v is a residual which is to be minimized.

(*b*) In the method of angles, if an *angle* 213 has been fallibly measured, the equation is

$$(a\delta\phi_1 + b\delta\lambda_1 + c\delta\phi_3 + d\delta\lambda_3) - (a\delta\phi_1 + b\delta\lambda_1 + c\delta\phi_2 + d\delta\lambda_2) = (\text{O}-\text{C}) + v, \tag{2.34}$$

in which a, b, c, and d in the first bracket refer to A_{13}, and those in the second to A_{12}. Rotation from 2 to 3 through the angle must be clockwise.

(*c*) If the adjustment is by *directions*, the following form replaces (2.34) for all recorded directions:

$$(a\delta\phi_1 + b\delta\lambda_1 + c\delta\phi_2 + d\delta\lambda_2) - Z_1 = (\text{O—C}) + v, \tag{2.35}$$

where Z_1 is the optimum correction to recorded directions at station 1, and (O—C) refers to the direction 1 to 2, etc. See § 2.22 and (2.31).

If observations at a station are made at significantly different dates, so that the stability of marks may be in question, each set of observations should be given different Z's, which constitute additional unknowns.

The reverse direction A_{21} will give rise to a similar equation with different (but similar) values of a, b, c, and d, and with Z_2 instead of Z_1.

(*d*) If the point 1 is to be *held fixed*, accept the required ϕ_1 and λ_1 as the provisional coordinates, and omit $\delta\phi_1$ and $\delta\lambda_1$ from all observation equations which would otherwise include them. Similarly for any other point. If two adjacent points 1 and 2 are both held fixed, the computed values of L_{12}, A_{12}, and A_{21} are automatically fixed also, and require no observation equations.

In the method of directions, all observed directions at 1 and 2 may be recorded as from these fixed azimuths as RMs, so that Z will be omitted from all equations (2.35) at 1 and 2. Alternatively, and better, Z_1 and Z_2 may be included as unknowns, but they will be the only unknowns in the direction equations 1 to 2 and 2 to 1.

The coordinates of several points, such as doppler fixes, can be treated as fallibly known by forming simple observation equations $\delta\phi$ or $\delta\lambda$ = (observed value) with appropriate weight. If the errors of the observed values of such stations are known to be correlated, and if their covariances can be estimated, these equations can collectively be given a full weight

† In all these equations observed distances, angles, and directions must first have been reduced to the spheroid. See § 2.18, first footnote.

matrix P instead of the usual diagonal matrix W, as in D.22 (c) formula (D.136).

In one or more of such stations are to be treated as infallible, their $\delta\phi$'s and $\delta\lambda$'s can be omitted as at the beginning of this sub-paragraph (d).

If every independent system there must be at least one station whose coordinates are either fixed or treated as fallibly observed. If there is only one, there must also be at least one measured distance and one azimuth. None of these three need necessarily be considered as infallible, but if there is only one fixed position, or one direction, or one azimuth, then that single one will be in fact be left unchanged by the process of adjustment.

(e) If a distance L_{12}, such as the length of an invar base, is to be accepted as errorless, the routine is to eliminate one of the unknowns, say $\delta\phi_2$† from the relevant equation (2.33) (without the v), and wherever that unknown occurs in any other equation, to replace it there by the resulting

$$(\text{O—C}—p\delta\phi_1—q\delta\lambda_1—s\delta\lambda_2)/r \tag{2.36}$$

wherever it occurs in any other equation. The (2.33) of L_{12} is then excluded from the summations which form the normal equations. $\delta\phi_2$ is eventually given by (2.36) after $\delta\phi_1$, $\delta\lambda_1$, and $\delta\lambda_2$ have been determined.

In preference to the above it will almost always be easier and more realistic to treat all observed distances (except those between pairs of fixed ϕ and λ) as fallible, as in (2.33), with suitable high weight. See also § D.19.

(f) A direction, such as a *Laplace azimuth*, which is to be accepted as an errorless azimuth, may be treated similarly, one of $\delta\phi_1$, etc. being eliminated from $a\delta\phi_1+b\delta\lambda_1+c\delta\phi_2+d\delta\lambda_2=(\text{O—C})$. But here there is the complication that the observed geodetic azimuth is obtained from an astronomical azimuth by applying the Laplace correction (Astro λ_1- Geod λ_1)sin ϕ_1, in which the provisional geodetic λ_1 may materially differ from the final $\lambda_1+\delta\lambda_1$. So the equation from which one unknown is to be eliminated is

$$a\delta\phi_1+(b-\sin\phi_1)\delta\lambda_1+c\delta\phi_2+d\delta\lambda_2$$
$$=\text{provisional Laplace } A_{12}-\text{Computed } A_{12}, \tag{2.37}$$

where the provisional Laplace A_{12}

$$=\text{astro } A_{12}-(\text{Astro } \lambda_1-\text{provisional Geod } \lambda_1)\sin\phi_1.$$

Note how easily the term $(b-\sin\phi_1)\delta\lambda_1$, used instead of $b\delta\lambda_1$ gets over the difficulty that a Laplace azimuth changes value if the accepted longitude changes.

If both the azimuth and the length of a line are to be accepted as

† If a line of known length lies north and south, eliminate a $\delta\phi$; if east and west, eliminate a $\delta\lambda$, and vice versa for a known azimuth.

errorless, but not the ϕ and λ at either end, avoid trying to eliminate the same unknown from both (2.33) and (2.37). See footnote to (*e*) above. It will probably be convenient to eliminate both ϕ and λ at the same end of the line.

(*g*) Instead of working as in (*f*) it will generally be easier and better to treat a Laplace azimuth as fallible, and to form an equation like (2.37) as an observation equation. Note that a Laplace azimuth is necessarily recorded as an observed direction, even though the method of angles may have been adopted.

(*h*) When the adjustment is by directions, the zero of recorded directions at a station where there is a Laplace azimuth, fallible or infallible, should be chosen so as to make the recorded direction of the RM agree exactly with its Laplace azimuth as computed with the provisional longitude. The Z of (2.35) should then be omitted, and for all the directions emanating from the Laplace station the observation equation becomes

$$a\delta\phi_1+(b-\sin\phi_1)\delta\lambda_1+c\delta\phi_2+d\delta\lambda_2=(\text{O as above})-\text{C}+v. \quad (2.38)$$

In other words, Z is replaced by $\delta\lambda_1 \sin\phi_1$.

In equations (2.32)–(2.38) the accuracy required in the coefficients a, $b, \ldots, s$ is not better than one part in 4000, § 2.21, and iteration will give correct results with larger errors. Four or at most five significant figures. will suffice.

(*i*) *The ratio of two measured distances*, $r=s_1/s_2$, may be formed to give an observation equation [169] and § 1.50. Let the two distances be AB of length s_1, and azimuth A_1, and AC of length s_2 and azimuth A_2. Then the observation equation is

$$r\left\{\left(\frac{\cos A_2}{s_2}-\frac{\cos A_1}{s_1}\right)\rho_A\delta\phi_A+\frac{\cos A_1}{s_1}\rho_B\delta\phi_B-\frac{\cos A_2}{s_2}\rho_C\delta\phi_C\right.$$
$$+\left(\frac{\sin A_2}{s_2}-\frac{\sin A_1}{s_1}\right)\nu_A\cos\phi_A\delta\lambda_A+\frac{\sin A_1}{s_1}\nu_B\cos\phi_B\delta\lambda_B$$
$$\left.-\frac{\sin A_2}{s_2}\nu_C\cos\phi_C\delta\lambda_C\right\}=(\text{O—C})+v. \quad (2.39)$$

As regards weights, the point of using distance ratios is that the s.e. of a ratio (in ppm) is smaller than that of a distance measurement. The s.e. of the ratio may be calculated using formula (D.19) in which the errors of the two measures will be highly correlated. [169] gives an example in which the coefficient of correlation is estimated to be 0.93, and the weight of the ratio in units of $(\text{ppm})^{-2}$ will be about 7 times that of a single measurement.

2.24. Weights

Before normal equations can be formed from the observation equations, each of the latter must be multiplied by the square root of the weight, w, of its observation O, or in matrix notation a weight matrix W must be assessed. This will generally be a diagonal matrix whose elements w are the estimated† weight of each observation O. Ideally the weight of an observation is the reciprocal of the square of its standard error. As they stand, the dimensions of equations (2.33) are length, while those of (2.34), (2.35), (2.37), and (2.38) are seconds of arc, so multiplication by $\sqrt{w}$ reduces all to zero dimensions.

It is, of course, generally unnecessary to estimate separate weights for every angle or direction. See § 2.13 (*c*). It may even happen that a uniform s.e. of 1″ is appropriate, which gives all w's = 1. An s.e. of one metre for all measured distances is unlikely to be appropriate, although those who express the observations O in feet, may not be far wrong if they adopt 1 foot as the s.e. of a 20 to 40-mile tellurometer line. The combination of 1 second of arc and one unit of length as s.e.'s, if it can be adopted, conveniently eliminates the numerical labour of weighting. Generally, however, the s.e. of a long measured distance is more nearly proportional to its length, and an s.e. of 5 ppm has sometimes been implied by putting $w = 1/(L \sin 1'')^2$. In other words, (2.33) has been weighted by multiplying it by $1/L \sin 1''$. Taken with unit weight for angles,‡ this may not be inappropriate, provided the above implications are understood and accepted. But it is little extra trouble to assess the s.e. of each separate distance by (1.55) or (1.57).

The values obtained for the $\delta\phi$'s and $\delta\lambda$'s will not be affected if the weights of *all* observations are taken to be some constant multiple of their (s.e.)$^{-2}$, such as (probable errors)$^{-2}$, but it may be necessary to remember what has been done, see § 2.27. On the other hand, if the observations comprise different fallible quantities, such as both angles and distances, it is not unlikely that the s.e.'s of members of the different classes may have been systematically over- or under-estimated by different factors. See § 2.28.

Conventionally, the s.e. of an angle is $\sqrt{2}$ times that of a direction, but this is unlikely to be correct since the errors of adjacent directions are likely to be to some extent correlated. In the absence of better evidence the weights of (60°) angles and of directions may be considered to be equal.

See Chapter 2, section 6 for methods of estimating the standard errors of distances, azimuths, and angles.

† Such weights may be described as *a priori* weights, as opposed to the revised weights given by (D.62) after the adjustment has been completed. See §§ 2.27 and 2.28.

‡ If the azimuths of all traverse lines are known with an s.e. of 1″, and the distances with one of 5 ppm, the error ellipse, § D.20, of the relative positions of the two ends will be circular. A circular ellipse is a reasonable aim, which observation programmes and instrument construction may be designed to secure.

2.25. The normal equations

The normal equations are formed as in (D.46), or in matrix notation in (D.51). Their number will be two or three times the number of unfixed stations, according to whether adjustment is by angles or by directions, less perhaps a few eliminated in connection with infallible bases or azimuths. In a large adjustment many of the coefficients will be zero, and if the observation equations have been suitably arranged, the non-zero coefficients will form a more or less narrow diagonal band. The solution of a large set of normal equations is much simplified if this band is kept as narrow as possible.

If there are n stations, let the unknowns be x_1 to x_{2n}, so that x_1 and x_2 are $\delta\phi$ and $\delta\lambda$ at station 1, and so on: with an extra x at each station for Z if the method of directions is used. Some of the x's will not appear, § 2.23 (d)–(h). Then in the matrix of the normal equations, it is clear from the form of (D.46) that the element in row i and column j will be zero unless there is at least one observation equation containing both x_i and x_j. The diagonal element ii will be non-zero provided x_i occurs in any observation equation. If it does not, x_i and the corresponding zero row and column must be excluded.

Fig. 2.11 shows a conventional net of 35 triangulation stations, in which 1 and 2 are to be held fixed and also constitute one infallible base and azimuth. Fig. 2.12 shows the resulting non-zero elements in part of the matrix of the normal equations if angles have been used, and Fig. 2.13 shows the corresponding part for directions.

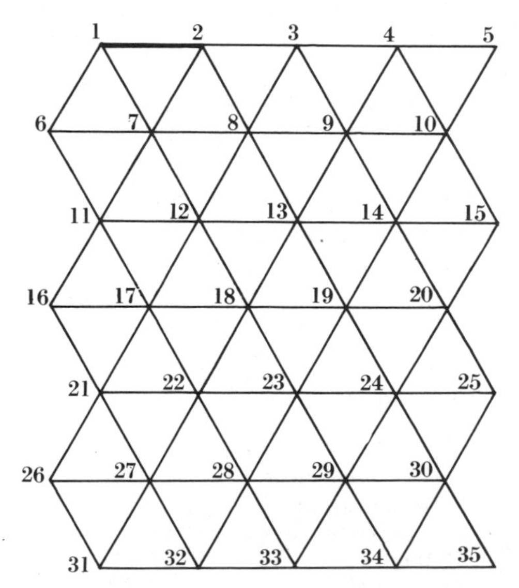

FIG. 2.11. Conventional triangulation net. Points 1 and 2 are held fixed.

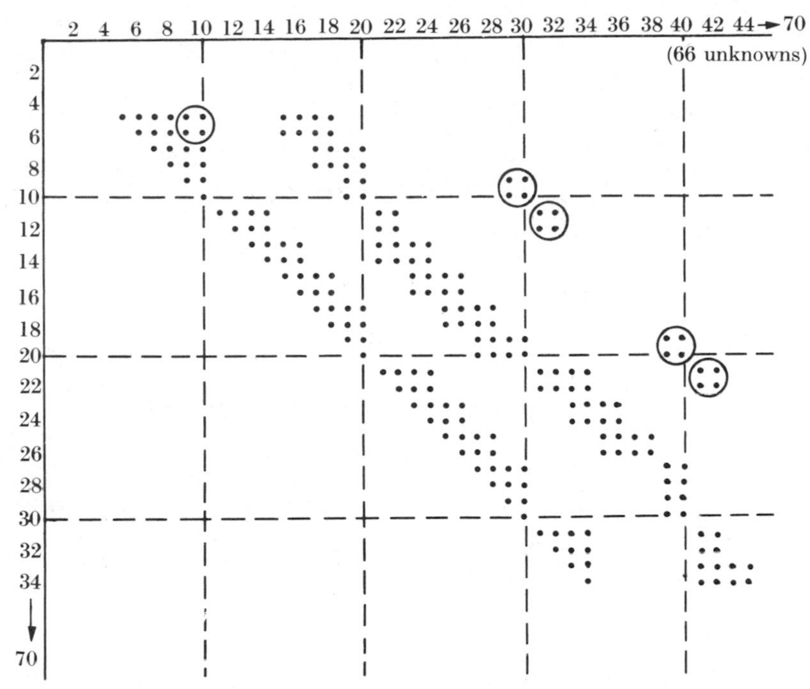

FIG. 2.12. Method of angles. Non-zero elements in the normal equations. Elements enclosed in circles occur only when external angles are included. Elements below the diagonal are omitted.

Fig. 2.11 shows pure triangulation, but the addition of any number of measured distances, fallible or infallible, between stations whose mutual angle or directions have already been included will add no non-zero elements to the normal equations. So far as the bandwidth of the normal equations is concerned, Fig. 2.11 may be triangulation or trilateration or both. Any number of Laplace azimuths may also be included, without introducing additional elements, provided they give the azimuth only over existing lines in the figure.

The observation of angles, directions, or distances over long lines such as 8 to 20 would add to the width of the band. From this point of view Fig. 2.11 is an optimal lay-out.

Table 2.1 gives some statistics for the computation of Fig. 2.11 by different methods.

In this example the method of angles has numerical advantage over that of directions, but the incorporation of exterior angles is a serious addition to it if the bandwidth is a critical matter. A computer, will be able to handle the maximum number of stations simultaneously if it uses the method of angles without exteriors, as is advised on other grounds in § 2.22.

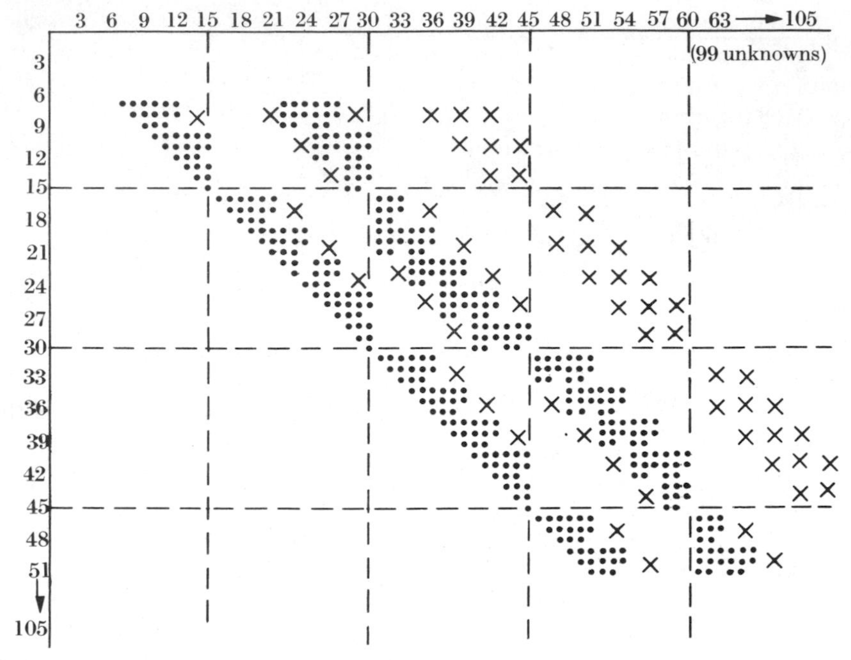

FIG. 2.13. Method of directions. Non-zero elements in the normal equations. Elements below the diagonal are omitted. If Schreiber's method is used, all blocks of 8 or 9 non-zero elements are reduced to 4 by the elimination of the Z's, but crosses indicate additional blocks of four non-zeros.

TABLE 2.1

	Angles excluding exterior	Angles including exterior	Directions without Schreiber	Directions with Schreiber
Number of observation equations	144	164	162	195
Number of unknowns, and normals	66	66	99	66
Number of non-zero elements in the normals, on or above the diagonal	425	481	850	1020
Maximum width of band, on and above the diagonal	14	22	21	24

At the start of the computation the bandwidth may be minimized by numbering the stations and unknowns in such order as will minimize the greatest difference between the serial numbers of the unknowns occurring in any one equation. Or, which is the same thing, by arranging that all the stations observed from any one station should bear the least possible range of numbers.†

† The width of the band is its width at its widest point. It is required to minimize the range at the stations where it is greatest.

In the method of directions, a routine known as *Schreiber's method* reduces the number of unknowns in the normal equations from 3 to 2 per station, by eliminating the Z's as follows.

After forming and weighting the observation equations (2.35) relating to all the directions emanating from any station (such as station 9 in Fig. 2.11), enter an additional equation

$$(i/\sqrt{m})\{[a\sqrt{w}]\delta\phi_9+[b\sqrt{w}]\delta\lambda_9+c_8(\sqrt{w})\delta\phi_8+d_8(\sqrt{w})\delta\lambda_8 \\ +c_3(\sqrt{w})\delta\phi_3+\ldots-[\sqrt{w}]Z_9\}(i\sqrt{m})][(O-C)\sqrt{w}], \quad (2.40)$$

where m is the number of these equations (six), $[a\sqrt{w}]$ and $[b\sqrt{w}]$ are the sums of the weighted coefficients of $\delta\phi_9$ and $\delta\lambda_9$ in them, $c_8\sqrt{w}$ and $d_8\sqrt{w}$ are the coefficients of $\delta\phi_8$ and $\delta\lambda_8$ in the equation referring to the direction 9 to 8, etc., $[\sqrt{w}]$ is the sum of the $\sqrt{w}$'s and $[(O-C]\sqrt{w}]$ is the sum of all the weighted (O−C)'s. In fact, (2.40) is the sum of all the direction equations emanating from 9, multiplied by $(i/\sqrt{m})$, [466], p. 175.

This equation and the others like it are then to be treated as additional observation equations when the normals are being formed, except that the marker i $(=\sqrt{-1})$ indicates that a product of any two of its terms is to enter the summations with reversed sign. The effect of this is as follows.

(*a*) All the Z terms are eliminated. Their coefficients in the normals are zero.

(*b*) The coefficients of all the other terms are what would have resulted from the elimination of the Z's from the normals in the ordinary way.

Schreiber's elimination may be worth adopting if the computation has to be by directions, but it leaves the bandwidth and the number of non-zero elements much greater than it would have been in the method of angles, because (2.40) contains all the stations which have been observed from the station whose Z is being eliminated. All pairs of these stations then produce non-zero elements; see Fig. 2.13.

2.26. The solution of the normal equations

The solution of n symmetrical simultaneous equations, without many zero elements, involves at least $n^3/6$ multiplications, [209], pp. 175–84, but if the non-zero elements are confined to a relatively narrow band of width k on and above the diagonal, this figure will be much reduced. Further, the ability of any particular computer to solve a very large system of equations by non-iterative methods will depend only on the width of the band. It must be able to hold rather more than k^2 elements in its high-speed store, [62], pp. 201–3.

The most suitable method of solution appears to be Cholesky's, § B.14 (*e*), if the available computer is large enough. If it is not, the solution must be carried out by partitioning as in § B.14 (*c*), or by other methods as described in Chapter 2, Section 5.

Iterative methods, § B.15, can be used, but non-iterative methods are currently (1978) preferred. The diagonal elements of the normal equations are generally formed by summing many more products than are most of the others. Typically, the diagonal terms and a few close to the diagonal are formed from 18 or 12 products (all positive in the case of the diagonals), compared with 6 to 2 products in the rest, although there are numerous exceptions. The diagonal terms thus tend to be the largest, and the equations are at least fairly suitable for iteration.

The normal equations may be far from orthogonal, especially if bases and Laplace stations are far apart, so that an error of (say) 10 cm in a measured base results in one of 100 cm in the coordinates of some station. But provided the layout of the system conforms to ordinary survey conventions, the normal equations in a computation by variation of coordinates will not suffer from unsuspected gross instability.

2.27. The precision of the results

If the allotted weights are the reciprocals of the squares of what are believed to be the standard errors of each observation, the s.e., $\sigma(x_m)$, of any unknown x_m will be given by

$$\sigma(x_m) = \sqrt{\alpha_{mm}}, \tag{2.41}$$

where α_{mm} is the diagonal element of row m in N^{-1}, the inverse of the matrix of the normal equations, § D.15. This is on the assumption that the s.e.'s assumed for the observations are correct. It is also assumed that the covariances of the observation errors are zero. Their omission may be serious, since the errors of measured distances (and less obviously satellite fixes and Laplace azimuths) are liable to be considerably correlated. Estimates of over-all errors of position are likely to be more or less too small on this account.

The average accuracy of the estimated s.e.'s can be checked by (D.62), which gives a factor σ_0 by which all the σ's and $\sigma(x)$'s may be multiplied. This will be an improvement, but if the factor is much greater than unity there will be doubt whether the estimates were all about equally optimistic, or whether some grossly inferior observations have been included by oversight. See §§ 2.28 for a routine for assessing different factors for angles, distances and azimuths.

The s.e.'s given for the coordinates of the stations are relative to the origin, or other point where coordinates have been assumed to be errorless. They will ordinarily increase with increasing distance from that point, and their variations may be exhibited by plotting contours of equal s.e. of position on a chart.

While such a chart is of value, it is not a means of assessing the relative error of pairs of points, because except when such pairs lie on opposite sides of the origin their errors are more or less closely correlated, especially if they are close together. It will then be proper to compute and record the s.e.'s of distance and azimuth between a number of selected pairs see § 2.49(*c*).

The s.e. of distance and azimuth between two stations m and n, where the corrections to ϕ and λ are (say) the unknowns x_1 and x_2 at m, and x_3 and x_4 at n, are given by

$$\sigma^2(\text{azimuth}) = a^2\alpha_{11}+b^2\alpha_{22}+c^2\alpha_{33}+d^2\alpha_{44}+2ab\alpha_{12}+2ac\alpha_{13}+ \\ +2ad\alpha_{14}+2bc\alpha_{23}+2bd\alpha_{24}+2cd\alpha_{34}, \qquad (2.42\text{A})$$

$$\sigma^2(\text{distance}) = p^2\alpha_{11}+q^2\alpha_{22}+r^2\alpha_{33}+s^2\alpha_{44}+2pq\alpha_{12}+2pr\alpha_{13}+ \\ +2ps\alpha_{14}+2qr\alpha_{23}+2qs\alpha_{24}+2rs\alpha_{34}, \qquad (2.42\text{B})$$

where a, b, c, d, p, q, r, and s are the coefficients in (2.32), and α_{ij} is the element in row i and column j in N^{-1}; see § D.15, (D.69). For this limited purpose it will not be necessary to invert the whole matrix N, but only to compute such columns of its inverse as contain any of the elements required for (2.42). For methods of computing† the elements of N^{-1} see § B.16. If the normals have been solved by iteration, the elements of N^{-1} are less easily got. Finally, multiply σ^2 by σ_0^2.

See further about error ellipses in §§ 2.49, D.20 and D.21.

If the weights given to the observations are based on probable errors instead of s.e.'s (2.42) will give $(\text{p.e.})^2$ instead of σ^2.

The variance–covariance matrix can be computed before the angles are observed. The right-hand sides of the observation equations do not enter into it. For the weights, which do matter, it is necessary to estimate the precision with which the angles are going to be observed. It is consequently possible to estimate the error of azimuth and distance between any two points before any observations are made.

2.28. Revision of weights

It may well happen that the *a priori* estimates of the σ's of angles (σ_α), of distances (σ_L), and of Laplace azimuths (σ_A) respectively may not be equally inaccurate. The estimate of σ_α may be too small and that of σ_L too large, or vice versa. In other words, different values of σ_0 may be appropriate to each of these classes, in preference to a single value for them all. The following routine enables separate values to be assessed, [68], pp. 22–23.

† If the s.e.'s of distance and azimuth between near points are required, (2.42 A and B) contain + and − terms of nearly equal magnitude. The σ^2 terms α_{11}, α_{22}, etc., may be as large as (5 metres)^2, while the required σ^2(distance) may be only (0.1 m)^2, and be required to (0.01 m)^2. So the elements of N^{-1} must be computed to at least 5 or 6 significant figures.

(*a*) A first adjustment is made using all the angles (or directions), but only one distance and one azimuth. This solution gives a value of σ_0, which is then used to correct the *a priori* weights of the angles by a factor of $1/(\sigma_0)^2$.

(*b*) A second adjustment is then made, using all the angles as before, but with their revised weights; all the distances with their *a priori* weights; and only one azimuth. If the solution gives a value of σ_0 other than 1, say $\sigma_0 = 1.3$, the blame is on the distances. A third adjustment is then made in which the distance weights are changed by such a factor as $1/(1.6)^2$, such as may be judged† likely to produce a σ_0 of 1. If in fact it produces a revised σ_0 of (say) 0.9, the position will be that the original σ_L gave $\sigma_0 = 1.3$, and weights based on $\sigma_L/(1.6)^2$ gave $\sigma_0 = 0.9$, so interpolation suggests that a fourth adjustment with $\sigma_L/(1.45)$ will give a σ_0 of 1.0. A fourth adjustment may be made to verify this.

(*c*) Finally an adjustment is made using revised weights for both angles and directions, and all the azimuths with their a priori weights. Revised weights for the azimuths are then obtained as in (*b*).

For numerical results in Great Britain, see § 2.44 (*a*) (ii).

The order in which angles, distances, and azimuths are considered is governed by the consideration that it must be possible to make an unambiguous computation using the whole of the first class with only one of the each of the other two. In a complete trilateration, supported by only some angles, distances would be taken first.

The pursuit of undue perfection in the determination of the three separate values of σ_0 may result in wasted effort. Errors of even 30 per cent in one of the σ_0's will probably cause unimportant changes in the adjusted coordinates, although the s.e.'s estimated for the results would be affected more seriously. See [168].

2.29. Tests for mistakes

The routine of least squares can easily be made self-checking, but tests as below are advised for proving the absence of blunders in the data, or in the formation of the observation equations.

(*a*) *To check the data.* Although the method of variation of coordinates does not in itself demand the setting up of condition equations, there is no better way of detecting mistakes in the data than by examining the closing errors of at least a large number of them, as follows.

(i) Triangular errors will have been examined in the field.

† If distances are less numerous than angles it may be expected that their weights may have to be changed by more than $1/(1.3)^2$ to reduce σ_0 to 1.

(ii) The side closures within triangulation figures are easily programmed, but it is not necessary to be certain that every independent side condition has been included. The side misclosures within an old base extension net are a valuable guide to its quality.

(iii) In a traverse the misclosures on Laplace stations must be examined, as they are the only check on the angles. They can be obtained from the computation itself, after it has been done, by summing the v's in the angle equations between Laplace stations. If a mistake is found, recomputation (on an electronic computer) is not a serious trouble.

(iv) In chains of triangulation, especially in a recomputation of old work, the examination of misclosures of scale and azimuth on bases and Laplace stations, and round circuits, is a valuable check. In a large and complicated piece of work there is a real risk that blunders will be incorporated unless this is done. A serious, but not enormous, mistake in a base or Laplace azimuth† may introduce very small adjustment errors into each of the possibly numerous angles between it and the next, and the computer may accept the error without producing abnormal residuals.

(v) With an electronic computer, circuit closures in position may be obtained by a preliminary computation in exactly the usual way except that a pair of stations in a circuit are given a double identity and are allowed to get two values of $\delta\phi$ and $\delta\lambda$, one from each arm of the circuit. If the differences are seen to be normal the computation is repeated, at slight cost, in the usual way.

(vi) All residuals, v, should be recorded and examined.

(*b*) *To check the setting up of the observation equations.* If a computer is used, a second approximation can be made with little trouble, and this should be done as a routine. Revised values of C are computed from the adjusted coordinates $\phi+\delta\phi$ and $\lambda+\delta\lambda$, and the solution is then repeated to give further corrections $\delta'\phi$, $\delta'\lambda$. If the provisional coordinates or the coefficients in the observation equations have been insufficiently accurate, the second solution will give changed results, but further repetitions should converge rapidly. Failure to converge very rapidly suggests either gross error in the left-hand sides of the observation equations or bad instability and lack of enough significant figures, rather than faulty observations.

Those who use non-electronic desk computers can verify that the $(\mathrm{O}+v)$'s of the solution satisfy the triangular and other conditions.

(*c*) A final check is to verify that (D.62) derives a proper value of σ_0

† See [113] 2 Edn, footnote to p. 126, for a mistake in a Laplace station. In the 1963 adjustment of Burma and Thailand there was a discrepancy between an old taped base and a geodimeter revision. Circiit and base misclosures suggested that the revision was the more likely to be wrong, and a re-check of the field notes showed doubt in the recovery of an old mark.

from the residuals. If the weights have been taken as $1/(\text{s.e.})^2$, and if the s.e.'s have been correctly estimated, σ_0 should be about 1. If w has been taken as $1/\kappa^2\,(\text{s.e.})^2$, the expected value of σ_0 will be $1/\kappa$. If it is greater than about $1\frac{1}{2}$ or $3/2\kappa$, see § D.15.

SECTION 4. ADJUSTMENT BY CONDITIONS

2.30. Utility of the method of conditions

Before 1955 it was usual to compute triangulation as described in Chapter 2, Section 2, adjusting separate figures by the method of conditions, and later adjusting the whole framework, probably in several sections, by the same method. See §§ D.16 and D.17 for the basic theory, and § 2.13 for details of simple cases. The following paragraphs describe the application of the method to more complicated cases.

For the reasons given in § 2.18, modern geodetic frameworks are likely to be computed by the method of variation of coordinates, and adjustment by conditions is obsolescent. Description of the method is nevertheless required for the proper understanding of the adjustments on which many existing surveys are based, and also because there may still be situations in which it is convenient to use it.† Even when the intention may be to adjust by variation of coordinates, it is proper to set up simple conditions such as triangular closures and easy side conditions, at an early stage, to give estimates of accuracy, and to confirm the absence of blunders in the data.

Condition equations may be of three kinds, as below.

(*a*) Between angles only, as in the triangular, central, and side conditions of § 2.13 (*a*).

(*b*) Between angles and measured distances, such as the misclosures on base lines in (2.43) in conventional triangulation.

(*c*) Between measured distances only, such as may occur in trilateration with redundant sides.

Existing triangulation frameworks have always contained sufficient observed angles to fix all stations with the help of one measured distance, although there may be more, and §§ 2.31 and 2.32 refer to such a situation.

In a pure trilateration, or one with the addition of only a few angles, the condition equations are not so easily formed. § 2.33 gives some guidance, but the easy procedure will be to compute by variation of coordinates.

† Such as in a chain of simple triangles, with a single base and Laplace azimuth, where the only conditions are the independent triangular closures. Even in this case, if a computer and variation of coordinates program are available, it is easier to use that program than to write another.

2.31. Types of condition equation

§ 2.13 (*a*) lists the types of condition equation which may occur in simple triangulation figures, namely triangle, side, and central conditions. In more complicated figures or nets or in systems of chains the following may also occur.

(*a*) *Between measured distances.* In Fig. 2.14 if AB and CD have been measured, a condition equation may be set up.

$$x_{AB}-x_{CD}+10^6 \sin 1''\{x_1 \cot 1-x_3 \cot 3+(x_{10}+x_{11})\cot(10+11)-$$
$$-x_9 \cot 9+x_5 \cot 5-(x_6+x_7)\cot(6+7)+$$
$$+(x_{18}+x_{19})\cot(18+19)-x_{17} \cot 17\}=\epsilon, \quad (2.43)$$

where x_{AB} and x_{CD} are the unknown errors of AB and CD in ppm, the other x's are the errors of the angles in seconds of arc, and ϵ is the distance CD as computed from AB minus its measured length, in parts per million.

If either of AB or CD, or both, are to be held errorless, the relevant x or x's are omitted.

The equation could equally well have been formed using some other angles. The best route involves the fewest angles, subject to the avoidance of any unusually small ones.

(*b*) *Between Laplace azimuths* such as at A and D in Fig. 2.14.

$$x_{A-B}-x_{D-C}-x_2-x_4-x_5-x_6-x_7-x_{12}=\epsilon, \quad (2.44)$$

where x_{A-B} and x_{D-C} are the errors of the Laplace azimuths, the other x's are the errors of a chain of intervening angles, and ϵ is the azimuth of C at D, as computed from A, minus the observed value, all in seconds of arc. In equations such as (2.44) the x's of the angles on the left-hand side of the selected route enter with positive sign.

(*c*) *Circuit closures.* In Fig. 2.14 starting from AB the length and azimuth of FG may be computed by two routes, clockwise and anticlockwise round the figure, which give equations in the form of (2.45) and (2.46).

(i) *Side*

$$10^6 \sin 1''\{x_1 \cot 1-x_3 \cot 3+(x_{10}+x_{11})\cot(10+11)-\ldots\}-$$
$$-10^6 \sin 1''\{x_2 \cot 2-x_3 \cot 3+(x_{20}+x_{21})\cot(20+21)-\ldots\}=\epsilon, \quad (2.45)$$

where ϵ is the distance FG computed clockwise minus its distance computed anticlockwise, in parts per million.

When there are intermediate measured sides, such as CD, this equation need only be formed between FG and those nearest to it on each side. If FG is itself a measured distance, (2.45) becomes two equations (2.43) formed with the nearest measured distances on each side of the

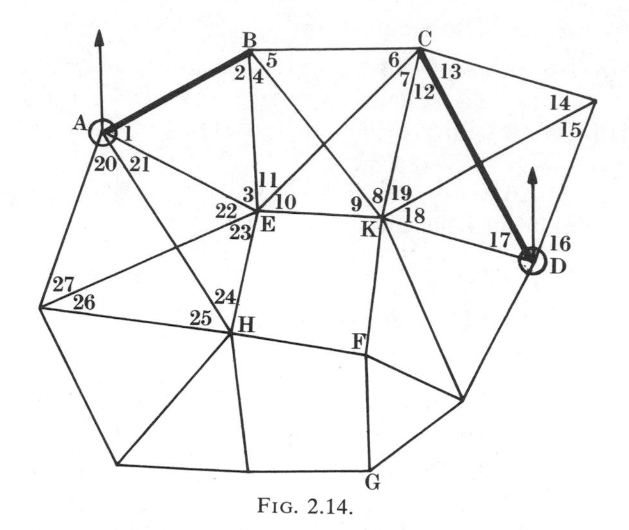

FIG. 2.14.

circuit. The same applies to azimuth equations

(ii) *Azimuth*

$$(x_1+x_3+x_{11}+x_{10}+\ldots)-(x_1-x_{22}-x_{23}+\ldots)=\epsilon. \qquad (2.46)$$

Note that the inclusion of x_1 is immaterial. It suffices to set up the equation round the inside of the chain. The x's on the left of the route have + signs inside the brackets, and ϵ is left-hand route minus right.

(iii) *Latitude and longitude.* Similarly, the ϕ and λ of F as computed from the fixed base AB should be the same via K as via H. Select two chains of simple triangles connecting F to AB by the two routes. Let ϵ_ϕ be the latitude of F computed by the route through K minus the other. Then

$$x_1(\partial\phi_F/\partial x_1)+\ldots+x_8(\partial\phi_F/\partial x_8)+\ldots \text{ via K}-$$
$$-x_1(\partial\phi_F/\partial x_1)-\ldots-x_{25}(\partial\phi_F/\partial x_{25})-\ldots \text{ via H}=\epsilon_\phi, \qquad (2.47)$$

where $\partial\phi_F/\partial x_1$ is the change in the latitude of F arising from a change of 1″ in x_1, and so on. There will be a similar equation for the closure in longitude.

To compute $(\partial\phi_F/\partial x_1)$, etc. see Fig. 2.15, which shows a chain of triangles between a base AB and a point F. In each triangle the angles marked 2 and 3 enter into the ratio of the sides, heavily drawn, through which scale is transmitted, while the angles marked 1 transmit only azimuth. At point D for instance, an error of x_1'' in 1 increases all forward azimuths by x_1'', and the effect of this on the position of F in what results from swinging DF through x_1''. At E the sign will be opposite. Similarly errors of x_2 and x_3 in angles 2 and 3 increase the scale of DE and thence of DF by $10^6 \sin 1''(x_2 \cot 2 - x_3 \cot 3)$ ppm. These changes in the direc-

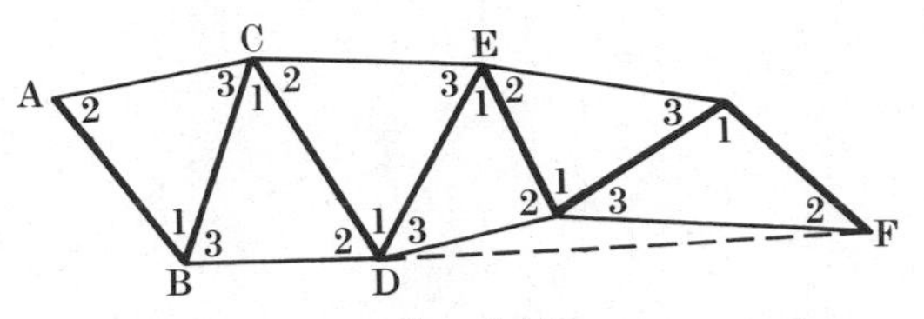

FIG. 2.15.

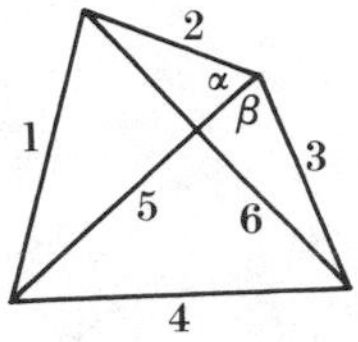

FIG. 2.16.

tion and scale of DF may be resolved into changes of ϕ and λ at F. The changes in ϕ_F are then summed for all the triangles whose angles 1 are at B, C, D,... (next before F). Further details are given in [466], pp. 238–44, and in [59], pp. 171–8. The angles 1 can conveniently be eliminated.

The labour of forming circuit equations may be very great, and in itself constitutes a good reason for not adjusting by conditions. The method of variation of coordinates avoids it.

In equations (2.43)–(2.47) the angles used when computing ϵ are ordinarily the observed, unadjusted angles, corrected as in § 2.09. One-third of the spherical excess should be deducted when side ratios are being computed, but not in other equations. Measured distances must be reduced to spheroid level.

2.32. The number of condition equations

§ 2.13 (*b*) lists the number of conditions included in the simple figures of which triangulation chains are usually composed. In more complicated figures or networks the correct number can be obtained as below. It is important, but not always easy, to set up exactly the correct number of equations, since if one is omitted the final computations will produce duplicate values for some ϕ's and λ's, while if a redundant equation is included the matrix of the normal equations will be singular, and no solution will be obtained. The routine for such a figure is as shown in Fig. 2.17 is as follows.

(*a*) Taking any side of the figure as base, draw a skeleton diagram consisting or just sufficient triangles (with at least two angles observed) to fix the position of all the stations. This diagram should be as simple as possible, should consist wholly of triangles without any redundant lines, and be such that each station is well fixed by lines from nearby points intersecting at a good angle. Fig. 2.18 illustrates such a diagram.

(*b*) Then each of the triangles in the diagram that have all three angles observed provides a triangular condition. Write them down.

(*c*) Next, taking the stations of the figure one at a time, enter on the diagram any other lines whose directions have been observed, either as a

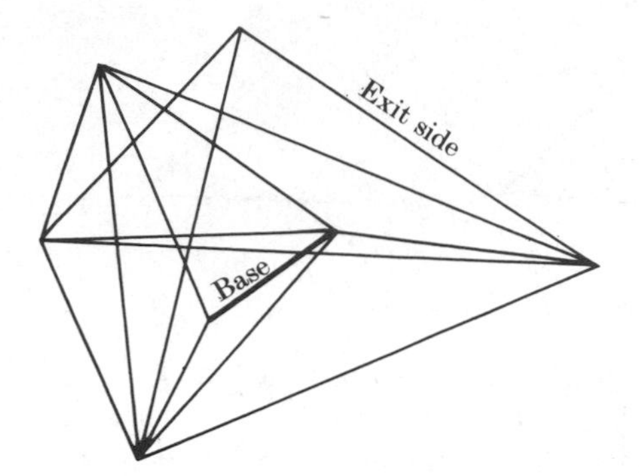

FIG. 2.17. Base extension figure.

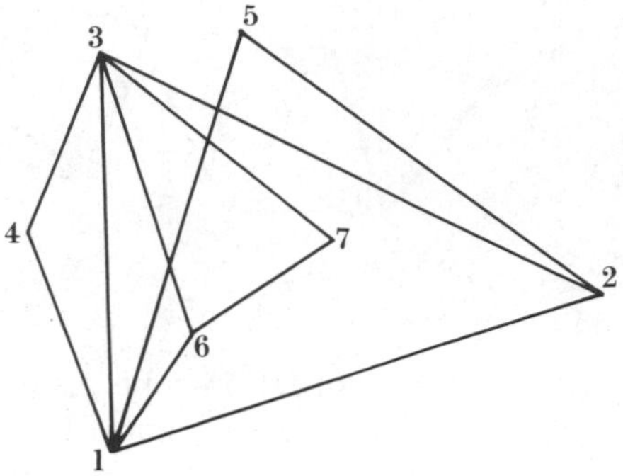

FIG. 2.18. Skeleton diagram. 1–2 is treated as starting base. Side 3–5 would have been included instead of 1–5, if it had been observed.

new line (which should be entered half full and half dotted, as if observed in one direction only), or recording the reciprocal observation along a line already introduced at a previous station, when the dotted half should be filled in solid.

(*d*) When such a line is introduced for the first time, a side equation must be written down. More than one may be possible, and § 2.32 (*l*) provides a guide to the best. This equation must not include lines not entered on the diagram, although it may include ones entered as observed in one direction only. Where the three angles of a triangle have not been observed, it is permissible to deduce the value of the third angle from them and the spherical excess, in order to form a side equation.

(*e*) When the reciprocal observation of a line that has been observed in both directions is entered, a triangular equation must be recorded. The triangle selected must include the line concerned, and must not include any side not yet shown in the diagram as observed in both directions. Where there is a choice see § 2.32 (*l*).

(*f*) When all observed directions have been included, the diagram will correctly represent the figure. Then write down a central equation for any station, not on the exterior perimeter of the figure, at which angles have been observed all round the horizon.

(*g*) Let S be the number of stations in the figure, N the number of observed angles, excluding exterior angles, and L the number of lines. Then the total number of equations will be $N-2S+4$, and of these $L-2S+3$ will be side equations. Thus in Fig. 2.17 there are 10 triangular equations, 5 side and 2 central, if all angles have been observed.

(*h*) In addition include an extra side equation like (2.43) for every measured distance additional to one, and an azimuth equation (2.44) for every extra measured azimuth. To avoid possible redundancies, introduce the measured lines in order, and form equations only between the new line and measured lines which have already been introduced. And the same for azimuths.

(*i*) Continuing (*d*), it sometimes happens that when a line is introduced, no side equation can be formed in the absence of other lines not yet on the diagram. Then postpone introducing that line until other stations have been dealt with, but in the meantime it cannot be entered on the diagram, and cannot be used for the formation of other equations. If it happens that no side equations can be written because each line has to wait for another which in turn has to wait for it, start again with a new basic diagram. Alternatively see [466], pp. 149–51 for the *fictitious angle* device, whereby a small net like Fig. 2.14, which contains a rectangle without any diagonals, can be adjusted without setting up circuit equations. When chains of triangles enclose such an area bounded by more than five or six sides, circuit equations will probably be unavoidable.

Another unusual situation is that in a quadrilateral ABCD the angles at B, C, and D may have been observed, and also BAC and CAD. No triangle has all three angles observed, but the angle at C has been observed which enters into no triangle. There is then one condition, namely the sum of the four angles of the quadrilateral, and this should be set up, as will be necessary to satisfy (g) above.

(*j*) See § 2.33 for a framework in which measured distances predominate.

(*k*) *Traverse.* The only conditions in an unclosed traverse are one of (2.44) per fixed or measured azimuth, less one. See § 2.16. A closed circuit contains one of (2.46) and a closed circuit or a traverse between fixed points also contains closures in ϕ and λ. Referring to Fig. 2.25, the effect on station 5 of an error x''_3, in the angle of 3 is to rotate the line 3–5 through an angle of x''_3, and the total error on the position of 5, due to angles, is the total of such changes. Errors x_{12},x_{23} etc., in the measured distances have no cumulative effect, but each changes ϕ_5 and λ_5 by an amount which is obvious in rectangular coordinates, and for which the spherical equivalent is easy to write down provided the area involved is small.

In principle an unclosed traverse is as ill-suited to variation of coordinates as anything could be, but if a computer and program exist, nothing is easier than using them. For a closed traverse the advantage of variation of coordinates is more considerable. For the adjustment of large traverse nets involving perhaps several hundred stations, see § 2.39.

(*l*) *Choice of conditions.* (i) *Side conditions.* There is generally a choice in the possible form of a side condition. See § 2.13 (*a*) for a braced quadrilateral. Which may be adopted is often immaterial, but it is usually best to compute round a central pole, such as O in Fig. 2.8 or the intersection of the diagonals in a braced quadrilateral, using as few angles as possible. See also [466], pp. 138–41, and [315], vol. i, pp. 274–86.

(ii) *Triangle conditions.* It is difficult to see why one triangle equation should be better than another, and no recommendation can be made.

(iii) For equations between measured distances and azimuths, and for circuit closures, see § 2.31 (*a*) and (*c*).

2.33. Trilateration

In a pure trilateration there will be one distance condition equation for every measured distance in excess of the minimum number necessary to fix all the stations. In a braced quadrilateral for instance, in which all sides but no angles have been measured, there is one condition.

These condition equations are not simple to form. As an example consider Fig. 2.16. The angle α can be computed from the measured sides 1, 2, and 5, and β from 3, 4, and 5, using equations of the form

$$A = \cos^{-1}\{(b^2+c^2-a^2)/2bc\}+\tfrac{1}{3}(\text{spherical excess}). \tag{2.48}$$

The side 6 can then be computed from 2, 3, and $(\alpha+\beta)$, and the difference between this and its measured value is the closing error of the condition equation. The equation itself is far from linear in the unknowns, and for practical work it would have to be linearized as in (D.47).

If any angles have been observed in a trilateration, each produces another condition, either a triangular or central condition in combination with other angles, or a relation between it and the measured lines. Thus in a triangle ABC in which a, b, c, and A have been measured,

$$x_A-(x_{BC}\sin A \operatorname{cosec} B \operatorname{cosec} C - x_{AC}\cot C - x_{AB}\cot B)10^{-6}\operatorname{cosec} 1'' = \epsilon_A, \tag{2.49}$$

where x_{BC}, etc., are in parts per million and ϵ_A is the observed value of A minus that computed by (2.48) in seconds. Only approximate values of sin A, cot C, etc., are required in (2.49).

A single trilaterated triangle with no observed angles contains no conditions. A quadrilateral with both diagonals contains one, and a centred triangle, etc., contains one. A triangle with all sides and angles observed contains three, either one triangular and two of (2.49), or three of (2.49).

These troublesome condition equations can be avoided by computing by the method of variation of coordinates. The number of normal equations may then be larger than it would have been using conditions, but that will probably not matter if a computer is used.

2.34. The least square solution

(*a*) Weights must be allotted to each observation. The weights should preferably equal $1/(\text{s.e.})^2$ of each observation, but $\propto 1/(\text{s.e.})^2$ is allowable. The remarks in § 2.24 about the weighting of distances and angles are relevant.

(*b*) The larger ϵ's in each type of equation should be of similar orders of magnitude, as they will be if angles are recorded in seconds, and errors of distance in parts per million or (less usually) in parts per 10^5.

(*c*) Before the normal equations are formed, the condition equations may be numbered and arranged in the most convenient order. In a small adjustment the order is immaterial, but if there are many equations they should be so arranged that the non-zero elements in the normal equations will lie near to the diagonal. See § 2.35, and compare with § 2.25.

(*d*) For solving the normal equations by desk computer the method of Cholesky, § B.14 (*e*), is better than Gauss. [466], pp. 75–96 gives practical examples. If large adjustments are done by conditions, the method of solution will depend on the available computer and the width of the band of non-zero elements. See §§ B.14 and B.15. The band may sometimes not be narrow, see § 2.35.

2.35. The normal equations for small frameworks. Conditions.

(*a*) *Continuous nets.* There will be a non-zero element m, n in the normal equations if conditions m and n have an unknown x in common. The conditions must therefore be numbered so as to minimize the greatest range of n's associated with any m.

Consider the net of 56 simple triangles shown in Fig. 2.11. The routine of § 2.32 (*a*)–(*g*) shows that there will be 48 normal equations. In Fig. 2.11 number the top row of triangle equations 1 to 8. Then number the central condition at station 7 as 9, and the side condition with 7 as pole as 10. Similarly conditions 11 and 12 at station 8, with 13 and 14 at station 9. Then number the second row triangles as 15 to 22, and so on. Fig. 2.19 shows the non-zero elements in the first 30 normal equations. The bandwidth is 18. The total number of non-zeros in all 78 equations is about 360. Compare variation of coordinates in Table 2.1, with 66 equations and 425 non-zeros in a band of 14. The difference is not great.

The addition of a redundant line crossing two triangles adds two normal equations, if it has been observed in both directions. A single extra line such as 9–12 in Fig. 2.11 would increase the bandwidth by 2, and 13–10 would add 2 more, but a more distant line such as 24–28 would widen the band further down, and not add to the increase produced by the earlier two.

The effect of measuring the length of side 34–35 is to produce a distance condition whose terms run right through the net from stations 1 to 35. In Fig. 2.19 it introduces a broken line of non-zero elements in column 79, which destroys the band. If this was the only extra measured distance, the normal equations could be partitioned as in § B.14 (*c*) as

$$\begin{bmatrix} N_{11} & N_{12} \\ N_{21} & N_{22} \end{bmatrix} \begin{bmatrix} x_1 \\ x_2 \end{bmatrix} = \begin{bmatrix} c_1 \\ c_2 \end{bmatrix} \tag{2.50}$$

where N_{22} is the single element 79, 79, N_{12} is column 79 without the element N_{22}, N_{21} is its transpose, and N_{11} is the rest, a band matrix. The solution is given by (B.61) and (B.62).

More frequent measured distances, at closer intervals, introduce shorter columns, adding to the bandwidth, but not so easily dealt with by partitioning. Fig. 2.19 shows the column arising from the measurement of 9–13. It increases the width to 25. If it is included, column 79 may be shortened. The effect of measured azimuths is similar.

(*b*) *Systems of chains.* The non-zero elements in the normals of a simple chain of braced quadrilaterals, each with four conditions, are as in Fig. 2.20, a series of 4×4 sub-matrices (or 8×8 for hexagons) all lying on the diagonal without entanglement. They can be adjusted separately as in § 2.13. An additional base or Laplace azimuth produces a column of non-zero elements running from the diagonal point in the matrix where it is introduced, up to the row of the preceding base or azimuth. The partitioned matrix then ceases to be diagonal.

If a chain forms a closed loop as in Fig. 2.21, the numbering of the conditions can proceed continuously from the point of closure, and the partitioned matrix will be diagonal except for four columns on the right,

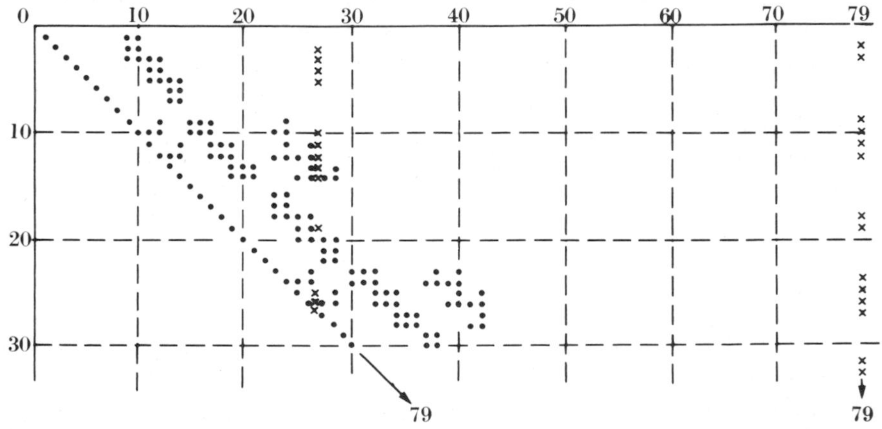

FIG. 2.19. Method of conditions. Dots are non-zero elements in the normal equations for the net shown in Fig. 2.11. Column 79 shows elements due to a measured distance equation between sides 1–2 and 34–35. Crosses between columns 26 and 27 show a similar equation between 1–2 and 9–13. Elements below the diagonal are omitted. The single dot in row 8 is the triangular equation of 4–5–10, which can be adjusted independently.

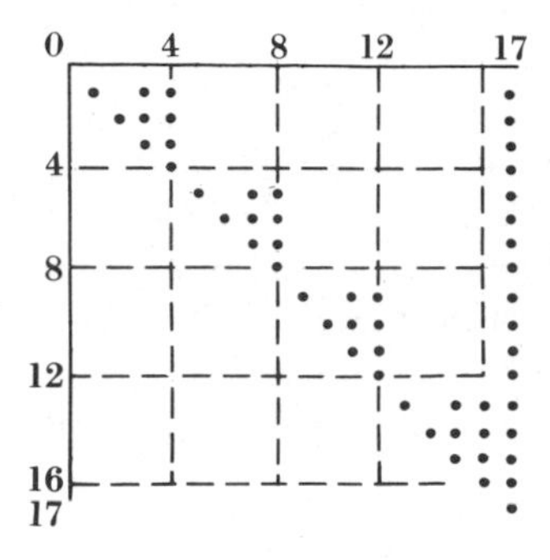

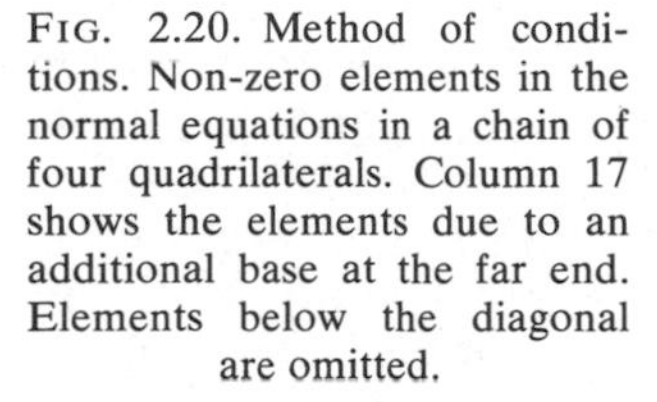

FIG. 2.20. Method of conditions. Non-zero elements in the normal equations in a chain of four quadrilaterals. Column 17 shows the elements due to an additional base at the far end. Elements below the diagonal are omitted.

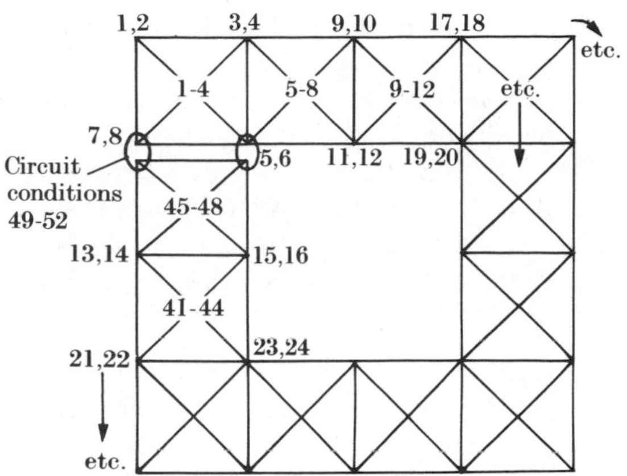

FIG. 2.21. Closed circuit of quadrilaterals. Numbers inside the quadrilaterals are the serial numbers of the figural condition equations, and 49–52 are circiut conditions. Numbers outside the quadrilaterals are unknown $\delta\phi$ and $\delta\lambda$'s for the method of variation of coordinates.

produced by the closures in ϕ, λ, side, and azimuth. If there are no more than the single base and Laplace station, the circuit can be dealt with as in (2.50), but otherwise the band is widened. A solution by variation of coordinates will give about the same number of normal equations, with a bandwidth of 20 (for quadrilaterals), but with no extra trouble from additional bases and Laplace azimuths.

(*c*) *Conclusion.* For a large network the difficulties in forming the condition equations almost inevitably result in variation of coordinates being adopted, apart from the fact that the normal equations of the latter are likely to be easier to solve.

SECTION 5. THE ADJUSTMENT OF A LARGE GEODETIC FRAMEWORK

2.36. Objects

The object of adjustment is to secure self-consistency, together with some improvement in accuracy both over-all and as between near points. The ideal is that the geodetic framework of a country, a continent, or eventually of the world, should be adjusted as a whole by the process of least squares. Until about 1950 the truly simultaneous adjustment of any but a very small country was impossible. Much more is now possible, but there are limitations as below.

(i) It may be inconvenient to postpone adjustment until all secondary in-filling, and perhaps the primary of some outlying areas, have been completed, even though they might have contributed something to the strength of the whole.

(ii) There must be some limit to the number of normal equations that can usefully be simultaneously solved. A net of secondary or even tertiary EDM lines may be as strong as a primary traverse enclosing it, but it is doubtfully practicable to make a simultaneous adjustment of the whole primary secondary, and tertiary framework of a continent. For most organizations a solution involving a few thousand normal equations is still a great piece of work.

Before the adjustment is begun it must be decided whether it is to be by variation of coordinates or by conditions, and whether the normal equations are to be solved by direct methods (§ B.14) or by iteration (§ B.15). Variation of coordinates and direct methods of solution are likely to be accepted.

It is also necessary to decide on the spheroid and datum as in § 2.02 (*c*) It is to be hoped that in future adjustments the centre of the adopted spheroid may be at the geocentre.

Other, less basic, points for decision are as below.

(i) Will adjustment be by angles or by directions? §§ 2.22 and 2.25. Angles are recommended.

(ii) If by directions, will Schreiber's method be used to eliminate the Z terms?

(iii) If by angles will exterior angles be excluded? § 2.22. Exclusion is recommended.

(iv) To what extent will station adjustment be considered as finalized before the main adjustment is begun? § 1.23.

2.37. Division into sections. Method of Helmert–Wolf

(*a*) *Description of the method* If it is possible, it is most convenient to form a normal equation matrix with a narrow band, as in § B.14 (g). If the band is too wide for the available computer, the method of Boltz, § B.14 (*c*), can be employed, or perhaps some simple form of partitioning as in § 2.35, (2.50). The following describes a method by which the labour of a large continental adjustment by truly simultaneous least squares may be broken up into separate parts, each of manageable size. [270], [65], and [610].

Fig. 2.22 shows four connected geodetic systems A, B, C, and D, such as those of four separate countries. Each country sets up its own observation equations for computation by variation of coordinates. In country A the vector of unknowns x_A includes the $\delta\phi$ and $\delta\lambda$'s which do not enter into any observation equation which includes the stations of another country, while vector y includes all the unknowns entering into equations

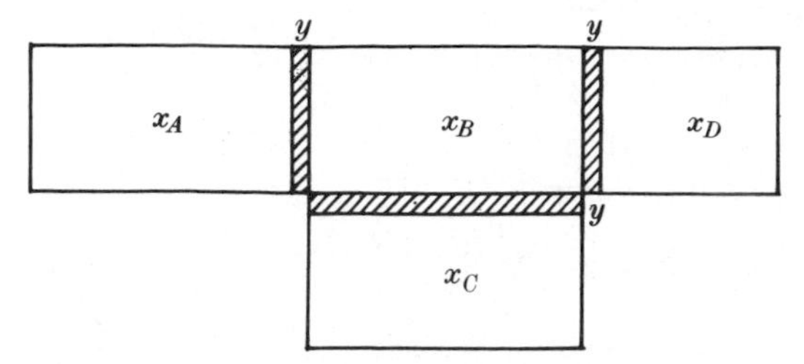

FIG. 2.22. Simultaneous adjustment of four national frameworks, A, B, C, and D. Stations whose unknowns are elements of vector y, in the shaded strips, are connected to stations of more than one country. Stations x_A are connected only to each other or to y stations, and similarly x_B, x_C, and x_D.

which do also include the stations of another country. Each country then allots weights, preferably the reciprocals of the estimated (s.e.)^2 of its observations,† and forms its normal equations. So A's normal equations may be partitioned thus

$$\begin{bmatrix} A_{11} & A_{12} \\ A_{12}^T & A_{22} \end{bmatrix} \begin{bmatrix} x_A \\ \hline y \end{bmatrix} = \begin{bmatrix} c_{A1} \\ c_{A2} \end{bmatrix}. \tag{2.51}$$

The vector y may be supposed to contain all the y's in all the junction strips between A and B, B and C, etc., with zero rows and columns in A_{22} and zero c's for those with which country A is not involved, but there will, of course, be no need to enter these zeros in the computations.

Then eliminating x_A from (2.51), just as in (B.60)–(B.62), gives

$$\begin{aligned} y &= (A_{22} - A_{12}^T A_{11}^{-1} A_{12})^{-1} (c_{A2} - A_{12}^T A_{11}^{-1} c_{A1}) \\ &= (\text{say}) E^{-1} C_E, \end{aligned} \tag{2.52}$$

and the variance–covariance matrix of this determination of the y's is $E^{-1} \equiv (A_{22} - A_{12}^T A_{11}^{-1} A_{12})^{-1}$. It may then be proper to compute an augmenting factor σ_E^2 from the consideration of the residuals, as in (D.62).‡

Countries B, C, and D then do the same, giving $y = F^{-1} C_F$, etc., and we then have 'observation' equations for the y's, on the lines of § D.22 (*d*) and (D.138)

$$\begin{bmatrix} I \\ I \\ I \\ I \end{bmatrix} [y] = \begin{bmatrix} E^{-1} C_E \\ F^{-1} C_F \\ G^{-1} C_G \\ H^{-1} C_H \end{bmatrix} \quad \begin{matrix} \text{with weights } E/\sigma_E^2 \\ F/\sigma_F^2 \\ G/\sigma_G^2 \\ H/\sigma_H^2. \end{matrix} \tag{2.53}$$

† See § 2.24. All observations need not have different weights.

‡ For this purpose a preliminary free adjustment should be made by each country, ignoring its connections with others.

Here y contains all the y unknowns, and E, F, C_E, C_F, etc. must contain the necessary zero rows and columns. The order of each I equals the number of y's.

As in (D.140) the solution is

$$y = (E/\sigma_E^2 + F/\sigma_F^2 + G/\sigma_G^2 + H/\sigma_H^2)^{-1}(C_E/\sigma_E^2 + C_F/\sigma_F^2 + C_G/\sigma_G^2 + C_H/\sigma_H^2). \tag{2.54}$$

The y's, namely the $\delta\phi$ and $\delta\lambda$'s of the frontier stations, having been determined, each country accepts them and substitutes them in its normal equations (2.51) to give $\delta\phi, \delta\lambda$ for its own internal stations. See [610].

(*b*) *The advantages* of this method are as follows.

(i) Each country is responsible for handling its own observations. This not only distributes the labour, but means that all the small intricacies and confusions which occur in every country's geodetic history are handled by those most likely to understand them.

(ii) No country has to solve a larger set of normal equations than its own system produces.

(iii) The number of y's only equals twice the number of frontier stations in the whole adjustment, and usually each frontier strip only includes a single line of stations, so the y matrix should have a narrow band and be no more troublesome than a single national matrix. But if it is inconveniently large, some of the smaller countries can first be joined together by combination on the same lines, leaving fewer y's for the final combination.

(*c*) *Notes*

(i) In the whole system there must be at least one 'observed' value of $\delta\phi$ and $\delta\lambda$. It may be an arbitrary value held fixed at some origin as in § 2.06, in which case the $\delta\phi$ and $\delta\lambda$ at the point can be omitted from all the observation equations. Alternatively at any *one* point observed or arbitrary values can be included as unknowns (if more convenient) with any reasonable weight. In this case, whatever the weight may be, the allotted values will be preserved unchanged by the adjustment process.

(ii) Not more than one arbitrary value may be included in the whole system. The matrices A, B, etc. of the other blocks will be singular, but no difficulty results. The equation for y, (2.53) will not be singular, and the normal equations of the separate blocks will case to be singular as soon as the y's are substituted in them.

(iii) If there are several (more than one) genuinely observed values, such as satellite fixes, they can be included wherever they occur as in § 2.23 (*d*).

(iv) All the countries connected to any particular y station must use the same provisional coordinates for it, so that they will all be seeking the same corrections $\delta\phi$ and $\delta\lambda$.

(v) Although the inverse of large matrices such as A_{11} occur in expressions such as $E = A_{22} - A_{12}^T A_{11}^{-1} A_{12}$, A_{11}^{-1} need not be explicitly computed. E is obtainable by the ordinary processes of solving equations, e.g. in Gauss's method, §§ B.14 (*b*) and 2.13 (*d*).

2.38. Less rigorous methods

Before about 1950 it was impossible to handle a least square adjustment with more than about 100 normal equations. Various approximate methods were adopted, and their results still survive. Some are noted below.

(*a*) The adjustment of chains of triangles, even where a continuous net may exist, enables a larger area to be adjusted. This was adopted for the 1950 European adjustment.

(*b*) Division into independent sections. A central block of triangulation, preferably one with the highest quality observations, can be adjusted first, and adjacent blocks can then be adjusted on to it, keeping the original block (and any other block which has been adjusted to it) unchanged. This method was used for the 1880 adjustment of India and for the 1935 adjustment of Great Britain. See [295] and [113], 3 Edn, p. 177.

(c) The Bowie method. For this method the triangulation is laid out in the form of intersecting north–south and east–west chains, with a base and Laplace station at or near each intersection. Each chain of triangles between intersections is first adjusted to the extension figure of the base and Laplace controls at its two ends, and it is then treated as a unit (rigid rod). Least-square changes in these units then provide unique optimum values of latitude and longitude at some point at each intersection. See more details in [1] and [113], 3 Edn, p. 179.

2.39. Systems of traverses

(*a*) *Approximate methods.* Traverses can be adjusted by rigorous least squares by variation of coordinates or by conditions, alone or in combination with triangulation. Unless traverse legs are very short, traverse stations will not be much more numerous than the stations in a system of triangulation chains enclosing rectangles of similar size, while traverse is likely to give the narrower bandwidth. The following simplified method has been used in Australia in 1966 [101].

See Fig. 1.4. Each traverse between junction points, averaging about 300–500 km in length, is treated as a unit, like the chains in the Bowie method. Each selected junction point is given a preliminary ϕ and λ, and each line of traverse is 'freely' computed by variation of coordinates using its angles, distances, and Laplace stations, with ϕ and λ fixed at only one of its ends. Closing errors result.

Circuit closing errors are then examined for abnormal error, which must if necessary be found and remedied. The mutual distance and azimuth† between the two ends of each traverse section, as given by these free computations, are then computed using § 2.15 (c) and constitute 'observations' for the next stage of the adjustment. Weights are allotted to them, dependent on the length and quality of the section.

For the final adjustment, the distances and mutual azimuths of the preliminary positions of adjacent junction points are computed using the same formula § 2.15(c), and each traverse section then produces two 'observation' equations, one like (2.33) comparing the length O given by the free adjustment with that C given by the preliminary positions: and the other‡ as in § 2.23 (g) comparing azimuths.

The coordinates of one junction point are held fixed, and the ordinary process of variation of coordinates gives the others. Finally, each traverse section is recomputed holding its two ends to the final positions of the junctions.

See [101], pp. 66–8 for details whereby undue violence to the angles at junction points is avoided.

This procedure does not produce exactly least square corrections to all observations, but it is not easy to say that it falls significantly short of doing so.

(b) *Rigorous adjustment of traverses.* [611], p. 262. A system of traverses can be adjusted in a way very similar to the Helmert–Wolf adjustment of a triangulation or other net, § 2.37. Let the vector x_2 be the coordinates of the nodal (junction) points, and let x_1 be the vector of the coordinates of all the remaining traverse points. Trial coordinates are allotted to the nodal points, and normal equations are formed for the adjustment of each separate traverse between its two terminal nodal points. These normal equations may be partitioned in the form

$$\begin{aligned} N_{11}x_1+N_{12}x_2&=c_1\\ N_{21}x_1+N_{22}x_2&=c_2. \end{aligned} \tag{2.55}$$

These are solved to the extent of eliminating the x_1's, leaving for each traverse a reduced equation $N'_{22}x_2=C_2$.

All these separate equations are combined to give

$$\sum N'_{22}x_2=\sum C_2, \text{ as in (2.54).} \tag{2.56}$$

These can be solved to give the coordinates of the nodal points, and

† The azimuths may either be of the normal section or of the geodesic, provided the same is used both here and when computing the mutual azimuths of the trial positions of the junction points.

‡ Each traverse section provides only one azimuth equation, not two like a triangulation line observed in both directions, because the azimuths O have been computed from the latitudes and longitudes. Apart from computational error, the forward and reverse azimuths must necessarily be in accord with each other.

back solutions of each traverse give the remaining points x_1. The least squares condition is fully satisfied.

2.40. Laplace azimuths and reduction of distances to spheroid level

(*a*) *Laplace azimuths.* An annoying complication has been that the Laplace correction depends on the geodetic longitude of the station, and this is not finally known until the adjustment has been completed. The method of variation of coordinates completely overcomes the trouble, § 2.23 (*f*) and (*g*), but adjustment by conditions demands that Laplace equations should be formed with an adequately accurate estimate of the final longitude. Otherwise a second approximation may be necessary.

It may even be asked whether such successive approximations will converge. See [113], 2 Edn., Appendix III, which concludes that there will be convergence, at any rate below latitude 70°, provided Laplace stations are not more than 300–400 km apart. Closer spacing is in any case advised for other reasons.

(*b*) *Distances.* Measured distances are involved in a similar difficulty. They cannot be reduced to spheroid level unless *N*, the height of the geoid above the spheroid, is known. Which is not possible until the deviations of the vertical are known. Which in turn are not finally known until the triangulation has been adjusted. An error of 6 metres in *N* corresponds to only 1 ppm in scale, so there is not likely to be trouble in an independent national survey of average size. But in a chain of transcontinental extent, such as when the African arc of the 30th meridian is computed on the European datum, the error can be 30 times as much. This accumulates to some seconds of latitude or longitude, which in turn vitiates the computed values of *N*. There are two methods of remedying this.

(i) To compute the deviations and the resulting *N* as the chain proceeds. As the computations reach each base, the value of *N* at that place is obtained and *N* at the next base is estimated, as it probably can be to 6 m or better. This estimated value is then used in the computations of the next section, and a final value results. If the estimate has been materially wrong, the necessary very small corrections can be made by slide-rule. The computations then proceed to the next section in the same way.

(ii) The Molodensky correction [417] and [419], pp. 29–37. Let the triangulation be computed in the ordinary way, but with bases reduced only to geoid level. As a result let the computed component of the deviation in the direction of the chain be ζ', instead of ζ as it would have been if the bases had all been reduced to the spheroid, and let the computed height of the geoid above the spheroid be N' instead of N. Then the Molodensky correction $N-N'$ is given by

$$N = N' + RA \sin s + B \cos s$$

and also

$$\zeta = \zeta' + A \cos s - (B/R) \sin s, \tag{2.57}$$

where s is distance along the chain (in arc) from a specified point,

$$A = \zeta_0 - \zeta_0' - (1/R)\int_0^s N' \cos s \, ds$$

and

$$B = N_0 - N_0' + \int_0^s N' \sin s \, ds.$$

and suffix zero refers to the point where s is zero. R = earth's radius. The solution is by iteration.

The integral $\int_s^0 N' \sin s \, ds$ (or $\cos s \, ds$) is not necessarily zero round a circuit. So, in a system which does not consist of a single chain it is necessary to apply the correction to each chain before circuits are adjusted. Only approximately correct results will be obtained if the correction is applied to an adjusted system.

2.41 Graphical adjustment

After a system of triangulation has been adjusted, it may be desired to incorporate new data, such as new or revised bases or Laplace stations. If the corrections are small a complete readjustment will hardly be worth while, especially if there is no intention of using the new values for mapping. It may then be convenient to make very simple computations of the changes in X and Y† at a few points, and to draw curves of equal changes in X and Y to be applied at all others. The curves of equal change should then so far as possible be such that the new positions of points should be an orthomorphic representation of the old, meaning that the angles between near points should be unchanged, and that at any point the changes of scale and azimuth should be the same in all directions. It is clear that at any point the separation of lines of equal change in X must equal the separation of the lines of equal change in Y, but it is shown below that if the representation is to be strictly orthomorphic these curves must follow two other rules. The extent to which this is found possible measures the success of the method. It will probably do well if the only imposed changes are of scale and azimuth (with a previously determined change of position at no more than one point), especially if the area covered is long and narrow. But if a certain area of

† X and Y. It is convenient to compute the changes of position on any rectangular system. Then unless the area is very large or in high latitudes $\delta\phi'' = \delta Y/31$ m and $\delta\lambda'' = \delta X \sec \phi/31$ m.

survey has to be fitted into a space bounded by other surveys which cannot be changed, a reasonably orthomorphic solution may be impossible to find.

A similar situation occurs when triangulation of a low order is to be adjusted to a higher order framework at a few points. The method of *divided differences*, § 2.64(*b*), may then be used, It will give results whose accuracy is intermediate between that of recomputation by variation of coordinates and that of graphical adjustment. See also [272].

Orthomorphic changes. Let P and Q in Fig. 2.23 be two points whose coordinates on the old system are (X, Y) and $(X+\Delta X, Y+\Delta Y)$, and on the new system let their coordinates be $(X+x, Y+y)$ and $(X+\Delta X+x+\delta x, Y+\Delta Y+y+\delta y)$, where ΔX and ΔY may be some tens of km, x and y tens of metres, and δx and δy of the order 10 cm (such as 1/200 000 of ΔX and ΔY). Let the original distance PQ $(=r)$ be changed to $r(1+\sigma)$ and let its azimuth θ be changed to $(\theta+\alpha)$. Then we wish to draw curves of constant x and of constant y subject to the condition that any point σ and α shall be independent of θ.

We have

$$\delta x = \left(\frac{\partial x}{\partial X}\right)\Delta X + \left(\frac{\partial x}{\partial Y}\right)\Delta Y = r\left(\frac{\partial x}{\partial X}\sin\theta + \frac{\partial x}{\partial Y}\cos\theta\right),$$

$$\delta y = \left(\frac{\partial y}{\partial X}\right)\Delta X + \left(\frac{\partial y}{\partial Y}\right)\Delta Y = r\left(\frac{\partial y}{\partial X}\sin\theta + \frac{\partial y}{\partial Y}\cos\theta\right),$$

provided ΔX and ΔY are small enough for second differentials to be neglected.

Also, from Fig. 2.23,

$$\delta x = r(\sigma\sin\theta + \alpha\cos\theta),$$
$$\delta y = r(\sigma\cos\theta - \alpha\sin\theta).$$

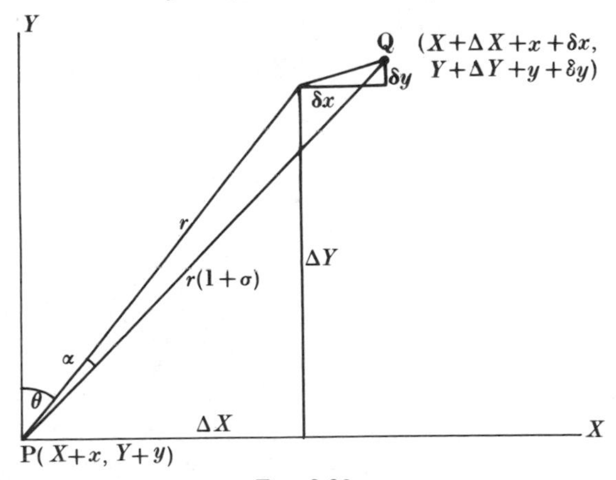

FIG. 2.23.

Equating the two expressions for δx and the two for δy gives two equations for σ and α in terms of θ and $\partial x/\partial X$, etc., solving which gives

$$\sigma = \sin^2\theta\frac{\partial x}{\partial X}+\sin\theta\cos\theta\left(\frac{\partial x}{\partial Y}+\frac{\partial y}{\partial X}\right)+\cos^2\theta\frac{\partial y}{\partial Y}$$

and

$$\alpha = \cos^2\theta\frac{\partial x}{\partial Y}+\sin\theta\cos\theta\left(\frac{\partial x}{\partial X}-\frac{\partial y}{\partial Y}\right)-\sin^2\theta\frac{\partial y}{\partial X}. \tag{2.58}$$

If these are to be independent of θ we must have

$$\frac{\partial x}{\partial X}-\frac{\partial y}{\partial Y}=0 \qquad \text{and} \qquad \frac{\partial x}{\partial Y}+\frac{\partial y}{\partial X}=0 \tag{2.59}$$

when (2.58) becomes

$$\sigma = \frac{\partial x}{\partial X}=\frac{\partial y}{\partial Y} \qquad \text{and} \qquad \alpha = \frac{\partial x}{\partial Y} = -\frac{\partial y}{\partial X}.$$

These equations occur in connection with map projections, see (F.31) which is the same in a different notation, and also § 6.72. Errors of survey

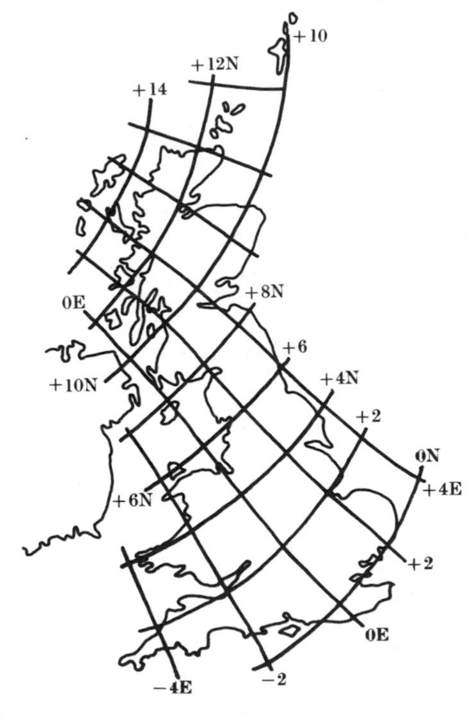

FIG. 2.24. Near-orthomorphic corrections to E and N resulting from new bases and Laplace stations. Metres. 1956.

tend to make orthomorphic changes, while tectonic movements do not.

The slope of a curve of constant value of x is $-(\partial x/\partial X)/(\partial x/\partial Y)$, and that of a curve of constant value of y is $-(\partial y/\partial X)/(\partial y/\partial Y)$, and (2.59) shows that their product is -1. The first rule which guides the drawing of these curves is then that the two sets should so far as possible be orthogonal.

As stated above, it is also clear that at any point the interval between lines of equal x must equal the interval between lines of equal y.

The second rule is that when the curves are straight,† they must for equal increments of x or of y be equally spaced, and when they are curved the inequality of their spacing (increasing outwards from the centre of curvature) must be inversely proportional to the radius of curvature. For proof see [113], 2 Edn. pp. 142–3.

Fig. 2.24 illustrates these rules. An example of orthomorphic correction curves is also in [272].

General references for Section 5. [65] and other papers in the same issue of the *Canadian Surveyor*, **28** (5), 1974.

SECTION 6. ESTIMATES OF ACCURACY

2.42. Summary and notation

(a) This section considers the following.

(i) Estimates of the accuracy of observed angles, distances and Laplace azimuths, §§ 2.43–2.45.

(ii) Estimates of the accuracy of triangulation chains and nets, and of traverses, §§ 2.46–2.48.

(iii) Methods of exhibiting the accuracy of geodetic nets. Error ellipses, § 2.49.

(b) The following notation is used in this section.

ppm = parts per million.
s.e. = standard error.
e = s.e. of an observed angle, in seconds. Also σ_α,
σ_D = s.e. of an observed direction, in seconds,
σ_L = s.e. of an observed distance, in ppm.
σ_A = s.e. of an observed Laplace azimuth, in seconds. Also t,
s = s.e. of a base measurement and extension, in ppm.
N_1 = s.e. of scale after 100 km in ppm.
N_2 = s.e. of azimuth after 100 km, in seconds.
$N = \surd(N_1^2 + 23.5N_2^2)$.
ϵ = Triangular error.
f = Number of figures in 100 km of chain.

† As they will be in an area in which there is a uniform change of scale and/or azimuth.

O = Origin from which errors are measured.
A = Terminal point at which s.e. is required.
$100S$ = Direct distance OA in km.
E_1 = s.e. of position at A in direction OA
E_2 = s.e. of position at A in direction perpendicular to OA.
$E = \surd(E_1^2+E_2^2)$ = total s.e. of position at A.
L = Length of a traverse in kilometres, § 2.47.
l = Length of a traverse leg in kilometres.

2.43. The accuracy of observed angles

This can be assessed in various ways.

(*a*) *By the accordance of the zero means* $m_1, m_2, \ldots, m_n$. If M is the general mean, its s.e. may be assessed as

$$1.25\sum|M-m| \div n\surd(n-1), \tag{2.60}$$

but this is unreliable, as all zero means are equally affected by most of the worst sources of error, § 1.24, and also by the omission of deviation corrections, § 2.09 (*b*). The result will generally be too small, although it may occasionally be too large, as when a theodolite has large periodic errors of graduation, or when lateral refraction is large but equal and opposite at different times of day.

(*b*) *From the triangular errors.* The s.e. of an observed angle is given by

$$e = \surd(\sum\epsilon^2/3n) = 0.72\epsilon_m, \tag{2.61}$$

where ϵ_m is the average triangular error, without regard to sign. The first expression is well known as General Ferrero's criterion. This is much better than (*a*), although certain types of error escape unnoticed, such as neglect of deviation corrections if the deviation is similar at all three angles of a triangle.†

(*c*) *From the adjustment of figures* more complex than simple triangles.

$$e = \surd(\sum wx^2/w_m c), \text{ [591], p. 344,‡} \tag{2.62}$$

$$= \surd(\sum x^2/c), \text{ if all } w\text{'s are more or less equal,} \tag{2.63}$$

where w is the weight of each angle, x the correction which each angle derives from the adjustment, w_m the mean of all the w's, and c the number of conditions. The summation may be taken over each figure separately, and the resulting values of e meaned, or it may be taken over all the figures other than simple triangles in any homogeneous piece of

† A more serious possibility is that the observer may have allowed his knowledge of the correct sum of the three angles to influence the rejection of discrepant zeros, or that he may have continued the re-observation of some zeros until a small triangular error has been obtained. Cases are on record where the triangular error has been no guide to actual accuracy.

‡ In this reference $[wx]^2$ in one place is a misprint for $[wx^2]$.

work. This method can only be used if a network contains a reasonably large number of side conditions. Fig. 2.11, for example contains 15 hexagons.

(*d*) *From closures on bases and Laplace stations.* § 2.46 (*d*) gives formulae for the probable misclosures on bases and Laplace stations or round closed circuits, given the quality of the intervening chains. Comparison between a number of actual and estimated closing errors, in the form of a table showing the frequency distribution of their ratios, will show whether the accepted s.e.'s of the observed angles are on the whole correct or not.

The quality of a triangulation net can be tested in this way by a provisional computation of side, azimuth, and position closures along selected chains of triangles or other simple figures. Table 2.2 is an example taken from [110].

TABLE 2.2

Distribution of ratios (actual closing errors) ÷ (their estimated s.e.'s)

	Number of cases in which ratio was between							Total number of cases
	0& 0.7	0.7& 1.3	1.3& 2.0	2.0& 2.7	2.7& 3.3	3.3& 4.0	4.0& 4.7	
Scale closures	32	14	4	2	—	—	—	52
Normal distribution of 52 cases	26	17	6	3	—	—	—	52
Azimuth closures	17	24	13	8	2	3	3	70
Normal distribution of 70 cases	35	23	9	2	1	—	—	70
Distribution of azimuth closures after multiplying estimated s.e.'s by $1\frac{1}{2}$	33	21	9	4	3	—	—	70

In Table 2.2 the actual distribution of scale closures agrees well with the normal, and the estimated s.e.'s have on average given correct results. With the azimuth closures, on the other hand, the estimates are too low. This might have been due to the estimated *e*'s being too small, or to error in the formulae which give their cumulative effects, or to unexpected inaccuracy in the Laplace stations themselves.† See sub-paragraph (*e*).

(*e*) *Large and small angles.* Sub-paragraphs (*b*) and (*c*) above, and the scale closures in (*d*) give the s.e.'s of a typical 60° angle, but large angles tend to be less accurate. Even though angles are measured not independently, but in rounds, instrumental errors, such as movements of the stand or tower, are likely to be greater in an angle of 180° than in one of

† If the last had been the chief cause, circuit errors analysed separately would have given a normal distribution, but actually they gave a distribution similar to that of the whole group.

60°, and so may be the effects of lateral refraction and of the neglect of deviation corrections. Internal evidence, the accordance of zero means, in an Indian sample, [110], p. 74, suggested that the s.e. of a 180° angle was 1.25 that of a 60° one, but internal evidence neglects many serious factors, and the distribution of azimuth closing errors given above is a better guide. The formula which was there used for the accumulation of azimuth error in a chain depended on the estimated s.e. of 180° angles, which in the absence of better evidence had been taken (as above) to be 1.25 times the s.e. given by figural adjustment. But the fact that scale closures followed the normal distribution, while azimuths only did so after adding 50 per cent to 1.25 times the s.e. of a 60° angle, suggests that a 180° angle is really 1.87 times as bad as a 60° one. By chance this is similar to the $e\sqrt{3}$ which is what would have resulted if three 60° angles had all been separately measured and treated as independent, as they are in § 2.46 (*b*) and (*c*).

This conclusion conflicts with the natural view that directions are the quantities which are actually observed, and that errors of adjacent angles will therefore be negatively correlated, as they clearly will be so far as they arise from random mis-pointing of the telescope or reading of the circle. These, however, are not major sources of error. The effect of lateral refraction on the other hand, and to some extent the effect of neglecting the deviation corrections, may be expected to give some positive correlation between the errors of adjacent 60° angles. It appears from Table 2.2 that the result of the two conflicting possibilities is that the coefficient of correlation is in fact near to zero. It is convenient to be able to treat adjacent angles of a round as uncorrelated, and it is satisfactory to have some justification for doing so.

(*f*) Estimates of *e*, however obtained, must be controlled by common sense. Thus first-order results cannot be expected from second-order instruments and programmes, and a small piece of work cannot give a reliable value for its *e*: it is better jointly assessed with some larger sample observed in similar conditions. Allowance is also sometimes necessary for unusual sources of error such as inexact recovery of markstones from one year to the next, as for instance in [110], pp. 77 § 4 (*e*).

(*g*) *Traverse angles.* The expected misclosure of a traverse between Laplace azimuths depends on the s.e. of each traverse angle (e) and on that of each Laplace azimuth (t). If there are n angles between the Laplace azimuths (four in Fig. 2.25),† the estimated (68:32) misclosure will be $(ne^2+2t^2)^{\frac{1}{2}}$. If m traverse sections with $n_1, n_2, \ldots, n_m$ intervening

† t is the s.e. of the angle between a star and a ground mark. If special marks are used as RM's at 0 and 5, in preference to stations 1 and 4, the number n will increase by 2, to 6 in Fig. 2.25.

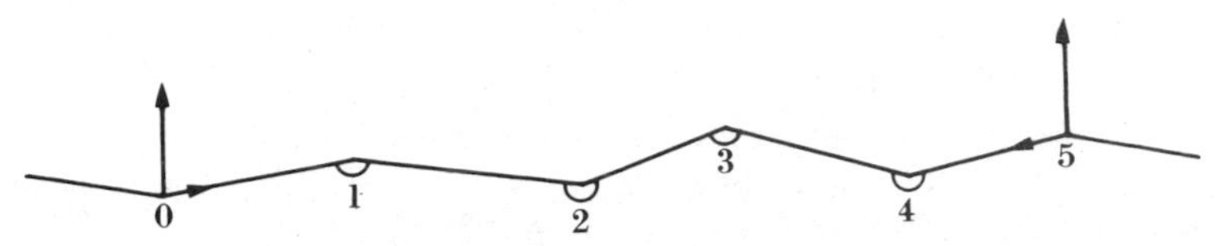

FIG. 2.25. Traverse. Closures on Laplace azimuths.

angles give misclosures of $\epsilon_1, \epsilon_2, \ldots, \epsilon_m$ the estimated value of e^2 will be

$$(1/m)\sum\{(\epsilon^2-2t^2)/n\}. \tag{2.64}$$

For values of t $(=\sigma_A)$ see § 2.45.

In Fig. 2.25 each traverse angle could perhaps be regarded as having been measured twice, the angle on one side being independent of the angle on the other. It must therefore be noted that (2.64) gives the s.e. of a traverse angle on the view that there is only one angle at each traverse station. If weights are derived from (2.64) and if both angles are treated as independent, angles are likely to be overweighted as compared with distances. If the two angles have been in any way independently measured, their total should be adjusted to 360° before computing (2.64), and thereafter only the angles on one side of the traverse should be considered. See also § 2.22, which recommends similar treatment for the exterior angles of a net or system of chains.

2.44. Accuracy of measured distances

(*a*) *EDM Random errors.* The s.e. of single EODM and MDM lines is given for different types of instrument in § 1.39 (*i*) and 1.43 (*h*) in the form $\pm p$ mm $\pm q$ ppm. Typically $p =$ about 15 mm and $q = 3$ ppm over lines of 25–50 km.

These figures are for random error in the sense that although the q ppm represents an error which is systematic over the length of any one line, the s.e.'s which are deduced are ordinarily treated as being random from one line to the next. This is a source of trouble, for which see (*c*) below.

Available figures for the accuracy of EDM are ultimately based on comparisons with invar baselines. They are good estimates of the accuracy to be expected in normal circumstances, but do nothing to confirm that the expected accuracy has actually been achieved in any particular piece of work. Figures for the latter can be got as follows.

(i) If a net containing measured sides, with or without measured angles as well, contains enough measured sides to fix all the stations with some redundancies, an adjustment by variation of coordinates, using the measured sides only, weighted in accord with the expected s.e.'s, will give an augmenting factor σ_0 of (D.62) by which the expected s.e.'s must on average be multiplied. It is unfortunate that pure trilateration generally contains few redundancies.

(ii) In a net in which most angles and many sides have been measured, the s.e.'s of the angles and distances may be separately determined as in § 2.28. For Great Britain [68] derives the following.

For directions	$\sigma_D = \pm 0.''63$
For MDM	$\sigma_L = \pm 2.54$ ppm
For Laplace azimuth pairs	$\sigma_A = \pm 0.''85$.

The average s.e. of an angle, σ_α, may doubtfully be supposed equal to that of a direction, § 2.22, in which case an s.e. of $\pm 0''.63$ implies an average triangular error of $0''.88$, which agrees reasonably well with the actual figure of $1''.1$.

(iii) In Great Britain, all the angles and sides of the Ridgeway and Caithness base extention figures have been measured. Then, working on the lines of (i) and (ii) above, [467] obtains the following for the s.e.'s of 20- to 50-km MDM measurements.

(A) From pure trilateration, ±2.4 ppm, agreeing well with (ii).
(B) Via the angles, ±1.8 ppm.

(iv) In Australia [365] gives the discrepancies between three sets each of three long traverses between pairs of points 4000 to 5000 km apart, as in Table 2.3. These traverses are all tolerably straight, not closed circuits.

TABLE 2.3

Set	Mean length	Difference (mean of 3) minus (each line)
1	4400 km	−1.4, +1.0, +0.4 ppm
2	4700 km	−0.6, +0.6, 0.0 ppm
3	5200 km	0.0, −1.4, +1.4 ppm

Table 2.3 suggests a random s.e. as small as 1 ppm for a single traverse over a distance of 4500 km, involving at least 100 traverse legs.

The figures quoted in the above examples mostly refer to MDM.

(*b*) *EODM. Random errors.* The accuracy of EODM may be expected to be better than that of MDM. In the US a system of zero-order traverses has been observed, § 1.06 (*c*), for which [406] gives the closing errors of five circuits as 5.5 m in 4000 km, 1.5 m in 4200 km, 1.2 m in 1200 km, 6.5 m in 8000 km, and 2.5 m in 3100 km. Average 0.8 ppm.

This has resulted from errors in both distance and direction, so the random error generated by distances is unlikely to have exceeded (say) ±0.6 ppm.

(*c*) *Systematic errors.* There is considerable evidence that in the average of a large sample, MDM is likely to measure 2 or 3 ppm too small, as follows.

(i) [251] reports 19 Tellurometer lines in Finland, of average length 30 km, which are on average 6 ppm shorter than the triangulated measures derived from several nearby invar bases.

(ii) In Canada [312] concludes that tellurometer measures (based on terminal meteorology) are 2 to 5 ppm shorter than geodimeter measures and that the latter are themselves (on average) 2 ppm too large during the day, and 2 ppm too small at night.

(iii) In Australia [379] gives figures for laser geodimenter measures minus tellurometer measures over the same lines as follows. The geodimeter was used in daylight.

(A) +2.4 ppm over 645 km of traverse along the east coast.

(B) +0.2 ppm over 1237 km in the dry central area.

(C) +4.6 ppm over 1298 km about 100 km from the sea in the Adelaide–Melbourne–Sydney area.

(iv) In the US, [406], satellite fixes made along the zero-order traverse lines referred to in (*b*) gave the following results. In 11 lines each of average length 1655 km, the doppler chord length averaged 1.0 ppm longer than the geodimeter measures. In 8 shorter lines of average length 706 km, the discrepancy was 1.3 ppm. The discrepancy was of the same sign in all 18 cases, suggesting that the geodimeter was systematically measuring too small, with random errors substantially smaller. Where only 1 ppm is involved, it might of course have been the doppler that was wrong. Both were indeed very good.

(v) In Great Britain [67] reports comparisons between EODM and MDM as giving the latter 2.6 ppm too small, and comparison of the triangulation net with doppler, after all MDM measures in the net had been increased by 2.6 ppm gave results which still appeared to be about 0.5 or 1.0 ppm too small. Here also it is possible that this small error could have been in the doppler.

(*d*) *Accuracy of long straight traverses.* Ignoring errors due to the angular measurements, the figures given in (*a*) and (*b*) above suggest that the length of a straight EDM traverse of 4000 km should be correct to ±1 ppm, but the possibility of systematic error changes the situation so far as MDM is concerned. On present evidence, the length of MDM lines based on terminal meteorological data should be increased by 2 or 3 ppm. As a rough guide, the s.e. of distance given by good MDM lines placed end to end in reasonably favourable conditions, after correction as above, may be expected to be

±4 ppm in a distance of 50 km
±3 ppm in a distance of 500 km
±2 ppm in a distance of 5000 km.

Better results may be expected in exceptionally arid areas.

For EODM the systematic error is unlikely to exceed 1 ppm.

All the above figures are given on the assumption that the measured distances are correctly reduced to spheroid level.

(*e*) *Invar base lines.* For good invar base lines measured since (say) 1920 the s.e. of measurement, including errors of standardization, should not exceed 1 ppm, but the lines are short and the triangulated extension is likely to have introduced more serious error. In any particular case this may be calculated by the method of § 2.27, estimating the s.e.'s of observed angles from the triangular errors and side closures within the extension, or from other work done in similar conditions.

By less elaborate methods [113] 2 Edn., p. 152 estimates an s.e. of 3 ppm for the extensions of Indian base lines, or $3\frac{1}{2}$ ppm after allowing something for the error of the base itself. These figures do not allow for doubt in the reduction to spheroid level, which is 1 ppm for an error of 6 metres in the accepted separation of geoid and spheroid.

2.45. Accuracy of Laplace azimuths

The s.e. of the astronomical azimuth is unlikely to be less than that of an ordinary horizontal angle observed on the same number of zeros with the same instrument, but provided the transverse bubble correction is varied and kept small, § 4.42 (*c*), it should not be much greater except in high latitudes. An independent estimate is obtained if half the zeros are observed on a west RM and half on an east one, or if Laplace azimuths are sometimes observed at both ends of a single line. See [589] for a full account of an unexpected 2″ discrepancy between different observers in work of apparently high accuracy. In Great Britain the r.m.s. discrepancies between the members of 15 pairs of reciprocal Laplace azimuths, using Black's method with a Wild T4 theodolite was 1″.29, from which the s.e. of the mean of a pair is 0″.64, in fair agreement with the 0″.85 quoted in § 2.44 (*a*), as obtained from the method of § 2.28.

[110] deduced 0″.75 as the probable error, or 1″.1 as the s.e., of a single (one-way) Laplace azimuth in India, which suggests 0″.8 for the mean of a pair.

2.46. The accumulation of error in a chain of triangles.

(*a*) § 2.27 provides the means for assessing the accuracy of any framework, but it is desirable to be able to form some estimate without having to go to the length of inverting the matrix of the normal equations, and this paragraph provides an approximate alternative. It is primarily applicable to a system of chains with rather infrequent base and azimuth control, but it can deal with a network by the selection of chains of triangles forming part of it, and at the end adding a little weight to allow for such lines as may have been ignored.

(*b*) *Scale error in a single figure.* For any figure, defined as in § 1.03 (*c*), given e the s.e. of an observed angle in seconds, the s.e. of scale and azimuth of the terminal side relative to the opening side can be got by the method of § 2.27.

In geodetic work such figures are often nearly regular, and for regular figures the formula gives the following for the s.e. of the side of exit, [591], p. 199.

For a pair of equilateral triangles	$5.6e$ ppm	
For a regular braced quadilateral	$4.8e$	
For a regular centred quadrilateral	$5.6e$	
For a regular centred pentagon	$5.8e$	
For a regular centred hexagon	$6.2e$	
For a regular centred heptagon	$6.8e$.	(2.65)

For simplicity, centred quadrilaterals may be treated as a pair of triangles, other single-centred figures as hexagons, and double-centred figures like Fig. 1.2 (*d*) as $1\frac{1}{2}$ hexagons.

The most usual departure from regularity is elongation in the direction of the chain. If a braced quadrilateral takes the form of a rectangle of length l (along the chain) and breadth b, the s.e. of the exit scale is increased in the ratio l/b, and the effect of elongating simple isosceles triangles is the same. § D.18 (*a*) and (*b*). This convenient rule is therefore applied to all other figures with some confidence, although the result will be too small if there is serious asymmetry as well as elongation.

(*c*) *Azimuth error in a single figure.* Similarly, assuming all angles to be independently measured, the s.e. of azimuth in the side of exit of regular figures is as below.

For a pair of equilateral triangles (and for centred quadrilaterals)	$1''.16e$
For a regular braced quadrilateral	$1.00e$
For a regular hexagon, etc.	$1.29e$.

(2.66)

(*d*) *Scale and azimuth in a chain.* Consider the accumulation of error in 100 km of chain. In this 100 km let N_1 be the accumulated s.e. of scale, in parts per million, and let N_2 seconds be the s.e. of azimuth. Let the chain be of uniform accuracy as regards the s.e. of observed angles, and let there be f triangulation figures in the 100 km. Then

$$N_1 = ABe\sqrt{f}, \qquad \text{and} \qquad N_2 = Ce\sqrt{f}, \tag{2.67}$$

where A has such a value as 5.6 from (2.65), or a mean weighted according to the number of different types of figure,

B is a mean value of l/b for the figures concerned,

and C is a weighted mean from (2.66).

Then in a chain of length $100S$ km the s.e.'s are

$$\text{In scale } N_1\sqrt{S} \text{ ppm, and in azimuth } N_2\sqrt{S} \text{ seconds.} \tag{2.68}$$

And after several consecutive chains of varying quality the s.e.'s are

$$\text{In scale } \sqrt{\{\sum(N_1^2S)\}}, \text{ and in azimuth } \sqrt{\{\sum(N_2^2S)\}}. \tag{2.69}$$

If the opening side has a s.e. of s ppm, and of t'' in azimuth, the terminal s.e.'s are

$$\text{In scale } \sqrt{\{s^2+\sum(N_1^2S)\}}, \text{ and in azimuth } \sqrt{\{t^2+\sum(N_2^2S)\}}. \tag{2.70}$$

The probable (68:32) misclosures round an uncontrolled circuit are given by (2.69), and those between base or Laplace controls by (2.70) with $2s^2$ and $2t^2$ substituted for s^2 and t^2.

It has been useful to tabulate N_1 and N_2 for all the chains of a national triangulation, and also

$$N \equiv \sqrt{N_1^2+23.5N_2^2),\dagger} \tag{2.71}$$

which combines N_1 and N_2 into a single criterion. In primary triangulation N varies between about 2.5 and 10.0, and in secondary from about 10.0 up to 40 or more.

(*e*) *Error of position.* Let a straight chain start at O, with an errorless base and Laplace station, and proceed to A, distant $100S$ km. Let E_1 be the s.e. of position of A in the direction OA, and E_2 the component at right angles. See § 2.42 (*b*) for notation. Then, see [110], Appendix VII for proof,

$$E_1 = 0.058N_1S\sqrt{S} \qquad E_2 = 0.28N_2S\sqrt{S},$$

and the total s.e. of position

$$E = 0.058NS\sqrt{S} \text{ metres.} \tag{2.72}$$

† $\sqrt{(23.5)} = 10^6 \sin 1''$, connecting seconds of arc with parts per million.

If the base and Laplace azimuth have s.e.'s of s ppm and t'', the resulting errors of $0.100Ss$ and $0.48St$ metres must be combined with E_1 and E_2 respectively, to give

$$E_1 = 0.058S\sqrt{(N_1^2S+3s^2)}, \qquad E_2 = 0.28S\sqrt{(N_2^2S+3t^2)},$$

and
$$E = 0.058S\sqrt{(N^2S+3s^2+70t^2)} \text{ metres.} \tag{2.73}$$

(*f*) *Errors in a chain with base and Laplace control.* In a straight chain OA let there be errorless bases and Laplace azimuths at both O and A. Then

$$E_1 = 0.029N_1S\sqrt{S}, \qquad E_2 = 0.14N_2S\sqrt{S},$$

and
$$E = 0.029NS\sqrt{S} \text{ metres.} \tag{2.74}$$

The two controls have thus halved the s.e. At B, the middle point of OA, the s.e. is 0.56 of the s.e. at A, and 0.79 of what it would be if OB stood alone without connection to the controls at A.

(*g*) Given the s.e. of a chain between base and azimuth controls, chains can be combined in parallel or in series by the ordinary rules for the accumulation of random error.

Further details, with more explicit formulae, are given in [113], 3 Edn, p. 194–8. Proofs are in [110].

2.47. Errors in a traverse

The error E_1 to be expected in the length of a straight EDM traverse is given rather doubtfully in § 2.44 (*a*) and (*d*). The following (*a*)–(*d*) refer to E_2, the s.e. of position at right angles to the line.

(*a*) In a straight traverse based on an errorless Laplace station at one end, and none at the other

$$E_2 = 0.0028eL\sqrt{(L/l)} \text{ metres,} \tag{2.75†}$$

where L is the length of the traverse in kilometres, l km is the average length of each leg, and e'' is the s.e. of each traverse angle, reckoning one angle at each station as in § 2.43 (g).

(*b*) If the s.e. of the Laplace station is t'',

$$E_2 = 0.0028L\sqrt{(e^2L/l+3t^2)}. \tag{2.76}$$

(*c*) If the traverse closes on to another Laplace station, also with s.e. $= t''$,

$$E_2 = 0.0014L\sqrt{(e^2L/l+6t^2)}. \tag{2.77}$$

(*d*) If a number of consecutive traverses lie end to end between Laplace

† The formula loses accuracy if $L/l < 4$.

stations, the total E_2 is approximately the square root of the sum of the squares of the E_2's given by (2.77), with the proviso that if e^2L/l is small compared with $6t^2$, the resulting s.e. will be 1.22 times too small, since the error of each Laplace azimuth (except the two terminal ones) enters equally into the two adjacent traverse sections.

[101] gives details of the closing errors of 59 loops of tellurometer traverse, with a little triangulation included, in Australia. The average is 2.2 ppm over 1440 km. The closing error of a circuit is of course unaffected by a constant scale error.

See § 2.44 (*b*) for the closing errors of circuits of the US zero-order geodimeter traverses, which averaged 0.8 ppm over 4000 km.

2.48. Effect on accuracy of varying numbers of base and azimuth controls.

In a triangulation net all the angles have usually been observed, but it has not yet become usual to measure the length of every side in addition, and still less is it usual to observe a Laplace azimuth at every station. It is necessary to enquire what the most useful proportion of measured distances and azimuths may be.

Considering Block VI of the European triangulation, Great Britain and Ireland, [69] analyses the strength of the existing triangulation net on the assumption that varying numbers of distances and azimuths have been included. It is assumed that the s.e. of a direction is $\pm 0''.65$, of a distance (MDM) ± 2.5 ppm, and of a Laplace pair $\pm 0''.82$, as have been deduced on the lines of § 2.28. The analysis is then carried out for a number of measured distances (between 1 and the total possible 1150), and for a number of azimuths (between 1 and the total possible of 350 stations, or of 175 pairs).

The criterion by which the resulting strength of the net is judged is the final s.e. of an adjusted distance (σ_L), and the final s.e. of an adjusted azimuth (σ_A). Selected figures are quoted in Table 2.4.

TABLE 2.4

15 azimuths plus No. of distances	σ_L/ppm	179 distances plus No. of azimuths	σ_A/arc seconds
1	7.49	1	1.15
12	3.27	15	0.52
45	2.91	33	0.46
90	2.75	51	0.44
179	2.44	350	0.36
251	2.07		
575	1.76		
1150	1.28		

It is clear that (in this example) increasing the number of measured lines beyond about 250, or the number of double Laplace azimuths to beyond 15 or 30, is relatively unrewarding. It may be noted that 200 is 17 per cent of the total number of distances and that 30 azimuths is 17 per cent of the total possible number of independent pairs.

These conclusions will depend in detail on the relative weights of angles, distances and Laplace azimuths, and they do not automatically apply in all circumstances. In particular, complete EODM trilateration might be expected to give substantially better results.

2.49. Error ellipses

(*a*) § 2.27 describes how the s.e. of the relative position of any two points, in distance or in azimuth, can be obtained from the elements of N^{-1}, the inverse of the matrix of the normal equations, and § D.20 describes how an *error ellipse* may be computed for every point in the net to give its s.e. of position, in any direction, relative to the fixed origin. § D.21 further describes three different definitions of the *standard error in any direction* on which three different error ellipses (or other figure) may be based. Here we consider only σ_D as defined in § D.21 (*b*), and the properties of the resulting ellipse as given in § D.21 (*d*).

In a medium-sized or large net the matrix N^{-1}, which is the basis of all estimates of accuracy, is an indigestible document, and something else is required to give a quick illustration of the accuracy of a net. Possible solutions are in (*b*) to (*d*) below.

(*b*) *Conventional position error ellipses.* A chart may be prepared showing the error ellipses, relative to the origin, at all the stations of a net, or at a selection of them only. The ellipse itself need not be drawn. It suffices to draw its two principal axes. The defects of this system are as below.

(i) The sizes of the ellipses vary from zero at the origin, increasing in size with distance from it.

(ii) The ellipses are in no way invariant with respect to change in the (arbitrary) location of the origin. A central origin will produce smaller ellipses than one situated at one corner of the net.

(iii) A pair of adjacent points may have large error ellipses which will provide no clue to their relative error of position, which may be quite small.

(*c*) *Free adjustment. Inner errors.* [415] and [408] describe an ingenious method of avoiding items (*b*) (i) and (ii). Instead of accepting arbitrary (or other) coordinates for a fixed origin by omitting $\delta\phi$ and $\delta\lambda$ at one point, all $\delta\phi$'s and $\delta\lambda$'s are included as unknowns. The resulting normal matrix is then singular, but a solution is obtained by imposing a condition

$$x^{\mathrm{T}}x = \sum(x_1^2+x_2^2+\ldots) \text{ is to be minimum,} \qquad (2.78)†$$

where x is the vector of unknowns $\delta\phi$ and $\delta\lambda$ in the adjustment by variation of coordinates. In other words, the net is to be located in such a way as will minimize its departures from the trial positions adopted before the adjustment. These trial positions may have been quite arbitrary, but the point of the method is to assess the internal accuracy of the net, and it does not have to be used for obtaining the coordinates for practical use.

The resulting variance–covariance matrix has smaller and more uniform diagonal elements than in the conventional method, but the relative standard errors of position of pairs of points, as given by (2.42) are unchanged.

The advantages of this method are as below.

(i) There is no point at which the ellipses are of zero size.‡ Given equal quality of layout and observation, the error ellipses will everywhere be equal. Variations indicate a change of quality.

(ii) If the ellipses are in fact tolerably equal and circular, a single-figure mean value of all their axes constitutes a reasonable measure of the quality of the whole net.

Disadvantages are.

(iii) As in (*b*) (iii) the ellipses are no indication of the accuracy of the relative position of near points.

(iv) It is difficult to give a precise statement indicating what is the distance (or azimuth or other parameter) whose accuracy is given by the radius vectors of these ellipses.

(*d*) *Relative position error ellipses.* Instead of drawing ellipses to show the standard errors of position of each separate point, ellipses may be drawn to illustrate the s.e. of distance and azimuth‖ in each separate side, or for a selection of sides, as given by (2.42). This meets the objections (*b*) (iii) and (*c*) (iii), and illustrates local accuracies very well. On the other hand it gives no indication of over-all accuracy, since local errors of scale

† This condition is in fact the same as minimizing the trace (the sum of the diagonal elements) of the variance–covariance matrix, N^{-1}.

‡ At first sight it might be thought that zero ellipses would be found near the centroid of the trial points where the changes are to be minimized, but it is impossible to locate this point exactly, except by the fallible observations of the net itself. There may be a station where $\delta\phi$ or $\delta\lambda$, or even both, may fortuitously be near zero, but that point does not become an origin of coordinates, because its $\delta\phi$ and $\delta\lambda$ are only zero plus or minus a non-zero standard error.

‖ Both expressed in metres. $L\times\sigma_{\mathrm{A}}$, in the case of azimuth.

and azimuth are likely to be correlated to some extent which is not indicated. To give a full picture it is therefore necessary to include the error ellipses of a selection of longer lines.

As an illustration, consider the small net of Fig. 2.11, for which error ellipses might be shown for the following.†

(i) Short sides. 1–6, 5–10, 16–17, 19–20, 26–31, 30–35.
(ii) Medium distances. 1–18, 5–18, 18–31, 18–35.
(iii) Long distances. 1–35, 5–31.

These lines involve 13 out of the 35 stations, so only 37 per cent of the columns of the inverse need to be computed. In a larger matrix the proportion would be much less.

(*e*) *Conclusion.* Method (*d*) appears to provide the most generally useful statement of the accuracy of a net. See also [64] and [70]. It must in any case be recognized that the (probable) exclusion of covariances from the weight matrix is likely to lead to the underestimation of over-all errors.

SECTION 7. CHANGE OF SPHEROID

2.50. Changes in the spheroidal elements

If triangulation has been computed on a certain spheroid, positions relative to a new spheroid may presently be required, as when connection with adjacent countries or the establishment of a world datum may make it possible to put the triangulation into international terms. It would be possible to do this by recomputing the whole triangulation and its adjustment using new tables of ρ and ν, new deviations of the vertical, new Laplace azimuths and new reductions of bases to spheroid level, but this would be laborious and it can be done more easily by the methods described in §§ 2.51–2.53.

It is first necessary to decide on the differences between the elements defining the old and new spheroids, S_1 and S_2, as listed in § 2.02 (*c*).

(i) It may generally be assumed that the minor axes of the two spheroids are parallel, both being defined as in § 2.02 (*c*) (i) by the direction of the CIO, § 4.09.

(ii) It is also to be expected that both will have accepted the same Greenwich meridian plane as their zero of longitude, but there is the possibility that non-satisfaction of Laplace's equation at the origin may have caused the longitudes of S_1 to need a constant correction, § 2.06 (*b*) (iii). In which case the correction should be applied to all points which are to be considered.

† In Fig. 2.11 the side 1–2 need not (in the present context) be the only measured distance.

(iii) Let the semi major axis of S_2 minus that of S_1 be δa. If the two values of a are doubtful they can be ascertained from the ρ's and ν's used in the computations, but it is necessary to confirm that the same units of length have been used in both cases. See § 1.29.

(iv) δf, the flattening of S_2 minus that of S_1 presents no difficulty. It will be such a figure as, for example,

$$(1/298.247)-(1/297.000)=-0.1408\times 10^{-4}$$

for conversion from the 1924 International figure to the reference ellipsoid of 1967.

(v) It remains to find the separation of the centres of the two spheroids, for which it is necessary to know ϕ, λ and h (above spheroid) relative to the two spheroids at one or more common points. For further details see §§ 2.51 and 2.53.

Notation

ϕ, λ, H = latitude, longitude, and height above spheroid. Suffix 1 refers to S_1 and 2 to S_2. Additional suffix 0, e.g. ϕ_{01}, refers to the common point O.

h = height above geoid, so $H = N + h$.

N = height of geoid above spheroid, with the same suffixes.

δN = height of S_1 above S_2. So $h_2 - h_1 = \delta N$.

a, f = semi major axis and flattening.

$\delta\phi, \delta\lambda, \delta h = \phi_2-\phi_1, \lambda_2-\lambda_1, h_2-h_1$.

x, y, z = cartesian coordinates with origin at the centre of the spheroid. See § 2.51.

$\delta x, \delta y, \delta z$ = S_2 coordinates $-$ S_1 coordinates.

ξ, η = deviations of the vertical.

$\delta\xi = \xi_2 - \xi_1 = -\delta\phi$.

$\delta\eta = \eta_2 - \eta_1 = -\delta\lambda \cos\phi$.

ρ, ν = principal radii of curvature.

$e = (1/a)\sqrt{(a^2-b^2)}$.

$\epsilon = e^2/(1-e^2)$.

L, A = spheroidal distance and azimuth.

See Fig. 2.26, in which P is any ground-level point and P_1 and P_2 are the corresponding spheroid points.

2.51. Change of spheroid via cartesian coordinates

With the centre of spheroid S_1 as origin take coordinate axes x_1, y_1, z_1 with z_1 lying along the minor axis of the spheroid, x_1 parallel to the zero meridian plane, and y_1 to the east. And take parallel axes x_2, y_2, z_2 with origin at the centre of S_2.

At the common point P_0 convert $\phi_{01}, \lambda_{01}, H_{01}$ to x, y, z using

$$x = (\nu + N + h)\cos\phi \cos\lambda$$

$$y = (\nu + N + h)\cos\phi \sin\lambda$$

$$z = \{(1-e^2)\nu + N + h\}\sin\phi, \tag{2.79}$$

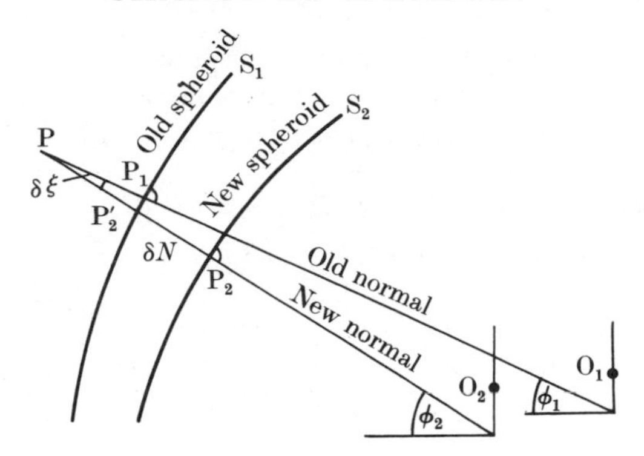

FIG. 2.26. Meridional section. $PP_1 = PP'_2 = h_1$. $P'_2P_2 = \delta N$ (drawn as positive, and much exaggerated). $\delta\phi = -\delta\xi = P'_2PP_1$ ($\delta\phi$ drawn as positive).

all with suffix 1 to give $x_1y_1z_1$. $N+h = H$. Then convert $\phi_{02}, \lambda_{02}, H_{02}$, using, $\nu_2, e_2, N_2, h_2, \phi_2$, and λ_2 to give x_2, y_2, z_2.

The differences $x_2 - x_1$, etc. give $\delta x, \delta y$, and δz, the separation of the two centres.

The ϕ, λ, H of any other point P are changed from S_1 to S_2 by converting their S_1 values to x_1, y_1, z_1, adding $\delta x, \delta y, \delta z$, and then converting them back to ϕ_2, λ_2, H_2, using

$$\tan\lambda = y/x \tag{2.80}$$

$$\tan\phi = \frac{z + \epsilon b \sin^3 u}{p - e^2 a \cos^3 u} \tag{2.81}$$

where $p = (x^2+y^2)^{\frac{1}{2}}$, a = major axis and b = minor axis of spheroid 2, and $\tan u = (z/p)(a/b)$. All of course with S_2 values. The expression for $\tan u$ is in principle only correct at spheroid level, but it involves an error of less than 0″.0000001 in ϕ_2, for any possible ground height H See [120].

Finally, $N+h = H = (x^2+y^2)^{\frac{1}{2}}(\sec\phi) - \nu$

$$\text{or} \qquad = z \operatorname{cosec}\phi - (1-e^2)\nu. \tag{2.82}$$

If there are several common points $\delta x, \delta y$, and δz can be got for each, and a suitably weighted mean accepted. But if the discrepancies are unacceptably large the only remedy is to readjust one or both of the two nets, so as to secure equal changes. The method of the adjustment will depend on the size of the discrepancies. It may be anything from a simple graphical adjustment as in § 2.41, to a complete combined adjustment on the lines of § 2.37.

The values of $\delta x, \delta y$, and δz required to convert many current national and continental datums to geocentric have now been obtained from satellite observations. See § 7.77, where they are described as $\Delta x, \Delta y$, and Δz.

This method involves no approximations, provided the computations are carried to a sufficient number of significant figures, such as 9-figures for 0.01 metres. It is the natural method to use with a computer.

2.52. Vincenty's method for desk computation

At the common poinf P_0 compute δx, etc., with 5 or 6 significant figures from

$$\delta x = (\delta N + \delta\nu)\cos\phi_0\cos\lambda_0 - (1+h/a) \times(\nu_0\delta\phi\sin\phi_0\cos\lambda_0 + \nu_0\delta\lambda\cos\phi_0\sin\lambda_0),$$

$$\delta y = (\delta N + \delta\nu)\cos\phi_0\sin\lambda_0 - (1+h/a) \times(\nu_0\delta\phi\sin\phi_0\sin\lambda_0 - \nu_0\delta\lambda\cos\phi_0\cos\lambda_0),\dagger$$

$$\delta z = \{\delta N + \delta\nu(1-e^2) - \nu_0\delta(e^2)\}\sin\phi_0 + \nu_0(1+h/a)(1-e^2)\delta\phi\cos\phi_0, \tag{2.83}$$

in which ϕ_0, λ_0, ν_0, h_0, and e have their S_1 values, and $\delta\nu$ is (ν in latitude ϕ_{02} on S_2) minus (ν in latitude ϕ_{01} on S_1).

Then at any other point

$$\delta\phi'' = (1/\nu)(1-h/a)\text{cosec }1''\{-\delta x\sin\phi\cos\lambda - \delta y\sin\phi\sin\lambda + \delta z(1+\epsilon)\cos\phi + \nu\delta(e^2)(1+\epsilon)\sin\phi\cos\phi - e^2\delta N\sin\phi\cos\phi\},$$

$$\delta\lambda'' = \{1/(\nu\cos\phi)\}(1-h/a)\text{cosec }1''(-\delta x\sin\lambda + \delta y\cos\lambda),$$

$$\delta N = (1-\epsilon\sin^2\phi)\{\delta x\cos\phi\cos\lambda + \delta y\cos\phi\sin\lambda + \delta z(1+\epsilon)\sin\phi + \nu(1+\epsilon)\delta(e^2)\sin^2\phi - \delta\nu\}. \tag{2.84}$$

As before, ϕ, λ, e, and ϵ have their S_1 values, and $\delta\nu$ must be ν for ϕ_2 on S_2 minus ν for ϕ_1 on S_1, but elsewhere ν can be taken to the nearest 100 metres from tables for either spheroid at the nearest 10′ of ϕ. In (2.84) δx, etc. should be as given by (2.83), in preference to using (2.79) as some compensation of small errors results.

Provided changes of δx, δy, and δz are less than about 300 m, as is usual, 5-figures will give a computational accuracy of 0.01 metres. There are approximations in the formulae which will amount to more than this at distances of 1000 km from P_0, but they will affect neighbouring points equally and are of no consequence. The terms $(1-h/a)$ are often very close to unity, but their omissions introduce errors which are not so closely the same at neighbouring points. [580] gives (2.84) in a rather more elaborate form, avoiding the use of tables for ν. It also gives smaller terms for higher accuracy, if required.

† The term $\nu_0\delta\lambda\cos\phi_0\cos\lambda_0$ was wrongly given with a + sign in [113] 3 Edn.

[519] and [419] give practically the same formulae in different form. If Δx, Δy, and Δz are known, as in § 7.77, changes in N, the height of the geoid above the spheroid, are adequately given by

$$\Delta N = \Delta x \cos\phi \cos\lambda + \Delta y \cos\phi \sin\lambda + \Delta z \sin\phi - \Delta a + a\Delta f \sin^2\phi. \tag{2.85}$$

which is substantially the same as the last line of (2.84).

2.53. Approximate formulae for slide–rule

The following formulae are for computing δN, $\delta\xi$, $\delta\eta$ by slide-rule to within 1 metre and 0″.1 or 0″.2. They are derived from [232], p. 139, or from [577], which are quoted in [113], 2 Edn. pp., 160–1. They ignore products of δN, $\delta\xi$, and $\delta\eta$ with f and with h/a. They are adequate for studies of the deviation of the vertical and of the form of the geoid, but not for the conversion of geodetic ϕ's and λ's.

First compute the following, of which the first three are constant for any particular pair of S_1 and S_2, while the last two also involve ϕ and λ.

$$\begin{aligned}
\delta\beta &= \delta a/a + \delta f \sin^2\phi_0,\\
A &= \delta\xi_0'' \cos\phi_0 - (\delta\beta + \delta N_0/a)\sin\phi_0 \operatorname{cosec} 1'',\\
C &= \delta\xi_0'' \sin\phi_0 + (\delta\beta + \delta N_0/a)\cos\phi_0 \operatorname{cosec} 1'',\\
B &= C\cos(\lambda-\lambda_0) - \delta\eta_0'' \sin(\lambda-\lambda_0),\\
D &= \sin\phi - \sin\phi_0.
\end{aligned}$$

Then

$$\begin{aligned}
\delta N &= a(B\cos\phi - A\sin\phi)\sin 1'' - \alpha\delta\beta + aD^2\delta f\\
&= k_1 + k_2\cos\phi\cos\lambda + k_3\sin\phi + k_4\sin^2\phi + k_5\cos\phi\sin\lambda,\\
\delta\xi'' &= (A - 2\delta f D \operatorname{cosec} 1'')\cos\phi + B\sin\phi\\
&= k_6\sin\phi\cos\lambda + k_7\sin\phi\sin\lambda + k_8\cos\phi + k_9\sin 2\phi,\\
\delta\eta'' &= C\sin(\lambda-\lambda_0) + \delta\eta_0''\cos(\lambda-\lambda_0)\\
&= k_{10}\sin\lambda + k_{11}\cos\lambda,
\end{aligned} \tag{2.86}$$

where the k's can be computed from $\delta\beta$, A, C, a, λ_0, and ϕ_0, and are constant for all values of ϕ and λ.

Conversely, if N_1 above a certain spheroid is given at a number of points, the changes δa, δf, δN_0, $\delta\xi_0$, and $\delta\eta_0$ to find another spheroid which will give minimum values of N_2 at these points, may be got by least squares. For forming the observation equations, (2.86) is more conveniently put into the form

$$-\delta N = P(U\sin\phi_0 + V\cos\phi_0) + Q\delta\eta_0 + R(V\sin\phi_0 - U\cos\phi_0) + \delta a - T\delta f, \tag{2.87}$$

where $P = \cos\phi \cos(\lambda - \lambda_0)$, $Q = a \cos\phi \sin(\lambda - \lambda_0)$, $R = \sin\phi$,

$$T = a \sin^2\phi,\ U = -a\delta\xi_0 - a\delta f \sin 2\phi_0,\ V = -\delta N_0 - \delta a + a\delta f \sin^2\phi_0,$$

$\delta\xi_0$ and $\delta\eta_0$ are in radians.

2.54. Changes in scale and azimuth

(*a*) *Scale.* If the ϕ's and λ's of two near points P and Q are converted from S_1 to S_2, the distance between them on S_2 will be approximately given by

$$L_2 = L_1(1 - \delta N/a).$$

More accurately,

$$L_2 = L_1(1 - \delta N/a) - (h_P - h_Q)(\delta\xi'' \cos A_{PQ} + \delta\eta'' \sin A_{PQ})\sin 1''. \tag{2.88}$$

These formulae can be used over distances such that mean values of δN, $\delta\xi$, and $\delta\eta$ are applicable.

For fully accurate results over long lines convert ϕ and λ to S_2 and compute L_2 by a formula from § 2.15.

(*b*) *Azimuth.* If A is the azimuth of Q at P,

approximately $$A_2 = A_1 + \delta\lambda \sin\phi.$$

More accurately

$$\delta A = \delta\lambda \sin\phi - (\delta\xi \sin A - \delta\eta \cos A)\tan\beta, \tag{2.89}$$

where β is the elevation of Q above the tangent at P. For short lines, such as 50 km or even more, $\tan\beta$ in this small term can be taken as $(h_Q - h_P)/L$. In (2.89) A, $\delta\lambda$, ϕ, $\delta\xi$, and $\delta\eta$ have their values at P.

For fully accurate results convert ϕ, λ to S_2 by § 2.51, and use Cunningham's (2.23).

SECTION 8. COMPUTATION IN PLANE RECTANGULAR COORDINATES

2.55. Definitions and notation

A *projection* is defined as any orderly system whereby the meridians and parallels of the spheroid† may be represented on a plane, and every surveyor knows the fundamental difficulty that except over an elementary area this cannot be done without some distortion or change of scale. As is also well known, certain projections are *orthomorphic*, by which is implied

† Or of a sphere for cartographical purposes. Here we only consider the spheroid.

that at any point the scale, whatever it may be, is the same in all directions so that there is no local distortion, although over large areas there is distortion due to the change of scale from place to place.

On the plane let there also be a system of rectangular coordinates N and E, which is often referred to as a *grid*. Then on the plane the *grid bearing*† of p_2 from p_1 is given by

$$\left.\begin{aligned} \tan\beta &= (E_2-E_1)\div(N_2-N_1) \\ \text{and} \quad p_1p_2 = l &= (N_2-N_1)\sec\beta = (E_2-E_1)\operatorname{cosec}\beta \\ \text{or} \quad N_2-N_1 &= l\cos\beta \quad \text{and} \quad E_2-E_1 = l\sin\beta \end{aligned}\right\}. \tag{2.90}$$

The study of projections from the cartographical point of view is not here considered as a branch of geodesy, but (especially where small areas are concerned) the simplicity of (2.90) often makes it convenient to compute and record triangulation and traverse directly in terms of rectangular coordinates. This is generally a topographical or military matter, since the permanent record of geodetic work is best kept in geographical coordinates and it might as well be computed in them, but it is one on which the local geodesist will be expected to advise.

A particular projection may be defined in various ways. (*a*) It may be, but generally is not, definable as the result of perspective projection of the meridians and parallels of the spheroid from a specified point on to a specified plane. Or (*b*), more usually, it is defined by a set of rules by which it can be geometrically constructed by ruler, compass, and scale. Whatever the definition, it must be possible to deduce from it a pair of formulae

$$N = f_1(\phi, \lambda) \quad \text{and} \quad E = f_2(\phi, \lambda), \tag{2.91}$$

whereby a point of given ϕ and λ can be plotted on the plane by its rectangular coordinates, and once these formulae are brought into use they constitute the effective definition of the projection.

The notation used in this section is as follows.

ϕ, λ = Latitude and longitude. Positive north and east.
$\zeta = 90° - \phi$, but not in § 2.62 (*a*).
α = Azimuth on spheroid. Clockwise from north.
β = Bearing on projection of straight line pq. Clockwise from N-axis.
L = Distance PQ on spheroid.
l = Straight line distance pq on projection.

† Geodesics and lines of normal section on the spheroid are represented by curved arcs on the projection. The words *grid bearing* and *grid distance* here refer to the bearing and length of the chords of such arcs, straight lines on the plane. Elsewhere, [46], these are described as *plane bearings* and *plane distances*, while the words grid bearing and grid distance have been used for the bearing of the end tangents to the arc and for the length of the projected curve, to which we here to find no need to give names.

N, E = Plane coordinates. N roughly north, and E (positive) roughly east.
γ = Convergence, clockwise from meridian to N-axis. $\beta \approx \alpha - \gamma$.
p = Length of perpendicular from P on a surface of revolution to the axis of rotation.
M = Meridian distance from origin on spheroid or other surface of revolution, positive north.
m = Scale of projection $= l/L$ when both are small.
δ = 'Arc to chord' correction. Sign defined by $\beta = \alpha - \gamma + \delta$.
σ = Radius of curvature of the plane projection of a geodesic.
t, n = Distances along, and normal to, a curve.
R = Radius of a sphere, except in § 2.62 (*c*) and (*d*).
a = Semi-major axis of spheroid.
ρ, ν = Principal radii of curvature of spheroid.
e = Eccentricity of spheroid. $\epsilon = e^2/(1-e^2) \approx 1/150$.
$r_0 = m_0\nu_0 \cot \phi_0$, on Lambert projection.
s = Modified meridian distance on Lambert projection.
$\mu = \log_{10} e = 0.4342945$.
ψ = Isometric latitude. $d\psi = dM/p$. (N.b. Not dM/ρ.)
$\omega = \lambda - \lambda_0$.

Suffix 0 refers to the origin of the projection, centrally situated in the area to be covered. Suffix m refers to the mean latitude of a line.

2.56. Convergence

At any point the *convergence* is the angle between the meridian as represented on the plane and the N grid line. It is positive when grid north lies east of true north, so $\beta = \alpha - \gamma$, but see § 2.59.

In (2.91) putting λ constant for any particular meridian gives the plane equation of the meridian in terms of the one variable ϕ, and thence

$$\tan \gamma = -\frac{\partial E}{\partial \phi} \Big/ \frac{\partial N}{\partial \phi}, \tag{2.92}$$

although, as in the Lambert projection, the value of γ is sometimes got more simply from the original definition.

2.57. Scale †

At any point the scale in meridian, i.e. the ratio of a short meridional distance on the plane to the distance between the corresponding points on the spheroid, is given by

$$m = \frac{1}{\rho} \sec \gamma \frac{\partial N}{\partial \phi},$$

and in parallel

$$m = \frac{1}{\nu} \sec \phi \sec \gamma' \frac{\partial E}{\partial \lambda}, \tag{2.93}$$

† When a projection is used cartographically, the scale m will be such a fraction as 1/100 000, but for the present purpose a projection is generally so defined by (2.91) that m departs from unity only by such amounts as its inevitable inconstancy necessitates.

where γ' is the angle between the parallel and the $E-W$ grid line. $\gamma' = \gamma$ if the projection is orthomorphic.

If the projection is orthomorphic both of (2.93) will be equal, and either then gives the local scale of the projection. For computing directly on the plane only orthomorphic projections can usefully be used, and henceforward m represents the local scale, identical in all directions, and obtained either from (2.93) or perhaps more simply from the original definition.†

2.58. Computation in rectangular coordinates

Let P_1 and P_2 be two points on the spheroid, and let p_1 and p_2 be the corresponding points on the plane. Let ϕ_1 and λ_1 (of P_1) be given, whence (2.91) gives N_1 and E_1. Let the spheroidal distance P_1P_2 $(=L)$ be given and also the azimuth α, and let it be required to compute N_2 and E_2. There are two ways of proceeding.

(*a*) Compute ϕ_2 and λ_2 as in Chapter 2, Sections 2 and 3, and then get N_2 and E_2 from (2.91). This is the natural method, but the present object is to explain a shorter one.

(*b*) Get l $(=p_1p_2)$ from L and the scale of the projection: get β the bearing of p_2 from p_1, by applying to α the convergence and also a small correction δ, for which see § 2.59 (*b*): and then compute N_2 and E_2 from (2.90). Provided P_1P_2 is not too large nor too remote from the central area of the projection, l and δ can often be got very easily, or can even be equated to L and zero for low-grade work, in which case the simplicity of (2.90) secures a great saving of labour.

When a projection is being designed for a certain area, an excellent check of the formulae and tables which are to be used is to compute a number of test lines by both these methods.

2.59. Scale and bearing over finite distances

(*a*) *Scale.* If l is small we have $l = mL$ while over a finite distance

$$l = \int_{P_1}^{P_2} m\,\mathrm{d}L \quad \text{or} \quad L = \int_{p_1}^{p_2} \mathrm{d}l/m, \tag{2.94}$$

where L is distance from P_1 measured along P_1P_2, and l is along p_1p_2.

(*b*) *Bearing.* If L is small $\beta_{12} = \alpha_{12} - \gamma_1$, but over a finite distance this does not exactly hold, since in the presence of varying scale the geodesic‡ (shortest line on the spheroid) does not project into a straight line

† The non-orthomorphic Cassini projection has been widely used for small areas or work of low accuracy, but it is not recommended.

‡ In the context of this section the distinction between the geodesic and the normal section is immaterial.

(shortest on the plane), but curves away somewhat so as to pass through a region of greater scale, as follows.†

See Fig. 2.27. Let ab $(=l')$ on the plane represent a short section of a geodesic of true spheroidal length L. Let the radius of curvature of ab be σ, and let the angles cab and cba be $\frac{1}{2}\theta$, ac and bc being tangents to ab. At

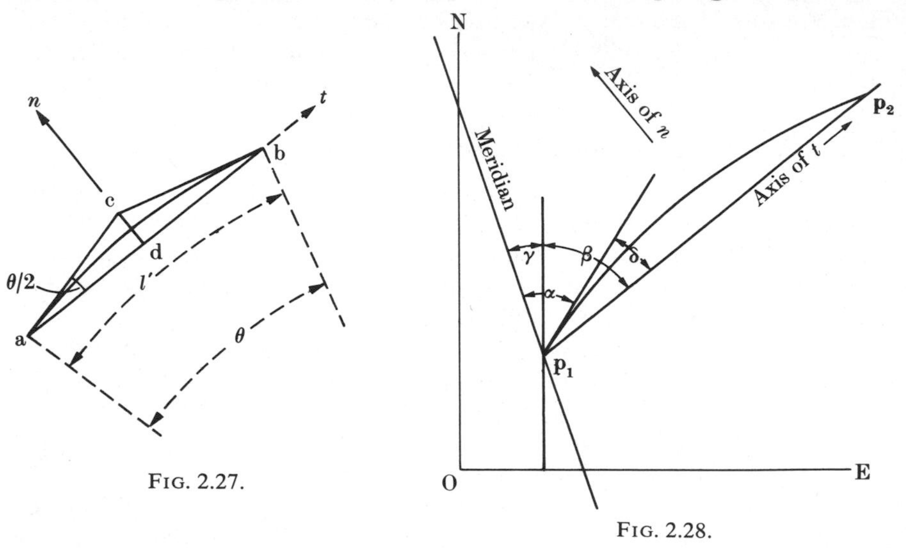

FIG. 2.27.

FIG. 2.28.

a and b, near points, let the scale be m, a variable. Let the gradient of m across ab, or $\partial m/\partial n$ where n is measured at right angles to ab, be locally equal to km, so that at a point distant n to one side of ab the scale is

$$m(1+kn), \tag{2.95}$$

n and $\partial m/\partial n$ both being positive in the same direction. k is not a constant.

On the plane,

$$\text{arc ab minus chord ab} = l'-l = \sigma\theta - 2\sigma \sin\theta/2 = \sigma\theta^3/24.$$

Then using (2.95) in conjunction with this and the fact that the mean value of n over the arc ab is $\frac{1}{3}$cd or $\frac{1}{12}\sigma\theta^2$, we have

$$\text{AB on the spheroid} = L = \frac{l}{m(1+k\sigma\theta^2/12)} + \frac{\sigma\theta^3}{24m}$$

$$= \frac{l}{m}(1-\tfrac{1}{12}kl\theta) + \frac{l\theta^2}{24m}, \text{ since } \sigma = l/\theta.$$

† A well-known example is that of 'great circle sailing' across the Atlantic, where the great circle, or geodesic, or shortest route, curves away to the north of the straight line on Mercator's projection. Note the resemblance between this problem and that of refraction, § 3.17 where a ray of light follows a path of minimum 'optical length'.

For this to be a minimum $dL/d\theta = 0 = -\frac{1}{2}kl^2/m + \frac{1}{12}l\theta/m$, so

$$\theta = kl, \quad \text{and} \quad 1/\sigma = k = \frac{1}{m}\frac{dm}{dn} = \frac{dm}{dn} \text{ if } m \approx 1. \tag{2.96}$$

Then over a finite distance p_1p_2, Fig. 2.28, we have $\beta_{12} = \alpha_{12} - \gamma_1 + \delta_{12}$, where δ is known as the *arc-to-chord* or *angle correction*,† and is given by

$$\delta'' = \frac{\operatorname{cosec} 1''}{l}\int^l \frac{l-t}{\sigma}dt, \; t \text{ being in the direction } p_1p_2. \tag{2.97}$$

And if k varies linearly along p_1p_2 so that $1/\sigma = k_1(1+ct)$, substitution in (2.97) gives

$$\delta'' = \frac{l}{2\sigma_3}\operatorname{cosec} 1'', \tag{2.98}$$

where σ_3 is the value of σ at the point on p_1p_2 distant $l/3$ from p_1. For many purposes this is an adequate approximation.

The correction required to convert a spheroidal angle to the corresponding plane angle is of course the difference of the δ's of its two arms. The formulae here given for scale and bearing over finite distances have been derived with reference to geodesic azimuths and distances on the spheroid, rather than to normal sections, since the geodesic is the shortest line and (2.96) is derived by minimizing L. In practice, however, when L is less than 100 km, as is most likely to be the case in rectangular computations, the difference between geodesic and normal section azimuths is less than 0".03, (2.10), and the formulae may consequently be used with normal section azimuths. The difference between geodesic and normal section distances is even more negligible, (2.13).

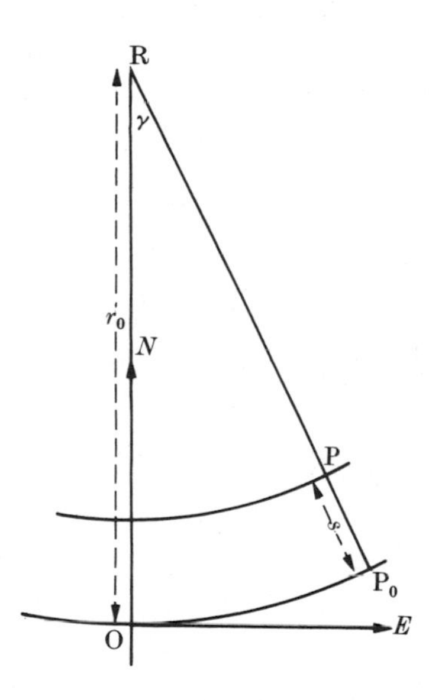

FIG. 2.29. Lambert projection.

§§ 2.60–2.63 give, without proof, the form of (2.91) to (2.94) and (2.97) for the four projections most often used.

2.60. Lambert's conical orthomorphic projection

(*a*) *Definition* (Fig. 2.29). Draw RO to represent a central meridian. With centre R and radius $r_0 = m_0\nu_0 \cot\phi_0$ draw an arc through O to

† Sometimes called the '$t-T$' correction.

represent a central parallel. Then O is known as the *origin.* All other meridians are straight lines through R making the angle

$$\text{PRO} = (\lambda_p - \lambda_0)\sin \phi_0.$$

All other parallels are arcs with centre R and radius $r_0 - s$, where s is the spheroidal meridian distance PP_0 so modified that the scale in meridian is everywhere equal to the slightly variable scale in parallel, thereby securing orthomorphism. At the origin the scale m_0, such a figure as 0.999, is generally so chosen that the scale over the area concerned varies between m_0 and $2 - m_0$.†

Then

$$\frac{s}{m_0} = M + \frac{M^3}{6\rho_0\nu_0} + \frac{M^4 \tan\phi_0(1-4\epsilon\cos^2\phi_0)}{24\rho_0\nu_0^2} + \\ + \frac{M^5(5+3\tan^2\phi_0 - 3\epsilon - \epsilon\cos^2\phi_0)}{120\rho_0^2\nu_0^2} + \frac{M^6\tan\phi_0(7+4\tan^2\phi_0)}{240\rho_0^2\nu_0^3} + \\ + \frac{M^7(60\tan^4\phi_0 + 180\tan^2\phi_0 + 61)}{5040\rho_0^3\nu_0^3} + \ldots, \quad (2.99)$$

where M is the length of the spheroidal meridian PP_0, and

$$\epsilon = e^2/(1-e^2).$$

[558], pp. 116–20 gives a proof with terms up to M^6. For any one projection, i.e. for any values of ϕ_0 and m_0, s can be tabulated against ϕ. The projection is orthomorphic. The scale is constant along any parallel, but increases to the north or south of ϕ_0. It is suitable for areas of large extent from east to west, but of limited extent from north to south.

(*b*) *Conversion.* Formulae (2.91) take the simple form

$$E = (r_0 - s)\sin\gamma, \quad \text{and} \quad N = s + E\tan\tfrac{1}{2}\gamma. \quad \gamma \text{ from (2.102) below.} \quad (2.100)$$

For the reverse process

$$\tan\gamma = E/(r_0 - N),\ \lambda - \lambda_0 = \gamma \operatorname{cosec}\phi_0,\ s = N - E\tan\tfrac{1}{2}\gamma. \quad (2.101)$$

Hence ϕ from the table.

In all conversions from spherical to rectangular coordinates, the formulae give N and E with reference to O, the defined *true origin* of the projection. As a convenience, some such figure as 1 000 000 m is then added to make all values of N and E positive. The point where these corrected values are zero is known as the *false origin*, a point of no special significance.

† If $m_0 = 1$ it is the projection known as the 'Conical orthomorphic with one standard parallel'. Otherwise it has two standard parallels.

(*c*) *Convergence.* From the definition, γ is constant along any meridian, and

$$\gamma = (\lambda - \lambda_0)\sin\phi_0. \tag{2.102}$$

(*d*) *Point scale.* The scale in any latitude is clearly ds/dM, which gives

$$m = m_0\left\{1 + \frac{M^2}{2\rho_0\nu_0} + \frac{M^3\tan\phi_0}{6\rho_0\nu_0^2} + \frac{M^4(5+3\tan^2\phi_0)}{24\rho_0^2\nu_0^2} + \ldots\right\}. \tag{2.103}$$

(*e*) *Finite distance. Scale.* A highly accurate formula for average scale over a long distance is complicated on this projection, since lines of constant scale are parallels, not approximately lines of constant N or E, and although (2.94) is simple in form it is not possible to express m simply in terms of either L or l. [156], p. 252,† gives

$$\begin{aligned}\log_{10}l = \log_{10}m_0L &+ \frac{\mu}{2\rho_0\nu_0}\left(\frac{M_1+M_2}{2}\right)^2 + \frac{\mu}{6\rho_0\nu_0}\left(\frac{M_1-M_2}{2}\right)^2 + \\ &+ \frac{\mu\tan\phi_0}{12\rho_0\nu_0^2}M_1M_2(M_1+M_2) + \frac{\mu\tan\phi_0}{24\rho_0\nu_0^2}L^2(M_1+M_2) - \\ &- \frac{2\mu\epsilon}{3}\frac{\tan\phi_0\cos^2\phi_0}{\rho_0\nu_0^2}M_1^3 + \frac{\mu(5+3\tan^2\phi_0)}{24\rho_0\nu_0^3}M_1^3(2M_2-M_1) + \ldots,\end{aligned} \tag{2.104}$$

where $\mu = 0.4342945$ is the modulus of common logs, and M_1 and M_2 are the spheroidal meridian distances of P_1 and P_2 from ϕ_0. Except in high latitudes, only the first and second terms are of topographical significance (>0.00002), provided $L < 160$ km and $\frac{1}{2}(M_1+M_2) < 450$. but all (or more) are needed for geodetic accuracy. [296], p. 161, gives another formula.

(*f*) *Finite distance. Direction.* The same difficulty occurs in using (2.97) to give an accurate formula for δ. For topographical accuracy with ϕ and λ taken from any reasonable map, [296], p. 158 gives‡

$$\delta_{12} = \tfrac{1}{2}(\sin\phi_3 - \sin\phi_0)(\lambda_2'' - \lambda_1''), \tag{2.105}$$

where ϕ_3 is $\frac{1}{3}(2\phi_1+\phi_2)$.

An accurate formula, but one that is difficult to apply without accurate knowledge of ϕ, α, and λ, is ([296], p. 158)

$$\delta_{12} = \frac{l\sin\alpha_3}{2m_3\nu_3\cos\phi_3}(\sin\phi_3 - \sin\phi_0)\operatorname{cosec}1''. \tag{2.106}$$

The following expression from [296], p. 160, in terms of rectangular coordinates gives geodetic accuracy, and with some tabulation is easy to

† In the last term [156] gives $(2+3\tan^2\phi_0)$, which is a misprint.

‡ In [296] (24.2), where ω is positive for west longitude, the sign of δ (here $-\delta$) should be +.

use, but while the Lambert projection is suitable enough for the direct computation of topographical triangulation, trouble will generally be saved by computing geodetic triangulation on the spheroid, and then converting to rectangular if required.

$$\delta_{12} = \frac{N_2E_1 - N_1E_2 + r_0(E_2 - E_1)}{2r^2 \sin 1''} \times \left\{A\left(\ln\frac{r}{r_0}\right) + B\left(\ln\frac{r}{r_0}\right)^2 + C\left(\ln\frac{r}{r_0}\right)^3 + D\left(\ln\frac{r}{r_0}\right)^4\right\}, \quad (2.107)$$

where $r^2 = (r_0 - N_3)^2 + E_3^2$, suffix 3 being for $\frac{1}{3}$ of the distance along p_1p_2,

$$A = -\cot^2\phi_0(1 + \epsilon\cos^2\phi_0),$$

$$B = -\cot^2\phi_0(1 + 3\epsilon\cos^2\phi_0 + 2\epsilon^2\cos^4\phi_0),$$

$$C = \tfrac{1}{3}\cot^4\phi_0(1 - 2\tan^2\phi_0 + 4\epsilon\cos^2\phi_0 - 14\epsilon\sin^2\phi_0),$$

$$D = \tfrac{1}{3}\cot^4\phi_0(2 - \tan^2\phi_0 + 15\epsilon\cos^2\phi_0 - 15\epsilon\sin^2\phi_0).$$

2.61. Mercator's projection

This is a Lambert conical projection orthomorphic projection in which the origin is on the equator, so $\phi_0 = 0$. Formulae (2.91) and (2.100) can be put into the form, [183], p. 115,

$$\begin{aligned}
N = m_0 a\psi &= m_0 a\left\{\tfrac{1}{2}\ln\frac{1+\sin\phi}{1-\sin\phi} - \tfrac{1}{2}e\ln\frac{1+e\sin\phi}{1-e\sin\phi}\right\} \\
&= m_0 a\{\ln\tan(\pi/4 + \phi/2) - e^2\sin\phi - (e^4/3)\sin^3\phi - (e^6/5)\sin^5\phi - \ldots\} \\
&= m_0 a\{\cosh^{-1}(\sec\phi) - e^2\sin\phi - (e^4/3)\sin^3\phi - (e^6/5)\sin^5\phi - \ldots\} \\
&= m_0 a\{\tanh^{-1}(\sin\phi) - e\tanh^{-1}(e\sin\phi)\}, \\
E &= m_0 a(\lambda'' - \lambda_0'')\sin 1'', \qquad (2.108)
\end{aligned}$$

where e is the eccentricity, logs are natural, and ψ is known as the *isometric latitude* defined by the equation

$$d\psi = dM/p, \text{ where } p = \nu\cos\phi, \quad (2.109)$$

p being the perpendicular from a point ϕ on to the axis. To prove (2.108) we have $dM = \rho\, d\phi$ and $\rho/\nu = (1-e^2)/(1-e^2\sin^2\phi)$ from (A.61) and (A.63). Whence

$$\psi = \int\frac{\rho\, d\phi}{\nu\cos\phi} = \int\frac{d\phi}{\cos\phi} - \int\frac{e^2\cos\phi\, d\phi}{1-e^2\sin^2\phi} \quad (2.110)$$

from which the first line of (2.108) follows.

The convergence is everywhere zero, and the local scale takes the form

$$m = m_0 a/(\nu\cos\phi). \quad (2.111)$$

Except near the equator, as a special case of Lambert, this projection is of no general use to surveyors, but it is the basis of other more useful projections, see § 2.62.

For the Mercator projection of the sphere $e=0$.

2.62. Transverse Mercator projection

(*a*) The *definition* of the transverse Mercator projection of the sphere (not spheroid) is easily visualized as follows. The projection is intended to cover an area of considerable extent in meridian, but little in parallel. Take a pole of coordinates Q on the equator in longitude 90° greater than that of the central meridian, ON in Fig. 2.30, of the area to be projected, use Mercator's projection to plot the resulting 'meridians' and 'parallels' on the plane, and then plot the true geographical meridians and parallels by reference to them, as below. The projection will be orthomorphic, of constant scale along the central meridian, and along small circles parallel to it, but the scale will increase with distance from the central meridian. The projection is a very useful one, and tables once made for any central meridian are applicable to any other.

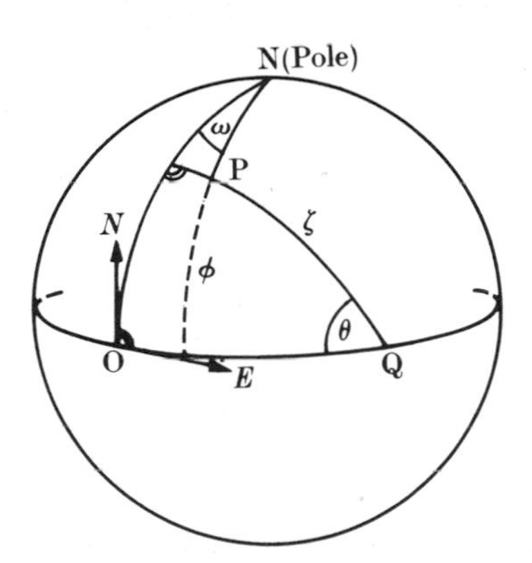

FIG. 2.30. Transverse Mercator projection. OQ = 90°.

The coordinates of any point P, Fig. 2.30, are given by

$$\cos\zeta = \cos\phi \sin\omega,$$

$$\sin\theta = \sin\phi \operatorname{cosec}\zeta,$$

$$N = m_0 R\theta \quad (\theta \text{ in radians}),$$

$$\begin{aligned} E &= m_0 R \ln \tan\{\tfrac{1}{4}\pi + \tfrac{1}{2}(\tfrac{1}{2}\pi - \zeta)\} \\ &= m_0 R \ln \cot \tfrac{1}{2}\zeta \\ &= m_0 R \tanh^{-1}(\cos\zeta), \end{aligned} \qquad (2.112)$$

where $\phi=\phi_P$, λ_0 is the longitude of thc projection's central meridian, $\omega=\lambda_P-\lambda_0$, R = earth's radius, ζ is the arc PQ, θ = OQP, and m_0 is the scale on the central meridian.

The scale error at a point is given by

$$m = m_0 \sec(90°-\zeta) = m_0 \operatorname{cosec}\zeta = m_0 \cosh(E/m_0R) \qquad (2.113)$$

since $\cos\zeta=\tanh(E/m_0R)$ from (2.112), whence $\sin\zeta=\operatorname{sech}(E/m_0R)$ from the properties of hyperbolic functions.

On the spheroid the strict definition is less clear, but it suffices to say that the projection is to be orthomorphic, and that the N-axis is to represent a meridian and to be of constant (minimum) scale. From this specification it is possible to deduce the conversion formulae, [374], p. 144, and [296], No. 63, p. 31†

$$N/m_0 = M + \tfrac{1}{2}(\lambda'' - \lambda_0'')^2 \nu \cos\phi \sin\phi \sin^2 1'' + \tfrac{1}{24} H(\lambda'' - \lambda_0'')^4 \nu \cos\phi \sin^4 1'',$$

$$E/(m_0 \nu \cos\phi) = (\lambda'' - \lambda_0'') \sin 1'' + \tfrac{1}{6} G(\lambda'' - \lambda_0'')^3 \sin^3 1'' + \tfrac{1}{120} J(\lambda'' - \lambda_0'')^5 \sin^5 1'', \quad (2.114)$$

where $G = \cos^2\phi(1 - \tan^2\phi + \epsilon \cos^2\phi)$,

$$H = \sin\phi \cos^2\phi(5 - \tan^2\phi + 9\epsilon \cos^2\phi + 4\epsilon^2 \cos^4\phi),$$

$$J = \cos^4\phi(5 - 18 \tan^2\phi + \tan^4\phi + 14\epsilon \cos^2\phi + 58\epsilon \sin^2\phi),$$

where M is the length of the spheroidal meridian from the equator to latitude ϕ, (A.68), less any constant which it may be convenient to subtract, and $\epsilon = e^2/(1-e^2)$. G, H, J, and K (2.115) can, if desired, be tabulated for different values of ϕ.

Reverse formulae for ϕ and λ, given N and E, are in [374], p. 149, and [296], p. 33.

(*b*) *Convergence and scale* at a point are given by

$$\frac{m}{m_0} \cos\gamma = 1 + \tfrac{1}{2} G(\lambda'' - \lambda_0'')^2 \sin^2 1'' + \tfrac{1}{24} J(\lambda'' - \lambda_0'')^4 \sin^4 1'',$$

$$\frac{m}{m_0} \sin\gamma = (\lambda'' - \lambda_0'') \sin\phi \sin 1'' + \tfrac{1}{6} H(\lambda'' - \lambda_0'')^3 \sin^3 1'' + \tfrac{1}{120} K(\lambda'' - \lambda_0'')^5 \sin^5 1'', \quad (2.115)$$

where G, H, and J are as in (2.114), and

$$K = \sin\phi \cos^4\phi(61 - 58 \tan^2\phi + \tan^4\phi + 270\epsilon \cos^2\phi - 330\epsilon \sin^2\phi).$$

Long formulae for N, E, m, and γ in terms of $(\lambda - \lambda_0)$ (in radians) and E as far as the 7th of 8th powers, for both spherical to rectangular and the reverse, are given in [479] and are reproduced in [44] and [113], 3 Edn. They are correct to 1 mm at a distance of 3^0 from the central meridian, and can be used as a standard by which approximate formulae can be judged.

(*c*) *Finite distance. Scale.* The scale at a point can be approximately expressed in rectangular coordinates as

$$m = m_0\left(1 + \frac{E^2}{2R^2} + \frac{E^4}{24R^4}\right). \quad (2.116)$$

where $R^2 = \rho\nu m_0^2$. This is correct to 1 in 10^7 if $E < 450$ km.

† There are two essentially different methods of projecting the spheroid, which are analogous to the projection of the sphere described above. The method here, and generally, adopted is the *Gauss–Kruger* projection. In the other [296], pp. 67–8, which has no widely recognized name, the scale along the central meridian is slightly variable. See [562].

This enters comfortably into (2.94), and with some simplification of the E^4 term [98] gives

$$m = \frac{l}{L} = m_0\left\{1+\frac{E_\mu^2}{6R_m^2}\left(1+\frac{E_\mu^2}{36R_m^2}\right)\right\}, \tag{2.117}$$

where $E_\mu^2 = E_1^2+E_1E_2+E_2^2$ and $R_m^2 = \rho_m\nu_m m_0^2$. This is correct to 1 in 10^7 if $l<100$ km and $E_m<450$ km. The small term is less than 1 in 10^6 when $E<400$ km, and can often be omitted. [98], § 20, quotes more terms. ρ_m and ν_m are at latitude $\frac{1}{2}(\phi_1+\phi_2)$.

For R_m in (2.117) and (2.119), and for R_3 in (2.118), ρ and ν may alternatively be taken for the 'foot-point' latitudes ϕ'.† This is often more convenient than the true latitude ϕ, which may not be known.

(*d*) *Finite distance. Direction.* [98] gives

$$\begin{aligned}\delta_{12}\sin 1'' = {} & \frac{(N_1-N_2)(2E_1+E_2)}{6R_3^2}+ \\ & +\frac{(E_1-E_2)(2E_1+E_2)^2\epsilon\sin\phi'\cos\phi'}{9R_3^3}-\frac{(N_1-N_2)(2E_1+E_2)}{72R_3^4}\times \\ & \times\left\{\frac{4(2E_1+E_2)^2(1-2\epsilon\cos^2\phi')}{9}-\frac{2(E_1-E_2)(E_1+2E_2)}{3}-(N_1-N_2)^2\right\},\end{aligned} \tag{2.118}$$

where $R_3^2 = \rho\nu m_0^2$ at $\frac{1}{3}(2\phi_1'+\phi_2')$. In $\sin\phi'$ and $\cos\phi'$ use ϕ if more convenient. This formula is likely to be correct to 0".001 if $l<100$ km and $E_m<450$ km. For an accuracy of 0".02 within 300 km of the central meridian this can be simplified to

$$\delta_{12}\sin 1'' = \frac{(N_1-N_2)(2E_1+E_2)}{6R_m^2}\left\{1-\frac{(2E_1+E_2)^2}{27R_m^2}\right\}. \tag{2.119}$$

The small term is less than 0.1 per cent of the main term and it can often be omitted also.

Many authors have given different formulae for finite distance scale and direction, using different notations. [98] gives a summary and criticism.

(*e*) *The Universal Transverse Mercator* is a world-wide system of Transverse Mercator projections. Each projection covers 6° of longitude, with the central meridian at 3°, 9°, etc., east of Greenwich, and extends from 84° N. to 80° S. The central scale factor $m_0 = 0.9996$, but there are local departures from that figure, notably in the USSR where it is 1.0000. The polar caps are covered by two Zenithal orthomorphic projections, § 2.63, with $m_0 = 0.994$.

† ϕ' is the latitude where $M = N/m_0$, namely the foot of the perpendicular (on the plane) from the point P on to the grid line $E = 0$, which represents the central meridian.

Projection tables necessarily depend on the spheroid used. [30] gives tables for the International (1924), Clarke 1880 and 1866, Bessel and Everest spheroids. [23] and [24] give tables and details of the Transverse Mercator used in Great Britain.

For the oblique Mercator projection, whose 'pole' is in any latitude, see [296] or [113], 2 Edn, pp. 182–5.

2.63. The polar zenithal orthomorphic projection

This is often called the *polar stereographic* projection. The projection of a sphere is very simply defined in polar coordinates (r, θ) by $\theta = \lambda$ (in the northern hemisphere: in the southern $\theta = -\lambda$), and

$$r = 2m_0R \tan \tfrac{1}{2}\zeta, \qquad \text{where } \zeta = 90° - \phi. \tag{2.120}$$

The scale at a point is

$$m = m_0 \sec^2 \tfrac{1}{2}\zeta \approx m_0(1 + \tfrac{1}{4}\zeta^2 + \ldots). \qquad \zeta \text{ in radians.} \tag{2.121}$$

For the spheroid, from [558], p. 129, which gives terms as far as ζ^{11},

$$r = \frac{m_0 a}{(1-e^2)^{\frac{1}{2}}} \left\{ \zeta + \frac{1-7e^2}{12(1-e^2)} \zeta^3 + \frac{1-2e^2+46e^4}{120(1-e^2)^2} \zeta^5 + \right.$$
$$\left. + \frac{17-93e^2-1335e^4-4889e^6}{20160(1-e^2)^3} \zeta^7 + \ldots \right\}. \tag{2.122}$$

And the scale at a point is given by

$$m = (r\lambda \text{ on the projection}) \div (\lambda\nu \cos\phi \text{ on the spheroid})$$
$$= \frac{r(1-e^2 \sin^2\phi)^{\frac{1}{2}}}{a \cos \phi}, \text{ from (A.63).} \tag{2.123}$$

For rectangular coordinates

$$x = r \cos \theta, \text{ towards Greenwich}$$

and

$$y = r \sin \theta, \text{ along meridian } 90° \text{ E.}$$

For finite distances, using Simpson's rule, (A.20),

$$m = \tfrac{1}{6}(m_1 + 4m_m + m_2), \tag{2.124}$$

where m_1 and m_2 are the point scales at the two ends, and m_m is the scale at the centre. [252], p. 45. And [252], p. 59,

$$\delta_{12} \sin 1'' = (l/4\nu_1)\cos\phi_1 \sin A_{12}, \tag{2.125}$$

where l is the length of the line. This and (2.124) will not give geodetic accuracy over long lines.

See [31] for projection tables on the international spheroid (1924).

2.64. Change of coordinate system

(*a*) It is possible to convert (x, y) in one plane projection to (X, Y) on another by passing through (ϕ, λ). [448] gives a shorter routine, applicable to adjacent zones of Universal Transverse Mercator, computed on the same spheroid and datum, for direct conversion from one to the other. This is easily carried out with a small calculator provided it can handle conversions between polar and rectangular coordinates. It has an accuracy of 1 mm at a distance of 200 km from the boundary.

(*b*) Interpolation. If the conversion of numerous points is required, a few may be computed directly and then the rest may be interpolated by the method of *divided differences.*

It will be convenient to write the pair of numbers x and y as the complex number $z_n \equiv x_n + iy_n$, see Appendix F. Let the coordinates z_1 to z_4 of (say) four points enclosing an area of perhaps 100×100 km be converted via ϕ and λ,† so that Z_1 to Z_4 are known. The remaining Z's are then required.

The routine is to tabulate the coordinate pairs z_1 to z_4 in the first two columns of a four-line table, and Z_1 to Z_4 in the next two. Then in the fifth and sixth columns enter the three first order *divided differences* $[Z_1Z_2]$, $[Z_2Z_3]$, and $[Z_3Z_4]$ defined by

$$[Z_1Z_2] = (Z_1 - Z_2) \div (z_1 - z_2), \text{ etc.} \tag{2.126}$$

or

$$= \frac{(X_1 - X_2) + i(Y_1 - Y_2)}{(x_1 - x_2) + i(y_1 - y_2)},$$

the quotient of two complex numbers such as $(A + iB) \div (a + ib)$ being

$$\frac{Aa + Bb}{a^2 + b^2} + i\frac{Ba - Ab}{a^2 + b^2}$$

as given in (F.15).

Then the seventh and eighth columns contain the two second-order differences $[Z_1Z_2Z_3]$ and $[Z_2Z_3Z_4]$ defined by

$$[Z_1Z_2Z_3] = \{[Z_1Z_2] - [Z_2Z_3]\} \div (z_1 - z_3), \text{ etc.}, \tag{2.127}$$

another complex quotient.

Finally, in the ninth and tenth columns enter the single third differencc

$$[Z_1Z_2Z_3Z_4] = \{[Z_1Z_2Z_3] - [Z_2Z_3Z_4]\} \div (z_1 - z_4). \tag{2.128}$$

† It may happen that the projection details of all the z's are lost in history, but the method can still be used provided the new Z's of the selected four points are known. The method can be used with three or five points instead of four as described below, in which case the divided differences are recorded as far as the second or fourth order, respectively, instead of to the third.

The coordinates of any other point in the area are then given by

$$Z = Z_1 + (z - z_1)[Z_1Z_2] + (z - z_1)(z - z_2)[Z_1Z_2Z_3] + \\ + (z - z_1)(z - z_2)(z - z_3)[Z_1Z_2Z_3Z_4], \quad (2.129)$$

where the products such as $(z - z_1)[Z_1Z_2]$ are products of complex numbers, given by

$$(a + ib)(A + iB) = (aA - bB) + i(aB + bA) \quad (2.130)$$

as in (F.8).

A check may be obtained by working backwards from Z_4. The method is convenient for electronic computation.

See [372], which gives four numerical examples.

Between *orthomorphic projections* over an area of 100×100 km the computation error may be of the order 2 cm.

This method may also be applied to the adjustment of a low-order triangulation net on to a few higher order common points. The z's are the unadjusted coordinates and the Z's are the adjusted. Z_1 to Z_3 or Z_4 will be the values given by the higher-order points enclosing a triangle or quadrilateral. The optimum adjustment is one that is so far as possible orthomorphic, involving smallest changes in the observed angles. Adjustment by variation of coordinates may be expected to give a better result.

General references for plane coordinates [296], [315], [356], [558], [562], and [620].

3

HEIGHTS ABOVE SEA-LEVEL

Section 1. Introductory

3.00. Definitions

There are two principal systems of height measurement, namely spirit levelling and vertical angles, of which details are given in Sections 2 and 3 of this chapter. Observations to artificial satellites may also give a three-dimensional fixation from which height above the spheroidal geodetic datum level can be computed, Chapter 7.

The position of points being defined by latitude and longitude on a prescribed spheroid, §§ 2.02 and 2.03, the rational definition of the height of any point is its distance above the spheroid measured along the spheroidal normal. Such heights are called *spheroidal heights.* Spheroidal heights are required for some purposes, but the geoid or mean sea-level surface has great significance, and a more generally useful height is the distance† above the geoid. Such heights are called *geoidal heights* and, unless otherwise stated, the height of a point implies its geoidal height. For common use it is essential that the zero height contour should lie close to mean sea-level, and spheroidal heights are not an acceptable basis for topographical contouring; see footnote to § 1.30(*f*).

The ordinary process of spirit levelling gives geoidal height with great accuracy. Theodolite vertical angles give it less accurately, but, especially if observed reciprocally between near stations, they give geoidal height rather than spheroidal. Spheroidal height will be directly given by various forms of satellite observations, but otherwise it can only be got by first measuring geoidal heights and then observing geoidal profiles, §§ 4.50–4.52, to give the separation of geoid and spheroid. This process may involve errors of many metres, § 4.52. In general the accuracy of geoidal heights, as given by spirit levelling, exceeds that of any other system.

The height of a point above the geoid, measured in metres or other linear units, is known as its *orthometric* (geoidal) height, and these are the heights generally quoted. But there is a complication because, at a height above sea-level, the level surfaces to which the local vertical is perpendicular are not parallel to the sea-level surface. The vertical is in fact curved. See §§ 6.24 and 6.55.

In Fig. 3.1 if A and B are two points at sea-level, while a and b are

† Whether it is measured along the vertical or the normal is immaterial.

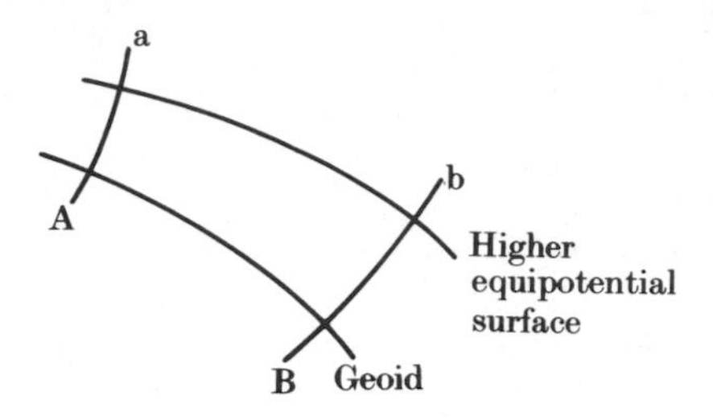

FIG. 3.1. The non-parallelism of level or equipotential surfaces. Meridional section.

vertically above them on the surface of a high-level lake, Aa will not generally equal Bb. The inequality is admittedly small, but other trouble results, notably significant circuit misclosures in otherwise errorless levelling.

Dynamic heights give identical height to all points on the same level surface. Let successive level surfaces be numbered 0 (for the geoid), 1, 2, etc., with decimal subdivisions, on some rational system. The height of any point is then the *number* of the level surface on which it lies. The most logical system is that the number of each surface should numerically equal the amount of work required to raise unit mass to it from sea-level, namely

$$\text{dynamic height} = \int_A^a g \, dh, \tag{3.1}$$

where g is the (variable) acceleration due to gravity. This is necessarily the same for all pairs of points Aa, Bb, etc., on the same pair of level surfaces. If g is measured in kilogals,† and h in metres, (3.1) will give dynamic heights which are numerically about 2 per cent less than orthometric heights. Heights given by (3.1) are known as *geopotential numbers.* The unit is known as a geopotential unit, GPU, and

$$1 \text{ GPU} = 1 \text{ kgal metre} = 100\,000 \text{ cm}^2 \text{ s}^{-2} = 10 \text{ m}^2 \text{ s}^{-2} \tag{3.2}$$

Its dimensions are those of potential, § 6.07, but note that whereas potential decreases with height, geopotential numbers are in this context defined to increase. The GPU is not a SI unit, [53].

3.01. Objects and accuracy required

Spheroidal heights are required for some purposes as follows.

(*a*) Reduction of measured distances to the spheroid for the computation of triangulation or traverse. An error of 1 m involves only 0·16 ppm.

(*b*) In connection with satellites, and baselines for the measurement of astronomical distances, positions are required in a defined coordinate system, such as spheroidal ϕ, λ, and spheroidal height. Here also, an error of one metre is probably tolerable.

† 1 gal = an acceleration of 1 cm s^{-2}. 1 kgal = 1000 gal = 10 ms^{-2}. Typically, gravity is about 980 gal.

For other purposes geoidal heights are either necessary or more convenient, as below.

(*c*) Topographical heights above sea-level. Errors of 0·5 metres seldom matter, or at great heights 0·5 m per 1000 m.

(*d*) Engineering. The head in a hydroelectric project is essentially a matter of potential (work), and the gradient of a canal is required with reference to level surfaces rather than to the spheroid (which may run uphill). For irrigation, in flat ground, heights may be required with an accuracy of $\sqrt{S}$ cm over distances of S km.

(*e*) Engineering. Tunnels cut from two ends need to meet in the middle to within a few centimetres. This may involve levelling over the surface for some tens of kilometres with height differences of 1000 m or more.

(*f*) To give heights to gravity stations. An error of 1 metre involves 0·3 mgal (3 μm s^{-2}).

(*g*) To measure the departure of mean sea-level from a single equipotential surface. A few centimetres are significant in distances of thousands of kilometres. The highest accuracy is required, but only in what amounts to the closure of a circuit, of which the sea surface itself forms a part.

(*h*) To record changes of height. The highest accuracy is required, except that errors are tolerable if they are identically repeated at different times and by different routes. This again implies that circuit errors are what must especially be avoided.

It is to be noted that the items which demand high accuracy, (*e*), (*g*), and (*h*) only require it in the closure of a circuit.

The object of a primary levelling system is to provide a framework on which topographers and engineers can base and adjust their heights with the expectation that, except quite locally, its errors will be less than theirs. It will also provide the principal evidence on the scientific questions (g) and (*h*) above.

3.02. Dynamic and orthometric heights

(*a*) *Normal orthometric heights.* The non-parallelism of the two level surfaces AB and ab in Fig. 3.1 derives from the fact that

$$\int_A^a g\,dh = \int_B^b g\,dh, \tag{3.3}$$

or

$$g_{mA}(\mathrm{Aa}) = g_{mB}(\mathrm{Bb}), \tag{3.4}$$

where g_{mA} and g_{mB} are the mean values of g along Aa and Bb.

Then θ the component in the direction AB of the inclination between the geoid and a level surface at height h is given by

$$\theta = -(h/g)(dg_m/dl) \text{ radians,}\dagger \tag{3.5}$$

where h and g have any approximate values, and dg_m/dl is the rate of change of g_m per unit distance in the direction A to B. θ is positive if Bb > Aa.

Fig. 3.2 shows a level set up between two staves from which it records the measured height difference $\delta\mathbf{M} = (s_1 - s_2)$. This is not the orthometric difference of height between the staff points S_2 and S_1, because the level line of sight is out of parallel to the geoid by the angle θ of (3.5). In fact, the difference of orthometric height is

$$\delta\mathbf{O} = \delta\mathbf{M} + \theta l = \delta\mathbf{M} - (hl/g)(dg_m/dl). \tag{3.6}$$

The average value of g between ground and geoid level is not observable, but for the small correction θl it is often sufficient to write $g = \gamma_h$, where

$$\gamma_h = 978(1 + 0.0053 \sin^2\phi - 2h/R) \text{ gals,} \tag{3.7}$$

ϕ is latitude and R is earth's radius. The exact values adopted for the constants 978 and 0.0053 are somewhat immaterial, provided one set is used consistently in any levelling net. Then

$$dg/d\phi \text{ (with } h \text{ constant)} = 978(0.0053 \sin 2\phi) \text{ gals per radian.} \tag{3.8}$$

$$= 9.78(0.0053 \sin 2\phi) \text{ ms}^{-2} \text{ rad}^{-1}.$$

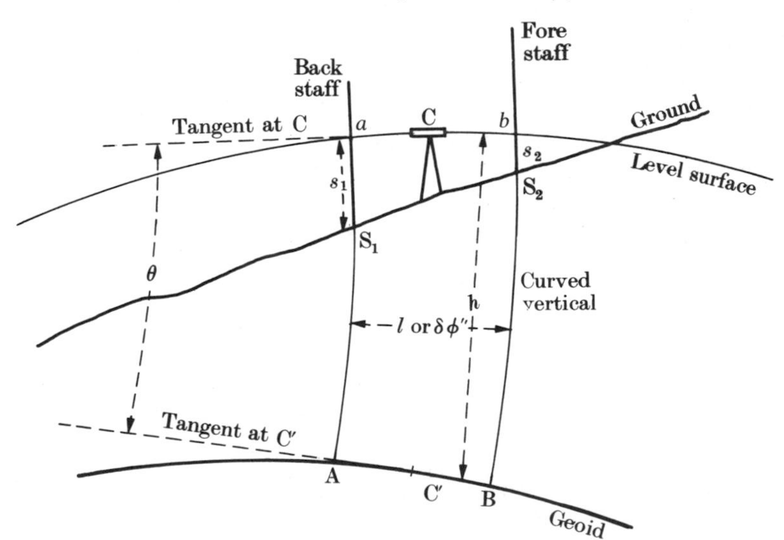

FIG. 3.2. From A to B dg_m/dl is negative.

† $\theta = (\text{Bb} - \text{Aa})/\text{AB}$, whence (3.5) follows from (3.4).

And for (3.6) we have

$$\delta \mathbf{O} = \delta \mathbf{M} - h(\delta\phi'')(0.0053 \sin 2\phi)\sin 1'', \qquad (3.9)$$

where $\delta\phi''$ is the latitude of S_2 (the fore staff) minus that of S_1, in seconds of arc. In Fig. 3.2 $\delta\phi''$ is negative.

Along any line of levelling (3.9) must be summed before circuit errors are found and adjusted.† The sum $\sum h(\delta\phi'')(0.0053 \sin 2\phi)\sin 1''$ will not in general be zero round a circuit. Thus in the circuit AabBA of Fig. 3.1, the sum is zero in the vertical climb between A and a, and also between b and B, where the $\delta\phi$'s $=0$. It is zero along BA where $h=0$, but it accumulates along ab. For example, if $h=1000$ m for a distance of 500 km along the meridian, and if $\phi=45°$, the total correction is 0.4 metres. If it is not applied it will make a significant contribution to the closing error of the circuit.

Orthometric heights computed with the normal value of gravity given by (3.7) are known as *normal orthometric heights*.‡

See (*d*) below for the errors which may arise from the use of (3.7).

(*b*) *Geopotential numbers.* With modern instruments gravity can be measured at frequent intervals along a levelling line. In Fig. 3.2 let g be observed at the instrument site. Then the dynamic height of S_2 minus that of S_1 will be

$$\delta \mathbf{C} = g\, dh = g(s_1 - s_2)\ \text{GPU}, \qquad (3.10)$$

if g is recorded in kilogals and $(s_1 - s_2)$ in metres. It is not necessary to observe g at every levelling station, since sufficiently accurate values can be obtained by interpolation, primarily with height. The appropriate distance between gravity stations depends on the roughness of the country, see [385] and [470]. In general the intervals may be 1.0 km in mountainous country (where slopes of the levelling line may average >8 per cent over distances of 1 km), $1\frac{1}{2}$ or 2 km in medium-sized hills (slopes of 4–7 per cent), and 5–10 km in flat ground (slopes <3 per cent), with a maximum of 100 metres difference of height between adjacent gravity stations in all cases. Gravity should especially be observed where the line crosses hill tops or valley bottoms, or at other abrupt changes of slope or of the direction of the route. An error of 10 mgal (100 μm s^{-2}) in interpolated g produces an error of only 1 mm in 100 metres of measured height difference.

Given sufficient values of g, this system in principle gives dynamic heights without error, and errorless observations should give zero closing error round a circuit, as required for § 3.01, objects (*e*), (g), and (h). If a

† The elements in the summation may be the intervals between BMs rather than those between staff points.

‡ They are not the same as the Normal Heights of § 6.47 (6.155).

circuit misclosure is required in linear units, as for a suspected crustal movement or an anomalous height of mean sea-level, it may be converted from GPUs to metres by adding 2 per cent.

Orthometric heights can be derived from geopotential numbers by dividing them by g_m, the mean value of g along the vertical from ground to geoid, in kilogals.

$$\mathbf{O} = \mathbf{C}/g_m. \tag{3.11}$$

This cannot be measured, so geopotential numbers are not an infallible source of orthometric heights, but g_m can be estimated as in § 6.42 (*a*).

(*c*) *Normal dynamic heights.* A system which has often been adopted [113] 2 Edn., §§ 4.00 and 4.03, is to compute dynamic height **D** from

$$\delta\mathbf{D} = \delta\mathbf{M} + \delta h(\gamma - \gamma_S)/\gamma_S, \tag{3.12}$$

where γ is given by (3.7), and γ_S is (3.7) with ϕ equal to ϕ_S, some arbitrary latitude central to the area concerned. These dynamic height differences are summed round the circuits, and misclosures are adjusted. The orthometric height of a BM in latitude ϕ is then given by

$$\mathbf{O} = \mathbf{D} - \int_0^h (\gamma - \gamma_S)\,\mathrm{d}h/\gamma_S = \mathbf{D} - 0.0053(\sin^2\phi - \sin^2\phi_0)h, \tag{3.13}$$

where the integral is taken along the vertical from geoid to ground level.

This procedure gives identically the same normal orthometric heights as the method of (*a*) above. It would give the same results as method (*b*) if g was actually equal to γ.

(*d*) *Errors resulting from the use of normal gravity.* In so far as $g - \gamma$ is not zero, an error of 10 mgal (100 μm s^{-2}) in γ makes an error of 1 mm in 100 metres of measured height difference. This is trivial unless it accumulates systematically. Any constant value of $g - \gamma$, or one which varies directly as h, makes $\int (g - \gamma)\,\mathrm{d}h = 0$ round a circuit, and does not affect closing errors. Similarly, such effects tend to cancel in the sum of a rise up one side of a hill and the descent on the other. The accumulation of error is thus limited.

[113] 2 Edn., pp. 207–9 provides some means of estimating the effect of using normal gravity on circuit closures, and quotes 44 circuits from [527] and 14 from [469], which have been computed with both g and γ. The average circuit-closing errors given by the two methods were the same.

So far as accuracy is concerned, normal orthometric heights are still acceptable substitutes for geopotential numbers for all the objects of § 3.01, but uniformity and continuity of method is most important. The primary levelling nets of Europe have now all been adjusted in GPUs, [526] and [529], and future work in Europe must clearly be done on the same system. If crustal movement, or the like, is to be deduced from the discrepancies between old and new levelling, it is essential that both must be computed on the same system. If necessary, the old levelling must be recomputed with the gravity data used for the new, or if the routes are different, old lines must now be given gravity data for their computation in GPUs.

[347] gives a full summary of the various methods which have been proposed and used in different countries for dealing with the dynamic-orthometric problem.

Section 2. Spirit Levelling

3.03. Layout of a primary level net

(*a*) *High precision levelling net.* In the past it has been customary to observe a primary levelling net in the form of rectangles or polygons with sides varying between 200 and 400 km in large or less developed countries, or of 100 km or less in well developed areas. Since 1912 [360] it has usually been specified that every such primary line should be independently levelled, fore and back, in conditions as widely different as possible. Subject to its meeting other specifications, such levelling has been known as *high precision levelling.* In recent years this form of primary net has been criticized, both as regards the density of the net, and on the extent to which the double levelling should be at widely different times.

(*b*) *The density of the net.* In all methods of survey it is traditional that detailed work should be based on, and controlled by, some preliminary framework of higher accuracy. In the case of levelling, its obvious advantages are as follows.

(i) The main, open, framework can be completed relatively rapidly over the whole country, so that the publication of detailed results on a preliminary basis, which will presently have to be corrected, is avoided. In an original survey, this argument is well founded, but it is less relevant when a second or later revised levelling is being considered.

(ii) Although later levelling lines may indeed be capable of adding to the strength of the primary net, in the past it has been impossible to incorporate them in a simultaneous adjustment. This consideration is now out-dated. The inclusion of large amounts of lower-order levelling in the primary adjustment is possible, if it is thought desirable, and if suitable weights are allotted.

(iii) A less obvious reason for having a control framework of specially high accuracy is that in the first-order levelling the instruments and system of work may be carefully adapted to the avoidance of systematic errors. It is then possible that the inclusion of a multiplicity of lower-order levelled lines, weighted in accordance with their random errors, may introduce systematic errors which might otherwise have been avoided.

The matter is one in which some compromise is now possible. In the US the problem has been carefully considered in [597], where it is recommended that a new revised primary net should consist of about 300 polygons with sides of 100 to 200 km.

(*c*) *Fore and back levelling.* The objects of double levelling (fore and back) have been as follows.

(i) To produce two independent results, with minimum risk of their sharing the same systematic errors or blunders.

(ii) To eliminate the serious systematic error due to the sinking or rising of the staff support, as described in § 3.06 (*a*) (iii).

(iii) To provide two independent measures of each section between well-founded bench marks (say 5 km), from which random accuracy can be assessed, systematic errors revealed, and blunders located.

On the whole, the system may be said to have succeeded in its objects, except perhaps in the difficult matter of systematic error. Criticism of the system is possible on the grounds that double levelling costs twice as much as single, and that money saved by single levelling might be put into the observation of a denser net, as in (*b*) above. It is also questionable whether the fore and back levellings should take place at widely separated dates, as advised in 1912, or whether the similarity of conditions resulting from close dates may be preferable.

In favour of close dates, the same instruments and similar working conditions, it may be said:

(i) This will largely cancel systematic error due to sinking or rising of the staff, § 3.06 (*a*) (iii), which has sometimes been a serious thing.

(ii) It is administratively convenient to have a single party levelling (say) 20–30 km in one direction, and then forthwith levelling the line back again in the other.

In favour of different dates and maximum change, it may be said:

(iii) If a single party is employed as in (ii), there is risk that a blunder made in the fore levelling may be repeated in the back. Such as some confusion about the bench mark at the end of a section.

(iv) Since errors which accumulate with height or latitude cannot, if constant, cause any discrepancy between fore and back, nearly simultaneous back levelling can do little to cancel them, and only serves to disguise them. If conditions are changed, there may at least be a warning.

(v) In the height difference between two low-lying points, such as two tide gauges, which may be separated by high ground, the cancellation of errors which accumulate with height will be helped if the ascent, and descent on the other side, are done with the same instruments at the same season. If the distance is not small, this is best secured by rapid work in one direction from one end to the other, rather than by doubling the time interval by observing short sections fore and back, the one immediately after the other.

The pros and cons are balanced, and it is not possible to make a general recommendation. On soft ground, where rising or sinking of the staff is to be feared, fore and back should be done in rapid succession as in (ii), before the condition of the ground changes. On the other hand, hills are best climbed and descended on the other side as in (v) without delaying to observe the ascent in the reverse direction. And the back levelling should similarly be uninterrupted.

(*d*) *US first-order class III levelling.* Going rather further than is suggested in (*c*) above, [597] recommends that the proposed relevelling of the US primary net should be carried out by single levelling, in which alternate daily sections are levelled in opposite directions. This goes a long way towards meeting the systematic error of 3.06 (*a*) (iii). The check on blunders is partly provided by the use of double-scaled levelling staves §3.04 (*d*), reinforced by the existence of many BM's of the previous

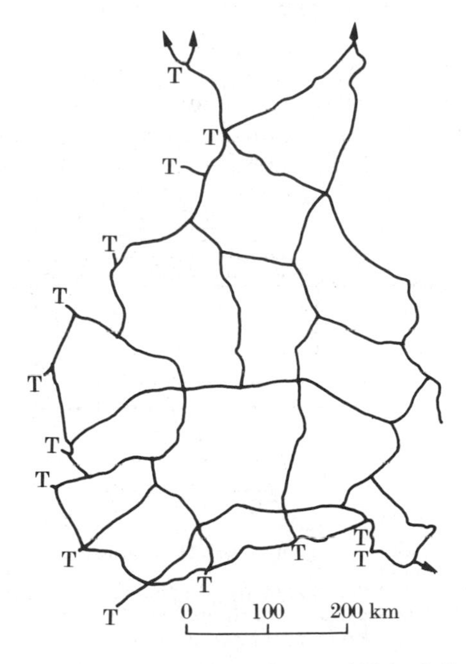

FIG. 3.3. High-precision levelling net in part of Finland. T = tide guage.

levelling, and by the electronic recording of readings and their differences, with automatic calls for verification when necessary. The estimation of accuracy can well be based on the very numerous and relatively small circuits. The proposal has been carefully costed, and has much to recommend it for the relevelment of a large country.

(*e*) *The location of primary levelling lines.* The lines should follow easy communications. The avoidance of railway lines and roads with heavy traffic has been advised, but railway lines with light traffic may be very convenient, and good work has been done on them. Great changes of height should if possible be avoided, unless heights are specifically required at the top. Other things being equal, lines should pass through areas where engineers are likely to use the bench marks (BM's). The existence of sites for stable BM's is also important, § 3.11.

Tidal stations should be established at least every few hundred kilometres along the coast, not up estuaries, but facing the open sea, and connected to the level net.

Fig. 3.3 shows part of the level net of Finland in 1962.

3.04. Levelling instruments

The level should of course be the best of its kind, typically with a focal length of 30–40 cm, aperture 50–60 mm, and magnification 40–45. Parts liable to introduce error through unequal heating should be made of invar [196], pp. 40–1. Two main types are in current use, as in (*a*) and (*b*) below.

(*a*) The well-established *tilting level*, Fig. 3.4, in which the axis is approximately levelled with a small circular bubble, and the main bubble is carefully centred with the tilting screw immediately before every reading. The sensitivity of the bubble is typically 3–5″ per mm.

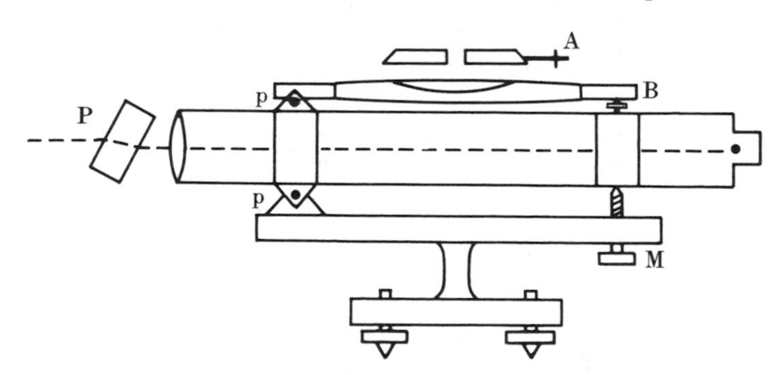

FIG. 3.4. Tilting level. Diagrammatic.

M = micrometer tilting screw. B = bubble adjustment.
A = longitudinal movement of prisms for fine adjustment of collimation.
P = parallel plate. p = pivots.

(*b*) *The newer type of automatic level*, which does not rely on any sensitive bubble. The axis is roughly levelled, and residual dislevelment of the telescope is automatically corrected by a system of compensating reflecting mirrors or prisms inside it, which are suspended under gravity.

Fig. 3.5 shows a form of compensator, diagrammatically. The line SA is horizontal, and the line of sight OC is inclined at a small angle α. Two mirrors (1) are suspended so that they always make angles of 45° with the vertical, while two others (2) are fixed to the telescope. A horizontal ray of light then follows the path SO1221C, during which its direction is changed by an angle of 4α. So if $BC = \frac{1}{4}OC$ it will fall on the horizontal wire.

There are of course many complications in the design. The suspension of the mirrors must be very frictionless, but strongly damped. The compensation must be insensitive to temperature and to lateral dislevelment. Changing focus, with varying distance of the staff is equivalent to changing OC, and the ratio OC/BC, which must be allowed for. See [277] and [514] for fuller descriptions.

If the ratio OC/BC is not equal to 4, or to whatever figure the system is designed for, the instrument is said to be over- (or under-) compensated, and the ray of light seen on the cross-wire will be non-horizontal by an amount proportional to the dislevelment of the telescope and to thc amount of over-compensation. [313] investigated this and other errors of compensation systems and for the best instruments, properly handled, came to the following conclusions.

(i) The effect of over-compensation may be limited to $\frac{1}{2}$ per cent of the dislevelment of the telescope. In so far as this is constant, as it will be if the vertical axis is truly vertical, this will cancel as between fore and back sights, but there will actually be an error of $\frac{1}{2}$ per cent of the error of verticality resulting from the use of the small circular level for the initial levelling. If this is 10″, which is a very small figure, the error of compensation will be 0″.05 of slope. A larger figure may well be tolerated, if it is random. But it will be systematic in so far as it is due to maladjustment of the circular bubble, if the initial levelling is always done

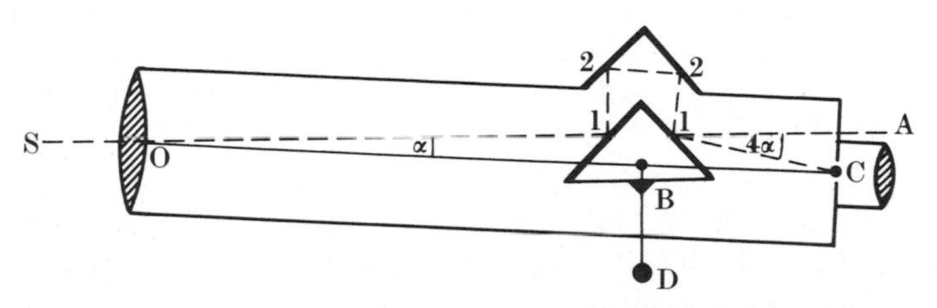

FIG. 3.5. Automatic compensator. SOA is horizontal and BD is vertical. O is the optical centre of the object glass, and C is the intersection of the cross wires. OC = 4BC.

with the telescope pointing towards (say) the back staff. The trouble is remedied by its being done with the telescope pointing fore and back at alternate stations. [313], p. 352, advises that it is also best to relevel with the circular bubble before readings are made to the second staff.

In modern instruments (1977) steps have been taken to reduce or eliminate these points. Thus, in the Zeiss Ni 002 level, collimation and compensation errors can be cancelled by reversing the compensator, and observing in both positions.

(ii) Lateral dislevelment of 1′ may dislevel the line of sight by 0″.1. Instruments should be checked to see that they are not abnormally sensitive to such trouble.

(iii) The compensation is strongly damped, and there may be an error of 1″ whose sign will depend on whether the compensator has last rested against its fore or back stop. This will be cancelled provided the same stop is involved in both the fore and back sights, as can be ensured by rotating the telescope about its vertical axis before observations. Centrifugal force will always throw it out in the same direction.

In some instruments, e.g. the Wild NA 2, a small button is pressed before reading the staff. This causes the compensator to swing off-level for a second or two. The observer watches this through the eyepiece and reads as soon as movement ceases.

(*c*) *Comparison of the two types of level.* The faster work of the automatic level is an advantage, not only financially, but because fast work may reduce errors due to changing refraction and subsidence of the staff, 3.06 (*a*) (iii).

It is possible that the best tilting levels are still the more accurate instruments over a short line, but the difference is probably more than offset by the saving of time and the associated hope of reduction in systematic error. Observers who have used automatic levels are certainly reluctant to change back to the old type.

The Zeiss Ni 002 is a good modern automatic level, and the Wild N3 is an equivalent tilting level.

(*d*) *Staves.* Geodetic levelling should be done with staves in which the divisions are marked on a strip of invar, generally about 3 metres long, not folding nor telescopic. The design of the staff should be such that its structure supports the invar strip without constraining its length. It is desirable that the invar strip should so far as possible take up the air temperature, and that its temperature should be insensitive to the direction from which sunlight falls on it, either when standing or being carried between staff points, § 3.06 (*c*) (ii). The painted divisions must not stand up above the surface, nor be recessed below them, § 3.06 (*c*) (i). The distance from the foot of the staff to the zero mark must be the same for both members of a pair, unless one only is designated as that which will be placed on all bench-marks.

Levelling staves are best made with two scales, side by side on the same face. Both scales are divided in either cm or 0.5 cm. The centimetre and decimetre divisions of the two scales are out of step, so that the reading of one introduces no bias into the reading of the other. The divisions may take the form of fine lines, or of alternate intervals in black or white (or yellow).

The length of the staff, and its coefficient of expansion, and the accuracy of its subdivisions must be certified, and the over-all length must be verified at least every few months. An error of 0.1 mm in 3 metres is just significant. Computations can be corrected to allow for known small errors.

Staves are held vertical by means of steadying poles or handles and circular bubbles. The adjustment of the bubble must be periodically checked, such as once a week. A constant, or average, error of $\frac{1}{2}^{\circ}$ out of plumb will introduce an error of one part in 26 000 in large height differences.

3.05. Field procedure

The general procedure for geodetic levelling is very similar to that for lower orders of accuracy, which need not be described. The usual specification for the best geodetic levelling, with details of some special precautions is given below.

(*a*) *Collimination.* The level should be set up halfway between the staves, and the collimation error should be kept adequately small. If the error is 2 mm on a staff at a distance of 40 m, a one-metre inequality between fore and back distances will introduce an error of 0.05 mm in height. If distances are taped, the error is easily kept to less than this. If larger errors are allowed, the stadia readings should be used to record and control the total of the distance differences (fore minus back), which should be kept below about 5 m. Lower standards of equality are allowable with an instrument like the Ni 002, in which the compensator is reversible.

Collimation, or the efficacy of reversing the compensator, should be verified every few days, as indicated by experience.

Additional reasons for the equality of fore and back sights are as below.

(i) Collimation is apt to be changed by change of focus.

(ii) Equality of sights eliminates error due to geoid curvature, and normal atmospheric refraction.

(*b*) *The parallel plate* should be used. Its readings can be made to 0.1 mm, with (doubtful) estimation to 0.01 mm. It gives much increased accuracy in short lines when atmospheric conditions are good, but cannot help with bad seeing ('boiling'), and does nothing to eliminate the systematic errors which are the chief source of trouble in long lines.

(*c*) *Length of sight.* Sights of more than 50 m from instrument to staff should be unusual, but on flat ground 50 m may be exceeded if the air is exceptionally steady. The resulting increased speed and reduction in the number of set-ups is itself beneficial. The line of sight should always be at least 50 cm above the ground.

(*d*) *Programme of readings.* Staves should move alternately. If staff A is back staff at the first set-up, it should be fore staff at the second, and back staff again at the third, and so on. Working on this system, one staff (say A) should always be read first, which ensures that the back staff is read first at alternate set-ups. This routine eliminates errors due to the following.

(i) Change in refraction with steadily rising or falling temperature.

(ii) Any constant tendency for the instrument to rise or fall during the observations.

(iii) Inequality of staff readings above their base plates. If there is any possibility of significant inequality, one particular staff should always be used on every bench mark.†

The sequence of readings may then be as below.

At one station	Staff	Scale	
	Back	left	
	Back	left	Stadia wire
	Fore	left	
	Fore	left	Stadia wire
	Fore	right	
	Back	right.	

At the next set-up interchange the words fore and back.

If the instrument has a reversible compensator, either the compensator may be reversed after the four left-scale readings, or alternatively the sequence may be

At one station	Staff	Scale	Compensator	
	Back	left	1	
	Back	left	1	Stadia wire
	Back	left	2	
	Fore	left	2	
	Fore	left	1	
	Fore	left	1	Stadia wire
	Fore	right	1	
	Fore	right	2	
	Back	right	2	
	Back	right	1	

Interchange the words fore and back at the next set-up.

† This involves an even number of set-ups between bench marks, which may sometimes mean that there is one more than might have been necessary. Inequality of staves is best avoided, if possible.

This second programme demands nearly twice as many readings. Random error is reduced, and there are additional numerical checks. On the other hand, time has to be paid for, and systematic errors may be increased.

A strict but not uncommon specification is that Back-minus-Fore on each scale should agree with the other scale within 0.4 mm. This should result in a standard random error (68:32) of 0.5 mm in one km of line, ignoring systematic errors. Actual s.e.'s of 0.3 mm in 1 km have been recorded, [514], pp. 158–9, and instrument-makers claim similar figures. If conditions are such that this criterion demands repetition more than once in about every ten stations, the 0.5 mm/km standard is not likely to be reached. The easy remedy is to shorten the lengths of sights, but abnormally short sights will probably lead to increased systematic error. If there is difficulty in meeting the 0.4-mm standard, either (*a*) further work must await better conditions, or (*b*) the expectation of a larger random error must be accepted. If repeat readings are made, it will generally be best to accept the mean of all readings except obvious blunders.

The instrument should be protected from the sun by an umbrella when set up, and by a cloth while being carried forward.

Staff readings should not be made before 20 seconds have elapsed since the setting up of either the instrument or the staff, since settlement is most to be expected immediately after setting up.

(*e*) *Staff points.*The flat base of the staff must rest on some sort of hemispherical knob, with a definite highest point. The stability of this support is most important. The ordinary three-cornered plate may be satisfactory in ground that is soft enough to prevent its moving sideways, but hard enough to prevent its sinking by 0.1 mm between its being used as a fore staff and later as a back staff. This condition may not be uncommon, but it is difficult to recognize it with certainty. The use of a spike or peg driven into the ground involves the risk that it or the ground may be rising elastically while observations are being made. From this point of view a sharp, light, iron spike is better than a wooden peg. In some types of ground it may only be possible to get good results if spikes are driven in well in advance of the levelling. See further in § 3.06 (*a*) (iii).

The following, based on [196], are observed figures for the average sinking of an iron spike† after a staff has been placed on it.

Locality	In 20 seconds	In 2 minutes
Road or rail	0.008 mm	0.010
Sandy ground	0.020	0.024
Turf	0.130	0.168

† Length not stated. In English 'spike' suggests 20–30 cm. [318], p. 22 reports 30–40 cm in Finland.

The sinking of triangular plates was 3 to 5 times as much.

Temperatures should be recorded, and the effect on staff lengths included in the computations if significant.

(*f*) *Check levelling.* The season's work must be started from a BM of proved stability. Even when a BM cut on solid rock is available, it should be checked (if only to prove correct recognition) by connection to two or three other old BM's. In general 4 or 5 km of old line, with as many BM's, should be relevelled whenever connection with old work is required, and if discrepancies are found, more may be necessary.

Each day's work should be closed and started on a single good mark cut on some solid structure, although connection to a second mark of some kind is a useful check. If nothing solid is available, work can be closed and started on three pegs. When more than one mark is used, care must be taken to avoid the inclusion of their height difference twice over (or not at all) in the running total of the line.

(*g*) *Relevelments.* Exceptional discrepancies between fore and back levelling between BM's will require relevelling in one or both directions. It is reasonable to relevel if the discrepancies exceed $2\frac{1}{2}$ times the s.e. of fore-minus-back as judged by what is usually being got in the line concerned. The criterion should not be too strict. It may sometimes be best to accept the mean of all four measures in preference to accepting only the second pair.

(*h*) *Motorized levelling.* In recent years equipment has been designed for the purpose of using motor transport for spirit levelling. One car carries the instrument, observer, booker, and an electronic recorder and computer, while two others carry the staff men and staves.

In the observer's car the instrument and stand are let down on to the road surface at the required distance from the last staff as measured by a specially sensitive cyclometer (trip-meter). The level is of the automatic type, being in fact a Zeiss Ni 002, which was originally designed for this purpose. Both fore and back sights are read from the side, without the observer having to move, and the compensation is reversible so that collimation errors are eliminated. The staves are similarly emplaced and set vertical without the staff man having to leave his vehicle, except when an established BM is to be connected. The observer and booker change places after every hour, as the rapid rate of work is tiring for the observer.

This system of work has been used in Sweden [84] for secondary and tertiary levelling, for which purpose it has been found satisfactory. The daily out-turn has been doubled as compared with work on foot or bicycle. Accuracy, as judged by circuit closing errors, has been $2\sqrt{K}$ mm, where K is distance in km. The economic advantage of course depends on the relative cost of labour and equipment.

3.06. Systematic errors

If levelling can be carried out with a standard error of e mm in 1 km of single levelling, and if the errors are random, the root mean square difference between the fore and back levelling of 1 km will be $e\sqrt{2}$, and for a distance of L km it will be $e\sqrt{(2L)}$. It is common experience that if e is determined from a number of single kilometre lines, the discrepancies in longer lines will be greater than $e\sqrt{(2L)}$. This is due to some sources of error having a constant effect on successive sections of the line, or an effect of predominantly constant sign. Such errors are described as systematic. The accumulated effect of such an error in L km of levelling is likely to be proportional to L rather than to $\sqrt{L}$.

It must be noted that distance is not necessarily the parameter to which the accumulation of error is proportional. It may be the number of set-ups, n, or the time spent on the work, t. Both of these are roughly proportional to L, and the distinction between L, n, and t can probably be ignored. On the other hand, errors which accumulate with the height, h, require quite distinct consideration, as may some less obvious errors which may accumulate with differences of latitude, ϕ. For instance, L, n, and t errors may be expected to show themselves in the closing errors of circuits, while h and ϕ errors, if constant, will not.

The following are examples of systematic errors of different kinds, and the steps taken to control them.

(*a*) *Errors accumulating with L, n, or t*

(i) The elementary procedure of having equal fore and back sights cancels errors due to collimation, earth's curvature, and symmetrical refraction. It also avoids change of focus.

(ii) The practice of sighting first on the fore staff at alternate stations, § 3.05 (*d*), cancels the effects of the following.

(A) Steadily decreasing refraction, as may occur every day from sunrise to 15.00 hours.

(B) Maladjustment of the circular bubble in an automatic level, § 3.04 (*b*) (i).

(C) Systematic sinking or rising of the instrument between observing fore and back staves.

(iii) Systematic sinking or rising of the staff between its use as a fore staff and as a back staff, § 3.05 (*e*). If on average, the staff sinks by 0.03 mm during this interval, the heights of forward BM's will be too great by about $0.3L$ mm, and too low if it rises. [429] illustrates persistent errors of 1 mm/km, as 2 mm/km discrepancies between fore and back levellings in Sind, when wooden pegs were driven into a dry clay surface which was apt to overlie water-logged sand. The sign of the error suggests elastic rising of the pegs. In other countries, [612] in Great Britain, the errors have tended to be of opposite sign, suggesting a sinking of the support as in § 3.05 (*e*).

This error will appear in the differences between fore and back levelling, and if it is the same in each, it will cancel in their mean. It constitutes an argument in favour of fore and back levellings being carried out in identical conditions.

(*b*) *Errors accumulating with height*

(i) An error of staff length, or a systematic error in its subdivision.

(ii) Non-verticality of the staff, § 3.04 (*d*).

(iii) Errors in the values of gravity used in the dynamic and orthometric corrections, § 3.02.

(iv) Errors due to unequal refraction. When the levelling is going uphill the foresight, being always nearer the ground than the back sight, passes through air where $\mathrm{d}T/\mathrm{d}h$ is numerically greater and refraction less. The higher BM thus tends to get too low a height. The correction depends on $\mathrm{d}^2T/\mathrm{d}h^2$, which is determinate if T is measured at three different heights above the ground, e.g. at the instrument and at the two staff readings, but it must be read to about 0.01 °C. This is barely possible, but mean values of $\mathrm{d}^2T/\mathrm{d}h^2$ taken over a section of several kilometres may be got from readings of lower accuracy. [104], pp. 32–7, gave a numerical investigation, and concluded that the error was of reduced consequence on steep hills where the sights are necessarily reduced to 30 m or less, and of course in very flat country, while in undulating country the error is soon reversed by the line going downhill. The worst error occurs in a long easy gradient, such as when a railway climbs to such a height as 1000 m, in which case it might be of the order 20 mm per 100 m rise. The remedy suggested is to shorten sights (level to staff) to 30 m† in such cases, even though the gradient may admit more. Sights on the bottom 50 cm of the staff should always be avoided.

An alternative method, [350] and [301], is to measure the temperature at only two heights to give $\mathrm{d}T/\mathrm{d}h$, and from it to deduce $\mathrm{d}^2T/\mathrm{d}h^2$ from the general experience of meteorologists according to the latitude and season. The apparatus takes the form of two nickel wire resistances mounted on a light rod at 0.5 and 2.5 m above the ground, and the difference of their resistances is measured by a Wheatstone bridge. Readings to 0.1 °C, suffice, and no delay to the work results. This method is probably as good as the first for giving an estimate of the magnitude of the error. [301] reports average errors of 1 mm per 17 metres of height difference in Finland, typically using 50-metre sights.

† With 30-metre sights on a slope of 1 in 20, the error was estimated to be 1 mm per 50 metres of height difference. On easier gradients 30-m sights would introduce less error. Stricter limits of error may now be demanded.

If they are constant, none of these h errors will affect the difference of fore and back levellings, nor circuit errors, nor the changes of level suggested by successive levellings. It is clear, however, that although they may be reasonably constant over distances of (say) 50 km, or for a period of months, their constancy cannot be relied on over different routes or in different years.

(*c*) *Errors accumulating with differences of latitude*, irrespective of the direction of work.

(i) If the black divisions on a staff (being thickly painted) stand up above the surface, or if (being recessed) they lie below it, a systematic bisection error will occur when the staff is lit by sunlight from above. In a line which is proceeding northwards (in northern latitude) this will happen more frequently to the fore staff than to the back. [192] and [439] show that this may have been a source of systematic error of 0.5 or 1.0 mm/km (in continuous bright sunlight) in north–south lines in Great Britain, when using staves of a particular make. Frequent absence of sun will, of course, have much reduced the average.

Provided there is no perceptible uneveness in the surface of the staves, significant error is unlikely to recur.

(ii) Similarly, when levelling in middle north latitudes is run from south to north or the opposite, the invar strip on a staff will more often be exposed to the sun when it is facing south than when it faces north. [249] reports that the rate of heating of a staff placed in bright sunshine may be as much as 0.8 °C/minute for the first 20 minutes, and that the rate of cooling may be the same when it is turned away from the sun. The remedy is fast work, and the avoidance of bright sunshine if the best results are required.

(iii) Tidal effects. The direction of gravity departs slightly from its mean according to the directions of the moon and sun. [310] and [528] Supplement give formulae and some examples, computed on the assumption of a rigid earth. For the effect of the moon on a north–south line the 14-day mean error is $0.042 \sin 2\phi(3 \cos \delta - 2)$† mm/km, where $+\delta$ to $-\delta$ is the range of the moon's declination, in the sense that the correction to forward heights is positive in a line going from north to south. The error will be $0.036 \sin 2\phi$ mm/km, or 230 mm from equator to pole, when the moon's declination varies between $+18°$ and $-18°$. And when the declination varies between + and $-28°$ the error will be $0.027 \sin 2\phi$ mm/km, or 170 mm from equator to pole. Since the moon's hour angle during normal hours of work varies from day to day, the average effect on an east–west line will be zero.

The average solar effect on a north–south line is about half that of the

† [528], p. 49, gives $0.021 \sin 2\phi(3 \cos \delta - 1)$, which is corrected in the Supplement.

moon, with the same sign. On an east–west line it would average zero if observations were made at all hours of the day, which they will not be. In examples given in [310] it does not amount to more than 0.003 mm/km. See §§ 3.23 and 3.24.

These average figures are little more than guides to the order of magnitude. They show that the lunar effect on a north–south line is not quite negligible. Approximations involved in calculating these average figures may be avoided by applying (3.73) to sections between BMs, using zenith distances and azimuths appropriate to the actual hours of work.

The problem is complicated by the yielding of the earth itself, and by ocean tides which affect the direction of gravity by the mass of their water, and also deform the adjacent crust by their weight. These considerations may multiply (3.73) by such a factor as 0.8, [310], p. 277. They add much to its uncertainty, especially within a few km of coasts where tides are abnormally large.

Like errors varying with height, these errors which depend on latitude, if constant, would not be visible as a difference between fore and back levelling, nor in the closing errors of circuits.

3.07. Measures of accuracy

(*a*) *Methods of representation.* If all errors were random the s.e. expected, or thought to have been achieved, in a line L km long, could be expressed in the form $\eta\sqrt{L}$. Two other sets of formulae have been proposed in order to express the existence of systematic error.

(i) *Lallemand's formulae* [360]. The error to be expected in a line of L km is given in the form

$$\text{s.e.} = (\eta^2 L + \sigma^2 L^2)^{\frac{1}{2}}, \tag{3.14}$$

where η is a random error in mm/km$^{\frac{1}{2}}$ accumulating as $\sqrt{L}$, while σ is systematic, accumulating in proportion to L. With large values of L the formula approximates to

$$\text{s.e.} = \sigma L, \tag{3.15}$$

the assumption being that σ is likely to continue unchanged to any distance.

(ii) *Vignal's formulae* [579]. It is recognized that the magnitude and sign of σ are liable to change after intervals of such as 10, 50, or 100 km and that over distances exceeding some figure Z, the systematic error itself becomes random.

Then for a short distance, such as 1 to 5 km,

$$\text{s.e.} = \eta\sqrt{L} \tag{3.16}$$

as before. Small systematic error may be present, but over a short distance it is of no consequence. And for distances exceeding Z

$$\text{s.e.} = \tau\sqrt{L}. \tag{3.17}$$

For intermediate distances $L < Z$

$$\text{s.e.} = \tau_L\sqrt{L}, \tag{3.18}$$

where τ_L is a figure which increases from η to τ as L increases from 0 to Z.

In the second geodetic levelling of Finland, [318], pp. 51–3, Lallemand's η and σ were found to be 0.35 mm/km$^{\frac{1}{2}}$ and 0.04 mm/km, while Vignal's η and τ were 0.33 and 0.60 mm/km$^{\frac{1}{2}}$, with Z taken as 50 km. Table 3.1 compares the resulting s.e.'s of height from (3.14), (3.16), and (3.17).

(iii) *Comparison.* For distances of up to about 250 km the results of the two systems are similar, but at 1000 km or more the differences are apparent, and Vignal's are logically the more likely to be correct. On the other hand, the figure 0.60 for τ is derived from the closing errors of circuits which average only 500 km in length, and it is not certain that $\tau\sqrt{L}$ will give correct expectations at 2000 km with the same τ as is appropriate at 500. Where a number of 1000- to 4000-km circuits are available it would be reasonable to compute a third figure from them. We would then have the following.

(A) η_1 computed from the discrepancies between fore and back levelling between adjacent BMs. So the s.e. over 0–5 km would be $\eta_1\sqrt{L}$.†

(B) η_2 computed from the closures of circuits of (say) 300–600 km. The s.e. over the mean of such distances will then be $\eta_2\sqrt{L}$.†

(C) η_3 from the closures of the largest circuits available, perhaps averaging 2000 km, and the s.e. over such distances will be $\eta_3\sqrt{L}$.

For intermediate distances use intermediate values of η.

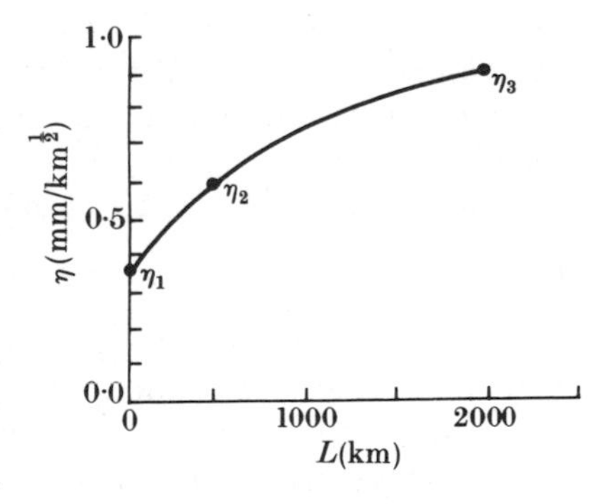

FIG. 3.6. Standard error of double levelling = $\eta\sqrt{L}$. η_1, η_2, and η_3 are (hypothetical) values of η appropriate for distances of 3,500, and 2000 km.

† In the 1971 printing of [113] 3 Edn., p. 245 $\eta_1\sqrt{L}$ and $\eta_2\sqrt{L}$ were misprinted as $\eta_1/\sqrt{L}$ and $\eta_2/\sqrt{L}$.

TABLE 3.1

Standard errors in millimetres, Finland

Distance (km)	Lallemand $(\eta^2 L + \sigma^2 L^2)^{\frac{1}{2}}$	Vignal	
		$\eta\sqrt{L}$	$\tau\sqrt{L}$
1	0.35 mm	0.33 mm	—
10	1.2	1.0	1.9
50	3.2	2.3	4.2
100	5.3	3.3	6.0
250	11.5	—	9.5
500	21	—	13
1000	40	—	19
2000	80	—	27
4000	160	—	38

The information obtainable from any levelling net will then be fully represented by these three, or more, values of η, with a statement of the distances to which each is applicable. A curve such as Fig. 3.6 results. If data are scant the accepted curve need not exactly pass through the plotted points. Negative values of $d\eta/dL$ are probably unacceptable.

(*b*) *Formulae for computation.* Formulae for computing Lallemand's η and σ are in [113] 2 Edn., [318], [360] and many others. Vignal's formulae are in [579], and more concisely in [318].

To compute the s.e. $\eta_1\sqrt{L}$ from discrepancies between fore and back,

$$\eta_1^2 = (1/4n)[\Delta^2/r]\ \text{mm}^2/\text{km}, \tag{3.19}$$

where Δ is the difference in mm between the fore and back levelling between BM's, r (1to 5 km) is the distance of each section between BM's in kilometres, the square brackets indicate summation, and n is the number of sections.

To compute the s.e. $\eta_2\sqrt{L}$ from the closing errors of circuits†

$$\eta_2^2 = [e^2/L]/n\ \text{mm}^2/\text{km}, \tag{3.20}$$

where e mm is the closing error of each circuit, L km is its length, the brackets indicate summation, and n is the number of circuits.

As suggested above η_2, η_3, etc., may be computed separately using homogeneous groups of circuits of comparable lengths.

The s.e.'s $\eta\sqrt{L}$ will partially or wholly ignore errors which tend to cancel round a circuit, such as those which depend on height or latitude, and the magnitude of such errors cannot be assessed from a comparison

† Vignal's formulae include the bounding circuit in the summation, since its lines are otherwise included only once, and all others twice, But since the bounding circuit is much longer than the others, we prefer to exclude it from the summation of the smaller circuits.

of numerical results.† They can only be eliminated if their causes are known. The technique of observation can then be designed to avoid them, or to apply corrections for them. And the magnitude of possible residual error can only be estimated in the same way. As a guide to this process, graphical comparison of the fore and back levelling of all lines should be made, and where the differences tend to increase systematically over considerable distances attempts should be made to correlate the errors with climatic, topographical, or (superficial) geological conditions, or with the instrumentation.

(*c*) *Accuracy attainable*. In 1912 the International Geodetic Association resolved that levelling should only be described as of high precision if the random probable error was <1 mm/km$^{\frac{1}{2}}$ and the systematic $<$ 0.2 mm/km, for the mean of fore and back levelling. The limit for s.e.'s would be $1\frac{1}{2}$ times as much.

In 1948 the Association redefined high precision levelling as having τ <2 mm/km$^{\frac{1}{2}}$ (probable) or <3 mm/km$^{\frac{1}{2}}$ (standard), i.e. the s.e. $<3L^{\frac{1}{2}}$ mm when L exceeds a distance of some tens of kilometres.

Values of τ^2 for west European level nets in 1959 are given in [526], and show τ varying between 0.6 and 2.5 mm/km$^{\frac{1}{2}}$. The larger figures are associated with mountainous country or old work.

3.08. The datum surface

Every survey organization, or group of such organizations, aims at recording the heights of the BM's above the geoid as defined in § 2.02 (*b*) as the equipotential surface which on average coincides with MSL in the open ocean. The method of securing this is to set up tide gauges at suitable places, and to accept one of them or the mean of a group of several, as the best available method of locating the geoid, and to define this MSL as being of zero geoidal height. There are, however, three further points to be considered.

(*a*) Although MSL in the open ocean is approximately an equipotential surface, it is not exactly so. See § 3.28. The difference may amount to 1 or 2 metres. Oceanographers can estimate the height of MSL above a world mean, i.e. above the geoid. Their estimates may be imperfect, but when MSL at a tide gauge is to be accepted as a levelling datum, its geoidal height should be taken as zero plus the oceanographers' correction. If several tide gauges are available, it will be questionable whether to accept each separate (corrected) MSL as of zero height and to adjust the connecting levelling accordingly, or whether the levelling should be accepted as errorless, so that the accepted height of each (corrected) MSL

† Except that large differences between repeat levellings of the same line in different years may suggest changing systematic error rather than changes of height.

departs from zero, but minimally so. Alternatively, of course, weights may be given to both the levelling and the estimated geoidal heights of MSL.

(*b*) As stated in § 3.06 (*c*) (iii) the direction of gravity, which is perpendicular to the equipotential surfaces, varies with the aspect of the sun and moon. In fact, any particular equipotential surface rises and falls, relative to a fixed point at the centre of the earth, through some tens of centimetres twice a day. The surface of the elastic solid earth moves similarly, but through only about half as much, § 6.78. The result is that the height of a surface point moves vertically relative to an equipotential surface through perhaps 10 cm every 12 hours. Recorded heights cannot be allowed to do the same. The remedy is, firstly, to insist that the geoid (and the datum of height) is an equipotential surface of the earth's attraction and centrifugal force only, not including the sun and moon, and secondly to define the height of a surface point to be that corresponding to its mean position with respect to the solid earth's tidal deformations. In principle, this demands that the tidal corrections of § 3.06 (*c*) (iii), modified to include the earth's deformation, should be applied to all levelling. Since the earth's deformation is imperfectly known, it is fortunate that the mean tidal height corrections are small, and negligible in their variation between near points.

(*c*) An important object of levelling is to record changes of height, of the land relative to the sea, and of one land area relative to another. For this purpose MSL is ill adapted to define an ultimate datum level, since it fluctuates through many causes which have no connection with deformations of the earth's crust. The best datum is constituted by a large number of well-constructed BM's, founded on solid rock, in locations which are likely to be free from disturbance from earthquakes, erosion, or local subsidence. Mean sea-level, as in (*a*) and (*b*) above can be used to give these BM's their geoidal heights, but crustal movement will best be initially studied on the preliminary assumption that on average the heights of the BM's remain unchanged.

3.09. River crossings

(*a*) *Bridges.* The easiest means of crossing a wide river is to use a bridge, provided the supporting piers are sufficiently close together for the level and staves to be placed always over a pier. Some increase in the usual maximum length of sight may have to be accepted, if only because any other course may be worse. The crossing of a bridge with spans of over (say) 100 metres calls for extra measurements and some thought may have to be given to what the errors are likely to be. Levelling along a suspension bridge with a span of (say) 300 m is clearly a questionable procedure.

Where a bridge cannot be used, wide rivers may be crossed either by levelling to special targets, as in (*b*) below, or by theodolites (*c*), or in some circumstances by the so-called tide gauge method, (*d*).

(*b*) *Levelling*, [308], [478], and [322]. When the staff is too distant to be read directly, two fixed targets are attached to the staff, one above the level line and the other below it. Then using the micrometer screw M in Fig. 3.4, the centre wire is first brought on to the top target and M read. M is then read with the bubble central, and finally with the centre wire on the lower target. The staff reading corresponding to a level target is then determinate. For a river 1 or 2 km wide about 100 sets of readings are advised. This is a good and convenient method.

The back sight can seldom be as long as the crossing. Even if it were possible to make it so, refraction conditions in the back sight and over the river would be so dissimilar as to make it undesirable. Reciprocal observations should therefore be made from both sides of the river, with back sights of the usual lengths, thereby cancelling curvature and normal refraction as well as collimation. Care must be taken not to jolt the level and change collimation while changing sides. A level which can be turned over about its long axis to eliminate collimation should be read in both positions.

Zeiss produce a special equipment for long horizontal lines crossing rivers or valleys, where fore and back sights are unequal. It consists of four Ni 2 levels, mounted in pairs for simultaneous use on either side. Absence of collimation error is verified by each pair being frequently turned in towards each other during the observations. Two targets in the form of horizonal vs are mounted on the staff at each side of the gap, 20, 40, or 60 cm apart. A rotatable wedge prism in one of each pair of levels enables the targets to be brought on to the graticule wire, and enables the recording of the angular elevation of each target to 2″, with estimation to 0″.2. An approximate knowledge of the distance converts these to height differences.

(*c*) *Theodolite angles.* Vertical angles are observed to lamps or helios, well stopped down, in the usual way. The line need not be exactly horizontal, but it is advantageous to have the slope less than the range of the micrometer. The micrometer run must be tested, and the error allowed for if significant. If the line is not level, its length must be known, but with low accuracy. [602] describes the method, and lists 51 crossings of between 0.4 and 8 km.

The best time of day is probably the first few hours after noon, but other times have been satisfactory.

(*d*) *Tide gauge method.* It is possible to cross long narrow lakes and deep slow-running rivers by levelling to the water's edge on each side and assuming equality of height. This will be very inaccurate if there are central shallows, or a strong wind along the crossing, but fair results may be possible. See [104], pp. 11–13, and § 3.29.

(*e*) *Refraction.* Using either the best levels or the best theodolites, the worst source of error is likely to be refraction. See (3.21) and (3.22). The line is likely to be sufficiently level and short for variations of P/T^2 to be insignificant, and the trouble lies in the variations of the temperature gradient $\mathrm{d}T/\mathrm{d}h$. Variations of the gradient with time can be got over by simultaneous reciprocal observations, and this is necessary in crossing of more than a few hundred metres by either level or theodolite, but there is no guarantee that $\mathrm{d}T/\mathrm{d}h$ is the same on both sides of the river, and its value in the half-distance nearest the instrument is what the refraction at each end mostly depends on. If one half is over dry sand and the other over water, the refraction at the two ends will be entirely different, and differences of water temperature, such as occur below a river junction or in water of different depths, may also be serious.

Unfavourable situations are (i) great width, (ii) low clearance above the ground or water, (iii) asymmetry of ground and water, (iv) clear skies, great heat or cold, and absence of wind. Items (i), (ii), and (iii) may conflict. The shortest and most symmetrical line may be got from the water's edge on either side, but the clearance will then be only $1\frac{1}{2}$ metres or less. It is difficult to say to what extent extra width and asymmetry should be accepted in order gain height, but the following from [104] may be some guide. In the mean of a long series of reciprocal sights in warm sunny weather for a line which is partly over dry sand and partly over water, the error due to asymmetry may be expected to be of the order of

$$\text{Error} = 16\mathrm{e}^{-0.21h}bc \times 10^{-8} \text{ metres,} \tag{3.21}$$

with an s.e. of

$$\pm 8.3L^2\mathrm{e}^{-0.21h}\{(b^2/L^2)(c^2/L^2) + (60/L)\}^{\frac{1}{2}} \times 10^{-8} \text{ metres,} \tag{3.22}$$

where L is the total width, b is the width of sand on one side, $c = L - b$ the width of water, h the height of the sight above water-level, all in metres, and e is the base of natural logs. If the crossing is all water, $b = 0$ and $L = c$ and (3.21) equals zero. The sign of the error in (3.21) is such that the sandy side will be given too great a height. Example: in a line over 600 m of sand followed by 900 m of water, 3 metres above water-level, expect an error of 0.05 ± 0.03 m.

These formulae have been obtained empirically from observations controlled by ordinary levelling across an adjacent bridge, and by direct measures of $\mathrm{d}T/\mathrm{d}h$. They cannot be precise, but are some guide to what may be expected. They are applicable to crossings between 1 and 4 km wide at heights of 1–10 m above water.

[320] describes corrections derived from measures of the vertical temperature gradient.

(*f*) *Accuracy.* For an apparently symmetrical site (3.22) reduces to

$$\text{s.e.} = 0.2e^{-0.21h}L^{\frac{3}{2}} \text{ metres}, \quad \text{if } L \text{ is in kilometres.} \tag{3.23}$$

If $h = 5$ metres, the s.e. becomes $\pm 7L\sqrt{L}$ mm. In Finland [352] gives s.e. $= 12L$ mm, for crossings of between 0.5 and 5 km between small islands in the Baltic, giving a reasonably similar result.

Repetitions on the same site, especially at the same time of day and in similar weather, may leave refraction errors unrevealed. A fair test will be made if observations are made both in the afternoon and late at night, or at different sites selected so that any apparent asymmetry may have opposite effects.

(*g*) The following are the relative advantages of using a level as in (*b*), or a theodolite as in (*c*).

In favour of the theodolite

(i) Not being held to a horizontal line, the theodolite gives a wider choice of sites, which may help to meet the need for greatest height in the shortest line.

(ii) The theodolite automatically eliminates collimation error by change of face. Not all levels can be used in both positions, and collimation may change while moving from one side of the crossing to the other.

(iii) In a very long line earth's curvature will cause a level sight from one end to pass high above the instrument at the other. The two level sights will not be truly reciprocal, but this is no more serious than having a theodolite line which is not horizontal.

In favour of the level

(iv) In a level there is no question of graduation error or micrometer run.

(v) The use of the level with practically no special equipment is more simple than bringing two theodolite parties to the site.

(*h*) *Asymmetrical geoid.* Reciprocal angles, or levels sight, cancel the correction for earth's curvature provided that the geoidal profile at the crossing is circular or otherwise symmetrical about the centre of the crossing. A rough guide to the fallibility of this assumption is given by the following. In a crossing AB, to the height of B above A as given by reciprocal level or theodolite sights there should be added

$$\tfrac{1}{3}(\psi_A + \psi_B - 2\psi_C)L \sin 1'', \tag{3.24}†$$

where ψ_A, ψ_B, and ψ_C are the components of the deviation of the vertical

† Simpson's rule gives $\frac{1}{6}(\psi_A + 4\psi_C + \psi_B)L$ for the difference of height of the geoid above the spheroid at the two ends. The assumption that the geoid is circular implies that the difference is $\frac{1}{2}(\psi_A + \psi_B)L$. Hence (3.24).

along AB at A, B, and C the mid-point in seconds, ψ_C being unknown: and as in (4.95)

$$\psi = -(\eta \sin A + \xi \cos A), \tag{3.25}$$

where A is the azimuth of AB.

Except among mountains $\frac{1}{3}(\psi_A + \psi_B - 2\psi_C)$ in a 5-km line is unlikely to be as much as 2″, giving an error of 17 mm, and in ordinary undulating country it may be expected to be very much less. This error will ordinarily be trivial in comparison with the uncertainty of refraction.

3.10. Hydrostatic levelling

A wide river or narrow sea may, in principle, be very simply crossed if a suitable pipe is available, since if the pipe is filled with water, the level of the water should in suitable conditions be the same at both ends. In the Netherlands, [584], a pipe of 10-mm inner diameter has been used to measure distances of up to 7 km across narrow sea or estuarine crossings. The results are believed to be correct to about 1 mm. Necessary conditions are listed below.

(*a*) The diameter of the pipe must be large enough to avoid friction and stoppage by bubbles, but small enough to damp the oscillation of the water and so to give stable readings. A 10-cm pipe has also been found satisfactory.

(*b*) The pipe must be free from contained air. This may be difficult, especially if there is a central rise. The absence of air may be tested by adding a known volume of water and seeing that the resulting rise at each end confirms that the contents of the pipe are incompressible.

(*c*) Corrections must be applied for differences of barometric pressure at the two ends, if appreciable.

(*d*) The density of the water in the rising parts of the pipe must be the same at both ends. If the pipe goes down to a great depth, this may be a serious difficulty. Unequal density may be caused by changes in temperature, small air bubbles, salinity, or suspended matter, This trouble is least in a floating pipe, but that is only possible in still water or on ice.

[52] reports the completion of a new level net in Holland consisting of 590 km of hydrostatic levelling. The average length of one measurement is 4.6 km. The lines follow canals. The equipment includes a lead pipe 8 km long, which is laid down and taken up by a specially equipped ship.

3.11. Bench marks

The number of BM's required for public purposes may vary from 3 per km in towns to one per 5 km, or less. From the geodetic point of view requirements are as follows.

(*a*) It is convenient to close the day's work (perhaps 5 km) on a solid mark, and to connect the same mark when relevelling in the opposite direction, see § 3.05 (*f*).

(*b*) A greater spacing is inconvenient where there is any chance of another line being started or closed, as check levelling to prove stability and right identification will then be laborious. Allowance must be made for future losses. From this and other points of view, one BM per 1 or 2 km is a good target.

(*c*) To avoid the complete loss of the levelling, such as may result from the reconstruction of a road, permanent protected BM's should be made every 50 km, and handed over to some local authority for care. These BM's and others specially constructed to monitor crustal movements, § 3.08 (*c*), are often known as *Fundamental* BM's (FBM's). It is desirable that they, and many other BM's (and triangulation stations) should have some legal protection.

(*d*) It is seldom possible to be sure about suspected tectonic movements, §§ 6.71 and 6.74, unless the BM's concerned have been cut on solid rock, nor even then if they are at the top of a cliff or on steeply sloping ground. If possible, rock-cut BM's should be made and permanently marked at least at every 100 km, and much more frequently in earthquake areas, and on either side of faults where recent movement is suspected.

(*e*) Triangulation stations should be connected where control of their heights is required, see § 3.16. It is important to record very clearly which of the (possibly several) station marks have been connected.

Permanent protected BM's such as in (*c*) above, if not cut on solid rock, should be based on about 3 cubic metres of concrete below ground, with a hard stone monument projecting. Two or three smaller witness marks should be set in the concrete, so distributed as to reveal tilting of the block, and a low wall should mark and protect the site. Recent alluvium, mining areas, steep slopes, massive buildings, railways, and heavy road traffic should all be avoided. These BM's should be built a year or more before connection.

Ordinary inscribed BM's, as in (*a*) and (*b*) above, can be on road (but preferably not railway) bridges. Milestones are bad, being liable to be relaid rather often, and the copings of wells are especially liable to subsidence. In soft ground (in Finland) driven pipes 3–5 metres long of 30–45 mm diameter have sunk less than 1 mm in 10 years, [318]. The record must contain full descriptions of all BM's to enable them to be found, and also to avoid confusion with witnesss marks, if any.

3.12. Computations and adjustments

(*a*) *The preliminary computations* are the same as for ordinary levelling, except that the height differences between adjacent BM's may be corrected for the following.

(i) Staff length, if significant.

(ii) Unequal refraction on hills, § 3.06 (*b*) (iv), if it can be significantly determined.

(iii) Tidal effects, § 3.06 (*c*) (iii).

Measured height differences are then either converted to orthometric by (3.9), if standard gravity is being used, or to geopotential numbers by (3.10) if preferred and if sufficient values of g are available. Or, as an alternative to geopotential numbers (3.12) may be used to convert measured differences to normal dynamic differences.

Where regular tectonic changes of height are known to be taking place, as in Finland, it may be necessary to express heights in terms of a selected epoch, and to allow for the movements of BM's from year to year when forming circuits. See [318], pp. 32–4 and 45–7.

(*b*) *The adjustment* may be carried out in terms of either orthometric heights, geopotential numbers, or normal dynamic heights. In Fig. 3.7, the basic data are the height differences (2, 1), . . . , (12, 11) between junction points, obtained by summing height differences obtained as in (*a*) above, and taking the mean of fore and back levellings.

The adjustment may be carried out either by the variation method, analogous to variation of coordinates in triangulation, or by conditions.

(*c*) *Adjustment by variation.* An approximate height is assumed for every junction point. Any value will do, such as zero, but (desk) arithmetic is simplified if it can be within a metre of the final result. Then from the approximate heights, record the corresponding height difference in each line (2, 1), etc., with the correct sign in accord with the (arbitrary) arrows in Fig. 3.7. Call these provisional differences C, and let the observed values be O. Then we require the height corrections at each junction such as will secure best agreement between the O's and the corrected C's. Comparing with § 2.23, the observation equations are

$$\delta h_1 - \delta h_2 = (\mathrm{O}-\mathrm{C})_{2,1} + v. \tag{3.26}$$

Weights must be allotted to each O. It will be convenient to take $w = 1/(\text{s.e.})^2$, with $\text{s.e.} = \eta\sqrt{L}$ as in § 3.07 (*a*) (iii),† but allowance can be made for river crossings, exceptional changes of altitude, or other unusual sources of error.

The height of at least one point, or tide gauge, must be held fixed and its δh omitted from the observation equations.

† Different values of η may be used for lines of different lengths, but it may be noted that this implies some correlation between the errors of adjacent lines, since e.g. $(\text{s.e.})^2_{6,4}$ will not equal $(\text{s.e.})^2_{6,5} + (\text{s.e.})^2_{5,4}$. § D.22 provides the means of handling such a situation, but it will be difficult to formulate the implied correlations. This point may well be ignored, or a single suitable η may be used for computing the weights of all the lines.

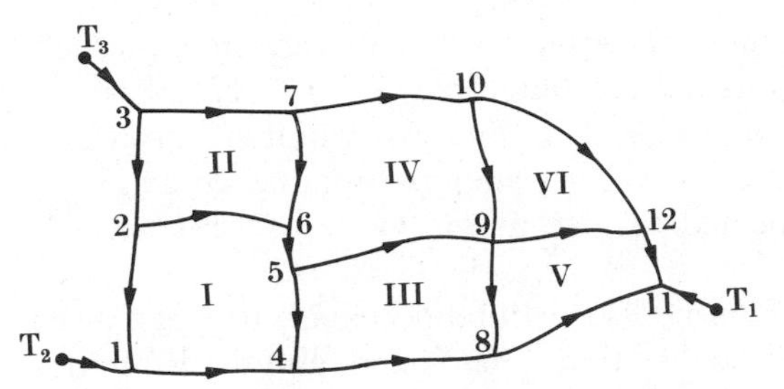

FIG. 3.7. Level net. Arrows show direction of computation. 1 to 12 are junction points. I to VI are circuits, computed clockwise. T_1 to T_3 are tidal stations.

The solution is carried out as in § D.14, and § D.15. (D.67) gives the s.e.'s and covariances of the junction point heights.

For an example see [466], pp. 218–21.

In this method there is one unknown per junction point and tidal station, less one or more held fixed. The solution is very much easier than that of a triangulation or traverse net covering a similar area. In a net of 100 circuits the bandwidth of the matrix of the normal equations will probably not much exceed 20 or 30.

When the heights of the junction points have been determined, the residuals in the observation equations must be distributed along the lines, evenly or unevenly according to the possibly varying quality of the work in each.

(*d*) *Adjustment by conditions.* Minimum corrections to the observed height differences between junctions are sought which will secure zero circuit-closing errors. Condition equations take the form

$$-x_{2,1}+x_{2,6}+x_{6,5}+x_{5,4}-x_{1,4}=\epsilon_{\text{I}}, \tag{3.27}$$

where $x_{2,1}$ is the correction to line (2.1).

There will be an extra condition for every tide gauge or other point (in excess of one in all), which is to be held fixed, in the form

$$x_{T_1,11}-x_{8,11}-x_{4,8}-x_{1,4}-x_{T_2,1}=\epsilon_{\text{VII}}. \tag{3.28}$$

Weights may be as for the variation method. The solution is carried out as in §§ D.16 and D.17. (D.87) gives the s.e.'s and covariance of the corrections.

For examples see [466], pp. 215–18.

The method of variation has not got the great advantages, as compared with conditions, that it generally has in triangulation or traverse. The method of conditions is the more self-checking, and it may well be preferred, [466], p. 221.

(*e*) *Orthometric heights.* If the adjustment has been done in terms of geopotential numbers or normal dynamic heights, the orthometric height of each BM is finally computed by (3.11) or (3.13) respectively.

(*f*) *Tidal stations.* It is a question whether (or which) tidal stations should be held fixed. Clearly stations situated up an estuary or tidal river cannot be expected to give a true value of MSL, or of the geoid. See §§ 3.08 and 3.28.

For scientific purposes it will be necessary to make a special solution in which only one tidal station is held fixed, and thence to obtain values (and their s.e.'s) for the height of MSL at all others. This solution may then help to indicate which tidal stations should be held fixed in the final solution on which the published values of the BM's will be based.

(*g*) *Acceptance of old work.* Once an adjustment has been completed another cannot be carried out and published whenever a new line is observed, and new lines have to be adjusted to the old, accepting all old BM's except any which have clearly been disturbed. After some decades there will probably be trouble from new lines having to accept significant errors for which they are not responsible, and a complete readjustment may then be considered. The question whether differences between old and new adjustments are due to real changes of ground-level, local movements of BM's, or errors in the levelling, will then be a problem calling for careful consideration

General references for levelling [318], [322], [478], and [597].

SECTION 3. HEIGHTS BY VERTICAL ANGLES, AND REFRACTION

3.13. Basic formulae

It is required to measure differences of height above the geoid. In Fig. 3.8 vertical angles are observed at P, at height h_1 above the geoid, and reciprocally from Q whose height h_2 is required. The section of the geoid is assumed to be the arc of a circle. PT is the horizontal at P, and Q′ is at the same height above the geoid as P. PQ subtends an angle of θ at the centre of the geoid's circular arc.

The vertical angle RPT ($=\beta$) is observed at P, as in § 1.26. The refraction angle Ω is computed as in § 3.18, and subtracted to give $QPT=\beta-\Omega$, of which a mean value may be used for computation.

Then $\beta' = QPQ' = \beta-\Omega+\frac{1}{2}\theta$

$$\Delta h = h_2-h_1 = QQ' = PQ' \sin\beta' \sec(\beta'+\tfrac{1}{2}\theta)$$
$$= L(1+h_1/R)\sin(\beta-\Omega+\tfrac{1}{2}\theta)\sec(\beta-\Omega+\theta), \tag{3.29}$$

or

$$\Delta h = L\tan\beta'\left(1+\frac{h_2+h_1}{2R}\right), \tag{3.30}$$

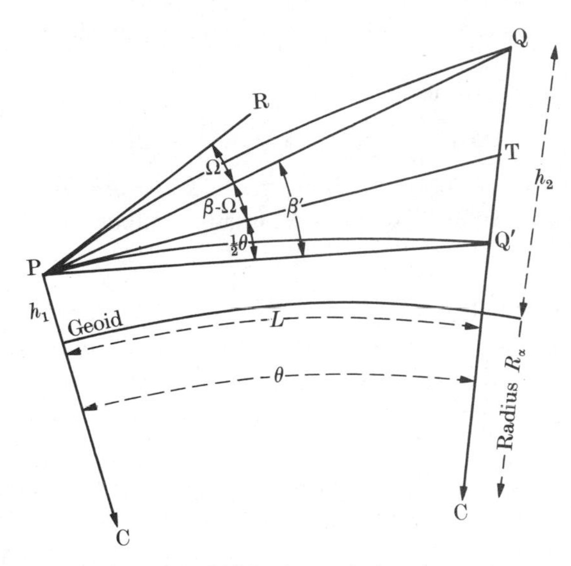

FIG. 3.8. Heights by vertical angles.

involving the approximations that chord $PQ' = L(1+h_1/R)$† and $L \tan \beta' = h_2 - h_1$ in the second term of (3.30), but the resulting errors are far smaller than those of refraction. The factor $1+(h_2+h_1)/2R$ is only material in lines rising to over 1500 m. For computing θ we have $\theta = L/R_\alpha$, where R_α may be taken to be the radius of curvature of the spheroid in the azimuth and latitude concerned, (A.67). If PQ has been directly measured, as by tellurometer,

$$h_2 - h_1 = PQ \sin \beta' \sec \theta/2. \tag{3.31}$$

It is best to compute the observations at P and Q separately, and to take the mean. Labour can be saved by a formula which combines the two, but the double computation provides a useful check, and gives some insight into the accuracy of the work. When the heights of P and Q are very unequal, separate computation enables allowance to be made for the normal variation of refraction with height, § 3.19.

The deduced height of Q must of course be increased by the height of the theodolite axis above the station mark at P, and decreased by the height of the signal at Q, and vice versa for observations from Q. To avoid gross error it is a convenient rule that when a station has more than one mark, § 1.12 (*a*), the accepted height should always refer to the one

† L being spheroidal distance, h should properly be spheroidal height when deducing PQ′, but the difference is generally trivial. If the ground–spheroid separation is known to be large, and if the line is steep, the estimated spheroidal height may be used.

which is approximately at ground level. This height can then be printed on maps.

The elementary formula

$$\Delta h = L \tan \beta + (L^2/2R_\alpha)(1-2k) \tag{3.32}$$

is sometimes a useful approximation, where $k = \sigma/\theta$ as in § 3.18, and R_α is the spheroidal radius of curvature in the latitude and azimuth concerned, (A.67).

3.14. Geoidal asymmetry

The treatment of §3.13 assumes that the geoidal section is circular and of known radius, although in the mean of reciprocal angles errors in this assumption cancel provided the geoidal section is symmetrical about its middle point.

(*a*) *Single observation from p to q.* See Fig. 3.9. A theodolite at p, h metres (orthometric) above the geoid, measures β the vertical angle to a point q. Ignoring refraction, the angle measured is qpx, where px is tangent to the equipotential through p, since the theodolite bubble lies parallel to px. The orthometric height of q is qQ$' = h_1$, and the elementary procedure is to write

$$h_1 - h = \{\text{distance} \times \tan(\beta - \Omega)\} + \text{curvature correction}, \tag{3.33}$$

where the curvature correction is TQ, the depression of the spheroid below its tangent PT.

More accurately, in Fig. 3.9 let pt be parallel to PT. Then

$$\text{xpt} = \psi = -(\eta \sin A + \xi \cos A)\dagger$$

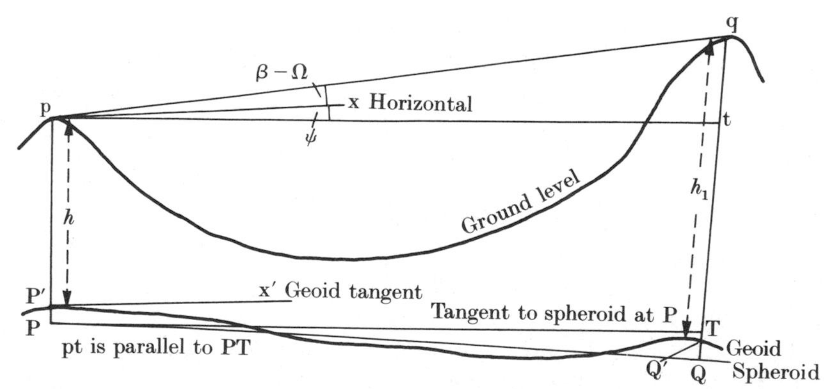

FIG. 3.9. PQ is spheroid and P′Q′ is geoid. PT is tangent to the spheroid, and pt is parallel to it.

† Note that the deviations of the vertical ξ and η refer to the level surface at p. They should not be reduced to geoid level as in §§ 6.24 or 6.55. A is the azimuth of Q at P.

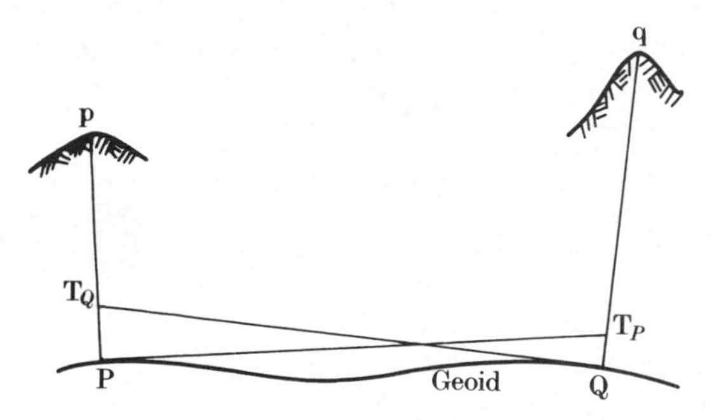

FIG. 3.10. PQ is the geoid. PT_P and QT_Q are tangent to it.

and $$h_1 - h = \text{qt} + \text{TQ}' + \text{P}'\text{P} = \text{qt} + \text{TQ} + (\text{PP}' - \text{QQ}'), \tag{3.34}$$

where $\text{qt} = L(1 + h_1/R)\sin(\beta - \Omega + \psi)\sec(\beta - \Omega + \psi + \theta)$.

This can be computed if ψ is given by astronomical observations at p, and if $\text{PP}' - \text{QQ}'$ is given by a geoidal section as in § 4.50. Vertical angles to an unvisited point, using only (3.32), give the orthometric difference of height above a spheroid whose axes are those of the reference spheroid, but which is tangent to the level surface at the point of observation. This is only a satisfactory approximation to the true orthometric difference over a short distance, such as 10 or 20 km.

(*b*) *Reciprocal observations.* In Fig. 3.10 the effect of a non-circular geoid will be cancelled if $T_Q P = T_P Q$. If this is not the case, the height of q above p as computed from reciprocal observations with conventional equal curvatures will require a correction of

$$\delta h = \tfrac{1}{2}(T_P Q - T_Q P). \tag{3.35}$$

Formula (3.24) offers a means of estimating the order of magnitude of δh, although in a long line (3.24) may be quite inapplicable. The error due to geoidal asymmetry is generally small compared with refraction error, except in high mountains, where one may be as large as the other. The criteria p and P of § 3.15 include the effects of both.

3.15. Accuracy of triangulated heights

The sum of the height differences in the three sides of a triangle should be zero. As observed, let it be ∇ metres. In any triangulation chain let

$$p = \sqrt{\textstyle\sum \nabla^2/3n}, \tag{3.36}$$

where n is the number of triangles involved, and p is then the s.e. of the unadjusted height difference in a single line. The s.e. of a line after triangular misclosure is $p \times \sqrt{(2/3)}$. When forming p ignore any redundant lines of exceptional length, both in the summation and in n. Then in a chain of triangles, quadrilaterals, etc., which provide three or four independent routes for carrying forward the height, let P be the s.e. of height after 100 km, and a good value for P will be

$$P = p\sqrt{(f/3)}, \tag{3.37}$$

where f is the average number of lines on each flank of the chain per 100 km. And after $100S$ km the s.e. will be $\pm P\sqrt{S}$.†

In India [103]‡ p, the s.e. of observed height in one side of a triangle, has been found to be 0.3 metres in favourable circumstances, but typically it is more like 0.5 m in flat ground with $L = 20$ km and height above ground 5 or 10 m; 0.5 m in hilly country with $L = 40$ km and height about 100 m; and 1.0 m in mountains with $L = 70$ km and height 1000 m. With these values of p and L (3.37) gives

$$P = 0.65,\ 0.45,\ 0.69 \text{ metres}, \tag{3.38}$$

for the s.e. of height after 100 km of triangulation in the three types of country. It is, of course, best to calculate the true value of P in any particular case. Approximate figures for a net can be obtained by treating it as an equivalent system of close-spaced parallel chains.

The error in a traverse is likely to be about twice as great.

When the heights of triangulation or traverse are being adjusted, each line should be weighted according to the inverse square of its s.e., which will vary according to the actual length of the line. Error mostly comes from the inequality of the refraction angles Ω at the two ends, and will be $\frac{1}{2}L(\Omega_P - \Omega_Q)$. Other circumstances such as height and weather being the same, $(\Omega_P - \Omega_Q)$ is likely to increase with L, perhaps in direct proportion to L or perhaps to $L^{\frac{1}{2}}$. It is better to take the less drastic $L^{\frac{1}{2}}$, so that the error of height varies as $L^{\frac{3}{2}}$, and to give the following, from [103] quoted above, as typical s.e.'s of height differences in lines of L km observed in both directions (on different days) during the hours of minimum refraction.

$$\begin{aligned} &\pm 0.5(L/20)^{\frac{3}{2}}\text{ m} = 0.0055L^{\frac{3}{2}}\text{ m in flat country, } h = 5 \text{ to } 10\text{ m}.\\ &\pm 0.5(L/40)^{\frac{3}{2}}\text{ m} = 0.0020L^{\frac{3}{2}}\text{ m in hills, } h = 100\ m.\\ &\pm 1.0(L/70)^{\frac{3}{2}}\text{ m} = 0.0017L^{\frac{3}{2}}\text{ m in mountains, } h = 1000\ m.\end{aligned} \tag{3.39}$$

The distinction between the last two can be ignored. The first line agrees well with § 3.09 (*f*) which gives $\pm 0.007L^{\frac{3}{2}}$ metres for river crossings at $h = 5$ m, from different data. In (3.39) h is the average height of the line above the ground.

† If the heights of a chain are adjusted between two errorless spirit-levelled values, at distances of $100S$ and $100KS$ km from a point, the s.e. of the adjusted height at the point will be $P\sqrt{S}\sqrt{\{K/(K+1)\}}$ m. If $K > 1$, i.e. if $100S$ km is the distance to the nearest connection, $\sqrt{\{K/(K+1)\}}$ is between 1.0 and 0.7 and is easily estimated.

‡ In [103] p is s.e. in feet, not metres. Its P is not the same as P is here.

Lines observed in one direction only. The error is the error in ΩL or in $k(L^2/R)$, where R is earth's radius and k is the coefficient of refraction. In the most favourable circumstances the s.e. of k may be about ± 0.005, and it will be perhaps as much as ± 0.3 in low lines observed on cold nights. Except between mountains, or over very short distances, such lines can have little or no weight compared with lines reciprocally observed.

3.16. Adjustment

Heights derived from vertical angles should be adjusted to spirit levelling or to mean sea-level, whenever possible.

(*a*) *Least squares*. Heights may be adjusted by least squares, either by the variation method or by conditions. Apart from the weighting, the situation is exactly as in § 3.12 (*c*) or (*d*), triangle sides (or whole chains of triangulation, or traverses) taking the place of level lines in a level net.

(*b*) *Weights*. In principle, weights should be $1/(\text{s.e.})^2$. If the adjustment is of a number of chains, each of which has proviously been made self-consistent, the s.e.'s are given by (3.37) using values of p given by the relevant triangle closures.

When individual lines have to be adjusted within a chain s.e.'s are similarly given by (3.39), using factors locally determined as above. If other conditions are equal, this weights lines as the inverse cube of their lengths. But it must be recognized that even in a single chain, or triangulation figure, other things may not be at all equal. Consideration must be given to such matters as below-average height clearance, wrong times of day, and abnormal discrepancies between reciprocal observations.

3.17. Refraction. The curvature of a ray of light

The path of light through a medium in which the refractive index n varies, is determined by the law

$$\int n \, dl \text{ along the path is a minimum.} \tag{3.40}$$

Light follows the shortest 'optical path'.

Consider an atmosphere is which the layers of equal n are horizontal planes,† and in which dn/dh is a constant, ordinarily negative, h being height above the ground-level plane. In Fig. 3.11, which lies in a vertical plane, the arc AB of length L_1 is a short element of the path of a ray of light PQ. Let the radius of curvature of AB be σ, which is large compared with L_1, so that the angle θ between the end tangents of AB is L_1/σ. Let

† More accurately spheres, but over the short element of distance AB the earth's curvature can be ignored.

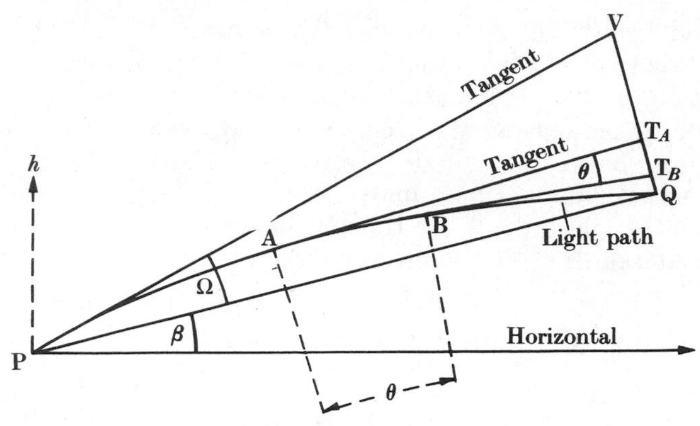

FIG. 3.11. $AB = L_1$, $PQ = L$. θ is the angle between tangents to the light path. $\cos\beta \approx 1$.

the inclination of PQ (and of AB) to the surfaces of equal n be β, so the gradient of n across AB is $(dn/dh)\cos\beta$. β is a small angle.

To determine σ is exactly the same problem as that of § 2.59 (*b*) with n here in the place of $1/m$ and h for n, and we have

$$\text{Arc AB} - \text{chord AB} = \sigma\theta^3/24. \tag{3.41}$$

The mean separation of arc and chord over the distance AB is $\sigma\theta^2/12$, and the mean value of

$$(n \text{ on arc}) - (n \text{ on chord}) = (1/12)\sigma\theta^2(dn/dh)\cos\beta. \tag{3.42}$$

Let $\Delta = \int n\,dl$ along arc $-\int n\,dl$ along chord.

Then $\Delta = n\sigma\theta^3/24$ from $(3.41) + L_1(\sigma\theta^2/12)(dn/dh)\cos\beta$ from (3.42)

$$= nL_1^3/(24\sigma^2) + (L_1^3/12\sigma)(dn/dh)\cos\beta. \tag{3.43}$$

For minimum $\int n\,dl$ along the arc $d\Delta/d\sigma = 0$, i.e.

$$nL_1^3/12\sigma^3 + (L_1^3/12\sigma^2)(dn/dh)\cos\beta = 0$$

$$\frac{1}{\sigma} = -\frac{1}{n}\frac{dn}{dh}\cos\beta. \tag{3.44}$$

In (3.44) $n = 1$ within 0.03 per cent, so

$$\frac{1}{\sigma} = -\frac{dn}{dh}\cos\beta, \quad h \text{ and } \sigma \text{ being in the same units.} \tag{3.45}$$

The refractive index of air is given by (1.10). In this context $(n_0 - 1)$ is generally taken as 293×10^{-6}, and $\alpha=1/273$. Differentiating (1.10) then gives

$$\frac{dn}{dh}=\frac{7.9}{T\times10^5}\left\{\left(\frac{dP}{dh}-0.14\frac{de}{dh}\right)-\frac{1}{T}\frac{dT}{dh}(P-0.14e)\right\}, \qquad (3.46)$$

in millibar, kelvin, and metre units†

In the dT/dh term the $0.14e$ is negligible, but de/dh in the first term is less so. At sea-level (1.25) gives typical $dP/dh=0.125$ mbar/m, while Table 1.2 gives $de/dh=0.013$ mbar/m as a maximum in normal conditions, and 0.06 at a height of 2 metres above damp grass. In the normal case $0.14\,de/dh$ will then be less than 2 per cent of dP/dh, which rises to 7 per cent in extreme conditions. The latter is not small, but in the conditions in which it may occur dT/dh is likely to be dominant and is likely to lead to much greater error. The normal 2 per cent maximum is reasonably admissible, and de/dh is virtually unobservable anyway, so in (3.46) de/dh and e are ignored. Then putting $dP/dh=-0.0342P/T$ from (1.26), (3.46) gives

$$\frac{1}{\sigma}=\frac{7.9P}{T^2\times10^5}\left(0.0342+\frac{dT}{dh}\right)\cos\beta \text{ metres}^{-1}. \qquad (3.47)$$

Here σ is the radius of the path in metres. Alternatively $1/\sigma$ may be regarded as the curvature of the path in radians per metre, or more conveniently

$$\frac{1}{\sigma}=16.3\frac{P}{T^2}\left(0.0342+\frac{dT}{dh}\right)\cos\beta \text{ seconds per metre}, \qquad (3.48)$$

in millibar, kelvin, and metre units‡. In this context $\cos\beta$ may ordinarily be taken as unity and ignored.

Now consider a longer line PQ in Fig. 3.11, along which σ is variable. The angle of refraction at P is given by

$$\Omega=(\text{VQ/PQ})\operatorname{cosec}1''=\frac{1}{L}\int_0^L\frac{L-l}{\sigma}dl \text{ seconds}, \qquad (3.49)$$

where $1/\sigma$ is expressed in seconds per metre as in (3.48), and L is in metres.

Then (3.48) and (3.49) are the fundamental formulae, and if P, T, and dT/dh are known at every point along the path, Ω is determinate. Actually P is generally known to 1 per cent or better. T can be measured at P, and the average value of T along the path in K can probably be

† If P and e are measured in Pa, substitute 0.079 for 7.9 here and in (3.47).

‡ If P is measured in Pa, substitute 0.163 for 16.3.

estimated within about 1 per cent too. The doubtful item is $\mathrm{d}T/\mathrm{d}h$. The adiabatic gradient for dry air, which might be expected to be the actual gradient in the afternoon, is −0.010 or 1 °C per 100 m, but apart from abnormal values near the ground −0.0055 is more usually observed. There is apt to be an uncertainty of as much as 0.003 in $\mathrm{d}T/\mathrm{d}h$, even in favourable circumstances, with a resulting uncertainty of 10 per cent in Ω.

Note the factor $(L-l)/L$ in (3.49). Values of $1/\sigma$ at the far end of a line have little effect on Ω.

3.18. The coefficient of refraction

(*a*) *Definition.* It is usual and convenient to express Ω in the form

$$\Omega = k\theta, \tag{3.50}$$

where $\theta = L/R$, the angle subtended by the line at the earth's centre, and k $(=\Omega/\theta)$ is the *coefficient of refraction.*†

(*b*) *Horizontal lines.* If the line is sufficiently horizontal for P and T to be treated as constant, such as if $h_P = h_Q$ within 100 or 200 metres, and if $\mathrm{d}T/\mathrm{d}h$ is also regarded as constant along the line, (3.49) gives

$$\Omega = L/2\sigma = R\theta/2\sigma. \tag{3.51}$$

So $\qquad k = R/2\sigma$, σ being in metres as in (3.47)

$$= 252\frac{P}{T^2}\left(0.0342+\frac{\mathrm{d}T}{\mathrm{d}h}\right), \tag{3.52}$$

in millibar, kelvin, and metre units.‡ At sea-level if $P=1000$, $T=300$ (27 °C) and $\mathrm{d}T/\mathrm{d}h = -0.0055$ °C/m (3 °F/1000 feet), k will be 0.080, while at a height of 3000 m where P may be 700 and $T=275$, k will be 0.066. These examples confirm the values of 0.075 or 0.07 which are generally used, but the best values (for horizontal lines) are got by substituting in (3.52) observed values of P and T, with −0.0055 or any better estimate for $\mathrm{d}T/\mathrm{d}h$.

In so far as $\mathrm{d}T/\mathrm{d}h$ can be assumed constant, its actual value is immaterial if reciprocal angles are observed at P and Q, for an error in the accepted value will produce identically opposite errors in the height differences computed from the two ends of the line, which will cancel in the mean. This fact, taken with § 3.14 (*b*), makes it essential to observe reciprocal angles.‖ One-way observations are of little value.

Observed reciprocal angles on horizontal lines provide a means of estimating k and hence $\mathrm{d}T/\mathrm{d}h$, which may then be used for computing the

† The coefficient of refraction is defined by many authors as R/σ. If σ is constant $R/\sigma = 2k$, and is described as K in (1.20). If σ is variable K has no unique value for a particular line.

‡ If P is measured in Pa, substitute 2.52 for 252.

‖ At the same (optimum) time of day, but not necessarily on the same day.

heights of unvisited intersected points. The formula is

$$\beta_P + \beta_Q + \psi_P - \psi_Q + \theta(1-2k) = 0, \qquad (3.53)\dagger$$

where β_P and β_Q (elevations + ve) are corrected for unequal heights of instrument and target, and ψ_P and ψ_Q are the components of the deviations at P and Q in the direction PQ, given by $-(\eta \sin A + \xi \cos A)$, where A is the azimuth from P to Q (for both ψs).

3.19. Inclined lines. Usual formula

If h_P and h_Q are so different that P and T cannot be treated as constant, P/T^2 may be considered to vary linearly with height, and a good value of k will be obtained substituting in (3.52) the estimated values of P and T at one-third of the distance along the line, i.e. at height $\frac{1}{3}(2h_P + h_Q)$ for observations at P and $\frac{1}{3}(h_P + 2h_Q)$ for Q. The most probable value of dT/dh remains -0.0055. The expressions $\frac{1}{3}(2h_P + h_Q)$ and $\frac{1}{3}(h_P + 2h_Q)$ ignore earth curvature, and in very long lines they will give both heights substantially (but equally) above the line of sight. In the mean of reciprocal angles the resulting errors of refraction will tend to cancel, provided the difference of altitude of the two ends of the line is not too great. In unusual circumstances the actual height of the line above sea-level at the two intermediate points may be calculated using (3.32).

In practice a table may be made, giving k for the normal values of P and T at different heights, with the differences resulting from departures from the stated normal of (say) 30 mbar in P and 10° in T, to enable the tabular values to be corrected by actual observations. Then k is computed for the appropriate height as above. This will give satisfactory results provided lines are from one steep-sided hill to another, not grazing the ground in between, and provided observations are made at both P and Q at the times of minimum refraction, noon to 16.00 hours.

There are of course circumstances in which this simple rule will give poor results, but it is seldom possible to suggest anything better. When it seems likely to be worth the labour, recourse may be had to (3.44) and (3.49). Let the line be divided into (say) five or ten equal sections, let the pressure, the temperature, and the temperature gradient be estimated at the mid-point of each section (if the necessary information exists) to give dn/dh, and numerical integration of (3.49) will give Ω at each end of the line.

3.20. Diurnal change in refraction

§ 1.36 (a) describes the temperature changes at a height h above the land surface in accordance with a theory of 'eddy conductivity', (1.42). In

† In [113], 3 Edn., (3.53) was wrongly given as $\beta_P + \beta_Q - \psi_P + \psi_Q + \theta(1-2k) = 0$.

rather more detail, if the diurnal variation of temperature at ground level can be expressed as a series of harmonic terms in the form

$$T_0 = T_m\left\{1 + \sum_{r=1} U_r \cos(rnt + u_r)\right\}, \tag{3.54}$$

the appropriate solution of (1.42) is

$$T_y = T_m\{1 + \lambda y/T_m + \sum U_r e^{-by\sqrt{r}} \cos(rnt + u_r - by\sqrt{r})\} \tag{3.55}$$

$$= T_m\{1 + \lambda y/T_m + f(y)\}, \quad \text{defining } f(y). \tag{3.56}$$

In (3.55) T_m is the diurnal mean at ground level, T_y is at time t and height y, λ is the mean temperature gradient, $n = 2\pi/(24 \text{ hours})$, t is the time in hours from midnight, and $b = \sqrt{(n/2\kappa)}$, see (1.42), whose dimensions are metres^{-1}. For an explanation of the simple case where the temperature variation can be expressed as a single harmonic see § 1.36, after (1.44), where $h' = 1/b$ of (3.55). For some recorded values of h' or $1/b$ see Table 1.3. Harmonic analysis of diurnal temperature variations is seldom possible, but [194] gives the first four harmonics at a wide range of stations, and shows that it is reasonably well representable by two harmonics, in which the amplitude of the second is one-third or one-quarter that of the first. This makes possible the approximations given in (3.58), (3.59), and (3.60).

The temperature gradient at height y above the ground is given by differentiating (3.56) with respect to y. Then σ is given by (3.48), and the refraction angle Ω by (3.49), which after various approximations and putting $\lambda = -0.0055$ gives

$$\Omega'' = \frac{0.24PL}{T_m^2}(1 + 15H \times 10^{-6}) + \frac{16.3P}{LT_m}\int_0^L (L - l)f'(y)\,dl = \Omega_c + \Omega_s, \tag{3.57}$$

where Ω_c is a constant term independent of the time of day, and Ω_s is periodic. The units are millibar, kelvin, and metres.† $H = h_Q - h_P$, y is height above the ground, and Ω'' is the angle of refraction in seconds. Note that in Ω_c, P, and T_m should have their values at one-third of the distance from P to Q, as in § 3.19, while in Ω_s they should be for the level at which diurnal variation is most occurring.

Before Ω_s can be integrated y must be expressed in terms of l, the distance from P. In general this cannot be done, but three special cases can be studied as in (a), (b), and (c) below.

See [113] 2 Edn., pp. 215–19 for fuller details.

(a) *Line from flat ground* P *to a distant high mountain* Q. If the height

† If P is in Pa, substitute 0.0024 and 0.163 for 0.24 and 16.3 in (3.57) and (3.58).

of Q is such that most of the line is above the part of the atmosphere where large diurnal variations occur, the difference of Ω between times t_1 and t_2 is given by

$$\Omega''_{t_1} - \Omega''_{t_2} = -\frac{16.3P\cot(\beta-\gamma)}{T_m^2}\left\{T_{t_1} - T_{t_2} - \frac{1}{mbH\sqrt{2}}(T_{(t_1-3)} - T_{(t_2-3)})\right\}, \tag{3.58}$$

where $t-3$ means 3 hours before t, H is the height of Q above the horizontal of P, β is the vertical angle at P, $m = (\beta-\gamma)/\beta$, where γ is the general slope of the foreground at P towards Q (+ if the ground rises).

In this situation refraction changes can be obtained from measured temperatures without harmonic analysis, and Fig. 3.12 shows the relation between Ω and T for different values of mbH. When mbH exceeds unity, the curve resembles an elongated ellipse, and when it is very large the term in $1/mbH\sqrt{2}$ may be neglected, and the relation is linear. This accords with the well-known fact that minimum refraction occurs at the hottest time of day, although see (*b*) and (*c*) below. The almost exact linear relationship is also clearly shown by long series of observations to Himalayan peaks made between sunrise and 16.00 hours in the afternoon, [230], p. 69.

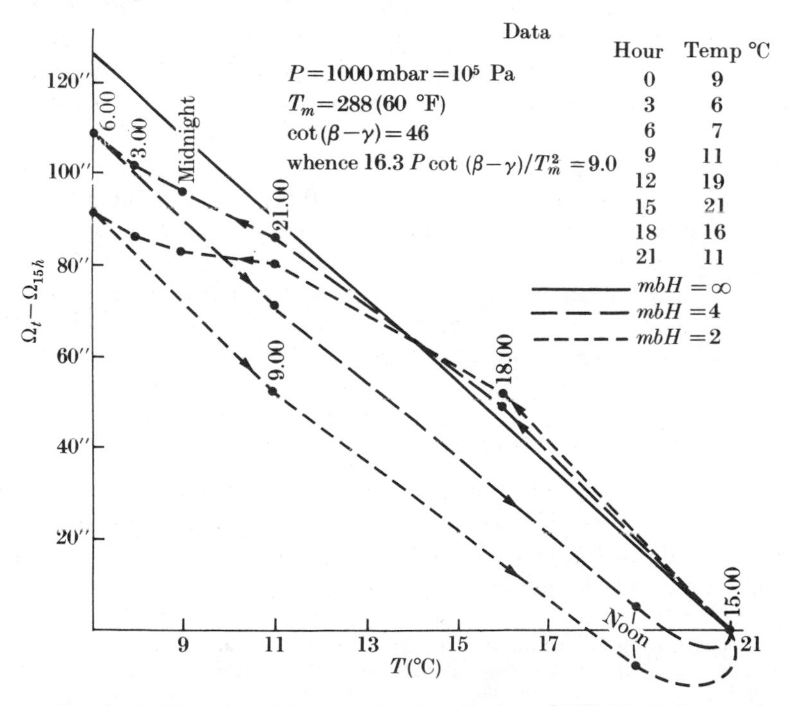

FIG. 3.12. Diurnal variation of refraction. Compare [387], Fig. 2. (inverted).

(3.58) only approximates to the truth if γ is small, for the mechanism of eddy conductivity will break down on steep slopes, but it does agree with fact in giving a small Ω_s if γ is a large negative, so that $\cot(\beta-\gamma)$ is comparatively small. There is little diurnal variation in a line where the ground falls steeply away in front of a high hill. Note also that Ω_s varies as $\cot(\beta-\gamma)$, not as the distance PQ. But the line considered is by hypothesis one that proceeds to some considerable height, such as at least $2/b$, and the factor $\cot(\beta-\gamma)$ is therefore a measure of the distance it travels in the periodically disturbed levels of the atmosphere. Note also that when Q is very high, the term in $1/mbH\sqrt{2}$ is very small, and the value of the eddy conductivity is immaterial, but when Q is lower so that the line does not proceed entirely beyond the disturbed levels, the value of b becomes more important. See (b) below, where the variation in a horizontal line is proportional to b, and to the length of the line.

(b) *Low horizontal lines.* (3.58) is inapplicable when the line is nearly parallel to the ground, as $\cot(\beta-\gamma)$ becomes very large, but the equation can be put into the form

$$\Omega''_{t_1}-\Omega''_{t_2}=\frac{16.3PLb}{T_m^2\sqrt{2}}(T_{(t_2+3)}-T_{(t_1+3)}). \qquad (3.59)\dagger$$

This shows that in a nearly horizontal line Ω_s varies as the length, as the change of temperature, and as b, and that the minimum refraction should occur 3 hours before maximum temperature. This last point is confirmed by observations in the flat plains of the Punjab, [591], Appendix 3.‡

(c) *From a steep hill* Q *to a plain* P. If the ground near Q is steep, little variation in the curvature of the ray occurs there. Bending near P occurs as in (a) above, but has less effect on Ω at Q than on Ω at P. Then in (3.57) put $y=m(h-H)$, where $m=(\beta-\gamma)/\beta$ at the distant station P. In this case h is height above the level of Q, and H is the value of h at P, both negative.

Then approximately

$$\Omega''_{t_1}-\Omega''_{t_2}=\frac{16.3P\cot(\beta-\gamma)}{mbHT_m^2\sqrt{2}}\{T_{(t_1-3)}-T_{(t_2-3)}\}. \qquad (3.60)$$

In this case the diurnal change is about $1/mbH\sqrt{2}$ times that in the line from P to Q, and minimum occurs 3 hours after maximum temperature at P. The reduced range is in accord with observation, and the late minimum is weakly confirmed by available data, so far as they go.

† If P is measured in Pa, substitute 0.163 for 16.3 in (3.59) and (3.60).

‡ The early minimum is not very apparent from the tables of figures, but in [591], Appendix 3, p. 89, it is stated that the objects observed were generally below the horizon between 10.00 and 13.00 hours, indicating minimum refraction at about 11.30.

[387] studies refraction on similar lines, as outlined in § 1.36, (1.45).

The preceding theory agrees with experience qualitatively, but it does not suffice for the computation of the refraction in any particular line. As the temperature increases on a hot day, the low density of the air near the ground results in convection and the break up of more regular eddy conductivity. This presumably tends to start when the adiabatic gradient is reached, but convection does not operate sufficiently vigorously to prevent the formation of much higher lapse rates, as in § 1.36, Table 1.1. At any one site, however, the limiting value of dT/dh is apparently somewhat constant, as witnessed by the observed considerable constancy in minimum refraction, although it must obviously vary with the wind speed if with nothing else.

There is no corresponding constancy in the maximum refraction at night, for the chilling of the lower air produces no unstable equilibrium, and increases steadily with the clearness of the sky and the stillness of the air.

(*d*) *Variations of k.* The coefficient of refraction decreases with height as shown in Table 1.1. In clear weather, especially in a dry continental climate, the diurnal variation of k in lines as low as 5 or 10 metres above flat ground may be very great. At midday k may fall to -0.10, while at night it may rise to as much as 1.2, [591]. The variations from day to day, and from place to place are much greater at night than by day, because the night values arise from cold air near the ground with warmer air above and a large positive dT/dh, which is a stable situation, while the midday values are caused by hot air near the ground with a large unstable negative gradient.

Over water the position is reversed, with rather high values of k at midday, due to hot air from the land lying over cooler water, and rather low values at night (cold air over less cold water).

At a height of a few hundred metres above the ground, the coefficient of refraction of light in a line of some kilometres in length may be expected to be within 0.01 of the value given by (3.52) using observed P and T, and dT/dh of -0.0055.

The conclusion is that for the computation of k it is not generally possible to improve on the system described in § 3.19, and to observe only in the early afternoon.

3.21. Direct measures of refraction

If a ray of white light from Q is bent by refraction, the bending of the red is less than that of the blue, by something of the order of 2 per cent, depending on the actual colours involved. Figure 3.13 shows a blue and a red ray from a target Q to the optical centre of a telescope object-glass O. Assuming that the two rays remain sufficiently close

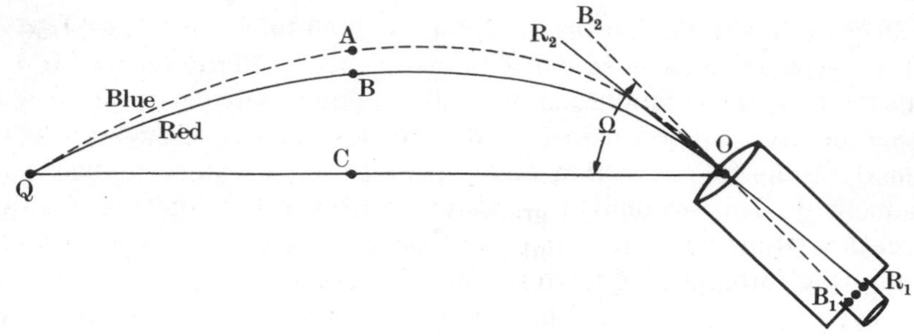

FIG. 3.13. Refraction measured by dispersion.

together throughout their course for dn/dh to have always the same value for each, the rato AB:BC in Fig. 3.13 will be the same, such as 1 to 50, at all points on the path whatever changes of curvature may occur, and the angle between the tangents at O, R_2OB_2, will be 2 per cent of the refraction angle Ω (=B_2OQ). So if R_2OB_2 can be measured, Ω is determined without any knowledge of the meteorological conditions.

In principle, incoming light at O can be separated by colour filters and the angular separation between two selected colours measured. The following are the basic difficulties.

(i) To measure the refraction to 1″ requires that the angle between the two colours should be measured to 0″.02.

(ii) Loss of light in the two filters. A two-colour laser is a possible remedy.

(iii) The oscillating movement of the images, especially in the middle of the day. But since the refraction is being actually measured, there is no need to do it at the time of minimum refraction. Night observations are permissible, if convenient.

[189] and [557] report ingenious means of getting over these difficulties. In 1977 neither is in more than experimental form, but [604] records some field tests.

3.22. Lateral refraction

From (3.48), taking P as constant, which implies the omission of the term 0.0342, the curvature of a ray of light in a horizontal plane is given as

$$16.3(P/T^2)(\mathrm{d}T/\mathrm{d}x) \quad \text{seconds per metre,} \qquad (3.61)\dagger$$

where P is in millibars, T in K, and dT/dx is measured horizontally at right angles to the line in °C per metre. Computation and correction will

† If P is measured in Pa substitute 0.163 for 16.3.

hardly be practicable, but the order of magnitude of possible lateral refraction can be illustrated as follows.

(*a*) For a distance of 3000 m let a line graze along a sloping hill-side, so that the ground is 1000 m from the line horizontally and 300 m vertically. If the hill is 5 °C hotter than the air through which the line passes, the average horizontal gradient will be 5 °C per 1000 m. On the line itself dT/dx will be less, but perhaps as much as 0.0015 per metre, and the total curvature in 3000 m will be about 1″ of arc.

(*b*) Two metres above the ground the vertical gradient may be 0.3 °C per metre, and the horizontal gradient might be the same at 2 metres to the side of a vertical rock in bright sunshine. A graze 15 metres long through such a gradient would produce a curvature of 1″.

See references in footnote to § 1.19 for some recorded examples.

(*c*) *Twist in a normally refracted ray*. The skewness of the spheroidal normals, to which the layers of equal atmospheric density may be supposed to be perpendicular, in theory may introduce a lateral component into terrestrial refraction. But [113], 2 Edn., p. 223 shows that the effect is negligible, such as $0''.004L^2$ in a line $100L$ km long.

SECTION 4. MEAN SEA-LEVEL AND THE TIDES

3.23. The tidal effect on equipotential surfaces

Fig. 3.14 represents the earth and moon, both revolving in the plane of the paper with angular velocity ω ($\approx 2\pi/28$ radians per day) about C their common centre of gravity, which is about 4720 km from the earth's centre. Let the mass of the earth, assumed spherical, be M and its radius R. Let the mass of the moon be m ($=M/81.30$), and let the distance between its centre and the earth's be d, assumed constant.

In Fig. 3.14 let AC $= c$. If P is a point on the earth's surface, let CP $= r$, and let PAB be z (approximately the moon's zenith distance at P). Let G be the gravitational constant, see § 6.06.

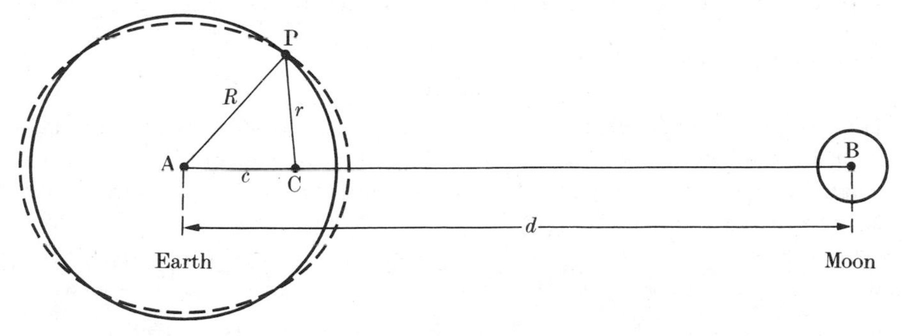

FIG. 3.14. The moon moves in the plane of the paper.

Then $\qquad Mc = m(d-c), \qquad r^2 = (R^2 - 2cR\cos z + c^2),$

and
$$\omega^2 = GM/\{d^3(1-c/d)\}, \tag{3.62}$$
since the moon's acceleration of $\omega^2(d-c)$ is due to a force GMm/d^2.

At P the potential of the earth's attraction, § 6.09 (*a*), is
$$V_1 = GM/R, \tag{3.63}$$
and that of the moon is
$$\begin{aligned} V_2 &= (Gm/d)\{1 - 2(R/d)\cos z + R^2/d^2\}^{-\frac{1}{2}} \\ &= \left(\frac{GM}{R}\right)\left(\frac{cR}{d^2}\right)\left(1 + \frac{R}{d}\cos z + \frac{3}{2}\frac{R^2}{d^2}\cos^2 z - \frac{1}{2}\frac{R^2}{d^2}\right) \div \left(1 - \frac{c}{d}\right). \end{aligned} \tag{3.64}$$

And at P the potential of the centrifugal force of the monthly rotation about C is
$$\begin{aligned} V_3 &= \tfrac{1}{2}\omega^2 r^2 = \tfrac{1}{2}\omega^2(R^2 - 2cR\cos z + c^2) \\ &= \left(\frac{GM}{R}\right)\left(\frac{1}{2}\frac{R}{d^3}\right)(R^2 - 2cR\cos z + c^2) \div \left(1 - \frac{c}{d}\right). \end{aligned} \tag{3.65}$$

Summing (3.63), (3.64), and (3.65) gives the total potential at P as
$$\begin{aligned} V &= V'_C + \left(\frac{3}{2}\frac{GMcR^2}{d^4}\cos^2 z\right) \div \left(1 - \frac{c}{d}\right) \\ &= V_C + V_M(\cos 2z + \tfrac{1}{3}), \end{aligned} \tag{3.66}$$
defining V'_C and V_C, which are constant, and V_M which is
$$V_M = \left(\frac{3}{4}\frac{GMcR^2}{d^4}\right) \div \left(1 - \frac{c}{d}\right). \tag{3.67}$$

The term $\frac{1}{3}$ in (3.66) is introduced so as to make V_C the mean value over the sphere.

Putting $GM = 398\,000 \times 10^9\ \text{m}^3\,\text{s}^{-2}$, $c = 4.72 \times 10^6$ m, $R = 6.37 \times 10^6$ m, and $d = 3.84 \times 10^8$ m, conforming to [601], gives
$$V_M = 2.62\ \text{m}^2\,\text{s}^{-2}. \tag{3.68}$$

The equipotential surface is thus a prolate spheroid with its long axis pointing towards the moon as shown in broken line in Fig. 3.14, and its height above the (spherical) earth at a point where the moon's zenith distance is z is
$$(V_M/g)(\cos 2z + \tfrac{1}{3}) = k_1(\cos 2z + \tfrac{1}{3}) = 0.267(\cos 2z + \tfrac{1}{3})\ \text{metres}, \tag{3.69}$$
defining k_1, and taking g as $9.81\ \text{m}\,\text{s}^{-2}$.

As the earth rotates daily about its axis, the equipotential surface at any point will have two high and two low tides. If the moon passes through the zenith and nadir, the height will vary between +35.6 and −17.8 cm.

The sun will produce a similar deformation of the equipotential surface, of

$$k_2(\cos 2z + \tfrac{1}{3}), \tag{3.70}$$

where $k_2 = 0.123$ metres, which produces two highs and two lows every solar day.

3.24. Effect on the direction of the vertical and on levelling

The direction of the downward vertical is deflected towards the moon (or sun) by

$$\frac{1}{R}\frac{\mathrm{d}}{\mathrm{d}z}\{k(\cos 2z + \tfrac{1}{3})\} \tag{3.71}$$

$$= \frac{2}{R} k \sin 2z \text{ radians} = \psi \sin 2z,$$

where $\psi_1 = 0''.017$, or 8.4 mm/100 km, for the moon,

and $\psi_2 = 0''.008$, or 3.9 mm/100 km, for the sun. (3.72)

The components of this deviation when the sun (or moon) is in azimuth A (from north) are

$\psi_s = -\psi \sin 2z \cos A$ towards the south

and $\psi_w = -\psi \sin 2z \sin A$ towards the west, (3.73)

where $\psi = \psi_1$ or ψ_2 for the moon or sun respectively.

Putting z and A into terms of the moon's (sun's) declination δ, hour angle t, and the latitude ϕ of the point considered, gives

$$\psi_s = (\psi/2)\{\cos 2t \cos^2\delta \sin 2\phi - 2 \cos t \sin 2\delta \cos 2\phi + \sin 2\phi(1 - 3\sin^2\delta)\},$$

$$\psi_w = \psi(\sin 2t \cos^2\delta \cos\phi + \sin t \sin 2\delta \sin\phi), \tag{3.74}$$

where ψ for the moon or the sun is given by (3.72).

For observations (e.g. levelling) regularly made at particular hours of the solar day, the moon's hour angle will vary from 0 to 24 h through the lunar month, and its declination varies harmonically from $+\delta_M$ to $-\delta_M$ and back in a similar period. δ_M varies from 18° to 28° during a 19-year period. For the sun, on the other hand, the hour angle is the same at any time of day, and the declination varies between $+23\tfrac{1}{2}^\circ$ and $-23\tfrac{1}{2}^\circ$ during the year. The figures given in § 3.06 (*c*) (iii) for the average tidal corrections to spirit levelling are appropriate mean values of (3.74).

3.25. Marine tides

If the whole earth was covered by deep water (more than 20 km deep), and if there were no meteorological or other such disturbances, the sea surface would conform to the equipotential surface given by the sum of (3.69) and (3.70). If in addition the moon and sun were always on the equator ($\delta = 0$), moving at constant speed and at constant distance, the tide† at any place on the equator would be expressible in the form

$$h = H_0 + H_1 \cos(2n_1 t + \zeta_1) + H_2 \cos(2n_2 t + \zeta_2), \tag{3.75}$$

where $H_1 = 0.267$ metres, $H_2 = 0.123$ m, $2\pi/n_1$ is a lunar day, $2\pi/n_2$ is a solar day, and ζ_1 and ζ_2 are the hour angles of the moon and sun at the epoch $t = 0$. In other latitudes H_1 and H_2 would be smaller. The result would be two highs and two lows daily, with fortnightly springs and neaps as the lunar and solar terms get in and out of phase.

This simple theory of the marine tides is invalidated by the existence of land and shallow water: by the varying declination, distances, and speeds of the moon and sun: and by meteorological disturbance. The consequence is that although there will, normally, everywhere be two high and two low tides daily, it is impossible for theory to give any approximation to the values of the H's or to the intervals between the moon's transit and high tide at any place. Many additional harmonic terms are also necessary, as below.

Overtides. Apart from the non-satisfaction of the other conditions (3.75) needs modification in shallow water, where the tidal variation cannot be represented by a simple cosine curve, but becomes asymmetrical as in Fig. H.2, the rise being more rapid than the fall. Such a curve can be represented by a series of cosine terms, § H.00. Thus for the moon

$$h = H_0 + H_{11} \cos(2n_1 t - \zeta_{11}) + H_{12} \cos(4n_1 t - \zeta_{12}) + \\ + H_{13} \cos(6n_1 t - \zeta_{13}) + \ldots. \tag{3.76}$$

The terms containing $4n_1$, $6n_1$, etc., are known as *overtides* or 2nd, 3rd, etc., harmonics. In shallow water and between converging coastlines, amplitudes can be increased fourfold, or more.

Other astronomical tides. The variations in the moon's and sun's declinations, distances, and speeds demand the addition of other harmonic terms whose n's depend on the periods of these disturbances and are given by rather complicated theory, [177]. Several have periods of approximately, but not exactly‡ half a day. Others are approximately diurnal, and there are also fortnightly and monthly terms.

† Assuming the earth itself to be rigid.

‡ A periodic variation in the amplitude, or in the phase, of a harmonic variation can be represented by the addition of *sidebands* namely harmonic terms of slightly different speeds. See §§ G.4–G.7.

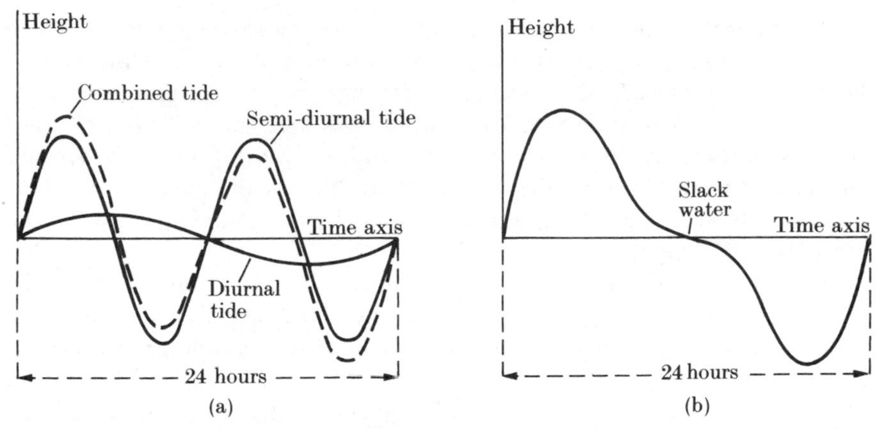

FIG. 3.15(a) Diurnal inequality. (b) Diurnal inequality when the diurnal tide is very large.

Diurnal inequality. Generally the most important of these extra terms are due to the sun and moon's declinations not being zero. Except on the equator this causes the zenith distances of their upper transits at any place to differ from their nadir distances at lower transit, and since their effect depends on their zenith distances, this is apt to cause considerable inequality in the range of the two tides in any one day. See Fig. 3.15. In an extreme case one of the two high and low tides may, for a day or two at about neap tide, become so small as to vanish, leaving only one high and one low tide per day, with a long period of slack water at the times of the others.

Compound tides. In shallow water the interaction of two simple harmonic tides is not accurately represented by summing the height of the water due to each, and other terms are needed whose speeds are the sum or difference of those of the two terms concerned. These are called *compound tides.* Their amplitudes are smaller than those of the original cosine terms, so it is only necessary to consider a few such combinations.

Meteorological effects. There are appreciable diurnal and annual variations of water-level, the latter with a 6-monthly 2nd harmonic, caused by periodic changes of wind and atmospheric pressure and the discharge of rivers.

3.26. Harmonic analysis and tidal prediction

The result of the above is that the tide at any place is given by the sum of 20 to 30 cosine terms whose speeds are given by theory, but whose amplitudes and phase angles are given by an analysis of observed water heights at the place concerned.

For a satisfactory analysis, observations must be made for a year, or better, several. The *tide gauge* used for this has traditionally consisted of a float controlling the movement of a pencil over a revolving drum, which gives a continuous record at a suitably reduced scale, from which hourly heights are measured. More modern instruments give a digital record on magnetic tape. A routine form of computation then produces the H's and the ζ's. Ordinarily these are true constants,† and if their values are substituted in the formula the height of the tide at the place of observation is quite accurately predictable several years ahead. The analysis also gives H_0, the height of mean sea-level.

With the analysis complete, the height of the tide at any time can be obtained by summing the harmonic terms. Traditionally, this has been done by a mechanical *tide-predicting machine*, but a computer now does it more easily.

At different places the H's (semi-ranges) of the different tides vary greatly. In the open ocean that of the lunar semi-diurnal tide known as M_2 is about 300 mm, while those of the overtides M_4, M_6, etc., are 3 mm or less. In estuaries on the other hand the H of M_2 may be over 3 m and that of M_4 0.5 m, although the other overtides are always much smaller. The solar semi-diurnal tide S_2 tends to be about one-third of M_2, while the H's of its overtides are even smaller in proportion. Of the diurnal tides the two largest, known as O (or O_1) and K_1, arise from the combination of the moon and sun's changes of declination, and their H's may amount to 0.3 to 0.6 m. Of the long period tides the largest is generally the annual *Sa*, whose H may be a metre at riverain ports like Calcutta, although it is ordinarily unlikely to exceed 300 mm. When *Sa* is large, its overtide *Ssa* with a 6-monthly period is often considerable too.

Except in stormy weather, and in riverain ports, harmonic analysis may be expected to give predictions correct to 15 or 30 minutes in time and to 5 or 10 per cent of the range in height.

3.27. The variations of mean sea-level

The height of mean sea-level is required as a basis for spirit levelling. The best method is to record the hourly height of the water for a year or more with a tide gauge as in § 3.26, and this should be done. The gauge is of course so adjusted that its zero is a known distance below a solidly built reference BM, and this adjustment is periodically checked, as also is the stability of the reference BM by comparison with others. The accepted MSL is then the mean of the hourly heights over a period (about a year) which should start and finish at (say) a spring tide superior‡ high water, so that it includes an exact whole number of the principal lunar

† Except that in the lunar tides the values of H vary slowly by a predictable amount depending on the angle between the plane of the lunar orbit and that of the earth's equator.

‡ 'Superior', with reference to the diurnal inequality.

and solar semi-diurnal and diurnal tides, whose mean values will then be zero.

A year's observations should give a value of MSL which agrees with any other year within some centimetres, except in riverain ports much affected by flood water, or where the range is exceptionally large, but such places are not suitable for a levelling datum, anyway. An approximate value can be got from a month's observations, but the result will ignore the effect of the annual tide, and may be even more seriously affected by the chance of abnormal weather conditions, and nothing less than one or preferably several years' work provides a sound datum for geodetic spirit levelling.

The mean sea-level at a place may vary from year to year for many reasons, as follows.

(*a*) The land may be rising or falling.

(*b*) Eustatic changes of sea-level due to an increase or decrease in the accumulation of polar ice, or to changes in the level of the sea bottom anywhere in the world.

(*c*) Barometric changes. The changes in annual mean sea-level at a place are clearly related to the changes of annual mean barometric pressure.

(*d*) The annual mean sea-level may similarly be related to the direction and magnitude of the annual total wind vector, which in turn is related to the annual mean barometric gradient. See [186].

(*e*) The 19-year astronomical tide. See [501], p. 296.

(*f*) Changes in water temperature, salinity, and currents.

The geodetic aim is to eliminate (*c*)–(*f*), and to determine (*a*) and (*b*). See [501] for an analysis of the variations of MSL in Europe.

The MSL above which heights on land are generally expressed must not be confused with the *datum of soundings*. The latter is selected to be a little below the lowest low water, and marine charts show depths below it, while tables of tidal predictions give heights above it. The actual depth of water at any time is then the depth given on the chart plus the figure given in the tide tables. This is convenient for mariners, but submarine contours based on it are out of terms with land contours by an amount rather greater than half the tidal range.

The *revised local reference* datum (RLR) above which the observed heights of the tide are now (1977) recorded, is defined to be at an integral number of decimetres (about 7 metres is advised) below the carefully selected stable *tide gauge bench mark* (TGBM). Oceanographers record the height of the water and the variations of MSL with reference to the RLR, while geodesists connect the TGBMs to each other and to other stable marks by spirit levelling.

3.28. Departure of MSL from a level surface

(*a*) In § 2.02 the geoid is defined as that equipotential of the earth's attraction, which on average coincides with MSL in the open ocean. The implication is that the ocean MSL is approximately an equipotential surface, but that is of course only approximately the case.

The free surface of a uniform liquid at rest is closely an equipotential surface. Apart from waves and tides which are eliminated in MSL, the ocean fails to meet these conditions in several respects.

(i) Its surface is overlain by air, whose pressure varies. It is not quite a free surface.

(ii) The wind applies a horizontal force to the surface.

(iii) The density of the water varies with its temperature and salinity.

(iv) The sources of water, rain, rivers, and melting ice, do not coincide with the areas where water is lost by evaporation.

(v) From these inequalities ocean currents result, acting towards the restoration of equilibrium, but with a time lag. The result is a mean ocean surface which departs from the equipotential by amounts which are more or less constant with time.

(*b*) Oceanographers can make estimates of the surface variation of potential at any suitable place on the basis of observed values of temperature, salinity, and ocean currents. In doing this they generally assume that below some such depth as 2 km the isobaric layer† (of equal pressure) is also equipotential, on the grounds that below this level conditions are relatively uniform and currents small.‡ The selected surface is known as the *deep sea isobaric reference surface*. They then calculate the potential difference between this reference surface and MSL at numerous points, and so construct charts showing the variations of potential at MSL all over the oceans.|| See [184], (i), p. 596 (Atlantic ocean), p. 657 (South Atlantic), and p. 670 (Southern Ocean), and [269] (world chart). These charts show the estimated departure of MSL from an equipotential surface to be varying through a range of about $2\frac{1}{2}$ to 3 m, generally with MSL above the geoid around the equator and below it at high latitudes in both hemispheres. But see (*c*) below.

Instead of giving results in the form of anomalies of potential, it is equally possible to give the variations of orthometric height (metres) of MSL above the deep sea reference surface, but results given in this form

† Pressures are commonly measured in decibars (db). 1 db = 10^4 Pa. In round figures the 2000-db surface is 1975–80 m below the ocean surface.

‡ The 2000-db surface is not necessarily selected. It may be 1000 or 4000.

|| These variations of potential are commonly described in *dynamic centimetres*, dyn cm. This unit, like the GPU of § 3.00 is a unit of potential, and is in fact equal to 10^{-2} GPU or 10^{-1} m^2 s^{-2}. It is not a SI unit [53]. A potential anomaly of 1 dyn cm is equivalent to a change of height of about 1.02 cm.

require careful interpretation. The deep sea reference and the geoid are not parallel, but (allowing for the density of the intervening water) they must be expected to converge from equator to pole by about 3 m for each km in the depth of the deep reference. This must be allowed for before comparing variations of orthometric heights above the deep reference with spirit-levelled variations of the heights of MSL above the geoid at tide gauges. See [61]. From this point of view it is better to work in potential units.

(*c*) *Comparison with spirit levelling*. Spirit levelling gives differences of geopotential between MSL at tidal stations, and these ought to agree with those computed by oceanographers as in (*b*) above, but there are of course discrepancies which may be due to the following causes.

(i) The potential difference between MSL and the deep reference surface in the open ocean may well not be applicable to a tide gauge site on an adjacent coast. Winds, tides, and currents in shallow water may produce abnormal results. See [378].

(ii) The assumption that the deep sea isobaric reference is an equipotential surface may also be imperfect, although [184] p. 686 gives estimates showing variations of only about 20 dyn cm at the 2000-db surface in the Atlantic Ocean over the area 50° N to 50° S.

(iii) Abnormal values of atmospheric pressure will not be included in the oceanographers' calculations. The effect of a 10-mbar increase in pressure is to depress MSL by about 0.1 m.

(iv) There is always some uncertainty about the accuracy of very long lines of geodetic spirit levelling. Systematic errors may be occuring, and it is in the nature of such errors that they are not easy to detect. See § 3.06.

Discrepancies amounting to about 0.5 m are found between latitudes 25° and 45° on the east coast of the US, and of 1 m between 33° and 45° on the west coast. See [61] which is based on [78] for oceanography and on [545] for spirit levelling. These discrepancies are of the same order of magnitude as the whole computed variations.

(*d*) *Satellite altimetry*, §§ 7.64 to 7.65, should be able to help in resolving the differences between oceanography and geodetic levelling, especially as it gives MSL heights in the open ocean. In 1977 it is too early to record results.

3.29. Computation of MSL across a narrow strait

Working on the general lines of § 3.28, it is possible to compute the difference of the height of MSL above the geoid at two points on either side of a narrow piece of water such as that between Dover (UK) and Calais (France), [147] and [148].

Take the axis of y across the gap, and x in the downstream direction, and at any point let v and u be the water velocity (ms^{-1}) along y and x, Fig. 3.16. Let h m be the height of the actual surface above a level surface. Then the difference of h between two points on the y axis is given by

$$\frac{\partial}{\partial y}(h+P)=\frac{1}{g}\left\{f(u)+\frac{F_S-F_B}{\rho D}-\frac{\partial v}{\partial t}-\frac{u\,\partial v}{\partial x}\right\}. \tag{3.77}$$

In this expression

P is atmospheric pressure in units of 100 mb or of 10^4 Pa, corresponding to the head of 1 metre of water. Easily measured.

$f(u)$ is the Coriolis acceleration, given by (I.10),

$$f(u)=\frac{4\pi u}{24\times 3600}\sin\phi=15\times 10^{-3}u\sin\phi,\ \text{ms}^{-2}$$

F_S is the drag of the wind, given by $F_S=1.8\times 10^{-3}v_S^2\,\text{Nm}^{-2}$, where v_S is the y-component of the wind.

F_B is bottom friction, given by $F_B=35\times 10^{-1}v_B^2\,\text{Nm}^{-2}$, where v_B is the transverse current velocity 1 metre above the bottom.

ρ is the density of water (kg m^{-3}), and D is the depth in metres.

$g=9.8\ \text{ms}^{-2}$.

v_S can be measured. It may be possible to measure v_B, but the site should be such that the effect of possible v_B can be neglected. In the sea this will be easy, but in a river this is likely to be the main source of error.

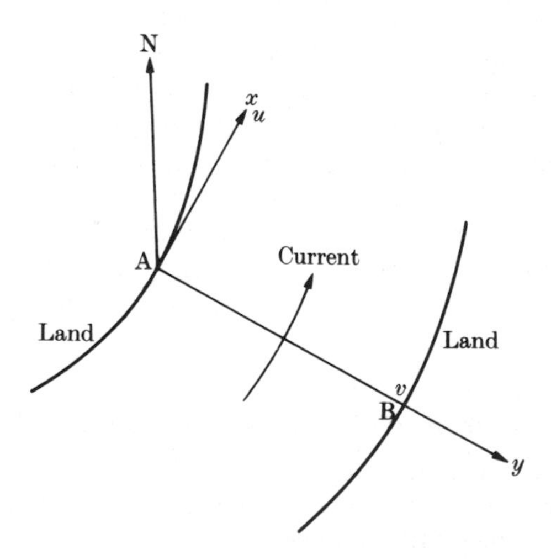

FIG. 3.16.

$\partial v/\partial t$, due to acceleration along y, must similarly be made negligible by choice of site.

$u\,\partial v/\partial x = u^2/R$, where R is the radius of curvature of the current, positive if the flow is concave towards A. An average value of R and u can be estimated well enough on a reasonably well-chosen site.

Neglecting terms whose effect can be kept small (3.77) can be written

$$h_B - h_A = (P_A - P_B) + \frac{L}{g}\left(\frac{F_S}{\rho D} + 15 \times 10^{-3}\,\bar{u}\sin\phi\right), \qquad (3.78)$$

where L is the width of the section, D is its mean depth, and $\bar{u}$ is the mean value (w.r.t. y) of u between A and B, in m and ms^{-1}. The value of $\bar{u}$ is important. If an electric telegraph wire exists across the gap, the variations of $\bar{u}$ can be derived from voltage fluctuations in the cable. This helps to reconcile separate measures of sea-level while the tide is flowing, but the long-time mean has to be determined by direct measures of the current.

The measured heights of A and B must be made on the surface of still water (as they will be in a tide-gauge, but not necessarily on a pole), since a velocity V (in any direction) is associated with a loss of head of $V^2/2g$.

[147] describes a transfer of levels across the Straits of Dover by this method.

General references for MSL and tides [177], [378], and [184].

4

GEODETIC ASTRONOMY

SECTION 1. INTRODUCTORY

4.00. Notation

THE following symbols are used in Sections 1 to 5 of this chapter.

ϕ, λ, A = Latitude, longitude, and azimuth. In this chapter generally astronomical, as defined in § 2.04. Suffixes $_A$ and $_G$ indicate astronomical or geodetic when distinction is necessary.
Suffixes $_0$ refer to the instantaneous pole and $_1$ to the CIO when distinction is necessary. ϕ_a and λ_a are also approximate trial values. In § 4.47 ϕ_0 is a calibration latitude.

t = Hour angle = LAST − RA. Also a whole number of years.

τ = A fraction of a year.

RA or α = Right ascension } Suffix $_0$ for mean place at epoch.

δ = Declination } Suffix t for epoch plus t years.

ζ, z = Apparent and true zenith distance, $z = \zeta + \psi$. Always positive except in (4.62).

h = Altitude above the horizon = $90° - z$.

q = Parallactic angle. Fig. 4.1.

ψ = Celestial refraction. Also luni-solar precession. Also a component of the deviation of the vertical, (4.95).

ξ, η = Deviations of the vertical in meridian and PV. Positive if downward vertical is south or west of the normal.

N = Height of geoid above spheroid. Also see (4.13),

P = Total atmospheric pressure.

e = Pressure of water vapour.

T = Temperature in K. Also time in centuries from some epoch.

n = Refractive index of air.

γ = The vernal equinox.

LAST = Local apparent sidereal time.

LMST = Local mean sideral time.

GAST, GMST = Greenwich ST.

UT = Universal time.

GAT, GMT = Greenwich apparent and mean solar time.

AT = Atomic time.

UTC = Coordinated universal time.

TAI = Temps atomique international.

AE = *Astronomical ephemeris.*

APFS = *Apparent places of fundamental stars.*

BIH = Bureau International de l'Heure.

PV = Prime vertical.

RM = Referring mark.

s.e. = Standard error. Probability level 68 : 32.

DT = Dynamic time. § 4.10 (c) (ii).

CIO = Conventional international origin.
CZM = Conventional zero meridian = BIH Greenwich.
E = Equation of time, (4.30).
EE = Equation of the equinoxes, (4.33). Apparent − mean ST.
x, y = Coordinates for polar motion, Fig. 4.3. Also coordinates in position line diagrams, (4.77).
P = Polar motion correction to LST, (4.57).
ϵ = Obliquity, (4.1).
ψ = Luni-solar precession, (4.2).
λ' = Planetary precession, (4.3).
p = General precession, (4.4).
m = Precession in RA, (4.5).
$n = \psi \sin \epsilon$ = Precession in declination, (4.10).
M, N, see (4.12).
μ = Proper motion of star.
FL, FR = Face left, face right.
LE, LW = Level east, level west. Two positions of the zenith telescope.
A, B, C, D, E = Besselian day numbers.
a, b, c, d, see (4.15), and also with different meanings in (4.60)–(4.63).
A, a, B, b, κ = Transit constants, (4.60)–(4.63).
c = Collimation error.
a = Azimuth error.
i = Inclination error.
v = velocity. Also least square residual.
c, k, A, i = Zenith telescope constants, § 4.18 (*d*). k is also in (4.70).
P, N = Rotation matrices, (4.19) and (4.24).
$\kappa, \omega, \nu = \zeta_0, z, \theta$. Constants in the precession matrix.
l, m, n = Direction cosines, (4.17), (4.18), and (4.23).
R_u, R_E, see § 4.10 (*d*) (i) and (4.32), respectively.
$R = R_u \pm 12$ hours.

4.01. Objects

The objects of geodetic astronomy have been as follows.

(*a*) To observe azimuth and longitude at Laplace stations § 2.05, or geodetic azimuth independently by Black's method § 4.47, for controlling the azimuths of geodetic triangulation, traverse, and trilateration. Also at satellite launching sites, if not connected to a geodetic framework.

(*b*) To measure the deviation of the vertical by observations of astronomical latitude, longitude, or (less suitably) azimuth, for the following purposes.

(i) Determining the local separation of the geoid and the reference spheroid, for the correct reduction of measured distances to spheroid level, §§ 1.30 (*f*) and 2.08. Geoid sections §§ 4.50–4.52.

(ii) The correction of horizontal angles, § 2.09.

(iii) In the combination with ground and satellite gravity surveys, and satellite altimetry, Chapter 7, Sections 8 and 10, to construct world-wide geoid charts.

(iv) The study of variations of density in the earth's crust and mantle.

(*c*) To observe latitude, longitude, and azimuth at the origins of independent surveys, or for the demarcation of astronomically defined boundaries.

(*d*) To observe changes of latitude and longitude with time, such as may arise from regularly periodic polar motion, § 4.9, or from continental drift and tectonic movements, §§ 6.71–6.75.

(*e*) To observe time. This is now done in fixed astronomical observatories, and the results are broadcast by wireless time signals.

(*f*) To photograph artificial satellites against a stellar background, Chapter 7, Sections 4 and 5.

These objectives have generally demanded an accuracy of ±1 second of arc, or better; much better for items (*d*) and (*e*).

The impact of artificial satellites, notably in the form of doppler fixes, Chapter 7, Section 7, has been considerable. It is necessary to assess the extent to which geodetic astronomy may be becoming obsolete. We take the items (*a*)–(*e*) above, in turn.

(*a*) As in § 1.09, satellite fixes accurate to 1 m and 200 km apart could effectively take the place of Laplace azimuths. Such fixes are not yet (1978) available, but they may not be far away.

(*b*) (i) Similarly, geocentric fixes correct to 1 or 2 m in the vertical component would suffice for the reduction of measured distances.

(*b*) (ii) There is no substitute for astronomical observations here, but the correction to horizontal angles has almost always been ignored hitherto, and with satellite fixes good to 1 m at intervals of 200 km, there would be no need to improve on current practice.

(*b*) (iii) This activity remains. Over short distances astro-geodetic sections are likely to be more accurate than satellite fixes for some time.

(*b*) (iv) This also remains, but other methods (gravity and seismological) contribute more than observations of the deviation.

(*c*) Origins of independent surveys can now be better and more easily fixed by satellite, see § 7.77, and modern boundaries are unlikely to be defined astronomically.

(*d*) Polar motion, see § 4.09. This is still observed astronomically, but satellite methods are beginning to do equally well. Tectonic crustal movements are likely to be measurable by satellite or VLBI methods before they can be measured by conventional astronomical observations.

(*e*) Time observatories are still necessary, but are doubtfully classed as geodesy, see § 4.10.

(*f*) Satellite photography is perhaps passing the zenith of its utility as a satellite technique, giving way to doppler, or to radio or laser ranging. It does however currently appear to be the best way of bringing the longitudes of satellite surveys into the conventional BIH Greenwich longitude system,† see § 7.59.

The conclusion is that geodetic astronomy is likely to become reduced in scope, but that it is not obsolete.

4.02. The celestial sphere and astronomical triangle. Summary

For fuller details see §§ 4.03–4.08. It is convenient to describe the stars as lying on the surface of a sphere of such large radius that it does not matter what point on the earth is considered to be the centre. The *poles* of this sphere are the points where it is cut by the prolongation of the earth's axis, and the *celestial equator* is the line in which the plane of the equator cuts the celestial sphere. The *ecliptic* is the path followed by the sun in its apparent annual motion round the earth, which intersects the plane of the equator at two points γ and γ'. The former, occupied by the sun at the vernal equinox is known as the *equinox*, or *first point of Aries.* The angle between the planes of the equator and of the ecliptic is known as the *obliquity*, ϵ, about $23\frac{1}{2}°$. The position of a star on the sphere is defined by its *right ascension*, RA or α, and its *declination* δ. The former (analogous to longitude) is measured from γ, see Fig. 4.1, and the latter, like latitude, is measured from the equator. The RA and declination of a star are approximately, but not exactly, constant. They are published as described in § 4.05 and 4.06.

The local sideral time is defined to be $00^h\,00^m\,00^s$ at the upper transit of γ, and the RA of a star, which is usually measured in hours rather than in degrees, is consequently the time elapsing between the transit of γ and the transit of the star. In other words the RA of a star is the local apparent sidereal time of its transit. The hour angle t of a star at any moment is the sidereal interval which has elapsed since its upper transit and is consequently the LAST minus its RA.

In Fig. 4.1 Z is the zenith, P the north pole of the celestial sphere, N the north point in the horizon of the station of observation, and S is a star. Then in the spherical triangle ZPS, if three of its six elements (three sides and three angles) are known, the other three can generally be determined.‡ In this triangle

(*a*) ZS is the zenith distance, which can be measured.

† VLBI, which is itself somewhat astronomical geodesy, provides an alternative method of getting BIH longitudes, § 7.67, but only if some of the RA sources used can be brought into terms of the equinox, γ.

‡ There are, of course, unfavourable circumstances in which a small change or error in a known element produces a large change in an unknown one, but the various practical systems of observation are designed to avoid such conditions.

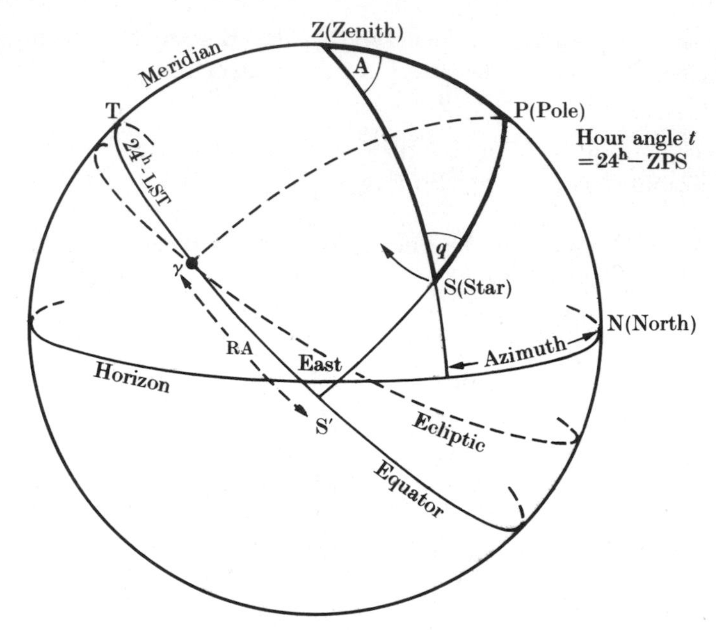

FIG. 4.1. The celestial sphere. LST = $24^h - \gamma$PT. Hour angle = $24^h -$ ZPS. RA(α) = γPS′. Declination (δ) = SS′. Azimuth (A) = PZS. Latitude (ϕ) = NP. Zenith distance (z) = ZS.
The celestial sphere is ordinarily drawn as above, as seen by an eye outside it. Patterns on the hemisphere nearest this eye are consequently seen by it in reverse as compared with their appearance when seen from the inside. The near concurrency of the horizon, the ecliptic, and the circle SS′ is fortuitous.

(*b*) ZP is the co-latitude, which may be approximately known, or may be one of the unknowns.

(*c*) PS, the north polar distance = $90° - \delta$, and is known.

(*d*) ZPS = $24^h - t$ for an east star or $+t$ for a west star, where t is the hour angle, namely LAST − RA. The RA is known, while the LAST may be approximately known or unknown.

(*e*) PZS = azimuth for an east star or $360° - A$ for a west star. Generally unknown.

(*f*) ZSP = the parallactic angle, q. It cannot be measured.

The celestial triangle can always be solved using the values of its internal angles, and carefully watching whether they are A or $360° - A$, t or $24^h - t$. In the southern hemisphere latitude is most conveniently considered negative. Different survey departments will generally have their own conventions and computation forms, and will not wish to change them.

The celestial sphere is comparable to the *unit sphere* of § C.01 (*b*). Its unit is large. A point on the celestial sphere does not define a single line in space, but the direction of all the members of a bundle of parallel lines.

Similarly, a great circle of the celestial sphere defines a set of parallel planes.

4.03. The ecliptic, axis of rotation, and equinox

(*a*) *The ecliptic.* See Fig. 4.2. If the planets caused no disturbance, the path round the sun of the earth-and-moon's centre of mass would lie in a fixed plane, the ecliptic, represented by a fixed great circle on the celestial sphere. The pole of this great circle, on its north side, K_1 in Fig. 4.2, is the *pole of the ecliptic.* In fact, the ecliptic and its pole are not fixed with reference to the inertial frame constituted by the distant stars, because the attraction of the planets moves the pole of the ecliptic $0''.47$ annually, to K_2.

Directions, or the positions of a star S, can be referred to the ecliptic and its pole by their *celestial latitude* SS″ positive north, and *celestial longitude* measured from γ in the same direction as the sun's annual motion. The celestial latitude and longitude of the sun are given in the *Astronomical Ephemeris* (AE).

(*b*) *The axis of rotation.* At any instant the motion of a rigid body may be described by two vectors, namely the linear velocity of any selected

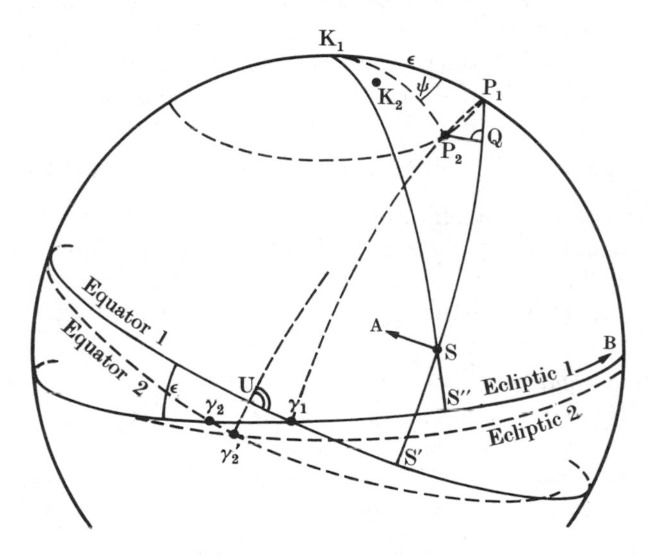

FIG. 4.2. The celestial sphere. K is the pole of the ecliptic. P is the celestial pole (earth's axis of rotation). S is a star. SS″ is its celestial latitude and γ_1S″ is its celestial longitude. Suffixes 1 and 2 indicate the two epochs 1967.0 and 1968.0. The changes are exaggerated. ϵ is the obliquity. $\gamma_1\gamma_2$ $(=P_1K_1P_2=\psi)$ is the luni-solar precession, and $\gamma_2\gamma_2'$ is the planetary precession. Arrow A indicates the direction of the apparent daily motion of the sun and stars. Arrow B indicates the sun's annual motion round the ecliptic, and the direction of increasing celestial longitudes. The angle $P_2P_1Q \approx \gamma_1P_1Q = \text{RA}$ of S, since P_1P_2 is very small.

point, and the angular velocity of the body [549], pp. 279–80. At different selected points in the body the linear velocity will differ, but the angular velocity is the same for all. Both may vary with the time. The instantaneous direction of the angular velocity vector defines the north celestial pole. In this context the earth's instantaneous axis of rotation is a *direction*, not especially associated with the earth's centre of mass or any other point.

If the earth was alone in space, inertia would hold its axis as a fixed direction relative to the distant stars, but the influence of the sun and moon, coupled with the earth's spheroidal form, causes the direction of the axis to change. The celestial Pole, P in Fig. 4.2, moves relative to the fixed stars. The principal change, precession, carries P roughly in a small circle round the pole of the ecliptic K. It goes round the whole circle once in 25 800 years. On this regular motion the moon imposes periodic motions, of which the principal terms are (i) *long-period nutation* with an amplitude of about 9″ and a period of 18.6 years, and (ii) *short-period nutation* with an amplitude of generally less than 0″.5 and a period of a fortnight. See (iii) below.

The *instantaneous* or *true axis* and equator are affected by nutation as above. The *mean axis* and equator are meaned with respect to both long- and short-period nutation.

Apart from nutation, the movement of the pole of the ecliptic results in the mean obliquity (Fig. 4.2) not being quite constant. At present its value is given by

$$\epsilon = 23°\,27'\,08''.26 - 46''.845T - 0''.0059T^2 + 0''.00181T^3, \qquad (4.1)$$

where T is measured in centuries from 1900.0.

(*c*) *The equinox*, being the intersection of the ecliptic and the equator, moves round the former, as the earth's axis and the celestial pole move round the pole of the ecliptic. Its movement is the sum of three elements, as below.

(i) *The luni-solar precession.* The regular precession of the axis causes γ to move fairly regularly round the ecliptic every 25 800 years. In Fig. 4.2 P_1 and P_2 are two successive positions of P, and γ_1 and γ_2 are the corresponding positions of γ. The annual motion of γ, known as the luni-solar precession ψ is given by

$$\psi = 50''.3708 + 0''.0050T, \qquad (4.2)$$

where, as before, T is in centuries from 1900.0. This and similar formulae are given in the annual *Astronomical Ephemeris* (AE) on or about p. 489. Its direction is against the sun's annual motion.

(ii) *The planetary precession.* The movement of the ecliptic itself causes a smaller movement of γ in the opposite direction, from γ_2 to γ_2' in Fig.

4.2. This is known as the planetary precession and is given by†

$$\lambda' = 0''.1247 - 0''.0188T. \tag{4.3}$$

Its direction is contrary to that of ψ.

The combination of the luni-solar and planetary precessions results in a total annual movement of γ amounting to

$$p = \psi - \lambda' \cos \epsilon = 50''.2564 + 0''.0222T. \tag{4.4}$$

This is known as the *general precession.*

The annual rotation of the meridian Pγ from which RAs are measured is $\gamma_1 U$, and is

$$m = \psi \cos \epsilon - \lambda' = 3^s.07234 + 0^s.00186T, \tag{4.5}$$

where T is in centuries from 1900.0. This is known as the *precession in RA*. See (4.10) for n, the precession in δ. Values of p, m, n, ϵ and other similar quantities are given on AE, p. 9.

(iii) *Nutation.* Long-period and short-period movements of γ, and in the obliquity ϵ, result from the nutation of the pole mentioned in (b) above. These are imposed on the general precession of the equinox. The complete expression of the nutation is very complicated, and currently consists of 69 significant harmonic terms. Those whose period are of more than 90 days are classed as long-period and can be linearly interpolated over periods of 10 days. Those with periods of less than 35 days are classed as short-period and have to be computed daily as in § 4.05. After the fortnightly and 18.6-year periods, the periods with the largest amplitudes are 6 months, 9.3 years, and one year. See [36] pp. 41–5.

(d) *The mean equinox* is a point whose movement is defined by (c) (i) and (ii) ignoring (iii). The movement of the *true equinox* is given by the sum of (i), (ii), and (iii). The separation of the mean and true equinoxes is known as the *equation of the equinoxes,* positive when the direction of the sun's annual motion is from true towards mean. It is given in *Apparent Places of Fundamental Stars,* APFS, (see § 4.05).

4.04. Star places

(a) The pattern formed by the distant stars changes very little during the course of a few years, but their positions (*star places*), as described by their RA and δ, vary with the changes in the equinox and obliquity described in § 4.03. In addition there are changes, generally very small, due to proper motion and parallax as in (b) and (c) below, and the direction in which a star is actually seen is affected by aberration, (d) below, and refraction § 4.11.

† λ' is smaller than the annual 0''.47 motion of K, for the chance reason that the direction of K's movement is somewhat closely towards P.

(*b*) *Proper motion.* Most stars are so distant that they maintain their relative positions on the celestial sphere without detectable change, but relative to the background of these 'fixed' stars many of the nearer stars show individual movements known as *proper motions*. Proper motions are small, 10″ a year being the largest known, and 1″ unusual. They are practically constant from year to year, although a few stars have perceptible orbital motion† in addition to their regular movement, on account of rotation in their own systems.

While the proper motions of all except orbital stars are constant in magnitude and direction from year to year if they are referred to the inertial frame defined by the distant stars which have no perceptible relative motion, the two components μ_α in RA and μ_δ in declination vary with the changes in the equinox and obliquity. If these components are $(\mu_0)_\alpha$ and $(\mu_0)_\delta$ in (say) 1950.0, the components in $1950+t$ years will be given by

$$(\mu_t^s - \mu_0^s)_\alpha = t\{\mu_\alpha^s \cos\alpha \tan\delta + (1/15)\mu_\delta'' \sin\alpha \sec^2\delta\} n'' \sin 1'', \quad (4.6)$$
$$(\mu_t'' - \mu_0'')_\delta = -t(15\mu_\alpha^s)\sin\alpha\,(n'' \sin 1''),$$

where n is defined in (4.10). These changes are small.

(*c*) *Annual parallax.* The earth's annual rotation round the sun causes a periodic change, with amplitude less than 1″ (<0″.25 for all but about 20),‡ in the positions of the nearest stars, [515]. The parallax due to the earth's radius is negligible for all objects outside the solar system. For the computation of annual parallax, if required, see [36], p. 155.

(*d*) *Aberration.* Since the speed of light is not quite infinite, the earth's annual movement round the sun, and its daily rotation on its axis, make the apparent direction of a star differ slightly from its true position. These effects are known as annual and diurnal aberration respectively. Aberration due to the sun's general motion through space has a constant effect on the α and δ of any star at all times and places, and can therefore be ignored.

(*e*) *Definitions.* For the *true place* of a star, the RA is measured from the true, or instantaneous, equinox of date, § 4.03 (*c*), and the declination refers to the instantaneous equator and pole. There is presumed to be neither aberration, refraction, nor parallax. The RA and δ are in fact what would be seen from the centre of mass of the solar system.

The *apparent place* only differs from the true place in that allowance is made for annual aberration and parallax. It is the direction of the star as it

† AP p. viii (in 1972) lists seven stars with significant orbital motion.

‡ Parallaxes are given in [515], but all stars of magnitude brighter than 6 with parallaxes known to be greater than 0″.25 are in AP, and so call for no computation.

would be seen from the geocentre if the mass of the earth and the atmosphere were concentrated there. Diurnal aberration is not included: it would be zero at the earth's centre.

The *mean place* only differs from the true place in that the RA and δ are measured from the mean equinox and equator for a specified date, generally but not necessarily the beginning of a Besselian year. It equals the true place corrected for both long- and short-period nutation.

4.05. Ephemerides

(*a*) Star places are given in two types of publications, ephemerides and catalogues. The former give star places at such intervals as 10 days, while the latter give them for a single epoch, such as 1950.0, from which their places at later dates can be computed. Catalogues are described in § 4.06. See also § 4.06 (*d*), for the relative advantages of ephemerides and catalogues.

(*b*) *Apparent places of fundamental stars.* (*APFS*). This is the best ephemeris for most geodetic purposes. It is published annually and gives the apparent places of 1483 stars at 10-day intervals, and of 52 circum-polar stars ($\delta > 81°$) daily. Places are given to 0″.01 for δ, to $0^s.01$ for RA of circum-polars, and to $0^s.001$ for RA of the rest. Nearly all the stars are of magnitude brighter than 6.5. APFS is based on the fundamental catalogue FK 4, § 4.06 (*a*), and its accuracy is the same as that of FK 4.

In AP the apparent places are as defined in § 4.04 (*e*), except that the short-period nutation is only included in the daily places of the circum-polar stars. For the 10-day stars it must be computed as in (*c*) (iii) below.

(*c*) *Use of APFS.* The introduction to APFS gives a worked example. The introduction to all such publications should be carefully studied before use. Old volumes may differ from new. The routine is as below.

(i) The recorded UT of the observation is converted to the decimal of a day, using AP Table V.

(ii) RA and δ are then interpolated between the 10-day places. Linear interpolation may introduce errors of a few units in the last figures, and second differences are required for the fullest accuracy. The introduction and Table VI give a convenient method.

(iii) Except for circum-polar stars, the correction for short-period nutation must be added, namely

$$\left.\begin{array}{ll} d\alpha(\psi)\,d\psi + d\alpha(\epsilon)\,d\epsilon & \text{to} \quad \alpha \\ \text{and} \qquad d\delta(\psi)\,d\psi + d\delta(\epsilon)\,d\epsilon & \text{to} \quad \delta, \end{array}\right\} \tag{4.7}$$

where $d\epsilon$ and $d\psi$ are changes in the obliquity and longitude caused by the nutation of the pole, and are dependent only on the date. They are given in AP, Table I. $d\alpha(\psi)$ means $d\alpha/d\psi$, etc. The values† of $d\alpha(\psi)$, etc. depend on α and δ, and are given in AP below the 10-day places.

(iv) If it is desired to include diurnal aberration at this stage, add

$$+0^s.0213 \cos\phi \cos t \sec\delta \quad \text{to} \quad \alpha$$

and

$$+0''.320 \cos\phi \sin t \sin\delta \quad \text{to} \quad \delta. \tag{4.8}$$

But for the best practical methods of correcting different observations see §§ 4.18 (*c*), 4.27 (*d*), 4.37 (*e*), and 4.42 (*e*).

(*d*) *Additional information in APFS.* AP includes other tables giving the following.

(i) Short-period nutation $d\psi$ and $d\epsilon$, daily.

(ii) The sidereal time (both apparent and mean) of 0^h UT 1.‡ See § 4.10.

(iii) Tables for the conversion of mean solar time to sidereal time and vice versa.

(iv) The equation of the equinoxes, § 4.03 (*d*).

(v) Besselian day numbers at 10-day intervals, excluding short-period nutation. See § 4.07 (*d*).

(*e*) *Other ephemerides.* APFS was started in 1941, before which 10-day star places were to be found in the *Nautical Almanac* and in the *American Ephemeris*. Since 1960 these two have been combined as the *Astronomical Ephemeris* (AE) which contains only a few star places. The *Star Almanac for Land Surveyors* gives 685 stars to $0^s.1$ and $1''$. It also gives R $(= R_u \pm 12^h)$ every 6 hours, and Pole star tables to $0'.1$. Where its accuracy suffices, it is very convenient. Its use is fully described in [492].

4.06. Star catalogues

See § 4.05 (*a*). Catalogues generally give a much greater number of stars than it is possible to include in any ephemeris.

(*a*) *FK 4 and its Supplement*, [211]. FK 4 contains the 1535 stars for which APFS is computed. It gives their mean places for 1950.0 and 1975.0, with their standard deviations, together with proper motions μ_α and μ_δ and their s.d.'s. It contains the best fixed stars, and in itself constitutes a reference system to which other catalogues should aim at conforming. It is used for the computation of time observations at all the principal observatories and by the BIH, § 4.10.

† $$d\delta/d\epsilon = \sin\alpha, \qquad d\alpha/d\epsilon = -\tfrac{1}{15}\cos\alpha\tan\delta, \qquad d\delta/d\psi = \cos\alpha\sin\epsilon$$

and

$$d\alpha/d\psi = \tfrac{1}{15}(\cos\epsilon + \sin\alpha\tan\delta\sin\epsilon),$$

see AP, Introduction, p. ix. These can be verified from Fig. 4.2.

‡ The heading of AP, Table II is 'Sidereal time at 0^h UT', without specifying UT 0, 1, or 2. If the UT is UT 1, the ST is that of the international Greenwich meridian as defined in § 4.09 (*c*), and using the CIO as pole. If the UT is UT 0 or UT 2, the ST is correspondingly different.

For the accuracy of FK 4 [437A] gives $\pm 0''.04$ as an average accidental standard error of either RA or δ at the mean epoch of its observations, generally between 1900 and 1930. To this must be added the s.e.'s of the two components of the proper motion since the mean epoch. Systematic errors are thought to be rather smaller than $\pm 0''.04$, except at high southern declinations.

The *Preliminary Supplement* to FK 4 contains the 1950.0 places and proper motions of 1987 additional stars, in the system of FK 4, mostly of magnitude between 5 and 7. Standard deviations are not given.

A revision of FK 4, namely FK 5, is being prepared for publication some years after 1980. It may contain as many as 5000 stars, some as faint as 9 magnitude. It will include recent observations, and so should have more accurate proper motions than those of FK 4. The epoch is likely to be the year 2000. It is hoped to provide a supplement giving the position of radio sources, and of faint optical sources which may be thought to coincide with radio sources. The two types may thus be brought into a common FK 4 system, see the last few lines of § 7.67.

(*b*) The *Smithsonian* (*SAO*) *Catalogue* of 1966. This catalogue gives the 1950.0 places of 258 997 stars. The principal data given are (i) RA and δ for 1950.0, and their combined standard deviations of position. (ii) The two components of the proper motion μ and their s.d.'s (iii) The RA and δ's at the mean epoch of the original observations, with their s.d.'s and mean dates. The latter being between 1875 and 1955, the digits 18 and 19 are omitted. The catalogue is available in four volumes, and also on magnetic tape.

The SAO catalogue does not give AV or SV for use in (4.9), so RA and δ at the start of later years must be computed by (4.13), or as in § 4.08.

Accuracy. The SAO catalogue is based on FK 4, and shares its systematic errors. Within that system its likely errors can be deduced from the standard deviations given in the catalogue.

(*c*) *Boss's General Catalogue* of 33 342 stars for 1950.0 was the catalogue most used for geodetic work demanding more stars than the 1535 of APFS, before the introduction of the SAO catalogue. For each star it gives α_{1950} and δ_{1950}, the mean places for 1950.0: AV, the *annual variation* of α and δ, namely the annual changes due to precession in 1950, (4.10) and (4.11), plus the annual proper motion: SV, the *secular variation* or change in AV per century: $3^{d}t$, the 'third term', for the change in SV: μ_α and μ_δ, the proper motions in α and δ: and in some cases the changes in μ per century.

The average s.d. of its places is probably about $\pm 0''.7$, but errors of $3''$ are known to occur. It is not likely to be used now, except where there may be an occasional need for a few stars which are not in APFS, in an organization which is not equipped to use the SAO catalogue.

(*d*) The relative advantages of catalogues and ephemerides are as below.

(i) The number of stars included in an ephemeris is necessarily rather limited.

(ii) The 10-day places of an ephemeris give a considerable saving in computation by traditional methods. On the other hand an electronic computer cannot conveniently store unlimited tabular material, and if star computations are being regularly undertaken, it will be easier to give the computer the catalogue RA and δ, and to let it compute the apparent places of date using basic formulae.

4.07. The use of star catalogues

(*a*) It is required to find apparent α and δ on some particular day on which observations have been made. This paragraph outlines methods that might be described as classical. The methods now used for satellite photography, are described in § 4.08 and in §§ 7.23 and 7.32.

(*b*) The first stage is to convert the 1950.0 catalogue places to mean places of $1950.0+t$, where t is the whole number of Besselian† years closest to the date for which places are required. This change involves only precession and proper motion. There are two methods.

(i) If the catalogue, like Boss, gives the annual variation (AV), the secular variation (SV) and the so-called 3d term,

$$\alpha'_t = \alpha_0 + t(\mathrm{AV}) + \tfrac{1}{2}t^2(\mathrm{SV})/100 + (t^3/10^6)(3^{\mathrm{d}}\ \mathrm{term}), \tag{4.9}$$

with a similar expression for δ'. Here α_0 is the mean place at 1950.0, and α'_t is the mean place at $1950.0+t$, including proper motion for the whole number of years.

(ii) With other catalogues, such as FK 4 and the SAO. In Fig. 4.2 the annual change in the declination of S is

$$\delta'_2 - \delta'_1 = \mathrm{P_1Q} = \mathrm{P_1P_2}\cos\alpha = \psi\sin\epsilon\cos\alpha = n\cos\alpha, \tag{4.10}$$

where $n = \psi\sin\epsilon = 20''.0468 - 0''.0085T$, and α is the RA of S. n is known as the *precession in declination.* Here δ' is mean declination as defined in § 4.04 (*e*).

Similarly, the annual change in RA is

$$\begin{aligned}\alpha'_2 - \alpha'_1 &= \gamma_1\mathrm{U} + \mathrm{P_2Q}\tan\delta \\ &= m + \psi\sin\epsilon\sin\alpha\tan\delta \\ &= m + n\sin\alpha\tan\delta,\end{aligned} \tag{4.11}$$

where m and n are as in (4.5) and (4.10). α' is mean RA.

† See § 4.10. 1950.0 is the beginning of the Besselian year 1950.

Formulae (4.11) and (4.10) give the annual changes in α and δ. For a fraction of a year τ the change is the annual change multiplied by τ, but for a longer period the variations of m and n begin to be of consequence, and m and n must be replaced by their mean values. AE, p. 9† for every year $1950.0+t$ gives

$$M=-t\times(\text{mean } m)$$

and

$$N=-t\times(\text{mean } n) \tag{4.12}$$

for the beginning of the year concerned, and its Table III gives their values between the year concerned and other years back to 1755.

So for an interval of many years,

$$\begin{aligned}\alpha_t'-\alpha_{1950}&=-M-N\sin\alpha_m\tan\delta_m,\\ \delta_t'-\delta_{1950}&=-N\cos\alpha_m,\end{aligned} \tag{4.13}$$

where α_{1950} and δ_{1950} are for 1950.0, and α_m and δ_m are for the middle of the period.

(*c*) To allow for proper motion, if not included in AV, the mean α and δ as given in the catalogue should be corrected by $t\mu_\alpha$ and $t\mu_\delta$, and the resulting 1950.0 place should then be corrected for precession as in (*b*) (ii) above, to give α_t and δ_t at $1950.0+t$.‡

(*d*) Having obtained α and δ for $1950.0+t$, the apparent α and δ (for $1950.0+t+\tau$), except for diurnal aberration and short-period nutation are given by

$$\begin{aligned}\alpha&=\alpha_t+Aa+Bb+Cc+Dd+E+J\tan^2\delta+\tau\mu_\alpha\\ \delta&=\delta_t+Aa'+Bb'+Cc'+Dd'+J'\tan\delta_t+\tau\mu_\delta,\end{aligned} \tag{4.14}$$

where A, B, C, D, E, J, and J' are *Besselian Day Numbers*, which are given in APFS at 10-day intervals. The year $1950+t$ or $1950+t-1$, should be used according to whether τ is positive or negative. These day numbers depend only on the date, while a, b, etc., depend on the α and δ of the star, as in (4.15) below, and τ is the *Fraction of the year* which is given in the day number table in APFS. τ should be between +1/2 and −1/2.

† Page 50 until 1972. Before 1972 there is room for confusion about signs. It is required to correct 1950.0 places to $1950.0+t$, not the opposite. Table III of AE gives values of M and N for correction from every fifth year (since 1755) to the date of publication.

‡ Alternatively α_t' and δ_t' as computed from the 1950.0 place uncorrected for proper motion, may then be corrected by applying $t\mu_\alpha$ and $t\mu_\delta$, where μ_α and μ_δ are as given by (4.6) for the year $1950+t$. Note that in this context the correction is $t\mu_{t\alpha}$, not $t\times$(mean of μ_0 and $\mu_t)_\alpha$. And similarly for δ.

$$\begin{aligned} a &= (1/15)\{(m/n)+\sin\alpha\tan\delta\} & a' &= \cos\alpha \\ b &= (1/15)\cos\alpha\tan\delta & b' &= -\sin\alpha \\ c &= (1/15)\cos\alpha\sec\delta & c' &= \tan\epsilon\cos\delta-\sin\alpha\sin\delta \\ d &= (1/15)\sin\alpha\sec\delta & d' &= \cos\alpha\sin\delta, \end{aligned} \tag{4.15}$$

where α_0 and δ_0 may be used for α and δ. (m/n) is given by

$$(m/n) = 2.29887+0.00237T. \tag{4.16}$$

For ϵ see (4.1) or AE, p. 9. The Explanation in AE includes an example.

In (4.14) A, B, and E provide the corrections for precession and long-period nutation; C and D correct for annual aberration. But it must be noted that the day numbers which are published in AE at one-day intervals, (but not those in APFS), also include short-period nutation as given in (4.7). So, when AE is used, (4.7) should not be applied in addition.†

(*e*) The diurnal aberration, and the short-period nutation if it has not been included already as above, must now be added as in (4.8) and (4.7) respectively.

4.08. Use of catalogues. Rotations

The methods of handling precession and nutation described here is slightly more accurate than that of § 4.07, and are now convenient for electronic computers. They are currently used for satellite work.

(*a*) *Precession.* The catalogue α and δ, corrected for proper motion to the required date, are first converted to direction cosines with reference to rectangular coordinates (w_1, w_2, w_3) defined by the equinox and equator of the catalogue (1950.0), w_3 being directed towards the pole, w_1 towards the equinox, and w_2 forming a right-handed triad. The direction cosines of a star are then given by

$$\begin{bmatrix} l_w \\ m_w \\ n_w \end{bmatrix} = \begin{bmatrix} \cos\delta\cos\alpha \\ \cos\delta\sin\alpha \\ \sin\delta \end{bmatrix} \tag{4.17}$$

Then, if rectangular coordinates in the mean sidereal system of the date of observation are (Z_1, Z_2, Z_3), the direction cosines of a star are given by

$$\begin{bmatrix} l_Z \\ m_Z \\ n_Z \end{bmatrix} = P \begin{bmatrix} l_w \\ m_w \\ n_w \end{bmatrix} \tag{4.18}$$

where P is the rotation matrix, [391], i, p. 29,

† As always, when using a strange catalogue on ephemeris, the Introduction or Explanation must be consulted.

$$P=\begin{bmatrix} -\sin\kappa\sin\omega+\cos\kappa\cos\omega\cos\nu & -\cos\kappa\sin\omega-\sin\kappa\cos\omega\cos\nu & -\cos\omega\sin\nu \\ \sin\kappa\cos\omega+\cos\kappa\sin\omega\cos\nu & \cos\kappa\cos\omega-\sin\kappa\sin\omega\cos\nu & -\sin\omega\sin\nu \\ \cos\kappa\sin\nu & -\sin\kappa\sin\nu & \cos\nu \end{bmatrix} \tag{4.19}$$

where, [36] p. 30,

$$\begin{aligned} \kappa &= (2304''.250+1.396T)t+0''.302t^2+0''.018t^3 \\ \omega &= \kappa+0''.791\,t^2 \\ \nu &= (2004''.682-0.853T)t-0''.426t^2-0''.042t^3. \end{aligned} \tag{4.20}$$

In (4.20) the epoch of the catalogue is $1900.0+T$, and the date of observation is $1900.0+T+t$, both T and t being measured in tropical ($\approx$Besselian) centuries, and t is not necessarily a whole number of years. In [36] and elsewhere κ, ω, and ν are known as ζ_0, z, and θ, but in geodesy these symbols are much preoccupied.

If $T=0.50$, as when the catalogue is for 1950.0, (4.20) simplifies to

$$\begin{aligned} \kappa &= 23''.0495t+0.''.30\times10^{-4}t^2 \\ \omega &= 23''.0495t+1''.09\times10^{-4}t^2 \\ \nu &= 20''.0426t-0''.43\times10^{-4}t^2, \end{aligned} \tag{4.21}$$

in which generally redundant small terms are neglected, [391] p. 35. In (4.21) t is measured in single tropical years from 1950.0. The mean α and δ of date can then be derived from (4.18), using

$$\begin{aligned} \tan\alpha &= m/l \\ \sin\delta &= n \\ \text{or}\quad \cos^2\delta &= l^2+m^2 \quad \text{if} \quad \delta\to 90°. \end{aligned} \tag{4.22}$$

(*b*) *Nutation.* The true α and δ of date are obtained by including nutation. Let (z_1, z_2, z_3) be the direction cosines of the true place. Then

$$\begin{bmatrix} l_z \\ m_z \\ n_z \end{bmatrix} = N \begin{bmatrix} l_Z \\ m_Z \\ n_Z \end{bmatrix} \tag{4.23}$$

where, [36] p. 43,

$$N=\begin{bmatrix} 1 & -\Delta\psi\cos\epsilon & -\Delta\psi\sin\epsilon \\ \Delta\psi\cos\epsilon & 1 & -\Delta\epsilon \\ \Delta\psi\sin\epsilon & \Delta\epsilon & 1 \end{bmatrix}. \tag{4.24}$$

In (4.24.) $\Delta\epsilon$ is the nutation in obliquity, which is given as $-B$ (the Besselian day number) in AE pp. 266–85, and $\Delta\psi$ is the nutation in longitude, which is tabulated in pp. 18–32. $\Delta\psi \cos \epsilon$ is known as the nutation in RA, or the equation of the equinoxes, and is given in APFS Table II. [391] refers to $\Delta\psi \sin \epsilon$ as the nutation in declination. In (4.24) second-order terms are neglected. They may amount to 10^{-8} in the direction cosines. [391] pp. 35–7 designates $\Delta\psi \cos \epsilon$ as $\Delta\mu$, and $\Delta\psi \sin \epsilon$ as $\Delta\nu$, and gives formulae for them and $\Delta\epsilon$ in terms of Julian days, using the four principal nutation terms only.

Then (4.22) derives the true α and δ from l_z, m_z, and n_z.

(*c*) *Other corrections.* Proper motion having already been applied, further corrections are necessary for the following.

(i) Parallax, if any. See § 4.04 (*c*).
(ii) Annual aberration, see § 7.22 (*e*).
(iii) Diurnal aberration as in (4.8).
(iv) Planetary aberration, for satellites only, see § 7.22 (*h*).
(v) Refraction, see § 4.11, or for satellites parallactic refraction, § 7.07.

4.09. Polar motion

(*a*) *Definitions.* The precession and nutation of the axis of rotation is shared by the body of the earth, and causes no change in astronomical latitudes or longitudes as defined in the first lines of § 2.04. There is, however, a movement of the body of the earth relative to the axis, known as *polar motion,* which does cause periodic changes in astronomical latitude and longitude. Conversely, relative to the body of the earth, but not to the stellar frame, the axis moves as in (i), (ii), and (iii) below.

(i) A periodic motion round a mean position with a period of one year and an amplitude which varies between about 0″.06 and 0″.10. It presumably results from meteorological causes, [305], 1959, p. 226.

(ii) A similar motion with a period of 14 months (the *Chandler period*) and amplitude of between 0″.08 and 0″.18, which can be ascribed to a minute non-parallelism of the axis of rotation and the principal axis of inertia. Theory gives 10 months as the correct period for a rigid earth, but elasticity lengthens the period, [305], 1959, pp. 211–12.

(iii) During the last 75 years the mean position has apparently moved through a total distance of 0″.25 along meridian 25° W.

The net result is a roughly circular movement about a slowly moving *mean pole of epoch,* with a radius which varies from about 0″.3 down to nearly zero and back again in a period of 7 years.

The movement of the pole is described with reference to axes of x along the Greenwich meridian and y along longitude 90° W., with an origin variously known as that of Cecchini's new system 1900–5, or as the mean pole of 1903.0, or (since 1967) as the *Conventional International Origin* (CIO), as shown in Fig. 4.3. The coordinates x'' and y'' are measured in seconds of arc and represent a direction, not a point on the surface of the earth. The CIO is a direction which is fixed in relation to the body of the earth. A rigidly mounted telescope, once directed at the CIO would (in principle) remain permanently pointed at it. On the other hand, the RA and polar distance of the CIO vary: the former through 24 hours every day, and the latter between 0″ and 0″.5 during the years.

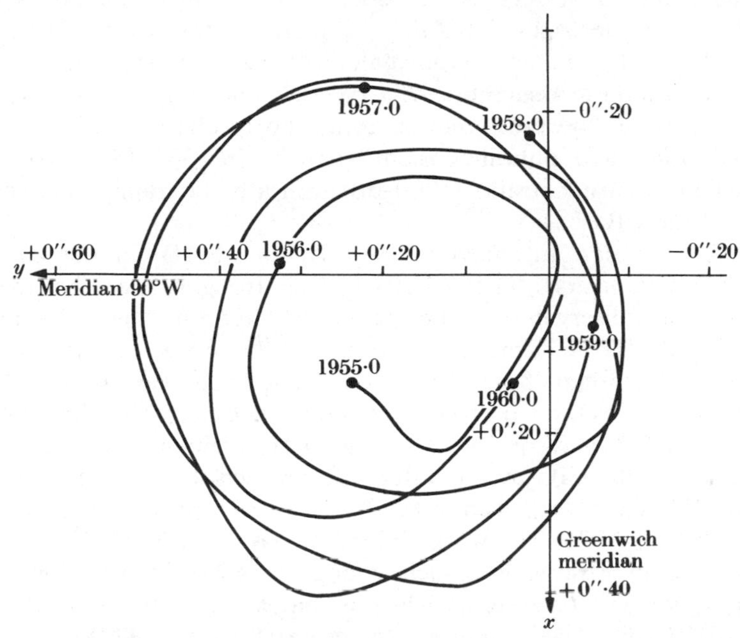

FIG. 4.3. Polar motion between 1955.0 and 1960.0. The origin of coordinates is the Conventional International Origin. This figure is drawn as viewed from the outside of the celestial sphere.

The precise definition† of the CIO is by the following values of the astronomical latitudes of the five observatories of the International Latitude Service, (ILS)

Mizusawa	+39°	08′	03″.602
Kitab			01.850
Carloforte			08.941
Gaithersburg			13.202
Ukiah			12.096

† The co-latitude is the angle between the pole and the vertical at a station. At any instant of time the verticals at the five ILS stations are five points on the celestial sphere, and any two of their latitudes then suffice to define the CIO. The five latitudes in the definition thus involve redundancies, but in so far as the published polar motion may be determined by a weighted mean of the five observatories, the origin from which variations are recorded is derivable from these five arbitrary latitudes using the same relative weights.

The CIO constitutes the mean axis of rotation to which the minor axes of geodetic reference spheroids should be defined to be parallel.

(*b*) *The measurement of polar motion.* The varying positions of the instantaneous pole relative to the CIO are (in 1978) determined by several sources, as below.

(i) The International Polar Motion Service (IPMS) uses old-type zenith telescopes at the five ILS stations, which define the CIO. It also receives reports from 50 or more other observatories which use photo zenith tubes, zenith telescopes, astrolabes, or meridian transit instruments.

The IPMS determines the movement of the pole as given by the five ILS stations on the assumption that no crustal movement has occurred. It also determines the polar motion using the results of the co-operating observatories, and publishes them annually in [38]. [446] contains a discussion of earlier results (1900–58) with a bibliography, and converts them to the CIO.

(ii) The Bureau International de l'Heure. (BIH), in the course of determining Universal time, § 4.10 (*d*), from (roughly speaking) the same co-operating observatories, uses them to determine the polar motion, which it publishes in its monthly circulars, only 4 to 8 weeks after the dates concerned. Its origin of coordinates was made to coincide with the CIO in 1968. In so far as the BIH results cannot agree exactly with those of the 5 ILS stations, it may be said that the BIH origin in subsequent years departs from the CIO by the difference between their results. The mean annual differences are given in the BIH annual reports, e.g. on p. A2 of the report for 1976. The worst difference has been 0″.06.

(iii) Satellite observations such as doppler, § 7.61, are now measuring polar motion with an accuracy which is comparable with that of the IPMS and BIH, while laser observations to satellites, § 7.43 (*e*), and lunar lasers, § 7.45 and VLBI §§ 7.66 to 7.72, hold out hopes of increased accuracy.

Some reorganization during the 1980s is probable, but for some time a certain amount of duplication will be necessary to ensure continuity. See [491].

(*c*) *Effect on astronomical observations.*

(i) Star places (RA and δ) are computed with reference to the instantaneous pole. Sidereal and solar time are defined in § 4.10 by the movement of the equinox and mean sun round the instantaneous equator. In general, astronomical computations of latitude, longitude, azimuth, and time are naturally carried out in terms of the reference system defined by the instantaneous axis and equator and the associated instantaneous Greenwich meridian.

On this system, the astronomical latitudes and longitudes of points fixed on the (supposedly) rigid earth are not constant. This is inconvenient, and the results of astronomical observations need to be converted to a reference system based on the CIO and its equator, together with a fixed zero of longitude.

(ii) *Conventional zero meridian* (*CZM*). In Fig. 4.4 P is the instantaneous pole and C is the CIO, with their equators E_0E_0 and E_1E_1 respectively. G is the direction defined by the vertical (not the spheroidal normal) at Greenwich, and Z is the vertical at some other observatory. The natural meridian to accept for the earth-fixed CIO system is CG produced to X_1 on the CIO equator,† thus preserving Greenwich as the zero of longitude on both systems. There are, however, complexities in matters of detail, and the actual definition of the point X_1 is provided by the BIHs accepted longitudes (on the CIO system) of a large number of observatories with which it co-operates. For a form of words see § 4.10 (*d*) (iii). It lies on the CIO equator, nearly but not exactly on CG produced.

(iii) *Formulae for change of pole.* Observations of astronomical latitude, etc., made and computed with reference to the instantaneous pole, must in principle be reduced to the CIO by the addition of the expressions given in (4.25)–(4.28), in which suffix 0 indicates reference to the instantaneous pole, and suffix 1 to the CIO. These results are clear from Fig. 4.4, in which x and y are measured in seconds of arc.

To *latitudes* add

$$\phi_1-\phi_0=ZZ_1-ZZ_0=-(x\cos\lambda-y\sin\lambda). \tag{4.25}$$

To astronomical *azimuths* add

$$A_1-A_0=-\theta=-(x\sin\lambda+y\cos\lambda)\sec\phi, \tag{4.26}$$

where θ is the angle PZC. This is not affected by any change in the zero of longitude.

To *longitudes*. Assuming that there is no change in the (equatorial) zero of longitude, add

$$-Z_1Z_0=-(x\sin\lambda+y\cos\lambda)\tan\phi. \tag{4.27}$$

To *Laplace azimuths* as computed in (2.4), combining the corrections to azimuth and longitude given by (4.26) and (4.27), add

$$A_1-A_0=-(x\sin\lambda+y\cos\lambda)\cos\phi. \tag{4.28‡}$$

† From the point of view of time, it is not material whether X_1, or any other such points, lie on the instantaneous or CIO equators. These two equators are separated by not more than 0″.6 in latitude, and the angle between the instantaneous and CIO meridians through any one such point also cannot exceed 0″.6. So if two meridians meet at a point on one equator they cannot be separated by more than 0″.6 tan 0″.6 ($=1''.8\times10^{-6}$) on the other. This can be ignored.

‡ $-\sec\phi+\sin\phi\tan\phi=-\cos\phi$.

It is assumed that in both systems the zero of astronomical longitudes is the same as that of geodetic longitudes.

The corrections of (4.25)–(4.28) seldom amount to 0″.5 except to longitude and azimuth in high latitudes. They have seldom been consistently applied in any national survey. Their small size makes it unlikely that any existing survey is any the worse for their omission, but they are easy to apply, and they should no longer be ignored.

(*d*) *The earth's north pole*, as opposed to the celestial pole, may be defined in many ways. The earth's *instantaneous astronomical*† *north pole* is a point on its surface where the vertical is parallel to the instantaneous axis of rotation. It moves relatively to the supposedly rigid earth as in Fig. 4.3.

The earth's *mean astronomical north pole* is a point where the vertical is parallel to the direction of the CIO. In so far as the earth is rigid, it is a fixed point. Its astronomical latitude as defined in § 2.04 is 90°.

A *geodetic north pole* is a point where the geodetic latitude is 90°. There are as many geodetic north poles as there are independent geodetic origins.

It would be possible to use the words 'north pole' to describe a point where the surface is cut by a line passing through the earth's centre of mass, and parallel to either the instantaneous axis of rotation or to the direction of the CIO. In the latter case it would be the geodetic pole of a datum centred on the earth's centre of mass.

4.10. Time

(*a*) *Definitions.* The definition of time is somewhat arbitrary. A clock is any object, natural or artificial, which moves with a regular cyclic rhythm, and two intervals of time may be described as equal if they contain an equal number of clock cycles. There are thus as many possible definitions of time as there are clocks. A particular clock and the associated time are, of course, only likely to be useful if a large number of other clocks agree with it in the definition of equal intervals of time,‡ and if reasonably simple natural laws (e.g. of motion and of gravity) can be associated with it.

The most natural clock is the sun's daily motion round the sky, which defines local *apparent solar time.* Local apparent solar time is the hour angle through which the sun has moved (relative to the earth) since its lower transit, converted into time by the ratio 1 hour = 15°. Judged by the

† *Astronomical* and *geodetic* here are as defined for ϕ and λ in §§ 2.03 and 2.04. They refer to the vertical and to the spheroidal normal respectively.

‡ What mostly matters is that if one clock records two time intervals as equal, others should do the same. It does not much matter if one clock beats (say) twice as fast as another, provided the ratio is constant.

movement of the stars, by artificial clocks, and by everything else, apparent time is irregular and inconvenient. It has therefore been replaced by local *mean solar time*, LMT, kept by a fictitious mean sun, which moves more regularly. Greenwich mean time has now in turn been replaced by *universal time* or UT, see (*d*) below.

Observations of the sun are imprecise, and the fictitious sun cannot be observed, so in practice time is obtained by observations to the stars. *Apparent sidereal time* is kept by the true equinox, § 4.03 (*c*) and (*d*). Local apparent sidereal time LAST is the angle through which the equinox has moved (relative to the earth) since its upper transit, converted by the ratio 1 hour = 15°. It is slightly irregular because of nutation, and *mean or uniform sidereal time* LMST is similarly kept by the mean equinox, § 4.03 (*d*).

Star observations give local apparent sidereal time. The regularity of mean sidereal time depends on the regularity of the earth's angular velocity.

In all the above, Greenwich time GAT, GMT, GAST, and GMST are zero at the appropriate transit of the sun, mean sun (lower transits), true equinox or mean equinox (upper transits) through the instantaneous Greenwich meridian.†

(*b*) *Solar and sidereal times. Conversions.* Conversions are effected as follows. In this sub-paragraph (*b*) meridians are those of the instantaneous pole. The effects of polar motion are considered in sub-paragraph (*d*).

(i) Greenwich time (of any kind)

$$= \text{Local time (of the same kind)} - \lambda \qquad (4.29)$$

where λ is the longitude of the local meridian east of Greenwich, all being expressed in hours, or in degrees, as convenient.

(ii) $\text{GMT} - \text{GAT} = 12^h - \text{E}$ (4.30)

where $12^h - \text{E}$ is the *equation of time.* E is tabulated to $0^s.1$ every 6 hours in The Star Almanac for Land Surveyors.

(iii) On any meridian

$$\text{LM sidereal time} - \text{LM solar time} = R_E + \text{ or } -12^h, \qquad (4.31)$$

where $$R_E = 18^h38^m45^s.836 + 8\,640\,184^s.542T_E + 0^s.0929T_E^2, \qquad (4.32)$$

and T_E is the number of Julian centuries, see (*i*) below, each of 36 525

† The Greenwich meridian is a plane parallel to the instantaneous polar axis, and to the (astronomical) vertical (not the geodetic normal) at Greenwich. See § 2.04. See also §§ 4.09 (*c*) (ii) and 4.10 (*d*) (iii) for the definition of *BIH Greenwich*, or the *Conventional Zero Meridian.*

days of ephemeris time which have elapsed since 12^h ET on 1900 Jan 0. R_E is the RA of the mean sun.

(iv) On any meridian

$$\text{LAST} - \text{LMST} = \text{EE}, \quad \text{the equation of the equinoxes.} \qquad (4.33)$$

The equation of the equinoxes is given daily in APFS Table II at 0^h UT, by subtraction of the columns Apparent minus Mean. It is to be noted that when the table is used to give LAST−LMST in longitudes other than that of Greenwich, the interpolation must be carried out for the correct instant of UT, not for the LMT of the observation.

(*c*) *Ephemeris and Atomic time.*

(i) *Ephemeris time.* So far, § 4.10 has considered time systems based on the clock constituted by the earth's rotation, but it is well understood that tidal friction will to some extent result in the rotation slowing down, and that any tendency for the earth to expand or contract will do the same or the reverse. Actually the length of the day has lately been increasing by about $0^s.00015$ every year, [165] p. 105. This may not signify much for geodetic purposes, but it upsets astronomical computations of the moon's orbit, and since 1950 astronomers have used *ephemeris time* for such purposes, based on the theory of the motion of the sun, moon, and planets, in accordance with Newton's laws of motion.

$$\text{ET} - \text{UT 1} = \text{between } 48^s \text{ and } 49^s \text{ in 1978}, \qquad (4.34)$$

increasing by about 1^s/year. See also (4.36). The exact value of ET is of no geodetic interest. See further under Atomic time, below.

(ii) *Atomic time.* For most purposes atomic time has superseded ephemeris time. Thus, the SI definition of the second is

The second is the duration of 9 192 631 770 periods of the radiation corresponding to the transition between the two hyperfine levels of the ground state of the caesium-133 atom.

The chosen number 9 192 631 770 was selected so as to make the atomic second accord so far as possible with the second of ephemeris time. On the other hand, the epoch is different.

$$\text{Atomic time} - \text{UT 1} = \text{between } 16^s \text{ and } 17^s \text{ in 1978}, \qquad (4.35)$$

increasing by about 1^s/year. The BIH have adopted a time system designated *International atomic time* (TAI), based on comparisons between atomic clocks throughout the world, and the monthly circulars of the BIH give UT 1−TAI for every fifth day of the preceding month.

In 1976 ET was replaced by a *dynamical time scale* (DT) in which one day equals exactly 86400 SI seconds, and

$$\text{DT} - \text{TAI} = 32.184 \text{ s}, \quad \text{exactly by definition.} \qquad (4.36)$$

The frequency of a caesium clock is thought to be stable to about 1 part in 10^{12}. Signals based on atomic time constitute a frequency standard, as well as being a source of time for astronomical purposes.

(*d*) *Universal time and the CIO.*

(i) *GMT and UT.* For a long time GMT was observed at Greenwich, and GMT equalled 0^h at the lower transit of the mean sun through the instantaneous Greenwich meridian, PG in Fig. 4.4. In practice, star observations gave GAST, whence GMST was given by (4.33), and from it GMT was given by definition as in § 4.10 (*b*) (iii).

In consequence of polar motion the Greenwich meridian PG is not fixed in relation to the earth's surface, and for practical purposes GMT has been replaced by *Universal Time.* UT 1 is defined as the hour angle measured from the conventional zero meridian, CZO, of a mean sun whose RA is given by $R_u \pm 12^h$. Here R_u is the same as R_E in (4.32) except that T_u (replacing T_E) is measured in centuries of 36 525 UT days elapsed since 12^h UT on 1900 Jan 0. [36] p. 73–4. R_u is given daily for 0^h UT in APFS, Table II.

UT 1 then differs from GMT as follows. (*a*) The small difference between the definitions of R_u and R_E. (*b*) The zero from which the hour angle UT 1 is measured is X_1 in Fig. 4.4, while the zero for GMT is X_0. And (*c*) UT 1 is determined by international co-operation from many observatories, not only (or now at all) at Greenwich.

(ii) *The measurement of UT* 1. UT 1 is measured in the sense that the difference between it and atomic time, UT 1 − TAI, at any instant is given by a large number of observatories, comprising 77 separate astronomical instruments in 1976, coordinated by the BIH. Each of these observatories is allotted an accepted longitude, on the CIO system, and these longitudes being substantially mutually self-consistent define the CZM, see (iii) below.

At any one of these observatories, Z in Fig. 4.4, let a star whose RA = α transit across the instantaneous meridian PZ at an instant of UT 1 as recorded by a clock, which may be assumed to be keeping universal time as disseminated by radio time signals.†

Then from Fig. 4.4, at the instant of transit

$$\begin{aligned} \text{UT 1} = \text{MX}_1 &= \alpha - \text{EE} - (R_u - 12^h) - \text{X}_1\text{Z}_0 \\ &= \alpha - \text{EE} - (R_u - 12^h) - \text{X}_1\text{Z}_1 - \text{Z}_1\text{Z}_0 \\ &= \{\alpha - \text{EE} - (R_u - 12^h) - \lambda_z\} - (1/15)(x \sin \lambda + y \cos \lambda) \tan \phi. \end{aligned} \tag{4.37}$$

† It is immaterial that neither the radio signals nor the clock may actually be keeping exact UT 1. The UT 1 error of the clock will be knowable from the BIH monthly circulars.

This is generally written

$$\mathrm{UT\,1} = \mathrm{UT\,0} - (1/15)(x \sin \lambda + y \cos \lambda)\tan \phi, \tag{4.38}$$

defining UT 0 for the particular observatory at the particular time.

If λ_z, x, and y are known, (4.37) gives UT 1, and the BIH clock can be corrected, or more properly the difference UT 1 − TAI is ascertained. In

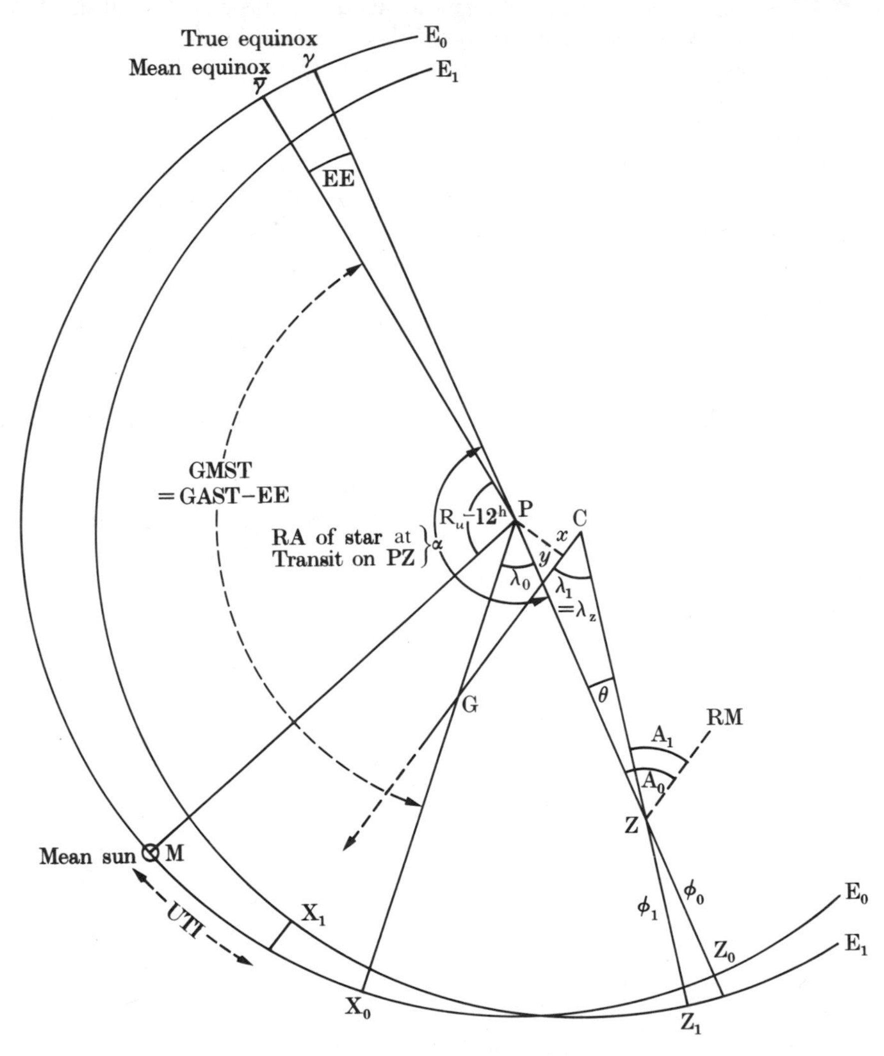

FIG. 4.4. The celestial sphere seen from outside above the north pole. P is the instantaneous pole and C is the CIO. CX_1 is the CZM, nearly coincident with CG. G and Z are the verticals at Greenwich and at another observatory. The instantaneous equator E_0E_0 and the CIO equator E_1E_1 are identical so far as concerns the definitions of longitude and time. α is the RA of star transiting over PZ_0.

practice, the corrections given by the 77 observatories are analysed at the BIH to give smoothed values of x and y and a smooth mean value of UT 1 − TAI.

(iii) It is now possible to give a formal definition of the CZM, or *BIH Greenwich*. It is a plane whose direction is the weighted mean of the directions of (in 1976) 77 planes, each of which is parallel to the direction of the CIO and makes a defined angle λ_i with a plane which is parallel to the CIO, and to the vertical at one of the 77 observatories.†

As the result of long experience, the accepted longitudes λ_i are closely self-consistent, so that there is little scatter among the 77 planes defining the CZM. They are also so chosen that so far as possible the CZM contains the vertical at Greenwich, i.e. so that X_1 lies close to CG produced.

When new observatories are introduced, or when old ones fall out, the accepted longitudes or their weights may be slightly changed by the BIH in such a way as will secure continuity in the position of X_1 and in the universal time based on it.

(iv) *The employment of UT* 1. If the point Z in Fig. 4.4 is not one of the BIH observatories, but a place whose longitude is required to be measured relative to the CIO and CZM, equation (4.37) gives its longitude on the CIO system if UT 1 time is observed and if x and y are known. See Section 3 for details.

(v) *UT* 2. UT 2 is a time system which allows for a conventionally defined annual fluctuation in the earth's rotation. The monthly circulars of the BIH give UT 2 − UTC, and UT 1 − UTC, whence UT 2 − UT 1 may be deduced. The difference varies between + and $-0^s.03$ during the year.

UT 2 is of little or no geodetic interest.

(*e*) *Co-ordinated Universal Time*, (UTC) is now transmitted by most of the world's time signals. Its aim is to combine the ordinary utility of UT 1 with the utility (to physicists) of a continuous atomic time frequency. It is transmitted as a series of pips at intervals of an exact atomic second, and consequently gets out of step with UT 1 by about 1^s every year. This is partially remedied by inserting a leap-second‡ when the atomic pips become $0^s.5$ or more fast on UT 1. Since it is desirable to predict an impending leap second some months in advance, and since the rate of change of TAI − UT is a little irregular, it can happen that the two can occasionally differ by as much as $0^s.9$. The position then is that

$$\text{TAI} - \text{UTC} = \text{an exact number of seconds } (= 17^s \text{ in Jan 1978}), \quad (4.39)$$

† See § C.01 (*b*) for the relation between the directions of lines in space and points on the unit or celestial sphere, and between planes in space and great circles on the sphere.

‡ i.e. an unnumbered pip, so about once a year 61 intervals are comprised in what the signal describes as one minute.

as given in the BIH monthly circulars. We have also

$$\text{UT 1} - \text{TAI} = \text{approximately} - 17^{\text{s}}, \tag{4.40}$$

the exact figure being given to $0^{\text{s}}.0001$ for every fifth day in the BIH circulars. And finally

$$|\text{UT 1} - \text{UTC}| \leqslant 0^{\text{s}}.9, \tag{4.41}$$

and the difference, known as DUT 1 is encoded into the signal, to the nearest $0^{\text{s}}.1$.

UTC thus supplies atomic frequency, UT to the nearest $0^{\text{s}}.1$ for ordinary use, and UT 1 correct to about $0^{\text{s}}.0001$ after waiting a month or two for the BIH circulars.

Between 1965 and 1971 UTC was defined differently. The frequency was altered as required in steps of 50×10^{-10}, and occasional abrupt changes of $0^{\text{s}}.100$ were made, which kept UT $1 - \text{UTC} < 0^{\text{s}}.100$. See [43], 1971 and 1972.

(*f*) *The earth's rotation.* The period of the earth's rotation in the inertial frame defined collectively by the distant stars is not exactly 360° or 24^{h} of sidereal time because the latter is measured from the transit of the equinox, which is itself moving as judged by the distant stars or by the natural laws of motion.

Let $\hat{\theta}$ be an angle (known as the *sidereal angle*) which at 1950.0 equalled the true sidereal time, and whose subsequent increase equals the earth's rotation, as measured from an equinox which is fixed in the inertial frame defined by the pole and equinox of 1950.0. Then [391], pp. 20–1, at any later date $\hat{\theta}$ is given by

$$\hat{\theta} = 100^\circ.075\,542 + 360^\circ.985\,612\,288(MJD - 33\,282.0) \tag{4.42}$$

where MJD is the modified Julian date, see (*i*) below. We have also

$$\hat{\theta} = \text{GMST} - (\mu + \Delta\mu) = \text{GAST} - \mu, \tag{4.43}$$

where μ is the precession in RA, which is designated M in (4.12), and $\Delta\mu$ is the nutation in RA, namely $\Delta\psi \cos \epsilon$ as in (4.24).

These formulae do not purport to describe the small periodic and secular irregularities in the earth's rotation which are mentioned in § 4.10 (*c*) and (*d*) (v) above, and which affect $\hat{\theta}$, GMST, and GAST equally.

(*g*) *The BIH monthly circulars*, have been issued regularly since 1967. In 1978 they give the following.

(i) For every fifth day, the coordinates of the pole to $0''.001$, UT 2 − UTC, UT 1 − UTC, UT 1 − TAI all to $0^{\text{s}}.0001$, and TAI − UTC.

(ii) A list of time signals which emit UTC within $\pm 0^{\text{s}}.0002$, and also one or two other signals in which the difference is a little larger, but known.

(iii) Comparisons between UTC as kept at the BIH and as received from other observatories by time signals, using previously deduced values of propagation delay. Comparisons are also made by the transport of atomic clocks. The differences between the two methods confirms, or suggests corrections to, the assumed velocities of propagation.

Details of the *Bulletin Horaire* before 1968 are given in [488], p. 17. [43] 1967 gives accepted longitudes of the contributing observatories. It also describes the effect of changes in the definitions of the mean pole and other matters, which were adopted in 1968, and gives corrections to UT 1 and UT 2 as published in 1955–67, to make them accord with the BIH system of 1968. Annual reports such as [43] are regularly published.

(*h*) *The day and year.*

(i) The day. The sidereal day, being measured from one upper transit of γ to the next, is not the true period of the earth's rotation relative to the stars. Conversion figures are:

Period of earth's rotation	23^h	56^m	$04^s.099$	mean solar time (UT),
One uniform sidereal day	23	56	04.091	mean solar time (UT),
One mean solar day	24	03	56.555	uniform sidereal time.

(ii) *The year.* The *tropical year* is the interval between passages of the fictitious mean sun through the mean vernal equinox.

The *Besselian solar year*, from the start of which Bessel day numbers are reckoned, begins when the RA of the mean sun, affected by aberration, and reckoned from the mean equinox, is $18^h 40^m$. This always happens on or about 1 January. It is a little shorter than the tropical year; see below. The start of the Besselian year is written as (e.g.) 1968.0.

The *sidereal year* is the time taken by the mean sun to make a circuit of the ecliptic in relation to the directions of distant stars. The precession of the equinox causes it to differ from the tropical year by $(50''.26 \sin 1'')$ $(365.25)/(2\pi)$ mean solar days.

One tropical year $= 365.2421\,9879 - 0.0000\,0614T$ mean solar days,

One Besselian year $=$ One tropical year $- 0^s.148T$,

One sidereal year $= 365.2563\,6042 + 0.0000\,0011T$ mean solar days,

where T is the interval in centuries from 1900.0. [36], pp. 30 and 99.

The *calendar year* is designed to keep in step with the tropical year, the vernal equinox being held at or about 21 March by the appropriate leap years.

(*i*) The *Julian date.* The Julian number of any day, reckoned from Greenwich mean noon, is the number of mean solar days which have elapsed since Greenwich noon on 1 January 4713 B.C., an arbitrarily selected date which avoids negative numbers for historical happenings. The fractional part of a Julian day is simply $\mathrm{UT}-12^{\mathrm{h}}$ expressed as a decimal of a mean solar day.

A *Julian year* consists of 365.25 mean solar days, and a Julian century contains 36525 Julian, or mean solar, days. It is sometimes a convenient unit for long periods of time, e.g. in (4.32).

The *Julian day number* is the integral part of the Julian date.

The *Modified Julian date* (MJD) is the Julian date minus 2 400 000.5. Unnecessarily large numbers are thus avoided, and the MJ day begins at 0^{h} UT.

4.11. Celestial refraction

(*a*) Refraction causes observed altitudes to be too great. On the assumption that the air is arranged in horizontal plane layers of equal density, but without assuming any specified density gradients, the refraction ψ is given by

$$\sin(\zeta+\psi)=n\sin\zeta, \tag{4.44}$$

where ζ is the apparent zenith distance, and n is the refractive index at the point of observation. Whence

$$\psi=16''.3(\tan\zeta)(P-0.14e)/T, \quad \text{if } P \text{ and } e \text{ are in mbar,}$$

$$\text{or} \quad \psi=0''.163(\tan\zeta)(P-0.14e)/T, \quad \text{if } P \text{ and } e \text{ are in Pa,} \tag{4.45}$$

from (1.9) for yellow light ($n=1.000293$), and (1.10). P is total pressure and e is water vapour pressure. T is in K. The effect of e is generally negligible. ψ is to be added to the apparent zenith distance. Conventionally adopting $n=1.000\,292\,7$ for typical starlight, and remarking that the humidity term is only 0.2 per cent if $e=15$ mbar (1500 Pa), [492] recommends the following for $\zeta<50°$,

$$\psi=0''.162\,76(P/T)\tan\zeta \tag{4.46}$$

with P in Pa. In this simple formula, neglected terms of longer formulae are less than $0''.02$ in the mean of a balanced pair (equal zenith distances north and south). This does not of course imply that this formula, or any other, will actually give the refraction with such accuracy.

More fully [511], Part 1, gives the following for zenith distances $<75°$.

$$\psi=0''.16271\tan\zeta\left[1+0.000\,000\,394\tan^2\zeta\left(\frac{P-0.156e}{T}\right)\right]\left(\frac{P-0.156e}{T}\right)$$
$$-0''.000\,749(\tan^3\zeta+\tan\zeta)(P/1000), \tag{4.47}$$

where ζ is the apparent zenith distance, P is the total pressure, and e the partial pressure of water vapour, both in Pa, and T is in degrees kelvin.

When $\zeta > 75°$ further terms are necessary, and these involve assumptions about the variations of density with height. The theory becomes complicated [609], and atmospheric anomalies are apt to cause inaccuracy. Such large zenith distances should not be measured in geodetic astronomy. [511] gives further terms for $\zeta \leqslant 80°$.

For ordinary zenith distances, refraction tables consist of a basic table giving the refraction for standard values of P and T, with correction tables to introduce observed values, [290], Tables V, VI, and VII. See § 7.07 for the parallactic refraction of artificial satellites.

(*b*) If the layers of equal density, instead of being horizontal, are uniformly inclined at an angle θ, the refraction tables will give correct results if entered with ζ measured from the normal to these layers instead of from the zenith. In practice this cannot be done, and the resulting error at $\zeta < 30°$ is $1''$ if $\theta = 1°$. In some form or other this is possibly the largest source of refraction error at altitudes of more than 60°, [471]. The tilting is sometimes apparently related to the wind direction. See also [107] where probably spurious variation of latitude is correlated with seasonal changes in the pressure gradient. Unequal ground temperatures on opposite sides of the observatory would produce a similar result, and [191], p. 615 shows that a horizontal temperature gradient of 1 °C per 30 metres would tilt the surfaces of equal density through 45°, and would change zenithal refraction by $0''.1$ if the gradient persisted to a height of only 15 metres.

Astronomical station sites should so far as possible be chosen so that ground radiation is symmetrical about them. Large buildings and camp fires are especially to be avoided. The best geodetic observatory is either a canvas tent or screen, or a light wooden structure which can be removed to a distance when observations are being made.

(*c*) The 'balancing' of stars is usually advocated, so that to every star in a programme there corresponds another at a similar altitude in an opposite azimuth. This will not eliminate error duc to uniform tilting of the layers of equal density, while if the layers are horizontal, carefully computed refraction can hardly be wrong by $1''$ if $\zeta < 30°$. But balancing will eliminate errors due to inaccurate values of P, as may be given by an aneroid, tilting of the layers symmetrically about the station, such as might occur on a hill-top, and bending of the instrument under its own weight. In high-class work approximate balance is essential.

(*d*) Table 4.1 gives values of ψ for $P = 1006$ mbar, $T = 10$ °C, and humidity 60 per cent, with Bessel's estimate of their probable error.

In round figures if $\zeta < 20°$ or 30°, ψ in seconds $\approx \zeta$ in degrees.

(*e*) *Instrument errors.* The question of what errors may be caused by

TABLE 4.1

ζ	ψ	p.e.	ζ	ψ	p.e.
0°	0″.0	—	70°	157″.8	0″.46
15°	15″.5	—	80°	317″.3	0″.92
30°	33″.4	—	85°	588″.4	1″.7
45°	57″.9	0″.27	87°	857″.6	3″.9
60°	100″.0	0″.34	89°	1452″	17″

unequal temperatures in different parts of the instrument is allied to that of abnormal refraction, in that both arise from somewhat similar circumstances, and both may result in errors of similar form.

(*f*) *Erratics.* Micro-turbulence, or some other source of small density variations in the atmosphere, introduces more or less periodic small variations in refraction, [363]. There appear to be two classes of such variations as below.

(i) Rapid variations with periods of less, or much less, than one second, and with amplitudes (semi-ranges) of a few seconds of arc when in the zenith. These are known as *erratics.*†

(ii) Slower variations with periods of up to one minute, and amplitudes of perhaps 0″.5 (seconds of arc). These are sometimes distinguished as *wanderings.*

The short-period erratics appear to be greater when seen through a telescope of apperture of under 10 cm, than when a larger instrument is used. The suggestion is that they arise from density abnormalities not more than about 10 cm across, possibly inside the telescope. The eye integrates them over their short period, so optical observations are little affected by them, but a short camera exposure is affected by their instantaneous magnitudes. See §7.28 (*c*).

The long-period wanderings presumably come from larger abnormalities of density, at greater distances. Neither the eye nor the camera can integrate them, but they are fortunately small, and a reasonable number of repetitions can adequately reduce them.

Both classes increase with zenith distance, perhaps in proportion to $\sec\zeta$, but $\sec^{\frac{1}{2}}\zeta$ and $\sec^2\zeta$ have also been suggested.

4.12. Spirit levels

Spirit levels or bubbles are essential features of conventional astronomical instruments, although they are not wholly satisfactory, and are sometimes replaced by a pool of mercury as in the astrolabe, or by some form of gravity suspension.

† Variations of brilliance are known as *scintillation.*

The value of one division of the bubble scale can be determined on an instrument known as a *bubble-tester*, which calibrates it against a micrometer screw of known pitch, [444], pp. 19–23. Alternatively a bubble may be calibrated against its own vertical circle by repeated intersection of a distant mark with various bubble readings. But with a very sensitive bubble, however accurately it may have been determined, the scale value cannot be relied on to be constant within (say) 50 per cent, [82]. In any set of observations the mean bubble correction should therefore be so small that such an error will not matter, and individual readings should be scattered on either side of zero, so that a grossly wrong scale value may be apparent.

The reading of the bubble when a mark is intersected with a particular circle reading may be found to depend on the direction of the last movement of its slow motion screw. It is therefore proper always to make this (say) clockwise, [100], p. 48.

A striding level, or a cross bubble on the lower plate of a theodolite, may be calibrated by the 'Wisconsin' method, [100], p. 49, or [492].

In large spirit levels the length of the bubble is adjustable. It should be adjusted so that when it is central its ends are approximately in the centres of the engraved scales.

A bubble tube is only ground to guaranteed accuracy along a narrow longitudinal strip of its length. It must not be rotated about its long axis so that the bubble ceases to move in this strip.

The reading of a bubble is sensitive to differences of temperature between its two ends, and care must be taken to avoid such changes as may arise from hot hands, breathing, artificial lighting, or unequal exposure to the night sky.

See § 4.26 for the adjustment of a striding level.

Bubbles should usually be read as soon as possible after an observation, before circles and micrometers, and so after minimum movement of body weight.

For the recording of bubble readings, see 1.26 (*d*), and also §§ 4.18 and 4.27.

General references for Section 1. For time [43], [488], and [512]. For star places [36], [391], and [533]. [492] deals very fully with all parts of Section 1, with worked examples.

SECTION 2. LATITUDE

4.13. Methods of observation

When a star is on the meridian

$$\phi = \delta \pm z, \tag{4.48}$$

the sign positive when it is south of the zenith: south declinations negative. When a transit is below the north pole substitute $180° - \delta$ for δ, and below the south pole substitute $-(180° + \delta)$. Methods of observing latitude are as follows.

(*a*) Meridian altitudes, using (4.48).

(*b*) Circum-meridian altitudes, in which the observations are continued before and after transit, and corrected to give the value of z on the meridian.

(*c*) Altitudes of Polaris, or of σ Octantis in the southern hemisphere, at any hour angle.

(*d*) The Talcott method. Often Known as Horrebow–Talcott.

(*e*) Equal altitudes. The timing of star passages in different azimuths across a constant altitude, such as 45° or 60°. Or with a really good theodolite, and with probably some loss of accuracy, the timing of passages across different measured altitudes. See §§ 4.37–4.40.

(*f*) Zenith photography. See §§ 4.34–4.36.

(*g*) Latitude can also be got as a by-product of the Black method of azimuth observation. See § 4.47.

Items (*e*) and (*f*) give latitude and longitude simultaneously, and are described in Section 4 of this chapter. Their use is advantageous when both latitude and longitude are required. Item (*g*) is in Section 5.

The most accurate methods are those used by the Photo zenith tube (PZT) in (*f*) and the Danjon astrolabe in (*e*), working in fixed observatories. More mobile and less precise instruments, using more or less similar principles are available, but are not necessarily as accurate as the Talcott or meridian altitude methods.

The relative merits of the Talcott and meridian altitude methods are arguable. In favour of Talcott is the fact that it is independent of accurate graduation of the vertical circle, and that currently available instruments provide more sensitive levels for the Talcott method than are attached to the vertical circles. In the past it has been used with a special instrument known as a Zenith Telescope, which has been designed for the purpose and has generally had a longer telescope than most theodolites, but now Astronomical theodolites, § 4.20, are used instead. These incorporate the special features required for field astronomical observations. In favour of the method of meridian altitudes: (i) its programme is much simpler; (ii) FK 4 and its supplement contain an ample number of stars, while the

Talcott method has to use less accurate stars from the Boss or SAO catalogues; and (iii) Talcott observations are much upset by a little cloud. The Talcott method is still in use, [447], but meridian altitudes will probably give as good a result, or a better one, for the expenditure of the same amount of time.

Circum-meridian altitudes may be preferred when a small instrument is being used, but it is generally better to get a single sight on each of many stars, that many sights on far fewer.

Polaris or σ Octantis may take the place of a meridian star in suitable circumstances.

Latitudes given by Black's azimuth observations (g) are generally of low accuracy, as the method is designed to concentrate on azimuth.

Throughout this section, δ and RA are apparent places. When latitude observations are made on or very close to the meridian, no further correction for diurnal aberration is required, since $t=0$ in (4.8).

All these methods give ϕ relative to the instantaneous pole, which must be corrected to the CIO by (4.25).

4.14. Talcott method and Zenith telescope

If two stars of declinations δ_1 and δ_2 transit at zenith distances of z and $z+\delta z$ north and south of the zenith respectively, (4.48) gives

$$\phi = \tfrac{1}{2}(\delta_1+\delta_2)+\tfrac{1}{2}\delta z. \qquad (4.49)$$

So z need not be measured. A theodolite with an eye-piece micrometer working in a vertical plane may be used to record the micrometer reading of the altitude of transit of the north star, and then with the vertical circle still clamped it may be swung to the south to await the transit of some south star which will pass within range of the micrometer. δz is thereby measured: δ_1 and δ_2 are known: and with suitable corrections for level, collimation, refraction, etc., ϕ is deduced without inaccuracy due to the graduation error of the theodolite's circle.

See [290] for many details.

In the past the Talcott method has generally been used with a special instrument, the Zenith telescope, but during the last 20 years several theodolites have been designed to include its special features, and they are now more generally used. See § 4.20.

The Zenith telescope is a telescope, usually 6- to 8-cm aperture and 75–100 cm focal length, mounted on horizontal and vertical axes as in Fig. 4.5(a). Stops are provided on the horizontal setting circle, so that the telescope can be turned into the meridian plane in either position 'level east' or 'level west', LE and LW for short, and provision is made for levelling the horizontal axis so that the telescope can turn in that plane.

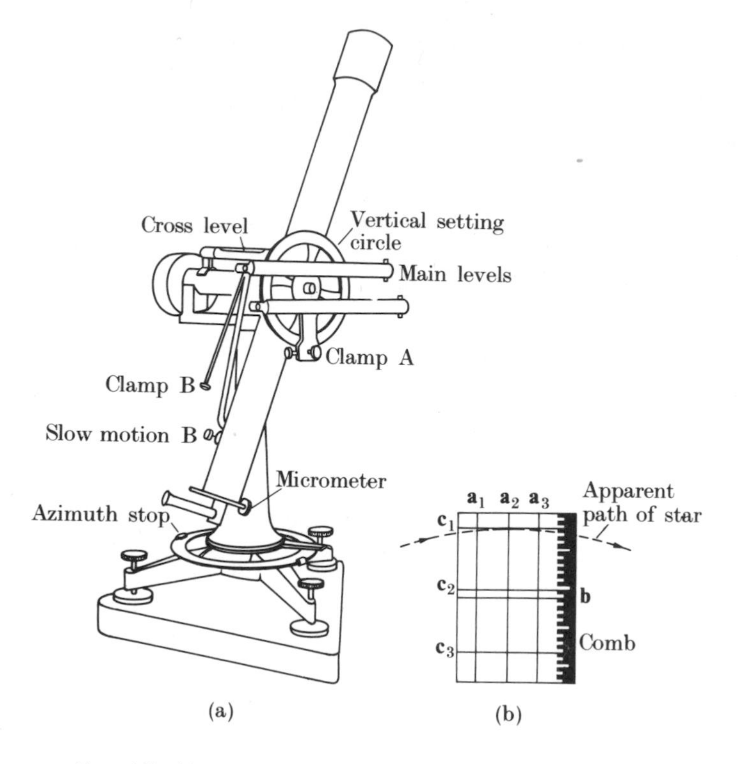

FIG. 4.5. (a). Zenith telescope. FIG. 4.5 (b).

Two very sensitive levels are provided ($\frac{1}{2}''$ or $1''$ per mm), which can be rigidly clamped to the telescope at any desired angle, as regulated by a vertical setting circle and the clamp and slow-motion screw marked A, while a second slow motion B can bodily rotate telescope and levels about the horizontal axis so as to centre the levels without disturbing the angle between them and the telescope. An eye-piece micrometer with its drum marked in 100 divisions of (say) $\frac{1}{2}''$ or $1''$ (second of arc) each, traverses the horizontal wires vertically across the field. Fig. 4.5(*b*) shows the usual arrangement of the diaphragm, $\mathbf{a}_1$, $\mathbf{a}_2$, and $\mathbf{a}_3$ being fixed vertical wires, **b** a fixed horizontal wire, and $\mathbf{c}_1$, $\mathbf{c}_2$, and $\mathbf{c}_3$ moving horizontal wires. A diagonal eye-piece is used.

The instrument is mounted on a firm base which for the best work is isolated from the weight of the observer's stool.

4.15. Adjustments

The vertical axis is made vertical by means of the foot-screws, and the horizontal axis is then made horizontal (if not already so) by an adjustment under one bearing, and its own level.† The telescope object glass is adjusted to bring the image of a star into the plane of the moving wires, and the eye-piece is adjusted to suit the eye. The vertical wires are made vertical, and the horizontal wires horizontal, giving preference to the horizontality of $\mathbf{c}_2$ if detectable imperfection is unavoidable. The micrometer drum and full-turn counter are made to read zero when wire $\mathbf{c}_2$ is on a long division of the comb, and the setting circle is adjusted to give true altitudes within 15″ when the two main levels are central.

Collimation. If a sharp distant object is available, wire $\mathbf{a}_2$ is easily brought on to the line of collimation by intersecting the object in both positions LE and LW, using the horizontal circle as with a theodolite, but unless the object is very distant, allowance must be made for the telescope being off the central axis, or else a double target may be provided. Polaris at elongation can also be used.

The collimation error of $\mathbf{a}_2$ should be less than 15″, see § 4.18 (*d*).

Azimuth. The stops on the horizontal circle must be so adjusted that in both positions the wire $\mathbf{a}_2$ lies in the meridian within 15″ or 40″ according to whether the outer wires $\mathbf{a}_1$ and $\mathbf{a}_3$ are or are not being used; see § 4.18 (*d*). The meridian must be determined by topographical methods, such as from Polaris. A distant azimuth mark is a convenience for checking the stability of the instrument in azimuth, but it can also be checked by recording the times of star transits, § 4.18 (*e*).

Determination of constants. The value of one division of the micrometer is determined by the star observations as in § 4.18 (*a*), but a preliminary value is easily got by timing the vertical movement of a star at elongation. Allow for change in refraction with altitude by subtracting about 1 part in 3500 from the resulting value of one turn of the screw, but see § 4.18 (*b*).

The interval, in terms of micro divisions, between the three horizontal wires at their intersections with the three vertical wires can then be got by comparison with the micrometer, and that between the three vertical wires by timing the transit of meridian stars across them.

For the bubble see § 4.12.

These constants are all reasonably invariable, and their values are best decided on after a fair volume of observations have been made.

4.16. Programme

A programme is necessary, comprising pairs of stars which satisfy the following conditions.

(*a*) Zenith distances, one north and one south, equal within the range

† Reverse the level end-for-end in each of the two positions LE and LW. Any dislevelment in the east–west plane is known as the *inclination*.

of the micrometer, (say) 20′, and not exceeding 45° or preferably 30°.

(*b*) RAs differing by not less than 2^m or 3^m,† nor preferably by more than 20^m, since no work can be done between the transits of the two members of a pair.

(*c*) RA of the first of a pair at least 3^m or 4^m† more than that of the second of the preceding pair.

(*d*) In a night's work the algebraic sum of the difference δz (south minus north) should be less than one micro turn per pair. This is to reduce the effect of error in the screw value.

(*e*) Magnitude brighter than (say) 6.5 (depending on the instrument and conditions), but the largest stars are preferably avoided.

The fulfilling of these conditions is quite troublesome. The ephemeris is not likely to have enough stars, and a catalogue must be used to supplement it. Even for making the programme, account must be taken of precession since the epoch of the catalogue. To make the programme, take each star of possible suitable RA, δ, and magnitude in succession, and look for another within the next 20^m of RA, whose declination is within 20′ of $2\phi-\delta$. All suitable pairs should be noted, so that the selected programme can easily be changed if cloud or other accident should cause a star to be missed.

For a field programme in varying latitudes it is convenient to plot suitable stars by RA and δ, in rectangular coordinates, on a long strip of transparent cloth. At any station the cloth is folded along the declination line corresponding to stars which pass through the zenith, and pairs can immediately be selected.

A programme should then be made showing for selected pairs: star numbers; magnitudes; RAs to 1^s; δ's to 1′; δz in micro turns; N or S; LE or LW (the first star of successive pairs should be taken E, W, W, E, E, W, etc.); micro readings of the stars' transits when the setting circle is set to read mean z (i.e. micro readings at which wire $\mathbf{c}_2$ gives true altitudes $\pm\frac{1}{2}\delta z$).

4.17. Observations

A reasonable night's programme is seven to fifteen pairs, which should be got in 3 to 5 hours. For each pair the setting circle is set to the mean z, LE or LW as shown in the programme, and the bubble is then carefully centred by slow-motion B. The star should then appear close to the micro setting given in the programme, the interval before transit depending on δ. A sidereal clock gives the time of transit (RA). As the star crosses the central wire $\mathbf{a}_2$, it is intersected by either $\mathbf{c}_1$, $\mathbf{c}_2$, or $\mathbf{c}_3$, these three wires

† This depends on the intervals between wires $\mathbf{a}_1$, $\mathbf{a}_2$, and $\mathbf{a}_3$, if the outer wires are used, and also on the polar distances of the stars.

being provided to avoid excessive turning of the screw. The moment of transit is (conveniently but not essentially) recorded to the nearest second by stop watch, or as convenient, and the micro reading is booked (whole turns from the comb: divisions and tenths from the drum), with a note of which wire has been used. Both main levels are read before the observation and immediately after. The telescope is then swung round to the other position ready for the other member of the pair, the levels are recentred by slow-motion B (A must not be touched), and the process is repeated. This completes the pair, and the setting circle is reset for the next. An assistant is required to book results, to keep an eye on the programme and read it out, and to record times. Recording times of transit checks the azimuth error.

The use of the outer vertical wires $\mathbf{a}_1$ and $\mathbf{a}_3$ is to allow of intersections being made with the moving wire as the star passes each of the three vertical wires, and so to divide random error by $\sqrt{3}$. But this reduces the permissible tolerance in the adjustments for azimuth, collimation, and inclination, § 4.18 (*d*). In field work they are probably best avoided.

Additional points to be recorded

(*a*) In low latitudes the time of transit of some close circumpolar star, as a strong check on azimuth error. In both positions, wire $\mathbf{a}_1$ or $\mathbf{a}_3$ being used for one of them.

(*b*) Readings on a distant mark, if available, as in § 4.15. Also of the cross-level for inclination. Twice a night should suffice, unless there is special fear of disturbance.

(*c*) The barometer. Temperature hourly. Weather notes.

(*d*) Radio check on chronometer. Not essential, but convenient as a check on azimuth, § 4.18 (*e*).

(*e*) Doubts about identity, with notes about which of any pair of very close stars have been observed. Record of any change of focus or other adjustment.

4.18. Computations

(*a*) *Basic formula.* If there is no error in collimation, azimuth, or inclination, the formula is

$$\phi = \tfrac{1}{2}(\delta_1+\delta_2) \pm \tfrac{1}{2}m(M_E - M_W) \pm \tfrac{1}{16}(d+d')[(n+n'+s+s')_E - {} \\ -(n+n'+s+s')_W] \pm \tfrac{1}{2}(\psi_1 - \psi_2), \tag{4.50}$$

where δ_1 and δ_2 are the declinations of the stars of a pair; m is the value in seconds of arc of one micro division; suffixes E and W refer to positions LE and LW; M_E and M_W are micro readings, in divisions of the drum, the interval between wire $\mathbf{c}_2$ and the upper or lower $\mathbf{c}_1$ or $\mathbf{c}_3$ being included in $M_E - M_W$ if either of those wires is used; d and d' are the values in seconds of arc of one division of each of the main levels, the level readings being n_E, s_E, etc. (numbered from one end of the level to the other), unprimed and primed for the two levels; and ψ_1 and ψ_2 are the astronomical refractions of the two stars. The signs of the micro and level terms in the formula depend on the direction of graduation of the scales concerned, and must be determined once and for all for any instrument. The sign of the refraction term will always be such as will numerically increase the micro term, since the tendency of refraction is to make uncorrected zenith distances a little too equal, see (*b*) below.

If m is known, each pair then gives a value of the latitude, and systematic comparisons between pairs with positive and negative values of $M_E - M_W$ show whether the accepted value of m is correct. It may be better to determine both ϕ and m by least squares, for which the observation equations take the simple form

$$\phi \pm \tfrac{1}{2}m(M_E - M_W) = K, \quad \text{etc.,} \tag{4.51}$$

K being the sum of the remaining terms in (4.50), all known.

The resulting values of ϕ and m may be accepted, or if changes of adjustment are not being made, values of m may be recorded in a register from which a final value is eventually decided on, with which to recompute ϕ. § 4.16 (*d*) ensures that ϕ is insensitive to changes in m.

(*b*) *Refraction.* Since the z's of a pair are equal within 20′, $\psi_1 - \psi_2$ is less than 0″.3, and is proportional to $M_E - M_W$ for all values of z up to 30°. The refraction term can then be omitted from (4.50) with the only result that a slightly fallacious value of m is obtained, without change in ϕ. No harm results, but values of m obtained by direct measurement as in § 4.15 must be increased by 1 in about 3500 to conform, which amounts to saying that the refraction correction there mentioned should be omitted, provided the star used for calibrating the micrometer is within 30° of the zenith. Otherwise the appropriate correction should be applied, and 1 in 3500 then added.

(*c*) *Diurnal aberration* involves no correction to δ or ϕ, since $t = 0$ in (4.8).

(*d*) *Collimation, inclination, and azimuth.* Let c be the collimation† error of the central vertical wire, $\mathbf{a}_2$, positive if the line of sight makes an angle of <90° with the easterly direction of the horizontal axis for north stars, and >90° for south stars. Let k be the additional error of the wire $\mathbf{a}_1$ or $\mathbf{a}_3$, making their collimation errors $c + k$ or $c - k$ respectively. Let A be the azimuth of the line of collimation, positive if the east end of the horizontal axis points south of true east. Let i be the inclination of the horizontal axis, positive if the east end is low. Let c, k, A, and i be

† For the definition of collimation see § 1.15 (i).

measured in sexagesimal seconds of arc. Then, [113], 2 Edn, pp. 277–8, if z is the true zenith distance at transit, and z' is recorded,

$$z - z' = \sin 1''[-\tfrac{1}{2}(c+k)^2 \tan\delta - \tfrac{1}{2}A^2 \sin z \cos\phi \sec\delta$$
$$-\tfrac{1}{2}i^2 \cos z \sin\phi \sec\delta - (c+k)(A + i\tan\phi)\cos\phi\sec\delta$$
$$- iA \cos z \cos\phi \sec\delta]''. \quad (4.52)\dagger$$

In (4.52) k may be 400″, and the only term for which a correction may conveniently be applied is

$$z - z' = \tfrac{1}{2}k^2 \tan\delta \sin 1'', \quad (4.53)$$

positive for stars which transit between the zenith and the equator, or below the elevated pole: otherwise negative. The values of c, A, and i must be small enough to make the rest of (4.52) negligible. Permissible tolerances will depend on circumstances as below. Suppose that $k = 400''$, and that each source of error may be allowed to contribute not more than 0″.1. Then examination of (4.52) shows that the critical terms are

$$ck \tan\delta \sin 1'' < 0''.1, \quad \text{or} \quad c < 50'' \cot\delta.$$
$$kA \cos\phi \sec\delta \sin 1'' < 0''.1, \quad \text{or} \quad A < 50'' \cos\delta \sec\phi.$$
$$ki \sin\phi \sec\delta \sin 1'' < 0''.1, \quad \text{or} \quad i < 50'' \cos\delta \operatorname{cosec}\phi.$$

All is well if c, A, and i are less than 15″, provided $\delta < 70°$. So when $\delta > 70°$, either this 15″ must be improved on, or only the centre wire $\mathbf{a}_2$ must be used. But note that if both $\mathbf{a}_1$ and $\mathbf{a}_3$ are used, the associated values of ck, kA, and ki will be equal and opposite and will cancel.

If $\mathbf{a}_1$ and $\mathbf{a}_3$ are never used, as may be more convenient in field work, so that $k = 0$, the critical terms in (4.52) are

$$\tfrac{1}{2}c^2 \tan\delta \sin 1'' < 0''.1, \quad \text{or} \quad c < 203'' \cot^{\frac{1}{2}}\delta.$$
$$iA \cos z \cos\phi \sec\delta \sin 1'' < 0''.1, \quad \text{or} \quad iA < 20\,600 \cos\delta \sec z \sec\phi.$$
$$cA \cos\phi \sec\delta \sin 1'' < 0''.1, \quad \text{or} \quad cA < 20\,600 \cos\delta \sec\phi.$$
$$\tfrac{1}{2}A^2 \sin z \cos\phi \sec\delta \sin 1'' < 0''.1, \quad \text{or} \quad A < 203'' \cos^{\frac{1}{2}}\delta \operatorname{cosec}^{\frac{1}{2}}z \sec^{\frac{1}{2}}\phi.$$

Provided $\delta < 88°$ all is well, if $A < 40''$ and c and i are each $< 15''$. The last two are probably easy, but a large tolerance in A may be convenient.

(*e*) *Checks on adjustment.* The timing of each transit across $\mathbf{a}_2$ checks adjustments. Azimuths error causes times of transits to be wrong by $(1/15)A'' \sin z \sec\delta$ seconds with different signs for south and north stars (upper transit). Collimation, on the other hand, introduces errors of $(1/15)c'' \sec\delta$ seconds of time which are of constant sign in zenith stars, and these two adjustment errors can thereby be detected and separated.

† The signs given are appropriate to north stars, above the pole in the northern hemisphere. [441], p. 71.

The effect of inclination, $(1/15)i'' \cos z \sec \delta$ seconds, is difficult to disentangle from that of collimation in the comparatively narrow range of declinations involved, but inclination is directly recorded by the readings of the cross-level, and can hardly occur unsuspected.

(*f*) *Accuracy*. The scatter of the values of ϕ given by six or more star pairs should give a reasonable value for the s.e. of their mean, using (D.23) or (D.24), although it may be a little too low if the accepted value of m is derived solely from the same observations. A programme of eight pairs will probably give an s.e. of $0''.5_n$ or better, in field conditions.

4.19. Variation of latitude

Since 1900 large Zenith telescopes have been used to measure the annual variation of latitude, §4.09, and they have been able to fix the monthly mean position of the pole to within a few times 0″.01. They have now been generally superseded for this purpose by PZTs, § 4.35, or by Danjon astrolabes, § 4.38 (*c*).

For this special purpose errors in star declinations have been eliminated by the programme of observations described in [113] 2 Edn., p. 280.

4.20. Astronomical theodolites

Several makes of geodetic theodolite, such as the Wild T4, the Zeiss 002, and the Kern DKM 3A, have been so designed that they can be used for Talcott latitudes, and for LST by meridian transits. They are tending to supersede special one-purpose instruments for geodetic field astronomy. They differ from ordinary theodolites as follows.

(*a*) They are provided with highly sensitive levels in the plane of the telescope's movement, with the necessary clamp and slow motion. These are sometimes called Horrebow levels.

(*b*) They have a similarly accurate striding or hanging level, parallel to the horizontal axis.

(*c*) They have a moving wire eye-piece micrometer for the impersonal observation of star transits, § 4.29 (*a*), which can be rotated through 90° and then serves for Talcott latitudes.

(*d*) The telescope is 'bent' as in Fig. 4.6 (p. 301), so that the eye-piece is conveniently situated for high altitude observations.

For Talcott latitudes, the use of an astronomical theodolite is the same as is described for the Zenith telescope in §§4.14–4.18, except as below.

(*a*) There are no stops on the horizontal circle for setting in azimuth. The circle divisions are used instead.

(*b*) Some of the extra diaphragm wires $\mathbf{a}_1$, $\mathbf{a}_3$, $\mathbf{c}_1$, or $\mathbf{c}_3$ may not be provided. They are not essential.

The Wild T4 is a heavy instrument, 122 kg including boxes and tripod. The horizontal and vertical circles are of 240 and 135 mm diameter respectively. The length of the horizontal axis is about 350 mm. The focal length of the telescope is 540 mm, and the eyepiece is situated at one end of the axis, as in Fig. 4.6. The weight may be inconvenient, but it has advantages in an instrument which has to remain unmoved in azimuth for some time, as when it is used for meridian transits. Its use and adjustment are fully described in [444] and [492].

The Kern DKM 3-A has 100-mm circles, focal length 523 mm, and weighs 33 kg. Its use is fully described in [434], [102], and [492].

4.21. Meridian altitudes, or the Sterneck method

(*a*) *The method.* This is the simplest of all latitude observations, using (4.48) with a minimum of corrections and complexities. In principle, stars are selected in pairs, one north and one south of the zenith. This goes some way towards eliminating refraction error, and also such graduation error as may be systematic over a few degrees, and if both observations are made on the same face, any small collimation error is also eliminated.

(*b*) The routine of the observation is as follows.

(i) *Azimuth mark.* The observations have to be made in the meridian, so it is necessary to lay out a fine illuminated mark, either in the meridian or in a known azimuth, so that the telescope can be set in the meridian by its horizontal circle. In most latitudes this can be laid out by means of Polaris or σ Octantis. The astronomical meridian is required, but if a geodetic azimuth is more easily available, it will ordinarily suffice.

(ii) The sequence of observations is that a group of three or four pairs of N and S stars are observed on (say) face left, followed by a similar group on face right. According to the accuracy required and obtained, two or more such pairs of groups may be observed. Within each group it is not essential that N and S stars should be observed alternately, but it is desirable that they should be well mixcd.

(iii) *Programme.* A programme must be prepared giving the declination, zenith distance N or S, magnitude, and chronometer time of transit of suitable stars in order of increasing RA. A minimum interval of $1\frac{1}{2}$ or 2 minutes between observed stars is necessary. Except in cloudless weather, failures must be expected, and the programme should include all stars in FK 4 and its supplement which may be useful if the preferred programme has to be changed. Stars should be paired with zenith distances equal within (say) 5°, and if possible with zenith distances between 5° and 30°. Magnitudes are best between 2.0 and 6.5, but a star which can be seen clearly is better than one which is difficult to see. In any one group the sum of the zenith distances of the north stars should equal the sum of the south within 5°.†

(iv) The observation of each star. Either the north or the south star of a pair may be observed first. The horizontal wire of the micrometer should be on the star as it passes across the vertical wire which is on the meridian. The time should be recorded as a check on azimuth setting. The north–south bubble should be read immediately after the star has been bisected, before the circle reading. The telescope is then swung about the vertical axis to the other side of the zenith for the second star of the pair. No change of face.

(*c*) *Tolerances.* Equation (4.52) applies to meridian altitudes, except that k can be ignored, since all observations are made on the central vertical wire. It is then desirable and possible to make A, c, and i so small that none of the terms of (4.52) are significant. Provided that δ is less than 86°, no term of (4.52) will exceed 0″.01 if $A < 15''$, $c < 10''$, and $i < 10''$. The collimation of the horizontal wire is immaterial, but it is easy to keep it down to (say) 5″.

(*d*) *Miscellaneous precautions.*

(i) The horizontal circle reading of the azimuth mark should be recorded on both faces at the beginning and end of each group.

(ii) The bubble readings of the stars within a group must be balanced so that $\sum N - \sum S$ is variable and small. See § 4.12 for other precautions.

(iii) The east–west bubble should be read on both faces at the beginning and end of each group, and the criterion $i < 10''$ should be satisfied.

(iv) A recorder is essential. A third man can advantageously be employed to read the bubble at the instant of transit, if there is room for him.

(*e*) *Computations.* The only corrections to (4.48) are for N–S bubble and for refraction. Each star gives a separate value for the latitude, but the mean values derived from N and S stars may differ significantly. Each group ought to contain an equal number of N and S, but difficulty in getting a balance, (*b*) (iii) above, may result in there being (say) 4 of one and 3 of the other, in which case the mean of the N and the mean of the S should be taken separately, and the mean of these two means should be accepted for the group. When the mean of two or more groups is being taken for the final mean, groups may be weighted in proportion to the number of complete pairs in each.

(*f*) *Accuracy.* The s.e. of the final mean may be deduced by (D.23) from the means of four or more groups. From experience in the USA [144] gives the s.e. of one night's latitudes as ±0″.22 as judged by such internal

† If this is difficult, a group may consist of 4 N and 3 S, or vice versa, to give the necessary balance.

considerations, but as ±0″.37 if the s.e. is deduced from the differences between different nights at the same place. Similarly in Australia [102], pp. 1–2 reports the internal s.e. of a 36-star programme as frequently less than ±0″.2, but the average difference between latitudes observed at the same place in different years was ±0″.43, suggesting 0″.38 as the s.e. of each.

4.22 Circum-meridian altitudes

(*a*) This is a well known variation of meridian altitudes. Instead of a single altitude at transit, a series of altitudes are measured during a few minutes on either side of transit. Face may be changed in such a sequence as RLLRRL . . . for each star, or (probably better) pairs of stars (one north and one south) should alternately be wholly observed on face left and face right.

The altitude of stars should not be below 30°, and above 60° may be inconvenient. Altitudes above 70° may call for some special care, see (*c*) below. The altitudes of the north and south members of a pair should agree within 5°. It is necessary to prepare a programme in advance, as for meridian altitudes.

The object of the method is to reduce the random errors of bisection and circle reading by increasing their number. This is done at the expense of using fewer stars in a given period of time, with correspondingly greater declination error. With a small theodolite this may be advantageous, but with a good instrument it is now generally thought better to observe single altitudes.

See [100] for a full description of the field routine. Some details are given below.

(*b*) *Level.* The setting of the vertical circle bubble is as important as the reading of the circle itself, and more liable to error. The bubble should be moved and re-centred before every star intersection.

(*c*) *Correction to meridian.* The observed zenith distance of a star observed off the meridian must be decreased (or increased if it is below the pole) by

$$\cos\phi\cos\delta\operatorname{cosec} z(2\sin^2\tfrac{1}{2}t)\operatorname{cosec} 1'', \tag{4.54†}$$

where t is the hour angle, before or after transit. In this formula $2\sin^2\frac{1}{2}t\operatorname{cosec} 1'' = 8''$ if $t = 2^{\mathrm{m}}$, and varies as t^2. If $t = 2^{\mathrm{m}}$ it changes by 1″ if t changes by 7^{s}, the rate of change being proportional to t, and changing sign with t. The factor $\cos\phi\cos\delta$ may be near 1. So if the time of transit

† There is another term

$$\cos^2\phi\cos^2\delta\cos z\operatorname{cosec}^3 z(2\sin^4\tfrac{1}{2}t)\operatorname{cosec} 1'', \tag{4.55}$$

which is negligible if the rules given below are followed.

has been wrongly recorded by 7^s, the error in a star whose hour angle is 2^m may be about $1''$ cosec z, but with opposite signs for observations made before and after transit. It is clear that there must be some limit on the size of t, on the errors in the recorded time of transit, and on the smallness of z. The following rules will ensure absence of significant error. They may be unnecessarily strict, but they are easy to follow. They can be relaxed provided examination of (4.54) shows that it is safe to do so.

(i) Avoid $z<20°$, altitude$>70°$.

(ii) Avoid $t>2^m$. This is easy if face is not being changed.

(iii) Determine t within 10^s. It may be possible to do much better, in which case the other rules can be relaxed.

(iv) Accepted observations should be well balanced on either side of transit.

The only difficulty is to get the right time of transit. This can be done in two ways.

(i) From the programme. An error of $1'$ in the longitude of the station causes an error of 4^s, which should be easy to avoid.

(ii) If a wireless is not carried, intersections of each star should be started 4 minutes before transit, and continued until 4 minutes after. The observed z's can then be plotted against time, and a smooth curve drawn through them. Those furthest from transit, where the curve is steep, will adequately give the time of transit, and only those near transit should be used for the latitude. If this is done, it is convenient to observe each star on one face only. This takes some time, but there is something to be said for having an observer study his results in the field.

(*d*) *Errors of collimation and inclination.* (4.52) *with* $k=0$ is applicable to circum-meridian altitudes, sin A being sin t cos δ cosec z, where t is the hour angle. The critical terms for collimation and inclination are then those in cA and iA, but large values will cancel provided positive and negative values of t are well balanced. If z and t are limited as in (*c*) above, errors of $15''$ in c and i will be harmless.

(*e*) *Final mean.* If observations to the members of a pair have all been on one face, the two stars may give very different values for ϕ, but the mean will be correct provided the collimation in altitude has been constant, as can be judged by the constancy of north minus south in successive pairs.

(*f*) *Use of pole star.* Polaris and σ Octantis never move more than $1''$ in altitude in 4 seconds of time, so if either of them is at a suitable altitude, and if LST is well enough known, it may be used as one member of a pair at any hour angle. Four to six intersections may be made spread over not more than 2 or 3 minutes of time, and the mean altitude may be ascribed to the mean of the times.

(*g*) *Accuracy*. From observations in the Sudan [436], pp. 113–4, gives a s.e. of $\pm 0''.3$ for the mean of eight pairs of N and S, and an average difference of $0''.5$ between such a programme and a good Talcott programme.

SECTION 3. LONGITUDE. MERIDIAN TRANSITS

4.23. Basic equation

$$\text{Longitude } \lambda = \text{LMST} - \text{GMST} \tag{4.56}$$

where LMST is the local mean sidereal time at which a star is at a certain altitude or azimuth at the place concerned, and GMST is the Greenwich mean ST at the same instant.

This may be written as

$$\lambda_a + \Delta\lambda = (\text{LAST} + \text{P} - \text{EE}) - (\text{UT } 1 + R_u + 12^{\text{h}}), \tag{4.57}$$

where λ_a is some preliminary value of λ, and $\lambda = \lambda_a + \Delta\lambda$.

LAST is the local apparent ST referred to the instantaneous pole.
$\text{P} = -(1/15)(x \sin \lambda + y \cos \lambda) \tan \phi$, as in (4.27).
EE is the equation of the equinoxes (Apparent − Mean) which is given in AP, Table II, columns 4 plus 5.
UT 1 and R_u are as in § 4.10 (*d*).

This again may be written as

$$\Delta\lambda = \text{LAST} + (\text{P} - \text{EE} - \lambda_a) - (\text{Clock time of transit}) - (\text{UT } 1 - \text{Clock} + R_u + 12^{\text{h}}). \tag{4.58}$$

In (4.58) the LAST can be computed from such data as the RA and declination, the latitude, and the altitude or azimuth of the star. The term $(\text{P} - \text{EE} - \lambda_a)$ is constant for any programme covering not more than a few hours. The clock time of transit is the mean recorded clock time of all the accepted intersections of any particular star, corrected as necessary for collimation, level, etc., as in (4.64).

The value of UT 1 minus Clock is got from wireless time signals as in (4.66), at least once before and once after star observations. At the clock time of any particular star it is got from these values by linear interpolation.

4.24. Methods of observing local time.

Methods are as follows

(*a*) Meridian transits, §§ 4.24–4.29.
(*b*) Altitudes of east and west stars, § 4.41.

(*c*) Simultaneously with latitude by equal altitudes, unequal altitudes, or zenith photography, §§ 4.34–4.40.

(*d*) With latitude, as a by-product of the Black azimuth method, § 4.47.

The *meridian transit* observation depends on the simple principle that the LAST of a star's transit equals the RA of the star, or its RA+12 hours if the transit is below the pole. Traditionally the observation has been made with a special instrument known as a *Transit telescope*, but for field work an astronomical theodolite, §§ 4.20, is now more generally used.

In fixed observatories meridian transit observations have now been superseded by the photo zenith tube or by the Danjon astrolabe, §§ 4.35 and 4.38 For field geodetic work meridian transits are generally less convenient than methods (*b*), (*c*), and (*d*), but they are still likely to be used, with astronomical theodolites, until superseded by portable zenith cameras or impersonal astrolabes.

4.25. The Transit telescope.

At its simplest, the Transit consists of a telescope, generally of from 50- to 125-cm focal length, and 6- to 10-cm aperture, mounted in the meridian on a horizontal axis about which its line of collimation can rotate through the zenith. Provision is made for the easy lifting of the telescope from its Y-bearings, turning it 180° about a vertical axis, and returning it to its Y's, so that observations can be made on both faces. Azimuth is adjustable through a degree or two to enable the line of collimation to be brought accurately into the meridian. The inclination of the horizontal axis is also adjustable, and is ascertained by means of a sensitive attached, or hanging, or striding level. The approximate setting in the meridian is got by observations to, for example, Polaris or σ Octantis.

In the old fixed-wire type of Transit the eye-piece is provided with a pair of horizontal wires and a large number of fixed vertical wires, the whole diaphragm being mounted, with the eye-piece, on a micrometer screw by which the centre wire can readily be adjusted on to the line of collimation. Two small setting circles are provided, by means of which the telescope can be so inclined that a star of known declination will transit in the field of view on either face.

The routine of observation is then that the telescope is set at the correct altitude, and approximately in the meridian, and the east–west level is recorded. When the star appears it is brought between the horizontal wires, and its time of passage over successive vertical wires is recorded by tappet and chronograph. Before the star reaches the centre

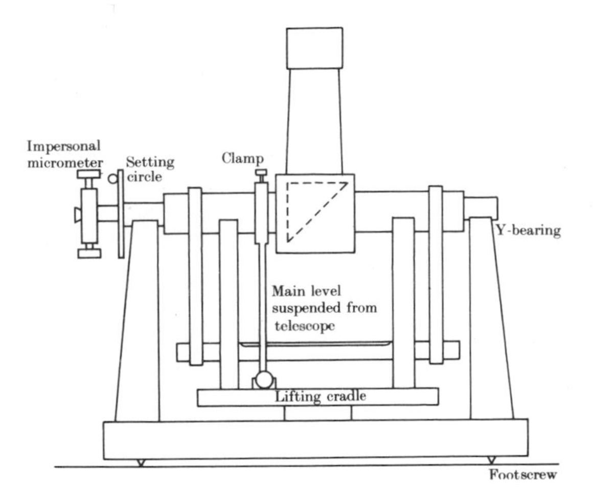

FIG. 4.6. Bent Transit. Diagrammatic.

wire the telescope is lifted and turned 180° in azimuth as above and reset in altitude so that the star passes again between the horizontal wires and back over the same vertical wires, its times of passage being again recorded. The level is then read again, and this completes the observation of one time star.

See § 4.29 (*a*) for the use of the impersonal micrometer. The following, and §§ 4.26–4.28, are applicable to both the micrometer and to the old fixed wire instruments.

The Transit instrument must be firmly mounted, preferably on an isolated brick pillar, in the hope that errors of azimuth and level will remain constant unless intentionally corrected. In the field a low rigid stand will suffice on hard ground, but on soft ground serious difficulty may be expected.

Modern instruments take the form of a *Bent Transit* in which a mirror reflects the light down the hollow horizontal axis to an eye-piece at one side. See Fig. 4.6. This makes a more portable instrument, avoids the use of a diagonal eye-piece, and makes the observer more comfortable.

Use of astronomical theodolite. Meridian transits can be observed with an astronomical theodolite, § 4.20, in exactly the same way, except that instead of the telescope being reversed by being lifted out of its Y-bearings, face is changed in the ordinary way. The level is either hanging or striding, not rigidly attached. An impersonal micrometer is always incorporated. The advantage of the Transit telescope lies in the weight and rigidity of the base, the absence of any vertical axis, and the fact that reversal places the telescope back in the meridian without any reading of

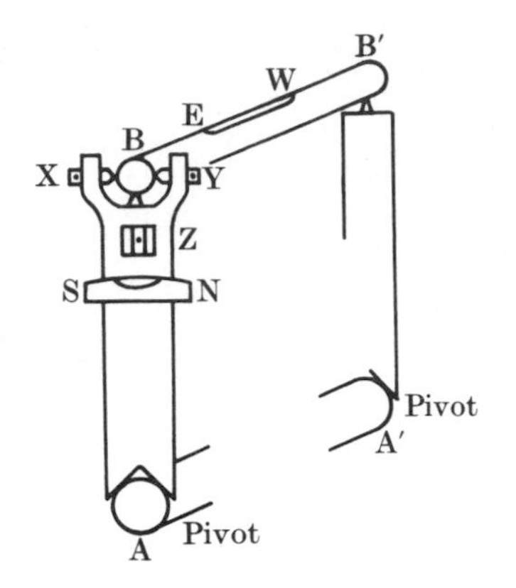

FIG. 4.7. Striding level. AB must be parallel to A'B'.

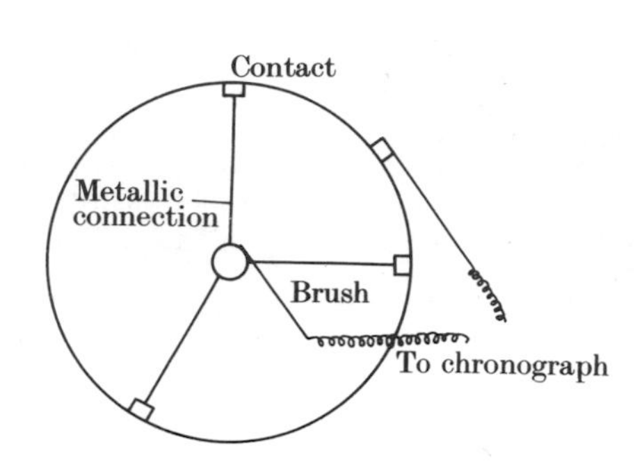

FIG. 4.8. Impersonal micrometer contact wheel.

a horizontal circle. The advantages of the theodolite are that changing face is a more gentle process than lifting the telescope and lowering it into the Y's, and of course the theodolite is of more general utility, such as for latitude and azimuth.

4.26 Adjustment of the east–west level

The correct recording of the inclination of the transit axis is most important, and assuming that the two pivots are equal co-axial cylinders it is given by the readings of the level (on both faces). This level may either be attached to the telescope, or may be a separate striding or hanging level. The latter can be (but is not necessarily) reversed end-for-end at each reading, while an attached level reverses only with the telescope. The striding level is used with the telescope pointing at the elevation of the relevant star, but for reading the fixed level the telescope must be pointing to the zenith.

The level itself can be adjusted, by a screw under one end, Z in Fig. 4.7, so as to lie close to the centre of its run when the axis is truly level. A small north–south bubble SN is generally attached to the main bubble, which must always be centred before the latter is read, and provision is also made (as in Fig. 4.7 for the striding level) to secure parallelism between bubble and transit axis as seen from above, so that when one support (of the striding level) is vertical, the other will be too. The test of this adjustment is that the bubble reading should not change when the whole striding level, or the whole telescope if the level is attached, is slightly tilted from north to south.

The effect of inequality of circular pivots is cancelled by change of face† and so is of little consequence. The effect of lack of true circularity, or the effect of rust or dirt, can be minimized by only using stars with very small z's, but cannot be positively cancelled. Modern precision is such that a clean new bearing should not be appreciably imperfect, but in practice serious error is possible.

4.27. Correction for level, azimuth, collimation, and aberration

When using (4.58) the computation is made for the instant of transit, not exactly the instant of observation, so the LAST equals the RA of the star. If the RA is got from AP, nutation must be added as in (4.7). The clock time in (4.58) is the mean recorded clock time of all accepted intersections, corrected for collimation (if necessary), level, azimuth, and diurnal aberration, as below, to give the time at which the star would have been on the true meridian by an instrument with no adjustment errors.

(*a*) *Collimation.* If every star is observed on both faces, and if the same wires or micrometer contacts are used on both, the mean of all the recorded times is free from collimation error. If a wire has been missed on one face it can be ignored on the other or, if observations are scanty, a knowledge of the interval between wires will enable a substitute for the missing reading to be deduced from its two neighbours, using the equation

$$\text{(time interval between two wires)} = \text{(equatorial interval)} \times \sec \delta. \quad (4.59)$$

If observations are made on one face only, the correction to the clock time of intersection by any particular wire is

$$+(c''/15)\sec \delta \text{ seconds for upper transits, or } -(c''/15)\sec \delta \text{ for lower,} \quad (4.60)$$

where c'' is the collimation error of the wire or contact concerned, positive if the line of sight points east of the line of collimation.

(*b*) *Level.* The correction to the recorded clock time of a star on account of dislevelment of the east–west axis is

$$\pm b \cos z \sec \delta \quad \text{or} \quad bB \text{ seconds, +for upper and } - \text{ for lower transits,} \quad (4.61)$$

where z is the zenith distance, and $b = (w_1 + w_2 - e_1 - e_2)d/60$, w_1, etc.,

† In this context 'FL' implies that a particular pivot is west of the meridian. Reversing the Transit in its Y's is not quite the same process as changing face on a theodolite.

being scale readings outward from the centre, w west and e east, and d'' the value of one division of the scale.† The inclination of the axis is thus $15b$ seconds of arc, considered positive if the west end is too high.

(c) *Azimuth.* The correction is

$$(a''/15)\sin z \sec \delta \quad \text{or} \quad (a''/15)A, \tag{4.62}$$

where a seconds of arc is the azimuth of the line of collimation when the telescope is horizontal, positive if east of the elevated pole, and z is the zenith distance as usual except that in this context it is considered positive between elevated pole and zenith and otherwise negative. Proofs of (4.59)–(4.62) are in [492] and [533].

(d) *Diurnal aberration.* From (4.8) with $t = 0°$ or $180°$ the correction to RA is

$$\kappa = \pm 0^{s}.021 \cos \phi \sec \delta, \tag{4.63}$$

but it is usually applied to the recorded time with opposite sign, namely negative for upper transits N or S, the usual case, and positive for transits below the pole.

(e) *To summarize.*

(clock time of true transit), for (4.58)

$$= (\text{mean recorded time}) \pm bB + (a''/15)A \mp \kappa, \tag{4.64}$$

to which add $\pm(c''/15)\sec \delta$ if observations have been on one face only. Signs are as above, and see (4.65) for micrometer contacts.

Tables for A, B, and κ are in [290] and for κ in AP, Table VII.

(f) *Tolerances.* In the correction for level, the factor B is $\cos z \sec \delta$ and is likely to be between 1 and 2 for near zenith stars in most latitudes, with the same sign for all stars. It follows that an error of 1″ in the bubble centring introduces errors of between $0^{s}.07$ and $0^{s}.13$ in the time.

For the azimuth correction the factor A is $\sin z \sec \delta$, which is smaller than B and has opposite signs for stars on opposite sides of the zenith. The azimuth error is easily made as small as 10″, and if north and south stars are balanced as in § 4.28 (a), this is harmless. Also see § 4.32.

If face is changed on each star, the collimation error on each separate face is necessarily large, such as 15′, but it is cancelled in the mean of both faces. It is convenient to know the micrometer reading at which it is zero, to within a few seconds, to facilitate the azimuth setting.

† If the scale is numbered from one end to the other, $b = \{(w_1 + e_1) - (w_2 + e_2)\}d/60$, where suffix 1 applies to readings made when the zero of the scale is east, and suffix 2 when it is west.

(*g*) *In high latitudes* the factor sec δ in all these corrections may be large, but compensating this is the factor cos ϕ in (2.2) which gives the deviation of the vertical. In high latitudes the longitude will be inaccurate, but the deviation need not be correspondingly so. § 4.28 (*c*) describes some necessary change in the usual star programme.

4.28. Programme of observations

(*a*) *Time stars.* It is usual to observe not less than 8 'time stars' with equal numbers either side of the zenith and all within at most 30° of it (or within 10 or 15° for the best work), and so balanced that the algebraic sum of the factors *A*, sin *z* sec δ, is between +1 and −1. This can probably be observed in 2 to 4 hours. The routine of observation, using an impersonal micrometer, is fully described in [290], [492], and [444].

(*b*) *Azimuth stars.* In addition to the time stars it has often been customary to observe one high declination star for every 4 to 6 time stars. Given an approximate azimuth, the balanced time stars give fair LST, with which the azimuth stars give a good azimuth. The time stars can then be corrected for the improved azimuth. The process converges rapidly. See § 4.32 (*b*). Owing to their slow movement, these high declination stars can only be observed on the most central wires.

(*c*) *In high latitudes* stars between the pole and the zenith are few in number and move slowly. Their *A* factors are large. The majority of the time stars will then have to be on the side of the zenith remote from the pole and at distances of up to perhaps 25° from it, so that a proper balance of the *A*'s can be secured. See [290], pp. 51–2. A few low-altitude stars can also be included as a substitute for good balancing of the time stars.

(*d*) *Preliminary programme.* A programme should be made giving star name, magnitude, declination, ZD north or south to 1′, RA = LST of transit to 1^s and a note of which face is to be observed first. Consecutive equatorial stars should be separated by not less than 4 minutes, which must be extended to 7 or 10 minutes for stars close to the pole. Stars of up to 6 or 6.5 magnitude can generally be used, and sufficient can be got from AP, or from FK 4 and its Supplement.

(*e*) *Change of face.* Reversing, or changing face, on each star eliminates collimation error, provided it is constant, but is liable to disturb the azimuth and level. An alternative system, [498] but not [444], is to observe four sets each of 4 to 6 time stars, with possibly one azimuth star in each set. The first and fourth sets are observed on FL, and the other two on FR. The level, striding or hanging, remains on the telescope unmoved, so that it is only reversed with the telescope. A least-squares solution, as in § 4.32 (*c*) (ii), is then made for the three azimuth errors (first set, last set, and the two middle ones together), the collimation† error (assumed constant throughout), the dislevelment of the axis when the bubble is central (also assumed constant), and the correction to the provisional longitude.

This system involves less disturbance of the instrument, and also enables stars to be taken more rapidly. On the other hand it assumes constancy of collimation and bubble for a longer period, and it may result in complete failure if there is trouble with cloud.

4.29. Personal equation

In the old system of recording the times of transit over fixed wires, the accuracy was much lessened by *personal equation*, a more or less constant tendency of any observer to press his tappet too early or too late. If this tendency was truly constant its effect could be eliminated by occasional comparative observations at some base station where longitude has been accurately determined. But in fact personal equation is one of those awkward errors which, while they may be fairly constant or at least of constant sign over a night or a week or more, vary slowly or suddenly but unpredictably. The accuracy deduced from the good mutual agreement of a series of stars may consequently be illusory, and the error suggested by the comparison of different observers and instruments may be disappointingly large. In magnitude it is generally less than $0^s.20$. Its constancy is likely to increase with experience, but its magnitude may increase with age. Devices which have been introduced to measure or eliminate it are described in (*a*)–(*g*) below. The sign is given by the rule that personal equation equals the correct east longitude minus the observed.

(*a*) *The impersonal or self-registering micrometer.* Instead of numerous vertical wires, there is a single wire which can be traversed across the field by a micrometer screw. A couple of minutes before transit this wire is moved so as to intersect the star, and thereafter kept on it by slowly turning the screw until about 30 seconds before transit. As the screw turns, three small metal contact strips let into the rim of a non-conducting wheel on the micro's axis (Fig. 4.8, p. 302) close the chronograph pen circuit for a short interval such as $0^s.1$.‡ The positions of the moving wire at the beginning of each of these contacts correspond to the positions of the fixed wires in the old type of transit, and §§ 4.25–4.28 apply.

Unless § 4.28 (*e*) is adopted, face is changed shortly before transit, and

† The same wires, or the same micrometer contacts § 4.29 (*a*), must be used for all stars, except that an interpolated reading may be provided in place of any occasional one that may be missed. See § 4.32 (*a*).

‡ Or non-conductors may cause a short break.

the star passes back over the same contacts. Collimation is thereby cancelled, and no determinations need be made of the distance from the line of collimation at which contacts occur, but on opposite faces contact is made on opposite sides of the metal strips, and their width must be determined and allowed for. To do this, turn the micro very slowly until a click is heard from the chronograph, and then read the micro head. Turn on past the contact, and then slowly turn back and read the micro as before when a click is heard. This gives the width of the contact in terms of micro revolutions. Alternatively the micrometer may be turned through a whole revolution during which the contact readings are noted, after which it is turned back again, to give the readings of the other sides of the contacts. Three such sets of measures may be made at every station. The value of one revolution is got by timing the movement of any star at transit. The time taken by an equatorial star to traverse one turn is $\cos\delta\times$(the time taken by a star of declination δ). Then the correction to the recorded time of transit of any other star (mean of all contacts on either or both faces) is

$$+\tfrac{1}{2}\sec\delta\,(\text{average width of micro contacts}) \times(\text{equatorial value of one turn}), \quad (4.65)$$

where δ is now the declination of the star concerned. The correction is positive. Contacts may need occasional cleaning.

The programme of § 4.28 (*e*), in which face is changed less often, is equally applicable when the impersonal micrometer is used.

Lost motion of the micrometer screw can be measured by reading the micrometer when a fixed wire is alternately bisected with an eastward and a westward movement. Then the difference is to be added to the width of the contacts in (4.65). Both corrections are always additive to the recorded time, and both are multiplied by $\frac{1}{2}\sec\delta$ as in (4.65). See [290], pp. 25–7, and [314].

The impersonal micrometer considerably reduces personal equation, but does not entirely eliminate it, as an observer may still have a tendency to keep the wire ahead of or behind the star. But with good observers it should not exceed $0^s.04$ or $0^s.05$. Sometimes modified as in (*b*) below, it is the generally accepted method of timing transits.

(*b*) *Impersonal micrometer. Motor drive.* Keeping the moving wire right on the star demands great concentration, and the blinking of an eye causes bad intersection for some seconds, during which a contact may occur. Comfort and accuracy are consequently increased if the micrometer is turned by an electric motor. The speed of the motor is regulated according to the declination of the star, and the action of the observer is to retard or accelerate the motor by pressing suitable electric contacts, or mechanically, as in [441], p. 44. Momentary lack of attention then has comparatively little effect. Rather unexpectedly, vibration can be eliminated and the apparatus does work but the device is not in common use.

(*c*) *Artificial moving star.* In principle it is easy to arrange a small artificial 'star' whose image can be passed across the field of view at the same speed as a real star. It can be halted on a vertical wire, and so adjusted as to close a recording pen circuit when it is exactly in that position. It is then moved across the field again at normal speed, and the observer records its passage by tappet. The difference between the tappet and the automatic contact then gives his personal equation. A similar procedure can be used to test or correct an impersonal micrometer.

Such an apparatus has been constructed from time to time, but no great use of the method has been reported.

(*d*) *Photo-electric methods.* When light falls on a photo-electric cell, a voltage is generated which is proportional to the intensity of the light. If a pen recorder, of a type in which the deflexion of the pen is proportional to the voltage, is attached to a photo-electric cell at the eye end of the telescope, the movement of the pen will record the passage of a star over the cross wire. A second pen marks the seconds of either a time signal or of the local clock. Personal equation is avoided, and with modern equipment there should be no instrumental lags. There are difficulties as below.

(i) The production of a sufficiently fine and sharp wire, or slit in a dark background, and

(ii) the scintillation, and erratic movement, of the star.

In Japan, [565] and [564], the sharp edge of a prism has been used instead of a fine slit. The light from a star is reflected from the two adjacent faces of the prism on to two cells, and when the difference of their outputs is zero, the star is deemed to be on the 'cross-wire'. Collimation error is eliminated by change of face. The prism edge is a great improvement on a slit. Using elaborate, but portable, amplifiers, and a pen recorder, excellent results have been obtained at about twenty stations with a small (70-cm focal length) Transit. Probable errors are reported as $0^s.01$ to $0^s.025$ for a single star, or $0^s.003$ to $0^s.005$ for a programme of ten to sixteen stars. The method has been adapted to latitude observations (by transits across the prime vertical), and also for use with an astrolabe. So far (1977) this type of instrument has not been reported as in use outside Japan.

(*e*) *Photographic methods.* Photographic recording eliminates any human personality except in the micrometric reading of the photographic plates, in which personality is cancelled by measurement in two reversed positions. Photographic recording is now being used for zenith photography to give both latitude and longitude simultaneously. See § 4.34.

(*f*) *Calibration.* Every field instrument, whether reputed to be impersonal or not, must occasionally be checked by a series of observations, in field conditions and with its field observer, at a place where the astronomical longitude has been determined by the better instruments used in fixed observatories. The series must, of course, be long enough to reduce random error to a negligible amount. If the error is shown to be substantially constant at different times, a correction can be applied to all field longitudes. More commonly with existing instruments, it will be found to vary. This will at least enable the true accuracy of the field observations to be judged. For a routine in connection with geoidal profiles, in which systematic error is especially unwelcome, see § 4.51 (*c*). For some results see § 4.33.

Since personal equation may vary with the speed of movement of the time stars, it is desirable that the latitudes of the fixed observatory and of the field stations should not differ by more than (say) 5°. Greater tolerance may be allowed in low latitudes.

4.30. Radio time signals

The rhythmic time signals on which geodetic longitude observations depended between 1925 and 1960 have now been discontinued, and in their place there are many, practically continuous, signals which emit short *pips* every (atomic time) second of UTC, as described in § 4.10 (*e*).

Radio transmissions are of two types.

(*a*) A1, in which the signal is given by a short interruption of the carrier wave, with no modulation.

(*b*) A2, in which the signal is given by a short modulation of the carrier.

Low frequency (LF) signals† are in many locations the easiest to receive, and they are frequently of A1 type, for the reception of which the receiver must incorporate a *beat frequency oscillator.*

Part C of the annual reports of the BIH [43] lists the most useful time signals, and gives their times of day, frequency, and type (A1 or A2), some description of the signal itself, and the latitude, longitude, and code name of the transmitter. The monthly circulars of the BIH list about 30 transmissions which give UTC correctly within $0^s.0002$.

† Time signals are classified as high frequency (HF) between 30 and 2.5 MHz, LF between 300 and 30 kHz, and VLF between 30 and 10 kHz.

All signals indicate exact minutes of UTC by such means as a lengthening of the pip. All are internationally requested to include in the signal a coded value of DUTI, § 4.10 (*e*). The recommended code is given in Part C of [43] and in [492]. Values of UT 1 − UTC are given to $0^s.0002$ for every fifth day in the BIH monthly circulars.

On account of the finite velocity of radio waves, the time of reception is a little later than the time of emission. The length of the path is not exactly that of the direct sea-level course, and varies a little with the frequency. If the distance is measured along a sea-level great circle (or from an atlas for short distances), the effective velocity may be taken as 285 000 km/s for HF transmissions and as 290 000 km/s for LF and VLF. The resulting figures for the delay are likely to be correct to $0^s.0005$ in a distance of up to 11 000 km.

The true UT 1 time of reception is then given by

$$(\text{reputed time of emission}) + (\text{distance} \div \text{velocity}) + (\text{UT 1} - \text{UTC}). \tag{4.66}$$

And (4.66) minus the local clock time of reception gives (UT 1 − Clock) for (4.58).

Radio receivers. The characteristics of a currently most useful radio receiver are that it should be fixed-tuned to frequencies of 2.5, 5, and 10 MHz for a variety of HF signals, 60 or 75 kHz for European LF signals, and one or two others for any other areas where it is expected to be used. It should have a beat frequency oscillator to enable it to receive A1 signals. See also [492] p. 19–5 to 8.

4.31. Clocks, chronographs, and recorders

The instruments described here are equally applicable to the observations described in Sections 4 and 5.

Until 1950–60 the standard equipment for comparing the times of star observations and UT 1 as given by the radio signals, consisted of the following.

(*a*) One or more spring chronometers, provided with break-circuit devices for the recording of seconds.

(*b*) A two-pen chronograph. One pen was actuated by the break-circuit of a clock, and the other by the contacts of an impersonal micrometer, or by a hand-held tappet for recording the pips of the radio signal or a star intersection. The second pen could also be used to compare the master clock with other clocks. For a description see [113], 3 Edn, or rather more fully in the 1 or 2 Edn.

This equipment is now obsolete. The spring chronometers are replaced by a crystal clock, whose rate can be regarded as perfectly constant during a few hours star observations, and the chronograph is replaced by a recorder which prints out, or provides a visual record of, the local clock time of any event fed into it, such as a radio pip, or a tappet break, or a micrometer contact. Such equipment can readily avoid lags and errors of as much as $0^s.001$. It is desirable to have three radio time receptions covering a night's star observations as a check on regularity.

The Chronocord, [492] and [489], is a typical modern combination of crystal clock and electronic time recorder. The clock keeps UT time with a rate which is uniform to about $0^s.001$ over the period of a day. Its printing unit prints out the clock time of either radio time signal pips, or of the closure of a tappet or impersonal micrometer. Its record is in binary coded decimals to $0^s.001$, and it can print the time of events separated by not less than one eighth of a second.

If radio reception is poor, with much noise, an alternative aural method is provided which is rather more troublesome, but involves no loss of accuracy provided the ear can recognize the pips.†

The dimensions of the Chronocord are $355 \times 300 \times 240$ mm; weight 10 kg; battery 12 volt. A frequency counter is available, as an attachment, for the calibration of EDM equipment, § 1.39 (*h*) (ii) and § 1.43 (*d*).

4.32. Final computation of longitude

(*a*) Equation (4.58) has to be formed separately for each star observed. If a mean time local clock is used, the term (UT 1 − clock) will either be constant or will vary regularly through a very small range. Its value for the middle time of each star observation is therefore accurately available from (4.66). The clock time of the observation is the mean‡ of the recorded times of all the accepted micrometer or other form of contacts,‖ and the clock time of transit will be given by (4.64) provided the azimuth error a'' is known, for which see (*b*) below, and provided each star is observed on both faces. The items in the term $(\text{P} - \text{EE} - \lambda_a)$ are all known. The true LAST of transit is equal to the star's RA, or $\text{RA} + 12^h$ for transits below the pole. All the terms in (4.58) are then known and one value of $\Delta\lambda$ results. No further correction to the CIO is required, see (4.58).

(*b*) There remains the question of azimuth error, which may not be adequately known. So let a value of a'' be assumed, and let (4.58) be

† The ear can separate signal from noise better than an ordinary recording apparatus.

‡ The recorded times should be examined for blunders or other unacceptable errors. If some contacts have been missed or rejected, corresponding rejections should be made on the other face to procure symmetry about the (reputed) line of collimation. If a star is observed only on one face, artificial figures (deduced from adjacent readings) should be entered as substitutes for the missing times, and recorded as such.

‖ After making allowance for width of contacts (4.65), and lost motion, if any.

computed to give provisional values of $\Delta\lambda$. The correctness of this assumption may be verified by separately listing the north time stars and south time stars. Then azimuth error δa will be revealed by systematic difference between them

$$\Delta\lambda\,(\text{by north}) - \Delta\lambda\,(\text{by south}) \approx (2/15)(\delta a)\sin z_m \sec \delta_m, \tag{4.67}$$

where z_m is the average zenith distance without regard to sign, and δ_m is the declination of a zenith star. If δa is significant, the azimuth stars may be recomputed using for the LAST the value of $\Delta\lambda$ given by the general mean of the time stars, and the time stars can then be recomputed with the revised azimuths. The process should converge rapidly. Finally, the general mean of $\Delta\lambda$ as given by the time stars can be accepted, except that in high latitudes, where north time stars may move slowly, such stars may be given reduced weight.

(*c*) *Least squares.* A least-squares solution may be made as in (i) or (ii) below.

(i) For $\Delta\lambda$ and δa as an alternative to (*b*). One observation equation is formed for each star

$$\Delta\lambda - (1/15)\delta a \sin z \sec \delta = 0, \tag{4.68}$$

where z and δ are the zenith distance and declination of the star concerned, z being positive between the elevated pole and the zenith, and otherwise negative. A least square solution may be especially desirable in high latitudes, § 4.28 (*c*) and [290] pp. 51–2.

[290] p. 61 recommends weighting each equation (4.68) in proportion to $(1+0.3\tan^2\delta)^{-1}$.

(ii) When stars have been observed in sets without change of face on each star, § 4.28 (*e*). Each star then gives an observation equation, from (4.58) in the form

$$\Delta\lambda + \text{Clock time from (4.64) and (4.65)} = (\text{LAST of transit})$$

$$+(\text{P}-\text{EE}-\lambda_a) - \{\text{UT 1 from (4.66)} - \text{Clock}\} - (R_u + 12^{\text{h}}) \tag{4.69}$$

or

$$\Delta\lambda + (a_n/15)A + (\delta b)B + (c/15)\sec\delta = k, \tag{4.70}$$

where A and B are as in (4.62) and (4.61),

a_n is a_1, a_2, or a_3, the azimuth error of the set concerned,
c is the collimation error,
δb is the bubble setting error,
and k is the sum of the other terms in (4.69), all being known quantities.

See (4.60)–(4.62) for the signs of the collimation, level, and azimuth terms. Weights may be as in (i).

4.33. Accuracy

The principal sources of error in meridian transit observations are as follows.

(*a*) The random error of star bisection.
(*b*) Systematic errors of bisection. Personal equation, including lost motion and error in contact width.
(*c*) The bubble. There should be no systematic error provided §§ 4.12 and 4.26 are followed.
(*d*) Star places. The errors of FK 4 and its supplement are small, and should not be significantly systematic over a range of several hours of RA. See § 4.06 (*a*).
(*e*) Refraction. This should be minimized by attention to § 4.11 (*b*). It is possible for refraction error to be systematic over long periods at any one place.
(*f*) Imperfectly cylindrical transit axis pivots. See § 4.26. The effect can be minimized by only using time stars which are close to the zenith.
(*g*) Electrical lags in recording star transits and radio signals. With modern equipment such errors should not amount to more than $0^s.001$.

Of these sources of errors some are clearly random and vary from star to star. The s.e. due to such random error will be determined by the scatter of the 6 to 8 stars in a set (after determining the optimum azimuth error). A figure of about $\pm 0^s.10 \sec \phi$ for a single star is likely to result.† See [102], p. 1 from Australia which gives $\pm 0''.9$ ($= 0^s.06$) for the s.e. of the mean of a pair of stars in low latitudes, giving $\pm 0^s.85 \sec \phi$ for a single star.

On the basis of $\pm 0^s.10 \sec \phi$ for each star, the s.e. of the mean of a set of 6 to 8 time stars is likely to be about $\pm 0^s.04 \sec \phi$. This can be confirmed by the scatter of the set means at each station,‡ which may be expected to produce a rather higher value.

The s.e. of the mean of two sets observed at one station may be expected to be about $\pm 0^s.03 \sec \phi$.

Analysing longitudes observed with the T4 in the US since 1962, [144] gives $\pm 0''.43$ ($0^s.03$) for the internal s.e. of the mean of a set, and $\pm 0''.61$ ($0^s.04$) for a set in older observations. These may be ascribed to latitude

† The latitude ϕ equals the average declination of near-zenith stars, and the stars' rate of movement varies as $\cos \delta$.

‡ There may be only 2, 3, or 4 sets at each station, but § D.08 describes how to get a reliable value for the s.e. of a single set.

40°. Comparing the results at stations which have been observed on two well-separated occasions, [144] gives $\pm 0''.77$ ($0^s.05$) as the s.e. of a single night of three sets, old or new. The older results were brought into modern terms by recomputation and the inclusion of polar motion and FK 4 star places.

The observations of [144] had been made without calibration at a base station, and the difference between the $0^s.03$ given by internal scatter, and the $0^s.05$ given after a period of years, would be accounted for by an r.m.s. value of $\pm 0^s.04$ for personal equation. This is a reasonable figure. [102] pp. 1–2 also confirms that the s.e. of a single station, as derived from stations which have been twice occupied, is $\pm 0^s.05$, in Australian latitudes.

Measures of personal equation. We are concerned to know the r.m.s. value of personal equations which have been ignored, and also the s.e. of those which may have been measured and accepted. The supply of data is not as plentiful as one might wish. Figures given below are accepted east longitude minus observed east longitude.

(*a*) *Old type fixed wire observations.* Reports of the Finnish Geodetic Institute [319] and [447] give a long series of calibrations (in latitude 60°) by six nights before and six nights after each of 10 field seasons. For observer EK (1961–66) the 6-night means of personal equation varied between $+0^s.042$ and $-0^s.068$, with (internal) s.e.'s of $\pm 0^s.004$ to $0^s.015$. For observer MO (1972–5) the figures were between $+0^s.244$ and $+0^s.143$, with s.e.'s of $\pm 0^s.007$ to $\pm 0^s.018$. In all the six seasons EK's personal equation decreased (or became more negative) during the field season. See Table 4.2.

(*b*) *Transit or T4, with hand driven impersonal micrometer.* The Ordnance Survey of Great Britain have provided unpublished details of calibrations made at Herstmonceux ($\phi = 51°$) with a T4 theodolite by a large number of observers between 1970 and 1976. See Table 4.3. Each item in the table is the mean of 4 to 9 sets of 6 to 10 stars. The standard errors of each item, judged by the accordance of its 6 to 10 sets, average $0^s.010$, with only 2 items out of 35 having s.e.'s greater than $\pm 0^s.018$.

The differences between observers are clearly significant. In the before and after-season pairs quoted the sign only changes in two, namely VQB 1970 +65 to −6 ms and 1972 −1 to +26. Judged by the internal errors, the s.e. of the difference between fore and after should average $\pm 0^s.015$ if there was no true change of personal equation. In the figures recorded the average difference (without regard to sign) is $0^s.025$.

(*c*) *Transit with motor driven impersonal micrometer.* [300] illustrates a T 4 theodolite with a motor-driven micrometer, and gives figures for

TABLE 4.2

Personal equation. Fixed wire transit. Finland

Observer		Personal equation/ms					
		1961	1962	1963	1964	1965	1966
EK	Before season	+19	+42	−16	−20	+15	+32
	After season	−10	−36	−68	−46	−34	−34
		1972	1973	1974	1975		
MO	Before season	+237	+196	+218	+143		
	After season	+185	+244	+231	+150		

TABLE 4.3

Personal equation. T4 hand driven micrometer. Great Britain

		Personal equation/ms			
		1970	1976		General mean
KI	Before season	+79	+88		
	After season	+49	+38		+63
		1970	1971	1972	
VQB	Before season	+65	−21	−1	
	After season	−6	−18	+26	+8
		1970	1971		
MP	Before season	+51	+69		
	After season	+86	+62		+67
		1973	1974		
MJH	Before season	+12	−3		
	After season	+15	−11		+3
		1974	1975		
HMK	Before season	00	+32		
	After season	−27	+4		+2
		1974	1975		
PET	Before season	−17	−10		
	After season	−27	−40		−23
Observer		RJF	PW	MDJ	IG
Year		1970	1970	1972	1973
	Before season	+40	+51	+63	+44
	After season	+56	+71	+22	+34
	Later	+25			

longitudes observed at a station in the Netherlands (latitude 52°) as obtained by three observers each observing 16 stars on each of two nights. Their discrepancies from a weighted mean are given in Table 4.4.

[349] gives the results of extensive experience in the USSR.

TABLE 4.4
Personal equation. Motor drive. Netherlands

Observer	Observed minus weighted mean/ms
B	−15 and +11
H	−31 and −13
V	+6 and +7

(*d*) *Conclusions.* Hand-driven impersonal micrometers give smaller personal equations than are got with fixed wires, as may be expected, but in the samples quoted they are not notably more constant for any particular observer. The motor-driven micrometer appears to reduce the figures further, but the sample quoted is a small one, and there is not much improvement in constancy.

There can be little doubt that photography is the current best answer to personal equation difficulties.

For bisection error see § 4.42 (*k*).

SECTION 4. TIME AND LATITUDE. SIMULTANEOUS OBSERVATIONS

This section includes the observation of LST by methods other than meridian transits. §§ 4.30–4.31, describing the use of clocks, recorders, and radio signals, are applicable.

4.34. Zenith photography

At the zenith in astronomical latitude ϕ and east longitude λ, both measured from the CIO, the declination and RA are given by

$$\phi = \delta + (x \cos\lambda - y \sin\lambda), \text{ from (4.48) and (4.25),}$$

$$\lambda = (\text{RA} - \text{EE}) - (1/15)(x \sin\lambda + y \cos\lambda)\tan\phi$$
$$- (\text{UT }1 + R_u + 12^{\text{h}}), \text{ from (4.57) and (4.27),} \qquad (4.71)$$

since the RA of a zenith star equals the LAST.

Then, if the position of the vertical can be located on a photograph which shows a number of stars surrounding the zenith, the apparent RA and δ of the zenith can in principle be derived by interpolation from the known places of the stars, whence (4.71) gives ϕ and λ. The plate position of the vertical can be located by rotating the plate about a vertical axis and making a second exposure.

The advantages of the method can be very great, as below.

(*a*) Latitude and longitude are obtained simultaneously.
(*b*) The field work is extremely rapid.
(*c*) Modern camera shutters and recording apparatus can give the UT 1 time of exposure to within $0^s.01$ or even $0^s.001$, and no personal equation is involved.
(*d*) In the zenith, ordinary refraction errors are much reduced. But erratics and wandering, § 4.11 (*f*), and refraction due to tilting of the isopicnics or to temperature abnormalities in or near the instrument remain.
(*e*) Since all observations are made with the telescope vertical, difficulties with non-circular pivots are eliminated.
(*f*) For the same reason, spirit levels are relatively easily replaced by a mercury bath, or some kind of automatic compensation. It is in this respect that different models are most likely to vary.

It is not necessarily easy to realize all these advantages fully in a field instrument.

4.35. The Photo Zenith Tube. PZT

The *Photo Zenith Tubes* used in fixed observatories aim at measuring latitude to $0''.01$ and time to $0^s.001$, and apart from errors in star places† actually appear to have s.e.'s not many times as much, in the mean of a night's work. As constructed, the PZT is far too elaborate for geodetic use in the field, but its principles may be adaptable.

The following description of the PZT at the Royal Greenwich Observatory is condensed from [458]. See Fig. 4.9. Light from the lens is reflected from a mercury bath back to a small photographic plate just below the lens.‡ The focal length is 3.5 m, and the maximum meridian zenith distance is 15′, so the plate is small, and provided it and its holder are symmetrical, it does little damage to the image. As a star transits, the plate is traversed at a constant speed such that the image is nearly, but not quite, stationary on it. A 20-second exposure is required for 9.5 magnitude stars. The position of the plate carriage on its track is automatically recorded at exact clock seconds. Between exposures (of one star), the lens, plate, and timing contacts are rotated about a vertical axis, and the plate is then traversed on and timed over the same contacts. This calls for careful construction, and might be difficult in a field instrument. Actually four exposures are made of each star, two before transit and two at approximately the same intervals after transit, the instrument being

† See the last 5 lines of § 4.35.
‡ The lens is designed to have its nodal point outside itself, and the plate is put there.

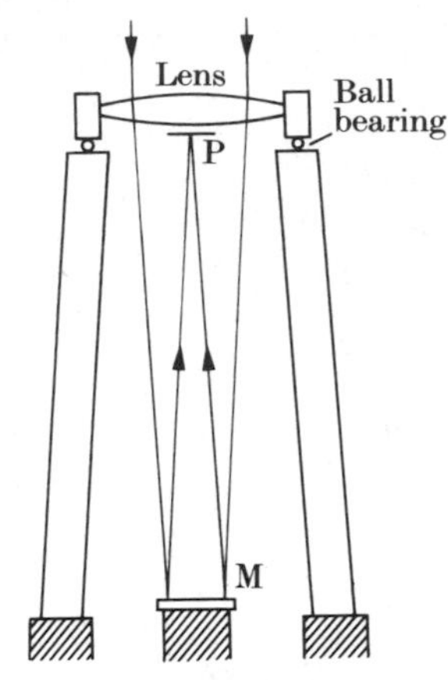

FIG. 4.9. Photo Zenith tube. P = Plate. M = Mercury.

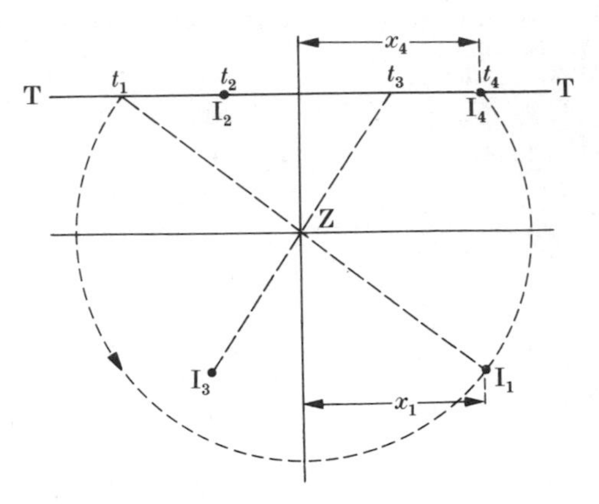

FIG. 4.10. PZT plate.

reversed as above after each. Fig. 4.10 represents the plate, in which Z is the image of the zenith, TT is the path of the star image moving east and west, but reversal has moved I_1 and I_3 to the opposite side of Z, so the four images (whose mid-exposure times are exactly known) form a rough rectangle on the plate. I_1 and I_4 give

$$t_m = \tfrac{1}{2}(t_1 + t_4) - \tfrac{1}{2}(x_4 - x_1)/v, \tag{4.72}$$

where t_m is the clock time of transit, x_4 and x_1 are scaled off the plate, and v is the rate of movement of the image. The 2nd and 3rd exposures have two uses.

(i) To determine the east–west line on the plate, to which x_4 and x_1 must be measured parallel. In Fig. 4.10, I_2I_4 and I_3I_1 lie east–west.

(ii) To determine the scale of the plate and v. For distances $(x_4 - x_2)$ and $(x_1 - x_3)$ are traversed in times $(t_4 - t_2)$ and $(t_3 - t_1)$.

For latitude, the meridian zenith distance is half the distance between I_2I_4 and I_3I_1, converted to arc by the determined scale.

With the PZT the accuracy of star places is very significant. When the PZT is used to measure polar motion, the inaccuracy of declinations is overcome by the programme of observations, as in § 4.19, but for the determination of continental drift (in longitude) it is necessary to place PZTs in similar latitudes, so that pairs can use the same stars.

4.36. Portable zenith cameras

A small zenith camera was used by a party from the US Coast and Geodetic Survey in Liberia in 1951 [122], of which a short description is given in [113] 3 Edn.

The following two instruments have recently been constructed and used in the field.

(*a*) An *Italian zenith camera* is described in [94]. It consists of an IGN satellite camera of focal length 30 cm, aperture 7 cm, and plate size 19×19 cm, which has been mounted on a 42-cm diameter base plate, on which the camera can be rotated about a vertical axis. Two orthogonal levels are attached to the camera, with sensitivities of 10″ per 2 mm. A single reading of a level is thought to have a standard error of 0″.7.

The blade shutter is opened at assigned instants of exactly 10^s on the clock. The latter is a quartz crystal, with a drift rate of not more than 50 ms/hour, operating through a relay with a calibrated delay of $0^s.11$. Radar signals are recorded with an error of about $0^s.001$, and the mean exposure time is considered to be known correctly to about $0^s.01$.

The observation procedure at each station is to make a set of six exposures at probably 30^s intervals, at each of azimuths 0°, 90°, 180°, and 270°. The levels are read before and after each set of six exposures, so that there are 8 readings of each level in all. Allowing for systematic error, the s.e. of the mean levelling is thought to be ±0″.5. All the 24 exposures are made on a single plate.

On the plate 16 stars are selected for reading, four in each quadrant within a radius of 6 to 8 cm of the principal point. There are thus 384 star images to be measured. The symmetrical distribution of the stars minimizes emulsion distortion, refraction, and comparator errors. Rotation through 180° eliminates error in the level adjustment, and of non-coincidence of the optical axis, the fiducial axis,† and the rotation axis, so the observations taken in four positions give two independent determinations of ϕ and λ. It is considered in [94] that the levelling, and to a lesser extent the timing, are the only significant sources of error, and that the total from a single plate is an s.e. of ±0″.5.

Radio time signals are taken before and after the observations at each station. Temperature and pressure are recorded for the refraction. Before the observations begin the instrument is set in azimuth (e.g. by Polaris) with an accuracy of 1°, but the rotations are made with the higher accuracy of 0.1°.

The apparatus is transportable by motor car, or in man-loads of less than 20 kg each. The time required for a full programme of six sets in four positions is about one hour, including setting up and dismantling, so if sites are accessible to motor transport, several closely separated stations can be observed in a single night. Alternatively, the accuracy of a single station may be increased by observing two or more plates at it.

Computation. For the selected 16 stars (all from FK 4) a computer

† The line joining the nodal point of the lens to the origin of plate coordinates.

computes the plate coordinates at the appropriate times, using provisional ϕ, λ, and camera calibration constants. These are compared with the measured coordinates, and the 384 pairs of coordinates give 384 pairs of equations for ϕ and λ and the six camera constants, (7.117). A least-squares solution, made using the 12 exposures in azimuths 0° and 180°, then gives values for ϕ and λ from which systematic error has been eliminated. The exposures at 90° and 270° give independent values.

Accuracy. As stated above the levelling is thought to be the largest source of error, and $\pm 0''.5$ is the expected s.e. of a single plate observed as above. In addition the assumed shutter opening lag may be wrong by $0^s.01$, which would introduce a systematic error of longitude of $0''.15$.

Comparison with results previously determined by Talcott latitudes and meridian transit longitudes at 20 old stations give average discrepancies (without regard to sign) of $0''.8$ in ϕ and $0''.7 \sec\phi$ in λ, which indicate (D.24) s.e.'s of $0''.64$ and $0''.55 \sec\phi$ respectively. These compare favourably with the internal estimates of error, even if the older observations are considered to have contributed no share of the differences.

It appears that an instrument of this type can give the accuracy required for deviations of the vertical which are to be used for astro-geodetic geoid sections.

(*b*) The *Transportable Zenith Camera* (*TZK*1), *Hanover*, is described in [224]. It is a rather more elaborate instrument than the Italian one described above. The focal length is 80 cm, aperture 16 cm, with 6×9 cm plates. An interior mirror 60 cm below the object glass, reduces the total height by deflecting the vertical rays to the horizontal. The camera can rotate on its base plate, and is used in two positions 180° apart. Dislevelment is indicated by two 'electronic levels' known as Talyvel, in which the inclination of the camera is indicated by a hanging pendulum. The two sides of the lower end of the pendulum each constitute one plate of each of two condensers, whose other two plates are attached to the frame of the instrument. If the instrument is level, the two condensers have equal capacitance, but if it is out of level, a detector circuit records the inequality, and the corresponding tilt. The pendulum is strongly damped (dead-beat in one second), and the sensitivity is $1''$ per small scale division. [46].

Observing programme. The beginning and end of the one-second exposure times are recorded by Chronocord, § 4.31. The levels are recorded at the times of exposure. The instrument is then turned through 180° for a second exposure. This constitutes the work done on a single plate, and gives a value of ϕ and λ. One or two such plates may suffice for one station, but to be on the safe side 4 to 6 are more usual. The instrument is transportable by motor car, and can be carried in man loads of up to 30 kg. The time spent on a station can be as little as 30 minutes.

Computation. About 15 stars are selected on each plate, and their coordinates are measured in both positions. Their magnitudes may have to be as faint as 10, because of the smaller plate. The computation uses standard coordinates as in (7.98) to (7.103), together with

$$\xi = (a_1 + b_1 x + c_1 y) \div (1 + b_2 x + c_2 y)$$

and
$$\eta = (a_2 + b_1 y - c_1 x) \div (1 + b_2 x + c_2 y), \tag{4.73}$$

with six unknowns in addition to the RA and δ of the zenith.

Accuracy. As compared with the Italian instrument the levelling and shutter timing arrangements may perhaps be better, but the instrument is heavier and the small plate involves the use of small stars not included in FK 4. In a test in the western Harz in 1974 at 15 stations where ϕ and λ had previously been observed with the NI2 astrolabe, the internal s.e. of a determination using 5 or 6 plates was between $\pm 0''.1$ and $\pm 0''.3$ in latitude, and between $\pm 0''.2$ and $\pm 0''.6$ in longitude. Comparison with the astrolabe gave differences of between $-0''.9$ and $+0''.5$ (mean $-0''.2 \pm 0''.1$) in latitude, and between $-3''.1$ and $+1''.5$ (mean $-0''.8 \pm 0''.3$) in longitude. In this area $\sec \phi = 1.5$.

Comparison with the PZT at Hamburg in a similar latitude gave differences as below

8/4/74 Mean of four plates. Lat. $+0''.4$, long. $0''.0$.

13/1/75 Mean of 12 plates. Lat. $+0''.7$, long. $-0''.5$.

This instrument also appears to give accuracy sufficient for geoid sections.

4.37. Position lines

(*a*) This is a method of computation which can be used for several methods of observation, such as are described in §§ 4.38–4.40.

At a certain instant of Greenwich apparent ST, a star of declination $= \delta$ and RA $= \alpha$ coincides on the celestial sphere with the zenith of a point on the earth whose astronomical latitude is δ and whose longitude is $(\alpha - \text{GAST})$. Its zenith distance when seen from another point on the earth, ϕ_a, λ_a at the same instant will then be given by

$$\cos z_a = \sin \delta \sin \phi_a + \cos \delta \cos \phi_a \cos t, \tag{4.74}$$

where t is the hour angle of the star, and

$$\lambda_a = t + \alpha - \text{GAST}. \tag{4.75}$$

See further in (4.81).

Let the zenith distance of a star of known δ and α be observed at a recorded time t at a point whose ϕ and λ are required. Then if ϕ is assumed to have the value ϕ_a (4.74) gives t, and then (4.75) gives a provisional λ_a. The azimuth of the star is given by

$$\sin A = -\cos\delta\sin t\operatorname{cosec} z. \qquad (4.76)$$

Now draw a chart in the form of Fig. 4.11, in which the line $\phi_a\phi_a$ represents the parallel ϕ_a divided into time seconds of longitude on such a scale as 5 or 10 cm $= 1^s$. For any star S_1 plot the longitude computed by (4.75) and draw S_1S_1' perpendicular to the azimuth of the star computed to the nearest 10′ by (4.76). Then it is clear that P, the true position of the point, lies on S_1S_1', since the locus of P is a circle centred on the distant point $\phi = \delta$, $\lambda = (\alpha - \text{GAST})$.

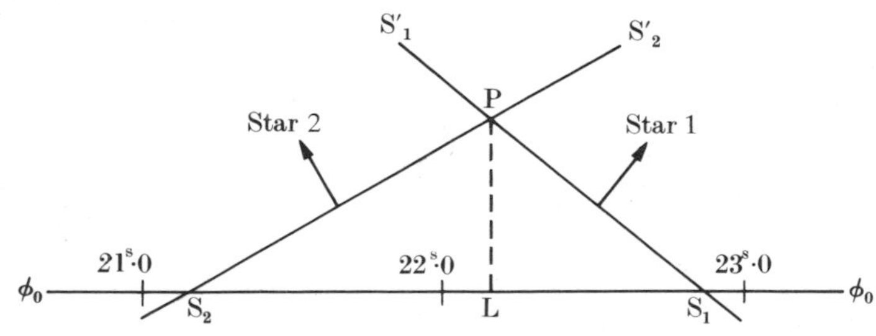

FIG. 4.11. Two position lines.

Observations to a second star at a different azimuth, and at the same or at a different z, give a second locus S_2S_2', which intersects S_1S_1' at the true position P. The longitude of P may then be scaled off the $\phi_a\phi_a$ line at L, and its latitude is $\phi_a + \text{PL}$, PL being measured on the λ scale and then multiplied by 15 cos ϕ_a.†

In practice numerous stars will be observed and a best position for P may be determined by least squares or graphically as below.

If the approximate latitude ϕ_a cannot be adequately‡ estimated by other means, a value can be got from a pair of stars provisionally computed with the best value available.

If the observed zenith distance is wrong by a small amount δz, the computed locus SS′ will be misplaced by δz (converted to distance on the diagram by the latitude scale) towards or away from the direction of the star. Then let four stars be observed, one in each azimuth quadrant, at recorded GAST's as they pass a constant, but not quite accurately known, zenith distance. This is advantageous, because it is often easier to keep z constant than to measure it. Then compute t from (4.74) as above with ϕ_a

† In [113], 3 Edn, a factor of (ν_a/ρ_a) was wrongly introduced. Fig. 4.11 is not a map of the ground, but a cylindrical orthomorphic projection of a unit sphere which is similar to the celestial sphere but rotating with the earth, see footnote on p. 681.

‡ If the lines on Fig. 4.11 are not to exceed 20 cm, and if 0″.1 of latitude is not to be less than 0.5 mm, ϕ_a must be correct to about 10″.

and z_a, and the diagram will be as in Fig. 4.12, in which perfect observations have resulted in all four loci being equally displaced from the correct point P, by a distance $\delta z = (\text{true } z) - z_a$.

In practice no circle will exactly touch the four lines, but P is selected as the intersection of the bisector of the north-west and south-east position lines with that of the north-east and south-west lines. If several sets of four stars are observed, the mean P of all sets can be accepted. Changes in refraction may result in changes in δz and in the size of the quadrilaterals, and stars should so far as possible be grouped in the order in which they have been observed.

To keep the diagram small, the true z and the refraction should be known within a few seconds of arc, or should be determined by a provisionally computed set of four stars.†

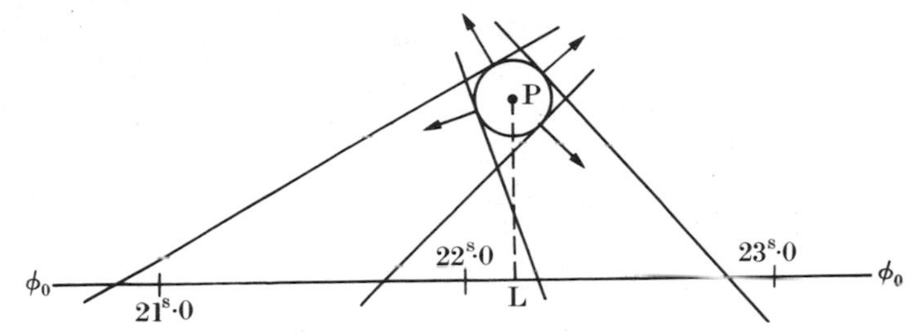

FIG. 4.12. Set of four position lines.

(*b*) *Least squares.* In Fig. 4.12 take axes of x along $\phi_a\phi_a$ and of y upwards, with origin at (say) the point $\lambda = 21^s.0 \equiv \lambda_a$. Let the unit of x and y be the same as the unit of longitude. So

$$x = (\lambda - \lambda_a), \quad \text{and} \quad y = \tfrac{1}{15}(\phi - \phi_a)\sec\phi_a.$$

Then the equation of any position line will be

$$y + x\tan A - \delta\lambda_c \tan A = 0, \tag{4.77}$$

where $\delta\lambda_c$ is the λ computed from (4.74) and (4.75) minus λ_a.

Let (x_P, y_P) be the required best position of P. Then the length of the perpendicular from P to the line (4.77) is

$$p = \delta\lambda_c \sin A - y_P \cos A - x_P \sin A.$$

We seek values of x_P, y_P, and δz such that $p = \frac{1}{15}\,\delta z \sec\phi_a - v$, with $\sum v^2$

† An alternative system for drawing the diagram is known as the Marc St. Hilaire method. In (4.74) and (4.75), the zenith distance z is computed using provisional values of both ϕ_a and λ_a. Then if this provisional position is taken as the origin in the diagram, each locus such as S_2PS_2' in Fig. 4.11 lies at a distance δz from it, where δz is z as computed by (4.74) minus the observed value (corrected for refraction), and the azimuth of the perpendicular from P to the position line will be as given by (4.76). The two methods, of course, give identical position lines.

minimum. So observation equations are

$$y_P \cos A + x_P \sin A + \tfrac{1}{15}\delta z \sec \phi_a = \delta\lambda_c \sin A + v.\dagger \qquad (4.78)$$

One equation for each star.

Other things being equal, especially if all altitudes are equal, these equations may all be given equal weight. The s.e.'s of deduced values of $\delta\lambda_c$ will vary as cosec A, so that the s.e. of $\delta\lambda_c \sin A$ is independent of A.

Least squares will be more convenient than the graphical method if a computer is used. The graphical method will probably be preferred by one who does his own computations in the field.

(*c*) *Curvature correction.* If a star is timed over several graticule wires, or by several contacts of a micrometer, or with a small change in the reading of the vertical circle, or with a change of face of a theodolite, it is natural (as a first approximation) to take the mean of the recorded times, and the mean of the zenith distances corresponding to the various contacts or graticule lines, and to use them to compute (4.74) and (4.75). To do this is to assume that dz/dt is constant, which is generally true enough over a single set of graticule lines or a couple of rotations of a micrometer, but not over the time involved by a change of face. The remedy when necessary is a so-called curvature correction.

Let z be the true zenith distance at the mean of the recorded times, and let z_m be the mean of the zenith distances, then

$$z - z_m = \cos^2\phi \cos A(\tan\phi - \cot z \cos A)m_0 \qquad (4.79)$$

where $m_0 = (1/n)\sum 2\sin^2(\frac{1}{2}\delta t)\text{cosec } 1''$, and δt is the interval between each recorded time and the mean of them all, in sidereal seconds. n is the number of recordings.

Alternatively, if t is the true time at which the zenith distance equals the mean of the actual z's, and if t_m is the mean of the recorded times

$$t - t_m = -\frac{\cot A \text{ cosec}^2 A}{30n \cos\phi}(\tan\phi - \cot z \cos A)\sum(\delta z)^2 \sin 1''. \qquad (4.80)$$

See [492], pp. 25–5 and 25–6 and also IV-5 to IV-8, where the subject of curvature corrections is treated in great detail.

(*d*) *Correction to mean pole.* The preceding sub-paragraphs (*a*), (*b*), and (*c*) have given the latitude $\phi_0 = ZZ_0$ (in Fig. 4.4) and longitude $\lambda_0 = X_0PZ_0$, in the system defined by the instantaneous pole P. It is clear that the latitude requires correction as in (4.25), to give ϕ_1 namely

$$\phi_1 = \phi_0 - (x \cos\lambda - y \sin\lambda).$$

The correction to longitude is less clear. The observation has given $\lambda_0 = X_0PZ_0$ in Fig. 4.4, while the required λ_1 (in the CIO system) is X_1CZ_1.

† In [113] 3 Edn the factor 1/15 was wrongly included in the term $x_P \sin A$.

We have $\lambda_1 = \lambda_0 + (X_1X_0 - Z_1Z_0)$

$$= t + \alpha - \text{GAST} + (X_1X_0 - Z_1Z_0) \text{ from (4.75).}$$

From Fig. 4.4. $\text{GAST} = \text{UT 1} + X_1X_0 + (R_u - 12^h)$,

so
$$\lambda_1 = t + \alpha - (\text{UT 1} + R_u - 12^h) - Z_1Z_0, \tag{4.81}$$

and the only necessary correction for polar motion is

$$-Z_1Z_0 = -(x \sin \lambda + y \cos \lambda)\tan \phi. \tag{4.82}$$

The correction X_1X_0 has disappeared because in the instantaneous system X_0 is the zero of both GAST and of instantaneous longitudes, and in the CIO system X_1 is the zero of both UT 1 and of the CIO longitudes.

(*e*) *Diurnal aberration.* No allowance should be made when computing *t*, but $0^s.021 \sin h$ should be added to the deduced (slow) LST error of the clock, or to the east longitude. Latitude is not affected.

4.38. Prismatic astrolabe

(*a*) The prismatic astrolabe is an instrument which is well adapted to the observation of ϕ and λ by timing the passage of stars through a prescribed (usually 60°) altitude. It was much used for this purpose between 1920 and 1940 in a form which did not incorporate any device to reduce or eliminate personal equation. Many attempts have been made to remedy this, and the Danjon astrolabe, described in (*c*) below has been very successful in fixed observatories, but it has not been used in the field. Other devices have been reported as promising, but they do not seem to have been widely used yet. The current tendency is to use either a zenith camera, § 4.36, or else to use an astronomical theodolite instead of an astrolabe.

(*b*) *The Claude and Driencourt astrolabe.* This is the simplest form of astrolabe. A full description of its use is given in [113], 3 Edn, pp. 341–4, which need not be repeated in full. The following is a short description.

The telescope is of 40-cm focal length and 5-cm aperture. Fig. 4.13

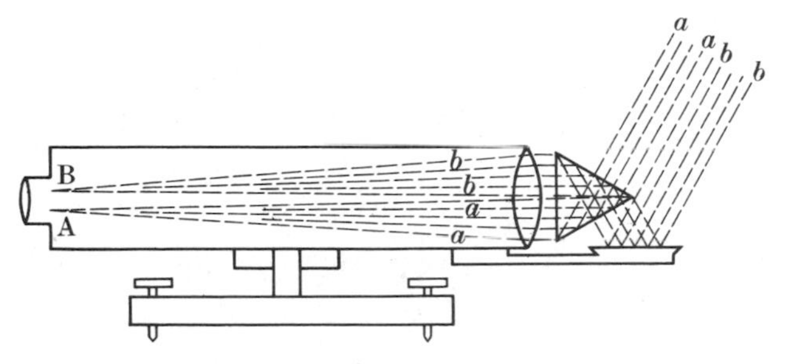

FIG. 4.13. Prismatic astrolabe.

shows the general arrangement. Light from a star falls partly on the upper surface of a 60° triangular prism, and partly on the horizontal surface of a mercury bath, and thence to the lower side of the prism. When the apparent altitude of a star is exactly equal to the front angle of the prism (60°) the two bundles of rays *aa* and *bb* are parallel as they emerge from the back of the prism, and the two star images seen in the eyepiece coincide, or in practice lie closely side by side horizontally, one moving up and the other down. The instant of coincidence is recorded by hand tappet, and the astrolabe is then turned on to the next star. The programme consists of a number of sets of four stars, with one star in each quadrant of azimuth, as near as may easily be to azimuths 45°, 135°, 225°, and 315°. To add strength to the longitude, a few pairs of nearly east and west stars may be inserted in gaps in the programme. With a good previously prepared programme it is possible to observe 15 FK 4 stars an hour in a cloudless sky in ordinary latitudes, and two hours work should produce 5 or 6 sets of four, with 6 to 8 extra longitude stars. From internal evidence this should give an apparent s.e. of $\pm 0''.5$ in latitude and $\pm 0^s.03$ in time. Calibration at a base station as in § 4.29 (*f*) is essential.

The essence of this instrument is that provided the mercury surface is level, the prism angle constant, and the atmospheric refraction unchanged during the observation of each set of four, a single set of four stars will give ϕ, λ, and the error of (prism + assumed refraction) with one redundancy.

(*c*) The *Danjon* astrolabe [27] and [175], 60°, contains a satisfactory self-registering micrometer, and gives results of the highest accuracy. It is used in fixed observatories for defining UT and for measuring variation of latitude, and is considered to be only slightly inferior to the large Photo Zenith Tubes. [175] reports the s.e. of a single star as $0''.09$ to $0''.26$ according to observing conditions, only exceeding $0''.26$ on 6 per cent of nights. Personal equation between observers over a year or two is reported as under $0^s.003$, and $0^s.007$ over a month is considered large.

The superiority of the instrument derives from a doubly refracting prism inserted just in front of the focal plane, see Fig. 4.14. This has two functions, as below.

(i) In Fig. 4.13, showing the old pattern astrolabe, the axes of the two pencils *aa* and *bb* which form the two images are not parallel, but this

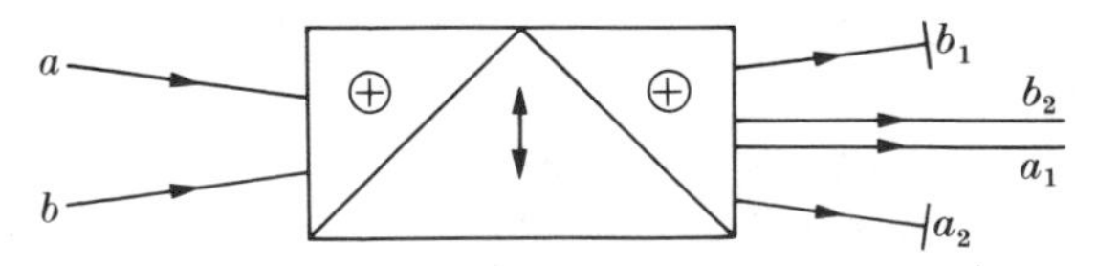

FIG. 4.14. Danjon astrolabe. Doubly refracting prism.

prism is so proportioned that two of the emergent pencils a_1 and b_2 are parallel, while the other two are screened off. The apparent coincidence is thus independent of eye-piece setting.

(ii) While the axes of pencils a_1 and b_2 are substantially parallel, the actual coincidence of the two images depends both on the zenith distance of the star and on the location of the prism along the beams. Coincidence can therefore be maintained for a short period by traversing the prism at a suitable speed, in the same way as the impersonal micrometer of the Transit. The prism is driven by an electric motor, subject to final control by the observer, and the clock time at which it passes contacts on its run is chronographically recorded.

The present model is too heavy for normal field use (175 kg).

(*d*) *Other impersonal devices.*

(i) In 1933 and 1938 the Nusl–Fric astrolabe was reported as having a self-registering impersonal device. It has not become generally available, and reports of its use in the field have not been received, but [490] reports further development. The new instrument is large, with focal length 100 cm, aperture 10 cm, and magnification 100. As with the earlier model, the impersonal micrometer consists of two thin prisms which can be rotated about the optical axis of the telescope, and times are recorded through contacts attached to these prisms. Rotating a prism changes the angle through which it deviates the star's apparent altitude, and a total deviation of between $+150''$ and $-150''$ activates 10 contacts. The instrument is said to be portable. The s.e. of a single star is about $0''.5$, but for the mean of a group of 15 stars it is about $\pm 0''.23$ in latitude or longitude.

(ii) *Zeiss Ni* 2 *astrolabe.* This instrument is a standard Ni 2 Zeiss level modified by the fitting of a 30°–60°–90° prism in front of the object glass, and the provision of an illuminated eye-end graticule with 10 pairs of fine horizontal lines, across which the passing of a star is timed.

Unlike the astrolabes described in (*b*), (*c*), and (*d*) (i) above, this instrument has no mercury bath, but depends for its levelling on the standard Zeiss automatic device used in their levelling instruments. Consequently there is only a single star image. The instrument has a horizontal circle for setting in azimuth on previously programmed stars, and a cross-level, which must be kept reasonably well centred. The automatic level is described by the makers as correct to $0''.5$. The aperture of the object glass is 40 mm diameter, and the magnification is 32.

The Ni 2 astrolabe has been used to measure ϕ and λ at 27 stations in Germany in 1966–1974 [323]. In 1966 and 67 internal evidence gave s.e.'s of about $\pm 0''.23$ in ϕ and $\lambda \cos \phi$, for the mean of a programme of

91 stars each observed only over a single graticule line. In 1974 the s.e.'s were $\pm 0''.10$ for the mean of 46 stars each observed over 8 graticule lines. When two nights were observed at each station their discrepancies gave s.e.'s of $\pm 0''.29$ in ϕ and $\pm 0''.33$ in $\lambda \cos \phi$, for the mean of two nights.

Calibration at a fixed station was carried out at the beginning and end of each season, and typically changed by 1″ or 2″ during a few months. [493] records experiments to provide an impersonal micrometer.

[300] reports three season's observations of ϕ and λ in the Netherlands, using a Ni 2 astrolabe, calibrated by observations at a base station. Corrections to λ, in about latitude 52°, were as below.

1974. $-3''.23$, $-2''.58$, and $-2''.03$
1975. -1.66, -1.46, and -1.58
1976. -1.74, -2.04.

Each figure is the mean of two nights. All refer to the same observer. The self-consistency is satisfactory.

(*e*) *Advantages of the prismatic astrolabe.* As compared with the zenith telescope and meridian Transit advantages are: (i) simultaneous determination of latitude and longitude if both are wanted; (ii) ease of setting up, adjustment and operation; (iii) no need for firm base or referring mark; (iv) sufficient stars can be got from an ephemeris; (v) absence of divided circles, moving parts, and trunion axes; (vi) Nothing to record except times of star passages.

Disadvantages are: (i) long computations; (ii) programme is troublesome if not already available; (iii) sensitive to wind; (iv) except in the Danjon model, no satisfactory personal equation apparatus is yet in general use for longitude, and it is less accurate than the zenith telescope for latitude.

If the sky is intermittently cloudy, it will be better to observe by theodolite at varying measured altitudes, as in § 4.40.

4.39. Equal altitudes with the theodolite

(*a*) A theodolite may be used in the same way as an astrolabe. The vertical circle is set to an altitude of 60°, 45°, or 30° or whatever is convenient, and all observations are made on one face. The essential thing is that during the whole evening's work no movement of the vertical slow-motion screw, nor of the (upper) bubble adjusting screw, should change the angle between the telescope and the bubble. Once observations have started any necessary movements or centring of the upper bubble must be done by the foot-screws.

No readings of the circle are necessary, except to give a value of the fixed altitude, and occasionally to confirm that it has not changed.

Observations can be made with fixed horizontal wires, or with an impersonal micrometer.

(*b*) *Fixed wires.* It may be convenient to have (say) 5 fixed horizontal wires, about 5′ apart, and to time the passage of each star over them. Between each pair of intersections the horizontal slow-motion screw should be turned so that the next intersection takes place near to the vertical wire. The upper bubble should be moved (by a foot-screw) and read, before each of the five intersections and after the last. The observation is not impersonal.

For full details, and an example of recording, see [100], pp. 20–6.

(*c*) *Impersonal micrometer.* If the theodolite has a moving wire impersonal micrometer, it should be turned so that the moving wire is horizontal. Each star should then be followed over the contact which takes place nearest the line of collimation and the 4 or 5 or more contacts on either side of it. It is essential to use the same contacts on all the stars observed. Face is not changed. The movement of the micrometer will be in opposite directions for east and west stars, so corrections must be made for width of contacts and lost motion. For the curvature correction see § 4.37 (*c*).

When the micrometer is being used, it is not possible to move the horizontal slow motion so that contacts are made when the star is near the vertical wire. Great care must therefore be taken to get the moving wire exactly horizontal, and to start observations at such a point on it that the star's course will pass close to the intersection of the fixed wires. The provision of radial lines in the outer part of the graticule may help with this.

For details see [100], pp. 27–35.

4.40 Unequal altitudes

When there is some cloud it is better to take stars at different measured altitudes. No programme is needed, and stars may be taken as they are seen in different parts of the sky. It is only necessary that they should be reasonably well balanced in altitude, and evenly distributed round the horizon, so that position lines can be plotted in groups of four with the hope that refraction errors will not be too different.† Altitudes should not be too low, with (say) 30° as a permissible minimum. Provided the diagonal eye-piece is used there is no maximum, but all members of a group of four should have equal altitudes within (say) five or ten degrees.

The observations, and measures of the vertical angles, should be on one face only . Collimation error of the horizontal wire, the disagreement between FL and FR vertical angles, should not be more than a few

† Refraction corrections are, of course, applied.

seconds, or should be known and allowed for in the computations. Otherwise the position line diagram will be too large.

Each star can advantageously be intersected on five fixed wires, or on a single wire with changed circle readings. The upper bubble should be recorded before all the intersections and after the last. Accuracy depends on bubble, intersection, angle-measurement, and timing, and all are the better for repetition.

Other things being the same, unequal altitudes with one intersection per star will give a larger s.e. than equal altitudes with a single intersection, but with five intersections unequal altitudes should be as good or better. Five intersections on each of sixteen well-placed stars should give a s.e. of $\pm 0''.5$ or $0''.7$ in latitude and in (longitude)$\times\cos\phi$.

Computation of azimuth for curvature correction and for plotting position lines, and for the recognition of misidentified stars, is avoided if the horizontal circle is read at the start and finish of each star, and on Polaris before and after work. The *Star Almanac* Polaris tables will then give the azimuth without computation.

Since face is not being changed, a curvature correction is unlikely to be material, as in § 4.37 (*c*).

4.41. Longitude by east and west stars

Instead of getting both ϕ and λ from observations to stars in all quadrants, it may be preferred to get ϕ from circum-meridian altitudes, and to get λ separately from equal (or unequal) altitudes of stars within (say) 10° of the prime vertical. Their observation is then exactly as in §§ 4.39 or 4.40, with or without the impersonal micrometer. The computation may be done by position lines as in § 4.37, the horizontal line $\phi_a - \phi_a$ being fixed by the separate latitude observations. Alternatively, ϕ and z being known, each star will give a value of the longitude using (4.74) and (4.75). Stars should finally be paired, one east and one west, both members of a pair having similar altitudes and not being far separated in time. The mean longitude from all pairs can then be accepted. As with all longitude observations, the scatter of the means of pairs may be small, but the real accuracy will depend on how personal equation and instrument lags have been eliminated.

SECTION 5. AZIMUTH

4.42. General principles

Astronomical azimuth is ordinarily observed by theodolite, preferably at the same time as the triangulation or traverse which it is to control. If the celestial pole was occupied by a visible object it would simply be included with other stations in the rounds of horizontal angles, but as this is not possible, observations are made between some station (or special referring mark which has been included in the usual rounds) and some star whose azimuth at the moments of intersection can be calculated from the triangle PZS. Then the azimuth of the RM immediately follows. §§ 4.43–4.48 detail the types of star best used in different circumstances, but the following considerations apply in all cases.

(*a*) *Sites of stations and RM's.* See §§ 1.03 (*c*), 1.05, 1.06, and 1.09 for frequency. Star observations cannot be spaced throughout the 24 hours, so a site which is likely to suffer from horizontal refraction, §§ 1.04, 1.19, and 3.22, is especially unwelcome.

There is much to be said for including two adjacent triangulation or traverse stations as RM's in the same rounds of angles as the stars. If the angles between these two RM's have also been observed by day, that will reveal abnormal horizontal refraction. And, especially in a traverse, it is good to have direct measures of the azimuths of both the adjacent lines.

In classical triangulation, where Laplace stations have been few, it has of late years been customary to observe them as reciprocal pairs at the two ends of one of the triangulation lines. This has provided a measure of their accuracy. Also, if (as a first approximation) lateral refraction is assumed to bend the ray into an arc of a circle, its effects will be eliminated by reciprocal observations. The latter result is to some extent (more easily) secured by selecting as RM's two adjacent stations at roughly equal distances and at azimuths differing by 180°.

In traverses with frequent Laplace stations, it is better to space the Laplace stations evenly, with RM's at both adjacent stations, than to distribute the same number of Laplace stations as more widely spaced pairs.

(*b*) To get a geodetic azimuth, the observed astronomical azimuth must be corrected for the east–west deviation of the vertical (η) as in (2.4).† This demands the observation of astronomical longitude at the azimuth station or so close to it that the error in an estimated value of η will not exceed $\cot\phi\times$(permissible error in the corrected azimuth), or (say) $0''.2\times\cot\phi$, although an accuracy of $1''\times\cot\phi$ will often be better than nothing.

(*c*) *Level.* Stars being highly elevated objects, it is impossible to get the vertical axis so accurately vertical that errors due to dislevelment can be ignored. See § 1.15, (vii). The cross-bubble, at right angles to the telescope, must therefore be read at each pointing to a star, and the following added to the zero-mean horizontal circle reading,

$$\frac{d}{n}\left(\sum L-\sum R\right)\cot z, \tag{4.83}$$

† But see § 4.47, where Black's method gets the geodetic azimuth directly, without the explicit determination of either η or the astronomical azimuth.

where d is the value of one division; n the number of scale readings, two per pointing; z = zenith distance; and $\sum L$ and $\sum R$ are the sum of the readings of the left- and right-hand ends of the bubble, seen from the eye-piece, the scale being numbered outwards from the centre. It must be noted that many theodolites have inferior cross-bubbles.

The final bubble correction must be kept small by periodical relevelment between zeros, in case calibration of the bubble is inaccurate, allowing for a possible error of 50 per cent. Similar corrections to the readings on the RM may be necessary if it is close and highly elevated, but careful levelling should ordinarily keep that negligible, as for ordinary stations.

(*d*) Observations must be made in pairs FL and FR with all the usual precautions of primary triangulation. For a Laplace azimuth the number of sets and zeros should be at least as many as for ordinary horizontal angles, § 1.18.

(*e*) *Diurnal aberration.* From observed horizontal circle readings to the star subtract $0''.32 \cos A \cos \phi \operatorname{cosec} z$ (arithmetically additive if $\cos A$ is negative, as when $90° < A < 270°$). If the star is near the pole this is constant, and $0''.32$ can be added to the final mean azimuth of the RM,† or subtracted in the southern hemisphere, but with other stars it is more variable. Put otherwise, the apparent position of the star is always east of its true position.

(*f*) *Mean pole.* As in (4.26) the observed astronomical azimuth requires reduction to mean pole by

$$\text{Mean} - \text{Observed} = -(x \sin \lambda + y \cos \lambda) \sec \phi. \tag{4.84}$$

For the Laplace azimuth the observed longitude will also have required correction as in (4.27). Alternatively, if both (4.27) and (4.84) are ignored, the correction to the Laplace geodetic azimuth is as in (4.28). For Black's method see § 4.47.

(*g*) When computing, the direction of a station used as RM requires correction for deviation, skew normals, and geodesic like any other station, if they are not negligibly small (§§ 2.09 and 2.12), but if a special RM is used its corrections immediately cancel when azimuth is transferred from it to the ordinary stations, and it is only necessary to see that they are either included or excluded at both stages. No such corrections are applicable to the star's readings except the Laplace correction itself as in (2.4) and except as in (4.92) if Black's method is used.

(*h*) *Curvature.* As in § 4.37 (*c*) the path of a star is not 'straight', in that dA/dt is not constant, and the mean of FL and FR horizontal readings, or of intersections on several wires, is not strictly applicable to the mean of

† Since Az of RM = Az of star + circle reading of RM − circle reading of star.

the observed times. The correction depends on the system of observation, and is detailed in the following paragraphs.

These curvature corrections are small, and can be computed by slide-rule. In the days before electronic computers, it was less trouble to compute (say) five curvature corrections by slide-rule and one solution of the astronomical triangle with seven-figure logarithms, than to compute five solutions of the triangle. The position is now different, and there is a choice of three courses.

(i) Let the electronic computer do the curvature corrections, and the solution of a single spherical triangle.

(ii) Let the computer compute every intersection separately in full. This is probably quicker than computing the curvature corrections, and less trouble to programme.

(iii) Let the field observer compute the curvature corrections as in the past.

(*i*) *Accuracy.* The s.e. of an astronomical azimuth cannot be expected to be less than that of an ordinary horizontal angle observed in similar circumstances, e in § 2.43, although in low latitudes there is no reason why it should be much worse, provided the bubble correction is kept small as in (*c*) above. In higher latitudes there are difficulties, and $\pm e \sec \phi$ is to be expected. See also § 2.45.

(*j*) *Personal equation.* Azimuth observations to stars very close to the pole, or to stars moving vertically, are free from personal equation, except of bisection as in § 1.27 (*f*), although the trouble enters into the determination of η, and thence into the geodetic azimuth (except on the equator). Black's method (§ 4.47) avoids determining η, but (except on the equator) involves intersecting stars which are moving in azimuth. See [485].

(*k*) *Error of bisection.* [145] analyses Polaris azimuths made in the US. No significant differences were found between one night and the next, nor between T3, T4, and DKM 3 theodolites, but a significant variation (s.d. $= 0''.82$) was found between nine observers. With Polaris errors of timing can hardly occur, so these figures are presumably the personal differences of bisection between Polaris and the referring lamp. It does not seem likely that the elevation of Polaris should be responsible for bisection error, so it appears that great care must be taken to equate the appearances of the lamp to that of the star. See also [485].

4.43. Polaris or σ Octantis at any hour angle

These stars of magnitudes 2.1 and 5.5 respectively, are about 1° from the north and south poles, and can conveniently be used between lats 55°

and 10°. It generally suffices to know LAST to the nearest second, since at the worst their azimuths change by 1″ sec ϕ in 4 seconds of time. The routine of a single pointing should be as follows. Intersect the star on the vertical wire and start the stop-watch; read cross-bubble; stop the stop-watch on some convenient reading of the chronometer,† and record as e.g. $12^h\,28^m\,30^s.0$ minus $28^s.7$; read horizontal circle; intersect and read RM (or this may come before the star). For this purpose LAST can be got well enough from any wireless time signal provided the astronomical‡ longitude is known within about 10″, or directly by ordinary topographical methods, such as the observation of east and west stars, the shortest programme being sufficient, provided there are enough stars to eliminate blunders.

The formula is (4.88), of which another form is

$$\tan A = -\cot \delta \sec \phi \sin t\left(\frac{1}{1-a}\right), \tag{4.85}$$

where $a = \cot \delta \tan \phi \cos t$, and $1/(1-a)$ can be tabulated, [290], Table XIII. Note that $\cos t$, and consequently a, is negative if t is between 06.00 and 18.00 hours. See [290] for an example. For signs it is probably best to record A as a small angle E or W of the elevated pole, correct it by (4.85), and then derive the azimuth of the RM with the help of a small diagram.

The correction for curvature is

$$-m_0 \tan A \sin^2\delta, \tag{4.86}$$

where $m_0 = (1/n)2 \sin^2\frac{1}{2}t \operatorname{cosec} 1''$, in which t is the interval between each intersection and the mean of all in sidereal seconds. A is the mean azimuth of the star during the set. The correction is applicable to the azimuth of the star as computed for the mean of the times; and its sign is such as to decrease the computed angle between the star and the elevated pole. The correction is small, for Polaris less than 1″ if t in $m_0 < 5$ minutes. [290], Tables XII and XIV, aid computations. For Polaris $\sin^2 \delta = 1$. Formula (4.86) is correct for any star at elongation. In general (4.89) and (4.90) apply. See § 4.42 (h).

For full details of the observation see [290], [444], and [100].

4.44. Circumpolar stars near elongation

When a star is at elongation $dA/dt = 0$, and in days when time-keeping was difficult it was usual to avoid time-keeping errors by observing high

† Where continuous time signals are easily heard, one hand of the stop-watch can act as chronometer, [100], pp. 15–17.

‡ Geodetic longitude may be known, but $A-G$ may not be.

declination stars at elongation. This is no longer necessary, but when fast-moving, low-declination stars are observed it is, of course, advantageous to have them near elongation if that is conveniently possible, to reduce the effects of personal equation and timing errors.

The hour angle of elongation is given by

$$\cos t = \cot \delta \tan \phi. \tag{4.87}$$

For computing, use (4.85). In *Apparent Places* circumpolar stars are listed, daily, after all others. For curvature at elongation (4.86) applies.

4.45. Meridian transits

In latitudes greater than about 55° the preceding methods become inaccurate, since any errors of time or intersection are multiplied by sec ϕ when the direction is transferred from the pole down to the horizon, and dislevelments of the transit axis are multiplied by tan ϕ. Inaccuracy can be reduced by using the Transit telescope, or a suitable astro-theodolite, such as the T4, although the programme outlined in § 4.28 is not suitable for azimuth as it stands. Only a few zenith time stars are needed, while as many high declination stars as possible should be included, and also stars of medium declination transiting below the pole. Time and azimuth may, of course, both be strongly determined by a suitable programme.

The US NGS (late USC & GS) use the Transit instrument for azimuth observations north of 50°, but on a different system, using the micrometer to make numerous measures of the angle between Polaris and the RM on both faces. [290], pp. 107–20. The range of the screw being about 25′, it is convenient to use Polaris near elongation, and the azimuth of the RM should be about $\pm$(NPD of Polaris $-$ 10′)sec ϕ, so that the micrometer readings are about equally positive and negative. The LAST is then got from a few zenith stars, or by wireless if the astronomical longitude is already known, and the value of the screw by timing the rate of movement of equatorial stars on the chronograph. For details see [290], p. 21. The screw value is needed accurately, although a good balance of positive and negative measures will minimize error. The value may change with the temperature. Note that micrometer readings of the star and of the RM (each reckoned from the line of collimation) must be multiplied by sec(altitude) to reduce them to the horizon.

The method of meridian Transits may be used in low latitudes if Polaris or σ Octantis are too near the horizon. The programme should then contain as many high and medium declination stars as can be got, and a considerable number of zenith stars, including some up to 100 degrees from the pole. The solution should be by least squares.

The RM should be at least 3 km away, unless very special care is taken.

A Transit in which the telescope is not vertically over the vertical axis will need (cancelling) corrections to centre on each face.

The use of the Transit telescope in the meridian requires a special RM very close to the meridian. In Canada (high latitudes) a similar method has been used with the instrument oriented in the vertical plane through the RM, which may be any adjacent station whose azimuth is within $\pm 30°$ of the meridian. See [500]. Stars are selected which will cut this vertical at zenith distances of between 40° and 70°, half on either side of the zenith, the north stars generally being below the pole. The method of observation is to follow the star with the impersonal micrometer as it crosses the selected vertical, just as in meridian transits. The movement of the star is not horizontal, so the verticality of the vertical wire is important. The instrument is reversed on each star, and the star's position relative to the horizontal wire should be the same at the start after reversal as it was at the finish before reversal. Correction for curvature is necessary, as in (4.90). The astronomical latitude must be known. If the longitude is also known, the azimuth is directly determined, but [500] considers the possibility of solving for longitude and azimuth simultaneously from the same observation programme.

4.46. East and west stars

Near the equator, if it is not convenient to use the Transit, azimuth can be got from stars near the prime vertical, knowing the time by time observations or by wireless if astronomical longitude is known. The formula is

$$\cot A = \sin \phi \cot t - \tan \delta \cos \phi \operatorname{cosec} t, \tag{4.88}$$

which is identically the same as (4.85).

The astronomical latitude must be known to the nearest 1″ or better, although error tends to cancel in the mean of east and west. The programme should include two east and two west stars, comprising between them the full number of primary sets and zeros, as in § 1.18.

Provided observations are reasonably rapidly made, corrections for curvature of the star's path between FL and FR are small and generally of opposite sign as between east and west, but they are not always negligible. The formula is

$$\Delta A'' = \tfrac{1}{8} \sin A \cos \phi \sec^2 h (\cos h \sin\delta - 2 \cos A \cos \phi)(15\Delta t^{s})^2 \sin 1'', \tag{4.89}$$

where Δt^{s} is the time interval between the two faces. For signs, h and $\cos \phi$ are essentially positive. Functions of δ and A should follow the usual rules, δ being negative south and A being clockwise from north.

Then $\Delta A''$ is the correction to the azimuth of the star as computed with the mean of the two times.

If several observations are made at short times t^s before or after the epoch to which all are to be reduced, the formula is

$$\Delta A'' = \sin A \cos \phi \sec^2 h(\cos h \sin \delta - 2 \cos A \cos \phi) \times m_0, \quad (4.90)$$

where m_0 is the mean value of $2 \sin^2 \frac{1}{2} t$ cosec 1″.

[293], 16, pp. 79–80, describes an application of this system. A star is selected whose azimuth at elongation is nearly the same as that of an adjacent triangulation station, and several measures of the small difference of azimuth are made by the eye-piece micrometer on both faces within a few minutes (of time) of elongation. Observations are repeated on several stars, using different stations as RM's. A star catalogue is probably necessary.

4.47. Black's method for azimuth

This method, [91], gives geodetic azimuth direct without explicit determination of η or ξ, although the programme may be arranged so that they also are given if only moderate accuracy is required. The method involves measuring horizontal angles between the RM and low altitude stars in all azimuths, at recorded times.

Fig. 4.15 represents the celestial sphere. Z_G and Z_A are the geodetic and astronomical zeniths, the points where the spheroidal normal and the ground-level vertical respectively meet the sphere. P is the pole, R is the referring mark whose geodetic azimuth is required, at a small altitude h_R above the horizon, and S is a star at altitude h_S. Then in the triangle PZ_GS, $PZ_G = 90° - \phi_G$, the geodetic co-latitude, PS is the polar distance

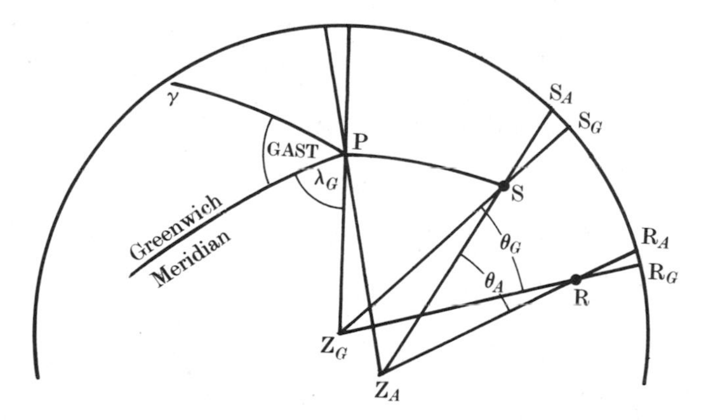

FIG. 4.15. Black's method for azimuth. Suffix G = geodetic. Suffix A = astronomical. PZ_GS = geodetic azimuth of star. PZ_AS = astro azimuth of star. SZ_AR = measured angle between star and RM.

of the star as usual, and

$$\begin{aligned} Z_G PS = -t &= RA - GAST - \lambda_G \\ &= RA - (UT\,1 + R_u - 12^h + EE) - \lambda_G, \end{aligned} \tag{4.91}$$

with the same notation as in (4.57). Then if ϕ_G, λ_G, and UT 1 are known, the geodetic azimuth of the star ($PZ_G S = A_S$) is given by (4.88). Multiple observations are corrected for curvature by (4.90) or (4.89). Correction to the CIO is still necessary.

To get the geodetic azimuth of R, we require the angle $RZ_G S$ ($= \theta_G$), which is added to A_S to give A_R. But the theodolite measures $RZ_A S$ ($= \theta_A$). From (2.5) we then have

$$\begin{aligned} \theta_G - \theta_A &= RZ_G S - RZ_A S = R_G S_G - R_A S_A \\ &= R_A R_G - S_A S_G \\ &= \tan h_R (\eta \cos A_R - \xi \sin A_R) - \tan h_S (\eta \cos A_S - \xi \sin A_S), \end{aligned} \tag{4.92}$$

where suffixes R and S refer to RM and star. If the RM is near the horizon, as will be usual, the first term will be small, and can probably be neglected. For its proper treatment see § 4.42 (g). Any approximate values of h and A will, of course, suffice.

If S also is on the horizon (4.92) = 0, and the geodetic azimuth is at once given by (4.88) using geodetic values of ϕ and λ as above. Otherwise three stars will give three equations for ξ, η, and the geodetic azimuth of the RM. If the stars' altitudes are low, ξ, and η will be weakly determined, and the azimuth will have maximum strength.

For a primary azimuth sixteen stars (on both faces) give a reasonable programme, from which ξ, η, and the azimuth are got by least squares as below. Altitudes should not be too low, for fear of lateral refraction, and 10° to 20° is a convenient range if azimuth without η or ξ is the primary requirement. Zero should be changed after each star.

The cross-level must be read before and after each star, and (4.83) must be applied to each reading of the horizontal circle. The time of passage across the centre of the vertical wire must be recorded by tappet and chronograph, and the UT is determined by wireless just as for longitude observations. Some laxity may be allowable near the equator where stars are moving nearly vertically, but elsewhere an error of $0^s.1$ in timing makes an error of the order of 1″ in azimuth. In medium or high latitudes it is convenient to have three or five vertical wires, and to record the time across each, or a moving wire micrometer may be used as in the T4 theodolite. The horizontal angle of the RM should be read before the star on the first face and after it on the second. The vertical angle of each star is wanted for (4.90) and (4.92). It should be recorded to within a few minutes, and also that of the RM.

As mentioned in §4.42 (*j*), the observation is not impersonal, and for the best work (except in low latitudes) a self-recording micrometer, as in the T4 theodolite, is necessary. Calibration by observing azimuth at the origin of the survey is a possible alternative. Note that the correction to an azimuth observation in lat ϕ will be $\delta A \sin\phi \operatorname{cosec}\phi_0$, where δA is the correction got by calibration in lat ϕ_0, [485].

This system of observing azimuth is especially accurate and convenient in high latitudes. The stars are seen comfortably without a diagonal eye-piece: no great accuracy is demanded of the cross-bubble: and the system avoids the loss of accuracy associated with the cutting down of an elevated point to the horizon.

The values obtained for azimuth, η, and ξ must be reduced to mean pole using (4.28), (4.27), and (4.25); the corrections to η and ξ being the same as those appropriate to $\lambda\cos\phi$ and ϕ respectively.†

The trouble in high latitudes where λ_G is only known with low accuracy, is dealt with in § 4.49.

Least squares. Each star gives an observation equation

$$A_R+\eta(\cos A_S\tan h_S-\cos A_R\tan h_R)-\xi(\sin A_S\tan h_S-\sin A_R\tan h_R)$$
$$=A_S+\theta_A,\quad (4.93)$$

in which the terms with suffix R (except the first term A_R itself) will generally be omitted.

High altitude stars are less reliably observed than low, and if h_S is high and variable it may be reasonable to weight the equations by multiplying them by $\cos h_S$, implying that the s.e. of A_S varies as $\sec h_S$. The s.e. of A_S will also depend on $\mathrm{d}A/\mathrm{d}t$, although not in direct proportion to it. We have

$$\mathrm{d}A/\mathrm{d}t=15(\sin\phi-\cos\phi\cos A_S\tan h_S).\qquad (4.94)$$

On the whole, unless h_S is very variable, all stars may reasonably be given equal weight.

For the strongest value of geodetic azimuth $\tan h_S$ should be small. If η and ξ are otherwise known, the best azimuth will then be got from stars with the smallest values of $\mathrm{d}A/\mathrm{d}t$, such as occur in the two quadrants nearest the elevated pole. But except in very low latitudes, $\mathrm{d}A/\mathrm{d}t$ at low altitudes will have the same sign for all stars.

If η and ξ are not otherwise known, their coefficients in (4.93), namely $\cos A_S\tan h_S$ and $\sin A_S\tan h_S$, must be of both signs, in order to

† So the correction to η is $-(x\cos\lambda+y\sin\lambda)\sin\phi$.

separate them from A_R in the solution, so the stars must be distributed through all four quadrants.

If the values of η and ξ are required for their own sakes tan h_S must not be too small. Altitudes of 35° or 40° make a reasonable compromise between their demands and that of azimuth. The stars observed should be equally distributed between the four quadrants, and the average altitudes should not differ between quadrants by more than (say) 5°.

Accuracy. A programme of 16 stars, evenly distributed as above and observed on FL and FR with a first-order theodolite, should give azimuth with an s.e. of about 0".7 sec h, and the same programme should give η and ξ with s.e.'s of 1".0 cot h, h being not more than 40°. Here h is an average value of h_S. These figures ignore personal equation.

Where these figures are not good enough, η and ξ must be determined by other methods, such as those of §§ 4.21 and 4.41, or 4.38–4.42. Black's method can then be used for azimuth with a minimum value of h.

[299] describes the careful observation of a Black azimuth using a T4 theodolite.

4.48. Summary

Methods recommended for azimuth are as follows.

(*a*) Black's method may be used in all latitudes, but it is especially advantageous in high latitudes. For the highest accuracy use low altitude stars, and determine η and ξ separately. It then becomes similar in principle to that of § 4.46.

(*b*) Between latitudes 10° and 35°, Polaris and σ Octantis are very convenient, especially if η is already known.

(*c*) Except in very high latitudes, the use of the transit telescope as in § 4.45, or of an astronomical theodolite in the same way, is acceptable if it is not inconvenient. Longitude and η will then, of course, be determined with the same instrument.

4.49. High latitudes

The computation of triangulation and traverse in very high latitudes presents various difficulties, which are considered below.

(*a*) The inequality of the second of latitude and of longitude. In latitude 89° 20', if ϕ is quoted to 0".01, the equivalent for λ is only the nearest 1". But this is no more than a nuisance.

(*b*) The computation formulae for latitudes, longitudes, and azimuths in § 2.14 are apt to be unstable in high latitudes. Any term including tan ϕ will cause trouble, other than a single power of tan ϕ in an expression for a difference of longitude. On the other hand, the reverse formulae of § 2.15 are stable (except azimuth in the USC & GS formula), so any possible computing difficulty can be got over by using the method of variation of coordinates. For Laplace azimuths see (*d*) below.

(*c*) Astronomical observations for azimuth by most methods become inaccurate. Thus, (i), close circum-polar stars are so high that errors of cross levelling are serious; (ii) the meridian, defined by pole and zenith is ill-defined when the two are close; and (iii) stars move nearly horizontally and their hour angles cannot be got from their altitudes. None of these difficulties affect Black's method, which can be used right up to the pole.

(*d*) A more fundamental difficulty is in the definition of azimuth as given in § 2.04, since the plane containing the pole and the vertical (or the normal) is ill defined when the pole and zenith are close. This definition may then be modified as follows.

The astronomical (or geodetic) azimuth from a point P to another point P_1 is the angle between two planes, both containing the vertical (or normal) at P, one of which contains the point P_1, and the other cuts the spheroid in a line which is locally parallel to a stated meridian plane λ_G, which approximates to the longitude of P.

In Fig. 4.16, the horizontal direction of PP_1 is defined by the two angles λ_G and A, irrespective of whether P is on the meridian λ_G or not. If λ_G is used in Black's method, errorless observations will give the angle A. If λ_G is not the longitude of P, η will be wrong, by the error of longitude multiplied by $\cos\phi$, as usual, a quantity which is not abnormally large in high latitudes.

The routine for forming Laplace equations in high latitudes when computing by variation of coordinates is then to assume trial values of ϕ_G and λ_G as usual, recording the trial λ_G to the full usual number of decimal places however close to the pole the point may be. Use this same λ_G when computing azimuth by Black's method. Use the geodetic azimuth given by Black for the 'Provisional Laplace azimuth' in the Laplace observation equation (2.37). The least-squares solution will then give a revised value of λ_G, for which the term

$$(b-\sin\phi)\delta\lambda$$

in (2.37) makes due allowance in the Laplace equation.

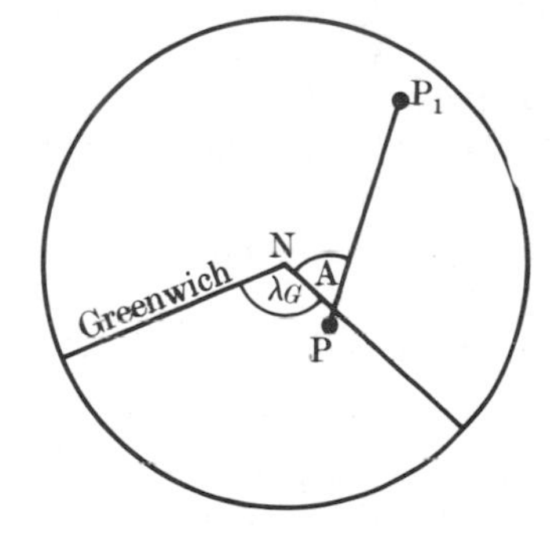

FIG. 4.16. The reference spheriod. N is geodetic latitude 90°.

General references for Sections 2, 3, 4, and 5. [100], [102], [441], [444], [492], and [533].

SECTION 6. GEOID SECTIONS

4.50. Geoid obtained by integration of ξ and η

Astronomical observations at a point fixed by triangulation or traverse give ξ and η, the two components of the deviation of the vertical, § 2.05. Then, if the deviation is known at all points along a line AB, the rise or fall of the geoid relative to the spheroid, between A and B, is given by

$$N_B - N_A = \int_A^B \psi \, dl, \qquad (4.95)\dagger$$

where $\psi = -(\eta \sin A + \xi \cos A)$, and is the component of the deviation in azimuth A along AB, positive if the geoid is rising towards the south and west, l = distance along AB, and N_A is the height of geoid above spheroid at A.

If A and B are close together and if ψ varies regularly, this may be written

$$N_B - N_A = \tfrac{1}{2}(\psi''_A + \psi''_B) l \sin 1'', \qquad (4.96A)\ddagger$$

or more conveniently

$$N_B - N_A = 0.90 \{\tfrac{1}{2}(\xi_A + \xi_B)\Delta\phi' + \tfrac{1}{2}(\eta_A + \eta_B)\Delta\lambda' \cos \phi\} \text{ cm}, \quad (4.96B)$$

where ξ and η are in seconds of arc, and $\Delta\phi'$ and $\Delta\lambda'$ are in minutes.

Except in mountainous country (4.96) is a good approximation if l is 15 or 30 km, see § 4.52 (a), and the form of the geoid can be accurately traced if the deviation is observed at such intervals along continuous section lines. An alternative is to observe at greater intervals, such as 100 or 200 km, and to construct the section with the help of a wide belt of gravity survey. See § 6.57 for this and other substitutes for closely spaced astronomical stations.

One of the aims of geodesy is to cover the land surface of the earth with a network of such sections, and thereby to get a direct measure of the form of the geoid.

4.51. Field routine

(a) *Geodetic fix.* This must be in terms of the highest class geodetic framework, but random error of 10 m at each station is harmless, and astronomical stations need not coincide with primary triangulation sta-

† (4.95) ignores changes in ϕ and λ and thence in ξ and η along the line, which may result from revised values of N. For the treatment of this, see § 2.40 (b).

‡ ψ_B, like ψ_A, must be reckoned in the direction AB not BA, and so must be calculated with the azimuth of B from A.

tions. They can be fixed by resection or any such expedient from the stations or intersected points of third-order triangulation, provided the latter is reliably and closely based on the primary. Resection from unvisited points generally involves some risk of blunder, and it is best done from at least five such points with a rough Polaris azimuth as well. Or a short base may be measured and the distance to one or two of the nearest points determined, as a further check. If there is a good cadastral map, which is known to be firmly in terms with the geodetic triangulation (a situation which is regrettably infrequent), the fix can of course be taken direct from it. The ideal line of section follows a motorable road through country where it is possible to get an easy geodetic fix correct to 10–15 metres every 15 or 25 km.

In mountains, formula (4.96) demands that stations should be sited either on mountain tops or in the middle of valley bottoms, where topographical anomalies will not be very large, or else symmetrically on the sides of ranges, so that the anomalies may be balanced. See §§ 4.52 and 6.59.

(*b*) *Astronomical fix.* The object is to get a standard error not much greater than $0''.7$ in latitude and in (Long)$\times \cos \phi$, free from systematic error of more than $0''.1$ or $0''.2$, or say $0^{s}.01$. In a meridional section it is only essential to observe ξ, although when the geoid chart is being drawn it is a pity not to have η as well, even if of rather lower accuracy, § 6.58 (*f*). Omitting η saves much time and equipment. In an east–west section no equipment and little time is saved by not observing ξ, and both components should be observed. In an oblique section both are necessary.

Suitable programmes are as below.

(i) Four pairs of circum-meridian latitudes and four pairs of east and west stars for time, each with 5 or 6 intersections, using a geodetic theodolite.

Or (ii) Sixteen quadrant stars at varying altitudes (§ 4.40) each with 5 or 6 intersections.

Or (iii) Sixteen to twenty-four quadrant stars at equal altitudes with a geodetic theodolite or preferably with an astrolabe: the higher figure if there is only one intersection on each.

Or (iv) A Black azimuth programme of 16 or 24 stars at an altitude of not less than 35°, § 4.47. This may be appropriate when the section is being observed along a line of primary traverse, where azimuth control is required, but it may be less accurate than (i), (ii), or (iii).

If an east–west section is being observed by methods (i) to (iii), two or three pairs of stars near the prime vertical can advantageously be added.

(*c*) *Personal equation.* In an east–west or oblique section personal equation demands serious attention. In the absence of effective and easily portable impersonal apparatus, recourse must be had to calibration as in § 4.29 (*f*). A reasonable programme is to observe four nights at a known longitude before the field programme and four at the end, with four more in the middle if the profile covers more than 750 km of easting (and two such sets for more than 1500 km, and so on). Such calibration must be carried out in field conditions, and the observer must follow his normal field living routine. Practice and training nights must not be included in the calibration programme.

Either the mean of 'before' and 'after' calibrations, or a smoothed set of values, may be applied to the field observations. One method is likely to be as correct as the other, and both will give the same over-all result in the section of the geoid concerned.

For calibration to be effective, the random error of observation must obviously be small. Also, but less certainly, calibration is likely to be more effective if the system of field observation is somewhat impersonal than if the personal equation which is to be eliminated is very large. In east–west sections it is consequently desirable to use the best available timing equipment, such as that described in § 4.31, even though a satisfactory standard (random) error may be obtained without it.

When a section has already been observed, without calibration or other elimination of personal equation, the only course is to reoccupy about a quarter of its stations with an instrument which (by calibration or otherwise) is impersonal. This system may be adopted in the first instance, see [482] where every fourth station of an east–west profile across the US was occupied by a Transit or T4 with impersonal micrometer, in addition to the T3 theodolite which was used at every station.

(*d*) *Rate of work.* With a motorable road, clear skies, and resection possible from camps on the road, thirty stations (600–800 km) can be observed in a month, but conditions are seldom ideal. Before leaving a station it is desirable to compute the geodetic fix, and to record the GST of all star intersections. This proves that everything is working well. Computation of t for each star can be left until later.

Some justification for the details of the recommended programme is given by (4.98), in which the errors due to the various main causes are all roughly equal.

(*e*) *Reduction to sea-level.* Formula (4.96) ignores the fact that ground-level observations of the angle between the vertical and the spheroidal normal have been assumed to be the same as the geoid-level values. This results in a systematic error and in a random one. The systematic error, which is in latitude only, arises from the progressive increase of gravity from equator to pole, and is removed by the application of (6.97). The effect is small, 5 metres in a geoidal profile from equator to pole at a height of 1000 m above sea-level, but (6.97) is easy to apply, and ξ should be corrected in this way before being used in (4.96).

The random error may be greater at individual stations, but it is largely self-cancelling. See § 6.55 (*b*) and (*c*) and the examples quoted at the end of 6.55 (*c*). It can be neglected. This source of error is included with random errors in the estimates of accuracy given in § 4.52.

This subject is continued in §§ 6.56–6.58.

4.52. Accuracy

The accuracy of geoidal profiles can be estimated as below. Details are in [111], [112], and [482]. L km is the total length of the section line (such as 4000 km) and l km is the interval between astronomical stations (such as 25 km).†

(*a*) *Interpolation error*, i.e. error in the assumption made in (4.96). This can be judged by comparing the observed value of the deviation at any station with the mean of the deviations at the two adjacent stations. This at once gives the error which would have been expected if the spacing had been $2l$, and it remains to find the power of l to which the error is proportional. From first principles, [111], the error in a total length L would be expected to vary as $l^2\sqrt{(lL)}$ in flat and geodetically featureless country in which the deviation varied quite smoothly from one station to the next. And at the other extreme, the error should vary as $\sqrt{(lL)}$ in sharply mountainous country in which the deviations at adjacent stations might be almost uncorrelated. Analysis of the Alpine section of [440], in which $l = 3\frac{1}{2}$ km, gives the accuracy for 7-, 14-, and 28-km spacing and confirms that error in L km varies as $\sqrt{(lL)}$. Indian (non-Himalayan) data suggest the first power of l. [445] investigates the problem using the closure of geoidal height round triangles in Finland, also using the first power. Results for the s.e. of geoidal height after L km can be summarized as:

$$\begin{aligned} \text{Alps} \quad & 0.012\sqrt{(lL)} \text{ metres,‡} \\ \text{India} \quad & 0.00052 l\sqrt{L} \text{ metres,} \\ \text{Finland} \quad & 0.00036 l\sqrt{L} \text{ metres.} \end{aligned} \tag{4.97}$$

These figures necessarily include the random error in the astronomical observations and geodetic fixes, and in the assumption that ground-level

† In this paragraph, as elsewhere, the figures given for the s.e.'s are such that the probability against their being exceeded is thought to be 68 : 32. Quoted references give probable errors (50 : 50), and these have been multiplied by $1\frac{1}{2}$.

‡ $0.008 l\sqrt{L}$ in [112] is a misprint for $0.008\sqrt{(lL)}$; p.e.'s.

deviations equal those at geoid level. They assume also that l is reasonably constant and between (say) 15 and 40 km.

[357] confirms (4.97).

(*b*) *Personal equation in longitude.* Suppose that adjacent sections 1000 km long are independently calibrated by observations made before and after the field work, and that the mean of the two calibrations is applied as a correction to all the stations of the section concerned, see § 4.51 (*c*). As an example, suppose that in a total distance of 4000 km successive calibration corrections of one observer (in units of $0^s.01$) were +8, +6, +7, and +9. Then a correction of +7.5 could be applied with confidence and with the expectation that no separate 1000 km would be wrong by more than (say) 2. But such clear corrections are not to be expected. A more likely set might be +2, +5, +3, and −2, whose mean is $+0^s.02$ with a conventional s.e. of $\pm 0^s.03$ for each. The significance of the $\pm 0^s.03$ is to suggest that personal equation is at any time liable to depart from its usual value by that amount, and (taking an unfavourable view) it may be assumed that the actual mean value of personal equation over any 1000-km section may depart from the value determined by any one calibration by $\pm 0^s.03$. Having accepted this, we may then at least expect that the remaining errors in each 1000 km are independent. Indian experience suggests that $\pm 0^s.03$ is a reasonable s.e. for a single calibration, or $\pm 0^s.020$ for the mean of two. The resulting s.e. of geoidal height after L km (when L is several thousand) from this cause in an east–west section is therefore $\pm 1.5\sqrt{(L/1000)}$ m.

(*c*) *Systematic error in geodetic position.* Chapter 2, Section 6, gives formulae for error in (horizontal) geodetic position. Based on this, [111] finds a s.e. of geoidal height of ± 2.5 m after 4000 km, or (say) $\pm 1.3\sqrt{(L/1000)}$ m, for work based on a first-class triangulation system.

(*d*) *Summary.* The s.e. of geoidal height after 4000 km with 25-km station spacing is then estimated as:

(i) Interpolation and other random error	± 1.5 m in typical country.	
(ii) Personal equation	± 3.0 m if east–west.	
(iii) Triangulation	± 2.5 m.†	
Total	± 4.2 m.	(4.98)

For comparison, [482] gives estimates of (i) = $\pm 1.0 \pm 1.3$ or ± 1.7 m, (ii) = ± 2.2 m, and (iii) = ± 2.9 m; total ± 4.0 m. Both estimates then agree on

$$\pm 2.0\sqrt{(L/1000)} \text{ m for east–west lines,}$$

or

$$\pm 1.5\sqrt{(L/1000)} \text{ m for north–south.} \quad (4.99)$$

† This figure is for a straight 4000 km section. In a 4000 km closed circuit the error should be much less.

These figures are in accord with the closing errors of nine Indian circuits quoted in [111] of lengths between 900 and 3200 km.

(4.99) is an empirical rule. When L is less than (say) 2000 km, item (i) which varies as $L^{\frac{1}{2}}$ is dominant. Over greater distances, such as 4000 km, item (iii) which varies as $L^{\frac{3}{2}}$ or L^2 is dominant. The numerical factor in (4.99) may be expected to give reasonable results for values of L between 500 and 4000 km, but it should not be used for greater distances.

4.53. Deviation of the vertical from azimuths

Formula (2.3) gives the east–west component of the deviation if the astronomical azimuth of a line of known geodetic azimuth is observed, but the method is seldom reliable. In low latitudes the factor $\cot\phi$ is adverse, and although it is less than unity in high latitudes, it does no more than compensate for the weakness of azimuth observations there. Further, unless triangulation or traverse has been satisfactorily adjusted on to frequent Laplace azimuths, its azimuth errors are liable to be a few seconds, which (especially after multiplication by $\cot\phi$ in low latitudes) is not negligible. Possible movement of station marks is also a source of serious error if the astronomical and geodetic observations are much separated in time.

Between (say) latitudes 40° and 65°, useful values of η may possibly be obtainable in this way, but it is not recommended, and if unexpected values of η result, they should be looked on with suspicion.

The subject of geoid sections, their combination and adjustment, and their control by other types of observations such as doppler satellite fixes, is continued in §§ 6.56–6.59.

Section 7. Gyro-theodolites

4.54. Introductory

(*a*) *Uses.* Gyro-theodolites provide a means of obtaining azimuth without having a clear view of the sky. They are of special use in mines and tunnels, and for military or other rapid surveys, especially where the sky is liable to be clouded. At present (1978) gyro-theodolites have not attained geodetic accuracy, but progress is being made, and geodetic accuracy may presently be obtainable. Except near the equator. It will, however, always be necessary to convert the astronomical azimuth, § 2.4, which is given by the gyro, to geodetic azimuth by some form of observation of the east–west component η of the deviation of the vertical. Gyro-theodolites cannot be expected to work accurately in high latitudes, since the precessional torque varies as $\cos\phi$, and is much reduced above (say) 75°.

The following §§ 4.55–4.62 describe the basic principles of the design and use of the original Wild GAK 1. § 4.63 outlines the directions in which progress has been made or attempted.

(*b*) *Working principles.* A rapidly rotating *spinner* is constrained to lie with its spin axis in a horizontal plane, but free to rotate about the vertical. The whole apparatus, and the direction of the vertical, rotates daily about the earth's axis. The spinner is then urged by precession to lay itself parallel to the earth's axis, but since it cannot rise above the horizontal, its urge is towards the north point, namely the horizontal line in the plane which contains the vertical and the celestial pole. Azimuths indicated by gyro-theodolites are consequently astronomical azimuths, as defined in § 2.04, except for the trivial difference that they refer to the instantaneous pole rather than to the mean pole.

(*c*) The following notation is used in this section.

- α Azimuth. Positive east of north.
- $\ddot{\alpha}$ Angular acceleration about the vertical.
- α_0 Azimuth of the V-notch, Fig. 4.17 (b).
- α_c Azimuth of spinner at mid-swing.
- ϕ Latitude.
- A, B Moments of inertia about the vertical and spin axes respectively.
- a Amplitude of swing, in arc.
- a' Amplitude of swing in scale divisions.
- m Value of one scale division. $a = a'm$.
- ω Angular velocity of spinner about its axle.
- Ω Angular velocity of the earth's rotation $\approx 2\pi/86\,164$ radians per second.
- T_U Period of swing. Spinner spinning. Theodolite following spinner.
- T_C Period of swing. Spinner spinning. Theodolite clamped.
- M_E $= B\omega\Omega \cos\phi$. $M_E \sin\alpha =$ torque due to precession.
- M_T = (Torque due to tape) ÷ (twist in tape).
- ϵ Angle between scale zero and position of zero twist. Positive east.
- t_0, t_1, etc. Recorded times.
- I Index correction, § 4.58.
- S Theodolite circle reading of mean swing. Theodolite following spinner.
- R Theodolite circle reading. Theodolite clamped.
- c $= 2\pi m T_U^2/T_C^3$, (4.113).
- δ_1 to δ_4 and α_1 to α_4 are scale readings of reversal points.
- β, δ See (4.116) and (4.115).
- C $= M_T/M_E$.

4.55. GAK 1 gyro-theodolite. Construction

See Fig. 4.17(a) and (b). The gyro attachment is mounted above an ordinary Wild T16 or T_2 theodolite. The spinner is suspended by a fine tape (2) from a column (3) which moves with the upper part of the theodolite. A mark (4) on the spinner mounting can be seen through the viewer (5), as enlarged in Fig. 4.17 (b). The axis of the spinner is designed to lie in the same meridional plane as the line of sight of the theodolite

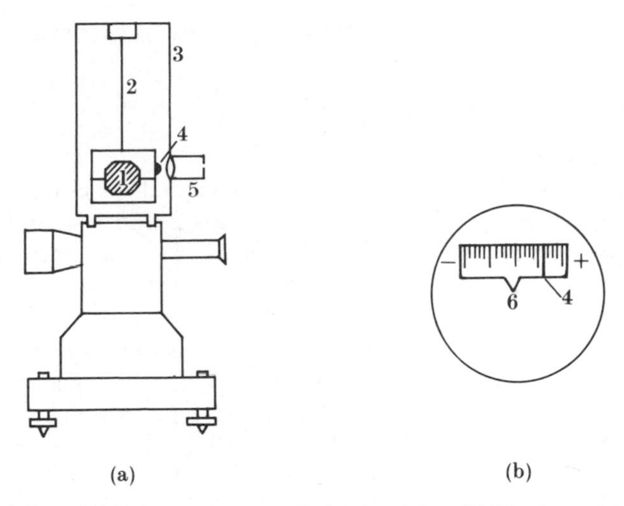

FIG. 4.17. (a) Gyro GAK 1 mounted on a T 16 theodolite. (b) The figure is drawn on the assumption that when the spinner points east of north, the mark (4) lies to the right of the V, positive on the scale. If the construction is otherwise, signs in the formulae may change.

when the mark (4) coincides with the V-notch (6) in the viewer, the zero of the scale. But see § 4.58.

The mounting of the spinner is attached to the bottom of the tape at a point above its mass centre, so the spinner axis is held substantially horizontal by gravity, and its only freedom is to rotate about a vertical axis.

4.56. Precession

If the spinner lies horizontal in azimuth α, the component of the earth's diurnal rotation (Ω) about the horizontal in azimuth $\alpha+90°$ is $\Omega\cos\phi\sin\alpha$, and this causes the spinner to precess about the vertical with angular acceleration towards the meridian of

$$\begin{aligned}\ddot{\alpha} &= -(B/A)\omega\Omega\cos\phi\sin\alpha \\ &= -(M_E/A)\sin\alpha,\end{aligned} \tag{4.100}$$

where $M_E = B\omega\Omega\cos\phi$, the gyroscopic torque when $\alpha = 90°$,† and B and A are the moments of inertia about the spin axis and vertical respectively. See [549], p. 399. The spinner will then swing to and fro about the meridian. If the azimuth range is small so that $\sin\alpha = \alpha$, and neglecting

† In the usual working position the spinner axis lies near north, in which position the northward urging torque $B\omega\Omega\cos\phi\sin\alpha$ is near zero, but where its rate of change with azimuth is greatest. The forced rotation is about a near east–west axis, which is called the *input axis*, and the precession is about the free vertical axis, which is called the *output axis*.

friction and torque in the suspending tape, this will be simple harmonic motion with a period of

$$T_U = 2\pi\surd(A/M_E), \tag{4.101}$$

which varies with latitude as $(\sec\phi)^{\frac{1}{2}}$.

Typically, ω may be $2\pi\times 370$ radians per second (22 000 rev/min in the GAK 1), kept constant by crystal control, Ω is $2\pi/86\,164$, and the period of the swing will be about $7(\sec\phi)^{\frac{1}{2}}$ minutes. For any particular instrument the equatorial period can be determined, and a table can be prepared giving T_U in all latitudes.

The levelling of the spinner axis is independent of the foot-screw levelling of the theodolite base, and slow changes in the base levelling, such as may happen on unstable ground will introduce no unusual precession. See § 4.63 (*c*).

4.57. Suspension torque

The torque in the suspension tape is far from negligible. At the upper end of the tape, adjustment is provided so that the torque may be approximately zero when the mark (4), Fig. 4.17 (b), is centred in the V. This may be adjusted by equating the eastward and westward swings of the spinner when it is not spinning, but there must always be some small residual error.

Fig. 4.18 shows the V zero of the viewer scale lying in azimuth α_0 east of true (astro) north, α_0 being a small unknown angle such as 10′ or 20′. Let the tape torsion be zero when the spinner points at T, at a small angle ϵ east of V. Let the theodolite be clamped so that α_0 and ϵ are constant. Let the torque due to the tape be $(M_T)\times$(angle between spinner and T).

Then if the swinging spinner lies in azimuth α, its angular acceleration, instead of being given by (4.100) will be given by

$$\begin{aligned} -A\ddot{\alpha} &= M_E\alpha + M_T(\alpha-\alpha_0-\epsilon) \\ &= M_E\{\alpha(1+M_T/M_E)-(M_T/M_E)(\alpha_0+\epsilon)\}. \end{aligned} \tag{4.102}$$

Typically M_T/M_E may be $\frac{1}{4}\sec\phi$, while $(\alpha_0+\epsilon)$ may be about 1/150 rad.

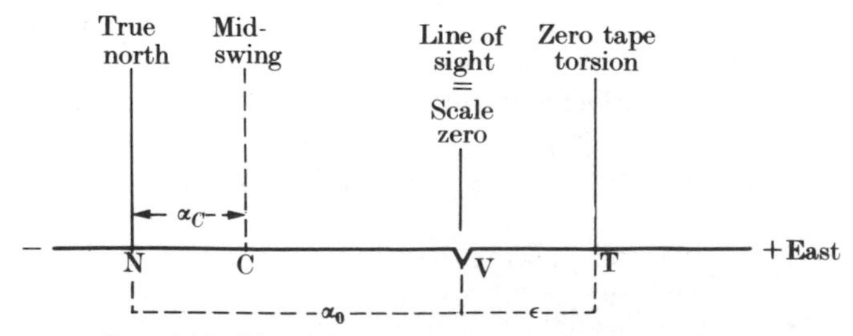

FIG. 4.18. Theodolite clamped, and spinner swinging.

Let T_C be the period of the swing. Then

$$T_C \approx 2\pi\sqrt{\left(\frac{A}{M_E}\right)}\sqrt{\left\{\frac{1}{1+(M_T/M_E)}\right\}}, \tag{4.103}$$

whence, using (4.101),

$$1+(M_T/M_E) = T_U^2/T_C^2. \tag{4.104}$$

The ratio M_T/M_E can thus be determined, since T_C is the period when the spinner swings normally, affected by both precession and tape torsion with the theodolite clamped, while T_U† is the period when the theodolite is unclamped and torsion is eliminated by the procedure described in § 4.59. For an alternative to (4.104) see (4.119).

The mid-point of the swing (the mean of the two extremes, if friction is ignored) will not be at N ($\alpha = 0$) in Fig. 4.18, nor at V ($\alpha = \alpha_0$), but at a point C where $\ddot{\alpha} = 0$. So, from (4.102) α_c is given by

$$\alpha_c(1+M_T/M_E)-(M_T/M_E)(\alpha_0+\epsilon) = 0, \tag{4.105}$$

whence NC in Fig. 4.18 $= \alpha_c = \dfrac{M_T}{M_E+M_T}(\alpha_0+\epsilon)$, (4.106)

$$\text{CT} = \frac{M_E}{M_E+M_T}(\alpha_0+\epsilon), \tag{4.107}$$

and

$$\text{CV} = \text{CT}-\epsilon = \frac{M_E\alpha_0}{M_E+M_T}-\frac{M_T\,\epsilon}{M_E+M_T}. \tag{4.108}$$

The angle ϵ can be determined by letting the spinner swing through a small angle, when it is not spinning, and noting the scale readings at the end of its swings, and taking a mean as in (4.109) below.

4.58. Index error

As in § 4.55, the line of sight of the theodolite (mean of FR and FL, free of collimation) is supposed to be in the same vertical plane as the spinner axis when the V-mark reads zero in Fig. 4.17 (b). Exact adjustment is not possible, and an *index correction*, *I*, must be determined and periodically checked by observing on lines of known astronomical azimuth. Apart from accidents, this correction should be constant.

There are four ways of determining the theodolite circle reading of astronomical north, which has to be subtracted from the FL and FR mean

† M_T/M_E is required for a small correction in (4.110) and (4.114). It is given with sufficient accuracy by a value of T_U determined as in § 4.59, but without the great care which the tracking method itself demands.

reading of the RM to give the latter's azimuth. These are described in §§ 4.59–4.62.

4.59. The tracking or turning point or reversal method

The theodolite is set to a few degrees off north, as may be got within a degree by magnetic compass or otherwise, and the spinner (spinning) is released. As it starts to swing, the theodolite is made to follow it by turning the upper horizontal slow-motion screw, so that the mark (4) remains on the zero of the scale in Fig. 4.17 (b). Tape torsion is then zero, provided $\epsilon = 0$. The theodolite circle is read at the end of each swing, and if there was no damping the mean of pairs of east and west swing readings would be the circle reading of true north. Allowing for damping, if $E_1 W_1 E_2 W_2$, etc., are successive circle readings, values for north are given by

$$\tfrac{1}{4}(E_1+2W_1+E_2), \qquad \tfrac{1}{4}(W_1+2E_2+W_2), \quad \text{etc.} \qquad (4.109)\dagger$$

These are known as *Schuler means.* They should be constant. Let their mean be S. Then the circle reading of astro north is

$$S+I-\epsilon(M_T/M_E), \qquad (4.110)$$

where I is the index correction from § 4.58 and $\epsilon(M_T/M_E)$ is the correction due to tape torque not being zero when the mark (4) is central. See Fig. 4.19. If ϵ is small, $\epsilon(M_T/M_E)$ will be constant over some degrees of latitude, and the correction can probably be treated as part of the index correction. If its value is required, it is given by (4.104).

This method can give good results, but the careful following with the slow motion screw calls for skill and concentration. The theodolite may also have to be fitted with a special long-threaded screw.

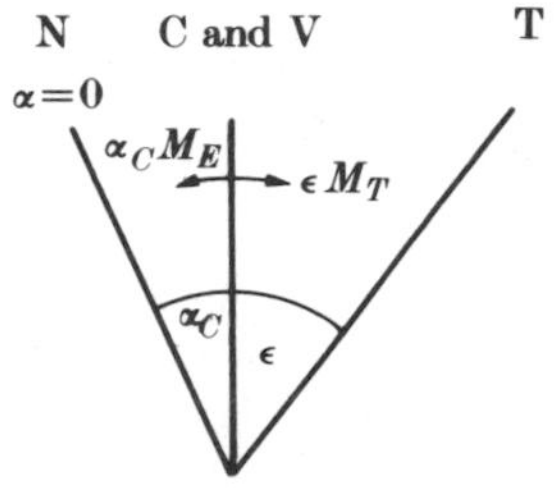

FIG. 4.19. Spinner swinging with mark held in V. N = north. C = mark when spinner is at mid-swing with total torque zero. T = tape torque zero. When total torque = 0, $\epsilon M_T = \alpha_c M_E$, and $\alpha_c = \epsilon M_T/M_E$.

4.60. Rapid approximate method

Let the spinner (spinning) be set swinging through a few degrees either side of north. Then if it is at the end of a swing at time t_0, and if the theodolite is turned as in § 4.59 with the mark held in the V until time $t_0+\tfrac{1}{4}T_U$, it will at that time be at mid-swing, C in Fig. 4.19, and after correction for index error the theodolite will be pointing close to north, since ϵ and α_c are small.

† See [606] for a discussion of the accuracy of these formulae.

In practice t_0 is hard to judge, but the following gets over the difficulty. As the theodolite is nearing the end of a swing, follow it with the mark held in the V, and stop following at a recorded time t_1 about 20 or 40 seconds before the expected end of the swing. The mark will then swing away from the V up to a maximum. When it returns to the V at time t_2 the holding of the mark in the V is begun again. Then $t_0 = \frac{1}{2}(t_1 + t_2)$.

This method will give a value of north correct to about 10′ or 15′, which is a sufficient approximation for starting the transit method of § 4.61. If the slow-motion screw is not long enough, it suffices for this purpose if the theodolite is carefully turned by hand without it.

4.61. The transit method

By the rapid method of § 4.60 (ignoring index error) the scale zero V is set to within 10′ or 15′ of north, and the spinner is set swinging through an arc of about 2° either way, with the theodolite clamped so that it swings under the influence of both precession and tape torsion. Let the mark moving east pass the V at time t_1, then moving west at time t_2, and again east at t_3. If $t_2 - t_1 = t_3 - t_2$, so that the east and west swings take equal times, the position of the mid-swing (C in Fig. 4.18) will be on the V. But there is likely to be a setting error, $\mathrm{NV} = \alpha_0$, of up to 20′, and C will not coincide with V.

Then if Δt is the time taken to swing from V to C,†

$$(t_2 + \Delta t) - (t_1 - \Delta t) = (t_3 - \Delta t) - (t_2 + \Delta t),$$

$$\Delta t = \tfrac{1}{4}(t_3 - 2t_2 + t_1). \tag{4.111}$$

If the amplitude of the swing is a degrees, the rate of change of α near mid-swing is $\dfrac{\pi}{2}\dfrac{a}{T_C/4}$,

so
$$\mathrm{CV} = \frac{\pi}{2}\frac{a}{T_C/4}\Delta t = 2\pi m a' \frac{\Delta t}{T_C}, \tag{4.112}$$

where a' is the amplitude (mean of east and west) in scale divisions, and m is the angular value of one division of the scale.

Then, from (4.108),

$$\begin{aligned}\alpha_0(=\mathrm{NV\ in\ Fig.\ 4.18}) &= \left(\frac{M_T + M_E}{M_E} \times \mathrm{CV}\right) + \epsilon\frac{M_T}{M_E}\\ &= \left(1 + \frac{M_T}{M_E}\right) 2\pi m a' \frac{\Delta t}{T_C} + \epsilon \frac{M_T}{M_E}\\ &= \frac{T_U^2}{T_C^3} 2\pi m a' \Delta t + \epsilon \frac{M_T}{M_E}, \quad \text{using (4.104)}\\ &= c a' \Delta t + \epsilon (M_T/M_E),\end{aligned} \tag{4.113}$$

† Here Δt is one quarter of Δt as defined in [544].

where $c = 2\pi m T_U^2/T_C^3$, which can be tabulated for different latitudes, and which actually varies very slowly between latitudes 0 and 75°, [544], p. 35.

Then the circle reading of astro north is

$$R + I + ca'(\Delta t) + \epsilon(M_T/M_E), \tag{4.114}$$

where R is the reading at which the theodolite is clamped, and I is the index correction from § 4.58. Note that Δt is negative when east swings take longer and C in Fig. 4.18 is east of V.

4.62. The amplitude method

This is the simplest method. It was not included among the methods originally recommended for GAK 1, as it depends on circle readings and the circle of the T16 can only be read to 1′, with estimation to 0′.1. Nevertheless, there is now some tendency to regard it as the best method, especially if the gyro is mounted on a larger theodolite.

The telescope is set within about 1° of north, and is clamped. The circle reading R is recorded. To determine the tape zero, the gyro is made to oscillate, with the spinner at rest, and the scale readings of four reversal points δ_1 to δ_4 are recorded.

Let
$$\delta = m(\delta_1 + 3\delta_2 + 3\delta_3 + \delta_4)/8. \tag{4.115}$$

The spinner is then started, and the gyro is again made to oscillate. When all is steady, four reversal points α_1 to α_4 are recorded, and

$$\beta = m(\alpha_1 + 3\alpha_2 + 3\alpha_3 + \alpha_4)/8. \tag{4.116}$$

Let
$$w = \beta(1 + C) - C\delta \tag{4.117}$$

where $C = M_T/M_E$, as given by (4.104).

$$\text{Then the circle reading of north} = R + w. \tag{4.118}$$

This programme is then repeated by changing the telescope reading R to R', with the gyro still suspended and the spinner spinning, where $R' = R \pm 2w$, the sign being such that R and R' are on opposite sides of north. The tape zero is redetermined, and reversal points α_1' to α_4' are recorded. δ', β', and w' result and give the new circle reading of north as $R' \mp w'$, which should of course be substantially the same as $R + w$. Some sources of error are cancelled in the mean.

An alternative formula for C, instead of (4.104) is

$$(1 + C) = (R' - R)/(\beta - \beta'), \tag{4.119}$$

which in the present context is more convenient than (4.104).

4.63. Subsequent improvements

The original GAK 1 mounted on a T16 theodolite was intended to have a standard error of about ±30″, and has generally been found to be better, with an s.e. of ±20″.

(a) The GAK 1 and T16 combination has been experimentally modified by the simple inclusion of a micrometer for interpolation between the circle divisions, which reads to 0′.1 more confidently than can be read from the T16's own reading scale, [537]. This is reported to have reduced the s.e. as given by the amplitude method to ±7″. With the GAK 1 mounted on a T2 theodolite ±4″ is reported.

(*b*) *In Canada* considerable development has taken place, [239], [241], [240], and [250].

(i) A gyro such as the GAK 1 has been mounted on a more accurate theodolite such as a Hungarian MOM Gi-B2 or a Kern DKM 3. This clearly gives expectations of better results if the amplitude method is being used.

(ii) An automatic electronic tracking device has been incorporated, which eliminates observer fatigue and reduces the handling of the instrument, when the tracking method is used. See [250] for a description.

(iii) New electronics for driving the spinner at even speed and temperature.

(iv) Automatic read-out.

(v) Careful specifications for the observing procedure and the rigidity of the support. The preliminary determination of true north, NV in Fig. 4.18, should be correct to 20–30″, and the point of no torsion (T in Fig. 4.18) should coincide within one division of the reading scale, [250], p. 220. These precautions eliminate drift in the apparent position of true north. It is also recommended that it is not useful to observe more than 2 or 3 full swings, 5–6 transits. It is better to make more numerous independent sets of swings, if required.

As a result of these improvements an internal s.e. of ±3″ or less is claimed for a single determination, in middle latitudes, [250].

(*c*) *Precision indicator of the meridian* (*PIM*). This has been a device acting on a different principle, see [113], 3 Edn, p. 444, and [437]. It is more elaborate than the GAK 1, but tests have shown it to be less reliable. The levelling of the spinner is not directly secured by the direction of gravity, but by the levelling of the instrument through the foot-screws. A progressive dislevelment due to temperature change or wet ground, amounting to as little as 4″ per hour, introduces a precessional torque which will correspond to an azimuth error of 15″ sec ϕ. Such a rate of change of level cannot be regarded as unexpected. Much larger errors may clearly be feared, and in fact have occurred from this or other (unknown) causes.

General references for gyro-theodolites. [544], [559], [250], and [157].

5

GRAVITY OBSERVATIONS

SECTION 1. INTRODUCTORY, AND ABSOLUTE VALUES OF GRAVITY

5.00. General principles

It is well known that two bodies A and B of mass m and M, whose centres of mass are separated by a distance r which is large compared with their own dimensions, attract each other with a force proportional to mM/r^2, so that the acceleration of A is proportional to M/r^2. This holds also even if B is large, provided it is spherical and of radius $<r$. If the earth was a sphere of uniform density, the acceleration towards it of small bodies on its surface would everywhere be the same, but its rotation and spheroidal form, the visible mountains and oceans, and the less obvious variations of density, combine to produce variations in g which are a fruitful source of study. The principal variation is with latitude between about 9.83 and 9.78 m s^{-2} (983 and 978 gals)† from pole to equator, and the next is with height, a decrease of about 10 μm s^{-2} (1 mGal) per 3 metres above sea-level. In addition there are random variations, which may amount 2000 or 3000 μm s^{-2} (200 or 300 mGal), but which are generally much smaller.

In geodesy gravity g means the vector sum of the acceleration due to the attraction of the earth, and that due to the centrifugal force arising from the use of rotating axes of coordinates. See § 6.08.

The value of g can be directly measured by the instruments described in § 5.04. These are known as *absolute observations*, and can currently be made correct to 0.1 μm s^{-2} (0.01 mGal) or better. The relative difference of g between one point and another on the earth's surface can be measured by *pendulum observations* which are described in §§ 5.06–5.11, or by *gravimeters* §§ 5.12–5.21. The latter are also used at sea, and with much loss of accuracy in aircraft, § 5.20. A generalized record of the variations of on the earth's surface is also obtainable from the orbits of artificial satellites, §§ 6.26 and 7.79.

Since about 1950 the great majority of gravity observations have been made by gravimeters, and the use of the pendulum has mostly been confined to the provision of national or regional base stations, and of the

† See 5.01.

calibration lines which are required for the effective use of gravimeters, §§ 5.13 and 5.05. As the mobility of absolute instruments increases, the use of the pendulum is likely to decrease further.

5.01. Units

Traditionally g has been measured in gals and milligals, where 1 gal = 1 cm s^{-2}. The word gal is derived from 'Galileo' but it is not an abbreviation, and it has been spelt with a small g. This has been a very convenient unit.

With the advent of the SI system the symbol for the gal, but not the name itself, has been given a capital letter, e.g. 2 Gal or 3 mGal. This is easily adopted. More seriously, the gal equals 0.01 m s^{-2} and is consequently not an SI unit†. The Gal and the mGal are on the list of symbols whose use it is hoped to abandon as quickly as possible, [53], p. 27. Like all change this involves temporary inconvenience, during the period when all records and recent literature are in mGals, while the SI μm s^{-2} (=0.1 mGal) will have been adopted for current uses. At the expense of some clumsiness, SI units are used here duplicated by Gals and mGals for readers who have not yet become accustomed to the new system.

The abbreviation GU (Gravity unit) has for some time been used for 0.1 mGal or 1 μm s^{-2}, especially in connection with observations whose expected errors are a few GU or less. It has not been accepted as the name of an SI unit, although both for speech and typewriting it has notable advantage over the μm s^{-2}.

5.02. Objects

The uses of gravity observations are as follows.

(*a*) To throw light on the constitution and strength of the earth, especially in the crust and upper mantle. Chapter 6, Section 11. This includes the prospecting for oil and other minerals.

(*b*) The prediction of artificial satellite orbits. Chapter 7.

(*c*) For use in Stokes's integral, Chapter 6, Section 6.

(*d*) For the computation of spirit-levelled heights in geopotential units, §§ 3.00 and 3.02.

(*e*) To measure earth tides and other crustal movements. Chapter 6, Sections 13 and 12. Also changes in mean sea-level, see § 6.75 (*b*).

(*f*) To detect possible changes in the gravitational constant, § 6.06. See [421], pp. 111–3.

For items (*b*), (*c*), and (*d*) and for general studies of the earth's interior, random errors of 20 μm s^{-2} (2 mGal) are harmless.

† cm s^{-2} is a legitimate SI symbol, but decimal multipliers of it, such as m cm s^{-2}, are not.

For local geophysical prospecting the accuracy needs to be 1 $\mu m\, s^{-2}$ (0.1 mGal) or better.

For (*e*) and (*f*) the greatest currently possible accuracy is required, such as 0.1 $\mu m\, s^{-2}$ (0.01 mGal) or better.

5.03. Records. Storage and retrieval of data

By 1978 gravity had been observed at at least one million stations [381], and the storage and retrieval of their details is a serious and increasing difficulty.

(*a*) *Records.* The record of a gravity observation needs to include the following.

(i) The deduced value of g.

(ii) The instrument used, the observer, and a reference to some publication giving full details.

(iii) The date, the latitude, longitude, and the height of the appropriate point in the apparatus above the geoid. The spheroid and datum used. The source of this information, if high accuracy is required in difficult places. See § 5.11.

(iv) In the case of a base or calibration station, which may later be revisited, a permanent mark must be made and described so that the exact site can be recovered. The height of the relevant part of the instrument above the mark must be recorded to the nearest cm.

(v) The base station from which differences purport to be recorded, and the value of g accepted for it. See also § 5.05 and § 6.27 (*b*). The reference should if possible be to stations of the IGSN.

(vi) Mention of the inclusion or exclusion of unusual small corrections, such as tidal, § 5.07 (*k*), or secular change in unstable localities, if relevant.

(vii) A statement of weather conditions, especially at sea, or of other factors affecting accuracy.

(viii) Internal and external estimates of accuracy, including some statement of the programme followed, the misclosures revealed by returns to base or other fixed stations, and the method of their adjustment. Scaling and drift of gravimeters.

(ix) If possible, the orographical correction O of § 6.41, and the values of any gravity anomalies which have been computed, with details of the systems adopted. Local geological information, if relevant.

(*b*) *Methods of distributing recorded information.* Of the items listed in (*a*) above, (ii), (v), (vi), and (viii) will probably be common to a large number of stations, and can be included in a separate report which can be published in the ordinary way. Such a report can also contain all the other items, if they are of sufficient local interest to justify it. Nevertheless, the recording of even the most basic facts about a million stations, in a form in which they can easily be retrieved, is a formidable matter. Possible methods are listed below.

(i) Many gravimeter observations are made for local geophysical exploration over limited areas. If all details are contained in a separate report, a world data bank may be content to record the area and mean height of the survey, and the mean value of $g-\gamma_A$ (§ 6.05), with a reference to the source of full details.

(ii) Data concerning important base stations, and national or international calibration bases can be published in full on the lines of [420], with further details such as descriptions of marks available on demand.

(iii) Contoured charts of gravity anomalies, on a scale of between 1/200 000 and 1/2M, according to the density of the stations, can present a great deal of information in convenient form. At sea, the contours can show $g-\gamma_A$, but on land (where $g-\gamma_A$ varies rapidly with height), $g-\gamma_B$ § 6.41, or an isostatic anomaly $g-\gamma_C$, are better. $g-\gamma_B$ is now most commonly used, § 6.60 (*c*), as it shows the local variations of abnormal density well, and is easy to compute.

(iv) Compilations of the mean values of $g-\gamma_A$ in areas of e.g. 1° Lat by 1° Long are useful for some purposes. See § 6.60 (*c*).

(vi) A spherical harmonic analysis of $g-\gamma_A$ can give a very generalized picture of the world-wide distribution. Such analyses are based on observations of satellite orbits, although information from surface observations may be included.

(vi) Finally, a national, continental, or world *Data Bank* can be prepared on magnetic tape from which information can be retrieved by computer. Detailed proposals are in [381] and [134].

(*c*) *Existing institutions.* The IAG Bureau Gravimètrique in Paris, and the US Defense Mapping Agency Aerospace Center (DMAAC) keep large compilations of gravity charts and data. Lists and charts giving details of new observations should be sent to both. In particular full details of observations at local base stations and at the stations of calibration lines § 5.05, should be sent to the IAG Bureau.

5.04. Absolute gravity observations

(*a*) *Reversible pendulums.* Originally, absolute values of g were measured by the Kater reversible pendulum, which has knife-edge supports near each end, so that it can be swung with either end downwards. Then if the pendulum is so adjusted, or constructed, that the period T is the same for both suspensions, and if l is the distance between the two knife edges, g is related to l and T by (5.5). There are, of course, practical difficulties, such as measuring to 1 ppm (or even 10 ppm) the distance between the two points about which the pendulum actually swings.

The value of g at Potsdam as given by this method in 1906 was internationally accepted as the basis of all gravity determinations until 1967, although it was becoming increasingly clear that it was in error by about 140 $\mu m\ s^{-2}$ (14.0 mGal), as has now been accepted, see §§ 5.05 and 6.27 (*b*). Reversible pendulums are no longer used.

[164] summarizes work done between 1938 and 1967.

(*b*) *The free fall method.* In principle this is the simplest of the modern methods, although it was not the first to be adopted. A small body is released and allowed to fall in a high vacuum. The times t_1, t_2, and t_3 at which it passes three fixed points on its line of fall are recorded, and the distances l_1 and l_2 between the fixed points are measured. Then

$$l_1 = u_0(t_2 - t_1) + \tfrac{1}{2}g(t_2 - t_1)^2$$

$$l_1 + l_2 = u_0(t_3 - t_1) + \tfrac{1}{2}g(t_3 - t_1)^2, \qquad (5.1)\dagger$$

where u_0 is the unknown velocity when the body passes the first fixed point. The equations give u_0 and g. No knowledge is required of the instant at which the body is released, nor of its initial position.

In detail the instrument is naturally somewhat elaborate. The Faller–Hammond model, which is described in [200] on which the following description is mostly based, is about 1 metre high, and weighs 300 kg. It is portable and should give g with an accuracy of 0.5 $\mu m\ s^{-2}$ (0.05 mGal) if a station is occupied for five days, including unpacking and packing up. Details will change with time, but the following illustrates some of the necessary elaborations.

(i) The falling body is enclosed in an inner vacuum case, which itself falls freely in a vacuum of 0.1 Pa ($=10^{-6}$ bar). So the small amount of air inside the inner case has a doubly small effect on the movement of the body.

(ii) The body is a corner reflector 2.5 cm in diameter, which constitutes one-half of a laser interferometer. The other half of the interferometer is fixed below the path of the body, and the changes of distance between the two are measured by fringe counters. In an early model two fringe counters were started simultaneously at time t_1, of which one stopped at t_2, and the other at t_3, but in a later model many more (50 to 500) separate fringe counts are made, which provide redundant equations in the form of (5.1). The counts are timed by a rubidium frequency standard. These numerous counts will help to detect and to some extent eliminate the effects of very short period seismic movements.

† Allowance must be made for the change in gravity with height, since (as they stand) g does not have the same mean value in the two equations.

(iii) The fringes are formed by a He–Ne laser, whose output is stabilized on an iodine absorption line. The fringe interval is thus extremely constant, and is estimated to contribute an error of no more than 0.01 μm s^{-2} to the deduced value of g.

(iv) The details of the design minimize vertical electrostatic forces, which in this instrument are considered to be negligible at the 0.1 μm s^{-2} level

(v) An in-built mini-computer processes the data.

(vi) Programme. Although one drop provides a great amount of data, it is considered necessary to make a large number of drops at every station. This is easily done, since the observation of a single drop takes less than one minute.

(*c*) *The up-and-down method.* In this method the body is thrown upwards, and having reached its maximum height it falls back. On the way up it is timed (t_1 and t_2) at two points separated by a distance l, and on the way down it is again timed (t_3 and t_4) at the same two points. Then, ignoring variations of g with height.

$$g = 8l\{(t_4 - t_1)^2 - (t_3 - t_2)^2\}^{-1}. \tag{5.2}$$

The fall must take place in a high vacuum, although the effect of air drag is largely cancelled out in (5.2) since it has the same sign as g on the up journey, but the opposite sign when going down.

Another advantage of the up-and-down method is that timing error may depend on whether the body is moving fast or slowly. In this system $(t_4 - t_1)$ and $(t_3 - t_2)$ are both differences of times recorded when the body is moving at the same speed, which is not the case in the free fall method. A disadvantage is that it is more difficult to throw a body up, without jolts, than it is to let it fall.

The up-and-down method was used for the *Cook* (*NPL*) instrument [161], the first of the modern absolute gravimeters to give results correct to less than 10 μm s^{-2} (1 mGal). The body was a glass ball, acting as a lens, whose passage past a pair of vertically separated slits was registered by a photomultiplier. The vertical separation of the slits was measured by interferometry between glass blocks on which the slits were mounted. The accuracy obtained has been about 1 μm s^{-2} (0.1 mGal). This instrument has not been made mobile.

The *Sakuma* (Sèvres) instrument [513] is an elaboration of the Cook apparatus. The body is a corner cube reflector, as described above for the free-fall method, so the distance l is measured directly by interferometer. The whole apparatus is sprung to eliminate seismic disturbance. These improvements increased the accuracy to 0.03 μm s^{-2} (0.003 mGal) in 1970, and to 0.01 μm s^{-2} or better in 1975, for a non-transportable model.

A portable Sakuma model of the Istituto di Metrologia, Torino, weighs 600 kg and is reported [421] to be able to complete a measurement correct to 0.03 or 0.05 $\mu m s^{-2}$ in a week.

Other instruments working on these general lines are being brought into use [513], and increases in convenience and accuracy are to be expected.

Absolute observations appear to have superseded the pendulum, but in the near future they are unlikely to supersede the gravimeter for the mass production required for detailed exploration surveys.

5.05. Standardization and calibration. IGSN

(*a*) *Historical.* When pendulum and gravimeter observations were based on Potsdam, the instrument was not of course taken to Potsdam as the opening station each season, but visited a local base station, such as Teddington for the UK, before and after work, and in an extended tour possibly other trustworthy stations in between. The values of g at base stations were determined by probably several pendulum interchanges between each base and Potsdam, and also between one base and another. The resulting redundancies could be reconciled by a least-squares adjustment. By 1965 this system was running into considerable difficulties, since every year's work made previous adjustments more or less out of date. Since 1970 the possibility of observing absolute values of g on a world-wide basis more accurately than differences could be measured by pendulums changed the situation, and made a comprehensive and more permanent adjustment possible.

Another factor was that the calibration of gravimeters requires a pair of accurate base stations with a rather wide difference of g, by which the readings of their scales can be calibrated. Since gravity varies primarily with latitude, this need can be met by a meridional line of calibration stations separated by about 15° of latitude or less, to give differences of 5000 $\mu m s^{-2}$ (500 mGal). Before 1970 such calibration lines had been observed in many countries, but they also needed to be incorporated in the general adjustment of the base stations.

(*b*) *The International Gravity Standardization Net*, 1971. In 1971 it was decided to carry out an adjustment to be known as the IGSN 1971, which is fully described in [420]. The data consisted of the following.

(i) Absolute observations at Teddington (UK), Sèvres (France), Washington, Middletown (Conn), Bedford (Mass), Fairbanks (Alaska), Denver (Col), and Bogota (Colombia). At Teddington the Cook (1969) and the portable Faller–Hammond observations differ by 0.5 $\mu m s^{-2}$ (0.05 mGal), and at Sèvres the Sakuma (1970) and Faller-Hammond observations differ by 0.3 $\mu m s^{-2}$. At the other stations only the Faller–Hammond apparatus was used with standard errors of about ± 0.4 $\mu m s^{-2}$.

(ii) A total of 1200 pendulum observations, mostly (82 per cent) made with the Gulf (USA) and Cambridge (UK) pendulums between 1952 and 1967. These pendulum observations were mostly concerned with the interconnection of national bases with each other and with the absolute stations. They also provide points to control three principal calibration bases, one through North and South America, one from Norway to South Africa, and a third from Alaska through Japan and Singapore to Australia.

(iii) A total of 12 000 gravimeter ties made with long-range LaCoste–Romberg instruments between 1961 and 1969, together with approximately 11 700 short-range connections between slightly separated sites, made with a number of different types of gravimeter. The 12 000 ties were made between places with similar values of g (within about 5000 $\mu m\, s^{-2}$), i.e. between points of similar latitude (without regard to sign).

The observations were weighted and simultaneously adjusted by least squares to give the value of g at 1854 station sites, together with 96 gravimeter scale factors and 26 pendulum and gravimeter drift rates. Of the 1854 sites many were in connected groups of several eccentric stations, and the total of independent stations was 473. The adjustment provides gravity values whose standard errors are in all cases thought to be less than 1 $\mu m\, s^{-2}$ (0.1 mGal), [420], p. 18.

The IGSN value of g at Potsdam (site A) is 981 260.19 mGal. The value at the old Potsdam base site, where g was accepted as 981 274.00, when transferred to site A is 981 274.20, so the change is −14.0 mGal, which is in accord with the Lucerne Gravity formula 1967, see § 6.27 (*b*).

(*c*) *National calibration lines and nets.* For internal convenience, some countries may have their own net, or meridional line, of accurately determined gravity values to provide locally a denser base and calibration framework than is given by the IGSN. Thus in 1966–7 the USA, [77], p. 23 and Fig. 10, observed and adjusted ties between about 60 evenly spaced stations covering the whole country. It was possible to incorporate all these in the IGSN, but it was not possible to postpone the adjustment of the IGSN until all other countries had done the same.

The observation of new national calibration lines and bases presents a problem. Clearly, they should include a number, say at least three, of IGSN stations, which will provide preliminary values and their weights, but the adjustment of the national net, with suitable weights for the new observations, will provide revised values for the IGSN stations. The IGSN cannot be recomputed whenever this occurs, and the recommendation [55], p. I–13, is that both the IGSN and the new local values should be included in the publication of the local net. If there is a difference of more than four times the standard error of the IGSN value, special investigation is advised, in the hope that one or the other can be corrected or excluded.

SECTION 2. PENDULUM OBSERVATIONS

5.06. Outline of method

Pendulums may be obsolescent, but until about 1970 they provided the basis, direct or indirect, for virtually all known values of gravity and it is still too soon to dismiss them as requiring no description. It is convenient to describe the use of a simple single pendulum as used 70 or more years ago, with mention of later improvements as they arise.

The principle is that a pendulum swinging through a small arc in a vertical plane swings from left to right and back again with a constant period T, given by

$$T = 2\pi(k_s^2/g\bar{y})^{\frac{1}{2}}, \tag{5.3}$$

where k_s is the radius of gyration about the point of support, and $\bar{y}$ is the distance from the point of support to the centre of mass. T is usually about 1 s. Writing

$$l = k_s^2/\bar{y}, \tag{5.4}$$

$$\text{(5.3) becomes } T = 2\pi(l/g)^{\frac{1}{2}}, \tag{5.5}$$

l being the *equivalent length* of the pendulum. We also have

$$k_s^2/\bar{y} = I_s/m\bar{y} = (I_0/m\bar{y}) + \bar{y} = (k_0^2/\bar{y}) + \bar{y}, \tag{5.6}$$

where I is the moment of inertia, suffix zero indicates moments about the centre of mass, and m is the mass of the pendulum.

Then if l is constant, T depends only upon g, and if T_1 and T_2 are the periods at two stations, the two values of g are related by

$$g_1/g_2 = T_2^2/T_1^2, \tag{5.7}$$

whence

$$g_2 - g_1 = \frac{2g_1}{T_1}(T_1 - T_2) + \frac{3g_1}{T_1^2}(T_1 - T_2)^2, \tag{5.8}$$

in which the second term is small. An error of 10^{-7} s in either T will cause an error of 2 μm s^{-2}, (0.2 mGal), in $g_2 - g_1$.

5.07. Use of the pendulum apparatus

(*a*) *Outline.* The simplest form of apparatus is shown in Fig. 5.1. A mirror is fixed to the top of the pendulum, so that a spot of light from A falls upon some form of recording device B when the pendulum is vertical, twice in each period T. It is required to record the number of times this happens in a given interval of time, with an accuracy sufficient to give T with an accuracy approaching 10^{-7} s.

(*b*) *Recording mechanism.* The early system [113], Edn 1 and 2, is of course obsolete. The light source was obscured except for a short interval

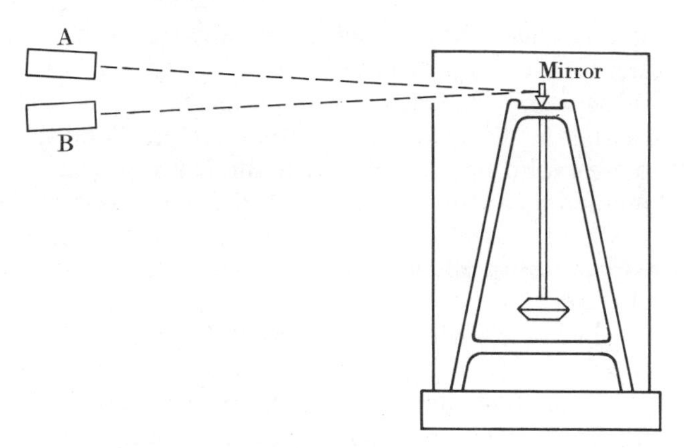

FIG. 5.1. Single pendulum. B is a lamp. A is a telescope, camera, or counter.

every second, when the clock opened a slit and a flash was reflected from the pendulum to the recording device. The period of the pendulum was 1.014 s. The counter B in Fig. 5.1 was a telescope with a graduated scale, placed so that when the pendulum was vertical the flash would fall on the zero of the scale. Consecutive flashes could thus be seen moving up and down the scale. A series of readings gave a comparison between clock time and pendulum swings, and another series an hour later gave another comparison, so that the two gave a good measure of the period, although many such determinations were needed to approach a 10^{-7} s accuracy.

In later models the swing of the pendulum has been recorded photographically, together with seconds markers introduced by a crystal oscillator, [304] and [303]. In a modern model photo-electric cells and electronic counters could be used over known time intervals. With modern equipment the timing of (say) 4000 one-second periods is sufficiently sensitive to provide the required accuracy, but there are, of course, other troubles.

(*c*) *Clock rate.* Between 1900 and 1920 this was a serious difficulty. The rate had to be determined by nightly star observations, the field clock was not very regular, and the pendulums (swinging at atmospheric pressure) could only swing for about an hour at a time. Between about 1920 and 1940 this situation was much improved, since radio time signals could be obtained more frequently, and the pendulums (at reduced pressure) could swing substantially from one signal to the next. Now, crystal or other oscillators, checked by wireless time signals as necessary, provide ample accuracy.

(*d*) *Lengths of the pendulums.* It is not easy to maintain the length of a pendulum to a fraction of 1 ppm, during work and journeys which may extend over months. Causes of change may be as below.

(i) *Accidental damage.* They should travel in hand travelling cases and should be watched over with care.

(ii) *Knife edge wear.* The effect of wear of the knife edge can be reduced by making the distance between the knife edge and the centre of mass equal to the radius of gyration about the centre of mass, as in Fig. 5.2. Such a pendulum cannot be made with the heavy bob at the end as shown in Fig. 5.1.

Pendulums should be gently lowered on to their supports by a suitably designed mechanism.

The knife edge need not necessarily be part of the pendulum. It can be fixed to the stand with its edge upward, bearing on a flat on the top side of a hole in the pendulum. But it is usually on the pendulum.

(iii) *Rust or corrosion.* The remedy is care, the avoidance of breathing on the pendulums, and handling them only with instruments. Quartz is good for this.

(iv) *Relaxation of internal stresses, or unexplained damage.* Probably the most serious source of error. Remedies are as follows.

(A) Good design.

(B) High thermal conductivity. Low coefficient of expansion. Avoidance of temperatures changes.

(C) 'Ageing' before use.

(D) Avoidance of invar, see § 1.30 (a), an otherwise very good material.

(E) The use of at least four pendulums at every station, so that damage or sudden change may be at once shown.

(F) Repeat swings at the base after a tour. For important work all stations can be visited in the reverse order on the return journey, as in [303]. One of the stations which have already been visited may be revisited after every few further stations, or a check can be obtained at any convenient station of the IGSN (1971). In this way changes can to some extent be located. But if most of the pendulums return to the original base with seriously changed lengths and without intermediate checks, a repetition of the whole work is clearly invited.

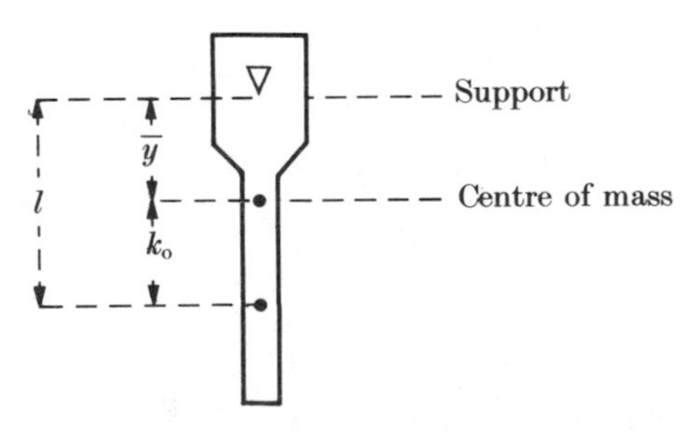

FIG. 5.2. $l = (k_0^2/\bar{y}) + \bar{y}$ from (5.6). For $dl/d\bar{y} = 0$ we have $k_0 = \bar{y}$, and $l = 2\bar{y}$.

[420] pp. 191–4 analyses the variations in the lengths of pendulums used for the IGSN. The mean drifts of different pendulums varied between 0.1 and 2.2 $\mu m\,s^{-2}$ (0.01 and 0.22 mGal) per station. The r.m.s. values of these figures varied between 0.8 and 3.3 $\mu m\,s^{-2}$. These figures suggest that, in a reasonably stable pendulum, future drift rates cannot well be forecast from their past history.

(*e*) *Temperature.* The temperature correction factor k_T, such that

$$\text{d(period)} = -k_T\ \text{d(temperature)}, \tag{5.9}$$

can be obtained by swinging a pendulum at different temperatures in otherwise identical conditions. It can be kept small by making the pendulums of invar or of fused quartz.†

Temperatures are best measured by a thermometer built into the stem of a dummy pendulum, of the same shape and material as the others, placed inside the vacuum chamber with them.

Brass or other stable material can only be used, if thermostatic apparatus enables the pendulums to be swung, stored, and transported at nearly constant and accurately ascertainable temperatures.

(*f*) *Pressure.* The atmospheric pressure affects the swing in two ways.

(i) By upward buoyancy acting against the gravity.

(ii) By its damping effect changing the period.

With modern apparatus using pressures of a few mbar (100 Pa) the formula takes the form

$$T_0 - T = -T_0(aD + bD^{\frac{1}{2}})\ \text{s}, \tag{5.10}$$

where D is the density expressed as a fraction of density at 0 °C and 1013 mbar (101.3 kPa), and a and b are empirical constants, such as 1200×10^{-7} and 180×10^{-7} respectively for the 'Cambridge' invar pendulums. T_0 is the period at 0 °C and 1013 mbar.

(*g*) *Arc.* Formula (5.5) is only correct when the pendulum swings through an arc of very small amplitude. The actual arc must be measured at the start and finish of the swing, and then

$$T_0 - T = -T\alpha_m^2/16 + T(\delta\alpha)^2/192\ \text{s}, \tag{5.11}$$

where $\alpha_m = \frac{1}{2}(\alpha_1 + \alpha_2)$, $\delta\alpha = \alpha_1 - \alpha_2$, α_1 is the initial amplitude of the pendulum (semi-arc), and α_2 is the final, both in radians. The second term is seldom of any consequence. Note that the amplitude of the pendulum is half that of the reflected beam in Fig. 5.1.

(*h*) *Levelling.* The knife edge and the flat on which it rests must be levelled so that the pendulum swings in a vertical plane. The error is $g(1-\cos\theta)$, or 1.7 $\mu m\,s^{-2}$ (0.17 mGal) if the dislevelment θ is 2 minutes of arc.

† Temperature coefficient = 0.4 ppm per °C.

(*i*) *Magnetization.* The period of an invar pendulum may be affected by either permanent magnetism or by magnetism induced by the earth's field. The presence of significant permanent magnetism can be detected by a compass or suitable magnetometer [548] before work at every station, and if it is present the pendulum must be demagnetized by a solenoid, which may be built into the carrying case. Other parts of the apparatus must be (and can be) non-magnetic. To avoid induced magnetism, the whole apparatus is surrounded by a pair of Helmholtz coils (one above and one below, with their axes vertical), 1 m in diameter, with the current adjusted so as to neutralize the vertical component of the earth's field. The pendulums are then swung in the magnetic east–west plane.

(*j*) *Flexure of the stand.*

(i) *Single pendulum.* The most rigid portable stand will swing a little with the pendulum, and will lengthen its period. Between about 1900 and 1925 this flexure correction was determined at every station by swinging a heavy pendulum, (isochronous with the working pendulums) on the usual mounting, while one of the working pendulums hung upon a rigidly attached auxiliary mounting. This latter started nearly motionless, but would soon begin to swing under the influence of the other. By measuring its changes of amplitude, the correction to the period of a working pendulum swinging on its normal mounting was deduced. See [76], pp. 625–38 or [113], previous editions.

(ii) *Two-pendulum apparatus.* Since about 1925, flexure has been eliminated by swinging two pendulums together 180° out of phase as in Fig. 5.3. Provided the phase difference remains between (say) 160° and 200° at the end of the swing, the flexure correction is negligible. For an

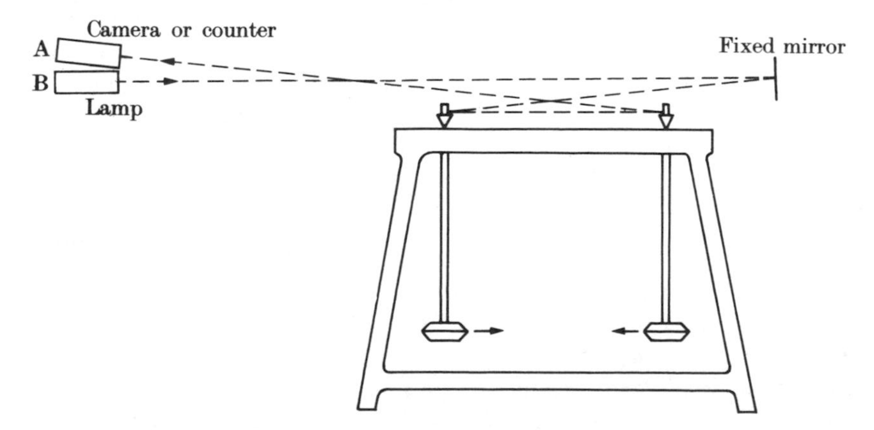

FIG. 5.3. Two pendulums swinging in anti-phase. The amplitude of the beam entering the camera or counter is twice the sum of the amplitudes of the two pendulums. Alternatively, the motion of each pendulum may be recorded separately.

8-hour swing this used to demand that the periods of the two pendulums should be equal within $30^s \times 10^{-7}$, which could only be achieved by much patient grinding. This trouble is saved with the 1-hour swings which are now possible.

This apparatus eliminates the effects of microseisms such as may be experienced in alluvial country near the coast in rough weather, as in Holland [132], p. 152.

(k) *Tidal effect.* If the earth was rigid, the passage of the sun or moon through the zenith or nadir would decrease g by about 0.08 and 0.16 mGal respectively relative to its value when they are on the horizon. Actually, owing to the yielding of the earth and sea, the effect is a little greater than it otherwise would be by a factor $(1+h-\frac{3}{2}k) \approx 1.15$, §§ 6.81 and 6.84, and its magnitude is not entirely predictable. See [345], p. 30.

It is generally ignored, but it is significant in modern absolute observations of g, and in good gravimeter observations on land. It is possibly just significant in the best pendulum results.

(l) *Programme of observations.* Details will of course vary with the perfection of the apparatus and the accuracy aimed at. The following has been adopted for the Cambridge 2-pendulum apparatus aiming at a standard error of about $\pm 2\ \mu m\ s^{-2}$ (0.2 mGal). Using pendulums in pairs as in Fig. 5.3, it is convenient to have at least four pendulums from which four pairs (such as AB, BC, CD, and DA) can be selected, although it is quite possible to work with a single pair. Then at every station each pair is swung two or three times for about 4000 seconds (1 hour and 7 minutes).

Each pair of pendulums, as above, is treated as a single inseparable entity whose period is determined. The two members of a pair are always swung in their appointed positions on the stand, A in the front and B at the back, or whatever it may be. There is no question of establishing the equality of the period of (say) A when swung in conjunction with B and when swung with D.

The corrections to the period of a combined pair for clock rate, temperature, pressure, and arc are the mean of the corrections applicable to each, regarded as single, as given by § 5.07 (c), (e), (f), and (g).

(m) *The pendulums.* Different organizations have their own pendulums. All have a period of about 1 s, and have sometimes been described as *half-second pendulums* as $\frac{1}{2}$ s is the interval between passages through the vertical. The following descriptions of the Cambridge and Gulf pendulums suffice to illustrate.

The Cambridge pendulums. These are of the type shown in Fig. 5.3, made of invar with stellite knife edges swinging on agate flats. There are two sets of three pendulums, known as I A, B, and C, and VI A, B, and C. Members of each set are closely isochronous with each other, but the two sets are not, so six pairs are available for swinging on the two-pendulum stand, but ordinarily four pairs have been selected.

The Gulf pendulums. These are as shown in Fig. 5.2. They are made of fused quartz and swing on pyrex flats. Two pairs, known as M and K, have been used, generally only the M pair, but sometimes both.

Very full descriptions of the use of the Cambridge and Gulf pendulum apparatuses are given in [132] and [548] respectively.

5.08. Accuracy of pendulum observations on land

(*a*) Observations made before 1900 without allowance for flexure may be wrong by 250 $\mu m\, s^{-2}$ (25 mGal) and are of no value.

(*b*) Observations made between 1900 and about 1925 with brass pendulums, swung singly at atmospheric pressure, and timed by local star transits, may be wrong by several tens of $\mu m\, s^{-2}$ (mGal), but generally give results of significant accuracy.

(*c*) Observations made on land between 1925 and about 1940, with invar pendulums in pairs, at reduced pressure, with wireless time signals, but without satisfactory demagnetization, will be seldom be wrong by more than 20 or 30 $\mu m\, s^{-2}$ (2 or 3 mGal), but unexpected discrepancies have occurred.

(*d*) Since 1950 accuracy on land has much improved, and the best results may suggest an s.e. of 2 $\mu m\, s^{-2}$ (0.2 mGal). Actual errors are unlikely to exceed 10 $\mu m\, s^{-2}$ (1 mGal).

(*e*) Observations made by pendulums in a submarine between 1920 and 1950, avoiding stormy weather, are liable to errors of perhaps 50 $\mu m\, s^{-2}$ (5 mGal).

Since 1950 the stability of the pendulums has probably been the most serious source of error. See § 5.07 (*d*)(iv).

5.09. Adjustment

(*a*) When pendulums or gravimeters returning to their base or to a previously visited station show a closing error, the discrepancy can be dispersed between the intervening stations by common sense. The changes may be considered to have progressed regularly, or consideration of the results given by different pendulums at different times may indicate an abrupt change in one, which can be treated accordingly.

(*b*) In more complicated cases a formal adjustment may be carried out by least squares. Between any two consecutive stations M and N, (5.8) gives the value of $(g_N - g_M)$ in terms of T_M and T_N. When several pendulum pairs have been used, separate equations (5.8) will be formed for each pair. If between stations M and N discrepancies between pairs suggest a sudden change in one of the pendulums, the measures involving that pendulum can be rejected, unless further observations are possible. Weights must be given to each measure of $(g_N - g_M)$, generally perhaps all equal, but some may be considered to be less reliable than others. A drift correction, δT per day, may be applied to each pendulum pair on the assumption that their period changes regularly with time. For this purpose a term $\delta T \times$Time interval (different for each pair) can then be included in (5.8) as an additional unknown. But see § 5.07 (*d*) (iv).

A value of g must be assumed for the base station, which must be entered numerically, and not treated as unknown. Other well fixed stations, such as those of the IGSN may be treated similarly, but if there is more than one such base, the preferred values at all may be included as separate observation equations in the simple form

$$g_{\text{Base}} = \text{IGSN value, with suitable weight.} \tag{5.12}$$

The observation equations are solved by the method of parametric method § D.14 (*a*). It will as usual be convenient to solve for Δg's as the unknowns, where $\Delta g = g -$(some provisional value).

The elaborate adjustments carried out for the IGSN by three different methods are described in [420].

5.10. Pendulum observations at sea

In about 1920 Vening Meinesz devised an apparatus suitable for the observation of g at sea in a submarine. Three closely isochronous pendulums, A, B, and C, of which the outer two A and B start 180° out of phase, swing together and the combined motions of the pairs are separately recorded to give periods T_{AC} and T_{BC}. The central pendulum C starts with nearly zero amplitude, which gradually increases as the result of outside disturbance. Its amplitude is recorded with reference to a strongly damped pendulum in the same plane, and a second damped pendulum records the tilting β at right angles to that plane. This apparatus eliminates the effect of moderate wave action, or enables it to be calculated as below, but except in the still water of a harbour it is necessary to observe in a submarine, usually at a depth of 20–40 m. The pendulum case is mounted in gimbals with the pendulums swinging fore and aft in the ship, so that β varies as the ship rolls, although less than the tilt of the ship itself. A photographic record is made of the movements of the fictitious combined pendulums AC and BC, of the amplitude of C, of β, and of the temperature.

This apparatus is no longer used, since gravimeters on a surface ship can now do as well as pendulums in a submarine, and with much less expense. Details of the pendulum observations are in [573]. Previous editions of [113] give a summary.

Before 1937 results were obtained without allowance being made for the *Browne* or *second-order* terms [131], but it was subsequently found possible to obtain their values from the photographic records.

See also § 5.18 (*a*) for the Eotvos correction. For the usual correction for the depth of the apparatus below the surface, see § 6.42 (*b*).

Vening Meinesz's work gave the first useful measurements of gravity at sea, and the geophysical benefits were great. A full record of the voyages undertaken by the Netherlands Geodetic Commission between 1923 and 1938 is in [575], with the emphasis on the geophysical aspect. This is continued in [133] for 1948 to 1958.

5.11. Supplementary field work

The height of a pendulum station above or below the geoid should if possible be measured within 0.2 m,† and the latitude within 0.12 cosec 2ϕ km, since either corresponds to a change of 1 μm s^{-2} (0.1 mGal). These are stringent limits, especially the height, and for many purposes and many observations, errors ten times as great may do little harm. The actual positions of base and calibration stations must be more accurately described, and marked, with a view to their exact sites being retrievable when required. § 5.03(*a*) (iv).

The problem is one of straightforward surveying. Greater accuracy in latitude (and longitude, for the record), will often be easy to get, but in unsurveyed country getting the height to even three metres may be a matter of serious difficulty, and may possibly be the limiting factor in the useful accuracy of the work. In remote areas, getting the height is more difficult than measuring *g*, especially if a gravimeter is used for the latter. The possible use of mercurial barometers or high-class aneroids should not be overlooked, especially in countries where the barometer fluctuates less than in western Europe, but comparative observations at a reasonably close station of known height, preferably distant not more than a few tens of kilometres, are essential. See [86].

The interpretation of gravity results calls for a contoured map, from which the disturbing effect of surrounding topography can be assessed. Inspection of the reduction tables referred to in §§ J.02–J.04 will indicate the accuracy required for the average height of annular zones at different distances from the station. If no contoured map as large as (say) 1 : 100 000 exists, the gravity observer can hardly be expected to make one, but he can sometimes usefully amplify an existing map up to a distance of 1 km or so, or alternatively he can site his stations in flat ground where the existing map may be good enough. Reliable information about the average rock densities may also be of value, but it is

† The free-air correction, § 6.23, suggests 0.3 m as equivalent to 1 μm s^{-2}, but here the height of the instrument above the ground is not in question; only the height of both above the geoid.

essential that rock specimens should be typical of the areas and depths to which their densities are going to be ascribed, and the observations of non-expert geologists must be treated with caution.

SECTION 3. GRAVIMETERS

5.12. Introductory

The principle of the gravimeter is that the length of a spring hung with a weight m at its lower end is given by

$$l = l_0 + mg/k, \tag{5.13}$$

where l_0 and k are constants. So the length changes with changes of g. It is, of course, important that the length should not change, or should only change predictably, on account of other circumstances.

Modern gravimeters are sensitive to as little as $g \times 10^{-8}$ or less ($0.1\ \mu\text{m s}^{-2}$). Subject to certain conditions they can measure differences of g correctly to $1\ \mu\text{m s}^{-2}$ (0.1 mGal). They are generally light and easily portable, and are much easier to operate than pendulums. Their limitation is the small range of gravity values which they can compare without calibration on pendulum stations. A gravimeter cannot usefully work in latitudes outside the range in which it has been calibrated.

Gravimeters can be used at sea on surface ships in reasonably fair weather, and they have begun to be used in aircraft.

Points which require consideration in a gravimeter are as follows.

(*a*) The delicacy and perfection of the spring. No hysteresis and no inexplicable irregular changes in its length.

(*b*) Drift. There is inevitably a slow but regular change in the length of the spring. The rate of change is measured by returning to a local base daily or weekly or as necessary. It can be as little as $1\ \mu\text{m s}^{-2}$ (0.1 mGal) per day.

(*c*) Calibration. The determination of the scale factor k in (5.13). See § 5.13.

(*d*) Range. Gravity varies by 5000 mGal between equator and pole, and it is not to be expected that a single value of k will give g correct to 0.1 mGal throughout this range. Nor that a delicate spring will be adequately free from hysteresis and the like, if exposed to such a change of load. A gravimeter used for the geophysical prospecting of a small area may have a small range, while a larger range is required for general geodetic purposes. Most gravimeters can easily be adjusted to change the belt of latitude within which they can operate, but that is not the same as being able to make a direct measure of the difference of g between two points where it greatly differs.

(*e*) Changes with temperature. See § 5.14.

(*f*) Changes of pressure. Gravimeters are enclosed at constant pressure, and so are not affected by air buoyancy. A very small change, of the order of 0.02 $\mu m\,s^{-2}$ per 10^4 Pa in the external pressure (0.002 mGal per 100 mbar) has been detected in some instruments, [345], pp. 18–19, presumably due to some minute deformation of the case. At any particular site the value of g itself will depend on the atmospheric pressure, since increased pressure arises from the presence of an increased mass of air above the site. From this cause g decreases by 0.0043 $\mu m\,s^{-2}$ (0.43 μ Gal)† per 100 Pa (1 mbar), but this figure may be reduced to something such as 0.0037 $\mu m\,s^{-2}$ or less by the elasticity of the earth, [345] p. 31 and [281] p. 148.

(*g*) The instrument must be correctly levelled.

(*h*) The movement of the weight must be damped.

(*i*) Readings to 0.1 $\mu m\,s^{-2}$ (0.01 mGal) call for extraordinary sensitivity.

(*j*) When the accuracy makes it relevant, a tidal correction should be applied, § 5.07 (*k*).

(*k*) Ground water. Changes in the water content of the underlying strata can change gravimeter readings in two ways, as below.

(i) By the gravitational attraction of the water.

(ii) By the expansion or contraction of loose sediments, and consequent change of height.

5.13. Calibration

Accurate observations at two places where g is already known give the value of *k* (or the value of one division of the reading dial) over the range covered. This range must not be smaller than the spread of the stations whose readings are to be based on it, nor so large that the scale value is materially variable over the interval. About 5000 $\mu m\,s^{-2}$ (500 mGal) is ordinarily a maximum. If the required range is large, the necessary difference of g can only be found at stations which are widely separated in latitude, but a small range can be covered by a moderate difference of height: 300 m for 1000 $\mu m\,s^{-2}$. The IGSN (1971), § 5.05, amplified by local calibration lines, provides for this.

Allowance must be made for instrumental drift between calibration stations, by returning to the first after visiting the second, and the second or first after visiting the third, and so on. For the most accurate calibration the atmospheric and tidal corrections should be included at both stations.

† Compare § 6.29, where the situation is different. § 6.29 considers the differences of g at different heights within a standard atmosphere. Here we consider the meteorological variation in the pressure at a fixed point, due to an influx of air from elsewhere.

The value of one dial division may have periodic irregularities due to eccentricities in the gearing, which may have amplitudes of 0.1 or 0.2 $\mu m\,s^{-2}$ (0.01 or 0.02 mGal). [345], p. 26.

Calibration in general, and of a Worden gravimeter in particular, is described in [85]. [345] describes the calibration and use of a Lacoste–Romberg gravimeter on land, from the point of view of obtaining the greatest accuracy.

5.14. Temperature

Temperature can affect a gravimeter in three ways.

(*a*) By changing the dimensions of its parts,

(*b*) By changing the elasticity of the spring,

(*c*) By changing the density of the air or liquid in which its weight is immersed.

Changes in the dimensions can be reduced by the use of fused quartz or invar. Changes of elasticity could be reduced by making springs of *elinvar*, a nickel alloy whose elastic modulus is relatively constant, but which is unstable in other respects. In most gravimeters changes are minimized either by thermostatic control or by the use of bimetallic or other such mechanical compensating devices, and in either case a residual temperature correction factor can be empirically determined.

5.15. Construction details

The necessary sensitivity of reading is got by a variety of ingenious devices, some of which are described in very general terms below. Photographs of some are in [266].

A basic form is shown in Fig. 5.4. A weight is carried on a pivoted beam, which is supported by the main spring. The beam is brought to its zero, horizontal, position by a variable auxiliary spring which operates near the pivot. The reading of this spring gives the value of gravity.

For maximum sensitivity it is necessary that a very small change of mg should result in relatively large movement of the mass supporting beam, and that the optical or other means of recording movements of the beam should provide a large magnification. Fig. 5.4 illustrates these two basic needs in the simplest way. Many different gravimeters have been used, with varying characteristics. The following sub-paragraphs (*a*), (*b*), and (*c*) give a little more detail about three which have been successfully used.

(a) *LaCoste–Romberg*. This instrument makes use of a *zero-length spring*.† Ordinarily the tension T in a spring of length l is

$$T = k(l - l_0) \tag{5.14}$$

† Instruments employing a zero-length spring are described as *astatic*.

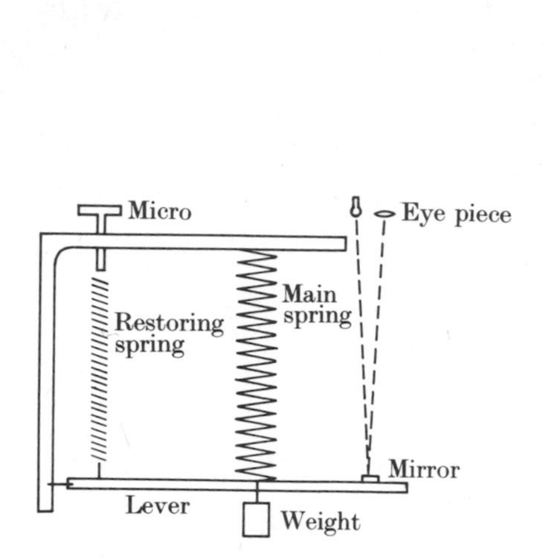

FIG. 5.4. Basic form of gravimeter.

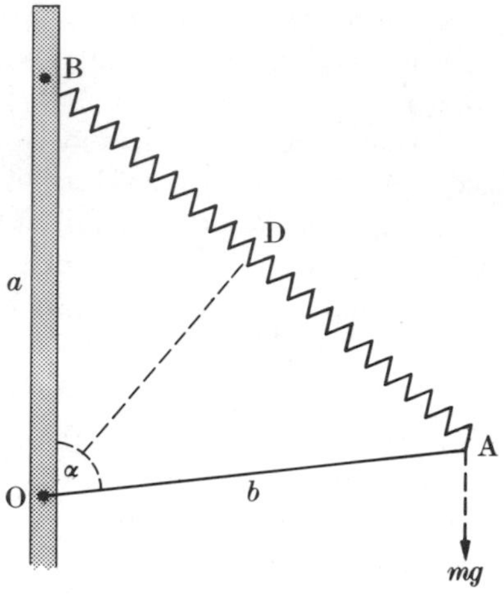

FIG. 5.5. The principle of the zero-length spring.

where k is a constant and $l = l_0$ when $T = 0$, and the weight of the spring is ignored. Then in Fig. 5.5 let a mass m be attached at A to a beam OA of length a, which is hinged at O and makes an angle α with the vertical OB. Let BA be a spring, whose windings are so proportioned that over its working range the tension in the spring equals $k(\mathrm{BA})$. If k is a constant, the effect on the mass m is the same as that of a hypothetical ordinary spring BA whose length is zero when $T = 0$. Hence the name zero-length spring.

Taking moments about O

$$mg \times (b \sin \alpha) = k(\mathrm{BA}) \times \mathrm{OD} = 2k(\text{Area BOA}) \qquad (5.15)$$
$$= k\,\mathrm{ab} \sin \alpha$$

So if there is equilibrium for any value of α, there is equilibrium for all. Then a slight alteration of the proportions will make the equilibrium stable, but very sensitive to small variations of g.

The rise and fall of A, or its adjustment to a null position, is read by a microscope or other more elaborate method of recording.

In the LaCoste–Romberg model G gravimeters the spring is metallic. The whole instrument is non-magnetic and is encased at low pressure with thermostatic control. The operating range is $70\,000\ \mu\mathrm{m\,s^{-2}}$ (7000 mGal), but they ordinarily work within a range of $5000\ \mu\mathrm{m\,s^{-2}}$ (500 mGal). They are also used at sea, § 5.16, and for recording earth tides and variations of g. A mobile model weighs 8 kg with another 4 kg for auxiliary equipment.

[187] describes methods of use required to give an accuracy of a few times $0.01\ \mu m\ s^{-2}$ ($1\ \mu Gal$) over a range of $500\ \mu m\ s^{-2}$. [345] gives a careful analysis of error sources which can be significant at the $0.01\ \mu m\ s^{-2}$ level. A model D *microgravimeter* is reported in [404] with a sensitivity of $0.01\ \mu m\ s^{-2}$ ($1\ \mu Gal$), and a standard error of $0.03\ \mu m\ s^{-2}$ over a range of $1\ \mu m\ s^{-2}$ (0.1 mGal). This model is suitable for the recording of earth tides and of variations of g.

(*b*) *Worden gravimeter*, This gravimeter uses a vertical zero-length spring operated as in Fig. 5.6., where the variations in the tension of the spring BA are proportional to BA over the working length. The spring is fastened to a member OA of length a which is rigidly attached to the weight carrying beam OM.

In Fig. 5.6 the line OX is horizontal and OM approximately coincides with it. β is the angle between OA and OM. BA is the spring, of length $d+a\sin\beta$, d being the constant distance above the horizontal through O. Then the moment of the spring tension about O is

$$k(d+a\sin\beta)\times a\cos\beta. \tag{5.16}$$

Now let β be changed by a small angle $d\beta$, such that $\cos d\beta=1$ and $\sin d\beta=d\beta$. The moment of the mass $mg\times OM$ will be unchanged, and the change in the moment of the tension of the spring will be

$$ak(a\cos 2\beta-d\sin\beta)d\beta. \tag{5.17}$$

For neutral equilibrium this must be zero, which it will be if

$$d=a\sin\beta,\ \text{when OM is horizontal} \tag{5.18}$$

and $\qquad \sin\beta=1/\sqrt{3}.$

A slight change in these dimensions will produce a sensitive stable equilibrium.

The general arrangement of the Worden gravimeter is as in Fig. 5.7. The reading of the beam is adjusted to zero by two springs and dials A and B. One is a coarse spring, to give a reading over a very wide range, and the other is a fine spring for use over a small range while A is left untouched. For the observations of the IGSN (1971) only small-range readings were accepted.

Worden gravimeters have been widely used for world-wide gravity ties. The model used in [613] had the following specification. The spring is made of fused silica. There is no thermostatic control, but there is a temperature compensating device with an empirical correction which has been given as $2.5\ \mu m\ s^{-2}$ per 30° C. It is barometrically sealed and is non-magnetic. Drift is less than $1\ \mu m\ s^{-2}$ per hour; weight, 10 kg with

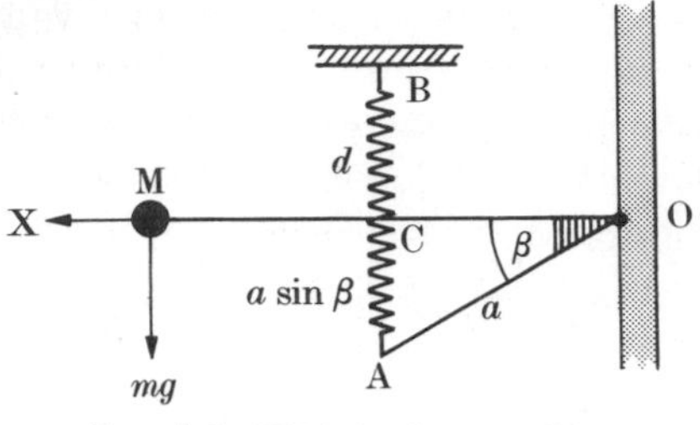

FIG. 5.6. OX is horizontal. OM approximately coincides with OX. AB is vertical. MOA = β and is a fixed angle.

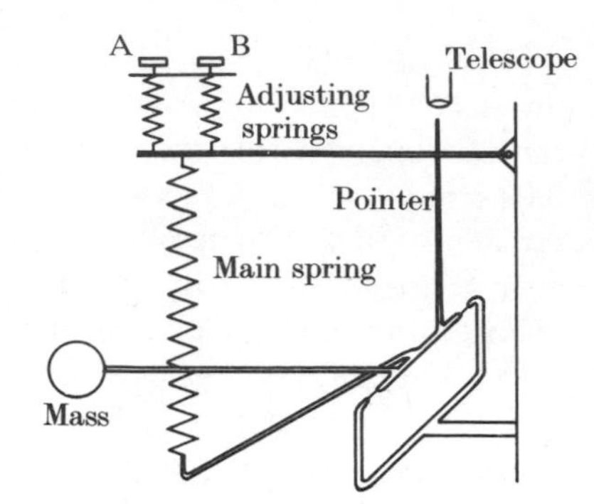

FIG. 5.7. Worden gravimeter.

case and tripod; time required to measure g, 5 minutes; sensitivity of reading, 2 μm s^{-2} (0.2 mGal) or less.

(*c*) *The Graf gravimeter.* In this instrument the weight-carrying beam is supported by the torsion of two horizontal springs as shown in Fig. 5.8. If a uniform spring has been twisted through an angle θ it will exert a torque of $k\theta$, k being a constant, so if m is the mass and l is the length of the beam

$$k\theta = mlg$$

and

$$d\theta = \theta\, dg/g \tag{5.19}$$

So high sensitivity is secured if θ is large, that is if the springs have been initially twisted through a large angle in order to hold the beam in equilibrium. There are two auxiliary vertical springs and dials, one to regulate the range through which the instrument will work, and a measuring spring to give a zero reading.

The Graf–Askania gravimeter has been much used at sea.

5.16. Shipborne gravimeters on stabilized platforms

(*a*) The possibility of measuring g in a surface ship instead of only in a submarine, has greatly increased the number of voyages during which g is determined. This can now be done using gravimeters, which have the added advantage that they can give a practically continuous record. Naval surveying ships frequently carry gravimeters and operate them continuously while at sea.

Accelerations arising from the rotations of the ship—pitching, yawing, and rolling—are minimized by putting the gravimeter near the ship's metacentre. The effect of waves is then that the instrument has a roughly circular motion imposed on its horizontal course. This circular motion takes place in a vertical plane, and results in accelerations of

$$\begin{aligned} \ddot{x} &= f \sin \omega t, \\ \ddot{z} &= f \cos \omega t, \end{aligned} \tag{5.20}$$

z being vertical, and x being horizontal in the direction of the wave movement. In moderately rough weather the period $2\pi/\omega$ may be of the order of 12 seconds, and the amplitude f may be 0.5 m s^{-2} (50 Gal). The corresponding velocity varies between ± 0.96 m s^{-1}, and the amplitude of the motion is ± 1.84 m. The object is to measure g to 10 μm s^{-2} (1 mGal) in these conditions.

There are two systems.

(i) The gravimeter is placed on a gyrosopically stabilized platform. This system is now preferred. It was originally adopted for the Graf–Askania.

(ii) Alternatively, the gravimeter is suspended in gimbals. See § 5.17. This system was originally adopted for the LaCoste–Romberg, but its use has been discontinued, and LaCoste–Romberg gravimeters are now used on stabilized platforms, as below.

For the general theory of these two methods see [358] and [359].

(*b*) *Stabilized platform.* The Graf–Askania equipment uses a Graf gravimeter, § 5.15 (*c*) and Fig. 5.8, with special modifications as follows. (i) It is constrained to respond only to accelerations, horizontal and vertical, in the vertical plane in which the weight-beam lies. (ii) It is very strongly damped. (iii) Instead of the beam being adjusted to zero for the reading, a continuous record of the deflexion of the beam is produced photo-electrically.

The gravimeter is mounted on a platform which remains level within about 1 minute of arc. If the ship was still, dislevelment of the platform would have the same effect as for the pendulum on land, § 5.07 (*h*), namely an error of 0.4 μm s^{-2} (0.04 mGal) for a dislevelment of 1′. This is trivial, but see (*d*) below.

The beam is mounted fore-and-aft in the ship. Take the axis of x horizontal in this direction, and we will assume accelerations as in (5.20), with the wave motion disadvantageously lying in the same plane as the beam.

(*c*) *Vertical accelerations.* The mean value of $\ddot{z}$ over a period of t seconds is $(1/t)(\dot{z}_t - \dot{z}_0)$, so if $\dot{z}$ starts as $+0.96$ m s^{-1} and finishes as -0.96, t will have to be 192 000 seconds to reduce the error to 10 μm s^{-2} (1 mGal). With a continuous record, the wave motion would of course be recognizable, and such a situation could be avoided by taking the mean over a whole number of periods. Actually, the difficulty is overcome by the very heavy damping which much reduces the effect of the periodic $\ddot{z}$ in the record.

(*d*) *Horizontal accelerations. Levelling error.* If the beam makes a small angle α with the horizontal, while experiencing an acceleration $\ddot{x}$, Fig. 5.9, the angle α will change, and the recorded value of g will change by

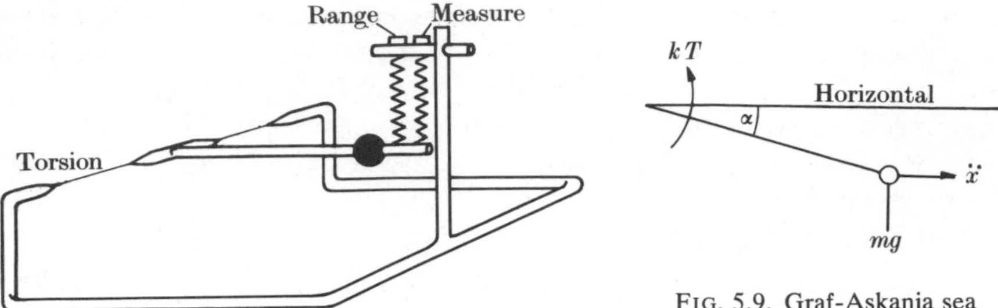

FIG. 5.8. Graf gravimeter.

FIG. 5.9. Graf-Askania sea gravimeter.

$\alpha\ddot{x}$. At sea $\ddot{x}$ will vary as in (5.20) and α will also vary periodically for two distinct reasons, namely (i) periodic error in the gyroscopic stabilization and (ii) the (much damped) recording of the vertical acceleration $\ddot{z}$. We now consider (i). For (ii) see (*e*) below.

Let the error in the platform vertical be $p\sin(\nu t+\lambda)$, and let $\ddot{x}=f\sin\omega t$ as in (5.20). Then the error in recorded g will be

$$fp\sin\omega t\sin(\nu t+\lambda). \tag{5.21†}$$

If $\omega=\nu$, as it may do, the mean value of this is $\frac{1}{2}fp\cos\lambda$, and if $f=0.5\ \mathrm{m\,s^{-2}}$ (50 gal) the error will be reduced to $10\ \mu\mathrm{m\,s^{-2}}$ (1 mGal) if p, the amplitude of the stabilization error, is $<8''\sec\lambda$. This cannot be assured. With $p=1'$ and $f=0.5\ \mathrm{m\,s^{-2}}$, errors of some tens of $\mu\mathrm{m\,s^{-2}}$ will no doubt sometimes occur, but these are extreme conditions [358], p. 91.

(*e*) *Cross-coupling error.* If $\ddot{x}=f\sin\omega t$ as before, and if α is proportional to $\ddot{z}$ (but 90° out of phase) namely $\alpha_0\sin\omega t$, the mean value of $\alpha\ddot{x}$ will not be zero, but will equal

$$\tfrac{1}{2}f\alpha_0. \tag{5.22}$$

In a typical apparatus, [19], p. 11, damping will reduce the amplitude α_0 to 7′ (1/500 radians) for $f=0.5\ \mathrm{m\,s^{-2}}$, but the error (5.22) will nevertheless be as much as $500\ \mu\mathrm{m\,s^{-2}}$ (50 mGal), a serious matter.

This is known as the *cross-coupling error*. There are various ways of reducing it, as follows.

(i) By increasing the damping, and so reducing α_0 in (5.22). But too much damping masks and distorts the true variations of g, and this is not practicable. See [450], pp. 182–3.

(ii) Instead of recording the varying deflexion of the beam, let it always be restored to near horizontal by a servo system. The cross-coupling

† Here λ and ν have no connection with longitude and the spheroidal normal.

effect can thus be reduced to 10 or 20 μm s^{-2} (1 or 2 mGal). The activity of the servo system will be proportional to the force it has to exert in keeping the beam level, and the value of g is obtained from a measure of this activity plus a record of the residual values of α. See [19], pp. 12 and 28–9.

(iii) Two gravimeters may be mounted back to back, so that their cross-coupling errors are equal and opposite, and cancel in the mean [253].

(iv) The acceleration $\ddot{x}$ and the deflexion α may be continuously recorded, and (5.22) may be continuously computed and applied as a correction to g, [117]. This is now (1978) the preferred method.

Methods (ii), (iii), or (iv) reduce the cross-coupling error to an s.e. of ± 10 or 20 μm s^{-2} (1 or 2 mGal) for $f = 0.5$ m s^{-2} (50 Gal). For the over-all accuracy obtained in an area of known g in 1966 see [450], p. 187.

(*f*) *Eötvös and tidal corrections*, § 5.18, are also applicable, except that the tidal correction will generally be negligible.

Currently Graf–Askania and LaCoste–Romberg gravimeters mounted on stabilized platforms probably give g correctly to within 10 or 20 μm s^{-2} (1 to 2 mGal) in reasonably good weather, apart from doubt in the Eötvös correction. A vibrating string accelerometer (VSA), § 5.21, has also given good results.

5.17. Shipborne gravimeters. In gimbals

A Lacoste–Romberg gravimeter has been used mounted in damped gimbals. [358], pp. 96–101 gives the theory of the resulting motion, and concludes that provided excessive swinging in the gimbals is avoided, recorded values of g need correction only by

$$-\tfrac{1}{2}g[\theta^2], \tag{5.23}$$

where $[\theta^2]$ is the time mean value of θ^2, θ being the deflexion of the gimbal system from the true vertical. If the period of swing of the gimbals and that of the waves is similar, resonance and excessive swinging is avoided by mounting the gimbal pivot in such a way that the pivot itself is free to swing with a suitable period, see [358], p. 101.

The gyro-stabilized platform is now preferred to the mounting in gimbals.

See [359] for details of both methods.

5.18. Eötvös and Tidal corrections

(*a*) *Eötvös*. Gravimeters whose supports are attached to an earth rotating with angular velocity ω will measure variations of g, the vector sum of the attraction and the centrifugal force, as in § 5.00. But if the support is moving relative to the earth as on a ship, submarine, or aircraft, in latitude ϕ with a velocity v the acceleration measured is

$$\text{(Attractive acceleration)} - \omega^2 \nu \cos^2\phi - 2\omega v \sin\alpha \cos\phi - v^2/R, \tag{5.24}$$

as in § I.03. The first two terms constitute g as ordinarily defined, but it is necessary to remove the last two by adding to recorded g

$$+4.0v \sin\alpha \cos\phi + 0.0012v^2 \text{ mGal}, \tag{5.25}$$

$\sin\alpha$ being positive if the velocity is towards the east, and v being in kilometres per hour.

The first term of (5.25) is known as the Eötvös correction. It can be a serious source of error, since an error of 4 mGal in g arises from one of 1 km/hour in the vehicle's ground speed.

The term v^2/R is almost negligible in a ship moving at less than 20 km/hour, but it is very important in an aircraft, § 5.20. In a free-falling satellite moving at (say) 8 km/s this term equals g.

Until recently the Eötvös correction has perhaps been the most serious source of error in shipborne gravimeter observations made in calm weather. Only accurate navigation can distinguish between a sea current of 1 km/hour and a gravity anomaly of 40 μm s^{-2} (4 mGal). But with the introduction of doppler fixes†, or other modern navigation aids, this difficulty is becoming less serious.

(*b*) *The tidal correction of* § 5.07 (*k*) is too small to be significant in shipborne observations, although the rise and fall of the tide, and of the ship with it, introduces a further correction. This is negligible in the open ocean ± 1 μm s^{-2} (0.1 mGal) but might be ± 10 μm s^{-2} in narrow waters with large tides.

5.19. Gravimeters on the sea bottom

Observations on the sea bottom at depths of up to 200 metres have for many years been made by a modified Gulf gravimeter, [266], pp. 113–15. The record is photographic, and the levelling is by remote control.

Observations at depths of 900 metres have been made with a modified LaCoste gravimeter [89], and [95] describes a vibrating string type of gravimeter which has been used sec § 5.21.

There is no fundamental difficulty in measuring g accurately on the sea bottom, but the geophysical interpretation of the results may be made a little doubtful by the difficulty of measuring the depth when it is very great.

† Using the navigation mode. § 7.47 (*c*).

5.20. Airborne gravimeters

An aircraft flying at high altitudes suffers smaller rotations and accelerations than a ship at sea, and on this account there is no basic impossibility about making accurate measures of g. There are, however, two difficulties which are much worse than they are at sea.

(a) The Eötvös correction. See § 5.18 (a). When an aircraft is moving at 300 km/hour in winds of 50 km/hour there is clearly serious difficulty in determining its instantaneous velocity to within a fraction of 1 km/hour. Especially if it is flying out of sight of land.

(b) The vertical acceleration. The term v^2/R in (5.24) is important, and the correct value of R must be used. But there is an additional difficulty.

The height of a ship above sea-level varies little, and in the long run abnormal vertical accelerations must average near zero. An aircraft is not so restricted. In a supposedly horizontal flight of L km, let the actual course of the aircraft in the vertical plane be a circle, such that at the mid-point of the L km the course is h cm above the horizontal circular course joining the two ends of the flight for which the correction v^2/R has been computed. Let the time over the L km be t seconds. Then the correction to the average recorded g is easily seen to be $8h/t^2$. So if $L = 20$ km, and $t = 200^s$ and h is as little as 1 metre, the correction to recorded g is as much as 200 $\mu m\ s^{-2}$ (20 mGal). The problem is one of navigation, and like that of the Eötvös correction it is more soluble in the mean of a long distance such as 50 or 100 km than for more instantaneous values.

As navigation aids, doppler, radar, and photo-theodolites have been used to give the course, hypsometers and radar altimeters for the height, and aerial photography for both, [605]. This problem may be solvable when the NAVSTAR system, § 7.44, provides position in three dimensions continuously with an s.d. of ±3 to 6 m, with hope of less in the mean over some distance.

[550] (1971) describes observations made by a LaCoste–Romberg and other gravimeters. The aircraft flew at an altitude of 8 km at a speed of 200 km/hour over an area of about 500×500 km in the south-central US, where there was exceptionally good surface gravity cover. The latter was extrapolated up to the height of 8 km with expected errors of not more than 10 $\mu m\ s^{-2}$ (1 mGal), or much less in the mean of a 100-km profile. Comparison between these extrapolated surface values and the values obtained in the aircraft gives the errors of the latter. After smoothing by various methods the aircraft values were combined to give $1° \times 1°$ and $5° \times 5°$ means, the former each being derived from one 500-km profile, and the latter from the mean of two such profiles. The standard error of the $1° \times 1°$ means was found to be 300 $\mu m\ s^{-2}$ (30 mGal) and that of the

$5° \times 5°$ means was 170 μm s^{-2} (17 mGal). The figure for the $5° \times 5°$ means as observed by a PIGA gravimeter flown in another aircraft at 800 km/hour was reduced to 118 μm s^{-2} (12 mGal).

[421] (1976) pp. 105–6 gives later reports with indications of a possible s.d. of 100 μm s^{-2} (10 mGal) for $\frac{1}{2}° \times \frac{1}{2}°$ means.

5.21. Vibrating string accelerometer (VSA)

Gravity may be measured by observing the period of transverse vibration of a thin wire tensioned by the weight of a mass M. The wire hangs between the pole pieces of a permanent magnet, and is made to oscillate by using it as a resonant element in an electronic oscillator. If the wire is flexible, of length L and of mass m per unit length, the frequency is given by

$$f = (1/2L)\sqrt{(Mg/m)}. \tag{5.26}$$

The method depends primarily on the constancy of M, m, and L, and not on constant elasticity. Corrections must be applied for the density of the surrounding air, the finite amplitude of the vibration, the end correction due to rigidity and the yielding of the supports, and for the effect of tilting. Hence differential values of g.

Such an instrument was used in a submarine in the English Channel in 1948 [226], giving results which agreed with pendulums with an s.d. of 22 μm s^{-2} (2.2 mGal), with a drift of 20 μm s^{-2} per day.

If, as in Fig. 5.10 (a), the mass is also attached to the bottom of the casing by another string whose tension is T_0 (5.26) becomes

$$f = (1/2L)\{(Mg + T_0)/m\}^{\frac{1}{2}} \tag{5.27}$$

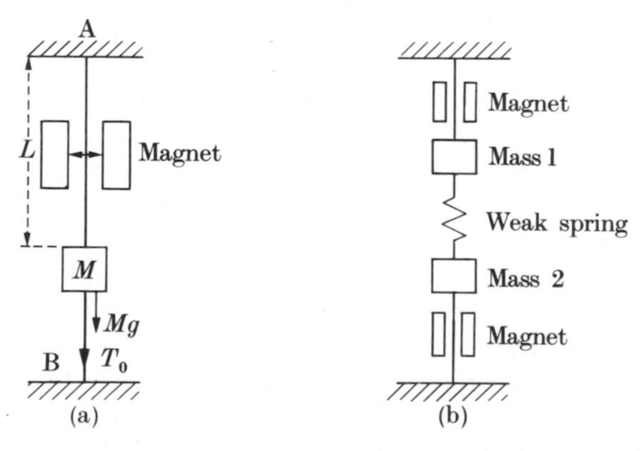

FIG. 5.10. Vibrating string accelerometer. (a) Simple instrument. (b) Invertible instrument with two strings.

which may be written as

$$f = K_0 + K_1 g + K_2 g^2 + \ldots \tag{5.28}$$

Other instruments have been made in which there are two masses as in Fig. 10 (*b*), connected by a weak spring. Then for the upper spring

$$f_1 = K_{01} + K_{11} g + K_{21} g^2 + \ldots$$

and for the lower

$$f_2 = K_{02} - K_{12} g + K_{22} g^2 + \ldots$$

and the difference

$$\Delta f_+ = (K_{01} - K_{02}) + (K_{11} + K_{12}) g + (K_{21} - K_{22}) g^2 + \ldots$$

If the instrument is now inverted, which is equivalent to changing the sign of g

$$\Delta f_- = (K_{02} - K_{01}) + (K_{12} + K_{11}) g + (K_{22} - K_{21}) g^2 + \ldots$$

and $\Delta f_+ + \Delta f_- = 2(K_{11} + K_{12}) g +$ odd powers of g. Calibration gives $(K_{11} + K_{12})$.

See [95], and for results in more detail see [119]. The sensitivity is $1\ \mu\text{m s}^{-2}$ (0.1 mGal) or less. When used in a ship in 1969, comparison with a LaCoste–Romberg gravimeter showed random errors of less than $10\ \mu\text{m s}^{-2}$, apart from drift and calibration errors. In the case quoted the latter were traced to a defect in the cross-coupling of the LaCoste–Romberg. On the other hand comparisons at track crossing points gave an r.m.s. difference of $32\ \mu\text{m s}^{-2}$ (3.2 mGal) in one area and of $60\ \mu\text{m s}^{-2}$ in another. This will have included Eötvös error, which was derived from satellite navigation.

6

PHYSICAL GEODESY

6.00. Introduction

PHYSICAL geodesy is primarily concerned with the determination of the spatial and temporal variations in the potential of the earth's gravitation and its derivatives. The data are provided by observations of the direction and intensity of gravity, Chapters 4 and 5, and of satellite orbits, Chapter 7.† The measurement of crustal movements is also included.

Geophysical problems are mentioned in this chapter only so far as may be required to illustrate the geophysical uses to which geodetic information may be put.

The following is a summary of the contents of this chapter.

Section 1 gives an outline of the earth's internal structure. This is not geodesy, but those who are concerned with the earth's potential field need to have some idea of the mass distributions which may cause anomalies in the potential.

Section 2 gives the potential field of simple standard bodies.

Sections 3 and 4 deal with the potential of solid spheroids, including spheroids and near-spheroids in which regularly distributed internal density is modified by density anomalies which can be described by spherical harmonics of fairly low degree. The potential fields of such solids approximate to that of the earth, although significant differences remain.

Section 5 describes the basis of currently accepted reference spheroids and gravity formulae.

The preceding Sections 3 to 5 deal primarily with the attraction of *earth models*, bodies whose form or potential fields are arbitrarily so defined as to approximate to those of the earth. Such bodies constitute reference systems, and the form and potential of the actual earth can be expressed as anomalies or small departures from the model, which have to be determined. The remaining sections deal more directly with the actual earth.

Section 6 deals with Stokes's integral, by which the form of a bounding equipotential surface, approximately the geoid, can be determined from measurements of gravity on it. The existence of external land masses constitutes a difficulty.

† Logically, Chapter 7 might well have been placed before Chapter 6. But satellite orbits depend on potential theory as well as providing data for it, and the present arrangement leads to fewer forward references than would result from the alternative.

Section 7 describes methods of gravity reduction, which, among other uses such as interpolation, eliminate the effects of the external masses and enable Stokes's integral to be applied to the actual earth.

Section 8 describes an alternative to the methods of Section 7, namely the direct application of Stokes's integral to gravity observations made on the earth's surface, without reduction to geoid level.

Section 9 deals with the deviation of the vertical, and describes its spatial variations, and the methods of reducing it to remove the local effects of the earth's topography according to different hypotheses.

Section 10 summarizes the methods which have been used to determine the form of the geoid, by means of astro-geodetic sections, gravity surveys, and satellite observations.

Section 11 describes the use of gravity anomalies as guides to the internal density of the earth.

Section 12 describes the measurement of crustal movements.

Section 13 gives an outline of the subject of earth tides, in so far as their effects are of geodetic significance.

6.01. Notation in Sections 1 to 10

x, y, z = Rectangular coordinates. The z-axis may be the axis of rotation, or the vertical, generally positive northwards or outwards respectively.

r, θ, λ = Polar coordinates. When appropriate, $\theta = 0$ is the axis of rotation or of symmetry. r is also the distance between two points.

ϕ, λ = Latitude and longitude + ve N and E. In § 6.14 ϕ is potential.

α = Azimuth clockwise from north.

ρ, ν = Principal radii of curvature of spheroid. ρ is also density.

p = Perpendicular from surface to axis of rotation.

r_x, r_y = Radii of curvature, § 6.23.

a, b, f = Major and minor semi-axes, and flattening, of spheroid.

$e^2 = (a^2 - b^2)/a^2$.

$e'^2 = (a^2 - b^2)/b^2 = \epsilon$.

a_1 = Semi-major axis of a higher equipotential surface.

R = Radius of a sphere. Approximate radius of the earth. Sphere of volume equal to that of spheroid, $a(1 - \frac{1}{3}f - \frac{1}{9}f^2)$.

$r_m = a(1 - \frac{1}{3}f - \frac{1}{5}f^2)$. § 6.18 (*b*).

ω = Angular velocity, especially of earth, for which $\omega = 7.292\,115\,1467 \times 10^{-5}$ rad s^{-1}, [423].

ξ, η = Deviations of the vertical. Signs as in § 2.05. Different in § 6.11.

$\xi_T, \eta_T, \xi_C, \eta_C$ = Deviations computed from topography, or compensated topography, § 6.54.

ζ = A component of the deviation. Also height of ground above telluroid, § 6.47.

G = Gravitational constant.

M = Mass of the earth, especially in GM. Any large mass.

m = A small mass. Also see m, m', etc., below.

σ = Mass per unit length, or per unit area.

ds, dS, dn, dv, dw = Elements of distance, area, along normal, volume, solid angle.

$m, m', \bar{m}, \widetilde{m}$ = See § 6.18.
h = Height above geoid.
h_S = Height above spheroid.
h' = Height of geoid above co-geoid.
H = Normal height, (6.155). Also precessional constant.
N, N' = Height of geoid, co-geoid, above spheroid.
V = The potential of the attraction of a body, § 6.07.
F = The acceleration due to the attraction of the body. No rotation.
W = The potential of the earth's gravitation. Attraction plus rotation.
g = The acceleration due to gravity, including rotation.
U = The potential of gravity for a model (near spheroidal) earth.
γ = standard gravity.
g_0 = g on the geoid, with no external masses.
g_0' = g on a co-geoid.
g_{00} = g on the geoid below ground level.
g_M = Mean g along a vertical.
γ_0 = γ on the surface of a spheroid.
γ_e, γ_p = γ_0 on the equator, at the pole.
$\gamma_A, \gamma_T, \gamma_C, \gamma_B$ = γ at an external point. Free air, topography removed, compensated topography removed, Bouguer convention, respectively.
γ_M = Mean γ along a vertical.
γ_m = Mean γ over the earth's surface.
γ_Q, etc. = γ at point Q, etc.
$\Delta g_0 = g - \gamma$ for inclusion in Stokes's integral (6.111).
$\Delta \bar{g}_0$ = Mean Δg_0 in an annular zone, or other area.
$\Delta g''$ = See (6.161).
$g - \gamma_E = \Delta g_E$. See (6.137).
O = Orographical correction. § 6.41.
B_2, B_4, μ_1, μ_2 = Constants in the gravity formula. (6.66).
χ = See (6.61). Also, differently in (6.161).
ν, K, E, λ, μ = Poison's ratio, Bulk modulus, Young's modulus, and two elastic constants respectively.
α, β = Velocity of P and S waves, § 6.03.
A, C = Moments of inertia about equatorial and polar axes.
$H = (C-A)/C$. The precessional constant. Also normal height.
ψ = Angle subtended at centre of spheroid. See (6.111).
$f(\psi)$ See (6.113).
N_1, N_2, N_3 See (6.128).
$F_e, \delta W$ = The changes in g and W when topography is moved. § 6.35.
CM, CV = Centre of mass, volume.
T = Potential anomaly = $\zeta\gamma_Q$, (6.152).
ζ = Height of ground above telluroid, § 6.47.
Δg = Anomaly of gravity defined by (6.153).
i See (6.147).
Y_n, u_n = General surface spherical harmonic of degree n. See § H.05.
Y_{nm} = Surface spherical harmonic of degree n and order m. See § H.05.
$C_{nm}\, S_{nm}$ = Coefficients in Y_{nm}.
$\bar{C}_{nm}\, \bar{S}_{nm}$ = Normalized coefficients in Y_{nm}, § H.06.
$P_n(\cos\theta) = P_n$ = Legendre function of degree n.
$-J_n$ = Coefficient of $(a/r)^n P_n(\cos\theta)$ in the earth's external attractive potential (7.1).

Gravity and attraction are forces, with dimensions $kg\,m\,s^{-2}$, while g and γ are accelerations due to gravity, with dimensions $m\,s^{-2}$. But where there is no ambiguity, g is generally spoken of as 'gravity'. It is numerically equal to the force acting on unit mass.

SECTION 1. THE EARTH'S INTERNAL STRUCTURE

6.02. Density, temperature, and strength

Geodesists measure the movements of the earth's crust, and the anomalies in the direction and intensity of the earth's gravitation, but the use of such data to discover the forces which cause the movements, and the density distribution which causes the anomalies, is the function of geophysicists, using the geodetic data in combination with data provided by seismology and other branches of geophysics. Geodesists, however, do need to have some slight knowledge of the earth's internal structure, if only to provide them with a common language when dealing with geophysicists.

(*a*) *Density*. Density increases towards the centre, but except within (say) 70 km of the surface, layers of equal density are closely (but not exactly) spheroidal equipotential surfaces. As a good first approximation the earth's interior below that depth may be said to be in hydrostatic equilibrium. In so far as this may not be the case, the hydrostatic assumption provides a useful model from which to reckon anomalies.

Evidence for the increase of density with depth is as in (i)–(iv) below.

(i) The earth's mean density as given by its volume and gravitation is about $5\frac{1}{2}\,g\,cm^{-3}$, compared with surface densities of $2\frac{1}{2}$ or $3\,g\,cm^{-3}$.

(ii) The known value of 1/298.25 for the flattening of the spheroidal form of the geoid, gives a figure of about $0.3307\,Ma^2$ for the polar moment of inertia of the earth, compared with the $0.4\,Ma^2$ which is correct if the density is uniform. § 6.20.

(iii) The natural expectation that density will increase with pressure.

(iv) The evidence of seismology, § 6.03.

Table 6.2 gives current estimates of the average densities at various depths.

Evidence about the extent to which surfaces of equal density depart from equipotential surfaces is as in (v)–(viii).

(v) It is visible at the surface. Geology and seismic sounding give direct evidence down to depths of (say) 10 km.

(vi) The value of the luni-solar precession, coupled with the known value of the flattening, § 6.20.

(vii) Anomalies in the intensity and direction of gravity at the surface, §§ 6.61–6.69.

(viii) Geophysical studies of the probable strength of rocks at the probable temperatures at different depths.

(*b*) *Temperature.* Near the surface the temperature increases with depth at a rate of between 10 and 40 °C/km, mean about 30 °C/km. The temperature at different depths between the surface and the centre has been variously estimated by different authors. The important thing in connection with density anomalies and possible movements of the crust, is the difference between the temperature at any depth and the melting point of the constituent rock at the pressure at that depth. Table 6.1, based on [541], p. 259, gives the estimated temperature at different depths, and also estimated melting points. Near the surface the temperature is naturally very well below the melting point, but between depths of 70 and 400 km they are much less separated.† Between a depth of 400 km and the top of the core, at the depth of 2900 km, temperatures appear to be well below the melting point.

(*c*) *Strength.* The engineer, working with ordinary materials, thinks of fracture or permanent deformation as occurring when tension or compression exceed certain limits, but at any depth inside the earth tension is impossible, as also is fracture due to uniform compression.‡ What leads to

TABLE 6.1

Temperatures, T, and melting points [541]

Depth/km	Convecting model T/K	Non-convecting model Continental T/K	Non-convecting model Oceanic T/K	Melting point T/K
0	300	300	300	1400
50	1200	1050	1350	1550
100	1500	1400	1900	1650
500	2200	2450	3050	3050
1000	2700	2900	3250	4500
2000	3300	3400	3550	5350
2900	3700	3700	3700	5900 Mantle
2900	3700	3700	3700	3150 Core
3500	3800	3800	3800	3400

† Between depths of 70 and 400 km [541] and Table 6.1 give temperatures beneath the oceans as being above the melting point, unless some convection in the mantle is accepted, but [541] accepts it as reasonably certain that the temperatures are actually below the melting point, implying that convection is probable.

‡ Although it is not impossible that changes of pressure may lead to abrupt changes of volume such as (e.g.) may be associated with a change from a vitreous to a crystalline state.

fracture or to gradual flow is *stress difference*, or inequality between (say) the vertical and horizontal compressions at a point. The reactions of materials to stress differences vary according to circumstances. A small change of form proportional to the stress difference may occur as soon as the stress is applied, and may remain unchanged until it is removed, when the body returns to its original form. The body is then described as *perfectly elastic*, and the tendency to return to the original form is *rigidity*. Alternatively, when a greater stress is applied or if the material is hotter, the final form may be reached more gradually, and although it tends to a definite limit there may be no complete return to the original state when the stress is removed. The body has then undergone *permanent set*. While a third state of affairs is that the change of shape may continue indefinitely so long as the stress difference is maintained, the behaviour of the body then being described as *plastic*.

The *strength* of a material at given temperature and pressure is an important property, defined as the stress difference above which the rate of change of shape does not decrease with time. Materials describable as liquids have zero strength, while most solids possess strength, although there are exceptions. Cold pitch, for instance, although in plain language a solid, and possessed of considerable rigidity in its reaction to rapidly changing stress, is of practically zero strength. If the melting point of a solid is defined, as in ordinary experiments, as the temperature at which rigidity vanishes and viscosity† is much reduced, the strength of a material at temperatures between the melting point and some hundreds of degrees below it is likely to be much less than its strength when cold.

From the estimates of temperature quoted in Table 6.1, it appears that below a depth of (say) 50 or 80 km the strength may be much reduced, with a minimum at around 100 km at which the strength may be very low. Below that, between 400 and 2900 km at the top of the core, the strength is apparently great. The core is liquid, with some uncertainty about an inner core with a radius of approximately 1250 km.

6.03. Seismological data

If (Fig. 6.1) an earthquake occurs at a point P a little below the surface of the earth, the shock is transmitted in all directions in the form of two kinds of waves, namely (*a*) a *compressional* or primary or *P* wave, in which every particle vibrates in the line of propagation (more or less radially from the point of origin), and (*b*) *a distortional* or secondary or *S* wave, in which the vibration is transverse. The velocity of the former wave is $\alpha = \sqrt{\{(\lambda + 2\mu)/\rho\}}$, and of the latter $\beta = \sqrt{(\mu/\rho)}$, where ρ is the

† Viscosity is defined as stress-difference $\div 2$(rate of shear).

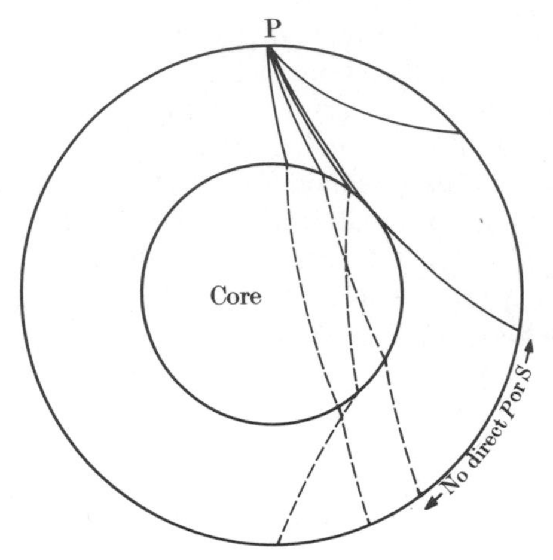

FIG. 6.1. Full lines represent both P and S waves: broken lines P only. Reflected waves, and waves refracted with change of type, are omitted.

density, and λ and μ are the two elastic constants of the rock through which the wave is passing.†

Seismographs on the earth's surface record the arrival of these waves, and the first wave of each type to arrive has clearly followed the path for which $\int ds/\alpha$ and $\int ds/\beta$ respectively is least. If the earth was homogeneous these paths would be straight, but if density varies with depth they are convex towards the side where velocity is greater. Reflections and refraction also occur at interfaces between layers of different constitution. Records at near observatories, reduced by well-established values for the velocity through surface rocks, give the time of the original shock, and in favourable circumstances some indication of its depth.‡ Records at more distant observatories then give average values of α and β for the paths of the waves reaching them, and consequently a relation between the density and elastic constants of the intervening rocks, or between ρ and the *bulk modulus* which is given by $K = \lambda + \frac{2}{3}\mu = \rho(\alpha^2 - \frac{4}{3}\beta^2)$. While a compressional wave can be transmitted by both solids and fluids, a distortional wave can only be transmitted by a solid, or something very like one.

† The modulus of rigidity μ is $E/2(1+\nu)$, where E is Young's modulus, ν is Poisson's ratio, which is typically about $\frac{1}{4}$ for most rocks, and λ is defined as $E\nu/(1+\nu)(1-2\nu)$. If $\nu = \frac{1}{4}$, $\lambda = \mu$ and $\alpha = \beta\sqrt{3}$. The *bulk modulus* is given by $K = E/3(1-2\nu)$.

‡ Most earthquakes originate at a depth of less than 50 km. It may be difficult to make a significant estimate of the depth of shallow earthquakes, but shocks also occur at much greater depths, down to several hundred kilometres, and figures for their depths can be deduced.

The core, or at least its outer 2000 km, transmits no transverse waves and is therefore describable as liquid.

The determination of the density and compressibility of different rocks at the temperatures and pressures existing at great depths is difficult, and results are necessarily open to some doubt. Table 6.2 gives current estimates of the density and possible constitution of the earth.

TABLE 6.2†

Depth/km	Density/g cm^{-3}	
0–30 variable	2.76 variable	Sedimentary, granitic, and basaltic Thickness varies between 6 and 70 km.
30–400	3.4–4.0	Silicates. Mantle
400–900	4.0–4.7	
900–2883	4.7–6.0	
2883–5120	10.0–12.0	Liquid iron. Outer core.
5120–6371	12.0–12.4	Solid. Probably iron. Inner core.

† Based on [541], p. 103. [137] gives very full treatment, and illustrates the doubts which still remain.

Seismology shows a very marked discontinuity at the base of the granite–basalt layer, which is known as the *Mohorovičić discontinuity*, known as *the Moho* for short. Its depth is variable, see § 6.04.

The rocks above the Moho are known as the *crust.* Those between the Moho and the core are the *mantle.*

6.04. The structure of the crust

The structure of the outer crust is irregular. At the surface, the interface between the air (density 0.0) and the rock (density 2–3 g cm^{-3}) generally slopes at angles of not more than a few degrees, but may be vertical. Under the sea, very steep slopes are probably rare, but the slope can average 3° or 4° over a distance of 100 km between the continental shelf and the deep ocean.

Within the crust there are, roughly speaking, three classes of rock.

(*a*) *Sedimentary rocks* with a thickness of between 0 and several kilometres. Their density is between 1.8 and 2.8, but the density of a large volume of sediment at a depth below the surface of more than 1 km will seldom be less than 2.4. In the deep ocean distant from land the sediments may be only a few hundred metres thick, but at the edge of the continental shelf there will be several kilometres of sediments.

(*b*) *Granitic rocks*, of average density 2.65 and of thickness probably 10–15 km in continental areas, but generally absent under the deep ocean. There may be no sharp discontinuity between this layer and the next.

(*c*) *Basaltic rocks*, of average density 2.87, continue down to the Moho, which may be at a depth of only 11 km below sea-level under the oceans, at about 35 km under ordinary land areas, and at 50 km or more under mountainous areas.

When the granitic and basaltic rocks are clearly separated, their interface is known as the *Conrad* discontinuity. The granitic layer (*b*) is then referred to as the upper layer, and the basaltic (*c*) as the intermediate layer. The density of the rocks, known as the lower layer, below the Moho is thought to be 3.32 g cm^{-3}.

[614], pp. iii, 7–35 lists the densities of many sedimentary and granitic rocks found at the surface and (of greater significance) at depth.

See [614] and [453] for fuller details of crustal structure.

Fig. 6.2 conventionally shows a typical section of the crust. A notable feature is that where the average land surface is high, the Moho is deep, and vice versa. The consequence (or cause) of this is that the total mass in vertical columns of unit cross-section, between the ground level (or the water level at sea) and a depth of perhaps 60 km or greater, tends to be constant. This tendency is known as *isostasy*. Where the masses of such columns above a certain depth are substantially equal, the rock above that depth stands in hydrostatic equilibrium on the lower layers, and the isostasy or *compensation* is said to be exact or perfect. The excess masses above sea-level, and the defective mass of the ocean, are said to be compensated by defects or excesses of mass below them. It is clear that columns whose cross-sections are say 250 × 250 km are likely to be more

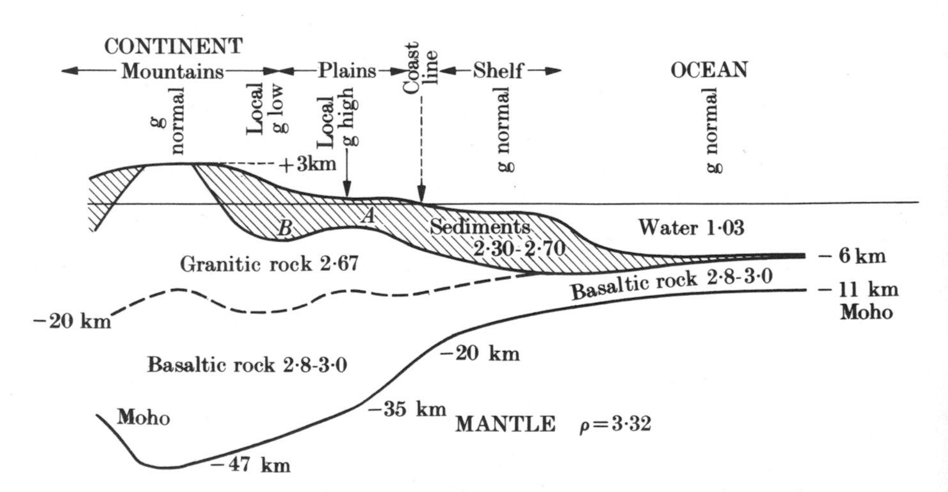

FIG. 6.2. Typical crustal structure. The undulations at A and B are to illustrate local abnormalities which are not necessarily typical of the situations in which they are drawn. The figure is not drawn to illustrate a coast at which a crustal plate is being forced forward and downwards under a continent.

closely compensated than columns of 100×100 km. Compensation tends to be *regional*.

The compensation of individual continents and oceans, taken as a whole, is reasonably nearly exact.† Some large mountain ranges are very closely compensated, such as the Alps, while others are not. Clearly a single mountain peak is unlikely to be individually compensated except by chance.

Departures from isostasy are to some extent measures of the earth's strength.

See further in Section 11.

6.05. Gravity anomalies

Variations in the intensity of gravity are the most direct geodetic guide to variations in the density and thickness of the underlying crust.

As a first approximation, gravity at sea-level varies with latitude from about 978 Gal (9.78 m s^{-2}) at the equator to about 983 Gal at the poles, and in the open air above the oceans it decreases with height by about 1 Gal (10 mm s^{-2}) per 3250 metres, § 6.23. The difference $g-(\gamma_0-h/3250)$, where g is observed gravity in Gal, γ_0 is as in (6.105), and h is height in metres, is then known as the *free-air gravity anomaly*, or $g-\gamma_A$, defining γ_A. This anomaly is the difference between g on the earth and gravity on or outside a standard spheroidal earth.

The most striking difference between the actual earth and such a standard body, when observations are made on land, is that the space between sea-level and the ground-level point of observation is not filled with air but with rock, which increases g. When h is large, $g-\gamma_A$ might be expected to indicate little but that fact, and over deep sea it might similarly be expected to reflect the relatively low density of water. The obvious remedy is then to calculate the attraction of all significant masses above sea-level and of the defective mass of the ocean, to add the total to γ_A, and to form the *topographical anomaly*, $g-\gamma_T$. This anomaly is the difference between g and gravity on a standard earth of the true superficial form, but of regular internal density. It is thus a guide to inequalities of density.

The utility of the anomaly $g-\gamma_T$ is, in some connections, limited by the fact that the earth's major surface features are approximately isostatically compensated. The anomaly may indicate no more than this already known fact. To remedy this, the compensation may be assumed to adopt

† Charts of the sea-level equipotential surface, and tables of the harmonic constants describing it as given by satellite orbits, show practically no correlation with the pattern of continents and oceans, Fig. 6.22. The 3-degree terms in Table 7.8 are perhaps the nearest approach to an exception.

one of various prescribed standard forms, §§ 6.63–6.66, and its attraction (positive or negative) can be calculated and added to γ_T, to give $g-\gamma_C$, an *isostatic anomaly*. This is the difference between g and gravity on a standard earth of the correct form, whose mountains and oceans are compensated in the prescribed way.

These anomalies $g-\gamma_A$, $g-\gamma_T$ and the various forms of $g-\gamma_C$ all have their uses in different circumstances. See Sections 7 and 11.

General references for geophysics. [305], [137], [165], [453], and [541]. Also [151] for Terrestrial Magnetism, [184] for Oceanography, and [409] for Earth Tides.

SECTION 2. THE ATTRACTION OF STANDARD BODIES

6.06. General formulae for attraction

The mutually attractive force of two particles of mass m and M separated by a distance r is GmM/r^2 along the line joining them. If F is the resulting acceleration of m, usually referred to as the *attraction* of M at the point occupied by m,

$$F = GM/r^2 = G\rho v/r^2, \tag{6.1}$$

where v is the volume of the particle M, ρ is its density, and G is the *gravitational constant*. In traditional units in which $\rho = 1\ \text{g cm}^{-3}$, $G = (6.672 \pm 0.004) \times 10^{-8}\ \text{cm}^3\ \text{g}^{-1}\ \text{s}^{-2}$ and (6.1) will give F in Gals. In S.I. units $G = 6.672 \times 10^{-11}\ \text{m}^3\ \text{kg}^{-1}\ \text{s}^{-2}$, [53].

The product GM, where M is the mass of the earth including the atmosphere is $3.986\ 005 \times 10^{14}\ \text{m}^3\ \text{s}^{-2} \pm 0.2$ ppm. Separately, G and M are less well known than their product. The figure given for G corresponds to $5.517 \pm 0.004\ \text{g cm}^{-3}$ for the earth's mean density.

At a point P the component, in any direction PS, of the attraction of a body of finite size is given by

$$F = G \iiint \rho \cos\theta \, dv/r^2, \tag{6.2}$$

where ρ is mass per unit volume and θ is the angle between PS and the direction of each element. When the form and density of the body cannot conveniently be expressed in simple mathematical terms, the integration must be carried out by quadratures.

It is sometimes convenient to regard a surface as covered by a thin layer of so many kg per square metre, or to regard the density of a line as so many kg per linear metre.

6.07. Potential

The attractive potential V at a point P of a number of particles m_1, m_2, etc., at distances r_1, r_2, etc., from P is defined to be

$$V = G\sum m/r, \tag{6.3}$$

and for bodies of finite size

$$V = G\iiint (\rho/r)\,\mathrm{d}v. \tag{6.4}$$

Note that potential is a scalar, not a vector.

Take a single particle m_1, Fig. 6.3 (a), and consider the potential V' at a point P′ distant ds from P in such a direction that APP′ = any angle ϕ. Then

$$\begin{aligned} V' - V &= \frac{\partial}{\partial s}\left(\frac{Gm_1}{r_1}\right)\mathrm{d}s = G\frac{m_1}{r_1^2}\cos\phi\,\mathrm{d}s \\ &= \mathrm{d}s \times (\text{component of attraction at P in direction PP}'). \end{aligned} \tag{6.5}$$

From this it follows that if δV is the difference of potential between two points P and P′, the work done by the attractive forces when a body of mass m is moved from P to P′ is $m\,\delta V$, positive if the potential at P′ is greater than at P. An alternative definition of the potential at P is then that the work done by the attractive forces is mV, when they move a mass to P from some point infinitely distant from all attracting matter, where $V = 0$.

Potential is not quite the same thing as potential energy. The sign is opposite, increasing downwards instead of upwards, and its dimensions are those of $\mathrm{m^2\,s^{-2}}$ instead of $\mathrm{kg\,m^2\,s^{-2}}$. But, see § 3.00, the GPU used for recording dynamic heights has the same sign as potential energy.

A surface on which V is constant is known as an *equipotential* or *level surface* of the attractive forces concerned. Formula (6.5) shows that on such a surface the tangential component of the attraction is zero, so the total resultant force is normal to the surface. The attraction is given by

$$F = -\partial V/\partial n, \tag{6.6}$$

where n is measured normal to the surface, positive outwards in the opposite direction to F. In other words, the separation between two near equipotential surfaces at different points varies inversely as the attraction. See Fig. 6.3 (b). It also follows that in equilibrium, any unconstrained liquid surface, such as mean sea-level, must be an equipotential (but see § 6.08), that gravity at sea-level is everywhere normal to the geoid, and

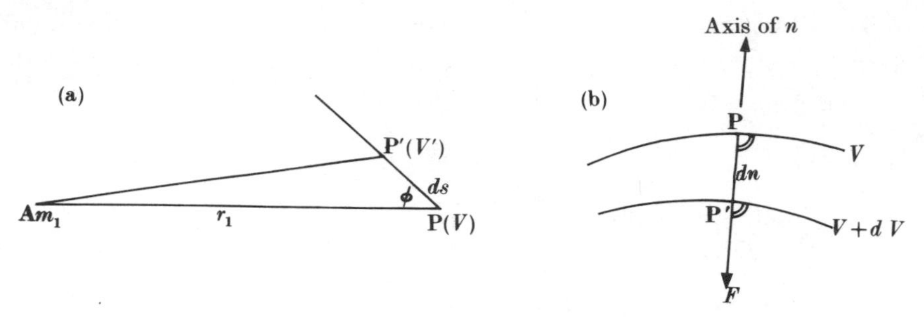

FIG. 6.3. In (b) The force along the normal $PP' = -\partial V/\partial n$, and is inversely proportional to the separation of the two surfaces.

that the separation of any two equipotential surfaces in different places is inversely proportional to the varying values of g.

In vector notation $\mathbf{F} = \text{grad } V$, see § E.11 where the axis of N is in the direction of increasing V, so $F = \partial V/\partial N$.

6.08. Rotating axes

When considering a rotating body such as the earth, it is convenient to employ axes which rotate with it, and so to be able to regard as stationary any point P which is fixed to the earth. This is possible, and the rotation can be ignored, provided a *centrifugal force* $\omega^2 p$, where ω is the angular velocity and p is the perpendicular from P to the axis of rotation, is considered to act on every particle of unit mass outwards along the perpendicular p.† When rotating axes are used this $\omega^2 p$ must then be vectorially added to (6.2). See §§ I.00 and I.01.

In these circumstances surfaces on which $V = \text{constant}$ cease to be the equilibrium surfaces of free liquids, which are now given by

$$U = V + \tfrac{1}{2}\omega^2 p^2 = \text{constant}, \tag{6.7}$$

since $\frac{1}{2}\omega^2 p^2 \equiv \int \omega^2 p \, dp$ is the work done by the centrifugal force when unit mass moves to P from the axis of rotation, where the centrifugal potential is zero. In what follows, when rotating axes are used (as they generally are on the earth, but not for artificial satellites), surfaces on which U is constant are referred to as equipotential or level surfaces, and the gravitational acceleration is

$$\gamma = -\partial U/\partial n \quad \text{and vector } \boldsymbol{\gamma} = \text{grad } U. \tag{6.8}$$

It is convenient to use the word *attraction*, F, for the acceleration

† If a body is moving relative to the rotating axes, a Coriolis force must also be added, see § I.02, but this does not arise in the present context.

resulting from the attractive force. When rotating axes are used, the word *gravity* implies the combined attractive and centrifugal forces, and γ is the resulting acceleration relative to the rotating frame. V and U are the corresponding potentials. When the reference is to the actual earth, rather than to a model, g and W take the place of γ and U.

6.09. Attraction and potential of bodies of simple form

The following cases can be derived from the formulae of §§ 6.06 and 6.07. See Fig. 6.4 (a)–(f). Total attraction at P $= F$, with components X, Y, Z. $V =$ potential. Unless otherwise stated the body is of uniform density, P is external to it, and axes are non-rotating.

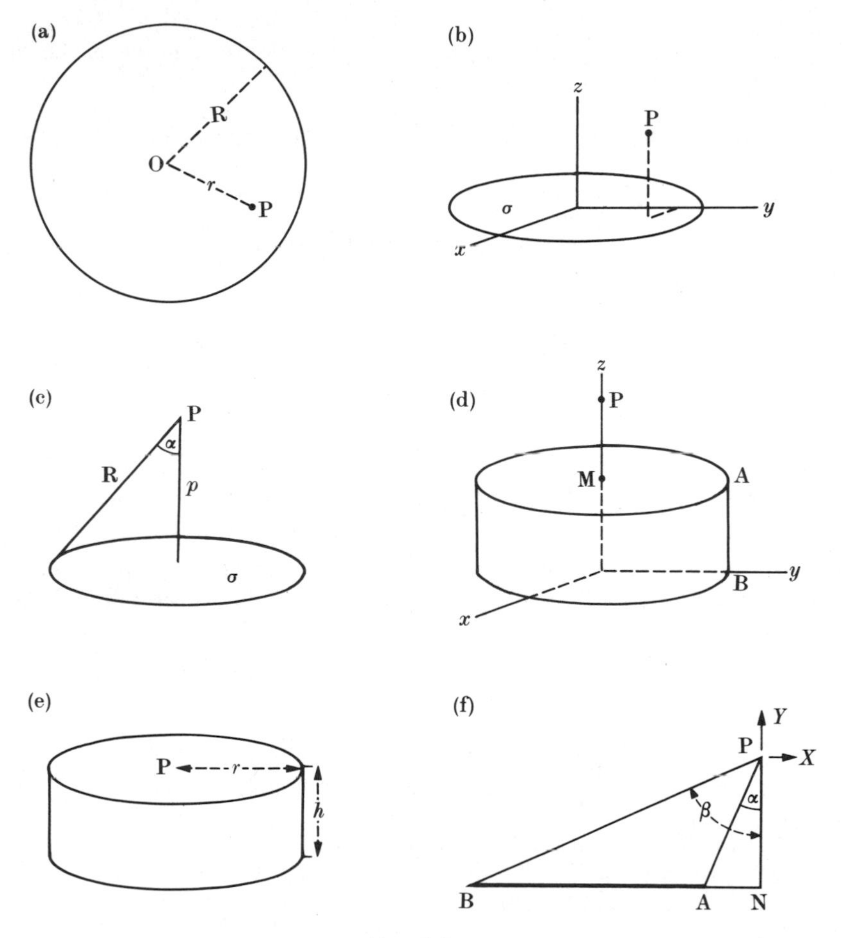

FIG. 6.4.

(*a*) *Spherical shell and solid sphere. External point.*

$$F = GM/r^2 \text{ towards the centre, and } V = GM/r, \quad (6.9)$$

where r is the distance from P to the centre, and M is the total mass. Note that at an external point uniform spheres and spherical shells, and spheres which are not uniform but whose density depends only on distance from the centre, can be regarded as concentrated at their centres.

(*b*) *Numerical example.* Consider the attraction of a (spherical) earth at a point on its surface. $r = 6.37 \times 10^6$ m, $\rho = 5.517 \times 10^3$ kg m^3, $G = 6.672 \times 10^{-11}$ kg^{-1} m^3 s^{-2} (§ 6.06). Whence $M = (4\pi/3)\rho r^3 = 5.98 \times 10^{24}$ kg and $F = 9.80$ m s^{-2}, as is approximately correct.

(*c*) *Spherical shell or sphere. Internal point.* Fig. 6.4 (a). The attraction of a uniform spherical shell at a point inside it is zero, so that the attraction of a uniform sphere at an internal point is GM'/r^2 towards the centre, where r is the distance from P to the centre, and M' is the mass inside the sphere of radius r.

$$\text{Inside a uniform shell of radius } R, \quad V = GM/R = \text{constant.} \quad (6.10)$$

$$\text{Inside a uniform sphere of radius } R, \quad V = \tfrac{2}{3}\pi G\rho(3R^2 - r^2). \quad (6.11)$$

(*d*) *Thin plane plate* of uniform surface density σ. Fig. 6.4 (b).

$$-Z = G\sigma w, \quad (6.12)$$

where w is the solid angle subtended by the plate at P. Hence for an infinite plate, or when P is very close to the centre part of a finite plate

$$F = -Z = 2\pi G\sigma. \quad (6.13)$$

This is an important result. Consider a plane layer of ordinary rock 9 m thick, of typical density 2.67 g cm^{-3}. Then $\sigma = 2.4 \times 10^4$ kg m^{-2}, and (6.13) gives the attraction as 10 μm s^{-2} (1 mGal), or 1.12 μm s^{-2} per metre of thickness. Note that this is independent of the distance of P from the layer, provided only that the extent of the layer in all directions is large compared with the distance of P from it.

It follows that over an infinite plate the equipotential surfaces are parallel planes equally spaced, and that at distance p from the plate

$$V = C - 2\pi G\sigma p, \quad \text{where } C \text{ is a large constant.} \quad (6.14)$$

(*e*) *Disc and cylinder.* For a thin disc (6.12) applies, but when P is on the axis it simplifies to

$$F = 2\pi G\sigma(1 - \cos\alpha) \text{ along the axis,}$$

$$V = 2\pi G\sigma(R - p), \quad (6.15)$$

R, p, and α being as in Fig. 6.4 (c).

For a *cylinder*, P external and on the axis, Fig. 6.4 (d),

$$F = 2\pi G\rho(\text{AB}+\text{PA}-\text{PB}) \text{ along the axis.} \tag{6.16}$$

And if P is at the centre of the upper surface of the cylinder, Fig. 6.4 (e),

$$F = 2\pi G\rho\{h + r - \sqrt{(r^2+h^2)}\} \text{ along the axis}$$
$$= \text{Attraction of an infinite plate} \times (1 - h/2r + \ldots).$$

If $h < r/5$, the factor in brackets exceeds 90 per cent.

$$V = \pi G\rho\left\{h\sqrt{(r^2+h^2)} - h^2 + r^2 \ln\left(\frac{h+\sqrt{(r^2+h^2)}}{r}\right)\right\}. \tag{6.17}$$

If P lies on the axis inside the cylinder, F is the difference of the attractions of the parts above and below, and V is the sum of their potentials.

(*f*) Straight rod of line density σ. AB in Fig. 6.4 (f).

$$-Y = (G\sigma/\text{PN})(\sin\beta - \sin\alpha),$$
$$-X = G\sigma(1/\text{PA} - 1/\text{PB}). \tag{6.18}$$

The resultant bisects the angle BPA. It follows that the equipotential surfaces are prolate spheroids with foci at B and A, and that

$$V = G\sigma \ln\left(\frac{a+l}{a-l}\right), \tag{6.19}$$

where $2a = \text{PB}+\text{PA} = \text{constant}$, and $2l$ is the length of the rod.

(*g*) *Rectangular blocks and plates.* The potential at P of a homogeneous rectangular block is simply expressed in terms of each of its faces considered as of unit surface density, and each of their potentials can be expressed in terms of the solid angles they subtend at P and the potentials of each of their sides, the latter being given by (6.19). See [502], pp. 130–3, for details. The final result is lengthy, and MacCullagh's theorem, (*i*) below, is often a good substitute.

(*h*) *Solid ellipsoid.* Expressions can be got for the attraction and potential of a uniform solid ellipsoid, but in general only in terms of elliptic functions or other integrals for which tables are not readily available. See [502], pp. 97–130, and [244], pp. 2–8.

A comparatively simple case is the potential at an internal point (x, y, z) of a solid homogeneous ellipsoid $x^2/a^2 + y^2/b^2 + z^2/c^2 = 1$, with small ellipticities $f_1 = (a-c)/a$ and $f' = (a-b)/a$, $a > b > c$. Then [244], pp. 2–3,

$$V = \pi Gabc\rho(A - A_1x^2 - A_2y^2 - A_3z^2), \tag{6.20}$$

where

$$\tfrac{1}{2}aA = 1+\tfrac{1}{3}(f_1+f')+\tfrac{2}{15}(f_1^2+f'^2)+\tfrac{1}{5}f_1f'+...,$$
$$\tfrac{1}{2}a^3A_1 = \tfrac{1}{3}+\tfrac{1}{5}(f_1+f')+\tfrac{4}{35}(f_1^2+f'^2)+\tfrac{1}{7}f_1f'+...,$$
$$\tfrac{1}{2}a^3A_2 = \tfrac{1}{3}+\tfrac{1}{5}(f_1+3f')+\tfrac{1}{35}(4f_1^2+27f'^2)+\tfrac{3}{7}f_1f'+...,$$
$$\tfrac{1}{2}a^3A_3 = \tfrac{1}{3}+\tfrac{1}{5}(3f_1+f')+\tfrac{1}{35}(27f_1^2+4f'^2)+\tfrac{3}{7}f_1f'+... .$$

The internal equipotential surfaces are then concentric ellipsoids with corresponding ellipticities of approximately $\tfrac{3}{5}f_1$ and $\tfrac{3}{5}f'$, and unless the formulae are modified to include rotation, the bounding surface is only an equipotential if $f_1 = f' = 0$, i.e. if the ellipsoid is a sphere.

Formula (6.20) is applicable to a point on the surface, and there the three components of the attraction will be

$$\left.\begin{aligned} X &= \partial V/\partial x = -2\pi Gabc\rho A_1 x \\ Y &= \partial V/\partial y = -2\pi Gabc\rho A_2 y \\ Z &= \partial V/\partial z = -2\pi Gabc\rho A_3 z \end{aligned}\right\}. \tag{6.21}$$

(*i*) *Any distant body, MacCullagh's theorem.* As a first approximation, attraction and potential at P are given by $F = GM/r^2$ and $V = GM/r$, where M is the total mass of the attracting body, O its centre of mass, and $r = \text{OP}$. For a closer approximation:

$$V = \frac{GM}{r} + \frac{G}{2r^3}(A+B+C-3I)+..., \tag{6.22}$$

where A, B, and C are the moments of inertia about any three mutually perpendicular lines through O, and I is the moment of inertia about OP. Note that the absence of a $1/r^2$ term in the external potential of a body indicates that r is measured from its centre of mass. [502], pp. 66–7.

Section 3. Green's and Clairaut's Theorems

6.10. Introductory

Given the external form of the earth and its internal density distribution in sufficient detail, the direction and intensity of gravity at any external point could clearly be calculated by some adaptation of (6.2). In fact, the internal density is not adequately known, but the external gravity field can (in principle) be determined without specific knowledge of it, by theorems due to Green and Clairaut.

In summary, Green's theorem states that if the intensity of gravity is known all over an equipotential surface such as the geoid, all matter within that surface may be replaced by a surface coating of density $g/4\pi G$

per unit area, without changing the external field. Such a coating is known as a *Green's equivalent layer.* See in more detail in § 6.14.

The second theorem, Clairaut's, states that if the form of an equipotential surface, which is external to all matter, is known, the intensity of gravity at all points on it is also known, and vice versa except for an unknown scale.†

The treatment given in §§ 6.15–6.18 demands that the irregularities of the equipotential surface should be expressed as a series of spherical harmonics. The corresponding variations of gravity are then obtained in similar form. For the converse process, determining the form of the surface from observations of the intensity of gravity on it, Section 6, §§ 6.30–6.31 gives alternative formulae due to Stokes, for numerical computation.

The consequence of Green's and Clairaut's theorems is that, apart from difficulties due to some of the earth's mass being external to the geoid, if g on the geoid is everywhere known, the form of the geoid is known and also the external field.

6.11. Laplace's theorem

In (6.3) let (ξ, η, ζ) be the coordinates of any particle m, and let (x, y, z) be those of P. Then

$$r^2=(x-\xi)^2+(y-\eta)^2+(z-\zeta)^2, \qquad r\,\partial r/\partial x = x-\xi,$$

$$\frac{\partial V}{\partial x}=-\frac{Gm}{r^2}\frac{\partial r}{\partial x}=-\frac{Gm}{r^3}(x-\xi), \quad \text{and} \quad \frac{\partial^2 V}{\partial x^2}=-\frac{Gm}{r^3}+\frac{3Gm}{r^5}(x-\xi)^2,$$

with similar expressions for $\partial^2 V/\partial y^2$ and $\partial^2 V/\partial z^2$.

Whence, by addition,

$$\frac{\partial^2 V}{\partial x^2}+\frac{\partial^2 V}{\partial y^2}+\frac{\partial^2 V}{\partial z^2}=0, \tag{6.23}$$

or in vector notation $\nabla^2 V=0$, as in § E.20.

If P coincides with any particle m, $\partial V/\partial x$ is indeterminate and the equation does not hold, see § 6.12, but otherwise, sincc it holds for each particle it holds for the whole of any attracting body. It is known as *Laplace's equation,* but must not be confused with his other equation in § 2.05.

With rotating axes as in § 6.08,

$$\nabla^2 U=\nabla^2 V+\nabla^2\{\tfrac{1}{2}\omega^2(x^2+y^2)\}=2\omega^2, \tag{6.24}$$

the z-axis being the axis of rotation, but the result is independent of the direction of the axes.

† The determination of the finer details of the gravity is unstable, see after (6.67).

6.12. Poisson's theorem

If in § 6.11, the point P is occupied by matter, describe a sphere containing P with centre near but not exactly at P, and with a very small radius. Within this sphere ρ can be considered constant. Then at P, $V = V_1 + V_2$, where V_1 is the potential of the matter inside the sphere, and V_2 that of matter outside. By (6.23) $\nabla^2 V_2 = 0$. Then differentiating (6.11) with respect to x, y, and z as in § 6.11 gives $\nabla^2 V_1 = -4\pi G\rho$, which is therefore the value of $\nabla^2 V$. [502], pp. 51–2.

Hence within matter of density ρ,

$$\nabla^2 V = -4\pi G\rho \tag{6.25}$$

and with rotating axes,

$$\nabla^2 U = -4\pi G\rho + 2\omega^2. \tag{6.26}$$

6.13. Gauss's theorem

Let S be a closed surface, within which there are various bodies of total mass M_1, with other bodies of mass M_2 outside the surface, and let F_n ($\equiv \mathbf{F} \cdot \mathbf{n}$) be the component of the attraction of all the bodies in the direction of the normal to the surface, positive outwards. Then, as proved in § E.15,

$$\int F_n \, \mathrm{d}S = -4\pi G M_1, \tag{6.27}$$

the integral being over the surface S. The result depends only on the total internal mass, and the integral is zero if all the masses are external.

6.14. Green's equivalent layer

A theorem due to Green, proved in § E.21, states that if S is any closed surface, and if ϕ_1 and ϕ_2 are any two scalar quantities which have varying values at all points inside and outside S,

$$\iint \left(\phi_1 \frac{\partial \phi_2}{\partial n} - \phi_2 \frac{\partial \phi_1}{\partial n}\right) \mathrm{d}S = \iiint (\phi_1 \nabla^2 \phi_2 - \phi_2 \nabla^2 \phi_1) \, \mathrm{d}v, \tag{6.28}$$

where n is distance along the outward normal. The first integral is taken over the surface S, and the second throughout the volume contained by it.

Now, as in Fig. 6.5, let S be an equipotential surface of the attraction of all matter inside it and outside it, such as the geoid (apart from the distinction between V and U, for which see later). Let P be a fixed point outside S, and let P′ be a (variable) point inside or on S. Let ρ be the variable density of matter inside S, and let PP′ $= r$.

Then in (6.28) let ϕ_1 be the potential at P′ of all matter both inside and

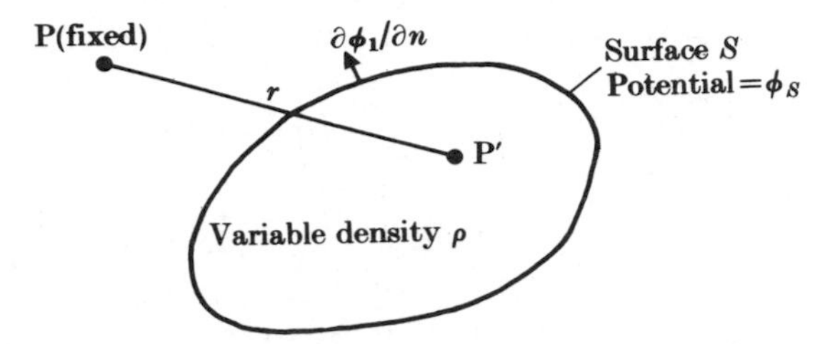

FIG. 6.5. ϕ_1 = potential at P'. $\phi_2 = 1/r$.

outside S. So on S, $\phi_1 = \phi_S$, a constant. Let $\phi_2 = 1/r$, namely the potential at P′ of a non-existent mass numerically equal to $1/G$ located at P.

We now examine the four terms of (6.28).

(i) $$\iint \phi_1 \frac{\partial \phi_2}{\partial n} \mathrm{d}S = \phi_S \iint \frac{\partial}{\partial n} (\text{potential of mass } 1/G \text{ at P})\, \mathrm{d}S$$

$$= 0, \quad \text{by (6.27) since P is external to } S. \qquad (6.29)$$

(ii) $\iiint \phi_1 \nabla^2 \phi_2 \,\mathrm{d}v$ is also zero, because ϕ_2 equals the potential at P′ of a mass at P. P′ is on or inside S and P is outside, so P′ cannot coincide with P. So $\nabla^2 \phi_2 = 0$ throughout the volume of the integration, § 6.11.

(iii) Inside S, $\nabla^2 \phi_1 = -4\pi G\rho$, where ρ is the variable density of the matter at the variable point P′, from (6.25). So

$$-\iiint \phi_2 \nabla^2 \phi_1 \,\mathrm{d}v = +\iiint \frac{1}{r} 4\pi\rho G \,\mathrm{d}v$$

$$= +4\pi(\text{potential at P of all matter within } S). \qquad (6.30)$$

(iv) $$-\iint \phi_2 \frac{\partial \phi_1}{\partial n} \mathrm{d}S = +\iint \frac{1}{r} F_0 \,\mathrm{d}S$$

$$= +\iint (\text{potential at P of matter of density } F_0/G \text{ per unit area})\, \mathrm{d}S, \qquad (6.31)$$

where F_0 is the total attractive force at each point on S.

Equating (iii) and (iv) then gives, at any external point P,

$$V_P = (\text{potential at P of all matter within } S)$$

$$= (\text{potential of a surface density of } F_0/4\pi G \text{ per unit area on } S). \quad (6.32)$$

The surface density is known as *Green's equivalent layer*.

With rotating axes, [233], pp. 382–3, ϕ_1 is taken to be U instead of V, as in (6.7), and S is an equipotential of U.

Then $\nabla^2\phi_1 = \nabla^2 U = -4\pi G\rho + 2\omega^2$ from (6.26), and (6.28) becomes

$$\iiint \frac{1}{r} 4\pi G\rho \, dv - 2\omega^2 \iiint \frac{1}{r} \, dv = \iint \frac{1}{r} \gamma_0 \, dS,$$

$$4\pi V_P = \iint \frac{\gamma_0}{r} dS + 2\omega^2 \iiint \frac{1}{r} dv, \quad (6.33)$$

or, in words, after dividing by 4π, at any external point P

(V of matter inside S)

$= (V$ at P of a surface density $\gamma_0/4\pi G$ on $S) +$

$+ (\omega^2/2\pi)$ (V at P of uniform density $1/G$ throughout the volume enclosed by S), (6.34)

where γ_0 is the vector sum of the centrifugal acceleration and the attraction of all masses internal and external, and V is attractive potential only.

Finally, U, the total potential at P, is given by

$$U = \{V \text{ of matter inside } S \text{ as in (6.34)}\} + (V \text{ of matter outside } S) + \tfrac{1}{2}\omega^2 p^2, \quad (6.35)$$

as in (6.7).

Equation (6.34) gives the potential at an external point. At a point actually on the surface the potential of the surface layer $\gamma_0/4\pi G$ also equals that of the internal masses, but its attraction is indeterminate. Just outside S the attraction of the small underlying element of density $\gamma_0/4\pi G$ is $\frac{1}{2}\gamma_0$, from (6.13), while the whole of the rest of the surface layer contributes the other $\frac{1}{2}\gamma_0$. Just inside S the local $\frac{1}{2}\gamma_0$ is reversed, so that the total attraction of the layer is zero.

6.15. Potential expressed in spherical harmonics

See Appendix H. In Fig. 6.6 (a) let P′ (r, θ, λ) be an element of attracting matter of mass m, and let P $(r_1, \theta_1, \lambda_1)$ be an external point at which the potential is required. Let $\text{POP}' = \theta'$. Then at P, $V = G \sum m/R'$, where $R' = \text{PP}' = \sqrt{(r^2 + r_1^2 - 2rr_1 \cos\theta')}$, and expanding $1/R'$ in terms of zonal harmonics, as in (H.17) gives

$$V = G \sum \frac{m}{r} \left\{1 + \left(\frac{r_1}{r}\right) P_1(\cos\theta') + \left(\frac{r_1}{r}\right)^2 P_2(\cos\theta') + \ldots\right\},$$

on account of matter whose $r > r_1$ (6.36)

and

$$V = G \sum \frac{m}{r_1} \left\{1 + \left(\frac{r}{r_1}\right) P_1(\cos\theta') + \left(\frac{r}{r_1}\right)^2 P_2(\cos\theta') + \ldots\right\},$$

on account of matter whose $r < r_1$, the more usual case. (6.37)

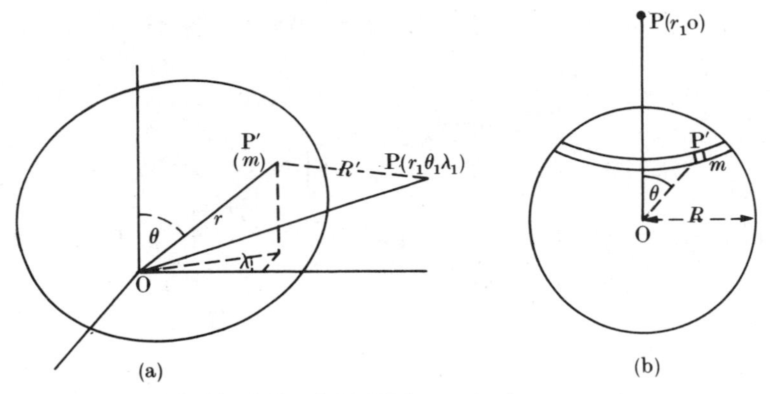

FIG. 6.6. (a) POP′ = θ'. (b) OP is an axis of symmetry for m.

If these are to be summed analytically, m must be expressed as a function of r, θ, λ, so that (6.37) can be integrated. Consider the following cases.

(*a*) *Axially symmetrical spherical shell.* Let P be on the axis of θ, Fig. 6.6 (b), so that $\theta' = \theta$. Let the masses m be confined to a spherical shell of radius R ($R < r_1$), and centre O, and let the density of the shell σ be independent of λ, so that m ($= \sigma R^2\,\mathrm{d}w$) can be expanded in a series of zonal harmonics

$$m = R^2\,\mathrm{d}w(a_0 + a_1P_1 + a_2P_2 + \ldots). \tag{6.38}$$

Substituting this in (6.37),

$$V \text{ at } P = \frac{GR^2}{r_1}\left\{\sum_0^n \int a_n(P_n)^2\left(\frac{R}{r_1}\right)^n \mathrm{d}w + \right.$$
$$\left. + \sum \text{ integrals involving } P_nP_m, \text{ which are zero by (H.21)}\right\}, \tag{6.39}$$

where the integrals are over the whole sphere, or between $\cos\theta = \pm 1$, and $\mathrm{d}w = 2\pi \sin\theta\,\mathrm{d}\theta$ or $-2\pi\,\mathrm{d}(\cos\theta)$, since σ is independent of λ.

So, using (H.23),

$$V = 4\pi GR \sum_0^n \frac{a_n}{2n+1}\left(\frac{R}{r_1}\right)^{n+1}. \tag{6.40}$$

This then gives the potential of an axially symmetrical spherical shell at an external point on its axis.

(*b*) *General spherical shell.* In Fig. 6.7 OZ is the axis of $\theta = 0$, and ZG is the meridian of $\lambda = 0$. Let P (r_0, θ_0, λ_0) be the point at which the potential is required, with OP cutting the sphere at Z_0. Let the density of the shell be expressed as a series of surface spherical harmonics with OZ

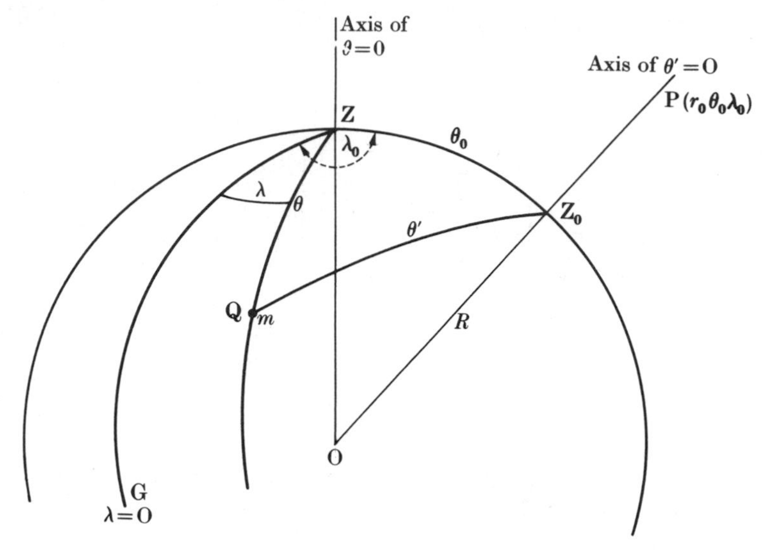

FIG. 6.7. ZQ $=\theta$ (variable). $Z_0Q=\theta'$ (variable). $ZZ_0=\theta_0$ (fixed).

as axis,

$$m = R^2\,\mathrm{d}w \sum_0^n Y_n(\cos\theta), \tag{6.41}$$

where each harmonic Y_n contains $(2n+1)$ constants, c_n, C_{nm}, and S_{nm} as in (H.32). Then, considering the mass at $Q(\theta, \lambda)$, writing $Z_0Q=\theta'$, and substituting (6.41) in (6.37), in which OZ_0 is the axis of $\theta'=0$, we have

$$V \text{ at } \mathrm{P} = \frac{GR^2}{r_0}\sum_0^n \left(\frac{R}{r_0}\right)^n Y_n(\cos\theta)P_n(\cos\theta')\,\mathrm{d}w$$

$$+\text{integrals involving } Y_m(\cos\theta)P_n(\cos\theta'). \tag{6.42}$$

From (H.60) and (H.62), in which θ' is written as γ, this is

$$4\pi GR \sum_0^n \frac{1}{2n+1}\left(\frac{R}{r_0}\right)^{n+1} Y_n', \tag{6.43}$$

where Y_n' is the value of $Y_n(\cos\theta)$ at Z_0, the point where OP cuts the sphere. Y_n' is thus a function of the constants C_{nm} etc., and of θ_0 and λ_0, the coordinates of Z_0.

Now let r_0, θ_0, and λ_0 become variables, so that V is the general expression for the external potential. Then the only necessary change in (6.43) is to write Y_n $(\cos\theta)$ instead of Y_n'.

So
$$V = 4\pi GR \sum_0^n \frac{Y_n(\cos\theta)}{2n+1}\left(\frac{R}{r}\right)^{n+1}, \tag{6.44}$$

and $$\text{Attraction} \approx -\frac{\partial V}{\partial r} = 4\pi G \sum_0^n \frac{n+1}{2n+1} Y_n(\cos\theta)\left(\frac{R}{r}\right)^{n+2}, \tag{6.45}$$

a good approximation provided the distribution of mass is such that the attraction is approximately radial.

Since R is a constant, (6.44) can be written

$$V = \sum_0^n \frac{Y_n(\cos\theta)}{r^{n+1}} = \left(\frac{Y_0}{r} + \frac{Y_1}{r^2} + \frac{Y_2}{r^3} + \ldots\right), \tag{6.46}$$

where the Y's now have different constants incorporating the fractions $(4\pi G R^{n+2})/(2n+1)$. So if the potential is known as a series of Y's all over a sphere of radius R, on a sphere of greater radius, r, the constant in Y_0 will be less in inverse proportion to the first power of r, and those in Y_n in proportion to $(1/r)^{n+1}$.

These results, giving the potential and attraction at an external point of any spherical shell whose surface density can be expressed as a series of spherical harmonics, illustrate the value of spherical harmonics in geodetic problems.

Numerical example. Consider the attraction at the earth's surface, supposed spherical, of a skin density, such as the topography supposed condensed at sea-level, represented by a series of spherical harmonics $\sum Y_n$, in which the constants represent mass per unit area, or a thickness of so many metres of rock of normal density. Then at sea-level $r = R$,

$$V = 4\pi G R \sum Y_n/(2n+1),$$

and $$\Delta g = -\partial V/\partial r = 4\pi G \sum Y_n(n+1)/(2n+1).$$

Now suppose that the topography takes the form of a series of parallel mountains and seas of amplitude (top to bottom) $2H$ metres† and 'wavelength' (crest to crest) L. Then $\sum Y_n$ can locally be approximately represented by a single Y_n, whose $n = 2\pi R/L$. Hence Δg will vary between $\pm 2\pi G H\rho$, since $(n+1)/(2n+1) \approx \frac{1}{2}$. And putting $\rho = 2.67$, Δg becomes $10\,\mu\text{m s}^{-2}$(1 mGal) per 9 metres of H, the same result as in § 6.09 (d).

The undulation of the geoid, $\pm N$, given by V/g, will be

$$\pm 4\pi G R H\rho/g(2n+1) = \pm 4\pi(6.67\times10^{-8})(6.4\times10^{8})(2.67)H/980(2n+1)$$
$$= \pm 1.45H/(2n+1),$$

which varies approximately inversely as n and directly as L. If $H = 300$ m

† Here H represents the mass deficiency of the sea. Allowing for the density of water the depth must be $1.6H$.

and $L = 180$ km, $n = 220$ and the undulations of the geoid will be of the order ± 1 m.

When an anomalous skin density can be expressed by a single harmonic of degree n,

$$\frac{N}{R} = \frac{1}{n+1} \frac{\Delta g}{g}, \tag{6.47}$$

where Δg is the change in g at a (near) external point.

(*c*) In the preceding sub-paragraph the attracting masses are assumed to lie on a spherical shell, but Green's equivalent layer, § 6.14, lies on an equipotential surface whose form is not exactly spherical but which depends on the masses which are assumed to lie on it. The relation between gravity (as opposed to attraction) on an equipotential surface and the form of the surface is a more complicated matter, and is dealt with in §§ 6.16 and 6.17.

The difference is apparent from the factor $(n+1)$ in (6.47), compared with the $(n-1)$ in (6.65) where gravity on an equipotential surface is being considered.

6.16. Clairaut's theorem†

Gravity on an external equipotential surface of a nearly spherical body. This important theorem gives the variations of gravity on a surface S, of known form, which is external to all matter under consideration, and which is an equipotential surface of the rotation and the attraction of the contained matter. Such a surface may be called a *bounding equipotential.* On the earth the geoid is not quite such a surface, as it is neither external to the whole earth, nor exactly an equipotential of the matter inside it, but the theorem is strictly applicable to the modified surfaces, known as co-geoids, described in § 6.35.

Let S be given by

$$r = R(1 + u_1 + u_2 + \ldots) \tag{6.48}$$

where R is the radius of the sphere of equal volume, and u_1, u_2, etc.‡ are surface spherical harmonics with small coefficients. Let the origin be at the centre of volume, so that $u_1 = 0$. The centrifugal acceleration is $\omega^2 r \sin\theta$.

† Originally Clairaut's [152], but later given in more general terms by Laplace and Stokes [542], ii, pp. 104–71. Second and higher powers of f are ignored. For second-order terms see § 6.17. The same result is obtained in [233] and in [159], the latter including terms in f^3.

‡ In this paragraph the notation u_n has been used for the spherical harmonics which describe the surface, while the Y's describe the external potential. In § 6.15 Y's were used to describe density variations. All these harmonics have the same form, as in § H.05, while the nature of the phenomenon which they describe is reflected in the dimensions of their numerical coefficients.

First, we need a formula to give the potential V of the attraction (only) at an external point. Such a formula must satisfy three conditions, namely: (i) It must be a solution of Laplace's equation $\nabla^2 V = 0$, (6.23); (ii) on the surface itself $U = V + \frac{1}{2}\omega^2 r^2 \sin^2\theta$ must be constant; and (iii) at great distances V must tend to zero. The first and third conditions demand that the formula must be a solid spherical harmonic of the form $V = \sum Y_n/r^{n+1}$, § H.09, in which the coefficients in the Y's are so determined that the formula will satisfy the boundary conditions. We write

$$V = \frac{Y_0}{r} + \frac{Y_1}{r^2} + \frac{Y_2}{r^3} + \ldots. \tag{6.49}$$

where the coefficients in the Y's, except Y_0 are small. Then if (6.48) is an equipotential surface of gravity

$$U = V + \tfrac{1}{2}\omega^2 r^2 \sin^2\theta = \text{constant on the surface,} \tag{6.50}$$

from (6.7), since p of that paragraph is $r \sin\theta$. So, using (H.24) to expand $\sin^2\theta$,

$$\frac{Y_0}{r} + \frac{Y_1}{r^2} + \ldots + \tfrac{1}{2}\omega^2 r^2 \{\tfrac{2}{3}P_0(\cos\theta) - \tfrac{2}{3}P_2(\cos\theta)\} = \text{constant.} \tag{6.51}$$

Substituting (6.48) in the first term, but being satisfied with $r = R$ in the smaller ones, gives

$$U = \frac{Y_0}{R}(1 - u_1 - u_2 - \ldots) + \frac{Y_1}{R^2} + \frac{Y_2}{R^3} + \ldots + \tfrac{1}{2}\omega^2 R^2 (\tfrac{2}{3}P_0 - \tfrac{2}{3}P_2) = \text{constant.} \tag{6.52}$$

Separately equating to zero the sums of all terms of degree greater than 0 gives

$$\begin{aligned} Y_1 &= RY_0 u_1 = 0, \text{ since } u_1 = 0 \text{ as above,} \\ Y_2 &= R^2 Y_0 u_2 + \tfrac{1}{3}\omega^2 R^5 P_2 = R^2 Y_0 u_2 - \tfrac{1}{2}\omega^2 R^5 (\tfrac{1}{3} - \cos^2\theta), \\ Y_3 &= R^3 Y_0 u_3, \text{ etc.} \end{aligned}$$

Whence, from (6.49),

V at any external point

$$= Y_0\left(\frac{1}{r} + \frac{R^2 u_2}{r^3} + \frac{R^3 u_3}{r^4} + \ldots\right) - \frac{\omega^2 R^5}{2r^3}(\tfrac{1}{3} - \cos^2\theta), \tag{6.53}$$

where $Y_0 = GM$, M being the total mass† of the earth, since GM/r is the value of V when r is very large. GM and ω are known.

† For the mass of the atmosphere see § 6.29.

To get γ_0, the intensity of gravity on the surface, we have at any external point†

$$\gamma \approx -(\partial U/\partial r) = -(\partial V/\partial r) - (\partial/\partial r)(\tfrac{1}{2}\omega^2 r^2 \sin^2\theta)$$
$$= GM\left(\frac{1}{r^2} + \frac{3R^2 u_2}{r^4} + \frac{4R^3 u_3}{r^5} + \ldots\right) - \frac{3\omega^2 R^5}{2r^4}(\tfrac{1}{3} - \cos^2\theta) - \omega^2 r \sin^2\theta. \tag{6.54}$$

Substituting r from (6.48) to give γ on the surface, and neglecting products of small quantities

$$\gamma_0 = \frac{GM}{R^2}\{1 + u_2 + 2u_3 + 3u_4 + \ldots + (n-1)u_n\} - \tfrac{2}{3}\omega^2 R - \tfrac{5}{2}\omega^2 R(\tfrac{1}{3} - \cos^2\theta). \tag{6.55}$$

If γ_m is the mean value of γ_0 over the surface,

$$\gamma_m = GM/R^2 - \tfrac{2}{3}\omega^2 R = (GM/R^2)(1 - \tfrac{2}{3}m'), \tag{6.56}$$

where $m' = \omega^2 R/\gamma_m$, since the mean of all harmonic terms is zero, (H.35). Then

$$\gamma_0 = \gamma_m\{1 - \tfrac{5}{2}m'(\tfrac{1}{3} - \cos^2\theta) + u_2 + 2u_3 + \ldots(n-1)u_n\}. \tag{6.57}$$

Numerical substitution. Let the earth be an oblate spheroid

$$r = a(1 - f\cos^2\theta),$$

ignoring higher powers of f, or $r = R\{1 + f(\tfrac{1}{3} - \cos^2\theta)\}$ for comparison with (6.48), u_2 then being $f(\tfrac{1}{3} - \cos^2\theta)$. Then

$$\gamma_0 = \gamma_m\{1 - (\tfrac{5}{2}m' - f)(\tfrac{1}{3} - \cos^2\theta)\}$$
$$= \gamma_e\{1 + (\tfrac{5}{2}m' - f)\cos^2\theta\} = \gamma_e\{1 + (\tfrac{5}{2}m' - f)\sin^2\phi\}, \tag{6.58}$$

where γ_e is the value of γ_0 at the equator ($\theta = 90°$), which is given by

$$\gamma_e = \gamma_m\{1 - \tfrac{1}{3}(\tfrac{5}{2}m' - f)\}. \tag{6.59}$$

When second-order terms are not considered, latitude $\phi = 90° - \theta$. Direct observation gives $f = 1/298$ and $m' = 1/289.20$, so

$$\gamma_0 = \gamma_e(1 + 0.00530\sin^2\phi), \tag{6.60}$$

which agrees well with observed values of gravity.

6.17. Second-order terms in Clairaut's theorem

§ 6.16 ignores terms in f^2 or fm', and the numerical substitution takes no account of harmonics higher than the second, their amplitudes being

† The approximation that the attractive force is the radial gradient of the attractive potential is a very close one, but see (6.64).

known to be of the order $5fu_2$ or less. Helmert and G. H. Darwin, [178] and [179], considered the term $Rf^2 \sin^2 2\theta$ in the formula for r, allowing also for a small departure from an exact elliptical section, as demanded by the earth's heavy centre, § 6.20 (b). See also [502], pp. 157–8.

Let the surface be

$$r = a\{1 - f\cos^2\theta - (\tfrac{3}{8}f^2 - \tfrac{1}{4}\chi)\sin^2 2\theta\} + a\sum_{2'} u_n, \tag{6.61}$$

where
$$\sum_{2'} u_n = u_2' + u_3 + u_4' + u_5 + \ldots,$$

u_2' and u_4' being the total second- and fourth-degree surface spherical harmonics less terms included in the first bracket. This will be an oblate spheroid if χ and $\sum u_n$ are zero.†

At an external point (r, θ) the potential of the attraction of the contained mass will be given by

$$V = \frac{GM}{r}\left\{1 - J_2\left(\frac{a}{r}\right)^2 P_2(\cos\theta) - J_4\left(\frac{a}{r}\right)^4 P_4(\cos\theta)\right\} + \frac{GM}{r}\sum_{2'}^{n}\left(\frac{a}{r}\right)^n u_n, \tag{6.62}$$

where J_2 and J_4 are to be determined. The last term is the same as before, the distinction between a and R being immaterial in these small terms.

On the surface U $(= V + \tfrac{1}{2}\omega^2 r^2 \sin^2\theta)$ is to be constant.

Substituting in U the values of r and V from (6.61) and (6.62), expressing all trigonometrical functions as powers of $\sin\theta$, and equating the coefficients of $\sin^2\theta$ and $\sin^4\theta$ to zero, gives

$$J_2 = \tfrac{2}{3}(f - \tfrac{1}{2}m + \tfrac{3}{4}m^2 + \tfrac{1}{7}mf - \tfrac{1}{2}f^2 - \tfrac{1}{7}\chi),$$
$$J_4 = \tfrac{8}{35}(\tfrac{5}{2}mf - \tfrac{7}{2}f^2 + \chi), \tag{6.63}$$

in which $\chi = 0$ if the section is an exact ellipse. Here $m = \omega^2 a/\gamma_e$, the ratio of the centrifugal acceleration to equatorial gravity, differing a little from the m' of § 6.16. $m' = m(1 - 5m/6)$. Then, m being adequately known, and, if $\chi = 0$, J_2, J_4, and f are known if either J_2 or f are defined. The constants B_2 and B_4 of (6.65) also follow.

On the surface

$$\gamma_0^2 = \left(\frac{\partial U}{\partial r}\right)^2 + \left(\frac{\partial U}{r\,\partial\theta}\right)^2, \qquad U \text{ being } V + \tfrac{1}{2}\omega^2 r^2 \sin^2\theta, \tag{6.64}$$

† The equation of an ellipse contains a small term in $f^3 u_6$, but it is negligible.

whence

$$\gamma_0 = \gamma_e\left\{1 + \mu_1 \cos^2\theta + (\mu_2/4)\sin^2 2\theta + \sum_{2'} (n-1)u_n\right\}$$

$$= \gamma_e\left\{1 + B_2 \sin^2\phi + B_4 \sin^2 2\phi + \sum_{2'} (n-1)u_n\right\}, \qquad (6.65)$$

using (A.56).

where

$$\mu_1 = B_2 = \tfrac{5}{2}m - f - \tfrac{17}{14}mf - \tfrac{2}{7}\chi,$$
$$\mu_2 = \tfrac{15}{2}mf - \tfrac{7}{2}f^2 + 3\chi,$$
$$B_4 = \tfrac{1}{4}\mu_2 - B_2 f = \tfrac{1}{8}(f^2 - 5mf + 6\chi),$$
$$\gamma_e = (GM/a^2)(1 - \tfrac{3}{2}m + f + \tfrac{9}{4}m^2 - \tfrac{27}{14}mf + f^2 - \tfrac{4}{7}\chi). \qquad (6.66)$$

If $\chi = 0$ for an elliptic section, the coefficient of $\sin^2 2\phi$ is $-5.9\gamma_e \times 10^{-6}$, and the term amounts to $-58\,\mu\mathrm{m\,s}^{-2}$ in latitudes 45° N and S.

Formula (6.65) constitutes a *standard gravity formula*, giving γ_0 the surface value of gravity in latitude ϕ, in terms of defined values of γ_e and f. χ is usually taken as zero and harmonic terms other than $\sin^2\phi$ and $\sin^2 2\phi$ are generally omitted. Currently accepted values of B_2 and B_4 are given by

$$\gamma_0 = 978.0318(1 + 0.005\,3024 \sin^2\phi - 0.000\,0059 \sin^2 2\phi). \text{ Gal.} \qquad (6.67)$$

Third-order terms can amount to about $1\,\mu\mathrm{m\,s}^{-2}$ (0.1 mGal). Also see (6.68), and §§ 6.27 and 6.28.

In (6.65) γ_0 is expressed as a series of harmonics in which (after the first two) the nth harmonic u_n in the formula for the surface is multiplied by a factor $(n-1)$ in the formula for gravity, which therefore becomes indeterminate if n is large. Short wavelength undulations in (r/R) of the surface are associated with variations in gravity (γ_0/γ_m) of the same wavelength but of much greater amplitude.

Considering only the first two terms of (6.65), involving $\sin^2\phi$ and $\sin^2 2\phi$, the following is an alternative. Instead of m it uses $\widetilde{m} = \omega^2 a^2 b/GM$, and the expansion includes a term of $\sin^4\phi$ instead of $\sin^2 2\phi$. See [262], p. 77 and [423], p. 60. Then

$$\gamma_0 = \gamma_e(1 + f_2 \sin^2\phi + f_4 \sin^4\phi), \qquad (6.68)$$

where

$$f_2 = -f + \tfrac{5}{2}\widetilde{m} + \tfrac{1}{2}f^2 - \tfrac{26}{7}f\widetilde{m} + \tfrac{15}{4}\widetilde{m}^2,$$
$$f_4 = -\tfrac{1}{2}f^2 + \tfrac{5}{2}f\widetilde{m}.$$

$$\gamma_e \text{ is given by } \frac{GM}{ab} = \gamma_e(1 + \tfrac{3}{2}\widetilde{m} + \tfrac{3}{7}f\widetilde{m} + \tfrac{9}{4}\widetilde{m}^2) \qquad (6.69)$$

also

$$\frac{\gamma_p - \gamma_e}{\gamma_e} = f_2 + f_4, \qquad (6.70)$$

where γ_p is standard gravity at the poles.

In (6.69) the omitted third-order terms amount to not more than 0.044 μm s^{-2} (0.0044 mGal).

Table 6.3, based on (6.65) and (6.66), gives the numerical relations between f, B_2, and B_4 (if $\chi = 0$), and also between B_4, J_4, and χ, if $\chi \neq 0$.

TABLE 6.3

f	$B_2 \times 10^6$	$B_4 \times 10^6$	
1/297	5288.4	−5.9	If $\chi = 0$
1/298	5299.6	−5.9	
1/299	5310.8	−5.9	
$\chi \times 10^6$	$B_4 \times 10^6$	$J_4 \times 10^6$	$\frac{1}{4}a\chi/$m. In (6.61)
−2.0	−7.4	−2.8	−3.2
0.0	−5.9	−2.4	0.0
+2.0	−4.4	−1.9	+3.2
+4.0	−2.9	−1.4	+6.4

Third-order terms are excluded.

6.18. Various forms of Clairaut's theorem

Formulae for γ_0, γ_e, and γ_m are expressed in terms of the parameters f, R, and m, but there is great variety in the exact definitions used for R and m. The following may be used for the radius.

(a) The semi-major axis a, as in § 6.17.

(b) The mean value of the geocentric distance over the surface of the spheroid, the mean being taken with respect to cos θ. $r_m = a(1 - \frac{1}{3}f - \frac{1}{5}f^2)$, (H.16) in which the P_2 and P_4 terms average zero.

(c) The mean value of the geocentric distance over the spheroid, the mean being taken with respect to the cosine of the eccentric angle u, Fig. A.1. $r'_m = a(1 - \frac{1}{3}f + \frac{1}{15}f^2 + \frac{1}{35}f^3)$, [159], p. 212.

(d) The radius of the sphere of equal volume. $R = a(1 - \frac{1}{3}f - \frac{1}{9}f^2)$.

Variations in the definitions of m are as below.

(i) $4m = \omega^2 a/\gamma_e = 0.003\,467\,8$. § 6.17. $m = \tilde{m} + \frac{3}{2}\tilde{m}^2$.

(ii) $m' = \omega^2 R/\gamma_m = 0.003\,457\,8$. $m' = m(1 - \frac{5}{6}m)$, § 6.17.

(iii) $\bar{m} = \omega^2 r_m^3/GM = 0.003\,449\,9 = m'(1 - \frac{2}{3}m + ...$[160], p. 227. where r_m is as in (b) above. Also in § 6.20 and [305], 1970, p. 191.

(iv) $\tilde{m} = \omega^2 a^2 b/GM = 0.003\,449\,80$, [159], p. 204, [262], p. 69, and § 6.21.

This variety of definitions is inconvenient, and the reconciliation of formulae based on different definitions is troublesome. Of the various m's (i) and (ii) have the disadvantage that they involve γ_e or γ_m, which are themselves being calculated. For the highest precision, iteration may be necessary. [159] gives expressions for γ_0, γ_e, and γ_m including third-order terms, f^3, etc., using (c) for the radius and (iv) for m.

6.19. The figure of a homogeneous rotating liquid

Putting $f'=0$, $b=a$, and $f_1=f$ in (6.20) and (6.21) gives the attraction on the surface of a homogeneous oblate spheroid correct to terms in f^2. If the resultant of this and the centrifugal acceleration, whose components are $\omega^2 x$, $\omega^2 y$, and 0, is to be normal to the spheroid

$$\frac{X+\omega^2 x}{x/a^2}=\frac{Z}{z/c^2}. \tag{6.71}$$

Substituting (6.21) and putting m' for $3\omega^2/4\pi G\rho$, gives

$$f=(5/4)m', \tag{6.72}$$

which for the earth gives $f=1/231$. But see § 6.20.

6.20. The figure of a rotating liquid with variable density

(a) *The flattening.* The difference between the figure 1/231 of § 6.19 and the earth's actual flattening of 1/298.25 is primarily caused by the increase of density towards the earth's centre.

Assuming layers of equal density to be equipotential surfaces, as they would be in a perfect liquid, and given all values of ρ (the density of any thin layer between surfaces whose semi-major axes† are r and $r+\delta r$, and whose flattenings are f and $f+\delta f$), f is given by

$$\frac{d^2 f}{dr^2}+\frac{6\rho}{r\rho_m}\frac{df}{dr}-6\left(1-\frac{\rho}{\rho_m}\right)\frac{f}{r^2}=0 \tag{6.73}$$

where ρ_m is the mean density of all matter inside each equipotential surface of semi-major axis r, which is given by

$$\int_0^r \rho r^2\,dr=\tfrac{1}{3}\rho_m r^3. \tag{6.74}$$

Equation (6.73) is also due to Clairaut. For a proof see [305], 1970, pp. 183–5.‡

Given figures for the variations of ρ, f can be computed for the external surface $r=a$, and any internal surface. For the method see [305], 1971, pp. 185–91, which shows that the f obtained for the external surface is

† ρ and f vary from the centre to the surface. ρ may be assumed to decrease outwards, and to be reasonably well in accord with the values given in Table 6.2.

‡ In [305], 1970, p. 185 (15), put $n=2$, $\rho_0=$ our ρ_m, and $\epsilon_2=\frac{2}{3}f$.

insensitive to any reasonable errors about the variations of ρ, provided they give a value of $C/(Ma^2)$ which accords with that of the earth. For a sphere of uniform density $C/(Ma^2)=0.4$. For the earth it is about 0.33. Here C is the moment of inertia about the polar axis, M is the mass of the earth, and a is the equatorial radius.

From (6.73), on the hydrostatic assumption (i.e. that the body is liquid) [305], 1971, derives

$$\frac{C}{Ma^2}=\frac{2}{3}\left\{1-\frac{2}{5}\left(\frac{5\bar{m}}{2f}-1\right)^{\frac{1}{2}}\right\}, \tag{6.75}$$

in which $\bar{m}=\omega^2 r_m^3/(GM)=0.0034499$, *see* § 6.19 item (iii). We also have†

$$J_2=(C-A)/(Ma^2), \tag{6.76}$$

in which A is the earth's moment of inertia about an equatorial axis.

Also, from (6.63)

$$J_2 \approx \tfrac{2}{3}(f-\tfrac{1}{2}\bar{m}). \tag{6.77}$$

Finally

$$(C-A)/C=H, \tag{6.78}$$

where H is the earth's *precessional constant*, whose value is known to be $0.003\,273 \pm 0.000\,000\,7$, [305], 1970, p. 191, as given by observations of the luni-solar precession.

Of equations (6.75) to (6.78) only the first depends on the hydrostatic assumption.

Then, combining (6.76) and (6.78) gives

$$J_2/H=C/(Ma^2). \tag{6.79}$$

TABLE 6.4.

$I/(MR^2)$ $\approx C/(Ma^2)$	$1/f$	$J_2\times 10^6$ $\chi=0$	$H\times 10^2$ See (6.78)
0.3200	311.75	984.9	0.30738
0.3267	304.28	1037.3	0.31715
0.3333	296.98	1091.3	0.32696
0.3400	289.81	1146.8	0.33686
0.3467	282.79	1203.9	0.34682

$C/(Ma^2)-I/(MR^2)=0.0001$. [305].

† At an external point on the polar axis, comparison of the J_2 term of (6.62) with (6.22) gives

$$-\frac{GM}{r}J_2\frac{a^2}{r^2}=\frac{G}{2r^3}(A+A+C-3C).$$

On the basis of (6.75), Table 6.4† gives the values of $I/(MR^2) \approx C/(Ma^2)$, $1/f$, J_2‡, and $H \times 100$ for different values of $I/(MR^2)$, in which R is the earth's mean radius and I is the mean moment of inertia. Here I is not the same as in (6.22). Then, given either J_2, f, or H, and accepting the hydrostatic assumption, the others are given by interpolation in the table.

The value of H has long been known, but before 1957 the value of J_2 was doubtful. Accepting 0.003 273 for H gives $C/(Ma^2) = 0.3336$, $f = 296.75$‖ and $J_2 = 1093 \times 10^{-6}$. These values are known to be wrong. They are actually 1/298.25 and 1082.63×10^{-6}. The error cannot be attributed to doubt in H or J_2, and it is consequently a measure of the failure of the hydrostatic assumption.

The following is another way of putting this result. Given H and J_2, (6.79) gives $C/(Ma^2) = 0.3307$. Entering Table 6.4 with this figure gives $f = 1/299.8$ and $J_2 = 1071 \times 10^{-6}$ on the hydrostatic assumption. The position then is that the flattening of the spheroid which best fits the geoid is 1/298.25, a situation which is sustained by the strength of the mantle, but if it was possible for the earth to relax and lose its strength without change of density, its flattening would be 1/299.8. Similarly, J_2 is 1082.63×10^{-6}, but it would become 1071. See also § 6.28.

For further details see also [335] and [262].

(*b*) *Depression in middle latitudes.* [178], p. 107. A further consequence of a heavy centre is that the meridional section of the bounding equipotential surface is not exactly an ellipse. On the hydrostatic assumption the earth's equipotential would be depressed by about 4 metres in latitudes 45° N and S. While the equation of an ellipse is

$$r = a(1 - f \cos^2\theta - \tfrac{3}{8}f^2 \sin^2 2\theta\}, \tag{6.80}$$

the equipotential surface, with a heavy centre, on the hydrostatic assumption is

$$r = a\{1 - f \cos^2\theta - (\tfrac{3}{8}f^2 - \tfrac{1}{4}\chi)\sin^2 2\theta\}, \tag{6.81}$$

where χ as well as f depends on the density distribution. For a liquid with the earth's actual densities, [135] gets $\chi = -2.8 \times 10^{-6}$, giving a depression of 4.5 m, corresponding to $J_4 = -3.0 \times 10^{-6}$, see Table 6.3. On the other hand (6.63) shows that χ can be determined by J_4 and f. J_4 is now known to be -1.6×10^{-6}, for which Table 6.3 gives $\chi = +3.2 \times 10^{-6}$, so that instead of being depressed by 4.5 metres in mid-latitudes the earth's

† From [305], 1970, p. 192. The values of f and H have here been adjusted to allow for second-order terms, as given below the Table in [305]. Figures may change a little if the values of H and J_2 are updated.

‡ J_2 is computed from (6.63), with $\chi = 0$.

‖ Using the density distribution derived from seismology, [305], 1970, p. 191 gets $f = 297.3$ by direct computation on the hydrostatic assumption.

surface equipotential is on average raised by 5.1 m. The difference must be ascribed to the inexactitude of the hydrostatic assumption. See also § 6.28.

6.21. Somigliana's formula

This formula [540] gives the variation of gravity on an equipotential surface which is an exact spheroid, no provision being made for any departure from the spheroidal form. The formula is closed, in terms of trigonometrical functions, without expansion in power series. [262], pp. 66–70 also gives a proof. The sequence of computation is as follows.

$$\tilde{m} = \omega^2 a^2 b/GM, \text{ as in § 6.18 (iv).} \tag{6.82}$$

$$e^2 = (a^2 - b^2)/a^2 \quad \text{and} \quad e'^2 = (a^2 - b^2)/b^2. \tag{6.83}$$

$$q_0 = \tfrac{1}{2}\{(\tan^{-1} e')(1 + 3/e'^2) - 3/e'\} \tag{6.84}$$

$$q_0' = 3(1 + 1/e'^2)\{1 - (1/e')\tan^{-1} e'\} - 1. \tag{6.85}$$

$$\gamma_e = \frac{GM}{ab}(1 - \tilde{m} - \tilde{m}e'q_0'/6q_0), \text{ at the equator.} \tag{6.86}$$

$$\gamma_p = \frac{GM}{a^2}(1 + \tilde{m}e'q_0'/3q_0), \text{ at the poles.} \tag{6.87}$$

$$\gamma_0 = \frac{a\gamma_e \cos^2\phi + b\gamma_p \sin^2\phi}{(a^2\cos^2\phi + b^2\sin^2\phi)^{\frac{1}{2}}}, \text{ Somigliana's formula.} \tag{6.88}$$

Alternatively
$$\gamma_0 = \gamma_e \frac{1 + k\sin^2\phi}{(1 - e^2\sin^2\phi)^{\frac{1}{2}}}, \text{ [423], p. 59,} \tag{6.89}$$

where
$$k = (b\gamma_p - a\gamma_e)/a\gamma_e.$$

It follows that

$$J_2 = \frac{e^2}{3}\left(1 - \frac{2}{15}\frac{\tilde{m}e'}{q_0}\right), \text{ [262], p. 78.} \tag{6.90}$$

This formula is of special value for the definition of a spheroidal reference surface which is to be used for both geometrical and gravitational purposes. See Section 5.

6.22. Significance of low-degree harmonics

Spherical harmonics of degree 0 to 2 in the form of a bounding equipotential S, or in the standard gravity formula, or in the actual value of g_0 (observed gravity reduced to geoid or co-geoid level), have special significance.†

† (H.05) gives the form of all harmonics of degrees 1 to 4.

(*a*) *Zero degree* constants represent the volume of *S*, or the average value of gravity on it, or the total mass, or average density.

(*b*) *First-degree* harmonics do not occur in the form of *S* if the origin of coordinates is the centre of volume of the surface. In these circumstances §§ 6.16 and 6.17 show that they are also absent from γ_0. It further follows that the centre of mass inside *S* must coincide with its centre of volume.†

(*c*) *The zonal harmonics* of the second and fourth degrees represent the spheroidal form and the consequent variation of γ_0 with latitude. Strictly, sixth and higher zonal harmonics are involved, but with negligible effect. A coefficient of P_4 other than $-(12/35)f^2$ introduces the depression in lats 45° referred to in § 6.20 (*b*).

(*d*) *A second-degree harmonic* involving $\sin^2\theta \cos 2(\lambda-\lambda_0)$, or P_{22}, represents an ellipticity of the equator. It could not occur if the earth was liquid throughout, but is present with measurable amplitude. It has no special significance.

(*e*) *A second-degree harmonic* involving $\sin 2\theta \cos(\lambda-\lambda_0)$, or P_{21}, would indicate that the axes of rotation and inertia, or figure, did not coincide. This might be the case, but in the earth it would result in the axis of rotation revolving round the axis of inertia with some such period as 1.2 years, depending on the earth's elasticity. This does in fact occur, and is partly responsible for the periodic variation of latitude, but the minute amplitude of the variation shows that the angle between the two axes is of the order 0.1 seconds of arc, so that the harmonic concerned is in fact virtually absent.

(*f*) The position of other low-degree harmonics is as in (*d*) above. They exist with small amplitudes, and their existence demands some strength in the crust, or mantle.

SECTION 4. THE VARIATION OF ATTRACTION AND GRAVITY WITH HEIGHT

In this section three cases have to be considered.

(*a*) Close to the surface. Gravity, with rotating axes, § 6.23.

(*b*) At satellite altitudes. Attraction, with fixed axes, § 6.25 (*a*).

(*c*) At satellite altitudes. Gravity, with rotating axes, § 6.25 (*b*). This is seldom required.

Formulae are given in terms of *V* or *U*, the potentials of the attraction *F*, or of the gravity γ of a standard body. With *W* and g replacing *U* and

†Because if $u_1 = 0$, i.e. if the origin is the CV, *V* or *U* as given by (6.53) will contain no term in $1/r^2$, and vice versa. Then see § 6.09 (*i*) last line. Note the general theorem that the centre of volume of an equipotential bounding surface must coincide with the centre of mass of the contained matter. [502], p. 157.

γ, and with the geoid taken as the bounding equipotential surface of the attracting body, the formulae are applicable to the earth, apart from the effects of masses outside the geoid, for which see Sections 7 and 8.

6.23. Gravity at low altitudes

Consider an approximately spheroidal surface S which is external to all matter. U is the potential of its gravity. Then, if the axis of z is taken tangent to the outward vertical at any point

$$\frac{\partial^2 U}{\partial z^2}=\frac{\partial}{\partial z}\frac{\partial U}{\partial z}=-\frac{\partial \gamma}{\partial z}. \tag{6.91}$$

Also, since the (x, y) plane is tangent to S at the same point

$$\frac{1}{r_x}=-\frac{1}{\gamma_0}\frac{\partial^2 U}{\partial x^2}, \quad \text{and} \quad \frac{1}{r_y}=-\frac{1}{\gamma_0}\frac{\partial^2 U}{\partial y^2}, \tag{6.92}$$

where r_x and r_y are the radii of curvature of S at the point, in the (x, z) and (y, z) planes respectively, and γ_0 is gravity on the surface. So from (6.24), outside matter

$$\frac{\partial \gamma}{\partial z}=-\gamma_0\left(\frac{1}{r_x}+\frac{1}{r_y}\right)-2\omega^2. \tag{6.93}$$

And for a spheroid $1/r_x+1/r_y=1/\nu+1/\rho$, ν and ρ being the principal radii of curvature, § A.03, so at a height h_s above a spheroid, standard gravity is

$$\gamma_A=\gamma_0-\left\{\gamma_M\left(\frac{1}{\rho_M}+\frac{1}{\nu_M}\right)+2\omega^2\right\}h_s, \tag{6.94}$$

where γ_M is the mean value of γ_A between heights 0 and h_s, and ρ_M and ν_M are similarly mean values of the radii of curvature of the intermediate equipotential surfaces. Then

$$\begin{aligned}\gamma_A &= \gamma_0-\gamma_M(1+m')(1/\rho_M+1/\nu_M)h_s\\ &= \gamma_0(1-2h_s/R')\\ &\approx \gamma_0-0.308h_s \text{ mGal, or } -3.08\ \mu\text{m s}^{-2}, \text{ if } h_s \text{ is in metres,}\end{aligned} \tag{6.95}$$

defining R', which can often be taken as equal to R, the earth's mean radius. The second term is about 1 mGal or 10 μm s^{-2} per 3.25 metres of h. Approximations involved in taking $R'=R$ are: from $\gamma_M=\gamma_0$, 0.1 per cent if $h_s=6000$ m; from ignoring m', 0.3 per cent; from $\rho_M=\nu_M=R$, up to 0.4 per cent for a spheroid without higher harmonics, but more on the actual earth.

Formulae (6.93)–(6.95) give the differences of γ between points in a vacuum, not separated by rock. For the gradient of γ or g in matter of density ρ see § 6.42. For the effect of the atmosphere see § 6.29. It is very small, less than 10 μm s^{-2}, or 1 mGal, over the whole thickness of the atmosphere. Its gradient is generally neglected.

The difference $\gamma_0 - \gamma_A$, as given by (6.95), or the corresponding $g_0 - g$, is known as the *free-air reduction*, by which observed gravity at geoidal height h is reduced to sea level, ignoring the density of the intervening mass.

This free-air reduction is similar in form to $g - \gamma_A$ as given in (6.95), but since it is desired to reduce the observation to sea-level, the station height must here be height above the geoid. See also §§ 6.37 and 6.62.

6.24. The curvature of the vertical

In Fig. 6.8 let S_0S_0' be a meridional section of a rotating oblate spheroid on which gravity at S_0 is given by $\gamma_0 = \gamma_e(1 + B_2 \sin^2\phi_s + \ldots)$ as in (6.65), and let S_1S_1' be a higher equipotential surface. Then from § 6.07 $S_0S_1/S_0'S_1' = \gamma_0'/\gamma_0$, and if S_0 and S_0' are close and $S_0S_1 = h$, the angle between the two surfaces is

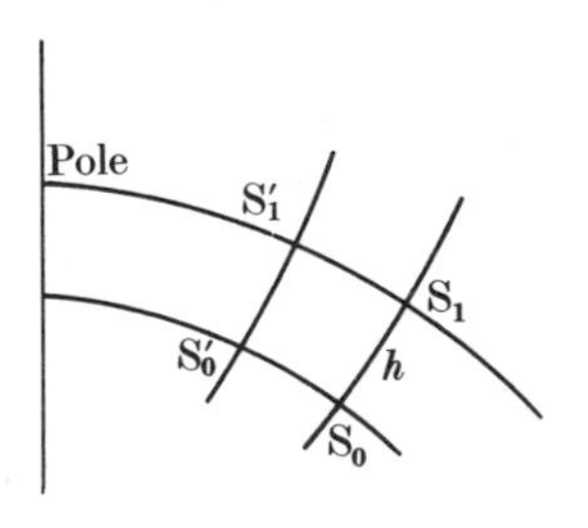

FIG. 6.8. Curvature of the vertical. $S_1'S_1$ and $S_0'S_0$ are equipotential surfaces. $S_1'S_0'$ and S_1S_0 are verticals, orthogonal to the equipotentials.

$$\frac{\gamma_0 - \gamma_0'}{\gamma_0}\frac{h}{R\,\mathrm{d}\phi} = \frac{h}{\gamma_0}\frac{\mathrm{d}\gamma_0}{R\,\mathrm{d}\phi} = \frac{hB_2 \sin 2\phi}{R}. \tag{6.96}$$

Putting $B_2 = 0.0053$, $R = 6.4 \times 10^6$, and expressing h in metres gives the inclination as

$$-0''.00017h \sin 2\phi, \ S_1S_0 \text{ being greater than } S_1'S_0'. \tag{6.97}$$

This is the correction conventionally applicable to an astronomical latitude in order to 'reduce it to sea-level'. See § 4.51 (*e*).

Put otherwise, the radius of curvature of the external verticals of the reference spheroid is $R/(0.0053 \sin 2\phi)$, concave to the axis of rotation, and confined to the meridian plane. The earth's actual upper equipotential surfaces are of course considerably modified by the topography and density anomalies, and these expressions are barely useful approximations. See § 6.55.

6.25. Attraction at satellite altitudes

(*a*) Let the bounding equipotential surface S of the attracting body be an oblate spheroid, modified by spherical harmonics of small amplitude as in (6.61) with $\chi=0$, namely

$$r=a\{1-f\cos^2\theta-(3/8)f^2\sin^2 2\theta\}+a\sum_{2'}^{n}u_n, \tag{6.98}$$

Then the potential at an external point (r, θ, λ) is given by (6.62), which in this context is more usually written in the form of (7.1).†

Given the potential, the attraction **F** equals grad V, whose magnitude is approximately, but not quite exactly, dV/dr. Precisely [578] gives its scalar magnitude as

$$F=\frac{GM}{r^2}\left\{1-3J_2\left(\frac{a}{r}\right)^2P_2-5J_4\left(\frac{a}{r}\right)^4P_4+\tfrac{1}{2}(f-\tfrac{1}{2}m)^2\left(\frac{a}{r}\right)^4\sin^2 2\theta+\right.$$
$$\left.+\sum_{2'}^{n}(n+1)u_n\left(\frac{a}{r}\right)^n\right\}. \tag{6.99}$$

For the angle ψ between the attractive force and the radial line [490] gives for first-order accuracy, ignoring u's,

$$\psi''=(f-\tfrac{1}{2}m)\frac{a^2}{r^2}\sin 2\theta\left\{1+(2f-m)\frac{a^2}{r^2}P_2\right\}\operatorname{cosec} 1'', \tag{6.100}$$

the sign being such that the inward direction of the force is on the side of the radial line remote from the minor axis.

For the equation of equipotential surfaces (of the attraction only), neglecting higher harmonics and with $\chi=0$, [578] gives

$$r=a_1\{1-(\alpha_2+\alpha_4)\cos^2\theta+\alpha_4\cos^4\theta\}, \tag{6.101}$$

where‡

$$\alpha_2=(f-\tfrac{1}{2}m+\tfrac{3}{4}m^2+\tfrac{1}{7}mf-\tfrac{1}{2}f^2)\left(\frac{a}{a_1}\right)^2-\tfrac{1}{2}(-f^2+\tfrac{9}{7}mf-\tfrac{1}{2}m^2)\left(\frac{a}{a_1}\right)^4, \tag{6.102}$$

$$\alpha_4=\tfrac{1}{2}(3f^2-mf-m^2)\left(\frac{a}{a_1}\right)^4. \tag{6.103}$$

In these formulae a_1 is the equatorial radius of the surface concerned, while f, m, and a refer to the bounding surface S (6.98). (6.102) then gives the flattening of these surfaces, and shows that at geoid level it is a little less than half that of the geoid since $(f-\frac{1}{2}m)\approx\frac{1}{2}f$, and that it decreases with height approximately as $(a/a_1)^2$.

† In this Chapter 6 and Appendix H, the degree of any harmonic is written as n, while in (7.1) and all Chapter 7 it is l.

‡ The first term of (6.102) is $\frac{3}{2}J_2a^2/a_1^2$.

In this treatment (6.99) derives from (6.98) on the assumption that there are no masses external to the surface S. In practice, if S is taken to be the geoid, the resulting error at satellite heights is very small, see § 6.44.

The orthometric height of a point (r, θ) above a spheroidal surface, such as (6.98) without the last term, is approximately $r-r_0$, where r_0 is the geocentric distance of the surface corresponding to θ in (6.98). For a spheroid with the earth's flattening, this is correct to 5 m at a height of 1000 km.†

(*b*) *Gravity at satellite altitudes.* If it is required to use rotating axes, the gravitational potential U is given by

$$U = V+\tfrac{1}{2}\omega^2 r^2 \sin^2\theta, \tag{6.104}$$

and the gravity vector is given by grad U. This is unlikely to be required.

6.26. Surface gravity deduced from high altitude potential

If satellite observations have given the potential of the attraction by determining all the harmonic constants in (6.62) or (7.1), it is required to find the form of the surface (6.98), and the intensity of gravity on it. Then the form of the surface is immediately given by (6.98) itself, using the values of a, f, and all the harmonic constants, as given by the satellites. See more explicitly, in § 6.34 (*b*) formula (6.129).

The intensity of gravity at the surface (including the centrifugal force) is given by (6.65) and (6.66)‡ using the same constants.

It must be remarked that in (6.62) the harmonic coefficients are multiplied by a factor of $(a/r)^n$, which may be a small fraction if n is large. Any error in the value of $u_n(a/r)^n$ as determined by satellite observations is then proportionately increased in the ratio $(r/a)^n$ when u_n is used in (6.98) to give the form of the geoid. High-altitude equipotential surfaces are smoother than sea-level ones. If the satellite altitude is 1000 km, the value of $(r/a)^n$ is 4.3 if $n=10$, or 8.9 if $n=15$.

[382], p. 101 estimates that the error in geoidal heights deduced from GEM 10 potential coefficients will have an r.m.s. error of between 1.5 and 1.9 m. These figures do not include error which will result from the truncation or omission of harmonic terms of higher degree than those included in GEM 10, i.e. terms generally higher than degree 22. The likely effect of the latter is not easy to estimate, but astro-geodetic

† Height is measured along the normal, not the radius, and is $(r-r_0)\cos(90°-\theta-\phi)$ plus the height of the tangent above the surface at a distance of $(r-r_0)\sin(90°-\theta-\phi)$. The angle $(90°-\theta-\phi)$ cannot exceed f, (A.56). The curvature correction is of opposite sign to the cosine correction, and their sum is numerically less than $(r-r_0)\cos f$ when $r-r_0$ is less than twice the earth's radius.

‡ In (6.66) the value of χ will ordinarily be taken as zero. If it is not, the effect on B_2 and B_4 will be cancelled by opposite changes in u_2' and u_4'.

surveys show that changes of 5 or even 10 metres of geoidal height may occur so locally that they could not possibly begin to be represented by (22×22) analysis.

In the harmonic expansion for gravity on the sea-level surface errors in the harmonic terms representing the surface are further multiplied by a factor of $(n-1)$ in the analysis of the variations of g, see (6.65).

SECTION 5. REFERENCE SPHEROIDS AND STANDARD GRAVITY FORMULAE

6.27. Historical

(*a*) *Reference spheroids for geometrical use.* It has long been known that the figure of the earth is an oblate spheroid†, and the last 150 years have provided successively more accurate values of the axes and flattening of the best fitting spheroid, obtained from arcs of meridian and parallel, from the mapping of the geoidal form over limited areas, from observations of the intensity of gravity, and finally from satellites. Such observations have been used to define reference spheroids as in § 2.02, but the point has been reached where change from the current (1967) spheroid would do little to improve it as a simple spheroidal representation of the irregular shape of the actual geoid.

In 1924 the figures 6 378 388 m and 1/297 were accepted as the *International spheroid*, which still bears that name (preferably with the date added), but in 1967 the IAG adopted the figures 6 378 160 and 1/298.247 as the *Reference ellipsoid 1967*, which was recommended for future use when it is convenient to change to a new reference spheroid, § 6.28.

(*b*) *Standard gravity formulae* in the form of (6.67), or less commonly of (6.68), are similarly used as standards with which observed gravity can be compared, and the differences recorded in the form of gravity anomalies, namely the difference between observed g and the standard γ for the same latitude and height. See more fully in Section 7.

Table 6.6 gives some of the constants which have been adopted from time to time.

The *International gravity formula* was adopted in 1930. In it and the other formulae in Table 6.6, except the last three, the equatorial value of gravity has been based on 981.274 cm s^{-2} as the value of g at Potsdam.

† The figure of the earth does not mean the figure of the irregular ground surface, but of the sea-level surface or geoid.

TABLE 6.5

Reference spheroids

		Equatorial semi-axis a		Polar semi-axis b		
	Date	a/feet	a/metres	b/feet	b/metres	$1/f$
Everest	1830	20 922 932		20 853 375		300.80
Bessel	1841	*20 923 600*	*6 377 397*	*20 853 656*	*6 356 079*	299.15
Clarke	1866	20 926 062	6 378 206	20 855 121	6 356 584	294.98
Clarke	1880	20 926 202	*6 378 301*	20 854 895	*6 356 566*	293.47
International (1924) (Hayford)	1909		6 378 388		6 356 912	297.00
Jeffreys	1948		6 378 099		6 356 631	297.10
Krassovsky	1948		6 378 245		6 356 863	298.3
Fischer	1960		6 378 155		6 356 773	298.3
Reference Ellipsoid 1967	1967		6 378 160		6 356 775	298.247
W.G.S. 72	1974		6 378 135		6 356 750	298.26
Best estimate [427]	1977		6 378 140		6 356 755	298.257

Note. Semi-axes are given here to the nearest foot or metre and flattening to one or two decimals, so consistency between f and axes is not quite perfect. Deduced figures are printed in italics.

References

Everest. [198]. Feet as given by Everest. In Indian feet as used in India.

Bessel. [254]. Defined by Bessel in toises. Feet as given by Clarke in [153], p. 734. Metres as used in US until 1880. [254], p. 172.

Clarke 1866. Feet and metres as given in [154], p. 287. Metres are as used in US since 1880, [21], p. 19. The foot-metre ratio used by Clarke was 1 m = 39.370 432 inches.

Clarke 1880. [155]. p. 319. Metres given by 1 m = 39.370 113 inches. 6 378 249 metres has also been used, e.g. in South Africa.

Hayford. [255]. The International Spheroid. [459]. Hayford gave minor axis as 6 356 909, but that is inconsistent with $f = 1/297$, and the international figure is as given.

Jeffreys. [306]. Flattening deduced from arcs, gravity, and astronomy.

Krassovsky. [302].

Fischer. Solution 3 of [204]. Flattening held as 1/298.3 from satellites.

Ref. ellipsoid 1967. See § 6.28 and [423].

WGS 72. [519].

This is now known to be about 0.014 cm s^{-2} too high, § 5.05. The *Gravity Formula 1967* is based on 981.260 for Potsdam and on the flattening of the 1967 reference ellipsoid.

The difference between the 1967 and 1930 formulae is

$$\gamma_{1967} - \gamma_{1930} = -(17.2 - 13.6 \sin^2\phi) \text{ mGal}$$

$$= -(172 - 136 \sin^2\phi) \ \mu\text{m s}^{-2}. \qquad (6.105)$$

TABLE 6.6

Standard gravity in terms of Potsdam 981.274 *except the last three entries.*

$$\gamma_0 = \gamma_e(1 + B_2 \sin^2\phi + B_4 \sin^2 2\phi). \quad (6.106)$$

	Date	γ_e/cm s^{-2}	B_2	B_4	Corresponding $1/f$
Helmert	1901	978.030	0.005 302	−0.000 007	298.3
International	1930	978.049	0.005 288 4	−0.000 005 9	297.0
Jeffreys	1948	978.051	0.005 289 1	−0.000 005 9	297.10
Gravity formula 1967	1970	978.032	0.005 302 4	−0.000 005 9	298.247

The following have been expressed in terms of $\sin^2\phi$ and $\sin^4\phi$, as in (6.68).

	Date	γ_e/cm s^{-2}	f_2	f_4	$1/f$
Gravity formula 1967	1970	978.03185	0.005 278 895	0.000 023 462	298.247 17
WGS 72	1974	978.03327	0.005 278 994	0.000 023 461	298.263 8

References

Helmert. [271] and [254], p. 172. Originally given as $\gamma_e = 978.046$ on the 'Vienna system'. Reduced to 978.030 on the system Potsdam = 981.274.

International [367].

Jeffreys. γ_e given as 978.037 3 in the original, in terms of Potsdam = 981.260 6. Flattening deduced from arcs and astronomy as well as from gravity. [306].

1967. See § 6.28 for fuller figures. Potsdam 981.260 [423].

WGS 72. [519]. *Potsdam* 981.260.

But the adoption of the new value of g at Potsdam changes all values of observed g by

$$g_{260} - g_{274} = -14 \text{ mGal} = -140 \ \mu\text{m s}^{-2}. \quad (6.107)$$

So the value of an anomaly $(g-\gamma)$ is changed by

$$(g-\gamma)_{1967} - (g-\gamma)_{1930} = 3.2 - 13.6 \sin^2\phi \text{ mGal}$$

$$= 32 - 136 \sin^2\phi \ \mu\text{m s}^{-2} \quad (6.108)$$

There is risk of confusion arising from the new and the old formulae, and the two values at Potsdam. When gravity values and anomalies are published it is essential to say which formula and which value at Potsdam has been used. If either has been changed, it is desirable to change both. [423] gives tables of γ_0 from the 1967 formula, and also of the changes (6.105) and (6.108).

6.28. Geodetic reference system, 1967

This system defines both the geometrical spheroid and the standard gravity, the latter being the gravity due to hypothetical masses lying within the spheroid, which is itself an equipotential surface of their gravity. The flattening f, which is required for the definition of the spheroid and the gravity formula, is replaced by the equivalent† J_2 as the actually defined quantity, while GM is chosen for definition instead of γ_e. The angular velocity of the earth's rotation ω in a Newtonian frame is also regarded as part of the definition, although its true value is so accurately known that the adopted value is in no way arbitrary.

The system is then defined by the following.

(i) Semi-major axis $a = 6\,378\,160$ m.
(ii) The coefficient $J_2 = 1082.7 \times 10^{-6}$.
(iii) GM, including the atmosphere, $= 398\,603 \times 10^9\ \text{m}^3\,\text{s}^{-2}$.
(iv) $\omega = 7.292\,115\ 1467 \times 10^{-5}\ \text{rad s}^{-1}$.

These defined quantities are accepted as exact. As compared with the actual earth, it is now (1977) thought probable that the true mean equatorial radius is more like 6 378 140 m, but 160 was chosen to avoid unnecessary change from a figure which had for some time been accepted as a standard. The current best value (1977) of J_2 is 1082.63×10^{-6}, and that of GM is $398\,600.5 \times 10^9\ \text{m}^3\,\text{s}^{-2}$ [427].‡ The value of ω is not very accurately required, and the figure accepted is more than amply accurate.

Given J_2 the constants f, J_4, B_2, and B_4 are given to second-order precision by (6.63) and (6.65) with $\chi = 0$, or with higher precision as in [423], p. 30 using

$$e^2 = 3J_2 + \frac{4}{15}\frac{\omega^2 a^3}{GM}\frac{e^3}{2q_0} \tag{6.109}$$

where e, e', m, and q_0 are as in (6.82)–(6.84). Since q_0 contains e', e and e' have to be obtained by iteration, using the defined values of a, J_2, GM, and ω. Given e, f is given by (A.41), and b follows. As an alternative to (6.63) and (6.65), γ_e and γ_p at the equator and poles are given by (6.85)–(6.87), [423], pp. 32–6, and then Somigliana's formula (6.88) gives γ_0, the standard value of gravity on the spheroid in latitude ϕ.

Very full details of the 1967 system are given in [423] using both Somigliana's formula, and also expansions on the lines of § 6.17, but with higher order terms. On the basis of the four defined quantities (i)–(iv)

† The relation between J_2 and f is given by (6.63) with $\chi = 0$, or with a higher degree of accuracy by (6.109).

‡ GEM 10 [382], p. 48, 1977, gives $GM = 398\,600.47$ including the atmosphere.

above, the following figures, among others, are deduced. The very large number of decimals are given to avoid any possible inconsistency.

$b =$ 6 356 774.516 1 m.
$e^2 =$ 0.006 694 605 328 56.
$f^{-1} =$ 298.247 167 427.
$U_0 =$ 6 263 703.052 3 kGal m. See (6.110). 1 kGal = 10 m s^{-2}.
$J_4 =$ −0.000 002 371 264 40.
$J_6 =$ 0.000 000 006 085 16.
$J_8 =$ −0.000 000 000 014 28.
$\widetilde{m} =$ 0.003 449 801 434 30. $\widetilde{m} = \omega^2 a^2 b/GM$.
$\gamma_e =$ 978.031 845 58 Gal. 1 Gal = 10 mm s^{-2}.
$\gamma_p =$ 983.217 727 92 Gal.
$k =$ 0.001 931 663 383 21, for (6.89)

The value of the gravitational potential U_0 on the surface of the reference spheroid is given by [423] and [262] as

$$U_0 = \frac{GM}{(a^2 - b^2)^{\frac{1}{2}}} (\arctan e') + \frac{\omega^2 a^2}{3}. \tag{6.110}$$

A possible criticism of the decision to adopt an exact spheroid, with $\chi = 0$, has been that no mass distribution with finite positive density within such a spheroid can make its surface an exact equipotential. On the other hand, a rational density distribution such as that of Table 6.2, together with a weak (negative) skin density on the surface will produce the desired result. Outside the surface, gravity would be insensibly changed if this skin density was replaced by a similar mass deficiency spread through (say) the upper 20 km of the crust. This would not be in any way unnatural, nor improper in a reference body, whether any such pattern of distribution actually exists or not. The point is well covered in [423], pp. 14–16.

There is something to be said for recording gravity and geoidal anomalies with reference to a strengthless solid spheroid with a flattening of 1/299.8 as is suggested at the end of § 6.20 (*a*). This has been done in (e.g.) [391], 2, p. 137. But in such charts the anomaly resulting from the second-degree harmonic may tend to swamp the pattern of the smaller features, and for practical purposes it is probably better to use 1/298.25. The strength demanded by the true flattening, and its associated J_4, can be separately assessed, see § 6.68 (*c*).

6.29. The atmosphere

The mass of the atmosphere is 0.89×10^{-6} times the mass of the earth. From the point of view of gravity measurements at or near sea level, the atmosphere is very closely an external spherical shell, whose effect on gravity at points inside it is zero. In these circumstances the value of GM in formulae such as (6.65) would conveniently include only the internal mass, without the mass of the atmosphere. On the other hand, satellites are substantially outside the atmosphere, and in formulae such as (7.1) GM should include the total mass. It is now most usual to include the atmosphere, in GM, as has been done in the definition of the reference system 1967, and in all figures for GM quoted here.

At a gravity station where the atmospheric pressure P is recorded in Pa, the value of standard gravity, γ_A in (6.94) and (6.95) must be decreased by $8.6\ P \times 10^{-5}\ \mu\text{m s}^{-2}$†, if γ_A is given by the 1967 or WGS 72 systems, for which GM includes the total mass. With older gravity formulae which exclude the mass of the atmosphere, its attraction may be included by adding $8.6\ (1-10^{-5}\ P)\ \mu\text{m s}^{-2}$ to γ_A. This has generally been ignored in the past, but since Potsdam and the great majority of other gravity stations are relatively close to sea level, the correction is likely to be similar at both, and the error in any resulting anomaly $g-\gamma_A$ will seldom exceed 1 or 2 $\mu\text{m s}^{-2}$. [423] gives a very full analysis, with tables for the necessary corrections at different heights using either the CIRA or US standard atmospheres.

SECTION 6. STOKES'S INTEGRAL

Stokes's integral has been an important factor in geodetic thinking for a long time, because it has appeared to be the only means of converting independent continental survey datums to a common geocentric reference frame. In the absence of sufficient world-wide gravity observations it has not yet been possible to use it for that purpose with satisfactory accuracy, and now the utility of trying to do so has become questionable, since satellite observations are taking its place with increasing accuracy. See § 6.32 (*c*) and §§ 7.74–7.77. There are, however, some other uses for Stokes's integral, see § 6.32 (*d*).

6.30. Stokes's integral

The converse of (6.65), a formula to give the form of a bounding equipotential surface S when the intensity of gravity all over it is known,

† If P is measured in mbar the correction to γ_A(1967) is $-0.86 \times 10^{-3} P$ mGal, and for older formulae $+0.86\ (1-10^{-3}P)$ mGal.

is easily got if the variations of gravity can be expressed as a series of spherical harmonics. For the $\sin^2\phi$ and $\sin^2 2\phi$ terms have only to be isolated to give f and χ, using (6.66) for B_2 and B_4, and the remaining harmonic terms each divided by $(n-1)$ give the u's of (6.61). These then give the form of S, except that the radius a has to be found by other means, such as by the measurement of arcs of meridian, or other geoid surveys.

Except for harmonics of fairly low degree, such as less than 10 or 20, it is not practicable to express observed g as a series of spherical harmonics. Significant variations would be lost, even in an analysis which went as far as degree 250 (involving 60 000 constants). The deduction of the form of the geoid from gravity must therefore be effected by quadratures. Stokes has given the following formula, [542].

Consider a spheroid or other similar surface of defined shape (here referred to as the spheroid) such as that given by (6.61), probably without the term $a\sum u_n$, and let γ_0 as in (6.65) be the value of gravity on its surface, on the assumption that it is an equipotential surface of the contained matter. In the actual earth there are internal† mass anomalies which result in actual g_0 (on the geoid) not being equal to γ_0, and which cause the actual equipotential surface of equal volume to lie at a height N (+ or −) above the spheroid. Let $g_0-\gamma_0=\Delta g_0$.

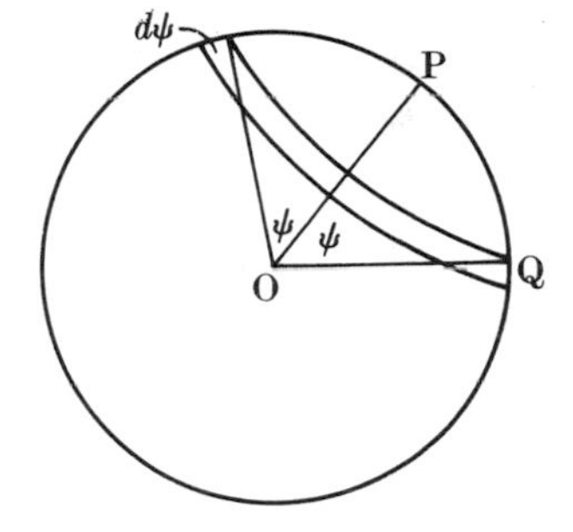

FIG. 6.9.

Then Stokes gives

$$N=\frac{R}{4\pi\gamma_m}\int_0^{2\pi}\int_0^{\pi}(\Delta g_0)f(\psi)\sin\psi\,d\psi\,d\alpha, \tag{6.111}$$

where ψ = the angle POQ in Fig. 6.9 between P, the point where N is to be calculated, and Q the location on the surface of an attracting element of a Green's equivalent layer of density $\Delta g_0/4\pi G$.

α is the azimuth of Q at P.

γ_m is a mean value of γ and R is a mean radius.

$$f(\psi)=\sum_2^\infty\frac{2n+1}{n-1}P_n(\cos\psi) \tag{6.112}$$

$$=1+\operatorname{cosec}\tfrac{1}{2}\psi-6\sin\tfrac{1}{2}\psi-5\cos\psi-3\cos\psi\ln(\sin\tfrac{1}{2}\psi+\sin^2\tfrac{1}{2}\psi). \tag{6.113}$$

† At this stage external masses such as mountains are treated as internal; see § 6.35.

N being small, ψ and α can be treated as if the earth was a sphere, see below. Further, the potential at P of any mass at Q depends only on the distance PQ, and is independent of α. So (6.111) can be written

$$N = \frac{R}{2\gamma_m} \int_0^{\pi} (\Delta \bar{g}_0) f(\psi) \sin \psi \, d\psi, \tag{6.114}$$

(where $\Delta \bar{g}_0$ is the mean value of Δg_0 in a surface ring between ψ and $\psi + d\psi$, as in Fig. 6.9.

Then $f(\psi)$ can be tabulated for all values of ψ between 0 and π, Δg_0 is averaged over annular rings of suitable width, § 6.33, and the integral is replaced by the summation of the effects of all the rings.

An alternative form of (6.111), for use if mean values of Δg_0 are estimated over areas bounded by regular lines of meridian and parallel, such as $1° \times 1°$ or $5° \times 5°$ 'rectangles', is

$$N = \frac{R}{4\pi\gamma_m} \sum_{\text{sphere}} \Delta \bar{g}_0 (\delta\phi' \, \delta\lambda' \cos \phi') f(\psi), \tag{6.115}$$

where N is required at the point (ϕ, λ).

$\Delta \bar{g}_0$ is the mean value of g over the area bounded by ϕ', $\phi' + \delta\phi'$, λ' and $\lambda' + \delta\lambda'$.

$$\cos \psi = \sin \phi \sin \phi' + \cos \phi \cos \phi' \cos(\lambda' - \lambda), \tag{6.116}$$

in which ϕ' and λ' are the latitude and longitude of the centre of each rectangle, or at some more appropriate point inside it, see § 6.34 (*c*) (ii).

Proof of (6.112). Treating the earth as a sphere rather than a spheroid, let $\Delta \bar{g}_0$ and N each be expressed as a series of zonal harmonics with P as pole and the earth's centre of mass as the point at which ψ is measured. Thus,

$$\Delta \bar{g}_0 = \gamma_m \sum_2^{\infty} a_n P_n(\cos \psi), \text{ the constants } a_n \text{ being presumed known,} \tag{6.117}$$

and (6.65) excluding spheroidal terms gives†

$$N = R \sum_2^{\infty} \frac{a_n}{n-1} P_n(\cos \psi). \tag{6.118}$$

† The distinction between R and a, and between γ_m and γ_e is immaterial here. The omission of P_0 makes the mean value of N zero, and the fact that ψ is measured at the earth's centre of volume eliminates P_1.

Then if $P_k(\cos\psi)$ is a particular harmonic, (6.117) gives

$$\frac{\Delta\bar{g}_0}{\gamma_m}P_k(\cos\psi)=\sum_2^\infty a_nP_k(\cos\psi)P_n(\cos\psi). \tag{6.119}$$

Integration over the sphere gives

$$\iint\frac{\Delta\bar{g}_0}{\gamma_m}P_k(\cos\psi)\,dw=\iint\sum_2^\infty a_nP_k(\cos\psi)P_n(\cos\psi)\,dw$$
$$=\iint a_kP_k(\cos\psi)P_k(\cos\psi)\,dw,$$

other products being zero, (H.21)

$$=\frac{4\pi}{2k+1}a_k \quad \text{from (H.23).} \tag{6.120}$$

So $$a_k=\frac{2k+1}{4\pi}\iint\frac{\Delta\bar{g}_0}{\gamma_m}P_k(\cos\psi)\,dw$$
$$=\frac{2k+1}{4\pi}\int_0^{2\pi}\int_0^{\pi}\frac{\Delta\bar{g}_0}{\gamma_m}P_k(\cos\psi)\sin\psi\,d\psi\,d\alpha. \tag{6.121}$$

Formula (6.118) gives N at $\psi=0$ to be $R\sum_2^\infty a_n/(n-1)$. (6.122)

Substituting (6.121) in (6.122), with n replacing k, gives

$$N=\frac{R}{4\pi\gamma_m}\int_0^{2\pi}\int_0^{\pi}\left\{\Delta\bar{g}_0\sum_2^\infty\frac{2n+1}{n-1}P_n(\cos\psi)\right\}\sin\psi\,d\psi\,d\alpha,$$

which is (6.111) with $f(\psi)$ as defined by (6.112).

It is perhaps surprising that $\sum_2^\infty\{(2n+1)/(n-1)\}P_n(\cos\psi)$ can be summed in terms of $\sin\psi$ and $\cos\psi$ to give the peculiar form of (6.113).

There is a looseness in this proof of Stokes's integral, namely that the anomalies Δg_0 are treated as being on a sphere, whereas they should be on a spheroid whose radius differs by between + and − one part in 600 from that of the best fitting sphere. But it is well known, Fig. 6.22, that the global extremes of N are about ±100 m, and it is surely clear that such a small error in the location of Δg is most unlikely to affect N by as much as ±1 m. [426], p. 23 quotes ±0.2 m as the global r.m.s. error arising if this point is ignored, and its next 20 pages give formulae for improved values.

If N is computed at a sufficient number of nearby points, the local form of the geoid with reference to a geocentric spheroid is given by interpolation, graphically or otherwise, between the computed values.

Since there is no first harmonic in N, the centre of the reference spheroid must automatically coincide with the centre of volume of the bounding equipotential surface.

For the practical application of Stokes's integral see §§ 6.33–to 6.38.

6.31. Formulae for the deviation of the vertical

Let ξ and η be the components of the inclination between the equipotential surface determined by Stokes's integral (6.111) and the (geocentric) reference spheroid. Then, considering the effect of an element $\Delta g_0\, dw$ located in azimuth α and at distance ψ from P in Fig. 6.9, the corresponding elements of ξ and η are

$$\xi = -\frac{dN}{R\, d\psi}\cos\alpha \text{ in meridian, and } \eta = -\frac{dN}{R\, d\psi}\sin\alpha \text{ in PV,}$$

where N is here the element of N arising from $\Delta g_0\, dw$.

So, [233], pp. 399–400, and [572A],

$$\xi'' = \frac{\operatorname{cosec} 1''}{4\pi\gamma_m}\iint \Delta g_0\left\{\frac{d}{d\psi}f(\psi)\right\}\cos\alpha \sin\psi\, d\psi\, d\alpha, \tag{6.123}$$

$$\eta'' = \frac{\operatorname{cosec} 1''}{4\pi\gamma_m}\iint \Delta g_0\left\{\frac{d}{d\psi}f(\psi)\right\}\sin\alpha \sin\psi\, d\psi\, d\alpha. \tag{6.124}$$

Positive values of ξ and η have the same significance as in § 2.05, but ξ and η are not identical with those of that paragraph, since the triangulator's reference spheroid (unlike Stokes's) is not centred on the earth's centre of mass.

Numerical integration is done by quadratures.

In (6.123) $\sin\psi \frac{d}{d\psi} f(\psi)$ is infinite when $\psi = 0$, but in practice no difficulty arises, for the deviations due to a small inner ring of radius r_0 km are

$$\xi'' = -(0.01051 r_0 + 0.12\times 10^{-5} r_0^2)\frac{d\,\Delta g_0}{dy}$$

$$\eta'' = -(0.01051 r_0 + 0.12\times 10^{-5} r_0^2)\frac{d\,\Delta g_0}{dx} \text{ seconds of arc,} \tag{6.125}$$

where the gravity gradients are in $\mu\text{m s}^{-2}$ per km, positive if increasing north (y) or east (x).† r_0 may be 200 m to 4 km. See [539], pp. 281–2. In

† Multiply the right-hand side of (6.125) by 10 if gradients are recorded in mGal per km.

[539] note that $f(\psi)$ is $\frac{1}{2}f(\psi)$ as given in (6.113): x and y are interchanged and signs of ξ and η are opposite, since [539] measures α from south. Terms in r_0^2 are very small, and are negligible if $r_0 < 10$ km.

[418], p. 20, remarks that these formulae give a better approximation to the ground-level values of the deviation than to the deviations at geoid level. This is convenient since the astronomical observations of ϕ and λ also give ground-level values.

[368], pp. 101–17, gives values of $\frac{1}{2}f(\psi)\sin\psi$, $\frac{1}{2}f(\psi)$, and $\frac{1}{2}\int_0^{\psi} f(\psi)\sin\psi\,d\psi$, where $f(\psi)$ is as in (6.113), at intervals of 0°.1 between 0° and 2°, and for every degree up to 180°. [539] gives values of

$$\frac{1}{2}\int_0^{\psi}\sin\psi\frac{d}{d\psi}f(\psi)\,d\psi, \qquad \tfrac{1}{2}\sin\psi\frac{d}{d\psi}f(\psi), \quad \text{and} \quad \frac{1}{2}\frac{d}{d\psi}f(\psi)$$

for (6.123) and (6.124).†

For the computation of deviations of the vertical using rectangular blocks, see [205] and [443].

6.32. Objects and limitations of Stokes's integral

(*a*) *Objects.* The expected uses of Stokes's integral and of the related formulae for the deviation of the vertical, (6.123) and (6.124), have been as follows.

(i) To compute N_0, ξ_0, and η_0 at survey origins and at selected control points of continental surveys, thereby putting all geodetic frameworks on to a common world geocentric reference spheroid. A reasonable standard to aim at (in 1978) is ± 1 m in N_0 and $\pm 0''.1$ in ξ_0 and η_0, although such accuracy is not yet possible.

(ii) To produce charts showing the undulations of the geoid, ideally for the whole earth, but more practicably over limited areas of perhaps 1000×1000 km, where it may be geophysically desirable, and where exceptionally dense gravity cover may make it possible. Here also an accuracy of ± 1 m in N, or possibly less, may be a reasonable target. One value of such surveys, in ocean areas, is that Stokes's integral gives the form of the geoid, while satellite altimetry, § 7.63 (*c*), gives the form of the water surface, and the difference is of oceanographical significance. But for this purpose errors of 1 m would be too great, and it will be necessary to aim at 10 cm.

(iii) As an aid to interpolation between astronomical stations in astro-geodetic geoidal profiles, if stations are unduly far apart, § 6.57 (*b*). This is a well established system of work in flat or gently undulating country.

† In [368] and [539] $f(\psi)$ means the same as $\frac{1}{2}f(\psi)$ in (6.113).

(*b*) *Difficulties and limitations.* The practical use of Stokes's integral involves three difficulties, as below.

(i) The integration covers the whole earth, and except for interpolation as in (*a*) (iii) above the effects of distant zones are not negligible. There is no early prospect of world-wide surface gravity observations providing adequate data. But see (*c*) (ii) below and § 6.34.

(ii) The theory on which Stokes's integral is based assumes that gravity is observed on an external equipotential surface, as it substantially is at sea and on low-lying land, but not elsewhere. § 6.35 describes the classical remedy, and Section 8 of this chapter describes a more recent approach.

(iii) Gravity is observed at a number of concrete points, often 100 km, or much more, apart. A single such observation does not provide a reliable mean value of Δg_0 over a surrounding area. From this point of view some systems for reducing observed gravity to geoid level may be better than others. They may be said to *interpolate* better, § 6.36 (*b*).

(*c*) *The impact of satellite geodesy* on the utility of Stokes's integral has been as in (i) and (ii) below.

(i) Satellite observations produce a world gravity survey in the form of a harmonic analysis of the geopotential, to about degree 16 or 20, from which a similar world gravity survey can be deduced. §§ 7.79–7.80 and 6.26. The amount of detail given by such an analysis is insufficient to replace the dense net of surface gravity observations required within (say) 20° of the place where N is to be computed, but it does constitute effective cover over the more distant parts of the earth. See §§ 6.33–6.34.

(ii) On the other hand, satellite observations are themselves producing good survey control, for object § 6.32 (*a*) (i) above. It cannot quite be said that the 1-m standard for N has yet been reached, but it does seem probable that satellite observations are, and will remain, better for this purpose than Stokes's integral, and with less effort. See §§ 7.77–7.78.

(*d*) It appears that the *current utility* of Stokes's integral is therefore limited to

(i) Local geoid surveys, as in § 6.32 (*a*) (ii).

(ii) Interpolation in aid of astro-geodetic geoid surveys on land, as in (*a*) (iii).

(iii) As an aid to introducing Greenwich longitude into satellite ephemerides, § 7.59.

6.33. Effects of error in $\Delta\bar{g}_0$

(*a*) *Accuracy required for computing N from* (6.114). In (6.114) consider a zone of width $\delta\psi = 2°$. Then the element of N arising from $\Delta\bar{g}_0$ in this zone is given by

$$\delta N = \tfrac{1}{2}R\frac{\Delta\bar{g}_0}{g}|f(\psi)\sin\psi|\frac{1}{28.7}\text{ metres,} \tag{6.126}$$

where $|f(\psi)\sin\psi|$ is its mean value in the zone, 1/28.7 is 2° in radians, and $R/g = 0.652\ \text{m}/\mu\text{m s}^{-2}$. Tables† show that $\tfrac{1}{2}f(\psi)\sin\psi$ is between +1.0 and +1.2 between $\psi = 0$ and 20°. It then changes sign and is about −1.0 between 60° and 90°, after which it changes sign again to become about +0.5 between 140° and 160°, decreasing to zero at 180°. See (A) in Fig. 6.10. Consequently, in round figures one metre of error in N arises from average errors of

40 μm s^{-2} of $\Delta\bar{g}_0$ in a 2° zone of $\delta\psi$, between $\psi = 0$ and 25°,
30 μm s^{-2} or more in a 5° zone between 25° and 55°,
20 μm s^{-2} in a 5° zone between 55° and 95°,
15 μm s^{-2} or more in a 10° zone between 95° and 180°.

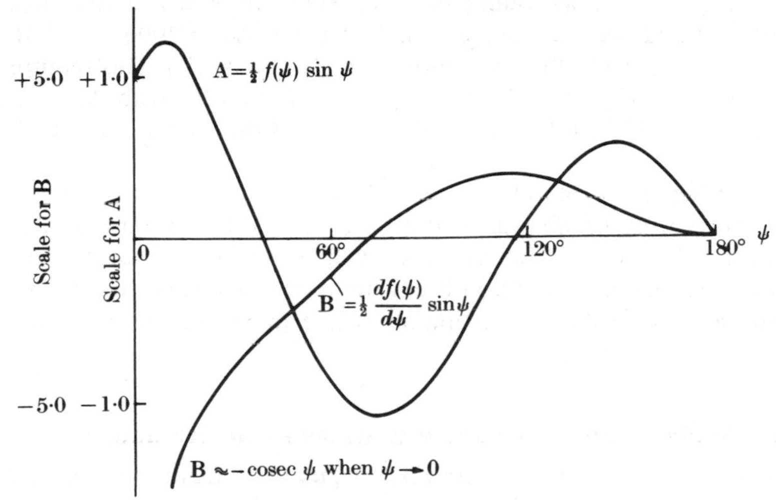

FIG. 6.10. Values of $A = \tfrac{1}{2}f(\psi)\sin\psi$, and of $B = \dfrac{1}{2}\dfrac{df(\psi)}{d\psi}\sin\psi$.

Then in the inner 24°, these figures show that if surface gravity in each of the twelve 2° annuli were estimated with an s.e. of ±20 μm s^{-2} (2 mGal), and if the errors of adjacent annuli were not correlated, the total error arising in the 24° cap would be about ±1.4 m.

In any particular area, careful analysis is necessary, special points of doubt arising from the following.

† As listed at the end of § 6.31.

(i) Random or systematic errors of observation at sea, and systematic errors due to the use of a common base station, § 5.05, or other errors of calibration.

(ii) Systematic error over large areas, due to a common method of reduction of observed gravity from ground-level to geoid level.

(*b*) *Accuracy required for computing* ξ *and* η, (6.123) *and* (6.124). In (6.123) and (6.124) consider an annular sector lying between radii ψ and $\psi+\delta\psi$, and subtending one radian of azimuth at P, where the deviation is to be computed. The contribution of this sector to the total deviation ξ is then

$$\delta\xi'' = 0.00167\left\{\frac{d}{d\psi}f(\psi)\right\}\Delta\bar{g}_0 \cos\bar{\alpha} \sin\psi\,\delta\psi, \text{ seconds of arc,} \quad (6.127)$$

if $\Delta\bar{g}_0$ is in $\mu\text{m s}^{-2}$, $\delta\psi$ is in radians, and $\bar{\alpha}$ is the mean azimuth of the sector. Fig. 6.10 shows values of $\frac{1}{2}f'(\psi)\sin\psi$, from which the effect of anomalous areas can be judged. For instance if $\Delta\bar{g}_0 = 200\ \mu\text{m s}^{-2}$, $\bar{\alpha}=0$, $\delta\psi=0.3$, and $\psi=40°$, the contribution to ξ will be 0″.26. This feature is a very large one. Closer to P, if $\psi=2°$ and $\delta\psi=1/120$, with $\Delta\bar{g}_0$ again $=200\ \mu\text{m s}^{-2}$, $f'(\psi)\sin\psi \approx -2\operatorname{cosec}\psi = -57$, and the contribution to ξ will be 0″.16.

See further in § 6.34 (*f*).

(*c*) *Areas with insufficient surface observations.* For distant areas, such as beyond 20° or 30°, satellite gravity may be used as in § 6.34. For nearer areas adequate surface observations are necessary. See § 6.43, for a guide to making the best estimates where there may be a shortage of observations in small areas.

6.34. Combination of surface gravity and satellite data

(*a*) *Size of surface gravity cap for computing N.* Continuing § 6.32 (*c*) (i). It is first necessary to decide on the area within which surface gravity will be used. This will probably be an annular cap of radius ψ_0. To decide on the value of ψ_0, as illustrated in [477] for an area in the North Atlantic, it is necessary to minimize the sum of three sources of error for caps of varying size, as follows.

(i) Error due to absence of adequate surface gravity data within the cap. This will be zero if $\psi_0=0$, increasing with ψ_0 fairly slowly so long as gravity cover is good, but more rapidly if the cap begins to include unsurveyed areas while ψ_0 is less than 20° or 30°, see Fig. 6.10. Estimates of the possible error must be based on the available data in any particular case. See § 6.33 (*a*).

(ii) Error in the coefficients in the satellite gravity formula. This clearly decreases as ψ_0 increases. [477], based on solution GEM 6, estimates s.e.'s of several metres if $\psi_0=0$, decreasing to ±1 or 2 m if $\psi_0=10°$, and to ±0.5 m if $\psi_0=30°$.

(iii) Truncation error, namely error arising from the omission of higher degree harmonics than are included in the satellite analysis. For a 16-degree solution [477] estimates ±4 m if $\psi_0=0$, $\pm1\frac{1}{2}$ m if $\psi_0=10°$, ±1 *m* if $\psi_0=20°$, and ±0.5 m if $\psi_0=30°$.

In [477], p. 24 the sum of these three errors is estimated to give s.e.'s of ±5 m for $\psi_0=0$, $\pm2\frac{1}{2}$ m for $\psi_0=10°$, ±2 m for $\psi_0=20°$, and $\pm1\frac{1}{2}$ m for $\psi_0=30°$. These figures suggest a cap radius of 30°, but 20° was accepted. The area selected for this test was intentionally a favourable one with good surface gravity cover out to 20°.

If it is required to use Stokes's integral to survey the geoid over an area of (say) 1000×1000 km, using a 20° cap, the surface gravity net will have to cover an area of about 5000×5000 km. Few such areas exist.

(*b*) *Combination of surface and satellite gravity.*

(i) In (6.111) or (6.115) it is desired to use surface values of gravity Δg_0 over a cap of perhaps 20° radius, designated area A_1, and to use satellite gravity Δg_s over the rest of the earth, area A_2. Then, following [477], the required value of N may be written as

$$N=N_1+N_2+N_3, \tag{6.128}$$

where N_1 would be the value of N if Δg_s was used for the whole earth.
N_2 is the difference between using Δg_0 and Δg_s in area A_1.
N_3 is the difference between using Δg_s and Δg_0 in selected parts of area A_2.

(*ii*) *Formula for* N_1.

$$N_1=\frac{GM}{r\gamma}\sum_{n=2}\left(\frac{a}{r}\right)^n\sum_{m=0}^{n}(\bar{C}'_{nm}\cos m\lambda+\bar{S}_{nm}\sin m\lambda)\bar{P}_{nm}(\cos\theta), \tag{6.129}$$

where r is the geocentric distance to P, Fig. 6.9.
a is the earth's equatorial radius,
θ, λ are polar coordinates of P.
γ is standard sea-level gravity at P.
$\bar{C}'$ and $\bar{S}$ are normalized coefficients, the prime against $\bar{C}'$ indicating that the potential coefficients implied by the adopted reference spheroid ($\bar{C}_{20}$ and $\bar{C}_{40}$) are subtracted from those given by the satellite analysis. Compare u' in (6.61).

Formula (6.129) is obtained from (7.1B), by adding $\frac{1}{2}\omega^2 r^2 \sin^2\theta$†, subtracting the value of the potential on the geoid, and using $\delta U = -\gamma\,\delta N$, as in (6.8).

(*iii*) *Formula for* N_2.

$$N_2 = \frac{R}{4\pi\gamma_m} \iint\limits_{A_1} (\Delta\bar{g}_0 - \Delta\bar{g}_S) f(\psi)\, d\sigma, \tag{6.130}$$

where R is the earth's mean radius (say) 6371 km.

γ_m is earth's standard gravity (say) 9.798 m s^{-2}.

σ is an element of area.

Integration is done numerically as in (6.115) and subparagraph (c) below. In any small area centred on θ', λ', Δg_S is given by

$$\Delta g_S = \frac{GM}{r^2} \sum_{n=2} (n-1)\left(\frac{a}{r}\right)^n \sum_{m=0}^{n} \bar{C}'_{nm} \cos m\lambda' + \bar{S}_{nm} \sin m\lambda')\bar{P}_{nm}(\cos\theta'), \tag{6.131}$$

as derived from (6.129) and (6.65).

(*iv*) *Formula for* N_3. Similarly

$$N_3 = \frac{R}{4\pi\gamma_m} \iint\limits_{A_2} (\Delta\bar{g}_0 - \Delta\bar{g}_S) f(\psi)\, d\sigma, \tag{6.132}$$

with notation as in (6.130).

In practice, the size of the cap A_1 is likely to be so chosen that Δg_0 only constitutes an improvement on Δg_S over such small areas, outside the cap, that N_3 is not worth computing.

(c) *Use of rectangular blocks instead of annular rings.*

Currently, and in future, Stokes's integral is likely to be used in the form of (6.115), and the same rectangular block limits will be used in (6.130). For each block we require (i), (ii), and (iii) below.

(i) The mean value of Δg_0. There is no difficulty in principle, other than lack of observations. But see §§ 6.35 and 6.37.

(ii) A mean value of $f(\psi)$. This can ordinarily be the value appropriate to the centre of the rectangle, but in the immediate neighbourhood of the

† To convert the potential of the attraction to the potential of gravity.

computation point P

$$f(\psi)=\frac{1}{\Delta\sigma}\iint\limits_{\Delta\sigma} f(\psi)\,d\sigma, \tag{6.133}$$

which may be evaluated by computing ψ by (6.116) at a number of suitably distributed points covering the rectangle. σ is an element of area.

(iii) A mean value of Δg_S. This is obtained by subdividing the standard size of block in the neighbourhood of P, using the smaller blocks for Δg_0 and $f(\psi)$ as well as for Δg_S. [477], on the basis of its trial area in the North Atlantic advises using $1°\times 1°$ blocks between $\psi=10°$ and $20°$, each divided into four between $\psi=5°$ and $10°$, into 16 between $2°$ and $5°$, and into 64 when $\psi<2°$. On this system [477], p. 12 estimates that the numerical integration will only contribute errors of the order 0.1 m, apart from errors in the analysis itself. [477] also estimates that if numerical integration were made using only the centre points of $1°\times 1°$ blocks, errors of 1 metre might result. In topographically or geophysically disturbed areas fine sub-division is of course more necessary than in more uniform areas.

(*d*) *Mass of the atmosphere.* Although the mass of the atmosphere is small, see § 6.29, the effect of ignoring it all over a cap of radius (say) 20° may be quite large. It is here assumed that the value of GM used for standard gravity treats the atmosphere as internal, as is correct for satellite data, but which will demand a positive correction to $g-\gamma_0$, if g is got from surface observations. [477], pp. 6–9 examines this and concludes that if the point P and its surroundings are at or near sea level, N should be increased by 0.56, 1.17, 1.75, 2.26, 2.67, or 2.97 metres if the cap radius is 5°, 10°, 15°, 20°, 25°, or 30°, respectively. If the average height of the cap is 400 m above sea-level, these figures should be reduced by about 5 per cent.

(*e*) *Sources of data.* For satellite data see §§ 7.79 and 7.80. For surface gravity currently available see § 6.60 (*c*). The effect of external masses above the geoid is considered in §§ 6.35–6.38.

(*f*) *Computation of ξ and η.* The routine of computation is outlined in [532]. The effect of an inner zone of radius about 1° is computed from the actually available point values of g, perhaps 200 values. In an intermediate zone values of g meaned over $1°\times 1°$ equal area 'squares' are employed. But within 10°, smaller subdivisions such as $10'\times 10'$ may be used if sufficient data is available. Over the rest of the world satellite-derived spherical harmonic constants are used.

The intermediate zone should extend to at least a radius of 30°. A radius of 40° is considered optimum, but the intermediate zone can clearly only go as far as detailed ground gravity surveys exist.

In a reasonably well gravitationally surveyed part of the world [532] hopes for an s.e. of between 1″.3 and 2″.2.

SECTION 7. MASSES OUTSIDE THE GEOID. GRAVITY ANOMALIES

Both Clairaut's theorem and Stokes's integral derive their results on the assumption that there are no masses such as mountains outside the equipotential surface, such as the geoid, on which the value of gravity is to be derived by Clairaut, or is accepted by Stokes. Actually on the earth there are external masses which are much too large to be ignored. Also, gravity observations in land areas are made at some height above the geoid, with rock intervening, and often with other external masses nearby, whose irregular shapes affect the local value of g. There are two ways of getting over this difficulty, namely the classical method which is described in § 6.35, and a newer 'non-classical' method which is described in Section 8.

6.35. Elimination of external masses. Co-geoids

(*a*) *Outline.* Referring to § 6.32 (*b*) (ii), the classical method of obtaining values of g on a bounding equipotential surface, without external masses, is as follows.

Gravity is observed at ground stations Q at known heights h above the geoid, Fig. 6.11. Then let all masses above the geoid be supposed to be

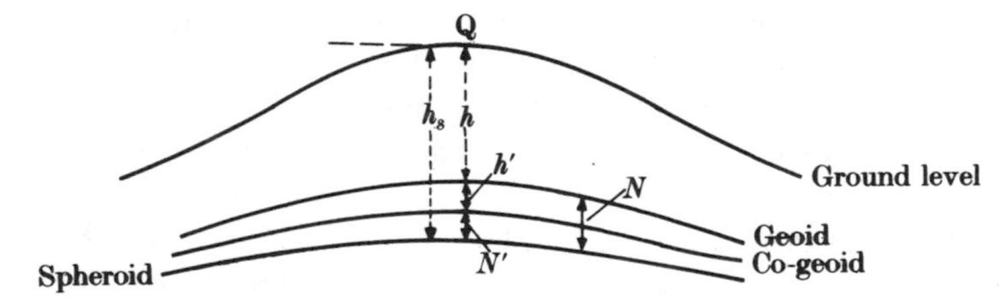

FIG. 6.11. $N = h' + N'$. $h_s = h + N$. The change of potential due to hypothetical mass movements is $-g_0 h'$.

removed, either to infinity, which will be inconvenient see § 6.38 (*a*), or to prescribed locations below the geoid. This will change gravity at Q from g to $g + F_e$, where F_e can be computed from topographical maps, assuming knowledge of the rock density.† See §§ J.02 and J.04. The potential on the geoid beneath Q will change from W to $W - \delta W$, where δW can also be calculated, § J.06. The equipotential surface will then be

† Estimates of density may be wrong by 10 per cent, but the total error at a station is likely to be a smaller percentage of F_e.

depressed by $h' = \delta W/g$, to form a new surface, which is external to all matter.† This surface is known as a *co-geoid.*‡ There are as many co-geoids as there are systems for removing the external masses and putting them in more convenient places.

The modified gravity at Q, $(g + F_e)$, can now be reduced to the co-geoid through a distance of $(h + h')$ by the free air reduction of §§ 6.23 and 6.37 to give g_0', the value of gravity on an equipotential surface with no external masses, and Stokes's integral can properly be used with

$$\begin{aligned}\Delta g_0 &= g_0' - \gamma_0\\ &= (g + F_e) + (\text{approximately})\ 0.308(h + h') - \gamma_0\ \text{mGal}\|\\ &= (g - \gamma_A) + F_e, \quad \text{see } (b) \text{ below,}\end{aligned} \qquad (6.134)$$

to give N_P' the height of the co-geoid above the reference spheroid at the required point P. The height of the geoid above the co-geoid at P, $h_P' = \delta W_P/g$, must then be calculated, and finally $N_P = N_P' + h_P'$.

The centre of the reference spheroid will now coincide with the centre of volume of the co-geoid.

While this method of solution is correct in principle, the calculation of F_e and δW at every gravity station which is to be used is an immense labour, especially in mountainous country. The arbitrary locations inside the geoid to which the external masses are moved must be chosen so as to minimize this labour.

(*b*) *The indirect effect.* When, as at the beginning of § 6.35 (*a*), masses between the ground level and the geoid are moved, the masses between the geoid and the co-geoid may remain external to the latter, in the form of a layer of thickness h' of perhaps some tens of metres. These masses must then be moved inside the co-geoid. Let this be done by putting them immediately inside it. No significant change in F_e will result, since h' varies slowly and the layer is virtually an infinite plate as in § 6.09 (*d*). Also the movement of this relatively small mass through this very small distance will produce no significant change in δW, and the co-geoid will remain an equipotential surface. The only necessary action is then to reduce g from ground level to the co-geoid by the free-air reduction through a height of $h_S = (h + h')$, as has already been done in (6.134). The

† Except for some matter between it and the geoid if h' is positive. For the remedy see the *indirect effect* in (*b*) below. h' is positive if the geoid is above the co-geoid.

‡ In the *Survey of India Geodetic Reports* and other papers, e.g. [238], such a surface has been called a *Compensated geoid,* the surface there used being derived from the geoid on the basis of the removal of topography and its isostatic compensation. But isostatic compensation is not an essential feature of the concept, and 'co-geoid' as introduced at the Oslo 1948 meeting of the International Geodetic Association is a better name.

‖ h is measured in metres. The factor is 3.08 if g is measured in $\mu\text{m s}^{-2}$.

term $2\gamma h'/R$ in this reduction, see (6.95), is sometimes known as the *Bowie correction.* For uniformity h' should be included whether it is positive or negative.

6.36. Gravity anomalies. Uses and characteristics

(*a*) *Uses.* § 6.05 outlines the use of gravity anomalies as giving a comparison between observed gravity and that of some specified model earth, for which the standard gravity γ_0 of a spheroid is modified to allow for the height of the gravity station, the attraction of the topography, and (optionally) of its isostatic compensation as defined by one of various systems. Then $g-(\gamma$ of the model) is a useful anomaly from the geophysical point of view.

For Stokes's integral we need to remove the topography. Here we need a model with no external topography, but probably with some corresponding internal irregularities of density. The attraction of the topography is then to be subtracted from observed g, and the attraction of the corresponding masses below the geoid is added. If the distribution of these internal masses is chosen so that they are equal and opposite to those postulated by some isostatic system, the resulting $g_0'-\gamma_0$, with g modified as above and reduced to co-geoid level, will have the same value as the $g-(\gamma$ of the model) of § 6.05. The routine for computing the latter for geophysical purposes will be the same as the routine for computing the former for Stokes's integral.

Another use for gravity anomalies is to aid the prediction of g where it has not been observed. With a good model, $g-\gamma$ will vary from place to place relatively slowly.

(*b*) *Characteristics of a useful anomaly.* The characteristics of a good system should so far as possible be as follows.

(i) Minimum labour in computing F_e, the effect of moving the topography from outside to inside the geoid, or of the effect of the topography and compensation (if any), if it is included in the model.

(ii) Minimum labour in computing δW. This is less serious than F_e.

(iii) The anomaly $g+F_e-\gamma_A$ must vary from place to place much less strongly than $g-\gamma_A$, so that widely separated values of it should be typical of the areas they have to represent. It must *interpolate well.*

(iv) The curvature of the co-geoid should be less irregular than that of the geoid, to reduce error in the free-air reduction, § 6.37 (*a*) (i).

(v) It is desirable, but not essential, that the computation of F_e should involve the same arithmetic as is required for computing anomalies used for geophysical purposes. The model should, in general, either be as similar as possible to the true earth, or to a model earth in which the mantle is strengthless.

6.37. The free-air reduction

(*a*) *From ground-level to geoid or co-geoid.* Continuing from § 6.23, the free-air reduction from Q at ground-level, Fig. 6.11, to the co-geoid $(h+h')$ metres below Q is given by

$$g_0' - (g+F_e) = g_M(1/r_x + 1/r_y)(1+m')(h+h'), \tag{6.135}$$

where g_M is the mean of $(g+F_e)$ and g_0', and is required with only low accuracy; $m' = 0.003\,46$; and r_x and r_y are the mean values of the radii of curvature of the equipotential surfaces between Q and the co-geoid after removal of the external masses, which will ordinarily be taken to equal ν and ρ.

For a gravity station 2000 m above sea-level this correction will be about 6.0 mm s^{-2} (600 mGal). On the surface of the actual earth the deviation of the vertical may change by 1″ per km, introducing an abnormality of 3 per cent in r_x or r_y, or of perhaps 2 per cent in the sum $1/r_x + 1/r_y = 2/r_M$, giving an error of 120 μm s^{-2} (12 mGal) in the reduction of a 2000-m station. This is significant, and even greater errors could sometimes occur. There are, however, two mitigating circumstances.

(i) In the present context, the topography which is responsible for the rapid variations of r_M will have been removed to some place where it will cause smaller, or perhaps very much smaller, abnormalities of curvature: leaving only density anomalies below the geoid, whose effect will also be much less.

(ii) For the practical application of Stokes's integral (6.114), we need average values of g_0 or g_0' over an area, and the mean curvature error over a large area is clearly likely to be less than the extreme error on (say) a mountain top. With the topography still in place (for instance) the upward verticals will tend to diverge more than average under a mountain, and less than average under a valley. Consider a square area 100 × 100 km. It would be a somewhat extreme case if the deviation of the (downward) vertical averaged 10″ inwards all round the perimeter, when the average r_M would be abnormal by 0.6 per cent corresponding to 40 μ m s^{-2} (4 mGal) for a 2000-metre station, and this is before the topography has been removed as in (i).

The approximation $r_x = \nu$ and $r_y = \rho$ is thus much less serious than it appears to be at first sight.

(*b*) *Use of free-air anomaly for Δg_0 in Stokes's integral.* The simplest of all gravity anomalies is $g - \gamma_A$, as in the last line of (6.95). Nothing could be more convenient in Stokes's integral than to be able to write $\Delta g_0 = g - \gamma_A$, but this would amount to ignoring the fact that the external topography has got to be removed as in § 6.35 (*a*). It simply takes F_e and δU as zero. Nevertheless, in country which is flat but not necessarily at a low altitude, the use of $\Delta g_0 = g - \gamma_A$ is in fact acceptable as a good approximation; see § 6.38 (b). It is thus necessary to consider to what extent $g-\gamma_A$ meets the necessity for accurate interpolation as demanded by § 6.36 (b) (iii). As it stands, $g-\gamma_A$ does not at all satisfy this criterion, since its local variations largely reflect the fact that the underlying rock is not of zero density. In a small area 10-μm s^{-2} contours of equal $g-\gamma_A$ are in fact very similar to 9-metre height contours, since 10 μm s^{-2} is the attraction of a 9-m layer of rock, § 6.09 (d). Figs. 6.12 and 6.24 (a) show $g-\gamma_A$ plotted against height of station in different types of country, and show that within distances of 100 or 200 km, $g-\gamma_A$ approximates to a variable sea-level value plus 1.12 μm s^{-2} (0.112 mGal) per metre of h. So if the mean height of such an area is known, the mean value of $g-\gamma_A$ in the area can be deduced from a single station. Many stations are of course better than only one, and if they strongly suggest a factor other than 1.12 for the area, it may be adopted. There is no need to do the calculation graphically.

In sea areas the factor 1.12 is replaced by 0.69, as in Fig. 6.24 (a).

It is important to note that while $g-\gamma_A$ is likely to vary linearly with height over a small area, it does not at all follow that it will on average be positive over a large area with a high average height. Over a large area,

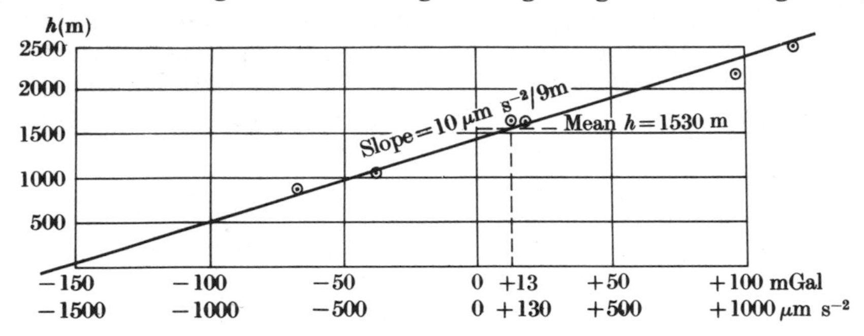

FIG. 6.12(a). Variation of $g-\gamma_A$ with height in Baluchistan between $\phi=30°$ and $31°$ N, $\lambda=67°$ to $68°$ E. The average height of the area is 1530 m, corresponding to mean $g-\gamma_A=130\ \mu$m s^{-2}.

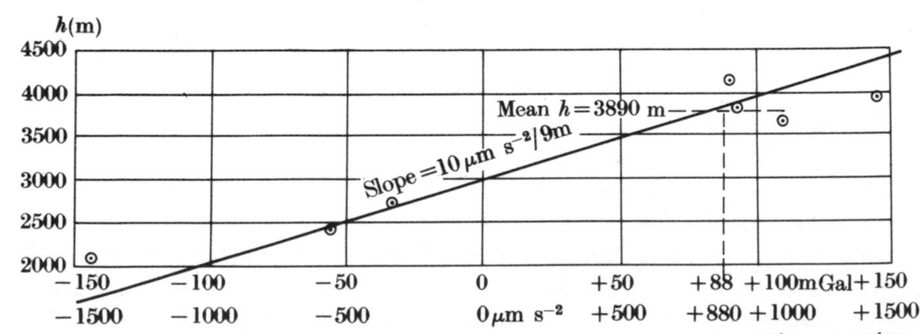

FIG. 6.12(b). Variation of $g-\gamma_A$ with height in Kashmir, between $\phi=34\frac{1}{2}°$ and $35\frac{1}{2}°$N, $\lambda=74\frac{1}{2}$ to $75\frac{1}{2}°$ E. The average height of the area is 3890 m, corresponding to $g-\gamma_A=+880\ \mu$m s^{-2}.

mountain or ocean, the most probable average value of $g-\gamma_A$ is not far from zero.

In this way the free-air anomaly can be made to interpolate well provided an *average height map* is available to give the average heights above the geoid of degree squares, or other similar areas, wherever there are gravity stations. Such maps have been prepared in many countries. [48] lists a number of average height maps known to exist in 1971. A world-wide compilation of $1°\times 1°$ means is maintained by the US Defense Mapping Agency Aerospace Center DMAAC. The desirable accuracy is ±25 m, but much larger errors have to be tolerated in many parts of the world, and are harmless at distances of more than (say) 1000 km.

6.38. Various co-geoids and associated gravity anomalies

The following systems have been considered for the removal of external masses.

(*a*) *Topography removed to infinity*. The obvious thing to do with the topography is to abolish it, but the resulting change of potential is such that the geoid and co-geoid may be separated by some hundreds of metres. F_e of (6.134) is the vertical component of the attraction at Q of all masses above sea-level, and $g-\gamma_A+F_e$ is written $g-\gamma_T$. It is very troublesome to compute. This system is not a practicable one.

(*b*) *External topography condensed to sea-level*. In Fig. 6.13, Q is a gravity station of height h_Q above the geoid. Let all masses outside the geoid be consensed vertically into a layer of varying skin density immediately below it. The column AA″B″B is replaced by equal mass on A″B″. The changes of potential are substantially zero,† so $\delta W=0=h'$.

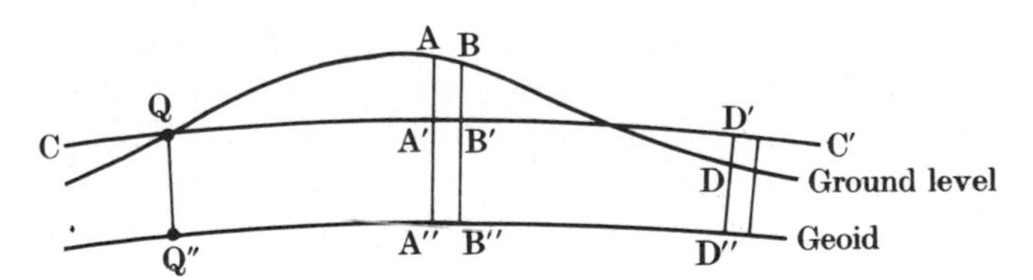

FIG. 6.13. Topography condensed to sea-level. Q is a gravity station.

To compute F_e, the resulting change in g at Q, topography may be regarded as

(i) a plateau CC′ of height h_Q and
(ii) excesses and defects AA′ and DD′.

Condensing the plateau to below the geoid causes no change in g, whether the plateau is regarded as in infinite plane or as a spherical shell. F_e is then the vertical component of the attraction at Q of the condensed

† $\delta W/g$ is less than 1 metre if $h=3000\ m$.

excesses and defects minus their attraction when in their true positions. When QA and QD are some thousands of kilometres, their contribution to F_e will clearly be small. It will also be small at a distance of a few hundred kilometres because the masses lie nearly on the horizon of Q both before and after moving. F therefore arises almost entirely from the excesses and defects within 50 km or less of Q. It can be computed with the tables mentioned in § J.04 using zero as the depth of compensation.

The condensation of mountains standing above Q contributes positive elements to F_e, while the contribution of valleys is negative. There is likely to be some cancellation in the total, but in mountainous country F_e can amount to some tens of mGal (100's of μm s^{-2}). Within this limit g_0' in (6.134) in this system is not very different from $g-\gamma_A$. It is relatively easy to compute, but it does not interpolate well.

(*c*) *Mirror image. Rudski reduction.* If every external element of mass is moved to a depth below the geoid equal† to its original height, the potential at Q will be unchanged.

The correction F_e can be calculated in the same way as in the method of sea-level condensation, but it will be rather larger and a little more troublesome. Interpolation will be better, but not good in mountainous country.

(*d*) *Isostatic reductions.* Let every item of external mass be moved inside the geoid to the point occupied by its compensation according to one of the formal hypotheses described in §§ 6.63–6.66. The resulting change of g at Q to $g+F_e$ is then the same as the 'isostatic reduction' which is commonly carried out for geophysical purposes. The change of potential will be such as will separate geoid and co-geoid by a maximum of 30 metres (in Tibet). That is a very great improvement on (*a*) above, although it cannot be neglected. It can be computed as in § J.06.

It is to be noted that the validity of the system is not dependent on the actual existence of compensation according to the hypothesis adopted, or in any other form, but that if the hypothesis was correct the modified crust would consist of layers of rock of uniform density, and conditions (iii) and (iv) of § 6.36 (*b*) (representative character of point values, and absence of abnormal curvature) would be satisfied. But even if compensation is wholly non-existent, the new location of the external masses will be sufficiently deep (or widespread if the hypothesis is a regional one) for these two conditions to be tolerably well satisfied in any case. See Fig. 6.24 (c) for examples of condition (iii).

Condition (v) is also satisfied, provided the deficient mass of the ocean is considered to be brought up from below, where the hypothesis places the compensation of the oceans. This is not necessary from the point of

† In the Rudski system, a very slight departure from equality of depth to original height allows for the spherical shape of the earth and secures identically zero change in *W*.

view of the validity of Stokes's integral, but it improves interpolation and is a convenience which is generally adopted.

The change from g to $g+F_e$ is apt to be large. The effects of topography may be appreciable up to distances of some hundreds of kilometres, but the effects of distant zones need only be calculated at a few stations, and then interpolated at others. See § J.07.

In general, the adoption of a large depth of compensation increases both F_e and δW and the labour of computing them, while it aids accurate interpolation. A depth of 25 or 30 km (Airy) is a good compromise.

(*e*) *Smoothed topography*. Let the earth's topographical features be supposed to be smoothed out by lateral transfers† of mass so that the slopes of the surface of the resulting earth model do not exceed some such figure as 0.01 as shown in Fig. 6.14. Then let the matter between the model and the geoid be condensed to just below the geoid as in § 6.38 (*b*). At a gravity station Q, the resulting total change of gravity from g to $g+F_e$ is then given by

$$F_e = \mathrm{O} - 1.12(h_Q - h_m) + \text{Zero}\ \mu\text{m s}^{-2}, \tag{6.136}$$

where h_Q is the height of Q above the geoid and h_m is the height in metres of the model above Q′ on the geoid. The last term, Zero, is the effect of condensing what is virtually an infinite plate, and O is the orographical correction of § 6.41. It is the change in g resulting from the smoothing out of nearby topography to the level of Q. It is relatively easily computed. The term $-1.12(h_Q - h_m)$ is the effect of laterally moving the remaining layer between the model and the level of Q. It also is substantially an infinite plate.

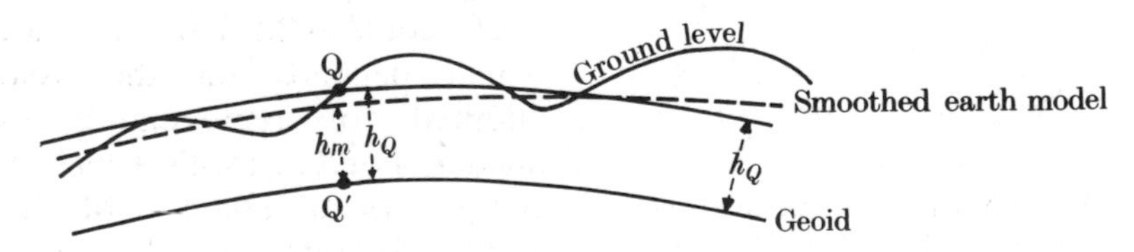

FIG. 6.14. Smoothed topography. The earth's curvature is exaggerated.

The changes of potential δW are small. The separation of the geoid and the co-geoid is estimated as 7 metres, [234], p. 5, in Himalayan areas and 2 metres in Alpine. It can ordinarily be ignored.

Then g_m, gravity on the bounding surface of the model is given by

$$g_m = g + \mathrm{O} - 1.12(h_Q - h_m) + (\text{free air from } h_Q \text{ to } h_m)\ \mu\text{m s}^{-2},$$

† This system of smoothing was proposed by de Graaff–Hunter in [234], in connection with the non-classical application of Stokes's integral to the earth's ground surface, §§ 6.47–6.49. It is here adapted to the classical treatment with substantially the same final result.

and $\Delta g_0 = g - \gamma_A + O - 1.12(h_Q - h_m)\ \mu m\ s^{-2} = g - \gamma_E = \Delta g_E$, (6.137)

where γ_A is computed for the height h_Q, defining γ_E and Δg_E. If the density between h_Q and h_m is known to differ from 2.67, write $(\rho/2.67) \times 1.12$ for 1.12. For anomalies in mGal use 0.112 instead of 1.12.

(6.137) shows that Δg_E is the Bouguer anomaly $g - \gamma_B$ of § 6.41 plus 1.12 $\mu m\ s^{-2}$ per metre of the local height of the model. In so far as h_Q and h_m tend to be somewhat equal, when large, the last term of Δg_E is not large, and Δg_E does not resemble $g - \gamma_B$ in being very large over high plateaux or deep oceans. In this it resembles an isostatic anomaly.†

This system involves less computation than isostatic systems. The basic requirement is an all-world average height map from which to compute the height of the model.

6.39. The earth's centre of mass

The following facts are conveniently collected here.

(a) Any surface whose equation is $r = R\left(1 + \sum_2 u_n\right)$ with no first harmonic u_1, has its centre of volume (CV) at the origin of coordinates.

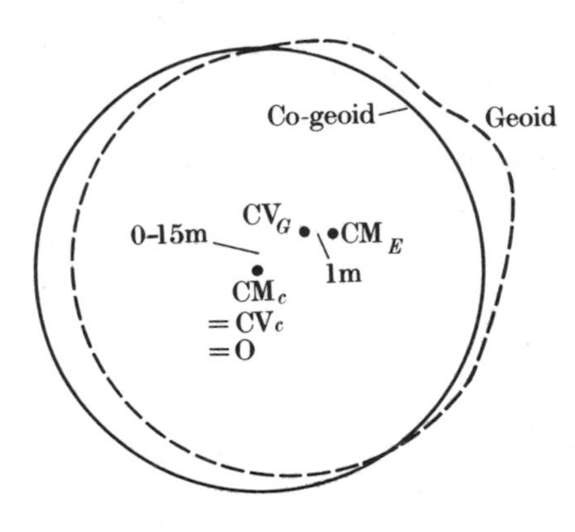

FIG. 6.15. CV_G is the centre of volume of the geoid, and CM_E is the centre of mass of the earth. CV_C is the centre of volume of a co-geoid, and CM_C is the centre of mass of the masses contained in it after the inward transfer of external masses. O is the centre of the spheroid above which Stokes's integral gives the heights of a bounding equipotential surface.

(b) The CV of a 'bounding equipotential surface' such as a co-geoid necessarily coincides with the centre of mass (CM) of the contained masses; see § 6.22 (b).

(c) The CM of the actual earth (CM_E) is separated from the CM of a co-geoid‡ (CM_C) by an amount which depends on the system adopted for removing external masses. For Hayford's equal mass compensation, § 6.64, CM_C is 5 metres distant from CM_E in the direction away from latitude 45° N, 30° E, [366], p. 169. For Hayford's equal pressure system the separation is 15 m. For more shallow systems of compensation the separation will be less.

†Moving a particle of external mass to some internal location, as when forming an isostatic anomaly, will have the same effect on the external field as spreading it out regionally over a more shallow surface, § 6.61 (c). The de Graaff–Hunter system of spread does not, of course, correspond exactly to any form of local isostasy, but its external effect differs only slightly from a rather shallow Airy isostatic system.

‡i.e. the CM of the masses contained inside the co-geoid after the inward transfer of external masses.

For the Rudski and sea-level condensation systems the separation is substantially zero. For topography removed to infinity it would be about 600 m.

(*d*) When the form of a co-geoid is determined by Stokes's integral, the centre of the reference spheroid coincides with CV_C, since there are no first harmonics, and consequently also with CM_C, and it is probably a few metres distant from CM_E as in (*c*) above.

(*e*) The CM of the earth and the CV of the geoid do not exactly coincide, because the geoid is not a bounding equipotential of the actual earth, but the separation is probably less than one metre.

(*f*) Since gravity on the co-geoid, g_0', contains no first degree harmonic, its mean value over any hemisphere of the Stokes reference spheroid must be the same as over any other hemisphere.

(*g*) The centres of reference spheroids used for survey frameworks, as defined in § 2.06, may be separated by some hundreds of metres from the earth's CM and from the centres of the spheroids with reference to which Stokes's integral gives N' and N.

6.40. The deviation of the vertical

In principle (6.123) and (6.124) are a possible means of computing the deviations of the vertical with reference to a spheroid centred on the earth's centre of mass, although sufficient gravity observations are not yet available. In mountainous country the deviation is mostly caused by the unevenness of the nearby air–rock interface, and if the integral is to give the effect of this correctly the values of $\Delta\bar{g}_0$ given to it must locally either be in very great detail, or else must be of a type which interpolates well: either $g-\gamma_A$ treated as in 6.37 (*b*) (in each separate annular sector), or an isostatic anomaly.† If one of the latter is used the integral will give the isostatic anomaly of the deviation, from which the deviation itself can be computed. See §§ 6.54 and J.01.

Survey framework origins, or control points, where it may be required to compute ξ and η in this way can generally be chosen so as to lie in flat country where local conditions are favourable.

See § 6.33 (*b*) for the accuracy required for Δg_0.

6.41. The Bouguer anomaly

The Bouguer anomaly, $g-\gamma_B$, is a simplified form of $g-\gamma_T$, § 6.38 (*a*). It is not suitable for use with Stokes's integral, but it is a useful form for

† $g-\gamma_T$ is not practicable, because the horizontal attraction of uncompensated mountain systems may be significant at a distance of some thousands of kilometres. For the deviation, an isostatic anomaly is not only probably closer to nature, but it is also much easier to compute. An anomaly based on the smoothed topography of § 6.38 (*e*) is a possible alternative.

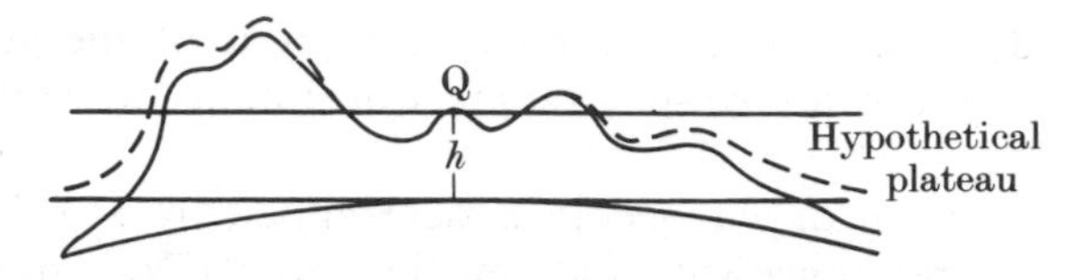

FIG. 6.16. Bouguer reduction. Full line is actual topography, and broken line is hypothetical.

geological and geophysical prospecting. It differs from $g-\gamma_T$ by ignoring the earth's curvature when the attraction of the topography is being computed, thus lessening the effect of distant features. The topography is assumed to take the form of a tangent plateau, as thick as the height of the gravity station, with the hills and valleys on its upper surface, as shown in Fig. 6.16. The result of this rather peculiar treatment is as follows.

(*a*) Within a radius of (say) 20 km of the gravity station the topographical effect is correctly computed.

(*b*) Beyond this there is a belt† in which the topographical effect is small, both for the actual earth and for the Bouguer model, because all masses are near the horizon.

(*c*) Beyond this belt, at distances of 1000 km or more, the curvature of the earth gives the attraction of the topography a substantial vertical component, which on the actual earth may be expected to be cancelled by nearby isostatic compensation. In the Bouguer model its effect is cancelled by the neglect of curvature.

The position thus is that $g-\gamma_B$ is a topographical anomaly as regards near masses, and an isostatic anomaly as regards distant masses, with an intervening belt in which the effect is small on both hypotheses. It is a theoretical monstrosity, and its unsuitability for Stokes's integral is clear, but it interpolates well and is easy to compute. A local irregularity in $g-\gamma_B$ correctly indicates a density anomaly in the underlying rocks.

We have

$$\begin{aligned}\gamma_B &= \gamma_A + (\text{attraction of plateau}) + (\text{attraction of hills and valleys})\\ &= \gamma_A + 2\pi G h_\rho - \mathrm{O}\\ &= \gamma_0 - 0.196h - \mathrm{O}\ \text{mGal, or } \gamma_0 - 1.96h - \mathrm{O}\ \mu\text{m s}^{-2},\end{aligned} \tag{6.138}$$

where h is the height of the gravity station Q in metres, and O is the upward attraction at Q of the surrounding hills and valleys, supposed imposed on the tangent plateau. O is known as the *orographical* or *terrain correction.*

† In high mountains this belt does not begin so near, but gravity is not a promising aid to geophysical prospecting in such country.

The correction O, which is subtracted† from γ, is always positive, since the removal of a hill above Q reduces γ, as also does the infilling of a valley below Q. There is thus no essential difference of sign between O at a station on a hill top, and one in a valley bottom. O is in fact a measure of the *roughness* of the topography.

In topography such as that of most of England, where hills seldom exceed 300 m, O seldom exceeds 2 or 3 mGal (20–30 μm s^{-2}). In the outer Himalayan ranges at heights of up to 2500 m it is commonly 20–25 mGal (200–250 μm s^{-2}). It is always much easier to compute than any isostatic anomaly. See § J.03.

Given a contoured map at such a scale as $1:10^6$, a few trial computations will enable types of areas to be marked.

(i) Areas in which O can be relied on to be not more than 2 or 3 mGal, so that it can be estimated within probably 1 mGal without computation.

(ii) Areas in which O is liable to be more than 10 mGal, and in which simple estimation may be wrong by 10 mGal or more. These areas will only cover a small fraction of the earth's surface.

(iii) Intermediate areas in which computation may be effective and not too difficult.

Geological survey departments, and others interested in prospecting for minerals, commonly prepare charts showing Bouguer anomalies.

6.42. Gravity below the surface

(*a*) *Below ground.* Let Q be a ground-level gravity station at height h. It is required to find g_{00}, gravity at Q′ on the geoid below Q with the intervening rock still in place. From (6.26)

$$\nabla^2 W = -4\pi G\rho + 2\omega^2, \tag{6.139}$$

whence, following (6.92)–(6.95),

$$g_{00} - g_Q = g_M(1/r_x + 1/r_y)(1+m')h - 4\pi G h\rho_M, \tag{6.140}$$

where r_x, r_y, and m' are as in (6.135), and g_M and ρ_M are mean values between ground level and geoid, ρ being density. Any approximate value will suffice for g_M.

Then if $r_x = r_y =$ the earth's mean radius, and if $\rho_M = 2.67$ g cm^{-3},

$$\begin{aligned} g_{00} - g_Q &= 0.308h - 0.224h \text{ mGal, if } h \text{ is in metres,} \\ &= 0.84h\ \mu\text{m s}^{-2}. \end{aligned} \tag{6.141}$$

The term $-4\pi G h\rho_M$ reflects the fact that the rock between Q and Q′ attracts downwards at Q and upwards at Q′.

† O might have been defined more rationally with opposite sign, but conventionally it is as stated. It is additive to $g - \gamma_B$.

Error may be due to either (i) or (ii) below.

(i) Error in ρ. Within 1 km of the surface ρ may be as little as 2.00 instead of 2.67, although that would be unusual. Estimates based on surface samples must be expected to be too low. An error of 0.27 (10 per cent) would cause an error of 22 mGal (220 μm s^{-2}) in g_{00} if $h = 1000$ m.

(ii) Error in assuming $r_x = r_y = R$. As in § 6.37 (*a*), an error of 2 per cent or of 6 mGal (60 μm s^{-2}) if $h = 1000$ m, is exceptional, but it may be exceeded in rough topography.

The mean value of g between ground and geoid levels, g_M, is given by

$$g_M = \tfrac{1}{2}(g_Q + g_{00})$$
$$= g_Q + 0.042 \text{ mGal } (0.42\ \mu\text{m s}^{-2}) \text{ per metre of } h_Q, \qquad (6.142)$$

using normal density and curvatures, with uncertainties equal to half those quoted for g_{00} above.

If the surface topography is flat, this formula is an accurate means of determining the density of the intervening rock, by observations at the top and bottom of a shaft.

(*b*) *Submarine gravity stations.* When observations are made in a submarine at a depth of d metres, it is more convenient to convert the observed g to its sea-level value g_0 than to form anomalies $g_d - \gamma_d$ at the submerged point of observation. Then, with $\rho = 1.03$ g cm^{-3},

$$g_0 = g_d - 2gd/R + 4\pi G d\rho \qquad (6.143)$$
$$= g_d - 0.308d + 0.086d$$
$$= g_d - 0.222d \text{ mGal}$$
$$= g_d - 2.22d\ \mu\text{m s}^{-2} \qquad (6.144)$$

Observations made on the sea bottom, at shallow depths, may conveniently be treated in the same way.

6.43. Estimates of g, where not observed

Stokes's integral demands a dense net of gravity stations in the neighbourhood of the point where N is to be computed, and it is generally probable that there will be gaps in the available cover, which must be filled by interpolation. Estimates of g at particular places may also be required for other purposes.

Within a small area such as a 100×100 km square, g at unobserved points can be obtained by computing an anomaly $g - \gamma$ at surrounding observation points, interpolating $g - \gamma$ from them at the required unobserved points or points, and thence obtaining g at the latter. For this purpose $g - \gamma_A$ may be used as in § 6.37 (*b*) and Fig. 6.12, or perhaps better, $g-\gamma_B$ § 6.41 may be used. An isostatic anomaly $g-\gamma_C$ will also interpolate well, but will be laborious to compute in mountainous country.

The interpolation may be made between two adjacent points if the requirement is only to compute along a profile, or by drawing an iso-anomaly chart if an area is involved. In the latter case the iso-anomaly contours may be drawn with an eye on the topography or known geology of the area. If the elaboration is considered worthwhile, the statistically most probable results are obtainable by the method of interpolation by least squares as described in § D.24, or by *collocation* as in § D.25, but the conclusions in §§ D.24 (*j*) and D.25 (*d*) are relevant.

At distances of some hundreds of kilometres from observed values, the best that can be done from the purely gravimetric point of view is to assume that some type of isostatic anomaly $g-\gamma_C$ is zero, except in so far as analogy with points situated in similar geological or geophysical conditions elsewhere may suggest some other value for $g-\gamma_C$.

Satellite observations, § 7.79, provide a harmonic analysis of the form of the geoid, whose s.e. may be ± a few metres, but it must be remarked that $\Delta g_0/g=(n-1)N/R$, where Δg_0 and N are the gravity and potential anomalies indicated by every harmonic term of degree n. A harmonic of degree 18 has constant sign over at least 10° of longitude, and for such a term $\Delta g_0/g=17\,N/R$. It is clear that satellite analysis cannot begin to describe the finer variations of g within an area of 10°× 10°. On the other hand, it can certainly give a significant figure for mean g in a square of 20°×20°. An estimate of g based on the assumption that $g-\gamma_C=0$ or any other geologically indicated figure, might well be improved by the addition of an area-mean value of $g-\gamma_A$ derived from satellites.

The accuracy of satellite analysis of gravity is discussed in [216], [587], [334], and § 7.80.

Kaula's covariance table [325] or [262], p. 254 gives the variance, the mean square value, of $g-\gamma_A$ derived from a world-wide sample. Also, for the same sample, it gives the covariance between the values of $g-\gamma_A$ at stations separated by varying distances ranging from 0°.5 to 150°. The r.m.s. value is 35 mGal (350 μm s^{-2}).

6.44. Effect of external masses on g given by satellites

In § 6.17 formula (6.62) gives the potential at external points of a body whose surface form is given by (6.61), and outside which there is no matter. On the earth there is matter external to the geoid, which in principle must be dealt with as in § 6.35, before the relationship between (6.61) and (6.62) is applicable. When that has been done, if observations of satellite orbits give the variations of the potential at satellite heights, the u's in (6.62), (6.61) will give the form of the co-geoid, and (6.65) will give g on it. Then if h' and F_e of § 6.35 are computed wherever required, the form of the geoid, and gravity on it, result. These computations would be a considerable labour if a world chart was required, unless the selected system of mass transfer was one which gave negligible values of h' and F_e.

The condensation of external masses to sea-level, § 6.38 (*b*), gives substantially zero values of h', and also of F_e in the oceans and other flat areas, but not in mountainous regions. It is hardly necessary to remark that satellite observations cannot be expected to record the sharply varying values of g which occur among mountains.

Alternatively, we may consider the system of smoothed topography described in § 6.38 (*e*). Here also h' is small (up to 2 m in the Alps and 7 m in the Himalayas), which can be neglected in a harmonic analysis which extends only to degree 20 or 30. It is also reasonably evident that although the F_e of this system will not be small in Alpine areas, a 20- or 30-degree analysis, being itself smoothed, is unlikely to differ much between the true earth and the smoothed model.

It may thus be concluded that, from this point of view and ignoring other difficulties, satellite observations can give a correct analysis of the form of the geoid and of gravity on it. Put otherwise, the movement of a satellite at 1000-km altitude will not be measurably affected by the smoothing postulated in § 6.38 (*e*).

Section 8. Stokes's Integral Applied to the Ground Surface

6.45. Introductory

The fact that Stokes's integral has to cover an inaccessible equipotential surface, separated from the ground surface by matter of inaccurately known density, is unsatisfactory, and this section considers the possibility of applying it to the actual ground surface. The theory of this process is easy enough, and in some respects is more satisfactory than that of Sections 6 and 7, but the practical difficulties caused by uneven topography and insufficient gravity data remain. Up to the present time (1978) the revised theory has made little or no contribution towards getting more accurate numerical results.

6.46. Green's theorem applied to the ground surface

Referring to § 6.14, in (6.28) let the surface S be the surface of the sea and land. This is not an equipotential surface. Let ϕ be the earth's gravitational potential†, W_Q, of all matter inside and outside‡ S at the variable point Q (P′ in § 6.14), and let $\phi_2 = 1/r$ as before, r being the distance between Q and a fixed point P which is either external to S or on S.

These changes affect the terms (i)–(iv) of § 6.14 as below.

(i) Since S is not an equipotential surface, ϕ in this term is not constant, and the term is not zero.

(ii) This term is zero as before.

(iii)
$$\nabla^2 W_Q = -4\pi G\rho + 2\omega^2, \qquad (6.145)$$

and
$$-\iiint \phi_1 \nabla^2 \phi_2 = \iiint \frac{1}{r}(4\pi G\rho)\,\mathrm{d}v - 2\omega^2 \iiint \frac{1}{r}\,\mathrm{d}v. \qquad (6.146)$$

(iv) In this term $(\partial\phi_1)/(\partial n) = g_0 \cos i$, where i is the angle between the vertical and the normal to the ground surface. Locally, i may vary between 0 and 90°.

Taking account of these changes, (6.33) becomes

$$4\pi V_P = \iint W_Q \frac{\mathrm{d}(1/r)}{\mathrm{d}n}\,\mathrm{d}S + \iint g_Q \cos i \frac{\mathrm{d}s}{r} + 2\omega^2 \iiint \frac{\mathrm{d}v}{r}. \qquad (6.147)$$

where V_P is the attractive potential at P of all matter within S. We have

$$V_P = W_P - \tfrac{1}{2}\omega^2 (x^2 + y^2)_P. \qquad (6.148)$$

Formula (6.147) applies when P is external to the surface S. It will later be necessary to place it actually on the surface, and this will result in infinite values of $1/r$. So let there be excluded from the volume of S a small hemisphere centred on P of radius d. As d vanishes, the effect on the second and third integrals also vanishes, but the effect on the first becomes

$$V_P \iint \frac{\partial(1/r)}{\partial n}\,\mathrm{d}S, \text{ over the hemisphere, } V_P \text{ being constant,}$$

$$= V_P(2\pi), \text{ positive since } r \text{ and } n \text{ are opposite.} \qquad (6.149)$$

So when P is on the surface, the first term $4\pi V_P$ of (6.147) is reduced by

† The notation is in § 6.01. g and W are the earth's gravity and gravitational potential (including rotation). γ and U are the gravity and its potential of some standard reference ellipsoid. F and V are the attraction (no rotation) and its potential, either of the earth or of some other body, as specified in the text. g_0 and W_0 are for points on the ground or water surface, while γ_0 and U_0 are on the reference surface.

‡ The only external masses are the sun and moon, whose effects are dealt with as earth tides, and the atmosphere. Recent earth models treat the atmosphere as internal. From the point of view of potential, this will give the correct result, (6.9) and (6.10). From the point of view of gravity see § 6.29.

subtraction to $2\pi V_P$, so (6.147) becomes

$$2\pi W_P = \iint W_Q \frac{d(1/r)}{dn} dS + \iint g_Q \cos i \frac{dS}{r} + 2\pi\omega^2(x^2+y^2)_P + 2\omega^2 \iiint \frac{dv}{r}. \tag{6.150}$$

6.47. Non-classical theory. The reference spheroid

The reference spheroid, with an ellipse as meridional section, is a solid rotating body centred on the earth's centre of mass. Its flattening is prescribed so as to match the gravity formula which is to be used for γ. Its mass equals that of the earth, including the atmosphere. Its internal density distribution is such that its surface is an equipotential of γ, its attraction and the centrifugal force. Gravity on the surface is γ_0 and its potential is U_0 which is equal to W_0, the potential of g on the geoid. At external points near the surface γ_A is given by (6.95), and

$$U = U_0 - \gamma_M h_s, \tag{6.151}$$

where $\gamma_M = \frac{1}{2}(\gamma_0 + \gamma_A)$ and h_s is height above the spheroid. Surfaces U = constant are known as *spherops*.

Fig. 6.17 shows the relation between the actual earth and the reference spheroid at any ground surface point Q. The geoid does not coincide with the spheroid, being separated by an unknown N. The ground station Q is at a height h above the geoid and $h+N$ above the spheroid. Gravity is g_0 on the geoid and g_Q at Q, and its potential at Q is W_Q. Surfaces of equal W are known as *geops*. W_Q does not equal U_Q, but there is a point q on the normal of Q where $U_q = W_Q$. Let $Qq = \zeta$ (positive when Q is above q) and let $W_Q - U_Q = T$, where T is the *potential anomaly* at Q.

Then
$$T = \zeta\gamma_Q. \tag{6.152}$$

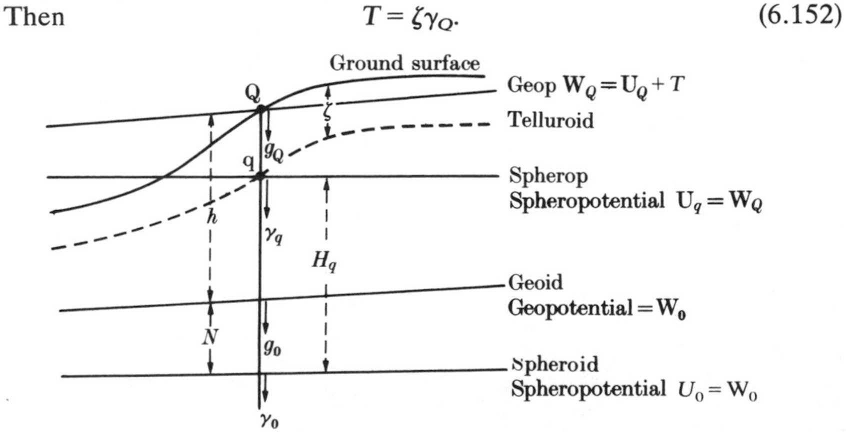

FIG. 6.17. W_Q is given by spirit levelling, and H_q the normal height of Q is derived from it by 6.155. i is the angle, not in the plane of the paper, between the ground surface and the geop at Q, and j is between the ground and the spherop.

The locus of points q, roughly parallel to the ground surface but departing from it by slowly varying amounts of perhaps + or − 50 metres, is known as the *Telluroid*, or Terroid in [235].

The scale of the spheroid is given by the condition that $\int \zeta \, dS = 0$, so that the volumes enclosed by the ground surface and the telluroid are equal.

The *anomaly of gravity* at Q is defined as

$$\Delta g_Q = g_Q - \gamma_q. \tag{6.153}$$

The difference of geopotential between Q and the geoid,

$$W_Q - W_0 = -\int_0^Q g \, dh, \tag{6.154}$$

is measured by spirit levelling supported by surface observations of gravity as in § 3.02 (*b*), so if W_0 is given any arbitrary value W_Q is known. Since $U_q = W_Q$, U_q is also known and thence the height of q above the spheroid is given by

$$H_q = (U_0 - U_q)/\tfrac{1}{2}(\gamma_0 + \gamma_q) = (W_0 - W_Q)/\tfrac{1}{2}(\gamma_0 + \gamma_q). \tag{6.155}$$

This is known as the *normal height* of the ground point Q. It is the height of the telluroid above the spheroid.

It follows from all the above that given the spirit-levelled height of Q, its height above the spheroid is known if ζ or T can be determined.

The word geoid has been used for the equipotential surface whose geopotential is defined to be W_0, as is proper if the body under consideration is the actual earth. On the other hand, if the earth is supposed to have undergone some form of smoothing as in §§ 6.38 (*e*) and 6.49, the new equipotential surface will differ from the geoid. It cannot be called a co-geoid since co-geoids are bounding equipotential surfaces. If confusion is possible it may be called the geoid of the modified earth.

The anomaly of gravity Δg and the anomaly of potential T are related as follows. Differentiating $T = W_Q - U_Q$ gives

$$-\frac{\partial T}{\partial h} = g_Q - \gamma_Q = g_Q - \left(\gamma_q + \zeta\frac{\partial \gamma}{\partial h}\right)$$

$$= \Delta g + \frac{T}{\gamma}\frac{2\gamma}{R}, \quad \text{using (6.95) and (6.152),}$$

so

$$\Delta g = -\frac{\partial T}{\partial h} - \frac{2T}{R}. \tag{6.156†}$$

† The term $2T/R = 2\zeta\gamma/R$, and is similar in form to the Bowie correction of § 6.35 (*b*).

6.48. de Graaff-Hunter's treatment

Formula (6.150) applies Green's theorem to the ground surface, taking ϕ_1 in (6.28) as W the geopotential. Let it now be rewritten with ϕ_1 taken as the spheropotential U, integrating over the ground surface as before. The result is

$$2\pi U_P = \iint U_Q \frac{d(1/r)}{dn}\, dS + \iint \gamma_0 \cos j \frac{dS}{r} + 2\pi\omega^2(x^2+y^2) + 2\omega^2 \iiint \frac{dv}{r}, \tag{6.157}$$

where j is the angle between the normals to the spherops and the ground surface.

Subtracting (6.157) from (6.150) gives

$$2\pi(W_P - U_P) = \iint (W_Q - U_Q) \frac{d(1/r)}{dn}\, dS + \iint (g_Q \cos i - \gamma_Q \cos j), \tag{6.158}$$

or

$$2\pi T_P = \iint T_Q \frac{\partial(1/r)}{\partial n}\, dS + \iint (g \cos i - \gamma \cos j) \frac{dS}{r}$$

$$= -\iint \frac{T_Q}{r} \frac{dr}{dn} \frac{dS}{r} + \iint (g \cos i - \gamma \cos j) \frac{dS}{r}, \tag{6.159}$$

where $T = W - U$. See Fig. 6.17. A little manipulation, which is given in full in [235], gives

$$(g \cos i - \gamma \cos j) = \left\{\Delta g_Q'' + 2T\left(\frac{1+m}{k}\right)\right\} \cos i, \tag{6.160}$$

where

$$m = 0.00346,$$

$$1/k = (\rho\nu)^{-\frac{1}{2}} = (1 + f \cos 2\phi)/a \approx 1/R,$$

$$\Delta g_Q'' = \Delta g_Q \text{ of } (6.153) + \gamma\chi \cos\alpha \tan i, \tag{6.161}$$

in which α is the azimuth difference between the direction of maximum ground slope and that of maximum deviation of the vertical χ.†

Formula (6.159) may then be written

$$2\pi T_P = \iint \left(\Delta g_Q'' + \frac{2T_Q}{R} - T_Q \frac{\sec i}{r} \frac{dr}{dn}\right) \frac{\cos i}{r}\, dS, \tag{6.162}$$

ignoring m and the difference between k and R with errors of about 0.3 per cent in the small correction terms in T on the right-hand side. This

† This χ, the total deviation of the vertical, has no connection with the χ of § 6.17 *et seq.* This χ is measured in radians.

equation gives the potential anomaly at a required point P in terms of the anomalies of gravity and potential at all other points on the surface. Some simplification is necessary before it can be solved.

The approximation $\Delta g'' = \Delta g$ in (6.162) would amount to ignoring $\gamma\chi \cos\alpha \tan i$, where $\chi \cos\alpha$ is a component of the deviation of the vertical which may be as much as 20″ or 1/10 000.† So the error in Δg would not exceed 1 mGal (10 μm s^{-2}) so long as $\tan i < 0.01$, i being the slope of the ground. The factor $\cos i$ may be put equal to unity, with errors of less than 0.3 per cent of Δg_Q if $i < 0.07$, a less stringent criterion. With these limitations $(1/r)(\mathrm{d}r/\mathrm{d}n) = (1/2) R$, [235], p. 196.‡ So if the slope of the earth's surface did not exceed 0.01, (6.162) could be written

$$T_P = \frac{1}{2\pi} \iint \left(\Delta g_Q + \frac{3}{2} \frac{T_Q}{R} \right) \frac{\mathrm{d}S}{r} \tag{6.163}$$

or

$$\zeta_P = \frac{1}{2\pi} \iint \left(\frac{\Delta g_Q}{\gamma} + \frac{3}{2} \frac{\zeta_Q}{R} \right) \frac{\mathrm{d}S}{r}, \tag{6.164}$$

in which T is potential anomaly (6.152), Δg is the anomaly of gravity (6.153), P is the ground point whose height above the spheriod is to be determined, and Q is the moving ground point where gravity is to be measured. $PQ = r = 2R \sin \frac{1}{2}\psi$. It is unfortunate that the slopes of the actual earth do much exceed 0.01. For the remedy see § 6.49.

Equation (6.163) is also given in [384] from a different approach. It is a fundamental equation from which the coefficients of the terms in u in (6.65) can be derived, and thence Stokes's integral (6.111), the only difference being that we are now concerned with Δg and ζ as defined in (6.153) and (6.152) instead of the Δg_0 and N of (6.111). We proceed as below.

Taking P as the pole of a series of Legendre functions let $\Delta \bar{g}/\gamma$ in the zones surrounding it be $\sum a_n P_n$, and let $\zeta = R \sum b_n P_n$. Then for any single value of n, ζ_P (at the pole) $= Rb_n$, and substitution in (6.164) gives

$$Rb_n = \frac{1}{2\pi} \int^{S} (a_n P_n + \tfrac{3}{2} b_n P_n)\, \mathrm{d}S/r$$

$$= R \int_{-1}^{+1} (a_n P_n + \tfrac{3}{2} b_n P_n)(1 + P_1 + P_2 + \ldots)\, \mathrm{d}(\cos\psi)$$

since

$$\frac{1}{r} = \frac{1}{R}(1 + P_1 + P_2 + \ldots) \quad \text{(from H.17)}$$

† Deviations of more than 20″ do occur, but gravity stations need not be on mountain sides where local topography results in large values.

‡ $r = PQ$. P and Q can be regarded as lying on a sphere of radius R.

and $$\mathrm{d}S = 2\pi R^2 \sin\psi \, \mathrm{d}\psi = -2\pi R^2 \, \mathrm{d}(\cos\psi).$$

Then from (H.22)

$$b_n = (a_n + \tfrac{3}{2}b_n)\frac{2}{2n+1},$$

or $$a_n = b_n(n-1),$$

or $$\frac{\Delta g}{\gamma} = \frac{\zeta}{R}(n-1). \qquad (6.165)$$

Compare (6.65) in which u's represent ζ/R, and which are the basis of the derivation of (6.112); see (6.118).

Summary. If the slopes of the earth's surface did not exceed 0.01, Stokes's integral (6.114) with $\Delta g_0 = \Delta g$ of (6.153) would give the potential anomalies $T = \zeta\gamma$ of (6.152), and the height of the ground level above the spheroid is then $H_q + \zeta$, where H_q is known from (6.155). The fact that the earth slopes exceed 0.01 is dealt with in § 6.49.

6.49. Smoothed topography

The topography may be supposed to be smoothed on the system described in § 6.38 (*e*). This causes changes of potential which are negligible at the point P provided it is a few hundred kilometres distant from mountains or continental slopes in the ocean, and which are of no consequence at gravity stations Q, since 1 or 2 metres of height have little significant effect on the value of Δg. The necessary change from g observed at ground level to g on the surface of the smoothed model is given by (6.137), and Δg_E of (6.137) is used for Δg_Q in (6.164) and (6.114).

6.50. The deviation of the vertical

Formulae (6.123) and (6.124) with $\Delta g_0 = \Delta g_E$ are applicable to the smoothed model, and give the deviations at its surface. It is then necessary to compute the difference between these and the ground level values on the actual earth. The system of computation may depend a little on the system used for defining the model. [234], pp. 2 and 15 describes a system. As with isostatic reductions, the effect of topographical transfers decreases greatly with distance. In the system described in [234], effects beyond 100 km are much decreased, while those beyond 1000 km are negligible.

It can be argued that if the deviation of the vertical is required in mountainous country, it may be easier to make the necessary astronomical observations. The geocentric geodetic fixation, to within about 0″.2 of latitude or longitude is obtainable by doppler satellite observations, if not otherwise available.

6.51. Molodenski's treatment

Molodenski's investigations on non-classical lines date from 1945 or earlier. His work is published in English in [419], and is summarized in [262], which see for details. In summary, using the same notation as in §§ 6.47 and 6.48, at a ground surface point P

$$T_P/\gamma = \zeta_P = \zeta_0 + \zeta_1 = \frac{R}{4\pi\gamma}\iint \Delta g_Q f(\psi)\,\mathrm{d}w + \frac{R}{4\pi\gamma}\iint G_1 f(\psi)\,\mathrm{d}w, \quad (6.166)$$

where Δg_Q is the anomaly of gravity defined by (6.153),

$f(\psi)$ is as in (6.113),

and
$$G_1 = \frac{R^2}{2\pi}\iint \frac{h_q - h_p}{r^3}\left(\Delta g_Q + \frac{3}{2}\frac{\gamma\zeta_0}{R}\right)\mathrm{d}w, \quad (6.167)$$

in which h_q and h_p are the spheroidal heights of Q and P, $r = \mathrm{PQ}$ and, $\mathrm{d}w = \mathrm{d(area)}/R^2$. To solve (6.166) ζ_0 is first computed from the first integral of (6.166) and then substituted in the second to give ζ_1.

Molodenski has not proposed using a smoothed model, but has preferred to rely on adequate gravity stations and interpolation using a topographical or other reduction, [419], p. 180. See also [425].

The corresponding formulae for the deviation of the vertical at a ground surface point P are, [262], p. 313,

$$\xi_P = \left\{\frac{1}{4\pi\gamma}\iint (\Delta g_Q + G_1)\frac{\partial f(\psi)}{\partial\psi}\cos\alpha\,\mathrm{d}w\right\} - \frac{\Delta g_P}{\gamma}\tan\beta_1 \quad (6.168)$$

$$\eta_P = \left\{\frac{1}{4\pi\gamma}\iint (\Delta g_Q + G_1)\frac{\partial f(\psi)}{\partial\psi}\sin\alpha\,\mathrm{d}w\right\} - \frac{\Delta g_P}{\gamma}\tan\beta_2, \quad (6.169)$$

where G_1 is as in (6.167), and β_1 and β_2 are the meridional and PV components of the slope of the ground, positive when it slopes upwards to the north-east.

Molodenski uses the expression *quasigeoid* for a surface lying above the spheroid by the distance ζ which separates the ground surface and the telluroid. It is not exactly a level surface. From Fig. 6.17 $h_Q + N = H_q + \zeta$, so the geoid and quasigeoid are separated by $H_q - h_Q$, the normal height of Q minus its orthometric height, or

$$\begin{aligned} N - \zeta &= (W_0 - W_Q)/\bar{\gamma} - (W_0 - W_Q)/\bar{g} \\ &= h_Q(\bar{g} - \bar{\gamma})/\bar{\gamma}, \end{aligned} \quad (6.170)$$

where $\bar{g}$ and $\bar{\gamma}$ are mean values between ground and geoid, and between spherop and spheroid. When the ground surface is at sea-level the geoid and the quasigeoid coincide. In Alpine country they may be separated by 1 or 2 metres.

References. [262], pp. 327–30 reviews the merits and limitations of the non-classical methods of applying Stokes's integral. Many other treatments are summarized in [422] and [276]. For USSR methods see [419].

6.52. The external field determined by integration

It may not be possible or convenient to expand the sea-level potential or gravitational anomalies as a series of spherical harmonics, as in § 6.25, especially when intense but local anomalies are under consideration. The following formulae give gravity at low altitudes in the form of integrals, like Stokes's, based on sea-level values of Δg over the whole earth. [262], p. 90 gives

$$\Delta g_P = \frac{1}{4\pi R_P} \iint \left(\frac{R_P^2 - R^2}{r^3} - \frac{1}{R_P} - \frac{3R}{R_P^2} \cos\psi \right) \Delta g_Q \, dS, \tag{6.171}$$

where

R is the earth's mean radius,
$R_P - R$ is the height of P above the ground surface,
r is the variable distance PQ,
Δg_P, Δg_Q is the anomaly of gravity, at P or Q, as in (6.153),
ψ is the angle POQ, where O is the centre of the spheroid.

If $R_P - R$ is small, and if the surface anomaly Δg_Q is constant, Δg_c, over a wide area below P, the integral naturally gives $\Delta g_P = \Delta g_c$.

For the external gravity gradient [262], p. 115 gives

$$\frac{\partial}{\partial h}(\Delta g_P) = -\frac{2\Delta g_P}{R} + \frac{1}{2\pi} \iint \frac{\Delta g_Q - \Delta g_P}{r^3} \, dS, \tag{6.172}$$

SECTION 9. THE DEVIATION OF THE VERTICAL

The deviation of the vertical is defined in § 2.05, and its uses are listed in § 4.01 (*b*). Its value depends on the reference spheroid employed, which should ideally be geocentric, but until quite lately this has not been possible. The necessary astronomical observations are described in Chapter 4, Section 6. The possible alternative method of computing the deviation from observations of g is described in §§ 6.31, 6.33 (*b*), 6.34 (*f*), and 6.51. The use of deviations to produce geoid charts is given in Chapter 6, Section 10.

The present section deals with the deviation itself, and with its variations with height and from place to place.

6.53. Variation from place to place

The deviation of the vertical, as derived from the difference between astronomical and geodetic fixations, results from the combination of many causes, as follows.

(*a*) The values of the deviation, as initially observed and published, have generally been with reference to spheroids which are not centred on the earth's centre of mass. The differences between current continental datums and geocentric datums may amount to a few seconds of arc, and to as much as 10″ for more local datums. Within any particular continental survey system, this difference may vary by 1″ or so in distances of 1000 km. The necessary corrections, by which positions given with reference to local datums can be converted to a geocentric datum, are now generally available, see §§ 7.77 and 7.78, and where they are not yet known they are easily obtainable.

(*b*) Deviations may have been referred to a spheroid which has unsuitable axes, notably the Everest or Bessel spheroids. Thus deviations in the prime vertical in Burma, 2000 km west of the origin of the Indian Everest spheroid, are abnormal by 12″ to 14″ on this account. Only these two spheroids are likely to have caused such large effects at the same distance. The remedy is to convert deviations and geoid charts to a modern spheroid, as in §§ 2.50–2.53. There is now no good reason why all deviations should not be expressed with reference to a common geocentric spheroid with up-to-date lengths for its axes.†

(*c*) Error may have accumulated in the triangulation or other framework between an origin and distant points. This should not amount to ±0″.5 of position in 4000 km of the best primary framework, but its effects will be systematic. Omission of the Molodensky correction, or an equivalent, § 2.40 (*b*), may introduce position errors of a few seconds in a very long section. This also can now be controlled to within 0.2″ by the use of doppler satellite observations to control the (geocentric) position of geodetic fixations.

(*d*) The most obvious cause of a deviation is irregular topography. Assuming normal densities, the effect can be computed as in § J.01. The effect may be as much as 20″ in Alpine topography, or 5″ in Scottish.

(*e*) If the preceding causes can be eliminated, what remains must be due to errors in the assumed densities of the crust or the mantle.

Large compilations of values of the deviation are [40] for Europe, Africa, Turkey, and Iran on the European datum; [188] for the United

† But for some purposes, such as the reduction of baselines and horizontal angles, deviations and the resulting geoid charts are still required with reference to the spheroid and datum on which the triangulation, etc. concerned is to be computed.

States; [245] for India, Pakistan, and Burma on the Indian datum. See § 6.60 (*b*) for geoid charts based on these compilations.

Such lists show that the deviation may predictably vary by such amounts as 20″ in 7 km in Alpine country; by 20″ in 15 km in Baluchistan; or by 6″ in 20 km in Scotland. These are the largest figures listed, but stations are naturally not located actually on steep hillsides where irregularities would be greater. Deviations of over 70″ have been measured near Mount Everest.

The variations in flat or gently undulating country, such as in the Punjab or in southern England, are not so very much smaller than those quoted for Scotland. Changes of 4″ in 20 km can occur, although changes of less than 2″ in 20 km are more usual. They are not necessarily related to the topography.

As an illustration Table 6.7 shows the deviations along a meridional profile in longitude 75° E, from the crest of the mountains separating Kashmir from the Punjab southwards into the Punjab plains.

The first ten entries in the column ξ of Table 6.7 show the strong northward deviations which are to be expected on the south side of a mountain range. The next seven show a southward deviation produced by no visible cause, although they are associated with positive gravity anomalies further south, see [108], Plate vi. The remaining entries mostly show the uniformity which would be expected in the topography of the Punjab plains, although changes of about 4″ do occur in two places. In this profile the s.e. of an observation is about ±0″.4.

TABLE 6.7

Deviations of the vertical. Indian International spheroid

$\lambda = 75°$ E, $\phi = 33°\ 30'$ to $28°\ 41'$ N. h = height in metres, d = kilometres south of preceding station. ξ = deviation in meridian in seconds. $\xi - \xi_c$ = Hayford isostatic anomaly, § 6.64.

h/m	d/km	ξ''	$\xi'' - \xi''_c$	h/m	d/km	ξ''	$\xi'' - \xi''_c$
2840	—	−5.1	−3.3	212	20	+2.7	+4.2
1540	15	−24.9	−5.4	224	15	+3.9	+5.2
740	15	−26.8	−6.1	234	18	+4.7	+5.8
1230	20	−28.5	−7.1	229	17	+5.8	+6.8
1170	4	−27.9	−6.3	226	17	+6.5	+7.4
770	9	−29.2	−9.1	225	17	+2.2	+2.9
650	18	−18.8	−7.1	225	17	+3.4	+4.1
630	15	−12.4	−1.5	224	17	+2.0	+2.7
325	20	−7.6	0.0	221	15	+3.8	+4.4
290	15	−1.3	+4.7	218	13	+3.1	+3.6
262	15	+2.7	+7.7	215	18	+4.1	+4.5
248	17	+6.6	+10.3	227	13	+3.4	+3.8
260	17	+7.2	+10.2	219	17	+3.2	+3.4
257	18	+8.5	+10.7	214	18	+3.9	+4.0
243	15	+8.6	+10.7	234	20	+4.5	+4.2
242	17	+7.5	+9.4	239	15	+5.1	+4.5
240	15	+6.7	+8.5	255	20	+3.9	+3.2

6.54. Deviation anomalies

Since the deviations ξ and η are largely caused by the topographical irregularities, it is sometimes useful to compute the effects of the topography as in § J.01, using an assumed density, to give ξ_T and η_T, § 6.05, and then to form the anomalies $\xi-\xi_T$ and $\eta-\eta_T$. It is to be expected that the anomalies will be smaller and less variable than ξ and η, and that their variations will indicate anomalies of mass.

A difficulty in computing this topographical effect is that it does not diminish rapidly with distance. The computed effect of a feature like the Tibet plateau may be a few seconds at a distance of 2000 km. This makes the computations laborious, and since such large features are in fact more or less isostatically compensated, the total effect of topography and its compensation is likely to be much nearer to zero than to the figure laboriously computed from the topography only. Instead of computing ξ_T and η_T it is for this reason actually easier to compute ξ_c and η_c on the basis of some standard form of compensation. The anomalies $\xi-\xi_c$ and $\eta-\eta_c$ have commonly been based on Hayford's system of compensation as described in § 6.64. They are then known as *Hayford deflection anomalies,* and ξ_c and η_c are *Hayford deflections.* [35] gives ξ_c and η_c at 3000 stations in many parts of the world, and [245] gives 900 stations in India.

To remove topography and compensation is exactly the same as to transfer the topography from above the geoid to locations inside the geoid, corresponding to the accepted isostatic hypothesis as in § 6.38 (*d*). So if geoid profiles are computed using $\xi-\xi_c$ and $\eta-\eta_c$, the resulting surface will be the co-geoid which would be given by Stokes's integral using $g-\gamma_c$ for Δg_0.† The geoid may then be computed from it by the method of § J.06. The early items in Table 6.7 illustrate the smooth variation of $\xi-\xi_c$ in moderately steep topography.

For the interpolation of the deviation at intermediate points in a geoid profile in places where the observations are unavoidably too widely spaced, it is sufficient to compute ξ_T and η_T to a radius of about twice the distance between observations. If observation stations are as little as 20 or 30 km apart the inclusion of compensation will save little labour and may not increase accuracy. See [543] for a very full treatment of interpolation using gravity.

6.55. The curvature of the vertical. Reduction to sea-level

(*a*) The deviation of the vertical as defined in §§ 2.04 and 2.05 and as observed astronomically is the angle between the vertical and the normal at *ground* level. For some purposes, such as for correcting horizontal angles, this is what is required, but for geoid profiles it should in principle

† It is the same surface, but the reference spheroid will be different, unless the deviations are with reference to a geocentric spheroid.

be reduced to *geoid* level. See § 6.24 for the conventional correction for a spheroidal earth, which is easily included, and see § 4.51 (*e*) which in general justifies the omission of anything more rigorous.

(*b*) The difference between ground level ξ or η and the geoid level ξ_0 or η_0, if required, is approximately given by the following formulae, [237],

$$\xi_0-\xi=-\sum \xi_T(1-F_0/F)-0''.000\,17h\sin 2\phi,$$
$$\eta_0-\eta=-\sum \eta_T(1-F_0/F), \qquad (6.173)$$

where h is the ground height in metres; ξ_T and η_T are the computed effect of the topography in each annular zone surrounding the station, § J.01; F is the horizontal attraction at ground level of an excess of mass in any particular zone; and F_0 is its horizontal attraction at geoid level. [237], p. 128 gives a table for $(1-F_0/F)$ for different zones and station heights. If $h=1000$, 2000, or 3000 metres, the factor is less than 0.1 beyond distances of 3, 7 and 15 km respectively.†

(*c*) The following shows a connection between ignoring the rigorous reduction to sea-level in geoid profiles, and the acceptance of standard gravity when computing levelling, §§ 3.02 (*c*) and (*d*). In Fig. 6.18 let

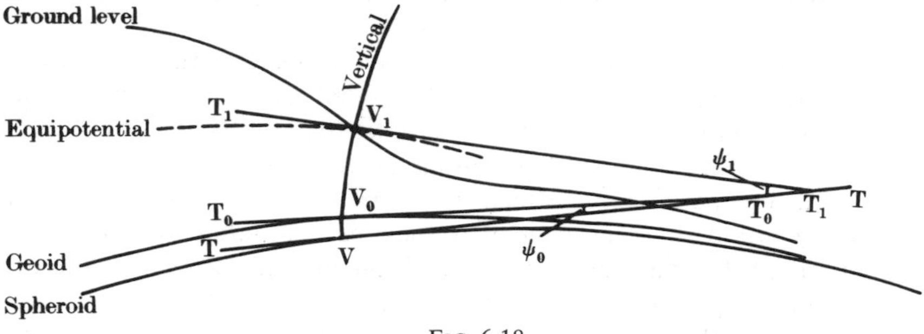

FIG. 6.18.

V_1V_0V be a curved vertical, and let VT, V_0T_0, and V_1T_1 be tangents to the spheroid, geoid, and ground-level equipotential at an astronomical station V_1. The two latter make angles ψ_0 and ψ_1 with VT, which are components of the sea-level and ground-level values of the deviation. Observation gives ψ_1, but (in principle) we want ψ_0 to put into (4.95) for getting the form of the geoid.

Now consider the geoidal rise U_0U-V_0V between V and U, Fig. 6.19. At an intermediate point P we have

$$(P'Q'-P_1Q_1)/P_1Q_1=-\frac{\partial g}{\partial s}ds/g,$$

† The basis of this simple tabulation is that close to the station, where the effect is greatest, the ground height is not unduly different from the station height.

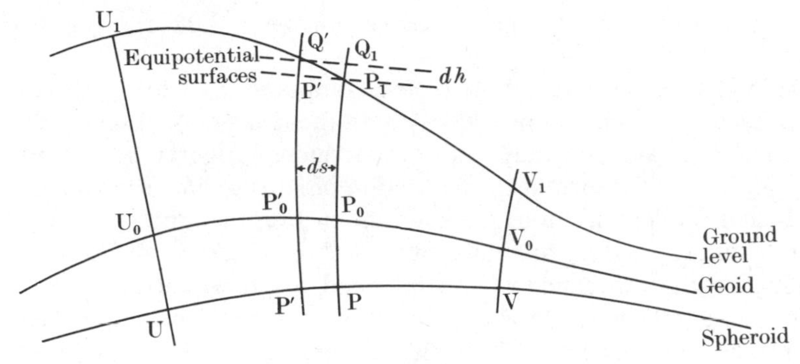

FIG. 6.19.

Since $g_{P'} \times P'Q' = g_P \times P_1Q_1$,

and at P
$$\psi_1 - \psi_0 = \int_{P_0}^{P_1} \frac{P'Q' - P_1Q_1}{P_1Q_1\, ds}\, dh = -\frac{1}{g}\int \frac{\partial g}{\partial s}\, dh, \tag{6.174}$$

whence the correction to $U_0U - V_0V$ is

$$-\frac{1}{g}\int_V^U \int_{P_0}^{P_1} \frac{\partial g}{\partial s}\, dh\, ds = -\frac{1}{g}\int_{\text{geoid}}^{\text{ground}} \Delta g\, dh, \tag{6.175}$$

where Δg is gravity at any point on the line $V_0V_1P_1U_1$ minus gravity at the same height on U_0U_1.

Referring to § 3.02 (*c*) and using observed g instead of γ, combining (3.12) and (3.13) gives

$$\mathbf{O} - \mathbf{M} = \int_{V_0}^{U_1} \frac{(g-\gamma_s)\, dh}{g} - \int_{U_0}^{U_1} \frac{(g-\gamma_s)\, dh}{g} = \frac{1}{g}\int_{\text{geoid}}^{\text{ground}} \Delta g\, dh \quad \text{as above,} \tag{6.176}$$

where **O** is the orthometric height of U_1 above U_0, and **M** is the height given by spirit levelling along $V_0V_1P_1U_1$ without any dynamic correction, the first $\int (g-\gamma_s)\, dh/g$ being along the line $V_0V_1P_1U_1$ and the second along the vertical U_0U_1. In the denominator g is indistinguishable from γ_s. The identity of (6.175) and (6.176) thus shows that to ignore reduction to sea-level when integrating deviations to get the form of the geoid results in identically the same error as does ignoring the dynamic and orthometric corrections in a line of levelling, and the use of the conventional sea-level reduction of § 6.24 takes the one problem as far as the conventional dynamic-orthometric correction of § 3.02 (*c*) takes the other.

[237], p. 124, gives an example of the computation of the geoidal rise under a 2000-m mountain range firstly with the conventional reduction to sea-level (6.97), and secondly with the reduction rigorously computed allowing for the difference of the attraction of surrounding topography at ground- and sea-levels. On a steep mountain side the corrections to the deviation differed by 5″, but only over a short distance and of course cancelling on the other side of the hill, and the total correction to the geoidal rise under the range was only 15 cm, the geoid being too high if the correction was ignored. [237] also computes $\mathbf{O}-\mathbf{M}$ by (6.176) and duly obtains a similar result.

Similarly [440], vol. 20 gives a section through the Swiss Alps, with observations at intervals of $3\frac{1}{2}$ km. The effect of rigorous reduction on the height of the geoid in the centre of the 185-km section is 0.4 m, decreasing to 0.2 m at the south end, with the suggestion that it probably decreases to 0.0 m in the plain of north Italy.

The labour of the rigorous reduction by either method is very great, and it is clear from these examples that (6.97) will generally suffice.

SECTION 10. GEOID SURVEYS

6.56. Methods of geoid survey

Geoid surveys, namely the preparation of charts which show the height N of the geoid above some defined reference spheroid, can be carried out by various methods, as in (*a*)–(*f*) below.

(*a*) Astro-geodetic sections are described in §§ 4.50–4.52, with further details in §§ 6.57–6.59. This method gives detailed surveys of land areas, but is not applicable at sea. Over great distances it is liable to accumulate significant error, which can now be controlled by other methods, such as by satellite doppler, in (*c*) below.

(*b*) Stokes's integral, § 6.30, can in principle give N at discrete points, and so can give a detailed geoid survey over such few areas as are surrounded by adequate surface gravity survey. §§ 6.32 (*a*) (ii) and 6.34.

(*c*) Ground stations from which satellites have been tracked similarly provide geocentric coordinates at discrete points, notably (in 1978) at doppler fixes, §§ 7.48 and 7.63. Also probably from Navstar fixes, § 7.44, after the year 1984. Such fixes can control astro-geodetic surveys, or if they are sufficiently numerous over a small area they can delineate the local geoid.

(*d*) The analysis of satellite orbits gives the anomaly of the earth's potential at external points (including points on the geoid†) in the form of a world-wide spherical harmonic analysis. This provides a geoid chart, § 6.26, which is locally considerably generalized as compared with charts given by methods (*a*) or (*b*).

(*e*) Satellite altimetry may be expected (in 1978) to give moderately detailed surveys over the oceans, § 7.64 and 7.65.

(*f*) Surface (land and sea) gravity observations can contribute to the satellite harmonic analysis of (*d*) above, § 7.79.

Geoid charts can of course be based on a combination of any two or more of the above methods.

§ 6.60 lists a few typical examples of geoid and world gravity charts obtained by these various methods.

Historically, the axis a and flattening f have been deduced from the varying curvature of arcs of meridian in different latitudes, later supported by arcs of longitude. Hayford [254] introduced the principle of finding values of a and f for a spheroid which most closely fits the geoid, or a co-geoid, over an area. Given values of N at a number of points, N being the height of the geoid or co-geoid above a defined spheroid, formula (2.87) gives an observation equation at each point from which the changes δa, δf, $\delta \xi_0$, $\delta \eta_0$, and δN_0 can be got by least squares, to give a spheroid and datum which will most closely fit the surface. Hayford used an isostatic co-geoid, on the reasonable grounds that over a limited area it would be more typical of the earth as a whole than the geoid itself would be.

In modern times interest has turned towards obtaining detailed surveys of the form of the geoid, rather than concentrating on trying to define its form in terms of two or three parameters.

6.57. Gaps in an astro-geodetic geoid section

It may happen that the accuracy of (4.96), namely

$$N_B - N_A = \tfrac{1}{2}(\psi''_A + \psi''_B) l \sin 1'' \tag{6.177}$$

is reduced, either by exceptionally rough topography or by the distance l being unavoidably increased by the presence of sea or inaccessible land. If the trouble is rough topography, the easiest remedy (in a good climate) is to reduce the intervals between stations. If this is not convenient, or when l is for other reasons unduly large, the situation can be improved by one of the following methods.

(*a*) By calculating the *Hayford* (or other) anomaly, § 6.54, at each station. These will run more smoothly, and intermediate values may be

† For the effect of matter external to the geoid, which in these circumstances in negligible, see § 6.44.

got by interpolation, which can be turned into deviations by calculating Hayford deflections at intermediate points. See [440], vol. 20. The validity of the process does not depend on the accurate truth of the isostatic hypothesis, but only on its giving smooth anomalies. It is to be noted that in this context the isostatic or other anomalies are used only as an aid to interpolation. Their use does not result in an isostatic or other co-geoid being obtained instead of the geoid.

(*b*) *Gravimetric interpolation.* If the deviations at two stations in a geoid profile, perhaps 100 km apart, have been observed astronomically, they can also be computed gravimetrically by (6.123) and (6.124) using data extending out to a limited radius (such as 200 km). If gravity data within this radius are adequate, the difference between the astro and gravimetric values will either be due to the centre of the spheroid not being at the earth's centre of gravity, or to the neglect of gravity outside the limited radius. If this latter error can be assumed to vary linearly between the two stations, the astro deviation at an intermediate point may be got by gravimetric computation (to the same radius) plus a correction linearly interpolated between the two astronomical stations.

The result is likely to be correct to within 1″, provided that

(i) the country is flat or gently undulating,

(ii) adequate gravity data extend to at least $1\frac{1}{2}$ times the distance between the two astronomical stations,

(iii) the corrections (astro deviation minus gravimetric) at the two stations are the same to within 1″, not only when the gravimetric deviation is computed to the limiting radius, but at all radii beyond about 75 per cent of the limit.

[481] gives examples of successful computations on these lines in the United States, using $\Delta g_0 = g - \gamma_A$. It gives computation details, and a table of radii for use with a template, such that a gravity anomaly of 1 mGal (10 μm s^{-2}) in a 10° sector of each zone contributes 0″.001 to the radial component of the deviation. There are fifty such zones between radii 0.1 km and 465 km.

[543] recommends the Bouguer anomaly, § 6.41, for Δg_0 when Stokes's integral is being used for interpolating deviations of the vertical, and gives details of the routine. The use of true local rock densities is also advised, if they are known.

If stations are spaced 100 or 200 km apart, they must be more accurately observed, since the random error of each will affect the profile over a greater distance. With ten times the normal spacing, the s.e. of each must be reduced to one-third of the usual figure or ±0″.25. This is not so easy to get, and to adopt the wider spacing voluntarily may result in little economy of effort, not to mention the cost of the necessary gravity survey (although it may, of course, be required for other purposes) and of the computation of its effects (which will not).

The elimination of personal equation, however, can be a factor in favour of this system, since it may be practicable to set up some rather elaborate impersonal instrument at a few stations, when it may be impossible to use it every 25 km.

(*c*) *Vertical theodolite angles* between triangulation stations, can in principle give accurate geoidal survey, if the refraction can be measured or accurately estimated. See § 6.59.

(*d*) *Satellite fixes.* A wide gap in a section may be bridged by the observation of satellite fixes at both ends of the gap, probably using translocation methods, § 7.50 (*b*)–(*d*). Errors in height of 1 or 2 metres may be feared, so this expedient is only applicable to rather wide gaps, until higher accuracy is assured. The gap itself will remain unsurveyed in detail, but the continuity of the section is preserved.

6.58. Adjustment of nets of geoid sections or of areal surveys

(*a*) *A net of geoid section lines.* When an astro-geodetic geoid survey is started in a continent or country, it is initially likely to take the form of one or two long lines of closely spaced stations, designed to connect with the geoid surveys of neighbouring countries, which may have already made such surveys or who hope to do so. Branch section lines may then be observed to provide geoid heights in out-lying areas. See Fig. 6.20 which shows the lines of geoid survey in Pakistan, India, and Burma, [245]. The main line is marked AB and other lines branch from it. The final stage is to connect the ends of the branch lines to form circuits, which not only extend the survey but whose closing errors give estimates of accuracy. Fig. 6.20 shows only a few such lines, as completed in 1955.

When a number of circuits have been observed, the net can be adjusted by least squares, using either the method of variations or of conditions, as is described for spirit levelling in § 3.12 (*c*) and (*d*). Weights may be allotted according to the lengths of sections, as in (4.97), but in practice there are likely to be weak places where the station spacing has been too great, either in mountainous country or for other reasons. It is difficult to assess the weakness of such occurrences, and a simple common-sense distribution of error may well be as good as a formal solution by least squares.

(*b*) *Conversion to geocentric spheroid.* Differences of geoidal height given by (6.177) are with reference to the spheroid and datum to which the geodetic coordinates in the values of (A–G) refer. If the geoid survey is required for the reduction of measured distances in the local geodetic

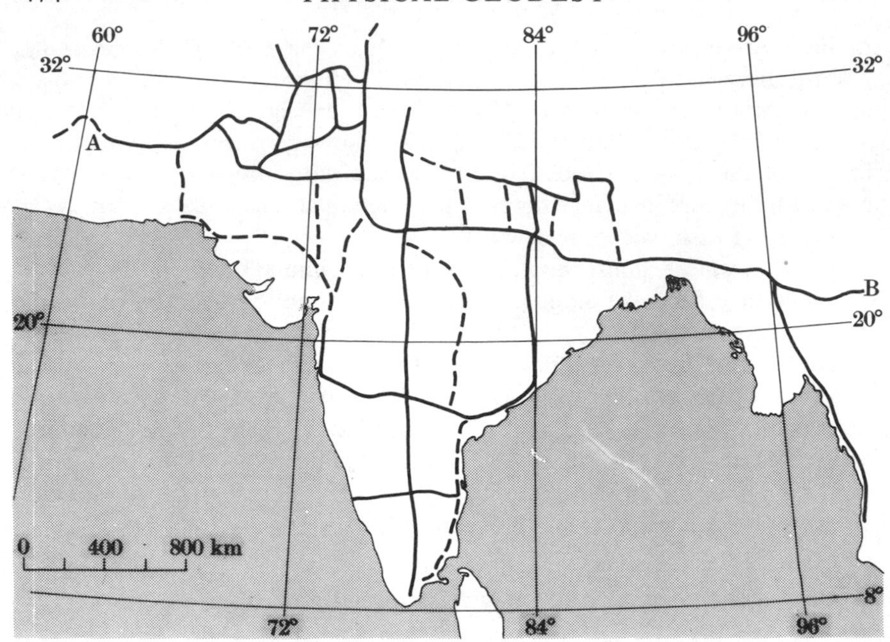

FIG. 6.20. India, Pakistan, and Burma. Geoid sections. Full lines are strong sections, spacing 20 km. Broken lines are weak sections, with gaps of up to 100 km. AB is the main east–west section line. There are also many observed deviations which do not form part of any section line.

framework, the geoidal heights must be above the spheroid on which the framework is to be computed. On the other hand, if the geoid survey is to be of geophysical significance it is best converted to a geocentric spheroid with currently accepted values of the axis and flattening. This conversion is easily carried out as in §§ 2.50–2.53, using formula (2.87) or otherwise This will only be possible if the connection between the geocentre and the centre of the local spheroid is known as in § 7.77.

Also see § 2.40 (*b*). If the existing framework has been computed with inaccurate values of N, its latitudes and longitudes may have to be corrected before it can produce sufficiently accurate values of N, Some simple iteration may be necessary.

(*c*) *Control by satellite fixes or Stokes's integral*, as in § 6.56 (*b*) or (*c*). If such control points are available, they can of course be included in the adjustment, with such weights as can best be estimated. Their values are likely to be known in geocentric terms, which must be converted to the datum on which the adjustment is being carried out. out. The problem may be complicated by covariance among groups of satellite stations, but it is very difficult to quantify such covariance and so to form a reliable covariance matrix for the adjustment.

(*d*) *Area surveys.* The treatment of an area over which observations of the deviation are more or less equally spaced is less obvious. The best and most elaborate treatment is by collocation. See (*e*) below. A simpler alternative is to write observation equations for the differences of N between each station and the surrounding nearest three or four stations, as given by § 4.50. Least squares will then give the heights of all stations, one being assumed known.

The simplest alternative is to form a number of N section lines containing most of the observations, some running (say) north and south, and some east and west, which are computed as in § 4.50. These will form a number of circuits, which can be adjusted as in (*a*) above.

(e) *Collocation.* The general principles of the method of collocation are described in § D.25. These principles can be applied to the determination of geoid heights, over an area, given its gradients $dN/d\phi$ and $dN/d\lambda \cos\phi$ at a number of discrete points. The method is fully described in [267] and [268], which apply it to the geoid survey of West Germany. See also [357].

Collocation methods essentially depend on the existence of a significant covariance function between values of the deviation, depending only on the distances between the discrete points concerned. Except in an area which is unusually devoid of moderately-sized topographical relief, such a

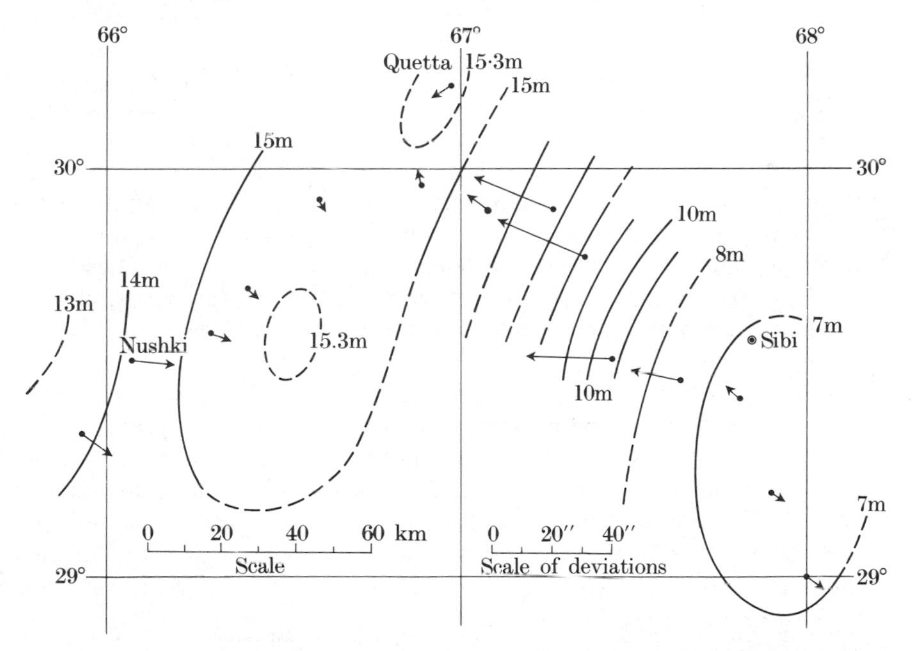

FIG. 6.21. The geoid in the Bolan Pass, Pakistan. Arrows show the deviation at the scale of 1 cm = 24″. Geoid contours are at 1 meter intervals. Broken lines indicate doubt.

function is unlikely to exist, since large changes and reversals of the signs of ξ and η can occur in quite short distances. On the other hand, isostatic anomalies ξ_C and η_C very much more regularly (§§ 6.53 and 6.54), and their covariance function may well be significant.

(*f*) *Drawing geoid contours.* See Fig. 6.21. Given the geoid heights of a number of points along a line or over an area, it is required to draw contours of equal *N*. Let the points (with their geoid heights) be plotted on such a scale as 1/2M, and at each station let an arrow represent the direction and magnitude of the deviation of the vertical†, at a scale of 1 or 2 mm for 1″. Contour lines will then be approximately perpendicular to the arrows, and the interval between contours will locally be inversely proportional to the lengths of the arrows. The contours can be sketched in accordingly, with an eye on the topography.

6.59. Geoid survey by vertical angles

In a triangulation net, if the vertical angles can be freed from refraction error, the heights of all stations above the spheroid can in principle be computed, provided the deviation of the vertical is known at one opening station. Then if the heights of all stations above the geoid are measured by spirit-levelling, the separation of geoid and spheroid below each station is known. The deviation of the vertical at each station is also determinate, from the reciprocal vertical angles. Errors will, of course, be controlled by the deviations being observed astronomically at a proportion of the stations. See § C.02.

This system is at present practicable in very mountainous country, where (*a*) formula (6.177) is liable to be inaccurate, and (*b*) the refraction can be computed with exceptional accuracy. See [615] for some results in the Alps. In more ordinary country the uncertainty of refraction (in 1978) introduces into this method greater errors than those caused by the use of (6.177). It is possible that the methods of measuring refraction described in § 3.21 may change this situation, and that geoid survey by vertical angles may become a practicable method everywhere.

6.60. Geoid and gravity charts

(*a*) This paragraph gives references to a selection of charts showing the form of the geoid as determined by different methods. Such charts have been drawn with reference to a great variety of spheroids and datums, and useful comparison between charts on different datums can only be made after converting one or both to a common reference system. See § 6.58 (*b*).

† It is convenient to draw these arrows with their points towards high areas on the geoid, i.e. towards what may be presumed to be areas of excess mass. The arrows then represent the deviation of the downward vertical relative to the spheroidal normal.

Sub-paragraph (*c*) below also lists various sources of gravity data and charts, such as may be used for determining the geoid as in § 6.56 (*b*) and (*f*). For average height maps see § 6.37.

(*b*) *Geoid charts.*

(i) *Astro-geodetic*

World charts on various geocentric datums. [207] 1968.
North America. NAD. [203] 1960.
Europe and South Asia. ED. [114] 1971.
Europe, S. Asia, and N. Africa. ED. [389] 1976.
India and adjacent. Indian International 1924. [243] 1951.
Australia. Australian datum. [399], 1971.
Africa (provisional). ED. [111], 1958.

The following are examples of a few detailed charts showing particular geoid sections.

North America. NAD. [482], 1962.
Switzerland. Swiss datum. [440], vol. 20, 1939. A remarkable section.
Great Britain. ED. [487], 1963.
India. Indian International 1924. [109], 1935.

(ii) *From Stokes's integral.* Geocentric datums.

Summary of early work. [261], 1965.
N. America, N. Atlantic, Europe, and Asia. [583], 1972.
North Atlantic. [477], 1975.

(iii) *From satellite orbits, excluding surface gravity.* Geocentric datums.

World. GEM 5. [383], Fig. 4-3, 1974.
World. GEM 7. [587], Fig. 35, 1976.
World. GEM 9. [382], Fig. 2, 1977.

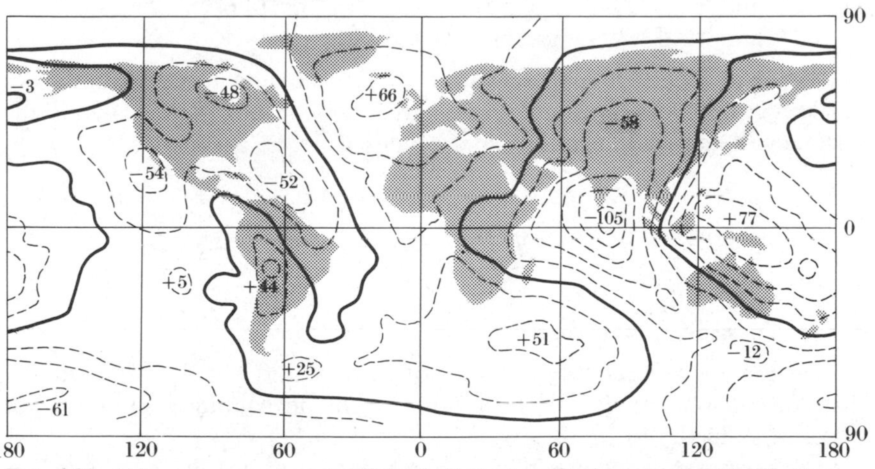

FIG. 6.22. *N* in metres as given by GEM 8. Flattening of spheroid is 1/298.255. Contour interval 20 m. Zero contour is full line, and others broken. Generalized from [588], Fig. 4.

(iv) *From satellite orbits, including surface gravity.* Geocentric datums.
World. GEM 6. [383], Fig. 4-4, 1974.
World. GEM 8. [588], Fig. 4, 1976.
World. GEM 10. [382], Fig. 10, 1977.
World. SAO III. [216], Fig. 6(a) and (c), 1973.
World. Grim 2. [79], 1976.
(v) *From satellite altimetry.*
North Atlantic. Geocentric datum. [505], 1977.

(*c*) *Gravity.*

(i) The International Gravity Bureau (IGB) has a great collection of gravity anomaly charts. Its holding of Bouguer anomaly charts is listed in [57], 33, 1973, pp. I-11 to I-127, with a supplement in 1977, pp. I-33 to I-56. It also holds many charts of free-air and isostatic anomalies. From these charts, and using the DMAAC lists of average heights of $1°\times 1°$ squares, § 6.37, the IGB is preparing a list of $1°\times 1°$ means of free-air gravity anomalies, hoping for an accuracy of 5 to 10 mGal (50–100 $\mu\text{m s}^{-2}$) in each $1°\times 1°$ square. DMAAC currently has a similar list of free-air mean anomalies, prepared earlier.

(ii) Lists of $5°\times 5°$ (or of 550×550 km equal area) mean values of the free-air anomaly are in [474], 1972, and in [608], 1977. $1°\times 1°$ (not equal area) world lists of observed free-air anomalies are in [475]: not wholly complete, but continuously revised and added to.

(iii) Examples of world charts of free-air gravity anomalies derived from satellite orbits are

SAO III. [216], Fig. 6(*b*), 1973.
GEM 5 and 6. [383], Figs. 4-1 and 4-2, 1974.
GEM 7 and 8. [587], Fig. 31, and [588] Fig. 5, 1976.
GEM 9 and 10. [382], Figs. 11 and 12, 1977.

(iv) [382], Fig. 1 is a world chart (1977) showing the distribution of surface gravity observations. [182] gives much information about the accuracy of the earth's gravity and potential fields, as known in 1972.

Section 11. Gravity as a Guide to Internal Densities

6.61. No unique solution

Given g at ground level, it is required to find the internal density distribution, but gravity by itself cannot provide a unique solution. For instance as in (*a*) to (*d*) below.

(*a*) Given a density distribution which does produce the known values of g, any sphere of uniform density may be expanded or contracted to fill a concentric sphere of any radius without affecting g at points outside the radii involved.

(*b*) Less obviously, regarding a small part of the earth as plane, let an anomaly $\sigma \sin px$ be added at depth z, x being a horizontal coordinate axis. Then the attraction at the surface will be $2\pi k\sigma e^{-pz} \sin px$, which is small if $1/p$ is small compared with z. It follows that surface observations can give no information about mass anomalies whose 'wavelength' $2\pi/p$ is small compared with 2π times their depth, [136], and also that such anomalies may be imposed on any solution which fits the data.

(*c*) Regarding the earth as a sphere of radius R, Fig. 6.23, consider a particle of unit mass at depth d_1. Then it can be replaced by a surface density of

$$\frac{(R-d_2)^2-(R-d_1)^2}{4\pi(R-d_2)r^3} \quad \text{or approximately} \quad \frac{d_1-d_2}{2\pi r^3}, \tag{6.178}$$

over a sphere of radius $R-d_2$, if $d_2<d_1$, where r is distance from the position of the original particle. See [305], (1929), p. 198. This amounts to saying that a concentrated particle at depth d_1 may be replaced by a regional distribution of equal total mass at any smaller depth d_2. Quite roughly, the regional distribution may be regarded as uniform over a radius of (say) d_1-d_2, since outside this radius $1/r^3<1(2\surd 2)$ times its maximum value, and it falls off rapidly.

(*d*) If g is known at n points, the observations can be satisfied by suitable masses $m_1,\ldots, m_n$ placed at any n arbitrarily selected locations, but if the selected locations are irrationally chosen, ridiculous values of g will be given at intermediate points between the gravity stations. A set of discrete gravity stations, at points not too widely separated, in fact determines g with reasonable accuracy at all points in the area covered.

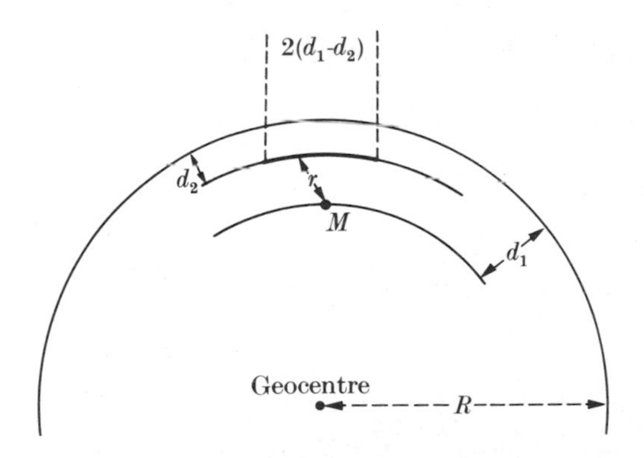

FIG. 6.23. A particle M at depth d_1 may be approximately replaced by an equal mass distributed over a radius of (d_1-d_2) at depth d_2. See § 6.61(*c*).

While gravity cannot solve the problem by itself, it does place considerable limitations on the possible solutions, and when gravity is considered together with other geophysical and geological data, a clear and significantly correct answer may be obtainable.

6.62. Gravity anomalies

It is convenient to compare g with γ, the attraction of some defined standard body, so that the differences $g-\gamma$ are the attraction of the masses which constitute the difference between the actual earth and the standard.

The simplest form of *standard earth* is a solid spheroid of defined major axis and flattening, whose internal density distribution is such as will make its surface an equipotential surface of its attraction and the centrifugal force, with its centre of mass coinciding with that of the earth. If the matter comprising this standard earth is to be of zero strength, its flattening must be 1/299.8 ($J_2=1071\times10^{-6}$), see § 6.20 (*a*), and its J_4 must be -3.0×10^{-6} with $\chi=-2.8$ in (6.31), see § 6.20 (*b*). As mentioned

---

Explanation of Fig. 6.24. Fig. 6.24 shows $g-\gamma_A$, $g-\gamma_T$, and $g-\gamma_C$ plotted against height of station or depth of ocean for eight pairs or groups of near stations of very different heights, as follows: A_1 and A_2, Skardu and Deosai II in Kashmir, about 40 km apart; B_1 and B_2, Colorado Springs and Pikes Peak, 20 km, in the U.S.A.; C_1 and C_2, Lalpur and Tosh Maidan, 20 km, in Kashmir; D_1 and D_2, Kurseong and Sandakphu, Sikkim, India, 35 km; E_1, E_2, and E_3, Vening Meinesz's station 111, Honolulu, and Mauna Kea, 180 and 300 km; F_1 and F_2, Vening Meinesz's 470 and 469 (*b*) in Madeira, 100 km; G_1, G_2, and G_3, Vening Meinesz's 562, 563 (Romanche deep), and 564 in the mid-Atlantic, 65 and 55 km; H_1 and H_2, Nero deep and Guam, 120 km. Also I_1, I_2, and I_3 are three ordinary low-lying places, New York, Paris, and Delhi for comparison (omitted from $g-\gamma_C$ to avoid overcrowding around the zero).

Fig. 6.24 (b) shows that adjacent stations have very similar topographical anomalies, but that high continental stations are negative and ocean stations positive. The line OJ represents a change of 1 mGal (10 μm s^{-2}) per 9 metres and OK one of 1 mGal per $14\frac{1}{2}$ metres, such as would be expected in $g-\gamma_T$ in uniform plateaux and oceans if perfect compensation actually existed. These lines represent the general run of the points except that ocean islands and isolated peaks naturally stand out above the line; ocean deeps and narrow valleys are below the line; and stations whose height or depth equals the local average are on the line.

Fig. 6.24 (c) shows that isostatic anomalies (Hayford 113 km) bring mountains and oceans together, and also maintain the close agreement between near stations of widely different heights, although there is a marked tendency for $g-\gamma_C$ to be rather more positive at the higher of a pair. The line OL represents 1 mGal (10 μm s^{-2}) per 45 metres, which purely empirical relation typifies the selected data. The cause may be that compensation is regional rather than local.

Fig. 6.24(a) shows that the free air anomalies are locally correlated with height, OM and ON showing the natural slopes of 1 mGal (10 μm s^{-2}) per 9 and $14\frac{1}{2}$ metres for comparison. But there is no systematic difference between continental and oceanic anomalies.

Note that except for the three plains stations I_1 to I_3, the stations shown here are all extraordinary in being on or above or near very abrupt topography. Such large anomalies are not typical of the earth as a whole.

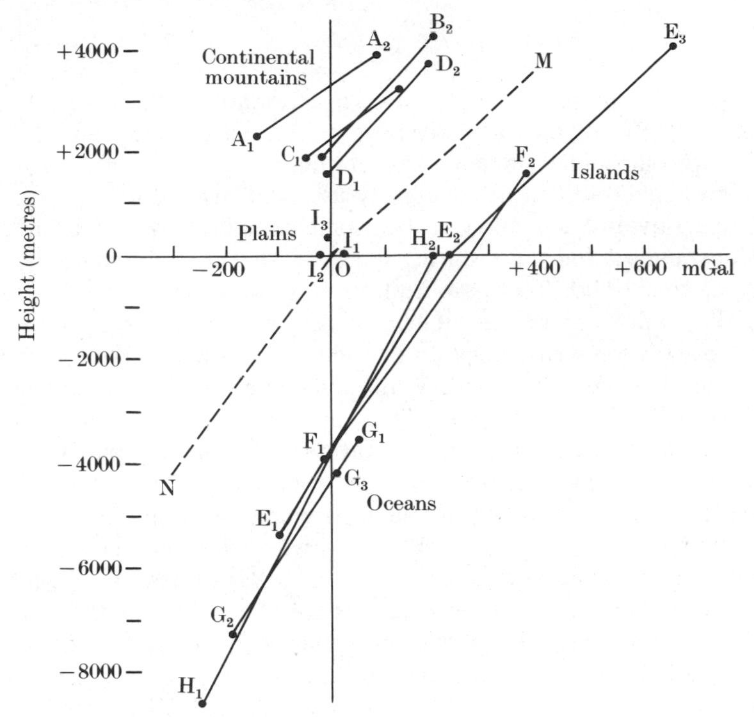

FIG. 6.24(a). $g-\gamma_A$. For explanation see facing page. 1 mgal = 10 μm s^{-2}.

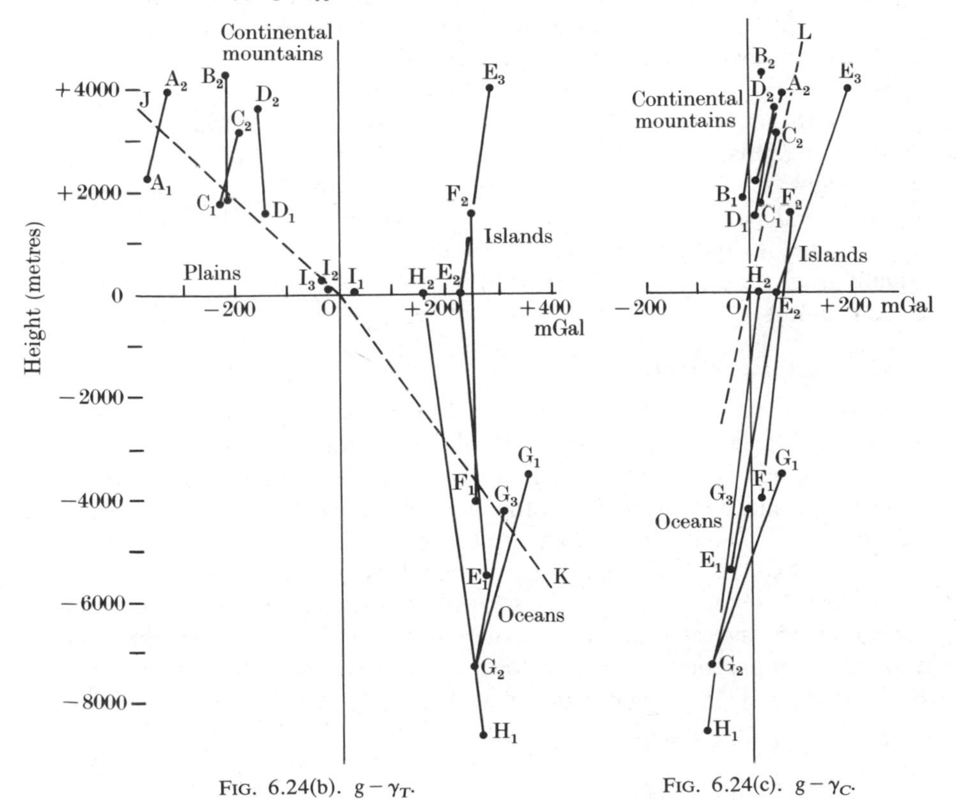

FIG. 6.24(b). $g-\gamma_T$.

FIG. 6.24(c). $g-\gamma_C$.

in § 6.28, it is more usual to use a standard with $f = 1/298.25$ ($J_2 = 1082.63 \times 10^{-6}$) which actually best fits the earth, and $J_4 = -2.4$ as for an exact ellipsoidal section. The minimum strength necessary to maintain this spheroid can be separately assessed, see § 6.68.

The most obvious deduction to be made from the use of a simple spheroidal standard and the resulting $g - \gamma_A$ is that the ground surface is not an exact spheroid. Topographical irregularities cannot be ignored, because they will disguise the other, smaller, effects which are sought. And similarly, in some situations, to include topography without isostatic compensation has the same result. While uncompensated topography can be, and is, used for local geophysical prospecting, more widespread studies may be better based on an isostatic model, namely the standard solid spheroid described above plus the mountains and oceans which actually exist, plus some system of isostatic compensation.

The anomalies $g - \gamma_A$, $g - \gamma_T$, and $g - \gamma_C$ which are now being considered are identical in form to those used for Stokes's integral in Sections 5 and 6 above, with the difference that here the height of the gravity station is (in principle) height above the spheroid, while in Section 7 the deduction (6.135) is through height above the co-geoid or geoid. On the other hand, it is generally convenient to use geoidal heights, and if the aim is to study the variations of crustal structure over a limited area (say 200×200 km), the differences between geoidal and spheroidal heights are likely to vary by only a few metres, the equivalent of even fewer mGal.

Fig. 6.24 (a), (b), and (c) illustrates the variations of $g - \gamma_A$, $g - \gamma_T$, and $g - \gamma_C$ with height, as between pairs of relatively near stations of greatly different heights. It clearly shows the strong local correlation of $g - \gamma_A$ with height, as is also shown in Fig. 6.12 (a) and (b). Fig. 6.24 (*b*) shows the local independence of $g - \gamma_T$ on height, but its strong dependence on the average height of large areas. In this respect $g - \gamma_B$ is the same. Finally, Fig. 6.24 (c) shows the relatively small variations of $g - \gamma_C$ in both circumstances.

6.63. Various isostatic systems

When anomalies are formed for use in Stokes's integral the object is to eliminate external masses and to obtain good interpolation, but in the present context the object is to compare g with a standard earth, which will differ from the true earth as little as possible. Such systems as the free air anomaly, condensation to sea-level as in § 6.38 (*b*), the mirror image system of § 6.38 (*c*), or that of smoothed topography (*e*) are therefore unsuitable for the present purpose, since they quite clearly do not represent any possible form of actual mass distribution. The Bouguer anomaly is in practice suitable for local structures, but its tangent plateau, constituting a different standard for each gravity station, is incompatible

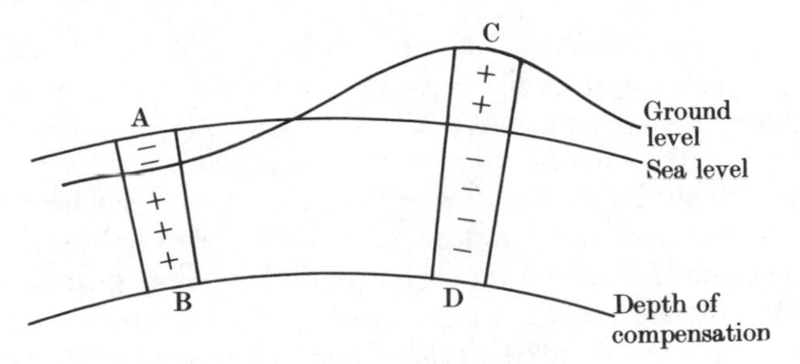

FIG. 6.25. Hayford compensation. If columns AB and CD are of equal cross-section, they are assumed to contain equal mass. Vertical distances are not drawn to scale.

with the present line of reasoning. The pure topographical anomaly $g-\gamma_T$, is more laborious to compute than an isostatic anomaly since its computation has to extend over the whole earth, and isostatic anomalies are also closer representations of the true mass distribution. Three main forms of isostatic standard remain, as described in §§ 6.64–6.66.

6.64. Hayford or Pratt compensation

The standard earth is a spheroid plus the actual topography, with every topographical excess or defect of mass compensated by an equal and opposite defect or excess, evenly distributed immediately below it between ground-level or sea-bottom level and a fixed depth D, commonly 113.7 km, known as the *depth of compensation*,† see Fig. 6.25. Every column of matter of unit cross-section based on this depth and extending up to ground- or (in the ocean) sea-level then contains equal mass.

At first sight equal mass in any two columns will ensure equal pressures on their bases, a hydrostatic state from which it is very natural to wish to reckon anomalies. But if allowance is made for the downward convergence of the vertical boundaries of the columns, the mass of the compensation must for this purpose be arithmetically reduced in the ratio $(R-D)/R$, so that the masses of the compensation and the corresponding topography are not equal but in the above ratio. If $D=113$ km, equal pressures will result if the mass of the compensation is everywhere arithmetically reduced by 1.8 per cent, and as its vertical attraction may

† In Hayford's original tables, [256], pp. 28–47, the compensation extends from ground or sea-bottom level to 113.7 km below it, a barely significant difference, which was adopted for computational convenience. If preferred, the compensation can of course be defined to lie between sea (or sea-bottom) level and the depth of compensation, but Hayford's original definition is as given in the text.

be a few hundred mGal the difference between the two systems may be appreciable. Either system is a legitimate standard, but it is necessary to state which is being used.

In the above, the increase of gravity with depth has been ignored, and this causes a slight inequality of pressure, allowance for which requires a further reduction of density in the ratio $(R-\frac{1}{4}D)/R$, which brings the total up to $2\frac{1}{4}$ per cent if $D = 113$ km. The point is usually neglected, illogically but with almost inappreciable result. See [376], pp. 103–9.

The theory originally behind this system was that the base of the crust is at a fixed depth, such as 113.7 km below sea-level, and that it everywhere exerts equal pressure on a weak layer below. Irregularities of ground level were presumed to exist because of variations in the density of the crust. This theory does not conform to current views, see Fig. 6.2, but the system has been much used, and in practice gives results very similar to what might be given by more modern theory.

6.65. Airy compensation

In the Airy system of compensation the earth's outer crust of constant density, usually 2.67, is assumed to float in a lower layer whose density is generally taken as 3.27. Density being constant, the thickness of the upper layer must vary as in Fig. 6.26. With the above figures, if D_1 is the normal thickness of the upper layer, it will extend deeper under a mountain of height h by an amount

$$h_C = h \times 2.67/(3.27-2.67) = 4.45h,$$

while under an ocean of depth d it will be less deep by

$$d(2.67-1.03)/0.6 = 2.74d.\dagger$$

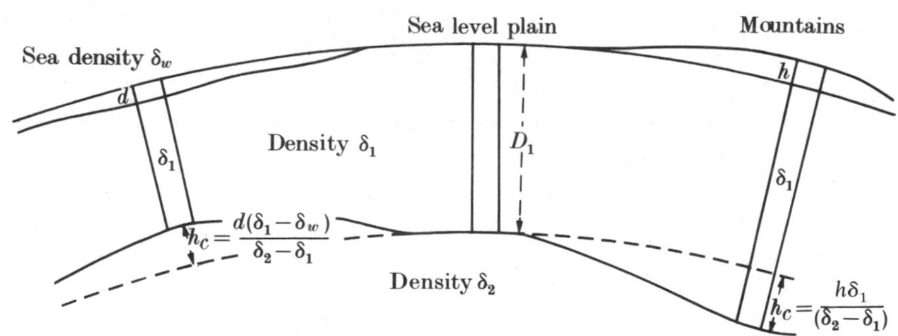

FIG. 6.26. Airy compensation.

† Allowance for the downward narrowing of prisms may introduce barely perceptible small terms, and the question of equal pressure or equal mass can also be a minor complication. [376], pp. 118–20.

Note that unless D_1 is very small, variations in the assumed density of the lower layer have little effect on computed γ, merely changing values of h_C, but not the mass displaced by it. On the other hand, variations in D_1 are important, since the vertical component of the attraction of medium-distant features will be proportional to it. Examination of anomalies is thus unlikely to suggest any change in the figure 3.27, but may give information about the most likely values of D_1.

Airy's system has a more probable physical basis than Hayford's but it is not noticeably more successful in its agreement with observed g: probably because the effect of departures from exact compensation in any form is apt to exceed and mask the differences between Hayford and Airy with $D_1 = \frac{1}{2}D$. For the method of computing see § J.05.

In so far as the thickness h_C in Fig. 6.26 may not be a large fraction of D_1, the layer h_C may be regarded as a skin density at depth D_1 of $-h \times 2670$ or of $d \times 1640$ kg m^{-2}, h and d being in metres. This approximation is convenient but not altogether accurate, see Fig. 6.2, which shows that the depth of the Mohorovičić discontinuity departs widely from its mean value.

6.66. Regional compensation

In the systems so far described the amount of compensation exactly follows the variations of the overlying topography. This may be convenient for computation, and possibly harmless as a standard, but it is obviously physically impossible since the strength of the upper crust must introduce some degree of smoothing.

In [574] Vening Meinesz described a system whereby calculations already made with local compensation may be modified to allow for regional compensation. If the strong crust is looked on as a thin plate, say 25 km thick, floating in a denser layer ($\rho = 3.27$), it will deflect under load as in Fig. 6.27. The proportions of the curve will be as there shown, but the linear dimension l will depend on the thickness and elasticity of the crust, and the maximum displacement d_m will depend on l and on the densities of crust and of the denser layer below. The depression of the light crust represents compensation whose attraction at any point P bears to that of exact local compensation a ratio which depends on l, on the distance from P, and on the depth D_1 (equal to the thickness of the light crust) at which the compensation is assumed concentrated. [574] gives these factors for concentration at $D_1 = 0$ or 25 km, and for an even distribution between the two depths, with $l = 50$ km, and [576] gives fuller tables for all depths to 60 km, for $l = 10$, 20, 40, 60, and 80 km corresponding to a distribution of the compensation through radii 2.9 times as great. (Not 3.89 times as in Fig. 6.27 and [574], but the

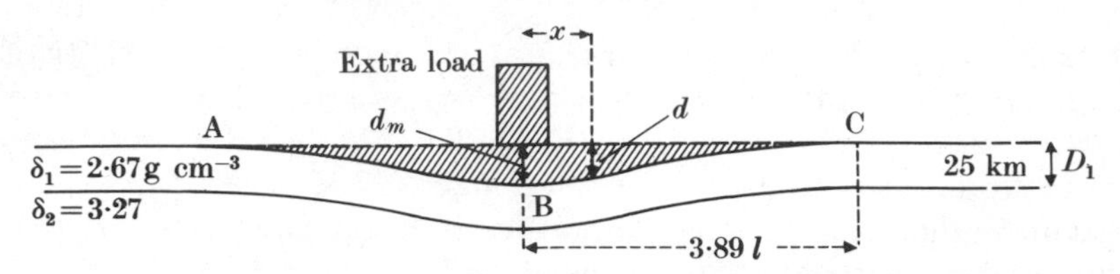

FIG. 6.27. Regional compensation. The form of the curve ABC is defined as follows. When $x = 0$, $d = d_m$, a quantity depending on the densities and other parameters. When $x = l$, $d = 0.646d_m$; $x = 2l$, $d = 0.258d_m$; $x = 3l$, $d = 0.066d_m$; and when $x = 3.89l$, $d = 0$.

distribution is not really very dissimilar.)

While the actual distribution of compensation must obviously be regional, there is no great reason to expect constancy in l or D_1 or in the form of the curve in Fig. 6.27, and the regional Hayford or Airy anomalies are not in general strikingly less than the ordinary Hayford, although evidence supporting them may be found in places.

If the standard earth is assumed to include local isostatic compensation, gravity anomalies $g - \gamma_C$ will generally have a tendency to be relatively positive in places such as mountain tops and crest lines which are notably higher than areas 10 or 20 km distant, with more negative anomalies in the lower ground nearby: and vice versa for depressions. A suitable form of regional compensation would remedy this. See also [81] and [592] for examples of apparent regional compensation.

Referring to § 6.61 (c), it is apparent that exact local compensation at a depth d_1 at which the skin density equals the topographic excess (with opposite sign) at any point, will have the same effect on surface g as will a system of regionally smoothed compensation at a shallower depth. So while Hayford compensation between depth 0 and 113 km is similar to Airy compensation at 56 km, both are in turn similar to regional compensation at a depth of (say) 15 or 30 km, at which local compensation is spread out over (roughly) a radius of 41 or 26 km respectively. These are reasonable figures.

6.67. Depth of compensation

Hayford's figure of 113.7 km [254] was selected from a variety of depths as the one which would produce the smallest anomalies in the deviations of the vertical in the United States, and Bowie [118] found a similar figure of 95 km from gravity in mountainous regions in the United States. Neither figure is applicable to ocean areas, in which Hayford had no data.

For the Airy system, where the depth would be expected to be half as much, Heiskanen [257] found 41 km in the Alps, 50 km in the United

States, and 77 km in the Caucasus, in good agreement with Hayford's figures. As mentioned in § 6.66, a regional anomaly at a shallower depth would do as well.

While these minimum-anomaly solutions do not necessarily rest on a very convincing basis, they do in fact agree very well with the depth of the Mohorivičić discontinuity, on land as given by seismology, when coupled with a reasonable amount of regional spread.

In so far as the undulations of the granite–basalt interface, where the density difference is 0.2, may follow those of the Moho (0.45), the average depth of compensation may appear to be somewhat less than the average depth of the Moho.

6.68. Stress differences caused by unequal loading

The determination of the stress differences near the surface which may result from unequal loading by visible topography, or at greater depths by its incomplete compensation, is a problem whose general solution is complicated. See [305] (1970 and earlier editions) for solutions in particular cases.

(*a*) *Thick strong layer*, i.e. a strong layer whose thickness is at least as great as the wavelength of the variations in the superficial load that is under consideration. Let the mass of this load be $\sigma\cos(2\pi x/l)$, lying on a strong layer of thickness D, which is supported by the uniform pressure of an underlying layer of zero strength. Take rectangular coordinates, x horizontally across the crests of the load, and z vertical (positive downwards). The wavelength of the load is l. This represents a succession of parallel mountain ranges distant l from crest to crest to crest, and if h is their height (crest above valley) and ρ their density, $\sigma=\frac{1}{2}h\rho$, [305], 1970, p. 253 shows that the maximum stress difference S_m will occur at depth Z below the surface of the strong layer, where

$$Z = l/2\pi$$

and

$$S_m = 2g\sigma/e, \quad \text{where} \quad e = 2.718$$

and at any depth z

$$S = 2g\sigma(2\pi z/l)\exp(-2\pi z/l)$$

Inserting figures let $\rho = 2670\ \text{kg m}^{-3}$, let $h = 5$ km so that $\sigma = \frac{1}{2}(13.4\times10^6)\ \text{kg m}^{-2}$, and let the ranges be 100 km from crest to crest, so that $l = 100\,000$ m. These are large mountains. We have $g = 9.8\ \text{m s}^{-2}$. Then $S_m = 4.8\times10^7\ \text{kg m}^{-1}\text{s}^{-2}$ (or $4.8\times10^8\ \text{dyn cm}^{-2}$), $Z = 16$ km, and

S at depth z is $8.2\times10^8\,(z/l)\exp(-2\pi z/l)$ kg m^{-1} s^{-2}. S_m is within the strength of basalt at 400 °C.† At 100-km depth the stress difference would be 1.5×10^6 kg m^{-1} s^{-2}, (or 1.5×10^7 dyn cm^{-2}) which may or may not be within the strength of the rocks at or above that depth. Such irregularities can in most places be supported by the strength of a 30-km crust, [305], 1959, p. 202. The assumed load has, however, been taken as alternatively positive and negative. While a system of several parallel mountain ranges and valleys of this size, all above sea-level, could probably be sustained without local isostatic support under each separate range, regional isostatic support in some form would be necessary.

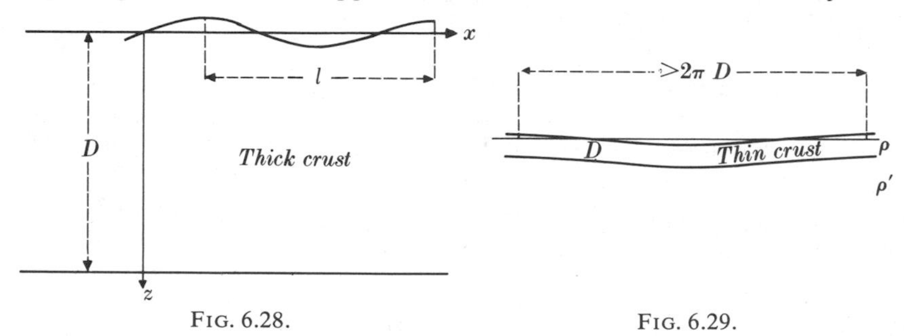

FIG. 6.28.

FIG. 6.29.

(b) Thin strong layer, Fig. 6.29, namely a strong layer (of density ρ), whose thickness is less than $1/2\pi$ times the wavelength of the imposed load. Then unless σ is very small, the variations of load can only be supported by the crust bending under it and thereby obtaining support from the increased hydrostatic pressure of the denser (ρ') underlying layer. And if $\rho'-\rho=1000$ kg m^{-3}, [305], (1929), pp. 175–6, shows (with certain simplifications) that this bending will involve a stress difference of $3g\sigma$ in the lower side of the crust, so that if the strength of the latter is 1.0×10^8 kg m^{-1} s^{-2}, fracture will occur when $\sigma=3.3\times10^6$ kg m^{-2}, corresponding to a loading of $\pm1\frac{1}{4}$ km of rock. But ρ/ρ' of the load will disappear from view, by reason of the crust's bending under it, so that the superficial maximum excess or defect will only be ±0.35 km. Greater irregularities can of course be supported, but the conclusion is that when such features persist with constant sign over widths greater than πD, the crust will have to fracture. Instead of bending as a whole as in Fig. 6.29, it will break into smaller pieces which will sink or rise independently. This places some limit on the possible regional spread of the compensation of large features.

† At the surface the strength of granite, the stress difference which it will support indefinitely, is 8×10^7 kg m^{-1} s^{-2}, and that of basalt is 12×10^7 kg m^{-1} s^{-2}, [305], 1929, p. 284.

The general conclusion of (*a*) and (*b*) above is that the earth's surface features, and anomalous masses which produce gravity anomalies of constant sign over widths of less than 100 km, could be supported by the strength of the crust, which in turn could be supported by 'floatation' on a strengthless mantle. It does not follow that they are actually so supported, nor that the mantle has no strength.

(*c*) *Widespread anomalies.* While more or less local irregularities of loading as in (*b*) demand no strength in the mantle, the widespread anomalies representable by low degree harmonics do. Thus [305], 1970, pp. 266–7 estimates that the anomaly of 10.6×10^{-6}(§ 6.20) in the value of J_2, with some small additions from the (2,2) and (3,1) harmonics, could be supported by a strength of 1.0×10^7 kg m^{-1} s^{-2} available from the surface to the top of the core, or of 3×10^7 kg m^{-1} s^{-2} in the top 600 km with no strength below. Making some allowance for other factors these figures are raised to 3.7×10^7 kg m^{-1} s^{-2} in the top 600 km alone, or a combination of 1.7×10^7 kg m^{-1} s^{-2} to 600 km followed by 1.0×10^7 kg m^{-1} s^{-2} down to the core. The higher figures in the outer 100 km for the support of local topography as quoted in (*a*) above, are of course additional to the figures here given for the support of very widespread anomalies.

6.69. Short-period anomalies must be caused by shallow masses

The gravity anomalies observed on the surface could be accounted for by a closely underlying skin density $\Delta g_0/2\pi G$, (6.13). Alternatively, subject to conditions, they could result from a suitable skin density, determined as below, on a lower equipotential surface, or by a combination of skin densities at different depths.

If a portion of the earth can be treated as flat, and if over this area Δg_0 can be represented by a single harmonic term $A\cos px$, it could be produced by a skin density of

$$(A/2\pi G)e^{pz}\cos px \qquad (6.179)$$

at depth z, and if Δg_0 is represented by a Fourier series, the distribution can be expressed by a series of such terms, provided that the series converges. [136] gives a formal method of determining the necessary skin density at depth z, which (*a*) duly applies to a sphere instead of to a plane as above, and (*b*) allows of numerical integration from the values of Δg_0 as exhibited on a contoured chart, instead of requiring a harmonic analysis. If this system is applied to determine possible mass distributions at successively greater depths, it will presently be found that the solution becomes irregular with large positive and negative densities closely adjacent. This is a sign that the depth reached is improbably great, or at any rate too great to account for the shorter wavelength variations of Δg_0. A routine is given whereby the shorter wavelengths can then be left behind, being accounted for a reasonable distribution at a shallow depth, while the determination of possible (but not necessarily more probable) solutions for the longer period irregularities proceeds to greater depths, until they in turn become unstable. As a general rule, the greatest depth at which a skin density can account for a change of surface gravity anomaly of $2g_m$ from maximum to minimum in x km is

$$(x/\pi)\ln(2\pi G\sigma/g_m)\text{ km},$$

where σ is the maximum acceptable value of the skin density, i.e. the maximum acceptable value of anomalous solid density multiplied by the maximum thickness of the layer, at mean depth z, in which it can occur [136], p. 342. If the maximum allowable σ is $g_m/2\pi G$, namely 9 metres of 2.67 density per mGal in g_m, the only possible depth is zero. This is then a minimum value of σ. If an acceptable maximum σ is eight times as much (720 metres of 0.27 density per mGal in g_m), which is a large figure, $\ln(2\pi G\sigma/g_m)\approx 2$, and the depth is $2x/\pi$. With figures such as $x = 100$ km and $g_m = 30$ mGal, 300 $\mu\text{m s}^{-2}$, involving a change of 60 mGal in a distance of 100 km, the maximum depth $2x/\pi$ is 64 km and the necessary σ is a density of 0.27 through a thickness of 21 km.

6.70. Details of mass anomalies by trial and error

The location of the anomalies of mass may be suggested by various considerations, such as seismic soundings or by other geological or geophysical surveys. Variations in the thickness and densities of the sedimentary layer, or in the depths of the Mohorovičić or Conrad discontinuities, are likely sources of gravity anomalies. The actual amount of anomalous mass in different places may either be solved for by least squares, so as to give best agreement with observed values of $g-\gamma_C$, or preliminary estimates may be made and the surface effects computed. If the correct surface g is not obtained, the estimates may be modified. Examples of such procedures are given in [221], pp. 63–139, [595], and [438].

For the detailed geophysical prospecting of small structures it is convenient to use the Bouguer anomaly $g-\gamma_B$, ignoring isostatic compensation. This anomaly interpolates well, and irregularities in it are likely to correspond to local anomalies of mass. Its more widespread variations, due to the existence of compensation and of low-degree harmonics, are for this purpose eliminated by removing its *regional trend* in the area.

General references for Section 11. [136], [221], [254], [266], [305], [575], and [614].

SECTION 12. CRUSTAL MOVEMENTS

6.71. Introductory

Geology reveals vertical crustal movements, by faulting and bending, which have amounted to 3000 metres or more in perhaps the last 25 or 50 million years, and the folding and overthrusting of mountains has involved horizontal movements of at least tens of kilometres. Much larger movements are involved in the possibility of continental drift. The contribution of geodesy to this subject is to detect and measure such movements as may currently be taking place.

The question of continental drift now appears to be settled, the general mobility of the earth's crust being accepted under the title of *plate tectonics.* It appears that crustal material emerges from the mid-ocean ridges, spreading outwards in the form of plates which are moving away at a rate of perhaps 3 to 10 cm/year. The total area of the earth's surface being relatively constant, the further edge of each plate has to disappear by *subduction* under an edge of some other plate. This occurs in the circum-Pacific and Alpine–Himalyan *collision zones*, where the leading edge of one plate dips forward and downwards under another. These zones are apt to be associated with marine trenches and narrow zones of negative gravity anomalies, with volcanic activity and deep-seated earthquakes over the deeper parts of the subducted plate. There are also lines where two plates are moving with different (or opposite) velocities in the same direction, on opposite sides of a long system of faults. About a dozen principal plates are currently recognized, but there is naturally a tendency for large plates to break up into smaller ones, and there is no limit to the number of separate plates which may eventually be recorded.

The subject of plate tectonics is nicely summarized in [453].

A large earthquake may often cause visible movements. Roads and fences may be put out of alignment, and vertical scarps a metre or two in height may appear. No geodetic instruments are required for measuring such changes, but to discover which side has moved, and how far the movement extends back from the visible rupture, demands more accurate measurement. The very practical question of whether measurable and recognizable strain takes place before an earthquake may also be a geodetic problem.

In addition to these more or less widespread and deep-seated movements, the surface of the earth is always apt to move in a more random way. From the geodetic point of view such matters as mining subsidence, alluvium shrinkage, soil creep, and landslides, as well as movements of particular survey marks due to no obvious cause, are simply sources of error which must be eliminated.

The causes of crustal movements are questions for geophysics. Geodesy is only concerned with their measurements. But geodetic programmes must be designed with some knowledge or where and what movements are to be expected, and the reality of any apparent changes recorded must partly be judged by the inherent probability of improbility of their occurrence, as well as by the accuracy of the survey which suggests them. Circumstances in which it will be relatively easy to believe in the tectonic reality of apparent changes are as below.

(*a*) If the recorded movements are large and have clear connection with earthquakes or volcanic activity.

(*b*) If observations at three or move epochs show progressive change.

(*c*) If movement is detected by two different methods of measurement.

(*d*) If geological evidence, such as raised beaches and the like, show that the geodetically suggested movement is a continuation of what has happened in the relatively recent past.

It must be recognized that movement is liable to be more or less intermittent. Rates of movement, or absence of movement recorded over a short period cannot necessarily be extrapolated over a period of centuries.

In some countries, notably the USSR, crustal movements are being studied continuously in *test areas*,† selected for the suitability and variety of their geological conditions. In these areas level lines and horizontal control frameworks are frequently remeasured, while tiltmeters, tide gauges, and strain gauges (when appropriate) are in continuous operation.

6.72. Horizontal movements. Revision of triangulation, trilateration, etc.

National control frameworks tend to be reobserved after periods of some decades, and the revision naturally does not give exactly the same coordinates as the original. The differences cannot lightly be ascribed to crustal movement. It is necessary to verify that both sets of coordinates are referred to the same reference spheroid and datum, and that the same standard of length is applicable to both. Both computations should also be based on the same geoid chart for the reduction of measured distances to spheroid level.‡ The possibility of the differences being due to ordinary random and other errors in the two sets of measurements, must then be assessed by estimating their accuracies by such methods as are described in § 2.27. The accuracy of a new survey will probably have been assessed

† These test areas are sometimes referred to as *polygons*. This is apparently a literal translation from the Russian, but it is not an apt term in English.

‡ Other matters in which identity is desirable are the inclusion or neglect of the small corrections listed in § 2.09, and also the relative weighting of measured distances and angles.

by modern methods, but older surveys will not have been so treated, and the necessary information, or funds, may not now be available. And whereas suspected errors in the new net can be remedied by further re-observation, those in the old net cannot. Unless the estimated accuracies strongly suggest that errors of observation cannot account for the differences, little more can usefully be said about them.

If it seems reasonable to suspect the existence of real tectonic changes, the suggested changes can be exhibited on charts† by either of the following alternatives.

(*a*) Arrows showing the magnitude and direction of the change at each common station, Fig. 6.30.

(*b*) Contouring showing equal changes in meridian and P.V.

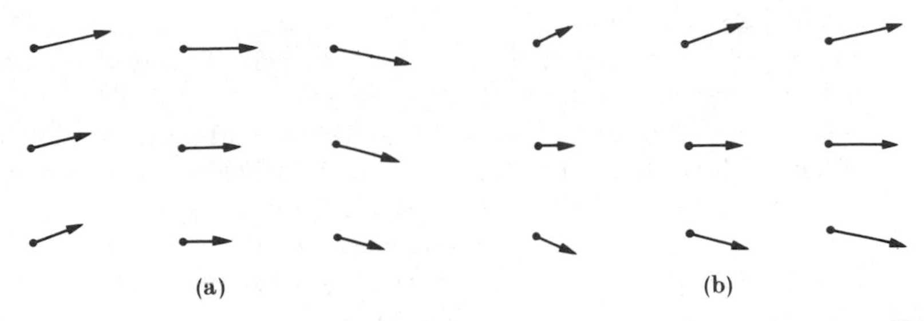

FIG. 6.30. The arrows represent changes of position of triangulation stations due to (a) a change of azimuth, and (b) a change of scale.

The chief source of error, at any rate in the older work, is likely to be the accumulation of error in scale and azimuth between widely spaced bases and Laplace stations. Such errors produce roughly orthomorphic errors of position, which can be recognized in charts giving contour lines of equal change, as in described in § 2.41 and shown in Fig. 2.24. Fig. 6.30 shows typical patterns for arrow charts. Errors of azimuth and scale may of course be combined. Crustal movements on the other hand are unlikely to produce orthomorphic changes except very locally. It may well happen that examination of the chart will at once show that the largest changes are unlikely to de due to crustal movement.

[26], 1957–9, Fig. 2, shows differences between the 1883–1909 triangulation of Japan and that of 1948–58. In that chart the arrows in Kyushu strongly suggest a change of scale, while those a little further north-east suggest, less strongly, a change of azimuth. Similarly [41], diagram 17, which shows changes in Great Britain, suggests scale changes in eastern England and azimuth changes in northerm Scotland. None of these

† Notably abnormal discrepancies at isolated stations may well arise from non-tectonic causes. Inspection of records may suggest non-identity, or movement of a station mark. Such stations can be eliminated from the comparison.

changes have the appearance of being tectonic. The charts of changes in faulted areas in the United States given in [599], [600], and [39], pp. 256–62, show the pattern which would be expected from tectonic movement along the faults.

Special projects. Where movement is most to be expected special observations of great accuracy may be undertaken and repeated at intervals of a few years. Movement along or across a known fault could be detected by measuring the sides and diagonals of a (reasonably square) quadrilateral, with two stations on each side of the fault, but it will generally be better to observe a more elaborate net, as insurance against an unstable mark, and to give better evidence of accuracy. Currently the accuracy of distance measurement is better than that of angles, so trilateration may be preferred to triangulation, although there are advantages in having both.

When possible, it is desirable to include at least two well separated stations in the net, on sites which are expected to be stable, or to have two such stations independently sited in the neighbourhood. If these stations show little change between the two epochs, confidence may be felt about the accuracy of the measures which do change.

It is of course desirable that the original and subsequent observations should be made with similar (or the same) instruments, and at similar times of year.

Station marks must be well constructed (sharp to 1 mm) on firm sites, preferably on rock. Where changes of a few mm are involved, it is not easy to be sure how far they extend beyond the rock on which the mark is built. A multiplicity of stations is thus an advantage.

If measured lines are steep, vertical angles must be measured, since a change of height might otherwise produce an apparent change of distance. Steep lines are generally best avoided.

6.73. Horizontal movements. Other methods

(*a*) *Astronomical.* Repeated observations of astronomical latitude and longitude at two stations will in principle give a measure of their relative movement during the interval, but even the best instruments (PZT or Danjon astrolabe, or the zenith telescope for latitude only) are only good enough to detect a relative movement of some metres, such as is only likely to arise after an interval of some decades or longer. No clear record exists of crustal movement, other than polar motion, having being measured astronomically.

(*b*) *Strain gauge.* Great accuracy is obtainable over short distances by specially devised instruments involving interference methods, or otherwise. See [566] for a laser strain gauge. To get the best from such methods, they should be used underground in disused mines or tunnels. When very short distances are measured, there will always be doubt about the extent to which the minute movements detected may extend to greater distances.

(*c*) *Satellite fixes.* Doppler observations, using translocation methods, § 7.50 (*b*), (*c*), and (*d*), can now probably give three-dimensional fixes with relative errors of less than one metre over distances of perhaps up to 100 km, but it is too early to claim that an apparent change of one metre between (say) 1978 and 1988 would be an undoubted proof of movement. For a review of satellite methods, present (1978) and future, see [58].

Laser tracking may give greater accuracy.

(*d*) *VLBI and lunar laser methods.* See §§ 7.66–7.72 and § 7.45. It is claimed that these methods will soon be able to measure the separation in all three dimensions between stations several thousands of km apart, with an accuracy of 10 cm or less. Such observations should be able to confirm or deny the movements postulated by plate tectonics in a period of ten years or less. But they have not yet (1978) been proved able to do it.

6.74. Vertical movements. Spirit levelling

Theodolite vertical angles are seldom accurate enough to reveal changes of height, but spirit levelling is a very accurate process which can often give significant results. When the re-observation of a levelling net has been completed the following points will call for special consideration.

(*a*) The accuracy of the old and new nets. Circuit closures and the discrepancies between fore- and back-levelling will have been good guides to the magnitudes of random errors, and will eliminate some sources of systematic error, but there is no easy way of estimating the success with which the systematic errors described in § 3.06 (*b*) and (*c*) have been eliminated. Levelling has sometimes been less good than it looks.

(*b*) No change can be considered as real, if there is a possibility of a single gross blunder in either of the two levellings, especially where wide river crossings are involved.

(*c*) Changes in the levelling will be more convincing if they are supported by accordant changes in the mean sea-level at a number of tide gauges.

(*d*) The re-observation of a net is likely to show changes in a large number of the common bench-marks, possibly in a large proportion of the whole, due to superficial causes such as the subsidence of buildings, consolidation of alluvium, mining subsidence, etc., and these will be more likely to be falls than rises. It is necessary to be sure that these bench-marks have been successfully excluded from the picture. If possible, conclusions should be based on bench-marks cut on solid rock as in § 3.11 (*d*).

The levelling of Finland is an example of deep-seated change which without doubt has been successfully measured. Confidence, [317], rests on the following.

(i) Geological and historical evidence of past changes.
(ii) Spirit levelling shows progressive changes.
(iii) Tide gauges agree.
(iv) The changes recorded are large, as much as 9 mm per year as a maximum, falling to zero change at a distance of 500 km.
(v) The favourable flat terrain and the high quality of the work.

Levelling in Japan also conclusively shows large changes connected with earthquakes, less regular in their nature than the changes in Finland. In contrast, [429] describes apparent variations of level in Pakistan amounting to 0.6 metres in 300 km, which cannot possibly be ascribed to tectonic causes. See § 3.06 (*a*) (iii).

6.75. Vertical movements. Other methods

(*a*) *Marine soundings.* Changes of level beneath the sea in coastal areas can be recorded by measuring the depth of the water and assessing the height of the tide from the readings of nearby tide gauges, or by trigonometrical fixing from land marks or artificial satellites. It may be difficult to trust the results to better than one metre in ordinary circumstances. It is necessary to be sure that apparent changes are not due to non-identity of horizontal position, which especially in the older survey may not have been well fixed. Changes of depth due to mud flows must not be confused with tectonic changes.

It will often be difficult to be sure about submarine movements unless an immovable target or transponder is placed on the sea bed.

(*b*) *Tide gauges* provide the most direct way of measuring changes in the height of the land. § 3.27 lists the various possible causes of an apparent change in the height of a tide gauge bench-mark above the annual mean sea-level. Before changes can be ascribed to tectonic causes other sources must be eliminated, and spirit levelling based on the tide gauge must show that the change is not entirely local.

If changes of height, as approximately revealed by changes of g as in (*c*) below, are compared with changes of height above mean sea-level as deduced from tide gauges and spirit levelling, the differences (within the limits of their errors) may be ascribed to real changes of MSL. It will of course still be necessary to distinguish between items (*b*) to (*f*) of § 3.27 before the differences can be ascribed to eustatic changes.

(*c*) *Variation of gravity.* A widespread elevation of the surface by 1 metre will decrease surface gravity by 0.3 mGal (6.95), if the elevation is due to expansion of the underlying crust without any addition of mass. But if the rise is due to the lateral intrusion of the equivalent of 1 metre of 2.67 g cm^{-3} density rock, the decrease will be only 0.2 mGal. See [280], pp. 139–41. Gravimeters are sensitive to perhaps one-tenth of these figures, but moving areas must be compared with stable areas where g is insignificantly different, as at the same height and latitude, to avoid calibration errors. Tidal effects and the effect of changes in ground water level must be eliminated.

At present (1978) secular change in the gravitational constant, or in the mean radius of the earth, must be assumed to have amounted to nothing in the short periods involved. See some examples in [421], pp. 111–113.

Modern gravity absolute observations at fixed sites are thought to be accurate to a few μGal (a few times 0.01 $\mu m\ s^{-2}$), corresponding to a few times 5 mm in height. Very small changes of height could thus be rapidly detected, but the changes of gravity cannot be accurately converted into changes of height unless the earth's internal mechanism is known.

(*d*) *Satellite fixes, VLBI, and lunar lasers.* See § 6.73 (*c*) and (*d*); these methods give both horizontal and vertical changes.

General references for crustal movements. Many reports of crustal movements, determined or suspected, and of the methods used, are to be found in the volumes of papers read at the IAG Crustal movement symposia, such as [39] and [45]. Lists of such occurrences cannot be included here.

SECTION 13. EARTH TIDES

6.76. Introductory

Earth tides are a subject which is on the border between geodesy and geophysics. They produce small periodic variations in gravity and levelling, and in astronomical latitudes and longitudes, which for most geodesists are no more than a trivial source of error, but for the geophysicist they provide information about the rigidity of the earth's solid substance. This section outlines their observable effects, but does not touch on their geophysical interpretation.

The subject is fully dealt with in [409], which describes special instruments used, gives numerical results, and includes a bibliography to 1966.

6.77. The equilibrium tide

The cause of earth tides is basically the same as that of the ocean tides, namely the changing potential of the moon and sun as their positions change. In fact, as shown in § 3.23, at any place and instant the earth's sea-level equipotential surface, the geoid, should be above its mean position by approximately

$$26.7(\cos 2z_1 + \tfrac{1}{3}) + 12.3(\cos 2z_2 + \tfrac{1}{3}) \text{ cm}, \qquad (6.180)$$

where z_1 and z_2 are the zenith distances of the moon and sun respectively. The approximations in (6.180) are that the moon and sun are assumed to be constantly at their mean distances from the earth's centre, in the equatorial plane, and that the solid earth and ocean are not deformed.

This tidal deformation of the geoid has its springs and neaps like those of the ocean tide, and its diurnal and other variations due to changes of declination and other factors. It can be expanded in terms of time as opposed to zenith distance in a series of harmonic terms, like those of the ocean tides, but of smaller amplitude and without the shallow water overtides or other distortions and phase lags caused by the topography of the ocean bed and coast-line. On the assumption that the earth is completely rigid, this geoidal tide is known as the *equilibrium tide.*

6.78. Love's numbers

The periodical change in potential results in various observable phenomena on the earth's surface, whose magnitude will depend on the extent to which the tide deforms the solid earth. As an example, consider the ocean tide as it might hypothetically be measured on a long tide pole set up on the bottom of a deep ocean which covered the whole earth.

In Fig. 6.31 the surface AA is the geoid as it would be in the absence of the moon and sun. BB is the geoid as modified by the equilibrium tide, so AB is given by (6.180). If the earth was wholly liquid, AA would be its surface without the moon and sun, and BB would be its surface with

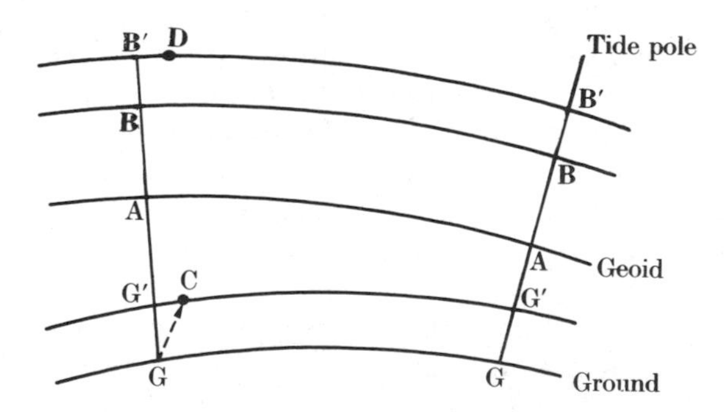

FIG. 6.31. AA is equipotential of the undeformed earth only. BB is equipotential of the moon, sun, and undeformed earth. GG is the undeformed ground level, and G′G′ is the deformed ground level. B′B′ is equipotential of the moon, sun, and deformed earth. G′C is the lateral movement of G. B′D = G′C.

them. But the movement of mass involved would in turn modify the potential, and would cause further deformation. If the earth was uniform and isotropic and of constant rigidity, or if its internal structure was fully known in these respects, the total change of form GG′ and the form of the resulting equipotential B′B′ could be computed. The actual earth is, of course, not uniform, but the tidal effects of simple model earths of defined structure can be computed and compared with observation to confirm or modify their details. As an intermediate aim, the deformation is expressed in terms of three ratios, *Love's numbers*, defined as follows, whose values can be determined.

(i) $h = \mathrm{GG'/AB}$, the ratio of the surface deformation to the height of the equilibrium tide. This is in fact about 0.6, see § 6.84.

(ii) $k = \mathrm{BB'/AB}$, the ratio of the additional tide, due to the earth's deformation, to the height of the equilibrium tide. This is about 0.3.

(iii) $l = \mathrm{G'C/AB}$, the ratio of the horizontal displacement to the height of the equilibrium tide. This is about 0.05.

Then, returning to the measurement of a hypothetical deep-ocean tide, the measured tide would be

$$\begin{aligned} \mathrm{G'B'} - \mathrm{GA} &= \mathrm{BB'} + \mathrm{AB} - \mathrm{GG'} \\ &= \mathrm{AB}(1 + k - h). \end{aligned} \qquad (6.181)$$

So the factor $(1 + k - h)$ is the ratio of the actual (hypothetical) tide to the equilibrium tide (6.180). While it is not practicable to measure this factor in this way, it and other combinations of h, k, and l can be measured as described in §§ 6.79–6.81 and § 6.83.

6.79. Deflexion of the vertical relative to the ground

We consider the variations with time of the angle θ between tangents to B′B′ and G′G′ in Fig. 6.31 at points on the same vertical: in other words, the variations in $\mathrm{d(G'B')/d}x$, where x is measured horizontally. We have

$$\begin{aligned} \mathrm{G'B'} &= \mathrm{B'B} + \mathrm{AB} + \mathrm{AG} - \mathrm{GG'} \\ &= \mathrm{AB}(1 + k - h) + \mathrm{AG}. \end{aligned}$$

Since AG does not change with time, it follows that the time changes of G′B′ are $(1 + k - h)$ times those of AB, and the ratio of the time changes of θ to those of θ_{AB} will be the same, where θ_{AB} is the angle between tangents to BB and AA on the same vertical.

Observed changes of θ are therefore $(1 + k - h)$ times those as calculated for the equilibrium tide. (6.182)

This is the same factor as that of (6.181).

Variations between the vertical or horizontal and the ground can be

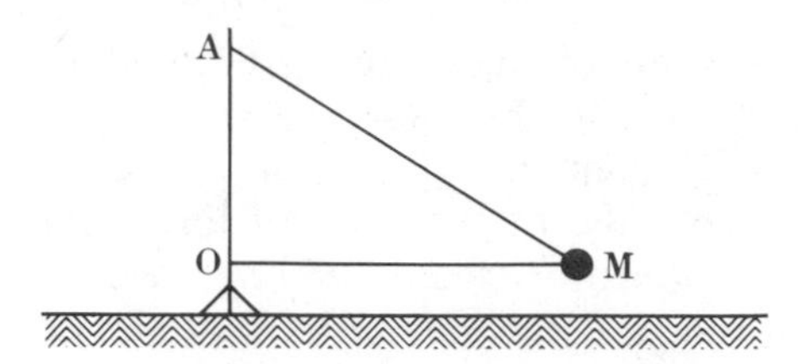

FIG. 6.32. Horizontal pendulum.

recorded by tiltmeters, of which one type is the *horizontal pendulum*, Fig. 6.32. In principle a weight at the end of a horizontal beam OM is supported by a wire MA, with A nearly exactly vertically above O. A small movement of the vertical out of the plane of the paper, not shared by the ground, will cause a much greater movement of M in the same direction. Two such instruments mounted in azimuths differing by 90° will record all variations of the vertical relative to the rock on which the instrument is mounted.

For the deflexions θ_{AB} due to the equilibrium tide see § 3.24, and for their effect on lines of spirit levelling see § 3.06 (*c*) (iii).

An alternative to the use of horizontal pendulums is to measure the variations of the slope of a free water surface relative to the ground. This may be done by means of tide gauges in a large enclosed body of water isolated from the ocean tides, or by measuring the changes in the height of the water at the ends of a long horizontal pipe attached to the ground, with short vertical ends. The pipe or other body of water, must be such that its natural period is short compared with 12 hours.

6.80. Deflexion of the vertical relative to the earth's axis of rotation

In Fig. 6.31 this is the time variation in the angle θ_{AD} between the tangent to AA at A and the tangent to B′B′ at D. As in § 6.79, $\theta_{AB'}/\theta_{AB} = (1+k)$, with no contribution from GG′ and h, because the ground level is not involved. But the lateral component of the ground movement G′C, which changes the direction of the vertical by G′C/(Earth's radius), or by $l(\text{AB})/R$, results in

$$\theta_{AD}/\theta_{AB} = (1+k-l). \tag{6.183}$$

This factor can be determined by harmonic analysis of continuous and highly accurate observations of latitude, or with more difficulty of longitude, at fixed stations such as those of the International Latitude Service.

The magnitude of θ_{AB}, the effect of the equilibrium tide in this paragraph, will be the same as in §§ 6.79 and 3.24.

6.81. The tidal variation of gravity

(*a*) *The equilibrium tide.* Differentiating (3.66) with respect to R shows that for the equilibrium tide

$$\Delta g = -\frac{\partial V}{\partial R} = -\frac{2V_M}{R}(\cos 2z + \tfrac{1}{3}) = -2(\text{AB})g/R, \qquad (6.184)$$

where Δg is the tidal effect on g, and V and z refer to either the moon or the sun. For the moon $\text{AB} = 26.7(\cos 2z + \tfrac{1}{3})$ cm, and for the sun the factor is 12.3 cm, as in (6.180), so

For the moon Δg varies from -0.110 to $+0.055$ mGal,

and for the sun Δg varies from -0.050 to $+0.026$ mGal.

(*b*) *On an elastic earth* gravity is changed

(i) On account of the equilibrium tide AB,
(ii) On account of the earth's deformation which produces the tide BB′, and
(iii) By the ground level being raised by an amount GG′.

The effects of these changes on gravity are as below.

(i) For the equilibrium tide, $-2(\text{AB})g/R$ from (6.184).
(ii) For BB′, which is a geoidal undulation represented by a second degree spherical harmonic, (6.47) gives

$$\Delta g/g = (n+1)(\text{BB}')/R$$
$$\Delta g = 3kg(\text{AB})/R.$$

(iii) For the free-air change through GG′,

$$\Delta g = -2g(\text{GG}')/R = -2hg(\text{AB})/R.$$

So the ratio between the total change (i)+(ii)+(iii) and the equilibrium Δg is

$$(1 + h - \tfrac{3}{2}k). \qquad (6.185)$$

This can be measured by sensitive gravimeters, carefully maintained.

6.82. The secondary effect of ocean tides

At places not too far from the sea, observations are affected by ocean tides in three ways.

(i) The potential and its horizontal gradient, and consequently the direction of the vertical, are affected by the mass movement of the sea-water. If the height of the tide is known, this can be computed by the method of § J.01.

Within a few kilometres of a coast where tides are large, this effect may be many times greater than the equilibrium tidal effect.

(ii) The extra weight of water at high tide depresses the solid earth and tilts the ground surface. This cannot be computed since the crustal elasticity is unknown.

(iii) The potential and the deviation of the vertical are further changed by the mass movement involved by (ii). This effect is likely to be small.

6.83. Other observations

The earth's elasticity affects other observable quantities as follows.

(*a*) *Strain gauges* or extensometers can measure the varying horizontal distance between near points in solid rock, which can in principle determine $(h-3l)$, [409], p. 115, and [566].

(*b*) The 14-month Chandler polar motion period, § 4.09 (*a*) (ii), is a measure of the earth's rigidity, and gives $k=0.28$, [409], p. 300.

(*c*) Periodic variations in the earth's speed of rotation also give $k=0.29$, [409], p. 300, or 0.303 ± 0.10 in [410], 1968, p. 370.

(*d*) Analysis of satellite orbits can include k as an unknown. The value currently considered the most probable from this source is 0.30, [10], p. 5324.

6.84. Numerical results

The following are a selection of numerical results.

(i) $k=0.28$ to 0.303, mean 0.29, from § 6.83 (*b*), (*c*), and (*d*).

(ii) $1+k-h=0.68\pm0.05$, from horizontal pendulums and water levels. [305], 1971, p. 276.
Also $1+k-h=0.68$ to 0.74†. [410], (1976), p. 336.

(iii) $1+h-\frac{3}{2}k=1.16\pm0.09$, from gravimeters. [410], 1968, p. 370.
Also $1+h-\frac{3}{2}k=1.15$ to 1.16.† [410], 1976, p. 337.

(iv) $1+k-l=1.20\pm0.10$, from International latitude stations. [305], 1971, p. 276.

(v) $h-3l=0.45$, from laser extensometer, [409], p. 300.

(vi) $l=0.04$, from strain meter. [305], 1971, p. 276.

From the above figures the following may be deduced.

$$k=0.29,$$

$$h=0.60,$$

$$l=0.05$$

† The figures vary according to the period of the tide analysed, sidereal day, solar day, or lunar day.

These give $1+k-h=0.69$: $1+h-\frac{3}{2}k=1.16$; $1+k-l=1.25$; and $h-3l=0.46$. These figures are all comfortably within the spread of the observations, but they could be wrong by a few times 0.01.†

General references for earth tides. [409] see § 6.76, [305], and [221] for the variation of gravity, and the IAG 4-yearly reports, [410].

† See [165] which concludes with the more conservative figures $k=0.3$, $h=0.7$, and $l=0.1$.

7

ARTIFICIAL SATELLITES

SECTION 1. INTRODUCTORY

7.00 Uses and notation

(*a*) The principal geodetic uses of artificial satellites are as follows.

(i) The fixing of ground stations in a geocentric frame, and thence the possibility of converting previously unconnected triangulation, or other systems of geodetic control, to one common world datum. With the improving accuracy of satellite fixings, it is becoming possible to use satellite results for the internal control of the more extensive survey systems.

(ii) The determination of the constants in a spherical harmonic analysis of the geopotential.

(iii) Other, less basic, results have a bearing on the following.

(A) Polar motion, and changes in the earth's angular velocity about its polar axis.

(B) Crustal movement.

(C) Earth tides.

(D) The departure of the ocean surface from an equipotential surface.

(E) The determination of *GM*, the product of the earth's mass and the gravitational constant.

The geodetic uses of artificial satellites cannot be described comprehensively in a single chapter. This chapter is intended to be an introduction which may be the basis of further study. The aim has been to indicate the lines on which problems can be solved, rather than to give the details of any particular routine. The emphasis is of course on the connection between satellite geodesy and classical methods.

Details of the construction of artificial satellites and of the associated ground equipment are not included.

(*b*) *Notation.* The following notation is used. Other lists as required for particular sections are in §§ 7.06, 7.16, 7.30, 7.46, 7.66, and 7.74.

$i, \Omega, p, e, \omega, t_0$ = Orbital elements. See § 7.01 (*c*).
$\Delta i, \Delta\Omega$, etc. = Changes in i, Ω etc. in the course of one orbit.
A = Semi-major axis of orbit.
t = Time.
n = Mean motion, see (7.3). Also a direction cosine.

M = Mean anomaly, see (7.7). Also mass of earth.
E = Eccentric anomaly, see (7.8).
v = True anomaly, see (7.9). Also velocity in various places.
T = Orbital period, node to node.
$u = v + \omega$.
α, δ = Right ascension and declination.
G = Gravitational constant.
M = Mass of earth, including the atmosphere.
$GM = 398\,600.5 \pm 0.2$ km^3 s^{-2}, [427].†
r = Scalar distance geocentre to satellite.
$\mathbf{r}$ = Vector from geocentre to satellite.
$\dot{\mathbf{r}}$ = Vector satellite velocity
x, y, z or r_1, r_2, r_3 = Components of $\mathbf{r}$.
S, T, W = Perturbing accelerations, see § 7.02.
h = see (7.6).
l, m, n = Direction cosines.
a = Earth's semi-major axis.
$\hat{\theta}$ = Sidereal angle, see § 4.10 (*f*).
ϕ, λ = Latitude and longitude.
θ = Angle between radius vector and polar axis. Also any other angle.
r, θ, λ = Polar coordinates.
Geocentre = Earth's centre of mass.

Throughout this chapter it is necessary to express the potential of the earth's attraction (excluding centrifugal force) at an external point (r, θ, λ) in the form of a series of spherical harmonics, thus

$$V = \frac{GM}{r}\left[1 - \sum_{l=2}^{\infty}\left(\frac{a}{r}\right)^l\left\{J_l P_l(\cos\theta) - \sum_{m=1}^{m=l}(\bar{C}_{lm}\cos m\lambda + \bar{S}_{lm}\sin m\lambda)\bar{P}_{lm}(\cos\theta)\right\}\right], \quad (7.1A)$$

or

$$V = \frac{GM}{r}\left[1 + \sum_{l=2}^{\infty}\left(\frac{a}{r}\right)^l \sum_{m=0}^{m=l}(\bar{C}_{lm}\cos m\lambda + \bar{S}_{lm}\sin m\lambda)\bar{P}_{lm}(\cos\theta)\right], \quad (7.1B)$$

where a is the earth's semi-major axis.
l, m are the degree and order of any particular harmonic.‡
J_l is the coefficient of $(a/r)^l P_l$ in a zonal harmonic.‖
$\bar{C}_{lm}, \bar{S}_{lm}$ are the coefficients of $(a/r)^l \cos m\lambda$ and $(a/r)^l \sin m\lambda$ in the normalized tesseral harmonic $\bar{P}_{lm}$. $\bar{C}_{l0} = -J_l\,(2l+1)^{-\frac{1}{2}}$. See § H.6.

† Later results are $398\,600.87 \pm 0.06$ km^3 s^{-2} from Mariner 9 [398] and $398\,600.46 \pm 0.02$ km^3 s^2 [383A], also from satellite laser ranging.

‡ In Chapter 6 and Appendix H, the letter n is used to signify the degree of any harmonic, as is usual, but in this chapter l is used in place of n, in order to avoid confusion with the mean motion of a satellite, which is always designated as n.

‖ In early literature $(3/2)J_2$ and $-(35/8)J_4$ were designated as J and D respectively.

(*c*) *Vectors.* Vectors are in bold type. Lower case does not here necessarily indicate unit vectors. Although vectors such as **r, S**, etc. are vectors in every sense of the word, in this chapter they are generally manipulated as matrix vectors, § B.03, i.e.

$$\mathbf{r}=\begin{bmatrix} r_1 \\ r_2 \\ r_3 \end{bmatrix},$$

and they are combined with other matrices according to the rules of matrix algebra given in Appendix B. In this chapter, they are seldom combined by the dot and cross products of Appendix E, although there are exceptions, such as in (7.14).

(*d*) The following abbreviations are used throughout this chapter.

Organizations

APL = Applied Physics Laboratory, Johns Hopkins University.
BIH = Bureau International de l'Heure.
CNES = Centre National d'études Spatiales.
DMAAC = Defense Mapping Agency Aerospace Center.
DOD = Department of Defense (US).
GSFC = Goddard Space Flight Center.
JPL = Jet Propulsion Laboratory.
NASA = National Aeronautics and Space Administration.
NGS = National Geodetic Survey.
NOS = National Ocean Survey.
} Previously US C & GS.
NSWL = Naval Surface Weapons Laboratory. (Previously NWL).
OSU = Ohio State University.
RAE = Royal Aircraft Establishment.
SAO = Smithsonian Astrophysical Observatory.

Observations and computations

AGD = Australian geodetic datum.
CIO = Conventional international origin. (Polar motion).
ED = European datum. Of 1950 unless otherwise stated.
GEM = Goddard earth model.
GRAAR = Goddard range and range-rate.
ISAGEX = International satellite geodesy experiment.
NAD = North American datum. Of 1927 unless otherwise stated.
NNSS = US Navy navigation satellite system.
SECOR = Sequential collation of range.
STADAN = Satellite tracking and data acquisition network.
SPEOPT = Special optical network.
VLBI = Very long base interferometry.
WGS = World geodetic system.
WN = World net (OSU).

7.01. Satellite orbits. Spherical earth

(*a*) A satellite orbiting a spherical earth moves with an acceleration of GM/r^2 towards the geocentre, subject to certain assumptions as below.

(i) The satellite must be small, both in size and weight. Artificial satellites very easily satisfy this condition.

(ii) The earth must not only be spherical, but its density must either be uniform or arranged in concentric uniform shells. The actual earth satisfies these conditions as a first approximation, but not at all exactly. See § 7.02.

(iii) There must be no disturbing forces, such as the attraction of the sun, moon, and planets, air drag, or radiation pressure. Actually, such forces do occur. They are small, but not negligible.

(*b*) On these assumptions a satellite will move in a fixed plane, in an unvarying ellipse with the geocentre at one focus. Its rate of movement round the ellipse is not constant, but is such that the radius vector from the geocentre sweeps out equal areas in equal times. The squares of the orbital periods (T) of different satellites are proportional to the cubes of their semi-major axes (A). In fact

$$T = 2\pi A^{\frac{3}{2}}(GM)^{-\frac{1}{2}}, \qquad (7.2)†$$

The value of GM is well known, see § 7.00.

These two sub-paragraphs (*a*) and (*b*) are elementary. Proofs are given in any textbook covering the dynamics of a particle such as [549] and [127].

(*c*) *Orbital elements.* The orbit and the position of the satellite at any instant can be described by six *orbital elements* i, Ω, and ω as shown in Fig. 7.1, with A, e, and t_0. They are defined as follows.

i is the *inclination* of the plane of the orbit to that of the equator.

Ω is the RA of the *ascending node*.‡ A node is the point of intersection on the celestial sphere of the orbit and the equator, the ascending node being when the satellite is passing from south to north. A progressive decrease of Ω is known as a regression of the node.

p is the semi-latus rectum of the orbit, and $p = A(1-e^2)$, where A is the semi-major axis.

e is the eccentricity of the orbit. Typically 0.03 to 0.2.

ω is the angle between the ascending node and the radius vector to the satellite at perigee, when its distance is minimum. It is known as the argument of perigee. The points of perigee and apogee are known as apses, so a progressive change in ω is a *rotation of the apse.*

† $T \approx 0.01A^{\frac{3}{2}}$ s, if A is recorded in km.

‡ Node, apse, argument of perigee, true anomaly, etc., are old-fashioned words whose relevance may not be obvious, but which are still in use.

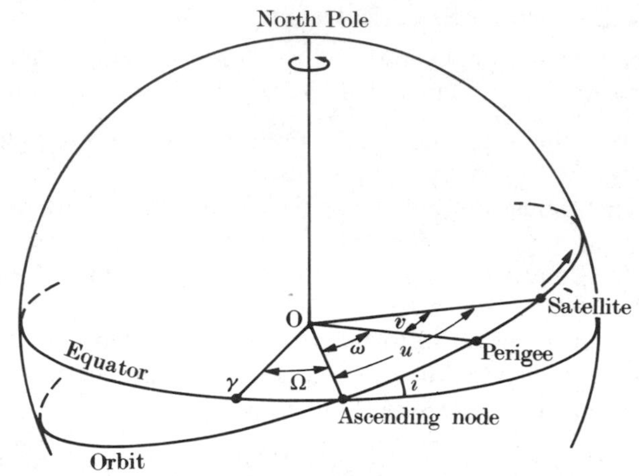

FIG. 7.1. The celestial sphere. Ω is measured in the plane of the equator, and ω and u in the plane of the orbit. γ is the equinox.

t_0 is the time when v, see Fig. 7.1, is zero and the satellite is at perigee. It may sometimes be convenient for t_0 to be at the time at the ascending node.

As an alternative to defining A it may be convenient to define, p, since $p = A(1-e^2)$. Another alternative is to define the *mean motion* n, namely

$$n = 2\pi/T = A^{-\frac{3}{2}}(GM)^{\frac{1}{2}}, \quad \text{from (7.2).} \tag{7.3}$$

Of these six elements i and Ω define the plane in which the orbit lies. The orbital ellipse is then described by p (or A or n), e, and ω. The position of the satellite in the orbit is given by the angle u in Fig. 7.1, where

$$u = \omega + v,$$

in which the angle v, known as the *true anomaly*, can be computed from n, e, and t_0 using (7.7)–(7.10).

(*d*) *Formulae for the ellipse.* The following useful formulae are given in textbooks.

(i) The equation of the ellipse, with one focus as origin, is

$$r = p(1+e\cos v)^{-1} = A(1-e\cos E)$$

in which $\quad p = A(1-e^2). \quad r_{\text{max}} = A(1+e). \quad r_{\text{min}} = A(1-e). \qquad (7.4)$

(ii) The velocity of the satellite in its orbit is given by

$$(\text{velocity})^2 = GM\left(\frac{2}{r} - \frac{1}{A}\right). \tag{7.5}$$

(iii) The area described by the radius vector (from the focus) in unit time is $\frac{1}{2}h$, where

$$h^2 = (GM) \times A(1-e^2) = (GM)p. \tag{7.6}$$

(*e*) *Reference system.* If constant orbital elements are to result from the absence of all forces other than the attraction GM/r^2 towards the centre, the reference frame must be a *Newtonial* or *inertial frame* as defined in I.00. The reference system used by the SAO, defined in § 7.32 (*a*) (i) adequately meets this requirement, as also does the slightly different system used by NSWL for the doppler system defined in § 7.55 (*b*). See 7.32 (*b*) for the conversion of coordinates between various reference systems.

The position of the satellite in space can be defined by i, Ω, u, and r, or by $\mathbf{r}$ with components $r_1 r_2 r_3$, or by α (RA), δ, and r. There is thus some choice about which six quantities should be used to define an orbit and the position of a satellite in it. A set which is often used is A, e, Ω, ω, i, and M, where M as in (7.7) takes the place of t_0 or u, by means of (7.7) to (7.12).

(*f*) *To obtain the geocentric α and δ at time t, from the orbital elements.* The mean anomaly† M is defined by

$$dM/dt = n, \text{ the mean motion,‡}$$

whence
$$M = n(t - t_0), \tag{7.7}$$

where t_0 is the time (at perigee) at which M is taken to be zero.

The eccentric anomaly E, defined geometrically as in Fig. 7.2 is given by

$$E - e \sin E = M. \tag{7.8}$$

Given e and M, E can be obtained by iteration without difficulty, since e is small or very small for most geodetically useful satellites.

Then the true anomaly v is given by

$$\tan(v/2) = \left(\frac{1+e}{1-e}\right)^{\frac{1}{2}} \tan(E/2), \tag{7.9}$$

whence u is given by $u = \omega + v$.

Alternatively

$$\left.\begin{aligned} \sin v &= (1-e^2)^{\frac{1}{2}}(1-e\cos E)^{-1}\sin E \\ \cos v &= (\cos E - e)(1-e\cos E)^{-1}. \end{aligned}\right\} \tag{7.10}$$

Finally,

$$\alpha = \Omega + \tan^{-1}(\cos i \tan u), \tag{7.11}$$

† M is an angle which, in simple elliptical motion, changes uniformly with time. E and v, on the other hand, do not.

‡ Either n is one of the orbital elements itself, or it is obtainable from A or p by (7.3).

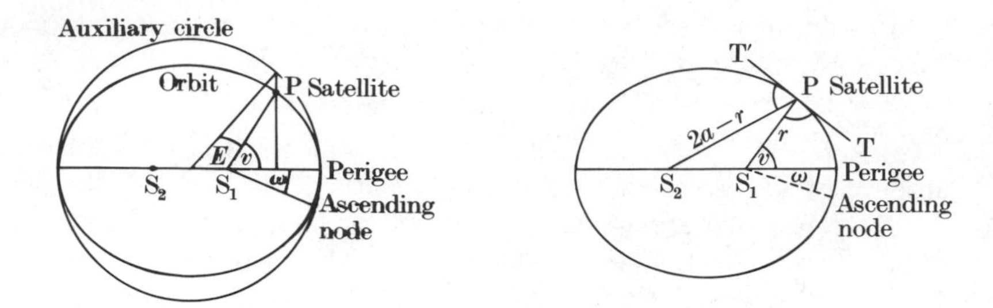

FIG. 7.2. In the plane of the orbit. S_1 and S_2 are foci. FIG. 7.3. In the plane of the orbit.

$$\sin \delta = \sin i \sin u, \tag{7.12}$$

$$r = A(1-e\cos E), \text{ as in (7.4).}$$

See [127], [391], and [461A].

(*g*) *To obtain rectangular coordinates* r_1, r_2, r_3.
Given the orbital elements, Fig. 7.1 shows that in the coordinate system defined by γ and the north pole, the three components of $\mathbf{r}$ are given by

$$\begin{aligned} r_1 &= r\{\cos \Omega \cos(\omega+v) - \sin \Omega \sin(\omega+v)\cos i\}, \\ r_2 &= r\{\sin \Omega \cos(\omega+v) + \cos \Omega \sin(\omega+v)\cos i\}, \\ r_3 &= r \sin(\omega+v)\sin i \end{aligned} \tag{7.13}$$

where the scalar $r = A(1-e^2)(1+e\cos v)^{-1}$, from (7.4).

(*h*) Conversely, it may be required to obtain the orbital elements from the rectangular coordinates and the velocity vector at any time. Formula (7.5) at once gives A. Let the components of $\mathbf{r}$ be $r_1 r_2 r_3$ at time t, and let $\dot{r}_1 \dot{r}_2 \dot{r}_3$ be the components of $\dot{\mathbf{r}}$, the velocity vector. Then the plane in which the orbit lies contains the geocentre and contains, or is parallel to, the vectors $\mathbf{r}$ and $\dot{\mathbf{r}}$. This determines† i and Ω.

† Let **i j k** represent the axes (**k** polar). Then the unit vector perpendicular to the plane of the orbit will be

$$\mathbf{r}\times\dot{\mathbf{r}} = \frac{1}{|r|\,|\dot{r}|}\{(r_2\dot{r}_3 - r_3\dot{r}_2)\mathbf{i} + (r_3\dot{r}_1 - r_1\dot{r}_3)\mathbf{j} + (r_1\dot{r}_2 - r_2\dot{r}_1)\mathbf{k}\} \tag{7.14}$$

as in (E.10), and its three components equal l, m, and n, the direction cosines of the perpendicular to the orbit. Then the inclination, i, which equals the angle between the perpendicular and the polar axis is given by

$$\cos i = n = (1/|r\,\dot{r}|)(r_1\dot{r}_2 - r_2\dot{r}_1), \tag{7.15}$$

and the direction of the ascending node is given by

$$\tan \Omega = -l/m = -(r_2\dot{r}_3 - r_3\dot{r}_2) \div (r_3\dot{r}_1 - r_1\dot{r}_3). \tag{7.16}$$

Then in Fig. 7.3 if S_1 is the geocentre and S_2 is the other focus, the direction PS_2 is given by $T'PS_2 = TPS_1$, and the distance $PS_2 = 2A - r$. This fixes S_2, and gives the length and direction of S_1S_2. Its length gives e, since $S_1S_2 = 2eA$, and its direction gives ω and v.

Given A and e, (7.3) and (7.4) give p, n, and dM/dt, and $(t - t_0)$ is then given by (7.7).

Since the orbital elements i, Ω, etc. can be determined from $\mathbf{r}$ and $\dot{\mathbf{r}}$, and vice versa, the latter may also be referred to as orbital elements. When necessary, the former set may be distinguished as *Keplerian elements.*

In so far as the Keplerian elements are constant, they and t_0 can in principle be determined by three suitable timed observations of the geocentric† RA and δ, although so few observations would be likely to lead to weak determinations, and a much longer set of observations has to be used, covering a period of time during which the variations of the elements must be allowed for, as in Sections 3, 5, and 7.

7.02. Satellite orbits. Actual earth

The actual earth is neither spherical nor even exactly spheroidal. The external potential V of its attraction is as described in §§ 6.14 to 6.16, whence V is expressible as a series of spherical harmonics in the form of (7.1A) or (7.1B), and the resulting acceleration of a satellite is grad V. There are also other significant perturbations as listed in § 7.01 (a) (iii). The largest harmonic term is the one whose coefficient is J_2, which arises from the earth's spheroidal shape. Its magnitude is of the order 10^{-3} GM/r. Other perturbations of the potential are of the order 10^{-6} GM/r or less. Terms as small as 10^{-8} GM/r may be significant. See also (7.222)

In consequence of these accelerations, the elements are constantly varying. At any moment the elements of the orbit which the satellite would thereafter follow, if all accelerations except GM/r^2 suddenly ceased, are known as the *osculating elements.*

Let a perturbing acceleration have components S along the radius vector (positive outwards), T in the orbital plane‡ and perpendicular to the radius vector, roughly tangential to the orbit (positive in the direction of motion), and W perpendicular to the orbital plane (forming a right-

† Observations of RA and δ made at the earth's surface can readily be converted to geocentric, provided the geocentric station coordinates of the observing station, and the distance of the satellite are known. The latter, and probably the station coordinates as well, are, of course, only approximately known. The procedures whereby these unknowns and the Keplerian elements can be simultaneously determined is outlined in Section 5.

‡ This T must not be confused with the orbital period.

handed system with S and T). Then the rates of change of the elements are as below.

$$\frac{dA}{dt}=\frac{2}{n(1-e^2)^{\frac{1}{2}}}\left\{\frac{AT}{r}(1-e^2)+Se\sin v\right\},$$

$$\frac{de}{dt}=\frac{(1-e^2)^{\frac{1}{2}}}{nA}\{T(\cos E+\cos v)+S\sin v\},$$

$$\frac{d\Omega}{dt}=\frac{r}{nA^2(1-e^2)^{\frac{1}{2}}}\times\frac{W\sin(\omega+v)}{\sin i},$$

$$\frac{d\omega}{dt}=-\frac{d\Omega}{dt}\cos i+\frac{(1-e^2)^{\frac{1}{2}}}{nAe}\left\{T\left(\frac{r}{p}+1\right)\sin v-S\cos v\right\},$$

$$\frac{di}{dt}=\frac{r}{nA^2(1-e^2)^{\frac{1}{2}}}\times W\cos(\omega+v),$$

$$\frac{dM}{dt}=-\frac{(1-e^2)}{nAe}\left(\frac{r}{p}+1\right)T\sin v-\frac{S}{nA}\left(\frac{2r}{A}-\frac{(1-e^2)}{e}\cos v\right) \quad (7.17)$$

in which $n=(GM/A^3)^{\frac{1}{2}}$, and $p=A(1-e^2)$.

These formulae are as given in [517], p. 43. They are also given in a variety of forms in [262], p. 345, [127], p. 301, and [273], p. 837.

In (7.17) dv/dt is also required. We have $dv/dt=(dv/dM)(dM/dt)$, and $(dv/dM)=(dv/dE)(dE/dM)$. Of these dE/dM is obtained from (7.8), and (dv/dE) from (7.10). With some manipulation, the result is

$$\frac{dv}{dt}=\frac{dM}{dt}\frac{(1-e^2)^{\frac{1}{2}}}{(1-e\cos E)^2} \quad (7.18)$$

The changes in the three components of $\mathbf{r}$ (r_1, r_2, r_3) resulting from changes in the elements A, e, Ω, ω, i, and v are obtainable by differentiating (7.13). Together with $\partial r/\partial M$, they are given in full in traditional form† in [391], i, pp. 105–6. [391], p. 107 also gives the changes in a more convenient vector form.

The total rate of change of r_1 resulting from the perturbing accelerations S, T, and W is then given by

$$\frac{dr_1}{dt}=\frac{\partial r_1}{\partial A}\frac{\partial A}{\partial t}+\frac{\partial r_1}{\partial e}\frac{\partial e}{\partial t}+\frac{\partial r_1}{\partial\Omega}\frac{\partial\Omega}{\partial t}+\frac{\partial r_1}{\partial\omega}\frac{\partial\omega}{\partial t}+\frac{\partial r_i}{\partial i}\frac{\partial i}{\partial t}+\frac{\partial r_1}{\partial v}\frac{\partial v}{\partial t}, \quad (7.20)$$

and similarly for r_2 and r_3.

The difficulty then is that the accelerations S, T, and W depend on the position of the satellite (r_1, r_2, r_3), and also independently on the time,

† e.g. $\frac{\partial r_1}{\partial\omega}=\frac{\partial r_1}{\partial v}=r\{-\cos\Omega\sin(\omega+v)-\sin\Omega\cos(\omega+v)\cos i\}.$ (7.19)

since irregularities in the earth's density and external potential rotate with it, while the satellite's orbit is computed in a non-rotating frame. There are also the (negative) accelerations due to air drag which depend on the satellite's velocity. There are two general methods of approaching this problem, which are outlined in §§ 7.34–7.38 and §§ 7.56–7.57. The second method is algebraically the more simple, but it demands a larger computer.

7.03. Summary of the remaining sections of Chapter 7

(*a*) Section 2. Formulae for corrections due to atmospheric refraction in satellite observations.

(*b*) Section 3. Determination of the coefficients J_l in (7.1), representing the spheroidal flattening and other axially symmetrical departures from a sphere of uniform density. They give rise to secular and long-period changes in the orbit, which are relatively easily measured. This section also includes § 7.15, which deals with resonant harmonics.

(*c*) Section 4. *Astro-triangulation*, a purely geometrical method, in which a satellite is photographed against a stellar background simultaneously from two stations, to determine the directions of straight chords joining camera stations separated by as much as 4000 km. The dynamics of the satellite orbit are not involved. A net of about 40 such stations surrounds the earth, but by itself this method gives their coordinates in a system which is not geocentric.

(*d*) Section 5. If the photographic observations are continued over a period of time at each station (but not necessarily simultaneously), the satellite's orbit will be determined relative to the camera stations, and vice versa, and since the geocentre must be at one focus of the orbit, the positions of the stations are obtained in a geocentric reference system. This is the *dynamic use* of satellite photography.

(*e*) Section 6. Laser or radar observations between ground stations and a satellite may provide the scale for astro-triangulation, or may constitute a geometrical *astro-trilateration*. By itself the latter will not be related to the geocentre.

(*f*) Section 7. If the orbit of a satellite is known, measurements of the *doppler shift* will determine the position of a ground station. If a number of such stations track the satellite regularly, the elements of the orbit will also be determinate. The reference system will be geocentric, but the longitude will not be directly determined.

(*g*) Section 8. *Satellite altimetry* can give a continuous record of the height of a satellite above the sea or other suitable surface. If the orbit of the satellite is known in geocentric terms, such altimetry will record the height of sea-level above a defined reference spheroid.

(*h*) Section 9. *Very long base interferometry* (VLBI) gives the lengths and directions of long lines with great accuracy. In principle an artificial satellite can be used, but astronomical sources (quasers) can also be used, and are preferable.

(*i*) Section 10 gives an outline of some of the principal observing programmes and computations, and of their geodetic results. It is a synthesis of the preceding sections.

7.04. Geodetically useful satellites

Geodetically useful satellites have commonly been at heights of between 500 and 4000 km, although a few are much higher. They are generally of low eccentricity. Many have high (near polar) inclinations. Satellites with low inclinations are relatively uncommon, but some are geodetically necessary.

The period of a satellite's movement round its orbit increases from about 90 minutes for a low satellite to about 180 minutes for a height of 4000 km, (7.2). The result of the earth's own rotation is that every point on the earth passes through the plane of the orbit twice a day. Then, provided the latitude of the point is not greater than the inclination of the satellite's orbit, the satellite will pass over the latitude of the point within at most 90 minutes of the time when the point was in its plane. So, provided the satellite is not very low, it will be above the point's horizon at least twice a day.

For a general discription of satellite movements see [337], pp. 1–33.

Lists of all satellites that have been launched are given in [56], published monthly, giving i, T, A, e, ω, and heights of apogee and perigee, at date of launch. Also size, shape, weight, and probable lifetime.

The following are some satellites which have been, or still are, of special geodetic importance. Figures given for heights are approximate. Dates are the date of launch.

(*a*) Sputnik 2 (Nov. 1957). Gave 298.3 ± 0.1 for the earth's flattening.

(*b*) Echo 1 (1960) and 2 (1964). Passive balloons 60 and 40 m in diameter. Used for astro-triangulation, at heights of 900–2100 and 1000–1300 km respectively. Now decayed.

(*c*) Explorer 9 (1961), and 19 (1963). 3.6 m passive balloons, at 600–2500 km. Now decayed.

(*d*) Anna-1 B (1962). Secor transponder, flashing lights, minitrack beacon, doppler transmitter. 1100 km.

(*e*) Pageos (1962). 30-m passive balloon. 4600 km (variable). Inclination 87°.

(*f*) Numerous Transit satellites since 1962, with transmitters for doppler tracking and ephemeris broadcasting. See § 7.48. Most of these satellites are at heights of about 1100 km, with inclinations within 5′ to 10′ of 90°.

(*g*) Secor satellites (1964–7), with radar transponders, in near circular orbits between 900 and 3700 km. Inclinations of 70° to 90°. See §.7.42.

(*h*) Geos 1 (Nov. 1965) Flashing lights, laser reflectors, radar for range and range rate, doppler, and Secor transponder. Height 1100–2270 km. Inclination 59°.

(*i*) Geos 2 (Jan 1968) Flashing lights, laser reflectors, radar for range and range rate, minitrack beacon, doppler, and Secor transponder. 1000–1570 km. Inclination 106°.

(*j*) Geos 3 (Apr. 1975). Doppler transmitter, C and S band radar, laser reflectors, altimeter (See § 7.64), satellite-to-satellite tracking (see below). 840–850 km. Inclination 115°.

(*k*) Starlett (Feb. 1975). 60 laser reflectors. High weight-to-surface ratio, 47 kg and 0.26-m diameter. Height 800–1100 km. Inclination 50°.

(*l*) Triad 2 (TIP) (Oct. 1975). Doppler transmitter. Discos, see below. Height 360–700 km. Inclination 90°.7.

(*m*) Lageos (May 1976). 426 laser reflectors. Weight 411 kg, 0.6-m diameter. Height 6000 km. Inclination 110°.

(*n*) ATS 6 (May 1974). Laser reflectors and S-band radar. 9-m diameter radar antenna. Height 35 800 m. Inclination 1°.6.

(*o*) Navstar satellites. See § 7.44.

(*p*) Seasat 1 (June 1978. Reported failed in Oct 78). Altimeter and associated meteorological sensor. S-band radar, doppler, and laser retroflectors for tracking. Height 800 km. Inclination 108°.

Examples of recent improvements in instrumentation are as in (*i*)–(*iii*) below.

(*i*) *Discos.* A solid 22-mm diameter sphere is enclosed in a larger hollow sphere of 40-mm internal diameter, which is firmly attached to the main structure of the satellite. The smaller sphere experiences no atmospheric drag or solar radiation, and will therefore follow an orbit which depends only on the various gravitational forces, provided the outer sphere keeps out of its way. This is ensured by very small jet engines which are activated by sensors detecting changes in the separation between the inner sphere and its container, [50].

Discos was included in Triad 1 and worked well, although for other reasons that satellite had a short life. The main effect which is to be eliminated is the along-track error, cross-track errors due to drag and solar radiation having relatively small effects. In Triad 2 the discos device only controls the along-track error.

(*ii*) *Satellite-to-satellite tracking.* The orbit of a low altitude satellite is much affected by air drag, and by the higher-degree harmonics in the geopotential, and is consequently difficult to predict accurately. A high orbit can be computed much more accurately, but observation of it provides no information about either the atmosphere or the high-degree geopotential harmonics† whose effect is much reduced at heights of 1000 or 2000 km, but whose determination is one of the main problems in satellite work. A remedy is that a high satellite such as ATS 6 at 35 800 km is accurately tracked from the ground, while it in turn tracks a low satellite such as Geos 3 at 850 km.

(*iii*) *Gravity gradiometers.* See [208], which summarizes various proposals for installing instruments in satellites to determine the gradients of the gravitational force along three orthogonal axes.

7.05. Accuracy aimed at

The accuracy currently aimed at (1978) for the positions of satellites in orbit, and for ground stations, may be said to be about 1 metre with respect to the geocentre. This has not yet been achieved with certainty, and errors of 5 m in the best work can still be expected.

For the relative fixing of ground stations a few hundred km apart, the one-metre standard is being met by the doppler system, and decimeter accuracy is the aim. Laser and VLBI may soon give equal or better results in favourable situations. It is also hoped that satellite altimetry may soon give decimetre accuracy in height in situations where the satellite can be well tracked by laser from nearby stable ground stations.

For the accuracy of current geopotential models see § 7.80. [13] gives a very full analysis of errors in the doppler system.

General references for sections 1 to 7. [18], [49], [56], [127], [216], [273], [327], [329], [337], [391], [519], and [572]. See also under separate sections.

Section 2. Atmospheric Refraction

7.06. Summary and notation

The refractive index of light and microwaves, with its dependence on pressure, temperature, humidity, and wavelength, and its effect on the velocity of propagation, is discussed in §§ 1.32 and 1.33. The curvature arising from the gradient of the refractive index is in §§ 1.35 and 3.17, and (for astronomical refraction) in § 4.11.

This section describes the effect of refraction on satellite observations, i.e. from the ground to objects outside the effective atmosphere.

Notation

p = total pressure } Suffix s = surface
T = temperature K } Suffix o = conventional sea-level.
e = partial pressure of water vapour. Suffix s = surface value.
ζ = apparent zenith distance.
z = true zenith distance.

† The amplitude of an l-degree harmonic term at sea level is reduced by a factor of $(r/a)^l$, where r is the distance from the geocentre and a is the earth's radius. See (7.1).

ζ^* = a modified zenith distance. See after (7.49).
ψ = refraction angle. True zenith distance − apparent.
$E = 90° - \zeta$.
n = refractive index of air. n_P, n_G = phase or group index, when confusion is possible.
$n_0 = n$ in standard conditions.
$N = (n-1)\times 10^6$
$N_d = N$ when $e = 0$. For dry air.
$N_w = N - N_d$. For humidity only.
r = slant distance to satellite.
h = height above sea level. Suffix s = ground surface.
H = height above the ground. $H_d = T_S/\alpha$, see below (7.32).
ρ = atmospheric density.
C_d = constant in $PV = CT$, for dry air.
S = ground-to-satellite distance along path.
τ = travel time (one way) of light or microwave.
ΔS = refraction correction. $S = c\tau - \Delta S$. Suffixes d and w for air and water vapour.
v_G, v_P = group velocity and phase velocity.
c = velocity of light *in vacuo* = 299 792 458 ± 1.2 m s^{-1}.
g_m = mean value of gravity in the atmosphere. See (7.25).
σ = radius of curvature of the path of light or microwaves.
θ = see (7.41) and (7.43), with different meanings.
B = see (7.29) and (7.43), with different meanings.
K_1 to K_4, N_e, e, m, ϵ_0, f, see (7.43) and (7.45).
h_S, H_d, H_{d0}, N_{ds}, α, μ, see (7.32). N_{ws}, H_w, see (7.35) and (7.36). a, r' z, see (7.42).

7.07. Parallactic refraction

When an object at a great height, such as an artificial satellite, is photographed against a background of stars, the refraction angle of the satellite is nearly but not exactly the same as that of a star in the same direction. In fact the tangent TQ to the ray at the satellite T (Fig. 7.4) passes above the point of observation P (if P is at sea-level) at a height of approximately

$$PQ = 2.3 \sec^2\zeta \text{ metres.} \qquad (7.21)\dagger$$

The refraction of the line to the satellite is then less than the normal celestial refraction, § 4.11, by $\Delta\zeta = (PQ/QT)\sin\zeta$ rad, plus any curvature of the path which may occur beyond T. For a satellite at an altitude of 500 km and with $\zeta = 45°$, $\Delta\zeta \approx 1''$. The curvature of the path beyond the satellite may be taken to be zero, if the altitude of the satellite exceeds 100 km.‡

More fully [567] gives

$$\Delta\zeta = 483''.8\,\frac{\tan\zeta}{r\cos\zeta}\left(1 - 0.00126\,\frac{2+\sin^2\zeta}{\cos^2\zeta}\right)\frac{P}{P_0} \qquad (7.22)$$

† The figure 2.3 m was originally estimated as 2.1 m in [568].

‡ Refraction varies as air density, which at 100 km is about 1/500 000 of its sea-level value.

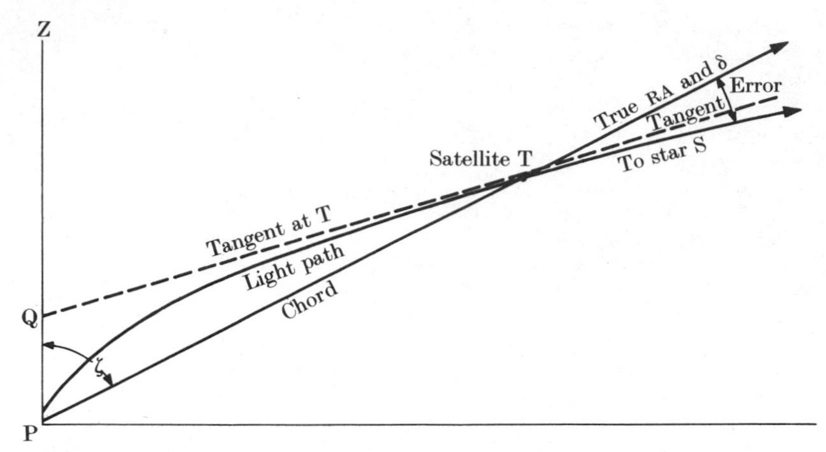

7.4. Parallactic refraction. QT is tangent to the light path STP.

where r is the slant range to the satellite in km, P_0 is the conventional standard pressure, 1013.25 mbar or 101.325 kPa, and P is the actual (total) pressure in the same units as P_0. The factor in brackets exceeds 0.94 if ζ is less than 75°, after which it decreases rather rapidly.

7.08. Refraction correction to measured distances. Light.

(a) Satellite distances measured by light (laser beams), or by radio or microwaves, require correction for the velocity of transmission. As in (1.8) $v = c/n$, where $c = 299\,792.458$ km/s, and n is the refractive index. The group index for light is given by (1.15) and (1.10).

With satellites most of the path lies outside the atmosphere, where n (for light) = 1. There is no question of the average n being given by the mean of its end values. Instead we have

$$S = c\tau - \Delta S,$$

where

$$\Delta S = \int_0^S (n-1)\,\mathrm{d}s$$

$$\approx \int_0^\infty (n-1)\,\mathrm{d}s, \qquad (7.23)$$

for any geodetically useful satellite. Here S is the distance measured along the path, τ is the travel time (for the one-way distance), and ds is an element of distance along the path. The group index must be used for n in (b) and (c) below.

(*b*) *Lines with small zenith distances.* Provided ζ is not too large, we have $\mathrm{d}s = \mathrm{d}h \sec \zeta$. Then (1.10) gives

$$(n-1) = 0.002696(n_0-1)(P-0.14e)/T$$
$$\approx 0.002696(n_0-1)(P/T), \; P \text{ and } e \text{ being in } Pa,$$

where the neglect† of $0.14e$ is unlikely to amount to 0.1 per cent.

Then $$(n-1) = 0.002696\,(n_0-1)C_d\rho \qquad (7.24)$$

where ρ is density‡, and $C_d = 2.8704\times10^2$ in SI units.

So $$\Delta S = \int_0^\infty (n-1)\,\mathrm{d}H = 0.002696(n_0-1)C_d\int_0^\infty \rho\,\mathrm{d}H$$
$$= 0.002696(n_0-1)(2.8704\times10^2)P_S/g_m \text{ metres.} \qquad (7.25)$$

where H is height above the ground, P_S is surface P in Pa, and g_m is a weighted mean value of gravity in m s^{-2}. This mean approximates to the value of g at 6 km above sea-level, and is conventionally taken to be 9.7877 m s^{-2} in latitude 45°, which is correct to 0.3 per cent in all latitudes. From (1.15), (n_0-1) is 298.0×10^{-6}, for the group velocity when $\Lambda = 0.6943\ \mu$m, as for ruby laser light.

Then $$\int_0^\infty (n-1)\,\mathrm{d}H \sec\zeta = 2.357P_S\times10^{-5}\sec\zeta \text{ m, with } P_S \text{ in Pa}$$
$$= 2.357P_S\times10^{-3}\sec\zeta \text{ m, with } P_S \text{ in mbar}$$
$$= 2.387(P_S/P_0)\sec\zeta \text{ m,} \qquad (7.26)$$

with P_S and P_0 both in any one unit. $P_0 = 101.325$ kPa or 1013.25 mbar.

Finally $$S = c\tau - 2.387(P_S/P_0)\sec\zeta \text{ metres,} \qquad (7.27)$$

for $\Lambda = 0.694\ \mu$m, provided ζ is less than 45°, see (*c*) below. In these circumstances it should be correct to about 1 cm.

(*c*) *Lines with large zenith distances.* For dry air [518] gives

$$S = c\tau - 2.387\sec z(1-0.00125\tan^2 z)(P_S/P_0) \text{ metres,} \qquad (7.28)$$

with P_S and P_0 both in the same units, as in (7.26).

† Not only is $0.14\,e$ small compared with P, but appreciable amounts of water vapour are confined to a relatively small range of height.

‡ $P/T = C_d\rho$, the perfect-gas law. The density of air is 1.276 kg/m^3 at 273.15 K and 1000 mbar or 10^5 Pa pressure.

Including an allowance for surface humidity, e_s, [511], p. 32 gives

$$S = c\tau - 2.387 \sec\zeta(P_S + 0.06e_S - B\tan^2\zeta)(1/P_0) - \delta \text{ metres,} \tag{7.29}$$

where P_S, P_0, and e_S are all in the same units, as in (7.26). B is tabulated in [511], which gives $B/\text{mbar} = (B/100\text{ Pa}) = 1.156$, 0.874, 0.654 for ground heights (h_S) of 0, 2, and 4 km respectively. δ is also tabulated, and is 0.003, 0.03, and 0.10 metres for $\zeta = 60°$, 75°, and 80° respectively for sea-level stations, and less for stations at heights of about 1 km or more.

(*d*) The *difference in length* between the path and the straight chord from station to satellite is given in [375], p. 43 as about $0.0014\tan^2\zeta\sec\zeta$ metres. It is ordinarily negligible.

7.09. Refraction correction to distance. Microwaves

(*a*) The refraction of microwaves differs from that of light as follows.

(i) The basic formula (1.16) starts with the factor 77.6 instead of 78.6 to 80.6 for light in (1.11).

(ii) The question of group velocity only arises in the ionosphere.

(iii) Humidity is important.

(iv) There is also a refractive effect in the ionosphere, see § 7.10. The present paragraph is concerned only with the tropospheric refraction, which includes refraction in the stratosphere.

Working on conventional lines [511] gives $S = c\tau - \Delta S$,

where $\Delta S = 0.002277\sec\zeta[P_S + (1255T_S^{-1} + 0.05)e_S - B\tan^2\zeta] - \delta$ metres,

if P_S and e_S are in mbar,

or $$\Delta S = 2.307\sec\zeta[P_S + (1255\,T_S^{-1} + 0.05)e_s - B\tan^2\zeta](1/P_0) - \delta \text{ metres,} \tag{7.30}$$†

where P_S, P_0, and e_S are all in the same units, with P_0 as given after (7.26). B and δ are as described after (7.29).

(*b*) *Hopfield's double quartic formula.* [284], [285], [283], and [288] approach the subject differently. Dry air is first considered, and an addition for humidity is made later. Remarking that the temperature gradient is nearly constant in the troposphere, which is responsible for most of the refraction, and initially considering the effect of

† Using (1.16) without the humidity terms gives the dry-air effect as

$$(n-1)10^6 = 77.6\,P_S/T, \quad \text{with } P \text{ in mbar.}$$

Then proceeding as in § 7.08(*b*), there results

$$S = c\tau - 2.304(P_S/P_0)\sec\zeta \text{ metres,} \tag{7.31}$$

which, so far as it goes, agrees well with (7.30).

$N_d = (n_d - 1) \times 10^6$ for dry air, [283] has shown that

$$N_d = N_{dS}\left(\frac{h_d - h}{h_d - h_S}\right)^{\mu} = N_{dS}\left(\frac{H_d - H}{H_d}\right)^{\mu}, \tag{7.32}$$

where h indicates height above the geoid.

H is height above the ground surface. $H = h - h_S$.

N_{dS} and T_S are surface values of N_d and T. N_{dS} is given by (1.16) as $77.6\ P_S/T_S$, if P is in mbar, or as $0.776\ P_S/T_S$ if p is in Pa.

$\alpha = -\mathrm{d}T/\mathrm{d}h$ or $-\mathrm{d}T/\mathrm{d}H$, assumed constant between $T = T_S$ and zero.

$H_d = T_S/\alpha$, by definition. The appropriate values of H_d and α are to be empirically determined.

$H_{d0} = H_d$ when $T_S = 273.15$ K. It is to be hoped that H_{d0} will be constant.

g_m is the weighted mean g in the atmosphere, namely $9.7877\ \mathrm{m\,s^{-2}}$, as in § 7.08(*b*).

C_d is the dry-air constant, 2.8704×10^2 in SI units.

$\mu = (g_m/C_d\alpha) - 1$.

Then if α is taken to be $0.00680\ \mathrm{K\,m^{-1}}$ as in [283], it follows that $\mu = 4$, and at a height H above the surface

$$N_d = N_{dS}\left(\frac{H_d - H}{H_d}\right)^4. \tag{7.33}$$

For a *vertical line*

$$\Delta S_d = 10^{-6} \int_0^{H_d} N_d\,\mathrm{d}H = (1/5)10^{-6}\, N_{dS} H_d \text{ metres.} \tag{7.34}$$

If $N_{dS} = 77.6\ P_S/T_S$ from (1.16), and if $H_d = T_S/0.00680$, and $P_S =$ 1013 mBar, (7.34) gives the dry-air correction as 2.312 metres. Compare this with (7.30) which gives 2.307, if ζ, e, and δ are all zero. In these circumstances, and if $T_S = 273.15$, $H_{d0} = 40\,169$ m.

As a test of these and other formulae [288] and [287]† report results from a year's daily balloon soundings of temperature, pressure, and humidity at 15 sites in North America and the Pacific. From these the dry-air effects can be separately computed by numerical integration, with results as given in (7.37) and (7.38) below.

While the dry component constitutes at least 90 per cent of the whole refraction correction, the wet component involves the greater doubt, since (7.32) does not hold for it.

For the wet component, N_w it may be provisionally assumed that

$$N_w = N_{ws}\left(\frac{H_w - H}{H_w}\right)^4, \tag{7.35}$$

† [284] and [285] give results of the earlier soundings.

where N_{wS} is the surface value of N_w, namely† $(3.76\times10^5\,e)/T_S^2$ from (1.16). Here the relation $H_w = T_S/\alpha$ does not hold, and it is only possible to write

$$\Delta S_w = 10^{-6}\int_0^{H_w} N_w\,dH = (1/5)\,10^{-6}\,N_{wS}\,H_w \text{ metres,} \tag{7.36}$$

after H_w has been determined empirically from balloon observations. These give values of H_w varying between 8 000 and 13 000 m.

Analysing available results, [288] remarks that for the dry component H_d is generally given by

$$H_d = (40.110\pm0.040)+(0.14881\pm0.00180)(T_S-273.15)\text{ km,} \tag{7.37}$$

above the surface, except that two of the 15 sites, which are on equatorial Pacific islands, are better fitted by

$$H_d = (40.402\pm0.097)+(0.14281\pm0.00330)(T_S-273.15)\text{ km} \tag{7.38}$$

These formulae are expected to give the dry component of a zenith line to within a few mm. The general constancy of $H_{d0} = 40\,110$ m is satisfactory.

For H_w, results are much more doubtful. [284] gives

$$H_w = 13.268-0.09796(T_S-273.15)\text{ km,} \tag{7.39}$$

but later papers are unable to justify such precision. [287] gives H_w $(=h_w-h_S)$ at 16 sites. For any other place the value of H_w may be selected (to the nearest 500 m) on the basis of climatic similarity. The resulting wet-air correction may be in error by 10 per cent, but this is likely to amount to 1 per cent or less of the total wet plus dry effect.

Finally, the total correction

$$\Delta S = \Delta S_d + \Delta S_w \tag{7.40}$$

It may be remarked that the tropospheric dry-air correction to the length of a vertical line, whose highest point is above the greatest height at which any significant tropospheric refraction occurs, depends only on the pressure and temperature, at the ground surface, and that the height of the ground station is only material in so far as it affects these factors. Variations in the temperature lapse rate also have no effect.

(*c*) *Non-vertical lines.* Hopfield's formula. See [286]. Sub-paragraph (*b*) has dealt only with vertical lines. Provided the zenith distance does

† (1.16) gives 3.76, but [288] and other papers use 3.73 from [536]. The difference is immaterial.

not exceed about 45°, a factor of sec ζ may be introduced into (7.34) and (7.36). For greater zenith distances the serious effect of water vapour, whose distribution within the atmosphere is irregular, inhibits precise theory such as has been applied to light waves, although (7.30) can be used, and [511], p. 34 estimates the standard error as not more than 10–20 cm if ζ is less than 80°.

For (7.34) and (7.36) [416] suggests replacing sec ζ by the empirical factor

$$\operatorname{cosec}[(E^2+\theta^2)^{\frac{1}{2}}] \tag{7.41}$$

where $E=90°-\zeta$, and is the (apparent) elevation of the satellite above the horizon, $\theta=2°.3$ for the dry-air term, and 1°.5 for the water vapour term.

For the most accurate work [618] gives

$$\Delta S=\frac{N_S\times 10^{-6}}{H^4}\int_{z=-H}^{z=0}\frac{(r'+z)z^4\,dz}{[(r'+z)^2-a^2]^{\frac{1}{2}}}, \tag{7.42}$$

with a suffix d or w on ΔS, N_S, H, and r'.
$H_d=T_S/\alpha$ as before $=40.1+0.149(T_S-273.15)$ km
H_w is determined as after (7.36).
$r'_S=$ distance from the geocentre to the ground station.
$r'=r'_S+H$
$a=r'_S\cos E$
$E=90°-\zeta$.
z is an integration dummy, between the given limits.

The integration of (7.42) involves numerical instability if E is near 0 or 90°. [618] gives an algorithm using series expansions for each of these two situations, with a wide intermediate range of E over which either series may be used.

(*d*) For doppler integrated counts, (S_2-S_1) receives the appropriate dry and wet corrections for S_2 minus the corrections for S_1, the difference arising from the change of ζ or E during the count.

7.10. Ionospheric refraction of radar and microwaves

(*a*) *The refractive index.* In the ionosphere radar and microwaves become involved in trouble which does not affect light, as a consequence of solar radiation detaching electrons from some of the atoms of the atmosphere. The resulting (phase) refractive index, n_P, is given by [400] in the form

$$(n_P-1)=(-K_1N_e/f^2)\pm(K_2N_eB\cos\theta)(1/f^3)$$
$$-[K_3N_e^2+K_4N_eB^2(1-\tfrac{1}{2}\sin^2\theta)](1/f^4), \tag{7.43}$$

where the K's are numerical parameters and [594] gives

$$(n_P - 1) = -(N_e e^2) \div (8\pi^2 m \epsilon_0 f^2) + \text{smaller terms similar to (7.43)}, \qquad (7.44)†$$

where N_e is the number of electrons per cubic metre, e is the electron charge, m is the electron mass, ϵ_0 is the permittivity, B is the earth's magnetic field, θ is the angle between the wave normal and the field, and f is the frequency of the wave. Using waves with $f =$ (say) 400 MHz, the first term (in $1/f^2$) is much the largest, although the terms in $1/f^3$ and $1/f^4$ are only marginally negligible.

In figures, [53] gives $e = 1.6022 \times 10^{-19}$ coulomb.

$$m = 9.1096 \times 10^{-31}\ \text{kg}$$
$$\epsilon_0 = 8.8542 \times 10^{-12}\ \text{F m}^{-1}.$$

Inserting these in the first term of (7.44) gives

$$(n_P - 1) = -40.3(N_e/f^2), \qquad (7.45)$$

So, for example, if $N_e = 10^{11}$ per m^3, and if $f = 449$ MHz as used for Secor, $(n-1)$ will equal -20 ppm.

From (7.45) the phase velocity is

$$v_P = c\{1 + 40.3(N_e/f^2)\}. \qquad (7.46)$$

Since $(n_P - 1)$ is negative, the path is convex towards areas of high electron density, and the phase velocity exceeds that of light in vacuum: a remarkable situation, but see the group velocity below.

Since n varies with f, the group velocity of radar and microwaves in the ionosphere is not the same as the phase velocity. As in (1.12)

$$\begin{aligned} v_G &= v_P - l(\mathrm{d}v/\mathrm{d}l), \quad \text{where } l \approx c/f \\ &= v_P(1 - 80.6 N_e/f^2), \quad \text{using (1.8) and (7.45).} \\ &= c(1 - 40.3 N_e/f^2), \quad \text{from (7.46).} \qquad (7.47) \end{aligned}$$

and

$$(n_G - 1) = +40.3\ N_e/f^2 \quad \text{from (1.13).} \qquad (7.48)$$

So the group velocity is less than c, as is proper. The curvature of the path is of course regulated by (7.45). See (f) below.

Then the length of a line is given by

$$S = c\tau - \Delta S,$$

where

$$\Delta S = \int_0^S (n-1)\,\mathrm{d}S \qquad (7.49)$$

† [594] p. 347 is in c.g.s. units, in which $\epsilon_0 = 1$, so ϵ_0 is omitted in its version of (7.44). In the SI system the unit value of ϵ_0 is replaced by $4\pi\epsilon_0$, [413] p. 335. Hence the difference of $4\pi\epsilon_0$ in the denominators of (7.44) and [594].

In (7.49) $(n-1)$ is given by (7.45) or (7.48). The group index (7.48) is applicable to Secor, which uses a modulated wave, but the phase index (7.45) will be used for doppler. The integration in (7.49) must be carried out numerically, taking account of the values of $(n-1)$ at different heights, and the corresponding values of dS/dh, which will depend on the zenith distance ζ. The approximation $dS = dh \sec \zeta$ may give tolerable results if $\zeta < 45°$, but when $\zeta = 45°$, 60°, 75°, and 90° it is better replaced by $dh \sec \zeta^*$, where $\zeta^* = 42°$, 55°, 65°, and 71° respectively. These figures result if the barycentre of the electron density is taken to be at an altitude of about 350 km. See [468], p. 42 for confirmation of the values given for 45° and 90°.

(*b*) The electron density varies greatly with time and other factors, and is not predictable. Fig. 7.5 illustrates average day maximum and night minimum densities at a time of little solar activity. At a time of high activity the density may be many times greater. The figure illustrates orders of magnitude, but little more.

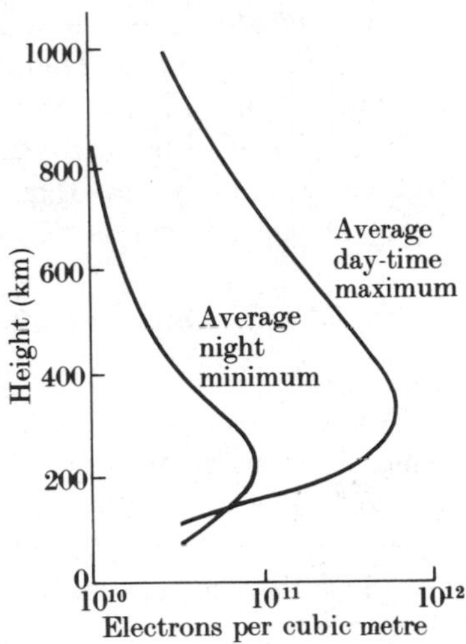

FIG. 7.5. Typical electron density profiles. Quiet solar activity. October 1964 to February 1965. From [201].

(*c*) *The ionospheric refraction correction. Main Term.* The integral in (7.49) demands a knowledge of n at all heights at the required time. To get an idea of the order of magnitude, consider the $1/f^2$ term only, as in (7.44), and let the N_e distribution be that of the average night minimum for quiet solar activity as shown in Fig. 7.5. Then, remembering that the figure is drawn with a log scale for density, we may estimate

$$\int_0^{100\,\text{km}} N_e \, dh \approx (3.5 \times 10^{10}) \times (8 \times 10^5)\ \text{m}^{-2} = 2.8 \times 10^{16}\ \text{m}^{-2} \tag{7.50}$$

Then, using the group index and $f = 449$ MHz, as for Secor, (7.48) and (7.49) give

$$\Delta S = (1/f^2)(40.3)(2.8 \times 10^{16}) \sec \zeta^* \text{ metres}$$

$$= (113 \times 10^{16} \sec \zeta^*)/f^2 \text{ metres,} \tag{7.51}$$

and

$$S = c\tau - 5.6 \sec \zeta^* \text{ metres,} \tag{7.52}$$

where ζ^* is a modified zenith distance as given after (7.49). For doppler, $f = 400$ MHz, and the factor 5.6 becomes 7.0.

If the instrumentation involves unmodulated observations, as in doppler, the phase velocity will be used, and the sign of the correction to $c\tau$ will be positive, as in (7.46).

This is a very considerable correction, especially at large zenith distances, and Fig. 7.5 suggests something of the order 10 times as great by day, and more again in active solar years. It is clearly impossible to get useful results by estimating the distribution of N_e. The remedy is either to increase f to 5 GHz or more, or to measure the distance simultaneously on two wavelengths, such as 449 and 224.5 for Secor, or 400 and 150 MHz for doppler. Then the corrections on both wavelengths can be computed from their difference, in so far as smaller terms can be ignored. VLBI, §§ 7.66–7.69, at present (1976) relies on a single very high frequency, namely 7.8 GHz.

(*d*) *Doppler.* Formula (7.170) gives†

$$\text{Doppler shift } \Delta f = -(f_t/c)\dot{S} = -\frac{f_t}{c}\left(\frac{S_2 - S_1}{t_2 - t_1}\right). \tag{7.53}$$

Consider a satellite at a height of 1100 km, with a velocity of 7 km s^{-1}, which has passed through the zenith and reached a ζ of 80°, for which $\zeta^* = 67°$. The rate of change of ζ^* will then be about 1/1800 rad s^{-1}, and d(sec ζ^*) = sec ζ^* tan ζ^* dζ^*, will be about 0.0034 s^{-1}. The correction to $S_2 - S_1$ will then be positive, and the correction to Δf will be

$$-(f_t/c)(7.0)(0.0034) = -0.031 \text{ Hz}. \tag{7.54}$$

At this time the value of Δf itself will be about 8000 Hz, so the correction is again seen to be very substantial (about 4 ppm in optimum conditions), unless it is eliminated by duplicating the wavelength.

(*e*) *The terms in* $1/f^3$ *and* $1/f^4$. [594] and [273], p. 98 give

$$\text{Doppler shift} = \Delta f = -(f/c)\dot{S} + (a_1/f) + (a_2/f^2) + (a_3/f^3), \tag{7.55}$$

where the last three terms correspond to the terms in $1/f^2$, $1/f^3$, and $1/f^4$ in (7.43) and (7,44). The first term a_1/f has been considered in (*c*) and (*d*) above.

† In (7.170) ΔS signifies $S_2 - S_1$, which must not be confused with the refraction correction ΔS of this section. To eliminate ionospheric refraction each of S_2 and S_1 would need to be numerically increased by the phase velocity value of ΔS, namely by approximately by +7.0 sec ζ^* m for optimum conditions, as computed from (7.51) with $f = 400$ MHz. The change in $S_2 - S_1$ thus arises from the change in ζ^*.

TABLE 7.1

Maximum values in any one pass. $\zeta < 80°$

	a_1/Hz^2 Average range	Max	a_3/Hz^4 Average range	Max
Sunspot minimum	0.1 to 5×10^8	10×10^8	0 to 2×10^{22}	2×10^{23}
Sunspot maximum	0.1 to 15×10^8	30×10^8	1 to 20×10^{22}	20×10^{23}

For a_1 and a_3 [594] gives Table 7.1, and states that the term a_2/f^2 is likely to be less important than a_3/f^3, and also that exceptionally large values of a_3 are likely to be associated with large values of a_1, which will be detected by the double frequency. At night only the smaller values are likely to occur. The figure of 0.031 obtained in (7.54) corresponds to $a_1 = 0.12\times10^8$, when $f = 400$ Hz. It is therefore not necessary to legislate for the undetected occurrence of the larger values of a given in the table. [594] also estimates that a 5-m error of position from a single pass will be caused by a value of 20×10^{22} for a_3, if the frequency pair is no higher than 250 and 500 MHz. Values of 20×10^{22} for a are no doubt unusual, and [13], p. 62 estimates that in ordinary circumstances, with doppler at 150 and 400 MHz, the slant range to the nearest point of approach may be wrong by less than 0.5 m for high elevation passes, and by 1 or 2 m for low elevations. Since the height of the station is mostly determined by high elevation passes, while longitude errors caused by ionospheric refraction change sign according to whether the pass is east or west, and since latitude is hardly affected at all (with satellites) in polar orbits), [13] estimates that the mean error from several passes should not be more than about 10 cm on account of ionospheric refraction.

(*f*) *The curvature of the path.* The curvature of the path in both the troposphere and the ionosphere must of course be considered as a whole when the difference between path length and chord is being considered.

As in (1.18) the curvature of the path is given by

$$\frac{1}{\sigma} = -\left(\frac{\mathrm{d}n}{\mathrm{d}h}\right)\sin\zeta. \tag{7.56}$$

For the troposphere, 4 lines below (1.37) suggests that $\sigma = 4R, 5R, 6R$, and $8R$ at heights of 0, 3, 4, and 6 km, respectively, where R is the earth's radius. At greater heights (1.36) can produce estimates.

In the ionosphere (7.45) gives

$$\frac{1}{\sigma} = \frac{40.3}{f^2}\frac{\mathrm{d}N_e}{\mathrm{d}h}\sin\zeta\ \mathrm{m}^{-1}, \tag{7.57}$$

where ζ is the local zenith distance of the tangent to the path. The gradient dN_e/dh will be very variable, but Fig. 7.5 suggests that at night, in a quiet period, it may average $-(3\times 10^{10})$ per 200 000 m at a height of more than 400 km. Using this figure, and taking 400 MHz for f, and putting $\sin\zeta = 0.8$, $\sigma = 34\times 10^6$ km. Compare this with the usual 5×10^4 km at a height of 6 km in the troposphere. With σ so large, the ionospheric path would be virtually straight, but σ varies as $1/(dN_e/dh)$, and could be 10 times smaller in less favourable circumstances. If estimates of σ are made for all parts of the path, the course of the path can be traced. The curvature in the troposphere is likely to be the most important, but the whole curvature correction is generally negligible. [286], p. 13 estimates that path curvature adds a few mm to the chord distance at elevations of about 20°, a few cm at 10°, and about 3 m at 1°. The sign of the curvature is such that in the troposphere the path is concave towards the ground, as usual. In the lower ionosphere up to the height of maximum N_e, the sign is the same. Above that height the path is convex towards the ground. Above 1000 km there is unlikely to be any significant curvature.

Section 3. Zero-order and Resonant Harmonics

7.11. Introductory

In this section, except in § 7.15, earth's potential field is assumed to be given by

$$V = \frac{GM}{r}\left\{1 - \sum_{2}^{\infty} J_l\left(\frac{a}{r}\right)^l P_l(\cos\theta)\right\}, \tag{7.58}$$

where a is the earth's semi-major axis, r is the geocentric distance to the satellite, P_l is the l-degree Legendre function, and J_l is its numerical coefficient. The field is thus assumed to be symmetrical about the earth's polar axis. In these conditions the orbit ceases to be a plane curve, and its osculating elements are continuously varying in several ways, as follows.

(i) Periodically during each orbit, as in § 7.12(h). This constitutes a considerable nuisance.

(ii) There are secular changes in Ω and ω.

(iii) There are long period changes in i, p, and e.

The secular and long-period changes are useful, since their magnitudes depend on the J's as in (7.59)–(7.72), and observations which give the changes in the elements enable the J's to be computed. Further, low accuracy observations extended over a fairly long period of time give the J's with a relatively high accuracy.

See § 7.00 for notation.

7.12. Formulae for $\Delta\Omega$, etc.

(a) The formulae quoted below give the first three terms, involving J_2, J_3, and J_4 in the expressions for $\Delta\Omega \ldots \Delta e$, these being the changes in each element in a single orbital period. Terms in $(J_2)^2$ are also included where they are significant. It must be remembered that J_2 to J_4 are only the first terms of a long and slowly convergent series. Attempts have been made to evaluate the J's as far as J_{26}, see Table 7.2.

The formulae given have been abstracted from [411], pp. 19–25, where terms as far as those of J_6 are included. All elements on the right-hand sides of these expressions are deemed to have their values at the preceding ascending node.† See also § 7.13.

The following approximate figures are guides to the order of magnitude of different terms in these expressions. $J_2 = 1082.6\times10^{-6}$, $J_3 = -2.5\times10^{-6}$, $J_4 = -1.6\times10^{-6}$. Other J's are less than 0.6×10^{-6}. $(J_2)^2 = 1.2\times10^{-6}$. For a satellite 1000 km above the earth $a/p = 0.87$, e is generally less than 0.1, and often less than 0.01. i, Ω, and ω may have any values.

(b) *The regression of the plane of the orbit*, or the rotation of the node in the equatorial plane. The most noticeable of the perturbations is that the plane of the orbit rotates about the polar axis. In fact

$$\begin{aligned}\Delta\Omega = &-3\pi J_2\left(\frac{a}{p}\right)^2 \cos i - \frac{3\pi}{4} J_3 e\left(\frac{a}{p}\right)^3 \cot i \sin\omega(15\sin^2 i - 4)\\ &-\frac{15\pi}{8} J_4\left(\frac{a}{p}\right)^4 \cos i(7\sin^2 i - 4) - \text{terms in } J_5, \text{ etc.}\\ &-9\pi J_2^2\left(\frac{a}{p}\right)^2 \cos i(\tfrac{5}{3}\sin^2 i - \tfrac{1}{4}) + (J_2^2 e, J_4 e^2).\end{aligned} \tag{7.59}$$

(7.59) can be extended to include higher numbered J's as follows, from [236] (1958),

$$\begin{aligned}\Delta\Omega = &-3\pi\cos i \sum_{n=1}^{\infty} C_{2n}\left(\frac{a}{p}\right)^{2n} J_{2n}\\ &-3\pi e\sin\omega\cot i \sum_{n=1}^{\infty} C_{2n+1}\left(\frac{a}{p}\right)^{2n+1} J_{2n+1} + (J_4 e^2, J_3 e^3, J_2^2),\end{aligned} \tag{7.60}$$

† Other formulae, [411], pp. 30–49, [332], and several others, give $\Delta\Omega$, etc. in terms of slightly differently defined elements, using values of the elements which are smoothed or meaned over one or more orbits, with a view to most conveniently dealing with the short-period perturbations.

where $$C_{2n}=\sum_{r=0}^{n-1}\frac{(-1)^r(4n-2r)!\,2^{2(r+1-2n)}}{3(r)!\,(2n-r)!\,(n-r)!\,(n-1-r)!}(\sin^2 i)^{(n-1-r)},$$

$$B_{2n+1}=\sum_{r=0}^{n}\frac{(-1)^r n(4n+2-2r)!\,2^{2(r-2n)}}{3(r)!\,(2n+1-r)!\,(n+1-r)!\,(n-r)!}(\sin^2 i)^{n-r},$$

$$C_{2n+1}=\sum_{r=0}^{n}(2n+1-2r)\times(\text{the expression following the summation sign in } B_{2n+1}). \tag{7.61}$$

B_{2n+1} is required for (7.64) for Δi.

For J_2 to J_6 these give

$$C_2=1.\qquad C_4=\tfrac{5}{8}(7\sin^2 i-4).\qquad C_6=\tfrac{35}{64}(33\sin^4 i-36\sin^2 i+8).$$

$$B_3=\tfrac{1}{4}(5\sin^2 i-4).\qquad B_5=\tfrac{5}{16}(21\sin^4 i-28\sin^2 i+8).$$

$$C_3=\tfrac{1}{4}(15\sin^2 i-4).\qquad C_5=\tfrac{5}{16}(105\sin^4 i-84\sin^2 i+8).$$

The terms with uneven J's in (7.60) contain $e\sin\omega$ as a factor,† and are therefore small if the orbit is nearly circular, and periodic unless ω is nearly constant, see (*c*) below.

An approximate formula, derived by combining (7.76) and (7.59) is

$$\Delta\Omega \text{ per day}=-10°\left(\frac{a}{p}\right)^{3\frac{1}{2}}\cos i. \tag{7.62}$$

(*c*) *The argument of perigee, or rotation of the apses.* The direction of the major axis of the orbit rotates in the plane of the orbit.

$$\Delta\omega=\frac{3\pi}{2}J_2\left(\frac{a}{p}\right)^2(4-5\sin^2 i)+\frac{3\pi}{4e}J_3\left(\frac{a}{p}\right)^3\sin i\sin\omega(4-5\sin^2 i)$$
$$-\frac{15\pi}{16}J_4\left(\frac{a}{p}\right)^4\{(49\sin^4 i-62\sin^2 i+16)$$
$$+\sin^2 i\cos 2\omega(6-7\sin^2 i)\}+(J_3e^2, J_4e^2, J_2^2). \tag{7.63}$$

Here $\Delta\omega$ is the rate of change relative to the (moving) node.

If i is $63\frac{1}{2}°$, $(4-5\sin^2 i)=0$, and changes in ω are small. When i is 30°, $\Delta\omega$ is about 15° per day and the apses make a complete rotation in about 24 days. When $e=0$ the J_3 term is infinite, but a circle has no apse. The changes of ω have a periodic effect on i, p, $r_{\min}$, and e, as below.

† The factor $\cot i$ does not imply instability when $i=0$. The direction of the node is not precisely determined when the planes of the orbit and equator are nearly parallel, so a small change in the plane of the orbit causes a large $\Delta\Omega$.

(*d*) *The inclination of the orbit*

$$\Delta i = \frac{3\pi}{4} eJ_3 \left(\frac{a}{p}\right)^3 \cos i \cos \omega (4-5\sin^2 i)$$
$$+\frac{15\pi}{32} e^2 J_4 \left(\frac{a}{p}\right)^4 \sin 2i \sin 2\omega (6-7\sin^2 i) - J_5 \text{ terms, etc.}$$
$$-\frac{3\pi}{4} eJ_2^2 \left(\frac{a}{p}\right)^4 \sin 2i \sin \omega (4-5\sin^2 i) + (eJ_5, e^2J_6, \ldots). \quad (7.64)$$

(7.64) can be extended to include higher odd-numbered J's using B_{2n+1} from (7.61), thus

$$\Delta i = -3\pi e \cos i \cos \omega \sum_{n=1}^{\infty} B_{2n+1} \left(\frac{a}{p}\right)^{2n+1} J_{2n+1}$$
$$+ J_2^2 \text{ as in (7.64)} + \text{terms with even } J\text{'s} \times e^2, \quad (7.65)$$

Unless $(4-5\sin^2 i)$ is small, i will vary periodically with ω. Taking $\Delta\omega$ from the first term of (7.63) and integrating the first term of (7.64) from $\omega = 0$ to $\frac{1}{2}\pi$ gives the amplitude (semi-range) of i as approximately

$$(eJ_3 \cos i)/(2J_2) \text{ radians} = 240'' e \cos i \quad (7.66)$$

(*e*) *Semi-latus rectum*

$$\Delta p = 2p(\Delta i)\tan i + (J_2)^2, \text{ where } \Delta i \text{ is as in (7.64)},$$
$$= \tfrac{3}{2}\pi p e J_3 \left(\frac{a}{p}\right)^3 \sin i \cos \omega (4-5\sin^2 i) + (e^2J_4, eJ_5, eJ_2^2). \quad (7.67)$$

(7.68)

(*f*) *Eccentricity*

$$\Delta e = -\left(\frac{1-e^2}{e}\right)\tan i \,\Delta i + (J_2^2/e), \quad (7.69)$$

where Δi is as in (7.64).

(*g*) *Perigee distance*

$$\Delta r_{\min} = \frac{3\pi A}{4} J_3 \left(\frac{a}{p}\right)^3 \sin i \cos \omega (4-5\sin^2 i) + (J_4 e, J_2^2). \quad (7.70)$$

Like i, $r_{\min}$ varies periodically with ω, and an approximation [337] to the amplitude of its variation is

$$\frac{aJ_3 \sin i}{2J_2}. \quad (7.71)$$

Variations of $r_{\min}$ are generally associated with changes of eccentricity, not with any general contraction of the orbit. In fact

$$\Delta A = 0 + (J_2^2, J_3^2). \tag{7.72}$$

(*h*) *Periodic changes during each orbit.* During each orbit Ω, p, and i vary approximately as follows. Suffix 0 indicates values at the ascending node.

$$\Omega = \Omega_0 - \tfrac{3}{2}J_2\left(\frac{a}{p}\right)^2 (u - \tfrac{1}{2}\sin 2u)\cos i + (eJ_2),$$
$$= \Omega_0 - 3\pi J_2(a/p)^2 \cos i, \quad \text{when } u = 2\pi \tag{7.73}$$

$$p = p_0 - 3pJ_2\left(\frac{a}{p}\right)^2 \sin^2 i \sin^2 u + (eJ_2), \tag{7.74}$$

$$i = i_0 - \tfrac{3}{4}J_2\left(\frac{a}{p}\right)^2 \sin 2i \sin^2 u + (eJ_2). \tag{7.75}$$

7.13. The nodal period

The period from one ascending node to the next is given in [411], p. 48, as

$$T = (2\pi/n')\left[1 - \left\{\frac{3J_2}{4}\left(\frac{a}{p_0}\right)^2 (1 + e_0\cos\omega_0)^{-2}(1 - e_0^2)^{\frac{3}{2}}(4 - 5\sin^2 i_0)\right\}\right]. \tag{7.76}$$

In this formula n' is not the value (n_0) of the mean motion at the node, but is the mean of the mean motion over the whole period, and is given by

$$n' = n_0(1 + 2A'F/GM) \tag{7.77}$$

where $F = (GM/r)\sum_2^\infty J_l(a/r)^l P_l(\cos\theta)$, namely $(GM)/r - V$ in (7.58), and A' is defined by $(n')^2(A')^3 = GM$.†

In this system the relation $(n')^2(A')^3 = GM$ is preserved, so p' and A' differ from the values of p and A at the node, and other elements are also affected. In this respect the RAE system of [411] does not exactly follow the SAO system [332] where the elements used are the time averages, $\bar{\Omega}$, etc., of the osculating elements taken over each revolution of the orbit. [412] shows that with the SAO definition $(\bar{n})^2(\bar{A})^3$ is not constant, but that results given by the two systems are precisely equivalent. For the sake of uniformity, the SAO system has generally been adopted.

† Writing v for the velocity of the satellite at any moment, we have

$$\begin{aligned}\text{Total energy per unit mass} &= \tfrac{1}{2}v^2 - V, \text{ since } -V = \text{potential energy per unit mass}\\ &= GM(\tfrac{1}{r} - \tfrac{1}{2A'}) - (GM/r - F), \text{ using (7,5)}\\ &= -(GM/2A') + F, \text{ which must therefore be constant.}\end{aligned} \tag{7.78}$$

7.14. The determination of the J's

(*a*) At the SAO [216], p. 257, the basic data are the observed mean orbital elements obtained from overlapping 4-day arcs, continuing over a period of weeks or months. Long-period perturbations arising from other causes, such as drag, radiation, and tidal effects, are carefully eliminated.

(*b*) *Numerical results. Even-numbered* J's. The most immediately useful of formulae (7.59)–(7.72) is (7.59) in its fuller form (7.60). In this equation the odd-numbered J's enter with e as a factor, which is generally small, so they are not well determined in this way. See (*c*) below. The determination of $\Delta\Omega$ for a number of satellites then gives a set of linear equations for an equal number of even J's, subject to the following considerations.

(i) The odd J's are ignored, as they probably can be. If any are known, even with low accuracy, they can be numerically substituted.

(ii) The coefficients of the even J's in (7.60) mostly depend on the inclination i, so it is necessary to select satellites with well-spaced inclinations. Satellites are now very numerous, but it is not practicable to solve for an unlimited number of even J's. The equations are liable to become unstable. It is best to solve by least squares for rather fewer J's than there are equations.

(iii) The effects of higher even J's are ignored. A finite number of equations for an infinite series of J's is being solved by assuming that the higher J's are zero. No doubt they are small, but there is no direct evidence of their being of no consequence. The total geoidal effect of high harmonics, including tesserals, can locally amount to ten metres or more. On the other hand, values of J_2 given in [339] (1966) only varied between 1082.70 and 1082.62×10^{-6} when given by solutions containing 2, 3, 4, 5, and 6 other even J's. The modern value of J_2, given below Table 7.2 is 1082.63, obtained in combination with other J's as high as J_{20}. It is probably correct to 0.01.

The value of f corresponding to $J_2=1082.6\times10^{-6}$ is approximately 1/298.25 with an uncertainty of less than 0.005, Table 6.4.

(*c*) *Numerical results. Odd-numbered* J's. Odd J's occur in $\Delta\omega$, the rate of rotation of the apses, with $1/e$ as a multiplier. A small e might appear to be helpful, but the apse of a nearly circular orbit is difficult to locate.

The rate of change of the inclination (7.65) includes e as a multiplier of the odd J's and e^2 of the even ones, so the latter can be ignored or numerically included if known† and a number of odd J's can be deduced from an equal or greater number of satellites with properly spaced

† The process of obtaining values of even J's with little knowledge of the odd J's, and then getting the odd J's from knowledge of the even, can of course be repeated if necessary, with improving results.

inclinations. High values of e, such as 0.1, and low values of i will be favourable. The inclination varies periodically with $\cos \omega$, with an amplitude of $240'' e \cos i$, see (7.66).

The latus rectum and eccentricity also vary with the same period as the inclination. In their formulae, the even J's have a factor e^2, and the odd ones only have e. Variations in the eccentricity, which are closely allied to those in the inclination, have been used by [338] and [341], and other sources, as the best guide to the odd-numbered J's.

For values obtained from various sources see Table 7.2.

(*d*) The determination of J_2 within a few weeks of the launch of Sputnik 2 was of great service to geodesy, the mean flattening of the geoid being given far more accurately than was previously possible, and differing significantly from the figure previously considered as the best.

TABLE 7.2.
Zonal harmonics. Values of $\bar{C}_{l,0} \times 10^6$

Degree l	Factor $-(2l+1)^{\frac{1}{2}}$	SAO iii [216] 1973	Wagner [586] 1973	King–Hele [341] 1974	GEM 8 [588] 1976	GEM 10 [382] 1978	s.e. of GEM 10 [382]
2	−2.236068	−484.170	−484.169		−484.165	−484.165	0.001
3	−2.6458	0.960	0.960	0.957	0.958	0.958	1
4	−3.0000	0.539	0.533		0.540	0.541	1
5	−3.3166	0.069	0.069	0.074	0.068	0.069	2
6	−3.606	−0.153	−0.147		−0.150	−0.151	2
7	−3.873	0.091	0.094	0.084	0.093	0.093	2
8	−4.123	0.050	0.049		0.050	0.051	2
9	−4.359	0.035	0.019	0.021	0.026	0.027	2
10	−4.58	0.052	0.049		0.054	0.053	2
11	−4.80	−0.065	−0.029	−0.033	−0.044	−0.049	3
12	−5.00	0.038	0.042		0.038	0.039	3
13	−5.20	0.065	0.019	0.025	0.036	0.044	4
14	−5.39	−0.020	−0.031		−0.025	−0.023	4
15	−5.57	−0.019	0.013	0.005	0.010	0.001	5
16	−5.74	−0.006	−0.001		−0.003	−0.007	4
17	−5.92	0.037	0.034	0.044	0.011	0.017	4
18	−6.08	0.017	0.014		0.008	0.010	3
19	−6.24	−0.016	−0.008		0.002	0.000	4
20	−6.40	0.019	0.013		0.018	0.023	4
21	−6.56	0.013	−0.002		0.002	0.000	4
22	−6.71	−0.014			0.001	−0.002	5
23	−6.86	−0.021			−0.024	−0.019	5
24					−0.003	−0.006	6
25					−0.002	−0.000	5
26					0.008	0.006	7

Note (1) $J_l = -(2l+1)^{\frac{1}{2}}\bar{C}_{l0}$.

(2) The values of $J_2 \times 10^6$ deduced from the above are 1082.637, 1082.635, 1082.627, and 1082.627.

(3) The values of J_2 to $J_6 \times 10^6$ from SAO [216] are $J_2 = 1082.637$, $J_3 = -2.541$, $J_4 = -1.618$, $J_5 = -0.228$, and $J_6 = +0.55$.

Knowledge of additional zero-order terms in the geopotential does not in itself add a great deal to knowledge of the geoidal form or of surface gravity, but when solutions for the tesseral harmonics are being made, it is most helpful that good values can be assigned for the J's, whereby error due to the large progressive variations which they cause can be eliminated.

7.15. Resonant harmonic terms

(*a*) *The* (15, 15) *harmonic.* In addition to the zonal harmonics (7.58) there are other terms in the expansion of V, (7.1), whose effect on the satellite is secular or of long period, notably some of the sectorial harmonics in certain circumstances.

Consider the effect of the (15, 15) sectorial harmonic, similar to Fig. H.7(*c*). If the orbital period of the satellite is about 95 minutes† the position of the satellite in relation to the positive and negative areas in this term of the geopotential will be unchanged after each orbit. Their effect is thus resonant and will accumulate constantly with time, as long as the period is unchanged. The effect of the (15, 15) term on a resonant satellite is thus much greater than that of any of the 15-degree tesserals, and although its numerical parameters $\bar{C}_{15,15}$ and $\bar{S}_{15,15}$ may be no larger than the others, they must (and can) be measured more accurately.

The special interest of the (15, 15) resonance is that the period of 95 minutes corresponds to a (circular) orbit at a height of about 500 km. There are many satellites in orbit at heights a little greater than this, but low enough to experience sufficient air drag for their orbits to decay after a number of years. As their orbits decay, they fall slowly through the height at which their period is 95 minutes, and so experience the effects of resonance and near-resonance for some months. For the (16, 16) resonance period of about 90 minutes, the height is about 200 km, where the effect of air drag is rather inconveniently large. For the 14 and lower harmonics the necessary height is 800 km or more, and although there are plenty of such satellites, the much lower air drag at such heights only introduces decay and consequent resonance after a much longer period.

(*b*) *Other resonant terms.*

(i) A condition for resonance is clearly

$$(\dot{\omega}+\dot{M}) = m(\dot{\theta}-\dot{\Omega}), \tag{7.79}$$

where $\dot{\omega}$, $\dot{M}$ $(=n)$, and $\dot{\Omega}$ are the time derivatives of the argument of perigee, the mean anomaly, and the RA of the ascending node respectively. $\dot{\theta}$ is the earth's rate of rotation relative to a fixed equinox, and m is the order of the harmonic concerned. $\dot{\theta}$ is 360.9856 degrees per Julian day of 86 400 seconds, (4.42) where θ is written as $\hat{\theta}$. See Fig. 7.1,

† 96 minutes = 24 hours/15. Actually 95 is correct, see (7.79).

where $(\dot{\omega}+\dot{M})$ is the velocity of the satellite in its orbit, and $(\dot{\theta}-\dot{\Omega})$ is the relative angular velocity of the orbital plane and the surface of the earth.

(ii) More generally, there will be resonance if

$$\alpha(\dot{\omega}+\dot{M})=\beta(\dot{\theta}-\dot{\Omega}), \tag{7.80}$$

where α and β are two mutually prime integers, such as 1 and 15, or 2 and 27, [6]. In practice α is most usually 1, sometimes 2, and rarely 3. β may typically be 13 to 16 with $\alpha=1$, 27, and 29, with $\alpha=2$, and possibly 44, 46, and 47 with $\alpha=3$. There will also be some long-period resonant effect if the equalities in (7.79) and (7.80) are only approximate.

Since $\dot{\Omega}$ and $\dot{\omega}$ depend on the inclination of the satellite, the exact height and period at which resonance occurs depends somewhat on the inclination. See [6], Fig. 1 for examples.

(iii) Finally, if the (15, 15) sectorial is resonant, the 15-order terms (16, 15) and (17, 15) etc. will be resonant too, since their tesserae are all separated by the same 24° of longitude, although their effect may be reduced by the changes in their sign which take place with latitude.

(*c*) *The effect of resonance on the satellite's orbit.* The resonant effect of a term for which $(l-m)$ is even, mostly appears as variations in the inclination of the orbit. As the orbital period shortens, the approach of resonance is shown by the inclination varying sinusoidally for several periods each of (say) one month's duration and of 0.003 degrees in amplitude. During approximately exact resonance the inclination falls steadily through 0°.02 or 0°.04 and then again varies sinusoidally for several more months. See [343] for examples. The theory is complicated, see [6] and [329]. As an example the following summarizes the theoretical effect of the 15-order resonance, [6].

$$\frac{\mathrm{d}i}{\mathrm{d}t}=n\left(\frac{15-\cos i}{\sin i}\right)(\bar{C}_{15}\sin\Phi_{1,15}-\bar{S}_{15}\cos\Phi_{1,15}) \tag{7.81}$$

$+$terms in cos and sin $2\Phi_{1,15}$ and $3\Phi_{1,15}$

$+$terms in eccentricity e, which can be neglected if $e<0.01$,

where $\Phi_{1,15}=\Phi_{\alpha\beta}=(\omega+M)+15(\Omega-\theta)$ in this case, (7.82)

θ is the sidereal hour angle.

$$\bar{C}_{15}=\bar{C}_{15,15}(a/A)^{15}F_{15,15,7}-\bar{C}_{17,15}(a/A)^{17}F_{17,15,8} + \bar{C}_{19,15}(a/A)^{19}F_{19,15,9}-\dots \tag{7.83}$$

$\bar{S}_{15}=$ The same with S instead of C,

$n=\mathrm{d}M/\mathrm{d}t=$ the mean motion of the satellite,

$A=$ the semi-major axis of the orbit,

a is the earth's mean equatorial radius.

$\bar{C}$ and $\bar{S}$ are normalized.

In (7.83) the functions F are known as F_{lmp}, in which summations are carried out over the ranges indicated by the figures given for l, m, and p in the suffix. As an example, for Ariel 3, with $i = 80° \, 18'$ and period 95 minutes, (7.83) took the form, [228], p. 7,

$$\bar{C}_{15} = \bar{C}_{15,15} + 0.347\bar{C}_{17,15} + 0.059\bar{C}_{19,15} - 0.097\bar{C}_{21,15} - 0.172\bar{C}_{23,15}. \tag{7.84}$$

$\bar{C}_{15}$ is known as the 15-order resonance *lumped coefficient.* It is a 'weighted' sum of the 15-order coefficients in the next few higher terms of odd degree, $(l-m)$ being even in all terms. If the lumped coefficients can be determined from several satellites with sufficiently different inclinations, the separate coefficients $\bar{C}_{15,15}$, $\bar{C}_{17,15}$, and $\bar{S}_{15,15}$, etc. can be separately determined on the assumption that those of higher degree can be neglected. [343] gives some results.

Resonant terms in which $(l-m)$ is odd are less easily determined. The principal effect of their resonance is on the eccentricity of the orbit. See [342].

Currently, a number of resonant coefficients are determined as a routine at the same time as the ordinary tesserals in the solutions for geopotential models, § 7.79. An outline of SAO methods is in [218], pp. 19–21.

General references for zonal harmonics: [411], [332], and [6]; for resonant harmonics: [326], [6], [228].

SECTION 4. SATELLITE PHOTOGRAPHY. ASTRO-TRIANGULATION

7.16. Summary and notation

For geodetic purposes the direction of a satellite at any instant cannot usefully be recorded by measuring its azimuth and zenith distance with a theodolite, if only because of the uncertainty in celestial refraction. The practical method is to photograph it against a background of stars and thereby to obtain the right ascension and declination of the line joining the ground station to the satellite at the instant of exposure. This is the basis of astro-triangulation.

In Fig. 7.6 A and B are two stations of an astro-triangulation network, separated by perhaps some thousands of kilometres and not intervisible. S_1 and S_2 are two positions of a satellite high enough to be seen from A and B and so located that the planes AS_1B and AS_2B intersect at an angle of between (say) 60° and 120°. When it is at S_1 the satellite is simultane-

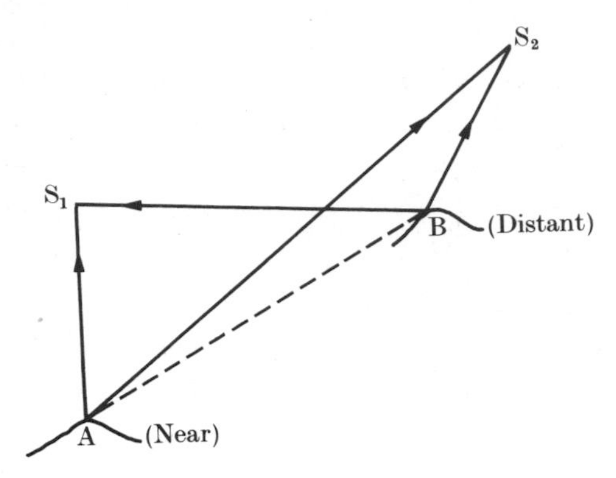

FIG. 7.6. A and B are two ground stations. S_1 and S_2 are satellite positions, one on either side of AB.

ously photographed from both A and B, a number of stars also being recorded on each plate, so that at a recorded time the RA and δ of the satellite as seen simultaneously from both A and B can be computed, § 7.22. The same is done when the satellite is at S_2, and the direction of the line AB in the geodetic reference system, § 7.21, can then be computed, § 7.24.

A world-wide net can be formed from triangles formed by such lines. With lines averaging 4000 km, sixty stations can cover the earth. If the direction of each line is known, the position of each point can be computed, in 3-D cartesian coordinates, relative to an arbitrary origin. At least one measured distance is, of course, required for scale, preferably several.

This section describes the purely geometric use of satellite photography, but the same system of photography can be used to determine the elements of the orbit, and thence the station positions, in geocentric terms, as described in Section 5.

§§ 7.17–7.22 and § 7.28 (*a*)–(*f*), which are applicable to both sections, describe the routine of satellite photography, and the use of the photographs to give the RA and δ of the satellite at recorded times. §§ 7.23–7.27 describe the system of astro-triangulation.

With the completion of the Pageos programme, § 7.75 (*c*), the future of astro-triangulation is perhaps a little doubtful. Greater accuracy than the 1″ or 2″ of photographic RA and δ is obtainable with laser distances or the doppler system. Further, astro-triangulation demands simultaneous good optical seeing from stations some thousands of km apart, and much time may be spend waiting for good weather. On the other hand, photography keeps good control of directions, and in particular it correctly locates the BIH Greenwich zero of longitude within the reference system.

Notation. The following notation is used in this section.

x, y, z = Cartesian coordinates, § 7.21 not § 7.22 (a).
$x, y,$ and z = Plate coordinates in § 7.22 (a).
X, Y, Z = Coordinates with lens as origin, § 7.22 (f) (iv).
X'', Y'' = Polar motion coordinates, § 4.09.
l, m, n = Direction cosines.
RA or α = Right ascension. δ = declination.
α_0, δ_0 = Direction of principal line of photographic plate.
α_P, δ_P = Approximate, reputed values of α_0, δ_0.
ξ, η = Standard coordinates, § 7.22 (c).
c = focal length of lens.
ρ, σ = See Fig. 7.9.
GAST = Greenwich apparent sidereal time.
ϕ, λ = Latitude and longitude.
h = Height above the geoid. Also constants in (7.118).
N = Height of geoid above spheroid.
ν = Normal terminated by minor axis.
θ, ψ = Direction angles, Fig. 7.7.
r = Radial distance.
ψ'' = Celestial refraction in seconds of arc.
ζ = Zenith distance.
q = See (7.101).
$\omega_1, \omega_2, \omega_3$ = Rotation angles, § 7.22 (f) (iv).
L = Distance between two points.
s.e. = Standard error, 68:32 probability.

7.17. Photographic satellites

From this point of view, satellites are of two kinds, passive and active. The former simply reflect an adequate amount of sunlight, while the latter carry equipment for emitting flashes.

(a) *Passive satellites.* Every satellite is to some exent capable of reflecting sunlight, but only spheres are likely to give continuous strong reflections to a ground station. Such satellites have been Echo 1 (1960–8), Echo 2 (1964–9), and Pageos 1 (1966–probably after 2000), spheres of 30, 41, and 30 metres diameter respectively.

The advantages of passive satellites have been the absence of risk of mechanical failure, and that suitable spheres are bright and easy to see and photograph. Refraction erratics can be averaged out by numerous exposures. There are also disadvantages, as below.

(i) Phase. If the point from which the sun is reflected is 10 metres off centre, the error is 1″ at a distance of 2000 km. If the sphere maintains its shape, this can be calculated and allowed for, but impact with meteorites may cause deformation.

(ii) They are visible only when they are in sunlight while the observer is in the dark. Unless the height is very great, this limits the opportunities for photography.

(iii) Synchronization of observations from different stations is not ensured, as it is by a flashing satellite.

(*b*) *Flashing satellites.* The necessity for accurate timing, and also the risk of phase error with large balloons, is eliminated if the satellite itself emits a series of flashes at appropriate times. Such flashes have been given by Anna (1962), Geos 1 (1965), and Geos 2 (1968), but the power supply limits the frequency of the flashes, and a limited number of short flashes will not well average out the effects of atmospheric erratics. There is also a limit to the life of the lamps.

7.18. Satellite photography

(*a*) *Light. Units and basic formulae.*

(i) The *candela* (cd) is the SI base unit of luminous intensity (brilliance), defined [53] as the luminous intensity, in the perpendicular direction, of a surface of 1/600 000 m^2 of a black body† at the temperature of freezing platinum under a pressure of 101 325 N m^{-2}.

To the uninitiated this definition may not seem very clear, but see remarks after the definition of the lumen in (ii) below.

(ii) The *lumen* (lm) is the unit of luminous flux, namely the amount of light in unit solid angle (sr) emitted by a body which emits light of intensity 1 cd in the direction concerned. Such a body emits a total of 4π lm if it is equally brilliant in all directions. The dimensions of the lm are cd $\times$ sr.

Light is a form of energy, and its amount (per second) can be measured in watts, but the amount of energy in one lumen depends on the wavelength. For average sunlight 1 lumen = 1/206 watts, [375], p. J1, and is approximately equivalent to 1.35×10^{16} photons per second.

The easy, but imprecise, way of understanding the definitions is to regard the lumen as a flow of light amounting to so many photons per second, or watts (depending on the wavelength). Then the candela is a measure of brilliance, namely the brilliance arising from the concentration of one lumen evenly over one steradian.

(iii) The *lux* (lx) is a measure of the illumination given by a source of light to a distant surface.

$$1\ \text{lux} = 1\ \text{lumen per square meter} = 1\ \text{lm m}^{-2},$$

the square metre being normal to the direction of the light. As an example, black print of ordinary size, on white paper, can be read if the illumination is 30 lx. On a clear day the illumination of a horizontal

† The dictionary definition of *black* in this context is 'a hypothetical perfect absorber and radiator'.

surface by a zenith sun is of the order 10^5 lx, [413], p. 319, and by the full zenith moon it is 0.2 lx.

The old unit of 1 foot-candle equals 1 lumen per square foot, or approximately 10.7 lm m^{-2}, or 10.7 lx. The old international candle is nearly the same as the candela.

(iv) The efficacy of a short flash of light of duration t seconds is, from the photographic point of view, proportional to its duration. Its intensity will be measurable in candela-seconds, and the number of protons received by a particular piece of apparatus will be proportional to lm $\times t$.

(v) From the above definitions, if a source emits C candelas in the direction of an eye or telescope which is pointed at it from a distance of r metres, the amount of light entering a lens of diameter d metres will be

$$C\pi d^2/(4r^2) \text{ lm}, \tag{7.85}$$ †

This must be multiplied by the factor k of (7.92) to allow for the extinction of light by the (apparently clear) atmosphere

(vi) The *magnitude*, m, of a star is related to its illumination of the earth by the formula

$$m = -(2.5 \log L) - 14.13, \tag{7.86}$$

where L is the illumination it produces, measured in lm m^{-2}, on a perpendicular surface. The figure 2.5 derives from the basic definition that a decrease of 5 magnitudes corresponds to a 100-fold increase in light.

(*b*) *Passive satellites.* When a star or satellite is photographed by a camera which is rotated so as to follow it, the power of the camera can be expressed by the statement that it requires an exposure of t_1 seconds to give a satisfactory image of a star of magnitude m_1. It then follows from the definition of stellar magnitudes that if t_2 is the minimum exposure for a star of magnitude m_2

$$(m_2 - m_1) = 2\tfrac{1}{2} \log_{10}(t_2/t_1). \tag{7.87}$$

When a fixed camera is used, if a star or satellite trails across the field of view with angular velocity ω, the amount of light falling on each part of the trail will be inversely proportional to ω. If m_1 is the magnitude which will just produce a clear trail when $\omega = 1°$/second, the magnitude which will produce a trail with velocity $\omega°$/second is given by

$$(m - m_1) = 2\tfrac{1}{2} \log_{10}(1/\omega),$$

or

$$m = m_1 - 2\tfrac{1}{2} \log_{10} \omega. \tag{7.88}$$

† C cd are distributed over 1 sr, i.e. over an area of r^2 at a distance r. The area of the lens being $\pi d^2/4$, the lens receives $(C/r^2)(\pi d^2/4)$ lm.

The magnitude m_1 which can just give a trail when $\omega = 1°$/second is known as the *tracking power*, *P*, of the camera. A large camera has a large *P*.

The *tracking capacity*, *Q*, of a satellite, represented by a large number if it is faint or fast moving, depends on its intrinsic luminosity, its distance, and its apparent angular velocity ω measured in degrees/second. We have

$$Q = m + 2\tfrac{1}{2}\log_{10}\omega, \tag{7.89}$$

where m is its magnitude. Geodetically useful satellites generally move with ω's which are less than 1°/second, and whose logs are consequently negative. If necessary, ω can be reduced by the camera following the satellite, see § 7.20(*a*)(ii).

A particular camera will then give a trail for a particular satellite if $P > Q$.

With present equipment the only source of continuous illumination, such as can give a trail, is reflected sunlight. The illumination given by a reflecting† sphere is given by

$$E = (ab^2/4r^2)E_0, \tag{7.90}$$

where a is the albedo of the surface, perhaps 0.8,
b is the radius of the satellite in metres,
r is the distance from satellite to camera in metres,
E_0 is the illumination incident on the satellite from the sun, namely 1.2×10^5 lumens per metre^2, [361], p. 15.

Then the apparent magnitude of the satellite is given by

$$m = -14.13 - 2\tfrac{1}{2}\log_{10} kE, \tag{7.91}$$

where k is a factor which allows for the reduction (extinction) of light passing through the atmosphere. For an apparently clear atmosphere

$$k = \exp\{-(0.0090\lambda^{-4} + 0.223)\sec z\}.$$

where z is the zenith distance, and λ is the wavelength in μm. For sunlight $\lambda = 0.52$, and

$$k = \exp(-0.345 \sec z). \tag{7.92}$$

See [361], which gives a nomogram for m on p. 17.

The magnitude of Pageos at a zenith distance of 40°, with $b =$ 15 metres, $a = 0.8$, and $r = 5500$ km is given by (7.80), (7.91), and (7.92) as +3.

(*c*) *Flashing satellites.* The flashes emitted by artificial satellites are very short and very intense. Thus Geos 1 carried four flash tubes, which

† i.e. a 'specular' sphere which reflects like a mirror. Not a matt surface which scatters the incident light, for which see [361], p. 15.

together gave a momentary brilliance of between 12×10^6 and 45×10^6cd in directions inclined at up to 60° to the vertical axis, with the maximum at about 50°, for observations at maximum distances. The duration of a flash was about $0^s.001$, and the significant figures are the products 12 000 to 45 000 candela seconds. See [333], p. 105.

If the light emitted in the direction of the camera is C cd s, (7.80) multiplied by (say) 0.6 for atmospheric extinction (7.92), and by another 0.7 for loss of light inside the camera, gives the total light falling on the emulsion per flash as

$$0.42C(\pi d^2)/(4r^2) \text{ lm s}$$

or
$$0.45C(d^2/r^2)\times10^{16} \text{ photons, from § 7.18 } (a) \text{ (ii)}, \tag{7.93}$$

where d is the lens diameter and r is the distance, both in the same units. So if a BC 4 camera with $d=11.7$ cm, § 7.20 (c), photographs Geos with $C=$ (say) 20 000 candela seconds at 2000 km, the number of photons reaching the emulsion will be 3×10^5. About 10^5 photons are required to produce a good image on the film used by the Baker–Nunn cameras, [375], p. 26. In these circumstances the margin is not very large.

[603] gives the following formula for the diameter p of the point image on sensitive emulsion

$$p=A_1q+A_2q^2+A_3q^3 \ \mu\text{m}, \tag{7.94}$$

where $A_1=7.47$, $A_2=0.112$, and $A_3=0.00083$, and

$$q=(d/r)(TC)^{\frac{1}{2}}P, \tag{7.95}$$

where d is the diameter of the lens in mm,
r is the slant range in km,
T is the factor for internal loss, say 0.7,
C is the intensity of the flash in cd s.
P is an extinction factor given by

$$P=\exp(-0.46\,\Delta m \sec z), \tag{7.96}$$

where z is zenith distance, and Δm is taken as 0.25 for clear conditions, and as 1.25 for moderate haze. Using the same figures as after (7.93) above, a BC 4 camera would get a 32-μm image of Geos at 2000 km.

7.19. Selection of orbits

(a) The *inclination* of the orbit must be such that the satellite can be seen in the desired azimuths and zenith distances from stations in the highest latitudes which the net is intended to cover. For a world-wide triangulation net a high inclination is necessary, but if observations in high latitudes are not envisaged, larger numbers of visible passages will be provided if the inclination is no greater than is necessary.

(*b*) *Altitude.* For astro-triangulation the optimum altitude is between 0.5 and 1.0 times the average length of triangle side, AB in Fig. 7.6, see [362]. The lower height probably gives maximum accuracy from each simultaneous pair of observations, but restricts the number of favourable opportunities. If the proposed use is for nets of varying side lengths, perigee height can be suited to the smallest and apogee to the largest, and the rotation of the apses will periodically bring perigee and apogee into all the latitudes covered by the inclination. The disadvantage is that for any particular line there may be considerable intervals of time during which conditions are not optimum.

7.20. Cameras

(*a*) The photographic tracking of a passive, continuously illuminated satellite involves the difficulty that the satellite and the stars move at different speeds. There are three possible systems.

(i) The camera may follow the stars. This is seldom done, except with the MOTS cameras.

(ii) To follow the satellite. A third rotation axis is necessary. It is in any case inapplicable to a flashing satellite, since its flashes would coalesce. The Baker–Nunn cameras can do this, and sometimes use the method with passive satellites.

(iii) To keep the camera still. This is the usual method. A flashing satellite is as well recorded by a still camera as by one which follows it. The resulting plate shows numerous short star trails with time breaks, and a much longer satellite trial with simultaneous breaks, or a line of flashes.

Sub-paragraphs (*b*) to (*e*) below give some details of cameras which have been much used. See also [390], pp. 103–28, with special reference to the photography of flashing satellites. It also describes the PC-1000, the K-50, and PTH-100 cameras.

(*b*) *The Baker–Nunn cameras* of the Smithsonian Astrophysical Observatory. Focal length 50 cm. Aperture 50 cm. Field of view $5° \times 30°$. Schmidt type. Not mobile, but installed at fifteen or more stations all over the world. Film is used in 300-metre rolls, pressed against a curved backing plate.

The camera is mounted on three axes, horizontal, vertical, and tracking. It can follow a satellite along any great circle at speeds of from 0° to 2° per second. It can record +12 magnitude stars or satellites with a $3^s.2$ exposure if it follows them. It will record a +8.5 mag satellite moving at 0.1°/second, a fairly typical speed. It is generally used still for satellites brighter than +7 mag.

Exposures can be of 0.2, 0.4, 0.8, 1.6, or 3.2 seconds, during which star or satellite trails are chopped into six equal segments by five fine breaks. The time of only the central break is recorded. In a typical satellite passage it is possible to observe three sets, each of ten such exposures (frames), giving a total of thirty timed breaks in the trails of any object.

The timing of the breaks is recorded to a few tenths of a millisecond. A single frame gives an apparent s.e. of ±2″ for the direction of the satellite.

See [391], i, p. 43, [216], pp. 17–26, and [430] for diagrams and photograph.

(*c*) *The BC* 4 *Wild camera.* Focal length 30 cm, later 45 cm for the Pageos programme. Aperture 11.7 cm. Field of view 33°×33° for the 30-cm lens. Can be carried with all associated electronics in a mobile trailer. Flat field, glass plates 18×18 cm or 19×25 cm.

The camera is mounted on the base of a T4 theodolite, with horizontal and vertical axes, and is only used still. It can record trails of +9 mag stars, and of a+6 mag satellite moving at 0.1°/second.

For satellite trails the chopper rotates 10, 5, or $2\frac{1}{2}$ times a second, to give exposures of 1/60, 1/30, or 1/15 seconds respectively. But the frequency of the exposures is reduced, either 2 or 5 times, by a capping shutter. A further reduction by 2, 4, 8, or 16 can be arranged by an auxiliary capping shutter mounted in front of the lens, but this is generally only used for star trails.

For star trails the high-speed chopper and the inner capping shutter are locked open, and sequences of five exposures are made with the auxiliary capping shutter, before and after the chopping of the satellite trails. These five exposures are ordinarily of different lengths, such as of 5, 2, 1, $\frac{1}{2}$, and $\frac{1}{4}$ second. The two exposures which give images of similar density to the satellite images are selected for measurement.

A single plate gives an apparent s.e. of ±2″.

See [430] and [553] for details and photographs.

(*d*) *The MOTS*-40 *camera.* Focal length 101.6 cm. Aperture 20 cm. Field of view 11°×14°. Flat field, with glass plates 20×25 cm. Housed in a fibre glass astronomical dome. The camera is equatorially mounted and is driven at sidereal speed, so that it follows the stars. It can record stars of +10 magnitude with a 35-s exposure, [273].

(*e*) *The PC*-1000. Focal length 100 cm. Aperture 20 cm. Field of view 10°×10°. Carried with its electronics in a mobile trailer. Flat field, glass plates 18×18 cm.

The camera is mounted on horizontal and vertical axes, and is used still. It can record stars of +7.0 mag, and satellites of +6 mag moving at 0.1°/second.

There is a between-lens iris shutter with a louvre shutter in front of it. The louvre is capable of making exposures of a few milliseconds, and is used for the satellite with the iris shutter open. The iris is used for the stars with the louvre open, and makes two sequences each of 2, 1, 0.5,

0.3, and 0.1 seconds before the satellite, and again after it, as described for the BC 4 above.

A single plate gives an apparent s.e. of ±2″. See [430].

(*f*) *The Hewitt camera* (Great Britain). Focal length 60 cm. Aperture 60 cm. Field of view 10° diameter. Schmidt type. Not mobile. Flat field and glass plates. There is significant radial displacement of the image, which is allowed for in the reduction, see 7.22(*f*)(v).

The camera is mounted on horizontal and vertical axes and is used still. Stars of +8 mag give an image of 42 μm diameter. It can record +6 or 6.5 mag satellites moving at 1°/second, or 8 mag at 0°.2/s.

Exposures with the capping shutter can be from $0^s.25$ upwards. Breaks of about 0.1 mm in the trails are made by an internal rotating sector shutter, and the times of the breaks are usually displayed to $0^s.0001$. The timing accuracy, in terms of the times of arrival of time signals, is $\pm 0^s.0002$. See [274]. The accuracy of direction is believed to be greater than that of other existing cameras.

(*g*) *Time-keeping.* With modern electronic equipment the measurement of short time differences between exposures at stations A and B, to within a few tenths of a millisecond, presents no difficulties. A crystal clock at each station records the times of reception of the same wireless signals, and also the movements of the camera chopping shutters.

7.21. Cartesian reference system

Having obtained a set of plates or films at a particular station showing the timed images of the satellite surrounded by simultaneous images of near-by stars, we require the direction cosines of the line from the station to the satellite with reference to some prescribed earth-based coordinate system. We also have a star catalogue, presumably the SAO catalogue, § 4.06 (*b*), which is likely to be the only one to include sufficient stars. The RA and δ of the stars being known, it is required to obtain the RA and δ of the satellite images (as seen from the ground station) by interpolation.

The *terrestrial reference system*, in which the final results are to be expressed, will clearly have its z-axis parallel to the direction defined by the CIO. The x-axis will be parallel to the corresponding equator, and also to the Greenwich meridian plane, the CZM, as defined by the BIH for recording UT 1, as in § 4.10 (*d*). The y-axis will be perpendicular to them in the meridian 90° E. These axes are parallel to those now generally adopted for geodetic reference systems, § 2.01 (*a*). Astro-triangulation cannot, by itself, locate the geocentre, and the origin of coordinates must be selected arbitrarily. It will be necessary to compute in the NAD, or other national cartesian coordinates† for any one of the

† i.e. cartesian coordinates derived from geodetic spherical coordinates, as in (C.1).

observing stations, which may (if desired) be provisionally converted to geocentric by using values of Δx, Δy, and Δz as derived by other satellite methods. See § 7.77.

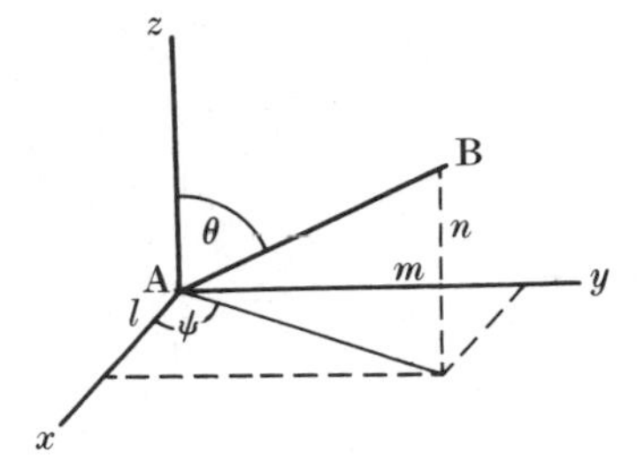

FIG. 7.7. If AB is of unit length, l, m, and n are its direction cosines.

Let l, m, n be the direction cosines of a certain line with reference to this system, so that $\cos\theta = n$ and $\tan\psi = m/l$, where θ and ψ are angles which also describe the direction of the line as in Fig. 7.7 and in § C.01 (b). Then, (C.19), (4.22), and (4.27), a line whose direction cosines are l, m, and n will meet the celestial sphere at the point given by

$$\begin{aligned}\text{RA} &= \text{GAST}+\psi+(1/15)(X''\sin\psi+Y''\cos\psi)\cot\theta\\ &= \text{UT }1+R_u+12\text{ hours}+(\text{nutation in RA})\\ &\quad+\psi+(1/15)(X''\sin\psi+Y''\cos\psi)\cot\theta,\text{ see Fig. 4.4.}\\ \delta &= (90^\circ-\theta)+(X''\cos\psi-Y''\sin\psi). \qquad (7.97)\dagger\end{aligned}$$

The description of computation methods given in § 7.22–7.24 gives some basic principles, but they are not an accurate account of the detailed handling of large masses of data in any particular computation programme. Such details vary from one programme to another. See, for example [391], [516], and [567].

7.22. Computation of satellite RA and δ from plate coordinates.

(a) *Plate coordinates.* A suitable camera, such as one of those described in § 7.20, photographs the satellite and various surrounding stars at a recorded instant of GAST. The RA and δ of the stars being known, it is required to find the RA and δ of the satellite at the instant of observation.

The *plate coordinates* of the star and satellite images on the photographic plate are measured in a *comparator* relative to arbitrary axes Mx and My,‡ the origin M being near the principal point, and My being approximately towards the image of the pole, Fig. 7.8. Let Mz, which is required later, be positive towards the lens. As an intermediary between these plate coordinates and the star positions (RA and δ) it is necessary to introduce a system of *standard coordinates* as below.

† The polar coordinates X'' and Y'' are written as x and y in § 4.09. X'' and Y'' are in seconds of arc. $\theta = 90-\phi$.

‡ These axes x, y have no connection with the world cartesian reference system x, y, z of § 7.21.

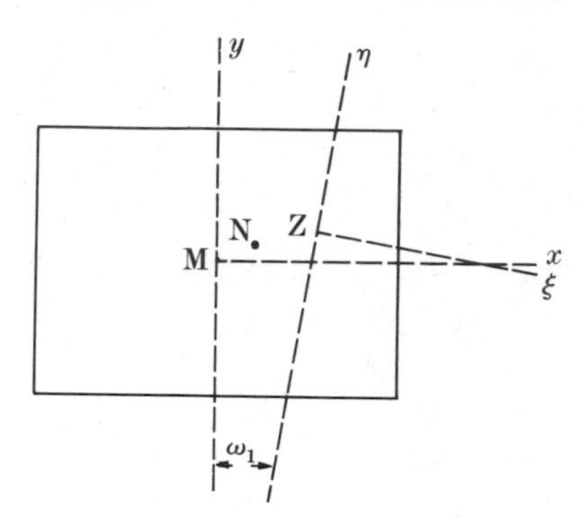

FIG. 7.8. The plate. The lens O is above the paper, and the axes of z and Z are positive upwards. M is the origin of plate coordinates x, y, z. N is the principal point. The direction of NO is α_0, δ_0, and the direction of ZO is α_P, δ_P, the provisional values of α_0, δ_0. The axes of x and y are close to those of ζ and η.

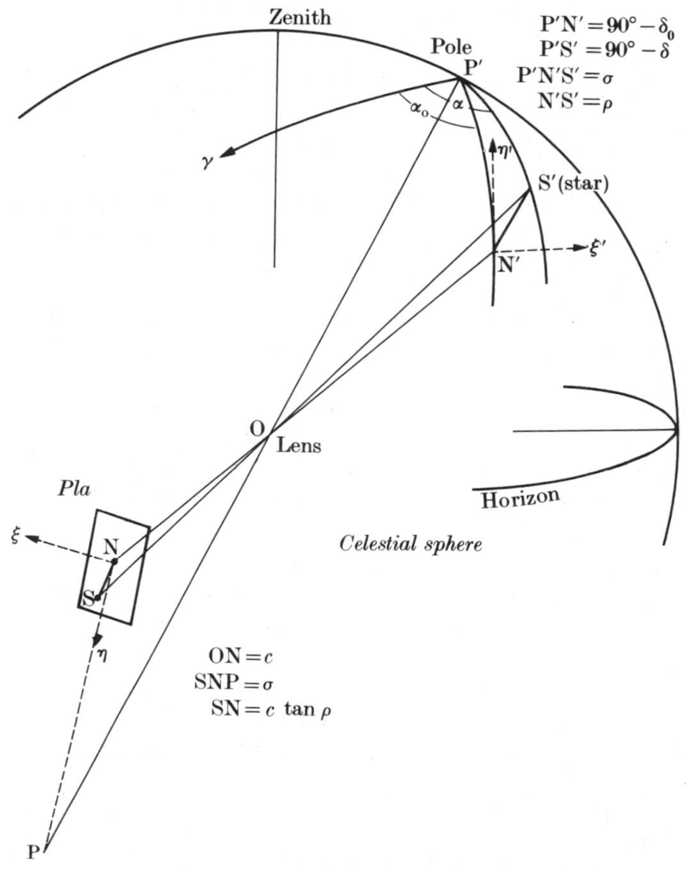

FIG. 7.9. Standard coordinates ξ, η. ON is perpendicular to the plate.

(*b*) Star places are taken from the SAO catalogue, which gives mean places for the epoch 1950.0. In the SAO routine [391] they are at this stage only brought up to date by applying the proper motion (in the 1950.0 frame) and annual parallax, if any. For refraction, see (*f*) (ii) below.

(*c*) *Standard coordinates*, [533], pp. 278–85. Fig. 7.9 shows the camera plate, O the nodal point of the lens, N the principal point of the plate, namely the foot of the perpendicular from O, and N′ the point on the celestial sphere towards which NO is directed. Let the RA and the declination of N′ be α_0 and δ_0. Let S′ be a star, α and δ, and let S be its image on the plate. Then the *standard coordinates* of S, ξ and η, are defined to have origin at N, the axis of η towards the image of the pole, and the axis of ξ in the direction shown in Fig. 7.9, so that stars with RA greater than α_0 have positive ξ's. Let the principal distance NO be taken as the unit of length. Let N′OS′ (=NOS) = ρ, and let PNS = σ. Then

$$\xi = \tan\rho \sin\sigma,$$

$$\eta = \tan\rho \cos\sigma. \tag{7.98}$$

We have $\cos\rho = \sin\delta \sin\delta_0 + \cos\delta \cos\delta_0 \cos(\alpha-\alpha_0)$,

$$\sin\rho \sin\sigma = \cos\delta \sin(\alpha-\alpha_0). \tag{7.99}$$

Also $\quad \sin\rho\cos\sigma = \sin\delta\cos\delta_0 - \cos\delta\sin\delta_0\cos(\alpha-\alpha_0)$,

from (A.30). Whence, dividing,

$$\eta = \frac{\cos\delta_0 - \cot\delta \sin\delta_0 \cos(\alpha-\alpha_0)}{\sin\delta_0 + \cot\delta\cos\delta_0\cos(\alpha-\alpha_0)} \tag{7.100}$$

or

$$\eta = \tan(q-\delta_0), \text{ where } \cot q = \cot\delta\cos(\alpha-\alpha_0). \tag{7.101}$$

and

$$\xi = \frac{\cot\delta\sin(\alpha-\alpha_0)}{\sin\delta_0 + \cot\delta\cos\delta_0\cos(\alpha-\alpha_0)} \tag{7.102}$$

or

$$\xi = \frac{\cos q \tan(\alpha-\alpha_0)}{\cos(q-\delta_0)}. \tag{7.103}$$

And conversely

$$\tan(\alpha-\alpha_0) = \frac{\xi\sec\delta_0}{1-\eta\tan\delta_0}, \tag{7.104}$$

and

$$\cot\delta\cos(\alpha-\alpha_0) = \frac{1-\eta\tan\delta_0}{\eta+\tan\delta_0}$$

or

$$\cot\delta\sin(\alpha-\alpha_0) = \frac{\xi\sec\delta_0}{\eta+\tan\delta_0}. \tag{7.105}$$

(*d*) *Reduction of plate coordinates to standard. Turner's method,* [533], pp. 285–97. The coordinates x, y which are measured on the plate are similar to ξ, η, but differ from them for various reasons as follows.

(i) OM, where M in Fig. 7.8 is the origin of the plate coordinates x, y, may not be perpendicular to the plate, so N will not coincide with M.

(ii) The supposed RA and declination of NO, as used to relate (α, δ) and (ξ, η) in (7.100)–(7.105) may not be the correct values α_0, δ_0 but approximate values α_P, δ_P. It is then generally convenient to accept the approximate values α_P and δ_P, and to regard the plate as being inadvertently inclined to the ξ, η axes at inclinations of $(\alpha_0-\alpha_P)\cos\delta_0$ and $(\delta_0-\delta_P)$. At Z, ξ and $\eta=0$.

(iii) The y-axis may not lie exactly in the meridian.

(iv) The principal distance may actually be $c_0+\delta c$ rather than the reputed length c_0, which is taken as the unit of length.

(v) The comparator axes x and y may not be mutually perpendicular, and their scales may be unequal.

(vi) The star and satellite images will be displaced by refraction, and the refraction of the satellite will be modified by parallactic refraction, § 7.07.

As a good approximation if the field of view is less than about 5°, and except for parallactic refraction and ordinary refraction at zenith distances of more than 60°, [533], pp. 287–97 shows that all the above sources of error, and also the neglect of annual and diurnal aberration, result in a linear relation between x, y and ξ, η. So we may write

$$\begin{aligned}\xi-x&=a_0+a_1x+a_2y,\\ \eta-y&=b_0+b_1x+b_2y,\end{aligned}\tag{7.106}$$

where the a's and b's are small constants, and the x's and y's are expressed in units of c_0. Then let (7.101), and (7.103) give the ξ and η of three or more stars of known RA and δ, whose plate coordinates have been measured, and we will have six or more equations for the six constants a_0, etc. These can be determined by least squares, and (7.106) will then give the ξ and η of any other object whose plate coordinates have been measured, such as a satellite. Equations (7.104) and (7.105) using α_P and δ_P then give its RA and declination at the moment of photography. This is Turner's method.

(*e*) *Aberration.* The satellite positions given by (*d*) now require correction for annual aberration. It could have been applied to the star positions earlier at the same time as proper motion, but it is more convenient at this stage. The corrections, [391], p. 56, are

$$\Delta\alpha=-(20''.47\sin\alpha\sin\odot+18''.87\cos\alpha\cos\odot)\sec\delta\tag{7.107}$$

$$\Delta\delta=-[20''.47\sin\delta\cos\alpha\sin\odot+18''.87\cos\odot(0.4337\cos\delta-\sin\delta\sin\alpha)],\tag{7.108}$$

where $\odot$ is the geocentric longitude of the sun, which is tabulated in AE

p. 20–35. For this purpose ⊙ is only required correct to one minute of arc.

(*f*) *Elaborations of Turner's method.* In some circumstances some of the following alternatives may be necessary, notably if the stars surrounding each satellite image are abnormally far from it.

(i) *Plate calibration*, Item (*d*) (v). Lack of perpendicularity and differences of scale between the x- and y-axes may be eliminated by calibration of the comparator.†

(ii) Item (*d*) (vi) above. Instead of relying on Turner's method to deal with differential refraction, the refraction for each star may be computed as in § 4.11 and applied to its RA and δ as below. The RA and δ as given for the satellite by Turner's method will then require corrections appropriate to its zenith distance, with opposite signs. In any case, the parallactic refraction of § 7.07 must be applied later.

The usual refraction correction ψ'' applicable to zenith distances can be resolved into corrections to RA and δ by

$$\text{Apparent declination} = \delta + \psi'' \cos q$$

$$\text{Apparent RA} = \alpha + (1/15)\,\psi'' \sin q \sec \delta, \tag{7.109}$$

where q is the parallactic angle ZSP in Fig. 4.1, which is given by

$$\sin q = \sin t \cos \phi \operatorname{cosec} \zeta$$

or

$$\sin \phi = \cos \zeta \sin \delta + \sin \zeta \cos \delta \cos q, \tag{7.110}$$

in which ϕ is latitude, ζ is zenith distance, and t is hour angle. When $t > 12$ hours, q is negative.

This leaves items (i) to (iv) of subparagraph (*d*) above.

(iii) Instead of (7.106) we may write

$$\begin{aligned} \xi - x &= a_0 + a_1 x + a_2 y + a_3 x^2 + a_4 xy + a_5 y^2, \\ \eta - y &= b_0 + b_1 x + b_2 y + b_3 x^2 + b_4 xy + b_5 y^2. \end{aligned} \tag{7.111}$$

Third-order terms may be added, as

$$a_6 x^3 + a_7 x^2 y + a_8 x y^2 + a_9 y^3, \tag{7.112}$$

and the corresponding b's.

Given sufficient stars, all the constants can be determined by least

† The comparator is periodically checked against a master grid. This grid can in turn be calibrated by the comparator itself, since without any knowledge of the comparator's accuracy, a pair of rigidly mounted microscopes can be used to verify
(*a*) that the four sides of a large grid square are equal,
(*b*) that its two diagonals are equal, and
(*c*) that its scale sub-divisions are equal.

squares. The utility of these extra terms depends on x and y being substantially smaller than the principal distance c.

(iv) *Photogrammetric method.* The following, based on [128] is a more rigorous treatment. Take X, Y, Z with origin at O, the lens, such that the direction OZ is α_P, δ_P the accepted values of α_0 and δ_0, and let the OX and OY axes be co-planar with the axes of ξ and η as so computed.

Fig. 7.8 shows the plate. M is the origin of plate coordinates x, y, and z, $N(x_N, y_N)$ is the principal point, and Z is the point where the Z-axis cuts the plate. If the accepted α_P and δ_P exactly equalled the true values α_0 and δ_0, the z- and Z-axes would be parallel, Z would coincide with N, and the OX and OY axes would be parallel to the plate, but actually this is not the case. Also, on account of adjustment error, the plate axis My is not parallel to the line in which the $Z\eta$ plane cuts the plate. The differences between the standard coordinates and the plate coordinates thus depend on the following six unknown quantities.

(A) x_N and y_N, the plate coordinates of the principal point N.

(B) The values of $(\alpha_0 - \alpha_P)$ and $(\delta_0 - \delta_P)$.

(C) δc, the error in the reputed principal distance.

(D) The angle between My and $Z\eta$ in Fig. 7.8.

Let $l_x m_x n_x$, $l_y m_y n_y$, and $l_z m_z n_z$ be the direction cosines of the plate axes x, y, z relative to X, Y, Z. Then [128] gives the following expressions for the x, y of an object S whose standard coordinates computed with α_P and δ_P are ξ, η.

$$x = x_N + c\frac{l_x\xi + m_x\eta + n_x}{l_z\xi + m_z\eta + n_z},$$
$$y = y_N + c\frac{l_y\xi + m_y\eta + n_y}{l_z\xi + m_z\eta + \eta_z}. \tag{7.113}$$

And conversely

$$\xi = \frac{l_x(x - x_N) + l_y(y - y_N) + l_z c}{n_x(x - x_N) + n_y(y - y_N) + n_z c},$$
$$\eta = \frac{m_x(x - x_N) + m_y(y - y_N) + m_z c}{n_x(x - x_N) + n_y(y - y_N) + n_z c}. \tag{7.114}$$

The nine direction cosines are not independent, being related by three identities $l^2 + m^2 + n^2 = 1$ and three others of the form

$$l_x l_y + m_x m_y + n_x n_y = 0.$$

They can be reduced into terms of three independent angles ω_1, ω_2, and ω_3, and are given in that form in (B.35), where $\omega_1\omega_2\omega_3$ are $\theta\phi\psi$.

If the ω's are small, so that their squares and products can be neglected, the direction cosines can be very simply expressed as

$$\begin{bmatrix} l_x & m_x & n_x \\ l_y & m_y & n_y \\ l_z & m_z & n_z \end{bmatrix} = \begin{bmatrix} 1 & \omega_1 & \omega_3 \\ -\omega_1 & 1 & \omega_2 \\ -\omega_3 & -\omega_2 & 1 \end{bmatrix}. \tag{7.115}$$

In fact

$\omega_1 =$ the angle between My and $Z\eta$ in Fig. 7.8,

$$\omega_2 = \delta_0 - \delta_P,$$

and $$\omega_3 = (\alpha_0 - \alpha_P)\cos\delta. \tag{7.116}$$

Substituting (7.115) in (7.113) and putting $c = c_0 + \delta c$ gives

$$\begin{aligned} x &= x_N + c_0\{\xi + \eta\omega_1 + \xi\eta\omega_2 + (1+\xi^2)\omega_3\} + \xi\delta c, \\ y &= y_N + c_0\{\eta - \xi\omega_1 + (1+\eta^2)\omega_2 + \xi\eta\omega_3\} + \eta\delta c. \end{aligned} \tag{7.117}$$

For a least-square solution equations (7.117) are the observation equations. The unknowns are x_N, y_N, ω_1, ω_2, ω_3, and δc. The fallible observed quantities whose precision determines the weights are x and y. The coefficients ξ and η are computed for each star by (7.101) and (7.103), using α_P and δ_P for α_0 and δ_0. If ω_1, ω_2, and ω_3 are less than $1'$, and δc less than (say) $c/3500$, ignored products will be of the order $c \times 10^{-7}$ and a second approximation will be unnecessary. One can be made if there is any doubt.

When the six unknowns have been determined, (7.115) and (7.114) give the standard coordinates of all satellite images, and (7.104) and (7.105) computed with α_P and δ_P, then give the satellite's RA and δ at the recorded times.

In all the observation equations (7.117) the coefficient of x_N is unity, and that of ω_3 is $c_0(1+\xi^2)$ which differs little from c_0 if ξ is small. The separate values of x_N and ω_3 are consequently ill-determined, as also are those of y_N and ω_2. But this does no damage, since the instability does not extend to the combinations of x_N and ω_3, or of y_N and ω_2, which occur in (7.114).

(v) *Radial distortion.* With some cameras [173] it is necessary to include in the solution an expression for radial distortion of the plate image. In (7.113) write

$$(x - x_N)\Big(1 + \sum_i h_i r^{2i}\Big) = c\frac{l_x\xi + m_x\eta + \eta_z}{l_z\xi + m_z\eta + n_z}$$

or

$$x = x_N - (1/r)(x - x_N)\sum_i h_i r^{2i+1} + c\frac{l_x\xi + m_x\eta + n_x}{l_z\xi + m_z\eta + n_z}, \tag{7.118}$$

and similarly for y. The h's are small coefficients, and r is radial distance from N in Fig. 7.8. As in (iv) above, x_N and y_N are probably sufficiently small or well known for a first approximation to give final results, and r can be taken as $(x^2+y^2)^{\frac{1}{2}}$, but a second approximation can be made if necessary.

In the Hewitt cameras, § 7.20 (*e*), h_1 is large and its inclusion is necessary, but the other h's are negligible.

(*g*) *Simultaneous observations.* Referring now to the process of astro-triangulation, at two stations A and B in Fig. 7.6, for a single passage of the satellite we have a series of plates taken at A containing a number of satellite images, and another series at B. Plates taken at the two stations overlap in time, but with exposures which are not exactly simultaneous, unless they arise from flashes sent out by the satellite. In the Baker–Nunn cameras there 4 to 10 satellite images on each plate, and in the BC 4 when observing Pageos there are 300. So a small number of 'fictitious' observations have to be interpolated from the actual timed images on each plate. These are obtained as follows.

An auxiliary plate coordinate system (x', y') is temporarily adopted, whose relationship to RA and δ is arbitrarily defined so that y' lies approximately along track and x' lies across it. Observed satellite RA's, and δ's are then converted to x' and y'. Then least squares determines the constants a and b in two polynomials such as

$$\begin{aligned} x' &= a_0 + a_1 t + a_2 t^2 + \ldots a_6 t^6 \\ y' &= b_0 + b_1 t + b_2 t^2 + \ldots b_6 t^6. \end{aligned} \qquad (7.119)$$

At selected times (typically seven) in each overlap, the polynomials for station A give x' and y', and thence RA and δ, while the polynomials for station B give similar information at the same instants. These simultaneous fictitious observations are then used for all later processes.

The SAO, using Baker–Nunn cameras with only seven satellite images on each plate, used 2-degree polynomials, but the Pageos programme with BC 4 cameras and 300 images, used 6-degree polynomials.

(*h*) *Final corrections to star places.* The following small corrections have to be applied, most conveniently at this stage.

(i) Diurnal aberration, as in (4.8).

(ii) Planetary aberration. If light takes t seconds to travel from the satellite to the camera, the plate will record the direction in which the satellite was at t seconds before the recorded time. The deduced RA and δ then require corrections of

$$(\mathrm{d}\alpha/\mathrm{d}t)d/299\,800 \text{ to RA, and } (\mathrm{d}\delta/\mathrm{d}t)d/299\,800 \text{ to } \delta, \qquad (7.120)$$

where d is its distance from the camera in km. The rates of change $\mathrm{d}\alpha/\mathrm{d}t$

and $d\delta/dt$ can be computed from the multiple images on the plate. This is called *planetary aberration.* Since it depends on the distance of the satellite, a preliminary rough knowledge of the orbit is necessary.

(iii) Parallactic refraction, see § 7.07.

7.23. Conversion to the terrestrial reference system.

The satellite directions so far obtained are up to date in respect of proper motion, annual parallax, aberration, and refraction, for which corrections have been applied, but they are still expressed in terms of 1950.0 equinox and equator. It is now necessary to convert them to the equinox and equator of date, using rotation matrices as below.

First, the values of RA and δ must be converted into direction cosines $l_1 m_1 n_1$ relative to the same frame (Fig. 7.7) by

$$\begin{aligned} l_1 &= \cos\delta\cos\alpha \\ m_1 &= \cos\delta\sin\alpha \\ n_1 &= \sin\delta \end{aligned} \tag{7.121}$$

The correction for precession then gives

$$\begin{bmatrix} l_2 \\ m_2 \\ n_2 \end{bmatrix} = P \begin{bmatrix} l_1 \\ m_1 \\ n_1 \end{bmatrix} \tag{7.122}$$

where P is as in § 4.08 (a).

$l_2 m_2 n_2$ are then similarly corrected for nutation to give

$$\begin{bmatrix} l_3 \\ m_3 \\ n_3 \end{bmatrix} = NP \begin{bmatrix} l_1 \\ m_1 \\ n_1 \end{bmatrix} \tag{7.123}$$

where N is as in § 4.08 (b).

It then remains to convert $l_3 m_3 n_3$ to the required terrestrial frame $l\,m\,n$, by a further rotation S, so that

$$\begin{bmatrix} l \\ m \\ n \end{bmatrix} = SNP \begin{bmatrix} l_1 \\ m_1 \\ n_1 \end{bmatrix} \tag{7.124}$$

$$\text{where } S = \begin{bmatrix} \cos\theta & \sin\theta & X'' \\ -\sin\theta & \cos\theta & -Y'' \\ -X''\cos\theta - Y''\sin\theta & -X''\sin\theta + Y''\cos\theta & 1 \end{bmatrix} \tag{7.125}$$

in which θ is the true (Greenwich) sidereal time, and X'' and Y'' are the coordinates of the instantaneous pole relative to the CIO, expressed in radians, [391], (i), p. 36.

Details of the procedure followed by the SAO are in [391]. [516] and [273] give details of the Pageos programme, and [567] describes observations in Holland in more detail.

7.24. Direction of the line joining adjacent stations

Referring to Fig. 7.6, it is required to find the direction cosines $l\,m\,n$ of the line AB in the cartesian system of § 7.21. Paragraphs 7.22 and 7.23 have given the direction cosines $l_1\,m_1\,n_1$ of the line AS_1, and $l_2\,m_2\,n_2$ of BS_1, at a series of identical instants, and from another passage of the satellite S_2 at different times we have a second series of simultaneous values $l_3\,m_3\,n_3$ for AS_2 and $l_4\,m_4\,n_4$, for BS_2, preferably with S_1 and S_2 on opposite sides of the line AB. Considering a single set of these four directions, the condition that the two lines AS_1 and BS_1 should meet is

$$\begin{vmatrix} l & m & n \\ l_1 & m_1 & n_1 \\ l_2 & m_2 & n_2 \end{vmatrix} = 0, \text{ as in (C.39)}, \tag{7.126}$$

where $l\,m\,n$ are the direction cosines of AB. Since AS_1 and BS_1 are known to intersect at S_1 (within the normal limits of error, while the lines AS_2 and BS_2 similarly also meet at S_2, we have

$$\begin{vmatrix} l & m & n \\ l_1 & m_1 & n_1 \\ l_2 & m_2 & n_2 \end{vmatrix} = \begin{vmatrix} l & m & n \\ l_3 & m_3 & n_3 \\ l_4 & m_4 & n_4 \end{vmatrix} = 0, \text{ as in (C.40)}, \tag{7.127}$$

which, taken with $l^2+m^2+n^2=1$, determines $l\,m\,n$. The solution is given in (C.41). Every set of four such directions gives another set of values for $l\,m\,n$.

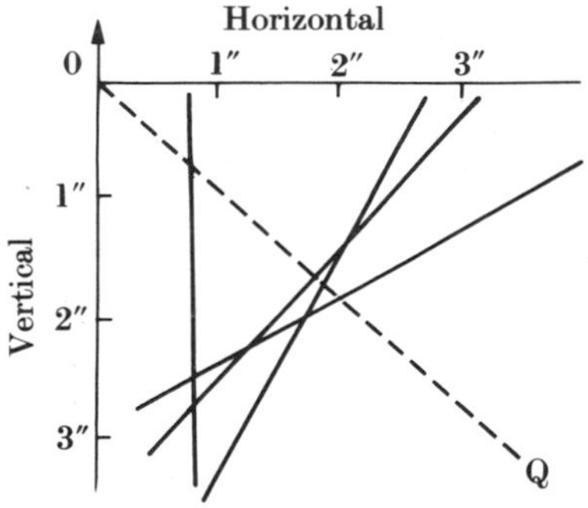

Necessary conditions for a firm solution are as below.

(*a*) The angles AS_1B and AS_2B must be neither small nor near 180°. 90° is the best, and anything between 60° and 120° is satisfactory.

(*b*) The planes AS_1B and AS_2B must similarly intersect at a good angle.

If more than two pairs of simultaneous observations AS and BS are made, as will be usual, the direction of AB will be got by least squares, with observation equations in the form of (C.42). Other

FIG. 7.10. View along AB. The diagram shows that further observations are required, such as will give planes roughly parallel to OQ.

things being equal, individual planes could be weighted in proportion to sin ASB. For a better routine see [570], and [391], i, pp. 67–71. The solution will give the error ellipse of the direction AB. The situation can also be exhibited graphically by a diagram. Fig. 7.10 is drawn in a plane perpendicular to AB, looking along AB. The oblique lines are the intersections of this plane by the various planes ASB. The range of angles between them, and their failure to be exactly concurrent, immediately shows whether the fixation is satisfactory, and in what directions the satellite should be further observed, if anything more is required.

[362] gives diagrams showing the optimum position of a satellite at the time of observation.

7.25. Scale

One or more of the sides of an astro-triangulation net must be measured. There are three possible systems as below.

(*a*) *Ground measurement.* Two or more of the stations of the astro net are connected by first or zero order triangulation or traverse, including the observation of geoid profiles between them as in Chapter 4, Section 6. If this is computed on any convenient spheroid and datum, the geodetic ϕ, λ and $(N+h)$ at these stations may be converted to cartesian coordinates, centred on the spheroid concerned, by C.1. The straight-line distance between them is then given by

$$L=\{(x_1-x_2)^2+(y_1-y_2)^2+(z_1-z_2)^2\}^{\frac{1}{2}}. \tag{7.128}$$

The direction cosines of the lines joining them are obtainable† from

$$l=(x_1-x_2)/L, \text{ etc.} \tag{7.129}$$

For this purpose existing frameworks may be used, if good enough, supplemented by additional field work if necessary, and probably recomputed. The Pageos programme was scaled by 5 or 6 special zero-order geodimeter traverses and geoid sections, as in § 1.06 (*c*).

(*b*) *Laser or microwave measurement of a single distance* such as AS_1 gives the lengths of all the lines in Fig. 7.6, if all directions have been measured as already described. The distance measurement must of course be made at substantially the same instant as the photography from A to B.

(*c*) If the distances to a satellite are simultaneously measured, by laser or microwave, from four connected astro-triangulation stations, the scale of the net is determined, even though the laser distances may not have

† Provided the spheroid has the CIO for the direction of its minor axis, and BIH Greenwich for its zero meridian.

been measured at the same time as the photography. Four observations determine four unknowns, namely the coordinates of the satellite, and the scale of the triangulation. Details of laser and of microwave distance measurements are in Section 6.

It is important to have a sufficient number of well-spaced scale controls in an extensive net of astro-triangulation. If such a net is not so controlled, the accumulation of errors of height (radial distance) through several triangles has the effect of systematically increasing or decreasing the differences of latitude and latitude between adjacent stations by 1 ppm per 6 metres of height error. Scale control will reduce this uncertainty by correcting the accumulated height error, as well as by controlling the errors of the 'horizontal' directions of the triangle sides.

7.26. Computation of the net. Conditions

(*a*) *Triangle conditions.* If the directions of the three sides of a triangle ABC have been determined in the cartesian frame as in §§ 7.21–7.24, the angles of the triangle are known, and if one side-length is known, the others follow. But it is first necessary to ascertain that the three directions are co-planar. If in Fig. 7.11 AC_1 and AB are considered to meet at A, while BC_2 and BA meet at B, the separation of AC_1 and BC_2 at C, perpendicular to the plane of the paper, is given by (C.33). This quantity should be computed as soon as possible, and should be seen to be satisfactorily small while there is still an opportunity to make more observations if necessary.

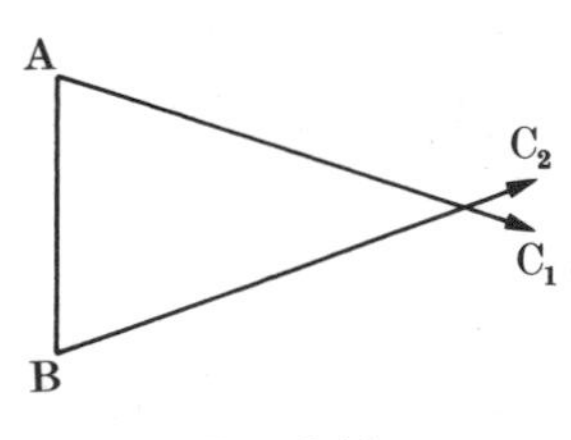

FIG. 7.11.

(*b*) *Side conditions.* Redundant lines, and measured distances in excess of one, introduce side conditions similar† to those of conventional triangulation.

(*c*) While it may be desirable to record these condition misclosures as above, if only as a guide to the accuracy achieved, there is no essential need to consider adjusting them, as the easiest method of adjusting the net is by variation of coordinates. On the other hand [391], (i), pp. 71–4 outlines a system for adjusting the triangle misclosures first, and then proceeding to the adjustment of the net using weight matrices derived from the triangular misclosures.

7.27. Computation of the net. Variation of coordinates

(*a*) The simplest, but not necessarily the best, procedure is as follows. The principle of the method is the same as that described in Chapter 2,

† But note that with these long sides it is not admissible to use the intersection of the diagonals as a pole. The diagonals will not, in general, intersect.

Section 3. The forms of possible observation equations in three dimensions are given in § C.02. In an astro-triangulation net as here described the computations can be very simple, and need only involve two kinds of observation equations as in (i) and (ii).

(i) Observed RA's and δ's, § (C.02(*f*), treating the deduced RA and δ of each surface line AB as if they were actually observed. It is also possible to form direction equations from ground surveys which may have been primarily intended for scale, either by using (7.129) and § C.02 (*b*), or by (C.21), (C.19), and § C.02 (*f*).

(ii) Observed lengths of sides, § C.02 (*a*). Weights will be the reciprocals of the squares of the estimated s.e.'s.

(*b*) In (*a*)(i) above, instead of treating the RA and δ of the lines AB as the basic observations, it is generally better to form observation equations with each set of observations of the RA and δ of AS_1, BS_1, AS_2, and BS_2 in Fig. 7.6. These are closer to what has been actually observed, and weights can be allotted more rationally.

(*c*) As usual in the method of variation of coordinates, changes to the coordinates of at least one station must be omitted. It is unlikely that more than one station will be held fixed, but if it were intended to hold both the length and direction of a side emanating from a fixed station, changes to the coordinates of the stations at each end would be omitted.

(*d*) It may be noted that the station coordinates resulting from observations made to artificial satellites, whether determined by the geometrical method, or by the photodynamic method, or by trilateration, or by doppler, are basically cartesian coordinates. If they are converted to latitude and longitude by (C.2)–(C.5), these will be spheroidal or geodetic (not astronomical) positions, in the sense of §§ 2.03–2.04. And their heights will be above the prescribed spheroid, not above the geoid. The local vertical, to which astronomical positions refer, does not enter into any of these processes.

7.28. Accuracy

If astro-triangulation is to be used to control continental primary frameworks, its standard error of position, horizontal or vertical, must not exceed 2 ppm over distances of more than (say) 2000 or 3000 km. See §§ 1.01 and 2.44. There are admittedly many large areas in which the best primary framework is of much lower accuracy, but to determine the direction of each astro-triangulated line with an s.e. of 1 ppm is a reasonable aim.

An s.e. of 1 ppm in a line such as AB in Fig. 7.6 in turn demands an s.e. of at most 0.5 ppm in each direction AS_1, BS_1, etc., or of $\sqrt{(n/8)}$ ppm in each, if n is the number of well-placed successful simultaneous observations of the satellite from A and B. In Fig. 7.6, $n = 2$. If $n = 32$, and if errors are uncorrelated, the s.e. required of each plate observed at A or B will have to be 2 ppm, or (say) $\frac{1}{2}''$ of arc, but the number n will have to be increased if the satellite is not always observed in optimum positions. In the necessary absence of horizontal planes in Fig. 7.10, vertical planes are of reduced value. See [362] for the merits of different configurations, and its Fig. 16 for the accuracy of the direction AB.

The sources of error are listed in (*a*)–(*f*) below.

(*a*) An error of 1″ results if *star exposure times* are wrong by $0^s.07 \cos \delta$. This easy to avoid.

(*b*) With a high (4000 km) satellite an error Of 1″ results from an error in the *satellite exposure time* of about $(d/1300)^s \sin \theta$, where d is the distance AS_1, etc., in thousands of kilometres, and θ is the angle between the line of sight and the satellite path. If the satellite is of the flashing type, no time-keeping is involved, but with a passive satellite tenths of a millesecond begin to be significant in the relative times of the exposures at A and B.

(*c*) *Refraction.* The parallactic refraction between the satellite and the stars is small and well determinate, except for the irregular stellar movements known as wanderings or erratics, § 4.11 (*f*). A satellite and a star exactly behind it would move together, and no harm would be done, but the apparent erratics of a satellite and a star one degree distant are likely to be quite unconnected. Over a period of some seconds these erratic movements will probably average out to within less than 1″, so if both stars and satellite had exposures of that length, the trouble would be much reduced. Unfortunately, one or the other is likely to have short exposures, and the relative error can then only be reduced to (say) ±1″ by having a large number of the short exposure images on each plate.

(*d*) *Plate measurement.* Given sharp images, plate measurements can probably be made with an s.e. of between 0.0015 and 0.003 mm, which is 0″.3 to 0″.6 if the focal length is 1 metre. It is to be expected that the mean error will be reduced when there are numerous star and satellite images, but there may be systematic error between the two different types of image.

(*e*) *Star places.* See § 4.06. There is likely to be some correlation between the errors in the places of stars appearing on the same small plate.

These sources of error (*a*)–(*e*), including those of timing, can be regarded as errors in the recorded plate coordinates at the reputed times.

(*f*) The accuracy of the world-wide BC 4 and Pageos programme is fully investigated in [516], from which the following figures are abstracted.

On any particular photographic plate there are five images of each of about 100 stars, and also about 300 satellite images, the latter producing the seven fictitious images for further use, as described in § 7.22 (g).

The self-consistency of the 500 star images in producing the six plate constants x_N, y_N, c_0, ω_1, ω_2, and ω_3 of § 7.22 (*f*) (iv) gives the figure of ±3.31 μm for the mean s.e. of each coordinate of a single star image, made up of†

Comparator error	s.e. = ±1.81 μm	
Emulsion shift	±1.00 μm	
Refraction erratics	±2.58 μm (1″.18)	(7.130)

The figures for the satellite images are similar, totalling ±3.52 μm (1″.61) as deduced from the polynomial smoothing. Timing errors were negligible.

The five images of each star were then treated as a single image with s.e. = ±3.31/√5 μm plus ±0.87 μm (0″.4) for error in the star catalogue, these combining to give ±1.72 μm (±0″.79) as the contribution of each coordinate to the orientation of the plate. The minimum number of stars necessary to compute six plate constants and 14 polynomial coefficients is 10 (two coordinates each). Since there are 100 stars on each plate, the plate may be regarded as providing 10 independent solutions, and the figure of 0″.79 is further divided by √10 to give for each coordinate

$$\text{Error of satellite direction given by one plate} = +0''.25 \qquad (7.131)$$

Equation (7.131) provides the weight for each 'observation' in the solution for the station positions, and the variance–covariance matrix in the solution shows that if there was one plate for each observed direction the average s.e. of the station coordinates would be ±7.0 m in ϕ and λ, and ±21.0 m in height, if there was no scale control. Actually, on average, there are about five plates for each direction, so these figures are divided by √5, and the height error is further reduced by the scale controls. Final figures given in [516] are

$$\text{s.e. in } \phi \text{ and } \lambda = \pm 3.1 \text{ m. s.e. in height} = \pm 5.6 \text{ m.}$$
$$\text{s.e. in position} = \pm[(3.1)^2+(3.1)^2+(5.6)^2]^{\frac{1}{2}} = \pm 6.7 \text{ m}\dagger \qquad (7.132)$$

These estimates must be considered in conjunction with the fact there are 37 doppler stations which are quasi-identical with BC 4 stations, the s.e.'s of which (apart from errors in their coordinate system) are quoted as ±1.5 m in ϕ, ±1.2 m in λ, and ±1.6 m in height, or ±2.5 m in position. [516], Table 8. The differences between doppler and BC 4 were analysed

† The figure ±3.31 μm was obtained by least squares, the comparator error by experiment, the emulsion shift by estimation, and the refraction erratics by subtraction, namely $(2.58)^2 = (3.31)^2 - (1.81)^2 - (1.00)^2$. The figure ±1″.18 for the erratics thus includes the effects of any unsuspected error from other causes. These figures are for single coordinates, x or y, each measured in two positions of the comparator.

† In [516] this figure is divided by √3.

to provide 3 translation parameters (to convert BC 4 results to the geocentre), a rotation about the polar axis (to bring doppler into terms with the Greenwich meridian), and also a scale difference, and two trivial rotations about the two equatorial axes. After applying these optimum corrections, the r.m.s. difference of position was 14.4 m, so some upgrading of the estimated ±6.7 m and/or of the ±2.5 m is clearly suggested.

7.29. Use of atmospheric sondes instead of satellites

In Finland, [321] and [51], a chain of 14 triangles with sides of about 200 km has been observed by photographing a flashing atmospheric sonde against a stellar background in exactly the same way as has been described for satellite photography.

The sonde is carried up to a height of about 35 km by a balloon, at which height the balloon either drifts slowly horizontally, or else bursts and releases the sonde on a parachute. The sonde carries radar and emits flashes on command. These are photographed by a camera with focal length 103 cm, aperture 34 cm, and a 5°×5° field, equatorially mounted and following the stars.

A satisfactory programme is considered to be the recording of 30 flashes simultaneously from each of the two stations concerned (each of AS_1 and BS_1 in Fig. 7.6). Such observations made to two suitably placed sondes give the direction of the line AB with an s.e. of ±0″.3 (seconds of arc) in each of RA and δ.

General references for Section 4. [49], [273], [391], [516], [567], and [568].

Section 5. Satellite Photography. Dynamic Use

7.30. Notation

$\mathbf{S}$ = Vector ground station to satellite.
$\mathbf{r}$ = Vector geocentre to satellite.
$\mathbf{a}$ = Vector geocentre to ground station.
$\mathbf{S}'$ = Observed value of $\mathbf{S}$.
$\mathbf{S}_0$ = $\mathbf{S}$ as computed from model.
$\mathbf{r}_0$ = $\mathbf{r}$ as computed from model.
S = Scalar distance ground station to satellite. Not related to S_1^{ω} in (7.158), nor to harmonic constant S_{lm} in (7.1).
v = Residual.
$p_1 \ldots p_n$ = Model parameters, including E_i, ψ_i, and $\mathbf{a}$.
E_i = Orbital parameters.
ψ_i = Perturbing parameters.
$\mathbf{a}_k$ = Vector $\mathbf{a}$, with components resolved in the terrestrial frame, in which they are presumed to be constant.
Δp_n = Unknown optimum changes in p_n.
y_1, y_2, y_3 = Vector components in the orbital system. See § 7.32.
w, x, Z, z See § 7.32.

N, P, R, S, T, U = Rotation matrices.
A = Rotation matrix, see § 7.33(*d*).
X'', Y'' = Coordinates of instantaneous pole relative to CIO.
$\hat{\theta}$ = Sidereal angle. See (7.144) and § 4.10 (*f*).
α, δ = Right ascension and declination.
θ, λ = Co-latitude and longitude.
$\bar{C}_{lm}\bar{S}_{lm}$ = Normalized harmonic constants, see (7.1).
B = Matrix of partial differential coefficients.
W = Weight matrix.

Vectors. (3×1) vectors are in clarendon type, e.g. $\mathbf{S}$. The scalar magnitude of $\mathbf{S}$ is in italics S, and its three components are S_1, S_2, S_3. See § 7.00 (*c*).

An equation between vectors, such as (7.145), comprises three equations implying the equality of the three components of the vectors concerned, in any one coordinate frame.

7.31. Introductory

Observations from one or more ground stations, made as in Section 4, will have provided the direction cosines, or the RA and declination, of the geometrical straight line joining the station to the satellite at recorded times, but in the dynamic method the observations at adjoining stations need not be at all simultaneous. Some direct distances may also have been measured, as in Section 6.

In principle, given the initial orbital elements of the satellite, or its initial position and velocity vector, and given the magnitudes and directions of the perturbing forces acting on the satellite at every place and time, the orbit can be calculated at all future times, with reference to the geocentre as origin. Then suitable timed observations of RA, δ, and/or distance from a ground station, will give that station's coordinates in the same reference frame.

In any particular situation one or all of (*a*) the orbital elements, (*b*) the parameters describing the perturbations, or (*c*) the coordinates of the ground stations, may be unknown. But, given an adequate series of observations, it is possible to determine them all by the method of variation of coordinates, in a way similar to that by which conventional ground surveys are computed by the method of variation of coordinates, § 2.18.

Every observation gives an equation in the form of

$$\sum \frac{\partial \mathbf{S}}{\partial p_n} \Delta_{p_n} = (\mathrm{O}-\mathrm{C}) + \mathbf{v} = (\mathbf{S}' - \mathbf{S}_0) + \mathbf{v} \tag{7.133}$$

where $p_1 \ldots . p_n$ are the various unknowns, orbital, perturbing, or station coordinates.

$\mathbf{S}'$ is the observed vector, ground to satellite, comprising observations of RA, δ, and distance S. For the present, it is assumed that all these have been simultaneously observed, although in practice this is very unusual. See further in § 7.33 (*d*).

$\mathbf{S}_0$ is the trial value of $\mathbf{S}$, computed for the appropriate time with the best available values of the p's. On the left-hand side of the equation the distinction between $\mathbf{S}'$ and $\mathbf{S}_0$ is assumed immaterial. If it is material, iteration is called for.

$\mathbf{v}$ is a residual.

$\Delta p_1 \ldots \Delta p_n$ are the required corrections to the trial values of $p_1 \ldots p_n$.

As here written, equation (7.133) is actually three equations, one for each of the cartesian components of $\mathbf{S}'$ and $\mathbf{S}_0$.

Then, the number of observations being much in excess of the number of unknowns, the observations are written in matrix form as

$$Bp = (\mathrm{O}-\mathrm{C}) + v = k + v, \text{ defining } k, \tag{7.134}$$

where B is the matrix formed by the coefficients $\partial\mathbf{S}/\partial p_n$, p is the vector matrix of the unknown Δp's, and k is the vector matrix of the (O−C)'s. For a least square solution the normal equations are

$$B'WBp = B'Wk \tag{7.135}$$

It is of course necessary to be assured that the equations are adequately well conditioned. The SAO routine for solving (7.135) is outlined in § 7.38.

7.32. Reference systems

(*a*) *Definitions.* The components of $\mathbf{S}'$ and $\mathbf{S}_0$ in (7.133) must of course be taken with reference to coordinate systems with parallel axes. This calls for some care, as several different systems are involved, as in (i) to (v) below.

(i) The *orbital system.* The orbit must be computed with reference to a Newtonian (inertial) system as defined by the distant stars (§ I.00). This has the inconvenience that the unknown station coordinates are with reference to a system which rotates with the earth, and whose pole precesses. The station positions also vary from an inertial frame, to a small extent, on account of polar motion.

The orbital system (y_1, y_2, y_3) adopted by the SAO, which is likely to involve minimum significant complexities, is defined as follows, [391], (i), pp. 9–42.

(A) The origin is at the geocentre.

(B) y_3 is directed towards the instantaneous pole.

(C) y_1 is directed towards a point $\hat{\gamma}$ on the instantaneous equator, which is east of the true vernal equinox of date by an angle $\mu + \Delta\mu$, namely the precession and nutation from 1950.0 to date.

(ii) The *celestial system* (w_1, w_2, w_3) in which the SAO catalogue gives RA and δ. Its axes are defined by the mean equinox and equator of 1950.0

(iii) The terrestrial system (x_1, x_2, x_3), in which the ground station coordinates $\mathbf{a}_k$ are assumed to be constant. The origin is the geocentre, the x_3 axis is towards the CIO, and the x_1 axis is parallel to the BIH Greenwich meridian plane (CZM).†

There are also two intermediate systems.

(iv) The *mean sidereal system* (Z_1, Z_2, Z_3), which is derived by rotating the celestial system by precession from 1950.0 to date. The origin is taken to be the geocentre.

(v) *The true sidereal system* (z_1, z_2, z_3), which is the same as the mean sidereal system, except that the celestial system is additionally rotated through the nutation. Its z_3 axis is directed towards the true pole of date, and z_1 towards the true vernal equinox.

(*b*) *Conversions.* [391], (i), pp. 29–39.

(i) Celestial to mean sidereal.

$$\mathbf{Z} = P\mathbf{w}, \tag{7.136}$$

where the rotation matrix P is as in (4.19). It is also in [391], (i), pp. 29 and 35.

(ii) Mean sidereal to true siderreal.

$$\mathbf{z} = N\mathbf{Z} = NP\mathbf{w}, \tag{7.137}$$

where N is $\begin{bmatrix} 1 & -\Delta\mu & -\Delta\nu \\ \Delta\mu & 1 & -\Delta\epsilon \\ \Delta\nu & \Delta\epsilon & 1 \end{bmatrix}$

and $\Delta\mu$, $\Delta\nu$, $\Delta\epsilon$ are the nutations in RA, δ, and obliquity as given in (4.24) and [391], (i), p. 36.

(iii) True sidereal to terrestrial.

$$\mathbf{x} = S\mathbf{z}, \tag{7.138}$$

where S is as in (7.125).

(iv) Celestial to terrestrial, combining (i), (ii), and (iii),

$$\mathbf{x} = SNP\mathbf{w}, \text{ as in (7.124).} \tag{7.139}$$

† As defined by the SAO there are some minor complications, [391], (i), p. 16. Their zero of longitude is 75° 03′ 55″.94 east of the mean meridian of the US naval observatory, which may not exactly coincide with the CZM.

(v) Celestial to orbital.

$$\mathbf{y} = R\mathbf{z} = RNP\mathbf{w}, \tag{7.140}$$

where $R = \begin{bmatrix} \cos(\mu+\Delta\mu) & \sin(\mu+\Delta\mu) & 0 \\ -\sin(\mu+\Delta\mu) & \cos(\mu+\Delta\mu) & 0 \\ 0 & 0 & 1 \end{bmatrix}$

in which $\mu = \kappa + \omega$, the precession in RA since 1950.0, and $\Delta\mu$ = the nutation in RA $= \Delta\psi \cos \epsilon$, which is given APFS Table II, as the equation of the equinoxes, and in [391], (i), pp. 36–8.

As an approximation, correct to 10^{-6} within about 15 years of 1950.0, the matrix RNP may have been written as

$$RNP = U = \begin{bmatrix} 1 & 0 & -(\nu+\Delta\nu) \\ 0 & 1 & -\Delta\epsilon \\ \nu+\Delta\nu & \Delta\epsilon & 1 \end{bmatrix} \tag{7.141}$$

where ν is as in (4.21), and $\Delta\nu$ is the nutation in declination, $\Delta\psi \sin \epsilon$.

Then $$\mathbf{y} = U\mathbf{w}. \tag{7.142}$$

(vi) Orbital to terrestrial.

$$\mathbf{x} = T\mathbf{y}, \tag{7.143}$$

where $$T = S \text{ of (7.125) with } \hat{\theta} \text{ in place of } \theta, \tag{7.144}$$

in which $\hat{\theta}$ is the *sidereal angle*, the rotation of the earth relative to an inertial frame since MJD 33282.0, see § 4.10 (f).

7.33. The observation equations

In (7.133) the components of the different terms must be expressed in the same reference system, Initially, but see (d) below, it is convenient to express them all in the orbital system.

(a) *The computed value*, $\mathbf{S}_0$. We have

$$\mathbf{S}_0 = \mathbf{r}_0 - \mathbf{a}, \tag{7.145}$$

where $\mathbf{r}_0$ is the computed geocentric position of the satellite, which will naturally be computed in the orbital system, and $\mathbf{a}$ is the geocentric position of the ground station. In the terrestrial system $\mathbf{a}$ and its three components are constant, and it may be written $\mathbf{a}_k$. In the orbital system

$$\mathbf{a} = T^{-1}\mathbf{a}_k, \text{ from (7.143)†}. \tag{7.146}$$

† $T^{-1} = T^T$, with the signs of $\hat{\theta}$, X'', and Y'' reversed.

So, written in full, (7.145) is

$$\begin{bmatrix} S_{01} \\ S_{02} \\ S_{03} \end{bmatrix} = \begin{bmatrix} r_{01} \\ r_{02} \\ r_{03} \end{bmatrix} - T^{-1} \begin{bmatrix} a_{k1} \\ a_{k2} \\ a_{k3} \end{bmatrix} \tag{7.147}$$

(*b*) *The observed* $\mathbf{S}'$.

(i) The plate coordinates are converted to RA and δ, and corrected for parallactic refraction, planetary aberration, and diurnal and annual aberration, as in § 7.22 (*e*) and (*h*).

(ii) Coordinates in the celestial system are given by

$$w_1 = S' \cos\delta \cos\alpha,\ w_2 = S' \cos\delta \sin\alpha,\ w_3 = S' \sin\delta, \tag{7.148}$$

where S is the scalar distance.

(iii) These are converted to the orbital system by

$$\mathbf{S}'(\text{orb}) = RNP\mathbf{S}'(\text{cel}) \tag{7.149}$$

(*c*) *The left-hand side of* (7.133), $(\partial\mathbf{S}/\partial p_n)\Delta p_n$. In this term only low accuracy is required, and there is no need to distinguish between $\mathbf{S}'$ and $\mathbf{S}_0$. As in (7.145) we have $\mathbf{S} = \mathbf{r} - \mathbf{a}$. The unknowns p_n are of three kinds,

(i) Orbital elements, E_i,
(ii) Other perturbing parameters ψ_i,
(iii) Station coordinates $\mathbf{a}_k$.

$$\text{Then } \sum \frac{\partial\mathbf{S}}{\partial p_n}\Delta p_n = \sum \frac{\partial\mathbf{r}}{\partial E_i}\Delta E_i + \sum \frac{\partial\mathbf{r}}{\partial\psi_i}\Delta\psi_i - (T^{-1}\Delta\mathbf{a}_k). \tag{7.150}$$

In (7.150) the simple form of the last term is due to the fact that $(\partial\mathbf{S}/\partial\mathbf{a}) = 1$, provided the components of $\mathbf{S}$ and $\mathbf{a}$ are expressed in the same coordinate system.

Equation (7.133) then reads

$$\sum \frac{\partial\mathbf{r}}{\partial E_i}\Delta E_1 + \sum \frac{\partial\mathbf{r}}{\partial\psi_i}\Delta\psi_i - (T^{-1}\Delta\mathbf{a}_k) = (RNP\mathbf{S}') - (\mathbf{r}_0) + (T^{-1}\mathbf{a}_k) + \mathbf{v}, \tag{7.151}$$

in which the components of all the terms are now expressed in the orbital system.

(d) As they stand, the three equations (7.151) are unsatisfactory, in that (7.148) contains S' the scalar distance as an observation, while actually it is seldom observed at the same time as RA (α) and δ. In fact, three equations have been formed when there are only two observations. Another weakness of (7.151) is that its observations are in the form of direction cosines, to which it is difficult to assign weights. Weights are much more easily assigned to α and δ, see § 7.37.

The remedy for both these troubles is to make a further transformation by multiplying both sides of (7.151) by a matrix A, [391], (i), p. 110, defined as

$$A = \frac{1}{S}\begin{bmatrix} -\frac{y_3 y_1}{Sk} & -\frac{y_3 y_2}{Sk} & \frac{k}{S} \\ -\frac{y_2}{k} & \frac{y_1}{k} & 0 \end{bmatrix} \tag{7.152}$$

where $k^2 = y_1^2 + y_2^2$, and $y_1\ y_2\ y_3$ are the components of $\mathbf{S}_0$ in the orbital system

The effect of this transformation is to make

$$A\mathbf{S}' = 0$$

and

$$A(T^{-1}a_k - r_0) = \begin{bmatrix} \partial\delta \\ \cos\delta\ \mathrm{d}\alpha \end{bmatrix}, \tag{7.153}$$

as can be verified by straight-forward algebra based on Fig. 7.12.†

So (7.151), becomes

$$A\left[\sum \frac{\partial \mathbf{r}}{\partial E_i}\Delta E_i + \sum \frac{\partial \mathbf{r}}{\partial \psi_i}\Delta\psi_i - (T^{-1}\Delta\mathbf{a}_k)\right] = \text{Zero} + \begin{bmatrix} \mathrm{d}\delta \\ \cos\delta\ \mathrm{d}\alpha \end{bmatrix} + \mathbf{v}, \tag{7.154)‡$$

In the left-hand sides (7.151) and (7.154) the three components of $\partial\mathbf{r}/\partial E_i$ (with $\mathbf{a}_k$ unchanged), namely $(\partial y_1/\partial E_i)$, $(\partial y_2/\partial E_i)$, $(\partial y_3/\partial E_i)$ are converted by A into $\partial\delta/\partial E_i$ and $\cos\delta(\partial\alpha/\partial E_i)$. And similarly for $\delta\mathbf{r}/\partial\psi_i$.

If it should happen that the distance S' is measured at the same time as α and δ, the expression

$$\frac{1}{S}\begin{bmatrix} \frac{y_1}{S} & \frac{y_2}{S} & \frac{y_3}{S} \end{bmatrix} \tag{7.155}$$

can be included as a third row in matrix A, (7.153). If S' has been measured by itself at a different time, the single resulting observation

† α and δ are by definition measured in the celestial system, while Fig. 7.12 shows the axes of the orbital system. But the directions of the axes of these two systems only differ by something of the order of 0.001 rad. The vector Q′Q is small (perhaps a few tens of seconds of arc), and has been accurately computed. In this context, it is immaterial whether its two components are recorded in the celestial or in the orbital system, since $\mathrm{d}\delta$ and $\mathrm{d}\alpha$ are very small. If greater accuracy is required, iteration will supply it.

‡ Note that the matrix vectors of (7.151) and (7.154), when pre-multiplied by the (2×3) matrix A, are converted into (2×1) matrices, as is proper for the two observations α and δ.

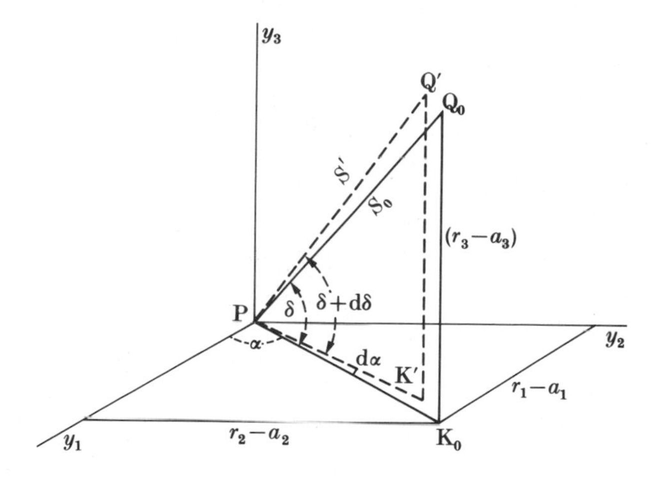

FIG. 7.12. P is the ground station. Q′ is the observed position of the satellite, and Q_0 is its position computed from the model. $Q'Q_0$ is small. y_1, y_2, y_3 are parallel to the axes of the inertial frame. (r_1-a_1), (r_2-a_2), and (r_3-a_3) are the three components of the computed geocentric satellite distance $\mathbf{S}_0$, which equals $(\mathbf{r}_0-\mathbf{a})$. The components of $\mathbf{a}$ in the inertial frame are given by $T^{-1}\mathbf{a}_K$. $d\delta$ is the component $(O-C)$ in the plane Q_0PK_0, and $d\alpha$ is the component in the plane y_1Py_2. $PK_0 = k$.

equation is formed by multiplying (7.153) by the (1×3) matrix (7.155).

(7.154) constitutes the final form of the observation equation (7.133).

It remains to compute the trial value of the geocentric vector $\mathbf{r}_0$, in §§ 7.34 and 7.35, the partial differential coefficients $\partial\mathbf{r}/\partial E_i$ and $\partial\mathbf{r}/\partial\psi_i$ in § 7.36 (*c*), to allot weights as in § 7.37, and to solve the observation equations by least squares in §§ 7.38 and 7.39.

7.34. Computation of $\mathbf{r}_0$.

(*a*) The best available values of all the unknowns E_i, ψ_i, and $\mathbf{a}_k$ constitute a *model* from which the orbital position of the satellite, $\mathbf{r}_0$, can be more or less accurately computed. Such a model is described in § 7.57, and further details of the geopotential part of the model are given in Table 7.2 and § 7.79. One object of further observations is to improve the model.

Given the initial position of a satellite, and a model which accurately specifies its accelerations at any place and time, its future is determinate, subject only to the difficulty that the acceleration at any instant is not known until the position has been determined. There is a second-order differential equation to be solved. There are two ways of approaching the problem.

(i) By numerical integration. This is currently used for observations made on the doppler system, although there is no reason why it should be confined to that. It is described in §§ 7.56 and 7.57.

(ii) In the method used by the SAO, approximate values of the unknowns are first obtained by analysis as far as it can go. Thus a satellite ephemeris can be computed which allows for the main term GM/r^2 in the acceleration, and for the J_l terms, § 7.11, which arise from an axially symmetrical earth. Approximate values having been obtained in this way, the remaining corrections due to the geopotential and other perturbations are at least relatively small, and are insensitive to small errors in the position of the satellite. See further in § 7.36 (*b*).

(*b*) The computation of the effect on $\mathbf{r}_0$ of the tesseral and sectorial terms of the potential is no small piece of work. The GM/r^2 term and the zonal J_l terms having been dealt with separately, the remaining terms of (7.1) are

$$\sum_{l=2} \left(\frac{a}{r}\right)^l \sum_{m=1}^{l} (\bar{C}_{lm} \cos m\lambda + \bar{S}_{lm} \sin m\lambda)\bar{P}_{lm}(\cos\theta), \tag{7.156}$$

with notation as in § 7.00. This gives the contribution of these terms to the geopotential at a point distant r from the geocentre in polar distance θ and longitude λ, which is constant in a frame which rotates with the earth. Equation (7.156) then has to be converted to an orbital frame (See § 7.39) in which it is time-dependent, and the resulting accelerations of the satellite are its gradients along the three components of $\mathbf{r}$ in that frame. The algebra is heavy. See [391], (i), pp. 94–104 and 156–184 for full details. For general principles see [329] and [326].

(*c*) Other perturbations included in the model are as below.

(i) Lunar and solar tides. See § 7.55 (*b*). Details are in [391], pp. 147–55, and [216], pp. 149–59.

(ii) Air drag. See § 7.57 (*c*) and [216], pp. 172–4.

(iii) Radiation pressure, See § 7.57 (*d*), and [216], pp. 168–72.

(iv) Earth, ocean, and atmospheric tides, see § 7.57 (*e*) and [216], pp. 149–159.

7.35. The unknown parameters

There are three classes of parameters for which improved values may be required.

(i) The orbital elements and their variations with time.

(ii) The geocentric coordinates of the observing stations.

(iii) The numerical parameters in the model describing the geopotential and other perturbations listed in § 7.34 (*c*).

Assuming that the parameters of the model are known, it is clear that a set of observation equations, in the form of (7.154) or (7.151) with matrix A included, can determine the orbital elements and the station coordinates, provided that the observations are sufficiently numerous and accurate, and distributed so as to avoid instability. On the other hand, to obtain information about item (iii) in order to improve the model, is more difficult. The SAO system has been to solve for items (i) and (ii) first, and then to consider what perturbing forces (of the correct theoretical form) could cause the outstanding residuals, § 7.39.

7.36. The orbital elements

(*a*) § 7.36, with §§ 7.37 and 7.38, considers the use of the observations to revise the orbital elements and the station coordinates on the assumption that the model requires no revision. The observation equations then take the form

$$A\sum \frac{\partial \mathbf{r}}{\partial E_i}\Delta E_i - A\sum T^{-1}\Delta \mathbf{a}_k = -A(\mathbf{r}_0 - T^{-1}\mathbf{a}_k) + \mathbf{v}$$

$$= \begin{bmatrix} d\delta \\ \cos\delta(d\alpha) \end{bmatrix} + \mathbf{v}, \qquad (7.157)$$

in which the numerical value of the right-hand side is known.

The six orbital elements employed by the SAO, [391], (i), p. 97, have been ω, Ω, i, e, n, and t_0, see § 7.01 (*c*).

(*b*) *The variation of the orbital elements with time.* These variations may be classed as

(i) Secular variations in ω and Ω, see §§ 7.11–7.15.

(ii) Periodic changes in all the elements, with periods generally greater than (say) 20 days, of the form $S\sin(\alpha+\beta t)$.

(iii) Short-period changes, which pass through one or more full periods in a single orbit.

In the SAO method the variations of the elements are represented by a series of simple functions containing numerical parameters which are to be determined, namely

(i) A polynomial in several powers of t.
(ii) A series of sine terms.
(iii) A series of so-called *hyperbolic terms.*

Thus, taking the element ω as an example

$$\omega = \omega_0 + \omega_1 t + \omega_2 t^2 + \ldots . + S_1^\omega \sin(\alpha_1^\omega + \beta_1^\omega t) + S_2^\omega \sin(\alpha_2^\omega + \beta_2^\omega t) + \ldots$$
$$+ H_1^\omega (L_1^\omega - T)^{k_1^\omega} + H_2^\omega (L_2^\omega - T)^{k_2^\omega} + \ldots \tag{7.158†}$$

where t is the time measured from the epoch of the orbit, and T is time in modified Julian days. The superscript ω indicates that the constants concerned relate to the variations of ω. Another equation in the form of (7.158) will give the variations of Ω, and so on. It is required to determine all the constants S, H, K, and L which have significant size, apart from any which may be already known with sufficient accuracy, such as perhaps some of the zonal harmonics. The α's and β's are given by theory, § 7.39.

(*c*) The partial differential coefficients, $\partial\mathbf{r}/\partial E_i$. These are required for the left-hand side of (7.157). Here E_i includes all the parameters in (7.158), namely $\omega_0, \omega_1 \ldots ., S_1^\omega, S_2^\omega$, etc., for ω and for each of the other elements.

Considering ω for example, we have

$$\frac{\partial\mathbf{r}}{\partial\omega_0}\Delta\omega_0 + \frac{\partial\mathbf{r}}{\partial\omega_1}\Delta\omega_1 + \ldots \frac{\partial\mathbf{r}}{\partial S_1^\omega}\Delta S_1^\omega + \ldots$$
$$= \frac{\partial\mathbf{r}}{\partial\omega}\cdot\frac{\partial\omega}{\partial\omega_0}\Delta\omega_0 + \frac{\partial\mathbf{r}}{\partial\omega}\cdot\frac{\partial\omega}{\partial\omega_1}\Delta\omega_1 + \ldots \tag{7.159}$$

in which $\partial\mathbf{r}/\partial\omega$ is obtained as in § 7.02 (7.19), and $\partial\omega/\partial\omega_0 = 1$, $\partial\omega/\partial\omega_1 = t \ldots . \partial\omega/\partial S_1^\omega = \sin(\alpha_1^\omega + \beta_1^\omega t) \ldots$ etc.

(7.160)‡

(*d*) If it is desired to improve the coordinates of any station, its $\Delta\mathbf{a}_k$ can be included in the left-hand side of (7.157). Any which are to be left unchanged are omitted.

7.37. Weights

When the normal equations are being formed, the weight of any observation should in principle be the inverse square of its estimated standard error. From this point of view, the use of direction cosines as 'observations' is bad. They are not only remote from the actual observations (plate coordinates), but the weight of a cosine depends very much on the magnitude of the angle.

† The parameters S of course have no connection with the S which denotes distance from ground to satellite. The hyperbolic terms with the parameters H, K, and L are required for drag and radiation pressure. In [391], (i), p. 114, they are given in the form $H_1^\omega \exp\{K_1^\omega \ln(L_1^\omega - T)\} \ldots$

‡ In the expression for $\partial\omega/\partial S_1^\omega$ the preliminary values of α_1, and β_1 must be used as previously determined. If they are not good enough, they must be dealt with in another iteration, but these partial differential coefficients are not required with very great accuracy.

To treat each individual plate coordinate as an independent observation would be impossible, but to treat measures of the satellite's RA and δ as independent observations is reasonable and convenient. A better method might be to regard along-track and across-track positions as independent, and consequently to allot a full (2×2) weight matrix to each pair of RA and δ, but this would be inconvenient and would probably make littlie difference.

Weighting is discussed at some length in [391], (i), pp. 115–26.

7.38. Solution of the equations. Differential orbit improvement

Once an initial series of observations has been used to give the unknown parameters in (perhaps) the usual direct way from (7.135), further observations can be added, and the resulting normal equations can be formed relatively easily, § D.14 (*e*). But although the normal equations are easily formed, their solution is not a small matter when the number of unknowns is very great. There may perhaps be 20 unknowns in the expressions for the orbital parameters describing each weekly or monthly arc of each satellite observed. [218], p. 23 mentions 100 orbital arcs with a total of 1500 unknowns in 1969, and later solutions will have included more. Further, there may be troublesome instability.

As an alternative to direct methods of solution the SAO, [216], p. 277, and [218], p. 23, have preferred to solve their normal equations (for the orbital elements and station coordinates) by an iterative process described as a Block Gauss–Seidel iteration, see § B.15 (*e*). In this method, iterations are carried out so as to improve the values of about 20 unknowns at a time. This is their *differential orbit improvement* programme (DOI). Judgement is required in the matter of which unknowns to include in each group of 20, and of the order in which groups should be brought forward for improvement. There can of course be advantage in this use of skill as opposed to mechanical routine.

The revision of the parameters of the model of the perturbations is outlined in § 7.39.

7.39. Revision of the geopotential model

In principle, changes $\Delta\bar{C}_{lm}$ and $\Delta\bar{S}_{lm}$ in the parameters of the geopotential (7.01) can be included as unknown ψ's in (7.157), provided the partial differential coefficients $\partial\mathbf{r}/\partial\psi_i$ can be computed for them. This is a complicated matter. It is required to find the changes in the position $\mathbf{r}$ at time t, which result from the varying accelerations (integrated over the time interval between the epoch t_0 and t) caused by each of the changes

$\Delta\bar{C}_{lm}$ and $\Delta\bar{S}_{lm}$. One method, which is used in conjunction with doppler observations, § 7.56 (*e*) (iii), is to compute these partials by numerical integration. The SAO has adopted an alternative method on the following lines.

The orbital elements and station coordinates having been obtained as in the preceding paragraphs, using the best available values of the geopotential parameters, the residuals in each observation equation (7.157) are recorded. The aim is then to find changes in the parameters which will reduce these residuals. The expression for the geopotential (7.01) is first transformed to a frame in which the pole of the spherical harmonics points towards the pole of the satellite's orbit, and in which the zero of λ in (H.26) coincides with the perigee of the orbit, instead of the Greenwich meridian. In this form the expression for the geopotential at the position of the satellite, where $\cos\theta = 0$, is much simplified. Being thus transformed, the harmonic terms (which are constant in the terrestrial frame) will of course be time-dependent, with periods which can be given by theory.

[218], pp. 16–18 and [216], p. 255 illustrate the periodic effects of various tesseral harmonics of degrees 3 to 8 on the element M (the mean anomaly) of a particular satellite at an altitude of about 1250 km. There are periods of 1.001, 0.971, 0.958, 0,497, 0.327, 0.091, 0.083, 0.071, 0.066, 0.041, and 0.040 days. Theory also gives the phase angles, the α's in (7.158). Analysis of the observed irregularities in the mean motion thus enables the amplitudes of these periodic perturbations to be determined, and thence the constants $\bar{C}_{lm}$ and $\bar{S}_{lm}$, always of course subject to the quantity, distribution, and accuracy of the observations being sufficient. There is a complication, namely that several tesseral terms may produce variations with the same periods and phase angles. Thus in the above sample, the amplitude of the 1.001-day period is a linear combination of $\bar{C}_{3,1}$, $\bar{C}_{5,1}$, and $\bar{C}_{7,1}$, and they cannot be separated by the data obtainable from a single satellite. The remedy lies in the fact that another satellite, with a different inclination, will give a similar combination of the same three constants, but with different factors. So three satellites will separate them.

Some outline of the methods used is in [391], (i), pp. 196–205, by which a solution is made for up to 100 constants at one time, by least squares, with frequent iterations. [216], pp. 277–308 gives further details. As with the DOI programme of § 7.38, this method of determining the geopotential parameters by analysis calls for judgment on the part of the operator, and the methods are continually being improved.

General references for Section 5. [216], [218], [273], [329], and [391].

SECTION 6. MEASUREMENT OF SATELLITE DISTANCES

7.40. Introductory

The distance between a ground station and a satellite may be measured by microwaves or by laser light with the objects of

(*a*) Controlling the scale errors of astro-triangulation, see § 7.25 (*b*) and (*c*).

(*b*) Adding strength to dynamically used satellite photography, or to doppler tracking.

(*c*) Constituting a pure trilateration system.

This section deals with the measurement of the distance. § 7.42 describes the Secor system, which has been used to give a world-wide trilateration, but which is now unlikely to be used again. § 7.43 describes laser ranging, which is very much more accurate than Secor. Its role is likely to be in co-operation with doppler or other dynamic systems, rather than pure trilateration.

7.41. C-band and S-band radar.

These systems have been primarily intended for spacecraft tracking and intercommunication. They operate with large ground antennas (3.7 to 26 metres in diameter). The C-band frequency is 5.400–5.900 GHz, and the S-band is from 2.100 to 2.300 GHz. [49], (i) gives a short description, with illustrations and charts showing locations. While they are not intended for geodetic work, they have produced data which have contributed to it, in the form of distances between some widely separated ground stations. The C- bandwidth is also being used for satellite altimetry, and the S-band was used for the Goddard range and range rate system (GRAAR).

7.42. Secor

(*a*) *General principles.* The distance between a ground station and a satellite is measured by the phase change in 420.9–MHz (67 cm) microwaves, modulated to 585.533 kHz, emitted by the ground station. They are transponded back by the satellite on both 449 and 224.5 MHz. The equipment allows four stations to work together, each obtaining a measure of distance from itself to the satellite every 0.050 s, between which interpolation gives a series of quasi-simultaneous values. As in instruments like the tellurometer, low modulation frequencies are provided to give coarse range readings.

In the usual routine three of the stations will have been fixed in x, y, and z, and it is required to fix the fourth, which will be in a position suitable for carrying the trilateration forward in the required direction.

Provided the geometry is suitable, the satellite is fixed by the intersection of three spheres of known radii centred on the three previously fixed stations, while the unfixed station lies on a sphere of known radius centred on the satellite. If this process is carried out with the satellite in three (or more for least squares) suitably placed positions, the unfixed station is fixed by the intersection of three (or more) spheres.

Secor transponders have been carried by the satellites Secor 1 to 9 and 13, with inclinations of 70° to 90°, in mostly near circular orbits of between 900 and 3700 km in height, and also by Geos 1 and 2.

The system has been used to form a 37-station trilateration net round the earth in equatorial and low latitudes, see § 7.75 (*f*). Further details are in [463], [171], [430], and [273], pp. 141–53 and 689.

(*b*) *Geometry*. The three spherical loci which fix a satellite must intersect at good angles. The pyramid whose base is formed by the three fixed stations, and whose apex is the satellite, must be neither too pointed nor too flat. The fix must be firm in altitude, as well as in latitude and longitude. And the same applies to the fixing of the fourth station from the satellite positions. In particular, neither the three fixed stations, nor the three satellite positions may be collinear, nor anything approaching it.

Accepting (say) 75° as a maximum zenith distance, and given the height of the satellite, a plan can be drawn with circular arcs centred on each fixed station which will enclose an area within which the satellite plumb point must lie when it is to be observed. The problem is then to find positions within this area at which the satellite will be strongly fixed as above, and which will collectively strongly fix the fourth station. Since the satellite cannot be expected to go to the exact optimum position when required, the whole area of common visibility will, in practice be covered by numerous (50 or 100) satellite fixes, from which the forward ground station will be fixed by least squares. But it must be said that numerous fixes with bad geometry are a poor substitute for a few fixes with good geometry.

The accuracy of the fix, especially in the vertical direction is helped by the use of satellites of widely different heights, such as between 1000 and 3000 km.

It is not possible to get strong geometry, and the system of pure trilateration is not likely to be used again, with any system of distance measurement.

As an alternative to simultaneous observations from four stations, simultaneous observations may be made at only three, to determine the orbit over short arcs in their neighbourhood, which may then be extrapolated to some distance, where a suitably placed fourth station can fix itself by three or more distances from itself to the satellite. Numerous such satellite passes will of course be used.

Another alternative, if the station clocks are well enough synchronized, is that no observations at all need to be simultaneous.

(*c*) *Calibration.* Delay in the satellite transponder is calibrated before launch. It may vary by a few metres according to the aspect of the ground station, but that should average out as the satellite spins. It is hoped that the delay will remain constant during launch and subsequently. But, see [273], p. 372 and Table 5.22, field tests against lasers have there indicated a 17-metre systematic error in the Secor.

The ground equipment is more liable to change, so it is calibrated against a ground transponder at a distance of 100 metres before and after every satellite passage. The correction, of the order of perhaps 10 metres, is different for the two frequencies.

(*d*) *Corrections to the recorded distances.* The following corrections have to be considered: (Correction = Final minus recorded).

(i) *Refraction.* See § 7.09 and 10. A formula which has been used for the tropospheric correction is

$$\text{Correction} = -\frac{2.6\{1-\exp(-S/6858)\}}{\sin\alpha + 0.0236\cos\alpha} \text{ metres,} \qquad (7.161)$$

where S is the slant range in metres, and α is the angle of elevation at the ground station. See [463], p. 325.

For the ionospheric correction, the main term is eliminated by the use of the two wavelengths at 449 and 229.5 MHz, and the smaller terms are neglected. The correction to the distance given by the 449-MHz signal is

$$\frac{f_2^{-2}+f_1^{-2}}{f_2^{-2}-f_3^{-2}}\Delta, \qquad (7.162)$$

where Δ is the distance given by the 224.5-MHz signal (f_3) minus that given by the 449-MHz signal (f_2) after applying calibration corrections. f_1 is the frequency of the station signal, 420.9 MHz. As written the correction should always be negative, but random error may result in apparent changes of sign, which should not be excluded from the general mean.

(ii) The *movement of the satellite* has a doppler effect on the frequency of the transponded signals, whose frequencies are consequently not exactly as given above. A device built into the equipment automatically applies the necessary correction.

7.43. Laser ranging

See § 1.49 for the characteristics of lasers and of the coherent light which they produce.

(*a*) *Equipment.* A short laser pulse is directed at the satellite, whence it is reflected back by corner cubes, and the elapsed time is recorded. The typical equipment of the first half of the 1970s† has been a ruby laser transmitting a pulse of about 20 ns (6 metres) in length about once a second, see Table 7.3. The principal source of error has been distortion in the shape of the returning pulse, so that the triggering of a timing device at the start of the outgoing and incoming pulses does not give a perfect measure of the travel time. With a 20-ns pulse the least r.m.s. error from this cause has been about ±0.5 m. Shortening the pulse has been a possible remedy, but since about 1974 the error has been much reduced (to perhaps 0.1 m) by recording the times of the centroids‡ of the two pulses. It is hoped to reduce the error further by using a Neodymium YAG laser with a pulse length of only 0.2 ns. [166], p. 457. outlines further improvements.

TABLE 7.3
Lasers for satellite ranging

	Early-1970s SAO [166], p. 432	Early-1970s Goddard [461]	Mid-1970s Goddard [166] p. 445	Late-1970s For lunar [166] p. 457
Laser type	Ruby	Ruby	Ruby	Neodymium YAG
Wavelength/nm	694	694	694	530
Pulse length/ns	25	8 to 12	4	0.2
Pulse rate/Hz	0.13	1	1	3
Energy per pulse/J	1 to 2	1	0.25	0.5
Receiver aperture/cm	38	30		

The Goddard mobile laser is carried on one large trailer, but four or five are required for the complete equipment.

(*b*) *Tracking.* The laser is automatically directed on to the satellite by a computer into which the satellite ephemeris is fed. The ground station need not be in darkness.

The beam width for the tracking of artificial satellites has varied between 5 and 0.5 mrad, but for lunar lasers this has been reduced to 0.01 mrad (2″).

(*c*) *Computation*

(i) *Refraction correction.* There is no ionospheric refraction delay. The delay in the troposphere (for the single distance) is given by (7.27) or (7.29), in which the figure 2.387 is for ruby light. For other colours (1.15) will give the necessary change. The group velocity is required.

† [217], p. 5395 estimates the standard errors of earlier lasers as 10 m for CNES lasers before 1970, 5 m for GSFC before 1970, and 2 m for the ISAGEX lasers.

‡ The receiver distributes the photons contained in each pulse into several time channels which cover the total length of the pulse. The distribution between the channels is more or less gaussian, and enables the time of the centroid to be determined.

(ii) *Velocity aberration.* A consequence of the satellite's movement relative to the ground is that a perfect corner cube, instead of returning the light to its source, returns it to a point distant $2Sv/c$ from the source, where S is the slant range, v is the component of the satellite's relative motion perpendicular to the line of sight, and c is the velocity of light, [375]. If $S = 1500$ km, this may be 75 m, equivalent to a deviation of 10″ at the satellite. The width of the beam returned by the corner cubes must of course be wider than this.

(iii) *Calibration.* Much calibration is necessary, on a short known distance such as 3 to 5 km, to determine the calibration corrections appropriate to different signal strengths. [457].

(*d*) *The advantages of laser light* as opposed to microwaves, are as follows.

(i) The simplicity and reliability of corner cubes, as compared with microwave transponders.

(ii) The great power which can momentarily be concentrated in a laser pulse.

(iii) Lasers can generate pulses of 20 ns, which are 100 times, or more, as short as the modulation wavelength of Secor.

(iv) Laser emission can be concentrated into a narrow beam by a 10-cm lens, while radar requires an antenna several metres wide.

(v) There are no anomalies of the refractive index of light in the ionosphere, and there is minimal effect of water vapour in the troposphere.

The disadvantage of light is its inability to penetrate cloud.

(*e*) *Uses of laser ranging.* As in § 7.25, laser ranges can be used to provide scale for astro-triangulation. Currently (1978) their principal use has been to be incorporated with measures of direction and range-rate (doppler) in general solutions for satellite tracking and for determining station coordinates and geopotential parameters.

Lasers which may be accurate to 10 cm, observing to a satellite such as STARLETTE or LAGEOS may be expected to produce significant results in connection with the following.

(i) Polar motion, which may be determinable to within 0″.03 in a period of 6 hours, [535] and [15].

(ii) Crustal movements, [3].

(iii) Irregularity in the rate of the earth's rotation.

(iv) Earth tides and Love's number.

The general principle for the determination of these variables is that distances, measured from two ground stations to at least three well-placed and well-tracked satellite positions, give the length and direction of the chord joining the two ground stations. If the two stations are not very far apart, the length and direction of the chord may be somewhat insensitive to errors in the satellite ephemeris.

7.44. Navstar. Global positioning system (GPS)

(*a*) *General description.* The Navstar system is intended to supersede the Transit doppler system in about 1984–5. The intention is to launch 24 satellites, in circular orbits at 20 000 km altitude, of which 8 will be in each of three planes at inclinations of 63° with their nodes equally spaced round the equator. The satellites will be tracked from ground stations, and will broadcast their ephemerides and their clock time. The geocentric coordinates of mobile ground stations will then be got by measuring the time delay of radar signals emitted from the satellite. At least six, and generally eight or nine satellites will be visible above the horizon at any time and place.

The first satellite was launched in 1977, and by the end of 1978 there should be six, which will provide cover over a test area. By 1982 there should be about 10, and the full 24 in 1984.

The following details are based on [167] and [454].

(*b*) *Tracking stations.* The first six satellites will be tracked from base stations in Guam, Alaska, Hawaii, and at Vandenburg (California), by one-way range measurement from the satellites' signals. Other tracking stations may be added later. Ephemerides are computed from the tracking observations, and once daily each satellite is sent a revised ephemeris (for perhaps the next 120 hours), and the error and predicted drift rate of its clock.

(*c*) *The satellites.* The satellite transmits on two L-band wavelengths, namely $L_1 = 1575.42$ MHz and $L_2 = 1227.60$ MHz. Each carrier is modulated to give the broadcast navigation message, consisting of the ephemeris, satellite clock time, and some ionospheric data. The signals of one satellite do not interfere with those of another.

Each satellite carries a very accurate caesium clock, with a drift rate which is expected to be of the order of 10^{-13} per day, and whose error may be expected to be known within a few nanoseconds. The instant at which any signal leaves the satellite is then known to this accuracy.

At the altitude of 20 000 km, air drag is virtually absent, and the effects of gravity anomalies are much minimized, leaving solar radiation as the largest source of error. It is to be hoped that the r.m.s. ephemeris error will not exceed ±1 or 2 metres.

The expected active life of one of these satellites is expected to be about 5 years.

(*d*) *Receiving stations.* The signals may be received at fixed ground stations, and on ships or aircraft. Different types of equipment will be available for different users. The receiving apparatus will predict the doppler shift to enable it to tune in on the satellite, and it will then decode the message, measure the range, apply atmospheric corrections, and compute the geocentric coordinates of the station.

An essential feature of the receiver is a clock of high stability. If the ground clock were perfectly synchronized with the satellite clock, the receipt of a satellite signal, after correction for atmospheric delays, would give the correct range. Simultaneous observations to three suitably placed satellites would then give the three coordinates of the station. Perfect synchronization is of course not possible, but the inclusion of a fourth (suitably placed) satellite provides an extra equation. This then gives the time difference between the two clocks, provided only that the difference is adequately constant during the short period of the observation.

(*e*) *Atmospheric corrections.*

(i) *Ionospheric refraction.* See § 7.10. If both wavelengths are used, the $1/f^2$ term is eliminated, and with $f = 1575$ MHz (as compared with the 400 MHz used for doppler) the $1/f^3$ and $1/f^4$ terms should be of the order of 0.1 metre or less in most conditions.

If only the 1575 MHz can be used (for instantaneous fixes), the figures given in § 7.10 (*c*) will be reduced, by the higher frequency, from $-5.6 \sec \zeta^*$ to $-0.46 \sec \zeta^*$ metres, for optimum conditions (quiet year, night), but which may be ten times or more as great by day.†

(ii) *Tropospheric refraction.* The correction to Navstar observations is the same as for any other radar, as described in § 7.09, with uncertainty arising from humidity.

(*f*) *Equipment and accuracy.* Various types of equipment will be available. For a fast-moving vehicle, such as an aircraft, which requires instantaneous results, the position is expected to be correct to 10 m, and the velocity of the vehicle correct to 0.1 m s^{-1}.

For geodetic purposes, where time can be spent at a stationary location, an s.e. of one metre is expected to be obtainable from a few hours of observation. It is hoped that the relative positions of ground stations within 1000 km of each other will be obtainable within perhaps 10 cm. [16] gives the latest available information (1978).

7.45. Lunar laser ranging

(*a*) *Equipment.* Five sets of corner cube reflectors have been placed at well distributed sites on the moon, three by Apollo missions and two by the USSR Lunakhod. Between 1971 and 1976 reflections had been regularly obtained from four of these sites, notably from lasers at

† Sec ζ^* varies from 1 to 3 for zero and 90° zenith distances. See after (7.49).

McDonald (Texas) and at the USSR Crimean observatory, [433], pp. 37–39. The lasers employed were ruby lasers with characteristics such as those in Table 7.3, columns (1)–(3). For the future, something like column (4) is probable. The sensitivity (as opposed to actual total accuracy) has generally been between ±0.1 and 1 m, but it is hoped to reduce this to ±0.03 m.

(*b*) *The geometry of the observations.* See Fig. 7.13. P is a laser station with earth coordinates *x*, *y*, *z* in the geocentric, CIO, and BIH (Greenwich) frame. Q is a reflector on the moon with coordinates *a*, *b*, *c* in the frame constituted by the moon's centre of mass, M, and the three orthogonal principal moments of inertia.

Computations are based on a model which gives the values of OM and PQ at any moment, based on the best current theory of the movements of the earth and moon, and on preliminary values of the coordinates of P and Q. As in other geodetic problems, the routine is to compare the observed value of PQ (corrected for refraction) with the value given by the model and, when sufficient observations have been obtained, to compute least-square corrections to the preliminary coordinates and to all the other parameters on which the model is based. If the model is insufficiently accurate, the variations may not be linear, but that can as usual be remedied by re-iteration of the solution.

The model is a complicated one. The coordinates of P are constant in a frame which rotates about the CIO axis, apart from earth tides, the effects of ocean tides, and secular or other crustal plate movements. For the highest accuracy the model must include parameters defining some or all of these, and the observations may provide corrections to the model. Further, the CIO is not exactly the earth's instantaneous axis of rotation, but differs from it by the polar motion, § 4.09, which may be allowed for

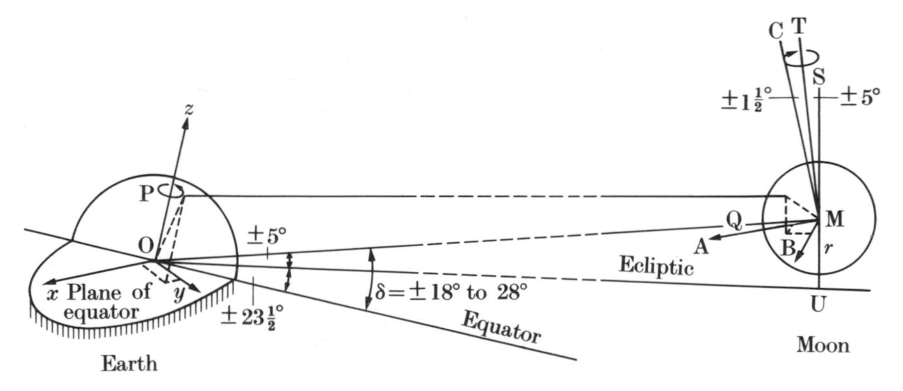

FIG. 7.13. OZ is perpendicular to the equator. SU is perpendicular to the ecliptic. TM is perpendicular to the moon's mean orbit. In the plane of the paper, the components of angles marked (e.g.) ±5° vary between the limits stated.

in the model and also treated as an unknown. Other complications are that the instantaneous axis is perpendicular to the equator, which makes a slightly variable angle of about $23\frac{1}{2}°$ to the ecliptic, in a direction which varies with precession and nutation (§ 4.03). The ecliptic itself is not quite exactly fixed with reference to a Newtonian frame, § I.00, as defined by the distant stars, and finally it is not the geocentre that moves round the sun in the plane of the ecliptic, but the joint centre of mass of the earth and moon, with some interference from the other planets.

The directions of the moon's principal axes of inertia, which define its coordinate axes in Fig. 7.13, also vary. The minimum axis A points approximately constantly towards the earth, but with a periodic variation, the libration, of amplitude 3° or 4°, or about 100 km on the moon's surface. The mean direction of the axis of maximum inertia, axis C, is inclined to the pole of the ecliptic at an angle of about 5°, and the instantaneous axis revolves about the mean with an amplitude of about $1\frac{1}{2}°$.

The relative movements of the earth's and moon's centres of gravity are given by Lunar Theory, which is currently based on the theory of E. W. Brown between 1896 and 1908, with some later modifications. Its accuracy has been sufficient to give the relative positions of points on the earth and moon to about 0.5 km, so if a model is required to be correct to (say) 0.05 m the accuracy of the existing theory will have to be increased about 10 000 fold. This is not considered to be impossible, [166], p. 463. The principal harmonic terms in the earth's gravitational field (which are now well enough known), and those of degrees 3 and 4 of the moon's field, must also be included in the model.

(*c*) *Geodetic uses.* The majority of the very numerous numerical parameters involved in the model are of no direct concern to geodesists, whose interests are confined to the following.

(i) The coordinates of the observing stations on the earth.
(ii) Polar motion.
(iii) Earth tides.
(iv) Crustal movements.
(v) Irregularities in the earth's rotation.

It is evident that provided the coordinates can be continuously and sufficiently accurately measured, the remaining items will follow. We are therefore only concerned that the observation equations should give stable solutions for the earth's station coordinates.

In Fig. 7.13, the distance PQ approximately differs from OM, the distance between the two centres of mass, by the sum of the projections of OP and MQ upon OM, since PQ and OM may for this present purpose be treated as parallel. The projection of OP varies with an amplitude of

$6370 \cos \phi$ km with a period of 24 hours. This is the principal source of variation in (OM − PQ), and a series of observations may clearly be expected to give the x- and y-coordinates with the same accuracy as the observations,† provided the effects of other unknowns can be eliminated.

The z-coordinate is determined by the variation of the projection of OP on to OM with the changes in the moon's declination, which varies with a period of one month and an amplitude of between 18° and 28°, according to the phase of the 19-year cycle. When the declination is changing between (say) +20° and −20°, and if the latitude of P is (say) 45° the length of the projection will vary between $6370 \cos 65°$ and $6370 \cos 25°$, i.e. with an amplitude of about 1500 km and a period of one month. This gives a good determination of z, but somewhat less strong than that of x and y, [433], p. 41.

The lunar coordinates of the reflectors are less accurately determined, since the relatively small libration causes much less variation in (OM − PQ) than is produced by the earth's daily rotation. [433], p. 34 suggests that the errors in lunar coordinates may be about 25 times as great as the errors in the earth station coordinates.

General references for Section 6.
Secor [171], [273], [432], and [508]; satellite laser ranging [461], [273]; and lunar laser ranging [166], [273], [393], and [433].

Section 7. Doppler

7.46. Notation

f_t = Frequency of transmitted signal.
f_r = Frequency of received signal.
f_0 = Frequency of receiver standard.
$f_b = (f_0 - f_r)$ = beat frequency.
$(f_0 - f_t)$ = Frequency offset.
$\Delta(f_0 - f_t)$ = Frequency bias.
$\Delta f = (f_r - f_t)$ = doppler shift.
N = Doppler count, (7.167)
c = Velocity of transmission in vacuum.
V = Velocity of satellite relative to ground.
S = Distance ground station to satellite.
$\dot{S}$ = Rate of change of S.
S_0 = Minimum value of S in any one pass.
t_0 = Time when $S = S_0$.
$\cos \theta = \dot{S}/V$, see Fig. 7.14. Also $\cos \theta = p$, the independent variable in spherical harmonics.

† Except that lunar ranging does not seem to connect directly with the earth's zero of longitude. The relative positions of two stations P_1 and P_2 will be accurately given, but the zero of longitude will have to be introduced by other means, as in the case of doppler, § 7.59. There is the same difficulty in VLBI, § 7.67.

$\Delta S = (S_2 - S_1)$ = change in S in time Δt. $\Delta t = (t_2 - t_1)$.
A,B,C,D = Rotation matrices for polar motion, earth rotation, nutation, and precession respectively, see § 7.55(*b*).
$E = (1/a_e)BCD$. (7.201)
v = Residual in an observation equation.
a_e = Semi-major axis of earth spheroid.
$\mathbf{a}$ = Matrix vector of components of distance from geocentre to station in inertial frame.
$\mathbf{a}'$ = Same as $\mathbf{a}$, but with components in terrestrial frame.
$x', y'.z'$ = Components of $\mathbf{a}'$.
$\Delta x', \Delta y', \Delta z'$ = Optimum corrections to x', y', z'.
$p = (f_0 - f_t)$, whose optimum correction is Δp. (7.175).
q = Standard correction for tropospheric refraction.
$C_r q$ = Optimum correction to q., (7.175). $C < 0.1$.
x_P, y_P = Coordinates on Guier plane.
$\mathbf{r}$ = Vector geocentre to satellite. Components in inertial frame, x, y, z.
$\dot{\mathbf{r}}$ = Satellite velocity vector. Components $\dot{x}, \dot{y}, \dot{z}$.
$\ddot{\mathbf{r}}$ = Satellite acceleration vector.
r = Scalar magnitude of $\mathbf{r}$.
$\mathbf{r}' = \mathbf{r}$, with components in the terrestrial frame. $\mathbf{r}' = [r_1' r_2' r_3']^T$.
p_k = one of the initial elements, $x, y, z, \dot{x}, \dot{y}, \dot{z}$.
p_j = A frequency, refraction, or atmospheric density bias.
$[\partial\ddot{\mathbf{r}}/\partial\mathbf{r}^T]$ = A (3×3) matrix, see (7.193).
$\boldsymbol{\xi} = \partial\mathbf{r}/\partial p_k$. $\dot{\boldsymbol{\xi}}$ and $\ddot{\boldsymbol{\xi}}$ are its first and second time derivatives.
ω = Earth's angular velocity in the inertial frame.
σ_0 = Standard error of an observation, or of a deduced unknown, based on internal consistency.
O, C = Observed (after processing) and computed values; the latter from an ephemeris.
U_l^m, V_l^m, see (7.199).

See also § 7.00, and the note about vectors after (7.1).

The word bias in American literature is used in the sense of systematic error, with no implication of prejudice, such as may be suggested by the use of the word in English. Thus $\Delta(f_0 - f_t)$ and p_j, as defined above, are errors which are common to all the observations of a single pass.

7.47. Basic principles

(*a*) In common with other dynamic methods, the doppler system determines the geocentric coordinates of the observing stations, and also provides information about the earth's geopotential field. From the geodetic point of view, its great use is that it constitutes a practical method of ground survey, whereby the geocentric positions (in all three coordinates) of numerous points, anywhere on the earth, can be accurately determined with relatively inexpensive field equipment. In different circumstances, the accuracy may be described (in 1978) as between ±1 and ±10 metres.

This § 7.47 describes the method in a very general way, with a view to illustrating how the system works, and how the most favourable geometry can be secured. Details of the observations and computations are given in later paragraphs.

(*b*) *The doppler shift.* Let a satellite S, Fig. 7.14, follow a polar orbit with constant velocity V at constant height above a spherical earth. Let its orbit be known, so that the cartesian coordinates of its position are

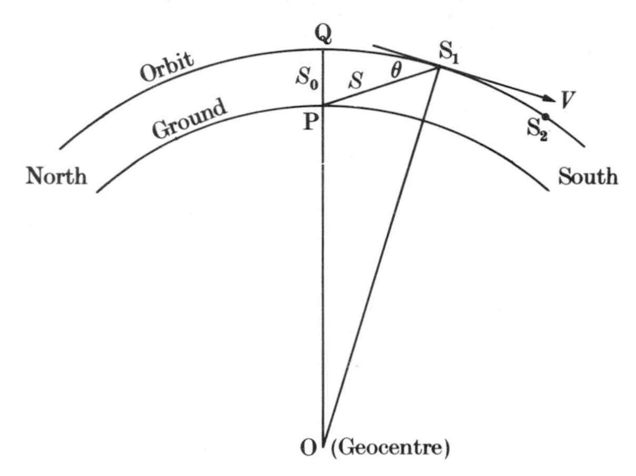

FIG. 7.14. Doppler geometry. Polar orbit, viewed from west. O and P and the ground section are in the plane of the paper. The short arc of the orbit is substantially parallel to the plane of the paper, but does not lie in it. Consequently, S, S_0, and θ are not measured in the plane of the paper.

known at any time, with reference to the geocentre O. Let the satellite emit a continuous signal with known frequency f_t, which is received at a ground station P with a frequency f_r. Let the velocity of the signal's transmission be provisionally assumed equal to c, the velocity in vacuum. Let PS $= S$, and let PS make an angle θ with the orbit at S_1, so that $\cos\theta = \dot{S}/V$. It is required to find the geocentric coordinates of P. For the moment, let the movement of P due to the earth's rotation be ignored. Then

$$f_r = f_t(1+\dot{S}/c)^{-1} \tag{7.163†}$$

Then the doppler shift is given by

$$\Delta f = (f_r - f_t) \approx -(f_t\dot{S}/c) \tag{7.164}$$

† Elementary considerations give (7.163) as for sound in air, but for electromagnetic waves relativity principles add a factor and give

$$f_r = f_t(1+\dot{S}/c)^{-1}(1-V^2/c^2)^{\frac{1}{2}}. \tag{7.165}$$

See [413], p. 321, in which $\theta = 180° - \theta$ as defined in Fig. 7.14. But see § 7.51. In practice the factor $(1-V^2/c^2)^{\frac{1}{2}}$ can be ignored.

We have $\dot{S}=V\cos\theta$, so with c, f_t, and V known and Δf measured, (7.164) gives θ, and the ground station must lie on a cone† whose axis is tangent to the orbit at S, and whose semi-apex angle is θ.

(*c*) *Geometry of the fix.* If, having been measured at S, Δf is again measured when the satellite is at S_2, another cone will be determined on which P must lie. These two cones will intersect in a nearly circular ellipse, in a plane approximately perpendicular to S_1S_2, on which P must lie. A series of such observations, made during a single pass, will give a series of cones intersecting in a series of ellipses all (in the absence of error) passing through P. These ellipses are not quite co-planar, because of the curvature of the orbit, but at P their tangents are substantially parallel. If the orbit passes through the zenith of P these tangents will all be horizontal, and if the point of the satellite's nearest approach is (say) 60° above the east or west horizon, the tangents will form a somewhat narrow bundle inclined to the horizontal at 30°. The consequence is that a good fix is not obtainable from a single pass unless other information is available, such as the distance OP. This is only known if we know the height of P above the geoid, and the height of the geoid above some geocentric spheroid. Given this, the geometry of the fix is good, provided

(i) The orbit is well removed (at least 30°) from the zenith at P.
(ii) The orbit, at its closest rises to at least 20° above the horizon.

If a fix is obtained in this way, the doppler system is said to have been used in the *Navigational mode.*

To obtain a fix in three dimensions without knowledge of the distance OP, observations must be made to two separate passes, preferably on opposite sides of the zenith. Fig. 7.15 shows two such passes, as seen from the south, with a pair of near-circular loci intersecting at P at a good angle, and giving a good three-dimensional fix with one redundancy, since the two circles should be co-planar. Actually, if each satellite is tracked throughout its pass, each of these two circles will (as above) be a bundle of circles all passing through P at narrowly divergent angles. When a fix is obtained in this way, preferably from several passes and subject to other conditions, the system is said to have been used in the *geodetic mode.* See § 7.49 and § 7.55.

The geodesist is only concerned with the geodetic mode.

An alternative view of the information obtainable from a single pass is shown in Fig. 7.16, which shows the variations in the doppler shift Δf in the course of the pass. It is clear that the shift is zero when the satellite is at Q, at its nearest to P, where $\dot{S}=0$. It is also clear that the slope of the

† In practice Δf is not measured, but its integral is recorded over a short interval of time, see § 7.51. The resulting locus is then a hyperboloid which is asymptotic to the cone which would arise from the measurement of Δf at the mid-point of the interval.

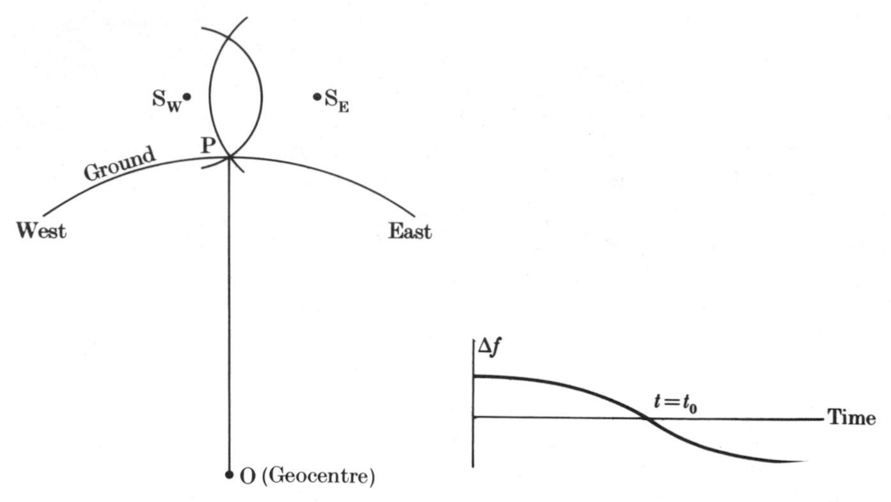

FIG. 7.15. Doppler geometry. Polar orbit, viewed from the south S_W and S_E are satellites whose orbits lie west and east of the zenith at P. Arcs centred on them are loci on which P must lie.

FIG. 7.16. Doppler shift Δf plotted against time. Time t_0 is when the satellite is closest to P.

curve at this point is an approximate measure of S_0, the minimum distance PQ. For when S is very close to Q, so that the orbit QS can be treated as straight, we have

$$S^2 = S_0^2 + V^2(t-t_0)^2$$

$$\dot{S} = (V^2/S)(t-t_0)$$

$$\ddot{S} = -(V^2/S^2)(t-t_0)\dot{S} + V^2/S$$

$$\text{and } \mathrm{d}(\Delta f)/\mathrm{d}t = -(f_t/c)\ddot{S}, \quad \text{from} \quad (7.164)$$

$$= -(f_t/c)(V^2/S_0), \quad \text{when} \quad t = t_0 \qquad (7.166)$$

Whence S_0 is obtainable.

It follows that a constant error in f_t (or in Δf) will cause an error in the time at which $\Delta f = 0$, and thence in the along-track positions of Q and P. But such an error will be revealed by unexpected inequality in the values of Δf before and after the erroneously deduced t_0. A possible error in f_t can therefore be included as an easily determinable unknown in the solution for the position of P.

An error which consistently makes Δf numerically too large or too small will introduce a constant proportional error in S_0, and thence in the longitude† and height deduced from a single pair of passes, although this

† Throughout this paragraph, the orbit is taken to be a polar orbit.

error will largely be eliminated in a solution based on many passes, with widely variable minimum zenith distances.

(*d*) In (*c*) above the earth's rotation has been ignored. While a polar satellite moves (north or south) at 7.3 km/s^{-1} at height 1100 km, with its sub-satellite point moving at 6.2 km/s^{-1}, a ground station in latitude ϕ moves eastwards at $0.46 \cos \phi$ km s^{-1}. It follows that the path of the sub-satellite point does not exactly coincide with a terrestrial meridian. It will in fact be inclined to the meridian by about 4 degrees at the equator, decreasing to zero at the poles. Consequently, instead of the point of closest approach being due east or west of the ground station, it departs from that by a few degrees to the north or south. This does not substantially affect the conclusions of (*c*) above from the point of view of considering the good or bad geometry of a fix, but the earth's rotation does of course enter very materially into the actual computations.

7.48. The Transit and Tranet systems

(*a*) *Satellites.* The doppler system, in the form known as the US Navy Navigation Satellite System, has been in operation since 1963. It was primarily intended for use by ships in the navigational mode, § 7.47 (*c*), to give positions correctly to about 100 m anywhere in the world. It originally employed five Transit satellites in polar orbits at altitudes of about 1100 km, with their nodes more or less evenly spaced round the equator. The orbital period is about 107 minutes, so every point on the earth is within range of at least two passes of each satellite every 12 hours. The original five satellites have been reinforced by another, Triad II, in 1975, with the Discos equipment described in § 7.04.

Each satellite continuously emits microwaves at frequencies slightly below 400 and 150 MHz,† controlled by a frequency standard carried in the satellite. These signals are received at the ship or ground station where they are compared with a local oscillator to measure the received frequency, and thence the doppler shift, or (actually) the change of the distance from ground to satellite in a finite short time. See § 7.51.

In addition, the satellite carries a core memory from which it emits an approximate *broadcast or predicted* ephemeris for a period of 16 hours, which is broadcast as a modulation of the 400- and 150-MHz transmissions. Each such emission takes two minutes, and is continuously repeated. It is updated every 12 hours. Suitable equipment at the ground station can decode this ephemeris, and will compute the geocentric coordinates of the satellite at the appropriate instants. See [93]. The satellite also broadcasts two-minute time marks and the amount in ppm by which its oscillator is at less than 400 MHz.

† Two frequencies to eliminate error due to ionospheric refraction.

Several dozen other satellites emit microwave signals which are suitable for doppler tracking. Some of them are used for the general geodetic solutions described in § 7.62.

(*b*) *Tracking stations.* For the original development of the system, ground equipment was located at 13 stations well distributed over the earth, which (by 1976) had been reinforced by 7 others. These stations continuously track one of the Transit satellites. From their observations covering consecutive periods of 48 hours, a *precise* or *tracked* ephemeris is prepared which is thought to give the geocentric position of the satellite within 2 or 3 metres, apart from slightly larger doubts about the scale and longitude of the reference system. See § 7.63 (*c*) and (*d*).

For the broadcast ephemeris the tracking of all the Transit satellites is carried out continuously from four stations only,† situated in Hawaii, California, Minnesota, and Maine. These stations report to a computing centre where the orbital elements of each satellite are updated, and an ephemeris for the next 16 hours is injected twice daily into each satellite's memory via one of the four tracking stations.

The broadcast ephemeris gives the geocentric coordinates of the satellite every two minutes, rounded to the nearest 10 metres. Its standard errors in 1976 were estimated, [129], p. 12, as ±24 m along track, ±17 m across track, and ±8 m in height.

(*c*) *Ground receiving equipment.* If the broadcast ephemeris is to be used to give rapid results the basic requirements are

(i) A clock and receiver to record the range-rate.

(ii) An element to demodulate and decode the ephemeris, and to compute the position of the satellite in the terrestrial frame.

(iii) A computer to process the doppler count, and to compute the geocentric position of the station.

The introduction of the *Geoceiver* in 1967 represented a great step forward from the geodetic point of view. It was a small portable instrument for use on land using the precise ephemeris, and simplified by the exclusion of facilities for receiving and decoding the broadcast ephemeris. An important improvement was the inclusion of its own high precision clock, whereas instruments intended for navigation only rely on the satellite clock.

Later instruments, such as the more recent Magnavox Geoceivers, the CMA 722B, and the JMR-1, have included the broadcast decoding and computing facilities, with a view to their using the broadcast ephemeris to

† These four stations form part of the NNSS or *Transit* system, a term which includes all observations and computations leading to the broadcast ephemeris. The word *Tranet* refers to the other tracking stations, and to the data which they produce, but not to the resulting ephemeris computations.

give results of geodetic accuracy over limited areas, using one of the Translocation or Short Arc modes of § 7.50 (*b*)–(*d*). Such instruments (in 1977) weigh about 30 kg in all, including antenna, battery, and computer, with a heaviest single package of 20 kg. The antenna is a single pole one or two metres in length. The 'electric centre' of the antenna is the point to which computed coordinates refer.

Needless to say, the doppler system works in any weather.

7.49. The relative advantages of the two ephemerides

The obvious advantage of the precise ephemeris is its accuracy. It is indeed possible for the predicted broadcast ephemeris to give the relative positions of nearby stations with an accuracy equal to that obtainable from the precise ephemeris, using translocation or short arc methods, but for world-wide fixations to connect the various classical ground survey systems, the precise ephemeris is necessary. On the other hand, the precise ephemeris has two disadvantages.

(*a*) It is only available for one or two satellites, so that the observation of (say) 30 passes may take four or five days, in mid-latitudes, while the six satellites operating the broadcast system can give 100 passes in six to eight days, [129], p. 14.

(*b*) Currently (1977) the precise ephemeris is only obtainable with some difficulty, with a delay of a few weeks.

7.50. Geodetic modes of observation

Geodetic fixes, i.e. those that determine all three coordinates from two or more passes, may be made singly to give absolute geocentric positions, or in combination with other fixes to give relative positions with high accuracy. Various methods are outlined in (*a*)–(*d*) below.

(*a*) *Point positioning.* This is defined as an isolated fixing, made on the assumption that the ephermeris is perfect. Point positioning using the broadcast ephemeris is not recommended for the best work, especially outside North America, even though a higher number of passes may be obtainable. The usual specification for a single fix is as below.

(i) 20 to 30 passes should be observed to satellites for which the precise ephemeris will be available.

(ii) Approximately equal numbers of passes should be north going and south going, and approximately equal numbers should pass east and west of the zenith.

(iii) Satellites should be observed from 10° above the horizon when rising, and down to 10° when setting. If this is not possible, the accepted counts should be balanced about the point of closest approach, leaving a total of not less than 12 30-second counts.

(iv) No pass should have a maximum altitude of less than 20°.

(v) See also §§ 7.53 and 7.63 (*a*) and (*b*) for internal standards of consistency, which should be satisfied.

For the resulting expected accuracy see § 7.63.

In (*b*)–(*d*) below, the same standards should be followed unless otherwise stated, except that the broadcast ephemeris is used.

(*b*) *Translocation.* If the intention is to obtain only the relative positions of two ground stations separated by some tens of km, or even perhaps a few hundreds, relative results which may well be as accurate as those given by the precise ephemeris are obtainable using the broadcast ephemeris.

A receiver is located at each of the two stations, and observations are made simultaneously from both to numerous (25 to 40) satellite passes. In the subsequent computations the doppler counts employed are restricted to such parts of these passes as have been satisfactorily observed at both stations. The separation of the two stations being small compared with the height of the satellite, ephemeris errors have a more or less similar effect at each station, especially along-track errors, which are generally the largest.

A possible routine is for one receiver to remain at a fixed central station, while a second successively occupies numerous surrounding stations. Relative accuracies of 3 metres or better may be expected from 25 passes, [129], p. 15, or even an s.e. of 0·5 m, in distances of 40 km, [72]. Alternatively, all possible pairs of stations may be independently observed, to give redundancies which can be adjusted in a way similar to the adjustment of ordinary triangulation and trilateration.

(*c*) *Short arc translocation.*† In a simple translocation as in (*b*) above, if the stations are rather widely separated, the use of only simultaneous counts looses a substantial amount of observations. Further, the assumption that ephemeris errors will cause equal errors in the two fixes will be less valid. A possible procedure is then to use all the data obtainable at each station, and to solve for a simple shift‡ in the broadcast ephemeris

† In satellite literature the expression *short arc* generally implies an arc extending over one or two days, as opposed to a long arc of 14 to 30 days, but in the present context it implies only a small fraction of a single orbit.

‡ Since 1976 the broadcast ephemeris has used the WGS 72 geopotential model, and its short-period perturbations may be expected to be well predicted. A parallel shift is consequently quite appropriate.

arc parallel to itself, such as will produce optimum agreement between the two stations, [129], p. 16.

(*d*) *Short arc geodetic adjustment.* [129], pp. 15–19. In this method simultaneous observations are made at a number of stations, and using the broadcast ephemeris a solution is made for the following.

(i) The coordinates of the stations.
(ii) Corrections to the orbital elements of the arcs employed.
(iii) For each arc five coefficients giving a zero bias, a timing bias, frequency bias, frequency drift, and a refraction correction.

As an example of the method [129] describes the observation of a small net of 16 stations, using five receivers. One receiver permanently occupies a central station. The other four first occupy four other stations forming a small quadrilateral ABCD, and observations are made. Two of the four receivers (say those at A and B) then move to stations E and F, such that CDEF forms another small quadrilateral interlocking with ABCD, and all five stations again observe. And so on, until the complete net has been incorporated in interlocking quadrilaterals. In this example 10 quadrilaterals were employed, and about 250 good simultaneous passes would have been observed in about 20 days. Solution by least squares would then give 48 station coordinates, 1500 orbital constants, and 5000 bias corrections, a total of 6548 unknowns. This is a formidable undertaking, made possible by suitable partitioning of the matrix, [129], pp. 22–4 and [130]. Relative accuracies of less than 1 metre were hoped for.

[130] describes several nets of satellite stations which have been observed in Canada on such lines as these, using the broadcast ephemeris. [72] reports s.e.'s of about ±0.5 m for the relative error of members of groups of stations about 40 km a part in Great Britain.

7.51. Integrated range-rate

The change of range ΔS or $(S_2 - S_1)$ over a period Δt or $(t_2 - t_1)$ is measured as follows. The frequency f_t of the satellite transmission is kept close to 399.968 MHz, and the ground station standard f_0 is 400.000 MHz, a difference of 32 kHz, which is substantially greater than any possible value of the doppler shift Δf, which can at most be about 8 kHz. The received signal of frequency f_r is compared with f_0, to give beats with a frequency of $f_b = f_0 - f_r$, which are accurately counted. The *integrated doppler count* N is the number of beats recorded between (UTC) times t_1 and t_2. The following assumes the velocity of transmission to equal c, the velocity of light in vacuum. Refraction corrections will of course have been applied.

Then

$$N = = \int_{t_1}^{t_2} f_b \, dt = \int_{t_1}^{t_2} (f_0 - f_r) \, dt \tag{7.167}$$

$$= \int_{t_1}^{t_2} \{f_0 - f_t(1 + \dot{S}/c)^{-1}(1 - V^2/c^2)^{\frac{1}{2}}\} \, dt, \quad \text{as in (7.165).}$$

$$= \int_{t_1}^{t_2} (f_0 - f_t) + f_t\dot{S}/c - f_t\dot{S}^2/c^2 + f_t V^2/2c^2\} \, dt$$

$$= (f_0 - f_t)\Delta t + f_t \Delta S/c - f_t \Delta S(\text{mean } \dot{S})/c^2 + f_t(V^2/2c^2)\Delta t. \tag{7.168}$$

Whence

$$\Delta S = (S_2 - S_1) = (c/f_t)\{N—(f_0 - f_t)\Delta t - f_t(V^2/2c^2)\Delta t\}\{1 - \text{mean } \dot{S}/c\}^{-1}, \tag{7.169}$$

where $c/f \approx 0.75$ m.

If Δt is very small, such as 1 second, the doppler shift at time $\frac{1}{2}(t_2 + t_1)$ is

$$\Delta f = -\left(\frac{f_t}{c}\right)\left(\frac{\Delta S}{\Delta t}\right). \tag{7.170}$$

To illustrate, insert figures $\Delta t = 30$ s, mean $\dot{S} = 3$ km/s. (It can vary between + and -7 km/s), and $f_0 - f_t = 32$ kHz. Then

$$\Delta S \approx (3/4)(1\,080\,000 - 960\,000 - 0.6)(1 - 10^{-5}) \approx 90\,000 \text{ m}. \tag{7.171}$$

The relativity term -0.6 is small and constant, and in practice it is omitted. It is indistinguishable from an error in f_t, which is treated as one of the unknowns in the solution, so the values of all other unknowns are unaffected.

In the foregoing, the times t_1 and t_2 between which the beats are counted are UTC or atomic times as recorded by the ground station clock. Alternatively it is possible to integrate between the instants at which the waves concerned have left the satellite, in which case the relevant Δt is less by a factor of $(1 - \text{mean } \dot{S}/c)$, so that this factor will disappear from (7.169). This is the way in which N has commonly been recorded (before 1976), and the formula used has been

$$\Delta S = (c/f_t)\{N - (f_0 - f_t)\Delta t\}, \tag{7.172}$$

in which Δt is the interval between the instants at which the first and last beat waves left the satellite.

The period Δt over which N is counted is approximately† 4.6 s, or some multiple of it, such as 6 or 7 such periods (known as 30-s counts) or (6+7+6+7) periods totalling about 120 s. The 4.6-s period is used at the original 13 tracking stations and produces about 200 counts in a typical 15-minute pass. The less numerous (20 to 40) counts produced by the 30-second periods are more convenient, and are usual in more modern equipment.

7.52. Data processing

Even when computations are to be postponed for the precise ephemeris, some processing of the data is usually done automatically by the receiving equipment, or by some small computer on the site. Such as the following.

(*a*) ΔS is computed from (7.169) or (7.172) for each count.

(*b*) The 400- and 150-MHz transmissions are compared in order to eliminate the effects of ionospheric refraction, §§ 7.54 and 7.10. The standard value of the tropospheric refraction is also applied.

(*c*) Some kind of filtering or preliminary computation is necessary to eliminate faulty or substandard values of ΔS, after determining and correcting for the frequency bias $\Delta(f_0 - f_t)$, and to confirm that normal consistency is being maintained.

(*d*) If the doppler count has been made over very many short periods, means may be taken to produce a smaller number of representative values.

(*e*) When the ground station is in a moving vehicle, such as a ship, allowance must be made for the motion. This does not concern the geodesist. See [93], p. 25 for some description of the errors involved.

7.53. Filtering. The Guier plane

Continuing from § 7.52(*c*), the following is a method of filtering the doppler counts observed in one pass. Preliminary computations as below of the station position are made in two dimensions on the Guier plane, which is defined as the sloping plane which contains the trial position of the ground station, and the satellite's position and velocity vector at the point of closest approach.

The approximate time of closest approach, and thence the ephemeris position and velocity of the satellite at that time, can be obtained by the inspection of successive doppler counts. At closest approach the doppler shift is zero, and for a short count centred on this point N should equal

† The exact value of the basic $(t_2 - t_1)$ at the satellite is 4.601016 s. N is ordinarily recorded as a whole number, with a possible error of up to $\pm\frac{1}{2}$. But in some instruments the beat frequency is internally magnified 100-fold, so that in (7.169) and (7.171) N is recorded to 0.01.

$(f_0 - f_t)\Delta t$, from (7.169), or approximately $32\,000\Delta t$. Two counts are found which bracket this figure, and interpolation between their middle times gives the time of closest approach.

With the position of the plane in the terrestrial reference system thus defined, the trial position of the ground station is taken as the origin of coordinates on it. One axis is taken to pass through the point of closest approach, and the third axis as perpendicular to the plane. Then satellite positions, as computed from the ephemeris transformed to the terrestrial frame, can be further transformed to the Guier frame, and each doppler count gives an equation for a revised position of the station (in that plane) in the form

$$\frac{\partial \Delta S}{\partial x_p}\Delta x_p + \frac{\partial \Delta S}{\partial y_p}\Delta y_p + \frac{\partial \Delta S}{\partial p}\Delta p = \mathrm{O} - \mathrm{C} + v, \tag{7.173}$$

where $p = (f_0 - f_t)$.

This is similar in form to the definitive solution (7.175), except that Δz and the correction to improve the tropospheric correction are omitted.

Normal equations are formed and solved to give Δx_p, Δy_p, and $\Delta(f_0 - f_t)$. A second solution can then be made, using $(x_p + \Delta x_p)$, $(y_p + \Delta y_p)$ and the new $(f_0 - f_t)$ to define a new Guier plane. Iteration on these lines continues until Δx_p and Δy_p are found to be negligible. Five iterations or less should suffice. The residuals v in the first solution should identify any observations which may contain gross error, and they can be rejected. Subsequent iterations may indicate some substandard observations whose rejection may be expected to improve the mean result. The values of v in the last solution will indicate σ_0, the s.e. of a single (satisfactory) observation, and also the s.e. of the pass. Doppler counts whose residuals exceed $3\sigma_0$ are commonly rejected. If the number of rejections in a pass exceeds 3 per cent, the whole pass is suspect.

The merit of this system of filtering are that, although the coordinates arrived at for the station will be seriously inaccurate, the self-consistency of the observations is not thereby much affected. In particular, errors in the ephemeris used, and in the trial position of the station, will have little effect from the point of view of consistency and rejections.

7.54. Refraction corrections

(*a*) *Tropospheric correction.* The correction resulting from the refractive index of microwaves is described in § 7.09. Hopfield's double quartic correction is currently (1976) used, and is applied as part of the data processing. Optionally, a refraction bias, $C_r \times$(Standard refraction correction), may be included as an unknown in the observation equations, see § 7.56 (*e*), (i) and (vi).

Zenith distances are currently limited to a maximum of 80°.

(*b*) *Ionospheric correction.* This is described in § 7.10. The first-order correction is applied by the ground receiver by analogue means. The smaller corrections a_2/f^2 and a_3/f^3 are ignored. See § 7.10(*e*), it may be hoped that the error on this account will not exceed 10 cm.

7.55. Computation of station coordinates. Geodetic mode

(*a*) In this paragraph and in those which follow, the emphasis is on the precise ephemeris with station coordinates computed in the geodetic mode.

Three separate types of computation have to be considered, as below.

(i) *General geodetic solutions*, which determine or revise the coordinates of the tracking stations and the values of the parameters of the geodetic model. § 7.62.

(ii) The computation of the *precise ephemeris*, accepting the results of the latest general solution for the station coordinates, and the best available model. § 7.56.

(iii) *Point positioning*, which gives the coordinates of a new ground station from several satellite passes, accepting the ephemeris as errorless. This is the subject of (*b*), (*c*), and (*d*) below.

(*b*) *Reference system.* The ephemeris is computed in an inertial reference system, centred on the geocentre, with axes defined by the mean equator and equinox (i.e. free of nutation) at zero hours on the first day of the 48 hours for which each ephemeris is computed. The station coordinates are required in a geocentric system, rotating with the earth, with axes defined by the CIO and BIH Greenwich.† So before publication the inertial coordinates of the ephemeris and velocity components of the satellite ($\mathbf{r}$ and $\dot{\mathbf{r}}$) are converted to the terrestrial frame to give $\mathbf{r}'$ and $\dot{\mathbf{r}}'$. We have

$$\mathbf{r}' = A\,B\,C\,D\,\mathbf{r} \qquad (7.174)$$

where A, B, C, and D are rotation matrices. D corrects for precession, C for nutation, B for the earth's rotation, and A for polar motion. They are given in detail in [451], pp. 117–33.

(*c*) *The observation equations.* One equation is set up for each value of $\Delta S(=S_2-S_1)$ which results from the processing of the observations. The equations take the form

$$\frac{\partial \Delta S}{\partial x'}\Delta x' + \frac{\partial \Delta S}{\partial y'}\Delta y' + \frac{\partial \Delta S}{\partial z'}\Delta z' + \frac{\partial \Delta S}{\partial p}\Delta p + (q_2 - q_1)C_r = \mathrm{O}-\mathrm{C}, \qquad (7.175)$$

† For longitude see further in § 7.59.

where x', y', z' are the station coordinates in the terrestrial frame, and $\Delta x'$, etc. are the required small corrections to some preliminary values.

p is $(f_0 - f_t)$, which is assumed to require a correction Δp, a frequency bias, which is constant throughout any one pass, but which may vary from one pass to the next.

q is the standard (model) tropospheric refraction, which is here assumed to require a correction $q \times C_r$, where C_r is constant throughout the pass, and is probably less than 0.1. The optimum value of the refraction is then $(1 + C_r)q$.

O is the observed value of ΔS, after processing.

C is the value of ΔS computed (in the terrestrial frame) from the precise ephemeris and the preliminary values x', y', z', between the instants of UTC at which the first and last waves of the count left the satellite.†

Partial differentials.

$$\text{(i)} \quad \frac{\partial \Delta S}{\partial x_1} = \frac{\partial (S_2 - S_1)}{\partial x_1} = \frac{x_1'}{S_1} - \frac{x_2'}{S_2} \tag{7.176}$$

$$\text{(ii)} \quad \frac{\partial \Delta S}{\partial p} = -\frac{\partial \Delta S}{\partial f_t} = \frac{c}{f_t} \Delta t, \quad \text{from (7.172)} \tag{7.177}$$

Weights. may either be the same for all the observations of any one pass, or may be reduced for observations made at low altitudes. The processing and filtering will indicate the relative weights of different passes.

(*d*) *The normal equations.* A single pass may produce (say) 30 observation equations, with three unknown station coordinates and two unknown biases. If 25 passes are observed there will be a total of 750 equations for 3 coordinates and 50 biases. Normal equations are formed in the usual way, and are partitioned in the form

$$\begin{bmatrix} N_{11} & N_{12} \\ N_{12}^T & N_{22} \end{bmatrix} \begin{bmatrix} X \\ P \end{bmatrix} = \begin{bmatrix} k_1 \\ k_2 \end{bmatrix} \tag{7.178}$$

where $X = \Delta[x' \; y' \; z']^T$
and $P = [\Delta p_1 \ldots \Delta p_n \, C_{r1} \ldots C_{rn}]^T$, for n passes.

Since the biases Δp and C_r are constant for each pass, but are assumed to be otherwise uncorrelated, N_{22} is block diagonal. So the normal equations can be formed and solved on a block-by-block basis, and improved values of x_0', y', z' are readily obtained as each new pass is introduced. Their convergence towards a stable value will be some indication that further passes would result in little improvement.

† Assuming (7.172) to be used.

7.56. The computation of a precise ephemeris

(*a*) *Object.* The object is to prepare a 48-hour ephemeris of a satellite, in the inertial frame defined in § 7.55(*b*), giving the three coordinates $\mathbf{r}=[x\ y\ z]^{\mathrm{T}}$ and $\dot{\mathbf{r}}=[\dot{x}\ \dot{y}\ \dot{z}]^{\mathrm{T}}$ at 60-s intervals. For publication, these coordinates and velocities are converted to the terrestrial frame, and provide the computed values (C) of ΔS in (7.175). The accuracy aimed at, but not as yet confidently achieved, is 1 metre.

(*b*) *Data.* The data on which the ephemeris is to be based are as in (i), (ii), and (iii) below.

(i) At time t_0, at the start of the 48 hours, we have the values x, y, z, and $\dot{x}$, $\dot{y}$, $\dot{z}$, at the end of the previous period. These constitute a set of orbital osculating elements, which are referred to as the *starting elements.* See §§ 7.01(*h*) and 7.02.

(ii) We have a model of the perturbing accelerations $\ddot{x}$, $\ddot{y}$, $\ddot{z}$, § 7.57, which is likely to give the perturbations of the satellite's motion correctly within a few metres throughout the 48 hours. It takes the form

$$\ddot{\mathbf{r}}=\mathbf{G}_E+\mathbf{G}_M+\mathbf{G}_S+\mathbf{D}+\mathbf{R}+\mathbf{T}_M+\mathbf{T}_S, \tag{7.179}$$

where $\mathbf{G}_E$ is the acceleration due to the earth's attractive potential, namely grad V, where V is as in (7.1).

$\mathbf{G}_M$, $\mathbf{G}_S$ are the differential accelerations due to the moon and sun.

$\mathbf{D}$ is the acceleration due to air drag.

$\mathbf{R}$ is the acceleration due to radiation pressure.

$\mathbf{T}_M$, $\mathbf{T}_S$ are the accelerations due to lunar and solar earth tides.

These are all vectors, and the model gives their three components.

(iii) Doppler observations made during the 48 hours at (in 1975) 21 tracking stations, whose earth fixed coordinates are treated as known. The use of these observations is to obtain a revised set of starting elements which will produce a best (least squares) fit between themselves and the ephemeris during the 48 hours.

(*c*) *Outline of the method used.* At its most simple, the procedure may be described as follows.

(i) First, the ephermeris of the previous 48 hours, expressed in the inertial frame, is expanded to cover the current 48 hours on the basis of their starting elements and the perturbation model. This done by a 12-order† Cowell second difference numerical integration at 60-s intervals with 14 significant figures. See (*d*) below.

† '12-order' because twelfth differences of $\ddot{\mathbf{r}}$ are recorded. It is 'second difference' because the data produced by the model are accelerations which have to be twice integrated to give the required positional ephemeris. The order and the 60-s interval can be varied to suit the available computer and the accuracy sought. A low order demands less storage, but a short interval demands more computer time.

(ii) The tracking observations are used to give corrections to the starting elements, as in (*e*) below.

(iii) Since (ii) will result in changes in the values of $\mathbf{r}$ and $\dot{\mathbf{r}}$ given by (i), the model will give changed accelerations, and if these are material, the numerical integration will have to be done again to produce a revised ephemeris. More than one such iteration may possibly be necessary.

(iv) When the new ephemeris is completed in the inertial frame, the values of $\mathbf{r}$ and $\dot{\mathbf{r}}$ are converted to the terrestrial frame, in which the positions of the field stations are to be computed.

(*d*) *Numerical integration. Cowell method.* See [127]. The description here given is worded as if it was for desk computation. The computer programme will of course differ in detail.

In the Cowell method each coordinate x, y, or z, of the satellite's position in the inertial system is treated separately. A table is prepared for each, as shown for x in Table 7.4.

The basic entries in Table 7.4 are the values of $\ddot{x}$, computed from (7.179) using the best available values of $x, \dot{x}, y, \dot{y}, z$, and $\dot{z}$. For this purpose the coordinates do not need to be as accurate as is required for the final ephemeris, but provisional values may have to be revised. In the table the tabular interval h is conveniently taken to be the unit of time,† but $\dot{x}$ is expressed in metres per second.

Table 7.4 starts with the last seven values of $\ddot{x}$ from the ephemeris of the previous 48 hours, followed by the first six of the current period. The difference columns $\delta^1\ddot{x}$ to $\delta^{12}\ddot{x}$ (written henceforward without the $\ddot{x}$) are then completed, but only provisionally, because the last six values of $\ddot{x}$ may require revision. For the further preparation of the table four formulae are required from the calculus of finite differences, as below, (7.180) to (7.183).‡ Suffix i implies time i in the first column of Table 7.4. See [127], p. 154.

$$\sum^2 \ddot{x}_i = x_i - (1/12)\ddot{x}_i + (1/240)\delta_i^2 - (31/60480)\delta_i^4 + \dots \quad (7.180)$$

$$\sum \ddot{x}_{i+\frac{1}{2}} = h\dot{x}_i + \tfrac{1}{2}\ddot{x}_i + (1/12)\mu\delta_i^1 - (11/720)\mu\delta_i^3 + (191/60480)\mu\delta_i^5 + \dots, \quad (7.181)$$

where h is the tabular interval, and $\mu\delta_i = \frac{1}{2}(\delta_{i-\frac{1}{2}} + \delta_{i+\frac{1}{2}})$, etc.

$$x_i = \sum^2 \ddot{x}_i + (1/12)\ddot{x}_{i-1} + (1/12)\delta^1_{i-1\frac{1}{2}} + (19/240)\delta^2_{i-2} + (18/240)\delta^3_{i-2\frac{1}{2}} + (1726/24192)\delta^4_{i-3} + (1650/24192)\delta^5_{i-3\frac{1}{2}} + 0.065\delta^6_{i-4} + 0.06\delta^7_{i-4\frac{1}{2}} + \dots \quad (7.182)$$

$$x_i = \sum^2 \ddot{x}_i + (1/12)\ddot{x}_{i-1} - (1/240)\delta^2_{i-1} - (1/240)\delta^3_{i-1\frac{1}{2}} - 0.00365\delta^4_{i-2} - 0.0031\delta^5_{i-2\frac{1}{2}} - 0.003\delta^6_{i-3} + \dots \quad (7.183)$$

† The units may be differently chosen if that is numerically more convenient.

‡ For a 12-order integration (7.180) to (7.183) are required with terms as far as δ^{12}.

TABLE 7.4
Cowell numerical integration. For x and $\dot{x}$

Time (minutes)	x $\sum^{2} \ddot{x}$	$\dot{x}$ $\sum \ddot{x}$	$\ddot{x}$	$\delta^1\ddot{x}$	$\delta^2\ddot{x}$	$\delta^3\ddot{x}$	$\delta^4\ddot{x}$	$\delta^5\ddot{x}$	$\delta^6\ddot{x}$	$\delta^7\ddot{x}$	$\delta^8\ddot{x}$	$\delta^9\ddot{x}$	$\delta^{10}\ddot{x}$	$\delta^{11}\ddot{x}$	$\delta^{12}\ddot{x}$
−6			–												
−5			–	–	–										
−4			–	–	–	–	–								
−3			–	–	–	–	–	–	–						
−2			–	–	–	–	–	–	–	–	–				
−1			–	–	–	–	–	–	–	–	–	–	–		
0	–		–	–	–	–	–	–	–	–	–	–	–	–	–
+1	–	–	–	–	–	–	–	–	–	–	–	–	–	–	×
+2	–	–	–	–	–	–	–	–	–	–	–	–	×	×	
+3	–	–	–	–	–	–	–	–	–	–	×	×			
+4	–	–	–	–	–	–	–	–	×	×					
+5	–	–	–	–	–	–	×	×							
+6	–	–	–	–	×	×									
+7	–	–	×	×											
+8	×	×													
+9															
+10															

$$\delta^p \ddot{x}_{i+\frac{1}{2}} = \delta^{p-1}\ddot{x}_{i+1} - \delta^{p-1}\ddot{x}_i$$

$$\sum \ddot{x}_{i+\frac{1}{2}} = \sum \ddot{x}_{i-\frac{1}{2}} + \ddot{x}_i$$

$$\sum^{2} \ddot{x}_{i+1} = \sum^{2} \ddot{x}_i + \sum \ddot{x}_{i+\frac{1}{2}}$$

Equation (7.182) is an approximation which only requires differences from a line one higher than is required for (7.183). It is based on the assumption that (if Table 7.4 is carried as far as δ^{12}), δ^{12}_{i-6} will equal δ^{12}_{i-7}.

Before the columns $\sum^{2} \ddot{x}$ and $\sum \ddot{x}$ can be completed, the two integration constants $\sum^{2} \ddot{x}_0$ and $\sum \ddot{x}_{\frac{1}{2}}$ must be computed by (7.180) and (7.181), using the x and $\dot{x}$ of the old ephemeris and the differences already entered. They are then entered in the table, and the two columns can be completed as far as $\sum \ddot{x}_{6\frac{1}{2}}$ and $\sum^{2} \ddot{x}_7$.

The provisional values used for x_1 to x_6 are then verified by (7.183), for which the additional diagonal of differences is now available. If their changes, or those in y or z, are large enough to affect $\ddot{x}$ (as is unlikely if the integration interval has been well chosen) $\ddot{x}$ and its differences must be revised, and possibly also the values of x itself.

All the entries so far mentioned are marked by a dash in Table 7.4. The tables for y and z must be completed concurrently with the x table, as $\ddot{x}$ depends on all three.

To extend the table to include $\sum \ddot{x}_{7\frac{1}{2}}$ and $\sum^{2} \ddot{x}_8$ a preliminary x_7 is computed by (7.182), together with y_7 and z_7 from their tables, and $\ddot{x}_7$ is computed from them and the model. Then $\sum \ddot{x}_{7\frac{1}{2}}$ and $\sum^{2} \ddot{x}_8$, and a diagonal row of differences, can be added to the table, and the preliminary x_7 is verified by (7.183). These entries are shown by crosses in Table 7.4, which can then be extended to give x_8, and so on.

When $\ddot{x}_7$ is being computed, no great accuracy is required for $\dot{x}_7$, since at this stage it is only required for the drag correction. Drag may be a serious matter, but no reasonably inaccurate value of $\dot{x}_7$ can add materially to its doubt. As soon as a few more values of x are available, $\dot{x}$ can be computed when required, using the differential formula, [127], p. 150.

$$h(\dot{x}_{i+\frac{1}{2}}) = \delta^1 - (1/24)\delta^3 + (3/640)\delta^5 - (5/7168)\delta^7 + (35/294912)\delta^9 + \dots \quad (7.184)$$

where the δ's are all differences of x (not $\ddot{x}$) with subscript $(i+\frac{1}{2})$.

(*e*) *The least-squares solution for revised starting elements.*

(i) The observation equations take the form

$$\frac{\partial N}{\partial p_k}\Delta p_k + \frac{\partial N}{\partial p_j}\Delta p_j = (\text{Observed } N\text{-Computed } N) + v. \text{ With suitable weight.} \quad (7.185)$$

Where p_k is one of the starting elements x_0, y_0, z_0, $\dot{x}_0$, $\dot{y}_0$, or $\dot{z}_0$.

N is the integrated doppler count, as in § 7.51, processed as in § 7.52.

p_j is one of several unknown parameters, namely (*A*) for each satellite pass, a frequency bias, and a tropospheric refraction factor C_r as in 7.55 (*c*) and (*d*), and (*B*) for the whole 48-hour period an atmospheric density factor affecting drag, and two polar motion components.

In (7.185) $\partial N/\partial p_k$ may be written as

$$\frac{\partial N}{\partial (S_2 - S_1)} \cdot \frac{\partial (S_2 - S_1)}{\partial p_k} \quad (7.186)$$

From (7.169) $\partial N/\partial (S_2 - S_1) = f_t/c$, which is constant, so the observation equations may be written

$$\frac{\partial (S_2 - S_1)}{\partial p_k}\Delta p_k + \frac{\partial (S_2 - S_1)}{\partial p_j}\Delta p_j = \text{Observed } (S_2 - S_1) - \text{Computed } (S_2 - S_1) + v. \quad (7.187)$$

(ii) In (7.187) the computed values of S_2 and S_1 are given by

$$\mathbf{S} = (\mathbf{r} \text{ at time } t_S) - (\mathbf{a} \text{ at time } t_R), \quad (7.188)$$

when t_S and t_R are the times of emission and reception, so $t_S = t_R - S/c$, and $\mathbf{r}$ is got by interpolation from the preliminary (inertial frame) ephemeris.

The station position $\mathbf{a}$ in the inertial frame is got as in § 7.55(*b*) by rotation of the earth-fixed coordinates $\mathbf{a}'$

$$\mathbf{a} = D^{\mathrm{T}}C^{\mathrm{T}}B^{\mathrm{T}}A^{\mathrm{T}}\mathbf{a}' \tag{7.189}$$

If the observed S_2 and S_1 have been corrected for ionospheric and standard tropospheric refraction as in § 7.54, the computed value will of course require no correction.

(iii) *Partial differential coefficients* $\partial(S_2 - S_1)\partial p_k$. In equation (7.187) $(S_2 - S_1)$ is a function of the satellite's position $\mathbf{r}$, $(=[x\ y\ z]^{\mathrm{T}})$, at times t_2 and t_1. So the single expression $\partial(S_2 - S_1)/\partial p_k$ may be written more fully as

$$\left(\frac{\partial S_2}{\partial x_2}\cdot\frac{\partial x_2}{\partial p_k} - \frac{\partial S_1}{\partial x_1}\cdot\frac{\partial x_1}{\partial p_k}\right) + \text{similar expressions for } y \text{ and } z, \tag{7.190}$$

in which, for each value of S there occur three partials $\partial S/\partial x$, $\partial S/\partial y$, and $\partial S/\partial z$, at times t_2 and t_1. There are also 18 other partials such as $\partial x/\partial x_0$, $\partial x/\partial y_0 \ldots \partial x/\partial \dot{z}_0$, all possible combinations of x, y, z with x_0, y_0, z_0, $\dot{x}_0$, $\dot{y}_0$, and $\dot{z}_0$. The significance of $\partial x/\partial y_0$, for example, is that it gives the change in x, at the time of observation t_2 or t_1, resulting from a change in y_0 at time t_0. The interval $t_1 - t_0$ may be as much as 48 hours, during which the satellite will have completed about 27 orbits, and have departed considerably from its starting orbit. These 18 partials thus vary with time, and are not easy to compute. See (v) below.

On the other hand, the changes Δx_0, $\Delta \dot{x}_0$, etc. in (7.187), which are required to optimize the starting orbit, will be very small, probably not more than 10 m in $\mathbf{r}$ and 3 cm s^{-1} in $\dot{\mathbf{r}}$, so the partials need not be very accurately known. Further, if the solution is to be iterated as in § 7.56(*c*)(iii), the necessary accuracy is no more than will secure reasonably quick convergence.

(iv) *The partials* $\partial(S_2 - S_1)/\partial x$, *etc.* These are simple.

$$\begin{aligned} \partial S_2/\partial x_2 &= (x\text{-component of } \mathbf{S}_2)/S_2, \\ \partial S_1/\partial x_1 &= (x\text{-component of } \mathbf{S}_1)/S_1, \end{aligned} \tag{7.191}$$

and similarly for y and z.

(v) There are two methods of computing the 18 partials $\partial x/\partial x_0$, etc. of (7.190).

(A) *Analytically*, assuming a simple form for the geopotential, namely the central acceleration, and the J_2 perturbations only. This is used in the APL for the broadcast ephemeris, and gives results which cause the solutions to converge reasonably quickly.

(B) By *numerical integration* of the perturbational equations.

Writing $\boldsymbol{\xi}_k = \partial\mathbf{r}/\partial p_k$, $\dot{\boldsymbol{\xi}}$ *for* $\partial\boldsymbol{\xi}/\partial t$, and $\ddot{\boldsymbol{\xi}}$ for $\partial^2\boldsymbol{\xi}/\partial t^2$ we have

$$\ddot{\boldsymbol{\xi}}_k = \left[\frac{\partial\ddot{\mathbf{r}}}{\partial\mathbf{r}^{\mathrm{T}}}\right]\boldsymbol{\xi}_k + \left[\frac{\partial\ddot{\mathbf{r}}}{\partial\dot{\mathbf{r}}^{\mathrm{T}}}\right]\dot{\boldsymbol{\xi}}_k + \frac{\partial\ddot{\mathbf{r}}}{\partial p_k}, \tag{7.192}$$

where $\ddot{\mathbf{r}}$ is the acceleration given by (7.179) with the full perturbation model. It is a function of $\mathbf{r}$ and $\dot{\mathbf{r}}$ (i.e. of $x, y, z, \dot{x}, \dot{y}, \dot{z}$), and the time. In (7.192) the expression $\partial\ddot{\mathbf{r}}/\partial\mathbf{r}^{\mathrm{T}}$† means the change in each of the three components of $\ddot{\mathbf{r}}$ resulting from instantaneous changes in each of the three components of $\mathbf{r}$. The expression $\partial\ddot{\mathbf{r}}/\partial\dot{\mathbf{r}}^{\mathrm{T}}$ is similarly the change in $\mathbf{r}$ resulting from an instantaneous change in velocity, as required for drag. And the last term $\partial\ddot{\mathbf{r}}/\partial p_k$ is the change of $\ddot{\mathbf{r}}$ at a fixed point caused by the rotation of the geopotential model in relation to the inertial frame. Equation (7.192) is thus a set of second-order differential equations which can be solved numerically by methods analogous to those of § 7.56(*d*), to give 18 values of $\partial x/\partial x_0$, etc. at times t_2 and t_1. This method is employed by NWL for its general geodetic solutions.

The derivation of (7.192) is as follows. Consider the x component of $\ddot{\mathbf{r}}$, namely $\ddot{x}$.

$$\delta\ddot{x} = \left(\frac{\partial\ddot{x}}{\partial x}\,\delta x + \frac{\partial\ddot{x}}{\partial y}\,\delta y + \frac{\partial\ddot{x}}{\partial z}\,\delta z\right) + \left(\frac{\partial\ddot{x}}{\partial\dot{x}}\,\delta\dot{x} + \frac{\partial\ddot{x}}{\partial\dot{y}}\,\delta\dot{y} + \frac{\partial\ddot{x}}{\partial\dot{z}}\,\delta\dot{z}\right) + \frac{\partial\ddot{x}}{\partial t}\,\delta t, \tag{7.194}$$

which may be symbolically abbreviated to

$$\delta\ddot{x} = \frac{\partial\ddot{x}}{\partial\mathbf{r}^{\mathrm{T}}}\,\delta\mathbf{r} + \frac{\partial\ddot{x}}{\partial\dot{\mathbf{r}}^{\mathrm{T}}}\,\delta\dot{\mathbf{r}} + \frac{\partial\ddot{x}}{\partial t}\,\delta t. \tag{7.195}$$

And the three equations for the three separate components of $\ddot{\mathbf{r}}$ may be combined and written as

$$\delta\ddot{\mathbf{r}} = \left[\frac{\partial\ddot{\mathbf{r}}}{\partial\mathbf{r}^{\mathrm{T}}}\right]\delta\mathbf{r} + \left[\frac{\partial\ddot{\mathbf{r}}}{\partial\dot{\mathbf{r}}}\right]\delta\dot{\mathbf{r}} + \frac{\partial\ddot{\mathbf{r}}}{\partial t}\,\delta t, \tag{7.196}$$

See footnote below (7.192).

† $[\partial\ddot{\mathbf{r}}/\partial\mathbf{r}^{\mathrm{T}}]\boldsymbol{\xi}_k$ is an abbreviation for the matrix product

$$\begin{bmatrix} \partial\ddot{r}_1/\partial r_1 & \partial\ddot{r}_1/\partial r_2 & \partial\ddot{r}_1/\partial r_3 \\ \partial\ddot{r}_2/\partial r_1 & \partial\ddot{r}_2/\partial r_2 & \partial\ddot{r}_2/\partial r_3 \\ \partial\ddot{r}_3/\partial r_1 & \partial\ddot{r}_3/\partial r_2 & \partial r_3/\partial r_3 \end{bmatrix} \begin{bmatrix} \partial r_1/\partial p_k \\ \partial r_2/\partial p_k \\ \partial r_3/\partial p_k \end{bmatrix} \tag{7.193}$$

where $r_1, r_2, r_3 = x, y, z$, the three components of $\mathbf{r}$.

We have

$$\ddot{\boldsymbol{\xi}}_K = \frac{d^2}{dt^2}\left(\frac{\partial \mathbf{r}}{\partial p_k}\right) = \frac{\partial}{\partial p_k}(\delta\ddot{\mathbf{r}}), \text{ by definition,}$$

$$\text{so } \ddot{\boldsymbol{\xi}}_K = \left[\frac{\partial \ddot{\mathbf{r}}}{\partial \dot{\mathbf{r}}^T}\right]\frac{\partial \mathbf{r}}{\partial p_k} + \left[\frac{\partial \ddot{\mathbf{r}}}{\partial \dot{\mathbf{r}}^T}\right]\frac{\partial \dot{\mathbf{r}}}{\partial p_k} + \frac{\partial \ddot{\mathbf{r}}}{\partial p_k}, \text{ from (7.196),} \tag{7.197}$$

which is the same as (7.192), since $\dfrac{\partial \dot{\mathbf{r}}}{\partial p_k} = \dfrac{\partial}{\partial t}\cdot\dfrac{\partial \mathbf{r}}{\partial p_k} = \dot{\boldsymbol{\xi}}$.

(vi) *The bias partials.* The frequency bias

$$\partial(S_2 - S_1)/\partial p_j = (c/f_t)(t_2 - t_1), \tag{7.198}$$

as in (7.177).

The tropospheric refraction factor can appear in the observation equations (7.187) in the same form, $(q_2 - q_1)C_r$, as in (7.175).

(vii). *The solution of the observation equations.* The values of the partial differentials appropriate to the times of each observation having been obtained as above, the equations are solved by least squares, weights being allotted as indicated by the data processing. The normal equations can advantageously be arranged as in § 7.55 (*d*). When the solution has been obtained, the ephemeris is recomputed using the new values of the starting elements instead of those provisionally accepted. The resulting ephemeris will then be such as will best accord with the tracking data.

This ephemeris in the inertial frame is finally converted to the terrestrial frame using (7.174). The velocity components are obtained from the values of the coordinates by subtraction.

7.57. The perturbation model

The separate items which constitute the model are listed in (7.179).

(*a*) *Earth's attraction.* The geopotential model used for the doppler, or any other method, need not be determined exclusively by the system concerned. The aim is to use the best model from all available sources. Currently (1977) the NWL accept the DOD *World geodetic System* 1972 model, [519], whose basis is described in §§ 7.76(*f*) and 7.79(*c*). It includes nearly all harmonic terms up to (19×19), together with selected resonant terms to degree 27.

The computation of the accelerations of a satellite at any place and time is complicated by the fact that they are required to be expressed in relation to the inertial frame, while the model gives the potential in terms of θ and λ as measured in the rotating terrestrial frame, in which it is not time-dependent.

In the terrestrial frame the geopotential is given in the form of (7.1), which for brevity may be written as

$$V = \sum_{l=0} \sum_{m=0}^{l} C_{lm} U_l^m + S_{lm} V_l^m \qquad (7.199)\dagger$$

where C_{lm} and S_{lm} are given by the model, and in this context are assumed to be correct.

$U_l^m = \dfrac{GM}{r}\left(\dfrac{a_e}{r}\right)^l P_{lm}(\cos\theta)\cos m\lambda$, and similarly for V with $\sin m\lambda$ instead of $\cos m\lambda$‡.

$\cos\theta = r_3'/r$, where r_1', r_2', r_3', are the components of $\mathbf{r}$ in the terrestrial frame, and $r = |\mathbf{r}|$.

$\cos\lambda = (r_1'/r)\sin\theta$, and $\sin\lambda = (r_2'/r)\sin\theta$. (7.200)

The required accelerations are the gradient of V, ∇V, whose components $\partial V/\partial x$, $\partial V/\partial y$, $\partial V/\partial z$ in the inertial frame are $\ddot{x}$, $\ddot{y}$, $\ddot{z}$.

The components of $\mathbf{r}'$ in the terrestrial frame are related to its components $(x\ y\ z)$ in the inertial frame by the rotations

$$\begin{bmatrix} r_1' \\ r_2' \\ r_3' \end{bmatrix} = BCD \begin{bmatrix} x \\ y \\ z \end{bmatrix} = a_e E \begin{bmatrix} x \\ y \\ z \end{bmatrix}, \qquad (7.201)$$

defining E. Here B, C, and D are rotation matrices as in (7.174), involving the earth's rotation, nutation, and precession respectively.‖ They are given in full in [451], pp, 128–32. a_e is the earth's semi-major axis.

In the terrestrial frame the components of $\ddot{\mathbf{r}}$ are given by

$$\ddot{\mathbf{r}}' = \nabla V = \begin{bmatrix} \dfrac{\partial V}{\partial r_1'} \\ \dfrac{\partial V}{\partial r_2'} \\ \dfrac{\partial V}{\partial r_3'} \end{bmatrix} = \sum_{l=0} \sum_{m=0}^{l} C_{lm} \nabla U_l^m + S_{lm} \nabla V_l^m, \qquad (7.202)$$

and in the inertial frame the three components are

$$\ddot{\mathbf{r}} = a_e E^{\mathrm{T}} \sum_{l=0} \sum_{m=0}^{l} C_{lm} \nabla U_l^m + S_{lm} \nabla V_l^m. \qquad (7.203)$$

† When $l=0$, $U_l^m = GM/r$, When $m=0$, $V_l^0 = 0$. When $l=1$, $C_{10} = C_{11} = S_{11} = 0$, since the geocentre is the origin of coordinates.

‡ Here in (7.199) C_{lm} and S_{lm} are not normalized, and P_{lm} is the corresponding unnormalized function. The reason for this is that the recurrence relations (7.204) to (7.206), as usually quoted, refer to unnormalized functions. [451] and [10].

‖ Of these three the rotations C and D are very small in this context. B is large. A of (7.174) is negligible.

In order to compute (7.203) for a given position (θ, λ) of the satellite in the terrestrial frame, there exist three pairs of recurrent relations for U_l^m and V_l^m, as in (7.204), (7.205), and (7.206).

$$\text{(i)}\quad U_{l+1}^m = (a_e/r)(l-m+1)^{-1}[(2l+1)U_l^m \cos\theta - (l+m)(a_e/r)U_{l-1}^m], \tag{7.204}$$

and the same for V_{l+1}^m with V's instead of U's. These relations are easily verified for any of the low harmonics given in (H.26).

$$\text{(ii)}\quad U_{l+1}^{l+1} = (a_e/r)(2l+1)[U_l^l \sin\theta\cos\lambda - V_l^l \sin\theta\sin\lambda]$$

and (7.205)

$$V_{l+1}^{l+1} = (a_e/r)(2l+1)[V_l^l \sin\theta\cos\lambda + U_l^l \sin\theta\sin\lambda]$$

Given starting values of U_l^m and V_l^m for some low values of l, such as (say) $l=2$ and 3 from (H.26), the values of all the U's and V's can be successively be obtained.†

(iii) Lastly, for the computation of (7.203) there are two more recurrence relations, see [451], pp. 138–9 and [10], p. 5322.

$$a_e \nabla U_l^m = \begin{bmatrix} \tfrac{1}{2}A_l^m U_{l+1}^{m-1} - \tfrac{1}{2}U_{l+1}^{m+1} \\ -\tfrac{1}{2}A_l^m V_{l+1}^{m-1} - \tfrac{1}{2}V_{l+1}^{m+1} \\ -(l-m+1)U_{l+1}^m \end{bmatrix}$$

these being its three components in the inertial frame.‡ And

$$a_e \nabla V_l^m = \begin{bmatrix} \tfrac{1}{2}A_l^m V_{l+1}^{m-1} - \tfrac{1}{2}V_{l+1}^{m+1} \\ \tfrac{1}{2}A_l^m U_{l+1}^{m-1} + \tfrac{1}{2}U_{l+1}^{m+1} \\ -(l-m+1)V_{l+1}^m \end{bmatrix} \tag{7.206}$$

where $A_l^m = (l-m+1)(l-m+2)$. These can be computed when the U's and V's have been got from (7.204) and (7.205).

(*b*) *Sun and moon's attraction.* The attraction of the sun is given by

$$\mathbf{G}_S = -GM_S\left(\frac{\mathbf{r}-\mathbf{r}_s}{|\mathbf{r}-\mathbf{r}_s|^3} + \frac{\mathbf{r}_s}{|\mathbf{r}_s|^3}\right) \tag{7.207}$$

where M_S is the mass of the sun, and $\mathbf{r}_S$ is the sun's vector distance from the geocentre. The second term in (7.207) is the earth's acceleration towards the sun, which is necessarily included because the frame emp-

† (7.204), described as *horizontal stepping*, is used with a constant value of m to give all the Us and Vs of order m and of degree greater than m. Starting with U_2^0 and U_3^0 from (H.26), (7.204) gives $U_4^0 \ldots U_l^0$ as far as may be required. Then $U_4^1 \ldots U_l^1$, $U_4^2 \ldots U_l^2$, and $U_4^3 \ldots U_l^3$ may be got in the same way.

(7.205), described as *diagonal stepping*, gives (say) U_4^4 when U_3^3 and V_3^3 are known. Horizontal stepping then gives $U_5^4 \ldots U_l^4$. The process may continue indefinitely.

‡ Like any other gradient of a scalar, ∇V is a vector.

loyed has the geocentre as its origin of coordinates. A similar formula applies to the moon, [10], p. 5323 and [451], p. 139.

(*c*) *Drag.* Atmospheric drag depends on the velocity of the satellite relative to the atmosphere. It also depends on the mass of the satellite m, and on its cross-sectional area s, and on the density of the air. At an altitude of 1100 km the air density is very low, but it has relatively large and unpredictable variations, and the drag correction is a significant source of error.

Assuming that the air is not moving relative to the terrestrial frame, the relative velocity is given by

$$\mathbf{v} = \begin{bmatrix} \dot{x} + \omega y \\ \dot{y} - \omega x \\ \dot{z} \end{bmatrix}, \tag{7.208}$$

where ω is the earth's angular velocity.

Unless a satellite is spherical, which the transits are not, its cross-section in this context varies between wide limits as the satellite tumbles about in its orbit, although the tumbling is probably rapid, and an average cross-section is probably predictable within 20 per cent. See [337], p. 110.

The air density ρ (measured in kg km^{-3}) is given by

$$\rho = \exp\{Ah - B - (Ch^2 + Dh - E)^{\frac{1}{2}}\}, \tag{7.209}$$

where h is height above sea level in km. $A = 0.01362\ \text{km}^{-1}$, $B = -8.3355$, $C = 0.0001018\ \text{km}^{-2}$, $D = 1.083\ \text{km}^{-1}$, and $E = 89.39$. See [530]. This value of ρ may be wrong by 25 per cent either way, [13], p. 69. In round figures, at 100, 200, and 300 km, $\rho = 7\times10^{-7}$, 4×10^{-10} and 3×10^{-11} respectively times its sea-level value.

Then the drag acceleration D, which has no connection with D in (7.209), is given by

$$D = -\gamma\rho(s/2m)v^2. \tag{7.210}$$

The drag factor γ, which is approximately 2.2 is given by an empirical table. See [10], p. 5323 and [451], p. 140.

In principle the satellite velocity should be corrected for the wind at satellite height, *wind* being the velocity of the air relative to the terrestrial frame. The only significant winds are either east or west. At all heights the atmosphere rotates in the same direction as the earth with a velocity which may be greater or less than that of the rotating frame by a multiplying factor Λ, [344]. This factor is variable, notably as between morning (04.00 to 12.00 LMT) and evening (18.00 to 24.00 LMT), but also depending on the latitude and solar activity. On average, the morning value of $\Lambda = 1.0$ at $h = 0$, and falls to about 0.8 at $h = 200$ km and above, as far as 700 km or more. The evening values increase from 1.0 at $h = 0$ to 1.4 at 300 km, falling back to near 1.0 at 500 km and above. The daily mean rises from 1.0 h at $h = 0$ to 1.3 at 350 km, falls to 1.0 at 400 km and to about 0.8 at 500 km and above.

In latitude 30° a value of $\Lambda = 1.2$ or 0.8 implies a west or east wind of 80 m s^{-1}. Since the velocity of a satellite is of the order of 7000 m s^{-1}, ignorance of the wind may not be very serious for satellites at 500 km or more, although winds of 500 m s^{-1} have been recorded during magnetic storms.

In the long run, drag causes the satellite to lose energy and fall to a lower orbit, where (at heights of more than about 160 km) its speed is necessarily increased by the increased gravity. The eccentricity is also reduced.

Of all the perturbations, drag (being consistently along-track) introduces the most uncertainty in a 48-hour uncontrolled orbit prediction. Tracking observations do, of course control the error, and as in § 7.56(*e*)(i) every computation includes an average correction factor, which is applicable to (7.208). [13], p. 69 computes that the effect of a single 50-per cent sudden change of drag in the middle of a 48-hour arc would be an r.m.s. correction of 0.64 m in the terminal position of the satellite.

A remedy for drag is the DISCOS system, § 7.04; also, of course, a high weight-to-cross-section ratio.

(*d*) *Solar radiation pressure.* The solar radiation pressure is of the order 1 dyn m^{-2} (10^{-5} Pa), [93], p. 53, acting along the vector from the sun to the satellite. For Triad 1 at about 750-km altitude, the radiation pressure is 4 or 5 times as great as the drag. It is zero in the earth's shadow, which makes it troublesome to program, although that difficulty can be overcome, [216], pp. 168–72 and [451], p. 140.

Some uncertainty arises from inconsistency in the solar radiation constant. As with air drag, the effect of solar radiation is eliminated by the DISCOS system, provided the system acts along all three axes.

(*e*) *Earth tides.* The deformation of the earth by the semi-diurnal lunar and solar tides is included as a pair of 2-degree zonal harmonics with their poles pointing towards the moon and sun respectively. Love's number, k in § 6.78, is generally taken as 0.30. Details are in [10], p. 5324.

(*f*) *Ocean tides.* are currently ignored. Their effect is small. They lack the regularity of the body tides, and no useful world-wide harmonic analysis is available. See [216], pp. 155–9, and [150].

7.58. Relativity

(*a*) *Special theory.* As in §§ 7.51 and 7.47(*b*) footnote, the special theory of relativity introduces a term $(V^2/2c)(t_2 - t_1)$ into the observed value of $(S_2 - S_1)$. This small term is indistinguishable from an error in the transmission frequency, which is in any case treated as an unknown. The term therefore can be, and is, neglected.

(*b*) *General theory.* There are two additional effects.

(i) [531] shows that while $(V^2/2c)(t_2-t_1)$ is correct for low altitudes, the general theory multiplies it by a varying factor which changes it to zero at a height of $\frac{1}{2}R$, (R = earth's radius), and which multiplies it by −0.5, −0.9, and −1.2 at heights of R, $2R$, and $3R$ respectively. This also can be dealt with as an error in frequency.

(ii) If the orbit is not circular [309] shows that there will be a small variation in the doppler shift, Δf, with a period of one orbit. It was measured in Geos 1 ($e = 0.072$), where the amplitude of the correction to Δf was about $f\times 0.8\times 10^{-10}$, agreeing well with theory. The magnitude of the correction is proportional to e, and for most purposes it is negligibly small. See also [13] where the effect on Δf is estimated to be 1 in 10^{-12} for the low eccentricity Transit satellites.

7.59. Longitude

(*a*) In common with other dynamic systems, the doppler system gives coordinates with the geocentre as origin, and with the z-axis aligned on the CIO, § 7.55(*b*), but the system makes no connection with the Greenwich meridian, and to enable it to give correct longitudes assistance must be sought from other sources. In principle, doppler longitudes can be corrected if the *geocentric* BIH geodetic longitude of any one doppler station (but of course preferably of many) can be obtained by any means. Three possible ways are in (*b*), (*c*), and (*d*) below.

(*b*) *Photographic methods.* Sections 4 and 5 of this chapter obtain the positions of their ground stations in frames whose axes are correctly aligned on the CIO and BIH meridian, although photo-geometric methods will not have the geocentre as origin. So if doppler observations are made at stations coinciding with at least two well-separated stations of any accurate photo-geometric net, common values of their coordinates can be obtained which will move the photo net to the geocentre, and the doppler zero of longitude to the BIH meridian. With a photo-dynamic net, whose origin is already at the geocentre, a single common station would suffice to correct the doppler longitudes. Least squares can be used when the number of common stations exceeds the minimum.

Many doppler stations are common to the SAO and BC 4 nets, and adjustment has been made, but of course with some small amount of error, since neither net can be errorless. See § 7.76 and 7.77.

(*c*) *Stokes's integral.* At a selected† doppler station let the deviation of

† The station should be in flat country with good gravity cover over the surrounding 20° or 30° or more in all directions, see § 6.33. In more distant areas satellite-determined values of gravity will have to be used to supplement sparse surface values.

the vertical and the geoid–spheroid separation be computed by Stokes's or Molodensky's integrals, (6.111), (6.123) and (6.124) or (6.168) and (6.169). Astronomical observations of latitude and longitude, and spirit-levelled heights, can then be corrected to give the geocentric coordinates with axes directed on the CIO and BIH Greenwich., and doppler longitudes can then be corrected. The accuracy is of course limited to that of the astronomical observations, but that can be increased by using several stations all within a small area. It is also likely to be limited, more seriously, by lack of sufficient gravity data. Sets of stations in several continents would be desirable, but the existing gravity cover limits the possibilities.

(*d*) *Comparison with astro-geodetic geoid sections.* §§ 4.50 to 4.52. Fig. 7.17 shows a line of points A ... F on a long zero-order west-to-east line of traverse and astro-geodetic section, as computed with some continental origin O (not geocentric) in terms of the CIO and BIH Greenwich. Let doppler observations have been made at a number of traverse points A ... F, giving coordinates referred to the geocentre and CIO, and to a provisional doppler 'Greenwich' meridian. Then if the two sets of coordinates are made to coincide as well as possible by least squares, the optimum relative positions of O and the geocentre, and of the BIH and doppler zero meridians, result. The traverse must be of reasonable length, such as may subtend an angle of (say) at least 30° or 40° at the geocentre for good geometry.

(*e*) Early doppler results were computed with a provisional longitude for the base station situated at the Johns Hopkins APL. Details have not been published, but the value used seems to agree very well with what would have been given if the NAD longitude of the station had been corrected by the best available value of the correction required to convert

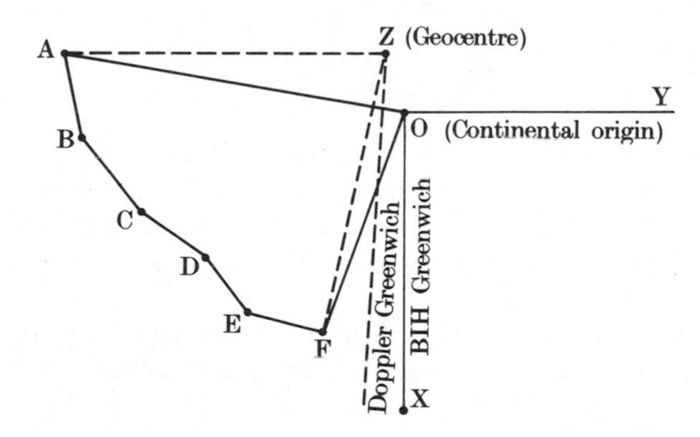

FIG. 7.17. Viewed down the Z-axis. ABCDEF is a zero-order traverse.

the NAD origin to the geocentre. The correction seems to have been well judged, since [222] reports that in 1970 the doppler longitude of the meridian mark at Greenwich was $-5''.64$, a figure which [223] in 1976 corrected to $-5''.69 \pm 0.17$ after recomputing with corrections for polar motion. The European Datum (1952) longitude of the mark is $-0''.37$, which can be converted to geocentric, using § 2.51, with $\Delta x = -85$ m, $\Delta y = -103$, and $\Delta z = -120$ m from Table 7.5. This change of datum gives a correction of $-5''.35$ to longitudes in the neighbourhood of Greenwich, whence the geocentric longitude of the mark is $-5''.72$, disagreeing with the broadcast doppler ephemeris (based on the WGS 72 geopotential model) by a fortuitously small $-0''.03$.† The relation between the broadcast ephemeris longitude datum and that of the precise ephemeris is not (in 1977) known.

(*f*) Of the three methods, comparison with photo-satellite nets seems likely to be the most accurate, but (in 1977) systematic errors of $\pm 0''.2$ in longitude may easily remain.

A correction of $0''.26$ was required to convert the doppler longitudes given by solution NWL-9D to the solution DOD WGS 72. The latter was constrained to agree with gravimetrically derived values of longitude in the area of the North American datum. See § 7.76 (*f*). After this correction, and a scale correction see § 7.60 NWL-9D became NWL-10F.

7.60. Scale

The scale of orbits and of ground station nets determined by doppler methods are in principle dependent on the value of earth's gravitational constant GM used in the general geodetic solution, § 7.62, which determines the coordinates of the tracking stations. Given the observed period of a satellite's orbit, (7.2) shows that if A is the semi-major axis of the orbit

$$\frac{\delta A}{A} = \frac{1}{3}\frac{\delta(GM)}{GM}. \tag{7.211}$$

Alternatively, the scale can be determined from outside sources, such as ground survey, or from laser or radar distance (dependent on the velocity of light) which may have been incorporated in nets of other satellite systems which may have some pairs of stations in common with the doppler net.

The doppler system itself can deduce a value of GM from its own observations, by treating GM as an unknown in its geodetic solutions, but

† This good agreement is sensitive to change in the figure -103 m accepted for Δy. A change of 1 m in Δy changes the discrepancy at Greenwich by $0''.05$.

in practice it is not very strongly determined by this means. Thus, the solution NWL-9D gave a value $GM = 398\,603\ \text{km}^3\ \text{s}^{-2}$ [11], while later values determined by the JPL are about 398 600.5, see § 7.00 (*b*).†

The figure 398 603 produced ground station distances which were over 2 ppm longer than the consensus of ground survey traverses. The value accepted for NWL-9D was 398 601 agreeing substantially with the JPL value, but still about 1 ppm longer than ground survey and VLBI values [11]. When NWL-9D was corrected for incorporation in WGS 72 as NWL-10F, its scale was further reduced by 0.8 ppm (−5.27 m) in the geocentric distance of its stations, so as to agree with the ground surveys. The scale is still a little open to question, possibly to the extent of about 1 ppm.

7.61. Polar motion

Polar motion, or variation of latitude, is a periodic motion of the earth's body relative to the axis of rotation. See § 4.09.

The satellite's orbit is computed in a reference frame which contains the instantaneous axis at the start of each 48-hour period. The ephemeris is then converted to the terrestrial frame, in which the z-axis is aligned on the CIO, using for the rotation matrix A of (7.174) the pole coordinates given by the BIH. In this terrestrial frame the coordinates of a fixed ground should remain unchanged with time. So (if the accuracy warrents it) repeated doppler based station coordinates should monitor or correct the BIH pole positions.

[13], p. 69 reports that the 48-hour mean values of polar motion derived from doppler appear to have a random error of about 0.7 m. At times there have been systematic differences of 0.5 to 1.0 m between BIH and doppler, lasting over some months, [9].

Doppler data have been brought into terms of the CIO by adjusting their results for 1964–7 into average agreement with the BIH over the same period, [14], p. 1.

It appears (in 1977) that doppler and conventional zenith telescopes give the polar motion with comparable accuracy, [551], [9], [14], [17], and [15].

7.62. General geodetic solutions

In the preparation of the ephemeris, § 7.56, the ground station coordinates and the geopotential constants are accepted as known. Every few years a *general geodetic* solution is made using satellites in non-polar orbits as well as the Transit satellites. The unknowns in the solution

† These figures are for the earth including its atmosphere, as is proper in this context.

NWL-9C of 1971 were as follows.

(a) The three coordinates of each of 75 ground stations.
(b) 479 gravity constants.
(c) One frequency parameter, and one frequency drift parameter, for each pass over each station.
(d) Six orbital elements, one solar radiation parameter, and eight drag parameters, for each of 39 arcs of between four and 16 days in length. The total number of doppler observations was 550 538, made to 10 non-polar satellites in addition to the Transit polar satellites.
(*e*) A single value of Love's number k, see § 6.78.

The equations of motion were numerically integrated using a 6-order Cowell procedure at 30-second intervals. Compare § 7.56 (*c*) and (*d*). The partial differential coefficients were obtained similarly, § 7.56 (*e*) (iii)–(vii). See [451], pp. 143–8 and [10], pp. 5321–3. Later solutions have used 6-order and 20-second intervals, or 12-order and 30-seconds.

The normal equations have been partitioned on the lines described in § 7.55 (*d*), so as to separate the instrumental bias parameters from the other unknowns, [10].

7.63. Accuracy of doppler fixes using the precise ephemeris

(*a*) *Within each pass.* Each pass produces 20 to 40 counts. Their self-consistency may be judged by computing a figure for each count N, using a best fitting solution for the ground position, and assuming an errorless orbit, § 7.53. Such a computation gives an r.m.s. residual σ_0 for observed N's amounting to only about 0.11 m, implying that 0.11 m is the standard error of range differences $(S_2 - S_1)$ deduced from each count. This small figure is obtained after rejecting a few counts (typically about 3 per cent of the total) whose residuals exceed $3\sigma_0$, [13], p. 45 and [10], p. 5326. Such a set should typically give the along-track position and slant range at the time of closest approach correctly to 0.50 m, from a single pass. [10], p. 5326 and [15], p. 379. [291] describes some early geoceiver passes in which σ_0 was much larger, such as 0.42 to 2.03 m, in which this large scatter was clearly associated with abnormal irregularity in the frequency biases observed for successive passes. Repeat observations changed the computed height of one station by 14 m, and three of the others gave heights differing by an abnormal 5 to 7 m from heights given by zero-order ground survey transcontinental traverse. It appears that if σ_0 is found to be greater than 20 or 30 cm, the resulting fix should not be accepted. Abnormal internal error seems to be associated with a great increase in total error.

(*b*) *At each station* 20 or 30 passes are observed. If the along-track position and slant range of the point of closest approach is computed for each pass from the final mean position accepted for the station, comparison with observation gives their standard errors in a single pass as 1.6 and 2.5 m respectively if the maximum elevation of the pass has been more than 30°, or 3 to 4 m if the maximum elevation has been as little as 10°, [13], p. 48. These figures compare adversly with the s.e. of 0.5 m obtained from internal evidence in (*a*) above, although passes with maximum elevations of less than 20° need not be considered. This is not unexpected, as within a single pass there will be much correlation between the ephemeris errors of the separate counts, and also between their atmospheric errors. [13], p. 43 estimates that the effect of errors in the geopotential model may be reduced to 0.7 m in the mean of 20 passes, but that figure will not be reduced by increasing their number to above 20.

(*c*) *Total station error.* At some stations observations have been more or less continuous over a period of 9 years (1964–72). If the possibility of ground movements can be ignored, the constancy of such observations is a good measure of their accuracy, [13] and [12]. Making the best of possible ground movements by computing a best fitting linear rate of change, [13] gives the r.m.s. residual of a 5- to 10-day fix as 1.7, 1.4, and 1.6 m in latitude, longitude, and height respectively. Working on the same lines over a period of a single year (1973), gave 0.7 m as the r.m.s. error in each component. It is reasonable for 1973 results to have been better than 1964–72.

It is perhaps possible to conclude that modern doppler fixes from 20 or more passes, in not unfavourable conditions, may be expected to be correct within 1 metre, except for possible systematic error in the reference frame, as in (*d*) below. See § 2.44 (*c*) (iv) for comparisons with zero-order traverses in the US. [380] gives comparisons with the Australian survey, and [67] and [72] in Great Britain.

(*d*) *Reference frame.* [13] considers that the origin of the current (1976) doppler frame coincides with the geocentre within one metre, but that the polar axis may be tilted by 3 to 6 m (in 6360 km) in relation to the axis on which star catalogues are based.† As stated in § 7.60, there is doubt of possibly 1 ppm in scale (i.e. in distances from the geocentre), and of perhaps 0″.02 in longitude.

General references for Section 7. [9], [10], [11], [13], [93], [129], [273], [406], and [451].

SECTION 8. SATELLITE ALTIMETRY

7.64. Observations

(*a*) *The principle of satellite altimetry* is that a satellite in a well-tracked orbit continuously records its height above the mean wave level of the sea

† The good agreement with the BIH figures for polar motion, § 7.61, suggests that the doppler *z*-axis is aligned with the CIO to within better than 1 metre.

(MWL) over an area of about one square km immediately below it. Ordinary waves are averaged out, but not the effects of tides nor of variations of current, temperature, salinity, or atmospheric pressure. In so far as this mean level coincides with the geoid, as it probably does within two or thre metres after correcting for tides, and apart from errors in tracking the satellite, altimetry enables the distance of the geoid from the geocentre to be computed. It provides a new method of surveying the geoid in sea areas, from which a rather generalized survey of the intensity of gravity can be computed. The apparatus is not yet designed for use over land, although large lakes would of course do as well as the sea.

(*b*) *Instrumentation.* Manually operated altimeters were installed in Skylab, but Geos 3 has given the first continuous results.

Geos 3, with an inclination of 115°, mean height 850 km, eccentricity 0.001, is provided with very full tracking facilities, namely laser reflectors, doppler transmitters, C- and S-band transponders, and satellite-to-satellite equipment for its tracking by ATS 6 ($i=1.^\circ 6$, mean height 35 800 km, $e=0.0001$). Exceptionally good tracking is thus possible, § 7.04 (ii), and it is to be hoped that its tracking will be good to a single metre or better. Its radar altimeter operates at 13.9 GHz (22 mm wavelength), or even at 27.8 GHz at reduced power. Long pulses are emitted, each 100 μs (30 km) in length, but the measurement is made at a modulated wavelength of 10 ns (3.0 m) in length, and the satellite's receiver responds only to returning signals at that wavelength.

The beam width of the antenna is 4°, and the satellite is sufficiently stabilized for this width to include the perpendicular from the satellite to the mean sea surface. The 4° beam will cover an area of about 60-km diameter at sea level, but the short 3-m working wavelength ensures that most of this area makes no contribution to the returning signal, as in (*c*) below.

(*c*) *Reflections from waves.* Fig. 7.18 illustrates the contact between mean wave level (MWL) and a 3-metre signal from a satellite at a height of 850 km. The curved arcs, centred on the satellite, are spaced at intervals of half a wavelength (1.5 m), and the horizontal lines represent MWL and surfaces 1.5, 3.0, etc. metres above and below it. Within the inner circle of radius 1.6 km, shown in plan below the diagram, the slant range from the satellite to MWL differs from the perpendicular distance by between 0 and 1.5 m, and within a smaller circle of radius 0.65 km, the variation will be between 0 and 0.25 m. In the first outer ring between 1.6 and 2.3 km, the variation will be between 1.5 and 3.0 m, with similar ranges of variation within each of the progressively narrower and more distant rings.

The reflecting surface of the sea may be regarded as consisting of a very large number of small (say 1 cm × 1 cm) reflecting elements, at varying inclinations, the great majority of which will fail to return any reflection

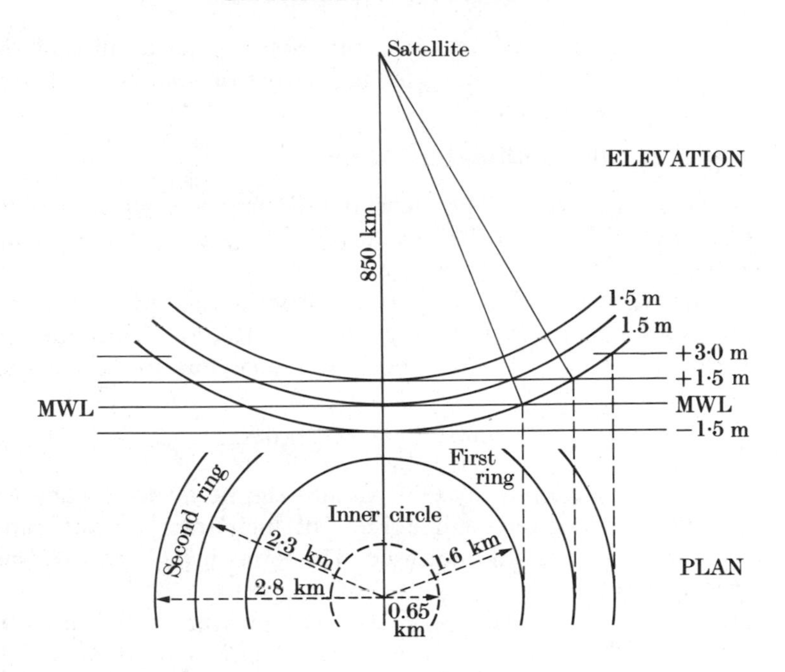

FIG. 7.18. Satellite altimetry. Vertical scale exaggerated.

to the satellite's antenna of diameter 0.4 m. Of those which do return a reflection, all reflectors which are at the same height within the inner 0.65-km ring will return signals whose half-delay will be the same within 0.25 m, a phase difference of 30°. In the outer part of the inner circle the phase angles of the signals from reflectors of the same height will vary between 30° and 180°. In each outer ring, reflectors of the same height will have a 180° range of phase difference. The demodulation of the received signal eliminates the effects of the outer rings as nothing but noise, leaving only the clear signal from the central area. On the assumption that the mean height of effective surface reflectors above MWL is zero, the delay given by the mean of the received signals will be the delay appropriate to the perpendicular distance from the satellite to the local MWL.

The pulse length of 30 km comprises 10 000 3-metre cycles. The object of having such a large number is to increase the total energy of the coherent return from the central 0.65 km of the target, as compared with the incoherent noise which emanates from the outer rings. The former builds up in proportion to the number of 3-metre cycles, while the latter increases as their square root. During the 100 μs of the long pulse the satellite will only travel through about 0.7 m.

[522], on which this paragraph is largely based, gives a full analysis of the reflection problem, and estimates that errors on that account should not exceed about 10 cm.

7.65. Computation and interpretation

(*a*) *Refraction correction.* The measured distance is vertical, so refraction errors are not enlarged by any factor sec ζ, and no path curvature needs to be considered.

(i) *Ionospheric refraction.* See § 7.10 (*c*). Considering the average night minimum in a quiet year, and taking $f = 13.9$ GHz, and using the group refractive index, (7.51) gives the ionospheric delay due to the main term as

$$\Delta S = (113 \times 10^{16}) \div \{(13.9)^2 \times 10^{18}\} = 0.6 \text{ cm}. \qquad (7.212)$$

This is virtually negligible. Day observations can be avoided when maximum accuracy is required, and periods of high solar activity can be detected by other means, and not used. The terms in $1/f^3$ and $1/f^4$ need not be considered.

(ii) *Tropospheric refraction.* See § 7.09. The pressure and temperature at sea level must be observed or estimated. An error of 10 mbar (1000 Pa) in pressure, or of 2 °C in temperature, will introduce an error of only 2.3 cm in the refraction correction. At the surface of the sea the humidity will always be 100 per cent.

The correction may be computed by (7.30), or by Hopfield's double quartic formula (7.40). For the latter a suitable value of H_W must be selected, on grounds of geographical similarity, from the list given in [287] or in any later such paper. It may be hoped that the error will not exceed an r.m.s. value of 5 cm.

(*b*) *Geometry.* In Fig. 7.19 S is the satellite, O is the geocentre, and P_M is the foot of the perpendicular from S to the mean wave level (MWL). This perpendicular meets the geoid at P_G which may be 2 or 3 metres plus tidal effects, totalling ΔN, above or below P_M. The reference spheroid at P lies N metres above or below P_G, where N may be 50 metres or more. In Fig. 7.19 $P_G P_S Q$ is the spheroidal normal, which makes an angle of ξ (the deviation of the vertical in meridian) with $P_G S$. ξ may be 20″ or more, although it is generally less than 10″. $P_S O$ is inclined to $P_S Q$ at an angle of up to 10′. It is apparent that the line SPO is substantially bent at P, although it is immaterial whether the bends are considered to occur at P_M, P_G, or P_S. The following places them both at P_S.

We then have

$$OS = OP_S \cos \beta + SP_S \cos \alpha \qquad (7.213)$$

where α and β are as in Fig. 7.19.

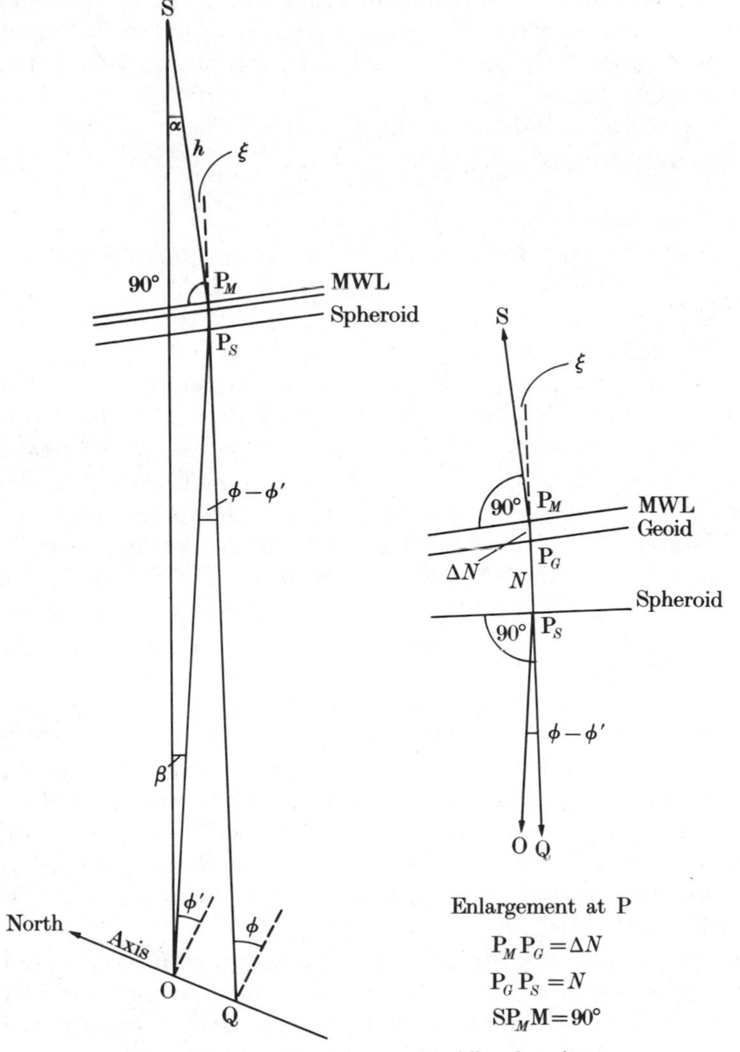

FIG. 7.19. Satellite altimetry. Meridional section.

Also $\alpha+\beta=\xi+\text{OPQ}=\xi+\phi-\phi'$

$$=\xi+694'' \sin 2\phi, \text{ from (A.56), if } f=1/298.25, \tag{7.214}$$

and $\quad \alpha/\beta=\text{OP}_S/\text{SP}_S.$ (7.215)

So α and β can be computed. The effect of a value of 20″ for ξ is less than 0.25 m. If its inclusion is necessary, it must be obtained from provisional computations of N made without it, along distances of some tens of km.

There is also the other component of the deviation, η, which may also be 20″ in the plane perpendicular to the paper in Fig. 7.19. Its effect is, however, negligible as it does not enter the formula in combination with the large bend $\phi - \phi'$.

(*c*) *Interpretation.* We have as in (7.213)

$$\mathrm{OS} = \mathrm{OP}_S \cos\beta + N + \Delta N + h\cos\alpha, \tag{7.216}$$

where OS is known from the ephemeris.

OP_S is known from the definition of the spheroid, given its geocentric angle $\phi' = \phi'_{\mathrm{sat}} - \beta$

h is the measured distance SP_M

$N = \mathrm{P}_S\mathrm{P}_G$, and $\Delta N = \mathrm{P}_G\mathrm{P}_M$.

So $N + \Delta N$ is known, but neither is given separately.

If the object is to determine the form of the geoid to within a couple of metres, as an end in itself, ΔN can be ignored, except for tidal effects which can be estimated or eliminated by repetition. (7.216) then gives N.

For other purposes, or if greater accuracy is required, either N or ΔN must be estimated by other means, always of course supposing that the accuracy of h and the satellite ephemeris justify it. Thus if it is desired to use N to obtain values of gravity in ocean areas, ΔN may perhaps be adequately estimated from meteorological and oceanographical considerations, as described in § 3.28 (*b*).

Alternatively, if the chief interest is in ΔN, the separation of mean sea-level from an equipotential surface, N must (if possible) be estimated from Stokes's integral, § 6.30, amplified by harmonic constants obtained by other satellite methods, § 6.34, with some control by altimetry or doppler fixes on adjacent coast lines or islands. This may be difficult. Changes of ΔN with time can of course be measured without knowing N, and changes of N over short distances can be obtained if ΔN can be assumed to be constant over the distance.

A use of satellite altimetry is that given N and thence the form of the geoid over sea areas, it is to some extent possible to compute more or less generalized average values of g. This is useful in areas like the Southern Ocean, where ship-based gravity values are still sparse. For results obtained up to 1977, see [476], and for the possible methods of computing g see [504].

General reference for altimetry. [522].

Section 9. Minitrack, and Very Long Baseline Interferometry

VLBI, which is described in §§ 7.66 to 7.72, is not usually practised with artificial satellites, but it is conveniently included in this section. § 7.73 describes the Minitrack system which, working on a similar method, has been used for tracking artificial satellites, but with relatively low accuracy.

7.66. VLBI. Introduction

Fig. 7.20 shows two ground stations A and B which may be some thousands of km apart. AQ and BQ point in the direction of a substantially infinitely distant radio source, a quasar, emitting radiation at a large range of high frequencies. AP is perpendicular to BQ, so (ignoring atmospheric refraction delays) a plane† wave front reaches A and P simultaneously, and reaches B T seconds later. The distance BP is then Tc, where c is the velocity of propagation.

Quasar radiation may cover a large frequency range, but the receiving apparatus may be tuned to receive only a relatively narrow band.

Given sufficiently accurate synchronized clocks at A and B, the delay T can be measured, and the length of BP obtained. Then if the length of AB is known, the angle QAB follows. This is the principle of Minitrack.

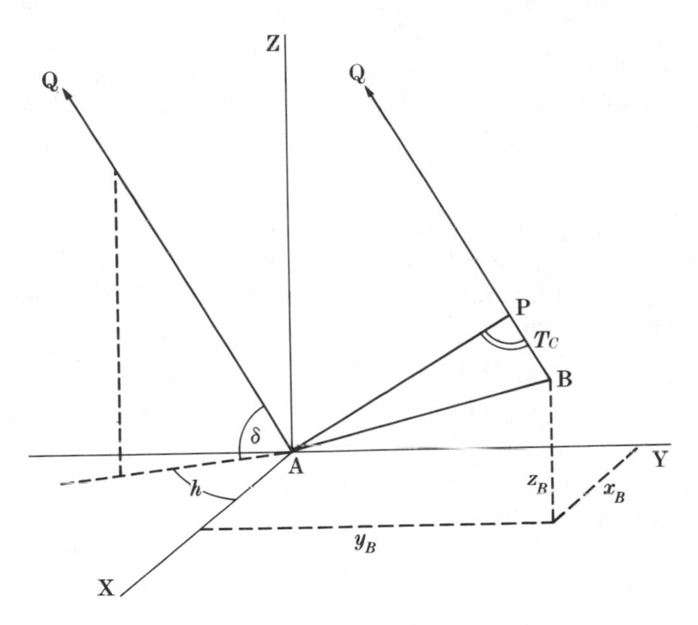

FIG. 7.20. The X-axis is upward from the paper. BQ is parallel to AQ.

† If observations are made to an earth satellite, the wave front will be spherical, and allowance for that must be made.

In VLBI, if the direction of the source is known, the length and direction of AB can be computed. Given suitably distributed observations, the relative declinations of different sources, their RA's (from an indeterminate zero) and other parameters such as the polar motion, can also be computed.

Notation.

c = Velocity of light in vacuum.
f = Frequency of radio reception.
δ = declination of source.
RA = Right ascension of source.
h = Greenwich hour angle of source.
T = Time delay between A and B, in SI seconds.
Tc = Distance BP.
t = sidereal time.
Ω = Earth's rotation (rad/sidereal seconds).
$\alpha, \beta, \gamma.$ = See (7.220).

7.67. Basic formulae

See Fig. 7.20. The arbitrary† origin of coordinates is taken to be at A. The z-axis is parallel to the instantaneous‡ axis of rotation. The x-axis is parallel to the instantaneous equator and to the BIH Greenwich meridian plane, CZM, and the y axis completes a right handed system. The coordinates of B are $x_B\ y_B\ z_B$. The declination of the source is δ, and its Greenwich hour angle is h. Over a short interval, such as a few minutes, we have

$$h = \text{GST} - \text{RA} = \Omega t \quad \text{if} \quad t = 0 \quad \text{when} \quad \text{GST} = \text{RA}. \tag{7.217}$$

RA and δ will be approximately known, although improvement will probably be necessary. GST will be available from the station clocks, so the direction of BQ and AQ in the frame of reference is at least approximately known.

$$\begin{aligned} \text{Then } T = \text{BP}/c &= (1/c)(\text{projection of AB upon BQ}) \\ &= (1/c)(\text{projection of AB upon AQ}) \\ &= (1/c)(-x_B \cos h \cos \delta + y_B \sin h \cos \delta - z_B \sin \delta), \end{aligned} \tag{7.218}$$

as is clear from Fig. 7.20. So if the time delay T could be instantaneously measured at each of three well separated times of day, and if RA and δ were known, and if the two station clocks were perfectly rated and

† The earth's centre of mass is not included in the observing process, so it is generally not a convenient origin. It, or any other point can be used as an origin, with suitable modifications to the formulae, if its position with reference to A is known.

‡ It is possible to take the z-axis as parallel to the direction defined by the CIO, but that complicates the formulae a little.

synchronized, the three equations (7.218) would give the three components of AB in the defined frame.

The earth's rotation. In practice, some minutes of observation are required to make a single determination of the delay T, during which time the hour angle of Q and the length BP are changing.

Differentiating (7.218)

$$dT/dt = (\Omega/c)(x_B \sin h + y_B \cos h) \cos \delta, \tag{7.219}$$

since $\Omega = dh/dt$.

Two measurements of dT/dt well spaced in h, would by themselves give values of x_B and y_B, and thence the length and direction of the projection of AB upon the equatorial plane, but the z component is not obtainable in that way.

In (2.218) let a term δt be included to allow for a lack of perfect synchronization between the two clocks, and let h be replaced by Ωt, as in (7.217). Then following [520]

$$\begin{aligned} T &= \delta t + (1/c)(-x_B \cos \Omega t \cos \delta + y_B \sin \Omega t \cos \delta - z_B \sin \delta) \\ &= \alpha + \beta \sin \Omega t + \gamma \cos \Omega t, \end{aligned} \tag{7.220}$$

where $\alpha = \delta t - (z_B/c) \sin \delta$,

$\beta = (y_B/c) \cos \delta$

$\gamma = -(x_B/c) \cos \delta$

Since δt and z_B cannot be separated, it follows that no number of observations of T from a single source can give a solution unless δt is ignored, although β and γ are determinate from three observations to give the length and direction of the projection of AB on to the equatorial plane, as before, if RA and δ are assumed to be known. If δt and z_B are to be separated, two suitably spaced observations must be made to two well separated sources.

Further, if the RA's and δ's of the sources are to be treated as unknown, at least three observations of T from each of three well separated sources will be necessary, to give

(*a*) The three components of AB.

(*b*) The clock synchronization error, δt.

(*c*) The three declinations δ_1, δ_2, and δ_3. [496], p. 103 reports that declinations have been measured to 0″.1, the equivalent of 3 metres, except that near-equatorial declinations are indeterminate. When higher accuracy is being considered, only $(\delta_1 - \delta_2)$ and $(\delta_2 - \delta_3)$ can be determined, unless the declination of one source is known, [142], p. 524.

(*d*) Two differences of RA, $(RA_1 - RA_2)$ and $(RA_2 - RA_3)$.

The RA of at least one source must be obtained by some other method, however many VLBI observations may be made.

For the computation of geodetic latitudes and longitudes, it is further necessary that the geodetic latitude, longitude, and N of at least one of the stations should be known.

7.68. The measurement of the time delay, T

The signals received at A and B are recorded on magnetic tapes together with time markers from their station clocks. The two tapes are then compared in a common laboratory. The object is to deduce the time delay in the form

(Observed difference between identical wave fronts)

$$= (T \text{ at time } t_0) + (t - t_0)\, \mathrm{d}T/\mathrm{d}t \qquad (7.221)$$

where $\mathrm{d}T/\mathrm{d}t$ is as given in (7.219), and $(t - t_0)$ covers the time length of the tapes, a few minutes.

There is the difficulty that the received signals contain no recognizable 'pips' or such-like markers. They are simply a mix of different frequencies with a bandwidth of perhaps 100 MHz centred on a mean frequency of something between 0.5 and 8 GHz (60 cm to 3.7 cm). If the received signal were absolutely monochromatic, observations of fractions of a wavelength would in principle be possible, but the number of whole wavelengths would be indeterminate.† The remedy is roughly as follows, [520] and [521]. Let the reception be confined to two relatively narrow bands each (e.g.) 0.36 MHz wide at frequencies of 23 MHz on either side of the mean frequency 7850 MHz (3.8 cm). If these bands were quite monochromatic, their combination would produce 'beats' or 'fringes' with a frequency of 46 MHz (6.5 m), fractions of which would resolve ambiguity in the number of carrier wavelengths. Perhaps fractions of the 6.5 m frequency might suffice for the final result for some purposes, but ambiguities in the number of 6.5-m wavelengths can in turn be resolved by using another pair of narrow bands separated by (say) 5 MHz, to give a fringe wavelength of 60 m, at which level existing surveys might be able to resolve further doubts. [520] gives a formula for the optimum choice of bands, and anticipates no difficulty in recording T correctly to 0.1 ns (3 cm) apart from error in clock synchronization.

The comparison of the tapes is carried out statistically by a process described as cross-correlation. Search is not made for any particular markers to be found in the signals, but the optimum value of T (at time t_0) is accepted to be that which secures the maximum cross-correlation

† The situation would be similar to that which arises in all EDM ground measurements, § 1.43 (*a*).

between the two tapes. [520] gives a description of the process which is not unduly technical.

7.69. Atmospheric refractive index

In the preceding paragraphs it has been assumed that the velocity of transmission is that of light in vacuum, but the times of reception at A and B have each to be corrected by the total time delay due to refraction between the outer atmosphere and the receivers. The two paths differ not only in their zenith distances, but also (especially if the distance AB is very great) in the atmospheric conditions affecting them, such as the time of day, latitude, topography, and weather. Uncertainty in the difference between the two corrections is likely to be the ultimate factor limiting accuracy.

The *tropospheric refraction* may be computed by the formulae of § 7.09, with reasonable hope of the error in the correction not exceeding 2 or 3 $\sec \zeta$ cm.

Ionospheric refraction. Currently (1977) the double wavelength system of observation has not been introduced for VLBI. Using 7.8 GHz, the magnitude of the main term, § 7.10 (*c*), is reduced to $2 \sec \zeta^*$ cm in the circumstances on which formula (7.51) is based, namely an average night minimum in a quiet solar year. Here ζ^* is a modified zenith distance, as after (7.49). This would be a satisfactory figure, but the conditions for which it is computed are more favourable than can be regularly expected. Much greater errors are possible, although the presence of larger errors would be recognizable from other phenomena. It is to be expected that the two-wavelength system will be introduced, and that the tropospheric refraction will then be the most serious source of error.

With a wavelength of 7.8 GHz the higher-order corrections of § 7.10 (*e*) are negligible.

It is to be expected that sources at large zenith distances will experience larger errors than those close to the zenith, and large zenith distances will not be avoidable if the three observations to each source are to be well spaced in hour angle, especially if the line AB is very long, so that sources near the zenith at A are near the horizon at B. The accuracy of very long lines would appear to be limited by this consideration, and a line of 10 000 km might best be measured in several shorter sections. On the other hand [496], p. 99 surprisingly reports no apparent decrease of accuracy with increasing zenith distance.

7.70. Accuracy obtainable

[521] reports the measurement of a line 3900 km long between Massachusetts and California. The observations were made on 9 different

dates in 1972–3. Each date covered 15 to 26 hours with, on average, 15 3-minute observations to each of 10 sources. Internal evidence gave an average r.m.s. of 0.40 m for the length of the line on the first five dates, and of 0.11 m for the last four, which had a better distribution of sources. The r.m.s. of the mean of the nine dates, judged by their mutual scatter, was 0.16 m. The final mean was 1.60 m greater than the distance given by conventional ground survey, equivalent to 0.4 ppm. Much, or all, of the difference could of course have been error in the ground survey.

In contrast to the length of the line, its direction was rather less well determined. The r.m.s. of the mean of the 9 dates judged by their scatter was 0″·06 (arc seconds) in hour angle and 0″·07 in declination, equivalent to 1.2 and 1.4 m in 3900 km. Compared with the survey these 2 components differed by 18 and 3.6 m, the former figure being substantially more than might be expected from the survey. The discrepancies with the survey were thought to be attributable to the non-identity of the terrestrial frame and the frame constituted by the quasar sources, as is indeed possible especially in hour angle, see last lines of § 7.67. Polar motion and rotation irregularities were not included as unknowns, and may have contributed to the increased values of internal scatter.

[143] similarly describes the measurement of the length and direction of a 5251-km line between Algonquin radio station in Canada and Chilbolton in England. The r.m.s. deviation of the mean determination of length was 1.05 m and of direction 0″·015.

7.71. Equipment

The equipment required is considerable, and VLBI is not (in 1976) an everyday aid to field surveying. Early experiments were made with 25- to 64-m diameter antennae, which existed for astronomical purposes, although [395] envisages the possibility of a smaller and portable antenna.

Hydrogen masers are desirable for time keeping and frequency stabilization at the two stations. The rates of the two clocks can then be relied on to be identical to within 1 in 10^{14}, which is adequate.

7.72. Other information obtainable from VLBI

(*a*) *Polar motion.* Variation in the relative coordinates at the ends of the line AB can of course be attributed partly to polar motion, the movement of the earth's crust relative to the axis of rotation adopted as the z-axis. Given sufficient observations, polar motion can be added to the list of unknowns, items (*a*) to (*d*) in § 7.67. See [59].

(*b*) Similarly, equation (7.220) can clearly provide information about variations in the earth's rate of rotation, Ω.

(*c*) Given sufficient accuracy, analysis of changes in the relative coordinates of A and B for variations of the correct periods may give figures for the lunar and solar earth tides.

(*d*) Repetition of observations after intervals of some years may be expected to give information about continental and plate movements.

7.73. Minitrack

Minitrack is a world-wide NASA system of about 10 interferometric tracking stations, installed in 1956–60. The system tracks artificial satellites in altitude and azimuth with an accuracy of about 10″ or 20″, [273], p. 376. The principle is as outlined in § 7.66, by which the time delay between the arrival of a radar wave at the two ends of a base of known length and direction, gives the direction of the source.

In the minitrack system there are two bases, one north–south and one east–west, each about 125 m long. They are designed for the reception of 136- to 138-MHz radio waves (2.2 m). There are four antennae on each base, to vary the sensitivity, and another five near the centre, where the two bases cross, to resolve ambiguities.

In its early years the system was useful for giving predictions, and for observations requiring low accuracy, such as for the long-period and secular changes arising from low-degree zonal harmonics in the geopotential field. Minitrack now has little direct geodetic significance.

Most minitrack sites have also included a MOTS camera, for calibration. Observations made with these cameras have been in general use for high accuracy orbital analysis, see §§ 7.75 and 7.76.

General references for VLBI. [520], [311], [521], [400], [142], [143], [496], and [58]; for Minitrack: [227] and [289].

SECTION 10. SYNTHESIS. GEODETIC RESULTS

7.74. Outline

The geodetic uses of artificial satellites are listed in § 7.00. This section gives an outline of the work which has so far been carried out by the various methods described in Sections 2–7. § 7.75 lists the larger programmes of observations, which generally involve a world-wide network of stations. § 7.76 lists the principal computations which have been carried out, generally combining data from more than one of the observing programmes. §§ 7.77 and 7.78 describe progress which has been made in the attempt to express all national classical surveys in terms of a geocentric coordinate system, while §§ 7.79 and 7.80 describe the principal analyses of the geopotential which have been made.

Notation

$\Delta x, \Delta y, \Delta z$. Coordinates of the centre of a local spheroid and datum referred to the geocentre.

ω, ψ, ϵ Right-handed rotations about the z, y, x axes respectively, viewed from the origin outward along the positive axes.

$\Delta\phi, \Delta\lambda, \Delta N$ Latitude, longitude, and height of geoid above spheroid on a geocentric spheroidal datum minus their values on a similar spheroid as centred for a local datum.

See also § 7.00, especially for abbreviations.

7.75. Principal satellite observing programmes

(*a*) With a few exceptions, most geodetically useful satellite observations have necessarily formed part of world-wide programmes, most of which have been initiated in the US. The currently (1977) most notable of such programmes are listed below.

(*b*) *Smithsonian Astrophysical Observatory* (*SAO*). Photographs were taken with Baker–Nunn cameras, originally at 12 sites from 1958 to 1964 or later. Since then about 20† other stations have been occupied for various periods, operated by various agencies. The distribution of the stations is world-wide. These observations are suitable for both photo-dynamic and photo-geometric computations, although the geometric net is not uniformly strong.

Since 1966–72 SAO lasers have been installed at nine stations, five of which coincide with Baker–Nunn camera stations.

(*c*) *National Ocean Survey. BC*-4 *Pageos programme.* This is a world-wide astro-triangulation net of 45 stations using the BC-4 cameras and the Pageos passive balloon satellite, (altitude 4000 km), 1966–70. The net and observing programme were designed for the photo-geometrical method only, and the geometry of the net is strong. Scale is provided by 5 or 6 first-order or zero-order traverses 3000–4000 km long, in the US, Europe, N. Africa, and Australia.

US Air Force PC-1000 cameras were used to densify the net in North and South America in 1969–70.

(*d*) *NASA satellite tracking and data acquisition network* (*STADAN*). This programme involved the Minitrack stations and a number of radar tracking antennae. It included 17 MOTS cameras at the minitrack stations, and four Goddard range and range-rate stations (GRARR), all observing on Geos 1 and 2, 1966–71.

(*e*) *NASA* (*GSFC*) *Special Optical network* (*SPEOPT*). An experimental net used for tracking Geos 1 and 2, mostly in the eastern US, comprising MOTS-40 cameras at 6 stations, BC-4 at one, Pth-100 at 6, and lasers at nine stations. This net was used for comparison with a variety of other methods.

(*f*) *US Army SECOR net.* 37 stations operated the SECOR system, using seven SECOR satellites, and Geos 1 and 2, and forming a net

† Such figures, here and elsewhere in this section, are only approximate. Exact figures involve explanations about such matters as the status of duplicate stations, where instruments may have been moved only a short distance, or of stations which have been occupied for abnormally short periods. The aim is to give a general description.

round the equator and low northern latitudes. The net consists of well shaped triangles, but by itself the system is geometrically weak. Thirteen of its stations coincide with those of the BC-4 Pageos net.

(g) *International satellite geodesy experiment* (*ISAGEX*). Coordinated by CNES and SAO. The programme incorporated nine lasers in a net with 34 cameras and a few other stations, about 50 stations in all. A number of these were collocated with Baker–Nunn and MOTS stations. About 64 000 laser ranges were observed to Geos 1 and 2, BE-B, BE-C, PEOLE, DI-C, and DI-D satellites in 1970–1.

(*h*) *US Transit doppler system.* For the tracking arrangements see § 7.48 (*b*). There is no limit to the number of position fixes which can be made by portable receivers, using the orbits computed in the US, and such field work does not need to be internationally organized. The results of such programmes are reported in [406], [596], [519] Fig. 1, and many later references.

(*i*) *US Navstar system.* This is not yet operative. See § 7.44.

7.76. Principal computations combining various observations.

The combination of various observation programmes is generally necessary in order to obtain ground fixes at the correct scale, and referred to the geocentre, the CIO and BIH Greenwich. The limitations of each method by itself are mentioned in § 7.03 (*c*) to (*f*).

Items (*a*)–(g) below list some of the principal computations which have recently been undertaken. Their results generally give the geocentric coordinates of ground stations, §§ 7.77 and 7.78, and a revised harmonic analysis of the earth's geopotential, § 7.79.

(*a*) *SAO Standard Earth I*, 1966 [391], *II*, 1969 [218], *and III*, 1973 [216]. The most recent, SAO III, was based on the most suitable observations selected from the following.

(i) Baker–Nunn photographs, used geometrically and dynamically.
(ii) SAO lasers.
(iii) NOS BC-4 Pageos programme.
(iv) NASA SPEOPT and lasers.
(v) ISAGEX lasers.
(vi) A few European camera stations.
(vii) Jet Propulsion Laboratory (JPL) deep space net.
(viii) Surface gravity. Estimated mean free-air anomalies in 19 328 1°× 1° squares (out of a possible 64 800).

The solution gives the geocentric (CIO and BIH) coordinates of 86 ground stations, and an 18×18 geopotential analysis, see § 7.79 (*a*).

The earlier solution SAO II of 1969 did not include the BC-4 or ISAGEX programmes, nor of course any other items after 1969. It included the early SAO, CNES, and GSFC lasers, and surface gravity in 935 5°×5° (equatorial degree) equal-area squares out of a possible 1650, based on [330], pp. 5303–14.

(*b*) *BC*-4 *Pageos programme.* [516]. The solution gives the coordinates of 45 camera stations, 37 of which were in the immediate vicinity of (collocated with) doppler stations, which enable the BC-4 net to be referred to the geocentre.

(*c*) *NASA*(*GSFC*). NASA has issued two long lists of tracking stations, with the object of recording their positions relative to both the local survey datum and to a geocentric datum. They also give exact details of the station locations, and describe the methods used for the satellite and ground survey fixes.

(i) [401] of 1971, which gives geocentric (CIO and BIH) positions relative to a 6 378 155-m and 1/298.255 spheroid.

(ii) [49] of 1973, in two volumes, giving positions relative to the modified Mercury datum of 1968 (6 378 150 m and 1/298.3). It also gives summaries of the various observation programmes with lists of stations involved in them, descriptions and photographs of the instruments, and world charts showing the stations which have contributed.

The 1971 list [401] is based on positions determined from 2-day arcs of Geos 1 and 2, from STADAN, MOTS-40, SPEOPT, SAO Baker–Nunn, and ISAGEX cameras (45 in all), with 3 lasers and GRAAR. It lists a number of other stations, whose positions can be brought into the same terms. The total is about 200.

The 1973 list [49] includes all the programmes listed in § 7.75 (except Navstar), together with JPL deep-space net, radio telescopes, and S-band radars, whose primary functions are not geodetic, but which have been able to contribute to the results. In includes more stations than [401], 113 in vol I and 243† in vol. II, and their positions have been updated.

(*d*) *Ohio* (*OSU*) *World Nets WN* 12, 14, *and* 16. [432] and [431]. These are geometric nets. The data used for WN 14 included.

(i) 23 Baker–Nunn, 14 PC-1000, 15 MOTS cameras, and a few others forming a previously adjusted net known as MPS, together with 7 C-band radar stations.

(ii) 9 PC-1000 cameras in South America, of which three were collocated with BC-4 stations.

(iii) 49 BC-4 stations.

(iv) 37 SECOR stations. of which 13 were collocated with the cameras of item (i).

The scale for WN 14 is provided by SECOR, by 5 or 6 ground survey

† Including duplication at collocated stations. A fourth edition has been issued in 1978.

traverses substantially the same as those of the BC-4 programme, 3 C-band radar distances, and by height control through heights above the spheroid (MSL heights + geoidal rise, N) computed from ground surveys, [432], pp. 106–16.

In WN 12 the ground survey height control is omitted. In WN 16 the ground survey traverses and C-band radar are omitted, but not the height controls.

[432] lists the coordinates of 175 stations based on WN 14 with an arbitrary origin, whose coordinates relative to the geocentre are believed to be, [432], p. 203, −21, −5, and +2 metres in x (Greenwich), y, and z (CIO).

(*e*) *Naval Surface Weapons Laboratory* (*NSWL*). The periodic doppler *general geodetic solutions*, § 7.62, give coordinates for the doppler tracking stations, and revise the geopotential model. The solution accepted for general use in 1975 is known as NWL-9D, [12], p. 3, which is based on the 18 semi-permanent Tranet stations and on about 100 more which have been temporarily occupied. An improved version, NWL-10 F, results from adding 0″·260 to the longitudes of 9D to agree with gravimetric data in North America, § 7.59, and by reducing all heights above the spheroid by 5.27 metres to make the scale accord with ground traverses, § 7.60. NWL-10E is an associated determination of the geopotential model, § 7.79 (*c*).

(*f*) *Department of Defense* (*DOD*) *World Geodetic system* '1972'

[519]. This solution is a combination of what were considered to be the best of the foregoing solutions, and uses the following.

(i) NSWL-10F doppler, see (*e*) above.
(ii) BC-4 Pageos programme, § 7.75 (*c*).
(iii) SECOR, § 7.75 (*f*).
(iv) SAO II, see (*a*) above.
(v) Surface values of gravity as in § 7.79 (*c*).
(vi) Astro-geodetic geoid charts [206].
(vii) Six 3000–4000 km first or zero-order ground survey traverses.

These data were in principle supplied in the form of normal equation matrices, which DOD combined to give their own solution. This gave the geocentric coordinates of the observing stations, a set of geopotential coefficients, and sets of datum shifts to convert 32 existing ground survey datums to geocentric. Coordinates are given on a 6 378 135 m and 1/298.26 spheroid.

In the final solution, the r.m.s. change from the original doppler, BC-4, SECOR, and SAO positions were 2.2, 5.4, 10.0, and 9.0 metres respectively, and the r.m.s. change in the length of the ground survey lines was 1.5 ppm.

The systematic corrections to longitude and scale, necessary to correct its four constituent solutions to WGS 72, are given as footnotes to Table 7.6. The longitudes of WGS 72 were defined by enforcing agreement between a selected set of doppler stations and gravimetrically derived values in the NAD area, [519], p. 13. A reduction in scale by (say) 1 ppm can be effected by reducing all computed heights by 6.4 metres.

(g) *Franco-German GRIM* 2, 1976. CNES and Technical University, Munich. [79]. This solution uses world-wide observations from 29 cameras and 14 lasers on sixty 10- to 14-day arcs distributed between 23 satellites. The object was to revise the geopotential analysis. Surface gravity was included. See § 7.79 (*d*).

7.77. Conversion of continental and local datums to geocentric

Let the coordinates of a ground station referred to the centre of its local geodetic datum or spheroid as origin be $x_1\ y_1\ z_1$, and let $x_2\ y_2\ z_2$ be its coordinates with reference to the geocentre as given by satellite observations, both sets of axes being aligned on the CIO and BIH Greenwich. Then $x_2 - x_1 = \Delta x$ etc. are the corrections required to convert local coordinates to geocentric. See § 2.51.

It is of course necessary that the satellite station should coincide with a primary station of the local ground survey, or have a trustworthy geodetic connection to one. It is further necessary that the height of such a station above the geoid should have been obtained by spirit levelling, and that the separation of the geoid and the local spheroid at the station should also be known.

If the survey based on a local datum is extensive, such as a thousand km or more, significantly different values of Δx, Δy, and Δz may be given by different satellite fixes in the area. In the present paragraph it is assumed that such differences arise from random error, and that a single mean value of Δx, etc. can consequently be accepted. See further in § 7.78.

Table 7.5 gives some values of the shifts which have been obtained for the conversion of NAD, ED, and AGD to geocentric. See also Table 7.7 for other solutions which include rotations. The most widely based entry in Table 7.5 is WGS 72 dated 1974, which to a great extent is a combination of the others. When accuracy is being assessed from the variations of the entries in Table 7.5, the following two considerations are relevant.

(*a*) The figures given are the means of the shifts at several stations within each datum, which are not necessarily the same stations in different lines of the table. Errors of ground survey are therefore contributing something to the discrepancies.

(*b*) On the other hand, many of the various satellite computations

TABLE 7.5
Three-parameter datum shifts.
Geocentric coordinates minus local datum coordinates (metres)

	NAD			ED			AGD		
	Δx	Δy	Δz	Δx	Δy	Δz	Δx	Δy	Δz
1. GSFC 1973 [402]†	−13	153	182	−85	−100	−110	−135	−42	138
2. BC-4+doppler. 1974 [516‡]	−15	166	174	−96	−93	−130	−122	−60	145
3. Doppler+BC-4. 1974 [10]	−27	161	181	−79	−105	−121	−120	−33	144
4. WGS 72. 1974 [519]	−22	157	176	−84	−103	−127	−122	−41	146
5. GRIM 2. 1976. [79]	−17	157	176	−83	−111	−124	–	–	–
6. Mean of 2 and 3.	−21	164	178	−88	−99	−126	−121	−46	144
7. Mean of 1, 5 and 6.	−17	158	179	−85	−103	−120	−129	−44	141
8. Range of 1, 5 and 6.	8	11	6	5	11	16	14	4	6

† Means from selected figures in Figs. 5 to 7.
‡ Means from selected stations.

contain a substantial amount of observations which are shared with others, and it is not easy to say by how much the scatter of the results is thereby reduced.

Of the entries in Table 7.5, the BC-4 and doppler combinations have little in common with GSFC 1973 or Grim 2, and these last two are also independent of each other. Table 7.5 then concludes with the general mean of (*a*) the mean of the two BC-4 and doppler combinations, (*b*) GSFC 1973, and (*c*) Grim 2. All means agree well with WGS 72, which is itself a combination solution.

The fact that doppler results, which are otherwise so strong, cannot by themselves give Greenwich longitudes, makes longitude a weakness. There is also room for a little doubt about scale, which is sometimes derived from ground survey, sometimes from SECOR or other radar net, and sometimes from satellite dynamics and the gravity constant *GM*, none of which (except now perhaps *GM*) is quite as perfectly known as might be desired. [11] gives a summary, reproduced in part in Table 7.6, in which the principal solutions† were brought into closer agreement with each other by the small changes of scale and longitude which were imposed on them when they were combined to form WGS 72.

In Table 7.6 the worst discrepancies are Δy for ED in WN-14, and Δz for Australia in SAO III. Not unnaturally, agreement is best in NAD,

† [11] also gives figures for the Tokyo and South American datums.

TABLE 7.6

Three-parameter datum shifts, after corrections to scale and longitude. Abstracted from [11]

	NAD			ED			AGD		
	Δx	Δy	Δz	Δx	Δy	Δz	Δx	Δy	Δz
1. WN-14	−23	154	173	−75	−146	−127	−122	−48	132
2. SAO III	−22	156	174	−85	−106	−130	−123	−47	122
3. GSFC 1973	−23	159	179	−84	−109	−116	−124	−37	141
4. NWL-10F	−21	164	178	−85	−101	−127	−122	−41	146
5. WGS-72	−22	157	176	−84	−103	−127	−122	−41	146

1. WN-14. Shift corrected as in § 7.76 (*d*), with in addition −0″.474 added to λ, and 1.03 ppm added to scale.
2. SAO III. 0.″371 added to λ, and −0.33 ppm added to scale.
3. GSFC 1973. −0.″439 added to λ, and 0.92 ppm added to scale.
4. NWL-9D. 0″.260 added to λ, and −0.82 ppm added to scale giving NWL-10F.

where observations are most numerous, and the geopotential model is probably most accurate.

[519] quotes 5 m, 10 m, and 15 m as the standard error of each component of the NAD, ED, and AGD shifts respectively. Tables 7.5 and 7.6 suggest that (in 1977) these figures may be over-estimates.

Lists of shifts for many other datums are given in [402], 19 datums; [516], 36 datums; [10], 26 datums; [432], 31 datums; [519], 9 datums; and [383], 11 datums.

7.78. Internal control of classical survey frameworks

(*a*) If the surveys based on a particular datum contain several well separated points which are fixed in the local geodetic framework, and also in a geocentric satellite frame the values of Δx, Δy, and Δz will naturally not be identical at the different points. The combination of the two systems can then be achieved in several ways, as in (i)–(v) below.

(i) If the geodetic survey is of limited extent, and if the variation of the Δx's etc. are not larger than might be expected to result from random error, in ground surveys or satelite, or both, a simple mean may be applied to the whole.

(ii) If there are at least three satellite stations suitably spaced, it is possible to solve for a single set of Δx, Δy, Δz, and for an over-all scale error for the survey (or the satellite orbits), and for three rotations: seven unknowns in all. These are known as 7-parameter solutions. Table 7.7 gives some examples. See further in (*b*) below.

(iii) Alternatively, if the extent of the survey is large, if there are many

TABLE 7.7

Seven-parameter datum shifts. Geocentric (CIO and BIH) coordinates minus local datum coordinates.
Scale correction and rotations required to correct local datum coordinates to satellite solution

	Δx (m)	Δy (m)	Δz (m)	Δ Scale (ppm)	ω''	ψ''	ϵ''
NAD GFSC 1973 [402], Table 19	−43	162	179	0.9	−1.1	−0.2	−0.0
GEM 6 [383], p. 84	−24	151	187	1.7	−0.8	0.1	−0.2
OSU 1973 [432], p. 187	−36	153	186	−0.8	−0.9	−0.2	−0.3
Doppler 1975 [291], p. 2	−24	153	174	0.6	+0.2	−0.1	+0.0
Doppler 1975 [291], p. 3	−22	166	179	1.7	+0.2	+0.1	−0.2
ED GFSC 1973	−149	−103	−92	5.0	0.6	−1.9	0.6
GEM 6	−83	−116	−120	−0.3	−0.6	0.4	0.6
OSU 1973	−112	−109	−154	7.3	1.8	0.0	0.4
AGD GFSC 1973	−137	−50	155	1.9	0.3	0.2	0.4
GEM 16	−135	−39	133	2.4	0.4	−1.2	−1.0
OSU 1973	−97	−36	119	1.2	−1.0	−1.0	0.2
				Scale	Rotation Az	Rotation E–W	Rotation N–S
NAD SAO III 1973 [216], p. 357	−31	154	176	1.8	0.1	−0.6	−0.2
ED SAO III	−85	−111	−132	2.6	0.6	−0.5	−0.2
AGD SAO III	−118	−39	120	2.3	0.2	0.8	−0.2

Rotations ω, ψ, and ϵ are positive for right-handed rotations about the z, y, and x axes respectively, viewed from the origin outward along the positive axes. In some references the signs of rotations are ambiguous, but the magnitudes quoted illustrate their possible values.

satellite stations, and if the changes in Δx, etc. progress somewhat regularly from one part of the survey to another, their changes may be exhibited by contour lines on small scale charts.† See for example charts for NAD in [519], Figs. 6–8. The changes can provisionally be ascribed to error in the ground survey, and on that hypothesis the survey's errors of scale, azimuth, and height of geoid can be deduced.

(iv) If a readjustment of the geodetic survey provides an opportunity, the satellite positions may be included in the adjustment with suitable weights (and perhaps covariances also), giving observed values of latitude, longitude, and of N, the height of the geoid above the spheroid.

† Instead of Δx, Δy, and Δz the charts may show $\Delta\phi$, $\Delta\lambda$, and ΔN as in (2.84).

(v) Astro-geodetic geoid surveys may readily be adjusted to conform to any equally or more reliable (especially doppler) values of N, either by simple interpolation or after alloting suitable weights. This should be done before an adjustment as in (iv), so that all measured distances can be best reduced to spheroid level before adjustment.

(*b*) *Seven-parameter solutions.* § 7.78 (*a*) (ii). Classical geodetic surveys have had no effective means of centering their spheroid on the geocentre. Satellites have made this possible, and three parameter solutions for Δx, Δy, and Δz constitute no conflict with the survey observations. On the other hand, rotations and corrections to scale, if more than trivial, are corrections to what purports to have been observed. It is proper to record differences between the results of satellite surveys and older methods, but rotations and scale corrections to ground surveys can only be accepted as correct, or even possible, so far as the weight of their determination by satellite may exceed the weight of the ground survey. Methods of assessing the accuracy of triangulation and traverse in scale and azimuth are described in §§ 2.27, 2.44, and 2.45, and liability to rotations in the definition of the datum in § 2.06. The accuracy of satellite observations is to some extent discussed in the description of the various methods in the earlier sections of this chapter. It is further necessary to remark that in areas extending to only 1000 or 2000 km from a central origin, it is not easy to distinguish between small changes Δx, Δy, Δz and rotations ω, ψ, ϵ, since throughout the area concerned both sets can produce very similar corrections to the coordinates of ground stations. In surveys covering a quadrant of the spheroid, or more, the two kinds of parameters are readily distinguishable, but first-class quality ground surveys of this extent do not exist†. Due account must be taken of the correlation between Δx, etc. and one or more of the rotations, when the reality of the latter is being considered. See [402], pp. 52–7 for some correlations. In the Australian solution for instance the correlation between rotation ψ and the three linear shifts is -0.988, 0.839, and 0.993 respectively. [432], pp. 189–201 gives correlations with less extreme values, although 0.7 is common and 0.9 does occur.

Comparison between Tables 7.5 and 7.7 shows that in Table 7.7 many of the linear shifts are quite abnormal, as the result of rotations being included. Seven parameters will no doubt fit the data with smaller residuals than 3 parameters will give, but they are liable to fail to locate the geocentre correctly, balancing that error by attributing non-existing rotations to the ground surveys.

† The European Datum can be extended to Vladivostock, Singapore, and Cape Town, but the density and quality of the connecting links are variable. It would be useless to obtain scale errors or rotations from one of these extensions, and to regard the result as applicable to Europe itself, or to the other two extensions. § 7.78 (*a*) (iii) would give more practical results.

7.79. Models of the geopotential

References to the zonal harmonics, and high-order resonant harmonics are in §§ 7.11–7.15.

During recent years numerous solutions have been madc for the constants in the tesseral harmonics, giving values for them up to some particular degree and order, together with a few of higher degree determined from resonant satellites. Some of these solutions are listed below.

(*a*) *SAO III.* 1973. [216] gives all constants to degree and order 18, (18×18), with a few selected resonant terms up to degree 24. It is based on the observations listed in § 7.76 (*a*). It has since been revised to produce SAO IV.I [219], 1975.

(*b*) *NASA* (*GSFC*) have madc a series of solutions known as GEM (Goddard earth models) 1 to 10, in pairs of which the odd-numbered models are based on satellite observations only, while the even numbers include observed surface free-air anomalies in their data.

GEM 5 [383], 1974, gives constants to 12×12, with some resonant terms to degree 22. It is based on camera observations to 23 satellites, including the low inclination PEOLE and SAS, covering 294 7-day arcs. Also radar, laser, and additional optical observations to 10 satellites in 68 7-day arcs, and 100 1- or 2-day arcs of Geos 1 and 2. Also doppler.

GEM 6 [383] gives constants as far as 16×16 with some resonant to degree 22, based on the same data as GEM 5 plus the BC-4 programme and surface gravity in the form of mean free-air anomalies in 1654 equal area 5°×5° squares, covering the whole earth. 1283 of these areas contained observations, and the rest were filled by interpolation based on geophysical correlations, [474].

GEM 7 and 8 [588], 1976. GEM 7 gives constants to 16×16 with some resonant to degree 29, and GEM 8 extends to 25×25 with some resonant to 30. GEM 7 is based on the same data as GEM 5, with the addition of laser data from ISAGEX, and GEM 8 is the same with rather more surface gravity than GEM 6.

GEM 9 and 10, [382]. GEM 9 is complete to 20×20, with zonals and some resonant harmonics to degree 29. GEM 10 is complete to 22×22 with zonals to 29 and some resonant to 30.

The data for GEM 9 are similar to those of GEM 7, with the addition of lasers to Geos 3, Starlette, Lageos, and BE-C, and doppler from Geos 2. GEM 10 has the same with the addition of gravity data. The latter is based on 1507 5°×5° equal-area squares containing gravity observations in 38 000 1°×1° squares, and on 147 5°×5° squares derived by interpolation.

The geocentric coordinates of 146 tracking stations are also listed in [382].

For the accuracy of the harmonic constants, see § 7.80 (*d*).

Table 7.8 lists the constants of the first five degrees of GEM 10. To illustrate the pattern of higher degree items, it also lists the contants of degrees 10 and 16.

(*c*) *NWL*-10*E and WGS* 72, 1974, [336] and [519]. These form a similar pair, with both based on the same satellite data, as listed under WGS 72 in § 7.76 (*f*). Additionally WGS 72 includes surface values of free-air anomalies in the form of means of equal area 10° × 10° squares, of which 45 per cent are based wholly on observations within the square, while figures for the rest are produced or reinforced by interpolation using geophysical correlations. Both are complete to 19 × 19, except for a few orders between degrees 11 and 19 in NWL-10E. The latter contains selected resonant terms to degree 27. Neither are yet (1977) published.

(*d*) *Grim* 2. 1976. [79] gives all constants to 30 × 30, using free-air anomalies of surface gravity as observed in 34 400 1° × 1° areas, from which 5° × 5° equatorial degree means were obtained from weighted 1° × 1° means and extrapolation. Satellite observations as in § 7.76 (g).

(*e*) *The inclusion of surface gravity.* The value of including surface gravity observations in satellite solutions for geopotential models is arguable. It undoubtedly improves the analysis in areas which are well

TABLE 7.8
Extracts from geopotential model GEM 10.
Some normalized coefficients from [382], *rounded to* 0.001

Degree	Order	$\bar{C}/10^{-6}$	$\bar{S}/10^{-6}$	Degree	Order	$\bar{C}/10^{-6}$	$\bar{S}/10^{-6}$	Degree	Order	$\bar{C}/10^{-6}$	$\bar{S}/10^{-6}$
2	0	−484.165	–	10	0	0.053	–	16	0	−0.007	–
2	2	2.434	−1.399	10	1	0.089	−0.130	16	1	0.020	0.004
3	0	0.958	–	10	2	−0.085	−0.013	16	2	−0.009	0.026
3	1	2.029	0.252	10	3	−0.019	−0.161	16	3	−0.011	−0.020
3	2	0.893	−0.623	10	4	−0.097	−0.077	16	4	0.032	0.032
3	3	0.700	1.412	10	5	−0.065	−0.032	16	5	−0.011	−0.001
4	0	0.541	–	10	6	−0.039	−0.085	16	6	−0.006	−0.029
4	1	−0.535	−0.469	10	7	0.004	0.018	16	7	−0.002	−0.008
4	2	0.352	0.664	10	8	0.043	−0.069	16	8	−0.019	0.008
4	3	0.988	−0.202	10	9	0.124	−0.049	16	9	−0.018	−0.045
4	4	−0.195	0.299	10	10	0.101	−0.022	16	10	0.009	−0.002
5	0	0.069	–					16	11	0.023	0.006
5	1	−0.051	−0.094					16	12	0.018	0.009
5	2	0.651	−0.328					16	13	0.012	−0.006
5	3	−0.467	−0.203					16	14	−0.019	−0.038
5	4	−0.288	0.050					16	15	−0.013	−0.026
5	5	0.156	−0.660					16	16	−0.026	0.008
Standard Errors											
Zonals		±0.001 to 0.002				±0.002				±0.004	
Tesserals		±0.002 to 0.012				±0.006 to 0.013				$m = 1-10 \pm 0.009-0.015$	
										$m = 11-10 \pm 0.002-0.008$	

Note $\bar{C}_{l0} = -J_l(2l+1)^{-\frac{1}{2}}$.

covered by surface gravity surveys, but having done that it imposes the resulting constants on areas in which there are no surveys. Harmonic analysis is a good means of interpolation, but it is notoriously bad for extrapolation. It would clearly be fallacious to suppose that (say) a 10-degree harmonic derived from the analysis of the world's main land areas would have any significance in the mid-Pacific. Surface observations made in one part of the earth can only be used to make significant predictions of g in an area where there is no gravity survey if the general topographical and geophysical situation in the latter matches that in particular areas which do have a gravity survey [158].

7.80. Accuracy of geopotential models

These models contain several hundred harmonic constants, $(l+1)^2$ constants for an $l \times l$ analysis†. It is not easy to assess the accuracy given by the model, nor the accuracy with which particular constants have been determined. Sub-paragraphs (*a*), (*b*), and (*c*) below consider the question from three different points of view.

(*a*) *Truncation. Kaula's rule.* The number of constants in the model is necessarily finite. It is relevant to enquire what error results from the neglect of the higher degree terms which are ignored.

As the degree l increases, the number of tesserals in each degree increases too, but the magnitude of each separate term decreases. *Kaula's rule* is an empirical rule, applicable when $2<l<15$, which gives the r.m.s. value of the normalized $\bar{C}_{lm}$ or $\bar{S}_{lm}$ (excluding zonals) in any one degree, as

$$\text{r.m.s.} = 10^{-5}/l^2 \tag{7.222}$$

For illustrations of the substantial accuracy of this rule see [216], p. 285 and [79], p. 11.

Accepting the rule, and assuming the magnitude of the $2l$ coefficients of degree l to be uncorrelated, it is possible to estimate the effect on an orbit of omitting (*truncating*) all terms, except zonal and resonant, of degree greater than some specified value of l. The following items (i)–(iv) quote various estimates. In all cases the satellite is assumed to be at a height of 1100 km. Unless otherwise stated the figures are applicable to a single pass, based on tracking facilities similar to those of the precise doppler ephemeris. In the mean of several passes the error is likely to be reduced.

(i) [13], p. 67 gives a chart showing the r.m.s. orbital position error likely to arise from the omission of all terms above $l=10$ to 20. For

† Less three since the first-degree constants are necessarily zero.

$l=12$ the r.m.s. error is 10 m, for $l=16$ it is 6 m, and for $l=20$ it is 2 m.

(ii) [13] also figures the result of a simulation of the effect of omitting the terms of degree >12 in a 24-hour arc of a satellite at 1100 km. It shows irregular sinusoidal variations with a period apparently equal to the orbital period (a little under 2 hours). The r.m.s. error is 1.1 m (maximum 2.5 m) vertically, 2.8 m (maximum 6 m) along track, and 2.7 m (maximum 7 m) across track. These figures are substantially smaller than those of (i).

(iii) [587], appendix C, considering only the vertical component gives $64/l$ metres as a reasonable upper limit for the amplitude of the sinusoidal variations arising from the truncation of all terms above degree l. This gives 5.3 m, 4.0 m, and 2.2 m for $l=12$, 16, and 20 respectively.

It may be thought that the mean of (i), (ii), and (iii) gives a significant indication of the errors arising from the truncation of the geopotential model in circumstances such as those of the doppler precise ephermeris.

(*b*) *Correlation between different independent solutions.* The mutual agreement or disagreement between any two solutions may be expressed by the differences between their two sets of $2l+1$ constants in each degree. The coefficient of correlation between two solutions 1 and 2 in their terms of degree l is

$$r_{12}=\frac{\sum(\bar{C}_1\bar{C}_2+\bar{S}_1\bar{S}_2)}{\sum[(\bar{C}_1^2+\bar{S}_1^2)\times(\bar{C}_2^2+\bar{S}_2^2)]^{\frac{1}{2}}}. \tag{7.223}$$

See (D.15) where in this context the v's are the normalized $\bar{C}$'s and $\bar{S}$'s in (7.1). Suffixes 1 and 2 refer to the two solutions. For every value of l, the summations are taken from $m=1$ to $m=l$.

If r_{12} is not exactly unity, both solutions cannot be exactly correct, but even high correlations such as 0.8 or 0.9 do not necessarily indicate that both are reliable, especially if both solutions derive from a high proportion of common data. Of the solutions listed as (*a*)–(*d*) in § 7.79, the SAO and the earlier GEM solutions have much in common, and so also do NWL and WGS 72, but these two pairs are fairly independent of each other, except for solutions using surface gravity, and GRIM 2 is independent of them both. Table 7.9, (abstracted from more numerous results in [334] and [336]), lists the coefficients for the fairly independent pairs (SAO and NWL-10E), (GEM 7 and NWL-10E), and (GEM 8 and NWL-10E). For comparison it also includes the pair (GEM 5 snd GEM 3) from [336], which have much basic data in common.

In Table 7.9 the first pair agree well with $r_{12}>0.95$ as far as $l=6$, and have definitely significant correlation with $r_{12}>0.7$ as far as $l=9$. In the next two pairs of solutions $r_{12}>0.95$ as far as $l=10$, and $r_{12}>0.7$ to $l=14$. In these three pairs high correlation may well suggest high

TABLE 7.9

Coefficients of correlation between harmonic constants of degree l in various pairs of geopotential models. From [334] *and* [336]

Degree	SAO III and NWL-10E	GEM 7 and NWL-10E	GEM 8 and NWL-10E	GEM 5 and GEM 3
2	1.0000	1.0000	1.0000	1.0000
3	0.9948	1.0000	1.0000	1.0000
4	0.9961	0.9950	0.9950	0.9999
5	0.9681	0.9993	0.9989	0.9995
6	0.9612	0.9977	0.9980	0.9990
7	0.9076	0.9895	0.9925	0.9966
8	0.6954	0.9828	0.9862	0.9943
9	0.8315	0.9305	0.9341	0.9909
10	0.5360	0.9585	0.9580	0.9796
11	0.5107	0.7661	0.8504	0.9653
12	0.3200	0.7403	0.7597	0.9046
13	0.4778	0.7844	0.8213	0.9498
14	0.4223	0.7145	0.7413	0.9424
15	0.2953	0.3469	0.4701	0.9419
16	0.3008	0.4089	0.4810	0.9561
17	0.4262	0.4397	0.3747	0.9105
18	0.1933	0.4116	0.2968	0.9509
19	0.2872	0.2028	0.3729	0.7493

accuracy. The last pair GEM 5 and 3, which have much common data, agree with other almost perfectly with $r > 0.94$ as far as $l = 16$, and with $r_{12} = 0.75$ for $l = 19$ as the only figure where r_{12} is less than 0.9. In this case the high correlation is no proof of high accuracy†.

(*c*) *Observed errors of the ephemeris.* A third approach is to note the improvement in the accuracy of an ephemeris (as judged by the consistancy of fixations made from it), when an improved geopotential model is brought into use. [92], p. 9 shows the r.m.s. of the NNSS broadcast ephemeris being reduced from 15 to 8 m when an older 14×14 model was superseded by the 19×19 WGS 72 model.

(*d*) [382] pp. 52–3 gives estimates of the s.e.'s of the normalized harmonic constants of GEM 9 and 10. Excluding the zonal harmonics (See Table 7.2), these estimates for GEM 10 are

(i) Between ±5 and 12×10^{-9} for $l < 7$.
(ii) Between ±8 and 17×10^{-9} for l between 7 and 22, and $m < 12$.
(iii) Between ±1 and 8×10^{-9} for $l = 12$ to 22, and $m = 12$ to 15.
(iv) Between ±4 and 15×10^{-9} for $l < 22$, and $m > 15$.

† Equally of course it is no indication of lack of accuracy. But the warning is worth noting.

(*e*) *Summary*. From the point of view of computing satellite orbits it may be said, as a general conclusion, that in comparison with other sources of error the best current (1977) solutions for the tesseral harmonics of the geopotential are accurate as far as $l<7$, and are probably good enough for $l<10$. As far as $l=14$ the larger coefficients are significantly determined, but not very accurately except for some resonant terms. Beyond $l=14$ most of the non-resonant coefficients are individually smaller than their estimated standard errors, although the figures ascribed to them may well have a beneficient collective effect on the prediction of orbits.

General references for Section 10. [18], [273], [334], [336], [401], [519], [587], and [588]. Of these references [273] includes a great deal of what is to be found in the others.

APPENDIX A

THE GEOMETRY OF THE SPHEROID

FOR notation in §§ A.02–A.04 see § 2.00. The word 'spheroid' is used to mean the surface generated by an ellipse rotated about its minor axis. Formulae are given without proof. They can either be found in elementary textbooks, or can be derived from standard formulae by expansion in power series.

A.00. Expansions

Geodetic formulae can generally be expanded in rapidly convergent series, either in powers of e^2 or of ϵ ($\approx 1/150$) or of L/R, which is generally less than 1/50. The following common expansions are given for reference.

$$f(x+h) = f(x) + hf'(x) + (h^2/2!)f''(x) + \ldots \quad \text{(Taylor's theorem).} \tag{A.1}$$

$$f(x) = f(0) + xf'(0) + x^2/2!)f''(0) + \ldots \quad \text{(Maclaurin's theorem).} \tag{A.2}$$

$$\sin\theta = \theta - \theta^3/3! + \theta^5/5! - \ldots . \tag{A.3}$$

$$\cos\theta = 1 - \theta^2/2! + \theta^4/4! - \ldots . \tag{A.4}$$

$$\tan\theta = \theta + \theta^3/3 + 2\theta^5/15 + 17\theta^7/315 + \ldots . \tag{A.5}$$

$$\sin^{-1}x = x + \frac{1}{2}\frac{x^3}{3} + \frac{1\cdot 3}{2\cdot 4}\frac{x^5}{5} + \frac{1\cdot 3\cdot 5}{2\cdot 4\cdot 6}\frac{x^7}{7} + \ldots . \tag{A.6}$$

$$\tan^{-1}x = x - x^3/3 + x^5/5 - x^7/7 + \ldots . \tag{A.7}$$

$$\sin^4\theta = (3 - 4\cos 2\theta + \cos 4\theta)/8. \tag{A.8}$$

$$\sin^6\theta = (10 - 15\cos 2\theta + 6\cos 4\theta - \cos 6\theta)/32. \tag{A.9}$$

$$e^x = \exp(x) = 1 + x + x^2/2! + x^3/3! + \ldots \tag{A.10}$$

$$\exp(\log_e x) = x \tag{A.11}$$

If $\cos x' - \cos x$ or $\sin(90° - x') - \sin(90° - x) = h$, which is small,

$$x' - x = -h \operatorname{cosec} x(1 + \tfrac{1}{2}h\cos x \operatorname{cosec}^2 x + \ldots). \tag{A.12}$$

If $\cot x' - \cot x$ or $\tan(90° - x') - \tan(90° - x) = h$,

$$x' - x = -h\sin^2 x(1 - \tfrac{1}{2}h\sin 2x + \ldots). \tag{A.13}$$

If $\tan x' = h\tan x$, h being near unity,

$$x' - x = \left(\frac{h-1}{h+1}\right)\sin 2x + \frac{1}{2}\left(\frac{h-1}{h+1}\right)^2\sin 4x + \ldots . \tag{A.14}$$

$$\ln(1+x) = x - \tfrac{1}{2}x^2 + \tfrac{1}{3}x^3 - \ldots, \tag{A.15}$$

$$(1+x)^n = 1 + nx + n(n-1)x^2/2! + n(n-1)(n-2)x^3/3! + \ldots . \tag{A.16}$$

A change of one unit in the seventh decimal of a common log (base 10) changes the anti-log by one part in 4 343 000. A change of 1 ppm in a number changes its common log by 4 · 343 in the seventh decimal.

When x and y are small the following approximations follow from (A.16).

$$1/(1+x)=1-x+\ldots,\quad 1/(1+x)^2=1-2x+\ldots,\ (1+x)(1+y)=1+x+y+\ldots. \tag{A.17}$$

If x_1 is small and x_2 very small

$$(1+x_1+x_2)^n=1+nx_1+nx_2+\tfrac{1}{2}n(n-1)x_1^2+\ldots. \tag{A.18}$$

If $y=x+a_2x^2+a_3x^3+\ldots$, then $x=y+b_2y^2+b_3.y^3+\ldots$,

$$\left.\begin{aligned}\text{where } b_2&=-a_2, \qquad b_3=-2a_2b_2-a_3=2a_2^2-a_3,\\ b_4&=-a_2(2b_3+b_2^2)-3a_3b_2-a_4\\ &=-5a_2^3+5a_2a_3-a_4.\end{aligned}\right\} \tag{A.19}$$

Numerical integration.

$$\text{Simpson's rule}\quad \int_{x_1}^{x_1+2h} f(x)\,dx=\frac{h}{3}(f_0+4f_1+f_2) \qquad \text{A.20}$$

$$\text{Gregory's formula}\quad \int_{x_1}^{x_1+4h} (f(x)\,dx=\frac{h}{45}\{14(f_0+f_4)+64(f_1+f_3)+24f_2\} \tag{A.21}$$

A.01. Spherical trigonometry

The following formulae are given for reference. Proofs are in [533]. They refer to a spherical triangle ABC with sides a, b, c.

$$\sin A/\sin a=\sin B/\sin b=\sin C/\sin c. \tag{A.22}$$

$$\text{Spherical excess } E=\text{Area of triangle} \div R^2 \tag{A.23}$$

$$\tan\tfrac{1}{4}E=\{\tan\tfrac{1}{2}s\,\tan\tfrac{1}{2}(s-a)\tan\tfrac{1}{2}(s-b)\tan\tfrac{1}{2}(s-c)\}^{\frac{1}{2}}, \tag{A.24}$$

where $2s=a+b+c$

$$\sin A/2=\sqrt{\left\{\frac{\sin(s-b)\sin(s-c)}{\sin b\,\sin c}\right\}}. \tag{A.25}$$

$$\cos A/2=\sqrt{\left\{\frac{\sin s\,\sin(s-a)}{\sin b\,\sin c}\right\}}. \tag{A.26}$$

$$\tan\tfrac{1}{2}(A-B)=\frac{\sin\tfrac{1}{2}(a-b)}{\sin\tfrac{1}{2}(a+b)}\cot\frac{C}{2}. \tag{A.27}$$

$$\tan\tfrac{1}{2}(A+B)=\frac{\cos\tfrac{1}{2}(a-b)}{\cos\tfrac{1}{2}(a+b)}\cot\frac{C}{2}. \tag{A.28}$$

$$\cos a=\cos b\,\cos c+\sin b\,\sin c\,\cos A. \tag{A.29}$$

$$\sin a\,\cos B=\cos b\,\sin c-\sin b\,\cos c\,\cos A. \tag{A.30}$$

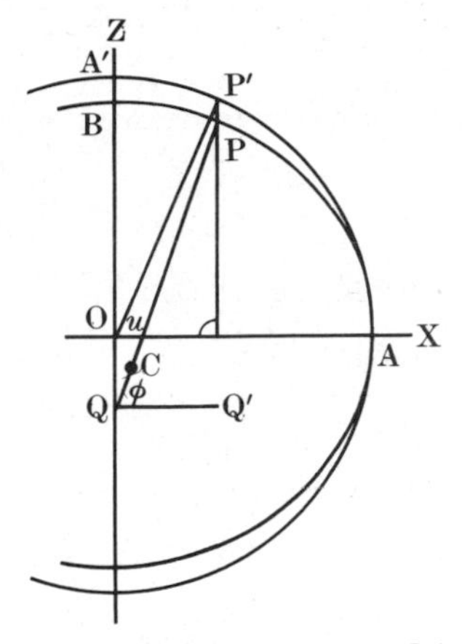

FIG. A.1. The meridional ellipse. OA = a. PQ = ν. OQ = $e^2\nu \sin \phi$. PC = ρ. OB = b.

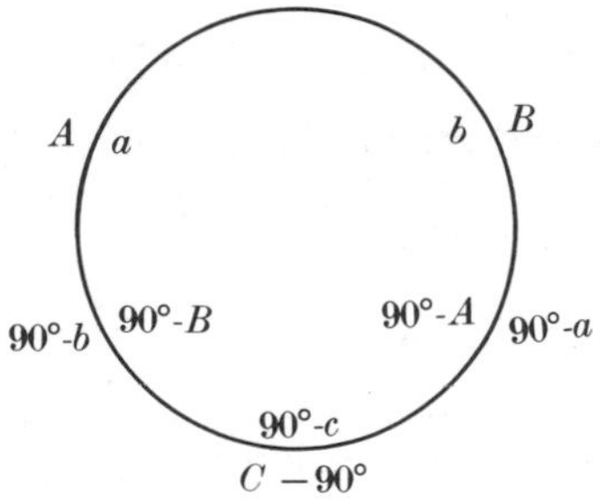

A a b B
90°-b 90°-B 90°-A 90°-a
90°-c
C − 90°

FIG. A.2. Napier's rules for right-angled and quadrantal triangles. Inside the circle $C = 90°$, outside it $c = 90°$.

In all these formulae A, B, and C may be interchanged, provided a, b, and c change with them. Further, excpet in (A.24) for the spherical excess, the formulae hold if $180° - a$ is written for A, $180° - b$ for B, etc. and $180° - A$ for a, etc. Thus (A.29) gives

$$-\cos A = \cos B \cos C - \sin B \sin C \cos a. \tag{A.31}$$

The '4-parts' formulae, [492], p. 5–3. Consider the sides and angles of any spherical triangle in order round the triangle, $AcBaCbAcB, \ldots$, and select any group of four, clockwise or anti-clockwise. Then, in that group,

$$\text{Cos(inner side)cos(inner angle)}$$
$$= \text{sin(inner side)cot(outer side)} - \text{sin(inner angle)cot(outer angle)}, \tag{A.32}$$

Taking $cAbC$ for example

$$\cos b \cos A = \sin b \cot c - \sin A \cot C. \tag{A.33}$$

Right-angled triangles. In the inner ring of Fig. A.2 $C = 90°$ Then taking any one of the five items in the figure as the 'middle part'

$$\text{sin(middle)} = \text{product of the tangents of the adjacent parts,}$$
$$= \text{product of cosines of opposite parts.} \tag{A.34}$$

Quadrantal triangles in which $c = 90°$. The same two rules hold, using the outer ring of Fig. A.2.

A.02. The spheroid and meridional ellipse

In Fig. A.1, AB is a quadrant of the meridional ellipse, AA′ is a circle, OBZ is the polar axis, and PQ is the normal at P.

The ellipse is $x^2/a^2 + z^2/b^2 = 1$. (A.35)

The spheroid is $x^2/a^2+y^2/a^2+z^2/b^2=1$. (A.36)

The flattening of the spheroid can be described by various parameters f, e, e', and n,

The flattening $f=(a-b)/a$. $b=a(1-f)$. (A.37)

The eccentricity e. $e^2=(a^2-b^2)/a^2$. $(1-e^2)=b^2/a^2$. (A.38)

$$e^2=2f-f^2, \text{ exactly.} \tag{A.39}$$

$$(1-e^2)^{\frac{1}{2}}=(1-f)=b/a. \tag{A.40}$$

$$f=e^2/2+e^4/8+e^6/16+\ldots. \tag{A.41}$$

The second eccentricity e'. $e'^2=\epsilon=(a^2-b^2)/b^2=e^2/(1-e^2)$. (A.42)

$$(1+\epsilon)^{\frac{1}{2}}=a/b. \tag{A.43}$$

$$1+\epsilon=1/(1-e^2). \tag{A.44}$$

$$\epsilon=e^2+e^4+e^6+\ldots\approx 2f+\ldots. \tag{A.45}$$

$$e^2=\epsilon/(1+\epsilon) \tag{A.46}$$

n, defined to be $(a-b)/(a+b)=f/(2-f)=\frac{1}{2}f+\frac{1}{4}f^2+\frac{1}{8}f^3+\ldots$

$$=e^2/4+e^4/8+5e^6/64+\ldots \tag{A.47}$$

The latus rectum $2p=2a(1-e^2)$. (A.48)

The polar equation of the ellipse is $r=p(1+e\cos\theta)^{-1}$, (A.49)
where r is the distance from one focus.

The distance from the focus to the centre $=ae$. (A.50)

The position of a point P on the meridional ellipse may be defined by various 'latitudes' ϕ, u, ϕ', and ψ.

$\phi=$ PQQ′ in Fig. A.1, is the geographical (goedetic) latitude.

$$\tan\phi=-\mathrm{d}x/\mathrm{d}z=(z/x)(1-e^2)^{-1} \tag{A.51}$$

$u=$ P′OA $=$ the reduced or parametric latitude or eccentric angle.

$\tan u=(1-e^2)^{\frac{1}{2}}\tan\phi=(b/a)\tan\phi=(1-f)\tan\phi$. And $a\cos u=\nu\cos\phi$. (A.52)

Note that this u is not the same as the u of satellite orbits in Fig. 7.1.

$$\left.\begin{aligned} u&=\phi-n\sin 2\phi+(n^2/2)\sin 4\phi-(n^3/3)\sin 6\phi+\ldots\text{ rad} \quad (A.53)\\ \phi&=u+n\sin 2u+(n^2/2)\sin 4u+(n^3/3)\sin 6u+\ldots\text{ rad} \quad (A.54)\end{aligned}\right\} n \text{ as in (A.47)}$$

$\phi'=$ POA $=$ geocentric latitude $=(90°-\theta)$. $\tan\phi'=z/x=(b/a)\tan u$

$$=(1-e^2)\tan\phi. \tag{A.55}$$

$$\phi'=\phi-n'\sin 2\phi+\tfrac{1}{2}n'^2\sin 4\phi-\ldots,\text{ rad}$$
$$\text{where } n'=e^2/(2-e^2)=f+\tfrac{1}{2}f^2-\tfrac{1}{4}f^4\ldots. \tag{A.56}$$

$\psi=$ the isometric latitude. See (2.109).

The differences between geodetic, reduced, geocentric, and various other 'latitudes' are tabulated at intervals of 30′ in [2].†

In the meridional ellipse

$$x = \nu \cos\phi = a\cos u,$$

$$z = (1-e^2)\nu\sin\phi = b\sin u, \text{ where } \nu = \mathrm{PQ}. \tag{A.57}$$

$$r = a(1 - f\sin^2\phi' - \tfrac{3}{8}f^2\sin^2 2\phi' + \ldots)$$
$$= a(1 - f\sin^2\phi + \tfrac{5}{8}f^2\sin^2 2\phi + \ldots\ \ ) \tag{A.58}$$

In the spheroid

$$x = \nu\cos\phi\cos\lambda,$$

$$y = \nu\cos\phi\sin\lambda,$$

$$z = (1-e^2)\nu\sin\phi. \tag{A.59}$$

In Fig. A.1 $\mathrm{OQ} = e^2\nu\sin\phi$ (exactly) $= \epsilon R\sin\phi + \ldots R(\epsilon^2)$. (A.60)

A.03. Radii of curvature

$$\rho = a(1-e^2)/(1-e^2\sin^2\phi)^{\frac{3}{2}} \tag{A.61}$$
$$= a(1-e^2)(1+\tfrac{3}{2}e^2\sin^2\phi + \tfrac{15}{8}e^4\sin^4\phi + \tfrac{35}{16}e^6\sin^6\phi + \ldots). \tag{A.62}$$

$$\nu = \mathrm{PQ} = a/(1-e^2\sin^2\phi)^{\frac{1}{2}} = a(1+\tfrac{1}{2}e^2\sin^2\phi + \tfrac{3}{8}e^4\sin^4\phi + \tfrac{5}{16}e^6\sin^6\phi + \ldots). \tag{A.63}$$

Radius of a parallel of latitude $= \nu\cos\phi$ (A.64)

K, defined to be $1/\rho\nu = (1-e^2\sin^2\phi)^2/a^2(1-e^2)$

$$= (1+\epsilon-2\epsilon\sin^2\phi)/a^2 + \ldots(\epsilon^2)/R^2$$
$$= (1+2f\cos 2\phi)/a^2 + \ldots(f^2)/R^2. \tag{A.65}$$

$$\nu/\rho = 1+\epsilon\cos^2\phi, \text{ exactly.} \tag{A.66}$$

Radius of curvature in azimuth A is given by

$$1/R_\alpha = (1/\nu)(1+\epsilon\cos^2\phi\cos^2 A)$$
$$= (\cos^2 A)/\rho + (\sin^2 A)/\nu = (\nu\cos^2 A + \rho\sin^2 A)/\rho\nu, \tag{A.67}$$

(Euler's theorem). [534], p. 214.

From (A.62), (A.8), and (A.9),

$$m = \text{meridional arc AP} = \int_0^\phi \rho\, d\phi = a(A_0\phi - A_2\sin 2\phi + A_4\sin 4\phi - \ldots), \tag{A.68‡}$$

† Note that 'isometric' latitude in [2] is not the same as in Chapter 2, Section 8.

‡ An approximate formula for the arc between ϕ_2 and ϕ_1, which is good enough for many purposes, is $(\phi_2-\phi_1)\rho_m + \frac{1}{8}\epsilon\rho_m(\phi_2-\phi_1)^3\cos(\phi_2+\phi_1) + \ldots$. ϕ's in radians. If $\phi_2-\phi_1 = 1°$ the second term is less than 0.03 m. [155], p. 112.

where

$$A_0 = 1 - \tfrac{1}{4}e^2 - \tfrac{3}{64}e^4 + \tfrac{5}{256}e^6 - \dots,$$

$$A_2 = \tfrac{3}{8}(e^2 + \tfrac{1}{4}e^4 + \tfrac{15}{128}e^6 + \dots),$$

$$A_4 = \tfrac{15}{256}(e^4 + \tfrac{3}{4}e^6 + \dots),$$

$$A_6 = \tfrac{35}{3072}e^6 + \dots,$$

$$\frac{d\rho}{d\phi} = \tfrac{3}{2}R\epsilon \sin 2\phi + \dots + R(\epsilon^2), \tag{A.69}$$

$$\frac{d\nu}{d\phi} = \tfrac{1}{2}R\epsilon \sin 2\phi + \dots + R(\epsilon^2), \tag{A.70}$$

$$\frac{dK}{d\phi} = -\frac{2\epsilon}{R^2}\sin 2\phi + \dots + (\epsilon^2)/R^2, \tag{A.71}$$

$$\frac{1}{\nu} = \frac{1}{\sqrt{(\rho\nu)}}(1 - \tfrac{1}{2}\epsilon \cos^2\phi) + \dots + \epsilon^2/R. \tag{A.72}$$

ρ and ν are the two principal radii of curvature of the spheroid at any point. In Fig. A.1 PQ $= \nu$, which is therefore known as the *normal terminated by the minor axis*. Q lies on the polar axis on the side of the equator remote from P. Except at the poles, where $\rho = \nu$, ρ is always less than ν, so that C the centre of curvature of the meridian lies on PQ, between Q and P.

At the pole ρ and ν have their maximum values $a/\sqrt{(1-e^2)} \approx a(1+f)$, and are equal. In about lat 55° $\rho = a$, and in about lat 35° $\rho = b$. On the equator $\nu = a$ and $\rho = a(1-e^2) \approx a(1-2f)$. The radius of a sphere of equal volume $= (a^2b)^{\frac{1}{3}} = a(1 - \frac{1}{3}f - \frac{1}{9}f^2 \dots)$.

Most large survey departments publish tables of ρ and ν for the spheroid used. Tables for the International spheroid are in [370] and [460]. Many others are in [22]. For the 1967 reference system, tables are in [423]. An electronic computer will compute ρ and ν as required in preference to referring to a table.

A.04. The geodesic

Defined as the shortest line on the spheroid, joining two points on it, the geodesic has two other properties.

(*a*) It is not a plane curve, but the plane which contains any three near points on it also contains the normal to the spheroid at the centre point of the three.

(*b*) Along any geodesic $\nu \cos\phi \sin\alpha$ is constant.†

Proof. From [296], No. 64, pp. 53–4. (*a*) Let a smooth, flexible, weightless string be stretched under tension between P_1 and P_2. It will clearly follow the geodesic or shortest line. Consider the forces acting on an element of its length. They are (i) the tensions at the two ends of the element, equal, but inclined to each other at a little under 180°, and both lying in the plane which locally contains the curved string; and (ii) the reaction of the smooth spheroid, which may be

† Note that this condition is also satisfied along a parallel of latitude, although it is not a geodesic.

considered to act along the normal at the centre point of the element. Then for equilibrium these forces must be co-planar.

Again consider the forces on any section CD of the string, not necessarily short, namely the end tensions and all the normal reactions, and take moments about the axis. The moments of the normal forces will be zero, since all normals intersect the minor axis, and so

$$T \times \nu_C \cos \phi_C \sin \alpha_C = T \times \nu_D \cos \phi_D \sin \alpha_D, \text{ i.e. } \nu \cos \phi \sin \alpha \text{ is constant.} \tag{A.73}$$

At any point on a geodesic

$$\begin{aligned} \mathrm{d}\phi/\mathrm{d}s &= (1/\rho)\cos \alpha \\ \mathrm{d}\lambda/\mathrm{d}s &= (1/\nu)\sec \phi \sin \alpha, \\ \mathrm{d}\alpha/\mathrm{d}s &= (1/\nu)\tan \phi \sin \alpha, \end{aligned} \tag{A.74}$$

where α is the azimuth of the geodesic at the point, and $\mathrm{d}s$ is an element of distance along it.

For the spheroidal geodesic triangle see § 2.12. Proofs of the formulae there quoted are in [563] and [113], 2 Edn., pp. 499–504. Textbooks such as [270], [620] and [315] give the geometry of the spheroid thorough treatment.

References. Tables of less usual mathematical functions, [174] and [414]. Lists of integrals, [414] and [413]. Mathematical and physical definitions and formulae, [413].

APPENDIX B

MATRIX ALGEBRA

B.00. Definitions

(*a*) *Scalars.* The ordinary numbers of arithmetic and elementary algebra are known as scalars. They can be represented by the positions of points on a scale. They include integers, fractions and decimals, positive, negative, and zero, real roots, and quantities such as π, e, and $\sin\theta$. They exclude (i) the square roots of negative numbers, (ii) vectors, which have a direction as well as a scalar magnitude, and (iii) the positions of points in a plane, or in space, which can only be described by two or three scalars.

Scalars can usefully be combined by addition, subtraction, multiplication, and division. Their manipulation is much facilitated by the following rules.

(i) Addition is *commutative.* $a+b=b+a$.
(ii) Addition is *associative.* $a+(b+c)=(a+b)+c$.
(iii) Multiplication is commutative. $a\times b=b\times a$.
(iv) Multiplication is associative. $(a\times b)\times c=a\times(b\times c)$.
(v) Multiplication is *distributive* with respect to addition.

$$a\times(b+c)=(a\times b)+(a\times c).$$

These rules do not apply to all processes. For instance, $a^b\neq b^a$. That they do apply to the addition and multiplication of scalars is very convenient.

Matrix algebra deals with the combination of blocks of scalars, by processes described as 'addition' and 'multiplication', whose utility is dependent on the extent to which these same rules can be applied.

(*b*) *Matrices.* A matrix consists of a number of scalars† arranged in one or more rows and columns. Each such scalar is known as an *element* of the matrix. Every row must contain the same number of elements. Some may be zero, but there may be no empty spaces. The elements of a matrix A may be written a_{ij}, indicating the element in row i and column j. Note 'Row before column' here and elsewhere. Thus A may be

$$\begin{bmatrix} a_{11} & a_{12} & a_{13} \\ a_{21} & a_{22} & a_{23} \end{bmatrix} \tag{B.1}$$

A may also be written $[a_{ij}]$, to show that its elements are a_{11}, etc.

The *order* of a matrix is the number of its rows and columns, rows before columns. Thus (B.1) is of order 2×3. If the number of rows and columns is equal, the matrix is *square.* A 1×1 matrix is a single scalar.

If the elements of a square matrix are symmetric about the *principal diagonal* (top left to bottom right), so that $a_{ij}=a_{ji}$, the matrix is *symmetric.*

(*c*) *Equality.* Two matrices are defined to be equal if they have the same

† Matrices with complex elements are sometimes considered, but not here.

number of rows and columns, and if their corresponding elements are all equal.

$$A = B \text{ and } B = A, \text{ if } a_{ij} = b_{ij}. \tag{B.2}$$

And conversely, if $A = B$ corresponding elements are equal.

(*d*) The *transpose* of an $m \times n$ matrix A is an $n \times m$ matrix A^T, such that each column of A^T is the same as the corresponding row of A. So $a_{ij}^T = a_{ji}$.

If
$$A = \begin{bmatrix} 2 & 3 & 4 \\ -1 & 0 & 3 \end{bmatrix}, A^T = \begin{bmatrix} 2 & -1 \\ 3 & 0 \\ 4 & 3 \end{bmatrix}.$$

The transpose of A^T, $(A^T)^T$ is A. Many authors write A', or $\widetilde{A}$, for A^T.

If A is symmetric, $A = A^T$.

(*e*) *Utility*. Any page of figures arranged in rows and columns is a matrix, but some are more usefully so described than others. The following are examples of useful matrices.

(i) The coefficients of the unknowns in the left-hand side of a set of linear simultaneous equations, § B.05.

(ii) The direction cosines of a set of rectangular axes X, Y, Z with reference to another set x, y, z, § B.08.

(iii) A tabular list of the variances and covariances of the expected errors of latitude and longitude of a number of fixed points, § D.15.

For the use of matrices in the method of least squares see §§ D.14–D.17.

The object of matrix algebra is to combine and minipulate matrices, while so far as possible treating them as entities, and ignoring the numerical values of their elements. When a problem has been reduced to its simplest form in this way, the elements of the resulting matrices may be given to an electronic computer to do the necessary arithmetic.

B.01. Addition of matrices

(*a*) Two matrices can be added if they are of the same order. If $C = A + B$, each element of C is defined to be the sum of the corresponding elements of A and B.

$$c_{ij} = a_{ij} + b_{ij}. \tag{B.3}$$

It follows that addition is commutative and associative as defined in § B.00 (*a*).

(*b*) *Multiplication by a scalar*. The product of a scalar k and a matrix A, written kA, is defined to be the matrix $[ka_{ij}]$, whose elements are ka_{ij}.

It follows that $A + A + \ldots . A$, k times, $= kA$ and that $k_1A + k_2A = (k_1 + k_2)A$. (B.4)

Also $k_1A + k_2B$, A and B being of the same order, is a matrix whose elements are $(k_1a_{ij} + k_2b_{ij})$.

(*c*) *Zero matrix*. If $A + B = A$, every element of B must be zero. Write $B = 0$, Note that 0 does not necessarily equal 0. There is a zero matrix of every order.

(*d*) *Subtraction.* If $A+B=0$, every element of B, b_{ij}, equals $-a_{ij}$. Write $B=-A$ and $A=-B$. The difference of two matrices B subtracted from A, is defined to be $A+(-B)$, written $A-B$, and its elements are $a_{ij}-b_{ij}$.

(*e*) Note that the matrix addition of two 1×1 matrices is the same as the ordinary addition of scalars.

B.02. Multiplication of matrices

(*a*) *Definition.* A matrix B of order $m_2\times n_2$ can be multiplied by another A of order $m_1\times n_1$ provided $n_1=m_2=$ (say) l. They are then described as conformable for multiplication in the order $A\times B$. Then the product $A\times B$ or $AB=$ (say) C is defined to be a matrix with as many rows as A and as many columns as B, and each element of C is defined to be

$$c_{ij}=a_{i1}b_{1j}+a_{i2}b_{2j}+\ldots+a_{il}b_{lj}. \tag{B.5}$$

$$\underset{m_1}{\overset{n_1=l}{[A]}}\times\underset{m_2=l}{\overset{n_2}{[B]}}=\underset{m_1}{\overset{n_2}{[C]}}$$

So every element of c_{ij} of C is the sum of l products, formed by multiplying successive elements of row i of A by the corresponding elements of column j of B. To form the product of two large matrices is a considerable piece of arithmetic, but one to which computers are fortunately well suited. It is not at all easy to visualize the pattern of the elements of a matrix product, as is often necessary in simple cases. As, for instance, when a number of elements in A and B are zero, and it is required to see how many in AB are zero too. (B.5) enables a particular element to be picked out quite readily, but the full product of two small matrices is more easily written out by the methods described in § B.05 (*b*) or § B.02 (g).

To the question 'Why adopt such an extraordinary definition for the product of two matrices?' the answer is 'Because it is useful'.

In (B.5) B is said to be *pre-multiplied* by A to give AB, *while A is post-multiplied by B.* If AB is possible, it does not follow that BA is possible. In fact, both will only be possible if $n_1=m_2$ and $m_1=n_2$. Any matrix can be both pre-multiplied and post-multiplied by its transpose.

(*b*) *Multiplication is not commutative.* It is easy to see from the definition that $(A\times B)\times C=A\times(B\times C)$, and that $A\times(B+C)=(A\times B)+(A\times C)$. Matrix multiplication is associative, and distributive with respect to addition.

On the other hand, AB does not in general equal BA, even if both are possible, and in general multiplication is not commutative. This limits the possibility of simplifying complicated expressions involving matrices. We cannot necessarily say $AB-BA=0$. Nor, if $AB=1$, can we say $ACB=C$. On the other hand, there are many common combinations of particular matrices which do commute, and the manipulation of matrix expressions depends on their recognition. They are listed in § B.06 (*a*).

If A is square, the product $A\times A$ is always possible and is written A^2. Similarly for A^3 and other positive integral powers. So $A^m\times A^n=A^{m+n}$.

(*c*) *Unit matrix*. The square matrix

$$\begin{bmatrix} 1 & 0 & . & . & . & 0 \\ 0 & 1 & . & . & . & 0 \\ . & & & & & \\ . & & & & & \\ . & & & & & \\ 0 & 0 & . & . & . & 1 \end{bmatrix} \equiv I \tag{B.6}$$

is known as the *unit matrix*, of order m if it has m rows. The pre- or post-multiplication of any matrix A by a unit matrix of appropriate order leaves A unchanged.

$$I_m A_{mn} = A_{mn}, \quad \text{and} \quad A_{mn} I_n = A_{mn}, \tag{B.7}$$

where the suffixes indicate orders. And conversely if BA or $AB = A$, $B = I$.

If A is square, I is of the same order for both pre- and post-multiplication, and

$$AI = IA. \tag{B.8}$$

The product $I \times I$ (both being of the same order) $= I^2 = I$, (B.9)

(*d*) *Scalar matrices*. The sum $I + I + \ldots I$ (k times, all of the same order) equals

$$\begin{bmatrix} k & 0 & . & . & . & 0 \\ 0 & k & . & . & . & 0 \\ . & & & & & \\ . & & & & & \\ . & & & & & \\ 0 & 0 & . & . & . & k \end{bmatrix} = kI, \tag{B.10}$$

as in § B.01 (*b*), and pre-multiplication or post-multiplication of A by kI, of the right order, is the same as multiplication by a scalar k. The matrix (B.10) is therefore known as a *scalar matrix*. We have

$$kIA = kA$$

and, if A is square,

$$kIA = AkI, \quad \text{or} \quad kA = Ak. \tag{B.11}$$

If A and B are 1×1 matrices, $AB = a_{11} b_{11} = BA$. See also § B.01 (*e*) for addition. Matrices of order 1×1 can be defined as equal to their single scalar element.

(*e*) *Multiplication by zero*. It is clear that $0 \times A = 0$ and $A \times 0 = 0$, but if $AB = 0$ it only follows that A or $B = 0$ if the other is non-singular, § B.05 (*d*).

(*f*) *Square roots*. If $B \times B = A$, B is the square root of A. Except for diagonal matrices, § B.07 (*a*), the square roots of matrices are best avoided. They are neither unique, nor necessarily finite in number. See [210], p. 38 for an example.

(*g*) *The cracovian product* of A and B, written $A \cdot B$ is defined to be $A^T B$. Column-by-column multiplication is easier to do mentally than row-by-column as defined in (*a*), and if the product of two matrices has to be written down by hand, it may be found helpful to transpose the first member and form the cracovian product $A^T \cdot B$, which equals AB. When the intention is that matrix products should always be cracovian, the matrices themselves are termed *cracovians*. This system is popular in eastern Europe.

(*h*) *The inverse, or reciprocal, of a matrix.* Let A^{-1} be defined to be a matrix such that

$$AA^{-1} = I. \tag{B.12}$$

It is not obvious that A^{-1} exists, but in fact it does, provided A is square and not *singular*, i.e. *provided the determinant of A is not zero.* See § B.05 (*d*).

A^{-1} is clearly of the same order as A. Pre-multiplying (B.12) by A^{-1} gives $A^{-1}(AA^{-1}) = A^{-1}I = A^{-1}$, whence from (B.7)

$$A^{-1}A = I = AA^{-1}. \tag{B.13}$$

Also

$$(AB)^{-1} = B^{-1}A^{-1}. \tag{B.14}$$

Proof. $(AB) \times (B^{-1}A^{-1}) = A(BB^{-1})A^{-1}$ (associative) $= AIA^{-1} = AA^{-1} = I$.

Multiplication by the inverse is the equivalent of division. $A^{-1}B$ is the result of dividing B by A. Note that it is only possible if A is square, non-singular, and conformable with B.

$$A^{-1}A^m = A^mA^{-1} = A^{m-1}.$$

If A is symmetrical, A^{-1} is symmetrical also.

(*i*) The trace of a square matrix is the algebraic sum of its diagonal elements.

B.03. Vectors

(*a*) *Definitions.* A matrix consisting of one column or one row is known as a *vector*. Column vectors are more common than row vectors, and are written in lower-case italic type as (e.g.) x.†

$$x = \begin{bmatrix} x_1 \\ x_2 \\ \cdot \\ \cdot \\ x_m \end{bmatrix}. \tag{B.15}$$

The transpose of a column vector is a row, so row vectors of x or c can be written as x^T or c^T. Column vectors take up much printing space, so (B.15) is conveniently printed‡ as $x = [x_1, x_2, \ldots, x_m]^T$. The elements of a vector are often called its *components*.

(*b*) *Products.* If a and b are two column vectors of the same order, the product a^Tb is a (1×1) matrix, namely a scalar.

$$a^Tb = a_1b_1 + a_2b_2 + \ldots + a_mb_m. \tag{B.16}$$

It is known as the *scalar product* of the two vectors. Compare § E.07. Note that

$$a^Tb = b^Ta, \tag{B.17}$$

† Compare the vector notation, using bold type, in § E.01. Bold type is not necessarily used for matrix vectors, although it is so used in Chapter 7. Lower case type is not used to distinguish matrix vectors such that $\sum x_i^2 = 1$.

‡ Some authors write a column vector as $\{x_1, x_2, \ldots, x_m\}$.

but a^Tb does not equal ba^T, which is an $m \times m$ matrix, whose elements comprise all combinations of a_i and b_i.

In particular $$x^Tx = x_1^2 + x_2^2 + \ldots + x_m^2, \tag{B.18}$$

So, in geometry x^Tx is the square of the distance from the origin to the point (x_1, x_2, x_3).

If $x^Tx = 1$, x is known as a *unit vector*, or as a *normalized vector*. The direction cosines l, m, n of a line are an example of a unit vector. The scalar product of two such vectors, $l_1l_2 + m_1m_2 + n_1n_2$ is the cosine of the angle between them.

B.04. Determinants

(*a*) *Definitions.* To every square matrix A there corresponds a scalar known as its determinant, written $|A|$ or $\det A$.

In full, $$|A| = \begin{vmatrix} a_{11} & a_{12} & \cdot & \cdot & \cdot & a_{1m} \\ \cdot & & & & & \\ \cdot & & & & & \\ \cdot & & & & & \\ a_{m1} & a_{m2} & \cdot & \cdot & \cdot & a_{mm} \end{vmatrix}. \tag{B.19}$$

The determinant of A is described as being of order m, the same as that of A. It is defined as follows.

For a first-order determinant $|A| = a_{11}$, the only element.

For the second-order

$$\begin{vmatrix} a_{11} & a_{12} \\ a_{21} & a_{22} \end{vmatrix} = a_{11}|a_{22}| - a_{12}|a_{21}| = a_{11}a_{22} - a_{12}a_{21}.$$

For the third-order

$$\begin{vmatrix} a_{11} & a_{12} & a_{13} \\ a_{21} & a_{22} & a_{23} \\ a_{31} & a_{32} & a_{33} \end{vmatrix} = a_{11} \times \begin{vmatrix} a_{22} & a_{23} \\ a_{32} & a_{33} \end{vmatrix} - a_{12} \times \begin{vmatrix} a_{21} & a_{23} \\ a_{31} & a_{33} \end{vmatrix} + a_{13} \times \begin{vmatrix} a_{21} & a_{22} \\ a_{31} & a_{32} \end{vmatrix}, \tag{B.20}$$

and so on for higher orders† with the signs of the terms alternatively + and −.

The *minor* of any element of a determinant is what is left in (B.19) after striking out the row and column which contains the element. The *co-factor* of the element a_{ij}, written A_{ij}, is defined as

$$A_{ij} = (-1)^{i+j} \times (\text{The minor of } a_{ij}). \tag{B.21}$$

So from (B.20), $$|A| = a_{11}A_{11} + a_{12}A_{12} + \ldots + a_{1m}A_{1m}. \tag{B.22}$$

Equally, a determinant may be *developed* as the sum of the products of the elements of any row or column and their co-factors.

(*b*) *Properties of determinants.* The following are stated without proof. See [616], pp. 6–12 for those which are not obvious.

† The 'diagonal method' of evaluating determinants, which is often quoted, is only correct for determinants of order 2 or 3, [616], p. 3.

(i) Determinants being scalars, their products commute.

$$|A|\times|B|=|B|\times|A|.$$

(ii) $|A|\times|B|=|AB|$ or $|BA|$, provided A and B are of the same order. This is not obvious.

(iii) $|A|+|B|$ does not in general equal $|A+B|$. If A and B are identical except for one row (or column), $|A+B|$ equals the determinant formed by entering the sum of these two rows (or columns) in their proper place, leaving the others unchanged.

(iv) $|A^{\mathrm{T}}|=|A|$.

(v) It follows from (ii) that $|A^{-1}|=(|A|)^{-1}$. And in general $|A^{p}|=(|A|)^{p}$.

(vi) If all the elements of any one row or column are multiplied by a constant k, the value of the determinant is multiplied by k.

(vii) The value of $|A|$ is unchanged if any multiple of one row (or column) is added to any other row (or column).

(viii) The sign of $|A|$ is changed if two rows (or columns) are interchanged. An even number of such interchanges restores the original sign.

(ix) $|A|=0$ if all the elements of any row (or column) are zero, or if any row (or column) is a multiple of another row (or column), or if one is a linear combination of the others, e.g.

$$\begin{vmatrix} 1 & 4 & 9 \\ 2 & 5 & 12 \\ 3 & 6 & 15 \end{vmatrix} = 0,$$

since the third column is the first plus twice the second.

(x) The sum of the products formed by multiplying the elements of one row (or column) by the co-factors of another is zero.

B.05. Linear equations

(*a*) A set of m linear equations for m unknowns,

$$\begin{aligned} a_{11}x_1+a_{12}x_2+\ldots+a_{1m}x_m &= k_1 \\ &\cdot \\ &\cdot \\ a_{m1}x_1+a_{m2}x_2+\ldots+a_{mm}x_m &= k_m, \end{aligned} \tag{B.23}$$

may be written $Ax=k$, where x and k are column vectors and A is a square matrix of order m, whose elements are a_{11}, etc. The product Ax is a column matrix whose elements are $a_{11}x_1+\ldots+a_{1m}x_m$, etc., as may be verified by multiplication.

Pre-multiplying both sides of $Ax=k$ by A^{-1} gives

$$A^{-1}Ax=A^{-1}k$$

or

$$x=A^{-1}k. \tag{B.24}$$

So if we can invert A, to give A^{-1}, the vector whose elements are the unknown x's is given by multiplication of A^{-1} and k.†

† While this statement is true, inversion of the matrix is not the easy way of solving equations. See §§ B.14 and B.15.

(b) The product of a matrix A (not necessarily square) and a vector x is easily recognizable as the left-hand side of a set of linear equations, another column vector. This may be used to facilitate writing the product AB of two matrices A and B as follows.

Pre-multiply the first column of B by A as above. This is the first column of AB. Then multiply the second column of B by A to give the second column of AB, and so on.

(c) *The adjugate matrix* of a square matrix A, written Adj A and sometimes called the *adjoint*, is defined by

$$\text{Adj } A = [A_{ij}]^{\mathrm{T}} = [A_{ji}]. \tag{B.25}$$

In other words, Adj A is the transpose of the matrix whose elements are the co-factors of the corresponding elements of A.

Direct multiplication, and reference to § B.04 (b) (x) shows that

$$A \times \text{Adj } A = \begin{bmatrix} |A| & 0 & . & . & . & 0 \\ 0 & |A| & . & . & . & 0 \\ \vdots & & & & & \\ 0 & 0 & . & . & . & |A| \end{bmatrix} = |A| \times I.$$

Whence

$$\frac{\text{Adj } A}{|A|} = A^{-1}. \tag{B.26}$$

Formula (B.24) then gives

$$x = \frac{1}{|A|}(\text{Adj } A)k, \tag{B.27}$$

and matrix multiplication of the right-hand side gives Cramer's rule for the solution of equations, [209], p. 34 and [598], p. 75.

(d) *Singular matrix.* If $|A| = 0$, (B.27) shows that the x's are either infinite or indeterminate, $(0 \div 0)$. A matrix whose determinant is zero has no inverse, by definition, and is said to be *singular.* The two sides of an equation must not be multiplied by the 'inverse' of a singular matrix. The result is meaningless and the identity ceases to hold.

(e) See §§ B.14 and B.15 for practical methods of solving equations, and § B.16 for computing the inverse of a matrix.

(f) *Degeneracy and rank.* A singular square matrix of order n is said to be degenerate. Its degeneracy is said to be of order 1, if there is at least one sub-matrix of order $(n-1)$ whose determinant is not zero. Similarly its degeneracy is of order m if there is at least one sub-matrix of order $(n-m)$ whose determinant is not zero, while the determinants of all sub-matrices of higher order are zero.

The *rank* of a square matrix is defined to be its order minus its degeneracy.

In a set of linear equations $Ax = k$, the order of A is the number of equations, while its rank is the number of equations which are independent.

B.06. Summary of rules of manipulation

(*a*) The following are situations in which matrix *multiplication is commutative.* Some references are to later paragraphs.

(i) Zero matrix. $A\times 0=0\times A$. § B.02 (*e*).

(ii) $AB=BA$ if both A and B are of order 1×1. § B.02 (*d*).

(iii) Unit. $A\times I=I\times A$. § B.02 (*c*).

(iv) Scalar matrices. $A\times k=k\times A$. § B.02 (*d*).

In (i), (iii), and (iv) it is implied that A is square. otherwise 0, I, and k will be of different orders on opposite sides of the equality.

(v) Scalars. $A(kB)=(kA)B$. A and B must be conformable. § B.01 (*b*).

(vi) Inverse. $A^{-1}\times A=A\times A^{-1}=I$. A being square. § B.02 (*h*).

(vii) $A^m\times A^n=A^n\times A^m=A^{m+n}$. A square. m and n whole numbers, positive or negative. § B.02 (*b*) and (*h*).

(viii) Transpose. If A is square and symmetrical $A=A^{\mathrm{T}}$ and $AA^{\mathrm{T}}=A^{\mathrm{T}}A$. § B.00 (*d*).

(ix) Diagonal matrix. $D_1D_2=D_2D_1$. § B.07 (*a*).

(x) Scalar product of two vectors. $b^{\mathrm{T}}a=a^{\mathrm{T}}b$, a and b being column vectors. § B.03 (*b*).

(xi) 2×2 rotation matrices. § B.08 (*b*).

(xii) 3×3 rotation matrices only commute if the rotation is small, as in (B.38).

(xiii) $|AB|=|BA|=|A|\times|B|$, A and B being square and conformable. § B.04 (*b*).

(*b*) *Transposition*

(xiv) $(AB)^{\mathrm{T}}=B^{\mathrm{T}}A^{\mathrm{T}}$. $(ABC)^{\mathrm{T}}=C^{\mathrm{T}}B^{\mathrm{T}}A^{\mathrm{T}}$, etc. [209], p. 30. So if $Ax=b$, $x^{\mathrm{T}}A^{\mathrm{T}}=b^{\mathrm{T}}$.

(xv) If A and B are square and symmetric $AB=(BA)^{\mathrm{T}}$, since both equal $A^{\mathrm{T}}B^{\mathrm{T}}$.

(xvi) If A is symmetric, A^2 is symmetric, but if A and B are both symmetric, it does not follow that AB is also. [209], p. 30.

(xvii) $(A+B)^{\mathrm{T}}=A^{\mathrm{T}}+B^{\mathrm{T}}$, as is clear from the definition.

(*c*) *Inverse*

(xviii) $(AB)^{-1}=B^{-1}A^{-1}$ and $(ABC)^{-1}=C^{-1}B^{-1}A^{-1}$, etc. [209], p. 30, A, B, and C being square.

(*d*) *Orthogonal matrices*

(xix) $A^{\mathrm{T}}=A^{-1}$, if A is orthogonal. § B.10.

(xx) $AA^{\mathrm{T}}=A^{\mathrm{T}}A=I$, if A is orthogonal. § B.10.

(*e*) *Cautions*

(xxi) Do not write a product AB or AI, etc. unless A and B, or A and I, etc., are conformable.

(xxii) Do not write the inverse of a matrix which is not both square and non-singular.

B.07. Simple matrices

(*a*) *Diagonal matrices* are square matrices, such as I, whose elements are all zero except those ($d_1, d_2, \ldots$) on the principal diagonal. If D_1, D_2, are diagonal matrices,

(i) D_1D_2 and D_2D_1 are diagonal, and their elements are the products of the corresponding elements of D_1 and D_2.

(ii) The elements of D^n are d^n, including D^{-1} and $D^{\frac{1}{2}}$.

(iii) $D \times A$ is a matrix whose rows are those of A each multiplied by the corresponding element of D. See the weight matrix $W^{\frac{1}{2}}$ in (D.50).

(iv) In $A \times D$ the columns of A are multiplied by the corresponding elements of D.

(v) $|D| = d_1 d_2 \ldots d_m$.

(*b*) *The matrix J is*

$$\begin{bmatrix} 0 & . & . & . & 0 & 1 \\ 0 & . & . & . & 1 & 0 \\ . & & & & & \\ . & & & & & \\ . & & & & & \\ 1 & . & . & . & 0 & 0 \end{bmatrix}.$$

Then $JA = A$ turned upside down.

$AJ = A$ reversed from side to side.

$J^2 = I.$

And in general, pre-multiplication of A by a matrix formed by interchanging any two rows of unit matrix I interchanges the corresponding rows of A. And post-multiplication by I with columns interchanged does the same for the columns of A.

(*c*) *Sparse matrices.* If A is pre-multiplied by a matrix B all of whose elements are zero except b_{ij}, all the elements of BA are zero except those of row i, whose elements are $b_{ij} \times$ those of row j of A. Post-multiplication of A by B has the same result except that column j of AB will be $b_{ij} \times$ column i of A.

(*d*) *Triangular matrices*

(i) *A lower triangular matrix*, L, is a square matrix all of whose elements above (excluding) the principal diagonal are zero. Similarly in an *upper triangular matrix*, U, all the elements below the diagonal are zero.

(ii) A set of equations which can be written $Lx = k$, is very easily solved, for the first line is $L_{11}x_1 = k_1$, from which $x_1 = k_1/L_{11}$. Substituting this in the next line $L_{21}x_1 + L_{22}x_2 = k_2$, then gives x_2, and so on. The process is known as *forward substitution*. Equations of the form $Ux = k$ are similarly solved by *back substitution*.

(iii) The determinant of a triangular matrix, like that of a diagonal matrix, is simply the product of its diagonal terms.

(iv) The product of two L matrices is an L, and that of two U's is a U. The transpose of an L is a U, but its inverse is an L.

(v) A triangular matrix in which all the diagonal elements are unity is called a *unit* (upper or lower) triangular matrix.

(vi) Products LU and UL are ordinary square matrices, and any square matrix A can (in general) be uniquely resolved into the product of two triangular matrices $L \times U$, of which one is a unit triangular. This is useful as a means of solving equations; see § B.14 (*d*).

B.08. Linear transformations and rotations

(*a*) *Transformations.* Let a set of dependent variables $y_1, y_2, \ldots, y_m$ be given in terms of $x_1, x_2, \ldots, x_m$ by the equations

$$y_1 = a_{11}x_1 + a_{12}x_2 + \ldots + a_{1m}x_m$$
$$\vdots$$
$$y_m = a_{m1}x_1 + a_{m2}x_2 + \ldots + a_{mm}x_m,$$

or

$$y = Ax.$$

And let a further set $z_1, z_2, \ldots, z_m$ be given in terms of the y's by

$$z = By.$$

Then it is easily verified that

$$z = BAx, \text{ in that order.} \tag{B.28}$$

(*b*) *Rotation.* As an example let rectangular plane axes x_1y_1 be inclined at an angle θ to axes x_0y_0 so that

$$x_1 = x_0 \cos\theta + y_0 \sin\theta,$$
$$y_1 = -x_0 \sin\theta + y_0 \cos\theta,$$

or

$$\begin{bmatrix} x_1 \\ y_1 \end{bmatrix} = \begin{bmatrix} \cos\theta & \sin\theta \\ -\sin\theta & \cos\theta \end{bmatrix} \begin{bmatrix} x_0 \\ y_0 \end{bmatrix}. \tag{B.29}$$

Now rotate axes to x_2y_2 through an angle ϕ, when

$$\begin{bmatrix} x_2 \\ y_2 \end{bmatrix} = \begin{bmatrix} \cos\phi & \sin\phi \\ -\sin\phi & \cos\phi \end{bmatrix} \begin{bmatrix} \cos\theta & \sin\theta \\ -\sin\theta & \cos\theta \end{bmatrix} \begin{bmatrix} x_0 \\ y_0 \end{bmatrix},$$

and matrix multiplication of the right-hand side gives

$$\begin{bmatrix} x_2 \\ y_2 \end{bmatrix} = \begin{bmatrix} \cos(\phi+\theta) & \sin(\phi+\theta) \\ -\sin(\phi+\theta) & \cos(\phi+\theta) \end{bmatrix} \begin{bmatrix} x_0 \\ y_0 \end{bmatrix}, \tag{B.30}$$

as is correct since the total rotation is $\phi + \theta$.

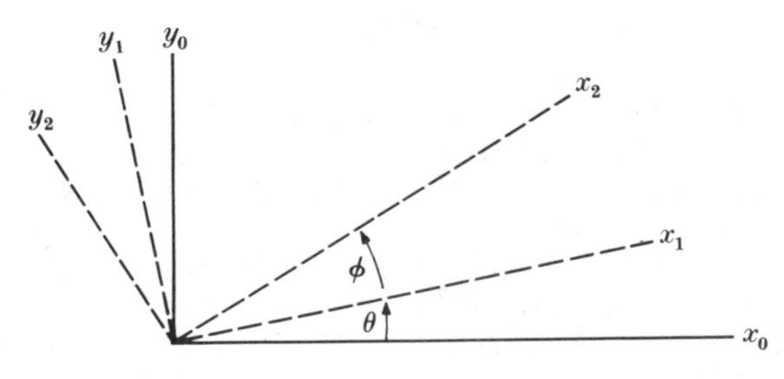

FIG. B.1.

Note that
$$\begin{bmatrix} \cos\theta & \sin\theta \\ -\sin\theta & \cos\theta \end{bmatrix}^n = \begin{bmatrix} \cos n\theta & \sin n\theta \\ -\sin n\theta & \cos n\theta \end{bmatrix}, \tag{B.31}$$

and in particular
$$\begin{bmatrix} \cos\theta & \sin\theta \\ -\sin\theta & \cos\theta \end{bmatrix}^{-1} = \begin{bmatrix} \cos\theta & -\sin\theta \\ \sin\theta & \cos\theta \end{bmatrix} = \begin{bmatrix} \cos\theta & \sin\theta \\ -\sin\theta & \cos\theta \end{bmatrix}^{\mathrm{T}}.$$

Powers and products of matrices in the form

$$\begin{bmatrix} a & b \\ -b & a \end{bmatrix}$$

are of the same form. Their multiplication is commutative. They are called *rotation matrices* (2×2).

When $\theta = 0$,
$$\begin{bmatrix} \cos\theta & \sin\theta \\ -\sin\theta & \cos\theta \end{bmatrix} = \begin{bmatrix} 1 & 0 \\ 0 & 1 \end{bmatrix} = I.$$

When $\theta = \pi/2$,
$$\begin{bmatrix} \cos\theta & \sin\theta \\ -\sin\theta & \cos\theta \end{bmatrix} = \begin{bmatrix} 0 & 1 \\ -1 & 0 \end{bmatrix}.$$

Note that
$$\begin{bmatrix} 0 & 1 \\ -1 & 0 \end{bmatrix}^2 = \begin{bmatrix} -1 & 0 \\ 0 & -1 \end{bmatrix} = -I.$$

We may write
$$\begin{bmatrix} 0 & 1 \\ -1 & 0 \end{bmatrix} \text{and} \begin{bmatrix} 0 & -1 \\ 1 & 0 \end{bmatrix} = + \text{ and } -\mathrm{i} = \sqrt{(-I)}, \tag{B.32}$$

since $\mathrm{i}\times\mathrm{i}$ does in fact equal $-I$. There is nothing 'imaginary' about this in matrix algebra.

Further,
$$\begin{bmatrix} \cos\theta & \sin\theta \\ -\sin\theta & \cos\theta \end{bmatrix} = \begin{bmatrix} \cos\theta & 0 \\ 0 & \cos\theta \end{bmatrix} + \begin{bmatrix} 0 & \sin\theta \\ -\sin\theta & 0 \end{bmatrix}$$
$$= I\cos\theta + \mathrm{i}\sin\theta.$$

Compare (B.31) with de Moivre's theorem.

(c) *Rotations in three dimensions.* If the direction cosines of rectangular axes $x_1y_1z_1$ relative to axes $x_0y_0z_0$ are $(l_xm_xn_x)$, $(l_ym_yn_y)$, $(l_zm_zn_z)$ we have

$$\begin{bmatrix} x_1 \\ y_1 \\ z_1 \end{bmatrix} = \begin{bmatrix} l_x & m_x & n_x \\ l_y & m_y & n_y \\ l_z & m_z & n_z \end{bmatrix} \begin{bmatrix} x_0 \\ y_0 \\ z_0 \end{bmatrix}, \text{[534], p. 30.} \tag{B.33}$$

The nine constants $l_x, \ldots, n_z$ are related by three equations of the form

$$(l_x^2 + m_x^2 + n_x^2) = 1,$$

and by three others $(l_xl_y + m_xm_y + n_xn_y) = 0$, etc., leaving only three independent relations expressing the three rotations which in general convert one rectangular set of axes into another. For the signs of rotation angles see (g) below.

Starting with axes $x_0y_0z_0$ let us change axes to $x_1y_1z_1$ by first rotating about z_0 through an angle θ, giving

$$\begin{bmatrix} x_1 \\ y_1 \\ z_1 \end{bmatrix} = \begin{bmatrix} \cos\theta & \sin\theta & 0 \\ -\sin\theta & \cos\theta & 0 \\ 0 & 0 & 1 \end{bmatrix} \begin{bmatrix} x_0 \\ y_0 \\ z_0 \end{bmatrix} = R_1 \begin{bmatrix} x_0 \\ y_0 \\ z_0 \end{bmatrix}. \tag{B.34}$$

Then rotate through ϕ about the axis x_1, giving

$$\begin{bmatrix} x_2 \\ y_2 \\ z_2 \end{bmatrix} = \begin{bmatrix} 1 & 0 & 0 \\ 0 & \cos\phi & \sin\phi \\ 0 & -\sin\phi & \cos\phi \end{bmatrix} \begin{bmatrix} x_1 \\ y_1 \\ z_1 \end{bmatrix} = R_2 \begin{bmatrix} x_1 \\ y_1 \\ z_1 \end{bmatrix}.$$

And finally rotate through ψ about y_2, giving

$$\begin{bmatrix} x_3 \\ y_3 \\ z_3 \end{bmatrix} = \begin{bmatrix} \cos\psi & 0 & -\sin\psi \\ 0 & 1 & 0 \\ \sin\psi & 0 & \cos\psi \end{bmatrix} \begin{bmatrix} x_2 \\ y_2 \\ z_2 \end{bmatrix} = R_3 \begin{bmatrix} x_2 \\ y_2 \\ z_2 \end{bmatrix}.$$

Combining these as in (B.28) gives

$$\begin{bmatrix} x_3 \\ y_3 \\ z_3 \end{bmatrix} = R_3R_2R_1 \begin{bmatrix} x_0 \\ y_0 \\ z_0 \end{bmatrix}$$

$$= \begin{bmatrix} \cos\psi\cos\theta - \sin\psi\sin\phi\sin\theta & \cos\psi\sin\theta + \sin\psi\sin\phi\cos\theta & -\sin\psi\cos\phi \\ -\cos\phi\sin\theta & \cos\phi\cos\theta & \sin\phi \\ \sin\psi\cos\theta + \cos\psi\sin\phi\sin\theta & \sin\psi\sin\theta - \cos\psi\sin\phi\cos\theta & \cos\psi\cos\phi \end{bmatrix} \times$$

$$\times \begin{bmatrix} x_0 \\ y_0 \\ z_0 \end{bmatrix} = R \begin{bmatrix} x_0 \\ y_0 \\ z_0 \end{bmatrix}, \text{ defining } R. \tag{B.35}$$

It follows that if $l_0m_0n_0$ are the direction cosines of any line relative to axes $x_0y_0z_0$, its direction cosines relative to axes $x_3y_3z_3$ are given by

$$\begin{bmatrix} l_3 \\ m_3 \\ n_3 \end{bmatrix} = R \begin{bmatrix} l_0 \\ m_0 \\ n_0 \end{bmatrix}. \tag{B.36}$$

so R is the matrix of the direction cosines $l_xm_xn_x$, etc. of the axes $x_3y_3z_3$ relative to the original axes $x_0y_0z_0$. Conversely, if these direction cosines are given, the values of the rotations θ, ϕ, and ψ, which will produce them, are easily found, since (from B.35),

$$\sin\phi = n_y$$
$$\tan\psi = -n_x/n_z$$
$$\cos\theta = m_y/\cos\phi$$
$$\text{or } \sin\theta = -l_y/\cos\phi. \tag{B.37}$$

(*d*) In the above, note that the value of matrix R ($=R_3R_2R_1$) depends on the order in which the three rotations θ, ϕ, ψ are carried out, unless they are so small that $\cos\theta$, etc., may be regarded as 1, $\sin\theta$ equal to θ, and $\sin\theta\sin\phi$, etc., as zero, in which case

$$R = \begin{bmatrix} 1 & \theta & -\psi \\ -\theta & 1 & \phi \\ +\psi & -\phi & 1 \end{bmatrix}, \tag{B.38}$$

and the (3×3) matrices R_3, R_2, and R_1 in these circumstances commute.

(*e*) The inverse of a single rotation matrix is obtained by changing the sign of the angle involved. Thus $[R_1(\theta)]^{-1} = R_1(-\theta) = [R_1(\theta)]^T$. In the inverse of products of rotation matrices the order of the separate rotations must be reversed. Thus $R^{-1} = R_1^{-1}R_2^{-1}R_3^{-1}$, as is geometrically obvious.

(*f*) *Euler's angles.* Instead of being treated as in (*c*) above, rotations are more usually expressed in terms of Euler's angles, [36], pp. 28–31 and [549], pp. 259–61, namely ϕ about z_0, θ about y_1, and ψ about z_2.

(B.35) then takes the form

$$\begin{bmatrix} x_3 \\ y_3 \\ z_3 \end{bmatrix} = \begin{bmatrix} \cos\psi & \sin\psi & 0 \\ -\sin\psi & \cos\psi & 0 \\ 0 & 0 & 1 \end{bmatrix} \begin{bmatrix} \cos\theta & 0 & -\sin\theta \\ 0 & 1 & 0 \\ \sin\theta & 0 & \cos\theta \end{bmatrix} \begin{bmatrix} \cos\phi & \sin\phi & 0 \\ -\sin\phi & \cos\phi & 0 \\ 0 & 0 & 1 \end{bmatrix} \begin{bmatrix} x_0 \\ y_0 \\ z_0 \end{bmatrix} \tag{B.39}$$

As in (*d*), when ϕ, θ, and ψ are small this simplifies to

$$\begin{bmatrix} x_3 \\ y_3 \\ z_3 \end{bmatrix} = \begin{bmatrix} 1 & (\phi+\psi) & -\theta \\ -(\phi+\psi) & 1 & 0 \\ \theta & 0 & 1 \end{bmatrix} \begin{bmatrix} x_0 \\ y_0 \\ z_0 \end{bmatrix}. \tag{B.40}$$

(B.40) is the same as (B.38) if the latter's θ is Euler's $(\phi+\psi)$, its ϕ is zero, and its ψ is Euler's θ.

(*g*) *Signs of rotation angles.* By convention axes x, y, z are always 'right-handed' as shown in Fig. B.2 (a) and (c). And rotations are considered positive if

(*a*) viewed outward along Oz, x turns clockwise towards y,
(*b*) viewed outward along Ox, y turns clockwise towards z, and
(*c*) viewed outward along Oy, z turns clockwise towards x.

As shown in Fig. B.2 (a).

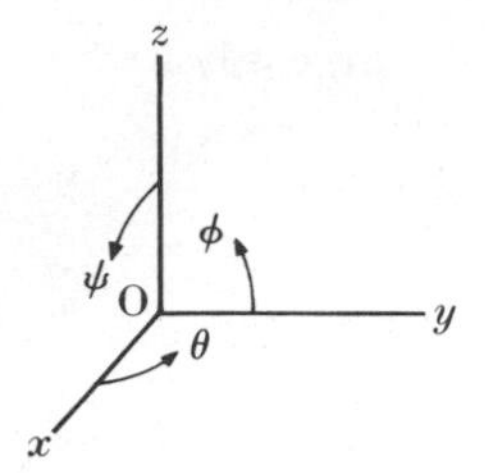

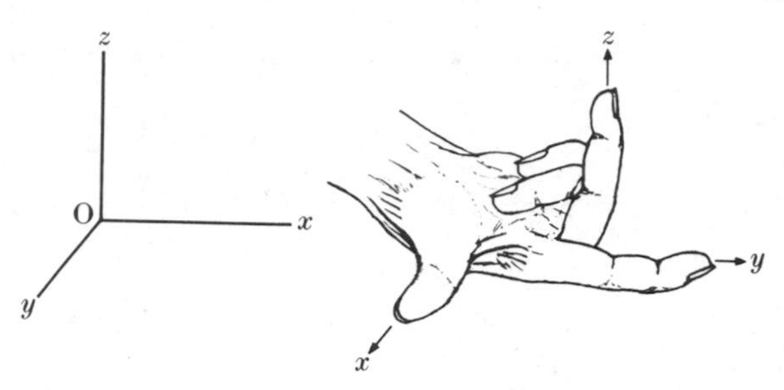

FIG. B.2 (a). Right-handed axes. Ox points upward from the paper. Positive rotations θ, ϕ, ψ are as shown.

FIG. B.2 (b). Left-handed axes. Oy is upward from the paper. Never used.

FIG. B.2 (c). A man's right hand indicating right-handed axes.

B.09. Partitioned matrices

(a) Any matrix may be *partitioned* by dividing it into *sub-matrices*, the divisions being horizontal, or vertical, or both. They may be as few or as numerous as required, but they must be straight. A sub-matrix may contain a single element. A matrix A whose elements are a_{11} to a_{mn} may, for example, be partitioned as below.

A_{11}	A_{12}	A_{13}	A_{14}
A_{21}	A_{22}	A_{23}	A_{24}

The sub-matrices are numbered as shown. Ordinarily there is no risk of confusing them with the co-factors in the determinant of A.

Partitioned matrices may be added and multiplied by the rules of matrix algebra, treating the sub-matrices as if they were single elements, subject to the following.

(i) *Addition and subtraction.* Not only must the two matrices be of the same order, but any two sub-matrices which have to be added must also be of the same order. Both matrices must be partitioned on exactly the same pattern.

(ii) *Mutliplication.* The two matrices must be conformable, and any two sub-matrices which have to be multiplied must be also. They will be so, provided the left-to-right vertical partitioning of the first matrix is of the same pattern as the downward horizontal partitioning of the second. A partitioned matrix and its transpose can always be multiplied, provided of course that the partitioning is done before the transposition; not after it with a different pattern. The final result is the same as if they had not been partitioned.

(b) *Inverse.* let a square matrix be partitioned into sub-matrices A, B, C, D, such that A and D are square but not necessarily of the same order as each other. Let the sub-matrices of its inverse be K, L, M, N, which will be of the same orders as A, B, C, and D respectively

$$\left[\begin{array}{c|c} A & B \\ \hline C & D \end{array}\right]^{-1} = \left[\begin{array}{c|c} K & L \\ \hline M & N \end{array}\right].$$

Then, [199], p. 103, $K. L, M, N$ are given by either of

$$\begin{aligned} N&=(D-CA^{-1}B)^{-1} & \text{or} \quad K&=(A-BD^{-1}C)^{-1} \\ M&=-NCA^{-1} & L&=KBD^{-1} \\ L&=-A^{-1}BN & M&=-D^{-1}CK \\ K&=A^{-1}-A^{-1}BM, & N&=D^{-1}-D^{-1}CL. \end{aligned} \tag{B.41}$$

(*c*) *Diagonal matrices.* If a square matrix can be partitioned so that all the sub-matrices on its diagonal are square, while the rest are zero, there results

$$\begin{bmatrix} D_1 & 0 & \cdot & \cdot & \cdot & 0 \\ 0 & D_2 & \cdot & \cdot & \cdot & 0 \\ \cdot & & \cdot & & & \\ \cdot & & & \cdot & & \\ \cdot & & & & \cdot & \\ 0 & 0 & \cdot & \cdot & \cdot & D_m \end{bmatrix}.$$

Then (i) its inverse is diagonal with elements $D_1^{-1}, D_2^{-1}, \ldots, D_m^{-1}$ and (ii) its determinant is $|D_1|\times|D_2|\times \ldots \times|D_m|$.

(*d*) Except for diagonal matrices as in (*c*), the determinant cannot be computed by applying the usual rules to sub-matrices, and then evaluating their determinants.

(*e*) For examples of useful partitioning see §§ B.14 (*c*) and D.22 (*d*).

B.10. Orthogonal matrices

(*a*) *Definition.* A square matrix is defined to be orthogonal if both

(i) the sum of the squares of the elements of every row and of every column equals 1,

(ii) the sum of the products of corresponding elements of every pair of columns is zero.

It follows that, if A is orthogonal, $AA^{\mathrm{T}}=A^{\mathrm{T}}A=I$.

(*b*) *Properties.* The properties of orthogonal matrices are as follows.

(i) $A^{-1}=A^{\mathrm{T}}$, since $AA^{\mathrm{T}}=I$. Orthogonal matrices are easily inverted.

(ii) The determinant $|AA^{\mathrm{T}}|=|I|=1$, and since $|A|$ always equals $|A^{\mathrm{T}}|$, § B.04 (*b*)(iv), it follows that $|A|=\pm 1$;

(iii) If A is orthogonal, A^{T} is also. So the definition in (*a*) (ii) above applies to rows as well as to columns.

(iv) If A and B are orthogonal AB and BA are orthogonal also, since

$$(AB)^{\mathrm{T}}=B^{\mathrm{T}}A^{\mathrm{T}}=B^{-1}A^{-1}=(AB)^{-1}.$$

(v) Orthogonal matrices are not necessarily symmetric.

(vi) Multiplication of orthogonal matrices does not necessarily commute.

(*c*) *Examples.* The matrix of direction cosines (B.33) clearly satisfies the definitions, and is orthogonal. So are the rotation matrices (B.34), (B.35), and (approximately) (B.38), and the modal matrix of § B.11 (*e*).

B.11. Latent roots

(*a*) The three equations

$$Lx \equiv \begin{bmatrix} l_1 & m_1 & n_1 \\ l_2 & m_2 & n_2 \\ l_3 & m_3 & n_3 \end{bmatrix} \begin{bmatrix} x_1 \\ x_2 \\ x_3 \end{bmatrix} = 0 \tag{B.42}$$

represent three planes through the origin, and their solution† is $x_1 = x_2 = x_3 = 0$.

If the determinant $|L|$ is zero, the line of intersection of two of the planes lies in the third, and so there is a line in which all three planes intersect. All points on this line are then additional solutions of the equations.

(*b*) *More generally*, let $Ax = 0$ be any set of m linear equations for m unknowns. If A is not singular, consider the matrix

$$A - \lambda I \equiv \begin{bmatrix} a_{11}-\lambda & a_{12} & \cdot & \cdot & \cdot & a_{1m} \\ a_{21} & a_{22}-\lambda & \cdot & \cdot & \cdot & a_{2m} \\ \vdots & & & & & \\ a_{m1} & a_{m2} & & & \cdot\ \cdot & a_{mm}-\lambda \end{bmatrix}. \tag{B.43}$$

There will be m values of λ for which the determinant $|A - \lambda I|$ will be zero, which may (in principle) be found by multiplying out the determinant and getting a linear equation of degree m in λ. The resulting values of λ are known as the *latent roots* or *eigenvalues* of A. They may not, of course, all be real nor all distinct, but they will all be real and positive if A is positive-definite, § B.12 (*b*), [616], pp. 42–4. If any one of them is zero, A is already singular.

The equation

$$(A - \lambda_i I)x = 0, \quad \text{or} \quad Ax = \lambda_i x, \tag{B.44}$$

where λ_i is one of the latent roots, in addition to the solution $x = 0$, will also have a solution

$$x = k \begin{bmatrix} u_1 \\ u_2 \\ \vdots \\ u_m \end{bmatrix} = ku. \tag{B.45}$$

Such a solution is called a *latent vector* or *eigenvector* of A. There are as many such solutions as there are latent roots. All may not be real, unless A is positive-definite. If for any latent vector k is so chosen that the sum of the squares of the elements of ku is unity, the vector is described as *normalized*. If $m = 3$, the elements of a normalized latent vector are the direction cosines of the line which is common to the three planes $(A - \lambda I)x = 0$, where λ is any one of the three latent roots.

(*c*) *Properties of latent roots*. Write $\lambda(A)$ to mean the latent roots of A.

(i) $\lambda(kA) = k\lambda(A)$, where k is any constant.

(ii) $\lambda(A - kI) = \lambda(A) - k$.

† This is often called the *trivial solution*, but it is the ordinary natural solution of the equation.

(iii) $\lambda(A^n) = \{\lambda(A)\}^n$ where n is a positive or negative integer, including $\lambda(A^{-1}) = \{\lambda(A)\}^{-1}$, [209], p. 42.

(iv) $\lambda(Y^{-1}AY) = \lambda(A)$ where Y is any conformable non-singular square matrix, [209], p. 43.

(v) $\lambda(A^T) = \lambda(A)$, and $\lambda(AB) = \lambda(BA)$, [209], p. 43.

(vi) If A is orthogonal $\lambda(A) = \lambda(A^{-1}) = \{\lambda(A)\}^{-1}$, so $\lambda = +$ or -1.

(vii) The latent roots of a diagonal matrix are its own elements.

(*d*) *Properties of latent vectors*

(i) The latent vectors, if distinct, are mutually orthogonal. $u_i^T u_j = 0$ $(i \neq j)$, [209], p. 44.

(ii) The normalized latent vectors of a diagonal matrix are the columns of I.

(*e*) *The modal matrix.* [209], pp. 42–3. Let M be the matrix whose columns are the normalized latent vectors of A. It is known as the modal matrix of A. Then the expression $Ax = \lambda_i u_i$, where λ_i is one of the latent roots of A and u_i is the corresponding latent vector, may be written to include all the λ's as

$$AM = MD, \tag{B.46}$$

where D is the diagonal matrix whose diagonal elements are the λ's. Whence

$$M^{-1}AM = D, \quad \text{and} \quad A = MDM^{-1}. \tag{B.47}$$

The modal matrix is orthogonal, so $M^{-1} = M^T$, [209], p. 45.

It follows that

$$\begin{aligned} A^n &= (MDM^T)^n \\ &= (MDM^T)(M^TDM^T)\ldots \\ &= MD^nM^T, \end{aligned} \tag{B.48}$$

where n is any positive integer. It is also true when $n = -1$.

In three dimensions, if in (B.42) one of λ_1, λ_2, or λ_3 is subtracted from the diagonal elements of L, the three resulting planes will have a common line. Then the columns of the modal matrix are the direction cosines of the three lines corresponding to λ_1, λ_2, and λ_3 respectively.

(*f*) *Use of latent roots.* (i) Latent roots and vectors occupy much of the space in textbooks on matrix algebra, and they have many applications in engineering. They have some use also in geodesy, as in (ii), (iii), and (iv) below.

(ii) It is clear from § B.11 (*b*) that the value of the smallest latent root is some indication of how close the matrix of a set of equations is to being singular. Near-singular matrices are ill-conditioned. A small change in a coefficient may make a large change in the solution. A better criterion sometimes is that the ratio of the largest root to the smallest should not be too great, see (*c*) (i) above.

(iii) The reciprocals of the latent roots of the normal equations in least squares are the squares of the lengths of the principal semi-axes of the error ellipsoid of the unknowns. See § D.20. The s.e.'s of all the unknowns therefore lie between the greatest and least latent roots.

(iv) The rate of convergence in the iterative method of solving equations depends on the latent roots of the matrix of the iteration, § B.15 (*c*).

(v) For the use of the modal matrix see §§ D.20 and D.22.

(vi) [534], Chapter III illustrates the use of (B.43), and the cubic equation for λ, in connection with the geometry of surfaces of the second degree in three dimensions.

[209] gives methods for computing latent roots and latent vectors, but see the footnote to (D.137). Their use does not necessarily demand the determination of their numerical values.

B.12. Quadratic forms

(*a*) *Definitions.* The expression

$$x^T A y = y^T A^T x = a_{11}x_1y_1 + a_{12}x_1y_2 + \ldots + a_{mm}x_my_m, \tag{B.49}$$

where A is any square matrix of order m, and x^T and y are row and column vectors of the same order, is a scalar. It is known as a *bilinear form.*†

If $x = y$ in (B.49), and if A is symmetric, the expression

$$x^T A x = a_{11}x_1^2 + a_{22}x_2^2 + \ldots + a_{mm}x_m^2 + 2a_{12}x_1x_2 + \ldots \text{ all combinations} \tag{B.50}$$

is known as a *quadratic form.* Such forms occur commonly in algebra, as in the following examples.

(i) If A is a unit matrix, $x^TAx = x^Tx$ or $\sum x_i^2$.

(ii) In 2 or 3 dimensions $x^TAx = k$ is the general equation of a conic or conicoid centred on the origin, and the criterion that the curve or surface shall be an ellipse or ellipsoid is that $x^TAx > 0$ for all values of x.

(*b*) A matrix is said to be *positive-definite* if x^TAx is positive for all real non-zero values of x. Compare (*a*) (ii) above. In least squares the matrix of the normal equations is of the form B^TB or BB^T and is always positive definite, since $x^TB^TBx = (Bx)^T(Bx)$, which is the sum of the squares of the elements of the vector Bx.

In matrix theory many theorems depend on a matrix being positive-definite, and it is fortunate that they apply to those with which we are most concerned.

(*c*) *Differentiation.* Bilinear and quadratic forms are scalars, so $\partial(x^TAy)/\partial y_i$ has its usual meaning, namely the change in the scalar x^TAy produced by a change in element y_i of y, all other elements being unchanged.

Then from (B.49)

$$\begin{aligned}\partial(x^TAy)/\partial y_i &= a_{1i}x_1 + a_{2i}x_2 + \ldots \\ &= (\text{column } i \text{ of } A)^T \times (\text{column vector } x).\end{aligned}$$

And the result for all values of i is included in the equality of the two column vectors

$$\partial(x^TAy)/\partial y = A^Tx. \tag{B.51}$$

Similarly,

$$\partial(x^TAy)/\partial x = Ay. \tag{B.52}$$

† The definition of bilinear forms does not exclude expressions in which A is not square, but we do not need to consider them here.

If $x = y$ and A is symmetric, so that $x^{\mathrm{T}}Ax$ is as in (B.50),

$$\partial(x^{\mathrm{T}}Ax)/\partial x_i = 2a_{ii}x_i + 2(a_{1i}x_1 + a_{2i}x_2 + \ldots \text{ excluding } x_i)$$
$$= 2(\text{row } i \text{ of } A) \times (\text{column vector } x).$$

And for all values of i,
$$\partial(x^{\mathrm{T}}Ax)/\partial x = 2Ax. \tag{B.53}$$

B.13. Differentiation of matrices

See [210], pp. 43–4. If the elements of a matrix A change by small quantities δa_{11}, etc., these changes may be arranged in matrix form as a matrix δA. If each element changes continuously with time or some such independent variable, there will be a matrix $\mathrm{d}A/\mathrm{d}t$, composed of the elements $\mathrm{d}a_{ij}/\mathrm{d}t$, and as the result of a small change δt, a scalar, A will become

$$A + \delta t(\mathrm{d}A/\mathrm{d}t). \tag{B.54}$$

It follows that
$$\mathrm{d}/\mathrm{d}t(A+B) = \mathrm{d}A/\mathrm{d}t + \mathrm{d}B/\mathrm{d}t, \tag{B.55}$$

and that
$$(\mathrm{d}/\mathrm{d}t)(AB) = A(\mathrm{d}B/\mathrm{d}t) + (\mathrm{d}A/\mathrm{d}t)B. \tag{B.56}$$

Note the order of multiplication in the second term.

Also, for instance, if A is a square matrix,

$$(\mathrm{d}/\mathrm{d}t)A^3 = (\mathrm{d}A/\mathrm{d}t)A^2 + A(\mathrm{d}A/\mathrm{d}t)A + A^2(\mathrm{d}A/\mathrm{d}t), \tag{B.57}$$

and
$$(\mathrm{d}/\mathrm{d}t)A^{-1} = -A^{-1}(\mathrm{d}A/\mathrm{d}t)A^{-1}. \tag{B.58}$$

If $Ax = b$, so that $x = A^{-1}b$,

$$\delta x = A^{-1}\delta b, \tag{B.59}$$

in which δx and δb are column vectors. If only one element of b changes, δb is of the form $[0, 0, \ldots, \delta b_i, \ldots, 0]^{\mathrm{T}}$.

B.14. Solution of equations. Direct methods.

(*a*) We consider solutions of equations $Ax = b$, where A is square, symmetrical, and positive-definite, as in the normal equations of least squares. The order of A may be very great, such as 1000×1000. There are two groups of methods.

(i) *Direct*, of which Gauss's method § 2.13 (*d*) is an example. These methods are described in (*b*)–(*g*) below.

(ii) *Iterative*, in which a trial solution is continuously tested by substitution, and improved until a sufficiently exact solution is found.† See § B.15.

The advantages of a direct solution are that it is generally quicker, it demands less attention from the operator, and the inversion of A, to give the variances and covariances of the unknowns, follows easily. Inversion by iterative methods is much more laborious.

The following are the advantages of an iterative solution. (*a*) If the non-zero

† The word 'iteration' is also used in connection with direct methods for the successive approximations referred to in §§ 2.21 and 2.29 (*b*). In the variation of coordinates at least one such repetition should always be made with the errors of the first solution as unknowns. With true iterative methods there are likely to be 100 iterations or more.

elements of A are few but distributed all over the matrix, solution by iteration may be possible on a computer on which convenient direct methods are not possible. (b) The programme is often more easily written. And (c) Unstable solutions are probably less liable to rounding-off errors.

Direct methods are advised when they are possible.

(b) *Gauss's method*, which is described in § 2.13 (d), can be expressed in matrix notation, but gains little in clarity thereby. It may, however, be noted for § 2.37 (c) (v) that at successive stages of the Gauss elimination the left-hand side of the solution is

$$A_{22}-A_{12}^{T}A_{11}^{-1}A_{12},$$

where A_{22}, A_{12}, etc., are sub-matrices. If at the stage concerned the first n unknowns have been eliminated, the partitioning is after the first n rows and columns, cf. (B.62). See [209], pp. 68–73 and [62], pp. 194–6 for details of this method and of those of (d) and (e) below.

(c) *Method of Boltz, or block elimination method.* See [315], vol i, pp. 300–6, which does not use matrix notation, and [62], pp. 199–201. To solve $AX=b$, let the equations be partitioned as

$$\left[\begin{array}{c|c} A_{11} & A_{12} \\ \hline A_{12}^{T} & A_{22} \end{array}\right]\left[\begin{array}{c} x_1 \\ \hline x_2 \end{array}\right]=\left[\begin{array}{c} b_1 \\ \hline b_2 \end{array}\right] \tag{B.60}$$

where A_{11} and A_{22} are square matrices, and A_{11} is small enough for (B.62) to be computed by any of the direct methods of solution.

Then
$$A_{11}x_1+A_{12}x_2=b_1 \quad A_{12}^{T}x_1+A_{22}x_2=b_2$$
$$x_1=A_{11}^{-1}(b_1-A_{12}x_2) \tag{B.61}$$

$$(A_{22}-A_{12}^{T}A_{11}^{-1}A_{12})x_2=(b_2-A_{12}^{T}A_{11}^{-1}b_1) \tag{B.62}$$

in which the order of the matrix $A_{22}-A_{12}^{T}A_{11}^{-1}A_{12}$ is less than that of A by the order of A_{11}. The process is then repeated by partitioning $A_{22}-A_{12}^{T}A_{11}^{-1}A_{12}$ in a similar way, until the last group of x's are directly determined. Previous groups of x's are then got by back substitution in the equations of the form of (B.61).

If A_{11} is taken as a single element, this is equivalent to Gauss's method. In principle quite a small computer can use Boltz's method to solve any number of equations, A_{11} being taken as a matrix of quite moderate order, such as 16. The method involves transfers of matrices and vectors between the fast and auxiliary stores, and these should be minimized by making A_{11} as large as possible.

(d) *Banachiewicz's method.* The solution of $Ax=b$ is carried out in three phases.

(i) First resolve A into the product of a unit lower triangular matrix, and an upper triangular, as in (B.66).

$$LU=A. \tag{B.63}$$

(ii) Then find a column vector f such that

$$Lf=b. \tag{B.64}$$

Given L, this is easily done by forward substitution as in § B.07 (d) (ii).

(iii) Finally the equation $Ax = b$ has become

$$LUx = Lf, \qquad Ux = f, \tag{B.65}$$

and x is given by back substitution.

The resolution of A into the product LU is done as follows. Write

$$\begin{bmatrix} 1 & 0 & . & . & . & 0 \\ l_{21} & 1 & . & . & . & 0 \\ \vdots & & & & & \\ l_{m1} & l_{m2} & . & . & . & 1 \end{bmatrix} \begin{bmatrix} u_{11} & u_{12} & . & . & . & u_{1m} \\ 0 & u_{22} & . & . & . & u_{2m} \\ \vdots & & & & & \\ 0 & 0 & . & . & . & u_{mm} \end{bmatrix} = \begin{bmatrix} a_{11} & a_{12} & . & . & . & a_{1m} \\ a_{21} & a_{22} & . & . & . & a_{2m} \\ \vdots & & & & & \\ a_{m1} & a_{m2} & . & . & . & a_{mm} \end{bmatrix} \tag{B.66}$$

Then multiply out the left-hand side and equate successive elements to those of A. Thus

$$u_{11} = a_{11}, \qquad l_{21}u_{11} = a_{21}, \text{ whence } l_{21} = a_{21}/u_{11},$$

and so on with increasing complexity. This is the most laborious part of the solution.

If A is symmetrical, U will be a scalar multiple of L^{T}.

If A is any square matrix, there will be difficulty if any of the diagonal elements of U are zero or very small, but this will not happen in the normal equations of least squares, [209], pp. 77–81.

(e) *Choleski's method*, [87]. If A is symmetrical, as the matrices of normal equations are, it may be decomposed so that $U = L^{\mathrm{T}}$, neither being unit triangular matrices. The solution then continues as in (d) (ii) and (iii) above. See [466], pp. 75–96 for full details of use with a desk computer, and also (g) below.

This method involves taking square roots for the diagonal elements of L, as may be seen by writing out $LL^{\mathrm{T}} = A$ in full, on the lines of (B.66), which gives $l_{11}^2 = a_{11}$, $l_{21}l_{11} = a_{21}$, and so on. There would be a difficulty if any of the squares of the diagonal elements of L were negative, although [209], p. 109 gives an easy way out of it. But in normal equations such diagonal elements will not occur.

Positive signs may be allotted to all square roots.

If the diagonal elements of A are very unequal, a useful preliminary to the solution may be to form

$$D^{-\frac{1}{2}}AD^{-\frac{1}{2}}(D^{\frac{1}{2}}x) = D^{-\frac{1}{2}}d, \tag{B.67}$$

where D is a diagonal matrix formed from the diagonal elements of A. The diagonal elements of $D^{-\frac{1}{2}}AD^{-\frac{1}{2}}$ are then all unity, and the solution gives values of $(D^{\frac{1}{2}}x)$. If A is symmetrical and positive definite, $D^{-\frac{1}{2}}AD^{\frac{1}{2}}$ will be also.

(f) *Capacity of the computer.* In the methods of Banachiewicz and Choleski, the number of products involved in the solution of m normal equations, if m is large, is of the order $m^3/3$ and $m^3/6$ respectively, [209], pp. 181–5. For a large computer this number is not overwhelming, even if m is 1000 or more, but if the non-zero elements are distributed all over the matrix, the fast store must be able to accommodate something more than m^2 elements. But see subparagraphs (c) and (g). It is possible to solve equations with m larger than 1000.

(g) *Band matrices.* If the non-zero elements of A (above the diagonal) are contained in a band of width k including the diagonal itself and the non-zero element furthest to the right of it in any row, the non-zero elements of L^{T} are confined to a band of the same width. See [62], p. 202 for a simple example, which shows that the available capacity of the fast store need not be much more than k^2, although a substantially larger capacity (up to a little more than km) is a convenience. See § 2.25 and § 2.34 for the securing of minimum bandwidth, and also [62], p. 204.

B.15. Solution of equations. Iterative methods

(a) *Jacobi's method.* Iterative methods can be illustrated by the following example. Let the equations $Ax = d$ be

$$\begin{bmatrix} 100 & 1 & 1 \\ 2 & 100 & 3 \\ 1 & 2 & 100 \end{bmatrix} \begin{bmatrix} x_1 \\ x_2 \\ x_3 \end{bmatrix} = \begin{bmatrix} 103 \\ 211 \\ 308 \end{bmatrix}.$$

The equations are arranged so that the largest coefficients are on the diagonal, as they will naturally be in the case of least-square normal equations. In this example an obvious first approximation is $x_1 = 103/100$, $x_2 = 211/100$, and $x_3 = 308/100$, which may be written $Dx^{(1)} = d$ or as $x^{(1)} = D^{-1}d$, where D is a diagonal matrix whose elements are the diagonal elements of A, and $x^{(1)}$ indicates the first approximation to the vector x.

Substituting these values of x in the original equations, and subtracting from each side of $Ax = d$, gives

$$\begin{bmatrix} 100 & 1 & 1 \\ 2 & 100 & 3 \\ 1 & 2 & 100 \end{bmatrix} \begin{bmatrix} x_1 - 1.03 \\ x_2 - 2.11 \\ x_3 - 3.08 \end{bmatrix} = \begin{bmatrix} 103 - (103 + 2.11 + 3.08) \\ 211 - (2.06 + 211 + 9.24) \\ 308 - (1.03 + 4.22 + 308) \end{bmatrix},$$

or

$$A(x - x^{(1)}) = d - Ax^{(1)} \tag{B.68}$$

For the next approximation we have

$$x_1^{(2)} = 1.03 - (2.11 + 3.08)/100 = 0.978,$$

$$x_2^{(2)} = 2.11 - (2.06 + 9.24)/100 = 1.997,$$

$$x_3^{(2)} = 3.08 - (1.03 + 4.22)/100 = 3.028,$$

or

$$D(x^{(2)} - x^{(1)}) = d - Ax^{(1)}, \tag{B.69}$$

or

$$x^{(2)} = (I - D^{-1}A)x^{(1)} + D^{-1}d. \tag{B.70}$$

In general, writing $x^{(r)}$ to indicate the result after r approximations

$$D(x^{(r+1)}-x^{(r)})=d-Ax^{(r)}, \tag{B.71}$$

$$x^{(r+1)}=(I-D^{-1}A)x^{(r)}+D^{-1}d \tag{B.72}$$

$$=-D^{-1}(L+U)x^{(r)}+D^{-1}d. \tag{B.73}$$

where A has been written as $L+D+U$, D being its diagonal terms, and L and U being triangular matrices with zeros on their diagonals.

Instead of taking $D^{-1}d$ as the first approximation, the equations $Ax=d$ could have been written

$$x_1=(d_1-a_{12}x_2-a_{13}x_3-\ldots)/a_{11},$$

$$x_2=(d_2-a_{21}x_1-a_{23}x_3-\ldots)/a_{22}, \text{ etc.} \tag{B.74}$$

Then if the initial approximation $x^{(0)}$ is taken as $[0,0,\ldots,0]^{\mathrm{T}}$, these equations give

$$x^{(1)}=[d_1/a_{11}, d_2/a_{22}\ldots]^{\mathrm{T}}$$

$$=D^{-1}d \text{ as before.} \tag{B.75}$$

(*b*) *Gauss–Seidel method.* In Jacobi's method every element of $x^{(r+1)}$ is obtained using the previous approximation $x^{(r)}$ for all the other elements. Clearly when computing $x_i^{(r+1)}$ there might be some advantage in using the already available $x_1^{(r+1)}, x_2^{(r+1)}, \ldots, x_{i-1}^{(r+1)}$. In the numerical example the second approximation for x_2 might have been $x_2^{(2)}=2.11-\{(2\times 0.978)+9.24\}/100$. This is the Gauss–Seidel method. Jacobi's method is known as that of *simultaneous displacements*, and that of Gauss–Seidel as of *successive displacements.*

In matrix notation

$$x^{(r+1)}-x^{(r)}=D^{-1}\{d-Lx^{(r+1)}-(U+D)x^{(r)}\},$$

$$x^{(r+1)}=-(D+L)^{-1}Ux^{(r)}+(D+L)^{-1}d. \tag{B.76}$$

(*c*) *Convergence.* The numerical example in (*a*) is exceptionally favourable. The large diagonal terms give a good first approximation, and ensure rapid convergence to an accurate solution. More typically, the first approximation $x=D^{-1}d$ may have no resemblance to the correct solution, and it is not obvious that the process will converge, however many iterations are made.

The normal equations of least squares, with which we are concerned, are more favourable than most. They are symmetric and positive-definite, and the solutions of all such equations must converge if the Gauss–Seidel method is used. Whether the Jacobi method will converge, and the actual rate of convergence of either method, depends on the dominance of the diagonal terms. In normal equations these are the sums of squares, and do tend to be larger than the other terms. For Jacobi to converge it is sufficient, but not necessary, that in any row the diagonal term should exceed the sum of the moduli of the rest of the row. This is unusual. But even when convergence is inevitable, a large set of equations may require hundreds of iterations to give a solution with four significant figures.

The convergence can be *accelerated* by the process known as *over-relaxation*, as below.

We define as follows.

(i) The *displacement vector* is the difference between successive approximations

$$\Delta^{(r)} = x^{(r+1)} - x^{(r)}. \tag{B.77}$$

From (B.71), for Jacobi

$$\Delta^{(r)} = D^{-1}(d - Ax^{(r)}). \tag{B.78}$$

(ii) *The residual vector* is the difference between the vector d and what results from substituting $x^{(r)}$ in Ax,

$$R^{(r)} = d - Ax^{(r)}. \tag{B.79}$$

From (B.71), for Jacobi

$$R^{(r)} = D(x^{(r+1)} - x^{(r)}),$$

or

$$\Delta^{(r)} = D^{-1}R^{(r)}. \tag{B.80}$$

Displacement vector $= D^{-1} \times$ residual vector.

(iii) The error vector is the error of $x^{(r)}$,

$$e^{(r)} = x^{(r)} - x. \tag{B.81}$$

$e^{(r)}$ is clearly unknown, but $e^{(r+1)}$ can be related to it. Thus (B.73) may be written

$$x^{(r+1)} = Cx^{(r)} + c, \tag{B.82}$$

where C is a matrix, $D^{-1}(L+U)$, and c is a constant vector, $D^{-1}d$. It follows from (B.82) that

$$e^{(r+1)} = Ce^{(r)} = C^2e^{(r-1)} = \text{etc.} \tag{B.83†}$$

If C is constant, independent of r, the iteration is said to be *stationary*. The matrix C is known as the *iteration matrix*. If the elements of $e^{(r)}$ are progressively reduced by multiplication by C, the iteration will clearly converge.

In more detail, it can be shown, [209], pp. 194–5 and [63], p. 81, that for any stationary iteration the error vector can be written in the form

$$e^{(r)} = \mu_1^r\{\alpha_1 u_1 + \alpha_2(\mu_2/\mu_1)^r u_2 + \ldots + \alpha_n(\mu_n/\mu_1)^r u_n\}, \tag{B.84‡}$$

where $\mu_1, \ldots, \mu_n$ are the latent roots of C in decreasing order of magnitude, so that $|\mu_1| > |\mu_2|$, etc.; $u_1, \ldots, u_n$ are the corresponding latent vectors of C; and $\alpha_1, \ldots, \alpha_n$ are unknown scalar constants.

Since μ_1 is the largest root, and provided $|\mu_2| \neq |\mu_1|$, it follows that when r is sufficiently large

$$e^{(r)} \approx \alpha_1 \mu_1^r u_1, \tag{B.85}$$

and

$$e^{(r+k)} \approx \mu_1^k e^{(r)}. \tag{B.86}$$

† Substituting (B.81) in (B.82) gives

$$e^{(r+1)} - e^{(r)} = x^{(r+1)} - x^{(r)} = Cx^{(r)} + c - x^{(r)} = Ce^{(r)} + Cx + c - e^{(r)} - x.$$

Whence $e^{(r+1)} = Ce^{(r)}$, since $Cx - x + c = 0$, from (B.82) provided $x^{(r+1)}$ converges to x when r is large.

‡ In (B.84) μ^r, with no brackets round the r, indicates that μ is raised to the power of r in the usual way.

Comparison with (B.83) shows that, when r is large, multiplication of $e^{(r)}$ by the matrix C approximates to multiplication by scalar $|\mu_1|$, and that the iteration will ultimately converge provided

$$|\mu_1| < 1. \tag{B.87}$$

Following from (B.86)

$$\Delta^{(r+k)} \approx \mu_1^k \Delta^{(r)}, \qquad \text{and} \qquad R^{(r+k)} \approx \mu_1^k R^{(r)} \tag{B.88}$$

and

$$\mu_1^k \approx \frac{\|\Delta^{(r+k)}\|}{\|\Delta^{(r)}\|} \approx \frac{\|R^{(r+k)}\|}{\|R^{(r)}\|}, \tag{B.89}$$

where $\|\Delta^{(r)}\|$, etc., indicates the square root of the sum of the squares of the elements of $\Delta^{(r)}$.†

(*d*) *Accelerated convergence.* If successive approximations regularly tend to reduce $\|\Delta^{(r)}\|$, the suggestion is that increasing the change (arithmetically) by a scalar *relaxation factor* w, somewhat greater than unity, may hasten the convergence. While if $\|\Delta^{(r)}\|$ varies irregularly, a smaller factor may be suggested. The two cases are known as *over-* or *under-relaxation.*

Further details for accelerating the Jacobi and Gauss–Seidel iterations are described in [63] and [113], 3 Edn., pp. 589–91. For iterative methods in general see also [199] and [209].

(*e*) *Block iteration. Gauss–Seidel.* In a large set of normal equations with (say) 1000 unknowns, it may be convenient to divide the unknowns into groups of (say) 20, and to iterate with these groups instead of with each of the 1000 unknowns separately, as follows.

Select the first group of 20 unknowns, and identify the 20 equations whose diagonal terms‡ refer to these unknowns. Then substitute the provisional values of the 980 other unknowns (mostly zero) in these 20 equations, so forming 20 equations for the 20 unknowns of the first group. These equations can be easily solved by any convenient method.

Then take the next group of 20. Identify the 20 equations with the new unknowns on the diagonal. Substitute the provisional values of the other 980 in them, using the values just determined for the members of the first group. Solve the resulting 20 equations. And so on. The whole process may have to be repeated many times.

The value of this process will depend on the wise selection of the members of each group, and on the order in which groups are brought forward for re-iteration. See § 7.38.

(*f*) *Gradient methods. Conjugate gradients.* As with previous methods, the object is to obtain the solution of $Ax = b$, where b is errorless, by successively substituting estimated values of the vector x, in the hope that the errors of these estimates will converge to near zero.

† If V is any vector, $\sqrt{(v'v)} \equiv \sqrt{(v_1^2 + \ldots + v_n^2)}$ is known as the *euclidean norm of* v.

‡ In normal equations there is a tendency for the diagonal terms to have the largest coefficients.

Any estimate of the vector x defines a point in n-dimensional space, and the function $f(x)=(x-A^{-1}b)^{\mathrm{T}}A(x-A^{-1}b)=$ constant is a quadratic form (B.50) which represents a family† of n-dimensional hyper-ellipsoids, whose centres are at $x=A^{-1}b$, the required solution. One member of this family will pass through the point defined by the first estimate $x^{(0)}$. To obtain a better approximation an obvious move would be to take the next estimate $x^{(1)}$ at a point on the inward normal to $f(x)=$ constant at $x^{(0)}$, at a distance from $x^{(0)}$ such that at $x^{(1)}$ this normal is tangent to an inner member of the family $f(x)=$ constant. We may write $p^{(0)}=$ the unit vector along the normal, and $\sigma^{(0)}=$ the (scalar) distance measured along it. As thus described the method is known as the *Optimum gradients* method. It clearly converges rapidly if the ellipsoid is nearly spherical, but much more slowly if the axes are very unequal.

The *conjugate gradients* method converges more quickly. Instead of each successive move being made along the normal, it is made along the n-dimensional hyper-plane which is conjugate to the tangent hyper-plane at the preceding estimate of x. For the properties of conjugate diametric planes in 3-D space, see [534] pp. 74–9. They contain the centre. It then follows that (apart from numerical rounding-off errors) the centre itself is bound to be reached after n moves along n different hyper-planes all of which contain the centre. The method might indeed be described as a method of direct solution, but the form of procedure is iterative, and in practice it may be possible to stop the iteration before the centre itself is reached.

The computational algorithm of the conjugate gradients method is as follows.

(i) Let the first estimate of x be $x^{(0)}$. Then $b-Ax^{(0)}=R^{(0)}$, the opening value of the residual vector, and initially $p^{(0)}=R^{(0)}$, the direction of the normal at $x^{(0)}$.

(ii) Compute $\sigma^{(0)}=\dfrac{(R^{(0)})^{\mathrm{T}}p^{(0)}}{(p^{(0)})^{\mathrm{T}}Ap^{(0)}}$, a scalar. (B.90)

(iii) Compute $x^{(1)}=x^{(0)}+\sigma^{(0)}p^{(0)}$ (B.91)

(iv) Compute $R^{(1)}=R^{(0)}-\sigma^{(0)}Ap^{(0)}$ (B.92)

$$T^{(0)}=-\frac{(R^{(1)})^{\mathrm{T}}Ap^{(0)}}{(p^{(0)})^{\mathrm{T}}Ap^{(0)}} \qquad \text{(B.93)}$$

$$p^{(1)}=R^{(1)}+T^{(0)}p^{(0)} \qquad \text{(B.94)}$$

The iteration can now be repeated from (ii), substituting superscript 1 for 0, and 2 for 1. In general write r for 0 and $(r+1)$ for 1.

The iteration can be summarized in the form

$$x^{(0)}\to p^{(0)}\to \underbrace{\sigma^{(0)}\to x^{(1)}\to R^{(1)}\to T^{(0)}\to p^{(1)}}_{\text{cycle}}\to \sigma^{(1)}\to \text{ etc.} \qquad \text{(B.95)}$$

This method is currently (1978) thought to be the most promising iterative method. The labour at each iteration is only marginally greater than that of the accelerated Jacobi or Gauss–Seidel methods, but fewer iterations are likely to be

† For different values of the constant. It is known that if A is positive-definite (as are the matrices of normal equations), the quadratic will be a closed ellipsoid, rather than a hyperboloid.

required, and the operator does not have to exercise judgement as for the over or under-relaxation of the latter methods.

For the conjugate gradient method see [73] and [195].

B.16. Computation of the inverse

The solution $Ax = \begin{bmatrix} 0 \\ \cdot \\ \cdot \\ \cdot \\ 1 \\ \cdot \\ 0 \end{bmatrix}$, where the 1 is element i, (B.96)

is $x = A^{-1}[0, \ldots, 1, \ldots, 0]^T$, which is clearly column i of A^{-1}. The columns of A^{-1} may thus be computed, one at a time.

If the equations $Ax = k$ have already been solved by any direct method, any column of A^{-1} is relatively easily computed, since all the 'left-hand side' of the computation remains unchanged. The inversion of a complete matrix of (large) order m involves m^3 multiplications as compared with $m^3/3$ for the solution of the equations. But the relative time on the computer is more than these figures suggest, since more transfers are required between the fast and auxiliary stores.

If the equations have been solved by iteration, each column of the inverse may be obtained by an iterative solution of a set of equations (B.96), but each column involves the same computer time as the solution of the equations.

B.17. Notes for students

Those unacquainted with matrix methods may on first reading omit §§ B.10–B.13 and most of §§ B.14–B.16. Familiarity may be acquired by illustrating and testing all statements such as those in §§ B.06 and B.07, with simple numerical examples using matrices of order 2 or 3. Matrix equations such as (D.49)–(D.54) and (D.76)–(D.80) may be visualized by drawing them as a series of squares and rectangles, as has been for $AB = C$ in (B.5).

General references for matrix algebra. [4], [199], [209], [210], [560], and [616].

APPENDIX C

CARTESIAN COORDINATES IN THREE DIMENSIONS

C.00. Summary and notation

THIS appendix collects formulae which are required when computations are done in rectangular or polar coordinates in three dimensions, as opposed to the more usual system of computing latitude and longitude on a spheroid and leaving the determination of height as a separate operation.

Computations in three dimensions are the natural method where satellites are involved. They might rationally have been used for the computation of long Shoran lines, but in practice they have not.

In case they should be required, § C.02 gives all necessary formulae for computing conventional triangulation and trilateration, as well as satellite triangulation, by variation of coordinates in three dimensions.

Notation. The following notation is used in this appendix.

x, y, z = Rectangular coordinates, § C.01 (a).
ϕ, λ, N = Latitude, longitude (+east) and height of geoid above spheroid.
ν = Normal terminated by minor axis.
e = Eccentricity of spheroid.
a, b = Semi-axes of spheroid. In (C.36)–(C.37), and again in (C.41), a, b, and c have different meanings.
h = Height above geoid.
H = Height above spheroid. $H = N + h$.
θ, ψ = Direction angles, § C.01 (b) (i).
l, m, n = Direction cosines.
A_{12} = Azimuth from 1 to 2.
L = Distance. Slant, not projected.
ξ, η = Deviations in meridian and in PV. Signs as in § 2.05.
ζ = Zenith distance, corrected for refraction, the same as z in § 4.00.
A_{213} = Horizontal angle at 1 between 2 and 3.
γ = Angle of a plane triangle.
RA, δ = Right ascension and declination.
O, C = Observed value, and value computed from trial coordinates.
v = Residual.

The *vertical* is the direction of gravity as opposed to the spheroidal *normal.*

C.01. Basic formulae

(a) *Conversion.* Take rectangular axes x, y, z. It will generally be convenient to take the x-axis parallel to the conventional zero meridian of Greenwich, the y-axis towards the east, and the z-axis parallel to the CIO (Conventional International Origin of polar motion), § 4.09, to which the minor axes of reference spheroids are defined to be parallel. The position of the origin of coordinates will be specified with reference either to the geocentre or to the vertical at some survey station, as in § 7.21.

With axes defined as above, if ϕ and λ are the geodetic latitude and longitude of a point, referred to a spheroid centred on the same origin as x, y, and z, and if the point lies $(N+h)$ above the spheroid, its rectangular coordinates are

$$x = (\nu+N+h)\cos\phi\cos\lambda,$$
$$y = (\nu+N+h)\cos\phi\sin\lambda,$$
$$z = \{(1-e^2)\nu+N+h\}\sin\phi. \tag{C.1}$$

The reverse process, getting ϕ, λ, and $N+h$ from x, y, z, is complicated by ν being a function of ϕ, (C.6).

$$\tan\lambda = y/x, \quad \text{giving } \lambda, \tag{C.2}$$

$$\tan\phi = \frac{z+e^2\nu\sin\phi}{(x^2+y^2)^{\frac{1}{2}}}, \tag{C.3}$$

using any approximate ϕ for $\sin\phi$ and ν, and re-iterating as necessary. Convergence will be rapid.

As an alternative to (C.3) use

$$\tan\phi = \frac{z+\epsilon b\sin^3 u}{p-e^2 a\cos^3 u}, \tag{C.4}$$

where $p = (x^2+y^2)^{\frac{1}{2}}$, and $\tan u = (z/p)(a/b)$, [120]. This formula is substantially exact without iteration, see (2.81).

Finally $$(N+h) = (x^2+y^2)^{\frac{1}{2}}\sec\phi - \nu = z\operatorname{cosec}\phi - (1-e^2)\nu. \tag{C.5}$$

Use the second expression in high latitudes.

$$\nu = a(1-e^2\sin^2\phi)^{-\frac{1}{2}}. \tag{C.6}$$

The equation of the spheroid is

$$x^2/a^2+y^2/a^2+z^2/b^2 = 1, \tag{C.7}$$

and the direction of the normal at (x, y, z) is given by

$$\tan\theta = \frac{b^2}{a^2}\frac{(x^2+y^2)^{\frac{1}{2}}}{z}, \quad \text{and} \quad \tan\psi = \frac{y}{x}, \tag{C.8}$$

where θ is the geodetic co-latitude, and ψ is the geodetic longitude.

The distance between $(x_1y_1z_1)$ and $(x_2y_2z_2)$ is

$$[(x_2-x_1)^2+(y_2-y_1)^2+(z_2-z_1)^2]^{\frac{1}{2}}. \tag{C.9}$$

The normal section azimuth A_{12} of $(x_2y_2z_2)$ from $(x_1y_1z_1)$ is given by

$$\cot A_{12} = [(z_2-z_1)\cos\phi-(x_2-x_1)\sin\phi] \div y_2, \tag{C.10}$$

if the axis of x is taken to lie in the meridian of $(x_1y_1z_1)$, so that $y_1 = 0$. In general,

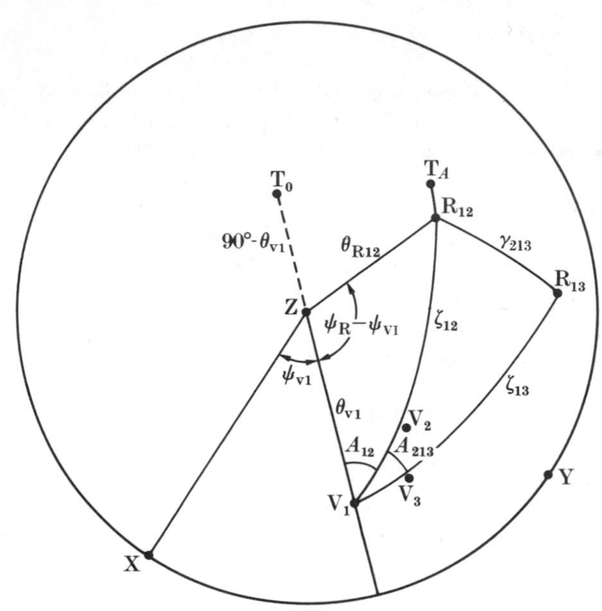

FIG. C.1. Directions and angles on the unit sphere. Points on the sphere define the directions of lines parallel to the radius at the point. Great circles on the sphere correspond to planes. The arcs of great circles correspond to angles between lines, and the angles between great circles correspond to the angles between planes.

if the axis of x lies in the zero meridian, in formula (C.10) substitute

$$-\sin\lambda(x_2-x_1)+\cos\lambda(y_2-y_1) \quad \text{in place of } y_2,$$

and

$$\cos\lambda(x_2-x_1)+\sin\lambda(y_2-y_1) \quad \text{in place of } (x_2-x_1). \tag{C.11}$$

where λ is the longitude of $(x_1y_1z_1)$.

(*b*) *Directions.* The *direction* of a line, which is the same as that of any parallel line, may be defined in three ways.

(i) By two angles θ and ψ, as in polar coordinates in solid geometry, Fig. 7.7. The axis of $\theta=0$ may be the CIO, the same as that of z in (*a*) above, and the Greenwich meridian (CZM) may be the zero of ψ.

(ii) By *direction cosines* l, m, and n, namely the cosines of the angles which the line makes with the x-, y-, and z-axes.

(iii) By the RA and declination on the celestial sphere to which the line points at a stated GAST and date. The RA and δ are ordinarily referred to the true instantaneous celestial pole and equator, but are open to definition as referring to either the true or mean (with regard to nutation) pole and equator at any stated epoch.

Direction cosines involve three numbers, compared with two for θ and ψ, but their formulae are generally more symmetrical and better for computation. On the other hand, polar coordinates lend themselves to a graphical figure, and may help with the solution of unfamiliar problems.

The unit sphere. Fig. C.1 represents a sphere of undefined (unit) radius. Take an axis $\theta = 0$ parallel to the axis of z, and let points on the sphere be defined by angles θ (0 to 180°) and ψ (0 to 360°), where ψ is measured eastwards from the positive direction of the x-axis. Then the direction in space of any line is defined by the θ and ψ of the point on the sphere at which the radius is parallel to the line. Points on the sphere thus correspond to systems of parallel lines. Great circles on the sphere correspond to systems of parallel planes, and the great circle joining any two points on the sphere corresponds to all the planes containing, or parallel to, the corresponding two lines. The angle between any two lines in space is the length of the arc between them on the unit sphere, and the angle between any two planes is the angle between the corresponding great circles at their point of intersection on the sphere.†

In Fig. C.1 this sphere is viewed down the θ-axis, and shows the following points.

(i) X, Y, Z the directions of the x-, y-, and z-axes.

(ii) V_1, V_2, V_3, the directions of the upward verticals at three points P_1, P_2, and P_3. At any V, $\theta_V =$ the astronomical‡ co-latitude, and $\psi_V =$ the astronomical longitude

$$ZV_1 = \theta_{V_1} = 90° - \phi_1,$$

and $\quad XZV_1 =$ the angle between meridian planes ZX and $ZV_1 = \psi_{V_1} = \lambda_1$. (C.12)

The V's of intervisible points are, of course, very close to each other; within 1° for 100 km. For the outward spheroidal normal, N

$$\theta_N = \theta_V + \xi, \quad \text{and} \quad \psi_N = \psi_V - \eta \sec \phi, \tag{C.13}$$

where ξ and η are the deviations of the vertical.

(iii) T_0 represents the horizontal line through P_1 in astronomical azimuth zero. $V_1T_0 = 90°$, so at T_0 $\theta = 90° - \theta_{V_1}$, $\psi = 180° + \psi_{V_1}$. Similarly T_A is the direction of the horizontal line in astronomical azimuth A_{12}. $V_1T_A = 90°$ as before, and $ZV_1T_A = A_{12}$, V_1T_A being a great circle.

(iv) R_{12} is the direction of the line from P_1 to P_2 in astronomical azimuth A_{12} and zenith distance $V_1R_{12} = \zeta_{12}$. Then

$$\cos \theta_R = \cos \theta_V \cos \zeta_{12} + \sin \theta_V \sin \zeta_{12} \cos A_{12},$$

and

$$\sin(\psi_R - \psi_V) = \sin A_{12} \sin \zeta_{12} \operatorname{cosec} \theta_R. \tag{C.14}$$

Conversely,

$$\cos \zeta_{12} = \cos \theta_V \cos \theta_R + \sin \theta_V \sin \theta_R \cos(\psi_R - \psi_V),$$

† The unit sphere is in fact very similar to the celestial sphere. The differences are (a) that the unit sphere is usually taken to rotate with the earth, and (b) its axis of $\theta = 0$ is fixed in the earth, parallel to the CIO, while the axis of $\delta = 90°$ on the celestial sphere is ordinarily either the true instantaneous axis of rotation, or else the axis meaned with regard to nutation (not polar motion).

‡ As defined in § 2.04, reduced to mean pole.

and $$\sin A_{12} = \sin \theta_R \sin(\psi_R - \psi_V)\operatorname{cosec} \zeta_{12}, \tag{C.15}$$

where θ_V and ψ_V refer to P_1, and suffix R refers to R_{12}.

Putting θ_N and ψ_N for θ_V and ψ_V in (C.15) will give the geodetic azimuth and geodetic zenith distance (as measured from the normal).

Note that V_2 necessarily lies very close to the arc V_1R_{12}, but not exactly on it unless the verticals at P_1 and P_2 happen to be co-planar. Note that V_1, R_{12}, and T_A are necessarily co-planar, so that these three points lie on one great circle in the diagram.

(v) Similarly R_{13} is the direction of the line P_1P_3, and the arc $R_{12}R_{13}$ is the angle γ_{213} in the plane triangle $P_1P_2P_3$. We have

$$\cos \gamma_{213} = \cos \zeta_{12} \cos \zeta_{13} + \sin \zeta_{12} \sin \zeta_{13} \cos A_{213}. \tag{C.16}$$

If the direction of P_1P_2 is (θ, ψ), that of P_2P_1 is

$$(180° - \theta, \psi + 180°). \tag{C.17}$$

In general the cosine of the angle between any two lines whose directions are (θ_1, ψ_1) and (θ_2, ψ_2) is

$$\cos \theta_1 \cos \theta_2 + \sin \theta_1 \sin \theta_2 \cos(\psi_1 - \psi_2). \tag{C.18}$$

The direction of the line, corrected for aberration and refraction, to a celestial point of known declination and RA is approximately

$$\theta = (90° - \delta), \quad \psi = (\mathrm{RA} - \mathrm{GAST}), \tag{C.19}$$

but if the axis of $\delta = 90°$ is the true axis of rotation, while that of θ is the CIO, (C.19) must be amplified as in (7.97).

The direction of the line from $(x_1y_1z_1)$ to $(x_2y_2z_2)$ is given by

$$\cos \theta = (z_2 - z_1)/L \quad \text{and} \quad \tan \psi = (y_2 - y_1)/(x_2 - x_1), \tag{C.20}$$

where $L^2 = (x_2 - x_1)^2 + (y_2 - y_1)^2 + (z_2 - z_1)^2$.

(*c*) *Direction cosines.* In Fig. C.1 the direction cosines of any line are the cosines of the angles it makes with the axes of x, y, and z. Thus, for V_1, they are the cosines of the arcs V_1X, V_1Y, and V_1Z.

Given the θ and ψ of any line,

$$l = \sin \theta \cos \psi, \quad m = \sin \theta \sin \psi, \quad n = \cos \theta.$$

And conversely, $$\tan \psi = m/l \quad \text{and} \quad \cos \theta = n. \tag{C.21}$$

If θ is small, use $\sin^2\theta = l^2 + m^2$ instead of $\cos \theta = n$.

Necessarily, $l^2 + m^2 + n^2 = 1$. The three cosines are not independent.

The direction cosines of the line joining $(x_1y_1z_1)$ and $(x_2y_2z_2)$, in the direction 1 to 2, are

$$(x_2 - x_1)/L, \quad (y_2 - y_1)/L \quad \text{and} \quad (z_2 - z_1)/L, \tag{C.22}$$

where $L^2 = (x_2 - x_1)^2 + (y_2 - y_1)^2 + (z_2 - z_1)^2$.

The equation of any line though $(x_1y_1z_1)$ is

$$(x - x_1)l = (y - y_1)/m = (z - z_1)/n = L, \tag{C.23}$$

where L is distance along the line from $(x_1y_1z_1)$. If $(x_1y_1z_1)$ is the origin $(0, 0, 0)$, we have

$$x/l = y/m = z/n = L,$$

and

$$l = x/L, \qquad m = y/L, \qquad n = z/L. \tag{C.24}$$

The cosine of the angle θ between two lines whose direction cosines are $l_1m_1n_1$ and $l_2m_2n_2$ is

$$l_1l_2 + m_1m_2 + n_1n_2, \quad [534], \text{p. } 17. \tag{C.25}$$

or

$$\cos(\theta/2) = \tfrac{1}{2}[(l_1+l_2)^2 + (m_1+m_2)^2 + (n_1+n_2)^2]^{\frac{1}{2}}$$

and

$$\sin(\theta/2) = \tfrac{1}{2}[(l_1-l_2)^2 + (m_1-m_2)^2 + (n_1-n_2)^2]^{\frac{1}{2}} \tag{C.26}$$

Formulae (C.19) or better (7.97), with (C.21), give the direction cosines of a line of given RA and δ.

Combining (C.12) and (C.21), the direction cosines of the upward vertical (or normal) at a point in astronomical (or geodetic) latitude and longitude ϕ, λ are

$$l = \cos\phi\cos\lambda, \qquad m = \cos\phi\sin\lambda, \qquad n = \sin\phi. \tag{C.27}$$

Given the astronomical azimuth A and zenith distance ζ of a line PR at a point P whose astronomical latitude and longitude are ϕ and λ, the direction cosines of PR are

$$\begin{aligned} l_R &= -\sin\lambda\sin A\sin\zeta - \sin\phi\cos\lambda\cos A\sin\zeta + \cos\phi\cos\lambda\cos\zeta, \\ m_R &= \cos\lambda\sin A\sin\zeta - \sin\phi\sin\lambda\cos A\sin\zeta + \cos\phi\sin\lambda\cos\zeta, \\ n_R &= \cos\phi\cos A\sin\zeta + \sin\phi\cos\zeta. \end{aligned} \tag{C.28}$$

Proof. Take axes X′, Y′, Z′ eastward in the horizontal plane, northward (along T_0), and outward along V respectively. Relative to these axes the direction cosines of R are

$$\sin A\sin\zeta, \qquad \cos A\sin\zeta, \qquad \cos\zeta.$$

Also, relative to these axes, the direction cosines of the axes X, Y, Z are

$$\begin{array}{llll} -\sin\lambda, & -\sin\phi\cos\lambda, & \cos\phi\cos\lambda, & \text{for X.} \\ \cos\lambda, & -\sin\phi\sin\lambda, & \cos\phi\sin\lambda, & \text{for Y.} \\ 0, & \cos\phi, & \sin\phi, & \text{for Z.} \end{array}$$

Whence (C.25) gives the result, since the direction cosines of R in the X, Y, Z system are the cosines of the angles it makes with those axes. In fact (C.28) is

$$\begin{bmatrix} l_R \\ m_R \\ n_R \end{bmatrix} = \begin{bmatrix} -\sin\lambda & -\sin\phi\cos\lambda & \cos\phi\cos\lambda \\ \cos\lambda & -\sin\phi\sin\lambda & \cos\phi\sin\lambda \\ 0 & \cos\phi & \sin\phi \end{bmatrix} \times \begin{bmatrix} \sin A\sin\zeta \\ \cos A\sin\zeta \\ \cos\zeta \end{bmatrix}. \tag{C.29}$$

Conversely, given l, m, n, the direction cosines of a line R, from $P_1(\phi, \lambda)$ to P_2, the azimuth and zenith distance of P_2 at P_1 are given by (C.21) and (C.15). Or alternatively

$$\begin{bmatrix} \sin A \sin \zeta \\ \cos A \sin \zeta \\ \cos \zeta \end{bmatrix} = \begin{bmatrix} -\sin \lambda & \cos \lambda & 0 \\ -\sin \phi \cos \lambda & -\sin \phi \sin \lambda & \cos \phi \\ \cos \phi \cos \lambda & \cos \phi \sin \lambda & \sin \phi \end{bmatrix} \times \begin{bmatrix} l \\ m \\ n \end{bmatrix}, \tag{C.30}$$

from (C.29), in which the rotation matrix is orthogonal, § B.10, and its inverse equals its transpose. The third row gives ζ directly, whence either the first or second gives A.

The perpendicular distance of a point (f, g, h) from the line (C.23) is given by the square root of

$$(f-x_1)^2+(g-y_1)^2+(h-z_1)^2-\{l(f-x_1)+m(g-y_1)+n(h-z_1)\}^2. \tag{C.31}$$

The condition that two lines with suffixes 1 and 2 in (C.23) shall meet is

$$\begin{vmatrix} x_1-x_2 & y_1-y_2 & z_1-z_2 \\ l_1 & m_1 & n_1 \\ l_2 & m_2 & n_2 \end{vmatrix} = 0, \quad \text{or} \quad L_0 \begin{vmatrix} l_0 & m_0 & n_0 \\ l_1 & m_1 & n_1 \\ l_2 & m_2 & n_2 \end{vmatrix} = 0 \tag{C.32}$$

where l_0, m_0, n_0 are the direction cosines of the line joining $(x_1 y_1 z_1)$ and $(x_2 y_2 z_2)$, and L_0 is the distance between them.

If they do not meet, the length of the common perpendicular between them is given by

$$L_0 \frac{l_0(m_1 n_2 - m_2 n_1) + m_0(n_1 l_2 - n_2 l_1) + n_0(l_1 m_2 - l_2 m_1)}{\sqrt{\{(m_1 n_2 - m_2 n_1)^2 + (n_1 l_2 - n_2 l_1)^2 + (l_1 m_2 - l_2 m_1)^2\}}} \tag{C.33}$$

$$= L_0 \frac{l_0 a_{12} + m_0 b_{12} + n_0 c_{12}}{\sqrt{(a_{12}^2 + b_{12}^2 + c_{12}^2)}}, \tag{C.34}$$

where the numerator is the same as (C.32), and

$$a_{12} = m_1 n_2 - m_2 n_1, \quad b_{12} = n_1 l_2 - n_2 l_1, \quad \text{and} \quad c_{12} = l_1 m_2 - l_2 m_1.$$

If a number of lines, suffixes 1 to n, are nearly but not quite concurrent, the position (f, g, h) from which the sum of the squares of the perpendiculars is least, is got by differentiating each (C.31) with respect to f, g, and h, multiplying by weights if necessary, summing, and equating each sum to zero, to give normal equations for f, g, and h, as below.

$$\begin{aligned} f[1-ll]-g[lm]-h[ln] &= [x(1-ll)]-[ylm]-[zln], \\ -f[ml]+g[1-mm]-h[mn] &= -[xml]+[y(1-mm)]-[zmn], \\ -f[nl]-g[nm]+h[1-nn] &= -[xnl]-[ynm]+[z(1-nn)]. \end{aligned} \tag{C.35}$$

(*d*) *Planes.* The 'direction' of a plane is defined by the direction cosines of its normals. The direction cosines of normals to the plane $ax+by+cz+d=0$ are

$$a/\sqrt{(a^2+b^2+c^2)}, \quad b/\sqrt{(a^2+b^2+c^2)}, \quad \text{and} \quad c/\sqrt{(a^2+b^2+c^2)}. \tag{C.36}$$

And p, the length of the perpendicular from the origin, is given by

$$-d/\sqrt{(a^2+b^2+c^2)}. \tag{C.37}$$

The equation of the plane may thus be written

$$lx+my+nz=p. \tag{C.38}$$

In astro-triangulation the condition that the three lines AB, AS_1, and BS_1, of Fig. 7.6, whose direction cosines are lmn, $l_1m_1n_1$, and $l_2m_2n_2$ respectively should be coplanar, i.e. that AS_1 and BS_1 should intersect, is

$$\begin{vmatrix} l & m & n \\ l_1 & m_1 & n_1 \\ l_2 & m_2 & n_2 \end{vmatrix} = 0, \quad \text{the same as (C.32).} \tag{C.39}$$

It follows that if AS_1 and BS_1 are known to intersect at S_1, and if two other lines $AS_2(l_3m_3n_3)$ and $BS_2(l_4m_4n_4)$ are known to intersect at S_2, the direction cosines of $AB(l\,m\,n)$ are given by three equations,

$$\begin{vmatrix} l & m & n \\ l_1 & m_1 & n_1 \\ l_2 & m_2 & n_2 \end{vmatrix} = 0, \quad \begin{vmatrix} l & m & n \\ l_3 & m_3 & n_3 \\ l_4 & m_4 & n_4 \end{vmatrix} = 0, \quad \text{and} \quad l^2+m^2+n^2=1. \tag{C.40}$$

Their solution is

$$l/(b_{12}c_{34}-b_{34}c_{12}) = m/(c_{12}a_{34}-c_{34}a_{12}) = n/(a_{12}b_{34}-a_{34}b_{12})$$
$$= 1/\{(b_{12}c_{34}-b_{34}c_{12})^2+(c_{12}a_{34}-c_{34}a_{12})^2+(a_{12}b_{34}-a_{34}b_{12})^2\}^{\frac{1}{2}}, \tag{C.41}$$

where a_{12}, b_{12}, and c_{12} are as in (C.34), in which a_{34}, etc., have suffixes 3 and 4, etc. There will be two solutions, one giving AB and the other BA, with opposite signs.

If there are more than two points S, to which simultaneous intersecting lines have been observed, the optimum values of l, m, n for AB are given by least squares with observation equations in the form

$$\begin{vmatrix} l & n & n \\ l_i & m_i & n_i \\ l_j & m_j & n_j \end{vmatrix} = 0,$$

or $$a_{ij}l+b_{ij}m+c_{ij}n=0, \quad \text{subject to} \quad l^2+m^2+n^2=1, \tag{C.42}$$

with a, b, and c as in (C.34).

The azimuth and zenith distance of the line AB are then given by (C.30), if required, for which ϕ and λ are derived from the x, y, z of A by (C.2)–(C.6).

C.02. Variation of coordinates

This paragraph gives formulae for the computation of triangulation and trilateration in three dimensions by variation of coordinates on the lines of Chapter 2, Section 3. In general, but not always, there are five unknowns at each unfixed station, namely δx, δy, and δz, the corrections to the trial coordinates, and also

$\delta\phi(=-\delta\theta_V)$ and $\delta\lambda(=\delta\psi_V)$, the corrections to the trial directions of the vertical. The last two will not aways arise. For instance, in the computation of astro-triangulation, § 7.23, in which directions are observed as RA and declination, the direction of the vertical at a station of observation is not involved. In ordinary triangulation, on the other hand, in which angles are measured in a horizontal plane, and in which altitudes are measured above the horizon, the direction of the vertical† is an essential unknown. There is no spheroid, and the vertical cannot as an approximation be taken as parallel to any spheroidal normal.

Accurate vertical angles are required in order to determine the direction of the vertical at a forward station, and these are not ordinarily available, because of refraction. Very frequent astronomical observations of ϕ and λ are a partial remedy, and it is possible that the instruments described in § 3.21 may provide a better one, but at present the accuracy of this method of computing conventional triangulation in three dimensions will not approach that of ordinary methods, except in mountains where the refraction is relatively regular, and the deviations of the vertical exceptionally irregular. But see § C.03 for a practicable modified system.

Possible forms of observation equations are in (*a*)–(*h*) below.

(*a*) Observed (slant) *distance* L between $(x_1y_1z_1)$ and $(x_2y_2z_2)$. The equation is

$$\frac{\partial L}{\partial x_1}\delta x_1+\frac{\partial L}{\partial y_1}\delta y_1+\frac{\partial L}{\partial z_1}\delta z_1+\frac{\partial L}{\partial x_2}\delta x_2+\frac{\partial L}{\partial y_2}\delta y_2+\frac{\partial L}{\delta z_2}\delta z_2=(\mathrm{O-C})_L+v, \quad (\mathrm{C.43})$$

where
$$\frac{\partial L}{\partial x_2}=(x_2-x_1)/L=l_{12}=-\frac{\partial L}{\partial x_1}.$$

And similarly for $\partial L/\partial y_2$, etc.

So (C.43) takes the form

$$(x_2-x_1)(\delta x_2-\delta x_1)/L+(y_2-y_1)(\delta y_2-\delta y_1)/L+(z_2-z_1)(\delta z_2-\delta z_1)/L=(\mathrm{O-C})_L+v$$

or
$$l_{12}(\delta x_2-\delta x_1)+m_{12}(\delta y_2-\delta y_1)+n_{12}(\delta z_2-\delta z_1)=(\mathrm{O-C})_L+v. \quad (\mathrm{C.44})$$

Note that $\phi_1\lambda_1$ and $\phi_2\lambda_2$, and the directions of the verticals, are not involved.

(*b*) 'Observed' *direction cosines* l, m, n, from $(x_1y_1z_1)$ to $(x_2y_2z_2)$ as given by (C.40). The equations are

$$\frac{1}{L}\{(1-l^2)(\delta x_2-\delta x_1)-lm(\delta y_2-\delta y_1)-ln(\delta z_2-\delta z_1)\}=(\mathrm{O-C})_l+v,$$

$$\frac{1}{L}\{-lm(\delta x_2-\delta x_1)+(1-m^2)(\delta y_2-\delta y_1)-mn(\delta z_2-\delta z_1)\}=(\mathrm{O-C})_m+v,$$

$$\frac{1}{L}\{-ln(\delta x_2-\delta x_1)-mn(\delta y_2-\delta y_1)+(1-n^2)(\delta z_2-\delta z_1)\}=(\mathrm{O-C})_n+v. \quad (\mathrm{C.45})$$

One equation is redundant. The one whose right-hand side refers to the largest of

† In principle, this unknown is the direction of the 'vertical' axis of the theodolite, rather than the true vertical. But in practice the only way of keeping the direction of the axis fixed is to use a spirit bubble to put it in the vertical and to keep it there.

l, m, or n may be omitted. Alternatively, and better, all three may be included with weights proportional to $1/(1-l^2)$, $1/(1-m^2)$, and $1/(1-n^2)$ respectively.

Since direction cosines are not directly observed, and are not independent, it is generally better to use (*c*) and (*e*), or (*f*), below.

Proof. From (C.22), $(x_2-x_1)=lL$. Differentiating with respect to x_2 gives

$$1=L\frac{\partial l}{\partial x_2}+l\frac{\partial L}{\partial x_2}=L\frac{\partial l}{\partial x_2}+l^2.$$

Whence

$$\frac{\partial l}{\partial x_2}=\frac{(1-l^2)}{L}, \qquad \text{(C.46)}$$

and similarly for the others. In short,

$$\begin{bmatrix} \dfrac{\partial}{\partial x_2} \\ \dfrac{\partial}{\partial y_2} \\ \dfrac{\partial}{\partial z_2} \end{bmatrix} [l \quad m \quad n]=\frac{1}{L}\begin{bmatrix} (1-l^2) & -lm & -ln \\ -lm & (1-m^2) & -mn \\ -ln & -mn & (1-n^2) \end{bmatrix}. \qquad \text{(C.47)}$$

Change sign for $\partial/\partial x_1$, etc.

(*c*) Observed *astronomical azimuth* A from $(x_1y_1z_1)$ to $(x_2y_2z_2)$. The equation is

$$\left(\frac{\partial A}{\partial x_1}\delta x_1+\frac{\partial A}{\partial y_1}\delta y_1+\frac{\partial A}{\partial z_1}\delta z_1\right)+\left(\frac{\partial A}{\partial x_2}\delta x_2+\frac{\partial A}{\partial y_2}\delta y_2+\frac{\partial A}{\partial z_2}\delta z_2\right)+\left(\frac{\partial A}{\partial \phi_1}\delta\phi_1+\frac{\partial A}{\partial \lambda_1}\delta\lambda_1\right)$$
$$=(\mathrm{O}-\mathrm{C})_A+v, \qquad \text{(C.48)}$$

where

$$\frac{\partial A}{\partial x_2}=-\frac{\partial A}{\partial x_1}=\frac{\operatorname{cosec}\zeta}{L}(-\cos A\sin\lambda+\sin A\sin\phi\cos\lambda)\operatorname{cosec}1'',$$

$$\frac{\partial A}{\partial y_2}=-\frac{\partial A}{\partial y_1}=\frac{\operatorname{cosec}\zeta}{L}(\cos A\cos\lambda+\sin A\sin\phi\sin\lambda)\operatorname{cosec}1'',$$

$$\frac{\partial A}{\partial z_2}=-\frac{\partial A}{\partial z_1}=-\frac{\operatorname{cosec}\zeta}{L}\sin A\cos\phi\operatorname{cosec}1'',$$

$$\frac{\partial A}{\partial \phi_1}=\sin A\cot\zeta \quad \text{and} \quad \frac{\partial A}{\partial \lambda_1}=-\cos\phi\cos A\cot\zeta+\sin\phi.$$

In the above, L, ζ, and A have suffix 12, and ϕ, λ have suffix 1. A, ϕ, and λ are in seconds. For the derivation of these differential coefficients substitute

$$(x_2-x_1)/L=l,$$

etc. from (C.22) in (C.30), giving

$$\begin{bmatrix} \sin A \sin \zeta \\ \cos A \sin \zeta \\ \cos \zeta \end{bmatrix} = \frac{1}{L} \begin{bmatrix} -\sin \lambda & \cos \lambda & 0 \\ -\sin \phi \cos \lambda & -\sin \phi \sin \lambda & \cos \phi \\ \cos \phi \cos \lambda & \cos \phi \sin \lambda & \sin \phi \end{bmatrix} \begin{bmatrix} x_2 - x_1 \\ y_2 - y_1 \\ z_2 - z_1 \end{bmatrix}. \tag{C.49}$$

Then from the third row of (C.49)

$$\cos \zeta = \frac{1}{L}\{\cos \phi \cos \lambda (x_2 - x_1) + \cos \phi \sin \lambda (y_2 - y_1) + \sin \phi (z_2 - z_1)\}, \tag{C.50}$$

whence differentiation (remembering that L is also variable) gives $\partial\zeta/\partial x_2$, $\partial\zeta/\partial y_2$, and $\partial\zeta/\partial z_2$ as required for (*e*) below.

And the first row of (C.49) is

$$\sin A \sin \zeta = (1/L)\{-\sin \lambda (x_2 - x_1) + \cos \lambda (y_2 - y_1)\}, \tag{C.51}$$

whence differentiation gives $\partial A/\partial x_2$, $\partial A/\partial y_2$, and $\partial A/\partial z_2$, using the expressions for $\partial\zeta/\partial x_2$, etc., already found. The algebra is heavy. For a geometrical derivation of (C.48) see [113], 3 Edn, § C.03.

The changes $\partial A/\partial\phi_1 (= -\partial A/\partial\theta_V)$, and $\partial A/\partial\lambda_1 (= \partial A/\partial\psi_V)$ are readily seen as the consequence of moving V_1 in Fig. C.1. The second term $\sin \phi_1$ in $\partial A/\partial\lambda_1$, which is recognizable as the Laplace correction $\eta \tan \phi$ of (2.4), is constant for all azimuths about any one vertical, and consequently disappears in a horizontal angle A_{213}, as in (*d*) below. Note that $\cot \zeta$ is generally small, so that $\partial A/\partial\phi_1$ and the first term in $\partial A/\partial\lambda_1$ are generally small also.

In (C.48) A_{12} is not affected by changes in ϕ_2 or λ_2.

(*d*) Observed *horizontal angle*, A_{213}, in which rotation from 2 to 3 through the angle is clockwise. The observation equation contains eight terms for the direction 1–3 as in (C.48), and eight more with changed signs, for the direction 1–2, involving eleven unknowns, $\delta x_1 \delta y_1 \delta z_1$, $\delta x_2 \delta y_2 \delta z_2$, $\delta x_3 \delta y_3 \delta z_3$, and $\delta\phi_1$, $\delta\lambda_1$. Not $\delta\phi_2$, $\delta\lambda_2$, $\delta\phi_3$, or $\delta\lambda_3$.

It will probably always be more convenient to compute with angles, than with directions and a Z-term as was used in § 2.22 (*b*).

(*e*) Observed vertical angle or *zenith distance*, ζ (corrected for refraction), from $(x_1 y_1 z_1)$ to $(x_2 y_2 z_2)$. The equation is of the same form as (C.48), writing ζ for A. The differential coefficients, derived as in (*c*) after (C.50), or as in [113], 3 Edn, § C.03, are

$$\frac{\partial\zeta}{\partial x_2} = -\frac{\partial\zeta}{\partial x_1} = -\frac{1}{L}(\cos A \cos \zeta \sin \phi \cos \lambda + \sin A \cos \zeta \sin \lambda + \sin \zeta \cos \phi \cos \lambda) \operatorname{cosec} 1'',$$

$$\frac{\partial\zeta}{\partial y_2} = -\frac{\partial\zeta}{\partial y_1} = -\frac{1}{L}(\cos A \cos \zeta \sin \phi \sin \lambda - \sin A \cos \zeta \cos \lambda + \sin \zeta \cos \phi \sin \lambda) \operatorname{cosec} 1'',$$

$$\frac{\partial\zeta}{\partial z_2} = -\frac{\partial\zeta}{\partial z_1} = \frac{1}{L}(\cos A \cos \zeta \cos \phi - \sin \zeta \sin \phi) \operatorname{cosec} 1'',$$

$$\frac{\partial\zeta}{\partial\phi_1} = -\cos A \quad \text{and} \quad \frac{\partial\zeta}{\partial\lambda_1} = -\sin A \cos \phi. \tag{C.52}$$

L, ζ, and A above have suffix 12, and ϕ, λ have suffix 1. ζ, ϕ, λ are in seconds.

(*f*) Observed *RA and declination* of the line from $(x_1y_1z_1)$ to $(x_2y_2z_2)$, as by stellar photography, §§ 7.23 and 7.24. Formula (C.19) gives $\theta = 90° - \delta$ and $\psi =$ RA − GAST, but polar motion terms should be included as in (7.97) if significant. These values of θ and ψ give the direction of the line.

From (C.22) and (C.21).

$$(y_2 - y_1) = (x_2 - x_1)\tan\psi. \tag{C.53}$$

Whence
$$\frac{\partial\psi}{\partial x_2} = -\frac{1}{L}\frac{\sin\psi}{\sin\theta}, \quad \frac{\partial\psi}{\partial y_2} = \frac{1}{L}\frac{\cos\psi}{\sin\theta}, \quad \text{and} \quad \frac{\partial\psi}{\partial z_2} = 0. \tag{C.54}$$

Similarly differentiating $(z_2 - z_1) = L\cos\theta$ gives

$$\frac{\partial\theta}{\partial x_2} = \frac{1}{L}\cos\theta\cos\psi, \quad \frac{\partial\theta}{\partial y_2} = \frac{1}{L}\cos\theta\sin\psi, \quad \text{and} \quad \frac{\partial\theta}{\partial z_2} = -\frac{1}{L}\sin\theta. \tag{C.55}$$

The observation equations are then

$$\cos\theta\cos\psi(\delta x_2 - \delta x_1)/(L\sin 1'') + \cos\theta\sin\psi(\delta y_2 - \delta y_1)/(L\sin 1'')$$
$$-\sin\theta(\delta z_2 - \delta z_1)/(L\sin 1'') = (\mathrm{O-C})_\theta + v \tag{C.56}$$

and

$$-\operatorname{cosec}\theta\sin\psi(\delta x_2 - \delta x_1)/(L\sin 1'') + \operatorname{cosec}\theta\cos\psi(\delta y_2 - \delta y_1)/(L\sin 1'')$$
$$= (\mathrm{O-C})_\psi + v. \tag{C.57}$$

(*g*) Observed direction of the vertical, *astronomical latitude or longitude*, ϕ, λ. The equations take the simple form

$$\begin{aligned} \delta\phi &= (\text{Observed} - \text{provisional}) + v, \\ \delta\lambda &= (\text{Observed} - \text{provisional}) + v. \end{aligned} \tag{C.58}$$

(*h*) Observed *height*. At a point $(x_1y_1z_1)$ the height H $(= N + h)$ above some reference spheroid may have been observed by independent means, such as by spirit levelling coupled with some estimate or determination of the separation of geoid and spheroid. Then, provided the spheroid concerned is centred on the origin of x, y, z, as in § C.01, the observation equation is

$$\delta x_1\cos\phi\cos\lambda + \delta y_1\cos\phi\sin\lambda + \delta z_1\sin\phi = (\mathrm{O-C})_H + v. \tag{C.59}$$

The values of ϕ and λ, and also the computed value of H, are got from (C.2) to (C.6), in which $N + h = H$.

Observations which are to be treated as infallible. If particular values of any of the unknowns x, y, z, ϕ, or λ are to be held fixed at any station or stations, it is only necessary to omit the relevant δx, δy, δz, $\delta\phi$, or $\delta\lambda$'s from any observation equations in which they might otherwise occur.

The length L, azimuth A or the direction (RA, δ), or (l, m, n) of any line, or the spheroidal height of any point, may be held fixed in various ways.

(i) If the x, y, and z at both ends of a line are fixed as above, its length, azimuth, and direction, and also the heights of its ends, are automatically fixed. Their observation equations should therefore be omitted.

(ii) From the observation equation whose observed value is to be treated as infallible, one unknown may be eliminated, as in § 2.23 (*e*) and (*f*), and replaced by the other unknowns in all other equations in which it occurs. Some care is required to avoid selecting an unsuitable unknown to eliminate.

(iii) In preference to (ii) it will generally be more convenient to treat the observation as fallible, but with a high weight.

Computation by conditions. As an alternative to variation of coordinates it is possible to compute in three dimensions by adjusting conditions. Details are given in [113], 2 Edn., pp. 195–7.

C.03. Variation of coordinates. Alternative method

As an alternative to the method of § C.02, [582] gives a method of computing, using the variation of cartesian coordinates x, y, z, which differs from § C.02 as follows.

(*a*) Heights above the spheroid are assumed to be known, as given by spirit levelling and geoidal section, or in some cases by vertical angles. This is assured by the elimination of one of the unknowns δx, δy, or δz.

(*b*) The direction of the vertical at every station (its astronomical latitude and longitude) is in the first instance assumed to be equal to the provisional geodetic latitude and longitude. This assumption can ordinarily be sufficiently accurate for the reduction of observed horizontal angles, for which purpose errors of several seconds of arc in the deviation are immaterial, and are customarily ignored in conventional methods of computation.

See also [121A].

APPENDIX D

THEORY OF ERRORS

D.00. Different types of error

ERRORS may be classified as follows.

(*a*) *Blunders.* These are generally due to carelessness. A blunder may be large and easily detectable, or smaller and more dangerous, or very small and indistinguishable from a random error. They are detected by repetition and by external checks, such as closing a traverse or substituting the solution of an equation in the original.

(*b*) *Systematic errors.* A systematic error is one that occurs with the same sign, and often with a similar magnitude, in a number of consecutive or otherwise related observations. Such as, for example, when a base is measured with a wrongly calibrated tape. There will, of course, be random errors in addition. Repetition does little or nothing to reduce the ill effect of systematic errors, which are therefore a most undesirable feature of any set of observations. Most of the care in making observations is directed towards eliminating or correcting systematic errors. See also § D.07.

(*c*) *Periodic errors* are such that in a complete set of observations there corresponds to every individual error another which is necessarily more or less equal and opposite. In a limited series the cancellation may not be quite exact, but the error of the mean of n observations may be expected to be $1/n$ of that of a single measure, or less. Example. The effect of the graduation error of a theodolite on an angle measured by repetition (§ 1.17) right round the 360° of the divided circle.

(*d*) *Random errors,* sometimes called accidental or casual. For practical purposes these can be described as all that remain. They are numerous, individually small, and each is as likely to be positive as negative. So far as they are concerned, the error of the mean of n observations is likely to be $1\surd/n$ that of a single observation. Their effect is consequently reduced by repetition, although a very long programme is needed to reduce it as much as ten times.

D.01. Frequency distribution of measurements

The theory of errors is based on the theory of statistics. Let the heights of a large number of men be measured. Let H_m be the mean, and let $h-H_m=v$ be the deviation of each separate measure from the mean. Then let the v's be grouped into classes such as -10 to -8 cm, -8 to -6 cm, etc., and let the number in each class be recorded. Ordinarily there will be many measures in the middle classes and few in the extremes. The distribution will be like Fig. D.1, which is known as a *frequency polygon.*

If the measures are very numerous, and the range of each class small, the polygon becomes, in the limit, a smooth curve as in Fig. D.2. The curve may not

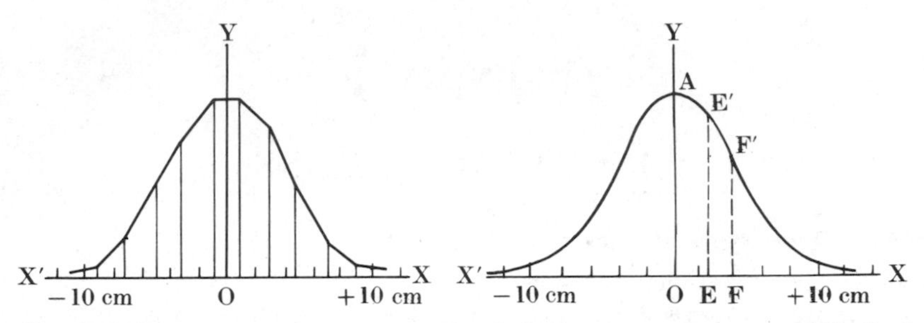

FIG. D.1. Frequency polygon. FIG. D.2. Frequency curve, normal distribution.

be symmetrical nor exactly of the form shown,† but Fig. D.2 shows a distribution which does often occur, and which is known as the *normal distribution.* If it can be assumed that the differences between the objects measured result only from a large number of small causes each as likely to cause an increase as a decrease, then it can be shown, [619], pp. 169–81 and [598], pp. 168–73, that this normal frequency will result, and that the equation of the curve will be

$$y = \frac{n}{\sigma\sqrt{(2\pi)}}\exp(-x^2/2\sigma^2), \tag{D.1}$$

where n is the total number of measures and σ is the standard deviation, see below. The multiplier $n/\sqrt{(2\pi)}$ is arbitrarily introduced so as to make the area under the curve numerically equal to n.

Some types of objects are more homogeneous than others. Men vary more than 2-metre survey rods. A measure of the homogeneity, or of the scatter, of a sample is given by

$$\sigma^2 = \sum v^2/n, \tag{D.2}$$

the mean of the squares of all the deviations. This is called the *variance,* and its positive‡ square root σ, which occurs in (D.1), is the *standard deviation* or *root-mean-square* (r.m.s.) *deviation.* So the heights of a group of men may be summarized as (say) 173 ± 6.3 cm, where 173 cm is the mean and 6.3 cm is the standard deviation.

In Fig. D.2 the number of deviations whose magnitude is between zero and such a value as +OE is

$$(\text{the total } n) \times \{(\text{area OEE'A}) \div (\text{total area XOX'AX})\}. \tag{D.3}$$

The probability that any particular deviation is numerically less than OE, i.e. between + and −OE, is therefore twice the area OEE′A divided by the whole

† For instance, if the men were of two very dissimilar races it might have two distinct maxima, or if children were included it would trail out further on the negative side than on the positive.

‡ Standard (and probable) errors are neither positive nor negative. They are conveniently preceded by the sign ±, which indicates their nature, but does not mean that their sign is ambiguous.

area under the curve, a proper fraction. If OF is the standard deviation it follows from (D.1) that 0.683 of the deviations will lie between + and $-\sigma$. The ordinate EE′ which bisects the area OAE′XO is such that half the deviations lie between +OE and −OE. Let this value of OE $= q$. Then in the normal distribution

$$q = 0.6745\sigma. \tag{D.4}$$

The probability that any particular deviation lies between + and $-k\sigma$ is

$$(2\sqrt{\pi}) \int_0^{k/\sqrt{2}} \exp(-t^2)\,\mathrm{d}t. \tag{D.5}$$

The probability that it lies between + and $-kq$ is

$$(2/\sqrt{\pi}) \int_0^{0.477k} \exp(-t^2)\,\mathrm{d}t. \tag{D.6}$$

And the probability that it lies between $x_1 + \frac{1}{2}\delta x$ and $x_1 - \frac{1}{2}\delta x$ is

$$\left(\frac{\delta x}{\sigma\sqrt{(2\pi)}}\right) \exp(-x_1^2/2\sigma^2). \tag{D.7}$$

Mathematical tables commonly include the *error function*,

$$\operatorname{erf}(x) \equiv (2/\sqrt{\pi}) \int_0^x \exp(-t^2)\,\mathrm{d}t, \tag{D.8}$$

from which (D.5) and (D.6) can be evaluated by putting $x = k/\sqrt{2}$ or $0.477k$ respectively. See [174], [466], and [414]. In Table D.1, column (2) gives the percentage probability that any particular deviation is between + and − the figure in column (1), and column (3) expresses the same thing in the form of odds against a particular deviation falling outside these limits.†

TABLE D.1

(1)	(2)	(3)
$\frac{1}{2}\sigma$	38%	0.61 : 1
σ	68.3‡	2.2 : 1
2σ	95.4	21 : 1
3σ	99.73	370 : 1
$\frac{1}{2}q$	26	0.35 : 1
q	50	1 : 1
$2q$	82	4.6 : 1
$3q$	95.7	22 : 1

† These figures are only correct when n is large. They are often good enough if n is more than about 7, but when $n = 5$ the 20 to 1 level is at $4q$ instead of at the normal $3q$. See [126], p. 164, ‘Student's t-table, in which for this simple example $(f+1)$ is the number of deviations.

‡ The similarity between this 0.683 and the 0.6745 of (D.4) is coincidence.

D.02. The standard deviation of the mean

Let the heights of 5000 normally distributed men be measured, with mean H and s.d. $= \sigma$. Then if they are divided at random into 50 groups of 100 each, the mean of each group will clearly be close to H, but not equal to it. In fact the frequency distribution of the 50-group means will form a normal curve like Fig. D.2, and experience supported by theory, (D.12) in § D.03, shows that the s.d. of the group means will in this case be $\sigma\sqrt{100}$. In general, the s.d. of the mean of p normally distributed measures will be $1/\sqrt{p}$ times the s.d. of a single measure. Provided the group means are reasonably numerous, their distribution is likely to be normal even though the distribution of the individual measures may not be.

D.03. Standard deviation of a function of several independent quantities

Let there be n quantities a whose mean is a_m and whose s.d. is σ_a, and let there be n others b with mean b_m and s.d. σ_b. Let each of the b's be associated with one of the a's, chosen at random, so that (for instance) large b's are not deliberately associated with large a's. Then the mean of $(a+b)$, the sum of the pairs, will be (a_m+b_m), and the deviation of each $(a+b)$ will be the sum of the deviations of the members of the pair, (v_a+v_b).

Then the variance of $(a+b)$ will be

$$(1/n)\sum (v_a+v_b)^2 = (1/n)\left(\sum v_a^2 + \sum v_b^2 + 2\sum v_a v_b\right)$$

$$= \sigma_a^2 + \sigma_b^2 + (2/n)\sum (v_a v_b). \tag{D.9}$$

In this expression the v_a^2 and v_b^2's are all positive, while products $v_a v_b$ are as likely to be positive as negative, since the association of the a's and b's is random. So, provided n is not too small, we have approximately,

the variance $\qquad \sigma_{(a+b)}^2 = \sigma_a^2 + \sigma_b^2$

and the standard deviation $\qquad \sigma_{(a+b)} = \sqrt{(\sigma_a^2+\sigma_b^2)} \qquad$ (D.10)

and, if $\sigma_a = \sigma_b$, we have $\sigma_{(a+b)} = \sigma_a\sqrt{2}$.

Contrast this with the obvious $\sigma_{2a} = 2\sigma_a$ (since each deviation is doubled).

Similarly the s.d. of the sum of p independent quantities, each with s.d. $= \sigma_a$, is

$$\sigma_{(a+b+\ldots+p)} = \sigma_a \times \sqrt{p}. \tag{D.11}$$

And the s.d. of the mean of these p quantities, their sum divided by their number, is

$$\sigma_{\text{mean}} = \sigma_a \div \sqrt{p}, \text{ as in § D.02.} \tag{D.12}$$

The s.d. of the difference of two independent quantities is clearly the same as that of their sum, since the term $2\sum v_a v_b$ in (D.9) is near zero, and

$$\sigma_{(a-b)} = \sqrt{(\sigma_a^2+\sigma_b^2)}. \tag{D.13}$$

And in general if F is a function of a number of independent measured quantities $a, b, \ldots,$ with s.d.'s $\sigma_a, \sigma_b, \ldots,$ the s.d. of F is†

$$\sigma_F = \sqrt{\{\sigma_a^2(\partial F/\partial a)^2 + \sigma_b^2(\partial F/\partial b)^2 + \ldots\}}. \tag{D.14}$$

D.04. Correlation

§ D.03 deals with pairs of measured quantities in which there is no reason to suppose that the deviation of one member of a pair is numerically related to that of the other. This may not always be so, and the measures are then said to be correlated.

Suppose two distances a and b are each measured six times, once with each of six different tapes as in Table D.2, in which the two measures shown in the same line have been made with the same tape. If the tapes had been well calibrated but the measures otherwise carelessly made, the two deviations in each pair would presumably be independent. On the other hand, if the measures had been carefully made, but with uncalibrated tapes, the pairs of deviations would probably be strongly correlated. We need a numerical means of calculating and expressing the amount of correlation in such cases.

The line below the table gives the variances σ_a^2 and σ_b^2 computed by (D.2). The last column contains the products of $v_a v_b$ and their mean $\sum v_a v_b/n$, which is known as the *covariance*, σ_{ab}. The *coefficient of correlation* is then defined as

$$r_{ab} = \frac{\sigma_{ab}}{\sigma_a \sigma_b} = \frac{\text{covariance } ab}{+\sqrt{(\text{variance } a \times \text{variance } b)}}. \tag{D.15}$$

TABLE D.2

Tape	Distance a	v_a	v_a^2	Distance b	v_b	v_b^2	$v_a v_b$
1	100.4	+0.1	0.01	150.3	0.0	0.00	0.00
2	100.0	−0.3	0.09	150.0	−0.3	0.09	+0.09
3	100.5	+0.2	0.04	150.7	+0.4	0.16	+0.08
4	100.2	−0.1	0.01	150.3	0.0	0.00	0.00
5	100.5	+0.2	0.04	150.5	+0.2	0.04	+0.04
6	100.2	−0.1	0.01	150.0	−0.3	0.09	+0.03

Mean $a = 100.3$. $\sum v_a^2/n = 0.033$. Mean b 150.3. $\sum v_b^2/n = 0.063$. $\sum v_a v_b/n = +0.04$.

The value of r necessarily lies between +1 and −1. In this example

$$r = (0.04) \div \sqrt{(0.033 \times 0.063)} = +0.87,$$

indicating that most of the variation between the different measures lies in the calibration of the tapes. This conclusion could have been more easily reached, qualitatively, by plotting the results as in Fig. D.3.

As in § D.03, if associated v_a's and v_b's are independent, the products $v_a v_b$ will tend to have equal numbers of + and − signs, and if n is large the covariance $\sum v_a v_b/n$ will tend to be small. The coefficient r is then nearly zero.

† *Proof.* On the lines of (D.9) it is easily seen to be true of a linear function. Then let $F = F_0 + \Delta F$, where F_0 is arbitrary and invariable, so that $\sigma_F^2 = \sigma_\Delta^2 F$ and $\sigma_a^2 = \Delta_{\Delta a}^2$, etc. We have $\Delta F = (\partial F/\partial a)\Delta a + (\partial F/\partial b)\Delta b + \ldots$, which is linear, so (D.14) gives $\sigma_{\Delta F}$.

On the other hand, if the v_a's and v_b's are actually linearly related, so that in any pair $v_b = kv_a$ where k is a constant (+ or −ve) for all pairs, we will have $\sigma_b = \pm k\sigma_a$ (both σ_a and σ_b being +), and $\sigma_{ab} = k\sigma_a^2$, whence $r = \pm 1$, the sign being that of k or of σ_{ab}. If $r = +$ or -1, the two sets of measures are said to be perfectly correlated, positively or negatively. Negative correlation implies that if any measure of a is too large the corresponding measure of b is likely to be too small. For instance, in measures of two angles AOB and BOC in a round of angles, an error in the bisection of station B tends to produce negative correlation in the measures of the two angles.

The variance of the sum $(a+b)$ will be, as in (D.9),

$$\begin{aligned}\sigma^2_{(a+b)} &= \sigma_a^2 + \sigma_b^2 + 2\sigma_{ab}\\ &= \sigma_a^2 + \sigma_b^2 + 2r\sigma_a\sigma_b, \end{aligned}\qquad \text{(D.16)}$$

and the rule $\sigma^2_{(a+b)} = \sigma_a^2 + \sigma_b^2$, (D.9), only holds when $r = 0$.

At the other extreme, if $r = 1$, $\sigma^2(a+b) = (\sigma_a + \sigma_b)^2$, and if $\sigma_a = \sigma_b$, $\sigma^2_{(a+b)} = 4\sigma_a^2$, instead of $2\sigma^2$ as it would be if they were uncorrelated.

Similarly,
$$\begin{aligned}\sigma^2_{(a-b)} &= \sigma_a^2 + \sigma_b^2 - 2\sigma_{ab}\\ &= \sigma_a^2 + \sigma_b^2 - 2r\sigma_a\sigma_b. \end{aligned}\qquad \text{(D.17)}$$

And
$$\begin{aligned}\sigma^2_{(a+b+c)} &= \sigma_a^2 + \sigma_b^2 + \sigma_c^2 + 2\sigma_{ab} + 2\sigma_{bc} + 2\sigma_{ca}\\ &= \sigma_a^2 + \sigma_b^2 + \sigma_c^2 + 2r_{ab}\sigma_a\sigma_b + 2r_{bc}\sigma_b\sigma_c + 2r_{ca}\sigma_c\sigma_a. \end{aligned}\qquad \text{(D.18)}$$

In general, if F is a function of a number of measured quantities a, b, etc.

$$\sigma_F^2 = \sigma_a^2(\partial F/\partial a)^2 + \sigma_b^2(\partial F/\partial b)^2 + \ldots + 2\sigma_{ab}\left(\frac{\partial F}{\partial a}\frac{\partial F}{\partial b}\right) + \ldots .\dagger \qquad \text{(D.19)}$$

The covariance of two linear functions of $a, b, \ldots, n$,

$$F \equiv f_1 a + \ldots + f_n n \qquad \text{and} \qquad G \equiv g_1 a + \ldots + g_n n$$

is given (by direct multiplication) as

$$\sigma_{FG} = f_1 g_1 \sigma_a^2 + f_2 g_2 \sigma_b^2 + \ldots + \sigma_{12}(f_1 g_2 + f_2 g_1) + \text{all pairs}. \qquad \text{(D.20)}$$

It follows from (D.18) that if σ_m is the s.d. of the mean of p positively correlated quantities each with s.d. σ, σ_m will vary between

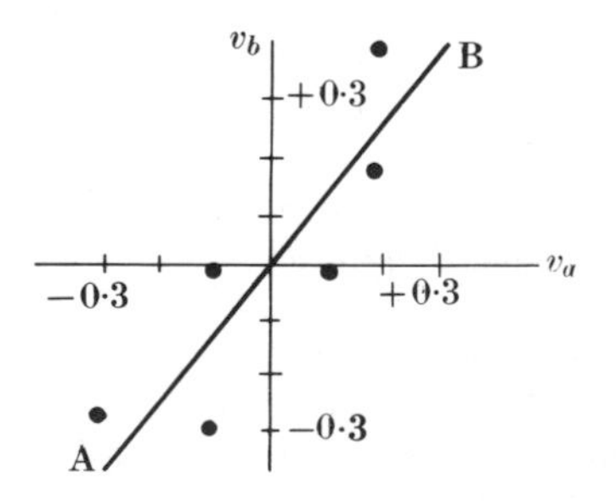

FIG. D.3. Regression line.

$\sigma \div \sqrt{p}$ as in (D.12), if all correlation coefficients are zero,

and σ, if all coefficients are +1.

Regression. When two quantities are correlated, like a and b in Table D.2, their correlation may be seen by plotting v_a and v_b on a diagram as in Fig. D.3.

If the coefficient of correlation is near 1 the points will fall roughly on a straight line, while if

† (D.19) comes from (D.18) in the same way as (D.14) comes from (D.9).

it is near zero, they will cover a square or rectangular area. In the former case a *regression line* AB can be drawn through them, by eye or by least squares, such that the sum of the squares of the vertical distances between each point and the line will be a minimum. Least squares gives the equation of the line as

$$v_b = \frac{\sigma_{ab}}{\sigma_a^2} v_a = r\frac{\sigma_b}{\sigma_a} v_a,$$

or

$$(b - \text{mean } b) = \frac{\sigma_{ab}}{\sigma_a^2}(a - \text{mean } a). \qquad \text{(D.21)}$$

The ratio σ_{ab}/σ_a^2 is known as the *coefficient of regression of b on a.*

Minimizing the horizontal distances between the points and the line will give another line of best fit

$$v_a = \frac{\sigma_{ab}}{\sigma_b^2} v_b \qquad \text{(D.22)}$$

in which σ_{ab}/σ_b^2 is the coefficient of regression of a on b.

From these lines of best fit, given one measure or deviation of a pair it is possible to estimate the other with more or less confidence, depending on the coefficient of correlation. Given a, (D.21) gives the estimated value of b with s.e. $= \sigma_b\sqrt{(1-r^2)}$, while the s.e. of an a deduced from a given b will be $\sigma_a\sqrt{(1-r^2)}$. The estimate is correct if $r = +$ or -1, while if $r = 0$ there is no reduction of the original σ_b or σ_a.

D.05. The frequency distribution of errors

Instead of measuring a number of similar objects, consider repeated measures of a single object. If n measures, h, are made of a distance or angle whose mean value is H_m, experience shows that the frequency distribution of the deviations $(h - H_m)$ tends to resemble the normal distribution of Fig. D.2. The formulae of §§ D.01–D.03 then apply. In particular the r.m.s. of the deviations is a measure of their scatter and is a guide to the precision of the observations.

If the measures are of something more complex than the simple length of a rod, there may be a few with abnormal errors very many times greater than the s.d., such as cannot occur in a normal distribution. They are described as blunders in § D.00 (a), and it must be assumed that they can be successfully eliminated.

There is, of course, no guarantee that the mean value H_m will equal H, the true value. If any systematic error has been made, H_m cannot be expected to equal H. Here also the theory of errors often has to work on the assumption that systematic errors have been avoided. But see § D.07.

Even if no systematic errors have been made, the mean of a finite number of observations cannot be expected to equal the true value exactly. If the s.d. of each measure is σ, that of H_m is $\sigma/\sqrt{n}$. The sum of the squares of the errors $(h - H)$ must necessarily be greater than $\sum (h - H_m)^2$, for it is a property of the mean that the sum of the squares of the deviations from it are a minimum. It follows that the r.m.s. error, known as the *standard error* (s.e.), is greater than the standard deviation. In fact, [598], p. 205,

$$\text{s.e. of a single measure} = \text{s.d.} \times \sqrt{\left\{\frac{n}{(n-1)}\right\}} = \sqrt{\{\textstyle\sum v^2/(n-1)\}},$$

$$\text{s.e. of the mean} = \sqrt{\{\textstyle\sum v^2/n(n-1)\}}. \tag{D.23}$$

The v's are deviations from the mean. These are known as *Bessel's formulae.*

The numerical difference between the standard error and the standard deviation is only the factor $\sqrt{n}/\sqrt{(n-1)}$ in (D.23), which is close to 1 unless n is very small. When errors are being considered, the symbol σ is used for the standard error.

Alternative formulae, known as *Peters*'s are as follows

$$\text{s.e. of a single measure} = 1.25\textstyle\sum |v| \div \sqrt{\{n(n-1)\}},$$

$$\text{s.e. of the mean} = 1.25\textstyle\sum |v| \div n\sqrt{(n-1)}, \tag{D.24}$$

where $|v|$ is the sum of the deviations, without regard to sign.

D.06. The probable error

The standard error σ is a measure of the scatter of the observations and is computed by (D.2), (D.8)–(D.14), (D.23) or (D.24). In certain circumstances—numerous observations, normal distribution, absence of systematic error—the chances are 68 : 32 that the actual error does not exceed the s.e., and this second aspect of the s.e., when it is reasonably true, is the more interesting. It would be convenient to have a different name for the measure of probability, but none exists.

The *probable error*, q in (D.4) is a term which also has two measings, namely:

(a) 0.6745σ. Regarded as a measure of scatter, this is redundant and useless.

(b) The magnitude which any error has a 50 : 50 chance of exceeding. This is a useful concept.

In the ideal circumstances mentioned above, the two definitions of the p.e. are the same, but it is the common experience of surveyors that if σ is calculated from the scatter of a set of observations, the magnitude which any error has a 50 : 50 chance of exceeding is substantially greater than 0.67σ.

Both the s.e. and the p.e. are recorded as (e.g.)± 0.75, and the two can be confused. If both were intended to indicate the same level of probability, the s.e. could very well be understood to refer to scatter and the p.e. to probability, but this is not the case and neither can have its current meaning changed.

It would be inconvenient to record the measure of scatter at any level other than σ's 68 : 32, because of the close association between variance and covariance, as in § D.04, (D.67), and (D.87), although for assessments of probability 50 : 50 may seem more natural than 68 : 32. Even more useful from the probability point of view is the 95 : 5 level, indicating an error which is 'unlikely' to be exceeded. This is given, in ideal circumstances, by either $3q$ or 2σ, Table D.1.

Formerly, surveyors always expressed their results in terms of p.e.'s, but almost everyone else has adopted the s.e. and the 68 : 32 level. A little reluctantly, we now do the same, and we use the expression standard error (σ) to indicate both the results derived from scatter, and our estimate of the 68 : 32 probability. The context must be a guide to which is intended. The words 'internal' or 'apparent' s.e. will emphasize simple scatter, while 'external' or 'estimated' or '68 : 32' will suggest truer indications of probability.

Bessel's formulae for the conventional probable error are

$$\text{p.e. of a single error} = 0.6745\sqrt{\{\sum v^2/(n-1)\}}.$$
$$\text{p.e. of the mean} = 0.6745\sqrt{\{\sum v^2/n(n-1)\}}. \quad \text{(D.25)}$$

Peters's formulae, D(24) gives

$$\text{p.e. of a single measure } = 0.85\sum |v| \div \sqrt{\{n(n-1)\}} \approx 0.85(\text{average error}). \quad \text{(D.26)}$$

The figure 0.6745 is only theoretically correct when n is very large. See [126], p. 164, 'Student's t-table. For $n=2$, 5, 10, and 30, better values are 1.00, 0.74, 0.70, and 0.68. For geodetic purposes 0.7 is precise enough, but it is sometimes convenient to print 0.6745 as a means of recognition.

D.07. Systematic errors

Continuing from § D.00 (*b*). if the members of a series of observations, instead of only being liable to random error, also share some single source of error of significant size, which has the same or similar effect on them all, they suffer from systematic error. The systematic error may not, of course, be constant. It may vary with time, or with the temperature, or with anything else. Various situations may be considered.

(*a*) The systematic error may be truly constant, in which case it is relatively easily dealt with by such methods as distributing the closing error round a levelling circuit, by summing the three angles of a triangle, or by calibration, namely the use of the apparatus to measure a quantity of known magnitude.

(*b*) The systematic error may reverse itself at frequent intervals, as does collimation error in a theodolite when face is changed, or it may vary periodically, with zero as its mean, at fairly rapid intervals. It then becomes a periodic error as in § D.00 (*c*), and it is only necessary to ensure that the programme contains an exact number of periods, or sufficient for the effect of an odd fraction to be negligible.

(*c*) The systematic error may remain substantially constant for a number of observations, and may then abruptly change, Fig. D.4. Let $\pm\sigma$ be the random error throughout. Let the systematic error have a mean value S. Let it change after every n observations, n being assumed constant, although there is no reason why it should be, and let the standard (68 : 32) deviation of the systematic error from S be $\pm s$.

The mean value S of the systematic error can perhaps be determined, and so eliminated, by calibration or otherwise, as in (*a*) above. Let it therefore be provisionally assumed to be zero.

The s.e. (68 : 32 probability) of the mean of each group of n observations will

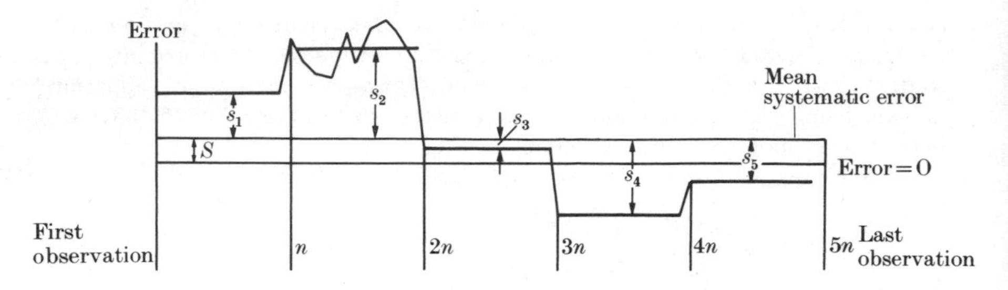

FIG. D.4. Systematic error. Systematic error is assumed to change abruptly after every n observations, having values $(S-s_1)\ldots(S-s_5)$. The r.m.s. value of $s_1\ldots s_5$ is $\pm s$. There is also a random error with r.m.s. value $\pm\sigma$, shown only between n and $2n$. As drawn, $n = 10$, and the number of groups, $m, = 5$.

clearly be $\pm\{(\sigma^2/n)+s^2\}^{\frac{1}{2}}$, which varies between $\pm s$, if s is dominant, and the usual $\pm\sigma/\sqrt{n}$ if s is negligible. Then if the whole series comprises m such groups, the s.e. of the general mean will be

$$\pm(1/\sqrt{m})\{(\sigma^2/n)+s^2\}^{\frac{1}{2}}, \tag{D.27}$$

which similarly varies between $\pm s/\sqrt{m}$ and $\pm\sigma/\sqrt{(mn)}$.

In the case where s is dominant, the actual variance would simple be s^2, and (D.23) would give approximately $s/\sqrt{(mn)}$ for the s.e. of the general mean. This is $1/\sqrt{n}$ times the true value $s/\sqrt{m}$ given above, which is a serious error if n is large.

(*d*) An even more intractable situation is that the systematic error may vary smoothly and unpredictably as it doubtless often does.

The only satisfactory remedy for systematic error is the careful consideration of every conceivable cause. The instrument design and the observation programme must then be so arranged that every significant cause is eliminated by either cancellation or calibration or correction. Classical examples are the measurement of invar bases, § 1.30, spirit levelling, §§ 3.06–3.07, and longitude observations, § 4.29. When the causes of systematic error are not clear, field observations may perhaps be carried out in a wide variety of conditions. Unexpectedly large variations will suggest that something is wrong, and analysis may suggest the cause. On the other hand, unnecessary variation of the conditions of work may upset a well-considered plan of calibration or of cancellation of error. There is no good substitute for a clear understanding of the causes.

Calibration. Let the apparatus be calibrated by measuring a known quantity after every p field observations, and let $\pm\sigma_c$ be the s.e. of the resulting calibration correction. The result of each calibration will (roughly speaking) be applied as a correction to the $\frac{1}{2}p$ observations made before and the $\frac{1}{2}p$ after the calibration, and its effect on them will be systematic. If $\pm\sigma$ is the s.e. of each field observation, the s.e. of the mean of the group of p field observations will be

$$\pm(\sigma^2/p+\sigma_c^2)^{\frac{1}{2}}. \tag{D.28}$$

This assumes that the systematic error of the apparatus is constant during the p observations and the associated calibration. In Fig. D.4 (a) a similar result would be got if a calibration was done in the middle of each of the m groups. This, of course, is not possible to arrange unless the causes of the abrupt changes are known and can be foreseen, which is unlikely. If the variation of (numerous) calibration corrections shows more scatter than can be attributed to the random errors of the calibration observations themselves, it will be clear that the systematic error is variable, and if the variation is too great, satisfactory results will not be got until the cause is discovered.

At the best, it may be hoped that the mean error of all the calibration corrections will tend to zero as their number increases.

D.08. Accuracy estimated from difference of two measures

Two observations are insufficient for (D.23) or (D.24) to give reliable results, but a figure for the s.e. can be got from the differences between a number of pairs of measures of different quantities, provided the circumstances are the same for all pairs. For if σ_1 is the r.m.s. of the differences between the two measures of each pair, the σ of the mean of each pair will be $\frac{1}{2}\sigma_1$.

Alternatively, if D is the average value of the differences of each pair, Peters's formula gives $0.63D$ for the s.e. of a mean.†

Similarly, [428], p. 155, and [246], p. 340, gives tables from which the standard deviation may be assessed from the average range (greatest minus least) of groups each of n measures, from which the following is abstracted.

$$\text{Average range of } n \text{ measures} = d \times \sigma \text{ of a single measure.}$$
$$\sigma \text{ of the mean of } n \text{ measures} = (\text{average range}) \div d\sqrt{n}.$$

n	d	$1/d\sqrt{n}$
2	1.128	0.63
3	1.693	0.34
4	2.059	0.24
5	2.326	0.19
10	3.078	0.10

When $n = 2$, the s.e. of the mean $= 0.63D$, as before.

D.09. Frequency distribution over an area or volume

Let the x and y coordinates of a point whose true position is $(0, 0)$ be determined with s.e.'s of σ_x and σ_y, and let the s.e. of r, the distance from $(0, 0)$, be σ_r. Then from (D.7), if errors of x and y are uncorrelated, the chance of any single determination falling within the limits x, $(x+\delta x)$, y and $(y+\delta y)$ is

$$(1/2\pi\sigma_x\sigma_y)[\exp(-x^2/2\sigma_x^2 - y^2/2\sigma_y^2)]\,dx\,dy. \tag{D.29}$$

Changing to polar coordinates, the chance of its falling within r, $(r+dr)$, θ, and $(\theta+d\theta)$ is

$$(1/2\pi\sigma_x\sigma_y)[\exp(-r^2\cos^2\theta/2\sigma_x^2 - r^2\sin^2\theta/2\sigma_y^2)]r\,dr\,d\theta. \tag{D.30}$$

† e.g. if a number of base lines are each measured twice, with an average difference of e ppm between the two measures, the apparent s.e. of each base is $0.63e$ ppm.

And if $\sigma_x = \sigma_y = \sigma$ this becomes

$$(1/2\pi\sigma^2)[\exp(-r^2/2\sigma^2)]r\,\mathrm{d}r\mathrm{d}\theta. \tag{D.31}$$

Then the chance of a determination being within r of the true position is

$$\frac{1}{\sigma^2}\int_0^r r[\exp(-r^2/2\sigma^2)]\,\mathrm{d}r = 1-\exp(-r^2/2\sigma^2)$$

$$= 1-\exp(-r^2/\sigma_r^2), \tag{D.32}$$

The distribution given by (D.31) is shown in Fig. D.5, and is not the same as the normal distribution of Fig. D.2. When r/σ is small the frequency is not a maximum, but near zero. The r.m.s. value of r, σ_r, is $\sigma\sqrt{2}$, as might be expected, but the value within which half the errors may be expected is $0.83\sigma_r$, not $0.67\sigma_r$, and 95 per cent may be expected within $1.73\sigma_r$, [225], p. 26.

When σ_x and σ_y are unequal (but still uncorrelated), σ_r^2 still equals $\sigma_x^2+\sigma_y^2$, but the frequency distribution is different, with a higher proportion of errors closer to $r/\sigma_r = 0$, and with the 50 : 50 probability nearer $0.67\sigma_r$.

Similarly in three dimensions, if $\sigma_x = \sigma_y = \sigma_z = \sigma = \sigma_r/\sqrt{3}$, and if errors of x, y, and z are uncorrelated, the chance of a determination being within a distance r of the true position is

$$\left\{\int_0^{r/\sigma} (r/\sigma)^2\exp(-r^2/2\sigma^2)\,\mathrm{d}(r/\sigma)\right\} \div \left\{\int_0^{\infty} (r/\sigma)^2\exp(-r^2/2\sigma^2)\mathrm{d}(r/\sigma)\right\}$$

$$= \sqrt{(2/\pi)}\left\{-\frac{r\sqrt{3}}{\sigma_r}\exp\left(-\frac{3r^2}{2\sigma_r^2}\right)+\sqrt{2}\int_0^{(r/\sigma_r)\sqrt{(3/2)}} \exp(-t^2)\,\mathrm{d}t\right\}. \tag{D.33}$$

Table D.3, amplifying Table D.1, compares the distributions in one, two, and three dimensions. See [37].

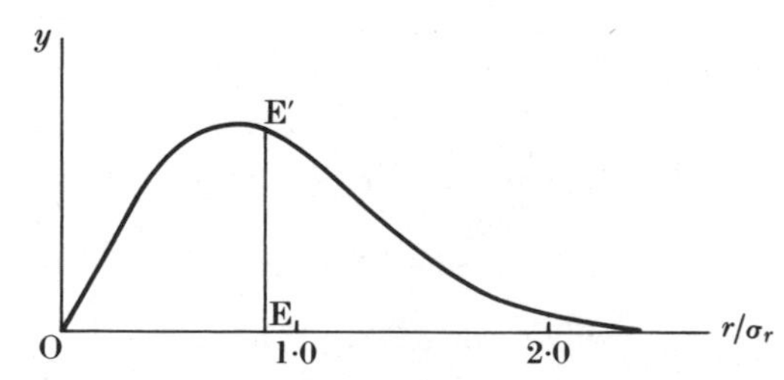

FIG. D.5. Distribution over an area. $y = (2/\sigma_r^2)r\exp(-r^2/\sigma_r^2)$. Total area = 1. EE′ bisects the area when OE equals 0.83.

TABLE D.3
Percentage deviations less than r/σ_r

r/σ_r	One dimension	Two dimensions	Three dimensions
0.5	38	23	14
1.0	68	63	61
1.5	87	89	91
2.0	95	98	99

In two dimensions $\sigma_r = \sigma\sqrt{2}$ if $\sigma_x = \sigma_y = \sigma$, and in three dimensions $\sigma_r = \sigma\sqrt{3}$ if $\sigma_x = \sigma_y = \sigma_z = \sigma$.

D.10. Weights

(*a*) *Definitions.* The weight of an observation is a measure of its reliability, and the best definition is

$$w = 1/(\text{s.e.})^2, \tag{D.34}$$

where the s.e. is the estimate of the 68 : 32 probability.

An alternative definition is

$$w = 1/\kappa^2(\text{s.e.})^2, \tag{D.35}$$

where κ^2 is any multiplier which has the same value for all the observations involved in any particular problem. Those who may still work with probable errors may put $\kappa = 0.67$. This or any other value of κ will give the same final results as when $\kappa = 1$, but see §§ D.15, D.17, and 2.27, where some care may be needed.

In both these definitions w is a dimensioned quantity, $(\text{length})^{-2}$ or (seconds of $\text{arc})^{-2}$ as the case may be. In principle, this is an essential property of weights used in least squares, § D.14, although the fact may be disguised, see § 2.24.

A third definition, which is effective in simple problems involving dimensionally similar quantities as in (*b*) and (*c*) below, is that weights may be pure numbers, expressing nothing but relative reliability. We measure an angle four times in the morning and twice in the afternoon. The instrument, observer and all other circumstances are thought to be the same, and the errors of all the measures are thought to be independent. So we ascribe weights of 4 to the morning mean and 2 to the afternoon: or 2 and 1 would do as well. Since the s.e. of a mean ordinarily varies as the square root of the number of observations, we are not in conflict with (D.35).

It is sometimes convenient to write u for $1/w$.

(*b*) *The weight of a function* of (similar) measured quantities. If a is measured with weight w_a and b with w_b, using the definition of either (D.34) or (D.35) and assuming no correlation, (D.10) gives

$$\sigma^2_{(a+b)} = \sigma^2_a + \sigma^2_b,$$

$$u_{(a+b)} = u_a + u_b, \tag{D.36}$$

$$w_{(a+b)} = \frac{w_a w_b}{w_a + w_b}.$$

The weight of their difference is the same, and (D.14) can be used to give the weight of any function of independent measures in terms of their separate weights.

(*c*) *The weighted mean.* Returning to the example of a measured angle in (*a*) above, it is perhaps obvious that the most probable value for the general mean is four times the morning mean plus twice the afternoon, divided by six. This is the same as taking the simple mean of all six measures.

In other words,

$$\text{weighted mean} = \frac{aw_a + bw_b}{w_a + w_b}. \tag{D.37}$$

And in general, if a quantity A is independently measured n times, with results $a_1, \ldots, a_n$ and weights $w_1, \ldots, w_n$, its most probable† value is given by

$$\begin{aligned} a_m &= (a_1 w_1 + \ldots + a_n w_n) \div (w_1 + \ldots + w_n) \\ &= \sum (aw) \div \sum w. \end{aligned} \tag{D.38}$$

And w_n, the weight of this mean, is $(w_1 + \ldots + w_n)$. (D.39)

If w_1, etc., are defined by (D.34) the s.e. of the mean is $1/\sqrt{w_m}$, and if they are defined by (D.35) it is $1/(\kappa\sqrt{w_m})$. If the w's are nothing but measures of relative reliability, the s.e. of the mean cannot be deduced from w_m.

Formula (D.39) is not the only guide to the s.e. of the mean. The estimated s.e.'s from which w_1, etc., are given by (D.34) may have been optimistic. Let $a_1 - a_m = v_1$, etc. Then the scatter of the residuals v provides a measure of the reliability of the a's. In fact,

$$\text{the s.e. of an observation of weight } w_1 = \sqrt{\left\{\frac{\sum wv^2}{(n-1)w_1}\right\}} \tag{D.40}$$

and

$$\text{the s.e. of the mean } a_m = \sqrt{\left\{\frac{\sum wv^2}{(n-1)\sum w}\right\}}. \tag{D.41}$$

In the absence of blunders, if (D.41) is greater than the s.e. given by (D.39), it suggests that the estimated s.e.'s of the measures a_1, etc., should (on average) have been greater in the same proportion.

D.11. Combination of bad observations with good

Consider two independent measurements, a and b, of the same object with s.e.'s σ and $\sigma\sqrt{k}$, so that their weights are in the ratio $1 : 1/k$. Let k be of the

† For proof, this is a very simple example of least squares, § D.14, the observation equations being $a_m = a_1$ with weight w_1, etc., with only one unknown a_m. See (D.48)

order 2, 3, or 4, so that b is markedly inferior to a. The two measurements could be combined to give a mean in three ways.

(i) Neglect the inferior b, and accept a. Its variance is σ^2.
(ii) Take the simple mean $(a+b)/2$. Its variance will be $\sigma^2(1+k)/4$.
(iii) Take the weighted mean $(a+b/k)\div(1+1/k)$. Its variance will be $k\sigma^2/(1+k)$.

If $k=2$, (i), (ii), and (iii) give σ^2, $0.75\sigma^2$, and $0.67\sigma^2$ respectively.
If $k=3$ (i), (ii), and (iii) give σ^2, σ^2, and $0.75\sigma^2$.
If $k=4$ (i), (ii), and (iii) give σ^2, $1.25\sigma^2$, and $0.80\sigma^2$.

The correctly weighted mean (iii), of course, always gives the best result. Note that if $k<3$, the simple mean is at least better than neglecting the weaker measurement, while if $k>3$ it is better to neglect it than to include it with unreduced weight. But see (a) below. In these three cases the range of the variance is not very great $(0.67-1.25)$ and the range of the corresponding s.e.'s will be only $0.82-1.12$. If $k<4$, wrong judgement is not very serious, although if the relative weights are known, their correct inclusion may be equivalent to an extra 25 per cent or more in the observation programme.

These are, however, two further considerations.

(a) How reliably have the relative weights been determined? If they are not properly known, their separate inclusion may be worse than treating them all as equal. If they are derived from internal evidence (the scatter of sets of observations), all the measures comprised in a may perhaps share a systematic error, while those in b may share a different one. If scatters gave s.e.'s of (say) 3″ for a and 1″.2 for b, and if we had reason to fear systematic errors of ±3″.0 in both, a and b should clearly be treated as of roughly equal weight. The proper course is to estimate s.e.'s as measures of 68 : 32 probability, § D.06, as best one can, and to weight the means accordingly.

(b) When something more than the measurement of a single object is concerned, the geometry of the problem may make a particular observation of special importance, in spite of its low weight.

As a general rule it may be concluded that there is little need to worry about whether the weight of one observation may be 2/3 or 3/2 times that of another, but outside this range it begins to be worth while to make estimates of s.e.'s and to allot corresponding weights.

D.12. Rejections

Various rules have been proposed as guides to whether one of a series of apparently good measures of the same thing should be rejected solely on the grounds that it differs too widely from the others. A common rule is to reject any observation which differs from the mean by more than 3 times the s.e. of a single observation, as deduced from the scatter of the whole series.

A point which needs to be stressed is that the scatter of any small series of observations is not all that is known about their accuracy. It is proper to draw on other experience. Let the recorded seconds of four measures of an angle be 3″, 5″, 4″, and 8″. If they have been measured over 40-km lines with a small single second theodolite, they would be accepted as an excellent set without hesitation.

On the other hand, if the same figures were given in laboratory conditions by a first-class modern geodetic theodolite of proven accuracy, the figure 8″ might be regarded as notably sub-standard.

Another point is that the errors of some types of observation have a tendency to be periodic. Theodolite angles may vary periodically with the circle zero-setting, or with the time of day, § 1.19. In such a case, the rejection of an outlying value may destroy the cancellation of error over the period.

The only satisfactory rule for rejections in the field is that the observer should know what scatter the proper handling of his instrument will produce in different conditions. If a single measure is grossly abnormal, an obvious blunder, it can be repeated and rejected, and if it is marginal it can be retained and swamped by a few repetitions. If trouble occurs more often than usual, either (*a*) a fault must be found and remedied, or (*b*) work must be postponed until conditions are better, or (*c*) the number of measures must be increased, or (*d*) a sub-standard result must be expected.

In some circumstances it may be said that somewhat numerous rejections will secure the most probable result, but that they are likely to falsify the estimate of its accuracy.

D.13. Least squares

If in any problem there are t unknown quantities, it is proper to make more than t observations to determine them. At its simplest, a single distance ($t=1$) may be measured twice, and the mean accepted. Less simply, the ϕ's and λ's of a 100-station geodetic framework ($t=200$) may be determined by the measurement of 450 angles and 150 distances. The object of the least squares routine is

(i) to produce unique values for the unknowns,

which (ii) will be of maximum probability,

and (iii) (optional extra) to indicate the precision with which the unknowns have been determined.

A least squares solution may be carried out in two different ways. With the same data and weights, both will give the same results, but there may be great differences in convenience. Taking a triangulation net as an example, the two methods are as below.

(*a*) The *Parametric* method, §§ D.14 and D.15, in which the unknown ϕ's and λ's are not directly observed, but are functions or combinations of the observed angles and distances. The method of Variation of Coordinates is an example, § 2.18, but the system of variation of preliminary values of the unknowns can be applied to many problems. It is also known as the method of *indirect observations*.

(*b*) The method of *Conditions*, §§D.16 and D.17, often described as the method of *Direct Observations* or of *Correlates*. The errors of all the observations are provisionally regarded as the unknowns, and when they have been determined the values of the ϕ, λ's, or whatever are ultimately required, can of course be computed. The number of observations is equal to the number of provisional unknowns, but redundancies arise from conditions between them, such as the known sum of the three angles of a triangle. An example is the adjustment of triangulation figures as described in § 2.13.

The nomenclature of the two methods is troublesome. The name parametric is current, but not very descriptive. The trouble with the names indirect and direct, as used in [113], 3 Edn., is that the normal equations can also be solved by two classes of methods known as *direct* and *iterative.* The use of the word direct in both situations is confusing. There is no easy alternative to the second use, so we here adopt parametric for the first and conditions for the second.

In all but the smallest problems most of the labour is in the solution of the simultaneous normal equations (D.46) and (D.74). In the parametric method their number equals the total number of unknown ϕ's and λ's, while in the condition method it equals the number of conditions, which is generally, but not necessarily, less. Before the introduction of computers the condition method was consequently almost invariably used, but now for the reasons listed in § 2.18 the parametric method is generally the easier for geodetic problems. See also Chapter 2, Section 4.

D.14. Least squares. Parametric method

(*a*) *Method.* The s observation equations† among the t unknowns $(s > t)$ may be written

$$
\begin{aligned}
a_1x_1 + b_1x_2 + \ldots + t_1x_t &= k_1 + v_1 \\
a_2x_1 + b_2x_2 + \ldots + t_2x_t &= k_2 + v_2 \\
\cdot \quad \cdot \quad \cdot \quad \cdot \quad \cdot \quad &\cdot \quad \cdot \quad \cdot \quad \cdot \quad \cdot \\
a_sx_1 + b_sx_2 + \ldots + t_sx_t &= k_s + v_s.
\end{aligned} \tag{D.42}
$$

Referring to § 2.23 for comparison, the x's are the unknown $\delta\phi$'s and $\delta\lambda$'s, the a, b's, etc., are the factors $a, b, \ldots, p, q, \ldots$ of (2.32)–(2.39), the k's are the (O − C)'s in which the O's are fallible measures, and the v's are the final corrections to the observations, namely the best estimates of (true − observed). When there are many unknowns, many of the factors $a, b, \ldots$ will be zero. The object is to find v's which will be as small as possible.

As in § D.05, it is assumed that the most probable value of any v is zero in accord with the observation, that the errors of all the observations are uncorrelated, and that the probability of any v being within any assigned small range $v + \frac{1}{2}\delta v$ to $v - \frac{1}{2}\delta v$ is given by the normal frequency distribution as in (D.7). For correlated observations see § D.22.

It is therefore assumed that the probability of the true value of the first v being within $\frac{1}{2}\delta v$ of any assigned value v_1 is proportional to $(\delta v)\exp(-v_1^2/2\sigma_1^2)$, where σ_1 is the standard (68 : 32) error of the observation in the right-hand side of the first equation. A similar expression refers to v_2, and the probability that all of any assigned set of values $v_1, \ldots, v_s$ are within δv of the truth is proportional to

$$(\delta v)^s\{\exp(-v_1^2/2\sigma_1^2) \times \ldots \times \exp(-v_s^2/2\sigma_s^2)\}.$$

The v's must therefore be chosen so as to maximize

$$-v_1^2/2\sigma_1^2 - \ldots - v_s^2/2\sigma_s^2,$$

† In the old literature these are sometimes called 'equations of condition', an unfortunate term since we also have 'condition equations' in D.16, which are not the same.

or to minimize
$$v_1^2/\sigma_1^2+\ldots+v_s^2/\sigma_s^2. \tag{D.43}$$

In words, the most probable solution gives a set of v's which satisfy the observation equations and minimize $[wv^2]$, where $w_i \equiv 1/\sigma_i^2$ is the *weight* of O_i, and the square brackets indicate summation from 1 to s.

$$[wv^2] \equiv w_1 v_1^2 + \ldots + w_s v_s^2. \tag{D.44}$$

We note

(i) that the maximum probability is also given if $w_i = 1/\kappa^2\sigma_i^2$, where κ^2 is any multiplier, the same for all the v's,
and (ii) that the definition of weight is the same as given in § D.10 (*a*).

Now multiply each of (D.42) by its own $\sqrt{w}$, and write

$$a_1x_1\sqrt{w_1}+b_1x_2\sqrt{w_1}+\ldots-k_1\sqrt{w_1}=v_1\sqrt{w_1}$$
$$\cdots\cdots\cdots$$
$$a_sx_1\sqrt{w_s}+b_sx_2\sqrt{w_s}+\ldots-k_s\sqrt{w_s}=v_s\sqrt{w_s}. \tag{D.45}$$

Substituting (D.45) in (D.44), differentiating with respect to $x_1, x_2, \ldots$, etc., and equating to zero for minimum, gives

$$a_1w_1(a_1x_1+b_1x_2+\ldots-k_1)+a_2w_2(a_2x_1+b_2x_2+\ldots-k_2)+\ldots=0,$$
$$b_1w_1(a_1x_1+b_1x_2+\ldots-k_1)+b_2w_2(a_2x_1+b_2x_2+\ldots-k_2)+\ldots=0, \text{ etc.,}$$

t equations, or

$$[aaw]x_1+[abw]x_2+\ldots+[atw]x_t=[akw],$$
$$[baw]x_1+[bbw]x_2+\ldots+[btw]x_t=[bkw], \text{ etc.,} \tag{D.46}$$

t equations, where

$$[aaw] \equiv a_1^2w_1+a_2^2w_2+\ldots+a_s^2w_s$$

and
$$[baw] \equiv [abw] \equiv a_1b_1w_1+a_2b_2w_2+\ldots+a_sb_sw_s, \text{ etc.}$$

Equations (D.46) are known as *normal equations.* They are ordinary simultaneous equations, and may be solved by any of the routines given in §§ B.14 or B.15. They are symmetrical about the diagonal $[aaw]x_1, [bbw]x_2, \ldots$.

If the k's are all dimensionally similar, identical results will be got if the w's are simple measures of relative reliability, but if both lengths and angles are involved, the weights must be dimensioned quantities. Then all equations (D.45) are of zero dimensions, and the summations $[aaw]$, etc., will not involve the addition of dissimilar quantities. See § 2.24.

(*b*) *Linearization.* The observation equations given in (D.42) are linear in the x's. More generally let them take the form $f_s(x_1, x_2, \ldots) = k_s$. By any means find an approximate solution X_1, X_2, etc., and tet $x_1 = X_1 + \delta x_1$. Then a set of linear equations for δx_1, δx_2 is

$$\frac{\partial}{\partial x_1} f_s(x_1, x_2, \ldots)\,\delta x_1 + \frac{\partial}{\partial x_2} f_s(x_1, x_2 \ldots)\,\delta x_2 + \ldots = k_s - f_s(x_1, x_2 \ldots), \tag{D.47}$$

with $X_1, X_2, \ldots$ substituted in $(\partial/\partial x_1)f_s(x_1, x_2 \ldots)$, etc.

When the observation equations are initially linear, arithmetical desk labour may sometimes be saved by finding an approximate solution as above, and then solving for $\delta x_1, \delta x_2, \ldots$ rather than for the larger $x_1, x_2, \ldots$.

(*c*) In (D.42) the coefficients $a, b, \ldots, t$ must be regarded as infallible. It may be remarked that in an observation equation such as $L_0 + \alpha t = L$, where the unknowns are L_0, the length of a bar at zero temperature, and α its coefficient of expansion, the doubtful measurement is that of the temperature t. Nevertheless, in this context t is infallible.† The fallible quantity is the observed length L, which purports to be the length at temperature t, but which may actually be the length at some other temperature.

It may be noted that if the observations are no more than repeated measures of (say) a distance L, the observation equations (D.42) are simply

$$L = l_1 + v_1, \text{ weight } w_1$$

$$L = l_2 + v_2, \text{ weight } w_2, \text{ etc.}$$

Then (D.45) is

$$L\sqrt{w_1} - l_1\sqrt{w_1} = v_1\sqrt{w_1}, \text{ etc.,}$$

and (D.46) is

$$[w]L = [wl], \text{ a single equation,} \tag{D.48}$$

whence

$$L = (w_1 l_1 + w_2 l_2 + \ldots) \div (w_1 + w_2 + \ldots),$$

as in (D.38), showing that subject to the assumptions on which the routine of least squares is based, the ordinary 'weighted mean' is the most probable value.

If the coefficients in the observation equations (D.42) or in (D.45) are very unequal, it is permissible to multiply the whole of one column (say that of x_i) by any convenient factor, such as a + or − power of 10. This amounts to changing the unknown x_i to $x_i \div$(the same factor), which must not be overlooked at the end. Rows, on the other hand may not arbitrarily be so treated, since multiplying a row by a factor amounts to a change of weight.

(*d*) *Matrix notation.* See Appendix B.

The observation equations (D.42) are $Ax = b + v$ (D.49)

The weighted equations (D.45) are $W^{\frac{1}{2}}Ax = W^{\frac{1}{2}}b + W^{\frac{1}{2}}v$, (D.50)‡
W being the diagonal matrix whose elements are the weights $w_1 \ldots w_s$.

The normal equations (D.46) which minimize $[wv^2]$ are $A^TWAx = A^TWb$
or $Nx = c$, (D.51)

† If the error of a coefficient is believed to vary according to some formula of known form, a limited number of constants in such a formula may be introduced as additional unknowns. For instance, the observed temperature t might, if appropriate, be replaced by $(t + et^2)$. The constant e is now an unknown. If the resulting equations are not linear use (D.47).

‡ There is no need to compute or use $W^{\frac{1}{2}}$, the square root of the weight matrix W. Only W itself occurs in the working computations. It is only included here to illustrate the connection with (D.45).

defining N, the matrix of the normal equations, and c. The solution is

$$\begin{aligned} x &= N^{-1}c \\ &= (A^{\mathrm{T}}WA)^{-1}A^{\mathrm{T}}Wb. \end{aligned} \tag{D.52}$$

The residuals are given by $v = Ax - b$ (D.53)

and the sum of the squares of the weighted residuals is

$$\begin{aligned} [wv^2] &= v^{\mathrm{T}}Wv \\ &= (Ax-b)^{\mathrm{T}}W(Ax-b). \end{aligned} \tag{D.54}$$

All the above may be directly verified by matrix multiplication. The brevity of the notation may be contrasted with the style of (D.42), (D.45), and (D.46)

(*e*) *Incorporation of later observations.*

(i) It may happen that after the normal equations have been formed from a set of observations $A_1x_1 = b_1$ with weight W_1, some more observations $A_2x_2 = b_2$ are made with weight W_2, which it is desired to incorporate into improved normal equations. Then, provided the vector of unknowns x_1 includes all the unknowns of x_2, the combined observation equations can be written as

$$\left[\frac{A_1}{A_2}\right] x_1 = \left[\frac{b_1}{b_2}\right], \text{ with diagonal weight matrix } \begin{bmatrix} W_1 & 0 \\ 0 & W_2 \end{bmatrix}, \tag{D.55}$$

and the normal equations are

$$\begin{aligned} [A_1^{\mathrm{T}}W_1A_1 + A_2^{\mathrm{T}}W_2A_2]x_1 &= A_1^{\mathrm{T}}W_1b_1 + A_2^{\mathrm{T}}W_2b_2 \\ \text{or } x_1 &= (P_1+P_2)^{-1}(P_1x_1+P_2x_2), \end{aligned} \tag{D.56}$$

where $P_1 = A_1^{\mathrm{T}}W_1A_1$ and $P_2 = A_2^{\mathrm{T}}W_2A_2$, and in which $A_1^{\mathrm{T}}W_1A_1$ and $A_1^{\mathrm{T}}W_1b_1$ are already known. This result may be compared with (D.140). The simplicity of both formulae results from the coefficients (zero or otherwise) of all the unknowns being included in A_1 and A_2.

(ii) If the new observation equations contain some new unknowns which were not included in the original equations, the new equations can be written

$$[A_{21} | A_{22}]\left[\frac{x_1}{x_2}\right] = b_2, \text{ with diagonal weight matrix } W_2, \tag{D.57}$$

where x_1 is the vector of the original unknowns, and x_2 is the vector of the unknowns which occur only in the new observations. The matrix A_{21} may consist largely of zeros.

Further, the original equations $A_1x_1 = b_1$ must be rewritten as

$$[A_{11} \,|\, 0]\left[\frac{x_1}{x_2}\right] = b_1, \text{ with diagonal weight matrix } W_1. \tag{D.58}$$

The combined set of observation equations is then

$$\begin{bmatrix} A_{11} & 0 \\ A_{21} & A_{22} \end{bmatrix} \begin{bmatrix} x_1 \\ x_2 \end{bmatrix} = \begin{bmatrix} b_1 \\ b_2 \end{bmatrix}, \tag{D.59}$$

and the normal equations are

$$\begin{bmatrix} A_{11}^{T}W_1A_{11}+A_{21}^{T}W_2A_{21} & A_{21}^{T}W_2A_{22} \\ A_{22}^{T}W_2A_{21} & A_{22}^{T}W_2A_{22} \end{bmatrix}\begin{bmatrix} x_1 \\ x_2 \end{bmatrix}=\begin{bmatrix} A_{11}^{T}W_1b_1+A_2^{T}W_2b_2 \\ A_{22}^{T}W_2b_2 \end{bmatrix}, \tag{D.60}$$

in which $A_{11}^{T}W_1A_{11}$ and $A_{11}^{T}W_1b_1$ are already known.

D.15. Standard errors and covariance. Parametric method.

Let the weights assigned to the observations be the reciprocals of the squares of their s.e.'s. Then it has for long been known, [598], pp. 239–41 following Gauss, that if these s.e.'s have been correctly estimated,† the s.e. of an unknown x_m will be

$$\sigma_m=(D_{mm}/D)^{\frac{1}{2}}, \tag{D.61}$$

where D is the $t\times t$ determinant formed by the coefficients $[aww]$, $[abw]$, etc., in the left-hand side of (D.46), and D_{mm} is the co-factor of $[mmw]$.

If the estimated s.e.'s on which the weights have been based are too large or too small, nothing in the solution will help to correct their relative values, but the sum of the squares of the residuals $v\sqrt{w}$ can be used to confirm or correct their average values, as follows.

Since the w of each observed k is 1/(its s.e.)^2, the estimated s.e. of each $k\sqrt{w}$ will be unity. The expected r.m.s. value of the $v\sqrt{w}$'s will be a little less, namely $\{(s-t)/s\}^{\frac{1}{2}}$, [598], pp. 243–5, or [560], p. 59.

Then if
$$\sigma_0=\{[wv^2]/(s-t)\}^{\frac{1}{2}}, \tag{D.62}$$

the expected value of σ_0 is unity, and if it is anything else a revised value for the s.e. of x_m is‡

$$\sigma_m=\sigma_0(D_{mm}/D)^{\frac{1}{2}}, \tag{D.63}$$

If σ_0 is significantly >1, the discrepancy may be due to various causes.

(*a*) All estimated s.e.'s being approximately $1/\sigma_0$ times their true values.
or (*b*) A few observations being much less accurate than estimated.
or (*c*) The inclusion of one or more blunders.

If (*a*) is the cause, the unknowns will have been rightly determined and (D.63) will rightly give their accuracies. If either (*b*) or (*c*) is the cause, and if the trouble can be located and remedied, recomputation will be necessary. The geographical position of large values of $v\sqrt{w}$ may reveal the cause.

It is unusual for σ_0 to be significantly less than its expected value.

It follows from (D.62) that if the s.e.'s have been correctly estimated, the sum of the squares of the weighted residuals, $[wv^2]$, will equal the number of observations s minus the number of unknowns t. And similarly in § D.16 (D.88) $[wx^2]$ should equal n, the number of conditions.

† And provided that the errors of different observations are uncorrelated. See § D.22.

‡ The revised estimate of the s.e. of an observation whose s.e. was originally estimated as unity, is σ_0. σ_0 has often been referred to as the s.e. of an observation of unit weight.

In all the above, if the weights have been taken as $1/\kappa^2(\text{s.e.})^2$ instead of $1/(\text{s.e.})^2$, (D.61) will read

$$\text{s.e. of } x_m = (1/\kappa)(D_{mm}/D)^{\frac{1}{2}} \tag{D.64}$$

and in (D.62) the expected value of σ_0 will be $1/\kappa$, as is evident from (D.63). If σ_0 is actually unequal to $1/\kappa$, (D.64) must be multiplied by a factor $\sigma_0 \div (1/\kappa)$, which gives it exactly the form of (D.63). It is clear that the value of κ does not enter into the final result. It is nevertheless desirable to have κ equal either to unity or to some other deliberately assigned value, so that the expected equality $\sigma_0 = 1/\kappa$ may be used to confirm, or dispute, the accuracy of the adopted weights. See (*a*), (*b*), and (*c*) below (D.63).

For those who use p.e.'s $\kappa = 0.67$.

In matrix notation. (D_{mm}/D) equals α_{mm} the diagonal element in row m of N^{-1}, the inverse of the matrix of the normal equations, as is immediately clear from (B.24), (B.26), and (B.25), so

$$\sigma_m^2 = \alpha_{mm}. \tag{D.65}$$

Further, as is shown in § D.23 (*a*), the covariance of any two unknowns x_m and x_n is given by

$$\sigma_{mn} = \alpha_{mn}, \tag{D.66}$$

where α_{mn} is the element in row m and column n of N^{-1}. The symmetric matrix N^{-1} is therefore known as the *variance–covariance matrix*, and on the basis that the weights have been based on correctly estimated s.e.'s, its diagonal elements are the variances of the unknowns, and its other elements are the covariances of the pairs of unknowns related to the row and column in which they lie.

We have
$$N^{-1} = \begin{bmatrix} \sigma_1^2 & \sigma_{12} & . & . & . & \sigma_{1t} \\ \sigma_{21} & \sigma_2^2 & . & . & . & \sigma_{2t} \\ . & . & . & . & . & . \\ \sigma_{t1} & \sigma_{t2} & . & . & . & \sigma_t^2 \end{bmatrix}. \tag{D.67}$$

If σ_0 in (D.62) is $\neq$ unity, the variance–covariance matrix is better given by

$$\sigma_0^2 N^{-1}. \tag{D.68}$$

Note that the coefficient of correlation (D.15) between any two unknowns is independent of σ_0.

The determination of the covariances is generally as important as that of the variances, since knowledge of the accuracy of fixation relative to the survey origin is likely to be of no more interest than knowledge of the relative accuracy of many other pairs of points. See § 2.27.

Then, if $F \equiv f_1x_1 + f_2x_2 + \ldots + f_tx_t$ is a linear function of the unknowns, such as the mutual distance and azimuth of two selected points, (D.19) gives

$$\sigma_F^2 = \sigma_1^2 f_1^2 + \ldots + \sigma_t^2 f_t^2 + 2\sigma_{12} f_1 f_2 + \ldots \text{all combinations } 2\sigma_{ij} f_i f_j. \tag{D.69}$$
$$= f' N^{-1} f, \text{ as may be directly verified by multiplication.}$$

If F is not linear, it can be made so by (D.47).

In particular, $$\sigma^2_{(x_m-x_n)} = \alpha_{mm} + \alpha_{nn} - 2\sigma_{mn}, \quad \text{(D.70)}$$

where the α's are elements of N^{-1}.

Formulae (2.41A) and (2.41B) follow from (D.69).

For methods of computing the elements of N^{-1} see § B.16.

D.16. Least squares. Method of conditions

Continuing from § D.13 (*b*). The unknowns are taken to be the errors of the s observations, so the observation equations are

$$x_1 = 0 + v_1 \text{ with weight } w_1$$

$$\cdot \quad \cdot \quad \cdot \quad \cdot$$

$$x_s = 0 + v_s \text{ with weight } w_s. \quad \text{(D.71)}$$

And the weighted observation equations are

$$x_1\sqrt{w_1} = 0 + v_1\sqrt{w_1}, \text{ etc.} \quad \text{(D.72)}$$

As before, we wish to minimize $[wv^2]$ or $[wx^2]$, which is the same thing, but subject to certain *condition equations*, n in number, n being $<s$, namely

$$a_1x_1 + a_2x_2 + \ldots + a_sx_s = \epsilon_a$$

$$b_1x_1 + b_2x_2 + \ldots + b_sx_s = \epsilon_b$$

$$\cdot \quad \cdot \quad \cdot \quad \cdot \quad \cdot \quad \cdot \quad \cdot \quad \cdot$$

$$n_1x_1 + n_2x_2 + \ldots + n_sx_s = \epsilon_n. \quad \text{(D.73)}$$

The routine is then to form from (D.73) n normal equations for n *correlates*†, or *Legendre coefficients*, $\lambda_a, \ldots, \lambda_n$

$$[aau]\lambda_a + [abu]\lambda_b + \ldots + [anu]\lambda_n = \epsilon_a$$

$$\cdot \quad \cdot \quad \cdot \quad \cdot \quad \cdot \quad \cdot \quad \cdot \quad \cdot \quad \cdot \quad \cdot \quad \cdot$$

$$[nau]\lambda_a + [nbu]\lambda_b + \ldots + [nnu]\lambda_n = \epsilon_n, \quad \text{(D.74)}$$

where $u = 1/w$, and $[aau] = \alpha_1^2u_1 + \ldots + a_s^2u_s$,

$$[abu] = [bau] = a_1b_1u_1 + \ldots + a_sb_su_s, \text{ etc.}$$

To form (D.74) each column of (D.73) is first multiplied by the $\sqrt{u}$ of its observation x. Ideally $\sqrt{u}$ is the s.e. of x.

These n equations are symmetrical, and like (D.46) they may be solved by any of the methods of §§ B.14 or B.15. Note that their number equals the number of conditions, in contrast to (D.46) whose number equals the number of unknowns.

Dimensions. In (D.73) the dimensions of a_m are those of (ϵ_a/x_m), which may be (seconds/seconds) or (ppm/seconds) or (ppm/ppm) according to whether the x's and ϵ's are errors of angle or of side. See § 2.31.

In (D.74) the dimensions of u_m are those of x_m^2, since $u = 1/w$ which should be σ_x^2, so the dimensions of $a_mb_mu_m$ are those of $\epsilon_a\epsilon_b$, and the dimensions of λ_b are those of $1/\epsilon_b$.

† An old name without obvious connection with 'correlation' as used in § D.04.

When the λ's are determined, the x's are given by

$$x_1 = u_1(\lambda_a a_1 + \lambda_b b_1 + \ldots + \lambda_n n_1)$$
$$\cdots\cdots\cdots$$
$$x_s = u_s(\lambda_a a_s + \lambda_b b_s + \ldots + \lambda_n n_s). \qquad \text{(D.75)}$$

For proof that the resulting $[wx^2]$ is minimum, see § D.23 (b).

Then the ϕ's and λ's are computed as in § 2.14 using observed angles and distances minus the x's. All computation routes should give identical results.

These λ's have no connection with the eigenvalue λs of § B.11.

In matrix notation, the condition equations (D.73) are

$$Bx = \epsilon \qquad \text{(D.76)}$$

The normal equations (D.74) are

$$BUB^{\mathrm{T}}\lambda = \epsilon$$

or $\quad N\lambda = \epsilon$, defining N, which is symmetrical, (D.77)

where U is a diagonal matrix whose elements are the u's. Contrast this with $A^{\mathrm{T}}WA$ in (D.51).

The solution is

$$\lambda = (BUB^{\mathrm{T}})^{-1}\epsilon$$
$$= N^{-1}\epsilon. \qquad \text{(D.78)}$$

Then as in (D.75) the x's are given by

$$x = UB^{\mathrm{T}}\lambda$$
$$= UB^{\mathrm{T}}N^{-1}\epsilon$$
$$= UB^{\mathrm{T}}(BUB^{\mathrm{T}})^{-1}\epsilon. \qquad \text{(D.79)}$$

The conditions are exactly satisfied. For the sum $[wx^2]$,

$$[wx^2] = [x^2/u] = \epsilon'(BUB^{\mathrm{T}})^{-1}\epsilon = \epsilon^{\mathrm{T}}N^{-1}\epsilon = \epsilon^{\mathrm{T}}\lambda. \qquad \text{(D.80)}$$

D.17. Standard errors and covariance. Method of conditions

As in § D.15 it is required to find the s.e., σ_F, of any function F of the adjusted observations $X_1, \ldots, X_s$, namely the observed values minus the x's. It is clear that the s.e. of each X is the same as that of its corresponding x.† If F is not linear, as in (D.47), take the observations as approximations to the adjusted values, and the expression

$$\Delta F \equiv (\partial F/\partial X_1)x_1 + \ldots + (\partial F/\partial X_s)x_s$$

is then linear in the x's, and the required $\sigma_F = \sigma_{\Delta F}$. it is therefore only necessary to consider linear functions.

† We measure a distance of exactly 100.00 m, and get the result 99.00 m. We proceed to find x, the error of that measure, and assess it as -0.85 m, giving 99.85 m as the adjusted result. The error of the figure 99.85 equals the error of $-x$, and the s.e. of one is the s.e. of the other.

We have $$F \equiv f_1X_1 + \ldots + f_sX_s$$

and $$\sigma_X = \sigma_x. \quad \text{(D.81)}$$

Then if the weights in (D.72) are $1/(\text{s.e.})^2$,† and if the s.e.'s have been correctly estimated

$$\sigma_F^2 = [ffu] - [afu]\mu_1 - [bfu]\mu_2 - \ldots - [nfu]\mu_n, \quad \text{(D.82)}$$

where $u = 1/w$, and $\mu_1, \ldots, \mu_n$ are given by

$$[aau]\mu_1 + [abu]\mu_2 + \ldots + [anu]\mu_n = [afu]$$

$$\cdot \quad \cdot \quad \cdot \quad \cdot \quad \cdot \quad \cdot \quad \cdot \quad \cdot$$

$$[nau]\mu_1 + [nbu]\mu_2 + \ldots + [nnu]\mu_n = [nfu]. \quad \text{(D.83)}$$

These equations have the same coefficients in the left-hand side as (D.74), with a different right-hand side, and if (D.74) has been solved by direct methods of solution, (D.83) can be solved with little fresh labour.

The s.e. of the adjusted value of any one X_m, σ_m, is got by putting $f_m = 1$, and all the other f's zero, in F.

In matrix notation (D.82) is

$$\sigma_F^2 = f^T U f - \mu^T B U f, \quad \text{(D.84)}$$

where f^T and μ^T are the row vectors $[f_1, \ldots, f_s]$ and $[\mu_1, \ldots, \mu_n]$, and μ is given by (D.83), which is

$$(BUB^T)\mu = BUf \text{ or } N\mu = BUf, \quad \text{(D.85)}$$

as may be verified by matrix multiplication. Then (D.84) states that

$$\sigma_F^2 = [f^T U - \{(BUB^T)^{-1}(BUf)\}^T BU]f$$

which simplifies to

$$\sigma_F^2 = f^T\{U - UB^T(BUB^T)^{-1}BU\}f$$
$$= f^T\sigma^2 f \quad \text{(D.86)}$$

where, corresponding to (D.67),

$$\sigma^2 = U - UB^T N^{-1} BU \quad \text{(D.87)}$$

where $N = BUB^T$ as in (D.77). In (D.87), σ^2 is then known as the *variance–covariance* matrix of the x's. It is symmetric. The diagonal term in row m of (D.87) is σ_m^2, and the element in row m and column n is σ_{mn}, the covariance of x_m and x_n.

As in § D.15, the values of $[wv^2] \equiv [wx^2]$ here also provide a check on the originally estimated s.e.'s. Let

$$\sigma_0^2 = [wx^2]/n = (x^T W x)/n. \quad \text{(D.88)}$$

Then if σ_0 differs significantly from its expected value of unity, the values of σ_F^2 given by (D.82), (D.84), and (D.86) may be multiplied by σ_0^2, and the variance–

† Then $u = 1/w = \sigma^2$.

covariance matrix (D.87) also. The remarks following (D.63) apply.† If the weights have been taken as $1/\kappa^2$ (s.e.)^2 the formulae for σ_F^2 must be multiplied by $1/\kappa^2$, and the expected value of σ_0 will be $1/\kappa$ instead of unity.

Note that σ_F is the s.e. of the observed quantities after they have received the benefit of adjustment. The ratio of the s.e. of each adjusted observation‡ to that of the observation itself is not the same for all, but a useful rule, [466], pp. 70–1, is

$$\text{average s.e. after adjustment} = (1 - n/s)^{\frac{1}{2}}(\text{average s.e. before adjustment}). \tag{D.89}$$

This rule applies to dimensionally similar observations of equal weight. In particular, it applies to the weighted observations $x\sqrt{w}$ of (D.72), which are of zero dimensions and of unit weight provided $w = 1/(\text{s.e.})^2$, as in (D.34).

D.18. Examples. Triangle and quadrilateral

(*a*) *Triangle.* In a triangle ABC, Fig. D.6, let AB be exactly known, and let the observed values of the angles be A, B, and C, each with equal weight (s.e. $= \sigma$), and unknown errors x_1, x_2, and x_3. Then there are three observation equations $x_1 = 0$, $x_2 = 0$, and $x_3 = 0$, subject to the one condition (D.73) $x_1 + x_2 + x_3 = \epsilon$ the known triangular error.

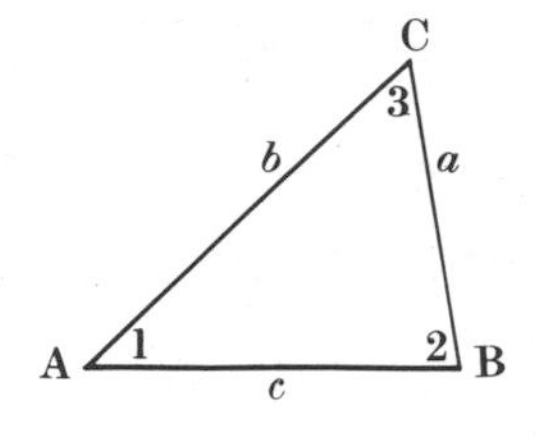

FIG. D.6.

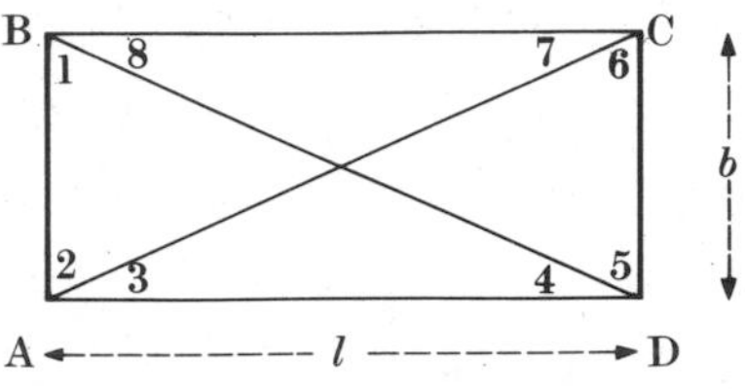

FIG. D.7.

Then (D.74) takes the simple form $3\lambda = \epsilon$, and (D.75) gives $x_1 = x_2 = x_3 = \epsilon/3$, confirming the obvious rule that if weights are equal the triangular error should be evenly distributed.

Now consider the length AC. We want the weight of the function

$$F = c \sin B / \sin C.$$

Putting this into linear form as in (D.47), dF (whose weight is the same as that of F) $= b[x_2 \cot B - x_3 \cot C]$.

† In the classical computation of a triangulation chain, § 2.13, $[wx^2]$ should not be computed separately for each figure, but from the wx^2's of as many figures as may appear to have been observed in similar conditions. In (D.88) n is then the total number of conditions in all the figures.

‡ (D.89) is applicable to single observations, not to functions of them. Distributing a triangular error improves the observations by a factor of 0.82. A base and Laplace station at the end of a long triangulation chain will improve each angle by a factor of perhaps 0.98, but each small improvement is cumulative, and the expected error of position at the end of the chain is halved. See 2.46 (*f*).

Then (D.83) takes the form

$$3\mu = b \cot B - b \cot C$$

and (D.82) gives

$$\sigma_F^2 = \sigma^2\{b^2 \cot^2 B + b^2 \cot^2 C - \tfrac{1}{3}(b \cot B - b \cot C)(b \cot B - b \cot C)\}$$
$$= \tfrac{2}{3}\sigma^2 b^2(\cot^2 B + \cot^2 C + \cot B \cot C). \quad \text{(D.90)}$$

The s.e. of AC is

$$\text{AC} \times \sigma\sqrt{\{\tfrac{2}{3}(\cot^2 B + \cot^2 C + \cot B \cot C)\}}, \quad \text{(D.91)}$$

σ being the s.e. of an observed angle in radians.

If the triangle is equilateral $\cot B = \cot C = 1/\sqrt{3}$, and if σ is in seconds and the s.e. of AC is expressed in ppm,

$$\text{s.e.} = 3.96\sigma \text{ for one triangle or } 5.60\sigma \text{ for a pair, as in (2.65).} \quad \text{(D.92)}$$

(*b*) *Braced quadrilateral.* In general, the formula for the s.e. of the side of exit cannot be written in simple form, and the four equations (D.83) must be written out and solved, but the special case of a rectangle of length l and breadth b, Fig. D.83, with all angles independently observed with equal weights is simply solved as follows.

The condition equations (D.73) are

$$x_1 + x_2 + x_3 + x_4 = \epsilon_1$$

$$x_5 + x_6 + x_7 + x_8 = \epsilon_2$$

$$x_3 + x_4 + x_5 + x_6 = \epsilon_3$$

$$x_1 \cot 1 + x_3 \cot 3 + x_5 \cot 5 + x_7 \cot 7 = x_2 \cot 2 + x_4 \cot 4 + x_6 \cot 6 + x_8 \cot 8 + \epsilon_4.$$

The last equation can be written in the form

$$x_1 - x_2 + rx_3 - rx_4 + x_5 - x_6 + rx_7 - rx_8 = \epsilon_4\sqrt{r}, \quad \text{where } r = l^2/b^2.$$

If AB and CD are the sides of entry and exit, the function whose weight is required, F, is CD = AB sin 3 sin(1 + 8)cosec 7 cosec(4 + 5), which can be put into linear form as

$$dF = \text{AB}\{x_3 \cot 3 + (x_1 + x_8)\cot 90° - x_7 \cot 7 - (x_4 + x_5)\cot 90°\}$$
$$= lx_3 - lx_7.$$

Then equations (D.83) are

$$4\mu_1 + 2\mu_3 = l$$

$$4\mu_2 + 2\mu_3 = -l$$

$$2\mu_1 + 2\mu_2 + 4\mu_3 = l$$

$$4(1 + r^2)\mu_4 = 0,$$

whence $\mu_1 = \mu_4 = 0$ and $\mu_2 = -\mu_3 = -l/2$.

And (D.82) gives

$$\sigma_F^2 = \sigma^2(2l^2 - l^2/2 - l^2/2) = \sigma^2 l^2 = \sigma^2 \,\mathrm{CD}^2 l^2/b^2.$$

The s.e. of CD is then

$$\sigma_F = (\mathrm{CD})\sigma l/b. \tag{D.93}$$

If σ is in seconds of arc, and the s.e. of CD is in parts per million,

$$\text{s.e.} = 4.85\sigma l/b. \tag{D.94}$$

This justifies the figure given in (2.65), and the rule that if the weights of all the separate angles are equal, the s.e. of CD is proportional to l/b.

D.19. Inclusion of infallible observations

In the method of indirect observations, it may happen that in addition to the fallible observations

$$A_1x = b_1 + v, \text{ with weight matrix } W, \tag{D.95}$$

in which $v^{\mathrm{T}}Wv$ is to be minimum, we may also have one or more equations

$$A_2x = b_2, \tag{D.96}$$

which are to be exactly satisfied. In (D.95) and (D.96) the vector x consists of all the unknowns, of which at least one must occur with non-zero coefficients in both sets of equations. This situation may arise in the adjustment of triangulation by variation of coordinates, as in § 2.23 (*e*) and (*f*). The unknowns x are the coordinates of the stations, and the observations b are the (observed–computed) values of angles or distances, some of which it may be desired to treat as errorless. § 2.23 (*e*) and (*g*) suggest simpler methods of treatment, which may generally be preferable to the method given here.

The procedure is as follows [66] and also [315], vol i, pp. 175–6 in non-matrix notation. We wish to minimize

$$\begin{aligned}\Phi &= v^{\mathrm{T}}Wv - 2(A_2x - b_2)^{\mathrm{T}}\lambda \\ &= (A_1x - b_1)^{\mathrm{T}}W(A_1X - b_1) - 2(A_2x - b_2)^{\mathrm{T}}\lambda,\end{aligned} \tag{D.97}$$

where λ is a vector of multipliers (independent of x) like those of (D.74) and (D.171), and W is symmetric.

Then
$$\partial\Phi/\partial x = 2A_1^{\mathrm{T}}WA_1x - 2A_1^{\mathrm{T}}Wb_1 - 2A_2^{\mathrm{T}}\lambda. \tag{D.98}$$

This has to be zero, so

$$x = (A_1^{\mathrm{T}}WA_1)^{-1}(A_1^{\mathrm{T}}Wb_1 - A_2^{\mathrm{T}}\lambda) \tag{D.99}$$

$$= N^{-1}(c + A_2^{\mathrm{T}}\lambda), \tag{D.100}$$

where $N = A_1^{\mathrm{T}}WA_1$, and $c = A_1^{\mathrm{T}}Wb_1$ as in (D.51)

But $A_2x = b_2$, so $A_2N^{-1}(c + A_2^{\mathrm{T}}\lambda) = b_2$,

whence
$$\lambda = (A_2N^{-1}A_2^{\mathrm{T}})^{-1}(b_2 - A_2N^{-1}c) \tag{D.101}$$

And from (D.100)
$$x = N^{-1}\{c + A_2^{\mathrm{T}}(A_2N^{-1}A_2^{\mathrm{T}})^{-1}(b_2 - A_2N^{-1}c)\}. \tag{D.102}$$

If there are no infallible observations $A_2x = b_2$, (D.102) becomes $x = N^{-1}c$, as in (D.52).

D.20. Error ellipses

Let the position of a point P have been fallibly found, relative to some other point such as the origin of the survey. At its true position O take axes Ox and Oy towards the east and north respectively, Fig. D.8. Let the position of P have been determined with s.e.'s of σ_x and σ_y in the directions Ox and Oy with covariance σ_{xy}, these having been determined as in §§ D.15 or D.17. It is required to find σ_M, the s.e. of the position of P in a direction OM which makes an angle ψ with Ox.

Take OM and ON as axes of x' and y', so $x' = x \cos\psi + y \sin\psi$. Then, from (D.19),

$$\sigma_M^2 = \sigma_{x'}^2 = \sigma_{x'}^2 \cos^2\psi + \sigma_y^2 \sin^2\psi + 2\sigma_{xy} \sin\psi \cos\psi. \qquad \text{(D.103)}$$

And in a perpendicular direction ON,

$$\sigma_N^2 = \sigma_x^2 \sin^2\psi + \sigma_y^2 \cos^2\psi - 2\sigma_{xy} \sin\psi \cos\psi. \qquad \text{(D.104)}$$

Note that $\sigma_M^2 + \sigma_N^2 = \sigma_x^2 + \sigma_y^2$ for all values of ψ.

For σ_M^2 to be maximum with respect to ψ, (D.103) or (D.104) gives

$$\tan 2\psi = 2\sigma_{xy}/(\sigma_x^2 - \sigma_y^2), \qquad \text{(D.105)}$$

whence there are two values of ψ, ψ_1 and ψ_2, differing by $\pi/2$, one for maximum σ and one for minimum.

Substituting these values of ψ in (D.103) or (D.104) gives the maximum and minimum values of σ^2, which may be reduced to

$$\text{maximum} = \sigma_X^2 = \tfrac{1}{2}[\sigma_x^2 + \sigma_y^2 + \{(\sigma_y^2 - \sigma_x^2)^2 + 4(\sigma_{xy})^2\}^{\frac{1}{2}}],$$

$$\text{mimimum} = \sigma_Y^2 = \tfrac{1}{2}[\sigma_x^2 + \sigma_y^2 - \{(\sigma_y^2 - \sigma_x^2)^2 + 4(\sigma_{xy})^2\}^{\frac{1}{2}}], \qquad \text{(D.106)}$$

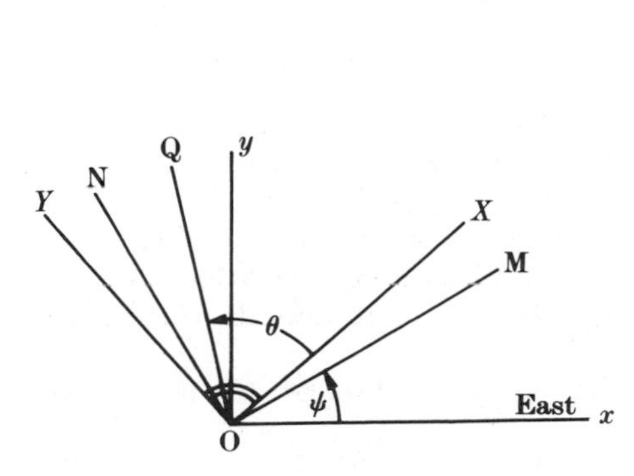

FIG. D.8. The axes Ox and Oy are east and north. OX and OY are the principal axes.

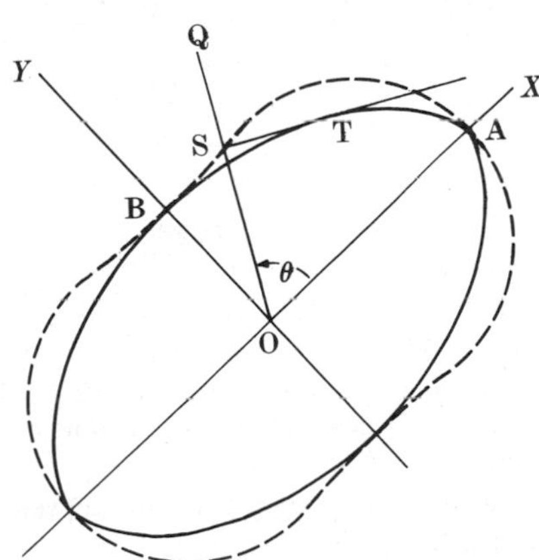

FIG. D.9. The error ellipse. OA = σ_X, maximum. OB = σ_Y, minimum. OS = σ_θ.

where X and Y are axes given by (D.105). These are known as the *principal axes*.

From (D.105) it is clear that if Ox, for which $\psi = 0$, happens to be a principal axis, σ_{xy} will be zero. So for the principal axes we have $\sigma_{XY} = 0$.

Then σ_θ^2 the variance of the position of P in any direction OQ which makes an angle θ with OX, will be given by (D.103) with θ instead of ψ, X and Y for x and y, and zero for σ_{xy},

$$\sigma_\theta^2 = \sigma_X^2 \cos^2 \theta + \sigma_Y^2 \sin^2 \theta, \tag{D.107}$$

and the total s.e. of the position of P is

$$(\sigma_\theta^2 + \sigma_{\theta+\pi/2}^2)^{\frac{1}{2}} = (\sigma_X^2 + \sigma_Y^2)^{\frac{1}{2}}. \tag{D.108}$$

Fig. D.9 gives a graphical construction for σ_θ. An ellipse is drawn with semi-axes σ_X and σ_Y. Then $\sigma_\theta = \text{OS}$, where $\text{QOX} = \theta$, and S a point on OQ such that the perpendicular ST is tangent to some point on the ellipse. The locus of S (D.107) is the *pedal curve* as shown in broken line. The ellipse is known as the *error ellipse* at the point P. See [315] vol i, pp. 438–45 and 555–8.

Formulae (D.103)–(D.107) are a simple case of a more general theorem. Let $x_1, \ldots, x_n$ be a set of quantities, such as the unknowns in a least-square solution, whose errors are correlated. Then we can find a set of linear functions of the x's $X_1, \ldots, X_n$, whose mutual covariances are zero. We can (if we like) visualize the x's as associated with n orthogonal axes in an n-dimensional space, and we can change variables to $X_1, \ldots, X_n$, by the matrix product

$$x = RX, \qquad \text{or} \qquad X = R^{-1}x = R^{\mathrm{T}}x, \tag{D.109}$$

where R^{T} is an orthogonal rotation matrix whose rows are the direction cosines of the X axes relative to the x axes. See § B.08 (c) for three dimensions.

In the present context, let the x's be the unknowns in the normal equations (D.51) $Nx = c$, so the variance–covariance matrix of the x's is N^{-1}, (D.67).

Then put $x = RX$, so that the observation equations are $ARX = k$ and the normals become

$$\begin{aligned} R^{\mathrm{T}}NRX &= R^{\mathrm{T}}c \\ X &= (R^{\mathrm{T}}NR)^{-1}R^{\mathrm{T}}c = (R^{\mathrm{T}}N^{-1}R)R^{\mathrm{T}}c, \end{aligned} \tag{D.110}$$

and the variance–covariance matrix of the X's is $(R^{\mathrm{T}}N^{-1}R)$.

If, now, R is chosen to be the *modal matrix* of N^{-1}, § B.11 (e) and [209], p. 42, namely the matrix M whose columns are the normalized latent vectors of N^{-1}, $M^{\mathrm{T}}N^{-1}M$ will be diagonal, so the variance–covariance matrix of the X's is diagonal and their covariances are zero. Their variances are the diagonal elements of $M^{\mathrm{T}}N^{-1}M$, which are the latent roots of N^{-1} or the reciprocals of the latent roots of N, § B.11 (f) (iii)

In the case of only two unknowns x and y as in (D.103)–(D.106), the rotation matrix M^{T} is

$$\begin{bmatrix} \cos\psi_1 & \sin\psi_1 \\ -\sin\psi_1 & \cos\psi_1 \end{bmatrix}, \tag{D.111}$$

where ψ_1 is given by (D.105). With some trouble it can be directly verified that this is the modal matrix of

$$N^{-1} \equiv \begin{bmatrix} \sigma_x^2 & \sigma_{xy} \\ \sigma_{xy} & \sigma_y^2 \end{bmatrix}, \qquad \text{(D.112)}$$

and that (D.106) gives the latent roots of N^{-1}.

If an error ellipsoid is required in a variation of coordinates solution in three-dimensional cartesian coordinates, the direction and magnitude of its principal axes can be obtained from N, the matrix of the normal equations, in the same way.

D.21. Standard error of position in a particular direction

There are three possible definitions of the s.e. in a particular direction, namely σ_A, σ_B, and σ_C.

(*a*) *Standard error* σ_A. Let the position of a point whose true position is O be fallibly determined. Then given the values of its variance, σ_X and σ_Y, along the two principal axes OX and OY, § D.20 obtains the variance and covariance of its surveyed position relative to axes OQ and OP, where QOX$=\theta$. See Fig. D.10. It is clear that in § D.20 the s.e. of position in the direction OQ implies the expected r.m.s. value of the perpendiculars from repeated surveyed positions on to the axis PP′, and (D.107) give it as

$$\sigma_A^2 = \sigma_X^2 \cos^2\theta + \sigma_Y^2 \sin^2\theta, \qquad \text{(D.113)}$$

where σ_A is the radius vector of the pedal curve, and σ_X and σ_Y are s.e.'s along the principal axes. In (D.107) σ_A is designated σ_θ, but here the notation is

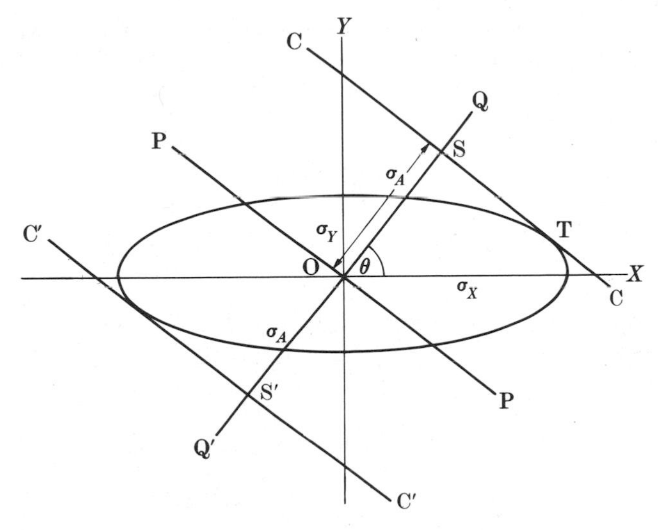

FIG. D.10. σ_X and σ_Y equal the semi-axes of the ellipse. σ_A = OS, as in Fig. D.9.

changed to σ_A for contrast with σ_B and σ_C, which are introduced below. The direction θ is implied in all cases unless otherwise stated.

Referring to Fig. D.10, the frequency distribution of perpendicular on to PP′ is normal, and 68 per cent of surveyed positions will lie between CC and C′C′, which are separated by a distance of $2\sigma_A$.

If σ_X is much greater than σ_Y, we remark that while this definition of σ_A is permissible, and indeed in some contexts useful, it does not (for some values of θ) well represent the distribution of errors lying along or close to OQ. It is an average over the whole plane, and near the line OQ the concentration will, in these circumstances, be much closer to O than the size of σ_A suggests.

(*b*) *Standard error* σ_B. See Fig. D.11. Consider a long, narrow rectangle Q′OQ of small width b, whose long axis makes an angle θ with OX. Let σ_B the r.m.s. value of distance from POP′ of errors within this rectangle.

Then, for proof see (*e*) (ii) below,

$$\sigma_B^2 = \frac{\sigma_X^2\sigma_Y^2}{\sigma_Y^2\cos^2\theta + \sigma_X^2\sin^2\theta}, \tag{D.114}$$

and the distribution along OQ is normal with 68 per cent errors lying between $+\sigma_B$ and $-\sigma_B$. Contrast this with the last line of (*a*) above.

Note. (i) The value of σ_B is the radius vector of the error ellipse whose axes are σ_X and σ_Y.

(ii) σ_B equals the conventional σ_A if $\sigma_X = \sigma_Y$, or if $\theta = 0°$ or $90°$.

From the descriptive point of view σ_B and the ellipse may be more convenient than σ_A and the pedal curve.

(*c*) *Standard error* σ_C. In Fig. D.12 let σ_C be the r.m.s. distance from O of all errors lying within the V-shaped area QOQ, bounded by directions $\theta + \frac{1}{2}\delta\theta$ and $\theta - \frac{1}{2}\delta\theta$. It is necessary to distinguish it from σ_r of § D.09, which is the r.m.s. value of radial distance all over the plane.

As with σ_r, the distribution of errors within QOQ is not the same as the normal univariate distribution of σ_A and σ_B, but is as in Fig. D.5. The frequency is zero at O, because it is in the narrow point of the V, and only 23 per cent are less than $\frac{1}{2}\sigma_C$, see Table D.3.

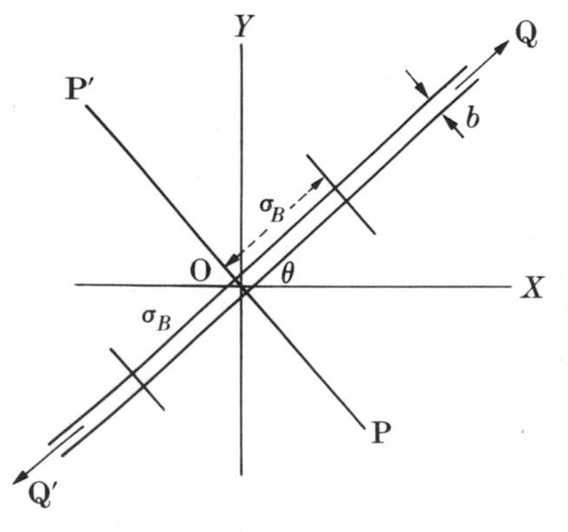

FIG. D.11.

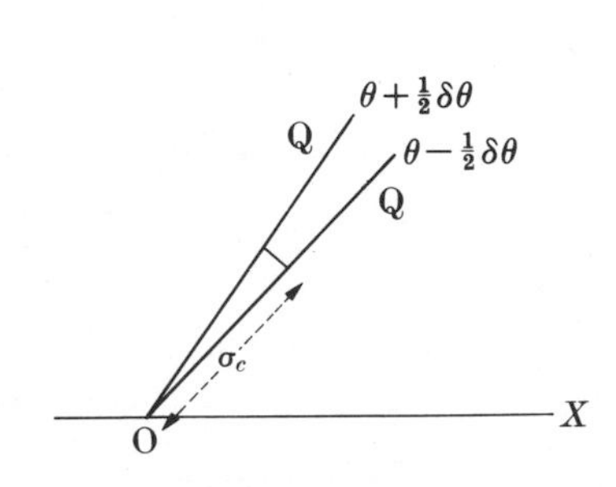

FIG. D.12.

This definition, and the use of the V-shaped area, is useful as a means of computing the proportion of errors lying within some closed area, such as an error ellipse, because integration with respect to θ sums up to the total area within the curve.

For the whole area of the plane, writing σ_X and σ_Y for the values of σ_A along OX and OY

$$\sigma_r^2 = \sigma_X^2 + \sigma_Y^2, \text{ as in § D.09.} \tag{D.115}$$

$$\sigma_r = \sigma_A\sqrt{2}, \text{ if } \sigma_X = \sigma_Y = \sigma_A. \tag{D.116}$$

$$\sigma_r = \sigma_X, \text{ if } \sigma_Y = 0, \text{ and } \sigma_R = \sigma_Y, \text{ if } \sigma_X = 0. \tag{D.117}$$

For the V-shaped area QOQ in the direction θ

$$\sigma_C^2 = \frac{2\sigma_X^2\sigma_Y^2}{\sigma_Y^2\cos^2\theta + \sigma_X^2\sin^2\theta} \tag{D.118}$$

$$= \frac{2}{(1/\sigma_X^2)\cos^2\theta + (1/\sigma_Y^2)\sin^2\theta} = 2\sigma_B^2 \tag{D.119}$$

For proof see (*e*) (i) below.

σ_r over the whole plane, is of value to give the probable magnitudes of the vector closing errors of circuits.

(*d*) *The properties of the error ellipse.*

(i) Its semi-axes are the principal (maximum and minimum) magnitudes of σ_A and σ_B, and equal those of σ_C divided by $\sqrt{2}$.

(ii) Its radius vector in any direction is σ_B or $(\sigma_C/\sqrt{2})$.

(iii) The radius vector of its pedal curve is σ_A.

(iv) It, and similar concentric ellipses, are loci or contours of equal density of errors.

(v) 39 per cent of all errors lie within the error ellipse whose principal axes are those of σ_A or σ_B. See [315], § 150 and [484], p. 111.

(*e*) *Two proofs.*

(i) *Proof of* (D.119) *for* σ_C. The chance that any particular error lies within a small rectangle with sides $r\,d\theta \times dr$ is (number of errors in $r\,d\theta \times dr$) divided by N, the total number of errors in the whole plane. As in (D.30), this chance is

$$\{1/(2\pi\sigma_X\sigma_Y)\}\{\exp(-r^2S)\}r\,d\theta\,dr, \tag{D.120}$$

where
$$S = \frac{\cos^2\theta}{2\sigma_X^2} + \frac{\sin^2\theta}{2\sigma_Y^2}.$$

So the number of errors in the area $r\,d\theta \times dr$ is

$$\{N\,d\theta/(2\pi\sigma_X\sigma_Y)\}\{\exp(-r^2S)\}r\,dr. \tag{D.121}$$

The required σ_C^2 = mean-square value of r for the errors within the infinite V-shaped area QOQ in Fig. D.12.

$$= \frac{\sum (\text{number in each element of } dr) \times r^2}{\text{Total number in the V}}$$

$$= \int_0^\infty \frac{N\,d\theta}{2\pi\sigma_X\sigma_Y} e^{-r^2S} r^3\,dr \div \int_0^\infty \frac{N\,d\theta}{2\pi\sigma_X\sigma_Y} e^{-r^2S} r\,dr \qquad \text{(D.122)}$$

We have $\int_0^\infty e^{-r^2S} r\,dr = (1/2S)\Big[1 - e^{-r^2S}\Big]_0^\infty = 1/2S$

and $\int_0^\infty e^{-r^2S} r^3\,dr = (1/2S)\Big[-r^2e^{-r^2S} - (1/S)e^{-r^2S}\Big]_0^\infty = 1/2S^2$

since in general [607], p. 123, $x^n \exp(-x) = 0$ for both $x = 0$ and $x = \infty$, if $n \neq 0$.

So, from (D.122) $\sigma_C^2 = 1/S = \dfrac{2\sigma_X^2\sigma_Y^2}{\sigma_Y^2\cos^2\theta + \sigma_X^2\sin^2\theta}$, as in (D.119). (D.123)

(ii) *Proof of* (D.114) for σ_B. Formula (D.121) gives the number of errors lying in a rectangle $r\,d\theta \times dr$. It is clear that the number lying in an area $b \times dr$ at the same position (r, θ) is

$$\{N/2\pi\sigma_X\sigma_Y\}\{\exp(-r^2S)\}b\,dr, \qquad \text{(D.124)}$$

where $S = \dfrac{\cos^2\theta}{2\sigma_X^2} + \dfrac{\sin^2\theta}{2\sigma_Y^2}$, as before.

In the rectangle QQ′ of Fig. D.11, b is constant, and the number of errors in the rectangle between O and r is

$$\frac{Nb}{2\pi\sigma_X\sigma_Y} \int_0^r e^{-r^2S}\,dr. \qquad \text{(D.125)}$$

This is a normal univariate distribution similar to (D.1), with $\sigma_B^2 = 1/(2S)$.

So $\qquad \sigma_B^2 = \dfrac{\sigma_X^2\sigma_Y^2}{\sigma_Y^2\cos^2\theta + \sigma_X^2\sin^2\theta}$, as in (D.114), (D.126)

and, as in (D.5), the proportion of errors within the rectangle lying between $+k\sigma_B$ and $-k\sigma_B$ is

$$\frac{2}{\sqrt{\pi}} \int_0^{k/\sqrt{2}} e^{-t^2}\,dt. \qquad \text{(D.127)}$$

D.22. Correlated observations

(*a*) In the method of least squares, using the parametric method, we seek a set of t unknowns $x_1, \ldots, x_t$ such that the s residuals $v_1, \ldots, v_s$ corresponding to observations $O_1, \ldots, O_S$, s being greater than t, will have minimum improbability.

If the errors of the observations, and consequently the expected values (s.v.'s)† of the v's, are uncorrelated, § D.14 shows that this is achieved by making $[v^2/\sigma^2]$ a minimum.

When errors of observation are correlated, this rule does not hold. In its place we assert that if we can find a set of uncorrelated quantities $V_1, \ldots, V_s$, linear functions of the v's, whose s.v.'s $\tau_1, \ldots, \tau_s$ are known in terms of the variances and covariances of the observations, the most probable values of the x's will be those which give rise to the least improbable values of the V's.

To illustrate this, imagine a set of orthoronal axes of $v_1, \ldots, v_s$ in an s-dimensional space, such as are shown for three dimensions in Fig. D.13 (a). We seek values of the x's which will give the vector of v's represented by the point A, which will have minimum improbability.

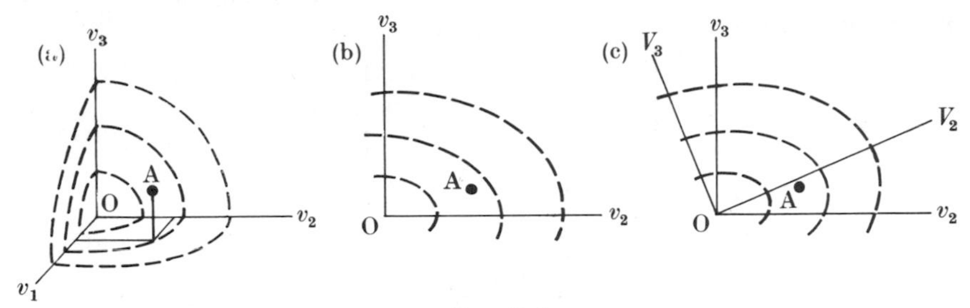

FIG. D.13.

If all the observations (and v's) are uncorrelated and have equal s.v.'s, the surfaces of equal probability for the point A will be hyperspheres centred on the origin O, at which all v's$=0$. In these circumstances the most probable position for A will be on the sphere with the smallest possible radius, so that

$$\sum^{s} v^2 = \text{minimum}. \tag{D.128}$$

If the s.v.'s of the v's are unequal, the hyperspheres will be replaced by hyper-ellipsoids. If the v's are uncorrelated, the axes of these ellipsoids will coincide with the axes of the v's, and their equations will be $(v_1/\sigma_1)^2 + \ldots\ldots (v_s/\sigma_s)^2 = k$. Then the most probable set of x's is that which places A on the ellipsoid with the smallest possible k, so that

$$\sum^{s} v^2/\sigma^2 = \text{minimum}. \tag{D.129}$$

Fig. D.13 (*b*) illustrates this in two dimensions.

† The expression 'standard error' is not applicable to the residuals v. Every observation O_n has a standard error σ_n at the 68 : 32 probability level, and the value of the corresponding v_n can be expected to be less than $\sigma_n(1-t/s)^{\frac{1}{2}}$ at the same level of probability. At the 50–50 level we can speak of the probable error of O_n and of the probable value of v_n. At the 68 : 32 level there is no recognized equivalent to 'probable value', but we can call it the 'standard value' s.v. We use the expression variance in connection with both (s.e.)^2 and (s.v.)^2.

If the observations are correlated, the axes of the ellipsoids will not coincide with the axes of the v's, but will lie along axes $V_1, \ldots, V_s$, as in Fig. D.13 (c).† The equations of the hyper-ellipsoids will now be $(V_1/\tau_1)^2 + \ldots . (V_s/\tau_s)^2 = k$, where $\tau_1, \ldots, \tau_s$ are the uncorrelated s.v.'s of $V_1, \ldots, V_s$, and the most probable position of A will now give

$$\sum^{s} V^2/\tau^2 = \text{minimum.} \tag{D.130}$$

(b) We now have to find the linear functions V, and their uncorrelated s.v.'s $\tau_1, \ldots, \tau_s$. Using matrix notation, as a trial let us take

$$V = M^{\mathrm{T}} v, \text{ and } v = MV, \tag{D.131}$$

where M is the modal matrix of Q, Q being the variance–covariance matrix of the O's, and that of the v's being $Q(1-t/s) \equiv Q_v$.

$$Q_v = \begin{bmatrix} \sigma_1^2 & \sigma_{12} & . & . & . & \sigma_{1s} \\ \sigma_{21} & \sigma_2^2 & . & . & . & \sigma_{2s} \\ & . & . & . & . & . \\ \sigma_{s1} & \sigma_{s2} & . & . & . & \sigma_s^2 \end{bmatrix} (1-t/s).) \tag{D.132}$$

So
$$M^{\mathrm{T}} Q_v M = D, \text{ a diagonal matrix, from (B.47).} \tag{D.133}$$

To get the (s.v.)^2 variances and covariances of the V's, p. 725 footnote,

$$(V_n)^2 = (u_{n1}v_1 + u_{n2}v_2 + \ldots)^2,$$

where the u's are the elements of M^{T}. So

$$\begin{aligned}(\tau_n)^2 &= \{\sigma_1^2 u_{n1}^2 + \sigma_2^2 u_{n2}^2 + \ldots + 2\sigma_{12} u_{n1} u_{n2} + \text{all pairs}\}(1-t/s) \\ &= [\text{row } n \text{ of } M^{\mathrm{T}}] Q_v [\text{column } n \text{ of } M].\end{aligned}$$

Similarly,

$$\begin{aligned}\tau_{nm} &= \{\sigma_1^2 u_{n1} u_{m1} + \sigma_2^2 u_{n2} u_{m2} + \ldots + \sigma_{12}(u_{n1}u_{m2} + u_{m1}u_{n2}) + \text{all pairs}\}(1-t/s) \\ &= [\text{row } n \text{ of } M^{\mathrm{T}}] Q_v [\text{column } m \text{ of } M].\end{aligned}$$

So the (s.v.)^2 variance–covariance matrix of V is

$$M^{\mathrm{T}} Q_v M. \tag{D.134}$$

As above, (D.133), this is diagonal, so the covariances of the V's are zero, and their variances τ^2 are the diagonal elements of D. So (D.131) is what is required.

It follows that
$$V_1^2/\tau_1^2 + \ldots + V_s^2/\tau_s^2 = V^{\mathrm{T}} D^{-1} V, \tag{D.135}$$

where $D = M^{\mathrm{T}} Q_v M$, as in (D.133). This is to be minimized.

† The situation has an appearance similar to that of § D.20, but in § D.20 the covariance involved is that of two of the unknown x's, namely the two coordinates of a survey station, and the reference axes are east and north. In this § D.22, we are concerned with the mutual covariance of all the observations and the associated v's, not with the covariance of the unknowns. The directions of the axes of the v's have no significance, except that they have to be orthogonal.

(*c*) We now assert, and prove later, that (D.135) will be minimized if, when following the usual least squares routine, instead of the diagonal weight matrix W, we use the full matrix

$$P \equiv Q^{-1}, \tag{D.136}$$

namely the reciprocal of the variance–covariance matrix of the observations, as follows.

$$\text{Observation equations.}\ Ax = b + v.$$
$$\text{Normal equations.}\ A^{\mathrm{T}}PAx = A^{\mathrm{T}}Pb,$$
$$x = (A^{\mathrm{T}}PA)^{-1}A^{\mathrm{T}}Pb,$$
$$\text{Variance–covariance of the } x\text{'s} = (A^{\mathrm{T}}PA)^{-1}. \qquad \text{(D.137)†}$$

Proof. We prove this assertion as follows.
We wish to minimize $V^{\mathrm{T}}D^{-1}V$.

We have
$$\begin{aligned}(1-t/s)V^{\mathrm{T}}D^{-1}V &= V^{\mathrm{T}}(M^{\mathrm{T}}QM)^{-1}V, \text{ from (D.133)}\\ &= V^{\mathrm{T}}(M^{\mathrm{T}}Q^{-1}M)V, \text{ from (B.45)}\\ &= (M^{\mathrm{T}}v)^{\mathrm{T}}(M^{\mathrm{T}}Q^{-1}M)(M^{\mathrm{T}}v)\\ &= v^{\mathrm{T}}Pv, \text{ since } M^{\mathrm{T}}M = I.\end{aligned}$$

It is finally required to prove that $v^{\mathrm{T}}Pv$ is minimized by the solution of

$$A^{\mathrm{T}}PAx = A^{\mathrm{T}}Pb, \text{ (D.137)}$$

That is to say that

$$(\partial/\partial x)(v^{\mathrm{T}}Pv) = 0 \text{ for all } x\text{'s given by } A^{\mathrm{T}}PAx = A^{\mathrm{T}}Pb.$$

P is symmetrical.

We have
$$\begin{aligned}v^{\mathrm{T}}Pv &= (Ax-b)^{\mathrm{T}}P(Ax-b)\\ &= (x^{\mathrm{T}}A^{\mathrm{T}} - b^{\mathrm{T}})P(Ax-b)\\ &= x^{\mathrm{T}}(A^{\mathrm{T}}PA)x - b^{\mathrm{T}}(PA)x - x^{\mathrm{T}}(A^{\mathrm{T}}P)b + b^{\mathrm{T}}Pb.\end{aligned}$$

Then, using (B.53) for the first term. (B.51) for the second, and (B.52) for the third,

$$\begin{aligned}\text{the column vector } (\partial/\partial x)(v^{\mathrm{T}}Pv) &= 2(A^{\mathrm{T}}PA)x - (PA)^{\mathrm{T}}b - A^{\mathrm{T}}Pb + 0\\ &= 2(A^{\mathrm{T}}PA)x - 2(A^{\mathrm{T}}Pb), \text{ since } P = P^{\mathrm{T}}\\ &= 0, \text{ if (D.137) is satisfied.}\end{aligned}$$

(*d*) *Applications.* If the x's are directly measured quantities, it will seldom be possible to estimate their covariances. We may hope that they are zero, or we may suspect that they are not, but if we cannot put a figure to our suspicions, we cannot use the rule (D.136).

† The facts that the columns of M are the latent vectors of Q, and that the elements of D are the latent roots of Q_v, are of interest, but neither M nor D has to be evaluated.

On the other hand, if the quantities x have been given by two or more least square solutions, their variances and covariances may be well known, and then their over-all most probable values may be got as below.

Let there be three sets of unknowns x_1, x_{12}, and x_2, such that x_1 and x_{12} are entangled in some observation equations, and x_{12} and x_2 in others, but x_1 and x_2 in none. We require the best values of the unknowns x_{12}.†

Then we can determine x_1 and x_{12} from their equations as in § D.14, and get

$$x_{12} = \text{(say) } x_A, \text{ with variance–covariance matrix } Q_A,$$

and the solution for x_2 and x_{12} gives

$$x_{12} = x_B \text{ with variance–covariance matrix } Q_B.$$

Let $\bar{x}$ be the required most probable values of x_{12}. Then the 'observation' equations for $\bar{x}$ can be set up as the partitioned matrices

$$\left[\frac{I}{I}\right][\bar{x}] = \left[\frac{x_A}{x_B}\right], \text{ where } I \text{ is a unit matrix.} \tag{D.138}$$

The normal equations are

$$[I|I]\left[\begin{array}{c|c} P_A & 0 \\ \hline 0 & P_B \end{array}\right]\left[\frac{I}{I}\right][\bar{x}] = [I|I]\left[\begin{array}{c|c} P_A & 0 \\ \hline 0 & P_B \end{array}\right]\left[\frac{x_A}{x_B}\right], \tag{D.139}$$

where the full matrix $P_A = Q_A^{-1}$, and $P_B = Q_B^{-1}$. Whence

$$P_A\bar{x} + P_B\bar{x} = P_Ax_A + P_Bx_B,$$

$$\bar{x} = (P_A + P_B)^{-1}(P_Ax_A + P_Bx_B). \tag{D.140}$$

The variance–covariance matrix of $\bar{x}$ is

$$(P_A + P_B)^{-1}. \tag{D.141}$$

And, if there was a third set of equations, $\bar{x} = x_C$, the solution would clearly be

$$\bar{x} = (P_A + P_B + P_C)^{-1}(P_Ax_A + P_Bx_B + P_Cx_C), \tag{D.142}$$

and so on.

This may be compared with § D.10 (*c*). In (D.140), (D.141) the full weight matrix of the initial solutions enters into the final result in the same form as do the weights of the separate measures of a single quantity into (D.37) and (D.39).

D.23. Two proofs

(*a*) *The variance–covariance matrix.* Proof of (D.66). Using the notation of (D.49)–(D.52), let the column vector δx be the change in the vector x resulting from errors $W^{\frac{1}{2}}\delta b$ in the weighted observations.

Then
$$\delta x = N^{-1}A^TW(\delta b) \tag{D.143}$$

† See § 2.37 for a practical illustration.

let δx^g be the changes in x resulting from the error in a single element $W^{\frac{1}{2}}_g b_g$, then

$$\delta x^g = N^{-1}A^T W^{\frac{1}{2}} \begin{bmatrix} 0 \\ \vdots \\ W_g^{\frac{1}{2}}\delta b_g \\ \vdots \\ 0 \end{bmatrix} \tag{D.144}$$

If all the errors, δb_g were exactly known, the total δx would simply be the sum of the δx^g's, but what is actually known is that the s.e.'s of the $(W^{\frac{1}{2}}b)$'s are unity, provided that weights have been taken as $1/(\text{s.e.})^2$.

So the r.m.s. value of $\delta x^g = \sigma^g = N^{-1}A^T W^{\frac{1}{2}} \begin{bmatrix} 0 \\ \vdots \\ 1 \\ \vdots \\ 0 \end{bmatrix}$

$$= \text{column } g \text{ of } N^{-1}A^T W^{\frac{1}{2}}. \tag{D.145}$$

And if σ_i^g is the r.m.s. value of a particular x, x_i, due to error in b_g

$$\sigma_i^g = \text{row } i \text{ of } N^{-1}A^T W^{\frac{1}{2}} \begin{bmatrix} 0 \\ \vdots \\ 1 \\ \vdots \\ 0 \end{bmatrix}$$

$$= \text{row } i \text{ of column } g \text{ of } N^{-1}A^T W^{\frac{1}{2}}, \tag{D.146}$$

and the total σ_i^2 due to errors in all the b's will be the sum of the squares of the elements in row i of $N^{-1}A^T W^{\frac{1}{2}}$, namely

$$\begin{aligned} \sigma_i^2 &= \text{the } (i, i) \text{ element of } (N^{-1}A^T W^{\frac{1}{2}})(N^{-1}A^T W^{\frac{1}{2}})^T \\ &= \text{the } (i, i) \text{ element of } N^{-1}, \text{ since } N = A^T WA \\ &= \alpha_{ii} \end{aligned} \tag{D.147}$$

Similarly for the covariance between x_i and x_j,

$$\begin{aligned} \sigma_{ij} &= \text{the sum of the products of corresponding elements in row } i \text{ and row } j \text{ of } N^{-1}A^T W^{\frac{1}{2}} \\ &= \text{the } (i, j) \text{ element of } (N^{-1}A^T W^{\frac{1}{2}})(N^{-1}A^T W^{\frac{1}{2}})^T \\ &= \alpha_{ij}, \text{ as above,} \end{aligned} \tag{D.148}$$

which proves the proposition.

(*b*) *Method of conditions. Proof of* § D.16. It is required to find values of $x_1, \ldots, x_s$, which

(i) satisfy the n conditions (D.73) exactly, and

(ii) subject to (i) make $X \equiv w_1 x_1^2 + \ldots + w_s x_s^2 = x^T W x$ minimum.

If the conditions imposed no restrictions on the possible values of the x's, X would be minimum when $x_1 = x_2 = \ldots = x_s = 0$, but the actual situation is less simple.

Consider the expression

$$\Phi \equiv (w_1x_1^2 + \ldots + w_sx_s^2) - 2\lambda_a(a_1x_1 + \ldots + a_sx_s - \epsilon_a) - \ldots$$
$$- 2\lambda_n(n_1x_1 + \ldots + n_sx_s - \epsilon_n)$$
$$\equiv X + \Lambda, \text{ defining } \Lambda, \qquad \text{(D.149)†}$$

where the λ's are independent of the x's and $(a_1x_1 + \ldots + a_sx_s)$, etc. are as in (D.73).

The conditions for Φ to be minimum, whether the variations of the x's are restricted or not, are

$$(\partial\Phi/\partial x_1) = \ldots = (\partial\Phi/\partial x_s) = 0,$$

i.e.

$$w_1x_1 = a_1\lambda_a + \ldots + n_1\lambda_n,$$
$$\cdots\cdots$$
$$w_sx_s = a_s\lambda_a + \ldots + n_s\lambda_n. \qquad \text{(D.150)}$$

X is required to be minimum with respect to variations of the x's, which are restricted by having to satisfy (D.73). This restriction permanently makes $\Lambda = 0$, so the required restricted minimum value of X is given by the permissible values of $x_1, \ldots, x_s$ and $\lambda_1, \ldots, \lambda_n$ which make Φ minimum.

To determine these $(n+s)$ unknowns we have $(n+s)$ equations, (D.73) and (D.150). Substituting (D.150) in the n conditions (D.73) gives n equations for the n λ's, namely (D.74). Then substituting the solution of (D.74) in (D.150), which is the same as (D.75), gives x's which satisfy the conditions, which make Λ zero, which make Φ minimum, and which therefore make $w_1x_1^2 + \ldots + w_sx_s^2$ as small as it can be while Λ is held to zero.

D.24. Interpolation by least squares

(*a*) *Interpolation.* A variable physical quantity U, such as the intensity of gravity or a component of the deviation of the vertical, or an anomaly of either, may be measured at a number of stations $A_1, \ldots, A_n$, whereas in theory continuous knowledge of U all over a certain area is probably required. Interpolation may be an acceptable means of densifying the observation net, or at any rate may be the best that can be done in particular circumstances. Interpolation may take many forms, as in (i)–(v) below, in all of which the observed values of U are presumed to be errorless.

(i) When observations are confined to points along a more or less straight line as in astro-geodetic geoid sections, § 4.50, simple interpolation between adjacent

† The terms $\epsilon_a \ldots \epsilon_n$ were omitted in [113], 3 Edn., but no error was carried forward to (D.150). The factors 2 preceding each λ in (D.149) are customarily included for slight numerical convenience, but their inclusion is immaterial, since Λ = zero.

stations amounts to weighting their observations inversely as their distance, and taking the weighted mean. Formulae (4.96) results.

(ii) When observations extend over an area, interpolated values may be obtained from the observed values at stations forming a triangle or quadrilateral surrounding the point where interpolation is required. As before, weighting may be inversely as the distance.

(iii) More elaborately, the interpolated value may be based on all available known points, or on all within a certain distance. And some different system of weighting may be preferred.

(iv) Even (iii) will not provide the best possible results. For instance, two of the observations may be rather close together. It is then questionable whether it would be best to treat the two as a single (mean) observation, or as two separate observations. A knowledge of C_{kk}, the variance of the values† of U, and of the covariance C_{kl} between observed values U_k and U_l is required before a suitable compromise can be arrived at.

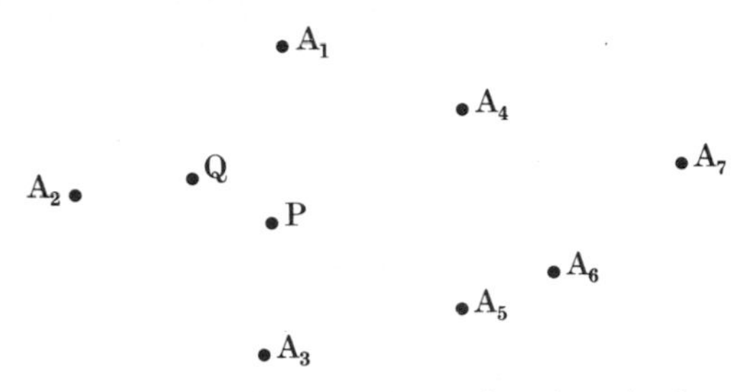

FIG. D.14. Observations of U have been made at points A_1 to A_7. Interpolated values of U are required at P and Q.

(v) Graphical interpolation by drawing contours of U on a chart is a simple intuitive alternative to all the above.

(*b*) *Weighting factors.* In general, following [262] pp. 266–70, let it be decided that the estimated interpolated value of U at a point P shall be given by

$$\bar{U}_P = \sum_k a_{kP} U_k = a_{kP}^T U_k \tag{D.151}$$

where a_{kP} is a weighting factor applicable to the value U_k observed at A_k. In the second equality in (D.151) a_{kP}^T is the vector $[a_{1P}, \ldots, a_{nP}]^T$. This matrix vector notation is more concise than the summation notation. In order to obtain the best values for the a's, the standard error of the deduced $\bar{U}_P$ must be expressed in terms of the a's and then minimized by least squares, as in (*e*) below..

It is first necessary to give numerical values to C_{kk} and C_{kl}. The former presents no difficulty, and the figure obtained from the observed values will in general be applicable to the required interpolated values. So $C_{PP} = C_{kk}$. For C_{kl} see (*c*) below.

(*c*) *The covariance function* C_{kl}. In practice the covariances are only computable if C_{kl} (between U_k and U_l) is assumed to be some simple function of the distance

† We require the variance (mean square) of the value of U, not of the error of U. We are here concerned with its magnitude, not with its precision.

S_{kl} between A_k and A_l. The form of the function can be empirically determined as below.

Consider all pairs of A's whose separations are between (say) 1 and 20 km, with mean S_1. If there are n_1 such pairs the covariance between them is $(1/n_1)\sum U_k U_l$, which can be computed. The covariance between points 20–40 km apart, with mean distance S_2, can be obtained similarly, and so on between suitable limits of distance. The results are then shown on a diagram, as in Fig. D.15, from which the covariance for any distance S_{kl} or S_{kP} can be read.

It may be convenient to replace the diagram by an algebraic expression designed to fit it, such as

$$C_{kl} \text{ or } C_{kP} = r_1 \exp(-s_1 S^2) + r_2 \exp(-s_2 S^2), \tag{D.152}$$

where r_1, r_2, s_1, and s_2 are determined from the data. A single exponential term may suffice. Equation (D.152) is the *covariance function of the observed U's*†. It is also applicable to the covariance between observations at points A and an unknown $\bar{U}_P$, but it is not applicable to a pair of unknown U_P and U_Q.

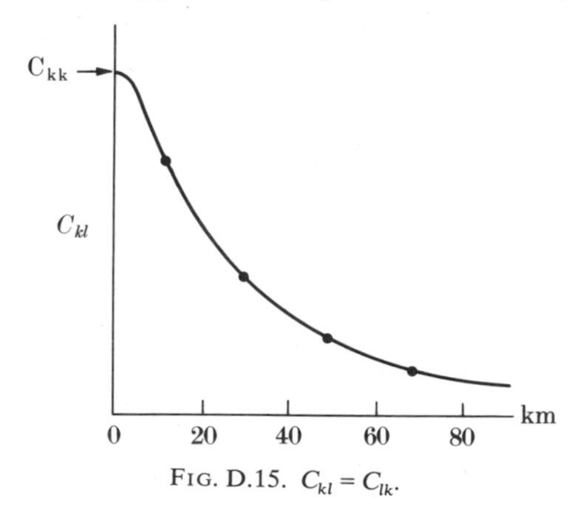

FIG. D.15. $C_{kl} = C_{lk}$.

(*d*) *The s.e. of the interpolated $\bar{U}$ at P.* Let U_P be the true value of U at P, and let $\bar{U}_P$ be the interpolated value. To evaluate σ_P, the s.e. of $\bar{U}_P$, we have

$$\text{Error of } \bar{U}_P = e_P = \bar{U}_P - U_P$$

$$= \sum_k (a_{kP} U_k) - U_P, \text{ from (D.151)}$$

$$\text{So } e_P^2 = U_P^2 - 2U_P \sum_k a_{kP} U_k + \sum_k \sum_l a_{kP} a_{lP} U_k U_l,$$

as may be verified by direct multiplication

and
$$\sigma_P^2 = \text{r.m.s. value of } e_P^2$$

$$= C_{kk} - 2\sum_k a_{kP} C_{kP} + \sum_k \sum_l a_{kP} a_{lP} C_{kl} \tag{D.153}$$

† It is the covariance between their magnitudes, not of their errors, which in the case of the observed U's are assumed to be zero.

since C_{kP} is the mean value of $U_k U_P$, and C_{kl} is the mean value of $U_k U_l$.

Given the a's by one of the rules suggested in (a) above, and given the C's from the diagram or (D.152), σ_P^2 can be computed.

(e) *Determination of the a's by minimizing* σ_P^2. To minimize σ_P^2 we have $\partial\sigma_P^2/\partial a_{mP} = 0$, n equations with $m = 1, \ldots, n$. Then (D.153) gives

$$-2C_{mP} + 2\sum_k a_{kP} C_{km} = 0, \tag{D.154}$$

since $\partial C_{kk}/\partial a_{mP} = 0$, $\partial\left(2\sum_k a_{kP}C_{kP}\right)\Big/\partial a_{mP} = 2C_{mP}$,

and $(\partial/\partial a_{mP})\sum_k\sum_l a_{kP}a_{lP}C_{kl} = 2\sum_k a_{kP}C_{km}$ †

As a matter of convenience the dummy m (which signifies all points $A_1, \ldots, A_n$) may be changed to l, which has the same meaning, and the n equations (D.154) are

$$\sum_k C_{kl} a_{kP} = C_{lP} \tag{D.155}$$

where l has values $1, \ldots, n$ in each of the n equations.

(f) In matrix notation, the n equations (D.155) may be written

$$C_{lk} a_{kP} = C_{lP} \tag{D.156}$$

where a_{kP} and C_{lP} are $(n \times 1)$ column vectors, and C_{lk} is a $(n \times n)$ symmetrical square matrix. The solution is

$$a_{kP} = (C_{lk})^{-1} C_{lP} \tag{D.157}$$

Then from (D.151) we have

$$\bar{U}_P = (C_{kl}^{-1} C_{lP})^{\mathrm{T}} U_k = C_{lP}^{\mathrm{T}} C_{kl}^{-1} U_k, \text{ since } C_{kl} \text{ is symmetrical.} \tag{D.158}$$

† In $(\partial/\partial a_{mP})\sum_k\sum_l a_{kP}a_{lP}C_{kl}$ only the terms which contain a_{mP} or a_{mP}^2 as a factor will contribute. First consider the second summation. When $l = m$, the expression becomes $\sum_k a_{kP}a_{mP}C_{km}$, whose contribution to $\partial/\partial a_{mP}$ is

$$\sum_k a_{kP} C_{km}, \tag{A}$$

except that when k also equals m, the expression is $a_{mP}^2 C_{mm}$ whose derivative is

$$2a_{mP}C_{mm}, \tag{B}$$

In the first summation, when $k = m$, the expression becomes $\sum_l a_{mP}a_{lP}C_{lm}$ whose contribution to $\partial/\partial a_{mP}$ is $\sum_l a_{lP}C_{lm}$, which is the same as

$$\sum_k a_{kP} C_{km} \tag{C}$$

But we note that although $a_{mP}^2 C_{mm}$ occurs in both summations, it actually only occurs once in the third term of (D.153), and its derivative must not be included twice.

Then summing (A), (B), and (C), but including $2a_{mP}C_{mm}$ only once, gives $(\partial/\partial a_{mP})\sum_k\sum_l a_{kP}a_{lP}C_{kl} = 2\sum_k a_{kP}C_{km}$.

More fully, as in [262], the value of $\bar{U}$ at P is given by

$$\bar{U}_P = [C_{1P} \quad C_{2P} \quad \cdots \quad C_{nP}] \begin{bmatrix} C_{11} & C_{12} & \cdots & C_{1n} \\ C_{21} & C_{22} & \cdots & C_{2n} \\ \vdots & \vdots & & \vdots \\ C_{n1} & C_{n2} & \cdots & C_{nn} \end{bmatrix}^{-1} \begin{bmatrix} U_1 \\ U_2 \\ \vdots \\ U_n \end{bmatrix} \tag{D.159}$$

(g) To determine σ_P^2 as given by (D.153), multiply the first equation of (D.155) by a_{1P}, the second by a_{2P}, etc., and add. On the left-hand side the sum equals the third term of (D.153), namely $\sum_k \sum_l a_{kP} a_{lP} C_{kl}$, while the sum of the right-hand side is $\sum_k a_{kP} C_{kP}$. Substituting this equality in (D.153) then gives

$$\sigma_P^2 = C_{kk} - \sum_k a_{kP} C_{kP}, \tag{D.160}$$

or, proceeding as in (D.157)–(D.159).

$$\begin{aligned} \sigma_P^2 &= C_{kk} - C_{lP}^{\mathrm{T}} C_{kl}^{-1} C_{kP} \\ &= C_{kk} - [C_{1P} \quad C_{2P} \quad \cdots \quad C_{nP}] \begin{bmatrix} C_{11} & C_{12} & \cdots & C_{1n} \\ C_{21} & C_{22} & \cdots & C_{2n} \\ \vdots & \vdots & & \vdots \\ C_{n1} & C_{n2} & \cdots & C_{nn} \end{bmatrix}^{-1} \begin{bmatrix} C_{1P} \\ C_{2P} \\ \vdots \\ C_{nP} \end{bmatrix} \end{aligned} \tag{D.161}$$

(h) If the interpolation is carried out at two points P and Q, the standard error of the interpolated value $\bar{U}$ at each is given by (D.161) but their errors will not be independent. The covariance of the errors of interpolation at P and Q is given by

$$\begin{aligned} \sigma_{PQ} &= C_{PQ} - C_{lP}^{\mathrm{T}} C_{kl}^{-1} C_{lQ} \\ &= C_{PQ} - [C_{1P} \quad C_{2P} \quad \cdots \quad C_{nP}] \begin{bmatrix} C_{11} & C_{12} & \cdots & C_{1n} \\ C_{21} & C_{22} & \cdots & C_{2n} \\ \vdots & \vdots & & \vdots \\ C_{n1} & C_{n2} & \cdots & C_{nn} \end{bmatrix}^{-1} \begin{bmatrix} C_{1Q} \\ C_{2Q} \\ \vdots \\ C_{nQ} \end{bmatrix}, \end{aligned} \tag{D.162}$$

which is similar in form to (D.159) and (D.161). See [262]. This covariance σ_{PQ} of the errors of the interpolated values is not to be confused with C_{kl}, the covariance function of the magnitudes of the U's.

(i) *Example.* As a very simple example, let it be required to compute the deviation of the vertical η at P in Fig. D.16, in which $A_1A_2 = A_2A_3 = A_3A_4 = 2$

FIG. D.16.

units of distance of (say) 10 km each, and the observed values of η at A_1, A_2, A_3, and A_4 are $-2''$, $-1''$, $+1''$, and $+3''$, as shown.

First C_{kl} is computed for $S=0$, 2, 4, and 6 units from the given data, as in (*c*) above, with interpolated values for $S=1$ and 3,

$S=0$	$C(0)=\frac{1}{4}(4+1+1+9)=+3.75$	
1	$C(1)$	$=+3.00$ or 2.00, by interpolation.
2	$C(2)=\frac{1}{3}(2-1+3)$	$=+1.33$
3	$C(3)$	$=-0.50$, by interpolation.
4	$C(4)=\frac{1}{2}(-2-3)$	$=-2.50$
5	$C(5)$	Not required
6	$C(6)=1(-6)$	$=-6.00$

In this example the value of $C(1)$ is important, but it is ill-determined because no pairs of observations are so close to each other. Let it first be taken as $+3.00$, and later as $+2.00$ for comparison.

As in (D.151) $\eta_P=-2a_{1P}-a_{2P}+a_{3P}+3a_{4P}$. Then (D.155) is

$$\begin{bmatrix} +3.75 & +1.33 & -2.50 & -6.00 \\ +1.33 & +3.75 & +1.33 & -2.50 \\ -2.50 & +1.33 & +3.75 & +1.33 \\ -6.00 & -2.50 & +1.33 & +3.75 \end{bmatrix} \begin{bmatrix} a_{1P} \\ a_{2P} \\ a_{3P} \\ a_{4P} \end{bmatrix} = \begin{bmatrix} -0.5 \\ +3.0 \\ +3.0 \\ -0.5 \end{bmatrix},$$

of which the solution is $a_{1P}=-0.09$, $a_{2P}=+0.59$, $a_{3P}=+0.55$, and $a_{4P}=-0.07$. Whence $\eta_P=-0''.07$.

From (D.160)

$$\sigma_P^2=3.75+0.09(-0.50)-0.59(3.00)-0.55(3.00)+0.07(-0.50)=0''.25.$$

So $\eta_P=-0''.07\pm0''.50$.

Computing with $C(1)=+2.00$ instead of $+3.00$ gives $a_{1P}=+0.02$, $a_{2P}=+0.40$, $a_{3P}=+0.39$, and $a_{4P}=+0.02$. Whence $\eta_P=+0''.01$

From (D.160)

$$\sigma_P^2=3.75-0.02(-0.50)-0.40(2.00)-0.39(2.00)-0.02(-0.50)=2''.19.$$

So $\quad \eta_P=+0''.01\pm1''.48$

In both cases the value of η_P is reasonable, differing insignificantly from the natural value of 0.00 derived by simple interpolation from the two nearest points. With a larger and less regular non-linear net of observations the utility of the least-square method would be more apparent. The smaller value of σ_P is perhaps not unreasonable, but it depends very largely on the value of $C(1)$ assumed for the covariance of the two nearest observations. In the more typical case of a non-linear net, there would probably be less uncertainty about this. In this example also the routine computation takes no account of the apparently regular increase of η from A_1 to A_4. If such regularity was typical in a larger sample, intuition would give a smaller s.e., on the given assumption that the observations were errorless.

(*j*) *Conclusions.* The effectiveness of the method depends on the significance of the covariance function C_{kl} deduced from the observations as in (*c*). If the assumption that the covariance depends exclusively on the distance is not reasonably valid, the method looses its utility. Notably, if gravity or deviation anomalies are not freed from all substantial topographical effects, this assumption is not valid, and it will probably be better to draw contours of U, or to interpolate simply by any other method, with an eye on the disturbing topography.

[262], p. 270 remarks that the main advantage of the method is that it can be done by computer. It also remarks that it is the most accurate method, although the improvement in accuracy is described as not striking. [267] agrees that the use of anomalies which have been freed from topographical effects is ordinarily essential.

D.25. Collocation†

(*a*) *Summary.* In the ordinary method of least squares, using indirect observations, § D.14, the observation equations are

$$Ax = b + v, \qquad \text{(D.163)}$$

where b is the vector of the observations, probably linearized, in the form of (observed value minus the value computed from some model).
x is the vector of the unknown corrections to the parameters of the model.
v is the vector of residuals. $v^{\mathrm{T}}Wv$ is to be minimized, where the weight matrix W may be diagonal, or a full matrix P, as in (D.136)

It may be remarked that there is no question of assigning a weight or covariance matrix to the unknown x's.

In the method of collocation, on the other hand, the vector $-v$ is replaced by two vectors $(s'+n)$. The vector s' is another set of unknowns, described as the *signal*, and n is the random error or *noise*. The essence of the method [424] is that by some means a covariance matrix can be assigned to the signal. For the noise it will generally be possible to assign a diagonal weight matrix in the usual way.

As an illustration, consider a group of gravity stations. Then b is the vector of observed g minus the value computed from some generalized model, such as a gravity formula applicable to the whole earth, or some empirical generalization intended to apply to some smaller area. The model may include parameters (such as density, or depth of isostatic compensation), to define some form of topographical or other reduction, and also instrumental constants and drift. Then the x's are the changes in the parameters of the model, the signal is the gravity anomaly at each station, and n is the random error of measurement. It may also be required to estimate the value of the signal at other, unvisited stations.

† The word collocation, as used here, has no connection with the more natural use of the word, by which two nearly but not quite identical observing sites are said to be collocated.

(D.163) is then written

$$b = Ax + s' + n \qquad \text{(D.164)}\dagger$$

$$= Ax + z, \text{ defining } z = s' + n, \qquad \text{(D.165)}$$

s' being written for the signal at observation stations, while s is reserved for the values of the signal at interpolated stations. Let there be q observations (and values of s'), p interpolated values of s, and m model parameters.

(*b*) *Covariance matrices.* We are concerned with the following.

(i) $C_{bb} = C_{zz}$, the expected covariance between the observed b's (or z's) for all pairs of the q observations. A $(q \times q)$ matrix.

(ii) $C_{s's'}$, the expected covariance between the signals (e.g. the gravity anomalies) for all pairs of the q observations. A $(q \times q)$ matrix.

(iii) C_{nn}, the expected value of n^2 at each station. A $(q \times q)$ diagonal matrix.

(iv) $C_{sb} = C_{sz} = C_{ss'}$, see [424], p. 12, the expected covariance between all pairs of mixed observed and interpolated signals . A $(p \times q)$ matrix.

(v) C_{ss}, the expected covariance between the signals at pairs of interpolated stations. A $(p \times p)$ matrix.

Values are got for these matrices as follows. The variance of the noise, C_{nn}, can be estimated in the usual way, according to the circumstances at each station, different types of instrument, land or sea, etc. The covariance C_{bb} between the observed b's at different stations can be got in the same way as described for C_{kl} in § D.24 (*c*). Then the elements of C_{bb} depend on the distances between the pair of stations concerned, and are given by a diagram or formula.

The elements of $C_{s's'}$ are then given by

$$C_{s's'} = C_{bb} - C_{nn}, \text{[424], p. 11}, \qquad \text{(D.166)}$$

and a similar diagram or formula may be prepared for $C_{s's'}$. This diagram will give the expected covariance of pairs of gravity anomalies, dependent on their separation. Clearly it is equally applicable to $C_{ss'}$ and C_{ss}. Figures for all the necessary covariances are thus available.

Let $\bar{C}_{bb}$ now be written for C_{bb}, the covariances of the observations, to distinguish it from C the covariances of the signals.

(*c*) *The least-squares solution.* It is required to obtain the m unknowns x and the signals s' at q observation stations, and also to predict the signals s at p unvisited stations. Then (D.165) may be written

$$Ax + Bv - b = 0, \; q \text{ equations}, \qquad \text{(D.167)}$$

where v is the column vector

$$[s_1 \ldots\ldots s_p \mid z_1 \ldots\ldots z_q]^{\mathrm{T}}, \qquad \text{(D.168)}$$

† The symbols x and b, used here to conform with other paragraphs, are more usually written in current collocation literature, e.g. [424], as X and x.

and B is the $q\times(p+q)$ matrix

$$\begin{bmatrix} 0 & . & . & 0 & 1 & 0 & . & . & 0 \\ . & & & . & 0 & 1 & . & . & 0 \\ . & & & . & . & . & . & . & \\ . & & & . & . & . & . & . & \\ 0 & . & . & 0 & 0 & 0 & . & . & 1 \end{bmatrix}, \tag{D.169}$$

a combination of a $(q\times p)$ zero matrix and a $(q\times q)$ unit matrix. Clearly $Bv = z$. It is now required to minimize $v^{\mathrm{T}}Pv$, where

$$P = Q^{-1} = \begin{bmatrix} C_{ss} & C_{ss'} \\ C_{s's} & \bar{C}_{bb} \end{bmatrix}^{-1}, \tag{D.170}$$

subject to the identities or conditions (D.167). For the solution, which is similar to that of § D.16, it is required to minimize Φ, where

$$\Phi = v^{\mathrm{T}}Pv - 2\lambda^{\mathrm{T}}(Ax + Bv - b), \tag{D.171}$$

in which λ is a column vector of Lagrangian multipliers, like those of (D.74).

Then

$$\partial\Phi/\partial v = 2v^{\mathrm{T}}P - 2\lambda^{\mathrm{T}}B, \text{ and is to equal zero.}$$

So
$$v^{\mathrm{T}} = \lambda^{\mathrm{T}}BP^{-1}, \text{ and } v = P^{-1}B^{\mathrm{T}}\lambda. \tag{D.172}$$

Also $\partial\Phi/\partial x = 2\lambda^{\mathrm{T}}A$ is to be zero, and the transpose $A^{\mathrm{T}}\lambda = 0$ also. (D.173)

Subtituting (D.172) into (D.167) gives

$$Ax + BP^{-1}B^{\mathrm{T}}\lambda - b = 0 \tag{D.174}$$

Then (D.173) and (D.174) are $q+m$ equations for the $q+m$ unknowns. To solve them, pre-multiply (D.174) by $A^{\mathrm{T}}(BP^{-1}B^{\mathrm{T}})^{-1}$ and subtract (D.173), to give

$$A^{\mathrm{T}}(BP^{-1}B^{\mathrm{T}})^{-1}Ax = A^{\mathrm{T}}(BP^{-1}B^{\mathrm{T}})^{-1}b \tag{D.175}$$

which determines x. Then v is given by (D.172) and (D.174)

as
$$v = P^{-1}B^{\mathrm{T}}(BP^{-1}B^{\mathrm{T}})^{-1}(b - Ax). \tag{D.176}$$

Substituting B from (D.169) and P^{-1} from (D.170) gives

$$BP^{-1}B^{\mathrm{T}} = \bar{C}_{bb}. \tag{D.177}$$

This gives (D.175) the simple form

$$A^{\mathrm{T}}\bar{C}_{bb}^{-1}Ax = A^{\mathrm{T}}\bar{C}_{bb}^{-1}b$$

whence
$$x = (A^{\mathrm{T}}\bar{C}_{bb}^{-1}A)^{-1}(A^{\mathrm{T}}\bar{C}_{bb}^{-1}b) \tag{D.178}$$

The vector x having been determined, substitution in (D.165) gives z, which is the best available value of s', since at any particular station the most probable value of n is zero.

Then, from (D.176)
$$v = QB^{\mathrm{T}}\bar{C}_{bb}^{-1}(b - Ax). \tag{D.179}$$

Partitioning (D.177) in accordance with (D.168) and (D.170) gives

$$\begin{bmatrix} s \\ z \end{bmatrix} = \begin{bmatrix} C_{ss} & C_{ss'} \\ C_{s's} & \bar{C}_{bb} \end{bmatrix} \begin{bmatrix} 0 \\ 1 \end{bmatrix} \bar{C}_{bb}^{-1}(b - Ax),$$

whence

$$s = C_{ss'}\bar{C}_{bb}^{-1}(b - Ax). \qquad \text{(D.180)†}$$

Formulae (D.178) and (D.180) are the solution of the problem. Fuller details are given in [424].

(*d*) *Conclusion.* This method of collocation combines the use of a covariance matrix to obtain the best value of the signal at the observation stations, with a system of interpolation (D.180) which is an improvement on (D.159) of § D.24, in that it makes a distinction between the matrices $\bar{C}$ and C.

The extent to which (D.180) can give a better value for the signal at interpolated points, depends on the magnitude of the noise n. If C_{nn} is small $\bar{C}$ will equal C, and (D.180) will be indistinguishable from (D.159).

In the case of gravity anomalies observed on land with modern instruments covering an area of (say) 100×100 km or more, the variance of the noise, is likely to be $(5\ \mu\text{m s}^{-2})^2$‡ or much less, while the variance of the anomaly may be $(100\ \mu\text{m s}^{-2})^2$ or more, and the difference between $\bar{C}$ and C will be immaterial. On the other hand, over a much wider area, if the 'observations' take the form of 1-degree or 5-degree means, of which some may depend on few actual observations, the variance of the random error of 'observation' may be relatively large, and it will be possible to form useful estimates of it.

General references for theory of errors. [7], [126], [262], [424], [560], [561], [598], and [619].

† If s is required at only one point P, $C_{ss'}$ is only a single row vector $[C_{P1}C_{P2}\ldots]$, which may be written as C_P^T. If s is also required at a second point Q, $C_{ss'}$ will be a two-row matrix, with C_P^T as its first row and C_Q^T as its second. And so on. The similarity to (D.159) of § D.24 may be noted.

‡ $5\ \mu\text{m s}^{-2} = 0.5$ mGal, and $100\ \mu\text{m s}^{-2} = 10$ mGal.

APPENDIX E

VECTOR ALGEBRA

E.00. Definitions

An ordinary number, such as can be geometrically represented by a point on a scale, is called a *scalar*. See § B.00. A *vector* is defined by a direction as well as a magnitude. It can be geometrically represented in three dimensions by a line of defined length and direction with an arrow at one end, or algebraically by three scalars, its components along three orthogonal axes. The length of the vector is known as its *modulus* or *magnitude.*

If the magnitude of a vector is zero, the vector is a *zero vector*, and its direction is not defined.

If the magnitude of a vector is unity, it is known as a *unit vector.*

The three components of a vector written horizontally constitute a 1×3 matrix vector, § B.03, but the treatment of vectors in vector algebra is not the same as the treatment of 1×3 matrices in matrix algebra.

As in matrix algebra, § B.00, the object of vector algebra is to deal with vectors as such, so far as possible, without resolving them into the three components which define each vector numerically.

E.01. Notation

Vectors are printed in bold type, and scalars in italic. Unit vectors are printed in lower-case type.

Scalars. A, B, C, R, x, y, z, p, etc.
Vectors.† $\mathbf{A}$, $\mathbf{B}$, $\mathbf{R}$, $\mathbf{F}$, etc.
Magnitude of $\mathbf{A} = A$ or more specifically $|\mathbf{A}|$.
Unit vectors.† $\mathbf{a}$, $\mathbf{b}$, $\mathbf{n}$, etc.
Unit vectors parallel to x, y, z axes. $\mathbf{i}$, $\mathbf{j}$, $\mathbf{k}$.
Product of vector and scalar. $p\mathbf{A}$.
Scalar product of two vectors, § E.07. $\mathbf{A} \cdot \mathbf{B}$.
Vector product of two vectors, § E.08. $\mathbf{A} \times \mathbf{B}$. Some authors use $\mathbf{A} \wedge \mathbf{B}$ or V$\mathbf{AB}$.
Element of distance. Scalar δL. Vector $\delta \mathbf{L}$.
Element of area. Scalar δS. Vector $\delta \mathbf{S} = \mathbf{n}\delta S$, where $\mathbf{n}$ is a unit vector normal to the surface, § E.13. For sign see § E.12.
Element of volume. δv.
Scalar components of $\mathbf{R}$ parallel to x, y, z axes. R_1, R_2, R_3. $R^2 = R_1^2 + R_2^2 + R_3^2$.

E.02. Examples of scalars and vectors

Examples of scalar quantities are density, temperature, and gravitational potential. Examples of vectors are velocity, acceleration, force, and angular velocity

† See footnote to § B.03 (a).
Some authors print non-unit vectors as $\mathbf{a}$, $\mathbf{b}$, $\mathbf{r}$ etc. and use $\hat{\mathbf{a}}$ for unit vectors.

about an axis which is the direction of the vector. A plane area can also be represented by a vector normal to the plane with magnitude equal to the area.

A space in which at every point (x, y, z) a certain scalar has a particular value is known as a *scalar field:* for example, the gravitational potential. And a space at every point of which some vector has a particular value is a *vector field.* For example, the velocity of the current of a moving liquid, or the gravitational force.

E.03. Equality of vectors

A vector is wholly defined by its magnitude and direction, without reference to the position of its end-point. Two parallel vectors of equal magnitude are identically equal, and the geometrical representation of a vector may be moved parallel to itself without changing it.

It follows that, when convenient, any two vectors can be regarded as co-planar, with one end of one coinciding with one end of the other, and their sum or product or other combination will not be affected by their being moved to such positions.†

E.04. Addition and subtraction of vectors

Vectors are added by the ordinary parallelogram rule. In Fig. E.1 the sum of OA and OB is defined to be OC, and in Fig. E.2 the sum of OA, OB, and OC is OR.

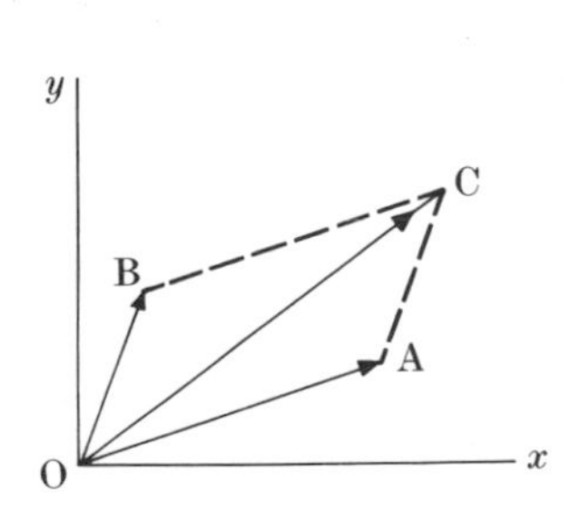

FIG. E.1.

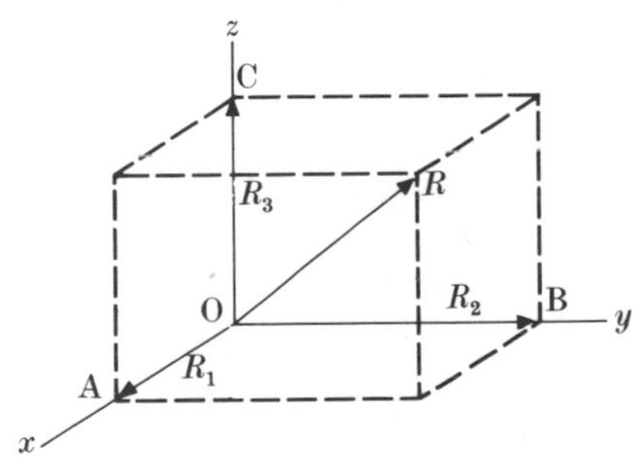

FIG. E.2.

It is clear that $\mathbf{A}+\mathbf{B} = \mathbf{B}+\mathbf{A}$, commutative,

and that $$\mathbf{A}+(\mathbf{B}+\mathbf{C}) = (\mathbf{A}+\mathbf{B})+\mathbf{C}, \quad \text{associative,} \tag{E.1}$$

† Although moving a vector in space does not affect it *as a vector*, any vector may be significantly associated with a particular point in space, as in a vector field, or as a force acting on a particular point of a rigid body. If two forces act on the same point of a body, their joint effect is given by their vector sum. If two forces which are otherwise vectorially equal to them, act on two different points of the body, their vector sum will be the same as that of the first two, but their effect on the body will not simply be that of a single force equal to their vector sum.

Similarly, the rotation of a rigid body through a finite angle may be represented by a vector **A**, and its subsequent rotation about some other axis by vector **B**. Then, as always, the sum $\mathbf{A}+\mathbf{B}$ is as defined in § E.04, but it does not follow that $\mathbf{A}+\mathbf{B}$ represents the result of first applying the vector **A** rotation, and subsequently vector **B**: as in fact it does not, unless both vectors are small.

the same as with scalars.

Subtraction. If $\mathbf{A}+\mathbf{B}=0$, a zero vector, their magnitudes must be equal ($B = A$), but their directions are opposite. Write $\mathbf{B}=-\mathbf{A}$ and $\mathbf{A}=-\mathbf{B}$. Then the difference between two vectors, $\mathbf{A}-\mathbf{B}$ is defined to be $\mathbf{A}+(-\mathbf{B})$.

E.05. Multiplication of a vector by a scalar

$\mathbf{A}+\mathbf{A}+\mathbf{A}+\ldots$ (p times) is defined to be $p\mathbf{A}$, a vector parallel to $\mathbf{A}$ whose magnitude is pA. The scalar need not be a whole number.

Similarly $\mathbf{A}/p$ ($\mathbf{A}$ divided by p) is such a vector that $p(\mathbf{A}/p)=\mathbf{A}$.

We have defined $p\mathbf{A}$. $\mathbf{A}p$ has no obvious meaning, so we may define $\mathbf{A}p=p\mathbf{A}$, so products of vectors and scalars are commutative. They are clearly also associative and distributive.

Thus far, vectors can be manipulated by the ordinary rules of scalar arithmetic, but note that if $\mathbf{C}=\mathbf{A}+\mathbf{B}$, it does not follow that the magnitude $C=A+B$. In fact $C=$ or $<(A+B)$.

E.06. Multiplication of vectors

The product of two vectors has no obvious meaning, and the expression can therefore be used for any useful combination of two vectors which is reasonably analogous to the multiplication of scalars. In fact there are two such combinations which are known as vector multiplication, namely the *scalar product* of two vectors, which is described in § E.07, and the *vector product* described in § E.08.

E.07. The scalar product

The scalar product, or *dot product*, of two vectors $\mathbf{A}$ and $\mathbf{B}$, is defined to be the scalar $AB\cos\theta$, where θ is the angle between the directions of $\mathbf{A}$ and $\mathbf{B}$. It is written $\mathbf{A}\cdot\mathbf{B}$, so

$$\mathbf{A}\cdot\mathbf{B}=AB\cos\theta. \tag{E.2}$$

This is a useful combination, as in the following examples.

(a) If $\mathbf{A}$ is a force and $\mathbf{B}$ is the movement of its point of application, $\mathbf{A}\cdot\mathbf{B}$ is the work done.

(b) If $\mathbf{A}$ is any vector, and $\mathbf{b}$ is a unit vector, $\mathbf{A}\cdot\mathbf{b}$ is the component of $\mathbf{A}$ in the direction of $\mathbf{b}$.

(c) Let $\mathbf{i}$, $\mathbf{j}$, and $\mathbf{k}$ be unit vectors in the positive directions of three right-handed orthogonal axes x, y, and z, § B.08 (g). Then if $\mathbf{R}$ is any vector, $\mathbf{R}\cdot\mathbf{i}$, $\mathbf{R}\cdot\mathbf{j}$, and $\mathbf{R}\cdot\mathbf{k}$ are its scalar components R_1, R_2, and R_3 along the three axes, and

$$\begin{aligned}\mathbf{R}&=R_1\mathbf{i}+R_2\mathbf{j}+R_3\mathbf{k}\\&=(\mathbf{R}\cdot\mathbf{i})\mathbf{i}+(\mathbf{R}\cdot\mathbf{j})\mathbf{j}+(\mathbf{R}\cdot\mathbf{k})\mathbf{k}\end{aligned} \tag{E.3}$$

and if $\mathbf{R}$ is a unit vector $\mathbf{r}$, whose direction cosines are l, m, and n,

$$\mathbf{r}=l\mathbf{i}+m\mathbf{j}+n\mathbf{k}. \tag{E.4}$$

It is clear that

$$\mathbf{A}\cdot\mathbf{B}=\mathbf{B}\cdot\mathbf{A}. \qquad \text{Commutative,}$$
$$\mathbf{A}\cdot(\mathbf{B}+\mathbf{C})=\mathbf{A}\cdot\mathbf{B}+\mathbf{A}\cdot\mathbf{C}. \qquad \text{Distributive.} \tag{E.5}$$

For triple products such as $(\mathbf{A}\cdot\mathbf{B})\mathbf{C}$ or $\mathbf{A}(\mathbf{B}\cdot\mathbf{C})$, note that the dot products are simply scalars, like p in § E.05. These two expressions are not equal. The triple product $\mathbf{A}\cdot\mathbf{B}\cdot\mathbf{C}$ is meaningless. There is no question of dot multiplication being associative.

$\mathbf{A}\cdot\mathbf{A}$ is written as $\mathbf{A}^2$ for short.

If $\mathbf{A}$ and $\mathbf{B}$ are parallel, $\mathbf{A}\cdot\mathbf{B}=AB$ and $\mathbf{A}\cdot\mathbf{A}$ or $\mathbf{A}^2=A^2$, while if they are perpendicular $\mathbf{A}\cdot\mathbf{B}=0$. In particular,

$$\mathbf{i}\cdot\mathbf{i}\,(\text{or }\mathbf{i}^2)=\mathbf{j}\cdot\mathbf{j}=\mathbf{k}\cdot\mathbf{k}=1$$

and

$$\mathbf{i}\cdot\mathbf{j}=\mathbf{j}\cdot\mathbf{k}=\mathbf{k}\cdot\mathbf{i}=\mathbf{j}\cdot\mathbf{i}, \text{ etc.}=0. \tag{E.6}$$

As further examples of the manipulation of scalar products,

$$(\mathbf{A}+\mathbf{B})\cdot(\mathbf{C}+\mathbf{D})=\mathbf{A}\cdot\mathbf{C}+\mathbf{A}\cdot\mathbf{D}+\mathbf{B}\cdot\mathbf{C}+\mathbf{B}\cdot\mathbf{D}.$$
$$(\mathbf{A}+\mathbf{B})\cdot(\mathbf{A}-\mathbf{B})=\mathbf{A}\cdot\mathbf{A}-\mathbf{B}\cdot\mathbf{B}=A^2-B^2.$$
$$(\mathbf{A}+\mathbf{B})^2=A^2+2\mathbf{A}\cdot\mathbf{B}+B^2=A^2+2AB\cos\theta+B^2,$$

and if $\mathbf{A}$ and $\mathbf{B}$ are perpendicular $(\mathbf{A}+\mathbf{B})^2=A^2+B^2$.

If $\mathbf{a}$ and $\mathbf{b}$ are unit vectors $\mathbf{a}\cdot\mathbf{b}=\cos\theta$. But also

$$\mathbf{a}\cdot\mathbf{b}=(a_1\mathbf{i}+a_2\mathbf{j}+a_3\mathbf{k})\cdot(b_1\mathbf{i}+b_2\mathbf{j}+b_3\mathbf{k})$$
$$=a_1b_1+a_2b_2+a_3b_3, \quad \text{using (E.6).}$$

So this expression equals $\cos\theta$, as given in ordinary analytical solid geometry, (C.25), since the direction cosines of $\mathbf{a}$ are a_1, a_2, a_3.

There is no process of division corresponding to the dot product, and no definition of the reciprocal of a vector.

E.08. The vector product

The vector product or *cross product* of two vectors $\mathbf{A}$ and $\mathbf{B}$ is defined to be a vector whose magnitude is $AB\sin\theta$, and whose direction is perpendicular to both $\mathbf{A}$ and $\mathbf{B}$ (i.e. to any plane to which both $\mathbf{A}$ and $\mathbf{B}$ are parallel). θ is the angle between the directions of $\mathbf{A}$ and $\mathbf{B}$.

We write

$$\mathbf{C}=\mathbf{A}\times\mathbf{B}, \quad \text{and the magnitude} \quad C=AB\sin\theta. \tag{E.7}$$

For the sign of $\mathbf{C}$, the rule is that a right-handed (or clockwise as viewed along $\mathbf{C}$) rotation carries $\mathbf{A}$ to $\mathbf{B}$ through an angle of less than 180°, making $\sin\theta$ positive. So in Fig. E.3 the vector $\mathbf{C}$ $(=\mathbf{A}\times\mathbf{B})$ is directed downwards into the paper.

This is a useful combination of two vectors. Examples:

(*a*) If $\mathbf{A}$ is the electric current in a conductor, and if $\mathbf{B}$ is the magnetic flux, $\mathbf{A}\times\mathbf{B}$ is the force acting on unit length of the conductor.

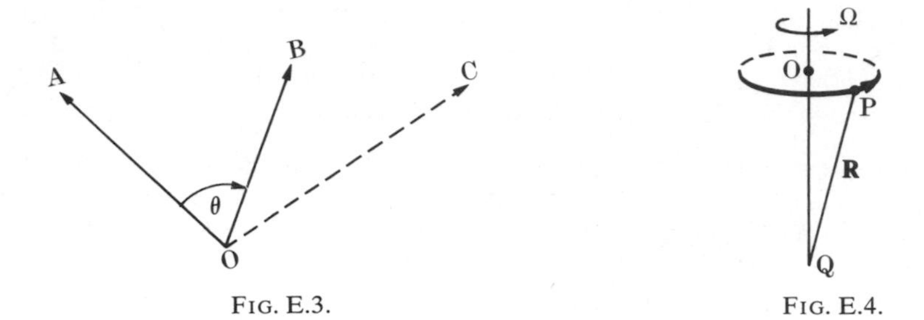

FIG. E.3. FIG. E.4.

(*b*) In Fig. E.4 the point P rotates about an axis QO with angular velocity Ω, which may be represented by a vector $\mathbf{\Omega}$ parallel to QO. Let the position of P relative to any point on QO be vector **R**. Then the vector linear velocity of P at any instant is $\mathbf{\Omega} \times \mathbf{R}$.

(*c*) If **A** is the vector representing the angular velocity of rotating axes x, y, z, and if **B** is the velocity of a moving point relative to this rotating frame, $2(\mathbf{A} \times \mathbf{B})$ is the Coriolis acceleration. See § I.02.

Vector multiplication is *not commutative.* In fact

$$\mathbf{A} \times \mathbf{B} = -\mathbf{B} \times \mathbf{A}. \tag{E.8}$$

Vector multiplication is distributive with respect to addition.

$$\mathbf{A} \times (\mathbf{B} + \mathbf{C}) = (\mathbf{A} \times \mathbf{B}) + (\mathbf{A} \times \mathbf{C}). \tag{E.9}$$

Hence $\quad (\mathbf{A} + \mathbf{B}) \times (\mathbf{C} + \mathbf{D}) = (\mathbf{A} \times \mathbf{C}) + (\mathbf{A} \times \mathbf{D}) + (\mathbf{B} \times \mathbf{C}) + (\mathbf{B} \times \mathbf{D}).$

Notes

(i) If two vectors are parallel, their vector product is zero. So $\mathbf{A} \times \mathbf{A} = 0$. This must not be written $\mathbf{A}^2$, which is reserved for $\mathbf{A} \cdot \mathbf{A}$.

(ii) If **A** and **B** are perpendicular, the magnitude of $\mathbf{A} \times \mathbf{B}$ is AB.

(iii) $\mathbf{i} \times \mathbf{i} = 0$, etc., $\mathbf{i} \times \mathbf{j} = \mathbf{k}$, $\mathbf{j} \times \mathbf{k} = \mathbf{i}$, $\mathbf{k} \times \mathbf{i} = \mathbf{j}$, and $\mathbf{j} \times \mathbf{i} = -\mathbf{k}$, etc.

(iv) There is no definition of division.

(v) Writing $\mathbf{A} = A_i\mathbf{i} + A_2\mathbf{j} + A_3\mathbf{k}$, and using (iii), we have

$$\mathbf{A} \times \mathbf{B} = (A_2B_3 - A_3B_2)\mathbf{i} + (A_3B_1 - A_1B_3)\mathbf{j} + (A_1B_2 - A_2B_1)\mathbf{k},$$

which may be written for short as

$$\begin{vmatrix} \mathbf{i} & \mathbf{j} & \mathbf{k} \\ A_1 & A_2 & A_3 \\ B_1 & B_2 & B_3 \end{vmatrix}. \tag{E.10}$$

But this is not a true determinant as defined in § B.04, since all its elements are not scalars.

E.09. Triple products

In addition to the triple products like $(\mathbf{A} \cdot \mathbf{B})\mathbf{C}$ mentioned in § E.07, which are simply products of a vector and a scalar, there are two other kinds of triple products, which involve the vector products of § E.08.

(*a*) The *scalar-vectorial* product, namely the scalar product of **A** with the vector product of **B** and **C**, $\mathbf{A}\cdot(\mathbf{B}\times\mathbf{C})$ or $\mathbf{A}\cdot\mathbf{B}\times\mathbf{C}$ for short. The latter cannot be confused with $(\mathbf{A}\cdot\mathbf{B})\times\mathbf{C}$, which has no meaning.

From the definitions

$$\mathbf{A}\cdot\mathbf{B}\times\mathbf{C}=\mathbf{C}\cdot\mathbf{A}\times\mathbf{B}=\mathbf{B}\cdot\mathbf{C}\times\mathbf{A}. \tag{E.11}$$

That is to say, this triple product is associative provided that the factors are permuted cyclicly. Otherwise, the sign changes. Further

$$\mathbf{A}\cdot(\mathbf{B}\times\mathbf{C})=(\mathbf{A}\times\mathbf{B})\cdot\mathbf{C}, \tag{E.12}$$

and all these expressions are scalars giving the volume of the parallelepiped defined by the three vectors. So in a scalar-vectorial product the dot and the cross can be interchanged, or for that matter omitted. Such a product may therefore be written $[\mathbf{ABC}]=[\mathbf{CAB}]=[\mathbf{BCA}]=-[\mathbf{ACB}]$, etc.

And putting $\mathbf{A}=A_1\mathbf{i}+A_2\mathbf{j}+A_3\mathbf{k}$, etc.,

$$[\mathbf{ABC}]=\begin{vmatrix} A_1 & A_2 & A_3 \\ B_1 & B_2 & B_3 \\ C_1 & C_2 & C_3 \end{vmatrix}. \tag{E.13}$$

If $[\mathbf{ABC}]=0$ the three vectors are co-planar, a condition which includes one of them being zero, or two of them being parallel.

(*b*) The *doubly vectorial* product of **A** with $(\mathbf{B}\times\mathbf{C})$. We have

$$\mathbf{A}\times(\mathbf{B}\times\mathbf{C})=(\mathbf{A}\cdot\mathbf{C})\mathbf{B}-(\mathbf{A}\cdot\mathbf{B})\mathbf{C}. \tag{E.14}$$

For proof see [616], pp. 470–2. This product is *not associative* even when permuted cyclicly, and $\mathbf{A}\times\mathbf{B}\times\mathbf{C}$ without brackets has no unique meaning.

Properties are:

$$\mathbf{A}\times(\mathbf{B}\times\mathbf{C})=(\mathbf{C}\times\mathbf{B})\times\mathbf{A} \tag{E.15}$$

and

$$\mathbf{A}\times(\mathbf{B}\times\mathbf{C})+\mathbf{B}\times(\mathbf{C}\times\mathbf{A})+\mathbf{C}\times(\mathbf{A}\times\mathbf{B})=0. \tag{E.16}$$

For various examples of the manipulation of vector products see [616], pp. 466–74 and [506], pp. 5–9. The utility of vector methods for constructive work naturally depends on the user being facile in their manipulation.

E.10. Vector functions of a single scalar variable

Let a vector **V** vary in magnitude and direction with the time t, or with any other scalar quantity with which it is associated. We may write $\mathbf{V}(t)$ for its value at time t. Then, decomposing it into its three components as in (E.3), we have

$$\mathbf{V}(t)=V_1(t)\mathbf{i}+V_2(t)\mathbf{j}+V_3(t)\mathbf{k}.$$

Write $\Delta\mathbf{V}=\mathbf{V}(t+\Delta t)-\mathbf{V}(t)=\Delta V_1\mathbf{i}+\Delta V_2\mathbf{j}+\Delta V_3\mathbf{k}$.

Then the derivative of **V** with respect to t is defined by

$$\frac{\mathrm{d}\mathbf{V}}{\mathrm{d}t}=\frac{\mathrm{d}V_1}{\mathrm{d}t}\mathbf{i}+\frac{\mathrm{d}V_2}{\mathrm{d}t}\mathbf{j}+\frac{\mathrm{d}V_3}{\mathrm{d}t}\mathbf{k}.$$

Let the changing vector **V** be placed with one end at the origin O, Fig. E.5. Then its other end P will move along a certain curve SS. At any instant the vector $\Delta\mathbf{V}$ is tangent to SS, and the vector velocity of P is $\Delta\mathbf{V}/\Delta t$ or $\mathrm{d}\mathbf{V}/\mathrm{d}t$, and its acceleration is $\mathrm{d}^2\mathbf{V}/\mathrm{d}t^2$, which does not in general lie along SS.

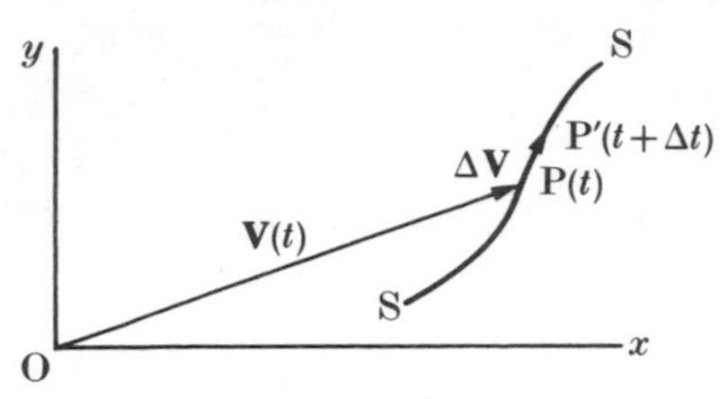

FIG. E.5. The curve SS need not be a plane curve.

If L is distance along SS, measured from some arbitrary point, the scalar velocity of P is $\mathrm{d}L/\mathrm{d}t$. So $\mathrm{d}L/\mathrm{d}t = |\mathrm{d}\mathbf{V}|/\mathrm{d}t$, whence $|\mathrm{d}\mathbf{V}/\mathrm{d}L| = 1$. So $\mathrm{d}\mathbf{V}/\mathrm{d}L$ is a unit vector tangent to SS.

As in the ordinary differential calculus, if **U** and **V** are two vectors, it follows from the definitions that

$$\frac{\mathrm{d}}{\mathrm{d}t}(\mathbf{U}\pm\mathbf{V}) = \frac{\mathrm{d}\mathbf{U}}{\mathrm{d}t} \pm \frac{\mathrm{d}\mathbf{V}}{\mathrm{d}t}. \tag{E.17}$$

$$\frac{\mathrm{d}}{\mathrm{d}t}\phi\mathbf{V} = \frac{\mathrm{d}\phi}{\mathrm{d}t}\mathbf{V} + \phi\frac{\mathrm{d}\mathbf{V}}{\mathrm{d}t}, \quad \text{where } \phi \text{ is a variable scalar.} \tag{E.18}$$

$$\frac{\mathrm{d}}{\mathrm{d}t}(\mathbf{U}\cdot\mathbf{V}) = \frac{\mathrm{d}\mathbf{U}}{\mathrm{d}t}\cdot\mathbf{V} + \mathbf{U}\cdot\frac{\mathrm{d}\mathbf{V}}{\mathrm{d}t}. \tag{E.19}$$

$$\frac{\mathrm{d}}{\mathrm{d}t}(\mathbf{U}\times\mathbf{V}) = \frac{\mathrm{d}\mathbf{U}}{\mathrm{d}t}\times\mathbf{V} + \mathbf{U}\times\frac{\mathrm{d}\mathbf{V}}{\mathrm{d}t}. \tag{E.20}$$

$$\frac{\mathrm{d}}{\mathrm{d}t}[\mathbf{UVW}] = \left[\frac{\mathrm{d}\mathbf{U}}{\mathrm{d}t}\mathbf{VW}\right] + \left[\mathbf{U}\frac{\mathrm{d}\mathbf{V}}{\mathrm{d}t}\mathbf{W}\right] + \left[\mathbf{UV}\frac{\mathrm{d}\mathbf{W}}{\mathrm{d}t}\right]. \tag{E.21}$$

$$\frac{\mathrm{d}}{\mathrm{d}t}\{\mathbf{U}\times(\mathbf{V}\times\mathbf{W})\} = \frac{\mathrm{d}\mathbf{U}}{\mathrm{d}t}\times(\mathbf{V}\times\mathbf{W}) + \mathbf{U}\times\left(\frac{\mathrm{d}\mathbf{V}}{\mathrm{d}t}\times\mathbf{W}\right) + \mathbf{U}\times\left(\mathbf{V}\times\frac{\mathrm{d}\mathbf{W}}{\mathrm{d}t}\right). \tag{E.22}$$

Note that in vector products the order of the factors must not be changed, and in (E.22) the brackets are essential. For applications of these formulae see [616], pp. 476–80.

E.11. The gradient of a scalar function. Grad ϕ

Consider a scalar field, § E.02, in which at every point the value of a scalar ϕ depends on the x, y, z of the point. So $\phi = f(x, y, z)$. For example, ϕ may be the gravitational potential at (x, y, z). Then there will be a family of *level surfaces* filling the space, over each of which ϕ has a constant value. The *gradient of* ϕ, or grad ϕ, at any point is defined to be a vector normal to the level surface at that point, whose magnitude is $\partial\phi/\partial N$, N being distance along the normal. If ϕ is the gravitational potential, grad ϕ is the gravitational attraction.

To evaluate grad ϕ, consider first a 2-dimensional case, Fig. E.6, in which let the scalar ϕ be h, the height of the ground above sea-level at (x, y). Then AA' and BB' are tangent to contours of h and $h+\Delta h$, and grad h is the slope of the ground (in the steepest direction) along the normal AB. In fact, grad h is the vector $(\partial h/\partial N)\mathbf{n}$, where **n** is unit vector in the direction of the normal, and $\partial h/\partial N$ is the scalar value of the slope.

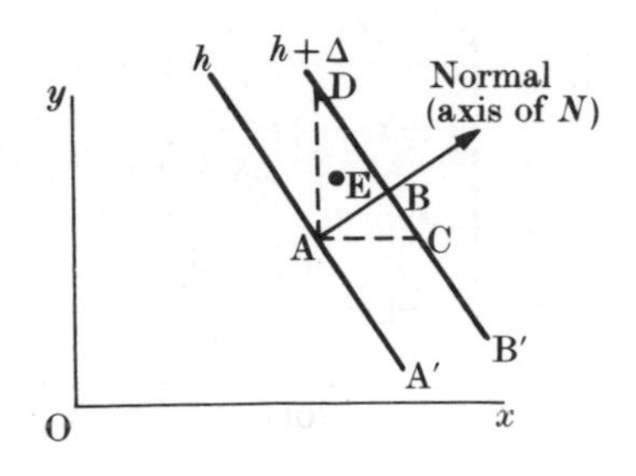

FIG. E.6. In general AA′ and BB′ are not straight, but the figure illustrates a small area over which they are substantially straight. AC is parallel to Ox.

If l and m are the direction cosines of AB, $\mathbf{n} = l\mathbf{i}+m\mathbf{j}$. Then

$$\begin{aligned}\text{grad}\,\phi &= \frac{\partial h}{\partial N}\mathbf{n} = \frac{\partial h}{\partial N}(l\mathbf{i}+m\mathbf{j}) \\ &= l\frac{\Delta h}{\text{AB}}\mathbf{i}+m\frac{\Delta h}{\text{AB}}\mathbf{j} \\ &= \frac{\text{AB}}{\text{AC}}\frac{\Delta h}{\text{AB}}\mathbf{i}+\frac{\text{AB}}{\text{AD}}\frac{\Delta h}{\text{AB}}\mathbf{j} \\ &= \frac{\Delta h}{\text{AC}}\mathbf{i}+\frac{\Delta h}{\text{AD}}\mathbf{j} \\ &= \frac{\partial h}{\partial x}\mathbf{i}+\frac{\partial h}{\partial y}\mathbf{j}.\end{aligned}$$

And in the general three-dimensional case we have

$$\text{grad}\,\phi = \frac{\partial \phi}{\partial N}\mathbf{n} = \frac{\partial \phi}{\partial x}\mathbf{i}+\frac{\partial \phi}{\partial y}\mathbf{j}+\frac{\partial \phi}{\partial z}\mathbf{k}. \tag{E.23}$$

For short this may be written

$$\text{grad}\,\phi = \left(\boldsymbol{i}\frac{\partial}{\partial x}+\mathbf{j}\frac{\partial}{\partial y}+\mathbf{k}\frac{\partial}{\partial z}\right)\phi, \tag{E.24}$$

or, shorter,

$$\text{grad}\,\phi = \nabla\phi, \tag{E.25}$$

where ∇ stands for $\left(\mathbf{i}\frac{\partial}{\partial x}+\mathbf{j}\frac{\partial}{\partial y}+\mathbf{k}\frac{\partial}{\partial z}\right)$. Read ∇ as *del* or *nabla*.

Although (E.24) and (E.25) have the form of a product of ϕ and ∇, ∇ is really an operator. This 'product' has no more connection with the product of ordinary arithmetical multiplication than has (say) the vector product of two vectors. On the other hand, if we like to so regard it, $\nabla\phi$ is no less a product of ∇ and ϕ than $\mathbf{A}\times\mathbf{B}$ is of $\mathbf{A}$ and $\mathbf{B}$. It is useful to regard it as a product if the operation of ∇ on ϕ obeys a reasonable number of rules resembling those of arithmetical or vector multiplication. See further §§ E.14, E.16, and E.18.

If in Fig. E.6 the position of A is given by the vector $\mathbf{R}$ (=OA), and that of another point E by the vector $\mathbf{R}+\Delta\mathbf{R}$, $\Delta\mathbf{R}$ being AE, it is clear that at $\mathbf{R}+\Delta\mathbf{R}$

$$\phi_{\text{R}+\Delta\text{R}} = \phi_{\text{R}} + \Delta\mathbf{R}\cdot\text{grad}\,\phi. \tag{E.26}$$

E.12. Integration of vectors. Line integral

Consider a vector field in which at every point (x, y, z) a vector $\mathbf{F}$ has a magnitude and direction depending on (x, y, z). The expression $\int_{\text{A}}^{\text{B}} \mathbf{F}\cdot d\mathbf{L}$ has no obvious meaning, so it may be defined in any useful way.

In Fig. E.7 let AB be a defined line in the field, not necessarily lying in one plane, and at any intermediate point P let $d\mathbf{L}$ be a small vector of magnitude dL

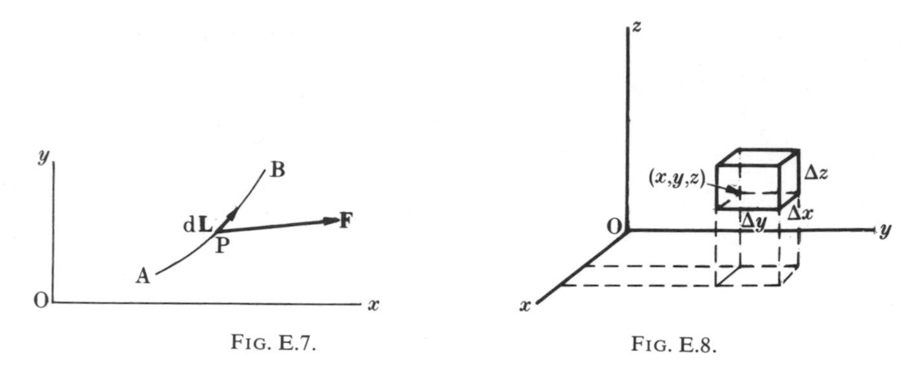

FIG. E.7. FIG. E.8.

tangent to the line at P. We now define $\int_A^B \mathbf{F}\cdot d\mathbf{L}$ to be the sum of all the scalar products of **F** and d**L** along the line from A to B.

Similarly $\int_C \mathbf{F}\cdot d\mathbf{L}$ is the integral round a closed circuit.

Every unit element of this integral is the component of **F** along the line, so (likening **F** to the current of a moving liquid) the integral round a closed circuit is called the *circulation* round the circuit.

If **F** is a force which moves its point of application along AB, $\int_A^B \mathbf{F}\cdot d\mathbf{L}$ is the work done.

The direction round a closed circuit is conventionally defined to be positive if it is clockwise when viewed along the positive direction of the normal to the surface. The outward normals to a closed surface are conventionally defined to be positive. With an open surface the choice is arbitrary, but the positive direction round closed circuits on the surface must conform.

E.13. Integration of vectors. Surface integral

Similarly, the surface integral of the variable vector **F** over a bounded or closed surface, not necessarily plane, is defined to be the sum of the elementary scalar products of **F** and d**S**, where d**S** at any point on the surface is a vector, normal to the surface, of magnitude equal to an elementary area surrounding the point. So $d\mathbf{S} = \mathbf{n}\,dS$, and the surface integral is $\iint \mathbf{F}\cdot\mathbf{n}\,dS$, where **n** is unit normal vector and dS is a scalar element of area.

E.14. Divergence

At any point in a vector field, the *divergence* of the vector **F**, written div **F**, is defined to be the surface integral of **F** over a small closed surface surrounding the point, divided by the volume dv enclosed by the surface.

$$\operatorname{div}\mathbf{F} = \text{the limit of } (1/dv)\iint_S \mathbf{F}\cdot\mathbf{n}\,dS, \quad \text{as } S \text{ and } dv \text{ tend to zero.} \qquad \text{(E.27)}$$

n is conventionally the outward normal.

For the elementary volume take a rectangular element situated between (x, y, z) and $(x+\Delta x, y+\Delta y, z+\Delta z)$, with sides parallel to the axes, as in Fig. E.8. Let the components of $\mathbf{F}$ be F_1, F_2, and F_3, so $\mathbf{F}=F_1\mathbf{i}+F_2\mathbf{j}+F_3\mathbf{k}$. Then considering the two sides perpendicular to the axis of x, their contribution to the surface integral is

$$\left(F_1+\frac{\partial F_1}{\partial x}\Delta x\right)\Delta y\Delta z-F_1\Delta y\Delta z=\frac{\partial F}{\partial x}\Delta x\Delta y\Delta z.$$

The other pairs of sides give similar contributions, with the result

$$\operatorname{div}\mathbf{F}=\frac{\partial F_1}{\partial x}+\frac{\partial F_2}{\partial y}+\frac{\partial F_3}{\partial z},\quad \text{a scalar.} \tag{E.28}$$

Remembering that $\mathbf{i}\cdot\mathbf{i}=1$ and $\mathbf{i}\cdot\mathbf{j}=0$, etc., this can be written

$$\operatorname{div}\mathbf{F}=\left(\mathbf{i}\frac{\partial}{\partial x}+\mathbf{j}\frac{\partial}{\partial y}+\mathbf{k}\frac{\partial}{\partial z}\right)\cdot(F_1\mathbf{i}+F_2\mathbf{j}+F_3\mathbf{k}),$$

or

$$\operatorname{div}\mathbf{F}=\boldsymbol{\nabla}\cdot\mathbf{F}. \tag{E.29}$$

In (E.29) operation by ∇ has again, as in (E.25), been treated as multiplication by a vector. In (E.25) this 'vector' multiplied a scalar to give a vector, and now in (E.29) it is multiplying another vector to give a scalar as in the dot multiplication of vectors. § E.16 will consider the 'vector product' of ∇ and a vector, which is written $\boldsymbol{\nabla}\times\mathbf{F}$.

The physical meaning of divergence. Let the vector $\mathbf{F}$ be the velocity of a moving liquid, varying from point to point. Then at any point the divergence is the volume of liquid leaving unit volume of space in unit time. If the liquid is incompressible, and if it cannot be created or destroyed, $\operatorname{div}\mathbf{F}$ must in this case equal zero. A point at which $\operatorname{div}\mathbf{F}$ is positive, as where a liquid is expanding or where an incompressible liquid is being created, is known as a *source*, and a point where $\operatorname{div}\mathbf{F}$ is negative is a *sink*. Fig. E.9 shows some examples, in two dimensions, the component perpendicular to the page being zero in all cases.

E.15. Gauss's theorem

Consider a closed space bounded by a surface of any shape. It is clear in the example quoted above that the total volume of (incompressible) liquid passing outwards through the surface must equal the total created inside it. In other words,

$$\iint_S \mathbf{F}\cdot\mathbf{n}\,\mathrm{d}S=\iiint_v \operatorname{div}\mathbf{F}\,\mathrm{d}v$$

$$=\iiint_v \boldsymbol{\nabla}\cdot\mathbf{F}\,\mathrm{d}v. \tag{E.30}$$

This is Gauss's theorem.

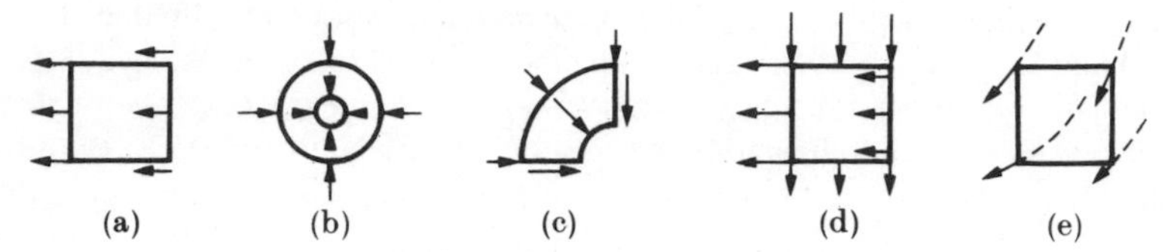

FIG. E.9. Divergence. In (a), (c), (d), and (e) the divergence is zero; (d) and (e) are identical; (b) represents a sink.

Example 1. Consider a small sphere of radius R. Its volume is $(4/3)\pi R^3$, and its area is $4\pi R^2$. Let it be a source from which a current of incompressible liquid flows radially outwards, equally in all directions, the velocity through the surface of the sphere being the vector $\mathbf{F}$. Then (E.30) takes the form

$$\begin{aligned} 4\pi R^2 F &= (4/3)\pi R^3 \operatorname{div} \mathbf{F}, \\ F &= (1/3) R \operatorname{div} \mathbf{F}. \end{aligned} \tag{E.31}$$

So the outward velocity through the surface is equal to $(R/3)$ times the rate of creation per unit volume.

Example 2. If $\mathbf{F}$ is the gravitational attraction of a small sphere of radius R and uniform density ρ, its magnitude F on the surface of the sphere is $-(4/3)\pi G\rho R$, where G is the gravitational constant, § 6.06. Then (E.30) takes the same form as (E.31), which gives

$$\operatorname{div} \mathbf{F} = -4\pi G\rho.$$

So over this small element

$$\iiint \operatorname{div} \mathbf{F} \, dv = -4\pi G \times (\text{contained mass}).$$

And (E.30) may be written

$$\iint_S \mathbf{F} \cdot \mathbf{n} \, dS = -4\pi G \times (\text{contained mass}) \tag{E.32}$$

$$= 0, \quad \text{if there is no mass within the surface.}$$

E.16. Curl

§§ E.11 and E.14 show that the operation of ∇ can be treated either as 'multiplication' of a scalar by this 'vector', or as the scalar multiplication of one vector by another. It is reasonable to inquire what might be a useful meaning for the vector product of a vector **F** by ∇. We would have

$$\begin{aligned}\nabla\times\mathbf{F} &= \left(\mathbf{i}\frac{\partial}{\partial x}+\mathbf{j}\frac{\partial}{\partial y}+\mathbf{k}\frac{\partial}{\partial z}\right)\times(F_1\mathbf{i}+F_2\mathbf{j}+F_3\mathbf{k})\\ &= \left(\frac{\partial F_3}{\partial y}-\frac{\partial F_2}{\partial z}\right)\mathbf{i}+\left(\frac{\partial F_1}{\partial z}-\frac{\partial F_3}{\partial x}\right)\mathbf{j}+\left(\frac{\partial F_2}{\partial x}-\frac{\partial F_1}{\partial y}\right)\mathbf{k}\\ &= \text{for short, }\begin{vmatrix}\mathbf{i} & \mathbf{j} & \mathbf{k}\\ \frac{\partial}{\partial x} & \frac{\partial}{\partial y} & \frac{\partial}{\partial z}\\ F_1 & F_2 & F_3\end{vmatrix}.\end{aligned} \tag{E.33}$$

If this expression has any physical meaning, it will clearly be convenient to denote it by $\nabla\times\mathbf{F}$.

Fig. E.10 shows the three components of the vector **F** at the point (x, y, z). Let ABCD be a small rectangle with sides δy and δz centred on (x, y, z) and lying in the yz plane. Consider the line integral of $\mathbf{F}\cdot dL$ (the circulation of **F**) round the sides of this rectangle. F_1 contributes nothing, and the circulation is

$$\left(F_2-\frac{1}{2}\frac{\partial F_2}{\partial z}\delta z\right)\delta y+\left(F_3+\frac{1}{2}\frac{\partial F_3}{\partial y}\delta y\right)\delta z-\left(F_2+\frac{1}{2}\frac{\partial F_2}{\partial z}\delta z\right)\delta y-\left(F_3-\frac{1}{2}\frac{\partial F_3}{\partial y}\delta y\right)\delta z$$
$$=\left(\frac{\partial F_3}{\partial y}-\frac{\partial F_2}{\partial z}\right)\delta y\delta z.$$

And the circulation per unit area of the surface enclosed by ABCD is

$$\frac{\partial F_3}{\partial y}-\frac{\partial F_2}{\partial z},\quad \text{a scalar.}$$

This is the circulation per unit area about an axis parallel to the x-axis, so it may well be represented by the vector

$$\mathbf{i}\left(\frac{\partial F_3}{\partial y}-\frac{\partial F_2}{\partial z}\right),$$

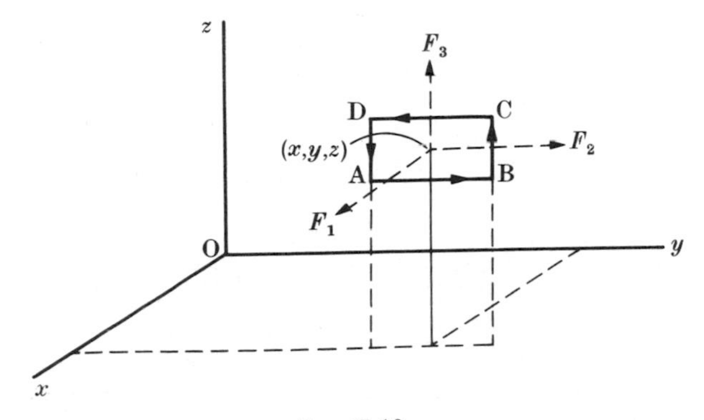

FIG. E.10.

the sign convention being that such a vector is positive if the circulation is clockwise when viewed along the positive direction of the vector, in this case outwards along the x-axis.

Two other similar vectors represent the circulation per unit area round small circuits parallel to the xz and xy planes, and the three vectors are the three components of one combined vector, which is called $\mathbf{C}$ or curl $\mathbf{F}$ and which is

$$\text{curl}\,\mathbf{F}=\mathbf{i}\left(\frac{\partial F_3}{\partial y}-\frac{\partial F_2}{\partial z}\right)+\mathbf{j}\left(\frac{\partial F_1}{\partial z}-\frac{\partial F_3}{\partial x}\right)+\mathbf{k}\left(\frac{\partial F_2}{\partial x}-\frac{\partial F_1}{\partial y}\right), \tag{E.34}$$

which, as in (E.33), may rationally be written as $\nabla\times\mathbf{F}$.

The physical meaning of curl. Round the circumference of a circle of radius R let the vector $\mathbf{F}$ be of constant magnitude ωR, and let its direction be tangent to the circle, such as is the velocity of points on the edge of a rotating disc. Then $|\text{curl}\,\mathbf{F}|$, the circulation per unit area, is clearly $(2\pi R\omega R)\div(\pi R^2)=2\omega$, namely twice the angular velocity about an axis parallel to the direction of the vector curl $\mathbf{F}$. And in general, if any rigid body is rotating about an axis with angular velocity ω, so that its rotation is a vector $\boldsymbol{\omega}$, with or without translation, and if the field vector $\mathbf{F}$ is the velocity of the different points of the body at any moment, varying from point to point, we have curl $\mathbf{F}=2\boldsymbol{\omega}$. See also [247], p. 47 and pp. 48–9 for hydrodynamic and electromagnetic applications.

Fig. E.11 gives two illustrations in two dimensions. In the examples of Fig. E.9 the curl is zero.

E.17. Stokes's theorem

This theorem about curl $\mathbf{F}$ is analogous to Gauss's theorem about div $\mathbf{F}$.

The components of the vector $\mathbf{C}$, or curl $\mathbf{F}$, in the directions of the three axes are $\mathbf{C}\cdot\mathbf{i}$, $\mathbf{C}\cdot\mathbf{j}$, and $\mathbf{C}\cdot\mathbf{k}$, each representing the circulation per unit area round circuits perpendicular to the three axes. Similarly, if $\mathbf{n}$ is any unit vector, the scalar product $\mathbf{C}\cdot\mathbf{n}$ is the circulation per unit area round a small circuit in a plane perpendicular to $\mathbf{n}$.

Fig. E.12 shows two elements of area S_1 and S_2, not necessarily quite co-planar, whose normal unit vectors are $\mathbf{n}_1$ and $\mathbf{n}_2$. In the two areas let the curl of a field vector $\mathbf{F}$ be $\mathbf{C}_1$ and $\mathbf{C}_2$, so the circulation $\int\mathbf{F}\cdot d\mathbf{L}$ round each is $\mathbf{C}_1\cdot\mathbf{n}_1\,dS_1$ and $\mathbf{C}_2\cdot\mathbf{n}_2\,dS_2$. Along the common boundary the integral of $\mathbf{F}\cdot d\mathbf{L}$ contributes equally

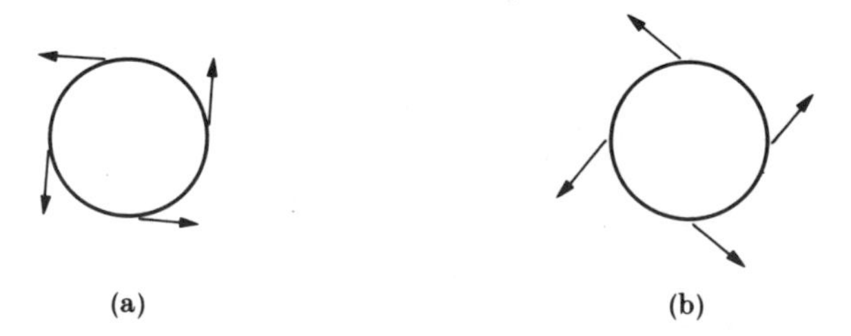

FIG. E.11. (a) Curl without divergence. (b) Curl with divergence.

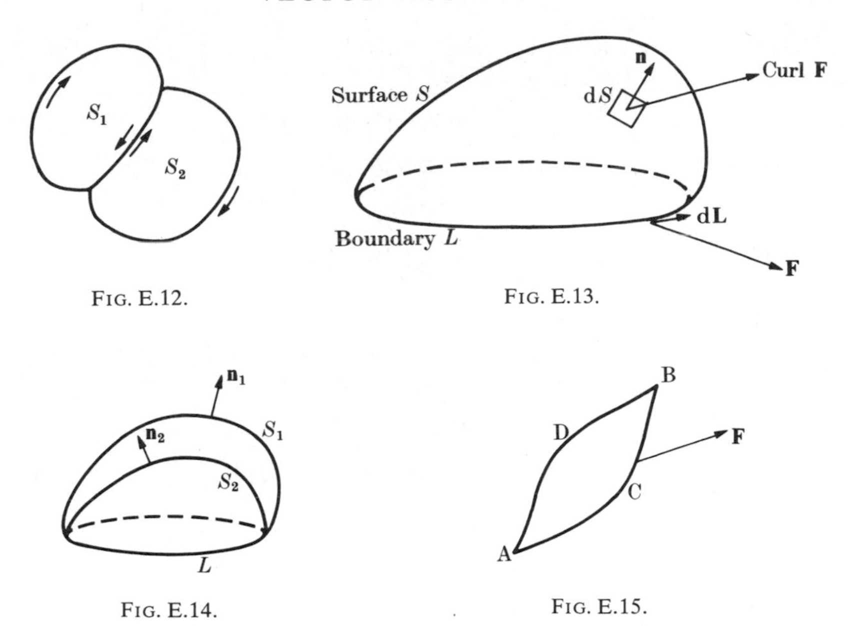

FIG. E.12. FIG. E.13.

FIG. E.14. FIG. E.15.

and oppositely to each circulation, so along the outer boundary which encloses both circuits

$$\int \mathbf{F} \cdot \mathrm{d}\mathbf{L} = (\mathbf{C}_1 \cdot \mathbf{n}_1\, \mathrm{d}S) + (\mathbf{C}_2 \cdot \mathbf{n}_2\, \mathrm{d}S).$$

Similarly, any number of elementary areas can be placed together to form a non-planar surface bounded by a closed (non-planar) line, Fig. E.13, and we have

$$\int \mathbf{F} \cdot \mathrm{d}\mathbf{L} \text{ round the boundary} = \iint \mathbf{n} \cdot \operatorname{curl} \mathbf{F}\, \mathrm{d}S \text{ over the surface.} \qquad \text{(E.35)}$$

This is Stokes's theorem, which states that if S is any open surface bounded by a closed curve L, the line integral of $\mathbf{F}$ round L (i.e. the circulation round L) equals the integral over the surface of the normal component of the curl.

Stokes's theorem must not be confused with Stokes's integral of § 6.30.

It follows that if two surfaces S_1 and S_2, whose (variable) unit normals are $\mathbf{n}_1$ and $\mathbf{n}_2$, Fig. E.14, are bounded by a common line L,

$$\iint \mathbf{n}_1 \cdot \operatorname{curl} \mathbf{F}\, \mathrm{d}S_1 = \iint \mathbf{n}_2 \cdot \operatorname{curl} \mathbf{F}\, \mathrm{d}S_2.$$

These two surfaces form one closed surface. Let $\mathbf{n}$ be the (variable) unit vector normal to this closed surface, reckoned positive outwards from it, so that in Fig. E.14 $\mathbf{n}_1 = \mathbf{n}$ but $\mathbf{n}_2 = -\mathbf{n}$. Then over the whole closed surface

$$\iint \mathbf{n} \cdot \operatorname{curl} \mathbf{F}\, \mathrm{d}S = 0. \qquad \text{(E.36)}$$

This is true of any closed surface, large or small, and of any vector **F**, so (referring to § E.14) we have always

$$\operatorname{div}\operatorname{curl}\mathbf{F}=0, \tag{E.37}$$

or, in symbols,

$$\boldsymbol{\nabla}\cdot\boldsymbol{\nabla}\times\mathbf{F}=0$$

or

$$[\boldsymbol{\nabla}\boldsymbol{\nabla}\mathbf{F}]=0,$$

as it would be if $\boldsymbol{\nabla}$ was a vector, see § E.09 (*a*).

E.18. Rules for manipulating $\boldsymbol{\nabla}$

The following may be established by writing out ∇ and the vectors **A**, **B**, etc. in the form of (E.24) and (E.3), when necessary.

$$\boldsymbol{\nabla}\cdot(\mathbf{A}+\mathbf{B})=\boldsymbol{\nabla}\cdot\mathbf{A}+\boldsymbol{\nabla}\cdot\mathbf{B}, \tag{E.38}$$

$$\boldsymbol{\nabla}\times(\mathbf{A}+\mathbf{B})=(\boldsymbol{\nabla}\times\mathbf{A})+(\boldsymbol{\nabla}\times\mathbf{B}), \tag{E.39}$$

$$\boldsymbol{\nabla}(\phi_1+\phi_2)=\boldsymbol{\nabla}\phi_1+\boldsymbol{\nabla}\phi_2, \tag{E.40}$$

$$\boldsymbol{\nabla}(\phi_1\phi_2)=\phi_1\boldsymbol{\nabla}\phi_2+\phi_2\boldsymbol{\nabla}\phi_1. \tag{E.41}$$

If ϕ is a variable scalar and **F** a variable vector,

$$\boldsymbol{\nabla}\cdot(\phi\mathbf{F})=\phi(\boldsymbol{\nabla}\cdot\mathbf{F})+(\boldsymbol{\nabla}\phi)\cdot\mathbf{F},$$

or

$$\operatorname{div}(\phi\mathbf{F})=\phi\operatorname{div}\mathbf{F}+(\operatorname{grad}\phi)\cdot\mathbf{F}, \tag{E.42}$$

$$\boldsymbol{\nabla}\times(\phi\mathbf{F})=\phi(\boldsymbol{\nabla}\times\mathbf{F})+(\boldsymbol{\nabla}\phi)\times\mathbf{F},$$

or

$$\operatorname{curl}(\phi\mathbf{F})=\phi\operatorname{curl}\mathbf{F}+(\operatorname{grad}\phi)\times\mathbf{F}. \tag{E.43}$$

The gradient of a scalar product

$$\boldsymbol{\nabla}(\mathbf{A}\cdot\mathbf{B})=(\mathbf{A}\cdot\boldsymbol{\nabla})\mathbf{B}+(\mathbf{B}\cdot\boldsymbol{\nabla})\mathbf{A}+\mathbf{A}\times(\boldsymbol{\nabla}\times\mathbf{B})+\mathbf{B}\times(\boldsymbol{\nabla}\times\mathbf{A}), \tag{E.44}$$

where $\mathbf{A}\cdot\boldsymbol{\nabla}$ is short for $A_1\dfrac{\partial}{\partial x}+A_2\dfrac{\partial}{\partial y}+A_3\dfrac{\partial}{\partial z}$.

The divergence of a vector product

$$\boldsymbol{\nabla}\cdot(\mathbf{A}\times\mathbf{B})=(\boldsymbol{\nabla}\times\mathbf{A})\cdot\mathbf{B}-\mathbf{A}\cdot(\boldsymbol{\nabla}\times\mathbf{B}),$$

or

$$\operatorname{div}(\mathbf{A}\times\mathbf{B})=\mathbf{B}\cdot\operatorname{curl}\mathbf{A}-\mathbf{A}\cdot\operatorname{curl}\mathbf{B}. \tag{E.45}$$

The curl of a vector product

$$\boldsymbol{\nabla}\times(\mathbf{A}\times\mathbf{B})=\mathbf{A}(\boldsymbol{\nabla}\cdot\mathbf{B})-(\boldsymbol{\nabla}\cdot\mathbf{A})\mathbf{B}+(\mathbf{B}\cdot\boldsymbol{\nabla})\mathbf{A}-(\mathbf{A}\cdot\boldsymbol{\nabla})\mathbf{B}. \tag{E.46}$$

For $\mathbf{A}\cdot\boldsymbol{\nabla}$ see below (E.44).

Note that (E.42)–(E.46) are not the same as would be given by §§ E.07 and E.09, if $\boldsymbol{\nabla}$ was simply a vector.

E.19. Curl grad ϕ

If a vector **F** is the gradient of a scalar ϕ, it is clear that between two points A and B in Fig. E.15 the difference of ϕ is the line integral of **F** along any route between A and B. In symbols

$$\phi_B - \phi_A = \int_A^B \mathbf{F} \cdot d\mathbf{L} \quad \text{along ACB.}$$

Also
$$\phi_B - \phi_A = \int_A^B \mathbf{F} \cdot d\mathbf{L} \quad \text{along any other route ADB.}$$

It follows that round any circuit such as ACBDA $\int \mathbf{F} \cdot d\mathbf{L} = 0$, if $\mathbf{F}$ is the gradient of a scalar. In other words,

$$\text{curl grad } \phi = 0 \quad \text{identically,}$$

or
$$\nabla \times (\nabla \phi) = 0 \tag{E.47}$$

as it would if ∇ was a vector, since $\mathbf{A} \times \mathbf{A} = 0$.

A vector field in which $\int \mathbf{F} \cdot d\mathbf{L}$ is independent of the route followed, and in which curl $\mathbf{F}$ is consequently everywhere zero, is known as a *conservative field.*

E.20. Div grad ϕ. Laplace's operator $\nabla^2\phi$

If ϕ is a variable scalar, we have grad $\phi = \nabla \phi$ as in § E.11. This is a vector, which will have a divergence.

$$\begin{aligned}\text{div grad } \phi &= \nabla \cdot (\nabla \phi) \text{ or } \nabla^2 \phi \text{ for short} \\ &= \left(\mathbf{i}\frac{\partial}{\partial x} + \mathbf{j}\frac{\partial}{\partial y} + \mathbf{k}\frac{\partial}{\partial z}\right) \cdot \left(\mathbf{i}\frac{\partial}{\partial x} + \mathbf{j}\frac{\partial}{\partial y} + \mathbf{k}\frac{\partial}{\partial z}\right)\phi \\ &= \left(\frac{\partial}{\partial x}\frac{\partial}{\partial x} + \frac{\partial}{\partial y}\frac{\partial}{\partial y} + \frac{\partial}{\partial z}\frac{\partial}{\partial z}\right)\phi, \quad \text{since } \mathbf{i} \cdot \mathbf{i} = \mathbf{1}, \text{ etc.} \\ &= \left(\frac{\partial^2}{\partial x^2} + \frac{\partial^2}{\partial y^2} + \frac{\partial^2}{\partial z^2}\right)\phi. \end{aligned} \tag{E.48}$$

If ϕ is the gravitational potential, grad ϕ is the attraction $\mathbf{F}$, and from § E.15, Example 2,

$$\begin{aligned}\text{div } \mathbf{F} \ (\equiv \text{div grad } \phi \text{ or } \nabla^2 \phi) &= 0 \text{ outside matter} \\ \text{and} &= -4\pi G\rho \text{ inside matter of density } \rho.\end{aligned} \tag{E.49}$$

That div grad ϕ should be zero except inside matter, when ϕ is the gravitational potential and grad ϕ is the attraction, is reasonable enough. Fig. E.9 (b) shows how divergence is non-zero when converging forces get smaller towards the centre, as they do inside a massive sphere, § 6.09 (*c*), while Fig. E.9 (c) suggests that (in three dimensions) the divergence is zero when the force decreases as the square of the distance.

Formulae (E.37), (E.47), and (E.48) give expressions for three of the nine possible combinations of the operators grad, div, and curl. The position of the other six is as below.

(*a*) curl div **F** and div div **F** have no meaning since div **F** is a scalar.

(*b*) Similarly grad curl **F** and grad grad ϕ have no meaning since curl **F** and grad ϕ are vectors.

(*c*) curl curl **F** and grad div **F** exist but are of less interest than the first three. See [247], p. 53.

E.21. Green's theorem

At every point (x, y, z) let there be two scalars ϕ_1 and ϕ_2. For instance, ϕ_1 might be the gravitational potential of all matter, while ϕ_2 might be the potential of some mass situated at a particular point. Then at every point in the space there is a vector $\phi_1\nabla\phi_2$, or ϕ_1 grad ϕ_2. Consider any closed surface S and apply Gauss's theorem (E.30) to the vector $\phi_1\nabla\phi_2$. It gives

$$\iint (\phi_1\nabla\phi_2)\cdot d\mathbf{S} = \iiint \nabla\cdot(\phi_1\nabla\phi_2)\,dv, \tag{E.50}$$

which from (E.42)

$$= \iiint \phi_1(\nabla\cdot\nabla\phi_2)\,dv + \iiint (\nabla\phi_1)\cdot\nabla\phi_2\,dv$$

or

$$= \iiint \phi_1\nabla^2\phi_2\,dv + \iiint (\nabla\phi_1)\cdot\nabla\phi_2\,dv. \tag{E.51}$$

There also exists the vector $\phi_2\nabla\phi_1$ or ϕ_2 grad ϕ_1, for which Gauss's theorem gives a similar expression, and subtracting it from (E.51) gives

$$\iint (\phi_1\nabla\phi_2 - \phi_2\nabla\phi_1)\cdot d\mathbf{S} = \iiint (\phi_1\nabla^2\phi_2 - \phi_2\nabla^2\phi_1)\,dv$$

or

$$\iint \left(\phi_1\frac{\partial\phi_2}{\partial n} - \phi_2\frac{\partial\phi_1}{\partial n}\right) dS = \iiint (\phi_1\nabla^2\phi_2 - \phi_2\nabla^2\phi_1)\,dv$$

or

$$\iint (\phi_1 \operatorname{grad}\phi_2 - \phi_2 \operatorname{grad}\phi_1)\cdot d\mathbf{S} = \iiint (\phi_1 \operatorname{div}\operatorname{grad}\phi_2 - \phi_2 \operatorname{div}\operatorname{grad}\phi_1)\,dv. \tag{E.52}$$

This is Green's theorem. In certain circumstances it takes a simpler form, e.g. if $\nabla^2\phi_1$ or $\nabla^2\phi_2 = 0$, or if ϕ_1 or ϕ_2 is constant. For a proof of Green's theorem without use of vector algebra, see [113], 2 Edn., pp. 395–6 or [502], pp. 74–6.

From Green's theorem can be derived the important theorem of *Green's equivalent layer*, see §§ 6.14 and 6.46.

General references for vector algebra. [247], [506], and [616], pp. 461–525.

APPENDIX F

COMPLEX NUMBERS AND CONFORMAL MAPPING

F.00. Definitions

JUST as matrix algebra deals with scalar numbers in blocks of m rows and n columns, so in the algebra of complex numbers they are dealt with in simple ordered† pairs. A pair of numbers is, of course, a 1×2 matrix, but complex numbers are not treated simply as small matrices. There is, for instance, a definition of the product of two complex numbers, such as does not exist for two 1×2 matrices.

Two ordered numbers define the position of a point in a plane with reference to rectangular axes, Fig. F.1, so to every point in the plane there corresponds a unique complex number, and vice versa. The subject can either be treated algebraically without diagrams, or geometrically. Here, the emphasis is on the geometrical aspect.‡

The position of the point P in Fig. F.1 may be described in seven ways, as below, all useful in different circumstances.

(*a*) (x, y), rectangular coordinates.

(*b*) $x + \mathrm{i}y$, where at this stage the $+\mathrm{i}$ means no more than the comma and brackets in (*a*). The marker i indicates the second member of the pair.

(*c*) For brevity we may write z or w for either of the above, and z_1 for (x_1, y_1), etc.

(*d*) (r, θ), polar coordinates.

(*e*) $(r \cos \theta, r \sin \theta)$ as in (*a*).

(*f*) $r \cos \theta + \mathrm{i}r \sin \theta$, or $r(\cos \theta + \mathrm{i} \sin \theta)$, as in (*b*).

(*g*) $re^{\mathrm{i}\theta}$ or $r \exp \mathrm{i}\theta$, see § F.03.

Since the point P defines the vector OP, the latter may also be defined in any of these ways.

In a complex number the first member of the pair, x or $r \cos \theta$, is known as the *real part*, and may be written $\operatorname{Re} z$ or $\operatorname{Re}(x + \mathrm{i}y)$, etc. The second member, y or $r \sin \theta$, is known as the *unreal* or *imaginary part*. These are traditional names, not very apt.

The *absolute value* or *modulus* of a complex number is defined to be $+(x^2 + y^2)^{\frac{1}{2}}$ or r, and is written $|z|$. The angle θ or $\tan^{-1} y/x$ is known as the *argument* of the complex number, and is written $\arg z$.

To any point z or $(x + \mathrm{i}y)$, etc., there corresponds its *opposite* $-x + \mathrm{i}(-y)$, written $-x - \mathrm{i}y$ and denoted by $-z$. There is also the point $x + \mathrm{i}(-y)$ or $x - \mathrm{i}y$, which is known as the *conjugate* of z and is written z^*.

The axis Ox is known as the *axis of reals*, and all points on it are of the form

† Ordered means that the order in which they are written is material. The pair 2 and 3 is not the same as 3 and 2.

‡ But see Appendix G, where complex numbers are used algebraically as an alternative to elementary geometry.

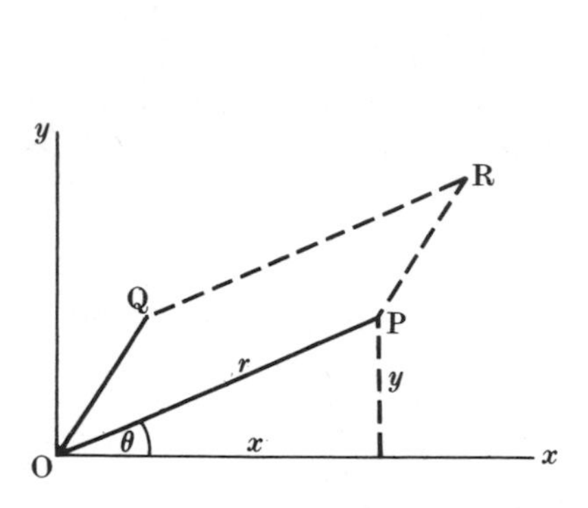

FIG. F.1. Addition. R = P + Q.

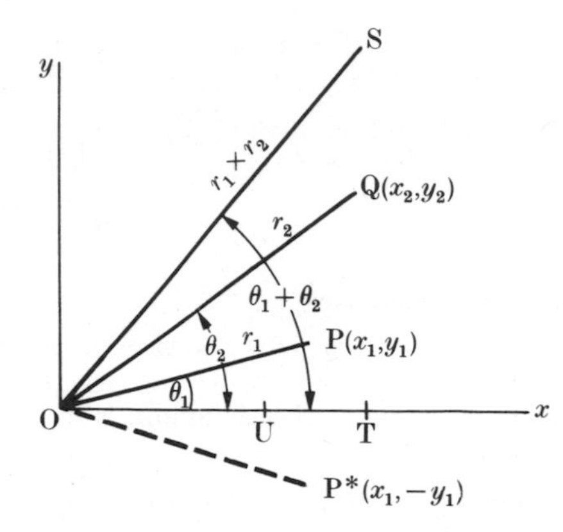

FIG. F.2. Multiplication. S = P × Q. U = (1, 0). P* is the conjugate of P. T = P* × P.

$x+i0$, or x for short. The axis Oy is the *axis of unreals* and all points on it are of the form $0+iy$, or iy for short. z^* is the mirror image of z in the axis of reals, and $-z^*$ is the image of z in the axis of unreals.

F.01. Addition of complex numbers

The sum of two complex numbers x_1+iy_1 and x_2+iy_2, such as P and Q in Fig. F.1, is defined to be

$$(x_1+x_2)+i(y_1+y_2), \tag{F.1}$$

the point R, such that OR is the vector sum of OP and OQ. Clearly

(*a*) $z_1+z_2 = z_2+z_1$, and addition is commutative. (F.2)

(*b*) $z_1+(z_2+z_3) = (z_1+z_2)+z_3$. Associative. (F.3)

(*c*) $z_1+z_1+z_1+\ldots k$ times $= kx_1+iky_1$, and may be written $k(x_1+iy_1)$ or kz_1. And the same if k is a fraction. (F.4)

The sum $z+(-z)$ or $(x+iy)+(-x-iy) = 0+i0$, and is the point O. It is called *zero*. A complex number is zero if and only if both its parts are zero.

Similarly, if $(x_1+iy_1) = (x_2+iy_2)$, both $x_1 = x_2$ and $y_1 = y_2$. The real and imaginary parts must both separately be equal.

Subtraction. $z_1+(-z_2)$ is written z_1-z_2, and z_2 is said to be subtracted from z_1.

$$(z_1-z_2) = (x_1-x_2)+i(y_1-y_2). \tag{F.5}$$

So far as concerns addition and subtraction, the ordinary rules of algebra can be unthinkingly applied, the marker i being treated as if it was a scalar like x and y. It is only necessary to note that actually it is not a scalar, and that it will never receive a numerical value, such as may presently be given to x and y.†

† When we write $x+y$ we have it in mind that presently we may say $x = 3$ and $y = 4$, so $x+y = 7$. But the sum $x+iy$ can be taken no further than $3+i4$, or $3+4i$ as it is more usually written.

F.02. Multiplication of complex numbers

(*a*) The product of the two complex numbers z_1 and z_2, represented by P and Q in Fig. F.2, is defined to be z_3, represented by the point S, such that

$$|z_3| = |z_1| \times |z_2|$$

and $$\arg z_3 = \arg z_1 + \arg z_2. \tag{F.6}$$

In polar coordinates

$$r_3 = r_1 r_2 \quad \text{and} \quad \theta_3 = \theta_1 + \theta_2. \tag{F.7}$$

The justification of this definition is that it is useful.

From the definition

$$\begin{aligned}(x_1+\mathrm{i}y_1)\times(x_2+\mathrm{i}y_2) &= (x_1^2+y_1^2)^{\frac{1}{2}}(x_2^2+y_2^2)^{\frac{1}{2}}\{\cos(\theta_1+\theta_2)+\mathrm{i}\sin(\theta_1+\theta_2)\}\\ &= (x_1x_2-y_1y_2)+\mathrm{i}(y_1x_2+x_1y_2).\end{aligned} \tag{F.8}$$

From the definition the following are clear.

(*a*) $z_1 \times z_2 = z_2 \times z_1$. Multiplication is commutative. (F.9)

(*b*) $(z_1 \times z_2) \times z_3 = z_1 \times (z_2 \times z_3)$. Associative. Both may be written $z_1 z_2 z_3$ in any order, and $z_1 z_1$ may be written z_1^2, etc. (F.10)

(*c*) $z_1 \times (z_2 + z_3) = z_1 z_2 + z_1 z_3$. Multiplication is distributive with respect to addition. (F.11)

If $z_1 \times z_2 = z_1$ it is clear that mod $z_2 = 1$ and $\arg z_2 = 0$. So $z_2 = 1 + \mathrm{i}0$, the point (1, 0) on the axis of reals, marked U in Fig. F.2. Also $zz^* = |z|^2 + \mathrm{i}0$, a point on the real axis.

Recalling that the rules of algebra are applicable to the addition and subtraction of complex numbers, we hopefully apply them to multiplication and get

$$(x_1+\mathrm{i}y_1)\times(x_2+\mathrm{i}y_2) = x_1x_2 + \mathrm{i}(y_1x_2+x_1y_2) + (\mathrm{i}\times\mathrm{i})(y_1y_2). \tag{F.12}$$

Compared with (F.8) the result is seen to be correct except that there is a term $(\mathrm{i}\times\mathrm{i})(y_1y_2)$ instead of $-y_1y_2$. It can be made correct if we make the rule that -1 is always to be written in the place of $\mathrm{i}\times\mathrm{i}$.

Note that $(0\times\mathrm{i}1)\times(0+\mathrm{i}1)$ or i^2 for short $=(0-1)=-1$, a point on the axis of reals, the opposite of U in Fig. F.2.†

(*b*) *Division.* The quotient $z_1 \div z_2$ is defined to be a complex number z_3 such that $z_2 \times z_3 = z_1$.

Then clearly $$|z_3| = |z_1| \div |z_2| \quad \text{from (F.6)} \tag{F.13}$$

and $$\arg z_3 = \arg z_1 - \arg z_2$$

or $$r_3 = r_1/r_2 \tag{F.14}$$

and $$\theta_3 = \theta_1 - \theta_2.$$

† To the question 'Does i^2 really equal -1?' the answer is 'Yes, if i stands for the complex number $0+\mathrm{i}1$, and if i^2 stands for $\mathrm{i}\times\mathrm{i}$, where the $\times$ indicates the multiplication of complex numbers'.

It follows that
$$\frac{x_1+iy_1}{x_2+iy_2}=(r_1/r_2)\{\cos(\theta_1-\theta_2)+i\sin(\theta_1-\theta_2)\}$$
$$=\frac{x_1x_2+y_1y_2}{x_2^2+y_2^2}+i\frac{y_1x_2-x_1y_2}{x_2^2+y_2^2}, \qquad (F.15)$$

and it is easily verified that $z_2\times z_3=z_1$ if -1 is substituted for $i\times i$.

The quotient z_1/z_2 is often most easily formed by writing it in the form $(z_1z_2^*)/(z_2z_2^*)$ in which the denominator is pure real.

(*c*) *Summary*. The general rule then is that addition, subtraction, multiplication, and division of complex numbers, as defined above, may be carried out by the ordinary rules of algebra, treating the marker i as an ordinary scalar, with the following provisos.

(i) It must be recognized that i will never have a numerical value.

And (ii) where $i\times i$ occurs in a product it is treated as equal to -1.

F.03. Exponentials

The expression $re^{i\theta}$, considered as $r\times(2.718\ldots)^{i\theta}$ has no clear meaning. On the other hand, if it is defined by the series

$$re^{i\theta}=r\{1+(i\theta)+(i\theta)^2/2!+(i\theta)^3/3!+\ldots\}, \qquad (F.16)$$

it can acquire a significant meaning in conjunction with the proviso of § F.02 (*c*) (ii). This definition is adopted, and

$$re^{i\theta}=r(1-\theta^2/2!+\theta^4/4!+\ldots)+ir(\theta-\theta^3/3!+\ldots)$$
$$=r(\cos\theta+i\sin\theta). \qquad (F.17)$$

As in § F.00, item (*g*), it is now an alternative way of describing the point (r,θ), P in Fig. F.1, and it is amenable to the rules of algebra subject to the provisos of § F.02 (*c*). For instance, if the sign $\times$ indicates the multiplication of ordinary algebra,

$$r_1e^{i\theta_1}\times r_2e^{i\theta_2}=r_1r_2e^{i(\theta_1+\theta_2)}, \qquad (F.18)$$

as is also correct if $\times$ indicates the multiplication of complex numbers.

We may write $\qquad x=r\cos\theta=\text{Re}\,re^{i\theta}\quad\text{or}\quad\text{Re}\,r\exp i\theta, \qquad (F.19)$

and the conjugate of $r\exp i\theta$ is

$$r\exp i(-\theta)\quad\text{or}\quad r\exp(-i\theta). \qquad (F.20)$$

From (F.17) there follows

(*a*) $\qquad(\cos\theta+i\sin\theta)\times(\cos\theta+i\sin\theta)=\exp(i2\theta)=\cos2\theta+i\sin2\theta, \qquad (F.21)$

and in general $\qquad(\cos\theta+i\sin\theta)^n=\cos n\theta+i\sin n\theta; \qquad (F.22)$

the well-known theorem of de Moivre.

(*b*) $\qquad e^{i\theta}+e^{-i\theta}=2\cos\theta+i0=2\cos\theta. \qquad (F.23)$

(*c*) $e^{i\theta}-e^{-i\theta}=0+2i\sin\theta$. Whence

$$\sin\theta=(1/2i)(e^{i\theta}-e^{-i\theta}). \qquad (F.24)$$

F.04. Uses of complex algebra

The complex notation for a pair of scalars, especially in the form

$$E = \text{Re } A \exp \text{i}(nt + \alpha)$$

is much used in electricity (where j is usually written for i, which is preoccupied). and other branches of engineering. See Appendix G for its application to the modulation of electromagnetic waves. It is also used to describe the motion of a pendulum on moving supports [573], and § 2.64 (*b*) uses (F.8) and (F.15) for the routine of interpolation in two dimensions. Treated as in §§ F.05–F.07, complex number theory has been much used in connection with map projections, [356].

F.05. Functions of a complex variable

As before, $z = x + \text{i}y$. Then let $w = u + \text{i}v$, where u and v are functions of x and y. For instance

(*a*) Let $w = (x + \text{i}y)^2 = (x^2 - y^2) + 2\text{i}xy$. So

$$u = (x^2 - y^2) \quad \text{and} \quad v = 2xy. \tag{F.25A}$$

or (*b*) Let $w = 4x + 2\text{i}y$. So

$$u = 4x \quad \text{and} \quad v = 2y. \tag{F.25B}$$

Whatever the form of u and v, some manipulation will enable w to be written as a function of z or of z and z^*. In these two examples

(*a*) $$w = z^2 \tag{F.26A}$$

and (*b*) $$w = 3z + z^*. \tag{F.26B}$$

Viewing this geometrically, to every point (x, y) there corresponds another, (u, v), and to every set of lines drawn with reference to the x, y axes 'on the z-plane', there corresponds a set of lines drawn with reference to the u, v axes 'on the w-plane'. Fig. F.3 shows a small square on the z-plane and the corresponding patterns on the w-plane when w is given by (F.25A) and (F.25B). The square on the z-plane is said to have been *mapped* on the w-plane. The figure suggests that (F.25A) has produced an orthomorphic or conformal map, while (F.25B) certainly

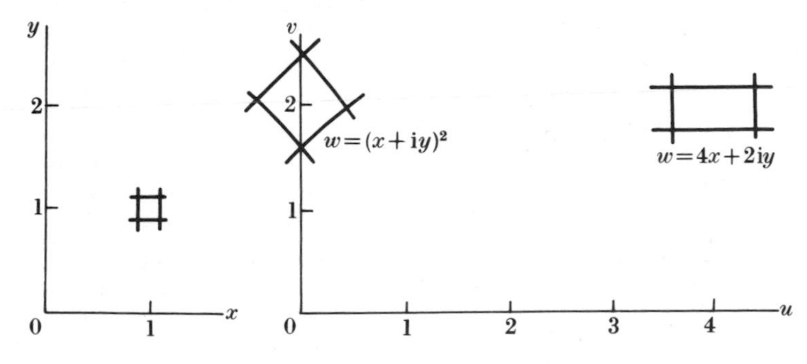

FIG. F.3 (a). The z-plane. FIG. 3 (b). The w-plane.

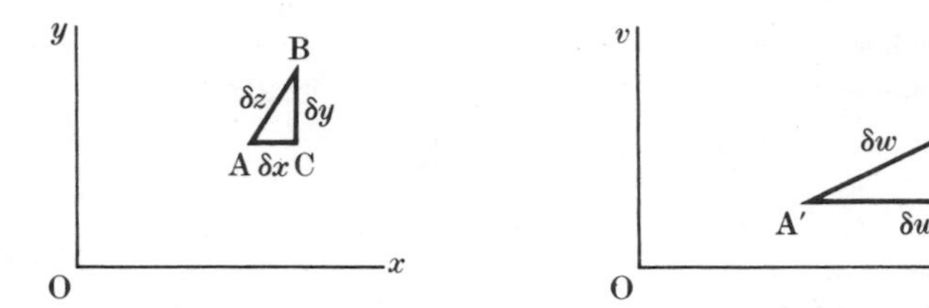

FIG. F.4. (a). The z-plane. $\delta z = \delta x + i\delta y$.

FIG. 4 (b). The w-plane. $\delta w = \delta u + i\delta v$.

has not. It is of interest to inquire what functions result in conformal maps, and what do not.

Note that if $f(z)$ is any function of z, $f(z^*)$ is the conjugate of $f(z)$. (F.27)

F.06. Differentiation of a function of a complex variable

The expression dw/dz may rationally be defined as the limit of the complex quotient

$$\frac{\delta u + i\delta v}{\delta x + i\delta y}, \tag{F.28}$$

where δu and δv are the changes in u and v resulting from small changes δx and δy. Geometrically, Fig. F.4 shows $A(x, y)$ and $B(x+\delta x, y+\delta y)$ on the z-plane, and the corresponding $A'(u, v)$ and $B'(u+\delta u, v+\delta v)$ on the w-plane. From (F.28) and (F.13) it is clear that

$$\left|\frac{dw}{dz}\right| = \frac{|\delta u + i\delta v|}{|\delta x + i\delta y|} = \frac{A'B'}{AB} \tag{F.29}$$

and

$$\arg\left(\frac{dw}{dz}\right) = B'A'C' - BAC = \tan^{-1}(\delta v/\delta u) - \tan^{-1}(\delta y/\delta x). \tag{F.30}$$

In other words, $|dw/dz|$ is the scale of the map of AB on the w-plane, and $\arg(dw/dz)$ is its change of orientation. Both, of course, depend on x and y, and are variable over the map. One point, however, remains to be considered.

In Fig. F.4 the direction of AB has been drawn making an angle of about 60° with the axis of x. If it had been drawn in some other direction, the numerical values of mod and $\arg(dw/dz)$ might well have been different, in which case the w-plane map would not have been conformal. If u and v are such functions of x and y that at any point (x, y) both mod and $\arg(dw/dz)$ have unique values independent of the direction AB (or of the ratio $\delta y/\delta x$), w is said to be an *analytic function* of z.

The condition that w should be an analytic function can be stated in two ways. Either (a), u and v must be such functions of x and y as satisfy the equations

$$\frac{\partial u}{\partial x} = \frac{\partial v}{\partial y} \quad \text{and} \quad \frac{\partial u}{\partial y} = -\frac{\partial v}{\partial x}. \tag{F.31}$$

These are known as the Cauchy–Riemann equations. For a proof see § 2.41.† Differentiating the first with respect to x and the second with respect to y shows that for an analytic function u must satisfy Laplace's equation

$$\partial^2 u/\partial x^2 + \partial^2 u/\partial y^2 = 0. \tag{F.32}$$

And the same for v.

Or (b) It must be possible to express w as a function of z, without involving z^*. See [616], pp. 549–50.

By either criterion (F.25A) is seen to be an analytic function, and it gives a conformal map as suggested by Fig. F.3. On the other hand, (F.25B) satisfies only the second equation of (F.31), and cannot be expressed as a function of z only. Fig. F.3 shows that its map is not conformal.

If w is an analytic function, so that dw/dz exists, dw/dx is given in terms of z by the ordinary routine of the differential calculus. Thus, if $w = z^2$,

$$dw/dz = 2z = 2x + 2iy. \tag{F.33}$$

So when $x = 1$ and $y = 1$ the scale of the conformal map is $|2z|$ or $2\sqrt{2}$, and its rotation relative to the z-plane is $\tan^{-1}2/2 = 45°$. Fig. F.3 confirms.

Alternatively,
$$\frac{dw}{dz} = \frac{\partial v}{\partial y} - i\frac{\partial u}{\partial y} = \frac{\partial u}{\partial x} + i\frac{\partial v}{\partial x}. \tag{F.34}$$

Both expressions give the same result, namely for $w = z^2$, $dw/dz = 2x + 2iy$ as before.

F.07. Map projections

On the z-plane of Fig. F.3 (a) let a map of some area be drawn on Mercator's projection, which is known to be orthomorphic. So $E = m_0 a\lambda$ and $N = m_0 a\psi$, where λ is longitude and ψ is isometric latitude as in (2.108). Then if

$$w = u + iv = f(E + iN) = m_0 af(\lambda + i\psi), \tag{F.35}$$

where f is any analytic function, the corresponding map on the w-plane, Fig. F.3 (b), will also be orthomorphic. By the choice of suitable functions, $f(\lambda + i\psi)$, to suit prescribed conditions (such as that a central meridian shall be straight and of constant scale) the algebraical expression of desired projections can be obtained. See [374] for a derivation of the Transverse Mercator projection in this way.

General references for Appendix F. [616], [356], and [562].

† X, Y, $(X+x)$ and $(Y+y)$ of § 2.41 are x, y, u, and v here, so (F.31) in the notation of § 2.41 is $\partial(X+x)/\partial X - \partial(Y+y)/\partial Y = 0$, or $\partial x/\partial X - \partial y/\partial Y = 0$, which § 2.41 gives as a condition for orthomorphic mapping, (2.59).

APPENDIX G

MODULATED WAVES AND TELLUROMETER GROUND SWING

G.00. Notation

l_C = wavelength of microwave carrier.
l_M = wavelength of microwave modulation.
f_C = frequency of carrier in hertz (cycles per second) $= v/l_C$.
f_M = frequency of modulation $= v/l_M$.
w = angular velocity of carrier $= 2\pi f_C$ radians per second.
W = angular velocity of modulation $= 2\pi f_M$ radians per second.
T = travel time over the direct route $= D/v$.
D = (single) distance along curved path.
v = velocity of microwaves along path.
Q = excess travel time of a reflected wave in single distance.
d = excess path of a reflected wave $= vQ$.
$z = \pi + wQ$.
$Z = WQ$.
$-WX$ = modulation phase error in single distance due to addition of reflected wave (G.21).
Δ = error of single distance due to reflected wave $= vX$.
E = instantaneous intensity of an electromagnetic wave.
X = error of transmission time (single distance) due to reflected wave.
A = amplitude (semi-range) of a wave.
α = phase angle.
t = time.
$\frac{1}{2}m$ = (amplitude of a sideband) ÷ (amplitude of carrier).
a = reflection factor.
$\beta = \tan^{-1}[(a \sin z)/(1 + a \cos z)]$. See (G.9).
$H = (1 + 2a \cos z + a^2)^{\frac{1}{2}}$. See (G.8).
$\psi = z - \beta$.
C = carrier wave.
S_1, S_2 = sidebands.
c = reflected carrier.
s_1, s_2 = reflected sidebands.
t_1, t_2, v_1, v_2 satisfy $s_1 = (t_1 + v_1)$ and $s_2 = (t_2 + v_2)$. See Fig. G.6.
j is used for the imaginary i in electrical contexts.
θ = the argument of a complex number as in $R \exp j\theta$.
n See (G.21).

G.01. Waves represented by a cosine formula

The instantaneous intensity of a microwave at a point M in Fig. G.1 (a) at time t may be represented as

$$E = A(\cos wt + \alpha), \tag{G.1}$$

where A is the amplitude (semi-range) of the intensity, $w = 2\pi \times$(frequency), and

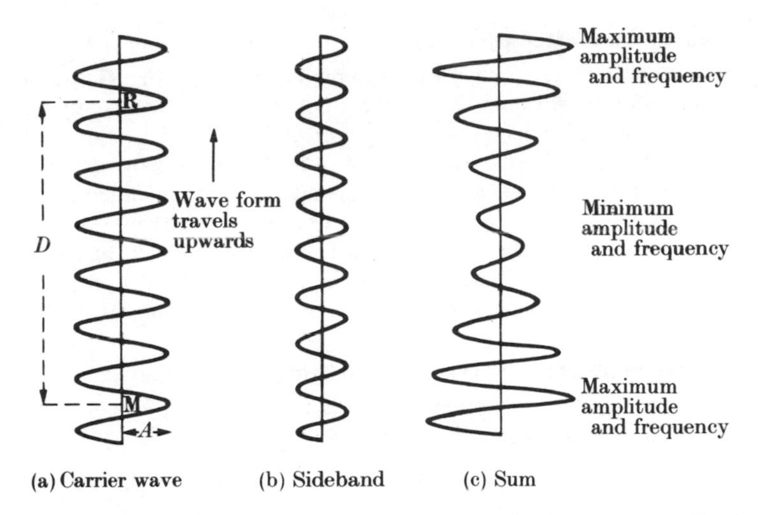

FIG. G.1. Modulation by one sideband. If M is the master and R the remote tellurometer, D in (a) will probably contain between 10^5 and 10^6 10-cm or 3-cm cycles, and there will be 300 cycles or more between maxima in (c). The changes of frequency in (c) can be seen by careful inspection of the intervals between zero crossings.

α is the phase angle. If the zero of t is taken to be when $E = A$, $\alpha = 0$ and

$$E \text{ at M} = A \cos wt. \tag{G.2}$$

If the wave travels from M to a point R at distance D with velocity v, the wave form arriving at R at time t will have left M at time $t - T$, where $T = D/v$. Then

$$E \text{ at R} = A \cos w(t - T). \tag{G.3}$$

If two waves with suffixes 1 and 2 are superposed, the intensity of their sum is clearly

$$A_1 \cos(w_1 t + \alpha_1) + A_2 \cos(w_2 t + \alpha_2). \tag{G.4}$$

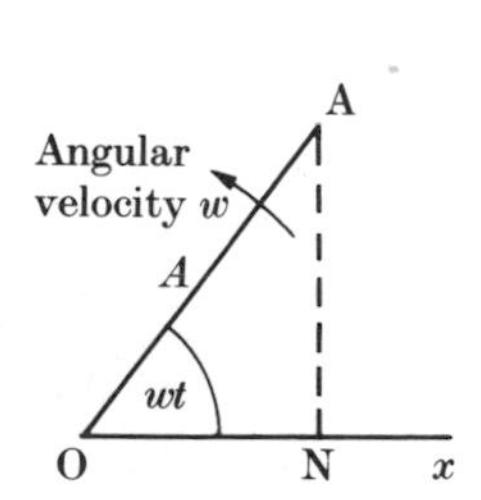

FIG. G.2 (a). Vector representation of a wave. ON $= A \cos wt$.

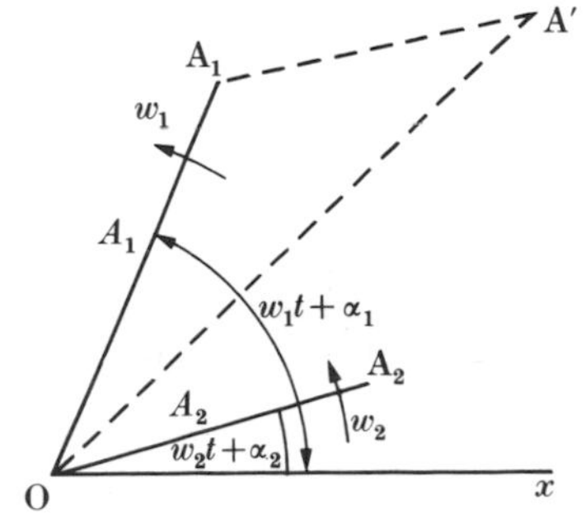

FIG. G.2 (b). Vector sum of two waves.

G.02. Waves represented by vectors

A wave may be represented graphically by a vector OA, Fig. G.2 (a), of length A rotating about O with velocity w radians per second and passing through Ox at time $t=0$. Then the projection of OA upon Ox is $A \cos wt$, as in (G.2). If this represents the wave at M, the wave at R may be represented by another vector of the same length and rotating at the same speed, but lagging behind it by an angle wT. The projection of this second vector on Ox will be $A \cos w(t-T)$ as in (G.3).

The sum of the two waves represented by OA_1 and OA_2 in Fig. G.2 (b) with different amplitudes, frequencies, and phases, will clearly be the projection on Ox of their instantaneous vector sum OA′, giving the same result as (G.4).

G.03. Waves represented by complex numbers

A third method of representing a wave is to remark that $A \cos(wt+\alpha)$ is the real part of $A \exp \mathrm{i}(wt+\alpha)$, as in (F.19). We may write

$$E = \mathrm{Re}\, A e^{\mathrm{i}(\omega t+\alpha)} \quad \text{or more usually} \quad E = A \exp \mathrm{j}(wt+\alpha), \tag{G.5}$$

omitting the Re as always understood, and in this subject customarily writing j instead of i.

The utility of this convention lies in the fact that from the definition of the multiplication of complex numbers given in § F.02,

$$(A_1 \exp \mathrm{j}\, \theta_1)\times(A_2 \exp \mathrm{j}\theta_2) = A_1 A_2 \exp \mathrm{j}(\theta_1+\theta_2). \tag{G.6}$$

So the wave $A \exp \mathrm{j}(wt+\alpha)$ may be written

$$A \exp \mathrm{j}(wt+\alpha) = (A \exp \mathrm{j}\alpha)\times(\exp \mathrm{j}wt). \tag{G.7}$$

A wave with phase angle $(\alpha+\beta)$ is then obtained by multiplying (G.7) by $\exp \mathrm{j}\beta$, and one with phase $(\alpha-\beta)$ by dividing it by $\exp \mathrm{j}\beta$.

If we have a vector $1 \exp \mathrm{j}0$, a unit vector lying on the real axis, and another vector $a \exp \mathrm{j}\alpha$, their sum is $(1+a \exp \mathrm{j}\alpha)$ as in Fig. G.3 (a), and the modulus of their sum is clearly given by

$$\mathrm{mod}(1+a \exp \mathrm{j}\alpha) = H = (1+2a \cos \alpha + a^2)^{\frac{1}{2}}, \tag{G.8}$$

and its argument by

$$\arg(1+a \exp \mathrm{j}\alpha) = \beta = \tan^{-1} \frac{a \sin \alpha}{1+a \cos \alpha}, \tag{G.9}$$

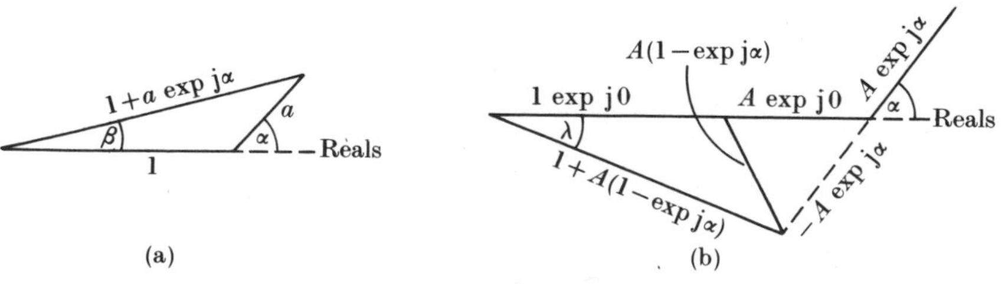

FIG. G.3.

defining H and β. So if a vector $A_0 \exp j\alpha_0$ is multiplied by $(1+a \exp j\alpha)$, the argument of the product is $\alpha_0+\beta$.

Fig. G.3 (b) shows a combination which will be required in § G.09. The vectors $A \exp j0$ and $-A \exp j\alpha$ are added to give $A(1-\exp j\alpha)$, and this is then added to the unit vector $1 \exp j0$. The argument of the total is then seen to be given by

$$\tan\lambda = \frac{-A\sin\alpha}{1+A-A\cos\alpha}. \tag{G.10}$$

Note that $\mathrm{Re}(A_1 \exp j\theta_1)(A_2 \exp j\theta_2)$ is not $A_1A_2\cos\theta_1\cos\theta_2$ but $A_1A_2\cos(\theta_1+\theta_2)$.

In the following paragraphs tellurometer ground swing is used to illustrate the vector and complex number methods.

G.04. Modulation with one sideband

To the wave $A \cos wt$ in Fig. G.1 (a) let there be added a wave $\frac{1}{2}mA\cos(w+W)t$,† in which m is a proper fraction, and W is much smaller than w. The second wave is known as a *sideband*. The sum will be as in Fig. G.1 (c) with the amplitude varying from maximum to minimum and back with a frequency of $W/2\pi$. The wave $A \cos wt$ is known as the *carrier wave* and is said to be modulated by the sideband. $W/2\pi$ is the *modulation frequency* and $2\pi v/W$ is its modulation wavelength. In as much as the amplitude is varying, the modulation is described as *amplitude modulation*, but Fig. G.1 (c) shows that with a single sideband the spacing of the zero points of the curve has also become variable, so that there is also some *frequency modulation*. This can be avoided as in § G.05.

Fig. G.4 (a) shows the vector representation of modulation by a single sideband.

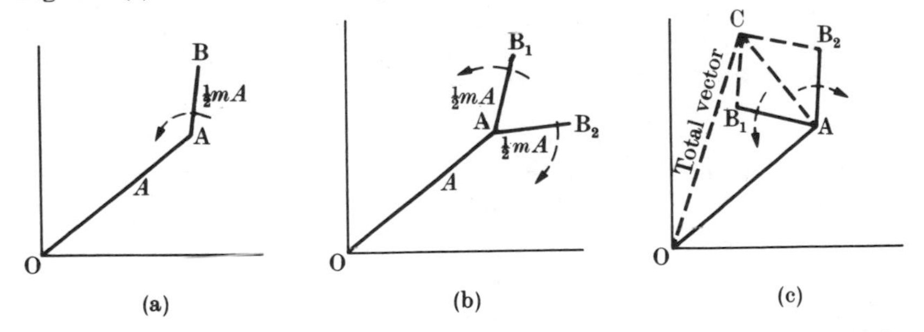

FIG. G.4. Modulation of carrier wave OA. The curved arrows show rotation of AB, AB_1, and AB_2 relative to OA with angular velocity W or $-W$. In addition, all rotate about O with much greater velocity w.

(a) Single sideband.

(b) Sidebands AB_1 and AB_2, rotating in opposite directions, coincide when they are parallel to OA. This is pure amplitude modulation. The total vector lies permanently along OA and gets longer and shorter.

(c) AB_1 and AB_2 coincide on AC. Approximate frequency modulation, if the sidebands are small. The total vector, of nearly constant length, swings backwards and forwards through OA.

† The zero of t is now one of the (relatively rare) instants when both waves are at maximum.

In the complex notation

$$E = A \exp jwt + \tfrac{1}{2}mA \exp j(w+W)t$$
$$= A \exp jwt(1 + \tfrac{1}{2}m \exp jWt). \qquad (G.11)$$

G.05. Amplitude modulation

Let two equal sidebands be added to the carrier with angular velocities differing from its w by $+$ and $-W$, phased so that they coincide with each other when they are in phase with the carrier. The total wave is then

$$E = A \cos wt + \tfrac{1}{2}mA \cos(w+W)t + \tfrac{1}{2}mA \cos(w-W)t$$
$$= A(1 + m \cos Wt)\cos wt, \qquad (G.12)$$

a wave with frequency $w/2\pi$ whose amplitude $A(1+m \cos Wt)$ is periodic with frequency $W/2\pi$.

The vector representation is as in Fig. G.4 (b). Note that the vector AB_1 is always the mirror image of AB_2 in OA.

In the complex notation

$$E = A \exp jwt\{1 + \tfrac{1}{2}m \exp jWt + \tfrac{1}{2}m \exp j(-Wt)\}, \qquad (G.13)$$

in which the term $\tfrac{1}{2}m \exp jWt$ is the conjugate of $\tfrac{1}{2}m \exp j(-Wt)$.

G.06. Frequency modulation

As in § G.05 let two equal sidebands be added, with velocities $w+W$ and $w-W$, but phased so that they coincide when 90° out of phase with the carrier. Then

$$E = A \cos wt + \tfrac{1}{2}mA \cos(w+W)t - \tfrac{1}{2}mA \cos(w-W)t$$
$$= A \cos wt - mA \sin wt \sin Wt$$
$$\approx A \cos(wt + m \sin Wt), \quad \text{if } m \text{ is small.} \qquad (G.14)$$

So if m is small, this gives frequency modulation with little variation of amplitude. To obtain pure frequency modulation requires a series of sidebands of diminishing amplitudes.

$$E = AJ_0(m)\cos wt + AJ_1(m)\{\cos(w+W)t - \cos(w-W)t\}$$
$$+ AJ_2(m)\{\cos(w+2Wt) + \cos(w-2Wt)\} + \ldots \qquad (G.15)$$

where the $J(m)$'s are Bessel functions of m, [174], Tables 30 and 31. Only the two first-order sidebands, as in (G.14) are used in the tellurometer.

The vector representation of Fig. G.4 (c) shows clearly how the periodic change of phase (and consequently of frequency) is associated with a change of amplitude which is small if m is small. Note that AB_1 is the mirror image of AB_2 in AC, which is perpendicular to OA.

In the complex notation

$$E = A \exp jwt\{1 + \tfrac{1}{2}m \exp jWt - \tfrac{1}{2}m \exp j(-Wt)\}. \qquad (G.16)$$

Compare this with (G.13) and note the difference in the sign of the last term.

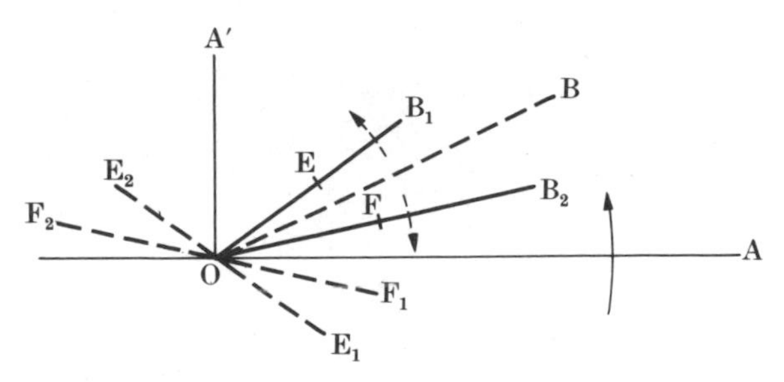

FIG. G.5. Mixed modulation.

G.07. Mixed modulation with two sidebands

If a carrier wave is modulated by two sidebands, with speeds $w+W$, and $w-W$ as before, but of unequal amplitudes or so phased that they do not coincide either when in phase with the carrier or when 90° out of phase with it, the result will be a combination of both amplitude and frequency modulation. The amplitude and phase of each type of modulation can then be determined as below.

Fig. G.5 shows the carrier wave OA modulated by OB_1 and OB_2, which are not necessarily of equal amplitude, and which rotate relative to it at equal speeds in opposite directions from the line OB on which they lie when coincident. OA′ is perpendicular to OA. Then add to the diagram OE_1, the mirror image of OE $(=\frac{1}{2}OB_1)$ in OA, and also add OE_2 the opposite of OE_1. The total addition is zero. Similarly add OF_1, the image of $\frac{1}{2}OB_2$, and also its opposite OF_2. It is now clear that OA is amplitude modulated by the combination of

(OE and OF) with (OE_1 and OF_1),

or by $\frac{1}{2}(OB_1+OB_2)$ with its image in the carrier OA, (G.17)

and that OA is (approximately) frequency modulated by

$\frac{1}{2}(OB_1+OB_2)$ with its image in OA′. (G.18)

In the complex notation the mirror image of any vector $A \exp j\theta$ in the axis of reals is its conjugate $A \exp j(-\theta)$, and its image in the axis of unreals is $-A \exp j(-\theta)$. So if the axis of reals is taken to rotate with the carrier vector OA, OE_1 and OF_1 are the conjugates of OE and OF, while OE_2 and OF_2 are their conjugates with reversed signs. See § G.09 for further illustrations.

G.08. Tellurometer ground swing. Vector representation

Consider the signal at Remote as transmitted along the direct path, combined with a single strong reflection with reflection factor a and phase change π, § 1.38 and Fig. 1.20. In the 10-cm tellurometer the carrier angular velocity $w = 2\pi\times 3000\times 10^6$ radians per second and the modulation velocity is $2\pi\times 10\times 10^6$. The travel time over the direct distance D is T $(=D/v)$, and the excess time taken by the reflected wave on account of the excess distance d is Q $(=d/v)$. Let the phase angle $z=\pi+wQ$, and let $Z=WQ$.

In Fig. G.6 C is the direct carrier wave. The axis of x, defined below, is taken to

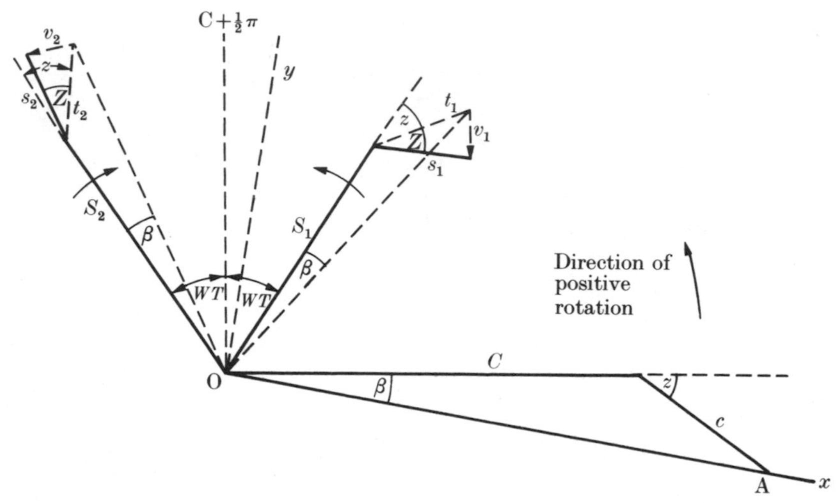

FIG. G.6. $|S_1| = |S_2| = 1$. $|C| = 2/m$. $|s_1| = |s_2| = |t_1| = |t_2| = a$. $|c| = 2a/m$. Angles are measured positive anti-clockwise.

rotate with the same angular velocity as C. S_1 and S_2 are two equal sidebands, whose angular velocities are $w+W$ and $w-W$, and which coincide in the direction $(C+\frac{1}{2}\pi)$, so their phase angles at remote will be $(C+\frac{1}{2}\pi)-WT$ and $(C+\frac{1}{2}\pi)+WT$. It is convenient to regard S_1 and S_2 as of unit length, so the length of C is $2/m$.†

The effect of the reflection is to add c to C, rotating at the same speed, with amplitude $2a/m$, and with phase behind C by $\pi + wQ = z$.‡ Then $(C+c)$ has amplitude $(2/m)(1+2a\cos z + a^2)^{\frac{1}{2}}$ and phase angle $-\beta$, given by

$$\tan\beta = a\sin z/(1+a\cos z),$$

as in (G.8) and (G.9). Let the axis Ox lie along $(C+c)$ and let Oy be perpendicular to it. $(C+c)$ is the total carrier wave, and it is modulated by two sidebands (S_1+s_1) and (S_2+s_2), in which s_1 and s_2 are each of amplitude a. s_1 having travelled a distance d more than S_1, and so having left M earlier, is behind it in phase by $\pi+(w+W)Q = z+Z$, and s_2 is behind S_2 by $\pi+(w-W)Q = z-Z$.

(S_1+s_1) and (S_2+s_2) have unequal amplitudes and they are not symmetrical about the axis Oy which is perpendicular to the carrier $(C+c)$, so the resulting modulation affects both amplitude and frequency. The circuits at Remote are designed to be sensitive only to frequency modulation, § 1.43 (a)‖, and as in

† $\frac{1}{2}m$ will be a fairly small fraction, but for clarity Fig. G.6 is drawn with $\frac{1}{2}m$ = about 0.6. Its magnitude does not affect the phase angles with which we are here concerned.

‡ In Fig. G.6 z is shown as a small angle, but it may be any size since it varies by 2π for every 10 cm in the excess path d. Z varies by 2π for every 30 metres in d. It may be large but it is very often small.

‖ This is the key to the solution.

§ G.07 this frequency modulation is produced by

$$\tfrac{1}{2}(S_1+s_1+S_2'+s_2')$$

with

$$\tfrac{1}{2}(S_2+s_2+S_1'+s_1'), \tag{G.19}$$

where S_2', s_2', etc., are the images of S_2, s_2, etc., in the y-axis.

The effect of summing the vector terms of (G.19) may be worked out geometrically as in [113], 3 Edn, pp. 661–3, but the same result is more conveniently obtained using complex notation as in § G.09.

Compared with the complex notation method, the geometrical vector method gives a much clearer picture to one who is not familiar with the subject, and it will always serve to illustrate a problem. On the other hand, the complex method avoids rather complicated diagrams and is easier to print. For those who know it well it is also easier to manipulate.

G.09. Tellurometer ground swing. Complex notation

The signal at Remote, including the reflection, $|C|$ now being taken to be of unit amplitude, is the sum of

$$\begin{aligned}
C &= \exp \mathrm{j}w(t-T),\\
S_1 &= \tfrac{1}{2}m \exp \mathrm{j}(w+W)(t-T),\\
S_2 &= -\tfrac{1}{2}m \exp \mathrm{j}(w-W)(t-T),\\
c &= a \exp \mathrm{j}\{w(t-T-Q)-\pi\} = a \exp \mathrm{j}\{w(t-T)-z\},\\
s_1 &= \tfrac{1}{2}am \exp \mathrm{j}\{(w+W)(t-T-Q)-\pi\} = \tfrac{1}{2}am \exp \mathrm{j}\{(w+W)(t-T)-(z+Z)\},\\
s_2 &= -\tfrac{1}{2}am \exp \mathrm{j}\{(w-W)(t-T-Q)-\pi\} = -\tfrac{1}{2}am \exp \mathrm{j}\{(w-W)(t-T)-(z-Z)\}.
\end{aligned}$$

If the axis of reals is rotated at angular velocity w, so as to coincide with C, the resulting signal is got by dividing all the above by $\exp \mathrm{j}w(t-T)$. And if the axis of reals is then turned through an angle $-\beta$, so that it coincides with $(C+c)$, they must be further divided by $1+a \exp \mathrm{j}(-z)$, as in (G.9).

The sum of the six vectors then simplifies† to

$$1+\tfrac{1}{2}m \exp \mathrm{j}W(t-T)\left\{\frac{1+a \exp \mathrm{j}(-z-Z)}{1+a \exp \mathrm{j}(-z)}\right\} - \tfrac{1}{2}m \exp \mathrm{j}\{-W(t-T)\}\left\{\frac{1+a \exp(-z+Z)}{1+a \exp \mathrm{j}(-z)}\right\}. \tag{G.20}$$

or $1+(S_1+s_1)+(S_2+s_2)$, where the initial 1 represents the (stationary) vector $(C+c)$. This is mixed amplitude and frequency modulation, and as in § G.07, the conjugate of (S_2+s_2) must be subtracted from (S_1+s_1) to give one sideband of the pure frequency modulation, namely

$$\tfrac{1}{2}m \exp \mathrm{j}W(t-T)\left\{\frac{1+a \exp \mathrm{j}(-z-Z)}{1+a \exp \mathrm{j}(-z)}+\frac{1+a \exp \mathrm{j}(z-Z)}{1+a \exp \mathrm{j}z}\right\}$$
$$= \text{(say)}\{\tfrac{1}{2}m \exp \mathrm{j}W(t-T)\}\times n, \quad \text{defining } n. \tag{G.21}$$

† The scale is also changed, so that the amplitude of $(C+c)$ becomes the unit of length.

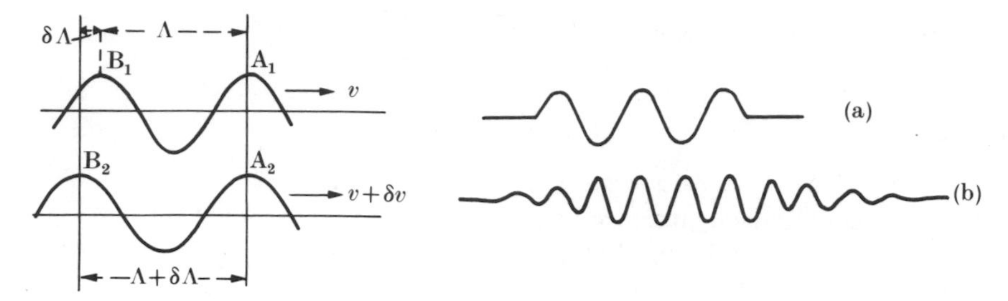

FIG. G.7. Group velocity.

FIG. G.8. Wave trains.

The other sideband will be

$$(S_2+s_2)-\text{the conjugate of } (S_1+s_1). \tag{G.22}$$

The modulation phase produced by the direct wave alone is $-WT$, so the error introduced by the reflection is $-\arg(n)$. We have

$$n=\frac{\{1+a\exp j(-z-Z)\}\{1+a\exp jz\}+\{1+a\exp j(z-Z)\}\{1+a\exp j(-z)\}}{\{1+a\exp j(-z)\}\{1+a\exp jz\}}. \tag{G.23}$$

Multiplying out the denominator gives $(1+2a\cos z+a^2)=H^2$, and the numerator simplifies to $H^2+a(a\cos z)\{\exp j(-Z)-1\}$, whence n may be put into the form

$$n=1-A\{1-\exp j(-Z)\} \tag{G.24}$$

where $A=a(a+\cos z)/H^2$.

From Fig. G.3 (b) and (G.10),† if n is in the form of (G.24), the error X is given by

$$WX=-\arg(n)=\tan^{-1}\left\{\frac{-A\sin(-Z)}{1-A+A\cos(-Z)}\right\}$$

$$=\tan^{-1}\left\{\frac{a(a-\cos wQ)\sin WQ}{1-a\cos wQ(1+\cos WQ)+a^2\cos WQ}\right\}, \tag{G.25}$$

G.10. Tellurometer ground swing. Conclusion

Formula (G.25) gives the modulation phase error in the single journey from M to R. The return journey is not quite identical, because its carrier frequency is different, 3033 MHz instead of 3000, § 1.43 (*a*). The combined error is then the mean of the values given by (G.25) using the two values of w, and z. In terms of distance the error is

$$\Delta=l_M(\text{mean } WX)/2\pi=v(\text{mean } X), \tag{G.26}$$

where l_M is the modulation wavelength.

† In (G.10) put $-Z$ for α, $-A$ for A and n for λ.

Weak reflections. If a is small, (G.25) becomes approximately

$$WX = -a \sin WQ \cos wQ. \tag{G.27}$$

As explained in §§ 1.43 (*b*) and 1.44, the error is then eliminated by varying wQ through a full cycle and making readings at equal intervals, so that the mean error is zero.

Strong reflections. If a is near unity, the ground swing curve takes the form shown in Fig. 1.22 (e), in which one of the two minima occurs in the outward path and one in the return.† The mean value of the curve is no longer exactly zero. Further, the narrow minima are not well defined by the spaced readings, and the signals at these wavelengths are weak.‡ A possible remedy is then to estimate the values of Q and a, and to use them to compute WX on the flat part of the curve. The observed values are then plotted as usual, and if the plot resembles the theoretical curve, the zero line can be set off from the flat part. See [462] and [220]. [355], pp. 30–4 also gives a routine. It must be recognized that the best treatment of lines with strong reflections is to avoid them.

In the special case where $a = 1$ and Q is small, the ground swing curve becomes a straight line giving a distance of $\frac{1}{2}d$ more than the direct path, with point indeterminacies corresponding to the minima of Fig. 1.22 (e).

Sharp minima in (G.25) and Fig. 1.22 (e) occur when the denominator in (G.25) is small. This is liable to occur when $\cos wQ = 1$, i.e. when wQ is a multiple of 2π or when d is a multiple of l_C. Such excess paths may be described as 'critical', [355]. In short lines with small d's it may be possible to chose instrument heights so as to avoid them: as when calibrating, § 1.43 (*c*).

Multiple reflections. In principle there is no difficulty in adding another reflection to Figs. G.6 and G.7, with given values of a, z, and Z. The resulting formula corresponding to (G.25) would be complicated, but the ground swing curve can be computed graphically, or otherwise, for different values of z. Fig. 1.22 (f) shows such a curve for one strong reflection and one weak, with the minima passing through infinity and reappearing as high positive values. See [617] and [462]. Such lines cannot be expected to be satisfactory.

G.11. The group velocity of light

This paragraph does not refer to microwaves, since their velocity is substantially independent of their wavelength. See § 1.32.

Fig. G.7 shows two light waves of wavelengths Λ and $\Lambda + \delta\Lambda$, with velocities v

† With weak reflections, instead of a curve with two sharp minima, the summation of two sine curves of different phases results in another sine curve of amplitude less than the sum of the two separate amplitudes. For small values of a (G.27) then gives a maximum value for the amplitude. See [355], p. 17.

‡ When z happens to be near 180°, the four vectors S_2', t_2', S_1, and t_1 will not lie in prolongation of each other, as drawn in Fig. G.7, but will be doubled back on themselves, so that if $a = 1$ their end-point is near O. The amplitude of the total will then be small, and its direction very sensitive to the exact values of z and Z, causing the sharp minima in the ground-swing curve.

and $v+\delta v$, the relation between δv and $\delta\Lambda$, derived from (1.9) and (1.8), being

$$\frac{\mathrm{d}v}{\mathrm{d}\Lambda}=c\left(\frac{3.26}{\Lambda^3}+\frac{0.05}{\Lambda^5}\right)10^{-6} \tag{G.28}$$

if Λ is in μm.

If two crests A_1 and A_2 coincide at time $t=0$, the difference of velocity will cause crests B_1 and B_2 to coincide at time $\delta t=\delta\Lambda/\delta v$. So in time δt the modulation maximum will have moved forward through a distance of $v\delta t-\Lambda$, and its velocity will therefore be

$$v-\Lambda/\delta t \quad \text{or} \quad v-\Lambda(\mathrm{d}v/\mathrm{d}\Lambda). \tag{G.29}$$

This is known as the group velocity, v_G, [392], pp. 106–8.

It is clear that the pattern resulting from the combination of any number of waves will move with this same group velocity, provided that the small term $\Lambda(\mathrm{d}v/\mathrm{d}\Lambda)$ is substantially constant over the range of wavelengths involved. In so far as $\Lambda(\mathrm{d}v/\mathrm{d}\Lambda)$ is not constant, the pattern is distorted as it travels, and the group velocity then has no more significance than has the velocity of a body whose shape is changing. The figures given after (1.12) show that a change of 1 ppm in the group velocity results from a change of about 0.02 μm or 200 Å in the wavelength.

That the beginning of a sharp pulse of light should travel with the group velocity is less obvious. It may be said that such a pattern as is shown in Fig. G.8 (a) can be expressed as a Fourier series of terms whose wavelengths are Λ, $\frac{1}{2}\Lambda$, $\frac{1}{3}\Lambda$, etc., but these wavelengths do not satisfy the condition that $\Lambda(\mathrm{d}v/\mathrm{d}\Lambda)$ should be constant. The following from [392], pp. 104–6 is perhaps more convincing.

Light emitted from a hot body consists of a very great number of waves, all of different wavelengths.† Each is too weak to be visible by itself, and in general they are out of phase and destroy each other by interference, but chance will continually be causing a sufficient number to be temporarily in phase with each other to produce a wave train which is sufficiently intense to be seen or recorded.

Since the group of waves forming each train will have different wavelengths, they will soon go out of phase, and the train will take the form of Fig. G.8 (b). The number of visible cycles in the train will depend on how monochromatic the members of the group are. A group comprising all wavelengths between 4000 and 7000 Å will be much reduced in intensity at a distance of a few wavelengths from the maximum, but a laser beam with a spectral bandwidth of 0.01 Å will produce a train of perhaps 10^6 waves whose length will be of the order of 50 cm. All such wave trains (of reasonably monochromatic light) will travel with the group velocity, and that velocity is therefore appropriate to the beginning of any visible pulse.

General references for modulated waves. [355] and [462].

† Each molecule emits a train of waves, and the Doppler effect causes its frequency and wavelength to depend on the momentary velocity of the molecule concerned.

APPENDIX H

SPHERICAL HARMONICS

H.00. Harmonic analysis

If $y = f(x)$ is any ordinary finite and single-valued curve, the $(2n+1)$ constants in the series

$$S = \tfrac{1}{2}b_0 + (b_1 \cos x + b_2 \cos 2x + \ldots + b_n \cos nx) + \\ + (a_1 \sin x + a_2 \sin 2x + \ldots + a_n \sin nx), \quad \text{(H.1)†}$$

or $$S = A_0 + A_1 \cos(x - \alpha_1) + A_2 \cos(2x - \alpha_2) + \ldots + A_n \cos(nx - \alpha_n), \quad \text{(H.2)}$$

can be so chosen as to make S and $f(x)$ identical at $(2n+1)$ values of x between $-\pi$ and $+\pi$, and it may ordinarily be expected that S and $f(x)$ will agree tolerably between these limits, but not necessarily at or beyond them. A series in the form of (H.1) or (H.2) is known as a *Fourier series.*

For example, Fig. H.1 shows the line $y = x$ fairly well fitted by the series

$$y = 2(\sin x - \tfrac{1}{2} \sin 2x + \tfrac{1}{3} \sin 3x - \tfrac{1}{4} \sin 4x) \quad \text{(H.3)}$$

between $\pm 4\pi/5$ with immediate divergence outside that range.

The series S is periodic with 2π as its period, and it is consequently very suitable for representing a periodic function of x. For if the unit of x is chosen so as to make 2π the period of $f(x)$, the good fit wlll persist to all values. For instance, Fig. H.2 shows an unsymmetrical wave which can be represented by the equation

$$y = \cos x - \tfrac{1}{4} \cos(2x - \tfrac{1}{2}\pi), \quad \text{(H.4)}$$

and Fig. H.3 shows an alternation of high and low waves represented by

$$y = \sin x + 2 \sin 2x. \quad \text{(H.5)}$$

To determine the constants $b_0, b_1, \ldots$ and $a_1, a_2, \ldots$, etc., in (H.1), m and n being integers, we have the elementary integrals

$$\int_{-\pi}^{+\pi} \cos mx \cos nx \, dx = 0, \text{ and } \int_{-\pi}^{+\pi} \sin mx \sin nx \, dx = 0, \text{ if } m \neq n,$$

$$\int_{-\pi}^{+\pi} \cos nx \sin mx \, dx = 0, \text{ and } \int_{-\pi}^{\pi} \cos nx \sin nx \, dx = 0, \quad \text{(H.6)}$$

$$\int_{-\pi}^{+\pi} \cos^2 nx \, dx = \int_{-\pi}^{+\pi} \sin^2 nx \, dx = \pi. \quad \text{(H.7)}$$

† The arbitrary $\tfrac{1}{2}$ before b_0 is to make (H.8) of general application.

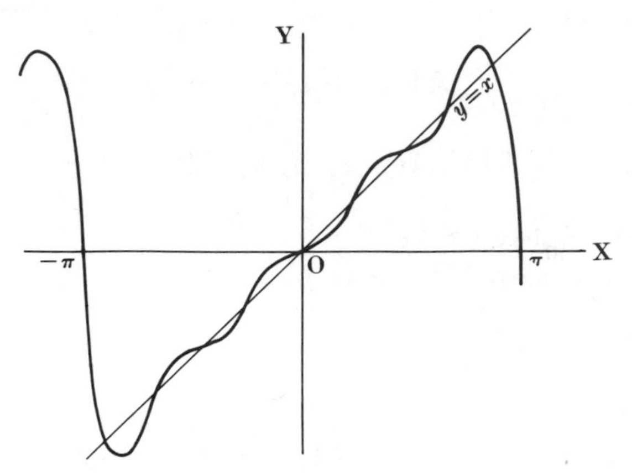

FIG. H.1.

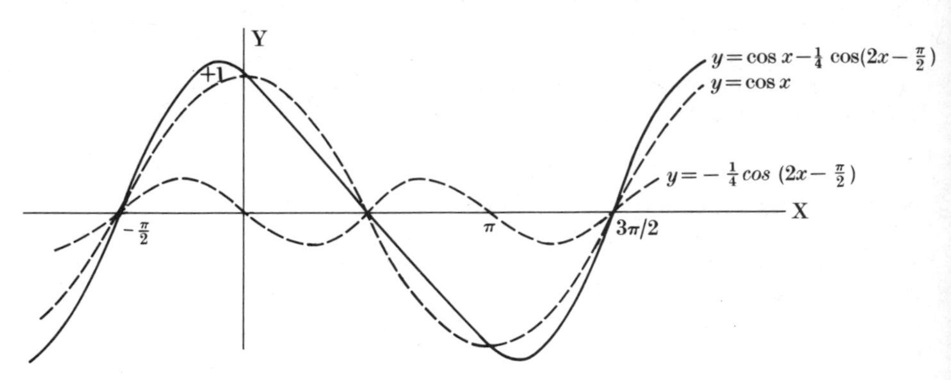

FIG. H.2.

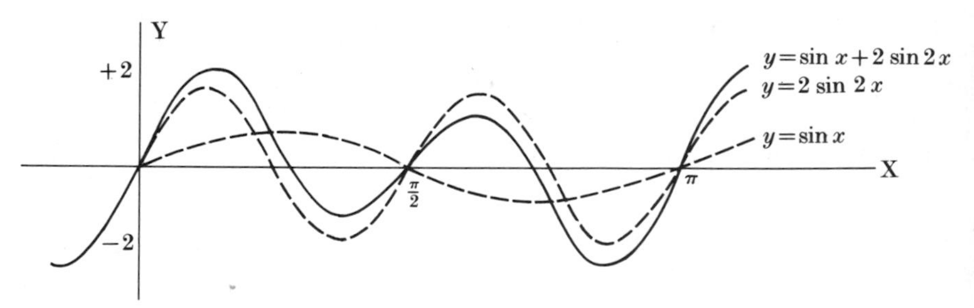

FIG. H.3.

Then multiplying (H.1) by $\cos x$, $\cos 2x, \ldots$, $\sin x$, $\sin 2x, \ldots$, and integrating each product $S \cos x \, dx$, $S \cos 2x \, dx, \ldots$ between $-\pi$ and $+\pi$ in turn, eliminates all but one term at a time and gives

$$b_n = \frac{1}{\pi} \int_{-\pi}^{+\pi} f(x) \cos nx \, dx, \quad a_n = \frac{1}{\pi} \int_{-\pi}^{+\pi} f(x) \sin nx \, dx. \tag{H.8}$$

These expressions can be integrated by quadrature or otherwise, the process being known as *harmonic analysis*. See [140], Chapter II or [396], Chapter I. For a numerical routine see [598], Chapter X.

H.01. Harmonic analysis by least squares

Another method of determining the constants in (H.1) is as follows. It may be required to determine a few terms so as to give the best possible representation of a large number of fallible observations. In Fig. H.4 for instance, it might be required to find the best a's and b's in

$$y = \tfrac{1}{2}b_0 + (b_1 \cos x + b_2 \cos 2x) + (a_1 \sin x + a_2 \sin 2x). \tag{H.9}$$

Then if there are T observations to determine t constants, the observation equations, weighted if necessary, take the form of (H.1) or (H.9), where the a's and b's are the unknowns, and the y's are the fallible observations corresponding to different values of x. In matrix notation

$$Ab = y. \tag{H.10}$$

where the elements of A are the appropriate numerical values of $\cos x$, $\sin x$, $\cos 2x$, etc., and the elements of the vector b are the a's and b's. The normal equations, if all observations are given unit weight, are

$$A^{\mathrm{T}}Ab = A^{\mathrm{T}}y. \tag{H.11}$$

In the matrix $A^{\mathrm{T}}A$ it is clear that the first diagonal term is $\frac{1}{4}T$ and that the other diagonal terms are sums of T values of $\cos^2 nx$ or $\sin^2 nx$. So if the observations have been equally spaced through the range 0 to 2π, these other diagonal terms will all equal $\frac{1}{2}T$, from (H.7). The non-diagonal terms will be sums of products such as those in the integrals of (H.6) and, in the same circumstances, they will be zero. The matrix is consequently diagonal and the a's and b's are each given independently by a single equation. The constant in any harmonic term can be got without solving for the others.

If the observations are not equally spaced, or if weights are unequal, the non-diagonal terms will not be zero, and the normal equations must either be solved in the usual way, or the data can be smoothed to give values at equal spacing.

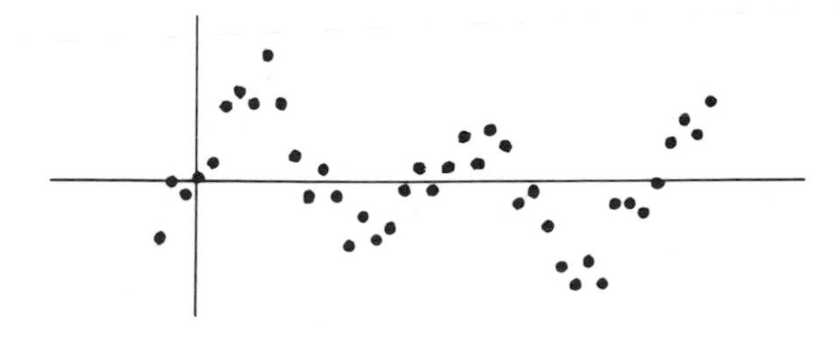

FIG. H.4.

The general use of Fourier series is to give a formal mathematical expression for an asymmetrical or more complex periodic curve, such as may represent an unpure musical note, or a wave in shallow water like that of Fig. H.2. For a more steep-fronted wave than that of Fig. H.2, more terms would be required.

We have so far considered the analysis of a curve into a series of terms whose periods are successively $\frac{1}{2}$, $\frac{1}{3}$, ... times that of the first harmonic term, such terms being known as overtones, overtides, or 2nd, 3rd, etc., harmonics, but such a phenomenon as the tidal wave can only be represented by numerous different terms of incommensurable periods, as explained in Chapter 3, Section 4, the larger of which may have to be represented by a short series like (H.1). For details of the procedure when the periods are given by theory, see the references given in Chapter 3, and for the routine when the periods themselves have to be empirically determined, see [598], Chapter XIII.

Harmonic analysis has little application to geodesy, unless tidal analysis is included in the subject, but its utility in many branches of physics is easily appreciated, and it is a good introduction to the more complicated subject of Spherical Harmonics, which is of great value in connection with the gravitational potential.

H.02. Spherical harmonics

§§ H.00 and H.01 have described the harmonic analysis of a function y which depends on a linear variable x, but since y is generally periodic it could equally well have been represented as a variable radial distance r, depending on an angle θ as in Fig. H.5, where the variations of r are small compared with the radius of the circle. Thus the equation of a nearly circular ellipse whose semi-major axis is a and whose flattening is f is

$$r = a\{(1-\tfrac{3}{8}f^2) - f\cos^2\theta + \tfrac{3}{8}f^2\cos^2 2\theta + \ldots + (f^3)\}$$
$$= a\{(1-\tfrac{1}{2}f-\tfrac{3}{16}f^2) - \tfrac{1}{2}f\cos 2\theta + \tfrac{3}{16}f^2\cos 4\theta + \ldots + (f^3)\}. \tag{H.12}$$

Here the angle θ is measured from the direction of the minor axis.

The use of *spherical harmonics* is to describe variations (of perhaps density or potential) over the surface of a sphere or throughout a volume. Three cases may be considered.

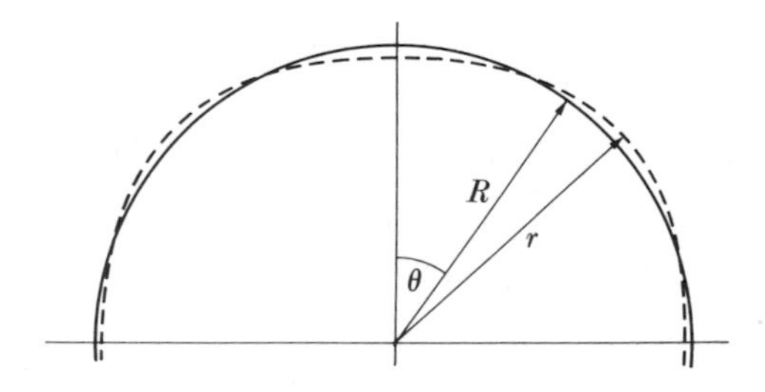

FIG. H.5. The broken line is $r = R - b_4 \cos 4\theta$.

(*a*) An axially symmetrical distribution over a sphere. This will depend only on a single angle θ, and a Fourier series like (H.12) is a possible solution, although in fact not generally the most useful, § H.03.

(*b*) A general distribution over a sphere, involving two angles, θ and λ, §§ H.05–H.08.

(*c*) A distribution through a volume involving r, θ, and λ, § H.09.

H.03. Legendre functions

For an axially symmetrical distribution over a sphere, instead of expanding in terms of $\cos\theta$, $\cos 2\theta$, $\sin\theta$, etc. it is convenient to expand in terms of the functions of θ listed below, writing $\cos\theta = p$, for short. On the axis of symmetry $\theta = 0$.

$$P_0 = 1,$$

$$P_1 = p,$$

$$P_2 = \tfrac{1}{4}(3\cos 2\theta + 1) = \tfrac{1}{2}(3p^2 - 1),$$

$$P_3 = \tfrac{1}{8}(5\cos 3\theta + 3\cos\theta) = \tfrac{1}{2}(5p^3 - 3p),$$

$$P_4 = \tfrac{1}{64}(35\cos 4\theta + 20\cos 2\theta + 9) = \tfrac{1}{8}(35p^4 - 30p^2 + 3).$$

And in general

$$P_n = \frac{1}{2^n}\sum_{k=0}^{k=r}(-1)^k\frac{(2n-2k)!(\cos\theta)^{n-2k}}{k!(n-k)!(n-2k)!}, \tag{H.13}$$

where r is the greatest integer $\leqslant n/2$.

An alternative definition is

$$P_n = \frac{1}{(2^n)n!}\frac{\mathrm{d}^n}{\mathrm{d}p^n}(p^2-1)^n. \tag{H.14}$$

These functions of $\cos\theta$ are known as *Legendre's functions* and are the simplest form of *surface zonal harmonics.* The fact that $\cos\theta$ is the variable is explicitly stated by writing $P_n(\cos\theta)$ or $P_n(p)$, but when nothing else is indicated P_n is taken to mean $P_n(\cos\theta)$. Since $p = \sin(90° - \theta)$, or $\sin\phi'$, ϕ' being geocentric latitude, P_n may also be written $P_n(\sin\phi')$ without change of meaning.

Then any axially symmetrical distribution over a sphere may be represented by a series such as

$$V = c_0P_0 + c_1P_1 + c_2P_2 + \ldots, \tag{H.15}$$

where the numerical coefficients c_0, c_1, etc., have the same physical dimensions as V. See for instance § 6.15. And the equation of an ellipse given in (H.12) can be written

$$r = a\{(1 - \tfrac{1}{3}f - \tfrac{1}{5}f^2) - \tfrac{2}{3}(f + \tfrac{3}{14}f^2)P_2(\cos\theta) + \tfrac{12}{35}f^2P_4(\cos\theta)\} + \ldots, \tag{H.16}$$

as can be directly verified.

The advantages of using Legendre functions instead of $\cos n\theta$ and $\sin n\theta$ are illustrated by the following.

(*a*) While the mean values of $\cos 2\theta$, $\cos 4\theta$, etc., over a circle are zero, their average values over the surface of a sphere are not. On the other hand, except for P_0 the mean value of P_n over a sphere is zero, and see § H.04 for integrals analogous to (H.6) and (H.7).

(*b*) If R' is the distance between two points $(r_1, 0)$ and (R, θ) as in Fig. 6.6(*b*), it is easily verified that

$$\frac{1}{R'} = (R^2 + r_1^2 - 2Rr_1 \cos\theta)^{-\frac{1}{2}} = \frac{1}{R}\left\{1 + P_1\left(\frac{r_1}{R}\right) + P_2\left(\frac{r_1}{R}\right)^2 + \ldots\right\} \text{ if } R > r_1$$

or

$$= \frac{1}{r_1}\left\{1 + P_1\left(\frac{R}{r_1}\right) + P_2\left(\frac{R}{r_1}\right)^2 + \ldots\right\} \text{ if } R < r_1, \tag{H.17}$$

and GM/R' is the potential at $(r_1, 0)$ of the mass M at (R, θ).

In (H.17) if $r_1 = R$, $R' = 2R \sin\frac{1}{2}\theta$, whence $\operatorname{cosec}\frac{1}{2}\theta = 2\sum_0^\infty P_n$. (H.18)

(*c*) The solid harmonics $r^n P_n(\cos\theta)$ are all solutions of Laplace's equation (H.67).

H.04. Properties of Legendre's functions

Fig. H.6 shows curves representing the first four harmonics. [174] gives tables for the first seven. $P_n = 1$ when $\theta = 0$, and $+1$ or -1 when $\theta = \pi$. It is zero at n slightly unequally spaced values of θ between 0 and π. If Fig. H.6 had been plotted with $\cos\theta$ instead of θ linearly spaced along OX the gradients at 0 and 1 would heve been steeper.

Fig. H.6 shows that

$$\int_0^\pi P_n(\cos\theta)\, d\theta \text{ is not zero, unless } n \text{ is odd.} \tag{H.19}$$

On the other hand, integrating over a sphere, $d\omega$ being an element of solid angle,

$$\int_S P_n(\cos\theta)\, d\omega = R^2 \int_0^{2\pi}\int_0^\pi P_n(\cos\theta)\sin\theta\, d\theta\, d\lambda$$

$$= 2\pi R^2 \int_{-1}^{+1} P_n(\cos\theta)\, d(\cos\theta), \tag{H.20}$$

which is zero unless $n = 0$.

Resembling (H.6) and (H.7),

$$\int_{-1}^{+1} P_n(p) P_t(p)\, dp = 0 \text{ if } t \neq n, \tag{H.21}$$

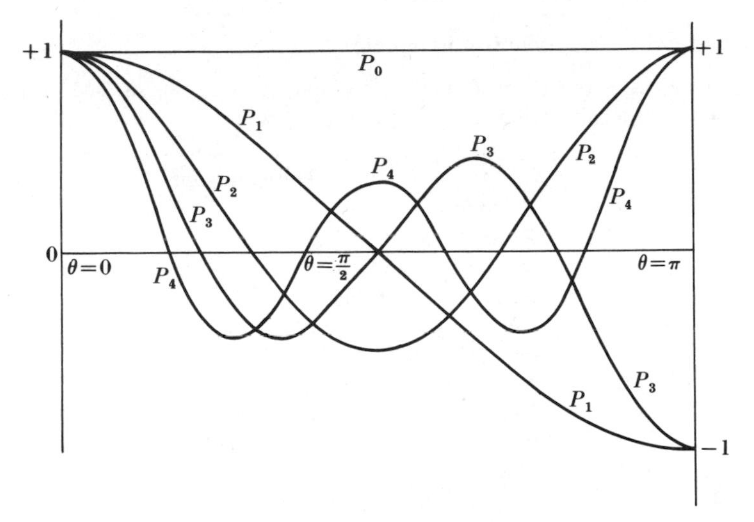

FIG. H.6. Legendre functions, P_1 to P_4.

where $dp = d(\cos\theta) = -\sin\theta\, d\theta$, which includes (H.20) since $P_t(p)$ includes $P_0(p)$ which is unity, and

$$\int_{-1}^{+1} \{P_n(p)\}^2\, dp = \frac{2}{2n+1} \tag{H.22}$$

or

$$\int_S \{P_n(p)\}^2\, d\omega = \frac{4\pi}{2n+1}. \tag{H.23}$$

For proof of (H.22) and (H.23) see [140], pp. 168–70, or other textbooks.

It follows from (H.23) that the r.m.s. value of any Legendre function over the sphere is $(2n+1)^{-\frac{1}{2}}$.

Trigonometrical functions can be expressed in terms of Legendre functions, thus

$$\begin{aligned}
&\cos\theta = P_1 && \cos^2\theta = \tfrac{2}{3}P_2 + \tfrac{1}{3}P_0 && \sin^2\theta = -\tfrac{2}{3}P_2 + \tfrac{2}{3}P_0\\
&\cos 2\theta = \tfrac{4}{3}P_2 - \tfrac{1}{3}P_0 && \cos^3\theta = \tfrac{2}{5}P_3 + \tfrac{3}{5}P_1 && \sin^4\theta = \tfrac{8}{35}P_4 - \tfrac{16}{21}P_2 + \tfrac{8}{15}P_0\\
&\cos 3\theta = \tfrac{8}{5}P_3 - \tfrac{3}{5}P_1 && \cos^4\theta = \tfrac{8}{35}P_4 + \tfrac{4}{7}P_2 + \tfrac{1}{5}P_0. && \sin^2 2\theta = -\tfrac{32}{35}P_4 + \tfrac{8}{21}P_2 + \tfrac{8}{15}P_0\\
&\cos 4\theta = \tfrac{64}{35}P_4 - \tfrac{16}{21}P_2 - \tfrac{1}{15}P_0.
\end{aligned} \tag{H.24}$$

Consecutive Legendre functions are connected by the relation

$$(2n+1)P_n = \frac{d}{dp}P_{n+1} - \frac{d}{dp}P_{n-1}$$

or

$$(n+1)P_{n+1} = (2n+1)pP_n - nP_{n-1}. \tag{H.25}$$

H.05. General surface spherical harmonic

An expansion to represent a general distribution over the surface of a sphere can be made in terms of the following expressions.

Degree	*Order*	*Harmonic term*
0	0	Constant
1	0	P_1 or $\cos\theta$
	1	$(\cos\lambda \text{ or } \sin\lambda)\times\sin\theta$
2	0	P_2 or $\frac{1}{2}(3\cos^2\theta-1)$
	1	$(\cos\lambda \text{ or } \sin\lambda)\times 3\sin\theta\cos\theta$
	2	$(\cos 2\lambda \text{ or } \sin 2\lambda)\times 3\sin^2\theta$
3	0	P_3 or $\frac{1}{2}(5\cos^3\theta-3\cos\theta)$
	1	$(\cos\lambda \text{ or } \sin\lambda)\times\frac{3}{2}\sin\theta(5\cos^2\theta-1)$
	2	$(\cos 2\lambda \text{ or } \sin 2\lambda)\times 15\sin^2\theta\cos\theta$
	3	$(\cos 3\lambda \text{ or } \sin 3\lambda)\times 15\sin^3\theta$
4	0	P_4 or $\frac{1}{8}(35\cos^4\theta-30\cos^2\theta+3)$
	1	$(\cos\lambda \text{ or } \sin\lambda)\times\frac{5}{2}\sin\theta(7\cos^3\theta-3\cos\theta)$
	2	$(\cos 2\lambda \text{ or } \sin 2\lambda)\times\frac{15}{2}\sin^2\theta(7\cos^2\theta-1)$
	3	$(\cos 3\lambda \text{ or } \sin 3\lambda)\times 105\sin^3\theta\cos\theta$
	4	$(\cos 4\lambda \text{ or } \sin 4\lambda)\times 105\sin^4\theta$, etc. (H.26)

The general expression, which the above are easily seen to satisfy, is

$$(\cos m\lambda \text{ or } \sin m\lambda)\sin^m\theta\frac{\mathrm{d}^m}{\mathrm{d}p^m}P_n, \text{ where } p\equiv\cos\theta, \tag{H.27}$$

in which n is known as the *degree* and m the *order* of each term, and $m\leqslant n$.†

λ is zero on some defined plane containing the axis $\theta=0$, such as the Greenwich meridian, or the XOZ plane if rectangular coordinates are in use.

These functions are known as *tesseral harmonics*‡, except those of order 0 which are the Legendre functions of § H.03. Those whose order m equals their degree n are known as *sectorial harmonics*. The reasons for their use are the same as those given for Legendre functions in § H.03 (a) and (c).

The lines on which surface spherical harmonics are zero divide the sphere into 'squares' or tesserae containing alternate positive and negative values, there being $(n-m)$ such parallels unequally spaced between $\theta=0$ and $\theta=\pi$, and $2m$ equally spaced meridians running from pole to pole, as in Fig. H.7(a)–(c), where shaded and unshaded areas are of opposite sign.

The part of the expression for a tesseral or sectorial harmonic which depends only on θ may be written as P_{nm}.‖

$$P_{nm}=\sin^m\theta(\mathrm{d}/\mathrm{d}p)^m P_n=\frac{\sin^m\theta}{(2^n)n!}(\mathrm{d}/\mathrm{d}p)^{m+n}(p^2-1)^n. \tag{H.28}$$

† In Chapter 7 the degree is referred to as l instead of n, since n is preoccupied.

‡ An older name is *Associated Legendre functions*.

‖ The notation has been variable. P_{nm} is in current use by most writers on geodetic subjects, see [262] and [391], but P_n^m has also been used. See also § H.06.

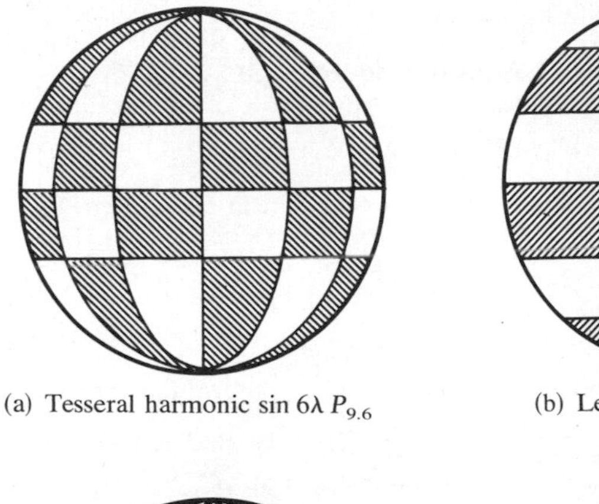

(a) Tesseral harmonic $\sin 6\lambda\, P_{9.6}$

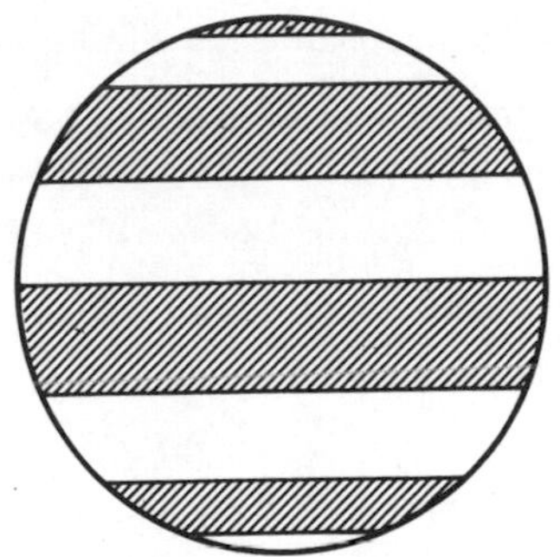

(b) Legendre function P_7

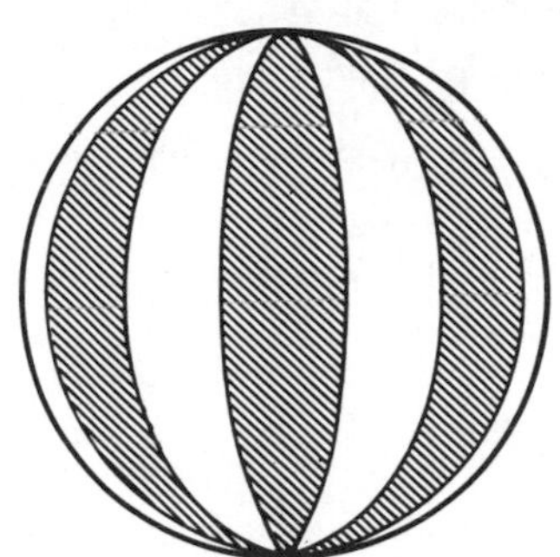

(c) Sectional harmonic $\sin 7\lambda\, P_{7.7}$

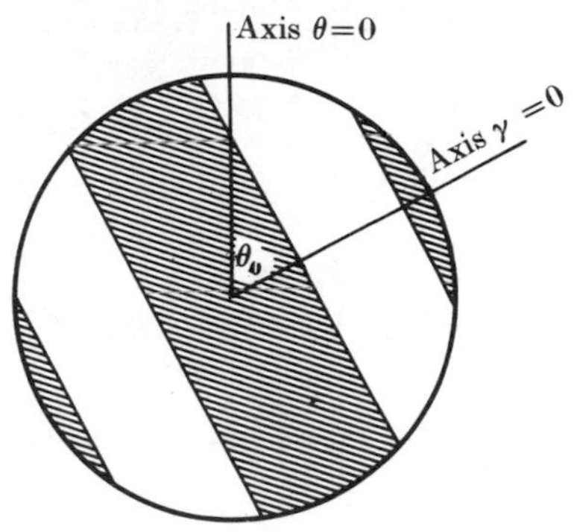

(d) Zonal harmonic $P_4(\cos\gamma)$

FIG. H.7. In (a), (b), and (c) the axis of θ is as shown in (d).

So the full harmonic terms as listed in (H.26) are

$$P_{nm}\cos m\lambda \quad \text{and} \quad P_{nm}\sin m\lambda \tag{H.29}$$

An alternative definition, [262], pp. 24–5, is

$$P_{nm}(\cos\theta)=\frac{1}{2^n}\sin^m\theta\sum_{k=0}^{r}(-1)^k\frac{(2n-2k)!}{k!(n-k)!(n-m-2k)!}(\cos\theta)^{n-m-2k}, \tag{H.30}$$

where r is the greatest integer $\leqslant\frac{1}{2}(n-m)$. This incorporates (H.13).

Recurrence relations similar to (H.25) exist between tesseral harmonics and the harmonics of adjacent degree and order. These are quoted in § 7.57 as (7.204)–(7.206).

The *general surface harmonic* of degree n can be written Y_n, and contains $(2n+1)$ numerical coefficients.

$$Y_n=\sum_{m=0}^{n}(C_{nm}\cos m\lambda+S_{nm}\sin m\lambda)P_{nm}. \tag{H.31}$$

As an example

$$Y_2 = c_2P_2 + 3C_{21}\cos\lambda\sin\theta\cos\theta + 3S_{21}\sin\lambda\sin\theta\cos\theta + 3C_{22}\cos 2\lambda\sin^2\theta + 3S_{22}\sin 2\lambda\sin^2\theta,$$

$$= c_2P_2 + (C_{21}\cos\lambda + S_{21}\sin\lambda)P_{21} + (C_{22}\cos 2\lambda + S_{22}\sin 2\lambda)P_{22} \quad \text{(H.32)†}$$

The single harmonic term of degree n and order m can be written Y_{nm}. For example

$$Y_{20} = c_2P_2$$

$$Y_{22} = (C_{22}\cos 2\lambda + S_{22}\sin 2\lambda)P_{22}. \quad \text{(H33)}$$

And if $C_{nm} = 1$ and $S_{mn} = 0$, $Y_{nm} = P_{nm}\cos m\lambda$, as in (H.26) and (H.29)

Every Y_{nm} $(m \neq 0)$ contains two numerical coefficients C_{nm} and S_{nm}. An alternative form is to write $A_{nm} = (C^2_{nm} + S^2_{nm})^{\frac{1}{2}}$ and $\tan m\lambda_{nm} = S_{nm}/C_{nm}$, giving

$$Y_{nm} = A_{nm}\cos m(\lambda - \lambda_{nm})P_{nm}. \quad \text{(H.34)}$$

Note that while Y_{n0}, Y_{nm} and Y_n contain numerical coefficients, P_n and P_{nm} do not.

Comparable with (H.21) and (H.22), for integrals over the whole surface of the sphere, Y_n and Y_s being of different degrees,

$$\int_S Y_nY_s\,d\omega = 0, \text{ including} \int_S Y_nP_s\,d\omega = 0 \quad \text{and} \quad \int_S Y_n\,d\omega = 0, \quad \text{(H.35)}$$

unless both n and s, or n only (in the last) $= 0$. And if $Y_n(\cos\theta)$ and $Y'_n(\cos\theta)$ are two general surface harmonics of the same degree, in which c'_n, C'_{nm}, and S'_{nm} are the coefficients in Y',

$$\int_S Y_nY'_n\,d\omega = \frac{4\pi}{2n+1}c_nc'_n + \frac{2\pi}{2n+1}\left\{\frac{(n+1)!}{(n-1)!}(C_{n1}C'_{n1} + S_{n1}S'_{n1}) + \dots\right.$$

$$\left.\dots + \frac{(n+m)!}{(n-m)!}(C_{nm}C'_{nm} + S_{nm}S'_{nm}) + \dots + (2n)!(C_{nn}C'_{nn} + S_{nn}S'_{nn})\right\}. \quad \text{(H.36)}$$

This equals zero unless there is at least one term for which both Y_n and Y'_n have a non-zero coefficient. For proof see [202], pp. 84–7. See also [151], pp. 610–11.

It follows from (H.36) that

$$\int_S Y_{nm}Y_{nt}\,d\omega = 0, \; m \neq t. \quad \text{(H.37)}$$

† See (6.62) and (6.99). When reference is to the earth's potential, the coefficient of P_n written as c_n above, has generally been written as $-J_n$. See also § H.06. It may also be written as C_{no} to conform to the other terms of (H.31).

This includes

$$\int_S Y_{nm} P_n \, d\omega = 0, \tag{H.38}$$

unless $m = 0$, when it is

$$4\pi c_n/(2n+1), \text{ from (H.23).} \tag{H.39}$$

And if Y_{nm} and Y'_{nm} have coefficients C, S, and C', S' respectively,

$$\int_S Y_{nm} Y'_{nm} \, d\omega = \frac{2\pi}{2n+1} \frac{(n+m)!}{(n-m)!} (CC' + SS') \text{ in general,} \tag{H.40}$$

or

$$= \frac{2\pi}{2n+1} (2n)!(CC' + SS') \quad \text{if } m = n \neq 0, \tag{H.41†}$$

$$= \frac{4\pi}{2n+1} c_n c'_n \quad \text{if } m = 0, \tag{H.42}$$

and

$$\int_S (Y_{nm})^2 \, d\omega = \frac{2\pi}{2n+1} \frac{(n+m)!}{(n-m)!} (C^2 + S^2) \quad \text{if } m \neq 0. \tag{H.43}$$

H.07. Normalization

Formula (H.43) shows that when n and m are large, the r.m.s. value of Y_{nm} over the sphere is $(C_{nm}^2 + S_{nm}^2)^{\frac{1}{2}}$ multiplied by a very large factor, namely

$$\left\{ \frac{1}{2(2n+1)} \frac{(n+m)!}{(n-m)!} \right\}^{\frac{1}{2}} \tag{H.44}$$

If $n = m = 10$, this of the order of 10^8. This is inconvenient, since if the value of the phenomenon represented by Y_{nm} is of the order 1 or 10, C_{nm} and S_{nm} will be of the order 10^{-7}. It is therefore often convenient to replace P_{nm} as defined by (H.28) by

$$\bar{P}_{nm} = \left\{ 2(2n+1) \frac{(n-m)!}{(n+m)!} \right\}^{\frac{1}{2}} P_{nm}, \text{ if } m \neq 0, \tag{H.45}$$

and to write the corresponding P_{nm}, C_{nm} and S_{nm} as $\bar{P}_{nm}$, $\bar{C}_{nm}$, and $\bar{S}_{nm}$. They are then described as *fully normalized*.‡ It follows that

$$\bar{C}_{nm} = C_{nm} \times \text{(H.44) and } \bar{S}_{nm} \times \text{(H.44), if } m \neq 0. \tag{H.46}$$

† $(n+m)! \div (n-m)!$ indicates $(n+m)!$ with the first $(n-m)$ terms omitted. So if $(n-m)$ equals 0 or 1, $(n+m) \div (n-m)$ is equal to $(n+m)$, since the omission of either no terms, or of the first term which is unity, makes no change.

‡ The notation has been variable. The symbols P_{nm} and P_n^m have been used for $\{2(n-m)!/(n+m)!\}^{\frac{1}{2}} P_{nm}$, and the resulting functions have been described as *normalized*. This is now seldom used, but it is still necessary to avoid confusion by using the words *fully* normalized for (H.45) and (H.46). The notation here given seems to be now most usually accepted, in connection with the earth's potential field.

An alternative way of writing (H.45) is

$$\bar{p}_{nm} = \left\{\frac{(2-\delta_{0m})(2n+1)(n-m)!}{(n+m)!}\right\}^{\frac{1}{2}} P_{nm}, \tag{H.47}$$

where δ_{0m} is the Kroneck delta, which equals 1 if $m=0$, and zero if $m \neq 0$. (H.47) thus includes the Legendre functions.

It follows that

$$\int_S (Y_{nm})^2 \, d\omega = 4\pi(\bar{C}^2_{nm}+\bar{S}^2_{nm}), \quad \text{if} \quad m \neq 0, \tag{H.48}$$

which includes

$$\int_S (\bar{P}_n)^2 \, d\omega = 4\pi, \text{ in which } m=0, \tag{H.49}$$

and the r.m.s. value of Y_{nm} $(m \neq 0)$ is $(\bar{C}^2_{nm}+\bar{S}^2_{nm})^{\frac{1}{2}}$. So the magnitudes of these normalized coefficients reflect the magnitude of the phenomenon they are describing.

Legendre functions $(m=0)$ are less generally normalized, although there is now a tendency towards it. The r.m.s. value of P_n is $(2n+1)^{-\frac{1}{2}}$, (H.23), so its normalized value is

$$\bar{P}_n = P_n(2n+1)^{\frac{1}{2}}. \tag{H.50}$$

its numerical coefficient may then be written as $\bar{C}_{no}$, and

$$\bar{C}_{n0} = c_n(2n+1)^{-\frac{1}{2}} \text{ or } -J_n(2n+1).^{-\frac{1}{2}}. \tag{H.51}$$

Y_{n0}, Y_{nm}, and Y_n are functions of θ and λ which also involve numerical coefficients C_{nm},S_{nm}, or c_n. P_n and P_{nm} are functions of θ or of θ and λ, which involve no numerical coefficients. P_n, P_{nm}, C_{nm}, S_{nm}, and c_n can be normalized, but their the products which form Y_{n0}, Y_{nm}, Y_n, and c_nP_n are unaffected by normalization.

H.07. Spherical harmonic analysis

If the value of some variable $f(\theta, \lambda)$, such as the height of the geoid above the spheroid, is known at a large number of points at well-distributed† values of θ and λ, the form of the surface can be expressed by a number of spherical harmonic terms in exactly the same way as described for a circular periodic function in §§ H.00 and H.01. If the variation is axially symmetrical, the expansion will be in terms of Legendre functions only, but otherwise it will be in terms of the general surface harmonics Y_n, and it will be more elaborate.

Since (H.21)–(H.23) and (H.37)–(H.43) have the same properties as (H.6) and (H.7), the methods of §§ H.00 and H.01 can be used for computing the numerical coefficients.

Then if $f(\theta, \lambda)$ is to be expressed in the form

$$f(\theta, \lambda) = \sum_{n=0}^{n} \sum_{m=0}^{n} \bar{C}_{nm}\bar{P}_{nm} \cos m\lambda + \bar{S}_{nm}\bar{P}_{nm} \sin m\lambda, \tag{H.52}$$

† The distribution of the points must be such that the integrals (H.54) can be adequately estimated.

$\bar{C}_{nm}$ is got by multiplying both sides by $\bar{P}_{nm} \cos m\lambda$, integrating over the sphere, and using (H.37) and (H.48).

$$\bar{C}_{nm}\int_S (\bar{P}_{nm}\cos m\lambda)^2\,\mathrm{d}\omega + \text{Zero terms} = \int_S f(\theta,\lambda)\bar{P}_{nm}\cos m\lambda\,\mathrm{d}\omega. \tag{H.53}$$

Then

$$\bar{C}_{nm} = \frac{1}{4\pi}\int_S f(\theta,\lambda)\bar{P}_{nm}\cos m\lambda\,\mathrm{d}\omega,$$

and similarly

$$\bar{S}_{nm} = \frac{1}{4\pi}\int_S f(\theta,\lambda)\bar{P}_{nm}\sin m\lambda\,\mathrm{d}\omega, \tag{H.54}$$

in which the right-hand sides can be integrated by quadratures.

H.08. Change of axis

A variable quantity $f(\theta, \lambda)$ on the surface of a sphere, such as an anomaly of the potential on its surface, may be represented by a series of spherical surface harmonics

$$f(\theta,\lambda) = \sum_{n=2}^{n}\sum_{m=0}^{m=n}(\bar{C}_{nm}\cos m\lambda + \bar{S}_{nm}\sin m\lambda)\bar{P}_{nm}(\cos\theta) = \sum_{2}^{n} \mathrm{Y}_n(\cos\theta), \tag{H.55}$$

in which the axis of $\theta = 0$ is OZ in Fig. H.8. It may be required to make an harmonic analysis of this distribution in the form of a series of zonal harmonics†

$$f(\theta,\lambda) = \sum_{2}^{n} \bar{c}P_n(\cos\gamma) \tag{H.56}$$

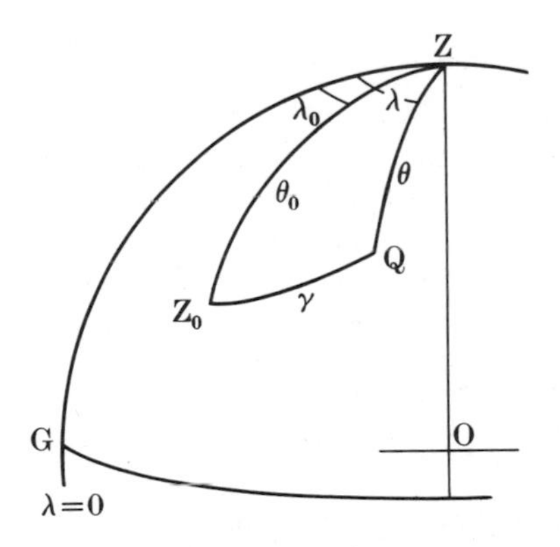

FIG. H.8. Change of axis from Z to Z_0.

† A series of zonal harmonics will be a poor representation of the original analysis which included tesserals, but it may be what is required, as for example in § 6.15.

in which the axis of $\gamma=0$ is OZ_0, where $ZZ_0=\theta_0$ and $GZZ_0=\lambda_0$.

We have

$$\cos\gamma=\cos\theta\cos\theta_0+\sin\theta\sin\theta_0\cos(\lambda-\lambda_0). \tag{H.57}$$

Then, [140], p. 211

$$P_n(\cos\gamma)=P_n(\cos\theta)P_n(\cos\theta_0)+2\sum_{m=1}^{n}\frac{(m-n)!}{(m+n)!}P_{nm}(\cos\theta_0)P_{nm}(\cos\theta)\cos m(\lambda-\lambda_0), \tag{H.58}$$

or, using fully normalized functions, and separating $\cos\lambda$ and $\sin\lambda$

$$\bar{P}_n(\cos\gamma)=(2n+1)^{-\frac{1}{2}}\sum_{m=0}^{n}\bar{P}_{nm}(\cos\theta)\bar{P}_{nm}(\cos\theta_0)(\cos m\lambda_0\cos m\lambda+\sin m\lambda_0\sin m\lambda). \tag{H.59}$$

Comparably with (H.38) and (H.39),

$$\int_S Y_n(\cos\theta)\bar{P}_s(\cos\gamma)\,d\omega=0,\text{ unless } s=n. \tag{H.60}$$

$$\int_S Y_n(\cos\theta)\bar{P}_n(\cos\gamma)\,d\omega=4\pi(2n+1)^{-\frac{1}{2}}Y'_n, \tag{H.61}$$

where Y'_n is the value of $Y_n(\cos\theta)$ at (θ_0, λ_0).

And if P_n is not normalized,

$$\int_S Y_n(\cos\theta)P_n(\cos\gamma)\,d\omega=4\pi(2n+1)^{-1}Y'_n. \tag{H.62}$$

To obtain the values of the coefficients $\bar{c}_n$ in (H.56), multiply the right-hand sides of both (H.55) and (H.56) by $\bar{P}_n(\cos\gamma)$, integrate over the sphere, and equate the two. We then have

$$\int_S \bar{c}_n\left\{\bar{P}_n(\cos\gamma)\right\}^2 d\omega=\int_S Y_n(\cos\theta)P_n(\cos\gamma)\,d\omega+\text{Zero terms, as in (H.60)}$$

$$4\pi\bar{c}_n=4\pi(2n+1)^{-\frac{1}{2}}Y'_n,\text{ as in (H.61)}$$

$$\bar{c}_n=Y'_n(2n+1)^{-\frac{1}{2}} \tag{H.63}$$

Then

$$\sum_{2}^{n}\bar{c}_n\bar{P}_n(\cos\gamma)=\sum_{2}^{n}Y'_n(2n+1)^{-\frac{1}{2}}\bar{P}_n(\cos\gamma), \tag{H.64}$$

using normalized P's. Or using unnormalized P's

$$\sum_{2}^{n}\bar{c}_nP_n(\cos\gamma)=\sum_{2}^{n}Y'_nP_n(\cos\gamma),\text{ from (H.50),} \tag{H.65}$$

where Y'_n is the value of Y_n $(\cos\theta)$ at (θ_0, λ_0).

H.09. Solid spherical harmonics

A distribution through three-dimensional space can be represented by a series of terms

$$r^n Y_n \text{ or } r^{-(n+1)} Y_n, \tag{H.66}$$

where r is distance from the origin and Y_n is a function of θ and λ as defined in § H.05. Such expressions are *solid spherical harmonics.* For example, (7.1) expresses the earth's potential at an external point in this form. On the surface of a sphere with $r = \text{constant}$, solid harmonics become surface harmonics.

If $V = \sum r^n Y_n$ or $\sum r^{-(n+1)} Y_n$, V is a solution of Laplace's equation $\nabla^2 V = 0$, in its polar form

$$\frac{1}{r^2}\left\{\frac{\partial}{\partial r}\left(\frac{r^2 \partial V}{\partial r}\right) + \frac{1}{\sin\theta}\frac{\partial}{\partial\theta}\left(\sin\theta\frac{\partial V}{\partial\theta}\right) + \frac{1}{\sin^2\theta}\frac{\partial^2 V}{\partial\lambda^2}\right\} = 0, \tag{H.67}$$

as can easily be verified for any low degree term as given in (H.26).

The subject of spherical harmonics can be, and conventionally is, approached by defining solid harmonics as homogeneous solutions of (H.67) and surface harmonics are then the simpler forms derived from them when r is constant.

General references to spherical harmonics. [140], [151], pp. 606–26, [202], [396], and [502].

APPENDIX I

ROTATING AXES. CORIOLIS FORCE

I.00. Rotating axes

NEWTON's law of motion, expressed as Force = Mass × Acceleration, only purports to be true if the acceleration is measured with reference to a *Newtonian* or *stellar* frame, namely a set of axes which neither linearly accelerates with reference to the centre of mass of the solar system, nor rotates with reference to the directions defined by the general body of distant stars.

Working on the surface of the earth, it is often convenient to uses axes which are fixed to the earth and rotate with it. The earth's linear accelerations towards the moon and sun produce tidal effects which are described in §§ 3.23–3.25 and 6.76–6.84. This appendix deals with rotating axes.

I.01. Centrifugal force

Fig. I.1 shows a point mass M moving on a smooth horizontal plane in a circle with angular velocity Ω relative to fixed axes OX and OY. Its acceleration relative to them is $\Omega^2 r$ inwards along MO, and the only horizontal force acting on it is tension in the string MO, which is $M\Omega^2 r$ towards O. Newton's law is satisfied.

Now let axes Ox and Oy rotate with the same velocity Ω. Relative to them the acceleration of M is zero, the tension MO remains the same as before, and the law is not satisfied. Although nothing is wrong, this is an inconvenience.

As is well known, the remedy is to introduce a fictitious *centrifugal force* $M\Omega^2 r$ acting outwards along OM. The total force is then zero, the acceleration relative to the coordinate axes is zero, and the law is satisfied. Two points require emphasis as below.

(*a*) The centrifugal force only arises if the reference frame rotates relative to the stellar frame, and its amount depends on the angular velocity of the frame.

(*b*) It only suffices to maintain Newton's law, if the masses concerned are stationary relative to the rotating axes. Masses which are moving relative to these axes require a further fictitious force, the *Coriolis force*, as in § I.02.

I.02. Coriolis force

In Fig. I.2 axes OX and OY are fixed in space, while Ox rotates with angular velocity Ω. A mass M is initially at A. It is moving, and is constrained to continue moving with constant velocity v relative to A along a straight line AB which rotates with Ox. For example, Ox and AB might be fixed to a horizontal rotating table, and M might be a miniature railway truck moving along a rail AB. Neglecting friction and vertical forces, the forces on M can be

(i) the pull of the engine or brake, forwards or backwards, along AB,
and (ii) lateral force from the rails, if necessary.

The acceleration relative to Ox and Oy has been stated to be zero. We have to find the accelerations relative to OX and OY. Let t be time, and let $\mathbf{R}$ be the position vector which locates M relative to OX, and let $\mathbf{V}$ be the velocity vector relative to OX. Then in vector notation, Appendix E,

$$\frac{d\mathbf{R}}{dt} = (\boldsymbol{\Omega} \times \mathbf{R}) + \mathbf{V}, \text{ as in Fig. I.2,} \tag{I.1}$$

$$\frac{d^2\mathbf{R}}{dt^2} = \left(\frac{d\boldsymbol{\Omega}}{dt} \times \mathbf{R}\right) + \left(\boldsymbol{\Omega} \times \frac{d\mathbf{R}}{dt}\right) + \frac{d\mathbf{V}}{dt}, \text{ using (E.20).}$$

The first bracket is zero. In the second bracket $d\mathbf{R}/dt$ is as in (I.1). So

$$\frac{d^2\mathbf{R}}{dt^2} = \boldsymbol{\Omega} \times (\boldsymbol{\Omega} \times \mathbf{R}) + (\boldsymbol{\Omega} \times \mathbf{V}) + (\boldsymbol{\Omega} \times \mathbf{V}), \text{ since } d\mathbf{V}/dt = (\boldsymbol{\Omega} \times \mathbf{V})$$

$$= (\boldsymbol{\Omega} \cdot \mathbf{R})\boldsymbol{\Omega} - (\boldsymbol{\Omega} \cdot \boldsymbol{\Omega})\mathbf{R} + 2(\boldsymbol{\Omega} \times \mathbf{V}), \text{ using (E.14).}$$

The first term is zero since $\boldsymbol{\Omega}$ and $\mathbf{R}$ are perpendicular.

So
$$\frac{d^2\mathbf{R}}{dt^2} = -|\boldsymbol{\Omega}|^2\mathbf{R} + 2(\boldsymbol{\Omega} \times \mathbf{V}). \tag{I.2}$$

In (I.2) the first term is a centripetal acceleration of $\Omega^2 R$ inwards, and the second term is an acceleration of $2\Omega v$ across $\mathbf{V}$ from right to left in Fig. I.2, since $\boldsymbol{\Omega}$ (up from the paper) is perpendicular to $\mathbf{V}$. This acceleration $2(\boldsymbol{\Omega} \times \mathbf{V})$ is known as the *Coriolis acceleration.*

Then to preserve Newton's law, so that zero acceleration relative to Ox is produced by zero force, the inward pull of the engine and the lateral pressure of the rails must be cancelled by the following two fictitious forces.

(a) A centrifugal force $M\Omega^2 R$ outwards, as before.

(b) A Coriolis force $2M\Omega v$ across AB from left to right in Fig. I.2, its sign being that of the vector product $(\mathbf{V} \times \boldsymbol{\Omega})$, as in § E.08. (I.4)

Put otherwise, if $\dot{x}$ and $\dot{y}$ are the components of v, the components of the Coriolis force along Ox and Oy are

$$2M\dot{y}\Omega \quad \text{and} \quad -2M\dot{x}\Omega \tag{I.5}$$

And those of the centrifugal force are

$$Mx\Omega^2 \quad \text{and} \quad My\Omega^2. \tag{I.6}$$

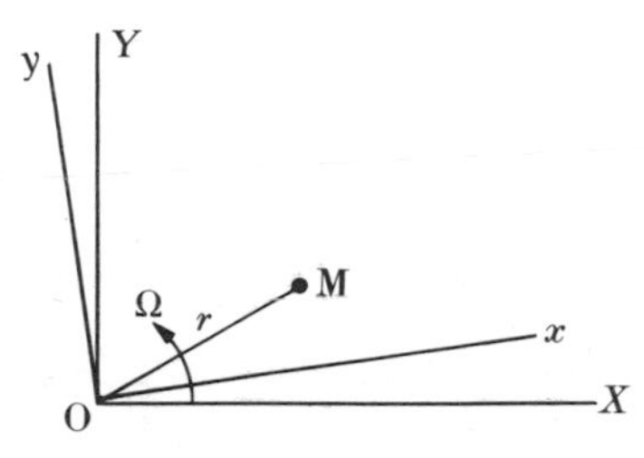

FIG. I.1 OX, OY are fixed axes. Ox, Oy rotate with velocity Ω.

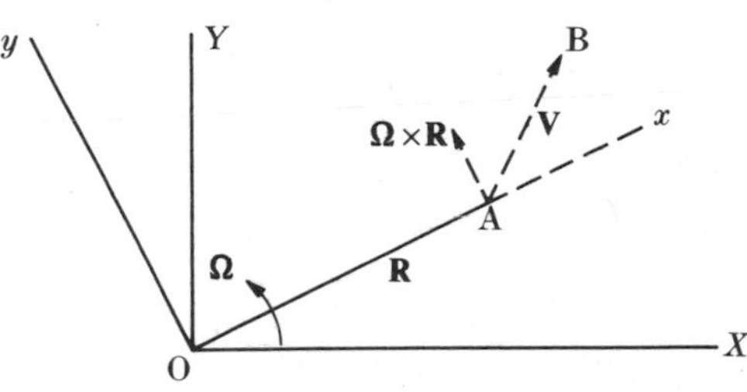

FIG. I.2. Axes OX and OY are fixed. Ox rotates with angular velocity Ω. The vector Ω is upwards out of the paper.

I.03. Coriolis force on the earth. The Eötvös gravity correction

When gravity is measured by pendulum or gravimeter on a moving vehicle, such as a ship or an aircraft, moving at constant height with velocity v in azimuth α (clockwise from north), the value deduced from the readings of the apparatus is the vector of the earth's attraction and of three fictitious forces (a), (b), and (c) below.

(a) The centrifugal force appropriate to axes rotating with the earth, namely $M\omega^2 \nu \cos\phi$ acting outwards from the axis of rotation, in a plane parallel to the equator. The vertical component is $m\omega^2 \nu \cos^2\phi$ upwards. The corresponding acceleration is conventionally included in the definition of g.

(b) The Coriolis force. To find the magnitude of its vertical component, the velocity of the vehicle may be resolved into $v \sin\alpha$ eastwards and $v \cos\alpha$ northwards, both in the horizontal plane.

The first component lies in the plane of the parallel ϕ, and demands a Coriolis force of $2M\omega v \sin\alpha$ acting radially outwards in this plane. This may be resolved into a vertical component, acting outwards, of

$$2M\omega v \sin\alpha \cos\phi, \tag{I.7}$$

and a horizontal southward component of

$$2M\omega v \sin\alpha \sin\phi. \tag{I.8}$$

The second component of v may be resolved into $v \cos\alpha \cos\phi$ parallel to the axis of rotation, from which no Coriolis force results, and $v \cos\alpha \sin\phi$ radial in the plane of parallel ϕ. This demands a horizontal eastward Coriolis force of

$$2M\omega v \cos\alpha \sin\phi. \tag{I.9}$$

Combining (I.8) and (I.9) gives a total horizontal Coriolis force of

$$2M\omega v \sin\phi \tag{I.10}$$

in the tangent plane, acting from left to right across v, together with a vertical force of

$$2M\omega v \sin\alpha \cos\phi \tag{I.11}$$

acting outwards if $\sin\alpha$ is positive. In the southern hemisphere ϕ is negative, so the horizontal force acts from right to left, but the sign of the vertical component is not affected. The rotation is measured in radians per second, so for the earth it is 2π in 86 400 sidereal seconds, and in figures the corresponding outward acceleration (I.11) is

$$2\omega v \sin\alpha \cos\phi = 4.0 v \sin\alpha \cos\phi \text{ mGal}, \tag{I.12}$$

where v is in kilometres per hour, and α is clockwise from north.

(c) If the vehicle maintains a constant height above sea-level, it is moving in a circular arc. So recorded g is diminished by

$$v^2/R_\alpha = 0.0012 v^2 \text{ mGal, if } v \text{ is in kilometres per hour,} \tag{I.13}$$

where R_α is the radius of curvature of the spheroid in latitude ϕ and azimuth α, as given by (A.67), increased by the height of the vehicle above sea-level, if material. On a ship (I.13) is generally quite negligible, but on an aircraft it is important. On a free falling satellite moving at (say) 8 km/s it equals g.

The accelerations (I.12) and (I.13) have to be reversed and added to g as recorded on moving vehicles. See § 5.18(a). (I.12) is known as the Eötvös correction.

General references for rotating axes. [507] and [549].

APPENDIX J

GRAVITY REDUCTION TABLES

J.00. Summary

THIS appendix outlines the systems used for computing the horizontal and vertical components of the attraction of the topography and its isostatic compensation (if any is assumed) at any point. In the past the general procedure has been that the country surrounding the station is divided into *zones* of specified radii, which are in turn radially divided into *compartments*. The average height of each compartment is estimated from maps with the help of celluloid templates on which the compartments are drawn, and suitable rules or tables then give the desired result. These methods are described in §§ J.01–J.06.

The effect of electronic computing is that templates, and rings centred on the point where g or N is required, are probably obsolescent (except in the close neighbourhood of the point), and that basic information about the topography and values of g is required in the form of means over 1° squares, or other convenient standard areas. See § J.07.

In [113], 3 Edn. the computation of stoke's integral was included in this Appendix. Now it is in §§ 6.31 to 6.34.

J.01. Deflexion tables

See [254]. In Fig. J.1 the horizontal northerly attraction at a station P of the portion of a zone bounded by radii r_1 and r_2 and by azimuth lines α_1 and α_2 is such as will cause a meridional deviation of the vertical at P of

$$7''.74\frac{\delta}{\rho}h(\sin\alpha_2-\sin\alpha_1)\ln\frac{r_2}{r_1}, \qquad (J.1)$$

where h is the average height of the zone in km, small compared with r_1 so that the whole mass effectively lies in the horizon of P, δ is its density, and ρ is the earth's mean density. See [155], pp. 294–6. Then if successive zones are chosen so that $r_2/r_1 = 1.426$, and if compartment divisions are such that $\sin\alpha_2 - \sin\alpha_1 = \frac{1}{4}$, and if $\delta/\rho = 1/2.09$ (Hayford's figures), there results the convenient rule that the required effect of any compartment is $0''.01$ for every 100 feet (30.5 m) of height, the height being reckoned above any datum such as sea-level, as the total effect is got by summing north and south compartments separately and subtracting. Fig. J.2 shows a suitable template. For estimating east and west components the same template is turned through 90°.

As the above stands, certain approximations are involved. Thus to allow for the earth's curvature, the factor 1.426 requires modification to the extent that between 670 and 4130 km the radii must be increased by between 0.3 and 170 km. [155], p. 296, and [254] give the formula. And if the ground-level in a compartment differs from the height of the station by an amount which is not

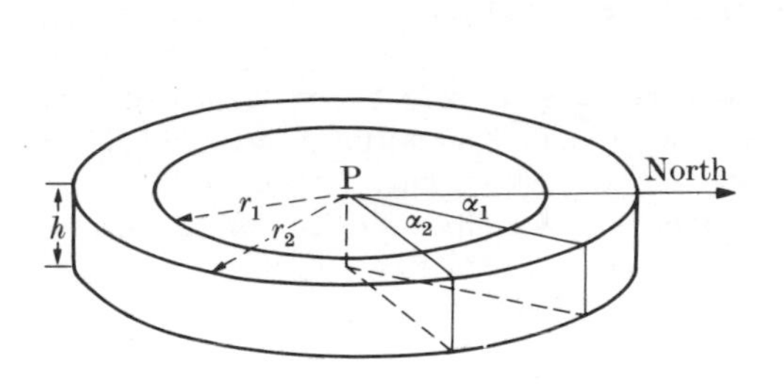

FIG. J.1.

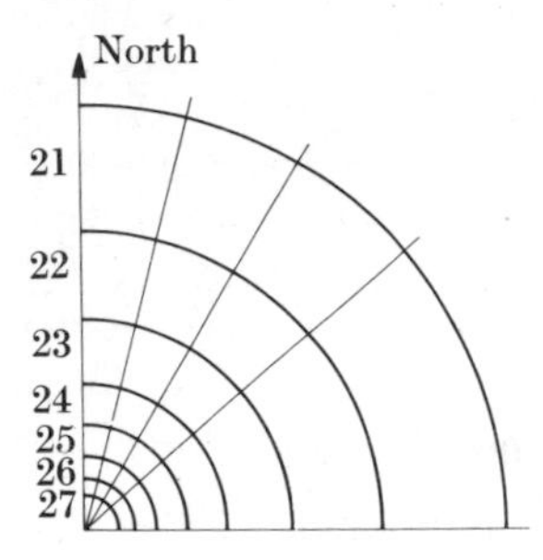

FIG. J.2. Part of template for scale 1/100 000.

sufficiently small compared with r_1, a small correction is required which is tabulated in [254].

Oceans constitute a defect, and so are recorded with negative sign, and their effect is 0″.00615 per 30.5 m (100 feet) of depth on account of their being filled with salt water.

Compensation. if the topography is compensated on Hayford's system, the attraction of the combined topography and compensation in any compartment depends on the radii of the zone and on the depth of compensation. In fact

$$\text{combined effect} = F \times (\text{topo effect}),$$

where

$$F = 1 - \left\{ \ln \frac{r_2 + \sqrt{(r_2^2 + D^2)}}{r_1 + \sqrt{(r_1^2 + D^2)}} \right\} \div \ln \frac{r_2}{r_1}. \tag{J.2}$$

This factor is tabulated in [254], p. 70, for each Hayford zone for eight different depths of compensation (D) between 56 and 330 km. Roughly speaking $1 > F > 0.9$ when $r < D/10$, $F < 0.1$ when $r > 2D$, and $F < 0.01$ when $r > 10D$. In other words, if $D = 100$ km compensation has little effect up to a radius of 10 km, but it makes the combined effect small beyond 200 km and practically nothing beyond 1000 km.

Hayford's zones extend to 4126 km, an arbitrary limit since topo effects beyond it are far from small. In some examples in fact, if no compensation is assumed, the last zone has as large an effect as all the inner zones combined. But, see § 6.54, accurate computation of uncompensated topo effects is almost impossible, and in any case of doubtful utility, while the limit of 4126 km is amply large enough if compensated topography is accepted as the standard.

Hayford's original tables are now superseded by [176]. For full formulae and explanation see [369].

J.02. Hayford's gravity tables

See [256]. The vertical component of the attraction of topography and compensation is obtained in a similar way, but with a different system of zones extending to the antipodes. In the nearer zones the tables are also more complex, since the effect depends so much on the relative height of zone and station. In these zones the attraction is the sum of (a) the attraction of the topography if the station is assumed to be of the same height as the average of the zone,† on which only the effect then depends, (*b*) attraction of compensation on the same assumption, and (*c*) a correction depending on the differences of height of station and zone. Allowance has to be made for water zones, and in some forms of the tables, [170], some allowance may be required for the earth's curvature. These 15 or 16 near zones A to O or P extend to 167 km, beyond which the combined effect of topography and compensation is almost exactly proportional to the height of the zone, and except for small corrections in the first five of these zones the work is simplified accordingly.

There is no essential division of zones into compartments at prescribed azimuths, as for the deviation, but subdivision into a number of compartments is usually made (*a*) to aid the estimation of average heights, and (*b*) because in the near zones the attraction is not at all proportional to height, and in a zone whose height is variable the tables must be entered separately with the mean heights of subdivisions in which the internal variation is not too large. Experience with the tables shows how much averaging may be done in different zones without serious error.

Before using gravity tables prepared by different authors it is essential to give careful attention to any introductory remarks about how to use them, and on what hypotheses they are based.

As with deviation, it is almost impossible to compute the effect of distant zones accurately unless compensation is included in the standard, although a better approximation is possible.

J.03. The Bouguer orographical correction

Curvature is ignored, and there is no compensation. Elaborate tables are not required.

The attraction of a cylindrical annulus of height h, measured above or below the level of the gravity station, of inner radius r_1 and outer radius r_2, and of density + or −2.67 is

$$Z = 0.115\{(r_2 - r_1) - (r_2^2 + h^2)^{\frac{1}{2}} + (r_1^2 + h^2)^{\frac{1}{2}}\}\ \text{mGal}$$

$$= 0.058\left\{h^2\left(\frac{1}{r_1} - \frac{1}{r_2}\right) - \tfrac{1}{4}h^4\left(\frac{1}{r_1^3} - \frac{1}{r_2^3}\right)\right\}\ \text{mGal}, \tag{J.3}$$

if h, r_1, and r_2 are in metres. If $r > 5h$, the term in h^4/r^3 is probably negligible.

The attraction is always upwards, whether h is positive or negative. When the height within an annulus is not substantially constant, the annulus must be divided into sectors of more constant height, since the mean of h^2 does not equal the mean of h.

With $h <$ about 2000 metres, the computation need seldom go beyond a radius of 50 km. Ordinarily a much smaller radius suffices.

J.04. Cassinis's gravity tables [149]

These tables follow Hayford's zones, but are of more fundamental character and can be used for any depth of compensation, including a varying depth such as

† Zone or subdivision of one. See below.

is prescribed by the Airy hypothesis. For each zone the tables give the vertical attraction of a complete cylindrical† annulus of unit density, whose bottom (or top) is at the same altitude as the station, and which extends upwards (or downwards) to all possible values of ground-level (or depths of compensation to 200 km). The attraction of topography or compensation in any sector of such an annulus of prescribed density and height or depth is then in simple proportion, and the total attraction at any station of any combination of topography and compensation can be obtained by subdivision and summation.

J.05. Airy and regional compensation

The vertical component of gravity on the Airy system (§ 6.65) can be got from Cassinis's tables, or from special tables of Heiskanen's for crustal thicknesses of 0, 20, 30, 40, 60, 80, and 100 km. [258] and [259].

For Vening Meinesz's system of regional compensation see § 6.66, and [574] and [576].

For a full summary of the different systems of reduction, and the formulae on which tables are based, see [376], pp. 63–125.

J.06. Geoidal rise

The modification $h' = \mathrm{d}W/g$ in § 6.35 (b) in the form of the geoid which would arise from the removal of the topography ($\rho = 2.67$), with or without compensation, can be calculated by estimating the heights of zones, and then entering suitable tables‡ An elaborate table, based on rigorous formulae and applicable to any depth of compensation between 80 and 130 km, is given in [368], and also the resulting Bowie correction, (§ 6.35(b)). [377] extends the Bowie correction tables to include all depths between 0 and 80 km.

[263] gives simple tables for the effect of removing topography and Hayford compensation (113.7 km) for zones whose radii are between 0 and 865 km, and a world chart giving the total Bowie correction arising in zones beyond 167 km. It gives a second world chart giving the total Bowie correction from zero radius, which can be used where the near topography (or ocean bottom) is not abrupt.

The correction for topography plus compensation will roughly be proportional to the mean depth of the compensation. So Airy 60 km will be about the same as Hayford 114 km, and Airy 30 km about half as much.

A generalized picture of the separation of the geoid and the co-geoid may be got from [464] or [325], which give harmonic analyses of the earth's topography to degree 16 and 8 respectively. Then if these excesses and defects of mass are treated as skin densities on a spherical earth, (6.44) with $R = r =$ earth's radius gives the reduction of potential when they are removed, and the same formula with a reduced R gives the increase when they are replaced at a prescribed mean depth inside the earth.

The change of potential at a fixed point associated with a change δN in the height of the adjacent geoid above the spheroid is $+g\delta N$.

† Actually conical, since the convergence of the verticals is allowed for.

‡ (6.17) gives W at the centre of one end of a cylinder, and subtraction gives that of an annulus. Then the geoidal rise or fall is W/g.

J.07. Rectangular areas instead of rings

(*a*) The labour of estimating average heights in rings or compartments around every gravity or deflection station is much reduced if average heights over 1-degree 'squares', or other such areas, can be estimated and recorded once and for all. Using an electronic computer, the labour of computation can also be much reduced by the direct computation of the effect of the topography in each such square, in preference to the effect being taken manually from a table. For the current availability of average height maps see § 6.37 (*b*), and [48] or an updated revision of it.

(*b*) *Rectangular areas.* The areas whose mean heights are recorded may be simply bounded by meridians and parallels at equal intervals, and so decreasing in size towards the poles, or they may be *equal area.* In the former case, the average height of each area will have to be weighted before it can be summed with others to give a total effect. Substantially equal areas may be obtained by maintaining a constant interval in latitude, while the longitude interval is suitably increased towards the poles, subject only to the condition that in any latitude belt each longitude interval must be a concrete subdivision of 360°. Equal-area 'squares' do not lend themselves to combination into larger squares, and the inconvenience of weighting is generally thought preferable.

The size of rectangle required to give adequately correct results varies with the distance from the station, and the ruggedness of the topography. At a distance of 500 km, $1° \times 1°$ (110×110 km) should be small enough, but $10' \times 10'$ may be too large at a distance of 50 km.

It can hardly be worthwhile to record the average heights of areas much less than $5' \times 5'$, all over a continent, so special estimates will have to be made for the area within perhaps 25 km of each station. Within this area the squares will have to become increasingly smaller as the station is approached, until the approximate equality of the height of the station and the topography produces effects which are clearly becoming negligible. In mountainous country the smallest rectangles may have to be as small as 100 m. The tables and formulae of §§ J.01–J.06 will be a guide to the necessary fineness of subdivision.

(*c*) *Computations.* Given the average height of a rectangular area, its effect at a station of given latitude and longitude is easily computed if the area is sufficiently small for the mass of its topography to be treated as concentrated at the centre of the rectangle at a height above the spheroid equal to half its average height. Similarly the isostatic compensation (if it is included in the model) can be placed at the hypothetical depth of compensation (for Airy), or at half its depth (for Hayford).

Then the attraction of the mass is given by (6.1). The attraction of the compensation is of course negative but not necessarily otherwise equal to that of the topography. Each acts along the chord joining the station to the point where the mass is concentrated. The earth being assumed to be spherical, the distance and direction are very simply programmed. The components of the attraction are thus computed (vertical, and horizontal northwards and eastwards), and thence the effect of each rectangle on g, ξ, and η at the station. Similarly, if the anomaly of potential is required, as in § J.06, it is given by (6.3).

J.08. Height estimation

This is the heavy labour. The following are aids.

(*a*) [442] gives world charts of the total vertical attraction, in zones beyond 865 km, of topography together with Hayford (113.7 km) compensation, or with Airy compensation at depths of 20, 30, 40, or 60 km.

(*b*) [324] gives similar charts for the total, in zones beyond 167 km, for topography with Airy compensation at 30 km.

(*c*) [264] gives similar charts for topography together with Airy compensation at 20 or 30 km, beyond 167 km, in Europe and the north Atlantic.

(*d*) When stations are near together some outer zones can often be got by interpolation, see [254] and [256], pp. 58–63.

(*e*) [265] describes an approximate method for dealing with topography within 18.8 km of the station.

Speed and accuracy only come with practice. Minute subdivision of compartments is the easy and accurate way, but is slow. The following are aids to speed.

(i) Subdivide into small areas of fairly even slope. Then the average height of each such area is that of the contour which bisects it. Take the mean of subdivisions weighted in proportion to area.

(ii) Where hills rise above a fairly level base, first estimate an average base level, and then an addition for what projects above. The latter may be unexpectedly small. Remember that the volume of a cone is $\frac{1}{3}\times$base$\times$height, and that a hill generally steepens towards the top, and that its sides are hollowed out by valleys. Also remember that the heighted points on a ridge are generally higher than the average crest line.

BIBLIOGRAPHY

This bibliography is not a comprehensive summary of geodetic literature, but includes only general works recommended for further reading, and specific references which amplify the text or acknowledge sources of information.

A full bibliography is published annually by the International Association of Geodesy.

1. Adams, O. S. The Bowie method of triangulation adjustment. *Spec. Publs U.S. Cst geod. Surv. No.* 159 (1930).
2. Adams. O. S. Latitude developments connected with geodesy and cartography. *Spec. Publs U.S. Cst geod. Surv. No.* 67 (1949).
3. Agreen, R. W., and Smith, D. E. A simulation of the San Andreas fault experiment. *J. geophys. Res.* **79** (29), 4413–17 (1974).
4. Aitken, A. C. *Determinants and matrices.* Oliver and Boyd (1958).
5. Allan, R. R. On the motion of nearly synchronous satellites. *Proc. R. Soc.* **A288,** 60–9 (1965).
6. Allan, R. R. Satellite resonance with longitude dependent Gravity, III, Inclination changes for close satellites. *Planet. Space Sci.* **21,** 205–25 (1973).
7. Allman, J. S. *Least square adjustment of observations.* University of New South Wales (1967).
8. Allman, J. S., and Bennett, G. C. Angles and directions. *Surv. Rev. No.* 139, pp. 219–28 (1966).
9. Anderle, R. J. Refined geodetic results based on doppler satellite observations. *NWL Tech. Rep.* 2889 (1971).
10. Anderle, R. J. Transformation of terrestrial survey data to doppler satellite datum. *J. geophys. Res.* **79** (35), 5319–31 (1974).
11. Anderle, R. J. Role of artificial earth satellites in redefinition of the North American datum. *Can. Surv.* **28** (5), 590–7 (1974).
12. Anderle, R. J. Long term consistency in position of sites determined from doppler satellite observations. *NSWC Dahlgreen, Tech. Rep.* 3433 (1975).
13. Anderle, R. J. Error model for geodetic positions derived from doppler satellite observations. *Bull. géod.* **50** (1), 43–77 (1976).
14. Anderle, R. J. Comparison of doppler and optical pole positions over twelve years. *NWL Tech. Rep.* 3464 (1976).
15. Anderle, R. J. Polar motion determined by doppler satellite observations. *Bull. géod.* **50** (4), 377–90 (1976).
16. Anderle, R. J. The global positioning system. *Phil. Trans. R. Soc.* (1979).
17. Anderle, R. J., and Beuglass, L. K. Doppler satellite observations of polar motion. *Bull. géod. No.* 96, pp. 125–41 (1970).

18. ANDERLE, R. J., *et al. J. geophs. Res.* **79** (35) (1974) includes [10], [217], [431], and [516] in one volume.

19. ANDERSEN, O. B. Surface-ship gravity measurement in the Skaggerak 1965–6. *Geodetic Institute Publication No.* 42, Copenhagen (1966).

20. ANGUS-LEPPAN, P. V. A study of refraction in the lower atmosphere. *Emp. Surv. Rev. Nos.* 120, pp. 62–9; 121, pp. 107–19; 122, pp. 166–77 (1961).

21. ANON. Formulas and tables for the computation of geodetic positions. Clarke, 1866. *Spec. publs US Cst geod. Surv. No.* 8 (1944).

22. ANON. *Latitude functions: Bessel spheroid, etc.* Separate volumes for each spheroid. Army Map Service, New York (1944).

23. ANON. *Projection tables for the Transverse Mercator projection of Great Britain.* H.M.S.O. (1950).

24. ANON. *Constants, formulae and methods used in the Transverse Mercator projection.* H.M.S.O. (1950).

25. ANON. *Tidal gravity effect tables.* Houston Technical Laboratories (1951).

26. ANON. *Japanese National Reports.* International Association of Geodesy (1954–6 and 1957–8).

27. ANON. Assemblée Générale de Rome 1954. Compte rendu, Section III. *Bull. géod. No.* 38, pp. 29–31 (1955).

28. ANON. *Final report on the Scotland–Norway tie* (1953), *Progress report on the North Atlantic tie,* U.S.A.F. (1955). and *Final report of results Crete–North Africa tie* (1957). U.S.Air Force.

29. ANON. *Smithsonian meteorological tables,* 6 edn (1958).

30. ANON. *Universal Transverse Mercator Grid Tables.* Military Survey, London (1958).

31. ANON. Universal polar stereographic grid tables for latitude 79° 30′ – 90°. International spheroid. *Army Map Service* TM 5-241-8 (1958).

32. ANON. Geodetic distance and azimuth computations. For lines over 500 miles. *ACIC Tech. Rep.* 80 (1959).

33. ANON. Geodetic distance and azimuth computations. For lines under 500 miles. *ACIC Tech. Rep.* 59 (1960).

34. ANON. *The refractive index of air for radio waves and microwaves.* National Physical Laboratory, London (1960).

35. ANON. Isostatically reduced deflections of the vertical at selected stations. *Army Map Service. Tech. Rep.* 30 (1960).

36. ANON. *Explanatory supplement to the Astronomical Ephemeris and the American Ephemeris and Nautical Almanac.* H.M.S.O. (1961).

37. ANON. Principles of error theory and cartographic applications. *ACIC Tech. Rep.* 96 (1962).

38. ANON. *Annual Reports of the polar motion service.* Mizusawa (from 1962).

39. ANON. *Proceedings* of *the Second International Symposium on Recent Crustal Movements, Academy of Science of Finland, Series AIII, No.* 90 (1966).

40. ANON. *Liste des stations de déviation de la verticale rattachées au réseau européen, No.* 4. International Association of Geodesy, Paris (1971).

41. ANON. *History of the retriangulation of Great Britain.* Ordnance Survey, H.M.S.O. (1967).

42. ANON. (Compilation). *Electromagnetic distance measurement.* Hilger and Watts (1967). Individual papers are listed under authors' names.
43. ANON. *Bureau international de l'heure. Rapports annuels.* Paris.
44. ANON. *The Australian map grid.* National Mapping Council, Canberra (1968).
45. ANON. *Problem of recent crustal movements of the earth. Third international symposium.* USSR Academy of sciences, Moscow (1969).
46. ANON. *Talyval.* Maker's handbook. Rank Precision Industries, Ltd. (c. 1970).
47. ANON. *Kingdom of Saudi Arabia. The traverse net. Final report of the National geodetic survey* (1970).
48. ANON. *Bull. Bureau Gravimètrique Int* **26** I, 26–34 (1971).
49. ANON. *NASA Directory of observation station locations*, 3 Edn, 2 vols. GSFC.
50. ANON. (Johns Hopkins APL). A satellite freed of all but gravitational forces; Triad I. *Use of artificial satellites for geodesy and geodynamics*, pp. 135–55. Athens (1973).
51. ANON. *Geodetic operations in Finland* 1971–4. National Rep. to AIG (1975).
52. ANON. *Geodetic work in the Netherlands* 1971–4. National Rep. to AIG (1975).
53. ANON. *Quantities, units and symbols.* Symbols committee of the R. Soc. London (1975).
54. ANON. Report on the symposium on scientific applications of lunar laser ranging. Selected abstracts. *Bull. géod.* **50**(3), 255–67 (1976).
55. ANON. *Bull. Bureau Gravimetrique Int.* **40** I, 13 (1977).
56. ANON. Table of earth satellites. Vol. i, 1959–68: Vol. ii, 1969–73: Vol. iii, 1974–8. RAE Farnborough.
57. ANON. *Bull. d'information. Bureau Gravimetrique International.* published triannually (1962–78).
58. ANON. *Applications of space technology to crustal dynamics and earthquake research*, NASA, Washington D.C. (1978).
59. ANON. The national geodetic survey project 'Polaris'. *Proc. of the* 1977 *Recent Crustal Movements symposium.* Elsevier (1978).
60. APPLETON, E., SHEPPARD, P. A., BOOKER, H. G., *et al.* Meteorological factors in radio wave propagation. *Joint report of the Physical and Royal Meteorological Societies*, London (1946).
61. ARUR, M. G., and MUELLER, I. I. Does mean sea level slope up or down towards north? *Bull. géod. No.* 117, pp. 289–97 (1975).
62. ASHKENAZI, V. Solution and error analysis of large geodetic networks. Direct methods. *Surv. Rev. Nos.* 146, pp. 166–73; 147, pp. 194–206 (1967).
63. ASHKENAZI, V. Solution and error analysis of large geodetic networks. Iterative methods. *Surv. Rev. Nos.* 151, pp. 34–9; 152, pp. 58–67 (1969).
64. ASHKENAZI, V. *Criteria for optimisation. A practical assessment of a free network adjustment.* IAG symposium, Oxford (1973).
65. ASHKENAZI, V. The proposed readjustment of the North American horizontal control networks. *Can. Surv.* **28** (5), 653–62 (1974).
66. ASHKENAZI, V. *Compendium of formulae on geodetic position networks.*

Nottingham (1976).
67. Ashkenazi, V., and Crane, B. A., *et al.* Terrestrial doppler adjustment and analysis of the primary triangulation of Great Britain. Preliminary report. *Phil. Trans. R Soc.* (1979).
68. Ashkenazi, V., and Cross, P. A., *et al.* The readjustment of the retriangulation of Great Britain, and its relationship to the European terrestrial and satellite networks. *Ordnance Survey Prof. Pap. NS* 24 (1972).
69. Ashkenazi, V., and Cross, P. A. Strength analysis of Block VI of the European Triangulation. *Bull. géod. No.* 103, pp. 5–24 (1972).
70. Ashkenazi, V., and Cross, P. A. *Strength of long lines in terrestrial geodetic control networks.* IAG, General Assembly (1975).
71. Ashkenazi, V., and Dodson, H. The Nottingham multi-pillar base line. *J. Assoc. Civil Engineers.* June 1977, 1–8(1977).
72. Ashkenazi, V., Gough, R. J., *et al. Preliminary analysis of the UK doppler campaign.* Council of Europe, Luxembourg (1977).
73. Ashkenazi, V., and Wood, R. C. Merits of the conjugate gradient method for solving geodetic normal equations. *African geodetic Journal* **1** (1), 75–94 (1974).
74. Asplund, L. Report on specifications for fundamental networks. *Trav. Ass. int. Géod.* **22,** 53 (1964).
75. Atia, K. A. The adjustment of trilateration networks observed in pairs. *Surv. Rev. No.* 181, 328–35 (1976).
76. Baeschlin, C. F. *Lehrbuch der geodäsie.* Zurich (1948).
77. Baker, L. S. *Geodetic operations in the US and other areas through international cooperation.* National Report to IAG 1967–70 (1971).
78. Balazs, E., Local mean sea level in relation to geodetic levelling along US coast lines. *Trans. Amer. Geophys. Union* **55** (3) (1974).
79. Balmino, G., Reigber, C. *et al.* The GRIM 2 earth gravity field model. *Dt. geod. Kommn.* A 86. Munich (1976).
80. Balmino, G., Lambeck, K. *et al.* A spherical harmonic analysis of the earth's topography. *J. geophys. Res.* **78** (2), 478–81 (1973).
81. Banks, R. J., Parker, R. L. *et al.* Isostatic compensation on a continental scale. Local versus regional mechanisms. *Geophys. J. R. astr. Soc.* **51** (2), 431–52 (1977).
82. Barnes, G. L., and Mueller, I. I. The dependence of the level sensibility on the position and length of the bubble in the Wild T4 theodolite. *Bull. géod. No.* 81, pp. 277–81 (1966).
83. Barrell, H., and Sears, J. E. The refraction and dispersion of air for the visible spectrum. *Phil. Trans. R. Soc.* **A238** (1939).
84. Becker, J-M. Experiences using motorised levelling techniques. *Proc. Fed. Int. Géomètres. Stockholm* (1977).
85. Bell, J. F. The field calibration of gravimeters. *Surv. Rev. No.* 144, pp. 72–9 (1967).
86. Bellamy, C. J., and Lodwick, G. D. The reduction of barometric networks and field gravity surveys. *Surv. Rev. No.* 147, pp. 216–27 (1968).
87. Benoit, E. Note sur une méthode de résolution des équations normales. *Bull. géod. No.* 2, pp. 66–77 (1924).
88. Best, A. C. Transfer of heat and momentum in the lower layers of the

atmosphere. *Met. Off. Geophys. Mem. No.* 65 (1935).

89. Beyer, L. A., von Huene, R. E. *et al.* Measuring gravity on the sea floor in deep water. *J. geophys. Res.* **71** 2091–100 (1966).
90. Bilby, J. S. The Bilby steel tower. *Spec. Publs U.S. Cst geod. Surv. No.* 158 (revised) (1940).
91. Black, A. N. Laplace points in moderate and high latitudes. *Emp. Surv. Rev. No.* 82, pp. 177–84 (1951).
92. Black, H. D. Position determination using the transit system. *Proc. Int. geod. Symposium, Las Cruces, Oct.* 1976 (1977).
93. Black, H. D., Jenkins, R. E., *et al. The Transit system,* 1975. AGU meeting, Washington, June 1975 (1975).
94. Biraldi, G. *Establishment of vertical deflection points by means of a photo-astronomic procedure.* IAG General Assembly, (1975).
95. Bock, R. O., and Wing, C. A new ocean floor gravimeter. *Proceedings of the First Marine Geodesy Symposium*, pp. 195–9. U.S. Government Printing Office (1967).
96. Böhn, J. Review of F. N. Krasovskij and V. V. Danilov *Rukovodstvo po vyschej geodesii,* Moscow (1938–42). *Bull. géod. No.* 8, pp. 178–86 (1948),
97. Bomford, A. G. Precise tellurometer traversing. *Emp. Surv. Rev. Nos.* 117, pp. 316–26; 118, pp. 342–50 (1960).
98. Bomford, A. G. Transverse Mercator arc-to-chord and finite distance scale factor formulae. *Emp. Surv. No.* 125, pp. 318–27 (1962).
99. Bomford, A. G. A least squares adjustment of angles and azimuths. *Cartography,* **5** (1 and 2) (1963).
100. Bomford, A. G. Astronomical observations with the Kern DKM 3A and Wild T3 theodolites. *National Mapping Tech. Rep. No.* 4. Canberra (1965).
101. Bomford, A. G. The geodetic adjustment of Australia. *Surv. Rev. No.* 144, pp. 52–71 (1967).
102. Bomford, A. G., Cook, D. P. *et al.* Astronomic observations in the Division of National Mapping 1966–70. *DNM Tech. Rep.* **10**. Canberra (1970).
103. Bomford, G. *Geod. Rep. Surv. India* **3**, 27–9 (1926–7).
104. Bomford, G. Three sources of error in precise levelling. *Prof. Pap. Surv. India, No.* 22 (928).
105. Bomford, G. *Geod. Rep. Surv. India,* **5,** 5, 6, 8–10 (1928–9); and *Geod. Rep. Surv. India,* 1934, pp. 16–22.
106. Bomford, G. *Geod. Rep. Surv. India,* 1933.
107. Bomford, G. *Geod. Rep. Surv. India,* 1933, pp. 44–6.
108. Bomford, G. Deviation of the vertical. *Geod. Rep. Surv. India,* **7** pp. 49–59 (1930–31).
109. Bomford, G. Deviation of the vertical. *Geod. Rep. Surv. India,* 1935, pp. 25–45.
110. Bomford, G. The readjustment of the Indian triangulation. *Prof. Pap. Surv. India, No.* 28 (1939).
111. Bomford, G. Deviations of the vertical 1954–6. *Trav. Ass. int. Géod.* **20** (5), (1958).

112. Bomford, G. The figure of the earth. Its departure from an exact spheroid. *Geophys. J. R. astr. Soc.* **3** (1), 83–95 (1960).
113. Bomford, G. *Geodesy*, 1 edn, 1952. 2 edn, 1962. 3 edn, 1971. Clarendon Press, Oxford.
114. Bomford, G. The geoid in Europe and connected countries. *Trav. Ass. int. Géod.* **24,** 357–71 and addendum (1971).
115. Bonsdorf, I. *Spec. Publs. Baltic geod. Commn No.* 5 (1935).
116. Bonsdorf, I. *Publication of the Finnish Geodetic Institute No.* 20. Helsinki (1934).
117. Bower, D. R. The determination of cross-coupling error in the measurement of gravity at sea. *Geophys. J. R. astr. Soc.* **71** (2), 487–93 (1966).
118. Bowie, W. Investigations of gravity and isostasy. *Spec. Publs U.S. Cst geod. Surv. No.* 40 (1917).
119. Bowin, C. O., Aldrich, T. C., *et al.* VSA gravity meter system. Tests and recent developments. *J. geophys. Res.* **77** 2018–33 (1972).
120. Bowring, B. R. Transformation from spatial to geographical coordinates. *Surv. Rev. No.* 181, pp. 323–27 (1976).
121. Bowring, B. R. Correspondence (1974–78).
121a. Bowring, B. R. The surface controlled spatial system for survey computations. *Surv. Rev. No.* 190, pp. 361–72 and *No.* 195, p. 240 (1978–9).
122. Boyer, J. O., and Sollins, A. D. Zenith camera observations in Liberia. *Jl. U.S. Cst geod. Surv.* **5** (1953).
123. Bradford, J. E. S. A method of observing primary horizontal angles. *Emp. Surv. Rev. No.* 67, pp. 222–30 (1948).
124. Bradsell, R. N. The electronic principles of the Mekometer III. *Surv. Rev. No.* 161, pp. 112–8 (1971).
125. Brazier, H. H., and Rainsford, H. F. Long lines on the earth. A new and easier solution. *Proceedings of the Commonwealth Survey Officers Conference*, London, 1951.
126. Brookes, B. C., and Dick, W. F. L. *Introduction to statistical method.* Heinemann (1951).
127. Brouwer, D., and Clemence, G. M. *Celestial mechanics.* Academic Press (1961).
128. Brown, D. C. Reduction of stellar plates for determination of direction of flashing light beacons. *The use of artificial satellites for geodesy*, vol. 1 (ed. G. Veis), pp. 163–86. First International Symposium, Washington, 1962. North-Holland, Amsterdam (1963).
129. Brown, D. C. Doppler surveying with the JMR-1 receiver. *Bull. géod.* **50** (1), 9–25 (1976).
130. Brown, D., and Trotter, J. SAGA. A computer program for short arc geodetic adjustment of satellite observations. *AFCRL Report* 69–0080 (1969).
131. Browne, B. C. The measurement of gravity at sea. *Mon. Not. R. astr. Soc. geophys. Suppl.* **4** (3), 271–9 (1937).
132. Browne, B. C. The technique of measuring gravity values on land by means of pendulums. *Bull. géod. No.* 64, pp. 145–68 (1962).
133. Bruins, G. J. (Ed.). *Gravity expeditions* 1948–54, vol. v. Delft (1960).

134. Buck, R. J., and Tanner, J. G. Storage and retrieval of gravity data. *Bull. géod. No.* 103, pp. 63–84 (1972).

135. Bullard, E. C. The figure of the earth. *Mon. Not. R. astr. Soc. geophys Suppl.* **5** (6), 186–92 (1948).

136. Bullard, E. C., and Cooper, R. I. B. The determination of masses necessary to produce a given gravitational field. *Proc. R. Soc.* **A194,** 332–47 (1948).

137. Bullen, K. E. *The earth's density.* Chapman and Hall (1975).

138. Burnside, C. D. *Electronic distance measurement.* Lockwood (1970).

139. Burrows, C. R., and Attwood, S. *Radio wave propagation.* Academic Press, New York (1949).

140. Byerley, W. E. *Fourier's series and spherical harmonics.* Dover Publications (1959).

141. Cabion, P. J. Principles and performance of an 8-mm wavelength tellurometer. *Electromagnetic distance measurement,* pp. 184–207. Hilger and Watts (1967).

142. Cannon, W. H. The classical analysis of the response of a long base line radio interferometer. *Geophys. J. R. astr. Soc.* **53** (3), 503–30. (1978).

143. Cannon, W. H., Langley, R. B., *et al.* Transatlantic geodesy by long base interferometry. *Symposium on geodesy and physics of the earth,* pp. 281–322. Weimar (1977).

144. Carroll, D. G. *Model (ii) analysis of variance of astronomical latitude and longitude. National Ocean Survey.* IAG General Assembly (1975).

145. Carter, W. E., Pettey, J. E., *et al.* The accuracy of astronomical azimuth determinations. *Bull. géod.* **52** (2), 107–13 (1978).

146. Carter, W. E., and Vincenty, T. Survey of the McDonald Observatory radial line scheme by relative lateration techniques. *NOAA Tech. Rep.* NOS 74, NGS 9 (1978).

147. Cartwright, D. E., and Crease, J. A comparison of the geodetic reference levels of England and France by means of the sea surface. *Proc. R. Soc* **A273,** 558–80 (1960).

148. Cartwright, D. E., and Crease, J. A proposed new method of levelling between France and England. *Bull. géod. No.* 55, pp. 97–9 (1960).

149. Cassinis, G., Dore, P., and Ballarin, S. *Fundamental tables for reducing observed gravity values.* Pavia (1937).

150. Cazenave, A., Daillet, S., *et al.* Tidal studies from the perturbations in satellite orbits. *Phil. Trans. R. Soc.* **A284,** 595–606 (1977).

151. Chapman, S., and Bartels, G. *Geomagnetism.* Oxford University Press (1940).

152. Clairaut, A. C. *Théorie de la figure de la terre.* Paris (1743).

153. Clarke, A. R. *Account of the principal triangulation.* London (1858).

154. Clarke, A. R. *Comparison of the standards of length of England, France, etc.* London (1866).

155. Clarke, A. R. *Geodesy.* Oxford University Press (1880).

156. Clendinning, J. Grid bearings and distances in the conical orthomorphic projection. *Emp. Surv. Rev. Nos.* 48, pp. 68–79; 51, pp. 211–21; 52, pp. 248–54 (1943–4).

157. Clendinning, J., and Olliver, J. G. *Principles and use of surveying instruments.* 3 edn. Blackie, London (1969).
158. Cochran, J. R., and Talwani, M. Free-air gravity anomalies in the World's oceans and their relationship to residual elevation. *Geophys. J. R. astr. Soc.* **50** (3), 495–552 (1977).
159. Cook, A. H. External gravity field of a rotating spheroid to the order of e^3. *Geophys. J. R. astr. Soc.* **2** (3), 199–214 (1959).
160. Cook, A. H. Developments in dynamical geodesy. *Geophys. J. R. astr. Soc.* **2** (3), 222–40 (1959).
161. Cook, A. H. A new absolute determination of the acceleration due to gravity at the NPL England. *Phil. Trans. R. Soc.* **A261,** 212–51 (1967).
162. Cook, A. H. Use of artificial satellites in the determination of the gravitational potential. *Trav. Ass. int. Géod.* **23** 245–56 (1968).
163. Cook, A. H. Determination of the earth's gravitational field from satellite orbits: methods and results. A discussion on orbital analysis. *Phil. Trans. R. Soc.* **A262,** 119–32 (1967).
164. Cook, A. H. Absolute measurement of gravity. *Trav. Ass. int. Géod.* **23** 271–8 (1968).
165. Cook, A. H. *Physics of the earth and planets.* Macmillan (1973).
166. Cook, A. H., *et al.* (Editors). A discussion on ranging to artificial satellites and the moon. *Phil. Trans. R. Soc.* **A284,** 421–619 (1978).
167. Cook, G. E., and Sudworth, J. R. An introduction to the Navstar global positioning system. *RAE Tech. Mem.* 37 (1976).
168. Cross, P. A. The effect of errors in weights. *Surv. Rev. No.* 165, pp. 319–25 (1972).
169. Cross, P. A., and Norris, J. C. The adjustment of length ratios. *Surv. Rev. No.* 181 (1976).
170. Couchman, H. J. The pendulum operations in India and Burma 1908–13. *Prof. Pap. Surv. India No.* 15 (1915).
171. Culley, F. L. Radio ranging on artificial satellites. *Electromagnetic distance measurement,* pp. 300–11. Hilger and Watts (1967).
172. Cunningham, L. B. C. Unpublished. Communicated by W. Rudoe.
173. Currie, I. F. The calibration of ballistic cameras and their use for the triangulation of satellite positions. *Réseau géodésique européen par observation de satellites,* pp. 235–54. IGN, Paris (1964).
174. Dale, J. B. *Five-figure tables of mathematical functions.* London (1903).
175. Danjon, A. The contribution of the impersonal astrolabe to fundamental astronomy. *Mon. Not. R. astr. Soc.* 118 (5), 411–31 (1958).
176. Darling, F. W. Fundamental tables for the deflection of the vertical. *Spec. Publs. U.S. Cst. geod. Surv. No.* 243 (1949).
177. Darwin, G. H. Harmonic analysis of tidal observations. *Report to the British Association* (1883).
178. Darwin, G. H. The theory of the figure of the earth. *Mon. Not. R. astr. Soc.* **60** 82–124 (1899).
179. Darwin, G. H. *Scientific papers,* vol. 3. Cambridge University Press (1910).
180. Davis, Q. V. Lasers and distance measurement. *Surv. Rev. Nos.* 139, pp. 194–207; 140, pp. 269–79 (1966).

181. Decae, A. E. Precision survey of the 28 BeV synchrotron. *Emp. Surv. Rev. Nos.* 121, pp. 98–106; 122, pp. 146–60; 123, pp. 194–208 (1961).

182. Decker, B. L. *Present day accuracy of the earth's gravitational field.* DMA St. Louis (1972).

183. Deetz, C. H., and Adams, O. S. Elements of map projections. *Spec. Publs U.S. Cst. geod. Surv. No.* 68 (1944).

184. Defant, A., *Physical Oceanography.* 2 vols. Pergamon Press (1961).

185. Denison, E. W. IAG Study Group 1.19 report on electronic distance measurement. *Trav. Ass int. Geod,* **24** 67–78 (1972).

186. Doodson, A. T. Mean sea level and geodesy. *Bull. géod. No.* 55, pp. 69–77 (1960).

187. Ducarme, B., Hosoyama, K., *et al.* An attempt to use the LaCoste–Romberg model G gravimeter at the microgal level. *Bull. Bureau Gravimetrique int.* **39** (1976).

188. Duerkson, J. A. Deflections of the vertical in the U.S. *Spec. Publs. U.S. Cst geod. Surv. No.* 229 (1941). With supplements (1954).

189. Dyson, J. Correction for atmospheric refraction in surveying and alignment. *Nature, Lond.* **216** 782 (1967).

190. Eaglesfield, C. G. Laser light. Macmillan (1967).

191. Eddington, A. S. Observations made with the Cookson floating telescope. *Mon. Not. R. astr. Soc.* **73** 605–16 (1913).

192. Edge, R. C. A. Some considerations arising from the results of the 2nd and 3rd geodetic levelling of England and Wales. *Bull. géod. No.* 52, pp. 28–36 (1959).

193. Edge, R. C. A., and Kenney, P. H. Electronic distance measurement. *Trav. Ass. int. Géod.* **22** 57–95 (1964).

194. Eliot, J., Blanford, H. F., *et al.* Daily variations of atmospheric conditions in India. *Memoirs of the Indian Meteorological Department* vols. 5 (1892–5) and 9 (1895–7).

195. Engeli, M., Ginsburg, Th., *et al.* Refined iterative methods for computation of the solution and the eigenvalues of self-adjoint boundary value problems. *Mitt. Inst. Ang. Mathematik.* Birkhäuser, Basle (1959).

196. Entin, I. I. Main systematic errors in precise levelling. *Bull. géod. No.* 52, pp. 37–45 (1959).

197. Essen, L., and Froome, K. D. The refractive index of air for radio waves and micro-waves. *Proc. phys. Soc.* **B54,** 862–73 (1951).

198. Everest, G. *An account of the measurement of an arc of the meridan between the parallels* 18° 3′ and 24° 7′. London (1830).

199. Faddeva, V. N. *Computational methods of linear algebra.* Dover Publications (1958).

200. Faller, J. E., and Hammond, J. A. A new portable absolute gravity instrument. *Bull. Bureau Gravimetrique int.* **35** I, 43–8 (1974).

201. Farley, D. J. Observations of the equatorial ionosphere. *Electron density profiles in ionosphere and exosphere.* Amsterdam (1966).

202. Ferrars, N. M. *Spherical harmonics.* Macmillan (1877).

203. Fischer, I. A map of geoidal contours in N. America. *Bull. géod. No.* 57, pp. 321–4 (1960).

204. Fischer, I. Present extent of the astro-geodetic geoid. *Bull. géod. No.* 61, pp. 245–64 (1961).
205. Fischer, I. Gravimetric interpolation of deflection of the vertical by electronic computer. *Bull. géod. No.* 81, pp. 267–75 (1966).
206. Fischer, I. Report of Study group V-29. Interpolation of deflections of the vertical. *Trav. Ass. int. Géod.* **24** (1972).
207. Fischer, I., Slutsky, M., Shirley, F. R., and Wyatt, P. Y. New pieces in the picture puzzle of an astro-geodetic geoid map of the world. *Bull. géod. No.* 88, pp. 199–221 (1968).
208. Forward, R. L. Review of artificial satellite gravity gradiometer techniques for geodesy. *Use of artificial satellites for geodesy and geodynamics*, pp. 157–92. Athens (1973).
209. Fox, L. *An introduction to numerical linear algebra.* Clarendon Press, Oxford (1964).
210. Frazer, R. A., Duncan, W. J., and Collar, A. R. *Elementary matrices.* Cambridge University Press (1963).
211. Fricke, W., and Kopff, A. *Fourth fundamental catalogue* (*FK*4). Astro. Rechen-Inst., Heidelberg, 10 (1963). With provisional *Supplement to the FK*4 (1963).
212. Froome, K. D. The refractive indices of water vapour, air, oxygen, nitrogen and argon at 72 kMc/s. *Proc. phys. Soc.* **B68,** 833–5 (1955).
213. Froome, K. D. Correspondence (1967).
214. Froome, K. D. Mekometer III. EDM with sub-mm resolution. *Surv. Rev. No.* 161, pp. 98–112 (1971).
215. Froome, K. D., and Bradsell, R. H. Long term stability of an NPL III wavelength standard. *Surv. Rev. No.* 186, pp. 166–71 (1977).
216. Gaposchkin, E. M. (ed.). 1973 Smithsonian standard earth (III). *SAO Special Report* 353 (1973).
217. Gaposchkin, E. M. Earth's gravity field to the 18 degree, and geocentric coordinates for 104 stations from satellite and terrestrial data. *J. geophys. Res.* **79** (35), 5377–411, (1974).
218. Gaposchkin, E. M., and Lambeck, K. 1969 Smithsonian standard earth (II). *SAO Special Report* 315 (1970).
219. Gaposchkin, E. M., and Williamson, M. R. *Revision of geodetic parameters.* IAG General Assembly (1975).
220. Gardiner-Hill, R. C. Refinements of precise distance measurement by tellurometer. *Proceedings of the Commonwealth Survey conference* 1963. H.M.S.O. (1964).
221. Garland, G. D. *The earth's shape and gravity.* Pergamon Press (1965).
222. Gebel, G., and Mathews, B. Navigation at the prime meridian. *Navigation,* **18** (2) (1971).
223. Gebel, G., and Pryor, L. L. Navigation at the prime meridian revisited. *Navigation,* **24** (3) (1977).
224. Gessler, J., and Seeber, G. Latitude and longitude determinations with a transportable zenith camera. *Dt. geod. Kommn.* B213, pp. 23–32. Munich (1975).
225. Gibbs, R. W. M. *The adjustment of errors in practical science.* London (1929).

226. GILBERT, R. L. G. A dynamic gravimeter of novel design. *Proc. Phys. Soc.* **62** 445–54 (1949).

227. GOODING, R. H. Orbit determination from Minitrack observations. *Phil. Trans. R. Soc.* **A262,** 1124, 79–88 (1967).

228. GOODING, R. H. Lumped geopotential coefficients $\bar{C}_{15,15}$ and $\bar{S}_{15,15}$ obtained from resonant variations in the orbit of Ariel 3. *RAE Tech. Rep. No.* 71068 (1971).

229. GOSSETT, F. R. Manual of geodetic triangulation. *Spec. publs. U.S. Cst geod. Surv. No.* 247 (1950).

230. GRAAFF-HUNTER, J. DE. Atmospheric refraction. *Prof. Pap. Surv. India No.* 14 (1913).

231. GRAAFF-HUNTER, J. DE. The earth's axes and triangulation. *Prof. Pap. Surv. India No.* 16 (1918).

232. GRAAFF-HUNTER, J. DE. Geodesy. *Survey of India Dept. Paper No.* 12 (1929).

233. GRAAF-HUNTER, J. DE. The figure of the earth from gravity observations. *Phil. Trans. R. Soc.* **A234,** 377–431 (1935).

234. GRAAF-HUNTER, J. DE Report of study-group 8, Reduction of observed gravity. *Bull. géod. No.* 50, pp. 1–16 (1958).

235. GRAAF-HUNTER, J. DE. The shape of the earth's surface expressed in terms of gravity at ground level. *Bull. géod. No.* 56, pp. 191–200 (1960).

236. GRAAF-HUNTER, J. DE. Unpublished (1958 and 1961).

237. GRAAF-HUNTER, J. DE, and BOMFORD, G. *Geod. Rep. Surv. India* **5,** 118–28 (1928–9).

238. GRAAF-HUNTER, J. DE, and BOMFORD, G. Construction of the geoid. *Bull. géod. No.* 29, pp. 22–6 (1931).

239. GREGERSON, L. F., *Report on experiments with a gyroscope equipped with electronic registration.* IAG General Assembly (1971).

240. GREGERSON, L. F. Use of gyroscopes in Canadian geodesy. *Proc. Int. Conf. Mining surveying.* Budapest (1972).

241. GREGERSON, L. F. Azimuth gyroscopy in North America. *Proc. Fed. Int. Géomètres,* Stockholm (1977).

242. GUILLAUME, C. E. INVAR and ELINVAR. *Dictionary of applied physics,* vol. 5, Macmillan, London (1923).

243. GULATEE, B. L. *Survey of India Tech. Rep.* 1950, Chart XXII (1951).

244. GULATEE, B. L. Gravity anomalies and the figure of the earth. *Prof. Pap. Surv. India No.* 30 (1940).

245. GULATEE, B. L. Deviations of the vertical in India. *Tech. pap. Surv. India No.* 9 (1955).

246. GUMBEL, E. J. Statistics of extremes (1958).

247. HAGUE, B. *An introduction to vector analysis.* Methuen (1951).

248. HALL, M. J. Tellurometer MRA 4. A survey of short range performance. *Proceedings of the Commonwealth Survey Conference 1967.* H.M.S.O. (1968).

249. HALLIDAY, E. X. Causes of systematic error in the geodetic levelling of England and Wales. *Proceedings of the Commonwealth Survey Conference 1967.* H.M.S.O. (1968).

250. Halmos, F. High precision measurement and evaluation method for azimuth determination with gyrotheodolites. *Manuscripta geodetica*, **3** 213–32 (1977).
251. Harmala, S., and Kukkamaki, T. J. Electromagnetic distance measurement in Finland. *Electromagnetic distance measurement*, pp. 22–4. Hilger and Watts (1967).
252. Hatfield, H. R. *Admiralty manual of hydrographic survey*, vol. 1 (1965).
253. Haworth, R. H. An analysis of the errors in gravity measurements at sea. *Bulletin du Bureau Gravimétrique International No.* 18, pp. 3–10. Paris (1968).
254. Hayford, J. H. *The figure of the earth and isostasy from measurements in the United States*. Washington (1909).
255. Hayford, J. H. *Supplementary investigation of the figure of the earth and isostasy*. Washington (1910).
256. Hayford, J. H., and Bowie, W. The effect of topography and isostatic compensation upon the intensity of gravity. *Spec. Publs U.S. Cst geod. Surv. No.* 10 (1912).
257. Heiskanen, W. Untersuchungen über Schwerkraft und Isostasie, *Publications of the Finnish Geodetic Institute No.* 4 (1924).
258. Heiskanen, W. Tables for the reduction of gravity on the basis of Airy's hypothesis. *Bull. géod. No.* 30, pp. 110–53 (1931).
259. Heiskanen, W. New Tables for the reduction of gravity on the Airy hypothesis. Helsinki (1938).
260. Heiskanen, W. The Nummela base measured with Vaisäla comparator. *Bull. géod. No.* 17, pp. 294–8 (1950).
261. Heiskanen, W. Present problems of physical geodesy. *Publs isostatic Inst. Ass. Geod. No.* 49 (1965).
262. Heiskanen, W., and Moritz, H. *Physical geodesy*. Freeman (1967).
263. Heiskanen, W., and Niskanen, E. World maps for the indirect effect of the undulations of the geoid on gravity anomalies. *Publs isostatic Inst. Ass. Geod. No.* 7 (1941).
264. Heiskanen, W., Niskanen, E., and Karki, P. Topo-isostatic reduction maps for Europe and the North Atlantic in the Hyaford zones 18 to 1 for the Airy-Heiskanen system. $T = 30$ km and 20 km. *Publs isostatic Inst. Ass. Geod. No.* 31 (1959).
265. Heiskanen, W., and Sala, I. The topographic-isostatic reduction of gravity anomalies by the aid of small-scale maps. *Publs isostatic Inst. Ass. Geod. No.* 21 (1949).
266. Heiskanen, W., and Vening-Meinesz, F. A. *The earth and its gravity field*. McGraw-Hill, New York (1958).
267. Heitz, S. Geoid determination by least-squares interpolation on the basis of measured and interpolated deflections of the vertical. *Dt. geod. Kommn.* C 124 (1968). ACIC Tech. translation No. 1643.
268. Heitz, S. An astro-geodetic determination of the geoid in West Germany. *Nachrichten aus dem Karten-und Vermessungswesen*, II, 24. Frankfurt (1969).
269. Hela, I., and Lisitzin, E. A world mean sea level, and marine geodesy.

First Marine geodesy symposium, 1966, 71–73. US Govt. Printing, Washington (1967).

270. HELMERT, F. R. *Theorieen der Höheren Geodäsie*. Leipzig (1880–4). Reprinted Minerva (1962).
271. HELMERT, F. R. Die normal Theil der Schwerkraft im Meeresniveau. *Sber. preuss. Akad. Wiss.* 1901, pp. 328–36.
272. HENDRIZK, D. R. The conformal transformation of a network of triangulation. *Emp. Surv. Rev. No.* 85, pp. 319–25 (1952).
273. HENRIKSEN, S. W. (Editor). *National geodetic satellite program.* 2 vols, 1030 pp. NASA SP-365 (1977).
274. HEWITT, J. An $f-1$ field flattened Schmidt system for precision measurements of satellite positions. *Photogr. Sci. Engng*, **9**. 10 (1965).
275. HIRVONEN, R. A. Tables for the computation of long geodetic lines. *Bull. géod. No.* 30, pp. 379–42 (1953).
276. HIRVONEN, R. A. New theory of gravimetric geodesy. *Publs isostatic Inst. Ass. Geod. No.* 32 (1960).
277. HOGG, F. B. R., and ARMSTRONG, J. A. Two new self-aligning levels. *Emp. Surv. Rev. No.* 111, pp. 2–6 (1959).
278. HONKOSALO, T. Measuring the 864 m long Nummela base line with the Vaisäla light interference comparator. *Finn. Geod. Inst. No.* 37 (1950).
279. HONKASALO, T. Measurement of standard base lines with the Vaisäla light-interference comparator. *J. geophys. Res.* **65** (2) (1960).
280. HONKASALO, T. Gravity and land upheaval in Fennoscandia. *Proceedings of the Second International Symposium on Recent Crustal Movements. Academy of Science of Finland, Series AIII, No.* 90, pp. 139–41 (1966).
281. HONKASALO, T. Special techniques of gravity measurement. *Trav. Ass. int. Geod.* **25** 145–50 (1976).
282. HOPCKE, W. The curvature of electromagnetic waves. *Surv. Rev. No.* 141, pp. 298–312 (1966).
283. HOPFIELD H. S. Two quartic tropospheric refraction profiles for correcting satellite data. *J. geophys. Res.* **74** (18), 4487–99 (1969).
284. HOPFIELD, H. S. Tropospheric effect on electromagnetically measured ranges. Prediction from surface weather data. *Radio Science* **6** (3), 357–67 (1971).
285. HOPFIELD, H. S. Tropospheric range error parameters. Further studies. *Goddard SFC* X–551–72–285 (1972).
286. HOPFIELD, H. S. Tropospheric effects on signals at very low elevation angles. *APL/JHU Tech. Memo.* TG 1291 (1976).
287. HOPFIELD, H. S. Tropospheric effects on low-elevation signals. Further studies. *APL/JHU.* SDO 4588 (1976).
288. HOPFIELD, H. S. Tropospheric correction of electromagnetic ranging signals to a satellite; study of parameters. *Electromagnetic distance symposium, Wageningen, Netherlands, May* 1977 (1977).
289. HOPKINS, H. G. Radio tracking at Winkfield. *Phil. Trans. R. Soc* **A262,** 1124, 46–49 (1967).
290. HOSKINSON, A. J., and DUERKSEN, J. A. Manual of geodetic astronomy. *Spec. Publs U.S. Cst geod. Surv. No.* 237 (1947).

291. HOTHEM, J. D. *Evaluation of precision and error sources associated with doppler positioning.* IAG General Assembly (1975).

292. HOTINE, M. The East African arc. Marks and beacons. *Emp. Surv. Rev. No.* 14, pp. 472–7 (1934).

293. HOTINE, M. The East African arc. *Emp. Surv. Rev. Nos.* 12, 14, and 16 (1934–5).

294. HOTINE, M. The East African arc. Base measurement. *Emp. Surv. Rev. No.* 18, pp. 203–18 (1935).

295. HOTINE, M. The re-triangulation of Great Britain. *Emp. Surv. Rev. Nos.* 25, 26, and 29 (1937–8).

296. HOTINE, M. The orthomorphic projection of the spheroid. *Emp. Surv. Rev. Nos.* 62, pp. 300–11; 63 pp. 25–35; 64, pp. 52–70; 65, pp. 112–23; 66, pp. 157–66 (1946–7).

297. HOTINE, M. *Mathematical geodesy.* ESSA Monograph. Washington (1969).

298. HUGGETT, G. R., and SLATER, L. E. *Recent advances in multi-wavelength distance measurement. Electronic distance measurement and atmospheric refraction,* pp. 141–52. Ned. Geod. Comm. (1977).

299. HUSTI, G. J. Twin Laplace point Ubachberg–Tongeren, applying the Black method. *Ned. Geod. Comm.* **4** (1) (1971).

300. HUSTI, G. J. Deviations of the vertical in the Netherlands. *Ned. Geod. Comm.* **6** (3) (1978).

301. HYTONEN, E. Measuring refraction in the second levelling of Finland. *Publication of the Finnish Geodetic Institute No.* 63 (1967).

302. ISOTOV, A. A. Reference ellipsoid and standard geodetic data adopted in USSR. *Bull. géod. No.* 53, pp. 1–6 (1959).

303. JACKSON, J. E. Observations with the Cambridge pendulum apparatus in North, Central and South America in 1958. *Geophys. J. R. astr. Soc.* **2** (4) 337–47 (1959).

304. JACKSON, J. E. The Cambridge pendulum apparatus. *Geophys. J. R. astr. Soc.* **4** 375–88 (1961).

305. JEFFREYS, H. *The earth.* Cambridge University Press. 2nd edn (1929); 4th edn (1959); 5th edn (1970).

306. JEFFREYS, H. The figures of the earth and moon. *Mon. Not. R. astr. Soc. geophys. Suppl.* **5** (7), 219–47 (1948).

307. JEFFREYS, H. On the hydrostatic theory of the earth. *Geophys. J. R. astr. Soc.* **8** (2), 196–202 (1963).

308. JELSTRUP, J. Crossing of fiords with precise levelling. *Bull. géod. No.* 38, pp. 55–63 (1955).

309. JENKINS, R. E. A satellite observation of the relativistic doppler shift. *Astronomical J.,* **74** (7), 960–3 (1969).

310. JENSEN, H. Formulas for the astronomical correction to the precise levelling. *Bull. géod. No.* 17, pp. 267–77 (1950).

311. JONES, H. E. Geodetic ties between continents by means of radio telescopes. *Can. Surv.* **23** (4), 377–88 (1969).

312. JONES H. E. Systematic errors in Tellurometer and Geodimeter measurements. *Can. Surv.* **25** (4) (1971).

313. JONES, P. BERTHON. Instrumental sources of error in levelling of high

precision by means of automatic levels. *Surv. Rev. Nos.* 132, pp. 276–86; 133, pp. 313–22; 134; pp. 346–54 (1964).

314. JONES, P. B. Correction for lost motion in theodolite eyepiece micrometers. *Bull. géod. No.* 107, pp. 5–11 (1973).

315. JORDAN, W., and EGGERT, O. *Handbuch der Vermessungskunde.* Revised 1939–41. Translated into English by M. W. Carta. Army Map Service, Washington (1962).

316. JORDAN, W., EGGERT, O., and KNEISSL, M. *Handbuch der Vermessungskunde.* In parts, 1958 and later years.

317. KAARIAINEN, E. On the recent uplift of the earth's crust in Finland. *Publication of the Finnish Geodetic Institute No.* 42 (1953).

318. KAARIAINEN, E. The second levelling of Finland in 1935–55. *Publication of the Finnish Geodetic Institute No.* 61 (1966).

319. KAARIAINEN, E. Astronomical determinations of latitude and longitude in 1961–66. *Publication of the Finnish Geodetic Institute No.* 71 (1971).

320. KAKKURI, J. Eliminating the refraction error from the long optical sights in water crossings, *Vermess Wes., Wien,* **25** 305–14 (1967).

321. KAKKURI, J. Stellar triangulation with balloon-borne beacons. *Publications of the Finnish Geodetic Institute No.* 76 (1973).

322. KAKKURI, J., and KAARIAINEN, J. The second levelling of Finland for the Aland Archipelago. *Publication of the Finnish Geodetic Institute No.* 82 (1977).

323. KANIUTH, K., and STUBER, K. Astrogeodatischen Lotabweichungsbestimmungen in 1966 bis 1977. *Dt. geod. Kommn.* B229 (1978).

324. KARKI, P., KIVIOJA, L. and HEISKANEN, W. Topo-isostatic reduction maps for the world for the Hayford zones 18 to 1, Airy–Heiskanen system, $T = 30$ km. *Publs isostatic Inst. Ass. Geod. No.* 35 (1961).

325. KAULA, W. M. Statistical and harmonic analysis of gravity. *Army Map Service Tech. Rep.* **24** Washington (1957).

326. KAULA, W. M. Determination of the earth's gravitational field. *Rev. Geophys.* **1** (4), 507–51 (1963).

327. KAULA, W. M. Analysis of gravitational and geometric aspects of geodetic utilisation of satellites. *Geophys. J. R. astr. Soc.* **5** 104–33 (1961).

328. KAULA, W. M. Review of tesseral harmonic determination from satellite orbit dynamics. *Use of artificial satellites for geodesy*, vol. 2 (ed. G. Veis), pp. 133–47. Athens (1965).

329. KAULA, W. M. *Theory of satellite geodesy.* Blaisdell Publishing Co. (1966).

330. KAULA, W. M. Tests and combinations of satellite determinations of the gravity field with gravimetry. *J. geophys. Res.*, **71** 5303–14 (1966).

331. KAULA, W. M., and LEE, W. A spherical harmonic analysis of the earth's topography. *J. geophys. Res.*, **72** 753–8 (1967).

332. KOZAI, Y. The motion of a close earth satellite. *Astron. J.*, **64** 367–77 (1959).

333. KERSHNER, R. B. The Geos satellite and its use in Geodesy. *The use of artificial satellites for geodesy*, vol. 2 (ed. G. Veis), pp. 97–110. Second International Symposium, Athens, 1965. National Technical University, Athens (1967).

334. KHAN, M. A. Comparative evaluation of recent global representations of the earth's gravity field. *Geophys. J. R. astr. Soc.* **46** (3), 535–53 (1976).

335. KHAN, M. A. Hydrostatic figure of the earth: theory and results. *Goddard SFC* X-592-73-105 (1973).

336. KHAN, M. A. Evaluation and comparisons of recent geopotential solutions. *Goddard SFC* X-921-74-275 (1974).

337. KING-HELE, D. G. *Satellites and scientific research.* (1960).

338. KING-HELE, D. G., COOK, G. E., and SCOTT, D. W. The odd zonal harmonics in the earth's gravitational potential. *Planet. Space Sci.* **13** 1213–32 (1965).

339. KING-HELE, D. G., COOK, G. E., and SCOTT, D. W. Even zonal harmonics in the earth's gravitational potential. *Planet. Space Sci.* **14** 49–52 (1966).

340. KING-HELE, D. G., COOK, G. E., and SCOTT, D. W. *The odd zonal harmonics in the geopotential of degree less than* 33, *from the analysis of* 22 *satellite orbits.* R.A.E., Farnborough (1968).

341. KING-HELE, D. G., and COOK, G. E. Analysis of 27 satellite orbits to determine odd zonal harmonics in the geopotential. *Planet. Space Sci.* **22** 645–72 (1974).

342. KING-HELE, D. G., WALKER, D. M. C., *et al.* Geopotential harmonics of order 15 and even degree from changes in orbital eccentricity at resonance. *Planet. Space Sci.* **23** 229–46 (1973).

343. KING-HELE, D. G., WALKER, D. M. C., *et al.* Geopotential harmonics of order 15 and odd degree from analysis of resonant orbits. *Planet. Space Sci.* **23** 1239–56 (1975).

344. KING-HELE, D. G., and WALKER, D. M. C. Upper atmosphere zonal winds from satellite orbit analysis. *Nature, Lond.* 264, 631–2 (1976).

345. KIVINIEMI, A. High precision measurements for studying the secular variation of gravity in Finland. *Publication of the Finnish Geodetic Institute No.* 78 (1974).

346. KNIGHT, D. J. E., and ROWLEY, W. R. C. Recent measurements of the speed of light. *Surv. Rev. No.* 185, pp. 131–4 (1977).

347. KRAKIWSKY, E. J. *Heights.* Ohio State University (1965).

348. KRASOVSKIJ, F. N., and DANILOV, V. V. *Rukovodstvo po vyschej geodesii.* Moscow (1938–42).

349. KRASNORYLOV, I. I. Longitude determinations. *Geodesy and Aero photography*, 1967, (5), 1969, (1), 1969 (3), and 1970, (6). English translation by Scripta Technica Inc., Amer. Geophys. Union, Washington, DC.

350. KUKKAMAKI, T. J. Über die nivellitische Refraktion. *And* Formeln und Tabellen zur Berechnung der nivellitischen Refraktion. *Publications of the Finnish Geodetic Institute Nos.* 25 and 27 (1938–9).

351. KUKKAMAKI, T. J. On lateral refraction in triangulation. *Bull. géod. No.* 11, pp. 78–80 (1949).

352. KUKKAMAKI, T. J. Levelling over the Turku and Åland Archipelago. *Det Tredie Nordiske Geodaetmode i Kobenhavn* 25–30 *Maj* 1959.

353. KUKKAMAKI, T. J., and HONKASALO, T. Measurement of the standard base line of Buenos Aires with Vaisäla comparator. *Bull. géod. No.* 34, pp. 355–62 (1954).

354. Kukkamaki, T. J. Vaisäla interference comparator. *Publications of the Finnish Geodetic Institute No.* 87 (1978).

355. Küpfer, H. P. *How to increase accuracy in EDM.* International Association of Geodesy, Lucerne (1967).

356. Laborde, J. *Traité des projections des cartes géographiques,* vol. 4. Paris (1937).

357. Lachapelle, G. Determination of the geoid using heterogeneous data. *Mitt. Geod. Inst. Graz,* **19** (1975).

358. LaCoste, L. J. B., and Harrison, J. C. Some theoretical considerations in the measurement of gravity at sea. *Geophys. J. R. astr. Soc.* **5** (2), 89–103 (1961).

359. LaCoste, L. J. B. Measurement of gravity at sea and in the air. *Rev. Geophys.* **5** (4), 477–526 (1967).

360. Lallemand, Ch. General report on levelling. *Proceedings of the International Geodetic Association,* Annexe B, viii, C. *Hamburg* (1912).

361. Lambeck, K. The probability of recording satellite images optically. *The use of artificial satellites for geodesy,* vol. 2 (ed. G. Veis), pp. 3–20. Second International Sympsoium, Athens, 1965. National Technical University, Athens (1967).

362. Lambeck, K. Optimum station-satellite configuration for simultaneous observations to satellites. *SAO Special Report 231* (1966).

363. Lambeck, K. Irregular atmospheric effects on satellite observations. *SAO Special Report* 269 (1968).

364. Lambert, B. P. The use of aerodist for filling in between tellurometer traverse loops. *Proceedings of the Commonwealth Survey Conference 1967.* H.M.S.O. (1968).

365. Lambert, B. P. The geodetic survey of Australia U.N. Cartographic Conference, Canberra, 1967.

366. Lambert, W. D. The reduction of observed values of gravity to sea-level. *Bull. géod. No.* 26, pp. 107–81 (1930).

367. Lambert, W. D. The international gravity formula. *Am. J. Sci.* **243A,** 360–92 (1945).

368. Lambert, W. D., and Darling, F. W. Tables for determining the form of the geoid and its indirect effect on gravity. *Spec. Publs. U.S. Cst geod. Surv. No.* 199 (1936).

369. Lambert, W. D., and Darling, F. W. Formulas and tables for the deflection of the vertical. *Bull. géod. No.* 57 (1938).

370. Lambert, W. D., and Swick, C. H. Formulas and tables for the computation of geodetic positions on the international spheroid. *Spec. Publs U.S. Cst geod. Surv. No.* 200 (1935).

371. Lane, J. A. Small-scale variation of radio refractive index in the troposphere. *Proc. Instn elect. Engrs.* **115** (9), 1227–29 (1968).

372. Lauf, G. B. and Young, F. Conformal transformation from one map projection to another using divided difference interpolation. *Bull. géod. No.* 61, pp. 191–212 (1961).

373. Laurila, S. H. *Electronic surveying and navigation.* Wiley (1976).

374. Lee, L. P. The Transverse Mercator projection of the spheroid. *Emp. Surv. Rev. No.* 38, pp. 142–52 (1945).

375. Lehr, C. G. Satellite tracking with a laser. *SAO Special Report* 215 (1966).

376. Lejay, P. *Développments modernes de la gravimétrie.* Paris (1947).

377. Lejay, P. Tables pour le calcul de l'effet indirect et la déformation du géoïde. *Bull. géod. No.* 8, pp. 99–163 (1948).

378. Lennon, G. W. Mean sea level as a reference for geodetic levelling. *Can. Surv.* **28** (5), 524–30 (1974).

379. Leppert, K. Two Australian baselines for the Pageos world triangulation. *Nat. Map. Tech. Rep. No.* 11 (1972).

380. Leppert, K. The Australian doppler satellite survey 1975–77. *Nat. Map. Tech. Rep. No.* 21 (1978).

381. Lepretre, J. P. Progress report for the creation of a world-wide gravimetric data bank. *Bull. Bureau Gravimètrique int.* **39** I-29 (1976).

382. Lerch, F. J., Klosko, S. M., *et al.* Gravity model improvement using Geos 3 (GEM 9 and 10). *Goddard SFC* X–921–77–246 (1977).

383. Lerch, F. J., Wagner, C. A., *et al.* Goddard earth models (5 and 6). *Goddard SFC* X-921-74-145 (1974).

383a. Lerch, F. J., Laubsher, R. E., *et al.* Determination of geocentric gravitational constant from near-earth laser ranging. Preprint (1978).

384. Levallois, J. J. Sur une équation intégrale très générale de la gravimétrie. *Bull. géod. No.* 50, pp. 36–49 (1958).

385. Levallois, J. J. Sur la fréquence des mesures de pesanteur dans les nivellements. *Bull. géod. No.* 74, pp. 317–25 (1964).

386. Levallois, J. J. *Géodésie générale.* Eyrolles (1970).

387. Levallois, J. J., and Masson d'Autume G, de. *Édude sur la réfraction géodésique et le nivellement barométrique.* Paris (1953). With *Annexe* of numerical tables printed separately.

388. Levallois, J. J. and Dupuy, M. Sur le calcul des grandes géodésiques. *Bull. géod. No.* 16, pp. 105–17 (1950).

389. Levallois, J. J., and Monge H. *Le géoid européen, version 1975.* IAG General Assembly (1975).

390. Lundquist, C. A. (ed.). Geodetic satellite results during 1967. *SAP Special Report* 264 (1967).

391. Lundquist, C. A., and Veis, G. (ed.). Geodetic parameters for a 1966 Smithsonian Institution standard earth. *SAO Special Report* 200 (1966).

392. Longhurst, R. S. *Geometrical and physical optics.* Longmans (1957).

393. Luck, J. M., Miller, M. J., *et al. The National Mapping lunar laser program.* Nat. Mapping, Canberra (1974).

394. Macdonald, A. S. An inexpensive tower for tellurometer traverses. *Emp. Surv. Rev. No.* 121, pp. 129–32 (1961).

395. Macduran, P. F. Very long base interferometry (VLBI) applications to secular geodynamics and earth strain. *Proc. Symposium on earth's gravitational field and secular variations in position*, pp. 380–94. Sydney (1973).

396. MacRobert, T. M. *Spherical harmonics.* London (1927).

397. Markowitz, W., and Guinot, B. *Continental drift, secular motion of the pole and rotation of the earth.* Reidel (1968).

398. Martin, C. F., Klosko, S. M., *et al.* Determination of the masses of the earth, moon, and sun, and the size of the earth from Mariner 9 range and doppler observations. *Goddard SFC* X–922–75–134 (1975).

399. Mather, R. S., Barlow, B. C., *et al. The earth's gravitational field in the Australian region.* IAG General Assembly (1971).

400. Mathur, N. C., Grossi, M. D., *et al.* Atmospheric effects in very long baseline interferometry. *Radio Sci.* **5** (10), 1253–61 (1970).

401. Marsh, J. G., Douglas, B. C., *et al.* A unified set of tracking station coordinates, derived from geodetic satellite tracking data. *Goddard SFC* X–553–71–370 (1971).

402. Marsh, J. G., Douglas, B. C., *et al.* A global station coordinate solution based upon camera and laser data. *Goddard SFC* X–592–73–171 (1973).

403. Marsh, J. C., Douglas, B. C., *et al.* Geodetic results from ISAGEX data. *Goddard SFC* X–921–74–250 (1974).

404. McConnel, R. K., Hearty, D. B., *et al.* An evaluation of the LaCoste–Romberg Model D microgravimeter. *Bull. Bureau Gravimètrique Int.* **36** I, 35–45 (1975).

405. Meade, B. K. High precision geodimeter traverse Surveys in the United States. *Bull. géod. No.* 90, pp. 371–85 (1968).

406. Meade, B. K. Doppler data versus results from high precision traverse. *Can. Surv.* **28** (5), 462–66 (1974).

407. Meier-Hirmer, B. Mekometer ME 3000. Theoretical aspects, frequency calibration, field tests. *Electronic distance measurement and atmospheric refraction*, pp. 21–40. Ned. Geod. Comm. (1977).

408. Meisl, P. Zusammenfassung und Ausbau der inneren Fehlertheorie eines Puncthaufens. *Dt. geod. Kommn.* A61 (1969).

409. Melchior, P. *The earth tides.* Pergamon Press (1966).

410. Melchior, P. Reports sur les marées terrestres. *Trav. Ass. int. Géod.* **23** 367–75 (1968); **24** 289–309 (1972); **25** 332–46 (1976).

411. Merson, R. H. The perturbation of a satellite orbit in an axi-symmetric gravitational field. *R.A.E. Tech. Note. Space* 26. Farnborough (1963).

412. Merson, R. H. A comparison of the satellite orbit theories of Kosai and Merson, and their application to Vanguard 2. *RAE Tech. Note*, Space 42 (1963).

413. Meyler, D. S., and Sutton, O. G. *A compendium of mathematics and physics.* English Universities Press (1958).

414. Milne-Thompson, L. M., and Comrie, L. J. *Standard four-figure mathematical tables.* Macmillan (1931).

415. Mittermayer, E. A generalisation of the least-square method for the adjustment of free networks. *Bull. Géod. No.* 104, pp. 139–57 (1972).

416. Moffett, J. B. (Editor) Program requirements for two-minute integrated doppler satellite navigation solutions. *APL Tech. Memo.* TG-819-1, revision 2, Appendix D. (1973).

417. Molodensky, M. S. *Principal problems in astro-gravimetric levelling in large areas* (in Russian). *GUGK*, 4, Moscow (1944).

418. Molodensky, M. S. New methods of studying the figure of the earth. *Bull. géod. No.* 50, pp. 17–21 (1958).

419. Molodensky, M. S., Eremeev, V. F., and Yurkina, M. I. *Methods for study of the external gravitational field and figure of the earth.* Translated from Russian. Israel programme for scientific translations (1962).

420. MORELLI, C. (Editor). The international gravity standardisation net, 1971. *AIG Publ. spécial No.* 4 (1974).

421. MORELLI, C. and HONKASALO, T. Gravimetry. General report. *Trav. Ass. int. Géod.* **25,** 94–113 (1976).

422. MORITZ, H. Linear solutions of the geodetic boundary value problem. *Dt. geod. Komm A*58. Munich (1968).

423. MORITZ, H., LEVALLOIS, J. J., *et al. Geodetic reference system,* 1967. Ass. int. Géod. (1970).

424. MORITZ, H. Advanced least–squares methods. *Ohio State Univ. Report No.* 175 (1972).

425. MORITZ, H. A new series solution of Molodensky's problem. *Bull. Géod. No.* 96, pp. 183–95 (1970).

426. MORITZ, H. Precise gravimetric geodesy. *Ohio State Univ. Report No.* 219 (1974).

427. MORITZ, H. Fundamental geodetic constants. Study Group 5.39. *Trav. Ass. int. Géod.* **25** 411–8 (1976).

428. MORONEY, M. J. *Facts from figures.* Pelican (1951).

429. MORTON, H. P. D. *Geod Rep. Surv. India* **5,** 94–5 (1928–9).

430. MUELLER, I. I. *Introduction to satellite geodesy.* Ungar, New York (1964).

431. MUELLER, I. I. Global satellite triangulation and trilateration results. *J. geophys. Res.* **79** (35), 5333–47 (1974).

432. MUELLER, I. I., KUMAR, M., *et al.* Global satellite triangulation and trilateration for the national geodetic satellite program. Solutions WN 12, 14, and 16. *Ohio State Univ. Report No.* 199 (1973).

433. MULHOLLAND, J. D. (ed.) *Scientific application of lunar laser ranging.* (Austin, Texas 1976). Reidel (1977).

434. MÜLLER, H. *Astronomical position, time and azimuth determinations with the Kern DKM* 3–A. Kern, Aarau (1973).

435. MUNSEY, D. F. Base measurement in the Sudan. *Emp. Surv. Rev. Nos.* 72 pp. 67–74; 73, pp. 98–105; 74, pp. 155–61 (1949).

436. MUNSEY, D. F. First and second order triangulation. 1943–52. *Sudan Survey departmental records,* vol. 2. Cambridge (1959).

437. MUNSEY, D. F., and PUGH, K. T. Observations with PIM. *Surv. Rev. No.* 161, pp. 119–31 (1971).

437A. MURRAY, C. A. The astrometric reference frame. *Observatory* **96** (1012), 90–7 (1976).

438. NEEDHAM, P. F. Detailed geopotential model based on point masses. *Ohio State Univ. Report No.* 149 (1970).

439. NIBLOCK, J. Graduations on geodetic levelling staves. *Surv. Rev. No.* 138, pp. 169–171 (1965).

440. NIETHAMMER, T. HUNZIKER, E., and ENGI, P. *Publications of the Swiss Geodetic Commission* 19, 20, and 22. Berne (1932, 1939, and 1944).

441. NIETHAMMER, T. *Die genauen Methoden der astronomisch-geographischen Ortsbestimmung.* Basel (1947).

442. NISKANEN, E., and KIVIOJA, L. Topo-isostatic world maps for the effect of the Hayford zones 10 to 1 for the Airy–Heiskanen and Pratt–Hayford systems. *Publs. isostatic Inst. Ass. Geod. No.* 27 (1951).

443. Obenson, G. Error analysis of deflection of the vertical and undulations from the accuracy of gravity anomalies. *Bull. Géod. No.* 108, pp. 141–56 (1973).

444. Odermatt, H. *Universal theodolite Wild T4. Instructions for the determination of geographic positions.* Heerbrugg (undated, about 1955).

445. Olander, V. R. The weight function in astronomical levelling. *Bull. géod. No.* 34, pp. 329–42 (1954).

446. Olander, V. R. Astronomical azimuths observed in 1920–59 in the primary triangulation net. *Publication of the Finnish Geodetic Institute No.* 60 (1965).

447. Ollikanen, M. Astronomical determination of latitude and longitude in 1972–5. *Publications of the Finnish Geodetic Institute No.* 81 (1977).

448. Olliver, J. G. Zone to zone transformation on the Australian map grid. *Aust. Surv.* **28** (3) (1976).

449. Olliver, J. G. Observation equations for observed directions and distances in spheroidal coordinates. *Surv. Rev. No.* 184, pp. 71–7 (1977).

450. Orlin, H. Marine gravity surveying instruments and practice. *Proceedings of the First Marine Geodesy Symposium*, pp. 181–7. U.S. Govt. Printing Office (1967).

451. O'Toole, J. W. Celeste computer program for computing satellite orbits. *Naval Surface Weapons Centre Tech. Rep. No.* 3565 (1976).

452. Owens, J. C. The use of atmospheric dispersion in optical distance measurement. *Bull. géod. No.* 89, pp. 277–91 (1968).

453. Oxburgh, E. R. The plain man's guide to plate tectonics. *Proc. Geologist's Ass.* **85** (3), 299–358 (1974).

454. Parkinson, B. W., Navstar global positioning system (GPS). *Proc. Nat. Telecommunication Conf.*, Dallas (1976), vol iii, 41.1–41.5 (1977).

455. Parm, T. High precision traverse of Finland. *Publications of the Finnish Geodetic Institute No.* 79, pp. 1–108 (1976).

456. Patterson, W. S. B. Atmospheric refraction above the inland ice in North Greenland. *Bull. géod. No.* 38, pp. 42–54 (1955).

457. Pearlman, M. R., Lehr, C. G., *et al. The Smithsonian satellite ranging laser system.* IAG General Assembly (1975).

458. Perfect, D. S. The photographic zenith tube of the Royal Greenwich Observatory, *Occ. Notes. R. astr. Soc. No.* 21 (1959).

459. Perrier, G. Comptes rendus de la Section de Géodésie, Madrid, 1924. *Bull. géod. No.* 7, pp. 552–6 (1925).

460. Perrier, G., and Hasse, E. Tables de l'ellipsoïde de référence internationale. *Special Publications of the International Association of Geodesy Nos.* 2 (sexagesimal) and 3 (centesimal) (1935 and 1938).

461. Plotkin, H. H. Laser technology for high precision satellite tracking. *Proc. Symposium on gravitational field and secular variations in position*, pp. 328–46. Sydney (1973).

461a. Plummer, H. C. *An introductory treatise on dynamical astronomy.* Cambridge (1918).

462. Poder, K., and Andersen O. B. Microwave reflection problems. *Electromagnetic distance measurement*, pp. 81–95. Hilger and Watts (1967).

463. Prescott, N. J. D. Experiences with Secor planning and data reduction. *Electromagnetic distance measurement*, pp. 312–38. Hilger and Watts (1967).

464. Prey, A. Darstellung der Höhen- und Tiefenverhältnisse der Erde durch eine Entwickelung nach Kugelfunctionen bis zur 16 Ordnung. *Abh. K. Ges. Wiss. Göttingen* **2** (1), (1922).

465. Rainsford, H. F. Long lines on the earth. *Emp. Surv. Rev. Nos.* 71, pp. 19–29; 72, pp. 74–82 (1949).

466. Rainsford, H. F. *Survey adjustments and least squares.* Constable (1957).

467. Rainsford, H. F. Combined adjustments of angles and distances. *Surv. Rev. No.* 150, pp. 348–63 (1968).

468. Ramastry, J., Rosenbaum, B., *et al.* Tracking of ATS 3 by VLBI technique. *Goddard SFC* X–553–72–290 (1972).

469. Ramsayer, K. Gravity reduction of the levelling network in Baden-Württemberg. *Bull. géod. No.* 52, pp. 76–9 (1959).

470. Ramsayer, K. Über den zulässigen Abstand der Schwerepunkte bei der Bestimmung geopotentieller Koten im Hochgebirge, Mittelgebirge und Flachland. *Dt. geod. Kommn* A44. Munich (1963).

471. Ramsayer, K. Errors in the determination of astronomical refraction *Vermess Wes., Wien* **25** 260–9 (1967).

472. Rannier, J. L., and Dennis, W. M., Variable personal equation of bisection. *Can. J. Res.* **10** 342–6 (1934).

473. Rannie, J. L., and Dennis, W. M. Improving the performance of Wild precision theodolites. *Emp. Surv. Rev. No.* 15, pp. 2–5 (1935).

474. Rapp, R. H. The formation and analysis of a 5° equal area block terrestrial gravity field. *Ohio State Univ. Report No.* 178 (1972).

475. Rapp, R. H. $1° \times 1°$ mean free air gravity anomalies. *DMAAC Reference publ.* 73.0002,100 pp. (1973).

476. Rapp, R. H. Mean gravity anomalies and sea surface heights derived from Geos 3 altimeter data. *Ohio State Univ. Report No.* 267 (1977).

477. Rapp, R. H., and Rummel, R. Methods for the computation of detailed geoids, and their accuracy. *Ohio State Univ. Report No.* 233 (1975).

478. Rappleye, H. S. Manual of geodetic levelling. *Spec. Publs U.S. Cst geod. Surv. No.* 239 (1948).

479. Redfearn, J. C. B. Transverse Mercator formulae. *Emp. Surv. Rev. No.* 69, pp. 318–22 (1948).

480. Reid, J. K. On the method of conjugate gradients for the solution of large sparse systems of linear equations. *Proc. Oxford Conference of the Inst. of mathematics and its applications* (1970).

481. Rice, D. A. Deflections of the vertical from gravity anomalies. *Bull. géod. No.* 25, pp. 285–312 (1952).

482. Rice, D. A. A geoidal section in the United States. *Bull. géod. No.* 65, pp. 243–51 (1962).

483. Richards, M. R. The use of thermometers for temperature measurement in EDM. *Surv. Rev. No.* 143, pp. 43–7 (1967).

484. Richardus, P., and Allman, J. S. *Project surveying.* North-Holland, Amsterdam (1966).

485. Robbins, A. R. Personal equation in the determination of geodetic azimuths. *Bull. géod. No.* 57 (1960).
486. Robbins, A. R. Long lines on the spheroid. *Emp. Surv. Rev. No. 125*, pp. 301–9 (1962).
487. Robbins, A. R. A geoidal section in Great Britain. *Surv. Rev. No.* 128, pp. 69–75; 129, pp. 121–32 (1963).
488. Robbins, A. R. Time in geodetic astronomy. *Surv. Rev. Nos.* 143, pp. 2–18; 144, pp. 94–5 (1967).
489. Robbins, A. R. The Chronocord Mk III. A portable recording crystal chronometer. *Bull. géod. No.* 87, pp. 87–94 (1968).
490. Robbins, A. R. Critical study of methods in geodetic astronomy. *Trav. Ass. int. Géod.* **24** 229–41 (1972).
491. Robbins, A. R. *A future international earth rotation service.* IAG General Assembly (1975).
492. Robbins, A. R. Field geodetic astronomy. *Military Engineering* XIII, part IX. MOD (1976).
493. Robbins, A. R. Problems in geodetic astronomy. *Trav. Ass. int. Géod.* **25** 23–30 (1976).
494. Robbins, A. R. Geodetic astronomy in the next decade. *Surv. Rev. No.* 185, pp. 99–108 (1977).
495. Robertson, K. D. A method for reducing the index of refraction errors in length measurement. *Survey and Mapping*, xxxv, pp. 115–29 (1975).
496. Robertson, D. S. Geodetic and Astronomic measurements with VLBI. *Goddard SFC* X–922–77–228 (1975).
497. Robinson, G. D. Some aspects of the meteorology and refractive index of the air near the earth's surface. *Electromagnetic distance measurement*, pp. 96–103. Hilger and Watts (1967).
498. Roelofs, R. Astronomical determination of longitude in the Netherlands, Appendix VI. *Trav. Ass. int. Géod.* **17** (1952).
499. Ross, J. E. R., *et al.* Geodetic application of Shoran. *Publs geod. Surv. Can. No.* 78 (1955).
500. Ross, J. E. R. Geodetic astronomy in Canada. *Festschrift C. F. Baeschlin*, pp. 225–38. Zurich (1957).
501. Rossiter J. R. An analysis of annual sea level variations in European waters. *Geophys. J. R. astr. Soc.* **12** (3), 259–99 (1967).
502. Routh, E. J. *Analytical statics*, vol. 2. Cambridge University Press (1932).
503. Rudoe, W. Manuscript notes, unpublished (1946).
504. Rummel, R., Sjaberg, L., *et al.* The determination of gravity anomalies from geoid heights. *Ohio State Univ. Report No.* 269 (1977).
505. Rummel, R., and Rapp, R. H. Undulations and anomaly estimates using Geos-3 altimeter data without precise satellite orbits. *Bull. Géod.* **51** (1), 73–88, (1977).
506. Rutherford, D. E. *Vector methods.* Oliver and Boyd (1948).
507. Rutherford, D. E. *Classical mechanics.* Oliver and Boyd (1951).
508. Rutscheidt, E. H. Preliminary results of the SECOR equatorial network. The use of artificial satellites for geodesy. *Geophysical Monographs, AGU*, **15** 49–58 (1972).

509. SAASTAMOINEN, J. The effect of path curvature of light waves on the refractive index. Applications to electronic distance measurement. *Can. Surv.* **16** (2) (1962).

510. SAASTAMOINEN, J. The path curvature of electromagnetic waves. *Electromagnetic distance measurement*, pp. 137–64. Hilger and Watts (1967).

511. SAASTAMOINEN, J. Contributions to the theory of atmospheric refraction. *Bull. Géod. No.* 105, pp. 279–98 (1972); *No.* 106, pp. 383–97 (1972); and *No.* 107, pp. 13–34 (1973).

512. SADLER, D. H. Astronomical measures of time. *Q. Jl R. astr. Soc.* **9** (3), 281–93 (1968).

513. SAKUMA, A. Absolute measurements of gravity. Study Group 3.18. *Trav. Ass int. Géod.* **25** (1976).

514. SCHELLENS, D. F. Design and application of automatic levels. *Can. Surveyor,* **19** (2) (1965).

515. SCHLESINGER, F., and JENKINS, L. F. *Catalogue of bright stars.* Yale (1940).

516. SCHMIDT, H. H. Worldwide geometric satellite triangulation. *J. geophys. Res.* **79** (35), 5549–76 (1974).

517. SCHOEPE, D. Über den einflus von Schwereanomalien auf die Positionen von Kunstlichen Erdsatelliten, *Zentral Inst. fur Physik der Erde,* **40** Potsdam (1976).

518. SCHURER, M. Improvements in satellite tracking. *The use of artificial satellites for geodesy and geodynamics* (ed. G. Veis), pp. 25–9. Athens (1974).

519. SEPPELIN, T. O. The department of defence world geodetic system, 1972. *Can. Surv.* **28** (5), 496–506 (1974)

520. SHAPIRO, I. I., and KNIGHT, C. A. Geophysical applications of long baseline radio interferometry. *Earthquake displacement fields and the rotation of the earth* (ed. N. Mansinha), pp. 284–301. Reidel (1970).

521. SHAPIRO, I. I., ROBERTSON, D. S., *et al.* Transcontinental baselines and the rotation of the earth measured by radio interferometry. *Science,* **186,** 920–2 (1974).

522. SHAPIRO, I. I., and YAPLEE, B. S. Potential of satellite radar altimetry for determination of short wavelength geoidal undulations. *Use of artificial satellites for geodesy and geodynamics* (ed. G. Veis), pp. 481–509. Athens (1973).

523. SHARMA, S. K. A note on geodesic lengths. *Surv. Rev. No.* 140, pp. 291–5 (1966).

524. SHIPLEY, G., BRADSELL, R. H., *et al.* A compact 2-colour EDM instrument. *Surv. Rev. No.* 179, pp. 210–33 (1976).

525. SIMONSEN, O. Triangulation between Denmark and Norway in 1945 by means of parachute flares. *Bull. Géod. No.* 11, pp. 33–52 (1949).

526. SIMONSEN, O. *Rapport sur l'établissement et la compensation de réseaux de nivellement européens.* International Association of Geodesy, Liverpool symposium. Copenhagen (1959).

527. SIMONSEN, O. *Closing errors and* Δc *and* $\Delta c'$ *from various countries.* International Association of Geodesy, Helsinki (1960).

528. SIMONSEN, O. *The astronomical correction for levelling of high precision when*

considering the definition of levelling datum. Danish Geodetic Institute (1965). With *Supplementary remarks* (1968).

529. Simonsen, O. *Report for the period Sept.* 1963 *to July* 1967 *on REUN.* Danish Geodetic Institute (1967).
530. Sims, T. NWL Precision ephemeris. *Naval Weapons Lab. Tech. Rep. No.* 2872 (1972).
531. Singer, S. F. Application of an artificial satellite to the measurement of the general relativistic red shift. *Phys. Rev.*, **104** (1), 11–14 (1976).
532. Sjaberg, L. The accuracy of gravimetric deflections of the vertical as derived from the GEM 7 potential coefficients and terrestrial gravity data. *Ohio State Univ. Report No.* 265 (1977).
533. Smart, W. M. *Spherical astronomy*. Cambridge University Press (1944).
534. Smith, C. *Solid geometry*. Macmillan (1920).
535. Smith, D. E., Kolonkiewicz, R., *et al.* polar motion from laser tracking of artificial satellites. *Goddard SFC* X–553–72–247 (1972).
536. Smith, E. K., and Weintraub, S. The constants in the equation for atmospheric refractive index at radio frequencies. *Proc. IRE*, **41** (8), 1035–37 (1953).
537. Smith, R. C. H. A modified GAK 1 gyro attachment. *Surv. Rev. No.* 183, pp. 3–24 (1977).
538. Smith, W. M. Angular measurements with the Lambert instrument tower. *Can. Surv.* **28** (3) (1974).
539. Sollins, A. D. Tables for the computation of the deflection of the vertical from gravity anomalies. *Bull. géod. No.* 6, pp. 279–300 (1947).
540. Somigliana, C. Teoria generale del campo gravitazionale dell'ellissoide di rotazione. *Memorie Soc. astr. ital.* **4** (1929).
541. Stacey, F. D. Physics of the earth. Space Science text series, Wiley (1969).
542. Stokes, G. G. On the variation of gravity at the surface of the earth. *Mathematical and physical papers*, vol. 2, pp. 104–71. Cambridge University Press (1883).
543. Strange, W. E., and Woolard, G. P. *Anomaly selection for deflection interpolation*, vols 1 and 2. Geophysics Institute, Hawaii (1964).
544. Strasser, G. J., and Schwendener, H. R. A north-seeking gyro attachment for the theodolite. *Bull. géod. No.* 79, pp. 23–38 (1966).
545. Sturges, W. Sea level slope along continental boundaries. *J. geophys. Res.* **79** (6), 826 (1974).
546. Sutton, O. G. *Atmospheric turbulence*. Methuen (1949).
547. Sutton, O. G. *Micrometeorology*. McGraw-Hill (1953).
548. Swick, C. H. Pendulum gravity measurements and isostatic reductions. *Spec. Publs U.S. Cst geod. Surv. No.* 232 (1942).
549. Synge, J. L., and Griffith, B. A. *Principles of mechanics*. McGraw-Hill (1959).
550. Szabo, B., and Anthony, D. Results of AFCRL's experimental aerial gravity measurements. *Bull. géod. No.* 100, pp. 179–202 (1971).
551. Takagi, S. *Comparison between results of astronomical and doppler satellite observation*. IAG General Assembly (1975).
552. Tarczy-Hornoch, A., and Eszto, P. The influence of wind on Jäderin wires. *Bull. géod. No.* 3, pp. 23–48 (1947).

553. TAYLOR, E. A. Optical tracking systems for space geodesy. *The use of artificial satellites for geodesy*, vol. 1 (ed. G. Veis), pp. 187–92. First International Symposium, Washington, 1962. North-Holland, Amsterdam (1963).

554. TAYLOR, G. I. Eddy motion in the atmosphere. *Phil. Trans. R. Soc.* **A215** (1915).

555. TAYLOR, G. I. Turbulence in the lower atmosphere. *Proc. R. Soc.* **A94,** 137–55 (1918).

556. TENGSTROM, E. *Research on methods of determining level surfaces of the earth's gravity field*, pp. 34–44. Uppsala (1964).

557. TENGSTROM, E. Elimination of refraction in vertical angle measurements, using lasers of different wavelength. *Z. Vermess Wes., Wien,* **25** 292–303 (1967).

558. THOMAS, P. D. Conformal projections in geodesy and cartography. *Spec. Publs U.S. Cst geod. Surv. No.* 251 (1952).

559. THOMAS, T. L. The suspended gyroscope. *Chartered Surv. Land, Hydro. and Minerals.* **2** (3), 39–49 (1975), and **3** (3), 33–43 (1976).

560. THOMPSON, E. H. The theory of the method of least squares. *Photogramm. Rec.* **4** (19) 53–65 (1962).

561. THOMPSON, E. H. *Introduction to the algebra of matrices with some applications.* Adam Hilger (1969).

562. THOMPSON, E. H. A note on conformal map projections. *Surv. Rev. No.* 175, pp. 17–28 (1975).

563. TOBEY, W. M. Geodesy. *Publs geod. Surv. Can. No.* 11 (1928).

564. TSUBOKAWA, I. *An electronic astrolabe.* International Association of Geodesy, Lucerne (1967).

565. TSUBOKAWA, I. and DAMBARA, T. *A method of photo-electric observation of a light-spot, and its application to geodesy.* Geodetic Society of Japan (1954 and 1957).

566. VALI, V., and BOSTROM, R. C. The use of a laser extensometer to observe strain in a large ground sample. *Bull. géod. No.* 88, pp. 151–5 (1968).

567. VAN LOON, D. L. F., and POELSTRA, T. J. Modified astronomic procedure of satellite plate reduction. *Ned. Geod. Comm.* **6** (2) (1976).

568. VEIS, G. Geodetic uses of artificial satellites. *Smithson. Contr. Astrophys.* **3** (9) (1960).

569. VEIS, G. (ed.). *The use of artificial satellites for geodesy*, vol. 1. First International Symposium, Washington, 1962. North-Holland, Amsterdam (1963).

570. VEIS, G. The determination of absolute directions in space with artificial satellites. *Bull. géod. No.* 72, pp. 147–66 (1964).

571. VEIS, G. (cd.). *The use of artificial satellites for geodesy*, vol. 2. Second International Symposium, Athens, 1965. National Technical University, Athens (1967).

572. VEIS, G. (ed.) *Use of artificial satellites for geodesy and geodynamics.* Athens (1973).

572A. VENING MEINESZ, F. A. A formula expressing the deviation of the plumbline in terms of the gravity field, and gravity potentials outside the geoid. *Proc. Sect. Sci. K. ned. Akad. Wet.* **31** (3) (1928).

573. Vening Meinesz, F. A. *Theory and practice of pendulum observations at sea*, Parts I and II. Delft (1929 and 1941).

574. Vening Meinesz, F. A. Méthode pour la réduction isostatique régionale. *Bull. géod. No.* 29, pp. 33–45 (1931).

575. Vening Meinesz, F. A. *Gravity expeditions at sea*, vols. 1–4. Delft (1932, 1934, 1941, and 1948).

596. Weightman, J. A. Doppler ties to the European datum and the European geoid. *Proc. Commonwealth. Surv. Conf.* (1975).

577. Vening Meinesz, F. A. Changes of deflections of the plumb-line brought about by a change of the reference ellipsoid. *Bull. géod. No.* 15, pp. 43–51 (1950).

578. Vening Meinesz, F. A. The outside gravity field up to a great distance from the earth. *Proc. Sect. Sci. ned. Akad. Wet.* **62B** (2) (1959).

579. Vignal, J. Evaluation de la précision d'une méthode de nivellement. *Bull. géod. No.* 49, pp. 1–159 (1936).

580. Vincenty, T. Transformation of geodetic data between reference ellipsoids. *J. geophys. Res.* **71** (10), 2619–24 (1966).

581. Vincenty, T. Direct and inverse solution of geodesics on the ellipsoid with application of nested equations. *Surv. Rev. No.* 176, pp. 88–93 (1975), and *No.* 180, pp. 294 (1976).

581a. Vincenty, T. The use of relative lateration for reducing errors in length measurements. *Surv. Rev. No.* 189, pp. 295–302 (1978).

581b. Vincenty, T. Correspondence, 1978.

582. Vincenty, T., and Bowring, B. R., Application of three-dimensional geodesy to adjustments of horizontal networks. *NOAA Tech. Memo.* NOS NGS–13 (1978).

583. Vincent, S., and Strange, W. E., et al. A detailed gravimetric geoid of N. America, the North Atlantic, Eurasia, and Australia. *Goddard SFC* X–553–72–331 (1972).

584. Waalewijn, I. A. Hydrostatic levelling in the Netherlands. *Surv. Rev. Nos.* 131, pp. 212–21; 132, pp. 267–76 (1964).

585. Wadley, T. L. The tellurometer system of distance measurement. *Emp. Surv. Rev. Nos.* 105, pp. 100–11; 106, pp. 146–60 (1957).

586. Wagner, C. A. Zonal gravity harmonics from long satellite arcs by a semi-numeric method. *J. Geophys. Res.* **78** (17), 3271–80 (1973).

587. Wagner, C. A. The accuracy of Goddard earth models. *Goddard SFC* X–921–76–187 (1976).

588. Wagner, C. A., Lerch, F. J., *et al.* Improvement in the geopotential derived from satellite and surface data. (GEM 7 and 8). *Goddard SFC* X–921–76–20 (1976).

589. Wakefield, R. C., and Munsey, D. F. Report on the arc of the 30th meridian in the Sudan (1935–40). *Sudan Survey Department Records*, vol. 1 (1950).

590. Walker, J. T. Survey of India. *Operations of the Great Trigonometrical Survey*, vol. 3, *The north-west quadrilateral* (1873).

591. Walker, J. T. Survey of India. *Operations of the Great Trigonometrical Survey*, vol. 2, *Reduction of the principal triangulation* (1879).

592. WATTS, A. B. and COCHRAN, J. R. Gravity anomalies and flexure of the lithosphere along the Hawaiian–Emperor seamount chain. *Geophys. J. R. astr. Soc.* **38** (1), 119–42 (1974).

593. WEBLEY, J. A. The tellurometer model MRA 101. *Electromagnetic distance measurement*, pp. 175–83. Hilger and Watts (1967).

594. WEIFFENBACH, G. C. Tropospheric and ionospheric propagation effects on satellite radio-Doppler geodesy. *Electromagnetic distance measurement*, pp. 339–52. Hilger and Watts (1967).

595. WEIGHTMAN, J. A. Gravity, geodesy and artificial satellites. *Use of artificial satellites for geodesy.* Athens (1963).

596. WEIGHTMAN, J. A. Doppler ties to the European datum and the European geoid. *Proc. Commonwealth Surv. Conf.* (1975).

597. WHALEN, C. T., and BALAZS, E. Test results of first-order class (iii) levelling. *NOAA Tech. Rep.* NOS 68 NGS 4 (1977).

598. WHITTAKER, E. T., and ROBINSON, G. *The calculus of observations.* Blackie (1926).

599. WHITTEN, C. A. Horizontal earth movement, San Francisco. *Trans. Am. geophys. Un.* **29** (3) (1948).

600. WHITTEN, C. A. Geodetic networks versus time. *Bull. géod. No.* 84, pp. 109–16 (1967).

601. WILKINS, G. A. The system of astronomical constants. *Q. Jl R. astr. Soc.* **5** (1), 23–31 (1964); **6** (1), 70–3 (1965).

602. WILLIAMS, J. W. Level transfers across water gaps by trigonometrical methods. *Proceedings of the Commonwealth Survey Conference* 1967. H.M.S.O. (1968).

603. WILLIAMS, O. W. Anna satellite yields photogrammetric parameters. *Photogramm. Engng*, **31** (2), 340–7 (1965).

604. WILLIAMS, D. C. First field tests of an angular dual wavelength instrument. *Electronic distance measurement and atmospheric refraction*, pp. 163–70. Ned. geod. Comm. (1977).

605. WILLIAMS, O. Airborne gravity measurements. *Bull. Bureau Gravimètrique. int.* **36** I–22 (1975).

606. WILLIAMS, H. S. and BELLING, G. E. The reduction of gyro-theodolite directions. *Surv. Rev. No.* 146, pp. 184–90 (1967).

607. WILLIAMSON, B. *An elementary treatise on the integral calculus.* Longmans (1918).

608. WILLIAMSON, M. R. Revised estimates of 550 km × 550 km mean gravity anomalies. *SAO Special Rep. No.* 377 (1977).

609. WILLIS, J. E. Determination of astronomical refraction from physical data. *Trans. Am. geophys. Un.* 1941, Pt. 2, pp. 324–36.

610. WOLF, H. Ideas and proposals for starting the adjustment of the European triangulation net. *Report of the Oct* 1962 *Munich symposium.* IAG (1963).

611. WOLF, H. Adjustment of a traverse network. *Surv. Rev. No.* 194, p. 190 (1979).

612. WOLFE A. J. and JOLLY, H. L. P. *The Second geodetic levelling of England and Wales*, p. 41. Ordnance Survey (1922).

613. WOOLARD, G. P. *World-wide gravity measurements with a gravity meter.* Woods Hole Oceanographic Institution (1949).

614. Woolard, G. P. *The relation of gravity anomalies to surface elevation, crustal structure and geology*. University of Wisconsin (1962).

615. Wunderlin, N. Lotabweichungen, Geoid und Meereshöhen in den Schweizer Alpen. *Publication of the Swiss Geodetic Commission No.* 26 (1967).

616. Wylie, C. R. *Advanced engineering mathematics*. McGraw-Hill (1960).

617. Yaskowich, S. A. Tellurometer ground swing on geodetic lines. *Can. Surveyor* **18** (1), 54–66 (1964).

618. Yionoulis, S. M. Algorithm to compute tropospheric refraction effects on range measurements. *J. geophys. Res.* **75** (36), 7636–7 (1970).

619. Yule, G. U., and Kendal, M. G. *An introduction to the theory of statistics*. Griffin (1957).

620. Zakatov, P. S. *A course in higher geodesy*. Moscow (1953). Translated into English by Israel translation program (1962).

INDEX

aberration 262, 268, 270
  annual, definition of 268
  corrections for 268, 270, 292, 304, 325, 550–1, 554
  diurnal, definition of 268
  planetary 554–5
  velocity 579
absolute gravity measurements 359–63
  accuracy of 362
  Cook 361
  Faller–Hammond 360
  Kater 359–60
  Sakuma 361–2
abstract, of angles 25, 27–8
accuracy
  of adjusted observations 712–13, 714–16
  of computations 105
  required of geodetic framework 2
  see also errors; least squares; standard errors; *under various instruments and methods*
acronyms 39, 506
adjustment (computational)
  by conditions 111–15, 144–53, 227, 713–18
  of correlated observations 154–7, 724–8
  by divided differences 161, 195
  figural 111–15, 716–18
  graphical 160–2
  of gravity observations 370–1, 374
  of large frameworks
    Bowie method 157
    using chains only 157
    division into independent sections 157
    Helmert–Wolf method 154–7
    objects of 2, 153–4
    transverse nets 157–8
  levelling 201, 226–8
  Molodensky correction 159–60
  satellite
    astro-triangulation 556–9
    general geodetic solutions 597, 613–14
  station 30–1, 131
  traverse nets 157–9
  traverses 124–5, 131, 149
  triangulated heights 233
  trilateration 126–44, 150
  by variation (parametric method)
    of coordinates 126–44, 707–13
    of heights 226–7
  see also theodolite, level, etc.; least squares; variation of coordinates
Aerodist 15, 87–8
Airy isostasy 484–7
  tables for 797
alluvium, effect on levelling 213–14, 491
Alps, geoidal profile 345, 470, 477
analytical function, definition 762
angle book 24, 27–8, 34, 36
  angles or directions 27–8
  description of station 28–9
  eccentric observation 29–30
angles exterior
  inclusion in adjustment 131, 137
  in traverse 131, 166–7
angles, horizontal
  accuracy of
    adjusted angles 141, 711–3
    observed angles 32–3, 164–7
  corrections for
    deviation 106–7
    geodesic 108–10
    skew normals 106–7
  observation of, *see* theodolite
angles or directions
  in abstract 27–8
  in figural adjustment 115
  in variation of coordinates 129–31, 137–9
angles, vertical, *see* vertical angles
Anna satellite 514, 540
anomaly
  of density or mass 478–90
  in deviation 466–7, 471–2
  of gravity (non-classical) 459, 463
  of potential 458
Apparent Places of Fundamental Stars 261, 263–4, 276
apse, orbital
  definition of 507–8
  rotation of 530
arc correction, pendulum 367
arc-to-chord correction 185–7
arcs, earth's figure deduced from 471
Aries, first point of, *see* equinox
astro-geodetic profiles, *see* geoid, sections
astrolabe, prismatic 325–8

astrolabe, prismatic (*contd.*)
  aberration 325
  accuracy 326–8
  advantages 328
  Claude and Driencourt 325–6
  computations by position lines 321–5
  Danjon 326–7
  impersonal 325–8
  Nusl–Fric 327
  programme 326
  Zeiss Ni2 327–8
astronomical azimuth, latitude, longitude
  definitions 97–8
  no use as framework control 95
  observations of, *see* azimuth; latitude; azimuth
  see also Laplace azimuth stations
Astronomical Ephemeris 264
astronomical theodolites 287, 294–5, 302, 328–9
astronomy, geodetic
  impact of satellites on 256–7
  objects of 255–7
astro-triangulation 537–62
  accuracy 559–62
  combined with ground survey 546–7, 557
  computation of directions 547–57
  computation of net 558–9
    conditions 558
    variation of coordinates 558–9
  definition 537–8
  net of 538
  notation 539
  optimum geometry 544, 556–7
  reference system 546, 555
  scale
    from ground survey 557
    from lasers 557–8
  selection of orbits 543–4
  suitable satellites 539–40
  use of atmospheric sondes 562
  utility of 537–8
  see also cameras; illumination; plate; satellite photography
atmosphere
  carbon dioxide 46
  composition 49–50
  density 50
  drag on satellite 608–9
  eddy conductivity 54–9, 237–41
  gravitational attraction 396, 422, 428, 430, 441
  pressure
    gradient of 50
    normal 51, 54, 60
  temperature
    abnormal 53–8, 60–1
    diurnal variation 55–7, 238
    gradient
      adiabatic 51
      near ground 53–7, 214
      normal 51–3, 238
    harmonic analysis of 238
    over water 58–9
  turbulence 54–9, 241
  water vapour pressure
    abnormal 53, 58–9, 62
    gradient of 51, 58–9
    normal 51
  see also ionosphere; meteorological observations; refractive index; refraction
atomic time, international (TAI) 276–9
ATS 6 satellite 515, 516
attraction
  of atmosphere 396, 422, 428, 430, 441
  definition of 396
    cf. gravity 398–9
  of earth at high altitudes 422–4, 511
  of plate 400
  of sphere 400
  of spherical shells 400
  of standard bodies 399–402
  see also gravity; potential
Australian traverses
  accuracy 168–9, 173–4
  adjustment 157–8
  MDM, humidity 62
  layout 9–10
  routine of observations 85–7
  use of mean pole 95
average height maps
  availability 447, 478
  uses 447, 798
axis of rotation, earth's
  instantaneous 259–60
  mean 260, 270–2
  see also polar motion; pole
azimuth, astronomical
  accuracy 170, 333
  Black's method 337–40
  circumpolar stars 334–5
  definition of 97–8
  to derive the deviation 100, 347
  east and west stars 336–7
  general rules 330–3
  in high latitudes 335, 340–1
  Laplace correction 4, 100, 331
  meridian transits 335–6
  personal equation 333
  pole star 333–4
  sigma Octantis 333–4
  summary of methods 340
azimuth, geodetic

computation of mutual 120–3
computation of reverse 117–20
definition of 96
derived from astro 4, 100
derived from satellites 13–15
error accumulated
in a chain 172–3
in a single figure 171
in a traverse 173
in high latitudes 341
observed by Black's method 337–41
see also Laplace azimuth

balloons as triangulation beacons
for photography 13, 562
for theodolite 13
see also astro-triangulation
Banachiewicz's method 670–1
band matrix 136–40, 151–3, 672
bar
definition 39
not SI unit, viii
barometric heights 33, 50–1, 61, 86, 372
basalt 393–4
baselines, invar 36, 40–2
accuracy 41, 170
extension 41, 170
frequency 4–5
measurement 40–1
reduction to spheroid 42, 105, 159–60
wind 41
see also invar; Vaisäla
beacons, opaque 17–18, 23
bearings, grid
defined 183
see also convergence; plane coordinates; projections
bench marks 224–5
fundamental 225
protected 225
tidal 248–9
see also movement of bench marks
Besselian
day numbers 267–68
solar year 281
Bessel's
formulae for s.e. and p.e. 698
refraction 284
spheroid 426
bias, meaning of 585; *see also* errors, systematic
BIH Greenwich, *see* Greenwich meridian
Bilby steel tower 18
bisection, personal error in 35, 333
Black's method for azimuth 337–41
in high latitudes 340–1
not impersonal 333

blunders
avoidance of 140, 711
definition of 691
rejection 32–3, 704–6
Boltz's method 154, 670
Boss's star catalogue 265–8, 286
use of 266–8
Bouguer gravity anomaly 451–3, 490
Bowie
correction 443–4, 797
method of adjustment 157
Browne terms 372
bubble, spirit
adjustment 21, 288–9, 302–3
calibration 22, 285
by Wisconsin method 283
correction to
horizontal angles and azimuths 22, 331–2
times of transit 303
vertical angles 34, 201–2
precautions when using 285
tester 22, 285
Bureau
Gravimetrique 359, 478
International de l'heure (BIH)
atomic time 276–8
BIH Greenwich or CZM 279
circulars 280–1
polar motion 272, 280
time of signal emission 280–1

calibration
in geoidal sections 344, 346
International gravity standardization net 362–3
lines, gravity 362–3, 374
see also errors, systematic; *and under the various instruments and methods*
cameras, satellite
Baker–Nunn 544–5, 628–31
BC 4 545, 628, 630
descriptions 544–6
Hewitt 546
modes of use 544
MOTS 40 545, 628, 630
PC 1000 545–6, 628, 630
timing of 546
tracking power 542
see also photo zenith tube; plate; zenith photography
cameras, zenith, *see* zenith cameras
Cassini's gravity tables 796–7
catalogues, star 264–70
accuracy of FK 4 265
Cauchy–Riemann equations 161–3, 762–3
Cecchini's 1900–5 origin 270–2

celestial
  equator 257
  latitude 259
  longitude 259
  pole 257
  refraction 282–4
    parallactic 517–8, 550–1
  sphere 257–9
  triangle 257–8
centrifugal force
  definition 790
  earth's equatorial, fraction of *g* 415
  inclusion in gravity 356, 398, 405–6, 410–15
  potential of 398
Chandler's period 270, 502
change of spheroid 177–82
  approximate formula 181–2
  changes in
    height above spheroid 181–2
    scale and azimuth 182
  via cartesian coordinates 179–80
  Vincenty's method 180–1
Choleski's method 151, 671
chord length, Clarke's formula 122–3
chronocord 311
  for MDM frequency calibration 69–70, 76
chronograph 310
chronometer, *see* clock
circum-meridian altitudes 297–9, 343
  accuracy 299
circumpolar stars, azimuth 334–5
Clairaut's theorem 402–3, 410–15, 442
  closed formulae 419
  Darwin's treatment 411–15
  various parameters in 415
Clarke's formulae
  for attraction of topography 794
  for latitude and longitude 118–19
  for length of chord 123
clock
  caesium 277
  crystal 311
  definition of 274
  for gravity observation 364–5
  hydrogen maser 626
  for longitude 310–11
  for satellite observations 546, 589–90, 624
clock comparisons
  with another clock 310
  with time signal 311
coefficient
  of correlation 695–6
  of expansion
    of air 44
    of invar 40
  of regression 696–7
  see also refraction, coefficient of
co-geoid
  centre of mass of contained mass 450
  centre of volume of 450
  curvature of 444
  definition of 442–4
  separation from geoid 443, 447–9
    tables for 797
  various systems 447–50
    comparison of 444
collimation
  definition of line of 21
  of level 208–9
  of theodolite 20–1, 36
  of Transit 303–4
  of zenith telescope 289, 292–3
collocation 736–9
  of gravity observations 454–5
  noise 736
  signal 736
  use of 739
compensated geoid 443
compensation 394–5; *see also* isostasy
complex numbers
  change of projection using 195–6
  for interpolation 195–6
  for projection of spheroid 763
  representation of waves by 766–74
  uses of 761
  see also para. headings in Appendix F
computation of triangulation, etc.
  checking of 125–6
  in plane coordinates 182–94
  of a single chain 107–26
  in three dimensions 678–90
  by variation of coordinates 126–44
  see also adjustment; triangulation, etc.
computers, electronic
  for change of spheroid 180
  checking data for 126, 159–61
  for gravity reductions 798
  for solution of equations 136–40, 670–2
  for star places 266, 268–70
  tables not required 107
  for variation of coordinates 127–8, 157–8
condition equations
  numbering of 169–72
  see also conditions
conditions
  accuracy
    of function of unknowns 714–15
    of unknowns 715–16
  adjustment by 111–15, 144–53, 713–14
  in astro-triangulation 558
  azimuth 145–6, 148
  between measured distances 112, 145
  central 113, 148

choice of 145, 149–50
circuit 145–7
normal equations 151–3
number of conditions 147–9
in quadrilateral 113, 149, 717–18
side 112–13, 145, 148
in station adjustment 30–1
in three dimensions 558, 684
in traverses 31, 149–50
triangular 111–13, 147–8
in trilateration 150
utility of 144, 153
cf. variation of coordinates 127–8, 144, 147, 157
weights 113–14, 151
see also least squares
conformal mapping 762–3
conical arthomorphic projection 187–90
conjugate gradients method 675–7
continental drift 491, 494–5
conventional international origin (CIO) 270–2; *see also* mean pole
conventional zero meridian (CZM) 273, 278–9; *see also* Greenwich meridian; zero of longitude
convergence, grid, definition of 184
coordinates
cartesian in 3-D 678–90
adjustment by conditions 690
basic formulae 678–85
for change of spheroid 178–82, 632–4
conversion from spheroidal 178–9, 679
direction cosines 680, 682–4
unit sphere 680–2
utility of 678
variation of 685–90
computation of
on plane 183, 185
on spheroid 115–26
summary 116–7
geocentric 95, 179, 632–6
photo plate 547–51
spheroidal, defined 93, 95–6
standard 549–54
variation of, *see* variation of coordinates
Coriolis force and acceleration
derivation of 790–2
on earth's surface 792
on pendulum or gravimeter 381–2
on sea level 252
corner cube reflectors, *see* reflectors
correlated observations 131, 135, 165–6, 695–7, 719–21, 724–8
correlates 114–15, 713–14
correlation
coefficient of 695–7
of errors in adjacent angles 165–6
covariance
definition of 695
function 731–2
uses of 719–21, 724–8, 730–9
see also variance–covariance matrix; correlation; regression
Cowell's method of numerical integration 599–602, 614
cracovians 653–4
crustal movement
astronomical observation of 318, 494
continental drift 491, 494-5
earth tides 215–16, 245, 497–503
effect on gravity 497
evidence for 492
exhibited on charts 493–4
in Finland 496
horizontal 494–5
in Great Britain 493–4
in Japan 493–4, 496
local strain 494–5
measured by
doppler 495
satellite lasers 495, 579
VLBI 495, 627
non-orthomorphic 493
plate tectonics 491
'polygons' 492
tectonic 492
test areas 492
in U.S. 494
vertical 248–9, 495-7
crustal structure 393–5, 478–90
Cunningham's formula 120
curl, definition 750–2
curvature 49
correction to path length
light 51–3, 63, 82–4, 520
microwaves 52–3, 63–4, 82–4, 527–8
of geodesic on plane projection 185–7
of geoid 223, 230–1, 444–5
in free air reduction 421, 445
related to gravity gradient 421
of star paths 324, 332–4, 336–7
of vertical
computed 467–70
conventional 97, 197–8, 344–5, 422
see also radius of curvature; refraction
curvature correction, star path
azimuth 332–3
by east and west 336–7
from Polaris 334
circum-meridian 297–8
position lines 324

Dalby's theorem 118
date, Julian 281–2

date, Julian (*contd.*)
  modified 282
datum
  levelling 219–20; *see also* mean sea level
  of soundings 249
  triangulation, etc. 102–3
    change of 177–82
datums, international
  separation determined by satellites 504, 632–6
day
  Julian 281–2
  solar and sidereal 281
declination
  annual change 266–7
  definition 257–8
  precession in 266
  see also star places
deflection of the vertical, *see* deviation
density
  of air 50
  earth's internal 389, 393–4
    determined from gravity 478–50
    mean 389–396
  of rock 393
  see also isostasy
determinants, properties of 655–6
development method of computation 105
deviation of the vertical
  accuracy required 256
  anomalies of 467
  causes of 465
  change of spheroid 181
  compilations 465–6
  computed
    from gravity 434–5, 462–3
    isostatic 467, 795
    from topography 467, 794–5
  correction to horizontal angles 106, 331
  definition 98–100
  derived
    from azimuths 100, 347
    by Black's method 337–40
    from Stokes's integral 434-5
    from vertical angles 476
  effect
    on river crossings 223–4
    on vertical angles 230–1
  geoid charts based on 470–5, 477
  in geoid profiles 342–7
  interpolated values of 342, 345, 471–6
    gravimetric 472–3
  in Laplace's equation 4, 100–1
  reduction to sea-level 97, 344–5, 422, 467–70
  summary of uses 255–6
  tables for computing 794–5
  in 3-D cartesians 681
  tidal variation of 500
  variations, spatial 465–6, 475–6
  see also geoid
differential orbit improvement 573
direction cosines
  of astro-triangulation lines 547, 552–6
  formulae 682–4
  in variation of coordinates (3D) 686–7
directions
  in adjustment by conditions 115
  corrections to
    arc-to-chord 185–7
    for deviation of the vertical 106, 132
    for dislevelment 22
    eccentric 29
    to geodesic 109
    for skew normals 106–7
    station adjustment 30–1
  on unit sphere 680–1
  in variation of coordinates 132, 136, 138–9
  weight of 135, 168
  Z term 130, 132, 136, 138–9
directions or angles
  in abstract 27
  in adjustment
    figural 115
    by variation of coordinates 129–30
  observation of 24
Discos 515, 609
dislevelment
  of automatic level 207–8
  of gyro-theodolite 350, 355
  of pendulum stand 367
  of theodolite 22, 331–2
  of Transit 303–4, 313
  of zenith telescope 292–4
distances, measured
  accuracy 167–70
  optimum frequency 174–5
  permanent record of 85
  reduction to spheroid 42, 82–5, 105, 159–60
  see also EODM; invar baselines; MDM
divided differences
  adjustment by 161
  interpolation by 195
doppler, *see* satellite doppler
duct, meteorological 63–4
DUT 1 280
dynamic
  centimetre viii, 250
  height 198, 226
  use of satellites 513, 562–74; *see also* doppler
dynamics
  centrifugal force 790

Coriolis force 790–3
see also centrifugal force; Coriolis force; satellite orbits

earth
figure of, *see* figure
models 386
standard 395, 480
earth tides 497–503
effect on levelling 215–16, 245
measured by
doppler 609
satellite lasers 579
VLBI 626
obscured by ocean tides 216
see also tidal effect
earthquakes 391–3, 491–2
earth's
axis, *see* axis of rotation
centre of mass 450–1
core 392–3
crustal structure 393–6
density distribution 389–95
from gravity 478–90
elasticity 391–2
figure, *see* figure of the earth
gravity, *see* gravity; potential
mantle 393
mean density 389, 393–4
moments of inertia 416–18
precessional constant 417
rotation 280–1, 428, 566
strength 390–1, 487–90
stress differences 391, 487–90
temperature 390–1
east and west stars
for azimuth 336–7
for longitude 330
eccentric stations
EDM 30
theodolite 5, 29–30
eccentricity of ellipse, formulae 646
Echo satellites 514, 539
ecliptic 259–61
pole of 259
eddy conductivity
definition 55
effect on vertical angles 237–41
measurement of 57
variation
diurnal 55
with height 55–7, 241
elastic constants 391–2
electromagnetic distance measurement (EDM) 60–91
eccentric stations 30, 85
general principles 37, 66–7, 74–5
length ratios 90–1, 134
reduction to spheroid 42, 82–5, 105, 159–60, 170
to satellites
C- and S-band 575
doppler 584–615
GPS 580–1
Secor 575–7
see also EODM; MDM; satellite distances; Shoran; traverse
electro-optical distance measurement (EODM)
accuracy 62, 68, 70–1, 167–70
calibration 69–71
curvature 51–3, 63, 82–4
cyclic error 70
computations 67–9, 82–5
favourable conditions 61
general description 66–9
georan 72
geodimeter 66–70
laser 68, 71–2
cf. MDM 87
mekometer 73–4
Nottingham calibration base 70
short range (infra-red) 72–3
statistics 68
two colour 61, 71–2
plus microwave 72
zero correction 69
see also atmosphere; light; satellite distances; traverse
elinvar 373
ellipse
equation of
in Legendre functions 778–9
in polar coordinates 646
geometry of 508, 645–8
orbital, dynamics 507–13
see also satellite orbits; spheroid; error ellipse
ellipsoid, triaxial 420; *see also* spheroid
elongation, hour angle of 335
engineering geodesy vi, 19
EDM for 72–3
heights 199
Eötvos correction 381–2, 790–2
ephemerides
astronomical 263–4
broadcast 591
doppler 589–91, 599–605
GPS 580
precise 590, 599–605
cf. broadcast 591
equation
of the equinoxes 261, 299
of time 275

equations, differential
  solution by numerical integration 600–2
equations, linear, solution of 656–7
  by direct methods 114–15, 139–40, 151–7, 669–72
  iteration of direct solutions 129, 143, 669
  by iterative methods 140, 572–7, 672–7
  see also normal equations
equator, celestial 257
equinox
  definition 257–60
  equation of 261, 299
  mean 261, 275
  nutation 260–1, 269–70
  precession 260–1, 268–9
    general 261
    luni-solar 260
    planetary 260–1
  true 261
equipotential surfaces 397
  earth's
    external 423–4, 505, 570, 637–42
    internal 416
    not parallel 97, 197–8, 422, 467–70
  Green's equivalent layer 404–6
  tidal effect on 220, 243–5, 497–9
  see also geoid; geopotential; co-geoid
erratics, refraction 284, 308
  cf. scintillation 284
  wanderings 284
error ellipse 175–7, 719–24
errors
  combination of 694–5
  correlated observations 724–8
  covariance 695, 719–21, 724–8
  ellipse 175–7, 719–24
  frequency distribution of 691–3
    over an area 701–3, 719–24
  inner 175–7
  periodic 691
  probable 698–9, 712
    cf. standard 698
  random 691
  rejection of observations 25, 32, 211–12, 595–6, 704–6
  standard 694, 697–8
    of a function 694–5
    from pairs of observations 701
  systematic 691, 699–701
    in EDM 62, 70–1, 169–70
    in invar baselines 170
    in levelling 213–6
    in longitude 306–9, 313–6, 333–4, 346–7, 610–2, 634
  variance–covariance matrix 140–1, 155
  see also least squares; standard error; accuracy *under various instruments and methods*
estimated values of gravity 454–6, 730–9
Euler's theorem 647
expansions, algebraic 643–4
Explorer satellites 514
external masses
  elimination of 436, 442–3, 447–50, 455–6
    equivalent to isostatic reduction 448
  smoothing of 449–50, 456, 462
    equivalent to Airy isostasy 450

faults, geological, movement of 225, 494
Ferrero's criterion 164
figure of rotating liquid 416–19
figure of the earth
  depression in mid-latitudes 412–15, 418
  from geoid profiles 425
  harmonic constants
    J's 413–15, 528–34
    tesserals 570–4, 637–9
  historical 425–6
  from luni-solar precession 412–15, 416–18
  from satellites 505, 528–34, 570–4, 613–4, 637–9
  statistics 426, 534, 638
  triaxial 420
  see also Clairaut's theorem; geopotential, harmonic analysis
figure, triangulation 5–6
  accumulation of azimuth error 171
  accumulation of scale error 172, 716–18
  adjustment of 111–15, 147–9
  definition of 5
  well conditioned 5
Finland
  atmospheric sondes 562
  crustal movement 496
  levelling 205, 218
  zero order traverse 11
flares, from aircraft 13; *see also* sondes, atmospheric
flattening
  of a spheroid, formulae 646
  of the earth, *see* figure
foot, as unit of distance 426; *see also* metre
formulae, standard
  algebraic 643–7
  geometry of spheroid 647–9, 678–85
  trigonometrical 644–5
Fourier series 775–8; *see also* harmonic analysis
framework, geodetic, *see* geodetic framework
free air
  gravity anomaly 421–2, 445–7, 480–3
    formulae 421–2

locally correlated with height 445–7, 480–3
for Stokes's integral 444–6
variations 446, 480–3
reduction to geoid 445
frequency distribution
of errors 697–9
normal 691–4
over an area or volume 701–3, 719–24
Fundamental Catalogue FK 4 264–5
accuracy of 265

gal
definition of 357
not SI unit viii, 357
Gauss's
conformal projection 192
method of solving equations 114–15, 157, 670
mid-latitude formula 123
theorem 404, 749–50
Gauss–Seidel method of iteration 673
by blocks 573, 675
general geodetic solutions, doppler 597, 613–14
geocentric coordinates 95, 179, 632–6
geodesic
definition of 109
formulae using 116, 120, 123–4
and normal section 108–10
not required for variation of coordinates 127
properties of 108–10, 124, 648–9
representation on plane 183, 185
triangle 109–10
geodesy
definition v, 1
impact of satellites on vi, vii, 13–15, 101, 179, 251, 256, 272, 436, 473, 477–8, 504, 613–14, 620, 632–9
objects v, vi, 1, 2, 198–9, 255–7, 357–8, 386, 435, 504
geodetic
azimuth, latitude and longitude, *see* azimuth etc.
framework
accuracy required 2
control by satellites 13–15, 632–6
defined 1
international connections vi, 12–13, 177–82, 632–4
layout of 3–9
objects of 1
primary, defined 1
zero order 9–11
reference system 1967 427–30
see also astronomy; levelling; traverse; triangulation; trilateration
geodimeter, *see* EODM
geoid
centre of volume of 451
charts of 473–8
compensated, *see* co-geoid
curvature of 223, 230–1, 421–2, 444–5
definition of 94–5
derived from
satellite altimetry 615–20
satellite surveys 424–5, 473; *see also* geopotential, harmonic analysis.
Stokes's integral 430–41
interpolation using
co-geoid 471–2
collocation 475, 730–9
gravity 472
sections 342–7, 470–4
accuracy of 345–7
adjustment 473
gaps in the section 471–3
in India 466, 473–5
layout of 473
observation of 342–5
use of v, 9, 42, 84, 342, 611–12
by vertical angles 476
from zero-order traverses 9–11
separation from mean sea-level 94, 248–53, 615–16, 620
separation from spheroid 2, 94–5, 342–7, 430–4, 470–8
change of spheroid 179–81
in conversion to 3-D 179, 678–9
correction of distances 42, 82–5, 105, 159–60, 170
geop 458
geophysical prospecting 490
geophysics, contact with geodesy v–vi, 386, 389
geopotential
height defined by 198
numbers 198
unit (GPU) viii, 198
see also equipotential surfaces; geopotential harmonic analysis; potential
geopotential, harmonic analysis of
combined solutions 438–42
accuracy 438–9, 442
from doppler 613–14
notation 505
numerical results 534, 638
from photo-dynamic 573–4
potential models 406–15, 477–8, 505, 605, 637–42
axially symmetrical 406–7, 528–34
see also Goddard earth models; harmonic analysis; resonant harmonics

Georan 72
Geos satellites 515–16, 540, 616, 628
Global positioning system (GPS) 580–1
Goddard earth models (GEM) 477–8, 637–42
Goddard space flight centre 628–42
de Graaff–Hunter's
  formula for coordinates 120
  model earth 449–50, 462
  treatment of Stokes's integral 460–2
gradient
  atmospheric
    of pressure 50
    of temperature 51–9, 214, 237–41
    of water vapour 51, 58–9
  of gravity
    below ground surface 453–4
    below sea surface 454
  method of solving equations
    conjugate 675–7
    optimum 676
  of scalar (grad ϕ) 746–7
gravimeters
  accuracy of 373, 381
  adjustment of circuits 373–5
  in aircraft 383–4
  astatic 375–6
  calibration 362–3, 374–5
  construction details 375–81
  cross-coupling errors 380–1
  drift 373–4
  effect of temperature 375
  Eütrös correction 381–2, 792
  Graf–Askania 378
  La Coste–Ramberg 375–6
  measurement of gravity by 373–5
  on sea-bottom 382
  in ship 378–82
    in gimbals 379–81
    on stabilized platform 379–81
  for standardization net 363
  vibration (VSA) 384–5
  Worden 377–8
  zero-length sprung 375–6
gravimetric interpolation of deviation 472–3
gravitation, constant of 396, 504–5, 612–13
gravity
  absolute 356, 359–62
  below sea-level 454
  between ground level and geoid 453–4
  definition of 356, 389, 398–9
  deduced from potential 411–12, 424–5
  dimensions of 389
  effect of crustal movement 497
  estimated values of 454–6
  harmonic analysis of 406–15, 431, 478, 620
    low degree harmonics, *see* low degree
  height of station 358, 372
  at high altitudes 422–4
    by integration 464
  local topography 372–3
  measurement of, *see* gravimeter; pendulum
  at Potsdam 360, 362, 425–7
  records and storage of data, 358–9, 478
  reduction tables 795–7
  reduction to sea-level 445, 454
  standard formulae 414, 427
  tidal variation 369, 382
  use of v, vi, 389, 478–80, 487–90
  variation
    with height 200, 421, 423–4, 445–7, 480–3
    with latitude 200, 397, 414–15, 427
    with time 497, 501
  see also gravimeter; gravity anomaly; pendulum; Stokes's theorem
gravity anomaly
  Bouguer 451–3, 472, 796
  charts and compilations 478
  definitions 395–6, 444, 480–7
  desirable features 444
  free air 421–2, 445–7, 480–3
  Hayford 480–4
  interpolation of 444, 454–5, 734–5
  isostatic 395, 448, 480–5
  in mountains and islands 480–2
  non-classical 459
  representative character 444
  Rudski 448
  smoothed topography 449–50
  for Stokes's integral 444
  strength demanded by 487–90
  topographical 395, 447, 480–1
  uses of 395–6, 444, 478–90
gravity measured in aircraft 383–4, 792–3
gravity unit (GU) 357
Green's
  equivalent layer 402–6
    with rotating axes 405–6
  theorem 756
    applied to equipotential surface 406
    applied to ground surface 456–8
Greenwich meridian
  conventional zero meridian 273, 278–9
  definition of 97, 103, 279
  for satellite surveys 546, 555, 565–6, 610–12
  for VLBI 623–4
Gregory's integration formula 644
grid
  definition of 183
  distance and bearing 183

see also plane coordinates; projection
GRIM 2 632–3, 638
ground swing, *see* tellurometer
group velocity
of laser beams 45, 89, 578
of light 44–6, 773–4
of microwaves
in air 45, 47
in ionosphere 523–4
Guier plane 595–6
gyro-theodolites 347–55
accuracy 355
measure astro azimuth 347
MOM 355
PIM 355
recent improvements 354–55
Wild GAK 1 348–54
amplitude method 354
approximate method 352–3
tracking method 352
transit method 353–4

harmonic analysis
of external potential 406–15
derived from satellites, *see* geopotential
of geoid 477–8, *and see* geopotential
of gravity 406–15, 431, 478
1st degree terms absent 411, 420
linear 775–8
significance of low degree harmonics 420
spherical 786–7
of temperature 238
tidal 246–8
Hayford's
deflection anomalies 466–7
deflection tables 794–5
gravity tables 795–6
isostasy 483–4, 486–7
spheroid 426
height
accuracy required 198–9, 372
average 447, 478, 798
barometric 33, 50–1, 61, 86, 372
changes of 220, 228, 491, 495–7, 498–9
correction
to deviation 344–5, 422, 467–70
to gravity 445, 454
to horizontal angles 106–7
to latitude 97–8
to measured distances 42, 82–5, 105, 159–60
datum of 197, 219–20
definition of
dynamic (geopotential numbers) 198, 226
geoidal 93, 197
geopotential unit 198
normal dynamic 202
normal non-classical 459
normal orthometric 199–203
orthometric 197, 202, 226
spheroidal 93, 197
effect on gravity 372, 421, 423–4, 445–7, 480–3
geoidal, when required 199
measurement, objects of 198–9
observation equation in 3-D 689–90
orthometric and dynamic 198–203, 467–70
spheroidal
methods for obtaining 197
when required 198–9
spirit levelled 203–223
traverse 86
triangulated 33–4, 197, 228–43
accuracy of 231–3
by vertical angles 33–4, 228–43
see also altimetry; levelling; refraction; vertical angles
heliotropes 23, 35
Helmert's
formula for distances 123
gravity formula 427
projection method 96
preferred to Pizzetti's 96
Hertz, definition of 37
high latitudes, observations in 305, 335, 340–1
Hiran 12–13
Hopfield's refraction formulae 520–3
horizontal, definition of 96
Horrebow levels 294; *see also* zenith telescope
hour angle 258, 275
humidity, *see* water vapour pressure
hydrostatic levelling 224
hyperbolic formula for distances 124

illumination
candela 540
lumen 540
lux 540
stellar magnitudes 541
tracking capacity of satellite 542
India
accuracy of triangulated heights 232
adjustment of triangulation 157
definition of longitude 102
deviations of the vertical 466, 474–5
geoidal profiles 474
levelling in Sind 213
reduction of baseline to spheroid 42
triangulation methods 3–4, 26
variation of gravity with height 446

indirect effect 442–4, 797
infallible observations
  azimuth 124–5, 133–4
  combined with fallible 718–9
  coordinates 132–3
  distance 133, 689–90
  height 689–90
  high weight preferred 133–4
infinite plate, attraction of 400
inner errors, free adjustments 175–7
interferometry
  for crustal movement 494–5
  minitrack 627
  Vaisäla comparator 41–2
  very long base (VLBI) 620–6
international
  Association of Geodesy vii, 19, 219, 359, 425
  connections, framework vi, 12–13, 153–7, 177–82, 504, 632–6
  Federation of Surveyors (FIG) 19
  framework adjustments 153–7, 177–82
  Gravity Bureau 359, 478
  gravity formula 425–7
  gravity standardization net 362–3
  polar motion service 272–80; *see also* polar motion
  scientific units viii, 198, 357
  spheroid (1924) 425–6
  see also Bureau international de l'heure
interpolation
  of the deviation 345, 471–6
  by divided differences 195–6
  of gravity anomalies 454–6
  by least squares 730–6
intervisibility, formula for 16
invar
  gravimeters 375
  levels 206
  levelling staves 208
  pendulums 360, 367
  properties of 40–1
  see also base lines, invar
inversion of matrix 139–41, 664, 677
ionosphere
  electron density 525
  refractive index
    light 523
    microwaves 45, 523–8
ISAGEX 629–3
isometric latitude 190
isostasy
  Airy 484–5
    equivalent to shallow regional 486
  definition of 394–5
  departures from 395–6, 481–3
  depth of compensation 486–7
  Hayford or Pratt 483–4
    equivalent to shallow Airy 485–6
  indirect effect 442–4
  location of compensation 485–7
  regional 485–6
  Rudski 448
  sea-level 447
isostatic
  anomalies
    deviation 466–7
    mass, location of 478–90
  gravity anomalies
    defined 395–6, 444, 480–7
    guide to densities 478–90
    for Stokes's integral 444
  reduction
    of deviation of the vertical 466–7
    equivalent to mass transfer 448–50, 461, 467
    good for Stokes's integral 444, 448
    tables 795–7
  see also isostasy
iteration
  method of solving equations 140, 572–7, 672–7
  of direct solutions 129, 143, 669

J's in geopotential
  accuracy 533–4
  definition of 413, 784
  determined by satellites 528–34
  normalized 534, 785–6
  numerical values of 533–4
  see also J_2; J_4
J_2
  actual value 417–18, 428, 533–4
  effect on satellite orbits 528–32
  internal strength required 489
  related to $C–A$ 417–18, 429–32
  in terms of flattening 413
J_4
  actual value 418, 429, 534
  effect on satellite orbits 528–32
  in terms of flattening 413
  resulting from heavy centre 415, 418
Jacobi's iteration method 672–5
Jet propulsion laboratory (JPL) 613, 629
Julian
  century 275, 282
  date 281–2
  modified date 280

Kater pendulum 359–60
Kaula's
  covariance function 455
  rule 639
Kern theodolites 20, 26, 295

Kerr cell 66
Krassovsky's spheroid 426

Lageos satellite 515, 579
Lallemande's formulae 216, 218
Lambert's projection 187–90
Lambert tower 18
lamps
  geodimeter 67–8
  in satellites 542–3
  triangulation 23–4, 35
    signals to 24
Laplace azimuth stations
  accuracy 170
  in adjustments 124–5, 133–4, 145–6, 148–9, 158
  definition 4, 100
  frequency 4–5, 8, 14–15, 174, 331
  in high latitudes 341
  in new chains 7–8, 17
  optimum number 174
  reciprocal 124, 170, 331
  satellite fixes in lieu 14–15
  in traverses 9, 26, 124, 149, 158, 331
  see also azimuth
Laplace's
  equation (azimuth) 100
    stability of 101, 133, 159
  operator (vector) 755
  theorem (potential) 403
    with rotating axes 403
  see also Laplace azimuth station
laser beams
  for EODM 68, 71–2, 168–9
  general description 88–9
  group index 45, 578
  lunar 581–4
  cf. microwaves 87
  cf. ordinary light 71, 88
  refractive index 45, 578
  safety 89
  for satellite ranging 577–9, 629
  statistics 89
  see also satellite distances
lateral refraction 7, 20, 27, 32, 242–3
latitude, astronomical
  accuracy 294, 296–7, 299, 320–1, 326–8, 330, 340
  definition of 97
  observations of, summary 285–6
    by Black's azimuth 337–40
    circum-meridian altitudes 297–9
    equal altitudes 325–9
    meridian altitudes (Sterneck) 295–7
    pole star 298
    Talcott method 287–94
    unequal altitudes 329–30
    zenith photography 316–21
  reduction
    to mean pole 97–8, 273, 324–5
    to sea level 97–8
  variation of, *see* polar motion
  see also coordinates
latitude, computation of geodetic, *see* coordinates
latitude, definitions
  astronomical 97–8
  celestial 259
  geocentric 646
  geodetic (spheroidal) 96
  isometric 190
  parametric 646
  reduced 646
layout
  astro-triangulation 518, 544, 556–7
  level net 203–6
  optimum number of controls 174–5
  Secor 575–6, 628–9
  Shoran 12
  traverse net 8–11, 15
  triangulation 3–7, 13–14
  trilateration 11–12, 15
least squares
  direct method (conditions) 111–15, 713–18
  given by weighted mean 704
  incorporation of later observations 710
  interpolation 730–6
  inclusion of infallible observations 718
  parametric (indirect) method 706–13
  residuals, r.m.s. of 711–13, 715–16
  tests for mistakes 114, 142–4, 711
  variance of functions of unknowns 141, 712–13, 715–16
  variation cf. conditions 127–8, 153
  with correlated observations 154–6, 724–8
  see also adjustment; normal equations; weights
Legendre
  coefficients 713–14
  functions 779–81; *see also* zonal harmonics; *J*'s
    normalized 785–6
    r.m.s. value 781
  theorem 108
length ratios in EDM 90–1, 134
Levallois and Dupuy's formula 120
level
  automatic
    design of 207–8
    utility of 208
  collimation of 209
  parallel plate for 209

level (*contd.*)
programme of observations 210–11
spirit, *see* bubble
surface, *see* equipotential
tilting 206
see also levelling
levelling
accuracy 199, 211, 216–19, 223
across rivers 220–4
geoidal asymmetry 223
across seas 250–53
adjustment 201, 226–8
bench marks
design 225
siting 224–5, 248–9; *see also* movement
check, at start of line 212
computations 225–6
datum surface 219–20
dynamic heights 198–203, 468–70
effect of sunlight 215
field procedure 209–12
fore and back 204–5, 218
geopotential
numbers viii, 198, 201–2
unit 198
of high precision 203
hydrostatic 224
instruments 206–8, 221
interval between gravity stations 201
length of sight 210, 214, 221
long slopes 214
mean sea-level 219–20, 228, 248–53
motorized 212
net
adjustment 226–8
density and layout 203–6
objects 199
primary 203–6, 209–12, 219
reduction to epoch 226
refraction 209, 213–15, 222–3
relevelments 212
revised local reference 249
revision 205, 228
secondary 203–4
staves, design and calibration 208–9, 214–15
supports of 208, 211–12, 213–14
systematic errors 213–16
tidal correction 215–16, 245
tidal stations 206, 221–2, 219, 228, 248–9
U.S. new system 204–6
see also height; level; satellite altimetry
light
coefficient of refraction 49, 51–3, 236–7, 241
curvature of path 51–3, 233–43
abnormal 52–3, 237–41
normal 52–3, 233–7
see also curvature; refraction
favourable conditions
EODM 61
vertical angles 32–3, 237, 241
infra-red 72–3
cf. microwaves 87
refractive index 44–5, 241–2
two-colour 71–2, 241–2
wavelength 44–5
see also atmosphere; illumination; laser; refraction
light, velocity of
average along line 49, 54, 60–4, 518
effect of
humidity 44
temperature and pressure 44
wave length 43–5
group 44–6, 89, 578, 773–4
when applicable 45, 89
in ionosphere 523
phase 43
see also refraction; satellite distances
lines in space 679–85
angle between 682
condition for intersection 684
directions of 681–4
plane containing two 685
longitude
celestial 259
computation of astronomical 299–300, 303–4, 311–13, 321–5
conventional zero meridian (CZM) 273, 278–9
definition
astronomical 97–8
geodetic 96
referred to Greenwich 103–4
observation of astronomical 300–30
photographic methods 316–21
variation of 273, 318
zero of 96–7, 103–4, 273, 278–9, 564–6, 610–12
see also coordinates; time local sidereal; transit telescope
lost motion (micrometer) 307, 313
Love's numbers 498–503, 609, 614
low degree harmonics in *g*
significance of 419–20
some inadmissible 411, 420
lumen, lux, *and* candela 540–2
lumped coefficients, *see* resonant
lunar laser ranging 581–4
uses of 583

MacCullagh's theorem 402, 417
magnetism of pendulums 368, 370

magnitude, stellar 541
mantle, earth's 393–4
matrix algebra, *see paragraph headings of Appendix B.*
  application to theory of errors 140–1, 154–7
  inverse, computation of 659, 663, 664, 677
  for least squares 709–12, 715, 718, 724–8
  for rotations 268–70, 555, 565–8, 597
  for solution of equations 669–77
mean pole
  astro latitude etc reduced to 97–8, 270–2, 299, 300, 312, 316, 325, 332
  conventional (CIO)
    defined 270–2
    defines spheroid axis 93, 95, 97, 272
  of epoch 270
  with respect to nutation 93, 260–1
  see also polar motion
mean sea-level 243–53
  deep sea reference surface 250
  determination of 248–9
  eustatic changes 249
  meteorological changes 249
  not equipotential 199, 219–20, 228, 248–53
  tectonic changes 249, 495–6
  see also geoid; satellite altimetry; tide
measure (of angle)
  for azimuth 338, 340
  definition 25
  number of 25–6
Mekometer 73–4
Mercator's projection 190
  transverse 191–4
  UTM 193–4
meridian
  altitudes 295–7
    accuracy 298
    advantages versus Talcott 286
  radius of curvature 647–8
  transits, *see* Transit telescope
  zero of, *see* Greenwich
meteorological
  notation 38
  observations 47–8
  units 39
  see also atmosphere *and* refraction
metre, international 39–40
foot–metre ratio 426
micrometer
  calibration
    for theodolite 21–2
    for zenith telescope 288–9, 292, 294
  impersonal
    for altitudes 294, 329
    in astrolabe 326–8
    in astro-theodolite 294, 329
    in Transit 306–7
    with motor drive 307–8
microwaves
  coefficient of refraction 52–4
  curvature
    abnormal 53, 63–4
    in ionosphere 527–8
    normal 52–4
  definition of 37
  8 mm 47, 77, 81
  favourable conditions 61–3
  frequency 77–9
  from Quasars, for VLBI 621
  ground swing, *see* MDM
  cf. light 87
  modulation 767–9
    of amplitude 768
    of frequency 74, 768
    mixed 769
    sidebands 767–9
  reflection of 64–5; *see also* MDM ground swing
  refractive index 46
  represented by
    complex numbers 760
    rotating vectors 760
  velocity 46–7
    average along line 60–3
    effect of
      rain 47
      temperature and pressure 46
      water vapour 46
    group 45, 47
    in ionosphere 523–8, 625
  wave length 77, 79
  see also MDM; satellite distances
microwave distance measurement (MDM)
  accuracy 78, 167–70
    systematic error 62, 78, 169–70
  calibration 74–6, 78
  computations 77, 82–5
  cyclic error 75–6
  8-mm 47, 77, 81
  cf. EODM 87
  favourable conditions 62–3, 78, 81–2
  general description 74–9
  ground swing 65, 75, 79–82
    complex notation 766–73
    double reflection 80–1
    strong reflection 80–1
    vector representation 766–71
    weak reflection 79–80
  ground swing curve 77, 79–82
  long lines 13, 63–4
  nomenclature 74

microwave distance measurement (*contd.*)
routine of measurement 74–5
statistics 77
tellurometer 74–82
traverse routine 85–7
various wavelengths 77–9
zero correction 75
see also microwaves and traverse
miniature geodesy, *see* engineering geodesy
Minitrack 627
modal matrix 667, 726–7
model earth (de G.-H.'s)
defined 449–50
for non-classical 462
for Stokes's integral 449–50
see also earth, models
modulation of microwaves 764–9; *see also* microwaves
Mohorovičić discontinuity 393–4, 486–7
Molodensky's
correction 159–60
treatment of Stokes's integral 463–4
moments of inertia
of earth, see earth
of gyroscope 349–51
see also MacCullagh's theorem
movement
of bench marks 206, 220, 228, 248–9
of triangulation marks 7, 17
see also crustal movement

N-unit (refractive index) definition 38, 44
Napier's rules 645
National aeronautics and space administration (NASA) 628–38
Naval Surface Weapons Laboratory (NSWL) 631, 634, 638
Navstar satellites, *see* global positioning system
Navy navigation system (NNSS) 590
Newtonian inertial frame 564, 597, 790
Newton's laws of motion 790–1
nodal period 507, 532
node, orbital, ascending
definition of 507
regression of 529
non-classical treatment of Stokes's integral 456–64
de Graaff–Hunter's 460–2
Molodensky's 463–4
normal equations, formation of
doppler positioning 598
doppler, general solutions 614
general 114–15
in method of conditions 151–3, 713
in variation of coordinates 136–40, 708–10
see also least squares
normal equations, solution of
Banachiewicz's method 670–1
Boltz's method 154, 670
Choleski's method 139–40, 671
with desk computers 114–15
Gauss's method 114–15, 157, 670
Helmert–Wolf method 154–6
iterative methods 140, 672–7
accelerated convergence 675
block Gauss–Seidel 573, 675
conjugate gradients 675–6
Gauss–Seidel 673–5
Jacobi 672–5
numerical checks 114, 142–4
partitioning 151–2, 154–7, 158, 664, 670
normal heights
dynamic 198, 202
misclosure of circuits 199–203
non-classical 459
orthometric 197–203
normal section
azimuth of 96
definition 105
cf. geodesic 108–10
'triangle' formed by 108–10
see also geodesic
normal, spheroidal
definition 96
terminated by minor axis
defined 107
formula for 647–8
normalization
of Legendre's functions 534, 786
of tesseral harmonics 785–6
of vectors 655
normals, skew
correction for 106–7
effect on lateral refraction 243
notation
accuracy of triangulation 163–4
astronomical 254
cartesian 3-D 678
change of spheroid 178
electromagnetic 38–9, 764
geometry of spheroid 92
gyro-theodolite 348
meteorological 38
physical geodesy 387–8
plane coordinates 183–4
refraction 38
satellite
abbreviations 506
doppler 584
orbital elements 504–5
photo-astro 529
photo-dynamic 562–3
refraction 516–17
synthesis 627–8

VLBI 622
trigonometrical computation 92
variation of coordinates 128
vector algebra 740
Nottingham calibration baseline 71
numerical integration
Cowell's method 600-2
Gregory's formula 644
Simpson's rule 194, 223, 644
Nusl–Fric astrolabe 327
nutation
definition of 260–1
in declination 270
effect on star places 263–4, 269–70
long period 260–1, 268
in obliquity 269
in RA 270
short period 260–1, 263–4

obliquity 257, 260
observation equations
doppler 602–9
in 3-D 686–90
linearization 708–9
satellite, dynamic 766–73
variation of coordinates 127, 131–4, 707–8
Ohio State University (OSU) 630, 633, 635
origin
of cartesian coordinates for
astro-triangulation 546–7
doppler 597
photo-dynamic 564
for polar motion 270–2
of plane projections 188
of survey
change of 177–82
definition of 101–3
errors of observation at 102
orographical correction
computation of 796
constancy of 452
definition of 452–3
magnitude 453
record of 358
orthometric height 198–203, 467–70
orthomorphic
adjustment, graphical 160–3
changes of position 492–4
conformal mapping 761–3
projections 182–3, 763

Pageos satellite 514, 538–9
parallactic angle 258
parallactic refraction 517–18
parallax
celestial, annual 262, 270
eye-piece 21–2, 35
parametric method, *see* least squares
pendulum
accuracy 370
arc correction 367
Browne terms 372
calibration 362–3, 366, 371
clock rate 365
distribution of closing error 370–1, 374
Eötvös correction 381–2, 792–3
flexure of stand 368–70
height of station 372
magnetisation 368, 370
measurement of gravity 364–5
on moving support 368–9
obsolescence 356–7
period 369
pressure correction 367
programme of observations 369
stability of length 365–7
in submarine 371–2
temperature correction 367
tidal correction 369
two or three on one support 368, 371
perigee, argument of 507, 530
personal equation
in astro azimuth 333, 339
in astrolabe 325–8
in geoidal sections 344, 346
in theodolite bisections 35, 333
in Transit times 306–9, 313–16
Peters's formulae 698–9
photo-electric cell 66, 308
photo zenith tube 317–18
accuracy 317
physical geodesy
definition and objects 386–7, 389
notation 387–8
pivots, imperfect 302–3, 313
plane bearing and distance 183
plane coordinates 93, 182–96
arc-to-chord correction 185–7
change of projection 195–6
convergence 184
conversion from and to spherical 183
corrections to bearings 185–6, 189–90, 193–4
Guier plane 595–6
notation 183–4
scale error
at point 184–5
finite distance 185, 189, 192–3, 194
t minus *T* correction 187
utility of 183
see also grid; projection; particular projections by name
planes

planes (*contd.*)
  directions of in 3-D 684
  elements of, as vectors 741–2
planetary aberration 554–5
plate, photographic
  coordinates 547–51
  measurement 547
    accuracy of 560–1
  standard coordinates 549–54
  see also satellite photography
plate tectonics 491
plumb line, *see* vertical
point positioning (doppler) 597–8
Poisson's
  ratio 392
  theorem 404
polar coordinates 93, 508, 648, 681, 778, 789
polar motion 270–4
  conventional international origin (CIO) 270–2, 279
  conventional zero meridian (CZM) 273, 278–9
  correction to
    azimuth 273, 332
    latitude 273
    longitude 273
    time 277–8
  mean pole 93, 95, 97–8, 271–2
    of epoch 270
  measured by
    conventional astronomy 272, 294
    lunar-laser 583
    satellite doppler 613
    satellite laser 579
    VLBI 626
  records of 272
  see also mean pole
pole
  celestial 257
  earth's north 274
  of ecliptic 259
  instantaneous 260
  mean, *see* mean pole
pole star
  for azimuth 333–4
  for latitude 286, 298
  for meridian setting 289, 300
position lines 321–5
  correction to mean pole 324–5
  curvature correction 324
  diurnal aberration 325
  least squares 323–4
  Marc St. Hilaire method 323
potential
  anomaly of (non-classical) 458
  definition of
    attractive 397–8
    gravitational 398
  earth's external 413, 423–4
  energy 397
  gradient 397–8
  height defined by 198
  at high altitudes 413, 423–4, 505
    by integration 464
  of rotating body 398
  in spherical harmonics 406–15
  of spherical shells 407–9
  of standard bodies 397–402
  see also, equipotential surface; geopotential; harmonic analysis
Potsdam, gravity at 360, 362–3, 427
Pratt isostasy, *see* Hayford
precession 260–9
  connection with earth's figure 416–18
  computed by rotation matrices 268–9, 555, 565
  in declination 266
  effect on star places 266–9
  general 261
  of gyroscope 348–50, 355
  luni-solar 260
  planetary 260–61
  in RA 261
  see also equinox
Prey's analysis 797
principal
  axes 719–20, 723, 725–6
  distance 552–3
  point 548, 552–2
  radii of curvature 647–8
probability; *see* errors, frequency distribution
probable error
  Bessel's formulae 699
  definition 698
  of least-square unknowns 712
  Peters's formulae 699
  cf. standard error 698–9
  see also standard error
projection
  change of 195–6
  computation on 185
  conical orthomorphic (Lambert's) 187–90
  definition of 182
  Mercator's 190–1
  orthomorphic, defined 182, 185
  transverse Mercator 191–4
  using complex variable 763
  UTM 193–4
  zenithal orthomorphic 194
  see also grid; plane coordinates
projection method
  Helmert's 96
  Pizzetti's 96

proper motion 262, 265, 267, 549
Puissant's formula 119–20

quadrilateral
  adjustment of 114–15, 717–18
  conditions in 113, 148
  error accumulated in 171, 717–18
  lay-out of 5
  trilaterated 150
quartz
  in gravimeters 375, 377
  in pendulums 367
quasars, used for VLBI 621
quasi-geoid 463

radar
  C- and S-band 575, 616, 631
  secor 515, 575–7, 628–31
radio receivers (time signals) 309–10
  transmissions 309–10
    velocity of 310
radio sources, RA and declination of
  inclusion in FK 5 265
  necessity for 621, 624
radius of curvature
  of spheroid 42, 648
  principal radii 648
    tables of 107, 120, 123
  see also curvature
ratios, distance 90–1, 134
reconnaissance
  traverse 16, 85
  triangulation 16–17
records of gravity observations 358–9
rectangular coordinates, *see* plane coordinates
reference ellipsoid 1967 426, 428–9
reference systems 93–4
  astro-triangulation 546–7
  cartesian 93–4, 179
  classical geodetic 93–6
  conversions 177–81
  doppler system 597
  geocentric 95, 179, 632–6
  inertial 564, 597, 790
  SAO satellite systems 564–5
    celestial 565
    mean sidereal 565
    orbital (inertial) 564
    terrestrial 565
    true sidereal 565
  see also coordinates; plane
referring mark
  for Laplace azimuth 331–3, 335–6
  for meridian altitudes 289, 295
  for triangulation 24–5, 30–1
reflection
  of light
    by corner cubes, *see* reflectors
    from spherical satellites 541–2
  of microwaves 64–5
  see also MDM ground swing
reflectors, corner cubes
  geodimeter (EODM) 68
  lunar lasers 581
  satellite lasers 515, 579
refraction, influence of, in geodesy 43; *see also following items*
refraction of light (effect on angles)
  angle of
    definition 228
    diurnal variation 237–41
    related to curvature 49, 51–3, 235–6
  celestial
    abnormal 283–4
    erratics 284
    normal 282–3
  coefficient of
    definition 49
    diurnal variations 53, 237–41
    in horizontal lines 236, 240
    in inclined lines 237–41
    from reciprocal observations 236–7
    related to dn/dh 51–3, 233–5
    variation
      with pressure 51–3, 237
      with temperature 51–3, 237
      with water vapour 51
  curvature of path 49, 51–3, 235–6
  direct measures of 241–2
  lateral 7, 20, 27, 30, 242–3
  in levelling 209, 213–15, 222–3
  parallactic 517–8
  vertical theodolite angles
    recommended normal procedure 237
  see also atmosphere; light; microwaves; refractive index; vertical angles
refraction, satellite distances
  altimetry 618–20
  curvature of path 520, 527–8
  distances along path
    laser light 518–20, 578
    microwave 520–7, 577
  doppler 526–7, 596–8
  GPS 581
  Hopfield's formulae 520–3
  ionospheric 520, 523–7, 577, 581, 597–8, 618, 625
  notation 516–7
  Secor 577
  VLBI 625
refractive index (ground stations)
  average along line 60–4
  dispersion 44, 71–2, 291–2

refractive index (ground stations) (*contd.*)
  group index 44–5
  of light 44–5
  measured by two colours 71–2
  of microwaves 46–7
    in ionosphere 523–4
  related to wavelength 44–6
  see also atmosphere; light; microwaves; refraction
regional isostasy 485–6
regression, coefficient of
rejection of observations 704–6
  angles 25, 32–3
  doppler filtering 595–6, 614
  levels 211,
relativity, effect on doppler 586, 594, 609–10
repetition of observations
  levelling 211–12
  theodolite 25, 32–3
resection, trigonometrical
  avoidance in triangulation 5
  in Secor 576
  in geoid profiles 342–3
resonant harmonics in geopotential 535–7, 637–8
  condition for resonance 535–6
  effect on orbit 536–7
  lumped coefficients 537
revised local reference (RLR), levelling 250
right ascension (RA) 257
  annual change 266–7
  nutation in 270
  precession in 261
  see also star places
rigidity 391–2
river crossings
  levelling 220–4
  theodolite 221–4
Robbins's formula 122
rotating axes
  centrifugal force 356, 790
  Coriolis force 252, 381–2, 790–3
  effect
    on attraction 356
    on green's theorem 405–6
    on Laplace's equation 403
    on Poisson's theorem 404
    on potential 398
rotating liquid
  with heavy centre 416–18
    depression in lat. 45° 418
  homogeneous, figure of 416
rotation, earth's
  angular velocity 280, 428
  period 281
  sideral angle 280, 566
rotations
  as matrix product 660–3
  right-handed, signs 664
Rudoe's formulae 117–18, 120–1

satellite altimetry 615–20
  accuracy 618
  general principle 615–16
  geometry 618–19
  instrumentation 616
  reflections from waves 616–18
  refraction correction 618
  uses of 620
satellite distances
  C- and S-band radar 575, 631
  laser 577–81, *see also* lunar laser
    accuracy 578
    calibration 579
    equipment 577–8
    refraction 578
    uses 579–80
    velocity aberration 579
  Secor 515, 575–7, 628–31
satellite doppler
  accuracy
    broadcast ephemeris 590–1
    geodetic mode 561–2, 614–15
    navigational mode 587
    precise ephemeris 591, 614–15
    translocation 592–3
  data filtering 595–6
  data processing 595
  doppler shift 586–7
  effect of relativity 586, 594, 609–10
  ephemeris, broadcast 589–91
  ephemeris, precise 540–1, 599–608
    least-square solution 599, 602–5
    numerical integration (Cowell) 600–2
    outline of method 599
    partial differentials 603–5
    starting elements 599
  frequencies 593–4
  general geodetic solutions 597, 613–4
  general principles 585–9
  geodetic mode 587–93
  geometry of the fix 587–9
  ground receiving equipment 590
  Guier plane 595–6
  integrated doppler count 593–5
  integrated range rate 593–5
  navigational mode 587
  notation 584–5
  perturbation model 605–9
  point positioning 591–2
  polar motion 613
  refraction correction 596, 602, 605
    ionospheric 597
  scale 612–13

station coordinates
normal equations 598
observation equations 597–8
partial differentials 598
reference system 597
tracking stations 590–1
tranet 589
translocation
short arc 592–3
short arc, geodetic 593
zero of longitude 610–2
satellite ephemeris, *see* satellite doppler
satellite geodesy, accuracy 516; *see also under various satellite methods*
satellite geodesy, general references
altimetry 615–20
for azimuth vii, 14–15
for control of triangulation vi, 13–15, 632–6
doppler 584–615
dynamics 507–13, 563–74, 599–610
geometric use, *see* astro-triangulation
for geopotential, *see* geopotential
impact on classical geodesy, *see* geodesy
inter-continental connections vi, 13, 632–6
observation programmes 628–32
photography, *see* satellite photography
refraction 516–28
satellite distances 575–84
VLBI 620–7
zonal harmonics 528–35
Details are indexed separately
satellite orbits
anomaly
eccentric 509
mean 509
true 509
apse
definition of 507–8
rotation of 530
argument of perigee 507–530
axially symmetrical earth 528–35
eccentricity 507, 531, 537
elements
Keplerian 507–8, 511
mean 532
osculating 511
starting 599, 602
variation with time 501–2, 528–32, 571–2
elliptic, elementary 508, 646–8, 778–9
inclination 507, 531, 536
latus rectum 507–8, 531
mean motion 509, 532
nodal period 507, 532
node
definition of 507
regression of 529
perigee distance distance 508, 531
perturbations 569–70, 605–9
air drag 608–9
earth tides 609
gravitational 605–7
models 569, 605
ocean tides 609
solar radiation 609
tidal sun, and moon 607–8
perturbing accelerations 599
selection of orbits 543–4, 576
spherical earth 507–11, 586–9
see also astro-triangulation; satellite doppler ephemeris; satellite photography dynamic use; zonal harmonics
satellite photography
aberration 554–5
accuracy 560–2
dynamic use
general principles 563–4
incroporation of distances 557–8
notation 562–3
observation equations 566–9
reference system 564–6
revision of potential model 573–4
unknown parameters 570–4
geometric use, *see* astro-triangulation
plate coordinates 547–51
calibration 551
refraction 551, 560
satellites
flashing 540, 542–3
passive 539, 541–2
standard coordinates 549–54
Turner's method 549–51
elaborations 551–54
see also cameras; illumination
satellite range rate (GRARR) 628, 630; *see also* satellite doppler
satellite stations, *see* eccentric
satellite to satellite tracking 515–16
satellites, artificial
flashing 540, 542–3
names and orbital details 514–15
passive 539, 541–2
see also under individual names, and under methods of use
scalar
definition of 650, 653, 740
field 741
scalars, ordered 757
scale error
accumulated in
chain of triangles 172–3
single figure 171, 716–18

scale error (*contd.*)
traverse 168–20
in plane projection 184–94
in satellite solutions 558, 612–3, 631–2, 634–5
Schreiber's method
for normal equations 138–9
for theodolite angles 24, 31
Seasat 1 satellite 515
secondary framework
Aerodist 15, 87–8
definition 1
traverse 13–15
triangulation 13–16
trilateration 13–15
Secor satellites 515; *see also* satellite distances
sectorial harmonics 535–7, 782–3
seismology 391–3
Sharma's formula 124
Shoran and Hiran 12–15
Simpson's rule 194, 223, 644
Smithsonian observatory (SAO)
Baker–Nunn cameras 544–5, 628
laser ranging 578
orbital reference systems 564–6
standard earths 629, 634–5
star catalogue 265
use of 268–70, 549
smoothed topography 449, 462
solution of triangles
Napier's rules 645
in spherical trigonometry 644–5
in triangulation 108–11
Somigliana's formula 419, 428
sondes, atmospheric 13, 562
spatial computations 678–90; *see also* coordinates, cartesian
SPEOPT 628–30
sphere
attraction of 400
celestial 257–9, 548–9
satellite orbiting 507–11
unit 680–2
spherical excess 108–10
spherical harmonics
applications of 406–14
change of axes 787–8
general surface 782–5
harmonic analysis 775–8, 786–7
Legendre's functions, *see* zonal harmonics
normalization 785–6
order and degree 782–3
properties 782–5
recurrence relations 607, 781
sectorial 782–3
solid 789
tesseral 782–5
zonal, *see* zonal harmonics
see also harmonic analysis; potential
spheroid, reference
centre of 101
as given by Stokes's integral 434, 451
separation from geocentre 451, 632–4
change of 177–82
definition of 95, 101–3
depression in latitude 45° 418, 429
direction of minor axis 95
ellipsoid 1967 426, 430
flattening 416–18, 429
geometry of 645–9
ground points projected onto 96
international 1967 426
non-classical 458–9
reduction of distances to 42, 82–5, 105, 159–60, 170
statistics 426
see also figure of the earth; spheroidal
spheroidal
azimuth, definition of 96
computations 108–26
coordinates 93, 96
conversion to rectangular 179, 678–9
latitude, definition 96
longitude, definition 96
triangles 108–10
triangular excess 108–9
spherop 458
spikes for levelling 211–12
Sputnik satellite 514, 534
STADAN 628–630
standard coordinates 549–54
Turner's method 549–54
standard deviation 692
of a function 694–5
of the mean 694
see also standard error
standard earth 395, 480
standard error
of adjusted observations 711–13
Bessel's formulae 698
definition of 697
of a function 696, 712, 715–16
of mean 698
from pairs of measures 701
Peters's formulae 698
cf. probable error 698
of observation of unit weight 711–13, 715–16
of unknowns (least squares) 711–13
see also errors; least squares
standard gravity 200, 202, 414, 425–30
effect of atmosphere 430
standard pressure 39, 50–1
standard value of residual 725
standards of length 39–40, 426

Star Almanac for Land Surveyors 264
star places
  accuracy 265
  apparent 262
  from catalogue 266–70
  from ephemeris 263–4
  mean 263
  nutation, *see* nutation
  precession, *see* precession
  proper motion 262, 265, 267, 549
  for satellite photography 549
  true 262
  see also aberration; declination; nutation; precession; right ascension
Starlett satellite 515, 579
station adjustment 30–1
  in traverse 31, 131, 166–7
stations
  descriptions of 28–9
  design of 17–18, 20
  eccentric 5, 29–30
  on highest points 7
  movement of marks 17, 491–4
  towers 17–18
  witnes marks 17, 28
  see also bench marks
staves, levelling 208–9
  calibration 209
  lighting 208, 215
  rising or sinking 211, 213–14
  temperatures 208, 214
  verticality 209
stellar
  frame (inertial) 564, 597, 790
  magnitude 541
  photography 540–54
Sterneck method, for latitude 295–7
Stokes's integral
  anomalies for 444–50
  data required, 436–9
  derivation 432 3
  derivatives 434
  elimination of external mass 436, 442–50
  geoid charts derived from 477
  impact of satellite geodesy vii, 430, 436, 438–42
  non-classical theory 456–64
    de Graaff–Hunter's treatment 460–2
    Molodensky's treatment 463–4
  omission of low-degree harmonics 434
  practical difficulties 436
  tables for computing 435
  use of free air anomaly 445–7
  use of isostatic anomaly 448–9
  use of rectangular blocks 440–2, 798
  use of satellite data 438–41
  utility of vii, 435
Stokes's theorem 752–4
strain gauges
  for crustal movement 494–5
  for earth tides 502
strength 390–1
  earth's 391, 487–9
  see also accuracy
striding level 21, 294, 302
Student's t-table 693, 699
submarine observations of g 371–2, 382

t minus T correction, *see* arc-to-chord
tables
  geodetic, for ρ and ν 107, 120, 123
    not for computers 107
  gravity reduction 794–7
  for Stokes's integral 435
Talcott method 287–95; *see also* zenith telescope
telluroid 459
tellurometer, *see* MDM
temperature
  absolute zero 39
  earth's internal 390–1
  gradient
    in atmosphere, *see* atmosphere
    in earth 390
  harmonic analysis of 238
  measurement of 47–8, 61
  see also atmosphere; thermometer
terrain correction, *see* orographical
terroid 459
tesseral harmonics
  of earth's gravity field
    numerical values 637–42
    from satellites 570, 573–4, 605–6, 613–14
theodolite
  adjustment 20–2
  astronomical 294–5
  effect of dislevelment 22, 32
  geodetic, characteristics 20
  micrometer run 22, 26
  observations (horizontal)
    abstract 25, 27–8
    broken rounds 25, 30
    closure of rounds 25, 30
    method of angles 24, 30
    precautions 35–6
    programme of observations 24–8
    rejections 32–3
    rounds and directions 24, 30
    Schreiber's method 24, 31
    sources of error 32, 35–6
    time of day 27
  optical plumbing 21
  recorder 36
  stand 20
  station adjustment 30–1

theodolite (*contd.*)
  see also vertical angles
thermometers
  meteorological 47, 61
  wet and dry 48
tidal effect
  on deviation of the vertical 245, 499–501
  on gravity 369, 382, 501
  on levelling 245
  on satellite altimetry 616, 620
  on strain gauge 502
tidal stations, in levelling 206, 221–2, 219, 228, 248–9
tide
  equilibrium 243–5, 497–9
  gauges 247–9
    for river crossings 221–2
  marine
    harmonic analysis 246–8
    prediction, accuracy of 248
  see also earth tides; mean sea-level
time
  and latitude, simultaneously 316–30
  definitions
    atomic (international), TAI 276–9
    dynamical time scale 276
    ephemeris 276
    Greenwich mean 275, 277
    sidereal
      apparent 257, 275
      mean 275
      uniform 275
    solar
      apparent 274
      mean 274–5
    time, in general 274
    universal (UT) 277
    UTC 279–80
    UT 0 278
    UT 1 277
    UT 2 279
  equation of 275
  as frequency standard 279
  local sidereal
    accuracy of 313–16, 317, 320–1, 326–8, 330
    by east and west stars 330
    by astrolabe 305–8
    by equal altitudes 328–9
    by meridian transits 299–316
    summary of methods 299–300
    by unequal altitudes 329–30
    by zenith photography 317–21
  signals
    corrections to 280–1, 310
    DUT 1 280
    lists of 309–10
    reception of 310–1
    rhythmic 309
    velocity of transmission 310
  see also clocks; longitude
topographical anomaly
  deviation 467
  gravity
    defined 395, 447
    large indirect effect 447
    use 395–6
    variation with average height 480–3
topographical formulae for coordinates 119, 123
towers, triangulation
  Bilby 18
  design of 17–18
  for azimuth observations 18
  Lambert 18
tracking
  capacity of satellite 542
  power of camera 542
tranet system 589–90
Transit satellites 514, 589–90
transit telescope
  for azimuth 335–6
  for longitude
    accuracy 313–16
    artificial star 308
    calibration 309, 314–16
    level adjustment 302–3
    in high latitudes 304–5
    impersonal micrometer 306–8
    personal equation 306–9, 314–16
    photo-electric 308–9
    photographic 309
    programme 301, 305–6
    reduction to meridian 303–4
Transit tracking system 589–90
transverse Mercator projection 191–4
  Universal T.M. 193–4
traverse
  accuracy 166–70, 173–4
  adjustment 124–5, 149, 157–9
  advantages 8
  azimuth control 8–9, 26, 86, 124–5, 149
  computation
    by direct method 124–5, 149
    of traverse net 157–8
    by variation of coordinates 124, 143, 149
  eccentric stations 30, 87
  layout 8–11, 14–15
  observation of angles 26, 86
  primary 8–11
  reconnaissance 16
  routine of observation 25–6, 28, 85–7
  secondary 15

station adjustment 31
use of only one angle 131, 166–7
see also EDM, EODM; MDM; zero order traverse
Triad (TIP) satellite 515
triangular error
adjustment of 108–13
as guide
to accuracy 164
to rejections 32–3
in height 231–2
usual value of 2, 32–3
triangulated heights, accuracy of 231–3
triangulation
abstract of angles 25, 27–8
accuracy 2, 32, 169–77
adjustment 2, 30–1, 111–15, 126–61
chains
accumulation of error 171–3, 716–18
adjustment 111–15, 126–61
advantages 3
computation 107–26
layout 3–7, 13–14
computation
checking of 125–6, 142–4
development method 105
of distance and and azimuth 120–4
of latitudes and longitudes 116–20
on plane 182–7
projection method (Helmerts) 96, 104–7
on spheroid 94–126
in three dimensions 678–90
by variation of coordinates 126–44
connection with old work 7–8, 17
control by satellite fixes 13–15
eccentric observations 5, 29–30
figural adjustment 111–15, 147–50
figure, definition of 4, 5
geodetic, definition 1, 2
grazing lines 7, 20, 27, 32, 242–3
in high latitudes 340–1
layout of 3–7, 13–15
length of sides 6, 7
observation systems 24–5
net 5–6
adjustment 126–60
advantages 3
layout 5–6, 15
optimum of base and Laplace 174–5
origin 101–3
primary, definition 1
programme of moves 23–4
reconnaisance 16–17
secondary 1, 13–16
stations 7, 17–18, 20, 28–30
towers 18
see also astro-triangulation; figure; theodolite; vertical angles
triaxial ellipsoid 401–2, 420
trigonometrical formulae 644–5
trilateration 11–12, 15
adjustment 126–44, 150, 154–6
Aerodist 15, 87–8
combined with triangulation 15, 126, 174
computed by variation of coordinates 126–44
conditions in 150
length ratios 90–91, 134
Laplace azimuth control 11, 13–15, 174
satellite 575–7
secondary 15, 87–88
Shoran 12
see also, EODM; MDM
truncation of harmonic analysis
effect on J_2 533
Kaula's rule 639
effect on satellite orbits 639–40
effect on Stokes 439
Turner's method 549–54
two colour laser EODM 71–2
two colours for vertical angles 241–2

unit sphere 680–5
unit weight, standard error of observation of 140–2, 711–13, 715–16
units
bar viii, 39
gal viii, 357
geopotential unit (GPU) viii, 198
gravity unit (GU) 357
international (SI) viii, 198, 357
metre 39–40
universal time (UT), *see* time
coordinated, *see* time, UTC
unvisited points
fixing of 5, 16
heights of 236–41
U.S. National geodetic survey (formerly USC and GS)
azimuths in high latitude 335
crustal movement 494
doppler control 169, 591–3, 632–6
formulae for coordinates 119–20, 123
frequency of
base lines 5
Laplace azimuths 5
geoid charts and sections 477–8
gravimetric interpolation 472
revision levelling 204–6
triangulation 5
see also zero-order traverse

Vaisäla comparator 41–2
variance
  definition 692
  variance–covariance matrix 141, 155, 712, 715, 726, 728–9, 737–8
  see also errors *and* standard errors
variation of coordinates
  accuracy
    of functions of unknowns 141, 712–13
    of unknowns 140–1, 712
  cf. adjustment by conditions 127–8, 144, 147, 153
  advantages 127–8, 153
  angles or directions 129–30, 137
  checks 142–4
  directions 130, 132, 136, 138–9
  exterior angles 131, 137
  fallible coefficients 709
  for traverses 149, 157
  general description 126–8
  infallible
    base or azimuth 133–4, 689–91
    coordinates 132–3, 690
  in three dimensions 685–90
  normal equations 136–40
    solution of, *see* normal equations
  observation equations 127, 131–4, 686–90
    numbering of 138
  provisional coordinates 126–7, 179
  Schreiber's method 138–9
  second approximation 129, 141–3
  station adjustment 30–1, 131
  weights 134–5
    revision of 141–2
  without station adjustment 31
variation of latitude, *see* polar motion
vector
  algebra, *see* paragraph headings of Appendix E
  applied to dynamics 790–1
  field 741
  in matrix algebra 654–5
  in satellite theory 506
velocity correction to gravity 792–3
Vening Meinesz
  derivatives of Stokes's integral 434–5
  pendulums 371–2
  regional isostasy 485–6
  submarine gravity results 372
vertical
  curvature of
    computed 467–70
    conventional 97, 197–8, 344–5, 422
  definition 96
  see also deviation of the vertical
vertical angles
  accuracy 33, 231–3
  bubble correction 21–2, 33–4, 284–5, 291
  geoid profiles by 476
  heights by
    accuracy of 33, 231–3
    adjustment of 233
    best time 34, 237–41
    computation of 228–31
    reciprocal 33–4, 221–3, 231, 237
    to unvisited points 230–1, 237–41
    weights 233
  observation equation in V of C 688
  observation of 33, 86
  refraction, *see* refraction (effect on angles)
  river crossings by 221–3
  see also refractions; meridian latitudes; zenith telescope *etc.*
very long base interferometry (VLBI) 620–7
  accuracy 625–6
  computation 622–4
  general principles 621–2
  measurement of delay 624
  notation 622
  reference system 622
  refraction correction 625
  uses vii, 623–4, 626–7
  zero of longitude 623–4
vibrating string accelerometer (VSA) 384–5
Vignal's formula 216–19
Vincenty's method for change of spheroid 180–1

water vapour pressure
  abnormal 58–60, 62
  effect of
    on refraction of light 44
    on microwaves 46, 52–3, 63–4
  Goff–Gratch formula 48
  gradient of 51, 53, 63–4
  measurement of 48
  normal 49–53
  over water 58, 63–4
  saturated 48, 58
  tables for 48
  see also refraction
wavelength
  of doppler 593
  and frequency 43
  of laser light 89, 578
  of light 37, 44
  of microwaves 37, 77, 79
  of radar 575
  of Secor 575
  of VLBI 624
weight
  of correlated observations 727

definition 135, 703–4
of directions 151
of dissimilar quantities 141–2, 150–1, 708, 713
of a function 141, 696
of height differences 233
of horizontal angles 113–4, 134–5, 141–2, 151, 155
in least squares 708, 713
in levelling 226–7
revision of 141–2
see also standard error; unit weight
weight matrix of correlated observations 724–5
weighted mean 704, 728
Wild
camera BC 4 545, 628, 630–1, 633
GAK gyro-theodolite 348–55
theodolites 20, 25, 295, 348–9, 545
wind, effect on
astrolabe 328
astronomical observations 283, 328
refractive index (EDM) 41, 55, 59, 61
on terrestrial refraction 61–2, 222, 241, 283
Wisconsin method
for testing bubbles 285
witness marks 17–28
World geodetic system, 1972 (WGS) 631–4, 638

year, tropical, etc., definitions 281

Zeiss Ni 002 level 208–9, 212
zenith
distance 257
photography, for time and latitude 316–21
advantages of 317
telescope 287–94
accuracy 294
adjustments 288–9, 292–4
computations 291–3
programme 289–3
refraction 292
utility 286
zenith cameras
accuracies 317, 320, 321
Italian 319
photo zenith tube 317
TZK 1 318
USC and GS 1951 321
zenithal projection 194
zero of longitude 96–7, 103–4, 273, 278–9, 564–6, 610–12
zero order traverse 1–2, 10–11, 14–15, 61
accuracy 168–9
azimuth control 10–11
comparison with, doppler 169, 612–13
extent of 11
provides geoid profile 10–11
provides scale for astro-triangulation 557–8, 628
zero settings, theodolite 26
zonal harmonics
change of axis 787–8
defined 779–81, 783
in geopotential, *see* *J*'s
normalized 785–6
properties 779–81